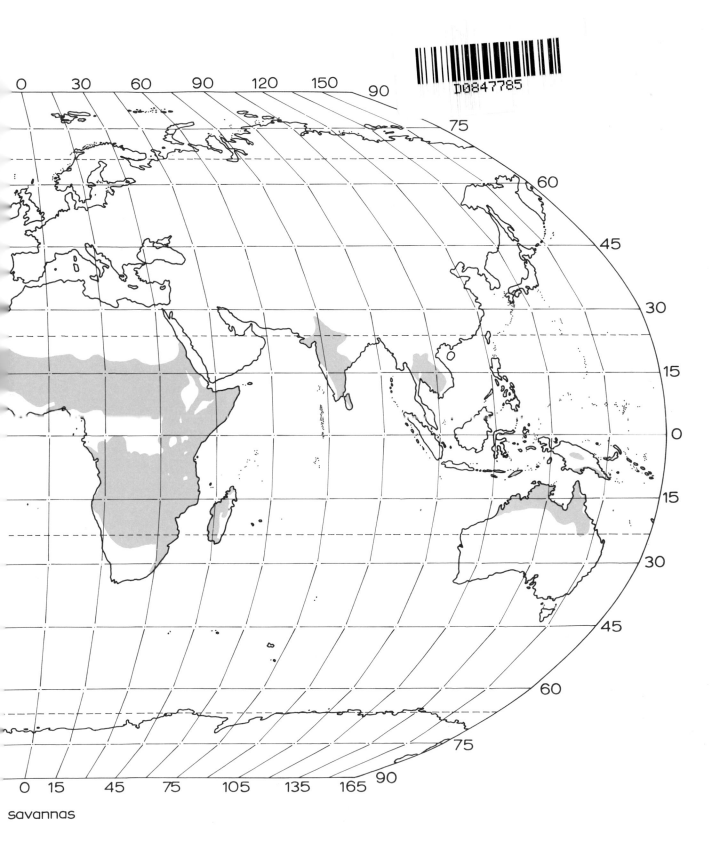

savannas

ECOSYSTEMS OF THE WORLD 13

TROPICAL SAVANNAS

ECOSYSTEMS OF THE WORLD

Editor in Chief:

David W. Goodall

CSIRO Division of Land Resources Management, Wembley, W.A. (Australia)

ECOSYSTEMS OF THE WORLD 13

TROPICAL SAVANNAS

Edited by

François Bourlière

Université René Descartes
45, rue des Saint-Pères
Paris (France)

ELSEVIER

Amsterdam — Oxford — New York 1983

ELSEVIER SCIENCE PUBLISHERS B.V.
Sara Burgerhartstraat 25
P.O. Box 211, 1000 AE Amsterdam, The Netherlands

Distributors for the United States and Canada:

ELSEVIER SCIENCE PUBLISHING COMPANY INC.
655, Avenue of the Americas
New York, NY 10010, U.S.A.

First edition 1983
Second impression 1990

ISBN 0-444-42035-5

PREFACE

This volume of *Ecosystems of the World* is concerned with the structure and dynamics of tropical savanna ecosystems. An underlying theme of the various contributors, botanists as well as zoologists, is to identify and describe the various strategies adopted by plants and animals to meet the constraints and opportunities of the savanna environment. All the contributors have done their best to consider the role of the particular taxonomic or functional group that they treat within the savanna system as a whole.

To avoid any duplication with the volumes in this series dealing with natural and artificial grasslands (the tropics included), little attention has been paid in the present volume to primary production processes. Many readers may feel, on the other hand, that too much emphasis has been laid in this book on vertebrate consumers in general, and mammals in particular. This is not merely due to my personal life-long interest in mammal ecology. Rightly or wrongly, more attention has been given during the past two decades to savanna mammals, large or small, than to any other class of animals. Their morphological, physiological and behavioural adaptations to this environment can nowadays be considered as reasonably well known, and some long-term studies of their population dynamics have been undertaken. Furthermore, the very different structure of the mammalian communities in Africa, the Neotropics and Australia paves the way for stimulating comparisons. It is to be hoped that field ecologists and granting agencies will, in the future, display the same interest for other "key" groups of savanna organisms, particularly grasshoppers and caterpillars among primary consumers, spiders and ants among secondary consumers, termites and soil micro-organisms among "decomposers".

I have tried to keep this book to a single volume and to a manageable size. Further information on tropical savannas is included in the comprehensive treatise recently issued by UNESCO (*Tropical Grazing Land Ecosystems*, Paris, 1979). As for the adaptation of man to the savanna environment it has been comprehensively discussed in the book edited by David R. Harris (*Human Ecology in Savanna Environments*, Academic Press, London, 1980). The reader is referred to these two volumes for further information on problems of applied ecology not treated in the present book.

As my editorial duties come to a close, I would like to extend my thanks to all of those who made the writing of this book possible. First of all, Dr David W. Goodall, the Editor in Chief of the series, for inviting me to contribute this volume and providing expert advice on many problems. The contributors have all been very receptive to my editorial suggestions and contributed to the project with enthusiasm. They did their very best to meet the deadlines we agreed upon initially, in spite of many other commitments. All but three of the final manuscripts were received during the second half of 1979, and bibliographies were updated to January 1, 1980. I took advantage of the delay imposed by the few "late comers" for translating and editing as carefully as possible the ten contributions originally written in French, a delicate and time-consuming task, in which I have been greatly helped by Mrs Felicia Silverman and Dr Malcolm Hadley. I heartily thank them for their cooperation. I do not want to forget either my secretaries, Mesdames Yannick Le Gall and Florence Petelle, who so carefully typed the successive drafts of many manuscripts and helped me in the preparation of the indexes.

FRANÇOIS BOURLIÈRE
Université René Descartes
Paris

LIST OF CONTRIBUTORS

G.P. ASKEW
Department of Soil Science
School of Agriculture
University of Newcastle upon Tyne
Newcastle upon Tyne, NE1 7RU (Great Britain)

R. BARBAULT
Laboratoire de Zoologie
Ecole Normale Supérieure
46, rue d'Ulm
75005 Paris (France)

F. BLASCO
Institut de la Carte Internationale
   du Tapis Végétal
39, allées Jules-Guesde
31400 Toulouse (France)

F. BOURLIÈRE
Université René Descartes
45, rue des Saint-Pères
75006 Paris (France)

C. FRY
Zoology Department
University of Aberdeen
Tillydrone Avenue
Aberdeen AB9 2TN (Great Britain)

A.N. GILLISON
Division of Land Use Research
CSIRO
Black Mountain, P.O. Box 1666
Canberra A.C.T. 2601 (Australia)

D. GILLON
Laboratoire de Zoologie
Ecole Normale Supérieure
46, rue d'Ulm
75005 Paris (France)

Y. GILLON
Laboratoire d'Entomologie
Université de Paris-Sud
Bâtiment 446
91405 Orsay (France)

M. HADLEY
Division des Sciences Ecologiques
UNESCO
7, Place de Fontenoy
75007 Paris (France)

D.C.D. HAPPOLD
Department of Zoology
The Australian National University
Box 4 P.O.
Canberra, A.C.T. 2600 (Australia)

B. HOPKINS
New England College
Tortington Park
Arundel, BN18 0DA (Great Britain)

I.G. HORAK
Tick Research Unit
Rhodes University
6140 Grahamstown
(South Africa)

G. JOSENS
Université Libre de Bruxelles
Laboratoire de Zoologie Systématique
C.P. 160
50, avenue Franklin D. Roosevelt
1050 Brussels (Belgium)

P. KAISER
Chaire de Microbiologie
Institut National Agronomique
9, rue de l'Arbalète
75231 Paris (France)

M. LAMOTTE
Laboratoire de Zoologie
Ecole Normale Supérieure
46, rue d'Ulm
75005 Paris (France)

H.F. LAMPREY
UNESCO Regional Office for
  Science and Technology for Africa
P.O. Box 30592
Nairobi (Kenya)

P. LAVELLE
Laboratoire de Zoologie
Ecole Normale Supérieure
46, rue d'Ulm
75005 Paris (France)

J. LEVIEUX
Laboratoire d'Ecologie Animale
Université d'Orléans
B.P. 6005
45045 Orléans (France)

J.C. MENAUT
Laboratoire de Zoologie
Ecole Normale Supérieure
46, rue d'Ulm
75005 Paris (France)

R. MISRA
International Society for Tropical Ecology
Department of Botany
Banaras Hindu University
Varanasi 221005 (India)

M. MONASTERIO
Facultad de Ciencias
Departamento de Biologia
Universidad de los Andes
Merida (Venezuela)

R.F. MONTGOMERY
Department of Soil Science
School of Agriculture
University of Newcastle upon Tyne
Newcastle upon Tyne, NE1 7RU (Great Britain)

A.E. NEWSOME
Division of Wildlife Research
CSIRO
P.O. Box 84
Lyneham, A.C.T. 2602 (Australia)

H.A. NIX
Institute of Earth Resources
Division of Land Use Research
CSIRO
P.O. Box 1666
Canberra, A.C.T. 2601 (Australia)

J. OJASTI
Instituto de Zoologia Tropical
Facultad de Ciencias
Universidad Central de Venezuela
Apartado 59058
Caracas 104 (Venezuela)

G. SARMIENTO
Facultad de Ciencias
Departamento de Biologia
Universidad de los Andes
Merida (Venezuela)

A.R.E. SINCLAIR
Institute of Animal Resource Ecology
2075 Wesbrook Mall
Vancouver, B.C., V6T 1W5 (Canada)

T. VAN DER HAMMEN
Hugo de Vries-Laboratorium
Universiteit van Amsterdam
Sarphatistraat 221
Amsterdam (The Netherlands)

CONTENTS

Chapter 1

PRESENT-DAY SAVANNAS: AN OVERVIEW

F. BOURLIÈRE and M. HADLEY

WHAT IS A SAVANNA?

There is no commonly agreed definition of the word "savanna" or its French (*savane*) or Spanish (*sabana*) equivalents. It is said to be derived from an old Carib word, but its etymology remains obscure. G. F. de Oviedo y Valdes (1535) stated that "this name *sabana* is applied to land without trees, but with much grass, short and tall". According to the *Shorter Oxford Dictionary* (3rd edition, 1944), the word savanna(h) has been known in English since 1555. It is derived from the sixteenth century Spanish word *zavana* and designates "a treeless plain". The French dictionaries of Littré (1863–72) and Robert (1964–65) also mentioned the use of the word *savane* since 1529 and its derivation from a Caribbean language, through the Spanish tongue. In these Western European languages, the word was obviously of creole origin. It was mostly used by the colonists of the Western Hemisphere, from Canada to Venezuela and Argentina, to designate what is now called a prairie landscape. As recently as 1942, the *Diccionario general de Americanismos* (Santa Maria, 1942) defined it as follows: "Llanura, en especial si es muy dilatada y sin vegetacion arborea, aun que suelo abundar en buenos pastos [A plain, particularly if it is very extensive, and without woody vegetation, though the soil carries plenty of good forage]".

For a long time, travelling naturalists and botanists seem to have followed common usage. In his *Tableaux de la Nature*, Alexander von Humboldt (1808/1828) wrote of "Missouri savannas", of "grassy steppes of the Apure and Rio de la Plata" or of "the South American steppes of the llanos". Referring again to the llanos, he did, however, mention that the treeless plains may be interspersed with *Mauritia* palms.

The first botanist to use savanna in its present meaning was apparently Grisebach (1872). In the second volume of his *Die Vegetation der Erde*, he mentioned that "savannas differ from temperate steppes in allowing arborescent vegetation" (quoted after the French edition, 1875, vol. 2, p. 177). The same view was held by Drude (1890), who considered that "the savannas are not only characterized by high grasses, but also by the presence of tropical woody plants which come into leaf during the rainy season ..." (French edition, p. 274). Schimper (1898) also concurred with the above definition. He proposed that "Grassland, when hygrophilous or tropophilous, is termed *meadow*; when xerophilous, *steppe*; and xerophilous grassland containing isolated trees is *savanna*" (English translation, 1903, p. 162).

Since that time, most authors have used the term savanna to designate a community having a continuous grass layer usually scattered with trees. Such is the case, for example, of Lanjouw (1936) and Beard (1953), both working in South America. Beard has defined a savanna as a "plant formation ... comprising a virtually continuous ecologically dominant stratum of more or less xeromorphic herbs, of which grasses and sedges are the principal components, with scattered shrubs, trees, or palms, sometimes present."

Most botanists working in Africa have also adopted a similar definition. The Scientific Council for Africa meeting on phytogeography held in Yangambi agreed to define savannas as follows: "Formations of grasses at least 80 cm high, forming a continuous layer dominating a lower stratum. Usually burnt annually. Leaves of grasses flat, basal and cauline. Woody plants usually present." (Conseil Scientifique pour l'Afrique, 1956). Within this physiognomic category, the Yangambi meeting

1

recognized four subcategories. First, savanna woodland, with trees and shrubs forming a canopy which is generally light. Second, tree savanna, with scattered trees and shrubs. Third, shrub savanna, with scattered shrubs. Fourth, grass savanna, where trees and shrubs are generally absent. High montane grasslands, aquatic grasslands and herb swamps were given the status of distinct formations.

A decade later, most of the members of the Caracas–Maracay International Geographical Union symposium also agreed that a savanna was a vegetation type characterized by scattered woody plants, in varying proportions, dispersed in a herbaceous grass layer (Hills, 1965; Hills and Randall, 1968). It was also emphasized, on that occasion, that in addition to "savanna ecosystem" there was a need for the term "savanna landscape". This term was taken to describe areas within which savanna vegetation dominates but is not necessarily exclusive, being interspersed with riparian or gallery forests, patches of woodland, swamps and/or marshes. This is a very frequent situation in the areas of forest/savanna mosaic, both in the Western and Eastern Hemisphere.

We recognize that the word "savanna" sometimes continues to be used where trees are not present. However, we have considered in a previous review (Bourlière and Hadley, 1970) that it should now be restricted to those tropical and subtropical formations: (1) where the grass stratum is continuous and important, occasionally interrupted by trees and shrubs; (2) where bush fires occur from time to time; and (3) where the main growth patterns are closely associated with alternating wet and dry seasons.

Most recent phytogeography and ecology textbooks agree with this or similar definitions (Polunin, 1960; Walter, 1964; Odum, 1971; Schnell, 1971; McNaughton and Wolf, 1973). Most of these authors also emphasize the fact that the gradient from savanna woodland to grass savanna forms a broad transition zone between the "closed" tropical forests and the "open" desertic steppes.

In proposing such a working use of the term "savanna", we are aware that some workers, especially in East Africa (Pratt et al., 1966), have rejected its use, as being a term seldom correctly applied outside its continent of origin. We would agree that the term, having been defined in so many

different ways, cannot be used satisfactorily in a precise classificatory sense (White, 1982). We nevertheless think that the term is a useful one, in that it groups different types of tropical vegetation which share a number of structural and functional characteristics. Thus, quoting White (1982), "in more general contexts, both popular and scientific, the antithesis 'forest and savanna' is a commonplace and will doubtless continue to be usefully used for certain tropical landscapes."

This is particularly true given that tropical savannas are, in a number of important respects, more similar to many temperate zone ecosystems than to tropical forests. We share with White the feeling that the value of the term "savanna" is greatly diminished if it is made to include, for instance, wooded *Sphagnum* bogs of temperate regions, as has been proposed by some authors.

In view of the different interpretations of the term "savanna", not surprisingly there exist several different systems of classification and nomenclature for tropical vegetation. Some proposed systems are regional in scope. Others are universal. A recent state-of-knowledge report on tropical grazing land ecosystems (UNESCO, 1979) includes a comparison of some of the principal systems of nomenclature that have been proposed. This publication also provides a listing of equivalent terms of tropical vegetation types in English, French and other languages.

NATURAL VERSUS SEMI-NATURAL ("ANTHROPOGENIC") SAVANNAS

The difficulty of clearly defining a tropical savanna arises from the fact that this physiognomic category of vegetation may be the end-result of either natural processes or of human activities — or a combination of the two. Unfortunately, it is often very difficult to assign many present-day savannas either to the category of so-called "natural" savannas or that of "anthropogenic" savannas. This is all the more difficult since human activities have probably more often contributed to extend considerably the area of already existing savannas than to create such an ecosystem "from scratch" in life zones where such a vegetation category did not exist previously.

That natural "climax" savannas must have existed for a long time is beyond question. The plants (the shade-intolerant C4 grasses, for example) and the animals which today live only in savannas must have come from somewhere, as very few can be considered as having originated either in closed tropical forests or in desertic steppes. Furthermore, human influences in some South American savannas are known to have been, and remain still, very slight.

The respective causal roles of a number of physical factors have been, and still are, the object of heated arguments, and are examined, from various angles, by many contributors to this volume. Suffice it to say at this stage that three categories of factors — climatic, edaphic and geomorphic — appear to be of major importance. Most authors stress the significance of moisture deficiency due to alternating short wet and long dry seasons, of low soil fertility and subsoil drainage conditions, and of the slope, elevation and degree of dissection, and of the presence or absence of a lateritic crust. However, many specialists disagree when they come to establish hierarchical relationships between these physical factors. Hills (1965) has accordingly stressed the importance of distinguishing more clearly between predisposing, causal, resulting, and maintaining factors.

Man-made and man-maintained savannas are particularly well represented in tropical Africa and Asia, where human activities have been carried out on a more or less large scale over many millennia. Fire has always been the major tool for clearing vegetation. Bush fires are still used by savanna hunter-gatherers for hunting — both in Africa (Lee and DeVore, 1976) and in Australia (Tindale, 1972; Mulvaney, 1975; Tonkinson, 1978) — as well as by African agriculturalists and cattle-herders (Klima, 1970).

There is little doubt that in many parts of West Africa, savannas were originally created and are still maintained by the regular burning of the grass cover for various human purposes. This is not to say that bush fires cannot be natural — that is caused by lightning; indisputable cases have been reported in Africa (Bourlière and Verschuren, 1960; Jones, 1963). But the impact of such fires upon the vegetation is far less extensive. In any case, when plots of such man-maintained savannas are protected from bush fires for a number of years, they very quickly turn into deciduous woodland.

The prevalence of "secondary" savannas in many parts of Africa has led a number of ecologists working in this continent to assign a major causal role to the anthropic factor in the origin and extension of savannas, all the more since the effects of man's actions have probably been accentuated by major climatic fluctuations in the past (Aubréville, 1949, 1962). That anthropic factors might have also played some role in the less densely populated South American savannas is shown by the 1579 document quoted by F. Tamayo (in Hills and Randall, 1968). Deforestation, mostly for the collection of firewood, and overgrazing by domestic cattle are also important factors which add to the destructive action of fire.

A "mixed" theory has been advocated by some West African ecologists. According to Adjanohoun (1963), for instance, present-day savannas at the northern edge of the Guinean forest block, and some savannas still presently "enclosed" within the rain forest, could not be the result of recent deforestation by man. They originated naturally during the last xeric phase of the Pleistocene, when the northern edge of the rain forest retreated south and the savannas reached the Gulf of Guinea shore in many places. Man entered the picture later on, acting to maintain the open savanna landscape when more humid conditions favoured a northern extension of the forested areas. Such a theory looks quite plausible, but still needs to be better documented by palynological, pedological and archeological studies.

SOME STRUCTURAL AND FUNCTIONAL CHARACTERISTICS

Whatever the category to which particular savanna areas belong, or their origin, all tropical savannas share a number of structural and functional characteristics which set them apart from other terrestrial ecosystems of the tropics. Several major characteristics, largely interlinked, can be recognized. Thus, tropical savannas form a transition zone between rain forest and desert. Their physiognomic structure changes with increasing aridity. Production patterns are markedly seasonal, with distinctive quantitative and qualitative characteristics. Savannas are dynamic systems; they have a relatively simple structure, and perturbation of one of the dominant species may have far-reaching repercussions on other system components. Each of these principal characteristics has a number of important consequences and manifestations (Fig. 1.1).

Physiognomic structure of savannas

The physiognomic structure of tropical savannas varies in space and time. In spatial terms, tropical savannas form a broad transition zone between the closed tropical forests and the open desertic steppes. The physiognomic gradient tends to follow the climatic gradient, with increasing density of trees with increasing rainfall. However, topographic features, and the effects of fire, herbivores and man, mean that the gradient is not a straightforward one. Fig. 1.2 illustrates four contrasting savanna areas, which have been the sites of intensive ecological studies. The figure highlights the within-site and between-site variability in the physiognomic structure of tropical savannas.

In a number of areas an apparently stable equilibrium has been reached between savanna trees and grasses. Walter (1964) has emphasized that these two life-forms, which are usually antagonists in other biomes, generally co-exist harmoniously in savanna ecosystems, even eventually benefiting each other (Jung, 1969; Gerakis and Tsangarakis, 1970). For example, at Fété Olé in Senegal, the mineralization of nitrogen has been shown to be much more important under *Acacia senegal* and even *Balanites aegyptiaca* than under pure grass cover (Bernhard-Reversat, 1977). Again, the recovery of herbaceous vegetation after the 1972–1973 drought in the Sahel was much quicker in areas where the trees and shrubs had not been cut down by the local pastoralists to feed their depleted herds of cattle and goats.

The nature of the balance between woody plants and perennial herbaceous vegetation, and the effects of various factors on this balance, is one of the most interesting aspects of the dynamics of savanna ecosystems. In the Neotropics, at least, there is evidence that particular areas have oscillated between rain forest and savanna during the Quaternary, with the forest undergoing repeated episodes of shrinkage and expansion (Haffer, 1969; Simpson Vuilleumier, 1971; Simpson and Haffer, 1978). Widespread invasion of savannas by rain

forest trees (as distinct from gradual advance of the forest edge) also appears to have recurred periodically during the Quaternary. Kellman (1979) has suggested that the key process involved in this invasion has been that of soil enrichment by savanna trees. This may itself be linked to the ability of trees gradually to establish an enlarged plant–litter–soil nutrient cycle within the savanna as a result of prolonged persistence at one site.

Analysis of the types of competition between trees and grasses (e.g. for soil moisture), and the effects of herbivory on the balance between them, would seem to indicate that the woody component of the ecosystem should generally have the advantage in tropical savanna areas (Noy-Meir, 1979). Thus, while trees compete with grasses for topsoil water, they have exclusive use of subsoil water. Grasses may become extinct in a sufficiently dense woody vegetation. Trees, on the other hand, can survive in the densest grassland, as long as rainfall is sufficient to reach the subsoil and there is no problem of recruitment from seedlings. Also, most trees are safe from extinction by herbivores, whatever the density of the latter, thanks to their "ungrazeable" upper branches. Grasses are susceptible to a collapse of reserves and to extinction, particularly if the herbivore is an efficient grazer.

Noy-Meir (1979), in presenting alternative models of plant–herbivore relations in savannas, suggested that moderately fixed herbivore pressure will shift the savanna equilibrium either in favour of the woody plants or in favour of the grasses, depending on the dietary preferences of the herbivore. He suggested that heavy sustained herbivore pressure would cause extinction of the perennial grasses (unless the herbivores are "absolute" browsers) and the conversion of a savanna into a woodland. The woodland at that stage may be dense or sparse (and reduced to the unreachable branches), depending on the tendency of the herbivore to browse. There may, however, be an intermediate range of herbivore densities in which the system can switch between a savanna and a woodland state, depending on initial conditions in relation to a critical threshold of grass biomass or grass reserves.

Another aspect of changing physiognomic structure in particular savanna areas concerns the relation between plant biomass and primary production. Within many savanna areas, there are large visible differences in standing tree and shrub bio-

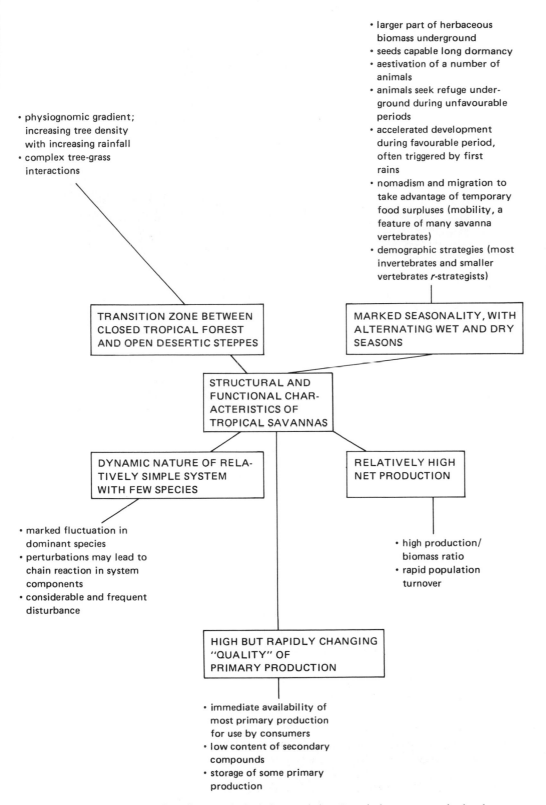

• larger part of herbaceous
  biomass underground
• seeds capable long dormancy
• aestivation of a number of
  animals
• animals seek refuge underground during unfavourable
  periods
• accelerated development
  during favourable period,
  often triggered by first
  rains
• nomadism and migration to
  take advantage of temporary
  food surpluses (mobility, a
  feature of many savanna
  vertebrates)
• demographic strategies (most
  invertebrates and smaller
  vertebrates *r*-strategists)

• physiognomic gradient;
  increasing tree density
  with increasing rainfall
• complex tree-grass
  interactions

TRANSITION ZONE BETWEEN
CLOSED TROPICAL FOREST
AND OPEN DESERTIC STEPPES

MARKED SEASONALITY, WITH
ALTERNATING WET AND DRY
SEASONS

STRUCTURAL AND
FUNCTIONAL CHAR-
ACTERISTICS OF
TROPICAL SAVANNAS

DYNAMIC NATURE OF RELA-
TIVELY SIMPLE SYSTEM
WITH FEW SPECIES

RELATIVELY HIGH
NET PRODUCTION

• marked fluctuation in
  dominant species
• perturbations may lead to
  chain reaction in system
  components
• considerable and frequent
  disturbance

• high production/
  biomass ratio
• rapid population
  turnover

HIGH BUT RAPIDLY CHANGING
"QUALITY" OF
PRIMARY PRODUCTION

• immediate availability of
  most primary production
  for use by consumers
• low content of secondary
  compounds
• storage of some primary
  production

Fig. 1.1. Schematic representation of some principal characteristics of tropical savannas, and related consequences and strategies.

Fig. 1.2A,B. For explanation see p. 9.

Fig. 1.2C,D. For explanation see p. 9.

mass, which are attributable to differences in topography, soil fertility, human use and grazing intensity. More heavily used land tends to support a lower tree and shrub biomass, but does not necessarily result in a correspondingly lower production of plants and animals. In eastern Uttar Pradesh (India), for example, similar levels of net production have been recorded in systems having very different tree densities, landscape physiognomies and biomass. Thus, in this region, the net primary production (in the absence of grazing) is 14, 10 and 13 t ha^{-1} yr^{-1} in forest, plantation and savanna, respectively (Table 1.1).

Seasonality in tropical savannas

Seasonal variations in climate and life cycles occur in all ecosystems, even apparently stable systems such as tropical rain forests. However, it is in tropical savannas that seasonality becomes marked, to such an extent that savannas are not only adapted to survive temporal variations but in all probability require such variations to maintain their resilience (Walker and Noy-Meir, 1979).

The seasonality of the phenological cycle of savanna plants and animals becomes increasingly marked as one proceeds from the edge of the rain forest to that of the desert. Along the same gradient, weather conditions tend to become more and more unpredictable, with large year-to-year variations in the amount and timing of rainfall, and the occurrence of long periods of drought. To cope with such conditions, savanna organisms have developed a wide range of morphological, physiological and behavioural adaptations (Bourlière and Hadley, 1970; Bourlière, 1978; See also the top right corner of Fig. 1.1).

In savanna regions, seasonality in organic production very frequently leads to temporary food surpluses which are used by nomadic secondary consumers and long-range migrants. Another feature is that the larger part of the living herbaceous

TABLE 1.1

Comparison of plant biomass and production in eastern Uttar Pradesh, India (after Singh and Misra, 1979)

Type of system	Biomass (t ha^{-1})	Production (t ha^{-1}yr^{-1})	Ratio of production/ biomass
Natural forest			
Protected from grazing	103.8	14.1	0.13
Open to grazing	94.5	12.6	0.13
Tectona plantation			
Protected from grazing	21.9	10.6	0.48
Open to grazing	10.0	4.5	0.44
Savanna			
Protected from grazing	8.9	13.2	1.48
Open to grazing	6.8	8.3	1.28

biomass tends to be found below ground. Thus, in the Sahelian savanna of Fété Olé in northern Senegal, the maximum standing crop biomass of grass is about 1.2 t ha^{-1} above ground, compared with 1.48 t ha^{-1} below ground (Bourlière, 1978). Much further south, at Lamto in the forest–savanna mosaic of the Ivory Coast, the average above-ground herbaceous biomass varies from 3.21 to 4.36 t ha^{-1} in various floristic categories of the savanna, whereas that of the underground biomass reaches 10.1 to 19.0 t ha^{-1} (Lamotte, 1978). In the Khirasara area of Gujarat State in India (mean annual rainfall 590 mm), Pandeya and Jain (1979) have reported a maximum above-ground biomass of 3.7 t ha^{-1}, with a maximum below-ground biomass figure of 5.0 t ha^{-1}. A similar trend is noticeable among trees and shrubs only in the driest savanna categories. Whereas at Lamto their root biomass represents 28.9 to 31.8% of the standing crop, it may reach 43.7% at Fété Olé.

Fig. 1.2. Between-site, year-to-year and long-term variability in physiognomic structure of tropical African savannas. Views of four areas which have been sites of intensive ecological studies during the past decades. A. The Lamto palm savanna in Ivory Coast (photo Y. Gillon). B. The short grass savanna of the Serengeti Plains, Tanzania, with a *kopje* in the background (photo J. Verschuren). C. The sahelian savanna of Fété-Olé, Senegal, at the end of the rains in a "normal" year, September 1969 (photo J.C. Bille). D. The same site in September, during the 1972 drought (photo J.C. Bille). E. The wooded savanna of the Rwindi Plain in the Virunga National Park, eastern Zaïre, in 1934 (photo G.F. de Witte). F. The same site, 25 years later in 1959. The wooded savanna has been transformed in a grass savanna with a few tree-euphorbias as a consequence of a steady increase of the elephant population (photo J. Verschuren).

Most of the savanna invertebrates also spend most of their life underground. Such is the case for termites, which are particularly numerous, as well as earthworms and soil arthropods, not to mention Bacteria, Protozoa and Fungi. In contrast, burrowing species are rather rare among vertebrates, though some amphibians, reptiles, shrews and rodents spend part of their life cycle in the soil or litter.

Levels and rates of turnover of primary production

Tropical savannas are characterized by relatively high levels of net primary production compared with the standing crop biomass. Though variable, levels of primary production can be high when water and soil fertility are not limiting, and can occasionally exceed that of forests growing under similar conditions. Thus, the total net production, above and below ground, of herbaceous and woody species in seven savanna faciès at Lamto, Ivory Coast, ranges between 21.5 and 35.8 t ha^{-1} yr^{-1} (César and Menaut, 1974).

A linked aspect of primary production is that the turnover of the savanna primary biomass — that is, the fraction of the plant material which is exchanged during a given time interval — is much more rapid than that of tropical forests. This stands out quite clearly when one calculates the $P/\bar{B}$ ratio, i.e. the ratio of annual primary production to the average plant biomass. For example, in the Banco rain forest, Ivory Coast, this ratio is 0.03, whereas it reaches 0.43 in the Lamto tree savanna a little further north, and 0.55 at Fété Olé, in the Sahelian part of Senegal. The turnover of organic biomass and nutrients is therefore much more rapid in savannas than in tropical rain forests.

Most of the savanna plants are also short-lived, as compared with those of the tropical forests. This is not only the case for the herbaceous species which make up a large percentage of the standing crop and most of the annual above-ground production (85% at Lamto, and 83.5% at Fété Olé), but of trees as well. With the exception of baobabs (Swart, 1963), very few savanna trees and bushes live for more than a few decades. At Fété Olé, acacias do not live for more than 30 to 35 years, and only *Balanites aegyptiaca* and *Grewia bicolor* sometimes attain 100 years of age (Poupon, 1979).

A rapid population turnover is also characteristic of many savanna animals of small size, even among vertebrates. Many savanna lizards, rodents and even passerines do not live for more than a year on average. One of the highest adult mortality rates so far known for a bird, 70% per year, is that of the Senegal fire finch, *Lagonosticta senegala* (Morel, 1969). High fecundity is needed to balance such high mortality rates, and many savanna plants and animals are clearly *r*-strategists.

Quality of primary production

From the qualitative angle, most of the primary production in savannas can immediately be used as food by secondary consumers. One of the major differences between tropical forest and savanna ecosystems lies in the differing proportions of the edible and inedible parts of their standing crops. While most of the biomass of a forest consists of wood — which remains inedible for animals, except for the larvae of xylophagous beetles and some termites — the bulk of a grass savanna biomass is made up of grass and other herbaceous plants. These can be consumed, in fresh or dead form, by a great number of invertebrates and vertebrates. In savanna woodlands and tree savannas, intermediate situations occur. A few figures can illustrate this point. Whereas leaf material only amounted to 1.60% of the above-ground plant biomass in the Banco forest, Ivory Coast (Huttel and Bernhard-Reversat, 1975), and 1.18% in the Pasoh forest, West Malaysia (Kira, 1978), grass made up 15% of the above-ground standing crop in the tree savanna of Lamto (Menaut, 1977) and 40% at Fété-Olé (Bourlière, 1978). In some parts of the Serengeti Plains (Tanzania), there are practically no trees, only short grass.

Furthermore, the leaves of grass and other herbaceous plants found in savannas are much more palatable for insects and vertebrates than those of rain forest trees. Rain forest trees appear to be rich in "chemical defences", and, in nutrient-poor tropical ecosystems, trees seem to synthesize relatively large amounts of secondary plant compounds such as alkaloids and polyphenols (Janzen, 1974, 1975). Janzen has argued that roots must expend more energy searching for nutrients in soils of low fertility than in soils of high fertility. In terms of energy economics, therefore, trees occurring along a gradient from rich to poor soils will tend to produce toxic secondary compounds which deter consumption by herbivores, rather than to replace leaves ingested by herbivores.

Support for this idea of reduction in herbivory through accumulation of secondary metabolic chemicals in leaves comes from the work of McKey et al. (1978) in two tropical rain forest sites in Africa. They found that leaves had a higher content of phenolic compounds on soils having lower nutrient contents. Again, in the tropical forest ecosystems of the Rio Negro area of Venezuela, the phenolic content of many tree species is high, and this may account for the relatively low (1–2%) amount of leaf area consumed by insects (Jordan and Medina, 1977; Herrera et al., 1978).

In contrast, grasses are remarkably free of secondary compounds compared with other plant groups (Bernays and Chapman, 1978; Kingsbury, 1978). The leaves of most savanna trees and shrubs are also eaten by many browsing mammals. Even some of their seeds (e.g. the pods of *Acacia tortilis* in Africa) are actively sought as food by many ungulates, which in turn contribute to their dispersal (Lamprey et al., 1974).

In spite of the immediate availability of most primary production for consumers, part of the primary production of the savannas can nevertheless be "stored", at least for a time, and constitutes an "energy reserve" for some consumers during the dry season. Such is the case for seeds, whose production is generally far in excess of that needed to ensure the maintenance of grass cover from year to year. This is in contrast to the situation in the rain forest, where high temperatures and humidity lead to a rapid destruction of organic material. Thus, granivorous animals, mostly ants, rodents and birds, can represent a major trophic category in tropical savannas. Dried grass, of low nutritive value for ruminants, is consumed by termites and decomposers, or destroyed by bush fires.

System dynamics

Compared to tropical forests, savannas have a relatively simple structure, with fewer species. Several groups may be dominant by their biomass. The system is a dynamic one, and a perturbation in the system may bring about rapid changes in the populations of dominant species. Changes in a dominant species may have far-reaching reper-

cussions on the rest of the system, leading to a type of chain reaction among system components. Sometimes, the perturbation may be abiotic in nature; other times it may be biotic.

Thus, in the north of Senegal, several authors have described how fluctuations in rainfall can have far-reaching and rapidly apparent effects on principal ecosystem components. For example, Galat and Galat-Luong (1977) have linked annual variations in reproductive success and mortality of the green monkey *Cercopithecus aethiops sabaeus* with different rainfall levels. The effect of rainfall is mediated through its impact on primary production and the food resources of the monkeys.

Again in northern Senegal, Poupon (1979) has described how climatic fluctuations can affect woody vegetation and subsequently other system components. The drought of 1972, for example, had major repercussions on the leaf stage (retarded leafing, restricted duration of active phase), on flowering (no flowering or late flowering, reduced number of woody plants in flower) and on fruiting (total absence of fruits or very low production). Drought caused a regression of the woody stratum during the period from 1970 to 1979, with important, and generally adverse, effects on microclimates, on soil erosion and soil fertility, and on the production of the herb layer.

An example of a chain reaction following a biotic perturbation in a savanna system is that of rinderpest. Sinclair (1979) has described the events following the first great rinderpest outbreak in East Africa at the end of the last century. The first noticeable effects appeared in 1890, and, by 1892, 95% of the cattle populations had died. The direct effects on man included famine and epidemic disease, particularly smallpox. The indirect consequences of the rinderpest outbreak were no less devastating. Reduction of populations of wild and domestic herbivores led to the appearance of maneating lions. This encouraged the human populations to leave rural areas. Bush encroachment, tsetse fly infestation and tryanosomiasis followed. Human populations retreated still further.

The removal of rinderpest in the Serengeti Plains in the 1960s has triggered an equally complex sequence of events (Sinclair, 1979). The immediate result of rinderpest removal was a doubling of yearling survival in the dominant herbivore, the wildebeest, from 25% to 50%. This allowed the

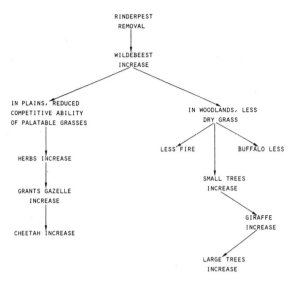

Fig. 1.3. Changes in ecosystem components following a perturbation (rinderpest removal) of a dominant species (wildebeest) in the Serengeti area of Tanzania (after Sinclair, 1979).

wildebeest population to increase from about 250 000 in 1961 to 500 000 in 1967. The buffalo population increased at about the same rate, from about 30 000 in 1961 to 50 000 in 1967. However, and most significantly, the zebra — which is a non-ruminant and therefore unaffected by rinderpest — showed no change in population during this period.

Removal of rinderpest thus acted as a comparatively sudden perturbation, allowing an eruption of the dominant grazer, wildebeest. The change in wildebeest numbers led to important changes in other ecosystem components (Fig. 1.3). Direct impact on the vegetation of the plains and of woodland areas in turn resulted, indirectly, in other changes in the vegetation and in populations of certain herbivores and carnivores.

The Serengeti Plains thus provide a good example of the strong interactions that can exist between a few components of a savanna ecosystem, and illustrate how changes in a few major components can have far-reaching effects on other components.

TOWARDS A CHARACTERIZATION OF SAVANNAS?

Tropical savannas exhibit marked variability, in both space and time. This variability is perhaps their most important feature. "There is no such thing as a typical savanna ecosystem. Rather there is a gradient of related ecosystems, ranging from open woodlands to almost treeless steppes" (Lamotte and Bourlière, Ch. 27).

In the light of these remarks, is there justification for trying to develop generalized models of tropical savannas? What are the types of information that might be included in such models? Most scientists would subscribe to the view expressed by many contributors to this book, about the need for more research. More data are indeed required — for many savanna components, processes and geographical regions — before a reasonably satisfactory picture of tropical savannas can be obtained. Caution is also called for in extrapolating the results of intensive site studies.

Nevertheless, some key features of tropical savannas can be described. Some patterns can be identified. Thus, tropical savannas might be characterized, simplistically, as follows (Walker and Noy-Meir, 1979): alternating wet and dry phases; structure primarily determined by competition between woody and grass plants for available soil moisture; fire, herbivores and soil nutrients as principal modifying factors.

Walker and Noy-Meir (1979) have proposed some simple models of tropical savannas, which provide a useful framework for discussion. Swift et al. (1979) have also prepared a number of charts which summarize the processes of decomposition in various ecosystem types in boreal, temperate and tropical zones; that for tropical savannas is reproduced as Appendix I. Both Appendix I and Fig. 1.2 are included here not as alternatives or as definitive statements. Rather they are intended to illustrate complementary ways and means by which the synthesis of information on tropical savannas might be approached and presented, in both verbal and figurative forms.

Many other structural and functional characteristics of savanna organisms are discussed by the contributors to this volume. Many data are presented. Nevertheless, a great deal remains to be done to identify and analyze the various strategies adopted by plants and animals to withstand, and sometimes to take advantage of, the constraints of the savanna environment.

Tropical savannas represent a young biome, on the geological time scale, as compared with the tropical forest biome. The new opportunities that they have offered to some tropical organisms, which were able to adapt successfully to this new environment, have had far-reaching consequences.

Thus, rapid and innovative change often characterizes taxa in savanna environments, as illustrated by ruminants among large herbivores (Webb, 1977, 1978) and by flowering plants (Stebbins, 1952, 1974). Stebbins has compared aspects of the evolution of angiosperms in mesic and xeric habitats. He has argued that tropical rain forests, far from being a cradle of origin of flowering plants as claimed by many authors, are more like museums — museums in which many archaic types can be preserved for millennia but which do not have the conditions favourable for radically new adaptive complexes. This contrasts with more xeric conditions, which promote rapid evolution.

During the Middle to Late Cenozoic cycle of increasing aridity, Stebbins has suggested that savanna-like formations may have acted as a "spe-cies pump" where phyla could shift first to more xeric and then back to more mesic situations. Stebbins has also remarked that evolution involves not only the origin of species, but also their migration and extinction. Thus, while savannas may support comparatively few species at the present or any one particular time, dry habitats may have supported more species than the mesic ones over the whole scale of geological time, because of a more rapid turnover in species composition.

Thus, savannas can be considered to be an important seat of ecological diversity, which have played a key role in the evolution of the biosphere and many of its component species. In particular, do not more and more primatologists and anthropologists nowadays favour the view that the colonization of savanna by protohominids has been the initial and decisive step taken in the history of our own species?

APPENDIX I: DECOMPOSITION PROCESSES IN TROPICAL SAVANNAS

Summary of the structure of the decomposition subsystem and its functional integration in tropical savanna ecosystems (after Swift et al., 1979, by permission of Blackwell). For visual comparison purposes, biomass pools are represented as π diagrams in which the size of circle is proportional to total biomass. Abbreviations: NPP = net primary production; SOM = soil organic matter; K = decay constant; tr = turnover constant.

Geographical distribution

The term "savanna" refers to the grasses which are a major component of the vegetation. Savanna embraces much of the tropical zone which is not forested and includes thorn scrub and coarse grass where drought is lengthy, through park-like grass-lands with scattered trees to savanna woodlands with tall elephant grasses. Vegetation forms are commonly affected by man, fire and grazing and many open areas are capable of regeneration to woodland. Soils lateritic.

Climate

Temperatures are high throughout the year particularly in the dry season when daytime maximum may exceed 40°C but fall to 15°C or less at night. Mean monthly temperatures show an annual range of 5 to 10°C. Amount and duration of precipitation varies widely from 1000 to 1500 mm, with only a short dry period, to central continental areas where less than 500 mm may fall over 1 to 2 months.

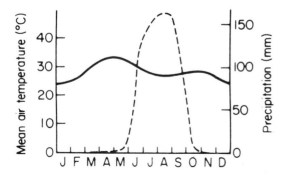

Plant biomass and production

Growth of woody plants occurs throughout the year but mainly during wet season. Growth of grasses is confined to the wet season and the large root biomass is efficient in nutrient uptake during periods of high rainfall and in moisture absorption at the ends of the dry season. A very rapid growth response of grasses to rain is characteristic of savanna as nutrients are released from materials accumulated during the dry season. Resource quality of grasses varies from sun-dried material at the end of the wet season, with high protein content, to highly silicious species.

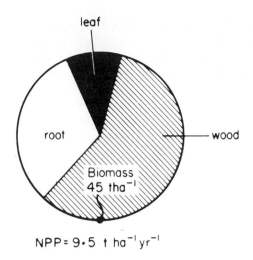

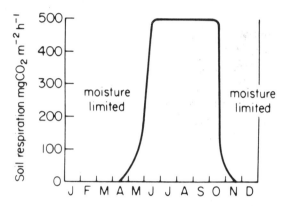

Soil respiration rates

Soil microbial activity is limited by moisture availability during the dry season but very high levels of CO_2 efflux are recorded during the rains when the non-woody NPP can be decomposed over a period of a few months. Soil respiration follows Q_{10} relationship 2. Thermophilic micro-organisms are important at midday when soil temperatures may exceed 35 C.

Major nutrient pools

Nitrogen standing crop is low and is largely held in the plant (and animal) biomass. Nutrient turnover rates are high and the system has low stability if perennial grasses and woody shrubs are cleared. At the beginning of the dry season the SOM content is less than 1% with a C:N 5:1, suggesting that residual organic matter is largely microbial tissues.

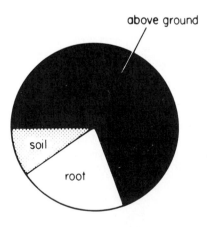

Decomposer biota

The soil fauna is dominated by the termites many of which can forage during the dry season by constructing sheets and galleries of soil over the resources. Almost all woody litter may be utilized by termites in West African savanna and the microbial component of wood decay is restricted to the fungus combs within termite mounds. Dung-beetle burrowing activities and termite foraging produce equivalent effects to earthworms in temperate grasslands for soil turnover and structuring. Microbial standing crop has a high turnover rate. Bacteria and Actinomycetes are important components of the soil microflora.

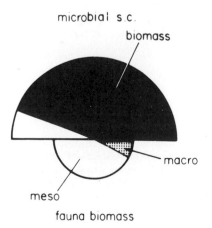

Decomposition rates

Decomposition is primarily moisture-limited and litter may accumulate on the soil surface during the dry season, although termite utilization of wood may be non-seasonal. When rain falls decomposition is rapid and temperature regulated. Grass litter has high nutrient content as a result of rapid drying at end of rains and non-woody NPP can be decomposed during rainy season. The effects of microclimate are similar but more extreme than temperate grassland and standing dead material has a long residence time in absence of termite attack.

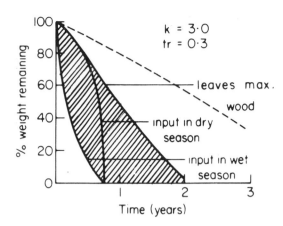

REFERENCES

Adjanohoun, E., 1963. *La Végétation des Savanes en Côte d'Ivoire Centrale*. Thesis, Université de Paris Sud, 178 pp.

Aubréville, A., 1949. *Climats, Forêts et Désertification en Afrique Tropicale*. Société d'Editions Géographiques, Maritimes et Coloniales, Paris, 351 pp.

Aubréville, A., 1962. Savanisation tropicale et glaciations quaternaires. *Adansonia*, 2: 16–84.

Beard, J.S., 1953. The savanna vegetation of northern tropical America. *Ecol. Mongr.*, 23: 149–215.

Bernays, E.A. and Chapman, R.F., 1978. Plant chemistry and acridoid feeding behaviour. In: J.B. Harborne (Editor), *Biochemical Aspects of Plant and Animal Coevolution*. Academic Press, London, pp. 99–141.

Bernhard-Reversat, F., 1977. Observations sur la minéralisation *in situ* de l'azote du sol en savane sahélienne (Sénégal). *Cah. ORSTOM, Sér. Biol.*, 12: 301–306.

Bourlière, F., 1978. La savane sahélienne de Fété Olé, Sénégal. In: M. Lamotte and F. Bourlière (Editors), *Structure et Fonctionnement des Ecosystèmes Terrestres*. Masson, Paris, pp. 187–229.

Bourlière, F. and Hadley, M., 1970. The ecology of tropical savannas. *Annu. Rev. Ecol. Syst.*, 1: 125–152.

Bourlière, F. and Verschuren, J., 1960. *Introduction à l'Ecologie des Ongulés du Parc National Albert*. Institut des Parcs Nationaux du Congo Belge, Bruxelles, 158 pp.

César, J. and Menaut, J.C., 1974. Le peuplement végétal. *Bull. Liaison Chercheurs Lamto*, Numéro Spéc., 2: 1–161.

Conseil Scientifique pour l'Afrique (CSA), 1956. *Réunion de Spécialistes du CSA en matière de Phytogéographie* [*CSA Specialist Meeting on Phytogeography*], *Yangambi 1956*. CCTA, London, Publ. No. 53, 35 pp.

Drude, O., 1890. *Handbuch der Pflanzengeographie*. Engelhorn, Stuttgart, 582 pp.

Galat, G. and Galat-Luong, A., 1977. Démographie et régime alimentaire d'une troupe de *Cercopithecus aethiops sabaeus* en habitat marginal au nord Sénégal. *Terre Vie*, 31: 557–577.

Gerakis, P.A. and Tsangarakis, C.E., 1970. The influence of *Acacia senegal* on fertility of sand sheet ("Goz") soil in the central Sudan. *Plant. Soil*, 33: 81–86.

Grisebach, A.H.R., 1872. *Die Vegetation der Erde nach ihrer klimatischen Anordnung*. Leipzig, 1335 pp.

Haffer, J., 1969. Speciation in Amazonian forest birds. *Science*, 165: 131–137.

Herrera, R., Jordan, C.F., Klinge, H. and Medina, E., 1978. Amazon ecosystems. Their structure and functioning with particular emphasis on nutrients. *Interciencia*, 3: 223–232.

Hills, T.L., 1965. Savannas: a review of a major research problem in tropical geography. *Can. Geogr.*, 9: 220–232.

Hills, T.L. and Randall, R.E., 1968. *The Ecology of the Forest/Savanna Boundary. Proceedings of the IGU Humid Tropics Commission Symposium, Venezuela, 1964*. Department of Geography, McGill University Montreal, Que., 128 pp.

Huttel, C. and Bernhard-Reversat, F., 1975. Recherches sur l'écosystème de la forêt subéquatoriale de Basse Côte d'Ivoire. V. Biomasse végétale et productivité primaire. Cycle de la matière organique. *Terre Vie*, 29: 203–228.

Humbold, A. de (von), 1808/1828. *Tableaux de la Nature*. Gide fils, Paris, 2 vols., 520 pp.

Janzen, D.H., 1974. Tropical blackwater rivers, animals, and mast fruiting by Dipterocarpaceae. *Biotropica*, 6: 69–103.

Janzen, D.H., 1975. *Ecology of Plants in the Tropics*. Studies in Biology, No. 58. Edward Arnold, London, 66 pp.

Jones, E.W., 1963. The forest outliers in the Guinea zone of northern Nigeria. *J. Ecol.*, 15: 415–434.

Jordan, C.F. and Medina, E., 1977. Ecosystem research in the tropics. *Ann. Mo. Bot. Gard.*, 64: 737–745.

Jung, G., 1969. Cycles biogéochimiques dans un écosystème de région tropicale sèche: *Acacia albida* (Del.) en sol ferrugineux tropical peu lessivé. *Oecol. Plant.*, 4: 195–210.

Kellman, M., 1979. Soil enrichment by Neotropical savanna trees. *J. Ecol.*, 67: 565–577.

Kingsbury, J.M., 1978. Ecology of poisoning. In: R.F. Keeler, K.R. van Kampen and L.F. James (Editors), *Effects of Poisonous Plants on Livestock*. Academic Press, New York, N.Y., pp. 81–91.

Kira, T., 1978. Community architecture and organic matter dynamics in tropical lowland rain forests of Southeast Asia with special reference to Pasoh Forest, West Malaysia. In: P.B. Tomlinson and M.H. Zimmermann (Editors), *Tropical Trees as Living Systems*. Cambridge University Press, Cambridge, pp. 560–590.

Klima, G.J., 1970. *The Barabaig. East African Cattle-Herders*. Holt, Rinehart and Winston, New York, N.Y., 114 pp.

Lamotte, M., 1978. La savane forestière de Lamto, Côte d'Ivoire. In: M. Lamotte and F. Bourlière (Editors), *Structure et Fonctionnement des Ecosystèmes Terrestres*. Masson, Paris, pp. 231–311.

Lamprey, H.F., Halevy, G. and Makacha, S., 1974. Interactions between *Acacia*, bruchid seed bettles and large herbivores. *E. Afr. Wildl. J.*, 12: 81–85.

Lanjouw, J., 1936. Studies on the vegetation of Suriname savannas and swamps. *Ned. Kruidkd. Arch.*, 46: 823–851.

Lee, R.B. and DeVore, I. (Editors), 1976. *Kalahari Hunter-Gatherers*. Harvard University Press, Cambridge, Mass., 408 pp.

Littré, M.P.E., 1863–1872. *Dictionnaire de la Langue Française*. Hachette, Paris, 4 vols.

McKey, D., Waterman, P.G., Mbi, C.N., Gartlan, J.S. and Struhsaker, T.T., 1978. Phenolic content of vegetation in two African rain forests: ecological implications. *Science*, 202: 61–64.

McNaughton, S.J. and Wolf, L.L., 1973. *General Ecology*. Holt, Rinehart and Winston, New York, N.Y., 710 pp.

Menaut, J.C., 1977. Analyse quantitative des ligneux dans une savane arbustive préforestière de Côte d'Ivoire. *Geo-Eco-Trop.*, 1: 77–94.

Morel, M.Y., 1969. Contribution à l'étude dynamique de la population de *Lagonosticta senegala* (Estrildidés) à Richard Toll (Sénégal). *Mem. Mus. Nat. Hist. Nat.*, A 78: 1–156.

Mulvaney, D.J., 1975. *The Prehistory of Australia*. Penguin Books, Ringwood, Vic., 2nd ed., 327 pp.

Noy-Meir, I. 1979. Stability of plant–herbivore models and possible applications to savanna. In: *Papers presented at the Workshop on Dynamic Changes in Savanna Ecosystems, Kruger National Park, May 1979*, in press.

Odum, E.P., 1971. *Fundamentals of Ecology*. Saunders, Philadelphia, Pa., 3rd ed., 574 pp.

Oviedo y Valdes, G.F. de, 1535. *Historia General y Natural de Las Indias*. [Quoted after Hills, 1965.]

Pandeya, S.C. and Jain, H.K., 1979. Tropical grazing land ecosystems of India. Description and functioning of arid to semi-arid grazing land ecosystem at Khirasara, near Rajkok (Gujarat). In: *Tropical Grazing Land Ecosystems*. A

state-of-knowledge report prepared by UNESCO/UNEP/ FAO. *Nat. Resour. Res.*, 16: 630–655.

Polunin, N., 1960 *Introduction to Plant Geography and Some Related Sciences*. Longmans, London, 640 pp.

Poupon, H., 1979. *Structure et Dynamique de la Strate Ligneuse d'une Steppe Sahélienne au Nord du Sénégal*. Thesis, Université de Paris Sud, Paris, 351 pp.

Pratt, D.J., Greenway, P.J. and Gwynne, M.D., 1966. A classification of East African rangeland, with an appendix on terminology. *J. Appl. Ecol.*, 3: 369–382.

Robert, P., 1964–1965. *Dictionnaire Alphabétique et Analogique de la Langue Française*. Société du Nouveau Littré, Paris, 6 vols.

Santa Maria, F.J., 1942. *Diccionario General de Americanismos*. Pedro Robredo, Mexico, 3 vols.

Schimper, A.F.W., 1898. *Pflanzen-Geographie auf physiologischer Grundlage*. Fischer, Jena, 898 pp. [Quoted after the English translation, Clarendon Press, Oxford, 1903, 839 pp.]

Schnell, R., 1971. *Introduction à la Phytogéographie des Pays Tropicaux. Les Problèmes Généraux. Vol. 1. Les Flores. Les Structures, Vol. 2. Les Milieux. Les Groupements Végétaux*. Gauthier-Villars, Paris, 951 pp.

Simpson Vuilleumier, B., 1971. Pleistocene changes in fauna and flora of South America. *Science*: 173, 771–780.

Simpson, B.B. and Haffer, J., 1978. Speciation patterns in the Amazonian forest biota. *Annu. Rev. Ecol. Syst.*, 9: 497–518.

Sinclair, A.R.E., 1979. Dynamics of the Serengeti ecosystem: process and pattern. In: A.E.R. Sinclair and M. Norton-Griffiths (Editors), *Serengeti: Dynamics of an Ecosystem*. Chicago University Press, Chicago, Ill., pp. 1–30.

Singh, K.P. and Misra, R., 1979. *Structure and Functioning of Natural, Modified and Silvicultural Ecosystems of Eastern Uttar Pradesh*. Indian MAB Research Committee Contribution to Research Project 1. Technical Report October 1975–October 1978. Banaras Hindu University, Varanasi, 161 pp.

Stebbins, G.L., Jr., 1952. Aridity as a stimulus to plant evolution. *Am. Nat.*, 86: 33–44.

Stebbins, G.L., Jr., 1974. *Flowering Plants: Evolution above the Species Level*. Edward Arnold, London, 399 pp.

Swart, E.R., 1963. Age of baobab tree. *Nature*, 198: 708–709.

Swift, M.J., Heal, O.W. and Anderson, J.M., 1979. *Decomposition in Terrestrial Ecosystems*. Studies in Ecology, 5. Blackwell, 371 pp.

Tindale, N.B., 1972. The Pitjandjara. In: M.G. Biccchieri (Editor), *Hunters and Gatherers Today*. Holt, Rinehart and Winston, New York, N.Y., pp. 217–268.

Tonkinson, R., 1978. *The Mardudjara Aborigines*. Holt, Rinehart and Winston, New York, N.Y., 149 pp.

UNESCO, 1979. *Tropical Grazing Land Ecosystems*. A state-of-knowledge report prepared by UNESCO/UNEP/ FAO. *Nat. Resour. Res.*, 16: 1–655.

Walker, B.H. and Noy-Meir, I., 1979. Aspects of the stability and resilience of savanna ecosystems. In: *Papers presented at the Workshop on Dynamic Changes in Savanna Ecosystems, Kruger National Park, May 1979*, in press.

Walter, H., 1964. *Die Vegetation der Erde, I. Die tropischen und subtropischen Zonen*. Gustav Fischer, Jena. [Quoted after the English translation *Ecology of Tropical and Sub-tropical Vegetation.* Translated by D. Mueller-Dombois. Edited by J.H. Burnett. Oliver and Boyd, Edinburgh, 1971, 539 pp.]

Webb, S.D., 1977. A history of savanna vertebrates in the New World. Part 1: North America. *Annu. Rev. Ecol. Syst.*, 8: 355–380.

Webb, S.D., 1978. A history of savanna vertebrates in the New World. Part II. South America and the great interchange. *Annu. Rev. Ecol. Syst.*, 9: 393–426.

White, F., 1982. *The Vegetation of Africa. A Descriptive Memoir to Accompany the UNESCO/AETFAT Vegetation Map of Africa.* UNESCO, Paris, in press.

Chapter 2

THE PALAEOECOLOGY AND PALAEOGEOGRAPHY OF SAVANNAS

T. VAN DER HAMMEN

INTRODUCTION

Many disciplines contribute to the knowledge of the relation of organisms (or communities of organisms) with their environment in the past, but palynology and in particular pollen analysis has supplied the most abundant and most relevant data. Such pollen studies enable one to reconstruct the history of vegetation and environment, and vegetation being the basic primary productive part of most terrestrial ecosystems, and forming their main visible part, it also defines the main features of their physiognomy. It is for these reasons that this chapter will be chiefly based on the results of palynological inquiry. Tropical palynology still being in its infancy, the history of savanna ecosystems is still insufficiently known, but the available data are very significant, because they show us that change and instability form important aspects of tropical savanna ecosystems.

PRE-QUATERNARY HISTORY

If savannas are defined as possessing a more or less continuous grass cover, they cannot have come into being before the Eocene, because the first records of pollen grains of grasses from the tropics date from the Middle Eocene. However, low, open vegetation types constituted by herbs other than grasses may have existed before that time. In his description of the present-day herb-savannas of tropical South America, which are practically devoid of grasses, Huber (1982) suggested that they are somehow derived from very old vegetation types. That herb-"savannas" of a somehow similar type (without grasses), may have been an important

vegetation type on the large African–South American land mass in the Cretaceous before the two continents became widely separated, seems to be a reasonable supposition. The Middle Cretaceous fossil flora of these continents contains very interesting sporomorphs, many of them of the *Ephedra* type or belonging to related fossil taxa; *Ephedra* is a genus at present mainly confined to dry habitats (Herngreen, 1974).

In Borneo the short-lived increases in grass pollen representations in the Tertiary seem to be related to the local destruction of the forest by volcanic activity with ash falls and not to climatic changes, because it seems that the climate was as humid during the Tertiary as it is now (Germeraad et al., 1968).

In Nigeria the major fluctuations in the percentage of grass pollen in Tertiary sediments most probably reflect the shifting boundary between forest and savanna in the lowlands (Germeraad et al., 1968).

In South America vegetation types with a grassy herb layer seem to have existed already by the Middle Eocene (the grasses were associated with Malvaceae and *Jussiaea* according to Gonzalez, 1969). In the Miocene and Pliocene, grass pollen became so abundant that open vegetation with a dominantly grassy ground layer must have been widespread (Van der Hammen and Wijmstra, 1964; Germeraad et al., 1968; Wijmstra, 1969, 1971). At the same time Compositae (Asteraceae) appear to have become locally abundant, and *Cuphea* entered upon the scene. Moreover, since the maxima of grass pollen often coincided with maxima of *Byrsonima* and later also with maxima of *Curatella*, there can (also in view of the composition of the pollen spectra of recent savannas) be very little doubt that savannas rather similar to the present ones existed already in the Miocene and Pliocene of northern South America, and constituted a rather important part of the total vegetation cover (Wijmstra, 1969, 1971; Van der Hammen, 1974). As man did not yet exist at that time, this provides (apart of course from the floristic arguments) one of the best arguments for the supposition that savanna vegetation is basically natural and not anthropogenic, quite independent of the question whether the present extension of the recent savannas is partly attributable to human activities.

As the base of the Lower Miocene is c. 25 million years old (and the base of the Middle Eocene even c. 50 m. y.), there has been ample time for the evolution and adaptation of tropical savanna floras. In this connexion it is interesting to note that in the Tertiary carbonized grass cuticles already occurred, so that fire (by natural causes) has presumably always been a factor of at least some importance.

SOME GENERAL ASPECTS OF THE QUATERNARY

Before the Quaternary palaeoecology and palaeogeography of savannas are discussed, it seems worthwhile to deal first with some general aspects of the Quaternary as established by modern research methods. The Quaternary was a period of strong climatic fluctuations resulting in colder and warmer periods called glacials and interglacials. Originally about four glacial periods were recognized, but later additional and earlier glacials and interglacials were found (see, for instance, Turekian, 1971; Van der Hammen et al., 1971; Van der Hammen, 1974), and there may have been up to twenty or more glacial–interglacial cycles.

Although many glacials do not represent simple climatic periods but consist of various glacial and (often rather warm) interstadial phases, and interglacials may show one or several cooler (or short, cold) intervals, a reasonable estimate of the average length of a glacial–interglacial cycle may be 100 000 years. Individual glacial or interglacial phases may last from about 10 000 to 20 000 years or more. Knowledge of the Pleistocene has increased considerably by the study of deep-sea cores (CLIMAP Project Members, 1976; Gates, 1976), and a reconstruction of the entire ice-age world seems to be within reach now because the correlation of data deduced from Quaternary events in both the oceans and the continents is now in principle possible (Van der Hammen, 1982).

As to the absolute age of the Quaternary and its subdivisions, radiocarbon (^{14}C) dating is only suitable for organic material dating from the last 50 000 (–70 000) years; and potassium/argon and fission-track dating are suitable for material such as volcanic ashes older than about 200 000 years. An indirect method of dating is palaeomagnetism (based on the recognition in the sediment series of palaeomagnetic reversals of polarity already dated by other methods elsewhere), which method is particularly suited for continuous series of fine-grained sediments such as deep-sea cores and clayey lake sediments.

Although the exact lower boundary of the Quaternary is still disputed and some workers defend an age of approximately 1.5 m.y., data are accumulating which favour the view that the age is at least 2.5 m.y. Recently Shackleton and Opdyke (1977) showed that in ocean sediments a significant change took place around 3 m.y. ago. Before that time the climate was more stable and warm until a cooling started at 3 m.y. B.P. (Before Present) which led to a glacial–interglacial type of rhythm around 2.5 m.y. B.P. As data from the tropics equally indicate about the same age for the first marked glacial, an age of 3 m.y. will be accepted here for the base of the Quaternary.

The Last Glacial lasted from c. 80 000 years ago to c. 10 000 B.P. The transition from the full glacial conditions of the Last Glacial (a period called Pleniglacial) to the interglacial conditions of our present period (the Holocene), is called the Late Glacial, which lasted from c. 13 000 (at most 14 000 and at least 12 500) B.P. to c. 10 000 B.P. Twenty years ago much information had already accumulated about the Quaternary history of the northern latitudes, where extensive glaciation took place, but next to nothing was known about the contemporaneous periods in the tropics. Glacials were thought to correspond with wet periods (pluvials) in the tropics and subtropics and interglacials to coincide with dry periods (interpluvials). In the last twenty years it has become clear that this view was too simple and even partly erroneous. A review of the palaeoecological and palaeogeographical data from the tropics by Livingstone and Van der Hammen (1978) listed the most relevant literature.

It seems indeed that an earlier part of the Last Glacial was quite wet in the tropics, but at the time when maximum glaciation was attained in the north about 20 000 B.P., the climate became extremely dry, to remain so till the beginning of the Late Glacial (*c.* 13 000 B.P.) when the effective precipitation became much higher again to decrease once more locally in the course of the Holocene. It needs to be pointed out that this sequence did not occur exactly as described here in all tropical regions and many local variations existed.

The present interglacial period, the Holocene, is likewise not a climatically stable period. Temperatures higher than the present ones prevailed during the middle part of the Holocene, a cooler phase began some 3000 years ago, and drier and wetter intervals alternated. This is the case not only in the northern latitudes, but also in the tropics.

THE BIOGEOGRAPHIC THEORY OF DIVERSIFICATION AND REFUGIA

For a long time the tropical rain forest was supposed to be a very old formation type developed without appreciable climatic changes in the same area for millions of years, and to represent a very stable ecosystem, but gradually an increasing amount of contradictory evidence accumulated. Fossil sand dunes now covered with rain forest, and certain geomorphological features and types of deposits found in the rain forest area, can only have been formed in a much drier climate with unstable soil conditions (compare, for instance, De Ploey, 1963, 1965; Journaux, 1975; and Ab'Saber, 1967, 1977). On the other hand, biogeographers found curious inhomogeneities in the distribution pattern of animals and plants that seemed to be quite incongruous in an area that is apparently so continuous and homogeneous as the tropical rain forest.

Detailed studies of selected groups of organisms led to the recognition of centres of diversity and endemism in the tropical rain forest, and of zones of contact and hybridization. This induced Haffer (1969, 1977), Vanzolini (1970, 1973), and others to postulate the existence in the past of much drier periods with a reduction of the continuous rain forest to "islands" (refugia) and an appreciable extension of savanna vegetation invading and oc-

cupying the areas between the forest refugia. The postulation of erstwhile refugia in Amazonia by Haffer was based on studies of the zoogeography of birds and Vanzolini came to the same conclusion on account of the distribution of lizards; they were followed by specialists in other groups such as butterflies (Brown, 1976) or certain plant families (Prance, 1973, 1978), who all arrived at the same result.

It appears, therefore, that the refugium theory is gaining ground; but one should bear in mind that more direct and more detailed historical evidence is required before the existence of forest refugia in the postulated places can be proved beyond reasonable doubt.

What is important for the study of the palaeoecology, palaeogeography and biogeography of savannas is that conceivably there was a considerable and repeated extension of savanna vegetation in the past during the drier climatic intervals, sometimes perhaps even connecting the areas of savannas north and south of the tropical rain forest zone. In the case of the Amazonian forest, this would mean that at one time the *llanos*, the *campos cerrados*, etc., and the isolated minor savannas (and campos) within this forest belt may have been linked at one time or another.

On the other hand one must reckon with the possibility that the present areas of savanna were considerably smaller during periods of a general increase of precipitation in the boundary areas leading to the invasion of savannas by (rain) forest. Conceivably such a situation could have led to savanna refugia. If the theory of forest refugia proves to be acceptable one might even regard some of the extant savanna areas as present-day refugia.

THE QUATERNARY PALAEOECOLOGY OF SAVANNAS

The Neotropics

Most palaeoecological data from tropical South America are from the northern Andes, especially from the area called "Sabana de Bogotá". This area is a high plain at *c.* 2600 m altitude with a (tropical) montane climate, and is not a tropical savanna as the name seems to suggest. However, the data from this area (see, for instance the compilation by Van der Hammen, 1974; and Livingstone and Van der Hammen, 1978) show that at one time the altitudinal forest limit in these tropical mountains was lowered by from 1200 to 1500 m and that the average annual temperature may have been $6°$ to $7°C$ lower than it is today. It was, moreover, established in one of the longest and most continuous sequences of lake sediments in the world that this was a recurrent phenomenon, so that more than twenty glacial–interglacial "cycles" could be recognized in the time-span of the last 3 m.y. Another fact of considerable importance is that during the period from c. 21 000 to c. 13 000 B.P. the climate was much drier than it is today, and the lake levels were much lower. The rainfall may have been less than half of the present amount.

The available data from the tropical lowlands will be discussed here, the Caribbean coast of Guyana and Surinam to be dealt with first.

The coastal lowlands of Guyana

The principal data from this area relevant to our subject were published by Van der Hammen (1963, 1974) and by Wijmstra (1969, 1971). The description is adapted from the summary given by Van der Hammen (1974).

Along the estuaries and the muddy shores of the sea and lagoons mangrove forest is found, of which the outermost zone is usually formed by *Rhizophora*, a band of *Avicennia* forest following behind it on still periodically inundated soil. Behind the mangrove belt there is an alternation of somewhat more elevated former beach ridges and lower-lying swampy areas. Part of the area behind the mangrove zone is known as "wet savanna"; the annual rainfall is here between 2000 and 3000 mm, and the edaphic conditions seem to be an important factor, so that this wet savanna seems to be closely related to open herbaceous swamps. The ridges are covered with a type of forest characteristic of drier soils, the swamps sustaining open herbaceous vegetation or swamp forest. Common elements in the open swamps are, for instance, *Cyperus*, *Typha* and *Acnida*; in the swamp forest such arborescent taxa as *Symphonia*, *Ilex*, *Virola* and *Tabebuia* abound, and palms may locally be very abundant. Farther inland, the relatively dry wallaba forest, or more humid tropical forest, is found. Where the edaphic conditions are extreme (as on the "white sands") the vegetation cover tends to approach the characteristics of a savanna, but true savannas are only found much farther in the interior, for instance, in the Rupununi and Sipaliwini savannas.

A 30-m section from a bore hole in the sediments of the coastal plain near Georgetown, Guyana, provided a pollen diagram (Fig. 2.1) that shows something of the history of this area during the last interglacial, the Last Glacial and the Holocene (Van der Hammen, 1963). The cumulative diagram shows the variation in time of the percentages of four ecological groups. The first two groups (left) represent elements of the mangrove forest (*Avicennia* and *Rhizophora*, respectively). The next groups contain the forest elements and grass-savanna elements. For a good understanding it must be borne in mind that *Avicennia* is a poor pollen producer and that its pollen is not widely dispersed, so that an appreciable pollen percentage could only accumulate within an erstwhile *Avicennia* belt. *Rhizophora*, on the other hand, produces much more pollen that is also readily transported seawards, so that a very high percentage accumulated not only within the mangrove belt *in situ*, but also in marine sediments on the shelf lying to seaward of this belt.

The lower part of the diagram, with a ^{14}C dating of $> 45 000$ B.P., indicates that at the time of deposition the site lay within the mangrove belt. In the diagram an extension of swamp forest elements follows, so that the sea must have retired and the site apparently came to lie behind the coastline. Mangrove elements disappear completely from the diagram in the next phase, open grass-savanna elements becoming completely dominant and the sediment showing clear signs of soil formation. The site was at that time well above sea level. This situation lasted until the beginning of the Holocene, when the area gradually became invaded by the sea. First the *Avicennia* belt passed across the site (c. 8600 B.P.) and subsequently *Rhizophora* dominated completely; the presence of microforaminifera in this part of the section indicates that at that time the coast line proper lay farther inland than at present. The later Holocene part of the diagram shows that the coast line moved northwards again and *Avicennia* forest, swamp forest and open swamps were frequent in the Georgetown area. The very last pollen spectra of the diagram show the present situation and the influence of man on the recent vegetation cover.

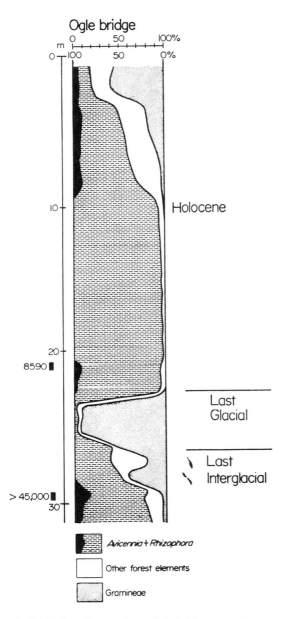

Fig. 2.1. Pollen diagram from Ogle Bridge, near Georgetown, Guyana (from Van der Hammen, 1974, by permission of Blackwell).

It is evident that the diagram not only reflects the glacial–interglacial eustatic sea-level fluctuations, but also shows that grass-savanna dominated in the area during at least a part of the Last Glacial.

Deep bore holes from Surinam and Guyana (Wijmstra, 1969, 1971; Van der Hammen, 1963) show that in the Quaternary sediments from the coastal plain the type of sequence as recorded from near Georgetown, is repeated many times, and

most probably reflects the sequence of Quaternary glacial–interglacial eustatic movements of the sea level. An example is given of a 120-m bore hole section from Surinam (Fig. 2.2). The Pleistocene (and Pliocene) age of the series can be deduced from the presence of a number of taxa such as *Alnus*. Although *Byrsonima* occurs locally in the Late Holocene part of the sections, the overall composition of the spectra indicates that at that time extensive grass-savannas were not developed in this area. On the other hand, many of the recorded older, low sea-level sites are similar to that from the Last Glacial in Georgetown. They show a dominance of grass-savanna with *Byrsonima* and *Curatella*. Therefore, one must conclude that the available data all point to the erstwhile dominance of real savanna vegetation in the present coastal area of Guyana and Surinam during glacial times. Although the influence of edaphic factors cannot be ruled out, it seems as if the dominance of grass-savannas over so large an area cannot be explained by edaphic factors alone and requires the existence, at times, of a zone with a "savanna climate" — that is, with a lower annual precipitation and/or more pronounced dry and wet seasons than at present.

The vegetational succession during a eustatic climatic cycle (see Wijmstra, 1971) is diagramatically indicated in Fig. 2.3 as being: stands of *Rhizophora* → stands of *Avicennia* → palm swamp forest → grass-savanna with *Byrsonima* and *Curatella* shrub, or *vice versa*.

The younger Holocene history of the coastal belt of open swamps and wet savannas was studied (partly in relation to prehistoric human settlements) by Laeijendecker-Roosenburg (1966). Cyperaceae, Amaranthaceae and *Typha* (and to a lesser extent also Poaceae) were well represented in the stand of vegetation as registered in the pollen diagrams, and minor relative fluctuations of sea level had a great influence on the vegetational changes.

The lower Magdalena Valley

The lower courses of the Colombian rivers Magdalena and Cauca form a very extensive, low-lying area subject to inundation stretching from approximately El Banco y Nechi to Plato. Vegetation types dominated by grasses are very common in the area; they are partly (more or less anthropogenic?) savannas, and partly form "floating mea-

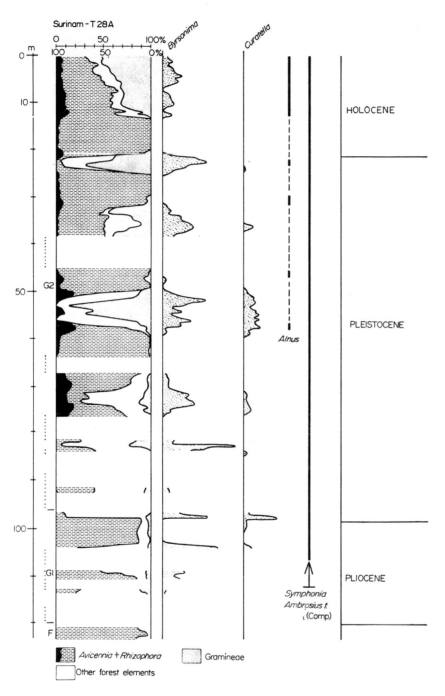

Fig. 2.2. Pollen diagrams from the bore hole Alliance, Surinam (from Van der Hammen, 1974, by permission of Blackwell; adapted from Wijmstra, 1971).

dows", which may in places rest on the ground during the dry season to start floating again during the wet season with higher water levels. Wijmstra (1967) studied the young Holocene history of these vegetation types and found them to be highly dynamic, and continuously adapting themselves to minor climatic changes. Periods of relatively low high-water levels resulted in "grounding" of floating meadows, an invasion of drier parts by *Byrsonima*, *Cecropia*, etc., and the formation of

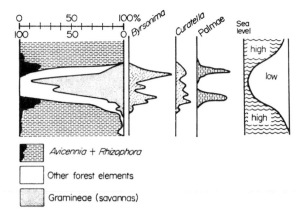

Fig. 2.3. Scheme of the succession of vegetation types (as reflected in the pollen diagrams) during an interglacial–glacial cycle in the present coastal plain of Guyana and Surinam (from Van der Hammen, by permission of Blackwell; adapted from Wijmstra, 1971).

peat layers. Later studies (A.E. Gonzalez, unpublished) showed that these changes played an important role during the entire Holocene. Radiocarbon datings of the peat layers led to the recognition of the main drier periods (i.e., of a lower effective precipitation in the Magdalena–Cauca River basin) of the Holocene. Approximate dates for the drier periods are: c. 7000 B.P., c. 5500 B.P., c. 4700 B.P., c. 4100 to c. 3500 B.P., c. 2700 to c. 2000 B.P., c. 500 A.D., c. 1250 A.D. and c. 1500 A.D. During these periods some of the areas occupied by floating meadows or the adjacent areas subjected to periodic inundation apparently assumed a more savanna-like character.

The inland savannas

A number of pollen diagrams are available from lakes in the savannas of the Llanos Orientales of Colombia and from the Rupununi savannas in Guyana (Wijmstra and Van der Hammen, 1966). Most of these are from the Holocene but one of them extends back to the Last Glacial. The following description is from a later compilation of the data and their interpretation (Van der Hammen, 1974). Two of the more informative diagrams reproduced in Fig. 2.4 show the variation in time of the percentage of pollen grains of three groups: forest elements; savanna woodland and *Byrsonima* and *Curatella* scrub; and open savanna (with Poaceae, Cyperaceae and other savanna herbs).

The first diagram is from Laguna de Agua Sucia,

south of San Martin and not far from the Ariari River in the Colombian Llanos Orientales. Nowadays the area is dominated by grass-savanna, with some swamp forest or gallery forest in low-lying places and in the Ariari Valley proper.

Two layers of peat, intercalated in the lake sediments, represent low lake levels and were dated at c. 4000 and c. 2200 years B.P. The lower part of the diagram, dating from possibly c. 6000–5000 to c. 4000 years ago, shows a complete dominance of grass-savanna elements; it seems that the lake became seasonally desiccated. After c. 3800 B.P. the lake level rose and open water remained in the lake all the year round. At the same time *Byrsonima* woodland invaded the open savanna and became dominant. There was also an increase of the other forest elements. All these changes point, generally speaking, to an increasingly wetter climate. Open savanna subsequently gradually increased again till about 2200 B.P. when the formation of peat indicates a very low lake level again; and by this time *Mauritia* swamp forests invaded part of the lake area. A little later the lake level rose again and a final sharp increase of grass-savanna elements can only be interpreted as signifying the effect of human influence (especially burning). From the data provided by this diagram one may conclude that, during the Holocene, changes in the rates of precipitation (in the annual total, and/or in the seasonal distribution) took place, which resulted in appreciable changes in the ratios of savanna and savanna woodland. A major period of open savanna, apparently of a "drier" type than the present stands, lasted from about 6000 or 5000 B.P. to c. 3800 B.P.

The second diagram of Fig. 2.4 is from Lake Moreiru in the Rupununi savannas of Guyana. Radiocarbon analyses of c. 7300 B.P. and c. 6000 B.P. date the upper part of the diagram. The lower part of the diagram shows approximately equal proportions of savanna woodland and open savanna, followed by a major extension of open savanna coinciding with a very low lake level. When the lake level rose again, the area around the lake was completely invaded by *Byrsonima* woodland, so that virtually no open grass-savanna remained. The age of the lower limit of this woodland period was not established by direct means. If the sedimentation rate between 7300 B.P. and the present is used as a yardstick, the age would be c. 13 000 B.P., but

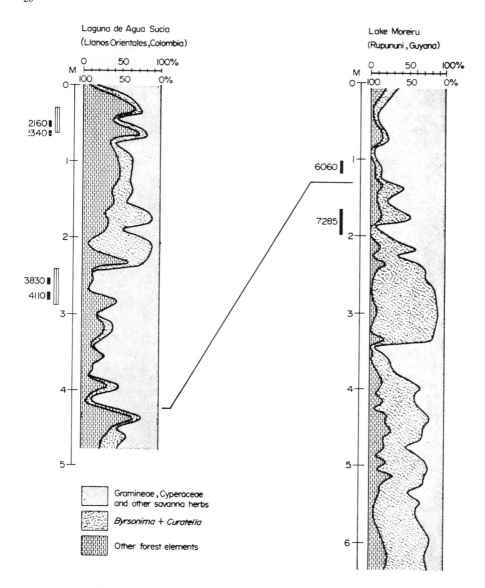

Fig. 2.4. Pollen diagrams from the Late Pleistocene and Holocene, from areas at present covered with savanna vegetation (Llanos Orientales, Colombia; Rupununi, Guyana) (from Van der Hammen, 1974, by permission of Blackwell; adapted from Wijmstra and Van der Hammen, 1966).

the sedimentation rate between c. 7300 and 6000 B.P. yields an extrapolated age of c. 10 000 B.P. There can, however, hardly be any doubt that the sediments below that limit (at c. 340 cm) date from the later part of the Last Glacial period. Towards the end of the above-mentioned savanna woodland period, open savanna increased again to become dominant around 7300 B.P. A subsequent minor increase of *Byrsonima* was in turn followed by an increase of open-savanna elements which began about c. 6000 B.P. Later, an ultimate, slight increase

of trees took place. If the calculated date of c. 13 000 B.P. is correct, this implies that the extreme grass-savanna period associated with very low lake levels immediately preceding it would be of Late Pleni-glacial age, and also that the effective precipitation was low, probably lower than at present, especially when one takes into consideration that the temperature was most probably also lower. The previous period must have been wetter, and the subsequent period, corresponding with the Late Glacial and possibly the Early Holocene, much wetter than the present. This implies that the curve for the effective precipitation corresponding with the lower part of the diagram is apparently in phase with that from Lake Fúquene in the Andes (Van Geel and Van der Hammen, 1973). This seems to be the most likely interpretation of the diagram[1].

There is no doubt that more sections and more radiocarbon datings are badly needed. However, the available data prove the existence of drier and wetter phases in the Holocene and in the Late Pleistocene beyond reasonable doubt and are highly suggestive of the prevalence of very dry conditions in the Late Pleniglacial and of a wet period in the Late Glacial.

Although palynological data on the Last Glacial and Early Holocene of the Colombian and Venezuelan Llanos (Llanos Orientales, Orinóco savannas) are still lacking, the study of the Quaternary geology, geomorphology and soils has yielded indications regarding the past climatic conditions. Aeolian sediments (dunes and loesses) are of common occurrence over large areas (see, for instance Goosen, 1971), which is indicative of climatic conditions different from the present ones. Apparently drier conditions prevailed during the latter part of the Last Glacial, and it is tempting to suppose that this interval corresponds in time with the "Fúquene stadial" dry period falling between c. 21 000 and c. 13 000 B.P. (see above).

Absy (1979) published a diagram of young Holocene sediments from Lago Galheiro in the savannas around Boa Vista, Território Roraima, Brasil. The diagram represents the last 1000 years and shows a continuous dominance of grasses. The sedimentary sequence suggests the presence of two dry phases, approximately around 1250 A.D. and 1500 A.D., respectively. The continuous representation of *Cuphea* pollen, also a characteristic element in the Colombian pollen diagram from Agua Sucia, is rather striking.

A number of pollen diagrams from the Meta Valley and bordering savannas (Colombian Llanos Orientales: Wijmstra and Van der Hammen, 1966) yield data concerning the recent Holocene history of gallery forests and savannas and a possible human influence.

These pollen diagrams suggest that both human activities and changing edaphic factors (inundations, changes of river courses) were important factors determining the local fluctuations of the

extension of the forests and of the open savannas. Open grass-savanna in the partly inundated valley (conceivably caused by felling and burning by man) may first have become invaded by savanna trees (*Byrsonima*), and subsequently replaced by a type of gallery forest. Two diagrams from the Meta Valley show an increase of forest elements and a decrease of open grass-savanna elements in recent time, whilst in the savanna areas proper the open grass type of savanna spread.

With the exclusion of the river valleys and savannas frequently subjected to river inundations, the available pollen-analytical records indicate that a dry forest or a closed-savanna woodland was at one time or another the climax vegetation in many of the extant savanna areas. Human activity was instrumental in the formation of an open grass-savanna out of this forest. Nevertheless, it seems that before human settlement took place natural causes, such as arid periods and lightning, exerted (and presumably locally may still exert) a similar, semi-permanent or temporary effect on this dry forest.

The (qualitative and quantitative) composition of the savanna woodland reflected in the pollen diagrams as high percentages of *Byrsonima* is not known in sufficient detail. It may have consisted of extensive, dense stands of *Byrsonima crassifolia*. It may also have contained a greater number of woody species which were co-dominant but produced very little pollen. Although *Byrsonima* cannot always be identified to the species level on the basis of the pollen morphology alone, it is certain that several species are represented in the pollen spectra. All pollen diagrams strongly suggest a very unstable equilibrium between the *Byrsonima* association and the open herbaceous savanna. This equilibrium was apparently influenced by changes of climate, lightning fires, and, in later time, by man-made fires (Wijmstra and Van der Hammen, 1966).

---

[1] If the date of 10 000 B.P. instead of 13 000 B.P. is accepted, the conclusion stands that the period corresponding in time with the El Abra stadial was dry, and was suceeded and preceded by wetter periods corresponding respectively with the Early Holocene and the Guantiva stadial. This also suggests that these changes may be in phase with those of Fúquene.

The Amazon Basin

Recently some palynological data became available that strongly favour the idea of the former existence of savannas in areas at present covered with tropical forest (Van der Hammen, 1972; Absy and Van der Hammen, 1976). The samples providing the evidence hail from Rondônia (Brazil) in the southern part of the Amazon Basin.

The general diagram presented here (Fig. 2.5) is composed of two sections from the same area. The uppermost part corresponds with 2 m of recent sediments of a river, and the lower part (representing c. 13 m of sediments) is from the bottom of a small valley. The cumulative diagram shows the variation in time of the percentages of three groups of pollen grains: those belonging to trees from the humid tropical forest; those belonging to *Mauritia* (mainly occurring in swamp forest); and those belonging to open grass-savanna (Poaceae, *Cuphea* and a few other herbs).

The lower section shows in its lower part a darker type of humic clay, and the pollen of species found in humid tropical and swamp forest dominates. In the upper part of the lower section the clays are of a lighter colour and show intercalations of reddish and sandy material, apparently representing slope deposits washed down from the sides of the valley during a period of instability. This part of the section shows a complete dominance of open-savanna elements (mainly grasses, but also such herbs as *Cuphea*), and is strikingly similar to the pollen diagram of, for instance, Lago Galheiro (see above).

In the upper section, of recent river valley sediments, elements of the humid tropical forest dominate completely. Here also savanna elements are much more abundant in the lower part.

The Holocene age of the uppermost, recent sediments seems to be certain; their pollen content is entirely in agreement with the present pollen rain. The age of the lower part of these river-valley sediments might still be Holocene or it is at most Late Glacial. Although the exact age of the other sections is not known, it is clear that they represent Plio-Pleistocene sediments from an earlier phase. It has so far not been ascertained whether they represent, for instance, the Last Glacial, but the data presented here show, beyond reasonable doubt, that there were periods during the Plio-Pleistocene when savannas locally replaced parts of the forest belt.

Recent studies by Absy (1979) have shown that, although there was no extension of savannas during the Holocene in the central Amazonian river-plains, there were periods of diminished effective precipitation in the Basin affecting the vegetation cover of this periodically inundated area (manifest from a local extension of floating meadows, etc.). The drier phases have been dated and rather closely correspond in age with those of the Llanos Orientales and the Lower Magdalena Cauca (*viz.*, around 4000 B.P., between c. 2700 and 2000 B.P. culminating around 2100 B.P., at c. 500 A.D., and at c. 1250 B.P.). It is not known what effect these drier phases had on *terra firme* vegetation; possibly the northern and southern boundary lines between the Amazonian forest and the savannas (or campos cerrados respectively) may repeatedly have shifted, but so far any confirmatory or negative historical data are wanting.

The data from Rondônia mentioned before are clearly indicative of rather extensive replacements of forest by savanna having taken place in the past in an area lying between postulated forest refugia (see also p. 21). This pleads in favour of the refugia theory, irrespective of the exact dating of the deposits in question (Absy, 1979; Van der Hammen, 1979). Much more palynological evidence gathered from the whole Amazon Basin is

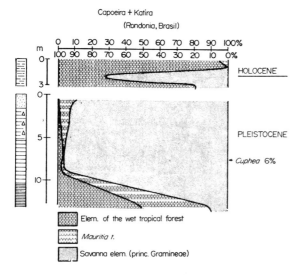

Fig. 2.5. Pollen diagrams from Capoeira and Katira (Rondônia, Brazil) (from Van der Hammen, 1974, by permission of Blackwell).

required, however, before the theory is confirmed or disproved in greater detail.

The Paleotropics

Africa

Many palaeoecological studies in Africa have been published or summarized in Van Zinderen Bakker's (1966–1979) series *Palaeoecology of Africa*, and summaries were also given by Kendall (1969), Hamilton (1973), and Livingstone (1975). The most recent summary is by Livingstone and Van der Hammen (1978).

From studies on the East African mountains it has become clear that the depression of the altitudinal forest limit during the Last Glacial must have been of the same order of magnitude as that in northern South America, which corresponds with an estimated lower mean annual temperature of between 5° and 8°C below the present figure. There are indications that the climate was also drier in the East African mountains during the Late Pleniglacial. Little is known from this area as regards the time before 33 000 B.P., but in view of the fact that the climatic sequence between 33 000 and the present closely resembles the Andean sequence (and partly also these of the northern temperate latitudes), there seems to be no reason to doubt that the area was subjected to a sequence of glacials and interglacials similar to that found elsewhere.

As mentioned before, grass-savannas may have existed in Africa from the Eocene onward and it has been established that major changes in the position of the forest/savanna boundary took place in the Neogene of Nigeria. The Tertiary age of savannas is also testified by the abundant fossil vertebrate faunas. A co-evolution of the large herbivores and (grass-) savannas is a reasonable assumption.

The fossil vertebrate faunas from the East African Quaternary sequences of sediments which locally yielded the famous fossils of early hominids (e.g., Bishop and Clark, 1967) indicate that savannas must have constituted an important part of the vegetation cover in that area during the Quaternary (i.e., during the last 2 to 3 m.y.). Bonnefille (e.g., 1972) published data on isolated pollen spectra from a number of "beds" from the Quaternary sequence in Ethiopia, from deposits in the Omo Valley between 2 and 3 m.y. old, and from a number from the last 1 to 1.5 m.y. collected

in the Awash Valley. Although it is difficult to draw more definite conclusions from these data, changes of climate are evident from the fluctuations in the proportion of arboreal and non-arboreal pollen. The changes in the lower parts of these areas may be explained in terms of increasing herbaceous or forest vegetation, and of fluctuations in the amount of effective precipitation, some intervals being drier than the present and others more humid.

As to the more recent palaeogeography of savannas in East Africa, the most informative pollen diagrams are those from Lake Victoria (at an altitude of 1100 m: Kendall, 1969).The best core of lake sediments is more than 17 m long and was radiocarbon-dated in 28 places so that apart from the normal percentage calculations (of which a diagrammatic representation is shown in Fig. 2.6), the number of pollen grains deposited per square centimetre annually could be calculated for all samples analyzed palynologically. The diagram covers the last 15 000 years. The vegetation stands prevailing from >15 000 to 12 500 B.P. produced mainly grass pollen, and the presence of a few pollen grains of Mimosaceae is highly suggestive of the erstwhile presence of a savanna at that site. Forest trees became much more abundant after 12 500 B.P., especially Oleaceae which appeared first (*Olea* being a pioneer). After a temporary decline of forest elements around 10 000 B.P., the forest expanded more extensively (pollen of Moraceae becoming very abundant in addition to that of many other taxa). This forest appears to have been evergreen, but at about 6000 B.P. it was replaced by a more seasonal, and drier, type of forest. Since 3000 B.P. this forest became gradually replaced by arable land.

The savanna period before 12 500 coincided with a very low lake level. Pollen of Cyperaceae and *Typha* became more abundant near the boring site, and there seems to be a hiatus in the sedimentation. The high concentration of calcium, magnesium and strontium in these sediments is explained by the fact that the water level became so low that Lake Victoria had no outlet at that time.

There is a considerable amount of data relating to East African lake levels (see, for instance, the summary in Livingstone and Van der Hammen, 1978), and most of them are indicative of low lake levels before *c*. 12 500 B.P. This period of low rainfall began at least 20 000 years ago. Before that time

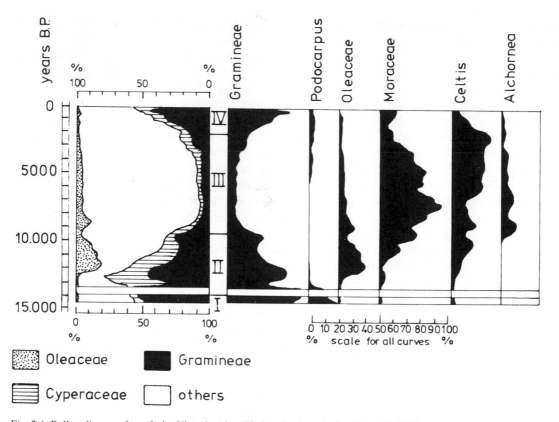

Fig. 2.6. Pollen diagram from Lake Victoria (simplified and adapted after Kendall, 1969).

there is evidence of alternatingly dry and more humid climates from c. 60 000 B.P. onwards. The driest period may have coincided with the period between the attainment of maximum glaciation and the beginning of the Late Glacial, when most lake levels started to rise again.

The past 12 000 years have been relatively humid, with intermittent short and dry phases at c. 10 000 and c. 6000 B.P. The period of greatest humidity most probably fell between 8000 and 7000 B.P., and the past 3000 years have (at least intermittently) been the driest during the last 12 000 years (Livingstone and Van der Hammen, 1978).

Other evidence from Central Africa, obtained from the study of soils and of the geomorphology (De Ploey, 1963, 1965), equally points to the prevalence of a dry climate in tropical Africa prior to c. 12 500 B.P. There can hardly be any doubt that this period of lower effective precipitation resulted in a considerable extension of the savannas (and possibly also of the dry forest vegetation), so that the parts under tropical rain forest decreased in area at that time.

As regards the effects of the changes in the amount of precipitation, the vast area south of the Tropic of Cancer is of special importance. Faure (1969) gave an excellent review covering the last 10 000 to 20 000 years in the Sahara area briefly summarized by Kendall (1969) as follows: "From Mauretania in the west to the plains south of Khartoum, south of the Tropic of Cancer, now *Acacia*-steppe or desert, was dotted in the past with lakes and active rivers. Several deposits left by these bodies of water have been dated; most of the dates are younger than 10 000 B.P. and tend to fall between 5000 and 8000 B.P. Most of the lakes had dried up by 3000 B.P., although the ones near Khartoum had apparently vanished by 7000 B.P., and the last high lake stage in the Chad Basin extended to at least 2450 B.P. This much of the record resembles the East African sequence". Diatomites in a caldera lake of the Tibesti were dated at c. 15 000 B.P. (upper part of the sequence), and may indicate that locally "pluvial" conditions already existed in the Sahara much earlier. The Chad Lake level was high c. 21 000 years ago, but subsequently dry conditions (evident from dune formation) apparently prevailed roughly between c. 20 000 and c. 10 000 B.P.

It will be clear that, under the influence of these changes in the amounts of effective precipitation, considerable vegetational changes took place in the "ecotone" between savanna and desert. During the wet phases the savanna moved northwards towards and penetrated into the present desert region to retract to the south again during the drier phases (see for instance, Quézel and Martinez, 1962).

South of the equator, namely from the wooded grasslands of Zambia, there is a 22 000 year old record from Ishiba Ngandu (Livingstone, 1971), but this record is hardly indicative of an appreciable change. This is conceivably attributable to the fact that the woody species produce very small amounts of pollen (as compared to Poaceae). However, the vegetation cover in this area must have contained a grass-dominated stratum all the time, and the forests were apparently never very dense.

Southeastern Asia

A single section from Taiwan representing 50 000 years of history (Tsukada, 1966), collected at an elevation of c. 750 m indicates that between 35 000 and 12 000 years B.P. the average annual temperatures may have been 2° to 6°C lower than they are today. During the earlier period (the Early Pleniglacial) the temperature may, according to Tsukada, have been 8° to 11°C lower than today. These data, in conjunction with records from New Guinea (see below), show that also in this part of the world temperature changes took place, which were, generally speaking, apparently contemporaneous with the last glacial–interglacial periods recorded elsewhere.

Palynological data from grass-dominated tropical vegetation in this part of the world are lacking. Vishnu-Mittre (1969) commented on some general aspects of the Indian flora, and pointed out that noticeable historical changes of this flora took the form of a gradual recession of tropical evergreen forests, the relics of which persist in the western Ghats and Assam, and their replacement by semievergreen forest and ultimately by deciduous forest and savanna vegetation. This process must have taken place after the Eocene.

At the northern limit of the tropical zone there are some data on more recent climatic changes from north-western India (Rajasthan: see, for instance Singh et al., 1972; and Singh 1977). The summary given by Livingstone and Van der Hammen (1978) is followed here. Pollen analysis, stratigraphy and radiocarbon dating revealed considerable changes in the environment, which is now semi-arid to arid. Before 10 000 B.P. there were severe arid environments and the now stabilized sand dunes were active. About 10 000 B.P. a deposition of lacustrine sediments began; the vegetation cover consisted of an open "steppe" rich in grasses, *Artemisia* and sedges, and poor in halophytes. Such species as *Typha angustifolia* and *Mimosa rubicaulis* (which now grow in areas with a higher rainfall) appear to have flourished in the semi-arid belt at that time. *Artemisia* and *Typha* even grew in what is now the arid belt. This suggests that a general westward shift of the rainfall belts took place. Between 5000 B.P. and 3000 B.P. an increase of rainfall is suggested by the increase of swamp vegetation elements, the increasing density of the vegetation cover inland, and the maximal pollen percentages of all mesophytic elements. Between 3800 B.P. and 3500 B.P. there was a short, relatively drier interval (correlated with the decline of the Indus Valley culture), and 3000 B.P. is the approximate date for the onset of aridity, which seems to have become widespread. By about 300 A.D. the climate ameliorated, which phase has lasted to the present. This type of climatic fluctuation, resembling that recorded from the zone immediately to the south of the Tropic of Cancer in Africa, may have been widespread in northern India, and have had a considerable influence on the local vegetation in general.

New Guinea and Australia

Much palynological and palaeoecological work has been carried out in this area during recent years by Walker and collaborators, and was summarized by Bowler et al. (1976). From the study of deposits from mountains in New Guinea it is known that a considerable lowering of the tree-line took place 25 000 to 15 000 years ago. The mean annual temperatures at that time must have been at least some 6°C lower than the present ones. Here again there is evidence of a considerable glacial lowering of the temperature in montane regions in the tropics.

From tropical Australia (northeastern Queensland), data relating to the history of vegetation and climate were gleaned from a pollen diagram from Lynch's Crater, representing the last 120 000 years (Kershaw, 1976, 1978). In Fig. 2.7 a simplified version of the diagram covering only the last 60 000 years is shown. The present site is within the rain forest area, but not too far from the border between this area and the sclerophyllous woodland. From c. 120 000 B.P. to c. 80 000 B.P. (including at least the last interglacial) vine forest dominated (with a calculated rainfall of about 3000 mm) gradually passing into araucarian vine forest (from c. 80 000 to c. 25 000 B.P.), and finally changing into sclero-

phyllous woodland (c. 25 000 to c. 10 000 B.P.), with a rainfall slightly above 500 mm. After 10 000 B.P. the rain forest re-established itself, after an estimated increase of the rainfall to over 3000 mm. These data show how important and how far-reaching such changes of precipitation were in this part of the tropics also.

SOME GENERAL CONCLUSIONS

Although the data presented here are still rather scanty, they all point to the occurrence in the past of similar climatic sequences in different parts of the tropics. Tropical highlands were affected by a considerable lowering of the annual temperature (by 5° to 8°C) during glacial periods. Although it is not known how much lower the temperatures were in the tropical lowlands, they presumably did not fluctuate to such an extent as they did in the highlands.

The changes in the amount of effective precipitation had nevertheless the greatest influence upon the type of formation covering the lowlands. Periods of relatively low rainfall apparently led to a shift of the savanna rain-forest boundary towards the equator, and may even have led to a more substantial reduction of the equatorial rain forest to isolated forest refugia, and to a concomitant local penetration of savannas into what is now the present rain-forest area. Although any direct evidence is wanting, it may be assumed that periods of high rainfall in the area of the present savanna/rain forest boundary led to a shift of this boundary away from the equator, and possibly savannas became reduced to smaller and partly isolated areas (savanna refugia). Similar movements resulting from changes in the amount of rainfall (effective precipitation) must also have taken place at the boundary between savanna and desert. This model is still partly conjectural, but there is an increasing amount of evidence pleading in its favour. It provides an explanation of many biogeographical problems, distribution patterns and centres of diversification and endemism, both as regards more hygrophilous and more xerophilous elements.

The climatic changes that took place within the present savanna areas must have resulted in vegetational changes ranging from the establishment of extreme, open grass-savannas without any forest or

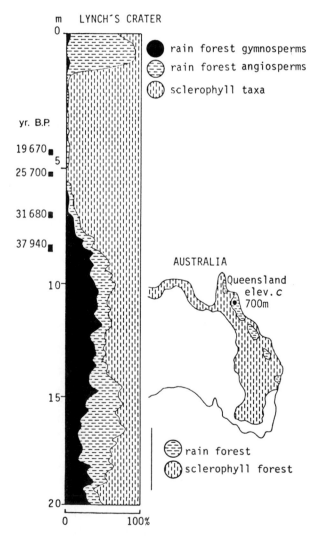

Fig. 2.7. Pollen diagram from Lynch's Crater, northeastern Queensland, Australia (simplified, after Kershaw, 1976).

scrub elements, to that of a closed savanna woodland or dry forest (nearer to the limits of the savanna belt, rain forests or semi-desert or desert vegetation may have occurred as pointed out above).

The most extreme situation as regards low amounts of rainfall in the tropics prevailed during the later part of the Last Glacial period (the Late Pleniglacial), that is, between about 25 000 or 20 000 B.P. and 12 500 B.P. This resulted in a considerable extension of the savannas towards the equator, and to a reduction of savanna woodland in the areas now covered with savannas (and locally to the formation of dunes and loess), while the limit of savanna and desert also seems to have moved towards the equator (at least in some places).

The Late Glacial is represented in the tropics by a period of increased rainfall (with some fluctuations), the forests gaining again upon the savannas in the equatorial area, and the savannas moving northward into the semi-desert zone. In the area of present savannas the proportion of savanna woodland locally may have increased considerably.

In the Holocene (from 10 000 B.P. onward) the period with a higher rainfall continued, and a maximum extension of forest (and a maximum extension of savanna in the boundary area with the desert) was attained. In the later part of the Holocene, sometime between 6000 B.P. and 3000 B.P., the climate became drier again, which led to the present situation. During the Holocene a number of minor but sometimes rather marked changes of effective precipitation affected lake levels and local stands of vegetation. This generalization is based on relatively few data; some of the trends indicated have only been established in one or a few places. Even if the generalization holds for the tropics as a whole, one must expect to find that there were alternative local developments and local deviations from the general scheme. Much more palynological and palaeoecological work will need to be carried out in the tropics before the local or regional validity of this scheme can be convincingly established.

Nevertheless, the recorded data and trends throw an altogether new light on the dynamics of tropical vegetation in general and of savanna vegetation in particular. Even over periods of a few centuries or even decades, there seems to be but little stability. Savanna ecosystems seem to react rapidly to every (even to a minor) change of effective precipitation (caused directly by a change in the amount of annual precipitation, by a change in the average annual temperature or by an increased or decreased seasonality), and to adjust themselves to the new circumstances by the incorporation or elimination of woody elements, the closing or opening of the herb stratum, the displacement of herds of herbivores, etc. As changes seem to occur continuously in savanna ecosystems, the latter, or some of their components in the course of their evolution seem to have developed adaptations to extreme changes (such as fire-resistance, specialized root systems, etc.). In this connexion it is important to bear in mind that the evolution of savanna ecosystems and their constitutent organisms may have started already in the Eocene and Oligocene, and continued during Miocene and Pliocene times when considerable changes took place already. The ultimate, extreme changes of the Quaternary extended over a period of about 3 m.y.

Fire has probably always been a natural, atmospheric factor in savanna regions. Man, however, increased the fire hazard tremendously, thus strongly reducing the woody component in many savannas and (together with the felling of trees) causing an extension of the open types of savanna to areas previously covered with scrub, woodland or forest.

REFERENCES

Ab'Saber, A.N., 1967. Problemas geomorfológicos da Amazônia Brasileira. In: *Atas do Sympósio sobre a Biota Amazônica, 1 (Geologia)*, pp. 35–67.

Ab'Saber, A.N., 1977. *Espaços ocupados pela espansão dos climas secos na América do Sul, por ocasião dos periodos glaciais Quaternários*. USP/IGEOG, São Paulo. 19 pp. (Paleoclimas 3).

Absy, M.L., 1979. *A Palynological Study of Holocene Sediments in the Amazon Basin*. Thesis, University of Amsterdam, Amsterdam, 105 pp.

Absy, M.L. and Van der Hammen, T., 1976. Some palaeoecological data from Rondônia, southern part of the Amazon basin. *Acta Amazônica*, 6: 293–299.

Bishop, W.W. and Clark J.D. (Editors), 1967. *Background to Evolution in Africa*. University of Chicago Press, Chicago. Ill., 935 pp.

Bonnefille, R., 1972. *Associations polliniques actuelles et quaternaires en Éthiopie (valées de l'Awash et de l'Omo)*. Thesis, Université de Paris VI, Paris (C.N.R.S. No. A07229) 513 pp.

Bonnefille, R., 1976. Implications of pollen assemblage from the Koobi Fora formation, East Rudolf, Kenya. *Nature* 264: 403–407.

Bowler, J.M., Hope, G.S., Jennings, J.N., Singh, G. and Walker, D., 1976. Late Quaternary climates of Australia and New Guinea. *Quaternary Res.*, 6: 359–394.

Brown Jr, K.S., 1976. Geographical patterns of evolution in Neotropical Lepidoptera. Systematics and derivation of known and new Heliconiini (Nymphalidae: Nymphalinae). *J. Entomol.*, B 44: 201–242.

CLIMAP Project Members, 1976. The surface of the ice-age earth. *Science* 191: 1131–1137.

De Ploey, J., 1963. Quelques indices sur l'évolution morphologique et paléoclimatique des environs du Stanley Pool (Congo). *Studia Univ. Lovanium, Fac. Sci.*, 13: 1–16.

De Ploey, J., 1965. Position géomorphologique, genèse et chronologie de certains dépots superficiels au Congo occidental. *Quaternaria*, 7: 131–154.

Faure, H., 1969. Les lacs quaternaires du Sahara. *Mitt. Int. Verein. Limnol.*, 17: 131–146.

Gates, W.L., 1976. Modeling the ice-age climate. *Science*, 191: 1138–1144.

Germeraad, J.H., Hopping, C.A. and Muller, J., 1968. Palynology of Tertiary sediments from tropical areas. *Rev. Palaeobot. Palynol.*, 6: 189–348.

Gonzalez Guzman, A.E., 1967. *A Palynological Study on the Upper Los Cuervos and Mirador Formations (Lower and Middle Eocene; Tibú Area, Colombia)*. Thesis, University of Amsterdam, Amsterdam; Brill, Leiden, 68 pp.

Goosen, D., 1971. Physiography and soils of the Llanos Orientales, Colombia. *Publ. Fys. Geogr. Bodemkd. Lab. Univ. Amsterdam*, 20: 198 pp.

Haffer, J., 1969. Speciation in Amazonian forest birds. *Science*, 165: 131–137.

Haffer, J., 1977. Pleistocene speciation in Amazonian birds. *Amazoniana*, 6: 161–192.

Hamilton, A.C., 1973. The history of vegetation. In: E.M. Lind and M.E.S. Morrison (Editors), *The Vegetation of East Africa*. Longman, London, pp. 188–209.

Herngreen, G.F.W., 1974. Middle Cretaceous palynomorphs from north-eastern Brazil. *Sci. Géol. Bull. (Strasbourg)*, 27: 101–116.

Huber, O., 1982. The ecological and phytogeographical significance of the actual savanna vegetation in the Amazon territory of Venezuela. In: G.T. Prance (Editor), *The Biological Model of Diversification in the Tropics*. Columbia University Press, New York, N.Y., in press.

Journaux, A., 1975. Recherche géomorphologiques en Amazonie brésilienne. *Bull. Centre Géomorphol. Caen (C.N.R.S.)* 20: 1–67.

Kendall, R.L., 1969. An ecological history of Lake Victoria basin. *Ecol. Monogr.*, 39: 121–176.

Kershaw, A.P., 1976. A Late Pleistocene and Holocene pollen diagram from Lynch's Crater, north-eastern Queensland. Australia. *New Phytol.*, 469–498.

Kershaw, A.P., 1978. Record of last interglacial–glacial cycle from north-eastern Queensland. *Nature*, 272: 159–161.

Laeyendecker-Roosenburg, D.M., 1966. A palynological investigation of some archeologically interesting sections in north-western Surinam. *Leidse Geol. Meded.*, 38: 31–36.

Livingstone, D.A., 1971. A 22,000-year pollen record from the plateau of Zambia. *Limnol. Oceanogr.*, 16: 349–356.

Livingstone, D.A., 1975. Late Quaternary climatic change in Africa. *Annu. Rev. Ecol. System.*, 6: 249–280.

Livingstone, D.A. and Van der Hammen, T., 1978. Palaeogeography and palaeoclimatology. In: *Tropical Forest Ecosystems; a State-of-Knowledge Report Prepared by UNESCO/UNEP/FAO*. UNESCO, Paris, pp. 61–90.

Peterson, G.M., Webb, T., Kutzbach, J.E., Van der Hammen, T., Wijmstra, T.A. and Street, F.A., 1979. The continental record of environmental conditions at 18,000 B.P.: an initial evaluation. *Quaternary Res.*, 12: 47–82.

Prance, G.T., 1973. Phytogeographic support for the theory of Pleistocene forest refugia in the Amazon basin based on evidence from distribution patterns in Caryocaraceae, Chrysobalanaceae, Dichapetalaceae and Lecythidaceae. *Acta Amazônica*, 3: 5–26.

Prance, G.T., 1978. The origin and evolution of the Amazon flora. *Interciencia*, 3: 207–222.

Quézel, P. and Martinez, M.C., 1962. Premiers résultats de l'analyse palynologique de sédiments recueillis au Sahara méridional a l'occasion de la mission Berliet-Tchad. In: *Missions Berliet Ténéré-Tchad*. Paris, pp. 313–327.

Shackleton, N.J. and Opdyke, N.D., 1977. Oxygen isotope and palaeomagnetic evidence for early northern hemisphere glaciation. *Nature*, 270: 216–219.

Singh, G., 1977. Climatic changes in the Indian desert. In: *Desertification and Its Control*. Indian Council of Agricultural Research, New Delhi, pp. 25–30.

Singh, G., Joshi, R.D. and Singh, A.B., 1972. Stratigraphic and radiocarbon evidence for the age and development of three salt lake deposits in Rajasthan, India. *Quaternary Res.*, 2: 496–505.

Tsukada, M., 1966. Late Pleistocene vegetation and climate in Taiwan (Formosa). *Proc. Natl. Acad. Sci. U.S.A.*, 55: 543–548.

Turekian, K.K. (Editor), 1971. *The Late Cenozoic Glacial Ages*. Yale University Press, New Haven, Conn., 606 pp.

Van der Hammen, T., 1963. A palynological study on the Quaternary of British Guaina. *Leidse Geol. Meded.*, 29: 125–180.

Van der Hammen, T., 1972. Changes in vegetation and climate in the Amazon Basin and surrounding areas during the Pleistocene. *Geol. Mijnb.*, 51: 641–643.

Van der Hammen, T., 1974. The Pleistocene changes of vegetation and climate in tropical South America. *J. Biogeogr.*, 1: 3–26.

Van der Hammen, T., 1982. Paleoecology of tropical South America. In: G.T. Prance (Editor), *The Biological Model of Diversification in the Tropics*. Columbia University Press, New York, N.Y., in press.

Van der Hammen, T. and Wijmstra, T.A., 1964. A palynological study on the Tertiary and Upper Cretaceous of British Guiana. *Leidse Geol. Meded.*, 30: 183–241.

Van der Hammen, T., Wijmstra, T.A. and Zagwijn, W.H., 1971. The floral record of the late Cenozoic of Europe. In: K.K. Turekian (Editor), *The Late Cenozoic Glacial Ages*. Yale University Press, New Haven, Conn., pp. 391–424.

Van Geel, B. and Van der Hammen, T., 1973. Upper Quaternary Vegetational and climatic sequence of the Fuquene area

(Eastern Cordillera), Colombia. *Palaeogeogr., Palaeoclimatol., Palaeoecol.*, 14: 9–92.

Van Zinderen Bakker, E.M., 1966–1979. *Palaeoecology of Africa*. Balkema, Cape Town, Rotterdam. Vols. 1–10.

Vanzolini, P.E., 1970. Zoología sistemática, geografia e a origem das espécies. *Inst. Geogr. Univ. São Paulo, Teses Monogr.*, 3: 56 pp.

Vanzolini, P., 1973. Paleoclimates, relief and species multiplication in equatorial forest. in: B.J. Meggers et al., *Tropical Forest Ecosystems in Africa and South America. A Comparative Review*. Smithsonian Press, Washington, D.C., pp. 225–258.

Vishnu-Mittre, 1969. Some evolutionary aspects of Indian flora. In: *J. Sen Memorial Volume*. Calcutta, pp. 385–394.

Wijmstra, T.A., 1967. A pollen diagram from the Upper Holocene of the lower Magdalena valley. *Leidse Geol. Meded.*, 39: 261–267.

Wijmstra, T.A., 1969. Palynology of the Alliance Well. *Geol. Mijnb.*, 48: 125–133.

Wijmstra, T.A., 1971. *The Palynology of thr Guiana Coastal Basin*. Thesis, University of Amsterdam; De Kempenaer, Oegstgeest, 62 pp.

Wijmstra, T.A. and Van der Hammen, T., 1966. Palynological data on the history of tropical savannas in northern South America. *Leidse Geol. Meded.*, 38: 71–90.

Chapter 3

CLIMATE OF TROPICAL SAVANNAS

HENRY A. NIX

INTRODUCTION

Problems of definition beset both the terms "tropical" and "savanna". It is not intended here to debate terminology, but some frame of reference is necessary before attempting to review the climates of tropical savanna ecosystems. The broad definition and description used by Bourlière and Hadley (1970) is accepted: "*a tropical formation where the grass stratum is continuous and important, but occasionally interrupted by trees and shrubs; the stratum is burnt from time to time and the main growth patterns are closely associated with alternating wet and dry seasons*". Consensus exists that major development of tropical savannas occurs in the seasonal wet-dry zones sandwiched between the humid equatorial zones and the arid zones in the mid-latitudes — that is, between about 10° and 30° north and south of the equator.

The usage of "tropical" *sensu strictu* refers to the geographic zone lying between the Tropics of Cancer and Capricorn (23.5°N to 23.5°S). But living organisms and vegetation formations do not recognize astronomically defined boundaries! Biologists, ecologists and climatologists prefer thermally defined boundaries for the tropical zone. Thus, Köppen (1931) differentiated his tropical (*A*) climates on the basis of mean temperature of the coldest month exceeding 18°C. A seemingly more meaningful boundary condition for living organisms is the frost-free limit, that is, the absence of freezing temperatures at any time. In truth, no single climatic boundary can adequately discriminate components of vegetation continua. The approach adopted in this chapter is to provide graphic illustration of *gradients* in selected climatic

parameters rather than to choose any particular boundary condition.

Usage of the term "savanna" has been undisciplined. Even where explicit and quantitative, definitions tend to be specific to continents and regions. Further confusion is added by argument as to the ultimate origins of savannas — whether they be natural or man-induced. The position adopted here is that, whether the tropical savanna ecosystem be natural, man-modified or man-made, climate and weather exert an overwhelming influence on its structure, processes and function. Ideally, what is required is a functional analysis, and synthesis, of the role of climate and weather in tropical savannas, but limits in ecophysiological understanding and in the available data base constrain this objective. However, careful selection of climatic parameters can illumine the more important controls on distribution and productivity. This chapter, then, aims at providing a brief conspectus of the general climatic characteristics of tropical savanna regions, the major climatic controls, spatial and temporal variation in major climatic elements, climatic classification and the use of climatic data in dynamic simulation of tropical savanna ecosystems.

GENERAL CLIMATIC CHARACTERISTICS OF TROPICAL SAVANNA REGIONS

As defined (cf. Bourlière and Hadley, 1970) savanna ecosystems are differentiated on the basis of vegetation structure. Similar structural formations may characterize a diverse array of climate, terrain, soil and biotic interactions. Although such interactions are extremely important in determining vegetation pattern at mesoscale, ultimate limits to vegetation development are set by macro-scale climate. There is general agreement that potential savanna environments are most extensive in the intermediate or transitional zone between the humid tropics and the dry mid-latitudes, with alternating wet and dry seasons.

The major weather systems of the tropical regions form part only of the global atmospheric circulation, but an important part. This is becoming clearer with a growing understanding of the dynamics of the total system (Sellers, 1969; Manabe, 1971; Wallace, 1971) and of the role of the tropical circulation within it (Riehl, 1969; Bates, 1972). More recently, developments in remote-sensing capability and operation of geostationary satellites are providing much-needed data for the development and testing of hypotheses concerning the general circulation and the mechanisms that drive it. For present purposes, it is sufficient to provide a brief, general, description of general climatic controls.

The climate of tropical regions is the outcome of a series of mechanisms that maintain the global energy balance through poleward transport of energy. Latent heat energy gained by the trade winds passing over the heated surface of the tropical oceans is released by convective mechanisms in the intertropical convergence zone (ITCZ) where the trade winds of summer and winter hemispheres converge. Poleward transport of this energy in the troposphere completes the meridional circulation known as the Hadley cell. Stable, subsiding air masses give rise to the high-pressure anticyclonic belts of the middle latitudes.

Very simply, the climates of tropical savanna regions are the outcome of sun-controlled shifts in the seasonal position of the sub-tropical high-pressure belts and the intertropical convergence zone. Tropical easterlies (including the trade winds) of low sun periods alternate with winds which have predominantly westerly components (equatorial westerlies or monsoon winds) during high sun periods. This maximum rainfall tends to occur at high sun, and minimum or total absence of rainfall at low sun. The relative length of wet and dry seasons is a function of proximity to the primary air-mass regions, whether this be the stable, dry air masses of the mid-latitude belt or the humid, unstable air masses of the ITCZ.

Although the equatorial westerlies or monsoon winds of the high sun period transport enormous quantities of evaporated moisture landwards from the tropical oceans, the rather homogenous air masses exhibit relatively weak atmospheric and temperature gradients. Accordingly, convective mechanisms play an important part in triggering rainfall. Typically, tropical savanna regions experience between 20 and 100 days per year with thunderstorms. Lightning strikes during such storms, particularly during the late dry season, contributed to natural fire regimes before the advent of more regular seasonal firing by man.

More organized weather systems such as weak cyclonic disturbances and depressions may characterize the main wet season. Major cyclonic depressions (cyclones, hurricanes, typhoons) are a hazard, but also may contribute significantly to wet-season rainfall and episodic recharge of drainage systems and aquifers in some tropical savanna regions. Areas where cyclones are important components of seasonal weather patterns are northern Australia, parts of peninsular India, Madagascar and the Caribbean.

Major climatic elements singled out for particular attention in discussion of the climates of tropical savanna ecosystems are solar radiation, temperature, evaporation and precipitation. This selection is due in part to data availability on a global basis, but also because of their direct use in specifying boundary conditions, in erection of classifications, in the calculation of energy and water balances and as input to simulations of plant response to the climatic environment.

Solar radiation

The global network of recording instruments for total solar radiation remains inadequate and imbalanced, but steady improvement has been made since the very comprehensive tabulations made by Löf et al. (1966). Continued development of empirical equations (cf. Ångström, 1924) for estimation of total solar radiation from sunshine-hour data has permitted more detailed mapping and analysis of spatial variations in many tropical regions. At a generalized global scale, the maps presented by Budyko (1974) are sufficient for the purpose of describing total annual values and the pattern of seasonal variation in tropical savanna environments.

Although one might expect highest radiation

totals in the tropical zone, peak values are attained at the edge of the arid tropics, mainly between 20 and 30° lat. on major land masses north and south of the equator. Clear skies and relatively high sun angles throughout the year yield annual totals in excess of 200 kcal cm^{-2} yr^{-1} ($\cong 9$ GJ m^{-2} yr^{-1}). Equatorwards there is a decline towards the humid zones, where cloud cover and suspended aerosols (mainly water vapour) reduce radiation receipts to values that can be less than 120 kcal cm^{-2} yr^{-1} ($\cong 5$ GJ m^{-2} yr^{-1}). The most extensive areas of tropical savannas are sandwiched between these two extremes and typically have annual totals in the range 140 to 190 kcal cm^{-2} yr^{-1} ($\cong 6$–8 GJ m^{-2} yr^{-1}). Tropical savanna zones in northern Australia, India and Africa south of the equator have somewhat higher totals at 160 to 190 kcal cm^{-2} yr^{-1} ($\cong 7$–8 GJ m^{-2} yr^{-1}) than do those in Brazil and Africa north of the equator where totals are 140 to 180 kcal cm^{-2} yr^{-1} ($\cong 6$–7.5 GJ m^{-2} yr^{-1}).

Latitudinal position dictates that seasonal amplitude in solar radiation should not be large. Greatest contrasts are imposed by the alternating wet and dry seasons. Seasonal variation is clearly related to rainfall patterns. During wet months, daily solar radiation receipts everywhere fall within a range of 13 to 17 MJ m^{-2} day^{-1}. In this, they parallel daily values in the humid tropics. During dry months and, in some cases, dry spells between rainfalls, daily values rise to 20 to 25 MJ m^{-2} day^{-1}. In this, they parallel daily values in the arid tropics. Commonly, in tropical savanna environments highest radiation receipts occur in the late dry season. Conversion of a higher proportion of this radiation to sensible heat, at this time, produces very high temperatures.

A global classification of solar radiation (Terjung, 1970) uses wave height (maximum input), wave fluctuation (amplitude between highest and lowest input), wavelength (for latitudes beyond 66.5°) and shape of curve (phase angle) to characterize distinctive seasonal radiation environments. Greatest development of tropical savannas occurs in the G radiation climates which have a high to very high radiation input and a low to very low seasonal amplitude. The G3 pattern predominates (peak radiation in spring, preceding zenith) but small areas of G2 pattern (bimodal peak) occur in West Africa and equatorial East Africa, in the Orinoco savanna of Venezuela, and much of the *cerrado* in Brazil. The G4 pattern (autumnal peak, or post-zenith) characterizes much of the Amazonian Basin, but also includes part of the cerrado region. Generally poleward of the G pattern are B patterns that characterize tropical deserts with extremely high input and low to very low seasonal amplitude. Savanna ecosystems in northern Australia, Burma, northern India, Africa south of the equator and Madagascar are characterized by a B3 pattern (vernal peak, preceding zenith). This accords with earlier statements that these areas of tropical savanna have higher radiation inputs. Finally, some tropical savannas occur in special edaphic or landscape situations in the more humid environments characterized by F patterns (vernal peak in northern Amazonia, Orinoco, East Africa, eastern islands of Indonesia and the Trans-Fly area of New Guinea; bimodal pattern in Matto Grosso, southern Zaire and coastal Nigeria).

Measurements and improved estimates of solar radiation provide essential input for calculations of photosynthesis, now used extensively in models of plant growth and biomass yields. In addition, radiation-based estimates of potential evaporation (e.g., Penman, 1963; Priestley and Taylor, 1972) are preferable to the more empirical methods based solely on temperature and humidity (e.g. Thornthwaite, 1948; Papadakis, 1961).

Precipitation

Annual rainfall totals may provide a useful frame of reference at regional and local scales, but add little to understanding of tropical savanna ecosystems at the global scale. Seasonality of rainfall pattern, effective rainfall during the growing season and the duration and severity of the dry season or seasons are of greater importance. Mean monthly rainfall data for some 4000 stations have been used in computation of selected attributes of the rainfall pattern that can be used to characterize tropical savanna environments.

In terms of mean annual rainfall, greatest development of tropical savanna ecosystems occurs in zones with between 1000 and 1500 mm (Fig. 3.1). However, local terrain, soil, groundwater and biotic influences (not to mention human disturbance) can modify macroclimate, so that savanna ecosys-

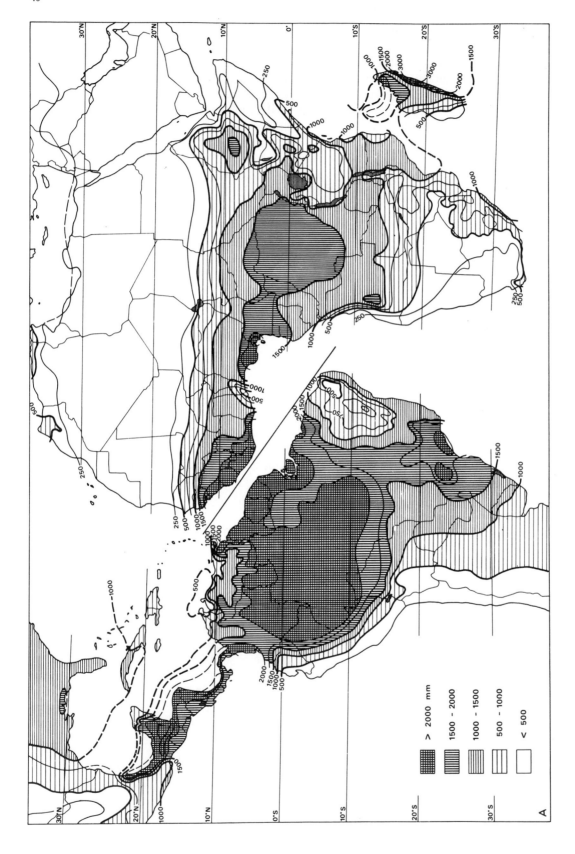

Fig. 3.1. Mean annual rainfall (mm) in the tropical savannas.

tems occur well into adjacent zones that would otherwise be too humid or too arid. Thus, towards the distal margins of the tropical zone, in slightly cooler uplands and inland from eastern coasts where rainfall may be less strongly seasonal, tropical savannas occur with annual rainfall values down to 500 mm and occasionally less. On shallow hard-pan soils and extremely porous soils, tropical savannas can occur in areas with greater than 3000 mm annual rainfall, providing that a short dry season is present. Extreme variants of tropical savannas occur on grossly nutrient-deficient sands in areas with no dry season and more than 3000 mm annual rainfall.

Seasonality of rainfall distribution is a marked feature of most tropical savanna environments. A useful comparative index of seasonality is the coefficient of variation (c.v., standard deviation of monthly rainfall expressed as a percentage of monthly mean rainfall). Values less than 50 indicate that rainfall is distributed throughout the year, although seasonal peaks may be present. Major occurrences in this category coincide closely with the equatorial humid zones and again with temperate humid zones at higher latitudes (Fig. 3.2).

Coefficients of variation greater than 75 indicate strong seasonal contrasts in precipitation. A surprisingly large proportion of the tropics falls within this category, and the zone so delineated includes virtually all major occurrences of savanna ecosystems. In all cases, with the notable exception of Australia, the arid limits to occurrence of tropical savanna fall within zones of extreme rainfall seasonality (i.e. C.V. exceeding 100). In northern Australia, increasing aridity southward coincides with decreasing seasonality. The probability of rainfall in both summer and winter seasons may help explain the apparent anomalous extension of woody trees and shrubs far into the arid zone.

Although useful, this index of seasonality is not a sufficient measure of the adequacy of wet-season rainfall or of the duration and severity of the dry season. Thus, areas with extremely high rainfall, such as the southern coastal areas of New Britain to the east of New Guinea, can have a high seasonality index, but rainfall during the trough of the seasonal rainfall cycle is still more than sufficient to meet evaporative demand. However, the patterns of rainfall seasonality shown in Fig. 2.4 provide an indication of the potential occurrence of savanna ecosystems.

Two further measures of seasonal rainfall distribution can provide insight into the water-limited boundaries of tropical savannas. Firstly, total rainfall during the wettest six months, irrespective of actual time of occurrence (Fig. 3.3) provides a measure of water supply during the period of peak growth. Major development of savanna ecosystems occurs where totals exceed 600 mm, but compensating factors such as deeper and more retentive soil, lower evaporative demands, or ground-water tables accessible to deeper-rooting trees and shrubs, may extend this to 400 mm and even beyond. No upper limit is indicated, since regions with more than 2000 mm during the wettest six months can have short but severe dry seasons, and this then becomes the limiting factor for development of a continuous tree cover.

Total rainfall during the driest three months, irrespective of actual time of occurrence, provides an excellent measure of the severity of the dry season. Within the tropics, potential evaporation rates for dry months can exceed 200 mm. With three months, a potential demand of 600 mm and a reasonable carry-over of stored soil water of, say, 150 mm, any rainfall totals less than 100 mm are likely to lead to severe water stress in plant communities. Again, it is immediately evident from Fig. 3.4, that a large proportion of all land in the tropics undergoes some degree of seasonal water stress. Evergreen closed-canopy forests are essentially restricted to those areas where the rainfall totals for the driest quarter exceed 100 mm. In Africa, Madagascar, India, Burma, Thailand and northern Australia, tropical savannas commonly occur where the driest-quarter values are less than 25 mm and approach 0 mm. In Latin America this is true in part, but substantial areas have driest-quarter totals exceeding 25 mm and even 50 mm.

Ideally, analysis of environmental water regimes should be based on a water balance that takes into account effective rooting depth, extractable water content of root zone, foliage cover and evaporative demand as well as precipitation. Simplified water-balance estimates at global scale have been made by Thornthwaite (1948) and Papadakis (1961) and, in each case, selected output values have provided a basis for climatic classification. At continental and regional scales, more detailed water balances have been computed, usually for agricultural or pastural purposes. Thus, for example, there is the work of

Hargreaves (1974) in Brazil, Cocheme and Franquin (1967), in the Sudanian–Sahelian zone in East Africa, Kayane (1971) in India and Southeast Asia, and Nix and Kalma (1972) for northern Australia and New Guinea.

Temperature

The powerful influence of the thermal environment on the zonation of vegetation was recognized at least three thousand years ago by the ancient Greeks. Most global climatic classifications since then have been based on some variant of their torrid, temperate and frigid zones. The choice of appropriate temperature parameters in defining boundaries has always presented problems for the simpler, empirical climatic classifications. Often, boundaries have been determined by direct comparisons of isothermal gradients and mapped vegetation patterns. Inevitably this involves circular reasoning and the product is as much a vegetation classification as it is a climate classification.

In describing the thermal environment of tropical savannas for this chapter, I have used monthly mean maximum and minimum screen temperature data for some 4000 stations lying between 40°N and 40°S. In order to integrate map output with that of previous workers, two maps have been generated: one showing gradients in annual mean air temperature (Fig. 3.5), and the other showing gradients in mean minimum temperature of the coldest month (Fig. 3.6). Both of these temperature parameters figure prominently in existing climatic classifications.

While there is broad agreement that the climatically defined tropical zone is distinguished by uniformly high air temperatures with little seasonal variation, and no or relatively few occurrences of freezing temperatures, there is considerable variation in the choice of temperature parameters and temperature values. Thus, Köppen (1931) differentiated tropical (A) climates on the basis of the mean temperature of the coldest month exceeding 18°C. Trewartha (1968) modified this to the mean temperature of the coldest month exceeding 65°F (18.3°C) in marine areas or the absence of killing frost. However, with the dry climates (B) Köppen used a mean annual temperature of 18°C to separate hot and cold variants. Thus, different criteria are used in differentiating "tropical" or "hot"

climates depending on whether they are rainy (A) or dry (B) climates. This separation is maintained in the Trewartha modification, but in this case the dry climates are differentiated by a more realistic criterion of eight months or more with an average temperature exceeding 50°F (10°C).

In terms of annual mean temperature (Fig. 3.5) values are high throughout the lowland tropical zone and generally exceed 24°C. Throughout this zone the average diurnal range of temperature usually exceeds the average annual range in mean temperature. Obviously, broad zonal patterns are perturbed by topography, elevation, proximity to water, presence of cold and warm ocean currents, relative distribution of land and water masses in insular regions, and relative dominance of major air masses in the general circulation. Many of these influences will be evident from the pattern of mean air temperature shown in Fig. 3.5, and it is not intended here to provide a detailed description of temperature variations in the tropical zone.

In Central and South America, the basically north–south alignment of the cordillera distorts broad zonal gradients, as does the more subdued elevation of the Brazilian Shield. The distribution of tropical savannas in Latin America as mapped by Sarmiento and Monasterio (1975) indicates that all occurrences have mean air temperatures exceeding 20°C, but that the most extensive occurrences have mean temperatures exceeding 24°C. In Africa north of the equator gradients are broadly zonal, but modified by the Ethiopian uplands and by less extensive uplands in West Africa. South of the equator, gradients are distorted by the north–south trending uplands of East Africa and Madagascar and, to a lesser extent, by the Angolan uplands. The influence of cold ocean currents is pronounced on the west coast of Latin America and in southern Africa. Existing maps of tropical savannas in Africa are conflicting, to say the least, but, as with Latin America virtually all occurrences have mean air temperatures exceeding 20°C. North of the equator, mean air temperatures exceed 24°C and, over large areas, 26°C. South of the equator the patterns are more complex and extensive areas occur with mean air temperatures between 20°C and 26°C.

In the remaining sector of southern Asia and Oceania, gradients are broadly zonal, but modified by the Himalayas and by lesser mountain ranges in

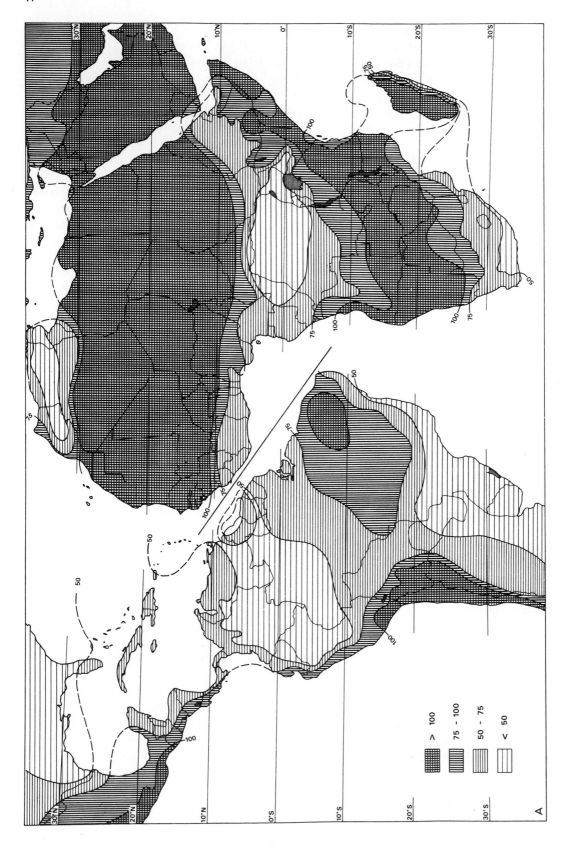

Fig. 3.2. Seasonality of precipitation (coefficient of variation of monthly variation) in the tropical savannas.

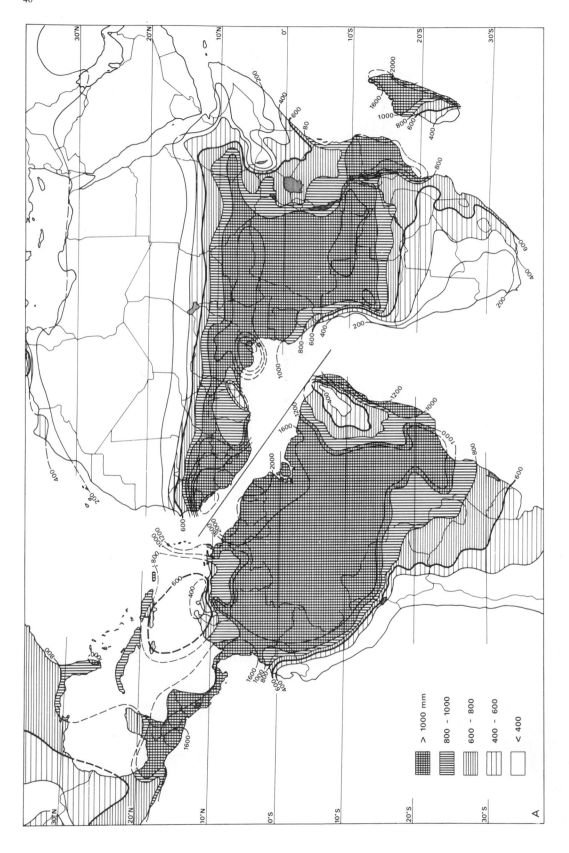

Fig. 3.3. Total rainfall (mm) of the wettest six months in the tropical savannas.

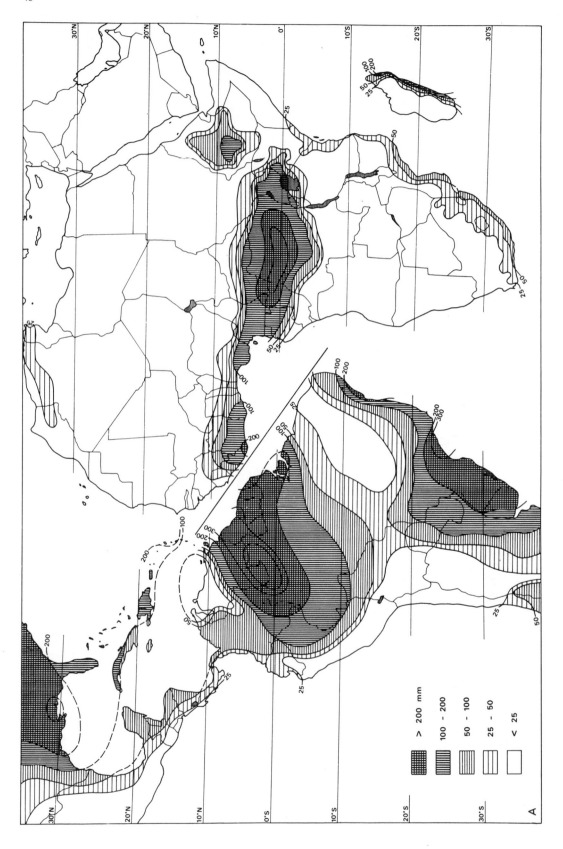

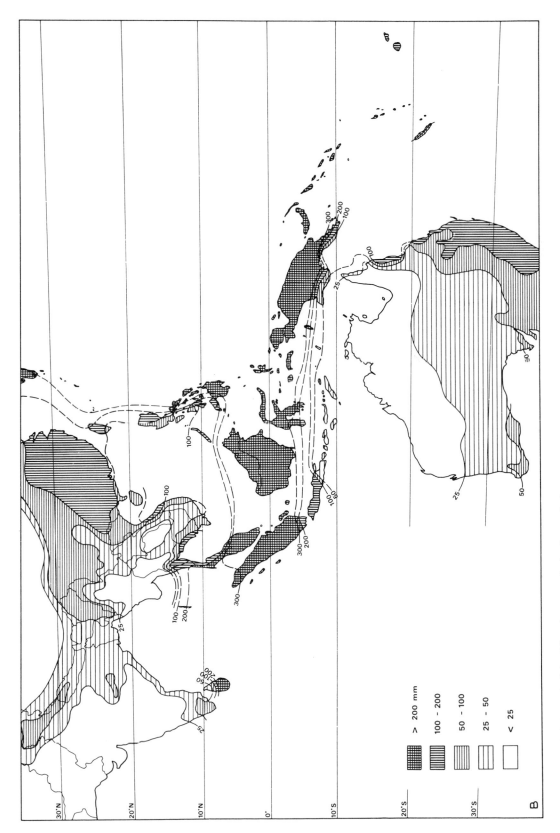

Fig. 3.4. Total rainfall (mm) of the driest three months in the tropical savannas.

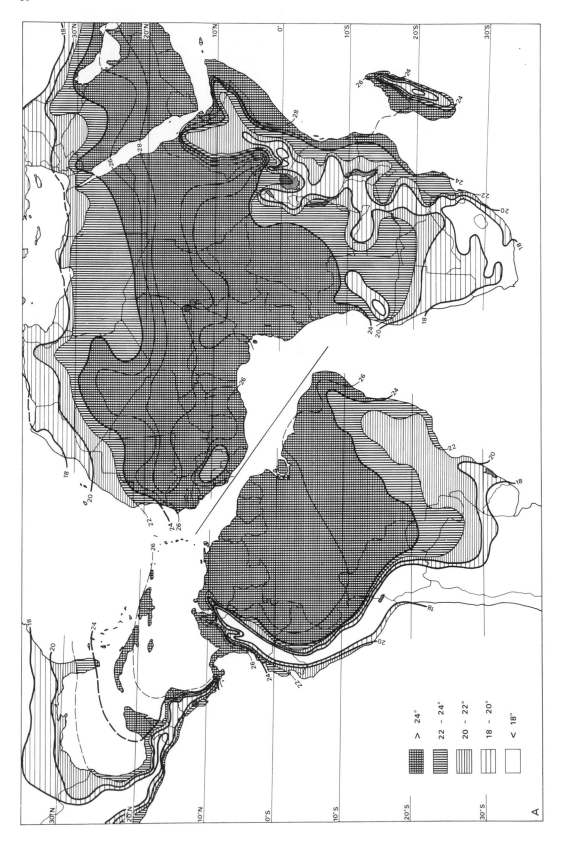

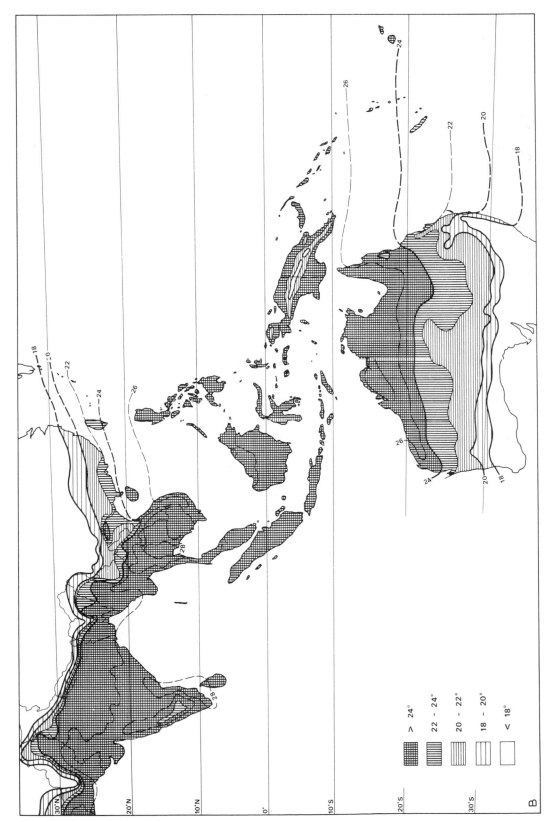

Fig. 3.5. Annual mean air temperature in the tropical savannas. (°C)

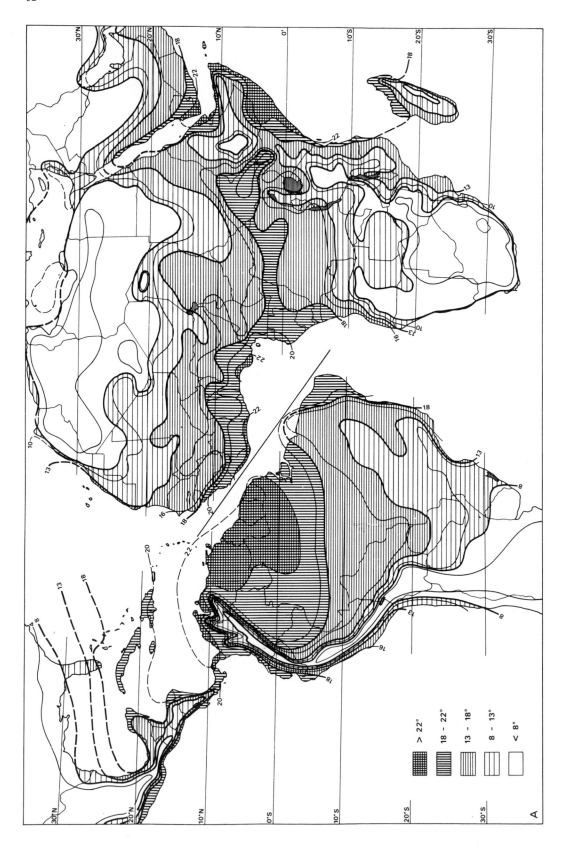

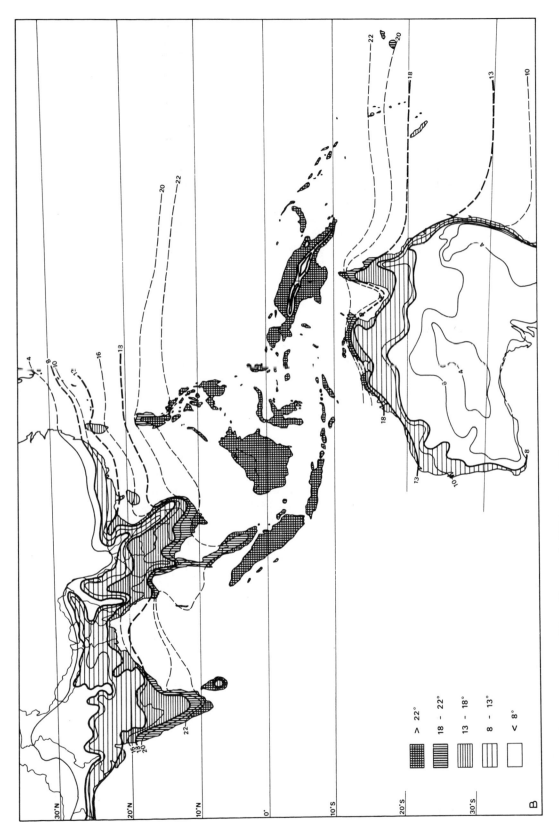

Fig. 3.6. Mean minimum temperature of the coldest month (°C) in the tropical savannas.

Southeast Asia and in New Guinea and eastern Australia. Most occurrences of tropical savannas reported in the literature for this region are found between 30°N and 30°S, and with mean air temperatures exceeding 20°C. However, greatest development of tropical savannas is in areas where the mean air temperatures exceed 24°C.

For what it is worth, the available evidence indicates that the 20°C isotherm encloses virtually all occurrences of tropical savanna, on all continents, and that major occurrences are enclosed by the 24°C isotherm. These two isotherms are shown in bold, together with the 18°C isotherm used by Köppen to differentiate his *B* climates, in Fig. 3.5. Mean air temperatures exceeding 28°C are most extensive in the broad Sahelian zone of Africa, with relatively localized occurrences in Latin America, in southern peninsular India, the central plain of Thailand and in north-western Australia.

While there is some agreement that presence/absence of tropically adapted vegetation is some function of minimum temperature and/or occurrence of freezing temperatures or frost, again there is no consensus on choice of parameter or parameter values. Where only mean monthly maximum and minimum temperatures are available, then the mean minimum temperature of the coldest month would seem the best index of low temperature limitation. Papadakis (1970) identifies an equatorial climate where the mean minimum temperature of the coldest month exceeds 18°C. His tropical-zone climates are bounded by the 8°C mean minimum of the coldest month. This isotherm coincides very approximately with the 50% probability of a freezing temperature occurring in any one year. A further subdivision of the tropical zone is based on the 13°C isotherm for the coldest month. In this case, there is broad agreement between this isotherm and the absence of freezing temperatures.

The 8°C, 13°C and 18°C isotherms are shown bold in Fig. 3.6. The major occurrences of tropical savannas are found where mean minimum temperatures of the coldest month range between 13°C and 18°C, but significant occurrences are found between the 8°C and 13°C isotherm. Few tropical savannas are found where mean minima of the coldest month exceed 18°C, since this is essentially a function of ever-humid equatorial conditions. While these selected isotherms appear to convey

some useful information, the limitations of such empirical, single-parameter approaches must be recognized. Taking all of the single-parameter maps used here, and broad correlations established with tropical savanna distribution, a general statement can be made that: in level terrain, on medium-textured soil of moderate depth (>100 cm), tropical savannas have the highest probability of occurrence where

(1) annual totals of solar radiation range between 6 and 8 GJ m^{-2} yr^{-1} (140–190 kcal cm^{-2} yr^{-1});
(2) annual mean rainfall values range between 1000 and 1500 mm;
(3) seasonality of rainfall is high, with a coefficient of variation greater than 75%;
(4) total rainfall during the wettest six months exceeds 600 mm;
(5) total rainfall during the driest three months is less than 50 mm;
(6) annual mean air temperature exceeds 24°C; and
(7) mean minimum temperature of coldest month is between 13° and 18°C.

Evaporation

Water is lost to the atmosphere from soil, plant and water surfaces by the physical process of evaporation. Evapotranspiration is the water lost by transpiration from the plant canopy and evaporation from the underlying soil surface. Because actual rates of evapotranspiration (E_a) are modified by soil, plant and atmospheric factors as the soil dries, the concept of potential evapotranspiration (E_t) is used for the case where soil-water supply is not limiting. The estimated ratio of actual to potential evapotranspiration (E_a/E_t) is a useful index of water stress in the vegetation studied and is calculated using a water balance equation.

Measurement of potential evapotranspiration is difficult and, commonly, this is a major source of error in computation of the soil-water balance. Direct measurement by lysimeter can be most accurate, but high costs of installation and maintenance normally restrict their use to research stations. Indirect methods involve computation using empirical formulae (e.g., Blaney and Criddle, 1942; Thornthwaite, 1948; Turc, 1954; Makkink,

1957; Papadakis, 1961; Fitzpatrick, 1963); an aerodynamic approach (e.g., Swinbank, 1951; Dyer and Maher, 1965); an energy budget approach (e.g., Penman, 1948; Priestley and Taylor, 1972); or an adjustment factor for observed water loss from a standard evaporimeter (e.g. U.S. Class A Pan).

Although the U.S. Class A pan has been accepted as a global standard, the network is as yet inadequate and data not readily available. The requirement for radiation and wind-speed data restricts use of the energy budget approach, and the aerodynamic approach requires further development. Only those empirical formulae requiring readily available data for temperature and humidity can draw upon a sufficiently extensive network of global scale. Published estimates for global networks are available in Thornthwaite (1948) and Papadakis (1961). Both formulae tend to underestimate potential evapotranspiration. In tropical savanna regions, wet-season daily values range between 2 and 4 mm and dry season values between 4 and 10 mm. Annual totals reflect relative lengths of wet and dry seasons; they everywhere exceed 1000 mm, but most commonly exceed 1500 mm with only the driest margins of the tropical savanna zone exceeding 2000 mm.

CLIMATE CLASSIFICATION

The search for pattern is a hallmark of scientific endeavour, and nowhere is this more in evidence than in the plethora of global classifications proposed for vegetation and climate. That broad patterns of climate and vegetation are indeed related was recognized at least as early as the ancient Greeks, but development of formal relationships had to wait until the full flowering of scientific exploration and measurement in the latter half of the nineteenth century. The historical development of vegetation mapping and classification, and the parallel development of climatic classification at global scale, have been ably summarized by De Laubenfels (1975).

General climatic classifications, particularly the more rational and systematic ones such as those of Köppen (1931) and Thornthwaite (1931, 1948), have made a major contribution to understanding and communication of climatic information at global and continental scales. Although such general-purpose classifications can be useful in the transfer of ecological information by analogy, they have limited utility for study of particular genotype/environment interactions or of ecosystem processes. Ultimately, the utility of any general multi-attribute classification depends on the relevance of the attributes used, the choice of class interval and the classification strategy adopted.

The most familiar and widely used of the general climatic classifications is that of Köppen (1931). Based upon annual and monthly means of temperature and precipitation, five major climatic groups are identified, four of them thermally defined and one by effectiveness of precipitation. Tropical (A) climates are defined as having mean temperature of the coolest month exceeding 18°C. Three major subdivisions based on seasonal rainfall distribution are:

Af Tropical wet climate: rainfall high throughout year and exceeding 60 mm in the driest month.

Am Tropical monsoon climate: rainfall high throughout year but with a short dry season.

Aw Tropical wet/dry climate: rainfall has a marked seasonal rhythm with a distinct dry season in the low-sun period.

All calculations are based on long-term means of temperature and rainfall.

Köppen arrived at his boundary conditions through consideration of global vegetation patterns; thus there is indeed reasonable agreement between his classes and major natural vegetation patterns. While tropical savanna formations are most frequently associated with the **Aw** climates, they do occur in **Am** climates and, to a much lesser extent, in **Af** climates, particularly where human disturbance is a factor. A more realistic assessment of year climate can be provided by calculating relative frequencies of Köppen climatic classes, over a series of years. Mizukoshi (1971) has calculated such frequencies for an extensive network in southern Asia and Oceania, in producing a map of dominant year climates. Significant shifts in boundaries are evident, with substantial increases in the area of **Aw** and **Am** climates, and encroachment of the dry **BSh** climate onto the margins of the **Aw** zones.

Another widely used global climate classification is that developed by Thornthwaite (1931, 1933) and resembling that of Köppen (1931) in that it is quantitative, uses readily available monthly mean temperature and precipitation data, uses a symbolic nomenclature in designating climatic types, and bases critical boundary conditions upon the then known distribution patterns of vegetation. However, Thornthwaite introduced two new concepts, that of precipitation effectiveness and that of temperature efficiency. The index of precipitation effectiveness is based on the precipitation/evaporation ratio of Transeau (1905), but involves summation of the P/E ratios for each month of the year. The temperature effectiveness index evaluates temperature on a linear basis, summing only those temperatures greater than 32°F (0°C). Empirical adjustments made to both formulae provide a corresponding range from zero with conditions totally limiting to 128 with conditions non-limiting. Five humidity and six thermal zones combine to yield thirty possible climatic regions. Additional subdivision is based on seasonal distribution of effective precipitation.

The primary division is based on precipitation effectiveness and uses capital letters from a maximum in **A** climate through to a minimum in **E**. Next, temperature efficiency classes similarly use capital letters primed (A^1 to F^1). The seasonality class uses lower case letters (r = rain at all seasons; s = summer dry, winter wet; w = winter dry, summer wet; d = dry at all seasons). Tropical savannas are well delineated by the CA^1w category on all continents, with minor occurrences in the higher rainfall BA^1w category of monsoon climate and in the somewhat cooler CB^1w category that characterizes extensive upland areas of Africa south of the equator.

Although these empirical and general classifications provide a useful basis for intercontinental comparison and have some value for didactic purposes, they are far too general and imprecise for ecological and bioclimatic analysis. An alternative to the use of indices and empirical formulae is that of graphical representation of important climatic elements in a climatic diagram. Such diagrams provide information directly on mean temperature and precipitation throughout the year, and can indicate the timing, duration and intensity of humid and arid seasons, hot and cold seasons and duration of frost season. The monumental work of Walter (1964, 1967, 1978) in developing such climatic diagrams for more than 8000 stations worldwide deserves recognition. Problems remain in the empirical adjustment of the temperature curve in relation to the precipitation curve (10°C = 20 mm rain) as a surrogate for potential evaporation. While this can provide a broad assessment of relative wetness and dryness, it is too imprecise for any detailed ecological and bioclimatic analysis. Despite this limitation, Walter's use of his climate diagrams reveals much ecological insight and understanding of ecophysiology and his evaluation of the role of the water balance in the relative dominance of grasses and woody plants in savanna ecosystems is masterly in its simplicity.

PLANT–ENVIRONMENT INTERACTIONS IN TROPICAL SAVANNA SYSTEMS

A major difficulty evident throughout all of the classificatory approaches is that all are based on a static approach to plant–environment interaction, and take little or no account of the dynamic, nonlinear responses of plants to their environment. Because of the wide range of plant adaptation and the need to take account of specific genotype/environment interactions the development of more dynamic approaches has been slow. Logically, such an approach requires a special classification for each genotype (or class of genotypes that have similar ecophysiological responses). Advances in computer technology and in ecophysiology now make such an approach entirely feasible.

The beginnings of such ecophysiological approaches are evident in the classical work of Livingstone (1916; Livingstone and Shreve, 1921) in North America. Introduction of the concept of a generalized water balance (Thornthwaite, 1948) that involved book-keeping transfers of water in six compartments — precipitation, potential evapotranspiration, actual evapotranspiration, soil moisture storage, soil moisture deficit and soil moisture surplus — represented another major step forward. Attempts to relate plant response to thermal and moisture regimes are evident in the works of Papadakis (1938, 1965, 1970) and Prescott (1943) among many others.

Direct coupling of the forcing functions of weather and climate with known ecophysiological response offers prospects of more relevant bioclimatic analysis and synthesis. Generalized models of plant response to the major light, temperature and moisture regimes were developed by Fitzpatrick and Nix (1970) for three broad groups of pasture plants — tropical grasses, tropical legumes, and temperate grasses and legumes. The non-linear responses of these groups to each of the light, temperature and moisture regimes were transformed to a linear, dimensionless scale, where zero represented completely limiting conditions, and unity non-limiting conditions for that factor. The separate indices together with a combinational multi-factor growth index were used in a bioclimatic analysis of the grassland ecology of the Australian continent. As an example, the gradients in the annual mean growth index for the tropical grass group are shown in Fig. 3.7.

Major differences between plant groups are most evident in patterns of temperature response. Five major groups can be identified, although further subdivision may be warranted when more response data became available. The terminology first used by De Candolle (1855) has been adopted and extended, with his megatherm type subdivided into two distinctive groups.

(1) *Megatherm*, C_4 and CAM photosynthetic pathway: optimum temperatures 30 to 32°C, lower temperature threshold 10°C, upper temperature threshold 46°C.

(2) *Megatherm*, C_3 pathway: optimum temperatures 26 to 28°C, lower temperature threshold 10°C, upper temperature threshold 36 to 38°C. 38°C.

(3) *Mesotherm*, C_3 pathway: optimum tempera-

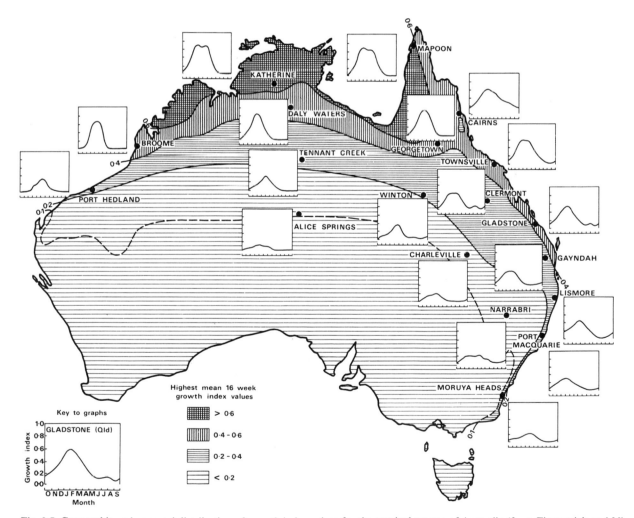

Fig. 3.7. Geographic and seasonal distribution of growth index values for the tropical grasses of Australia (from Fitzpatrick and Nix, 1970).

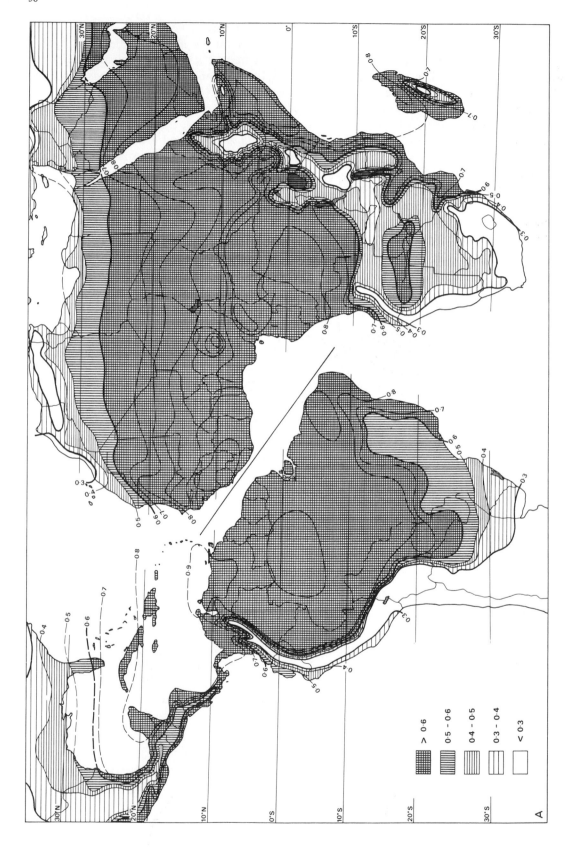

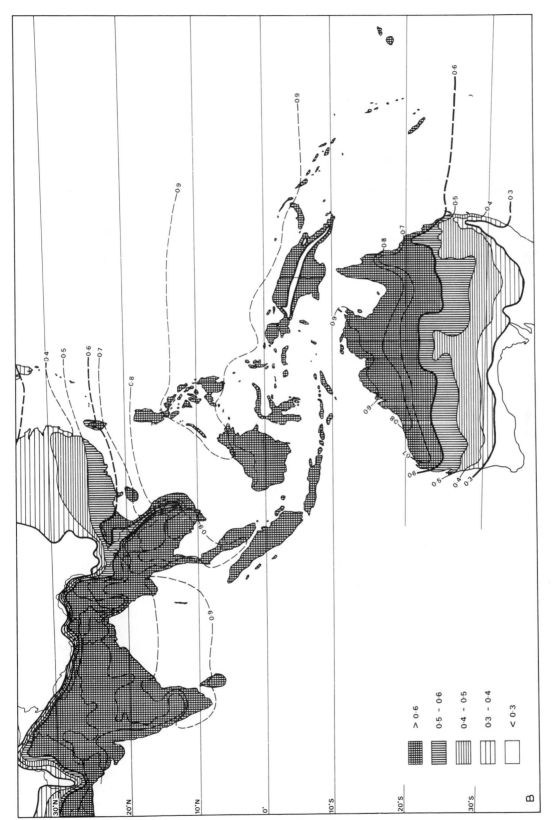

Fig. 3.8. Year-mean thermal index for Megatherm C$_4$ grasses, in the tropical savannas.

ture 19 to 22°C, lower temperature threshold 5°C, upper temperature threshold 33°C.

(4) *Microtherm*, C_3 pathway: optimum temperature 12 to 14°C, lower temperature threshold 0°C, upper temperature threshold 25°C.

(5) *Hekistotherm*, C_3 pathway: optimum temperature 6 to 8°C, lower temperature threshold −10°C, upper temperature threshold 25°C.

Only the first four types are of areal significance in Australia, and calculations of growth index values for these, on a weekly timestep, for over 1000 stations have provided a useful basis for bioclimatic analysis at the continental scale. Annual mean values, seasonal mean values and an index of seasonality (coefficient of variation of weekly mean values) have proven to be of greatest value (Nix, 1976). These values have provided a basis for bioclimatic classification in the study of woodlands and savanna ecosystems in Australia (Walker and Gillison, 1981; Gillison and Walker, 1981).

In terms of the present review of the climate of tropical savannas, and the emphasis given to the grass stratum in the accepted definition of Bourlière and Hadley (1970), it is clear that any functional interpretation of climate must be based on the Megatherm C_4 group of plants — the tropical grasses. Since the initial discovery of the C_4 photosynthetic pathway (Kortshack et al., 1965; Hatch and Slack, 1966) many physiological and biochemical studies have shown that C_4 photosynthesis is favoured under conditions of high light intensity, high temperature and high evaporation rates that are characteristic of the seasonally humid/arid tropics.

Ehleringer (1978) examined the implications of quantum yield differences on distribution patterns of C_4 and C_3 grasses, and found good agreement between simulations of photosynthetic response for each group and observed ratios of C_4 to C_3 grasses on the Great Plains of the U.S.A. (Teeri and Stowe, 1976). Analysis of the ecological distribution of C_4 and C_3 grasses along elevational gradients in Hawaii (Rundel, 1980) showed that temperatures at the point of floristic balance were considerably lower than those reported for the latitudinal point of floristic balance in North America. This is a reflection of near-continuous growth patterns in Hawaii versus a restricted summer seasonal pattern on the Great Plains. In all of the studies relating to spatial and temporal patterning of C_4 and C_3

grasses, it is clear that their different temperature responses are a major factor.

Formulation of thermal response functions has been hampered by lack of growth studies in controlled environment conditions, but data for a number of C_4 grass species were generated by Ivory and Whiteman (1978), and this has permitted recalibration of the thermal-response function for C_4 grasses as originally used by Fitzpatrick and Nix (1970). Optimum, lower and upper temperatures using daily mean values were set at 32°C, 10°C and 46°C. Thermal index (*TI*) values were estimated, from the response curve, for each month. Monthly values were summed and this total divided by 12 (a completely optimum thermal environment would have 12 monthly *TI* values of unity) to provide a comparative estimate of year-round thermal environment for C_4 grasses on a scale from zero to unity. These year-mean thermal indices were calculated for some 4000 stations and the resultant patterns are shown in Fig. 3.8.

Calculation of year-mean thermal indices for locations on the Great Plains used by Teeri and Stowe (1976) enabled calculation of correlations between these values and the ratio of C_3/C_4 grasses. A functional relationship between year-mean *TI* and the ratio of C_3/C_4 grasses was developed:

$$R = 206.90*TI - 123.03 TI^2 \quad (r^2 = 0.983)$$

where $R = C_4/C_3$ and TI = year-mean *TI*. At least for the Great Plains of the U.S.A., the point of floristic balance (i.e., 50% C_4 and 50% C_3 grasses) occurs close to a year-mean thermal index value of 0.3. An arbitrarily selected level of floristic dominance (i.e., 80% C_4 and 20% C_3 grasses) occurs close to year-mean thermal index values of 0.6. These two levels of 0.3 and 0.6 are shown bold in Fig. 3.8, solely as a basis for comparison between continents. This is not intended to imply that a relationship developed on one continent need necessarily hold for other continents. As far as the reported occurrences of tropical savannas are concerned practically all occur where year-mean *TI* values exceed 0.6, but limited occurrences are found between the 0.3 and 0.6 isolines.

The patterns in Fig. 3.8 show subtle differences from those of annual mean air temperature shown in Fig. 3.5, and this reflects the non-linear response of plants to temperature. Major northward excursions of 0.3 values in North America and in China reflect the high summer-season temperatures and help to explain the success of short-term C_4 crop plants (maize, sorghum) in those environments. Maximum TI values (>0.9) are attained most extensively in the Sahelian zone of Africa, with other lesser occurrences in Madagascar, in the Orinoco and Amazon Basins, southern India and northern Sri Lanka, the Chao Phraya and Mekong flood plains in Southeast Asia, and a segment of northwestern Australia. A substantial proportion of all land areas, excepting uplands, between 35°N and 35°S has values exceeding 0.3, and often exceeding 0.6. The influence of upland areas and of cold ocean currents lowering temperatures on the western side of South America and South Africa is evident.

While the patterns shown in Fig. 3.8 reflect the major thermal controls of C_4 grass distribution, obviously these will be distorted when interactions with light and moisture regimes are considered. Thus, continuously humid tropical environments favour development of closed-canopy C_3 tree species. Low light below the canopy effectively excludes C_4 grass species. When such forests are cleared and the regrowth controlled then C_4 grasses can be enormously productive. In certain continuously humid environments in Latin America, extremely nutrient-deficient soils and seasonally waterlogged soils can inhibit tree development and favour C_4 grasses of the tropical savanna (Sarmiento and Monasterio, 1975).

The lack of acceptable estimates of potential evaporation for the extensive global network used in this study has prevented an integrated analysis of the responses of C_4 grasses to light, temperature and water regimes. Such an analysis is available for Australia (Fitzpatrick and Nix, 1970), where the resultant patterns of simulated growth response (Fig. 3.7) indicate significant differences to the single, temperature-controlled pattern as shown in Fig. 3.8. Across northern Australia gradients remain much the same but favourable environments for C_4 grasses extend far down the eastern coast of Australia, due to a combination of high summer temperatures and summer rainfall. Similar extensions can be expected in southern Africa, southern South America, the Great Plains of the U.S.A. and the lowland basins of China. Otherwise highly favourable light and temperature conditions in the

tropical arid regions are of course offset by limitations in the water regime.

REFERENCES

Ångström, A., 1924. Solar and terrestrial radiation. *Q. J. R. Meteorol. Soc.*, 50: 121–126.

Bates, J.R., 1972. Tropical disturbances and the general circulation. *Q. J. R. Meteorol. Soc.*, 98: 1–16.

Blaney, H.F. and Criddle, W.D., 1950. *Determining Water Requirements in Irrigated Areas from Climatological Data.* Soil Conservation Service, Tech. Publ., U.S. Dep. Agric., Washington, D.C.

Bourlière, F. and Hadley, M., 1970. The ecology of tropical savannas. *Annu. Rev. Ecol. Syst.*, 1: 125–152.

Budyko, M.I., 1974. *Climate and Life.* Edited and translated by D.H. Miller. Academic Press, New York, N.Y.

Cocheme, J. and Franquin, P., 1967. *An Agroclimatology Survey of a Semi-arid Area in Africa South of the Sahara.* WMO Tech. Note No. 86, Geneva.

De Candolle, A., 1855. *Geographie botanique raisonnée.* Paris, Genève, 2 vols. 1366 pp.

De Laubenfels, D.J., 1975. *Mapping the World's Vegetation: Regionalization of Formations and Flora.* Syracuse University Press, Syracuse, N.Y.

Dyer, A.J. and Maher, F.J., 1965. The evapotron: an instrument for the measurement of eddy fluxes in the lower atmosphere. *CSIRO Div. Meteorol. Phys. Tech. Pap.*, 15.

Ehleringer, J., 1978. Implications of quantum yield differences on the distribution of C_3 and C_4 grasses. *Oecologia*, 31: 255–267.

Fitzpatrick, E.A. 1963. Estimates of pan evaporation from mean maximum temperature and vapour pressure. *J. Appl. Meteorol.*, 2: 780–792.

Fitzpatrick, E.A. and Nix, H.A., 1970. The climatic factor in Australian grassland ecology. In: R.M. Moore (Editor), *Australian Grasslands.* ANU Press, Canberra, A.C.T.

Gillison, A.N. and Walker, J., 1981. Woodlands of Australia. In: R.H. Groves (Editor), *Australian Vegetation*, Cambridge University Press, Cambridge, pp. 177–197.

Hargreaves, G.H., 1974. *Precipitation Dependability and Potential for Agricultural Production in Northeast Brazil.* EMBRAPA and Utah State University, Logan, Utah, 74-D155, 123 pp.

Hatch, M.D. and Slack, C.R., 1966. Photosynthesis by sugarcane leaves. *Biochem. J.*, 101: 103–111.

Ivory, D.A. and Whiteman, P.C., 1978. Effect of temperature on growth of five subtropical grasses I. Effect of the day and night temperature on growth and morphological development. *Aust. J. Plant Physiol.*, 5: 131–148.

Kayane, I., 1971. Hydrological regions in monsoon Asia. In: M.M. Yoshimo (Editor), *Water Balance of Monsoon Asia — A Climatological Approach.* University of Hawaii Press, Honolulu, Hawaii.

Köppen, W., 1931. *Grundriss der Klimakunde.* De Gruyter, Berlin.

Kortshack, H.P., Hartt, C.E. and Burr, G.O., 1965. Carbon dioxide fixation in sugar cane leaves. *Plant Physiol.*, 40: 209–213.

Livingstone, B.E., 1916a. Physiological temperature indices for the study of plant growth in relation to climatic conditions. *Phys. Res.*, 1: 399–420.

Livingstone, B.E. and Shreve, F., 1921. The distribution of vegetation in the United States as related to climatic conditions. *Carnegie Inst. Publ.*, No. 284.

Löf, G.O.G., Duffie, J.A. and Smith, C.O., 1966. *World Distribution of Solar Energy.* Solar Energy Lab, University of Wisconsin, Madison, Wis.

Makkink, G.F., 1957. Examining the Penman formula. *Neth. J. Agric. Sci.*, 5: 290–305.

Manabe, S., 1971. General circulation of the atmosphere. *Trans. Am. Geophys. Union*, 52: 313–320.

Mizukoshi, M., 1971. Regional divisions of monsoon Asia by Köppen's classification of climate. In: M.M. Yoshino (Editor), *Water Balance of Monsoon Asia — A Climatological Approach.* University of Hawaii Press, Honolulu, Hawaii.

Nix, H.A. and Kalma, J.D., 1972. Climate as a dominant control in the biogeography of northern Australia and New Guinea. In: D. Walker (Editor), *Torres Strait — Bridge and Barrier.* ANU Press, Canberra, A.C.T.

Papadakis, J., 1938. *Ecologie agricole.* Jules Ducolot, Gembloux.

Papadakis, J., 1961. *Climatic Tables for the World.* Published by the author, Buenos Aires.

Papadakis, J., 1970. *Climates of the World: Their Classification, Similitudes, Differences and Geographic Distribution.* Published by the author, Buenos Aires, 174 pp.

Penman, H.L., 1948. Natural evaporation from open water, bare soil and grass. *Proc. R. Soc. Lond.*, A, 193: 120–145.

Prescott, J.A., 1943. The Australian homoclimate of the zone of natural occurrence of *Parthenium argentatum. Trans. R. Soc. S. Aust.*, 67 (2).

Priestley, C.H.B. and Taylor, R.J., 1972. On the assessment of surface heat flux and evaporation using large-scale parameters. *Mon. Weather Rev.*, 100: 81–92.

Riehl, H., 1969. On the role of the tropics in the general circulation of the atmosphere. *Weather*, 24: 288 – 308.

Rundel, P.W., 1980. The ecological distribution of C_3 and C_4 grasses in the Hawaiian Islands. *Oecologia*, 45: 354–359.

Sarmiento, G. and Monasterio, M., 1975. A critical consideration of the environmental conditions associated with the occurrence of savanna ecosystems in tropical America. In: F. Golley and E. Medina (Editors), *Tropical Ecological Systems.* Springer-Verlag, Berlin, pp. 223–250.

Sellers, W.D., 1969. A global climatic model based on the energy balance of the earth–atmosphere system. *J. Appl. Meteorol.*, 8: 392–400.

Swinbank, W.C., 1951. Turbulent transfer in the lower atmosphere. In: *Climatology and Microclimatology.* UNESCO, Paris.

Teeri, J. and Stowe, L., 1976. Climatic patterns and the distribution of C_4 grasses in North America. *Oecologia*, 23: 1–12.

Terjung, W.H., 1970. A global classification of solar radiation. *Solar Energy*, 13: 67–81.

Thornthwaite, C.W., 1931. The climates of North America according to a new classification. *Geogr. Rev.*, 21: 633–655.

Thornthwaite, C.W., 1948. An approach toward a rational classification of climate. *Geogr. Rev.*, 38: 55–94.

Trewartha, G. T., 1968. *An Introduction to Climate.* McGraw-Hill, New York, N.Y., 4th ed.

Turc, L., 1954. Le bilan d'eau des sols: relations entre précipitations, l'évaporation et l'écoulement. *Sols Afr.*, 3: 138–72.

Walker, J. and Gillison, A.N., 1981. Australian savannas. In: B.J. Huntley and B.H. Walker (Editors), *Dynamics of Savanna Ecosystems.* Elsevier, Amsterdam, in press.

Wallace, J.M., 1971. Tropical meteorology. *Trans. Am. Geophys. Union*, 52: 395–399.

Walter, H., 1964. *Vegetation der Erde in öko-physiologischer Betrachtung, 1.* Gustav Fischer-Verlag, Jena.

Walter, H., 1978. *Vegetation of the Earth: in Relation to Climate and the Ecophysiological Conditions.* Translated by Joy Weiser, Springer-Verlag, Berlin, 2nd ed., 274 pp.

Walter, H. and Lieth, H., 1967. *Klimadiagramm-Weltatlas.* Gustav Fischer-Verlag, Jena, 245 pp.

Chapter 4

SOILS OF TROPICAL SAVANNAS

R.F. MONTGOMERY and G.P. ASKEW

INTRODUCTION

Many of the explanations for the occurrence of savannas involve the soil either as a primary cause or as an indirect factor. Edaphic explanations have included soil moisture and soil nutrient status whilst non-edaphic explanations also often have a soil aspect. Climatic theories, for instance, which relate savannas to wet and dry season conditions are postulating, in effect, that the savanna ecosystem is better adapted than any other plant community to stand the alternating soil moisture conditions in the soil which could vary from drought to waterlogging. Explanations for the occurrence of savannas which centre on clearing and burning of forest, with or without a cultivation stage, as a precursor to the invasion of savanna species are also closely related to soil conditions as it is argued that savanna is more likely to follow forest if the soils are nutrient poor. A knowledge of soils is essential, therefore, if the occurrence and ecology of savannas are to be understood.

As a variety of different soil types are found under tropical savannas it is impossible to make generalizations about a typical "savanna soil". Not only are different soil types to be found under climax savanna but in those areas where the savannas are only recently derived from another vegetation type, for example, forest, the soils could be regarded not as "savanna soils" but as "forest soils". The existence of such problems as relic soil features at the boundaries of vegetation formations brings to light the different emphasis given by the ecologist and the soil scientist to the problems of the ecosystem. The ecologist is interested in the underlying soil as part of the ecosystem and especially the extent to which it is a factor determining the occurrence and physiognomy of the vegetation, but he is not usually as interested as the soil scientist in the influence of the vegetation on soil properties and the extent to which the vegetation should be regarded as a major soil-forming factor and hence taken into account in classifying the soil. This problem of the interrelationships between soil and vegetation is important in the dynamic zone at the regional forest–savanna boundary, where, as discussed later, soil properties thought to be primary and hence, perhaps, affecting the vegetation can be separated from soil properties thought to be influenced, at least partly, by the vegetation.

It is clearly an advantage when discussing soils to be able to place them within the categories of a classification system but, unfortunately, no single system has gained universal acceptance and it has proved necessary to preface this review by a short account of the different systems in use for savanna soils. The subsequent description and discussion of the soils is arranged within a climatological framework and is followed by short accounts of the effect of termites on these soils and of their fertility status and utilization.

CLASSIFICATION OF SAVANNA SOILS

Three soil classification systems, each of which has some advantages and disadvantages, are widely used for tropical soils. D'Hoore's classification in the *Soil Map of Africa* (1964), as developed by Young (1976) has advantages in that its nomenclature is eclectic and largely based on well-established names. In this scheme each soil type is defined in terms of parent material, profile morphology and broad analytical features and reference is made to

63

the general environment in which the soil occurs. Young did not claim to have achieved a satisfactory formal classification but rather to offer a scheme that helps in the identification of the natural soil types of the tropics. Young's system defines soil units in a broader descriptive way than any other widely used scheme, it has flexible limits to the soil classes and its use requires less knowledge of detailed pedogenesis and horizon nomenclature than other modern systems.

The two other widely used classification schemes are firstly the official soil classification system of the United States Department of Agriculture (Soil Survey Staff, 1975) and secondly, the FAO Soil Classification system (FAO–UNESCO, 1974) used in the *Soil Map of the World*. The U.S.D.A. system is now widely used but it has extremely detailed and complicated definitions of diagnostic horizons and a completely new terminology, both for horizons and class names, derived from Greek and Latin roots. The FAO scheme is largely dependent on the U.S.D.A. scheme for many of its horizon concepts but its definitions are simpler and therefore easier to use and it retains many of the traditional names for soil types. An additional reason for making reference to the FAO scheme is the existence of small-scale maps for all regions of the world based on the FAO terminology. In the following account only outline descriptions of the soils are given and no attempt is made to describe in detail the features of each soil type. The reader who is not familiar with the soil types of the tropics and requires further information is referred to the classification schemes mentioned above.

The most widely occurring freely drained soil of the tropical forests is the Ferrallitic Soil (Young–D'Hoore system) which can be correlated with the Oxisols of the American classification and the Ferralsols of the FAO soil classification. Such soils have an oxic horizon[1] and show virtually no horizonation below the organo-mineral surface A horizon. By contrast one of the most widely occurring freely drained soils of the savanna regions is the Ferruginous Soil (Young–D'Hoore system) which can be broadly correlated with the Ultisols and some Alfisols of the U.S.D.A. system and with the Acrisols and Luvisols of the FAO scheme. Ferruginous Soils have either an argillic horizon (i.e. a B horizon of illuvial clay accumulation) or a well-developed structural B horizon or both. The clay

fraction of Ferruginous Soils, like that of the Ferrallitic Soils, is dominated by kaolinite and hydrous oxides of iron and aluminium but, unlike Ferrallitic Soils, 2:1 lattice clay minerals may also be present. The cation exchange capacity and the base saturation of Ferruginous Soils are higher, in general, than in Ferrallitic Soils. The differentiation of these two soil types is one of the most important in tropical soil classification and is further accentuated by the fact that Ferruginous Soils have a significant reserve of bases and other mineral elements whereas Ferrallitic Soils contain no significant reserve of bases and nutrient elements. In addition Ferruginous Soils are often characterized by the presence of concretionary segregations of iron oxides.

The correlation of freely drained tropical soils is complicated by the recognition in both the Young–D'Hoore system and in the FAO system of separate soil types over basic rocks and by the separation by Young of the Ferrallitic Soils into two groups depending on whether they occur under forest or savanna. Thus Young, after D'Hoore, uses the names Ferrisols and Eutrophic Brown Soils and the FAO system the names Dystric Nitosols and Eutric Nitosols to cover respectively the moderately nutrient-rich and the nutrient-rich soils over basic rocks. Both these soil types exhibit an argillic B horizon with clay skins on ped surfaces and they usually have an overall dark colour in reds or browns. Young, himself, subdivides Ferrallitic Soils into the Leached Ferrallitic Soils of the forests and the Weathered Ferrallitic Soils of the savannas. The structural condition of the latter subgroup is more weakly developed and their base saturation and pH is generally higher than in forest soils. Weathered Ferrallitic Soils are often paler and rather more likely to contain massive laterite at depth than the Leached Ferrallitic Soils though neither of these criteria is diagnostic. The relationships between the soil types described so far are shown in Table 4.1. Other soils which may be encountered under savanna are briefly described, as they are discussed, later in the chapter.

---

[1]Oxic horizon — a mineral subsurface horizon in an advanced state of weathering which consists of a mixture of hydrated oxides of iron or aluminium, or both, with variable amounts of 1:1 lattice clay. For full description see Soil Survey Staff (1975, pp. 36–41).

TABLE 4.1

Interrelationships of rocks, vegetation and soil types in Young's (1976) and the FAO (1974) classification schemes and some comparisons with the U.S.D.A. (1975) classification scheme

Rock type / Vegetation	Young's (1976) classification scheme		FAO (1974) classification scheme		U.S.D.A. (1975) classification scheme
	Felsic to intermediate rocks	Basic rocks	Felsic to intermediate rocks	Basic rocks	No subdivisions according to rock type
Rain forest	Leached Ferrallitic Soils	Leached Ferrisols	Ferralsols		Oxisols
Forest/savanna boundary	Weakly Ferrallitic Soils — Ferruginous Soils	Ferrisols	Ferralsols — Acrisols	Dystric Nitosols	Oxisols — Ultisols
Savanna	Weathered Ferrallitic Soils	Eutrophic Brown Soils	Ferralsols	Eutric Nitosols	Oxisols
	Ferruginous Soils		Acrisols — Luvisols		Ultisols — Alfisols

THE SOILS OF WORLD SAVANNAS

The causal factors which determine the occurrence of savannas are many and their relative importance will differ from region to region. The importance of climate as one factor either directly or through its influence on soil moisture is not disputed and it is convenient to discuss the soils of savanna landscapes within a climatic framework.

Whilst there are many climatic classifications of the savanna lands (e.g., Blydenstein, 1967; Hills, 1969; Sarmiento and Monasterio, 1975; and Young, 1976; see also Chapter 3) there is considerable agreement between them. All exhibit a climatic progression of savanna types from alternating dry and moist savannas, where there is a shortage of water in the dry season and just adequate or fully adequate water in the wet season, through wetter savannas where there is again deficiency of water in the dry season but excessive water, usually in the form of intermittent inundations, in the wet season, to swampy savannas where there is profile saturation by water in both wet and dry seasons. The differences in detail between the schemes do not detract from the fact that there is a widely accepted savanna classification which, because of its basis on climate, has considerable pedological significance. The groupings of soils in the following discussion are based on the climatic classes described by the authors mentioned above.

Well-drained soils of savannas in seasonal climatic zones

The savannas on well-drained soils in seasonal climates are considered by many as the most widespread of the savanna ecosystems. Their climate is such that the soils are never wet throughout the year, so that even in the wet season the soils only have "moist", "adequate" or "intermediate" water contents. Within this broad group it is possible to distinguish between the soils of the dry seasonal and of the moist seasonal savannas but it must be emphasized that there is no clear boundary between them, both categories having a noticeable wet and dry season. The difference is one of degree, the moister savannas having a shorter dry season and a higher mean annual rainfall.

The occurrence of savannas on freely drained soils in seasonal climates includes vast areas of South America, amongst which the *cerrados* of central Brazil are paramount, together with smaller areas in Surinam, the Guianas and Venezuela. Large areas of these seasonal savannas are present in Africa too, where they range in West Africa from very dry types in the north to moister seasonal savannas in the south, and where a wide range also occurs in Central and East Africa. The Deccan of India too carries some savannas of this nature.

The soils of the moist seasonal savannas

The most extensive well-drained soils of this group are the Ferruginous Soils and the Weathered and Weakly Ferrallitic Soils: the former, compared with the Ferrallitic Soils, are only moderately weathered and leached. At times downward leaching takes place but during other parts of the year the upper parts of the profile dry to wilting point. However, even in the dry season, a typical profile would be moist below about a metre from the surface. Young (1976) described examples of upper slope Weathered Ferrallitic Soils giving place to either Weakly Ferrallitic Soil or Ferruginous Soil on lower valley sides. Where intermediate or basic rocks are present Ferrisols or Eutrophic Brown Soils will be found.

Blydenstein (1967) working in Colombia where the mean annual rainfall is about 1700 to 2000 mm, found that together Entisols (soils with little horizon differentiation) and Inceptisols (soils with altered horizons that have lost bases or iron and aluminium but lack an illuvial horizon) occupied 75% of the 116 sites he examined in this ecoclimate. Oxisols only occupied around 20%.

Goodland (1971) identified Oxisols under cerrado in the Triangulo Mineiro region of central Brazil, where the annual rainfall is 1600 mm and, correlating them with the widespread Latosols of Brazil (Bennema et al., 1959; Jacomine, 1969; and Ranzani, 1971), he considered that Oxisols were the principal soils associated with cerrado vegetation. Askew and his colleagues (1970a) working in Mato Grosso, central Brazil, in an area lying to the south of the regional Dry Forest (Ratter et al., 1973) where the rainfall was thought to be about 1370 mm per annum, also classified the soils underlying the cerrado as Oxisols (Ferrallitic Soils) and oxic Entisols. On the other hand, soils in Nigeria in a rather similar rainfall zone were classified by Jones and Wild (1975) as mainly Ferruginous Soils, although some Ferrallitic Soils were recorded. Murdoch and associates (1976) working in the north of western Nigeria, within the area reported by Jones and Wild, found that Luvisols occupied more than 60% of the landscape and Fluvisols (alluvial soils), Arenosols and Cambisols (correla-

ting approximately with Inceptisols) together occupied about 25%. Since Luvisols can be equated with Ferruginous Soils this is support for the work of Jones and Wild.

Soil moisture. It has been suggested in a review by Hills (1965) that the moisture content of freely drained soils is thought by many workers to be an important factor in determining whether or not savanna exists in a particular landscape. Ferri (1962) and Arens (1963), however, have expressed the view that, in the case of one type of savanna vegetation, the deeply rooting cerrado plant species are rarely subject to serious water stress even in the dry season and have concluded that their xeromorphy is more likely to be oligotrophic scleromorphy. Now it may be that non-cerrado species in the same soil conditions would have experienced serious stress. Nevertheless, it is interesting to note the work of Askew et al. (1971) who found in Mato Grosso that the water table under both cerrado and adjacent forest is generally far too deep to measure in soil dip wells. They offered some evidence of water-table level at a depth of greater than 30 m in the dry season.

Texture and structure. Many authors [e.g., Askew et al. (1970a) and Goodland (1971) in central Brazil; Jones and Wild (1975) in West Africa] have referred to the generally sandy nature of the soils under savanna in this climatic zone. Askew and his colleagues reported that the savanna soils in the area they studied in northeastern Mato Grosso had loamy sand and sandy loam topsoils. Further to the south in Mato Grosso they found savannas over a wide range of textures including clay. Goodland described the Triangulo Mineiro soils as sandy loams with subsidiary loamy sands and sands but commented that elsewhere cerrado soils were less sandy and were generally sandy clays. Jones and Wild also emphasized the sandiness of the savanna soils in West Africa but they reported a significant siltiness and a fineness in the sand fraction, unlike the Brazilian soils, which have silt contents of less than 3%. According to Jones and Wild the siltiness causes structural collapse and therefore capping of the surface soil in the West African environment. The soil texture also affects water holding capacity; whilst this can be attributed to the sandy textures already mentioned, shallowness over coarse rocks, such as sandstones or quartzites, may also be a factor.

Both Brazilian and West African savanna soils are generally weakly structured and have a loose consistency due partly to their low clay content and partly to kaolinite being the dominant clay mineral. Lower horizons, however, are generally more compact and often rather structureless.

Organic matter. The soils of savannas in this climatic zone do not normally have a continuous litter layer, and organic matter is distributed through the top few centimetres of the profile. Typical organic carbon figures would be of a generally low level; Jones and Wild (1975) quoted 1.2% as a mean of 245 West African samples whilst Askew et al. (1970a) quoted about 1.7% and Goodland (1971) 2.0% in Brazil.

Cation exchange properties. The fact that the soils are low in clay and that this is of kaolinitic structure means that the contribution of the mineral part of the soil to the cation exchange capacity is not great. For cation exchange capacity figures less than 40 $\mu eq\ g^{-1}$ in the topsoil, and not infrequently less than 20 μeq, would be common and most of this is attributable to organic matter.

The pH and saturation of the exchange complexes of soils in this category, more than the properties already discussed, emphasize the differences between Africa and South America alluded to earlier. The African soils, mainly Ferruginous Soils, have pH in water of around 6.2, even the Ferrallitic Soils having mean values according to Jones and Wild (1975) of about 6. On the other hand, the well-drained Brazilian cerrado soils, often Ferrallitic Soils (Oxisols or Ferralsols) or Arenosols, have pH values in water which are very much lower. Ranzani (1963) for instance, quoted topsoil figures for 36 sites which, apart from 2 with a pH value just over 6, have a mean value of 4.9, and Goodland (1971) gave a mean pH of 5.0 for 110 samples. Saturation, as is usual, goes in line with the pH, and Askew et al (1970a) quoted under 5% for soils in their studies in Brazil. Jones and Wild gave examples of around 70% for Ferruginous Tropical Soils, from 25 to 70% for Ferrallitic Soils and around 25% for Ferrisols. Individual ions are low in all these soils; the exceptionally low figures for central Brazil (e.g., calcium 0.2, magnesium 1.0,

potassium 0.3 $\mu eq\ g^{-1}$) emphasized by both Askew et al. and Goodland are worthy of mention.

It is important to appreciate that in the past many of the measurements of cation exchange capacity and saturation have been made using molar ammonium acetate at pH 7. This method is now considered rather inappropriate for analyzing soils with low pH, especially as a high proportion of the charges on the clay complexes are pH dependent, but it still has some value for comparative studies. Jones and Wild suggested that measurements of this nature should be taken at the pH of the soil under study.

The soils of the drier seasonal savannas

The soils in the drier seasonal savannas are always well-drained, usually coarse-textured and sometimes gravelly or shallow. During the dry season, which may last about five months, the soils are dried to wilting point to great depth, possibly deeper than 2 m from the surface. The chemical properties of the soils reflect this particular moisture regime; leaching is not pronounced, so that whilst calcium carbonates are removed from the profile the soils are of medium base status and pH. Young (1976) suggested that base saturation in these soils might be about 60 to 90% and postulated that the mean annual rainfall would be about 600 to 900 mm.

In this zone the Ferruginous Soils and Weathered Ferrallitic Soils of Young are common. There is some evidence, as discussed in the preceding section, that Ferruginous Soils are more widespread in Africa than South America. D'Hoore (1964) was of the opinion that the relatively dry types of savanna in Africa occupy less important and more localized areas, often where there are special climates. His map shows a wide range of soil types under drier savanna areas including Ferruginous Tropical Soils, Lithosols (shallow soils over rock), Vertisols (clayey soils that have deep wide cracks at some time of the year, and are more fully described later on p. 69), Juvenile Soils on recent alluvia and even some Halomorphic Soils. Where there are underlying basic rocks Eutrophic Brown Soils are common and Ferrisols would also be present.

Soil influence on differences within savanna vegetation

Cole (1968) has related the occurrence of savanna species in Africa and Australia with bedrock geology operating through soil composition (particularly texture and nutrient status). In Central Africa she related some species of *Brachystegia* and some species of *Isoberlinia* with sandy soils formed from quartzites or granites whereas another *Brachystegia* species and *Uapaca* are more likely to be found on clayey soils derived from shales or gabbro. She classified all these soils as old lateritic soils on pediplain surfaces (probably Ferruginous Soils in current classifications) and such soils in the Darwin district of Australia she related to particular species of *Eucalyptus*. Young (1976) reported that short tree savanna occurs on Weathered Ferrallitic Soils and tall dense savanna on Ferruginous Soils. Goodland and Pollard (1973) have described the parallel relationship of a soil fertility gradient as measured by the levels of phosphorus, nitrogen and potassium in the soil to a physiognomic gradient of cerrado vegetation from small widely scattered trees through orchard vegetation to well-developed woodland (*cerradão*). The richer the soils are in these nutrients the greater is the standing vegetation. Whilst the authors stated that their results would suggest that the nutrient status of the soils is especially influential in determining the physiognomic gradient, they emphasized that soil fertility is not necessarily the cause of this gradient, which may have arisen from differential burning or moisture conditions. The soil nutrient gradient could well be the effect rather than the cause of the physiognomic gradient. Also working in Brazil, Ratter (1971) reported the correlation of *Hirtella glandulosa* cerradão with poor dystrophic soils and *Magonia pubescens* cerradão with comparatively nutrient rich mesotrophic soils. A mesotrophic community dominated by the latter species was studied by Ratter et al. (1977) in Mato Grosso, Goias and Minas Gerais states of Brazil and found to be associated with soils of higher pH and exchangeable calcium than other types of cerradão.

Hydromorphic soils in seasonal climatic zones

In the wetter of the two savanna areas described in the previous section there are many soils which are subjected to alternating moisture conditions of a much more serious nature than those described for the zonal soils. Hills (1969) described such areas as having wet savanna ecoclimates with alternating dry and *excessively* wet phases. The soils are deficient in water during the dry season but in the wet season there is an excess of water often caused by intermittent flooding and the soils are saturated for significant periods. It is also possible to have soils which are waterlogged for long periods in both the wet and dry season. Surface water to a depth of over 50 cm may lie on the surface for several months in the year and shallower surface water could still be significant well into the dry season. The long saturation of these soils results in the prevalence of reducing conditions, at least, in the lower horizons with consequent gleying of the soils.

Gleys, or Gleysols as they are named in the F.A.O. Classification, occupy the poorly drained sites in these savanna lands. In narrow valleys much of the parent material from which they are formed is colluvial and takes the character of the immediately adjoining materials but in broader low-lying areas the alluvial contribution will be greater and the character more variable. Particularly in savanna lands near the regional forests ribbon forests may occupy the narrow valleys but the wider the poorly drained sites are, or the further away from the forests, the less this is so. Savanna gleys have much in common with their counterparts in temperate countries and the grey, bluish grey or brownish grey gleyed horizons with or without prominent mottling, will be present within 50 cm from the surface. The tropical gley soils cover the whole range of textural types; but undoubtedly the most common are dark-grey or black clays. Such clays are often tenaciously sticky when wet, plastic when moist, and hard like concrete when dry. A few concretions of iron are likely to be present and calcium carbonate may occur as a minor feature.

Young (1976) stated that savanna Gleysols usually have weakly acid topsoils but are nearer neutrality in depth although in the cerrados of central Brazil the present authors found that the gleys on the valley floors and lower valley sides were only slightly less acid and base desaturated than the adjacent very acid freely drained soils.

Most savanna hydromorphic soils are bounded by mottled soils and on the borders of the low-lying hydromorphic soils it is usual, in the savanna zones, to find soils which show mottling rather than a well-developed gley horizon. The texture of such soils is variable but they are usually more sandy than the Gleysols. Topographically such soils lie just a little higher in the catena than the Gleysols, a situation which may also be occupied by laterite or profiles containing large and small iron concretions. Where the profiles contain laterite there is often a noticeable break of slope.

Groundwater Laterites, which are poorly drained soils, are much more widespread than the break-of-slope laterites mentioned in the previous paragraph. In Ghana, for instance, extensive upland sites of gentle relief under savanna have profiles which contain significant mottling, but also either noticeable concretions or iron pan at a shallow depth from the surface. These soils are saturated in the wet season with scattered pools of standing water but dry in the dry season. Their adverse profile, morphological features and low nutrient status cause these to be amongst the poorest of savanna soils and it is not surprising that they have been utilized only for low-intensity nomadic grazing. Although the Groundwater Laterite soils of Ghana are formed in a rainfall of from 1000 to 1200 mm it must be emphasized that the critical factor is the impermeability of the parent material rather than the mean annual rainfall and the same kind of soils have been reported in the drier savannas.

Impermeability may also be an important characteristic of the Vertisols, a distinctive soil type on which savanna growth is common. Vertisols, or black cotton soils as they are also known, are found in climates ranging from 250 to 1500 mm annual rainfall, but they can conveniently be described in this section because a dry season of over four months duration is a diagnostic feature.

Vertisols are dark cracking soils with a high content of fine-sized clay of montmorillonitic composition. Drainage ranges from virtually free to poor depending on a number of factors of which the nature and extent of the usually well-developed soil structure would be a main one. The high montmorillonite content causes shrinkage in the dry season and swelling in the wet season and the associated churning and mixing gives the well-known self-mulching effect.

Permeability is usually very low and bulk density high in these soils, and Young (1976) related these features to very fine pore size and compaction due to lateral pressure on swelling. Associated with high montmorillonite contents are high cation exchange capacities, high saturation, particularly in calcium and magnesium, and high pH (6–7.5).

Vertisols are a major feature in eastern Australia, the Deccan of India and the Sudan, but they are also to be found in many West African countries. They are not a common feature of South American landscapes. In India these soils are commonly under non-irrigated cereals or irrigated rice, but in West Africa it is more usual to see their natural savanna vegetation used for low intensity grazing by cattle.

Other intrazonal soil areas under savanna have been described by Blydenstein (1967) and Vesey-Fitzgerald (1963). The former working in Colombia found that, of the 10% of intrazonal soil sites he examined, more than half were hydromorphic Oxisols dominantly Aquox, whereas Inceptisols and Entisols were only about a third (the figures should be compared with breakdown of those for the well-drained soils of Colombia already discussed). Vesey-Fitzgerald, working in Central Africa where the annual rainfall was in the range from to 750 to 1270 mm, and the upland vegetation was *Brachystegia–Isoberlinia* woodland, was able to subdivide the edaphic grasslands of the sluggish drainage bottom lands into different vegetation classes associated with different types of hydromorphic soils including Vertisols.

Budowski (1968) and Tamayo (1968) examined soils in different parts of the world with savanna vegetation. The former described swamp savannas on permanently or semi-permanently flooded areas such as those close to estuaries and the latter recognized hygrophilous savannas on soils which remain wet throughout the year. The best known examples of savannas of this kind are in South America. The Llanos de Mojos, some 200 000 km^{-2} between western Brazil and the Andes, and the Pantanal de Mato Grosso, for instance, have depressions that are under water for periods ranging from five months to over ten months in the year. Gleysols of varying textures and nutrient status and Groundwater Laterites are common. There are no similar vast areas in Africa, although plenty of the small *dambos* and *mbugas* might have some very similar features on a small scale. These small-scale African depressions usually are not as badly drained as those in the Ivory Coast, described by Adjanohoun (1968) as "marshy savannas'.

It is not uncommon for small changes in relief to make very significant differences in these areas of hydromorphic savannas. Sometimes, as in Llanos de Mojos, there may be topographic changes of a metre or two over a short area and it is possible in these circumstances, in the flood period, to see Gleysols under water supporting savanna and islands supporting forest on relatively well-drained soils.

Blydenstein (1967) working in the frequently flooded savanna areas in Colombia to the north and west of the Orinoco found that of the 79 sites he examined over 60% were Inceptisols (Aquepts) and about 30% Aquox.

Soils of savannas in non-seasonal equatorial climates

Savannas are present in the areas having climates classified as non-seasonal equatorial even though the climatic climax vegetation there is undoubtedly tropical rain forest. Hills (1969), for instance, noted that a close examination of the Amazon Basin showed a "motheaten cover" in that there are discrete savanna landscapes breaking up a blanket of rain forest. Before these savannas and their soils are discussed it is necessary to appreciate that there may be a special reason for their existence, local climate or a particular anthropic history for example, and their classification in an equatorial climate does not necessarily mean that the climate is causal. There is no doubt, however, that there is a group of "island" savannas in this climatic zone, particularly in the north of South America, and it is convenient to discuss them together. Some of the savanna areas are well known and well documented. Probably the best known is the Rupununi savanna of Guyana although it is really only one of a number including the savannas of Rio Branco and Gran Sabana. All these savannas, as has been suggested by Sarmiento and Monasterio (1975), are either formed on white quartzose sands, when the closed nutrient cycle which would normally give rise to scleromorphic forest is broken, or on senile Oxisols formed on ancient sediments of the Pleistocene. Savannas do not normally occur on the podzols that form in this area. The excessively drained, extremely nutrient-leached podzols of the white sands give rise to scleromorphic forests whereas the less extreme Latosols (Oxisols) which can also form on these sands give rise to savannas. Sarmiento and Monasterio suggested that soil mineral deficiency may be a determining factor for open forest but a predisposing factor for savanna: these authors felt that fire might be a more direct cause of savanna establishment and permanence. Hills (1969) attributed the Rupununi and Rio Branco savannas to imperfect drainage, impermeable subsoil and/or sandy and bleached topsoils. Eden (1970) has emphasized the wide range of soil types [Latosols (Oxisols), Regosols, hydromorphic soils and Groundwater Laterites] present in the northern Rupununi area and considered that oligotrophism and, for other nearby sites, hydromorphism were probably secondary factors favouring the persistence of savannas as climatic relics. The physiognomic differences within the Rupununi savannas were related by Eden to soil types: the *campo limpo* (treeless savannas) occurred on the hydromorphic soils and the *cerrado* on the freely drained Latosols. The same author (1971) suggested that the Venezuelan savannas also are probably climatic relics maintained by fire.

Soils and the savanna/forest boundary

Whilst the overall distribution of regional forest and savanna would seem to be a function of climate, several authors have proposed that soil also is an independent factor influencing savanna occurrence. Cole (1960) considered the cerrado in parts of Brazil an edaphic climax on level plateaux and forest a climatic climax on dissected landscapes of better drained and relatively younger, more nutrient-rich, soils. Lemos and his colleagues (1960) also associated cerrado with nutrient-poor soils on uplifted plateaux. On the other hand, Askew et al. (1970a), who studied the soils across the forest savanna boundary in northeastern Mato Grosso, found that in this area cerrado was associated with the more dissected and younger surfaces of the Araguaia drainage system and Dry Forest (a type of Seasonal Evergreen Forest[1]) with the older surface related to the Xingu drainage system. Goodland (1970) and Denevan (1965) also

---

[1] The vegetation types are those defined by Ratter et al. (1973).

have described cerrado forest occurrences which do not fit Cole's geomorphological hypothesis, though there is no doubt that it holds for many situations in Brazil, especially where dissected slopes are associated with more fertile soils supporting a deciduous type of forest.

In their comparative study of the properties of soils under cerrado and forest Askew et al. (1970b, 1971) found that the soils on the interfluves under the two types of vegetation differed in texture, the cerrado soils being mainly sands or loamy sands and the forest soils sandy clay loams. The soils under Dry Forest have thick litter and fermentation layers and well-defined granular and sub-angular blocky peds in the surface horizons, whereas the soils under cerrado have virtually no litter layer, the organic matter is distributed diffusely throughout the top few centimetres and structure is very weakly developed. These finding appear to be in general agreement with the work of Waibel (1948) who, working in the Planalto Central, wrote of cerrados: "they are generally sandy and definitely less fertile and drier than forest soils and humus is absent". It must be stressed, however, that Askew et al. only found a correlation between textures and the occurrence of cerrados in the tension zone at the boundary between the two vegetation types: away from the boundary they found cerrado on a range of soil textural classes including clay and forest on a range of soil textural classes but not generally on very sandy soils.

Ahn (1970) found that the most important difference between West African forests and savannas was the much lower organic matter content of the latter compared to the former. Such differences in organic matter status, as discussed below, are likely to be the result of, rather than the cause of vegetational differences. The higher organic matter and clay content of the soils under the Dry Forest, relative to cerrado, reported by Askew et al. might suggest that forest soils would be more "fertile" than savanna soils — as indeed has been suggested by Waibel (1948), Arens (1963) and Ranzani (1963). But the English group, after studying the analyses of volume samples across this particular boundary, found that although this is so in terms of structure, organic matter and phosphorus, it is not so in terms of pH and calcium content. Their conclusions were that "soil fertility" has to be clearly defined; that general themes regarding the

regional savanna forest boundary cannot be put forward on the evidence of one particular situation; and that a distinction has to be made between primary soil properties, which may be of causal importance in influencing the occurrence and characteristics of vegetation types, and secondary soil properties, which have been determined largely by the vegetation and associated organisms in their role as pedogenetic factors. Of the many factors they considered, only texture was thought to be primary and therefore possibly causal.

Blydenstein (1967) working in Colombia reported that in some instances the same soil series was found under both savanna and forest with only minor differences such as a lower pH, darker colours at depth, deeper root penetration and slightly higher soil mineral content (not specified) under the forest. These findings, which are in general agreement with the studies in Mato Grosso, were interpreted by him in a similar manner — that is, the soil differences under savanna and forest are the *result* of the influence of the vegetation on the soil.

In Western Nigeria the Soil Survey (Ojo-Atere and Murdoch 1971) named the same series on both sides of the forest/savanna boundary. Since the soils had the same horizon sequence, differing only in organic matter, nitrogen and associated carbon/nitrogen ratio, it appeared to these authors that they could regard the savanna soils as low-humic variants rather than different soil series. It is important to appreciate, however, that the savanna here is a derived community, the savanna existing on land previously occupied by forest (Keay, 1953; Clayton, 1962; Murdoch, quoted by Young, 1976). Keay (1953) and Smyth and Montgomery (1962) related the advance of this savanna, which may extend to over 50 km in width, to the fact that the forest, over rather sandy soils, is repeatedly cut and burnt by farmers for cultivation and that the forest species eventually do not regenerate on the poor and exhausted soils after cultivation has ceased. The findings of Denevan (1968) in Nicaragua were similar in that he suggested that these savannas which have been created from forest by man's clearing and burning can be related to soils derived from siliceous rocks. Forest it seems has difficulty in re-establishing itself in that area, after clearing and burning, unless the soils are derived from basalt. Moss and Morgan (1970), working like

Keay, Smyth and Montgomery in the forest/savanna boundary zone of Western Nigeria, related the occurrence of forest communities to clayey soils of better moisture status and of savanna to more sandy soils with poorer moisture conditions.

The conflicting conclusions reached on the relationship of the savanna/forest boundary to soil conditions is due, in part at least, to the existence of different types of forest with particular soil requirements. For example, whilst in central Brazil both Dry Forest and cerrado occur on similarly dystrophic Oxisols (Askew et al., 1971), the floristically distinct mesophytic Deciduous Seasonal Forest present in the same area is invariably associated with soils of much higher nutrient status (Ratter et al., 1977), and the distribution of this and related mesophytic forest types as "islands" within the regional cerrado landscape coincides with the occurrence of less highly leached and weathered soils (Waibel, 1948; Rizzini and Heringer, 1962; Eiten, 1978) such as Alfisols and Inceptisols on the outcrops of more basic rocks (Askew et al., 1970b).

Laterite and vegetation

Iron segregation can be an important feature in drier forest and savanna soils, and small pisolithic concretions, for instance, may be found in both. On the other hand, indurated ferruginous crusts (laterite) are far more widely distributed under savanna than forest and deserve a mention in any chapter on savanna soil landscapes. There may, moreover, be a link between the two vegetation formations and laterite in that Maignien (1966) suggested that often the material is formed under forest vegetation but indurated after the forest disappears. This proposal fits well with a theory that, although laterites may form in forests, the hardening off process takes place under savanna and is therefore common at the boundary. Alexander and Cady (1962) introduced an anthropic factor by reporting the change from soft laterite (i.e. plinthite) to hard ironstone in a few years on a site where forest was denuded as a result of man's activity. Phillips (1959) also stated that concretionary ironstone associated with laterization occurs widely in the *derived* savanna zone of Africa but is present to a lesser extent and to a much less advanced degree in the rain forest. The implication of the degradation of forest to savanna

accelerating laterite formation is again clear. The present authors noted in Mato Grosso the frequent abundance of hard laterite in the savanna landscapes and its comparative paucity in the Dry Forests, except in the forest regions which according to Ratter et al. (1973) have advanced at the expense of cerrado. There is one report that in the course of time hard laterite may soften under forest (Roscvcar, 1942).

Soils of the savannas of semi-arid climates

Tricart (1972) has remarked that the savanna of Australia, which he equates with the Brazilian cerrado, has long been adapted to drought and it therefore tolerates a rainfall of approximately 400 to 500 mm which is much lower than on other continents. A comparison of the areas of Australia covered by savanna as depicted by Hills (1965) with the FAO *Soil Map of the World* shows that significant areas of Australian savanna soils as in the Barkly region and around the Flinders and Darling Rivers, can be classified as Vertisols. The other soils which are common are Yermosols (sub-desert soils), some of which are calcic and some of which are transitional to Luvisols (Luvic Yermosols). Regosols (weakly developed soils, lacking prominent horizonation, often formed over deep rather unconsolidated material), Lithosols, and Arenosols are also present.

Soil variability in savannas

The above review has shown that an unusually wide range of soil conditions and types are present in savanna landscapes. In terms of soil properties the soils range in particle size distribution from gravels and sands to the finest clays; in drainage, from soils which are excessively drained and droughty to those which are very poorly drained and, at least seasonally, waterlogged or even inundated; in depth, from thin soils over rock to very thick soils in deeply weathered regoliths; in mineralogy, from montmorillonitic soils to those composed of hydrous oxides and kaolinite; and in base and nutrient status from strongly leached and acid soils with the lowest recorded levels of bases through base-saturated and calcareous soils to even occasionally soils with soluble salts and alkalies. In terms of soil taxonomic units, savanna vegetation grows, for example, on soils from most of the major Orders of the U.S.D.A. classification system. Whilst Oxisols (Ferrallitic Soils) and Ultisols (Ferruginous Soils in part) are especially widespread in savanna lands, various types of Vertisols, Alfisols, Entisols, Inceptisols and Aridosols are present in particular environments. The authors are unaware of any occurrences of undoubted Mollisols (chernozemic soils) associated with savanna vegetation, whilst references to the occurrence of Spodosols (podzols) such as that of Jacomine (1969) are infrequent. The wide range of soil conditions associated with savannas reflects the variability of this vegetation formation from near forest-like communities to treeless grasslands and it is this variability of both soil and vegetation which explains the apparently contradictory reports of vegetation–soil relationships.

Termites and savanna soils

The savanna traveller cannot but be impressed by the termite (Isoptera) earth mounds which are everywhere scattered over the land surface (see Ch. 22). Often mounds are several metres high and Nye (1955) thought that there could be on average five per hectare in many parts of central Western Nigeria. The diameter of a typical mound at ground level may be much larger than the height and when this is appreciated it is not difficult to realize the great weight of soil material (may be up to 3 t per mound) that is moulded into these nests and that the activities of termites constitutes a major pedological factor. Calculations by Nye in Western Nigeria suggest that a layer of soil 30 cm thick could be deposited at the surface of the land by the breakdown of the mounds by natural processes in about 12 000 years. This figure is broadly supported by Meyer (1960) who, in a Zaïre study area, suggested that the amount of material in the scattered mounds would be equivalent to a surface layer of about 20 cm depth over the whole area.

The main morphological effect of the mound is the development of a non-stony, non-gravelly top part of the normal soil profile. In Nigeria where the profiles normally contain a gravel concretion horizon this upper non-gravelly horizon is very noticeable. Nye suggested that there might be need for a new nomenclature to indicate a creep horizon, which he designated Cr.T, due to termites but

despite the attractiveness of his ideas for Western Nigerian conditions they have not been taken up elsewhere.

The effect of termites on soil profiles is not restricted to the actual movement of soil. There is a considerable disruption of the fabric of soils as a result of the underground passageways made by the termite workers. It is also true that although the main effect is by the mound-building termite, the activity of the grass-eating termites which make small mounds only a few centimetres across and the tree-nesting termite are also important. The latter use soil, often red subsoil, to build noticeable ball-like nests, which with their soil covered runs on ground surface and tree trunks are such a prominent feature of savanna trees.

Termite mounds in Africa, South America and Australia are interrelated with vegetation patterns. In South America for instance they form the core of each of the cerrado "islands" in the broad treeless savannas (*pantanal*) associated with some of the flood plains of large rivers. In a like manner they form the core of the "islands" in many of the stream side *campos* (grasslands) which typically cover the lower valley sides in the cerrado landscapes. It is obvious that the mounds are better drained sites in otherwise poorly drained intermittently inundated soils, and that the savanna vegetation is to be expected on them just as grass is to be expected on the lower-lying areas. What is not clear is the origin of the islands, and one can argue about whether the termites predate or postdate the vegetation. The most likely theory on the origin is that the mounds are erosional features left after lateral surface wash (Paton, 1978). This explanation seems to fit the authors' experience in central Brazil, and is in general accord with the work of Glover et al. (1964) who attribute much of the vegetation pattern in Central Africa to lateral surface effects in areas where termite mounds may have a frequency of 650 km^{-2}.

Paton also discussed the question of texture change due to the transfer of material by termites from low in the profile to the surface. He quoted cases where there is preferential accumulation of clay in sand-rich materials and vice versa, although there are also cases where termite mound material has exactly the same particle size distribution as the subsoil material used in its formation.

Clay or sand transference by termites may of

itself have an important effect on soil fertility, but this becomes much more important if the material being moved by the termites has a significantly different chemical status to the existing upper part of the profile, as might be the case in a duplex profile. Even apart from this rather special situation there is evidence from the work of Hesse (1955), Watson (1962) and Sys (1955) in Central Africa which suggests that pH is nearly always higher within the mound (up to a unit higher has been quoted) and that the content of carbon, nitrogen and exchangeable bases, especially calcium and sodium are higher in the mounds than in surrounding soil. Calcium carbonate concretions in mounds have been reported and it seems not unlikely that the selectivity of the termite applies not only to particle size but to nutrients as well. Pomeroy (1976) however, working in Uganda argued that "termites only slightly affect the physical and chemical properties of Ugandan soils, even where mounds are comparatively abundant".

THE CAPABILITY AND UTILIZATION OF SAVANNA SOILS

It has been shown that there are many different soil types on which savanna grows and it is therefore difficult to make general statements about their fertility — that is, their capacity for cropping. Many workers in writing about soil fertility refer particularly to chemical status; but clearly in the poorly drained savannas the chemical status is of little importance until the drainage is corrected. Beard (1953) expressed this quite succinctly for the South American *llanos* soils. The statements which are made about the fertility of savanna soils in general usually refer to the well-drained soils as exemplified by large areas under cerrado in South America and large areas of Africa and Australia. Most workers seem to regard these soils as of low inherent chemical fertility or at least of low fertility relative to the fertility of adjoining forest soils. Bennema (1963) commented on this point at a conference as early as 1961 and Hills mentioned it in a review article in 1965. Ahmad and Jones (1969), Lobato and Goedert (1977), Jones and Wild (1975) and Beadle (1966) made the same point in particular studies in Trinidad, Brazil, West Africa and Australia, respectively. Some workers, however,

notably Askew et al. (1970b) and Blydenstein (1967), only accept it with reservation. In any comparison of the relative fertility of savanna and forest a distinction is needed between the mineralogical status of the soils and their organic matter status. The characteristic freely drained soil of the tropical forest is the Ferrallitic Soil, which has no reserve of weatherable minerals and in which bases and other nutrients are solely contained within the organic matter cycle, whereas the predominant freely drained soil of the savanna lands, in Africa at least, is the Ferruginous Soil which still contains some weatherable minerals in addition to the nutrients present in the organic matter cycle. Consequently, in terms of inherent mineral reserves the Ferruginous Soil is more "fertile" than the Ferrallitic Soil, even if the rate of release of minerals in weathering is very slow. By contrast forest soils contain generally a higher content of organic matter and humus than savanna soils so that following the initial burning and cultivation the forest soils will be more fertile, at least temporarily, in that they will have a higher level of nutrients and a more satisfactory structural condition. Apart from eutrophic soils under mesophytic forests, the higher organic status and better physical condition of forest soils is a function rather than the cause of the forest vegetation.

Nutrient status survey

The information on carbon/nitrogen ratios of savanna topsoils is interesting in that the ratio is linked to the release and availability of nitrogen during mineralization. Odu and Vine (1968) reported high figures of 15 to 20 for the Derived Savanna and Southern Guinean Zone[1] of West Africa but normal figures of 10 to 12 in the drier (500–950 mm mean annual rainfall) Northern Guinean and Sudanian Zones. The high carbon/nitrogen figures for the former zones are associated with a great deficiency of available nitrogen when bringing land into cultivation. Sys (1976) however, reported 11.2 and 12.7 respectively as the ratios for the Sudanian and Guinean Zones. Askew et al. (1970b) reported a mean carbon/nitrogen ratio of

---

[1]These zones were defined by Keay (1953).

11.2 for soils they described under savanna in Mato Grosso where the rainfall was 1370 mm in 1968.

Low phosphorus content of Australian soils has been discussed by Wild (1958); for instance, it is regarded by Beadle (1962) as a major factor in the delimitation of forest, woodland and scrub around Sydney. Phosphorus deficiency is also reported in Brazil (McClung et al., 1958), in Zimbabwe (Rhodesia) where it is claimed to affect nitrification (Purchase, 1974), and in Ghana (Stephens, 1960). Trace element deficiencies are reported in savanna soils in Burma (Obukhov, 1968), Mali (Peive, 1963) and parts of Brazil (Verdade, 1971). A number of trace elements are usually mentioned but zinc, boron, molybdenum and cobalt are prominent. Potassium does not appear to be low in savanna soils, and the relatively high content of potassium in Nigerian savanna soils is regarded by Lombin and Fayemi (1976) as a factor in the magnesium deficiency which they visualize as likely under cropping. Low base status and high exchangeable aluminium are often reported in South American conditions, and poor sulphur supply in savanna soils has been reported both from Nigeria (Enwezor, 1976) and from Brazil (McClung et al., 1959).

In the light of this survey it is not surprising that crop yield increases to macro and minor nutrients are widely reported throughout the savanna zone. Moderate or high responses were obtained in Ghana to the application of nitrogen and phosphorus (Nye, 1952; Koli, 1973). Examples of the improvement of cropping following fertilization in Brazil are: responses to nitrogen, phosphorus, potassium, sulphur, dolomitic lime and trace elements, especially boron and zinc, on coffee (Lazzarini et al., 1975); responses to nitrogen, phosphorus, potassium, zinc and lime (Freitas et al., 1960); responses to nitrogen, phosphorus, potassium and zinc (Britto et al., 1971) for maize, cotton and beans; and responses to nitrogen, phosphorus, potassium, calcium, magnesium, sulphur, zinc and molybdenum (Freitas et al., 1971). Zinc was given special significance in the work by Osiname et al. (1973) in Nigeria and Pereira et al. (1973) in Brazil, both studies using maize. Lime also is regarded as especially important in a review for South American conditions reported by Soares and colleagues (1975).

Land use in the savannas

At the present time large regions of the savanna are used for grazing. Often the grazing system is part of a nomadic way of life and it is not harmful to the soil: occasionally, however, when the system is static, overgrazing may occur and cause erosion. In many other parts of the savanna arable land use is much more common and distinctive farming systems have evolved. Three common agricultural systems have been described by Jones and Wild (1975) and although their descriptions refer to West Africa the systems are widespread throughout savanna lands. Where population is very low, a haphazard system of shifting cultivation is practised, often with movement of village sites; where population is low to medium, a deliberate and regular system of rotational bush fallowing is common; and where population density is high, as around large cities, virtually continuous cropping using fertilizers may be practised. It must be emphasized that there is a continuous transition from one system to the next, and, although a run-down in soil fertility is one of the critical factors in the decision of the peasant farmer on the timing and duration of the various parts of the cycle in the two non-continuous cropping systems, other factors, such as weed infestation, also influence his decision. Prior to using the land in any of the systems mentioned, clearing and burning take place.

The role of the savanna fallow in maintaining soil fertility has been fully described by Jones and Wild (1975) and by Nye and Greenland (1960) and will not be further discussed here, especially as the fallow is essentially part of the cropping system rather than the natural savanna ecosystem. The role of fire, however, is different in that it may start accidentally due to lightning or it may be started deliberately for hunting rather than cropping; even when started deliberately with cultivation in view its effect may be much more widespread than intended by the farmer. Thus, although fire may be part of the cultivation cycle it is just as likely to be part of the natural (non-anthropogenic) cycle of events in the savanna ecosystem. Many workers, as emphasized elsewhere in this book regard fire, if not as a causative factor in savanna, at least as a maintaining factor. The intensity of the burn is closely related to whether it occurs in the early or late dry season. The effect on the soil fertility, however, can be described in terms of general principles, as long as it is remembered that the effects will be much more pronounced if the burn is carried out late in the dry season.

It will be easily appreciated that there are losses as well as gains during the burning process. Nitrogen and sulphur are lost as gases. Phosphorus may be lost as well but this is less likely. The litter, but not the intimate humus, may be destroyed. Woody species may be replaced by grasses which are less effective in accumulating and storing nutrients and building up organic matter. The beneficial side of burning is the nutrient addition in the ash. Calcium, magnesium and potassium are all added to the soil and the addition of the alkaline ash will raise pH values by from 0.1 to about 0.3 pH units. Unfortunately this increase can be quickly lost during any subsequent cultivation cycle.

The extent and severity of erosion in savannas generally is not known although maps, such as that produced by Fournier in 1962, for Africa give a good indication of potential hazards. The FAO (1965) estimated that 70% of the soil of Northern Nigeria was eroded to some extent. Most soil erosion problems, for example, sheet wash and gullying, can be related to overgrazing or lack of proper management techniques during cropping. Leaving the soil bare for the minimum period of time, after clearing and burning, using mulches, if available, and using contour ridges and strip cropping, are some of the techniques that are available to minimize erosion.

REFERENCES

Adjanohoun, E., 1968. Some ecological considerations of the forest/savanna boundary in the Ivory Coast. In: T.L. Hills and R.E. Randall (Editors), *The Ecology of the Forest/Savanna Boundary*. Proc. I.G.U. Symposium, Venezuela 1964, pp. 93–96.

Ahmad, N. and Jones, R.L., 1969. A plinthaquult of the Aripo savannas, north Trinidad. *Proc. Soil. Sci. Soc. Am.*, 33: 762–768.

Ahn, P., 1970. *West African Soils*. Oxford University Press, Oxford 332 pp.

Alexander, L.T. and Cady, J.G., 1962. Genesis and hardening of laterite in soils. *U.S. Dep. Agric. Tech. Bull.*, No. 1282: 90 pp.

Arens, K., 1963. As plantas lenhosas dos campos cerrados como flora adaptada as deficinecias minerais do solo. In: M.G. Ferri (Editor), *Simpósio sôbre o Cerrado*. University of São Paulo, São Paulo, pp. 285–289.

Askew, G.P., Moffatt, D.J., Montgomery, R.F. and Searl, P.L., 1970a. Soil landscapes in north-eastern Mato Grossó. *Geogr. J.*, 136: 211–227.

Askew, G.P., Moffatt, D.J., Montgomery, R.F. and Searl, P.L., 1970b. Interrelationships of soils and vegetation in the savanna–forest boundary zone of north-eastern Mato Grosso. *Geogr. J.*, 136: 370–376.

Askew, G.P., Moffatt, D.J., Montgomery, R.F. and Searl, P.L., 1971. Soils and soil moisture as factors influencing the distribution of the vegetation formations of the Serra do Roncador, Mato Grosso. In M.G. Ferri (Editor), *III Simpósio sôbre o Cerrado*. University of São Paulo, São Paulo, pp. 150–160.

Beadle, N.C.W., 1962. Soil phosphate and the delimitation of plant communities in Eastern Australia II. *Ecology*, 43: 281–288.

Beadle, N.C.W., 1966. Soil phosphate and its role in moulding segments of the Australian flora and vegetation with special reference to xeromorphy and sclerophylly. *Ecology*, 47: 992–1007.

Beard, J.S., 1953. The savanna vegetation of northern tropical America. *Ecol. Monogr.*, 23: 149–215.

Bennema, J., 1963. Caracteristicas quimicas e fisicas de latossolos sob vegetação de cerrado. *Bol. Dep. P.E.A.*, 15: 137–143.

Bennema, J., Lemos, R.C. and Vettori, L., 1959. Latosols in Brazil. In: *Proc. 3rd Inter-Afr. Soils Conf., Dalaba*, pp. 273–281.

Blydenstein, J., 1967. Tropical savanna vegetation of the llanos of Colombia. *Ecology*, 48: 1–15.

Britto, D.P.P. de S., Castro, A.F. de, Mendes, W., Jaccoud, A., Ramos, D.P. and Costa, F.A., 1971. Response to micronutrients of a dark red latosol under cerrado vegetation. *Pesqui. Agropecu. Brasil.*, 6: 17–22.

Budowski, G., 1968. Classification and origin of savannas in the light of a world vegetation classification. In: T.L. Hills and R.E. Randall (Editors), *The Ecology of the Forest/Savanna Boundary*. Proc. I.G.U. Symposium, Venezuela 1964, pp. 1–4.

Clayton, W.D., 1962. Derived savanna in Kabba Province, Nigeria. *Samaru Res. Bull.*, 15: 595–604.

Cole, M.M., 1960. Cerrado, caatinga and pantanal: the distribution and origin of the savanna vegetation of Brazil. *Geogr. J.*, 126: 168–179.

Cole, M.M., 1968. Categories of savanna vegetation. Their distribution in relation to soils and geomorphology. In: T.L. Hills and R.E. Randall (Editors), *The Ecology of the Forest/Savanna Boundary*, Proc. I.G.U. Symposium, Venezuela 1964, pp. 57–61.

Denevan, W.M., 1965. The campo cerrado vegetation of Central Brazil. *Geogr. Rev.*, 55: 112–115.

Denevan, W.M., 1968. Observations on savanna/forest boundaries in tropical America. In: T.L. Hills and R.E. Randall (Editors), *The Ecology of the Forest/Savanna Boundary*. Proc. I.G.U. Symposium, Venezuela 1964, pp. 66–67.

D'Hoore, J.L., 1964. *Soil Map of Africa Scale 1 to 5,000,000. Explanatory Monograph*. C.C.T.A., Lagos, 205 pp.

Eden, M.J., 1970. Savanna vegetation in the northern Rupununi, Guyana. *J. Trop. Geogr.*, 30: 17–28.

Eden, M.J., 1971. Scientific exploration in Venezuelan Amazonas. *Geogr. J.*, 137: 149–156.

Eiten, G., 1978. Delimitation of the cerrado concept. *Vegetatio*, 36: 169–178.

Enwezor, W.O., 1976. Sulphur deficiencies in soils of southeastern Nigera. *Geoderma*, 15: 401–411.

FAO Staff, 1965. *Agricultural Development in Nigeria 1965–80*. FAO, Rome, 512 pp.

FAO–UNESCO, 1974. *Soil Map of the World 1:5,000,000. Vol. 1. Legend*. FAO–UNESCO, Paris, 59 pp.

Ferri, M.G., 1962. Problems of water relations of some Brazilian vegetation types, with special consideration of the concepts of xeromorphy and xerophytism. In: *Proc. Madrid Symp. on Plant–Water Relationships in Arid and Semi-arid Conditions. Arid Zone Res.*, 16: 191–197.

Fournier, F., 1962. *Carte du danger d'érosion en Afrique au sud du Sahara*. CEE-CCTA.

Freitas, L.M.M. de, McClung, A.C. and Lott, W.L., 1960. Field studies on fertility problems of two Brazilian campos-cerrados — 1958–1959. *I.B.E.C. Res. Inst. Bull*, 21: 31 pp.

Freitas, L.M.M. de, Lobato, E. and Soares, W.V., 1971. Lime and fertilizer experiments on soils under cerrado vegetation in the Federal District 1971. *Pesqui. Agropecu. Brasil.*, 6: 81–89.

Glover, P.E., Trump, E.C. and Wateridge, L.E.D., 1964. Termitaria and vegetation patterns on the Loita Plains of Kenya. *J. Ecol.*, 52: 365–377.

Goodland, R., 1970. The savanna controversy: background information on the Brasilian cerrado vegetation. *McGill Univ. Sav. Res. Ser.*, 15: 66 pp.

Goodland, R., 1971. The cerrado oxisols of the Triangulo Mineiro, Central Brazil. *An. Acad. Bras. Ciênc.*, 43: 407–414.

Goodland, R. and Pollard, R., 1973. The Brazilian cerrado vegetation: a fertility gradient. *J. Ecol.*, 61: 219–224.

Hesse, P.R., 1955. A chemical and physical study of the soils of termite mounds in East Africa. *J. Ecol.*, 43: 449–461.

Hills, T.L., 1965. Savannas: a review of a major research problem in tropical geography. *Can. Geogr.*, 9: 216–228.

Hills, T.L., 1969. The savanna landscapes of the Amazon basin, *McGill Univ. Sav. Res. Series*, 14: 38 pp.

Jacomine, P.K.T., 1969. Descricão das caracteristicas morfologicas, fisicas, quimicas e mineralogicas de alguns perfis de solos sob vegetação de cerrado. In: *Bol. 11 Escritorio de Pesquisas e Experimentacão, M.A., Rio de Janeiro*, 126 pp.

Jones, M.J. and Wild, A., 1975. Soils of the West African Savanna. *C.A.B. Tech. Comm.*, No. 55: 246 pp.

Keay, R.W.J., 1953. *An Outline of Nigerian Vegetation*. Government Printer, Ibadan, 55 pp.

Koli, S.E., 1973. The response of Yam (*Dioscorea rotundata*) to fertilizer application in Northern Ghana. *J. Agric. Sci.*, 80: 245–249.

Lazzarini, W., Moraes, F.R.P. de, Cervellini, G.S., Toledo, S.V. de, Figueiredo, J.I., Reis, A.J., Conagin, A. and Franco, C.M., 1975. Coffee production on a red yellow latosol in the Batatais region São Paulo. *Bragantia*, 34: 229–239.

Lemos, R.C. de (Editor), 1960. Levantamento de recohecimento dos solos do Estado de São Paulo. *C.N.E.P.A. Brazil, S.N.P.A., Bol.* No. 12; 633 pp.

Lobato, E. and Goedert, W.J., 1977. Increasing the productivity of Brazilian cerrado soils. In: *Proc. Int. Seminar on Soil Environment and Fertility Management in Intensive Agriculture, Tokyo*, pp. 462–471.

Lombin, G. and Fayemi, A.A.A., 1976. Magnesium status and availability in soils of western Nigeria. *Soil Sci.*, 122: 91–99.

McClung, A.C., Freitas, L.M.M. de, Gallo, J.R., Quinn, L.R. and Mott, G., 1958. Preliminary fertility studies on 'campos cerrados' soils in Brazil. *I.B.E.C. Res. Inst. Bull.*, 13: 19 pp.

McClung, A.C., Freitas, L.M.M. de, Lott, W.L., 1959. Analyses of several Brazilian soils in relation to plant responses to sulphur. *Soil Sci. Soc. Am. Proc.*, 23: 221–224.

Magnien, R., 1966. *Review of Research on Laterites*. UNESCO, Paris, 148 pp.

Meyer, J.A., 1960. Resultats agronomiques d'un essai de nivellement des termitieres realisé dans la Cuvette centrale Congolaise. *Bull. Agric. Congo Belge*, 51: 1047–1059.

Moss, R.P. and Morgan, W.B., 1970. Soils, plants and farmers in West Africa. In: J.P. Garlick (Editor), *Human Ecology in the Tropics*. Pergamon Press, London, pp. 1–31.

Murdoch, G., Ojo-Atere, J., Colborne, G., Olomu, E.I. and Odugbesan, E.M., 1976. Soils of the Western State savanna in Nigeria, 1. The Environment. *Min. Overseas Devel., Land Res. Stud.*, 23: 102 pp.

Nye, P.H., 1952. Studies on the fertility of Gold Coast soils. III. The phosphate status of the soils. *Emp. J. Exper. Agric.*, 20: 47–55.

Nye, P.H., 1955. Some soil-forming processes in the humid tropics IV. The action of the soil fauna. *J. Soil Sci.*, 6: 73–83.

Nye, P.H. and Grenland, D., 1960. The soil under shifting cultivation. *C.A.B., Tech. Comm.*, No. 51: 156 pp.

Obukhov, A.I., 1968. Content and distribution of trace elements in soils of the dry tropical zone of Burma. *Pochvovedenie*, 22: 93–102 (in Russian; Abstract 5 Annotated Bibliography S1414R on Savanna Soils 1959–1970, C.A.B., England).

Odu, C.T.I. and Vine, H., 1968. Transformations of ^{15}N-tagged ammonium and nitrate in some savanna soil samples. In: *Isotopes and Radiation in Soil Organic Matter Studies*. International Atomic Energy Agency, Vienna, pp. 351–361.

Ojo-Atere, J. and Murdoch, G., 1971. Soil series straddling the forest/savanna boundary in the Western State of Nigeria. *Land Resour. Div., Misc. Rep.*, 110.

Osiname, O.A., Kang, B.T., Schulte, E.E. and Corey, R.B., 1973. Zinc responses of maize (*Zea mays*) grown on sandy inceptisols in Western Nigeria. *Agron. J.*, 875–877.

Paton, T.R., 1978. *The Formation of Soil Material*. Allen and Unwin, London, 143 pp.

Peive, Ya, V., 1963. Content of trace elements in soils of the savanna zone of the Mali Republic. *Pochvovedenie*, 11: 47–50 (in Russian; Abstract 25 Annotated Bibliography S1414R on Savanna Soils 1959–1970, C.A.B., England).

Pereira, J., Vieira, I.F., Moraes, E.A. and Rego, A.S., 1973. Levels of zinc sulphate in maize (*Zea mays*) grown in cerrado soils. *Pesqui. Agropecu. Brasil.*, 8: 187–189.

Phillips, J.F.V., 1959. *Agriculture and Ecology in Africa*. Faber and Faber, London, 424 pp.

Pomeroy, D.E., 1976. Some effects of mound-building termites on soils in Uganda. *J. Soil Sci.*, 27: 377–394.

Purchase, B.S., 1974. The influence of phosphate deficiency on nitrification. *Plant Soil*, 41: 541–547.

Ranzani, G., 1963. Solos do cerrado. In: M.G. Ferri (Editor), *Simpósio sôbre o Cerrado*. University of São Paulo, São Paulo, pp. 51–92.

Ranzani, G., 1971. Solos do cerrado no Brasil. In: M.G. Ferri (Editor), *III Simpósio sobre o Cerrado*. University of São Paulo, São Paulo, pp. 26–43.

Ratter, J.A., 1971. Some notes on two types of cerradão occurring in north eastern Mato Grosso. In: M.G. Ferri (Editor), *III Simpósio sôbre o Cerrado*. University of São Paulo, São Paulo, pp. 100–102.

Ratter, J.A., Richards, P.W., Argent, G. and Gifford, D.R., 1973. Observations on the vegetation of north eastern Mato Grosso I. The woody vegetation types of the Xavantina–Cachimbo Expedition area. *Philos. Trans. R. Soc. Lond.*, B 266: 449–492.

Ratter, J.A., Askew, G.P., Montgomery, R.F. and Gifford, D.R., 1977. Observacoes adicionas sobre o cerradão de solos mesotroficos no Brasil Central. In: M.G. Ferri (Editor), *IV Simpósio sôbre o Cerrado*. University of São Paulo, São Paulo, pp. 303–316.

Rizzini, C.T. and Heringer, E.P., 1962. *Preliminares acêrca das formações vegetais e do reflorestamento no Brasil Central*. Ministerio da Agricultura, Rio de Janeiro, 79 pp.

Rosevear, R.D., 1942. Soil changes in Enugu Plantation. *Farm For.*, 5(1).

Sarmiento, G. and Monasterio, M., 1975. A critical consideration of the environmental conditions associated with the occurrence of savanna ecosystems in tropical America. In: F.B. Golley and E. Medina (Editors), *Tropical Ecological Systems*. Springer-Verlag, New York, N.Y., 223–250.

Smyth, A.J. and Montgomery, R.F., 1962. *Soils and Land Use in Central Western Nigeria*. Government Printer, Ibadan, 265 pp.

Soares, W.V., Lobato, E., Gonzales, E. and Naderman, G.C. Jr.,

1975. Liming soils of the Brazilian cerrado. In: *Proc. Seminar on Soil Management in Tropical America, Colombia 1974*. Soil Sciences Department, University of North Carolina, Chapel Hill, N.C., pp. 283–299.

Soil Survey Staff, 1975. *Soil Taxonomy*. U.S. Dep. Agric., Washington, D.C., 754 pp.

Stephens, D., 1960. Three rotation experiments with grass fallows and fertilizers. *Emp. J. Exper. Agric.*, 28: 165–178.

Sys, C., 1955. The importance of termites in the formation of latosols in the region of Elizabethville. *Soils Afr.*, 3: 393–395.

Sys, C., 1976. Influence of ecological conditions on the nutrition status of tropical soils. *Pédologie*, 26: 179–190.

Tamayo, F., 1968. Classification of the Venezuelan Savannas. In: T.L. Hills and R.E. Randall (Editors), *The Ecology of the Forest/Savanna Boundary*. Proc. I.G.U. Symposium, Venezuela 1964, pp. 8–9.

Tricart, J., 1972. *The Landforms of the Humid Tropics, Forests and Savannas*. Longman, London, 306 pp.

Verdade, F.C., 1971. Agricultura e silvicultura no cerrado. In: M.G. Ferri (Editor), *III Simpósio sôbre o cerrado*. University of São Paulo, São Paulo, pp. 65–76.

Vesey-Fitzgerald, D.F., 1963. Central African grasslands. *J. Ecol.*, 51: 243–274.

Waibel, L., 1948. Vegetation and land use in the Planalto Central of Brazil. *Geogr. Rev.*, 38: 529–554.

Watson, J.P., 1962. The soil below a termite mound. *J. Soil Sci.*, 13: 46–51.

Wild, A., 1958. The phosphate content of Australian soils. *Aust. J. Agric. Res.*, 9: 193–204.

Young, A., 1976. *Tropical Soils and Soil Survey*. Cambridge Geographical Studies, 9. Cambridge University Press, London, 468 pp.

Chapter 5

LIFE FORMS AND PHENOLOGY

G. SARMIENTO and M. MONASTERIO

INTRODUCTION

One of the most enlightening approaches to the study of terrestrial ecosystems is that focused on the careful consideration of the morphological features and the phenological behavior of their member species, as expressed both by peculiar traits of external morphology and overall organization, and by sequential development of structures during each annual cycle. In ecosystems evolving in extreme environments, the external constraints to plant survival are so strong that just a few possibilities of morphological and phenological expression remain available, narrowing the range of possible plant responses to such unfavorable conditions; in these cases, then, species follow only a few different behavioral patterns. Warm humid tropical ecosystems, on the other hand, occupy the other extreme of the range of natural environments; here the external constraints are much more gentle and less varied, imposing only some broad limitations, mainly as a need to overcome some kind of seasonal water stress. Under such circumstances the possibilities open to the flora are so numerous that a wide range of life-cycle and life-form strategies becomes compatible with the successful performance of plant species.

According to this reasoning, warm humid tropical ecosystems such as rain forests and savannas will exhibit a diversity of specific solutions to overcome the obstacles set by the biotic and physical environment. A critical analysis of these plant responses may accomplish in this case a dual purpose: first, to penetrate into the very essence of the functioning of the ecosystems by recognizing patterns of spatial and temporal division in the allocation of resources among their populations;

and, second, to grasp the fantastic evolutionary skill displayed by plant life that led to successful survival under such a set of external constraints, including among the restrictive factors both abiotic strains and the biotic pressures derived from competition and coevolution.

The rich and taxonomically diversified floristic stocks of tropical savannas have evolved under environmental pressures that, though not extremely harsh, represent a serious handicap to the performance of flowering plants. Species with a long evolutive acquaintance with these ecosystems have found adaptive answers to these challenges through different architectural designs and developmental patterns. Furthermore, the type of seasonal stress imposed by the humidity regime in savanna ecosystems has allowed a wider spectrum of plant responses than in the case of temperate ecosystems synchronized mainly with thermal seasonality, where low temperatures permit only a few strategies of survival through a cold season dormancy.

Within the whole tropical life world, savannas appear to be one of the most marked examples of seasonal ecosystems, the changes during each annual cycle being perhaps one of their most striking observable features. The woody and the herbaceous savanna species have to overcome the same major environmental risks: seasonal drought and/or flood, and periodic burning; they attain these goals by means of a precise synchronization of their successive phases with the rhythmically changing environment. The allocation of resources to the various plant structures must become compatible with the periodic consumption of almost the whole aerial standing crop by fire, and this sole drastic restriction has necessarily imposed certain architectural patterns, while other solutions have had to be discarded as less efficient. The same may be said about the filtering effect of an extended dry season, operating as a strong selective sieve against plant forms and developmental strategies not suited to avoid, resist or escape this major environmental stress.

Besides the intrinsic importance of knowing the forms and rhythms of savanna species, this knowledge also appears as a necessary precondition to implement efficient procedures of ecosystem management, particularly in those ecosystems whose main role in the near future seem to be as natural rangelands. The desirability of improving the quality of this resource makes still more urgent an understanding of every aspect related to biomass allocation and to the seasonality of the species.

After the pioneer work of some of the founders of plant ecology (Warming, 1909; Clements, 1920; Braun-Blanquet, 1932; Du Rietz, 1931; Raunkiaer, 1934), the morphophenological approach to the study of plant populations and ecosystems fell into relative disregard, overwhelmed by an avalanche of floristic works of an increasingly quantitative nature. A more balanced treatment of the various aspects of plant ecology has been attained many decades later, following the ambitious projects of the International Biological Programme between 1968 and 1974, partly because of the obvious implications of growth forms and rhythms for an understanding of primary production processes. Furthermore, evolutionary plant ecology regained its lost position as an important biological subject, highly deserving recognition in its own right as a growing and powerful field of inquiry. The few far-sighted publications arguing for a more Darwinian approach to plant ecological problems (Harper, 1967) bore fruit in several convergent research lines which tackled the problems of plant populations within their respective ecosystems (Harper, 1977). One of these new research lines leads to a more dynamic approach to plant form and organization, viewed not just as ready-made structures, but as the accomplishment of an architectural design, coded in the genotype, but permitting varied phenotypic expression in accordance with the hazards of life cycles under the influences of diverse external modelling agents (Hallé and Oldeman, 1970). This integrated approach contributed to the renewed interest in plant form and in developmental pat-

terns, not only by linking together both kinds of phenomena, but by putting them within a common frame of the natural ecosystems in which these structures and patterns have evolved (Oldeman, 1974).

However, in spite of all these recent advances in the ecology of plant populations, it is still too early to see them reflected in ecological studies of savanna species. Thus, for example, the architectural approach to plant form as defined by Hallé and Oldeman (1970) implies a detailed knowledge of the successive phases of plant development that is still far from being reached in savanna species, which by the very nature of the external stresses to which they are subjected, have a twisted appearance of quite difficult interpretation. Furthermore, the disentangling of architectural patterns in herbaceous plants is still only in its very beginning (Jeannoda-Robinson, 1977) and almost nothing is known about the architecture of perennial grasses. Neither has knowledge of the demography of perennial grasses advanced beyond its first steps, although one may hope for a vigorous development of these subjects during the next decade. We will be obliged, then, to restrict our treatment to a more classic approach, but on any possible occasion we will risk discussion of implications concerning architectural models and considerations regarding the evolutionary advantages of plant forms and developmental patterns.

The present situation concerning the knowledge of phenological strategies in savanna species seems to be somewhat better, since relevant information is already available at least for a few well-known ecosystems, such as the Lamto savannas in the Ivory Coast and the *llanos* savannas in Venezuela. Here, our task has mostly been limited to an analysis of this information in order to point out the main phenological patterns found so far in tropical savannas, and afterwards to discuss their possible significance as responses to the major environmental impulses and competitive pressures characteristic of these ecosystems.

Though both aspects of form and development cannot be isolated from each other if they are to provide a basis for the comprehension of adaptive strategies, in order to attain a clearer picture of these two aspects in our limited space we will proceed first to discuss life forms afterwards to enter into the consideration of seasonal cycles, and to end the chapter with a joint treatment of the evolutionary implications of forms and rhythms, as well as of their relevance to understanding of the major characteristics of tropical savanna ecosystems.

LIFE FORMS

The Raunkiaer system of life forms and biological spectra (Raunkiaer, 1934), with various later improvements, has been widely employed to compare numerous floras from everywhere in the world. Table 5.1 shows some results for the floras of several savannas from different tropical regions.

Compared with Raunkiaer's normal spectrum, it is clear from Table 5.1 that phanerophytes are very much under-represented in the savanna flora, while therophytes appear in a much greater proportion than in the average world flora. The ensemble of chamaephytes, hemicryptophytes and geophytes is over-represented, though the participation of each group varies among the different savannas — perhaps partly because of differences in criteria among the authors about what life form should be ascribed to some species of subshrubs and perennial herbs. In general, hemicryptophytes and geophytes together seem to be the dominant group of life forms in the savanna flora. But, according to Raunkiaer's principles and conclusions, a "hemi-cryptophtic–geophytic phytoclimate" corresponds to a cool- or cold-humid climate proper to high latitudes or altitudes, quite unrelated ecologically to the tropical savanna environment.

This sort of result has led to the almost unanimous opinion among tropical ecologists that the applicability of Raunkiaer's life form system to savannas in particular, and to the whole tropical plant world in general, provides information of very dubious interpretation. On the one hand, this system classifies life forms primarily according to the position of buds during the unfavorable season, on the supposition that the limiting factor for plant growth was the low winter temperature. In tropical lowlands, obviously, low temperatures do not represent an ecological factor to be taken into account, the crucial environmental limitations being quite different, such as an extended drought as well as periodic burnings in seasonal savannas, plus a period of waterlogging in hyperseasonal savannas.[1] Bud height above ground level does not

[1] The phrase "hyperseasonal savannas" was introduced by the authors (Sarmiento and Monasterio, 1975) to describe savannas subject to both seasonal drought and seasonal waterlogging.

TABLE 5.1

Life-form spectra for several savannas, according to the Raunkiaer system

Locality	Phanero-phytes	Chamae-phytes	Hemicrypto-phytes	Geophytes	Thero-phytes	Reference
Zaire, Lake Edward plains; mean of 2 savannas	5	38	22	5	29	Lebrun (1947)
Nigeria, Olokemeji savanna site	30	0	23	21	25	Hopkins (1962)
Ivory Coast, Lamto; mean of 8 savannas	9	1	62	9	19	César (1971)
Southwest Madagascar: mean of 11 savannas	21	18	26	3	32	Morat (1973)
Northern Surinam; total savanna flora	8	3	38[1]	28	23	Van Donselaar-Ten Bokkel Huinink (1966)
Central Venezuelan llanos, savanna flora of Calabozo	28	7	31	5	29	Aristeguieta (1966)
Western Venezuelan llanos, savannas of Barinas	11	3	18	40	28	Sarmiento (unpubl.)

[1] Including Geophyta geopodiosa.

influence the response of plants to these factors as it does to cold, which therefore renders the Raunkiaer system of low interpretative value in comparing tropical with extratropical plant communities.

On the other hand, a feature often noticed in tropical savannas is the occurrence of many species which are difficult to classify into one or another life form, since they behave as phanerophytes under certain conditions, and as hemicryptophytes under another environmental stress; there are even certain species that pass through three or even four different life forms during their development. In many tropical savannas, particularly those subjected to a severe dry season and to frequent fires, the growth patterns of certain perennial herbs and subshrubs show features that make their classification as geophytes, hemicryptophytes or chamaephytes so arbitrary as to lack any ecological significance; examples of this phenomenon will be mentioned later.

Due then to these inherent limitations of the Raunkiaer life form system, we thought it more fruitful to consider the most conspicuous forms occurring in savanna ecosystems without pretending to follow any formal system, focusing our attention on those morphological features that seemed to us to be of adaptive and evolutionary significance for plant fitness under the environmental conditions where these populations have been successful.

To start with, savanna species will be divided into three morphofunctional groups according to the degree of perennation of their vegetative structures. The first group comprises those species that have certain permanent above-ground structures, that is, some organs that live more than one year and receive therefore the full impact of a whole cycle of environmental stresses. The aerial plant parts living more than one year may, in a few cases, be the entire shoot including leaves and young branches, such as in palms for instance, more commonly the single persistent structure is the trunk together with old branches. This group includes all woody species, such as trees and shrubs, as well as caulescent palms and a few succulent rosettes, like the aloes in African savannas and the bromeliads in America. In Raunkiaer terminology all phanerophytes and chamaeophytes, and certain hemicryptophytes are assembled here.

Our second group is formed by perennial species in which all above-ground parts are entirely seasonal — the aerial organs do not survive from one annual cycle to the next, but the plants have perennating underground structures of different types. To this group belong the geoxyles or hemixyles, that is, species with woody underground organs, as well as perennial herbs with non-woody rhizomes or bulbs. All geophytes, most hemicryptophytes and a group of species hard to classify into one or another category, have been assembled into this group.

A third type of growth form includes annuals, all of them lacking any perennating vegetative structure, of which only the seed bank persists during the unfavourable season. It corresponds to the therophytes of the Raunkiaer system.

We consider that this primary differentiation of three growth forms is more meaningful from an ecological viewpoint that the traditional division between woody and herbaceous species — firstly because it gives a clearer idea of the structural behavior of each species in face of the environmental constraints, and secondly because the strict anatomical differentiation of herbaceous and woody plants is not so straightforward as it might appear at first sight. Separation of primary and secondary tissues and structures is often difficult since in dicotyledons there are varying degrees of lignification (Esau, 1953), besides the great dissimilarities between "woody" monocotyledons and dicotyledons.

However, there is a difficulty with our proposed system: often enough, the ground level is not a sharp boundary for the separation of perennating structures, since in many species a woody organ may persist almost at ground level or immediately above or below it. Under other circumstances, the ground level may change a few centimeters either by sheet erosion, deposition or some biological artifacts, without affecting the survival of these species. Our criterion in these cases has been to consider a given structure as aerial or subterranean according to where most of its living biomass is located.

SPECIES WITH PERENNIAL ABOVE-GROUND ORGANS

In savannas, the tree form constitutes the major architectural type within this group, while palms, shrubs, woody vines, succulents, caulescent rosettes, epiphytes and so forth are much less frequent types. We consider first, then, the main morphological features of savanna trees.

It is convenient to keep in mind that there has not been unanimity as to whether woody savanna species should be considered as trees or as shrubs, therefore we will regard as trees all woody plants with a well-defined trunk, and also those forms branching close to the ground or even bifurcating from the ground level itself. In fact, the branching patterns and the existence of a well-defined bole, or of several suckers sprouting from the stem base or from roots, are dependent on the past events acting upon any particular population; hence, the distinction between forms with one or with two or three main axes does not seem very important. Trees branching from ground level represent less frequent forms originated by suckering after fire damage, but under suitable circumstances these suckers will be able to become trees.

Once having stated what we call a tree, the first feature of savanna trees to note is the relatively modest development attained by their aerial biomass. In fact, in most savanna communities an overwhelming proportion of the tree flora have a mean height ranging from 2 to 6 m, the tallest individuals rarely exceed 12 m. When they reach their maximum development, most trees are just microphanerophytes or at the best some of them slightly surpass the upper limit of this class (8 m). This most obvious trait neatly separates the forest canopy species (in rain forests or in tropical deciduous forests) from the tallest savanna trees.

In accordance with their limited height, the arboreal savanna species also have rather small stem diameters; in any species, when an individual has already formed a single main bole and attained its maximum development, its diameter will usually be well below 40 or 50 cm, the average for mature trees being of the order of 20 to 30 cm. Lawson et al. (1968) gave the girth class distribution of all tree species (34 species) in four sample plots (50 × 50 m) in the Guinean savanna of northern Ghana. Their figures show a modal class at 20 to 30 cm, and a sharp decline in frequency above 70 cm, with just a few trees reaching maximum girth values slightly surpassing 100 cm. Menaut (1971) sampled the woody populations in the Lamto savannas with five 50 × 50 m quadrats, measuring tree height and stem girth at the base. His diagrams show that all trees were lower than 10 m, except the palm *Borassus aethiopum* which might attain 20 m; stem girths very rarely surpassed 150 cm, with the mean in the classes from 20 to 40 cm, while the histograms of frequency sharply declined above 50 cm. Ataroff (1975) gave figures for 200 randomly chosen individuals of the two most frequent tree species (*Curatella americana* and *Byrsonima crassifolia*) in the seasonal savannas of the western Venezuelan llanos. The mean stem girth at ground level calculated on her data are 26 and 23 cm, respectively; the largest girth recorded was 125 cm. These modest dimensions, both in height and trunk girth, together with a typically tortuous branching pattern, make the gnarled shape[1] the dominant tree form in most savannas. As a natural corollary of these facts, one may suppose that these low trees will show much less longevity than forest species. Through counts of annual rings, Warming (1892) found ages of about 30 to 40 years for the oldest individuals of several species of cerrado trees.

Taller trees, well above 12 m, occur in ecosystems transitional between savannas and deciduous forests as well as in small forest outliers within a more or less continuous savanna matrix, as has been reported for the central Venezuelan llanos by Sarmiento and Monasterio (1971). The same may be said about the "derived savannas" of the Guinean zone, where mesophanerophytes dominate the savanna tree layer (Hopkins, 1962; Menaut, 1971). These cases have to be considered rather as ecotones, mosaics or seral stages, where certain pioneering forest species may become established. Likewise the *miombo* woodlands in the Zambezian region have a taller tree canopy up to 20 or 25 m high; but most authors consider these ecosystems rather as open forests than as savannas, though when they have been cleared the resulting open woodlands have been often referred to as savannas.

Other morphological features shared by a high proportion of the arboreal savanna flora relate to leaf size, shape, texture and life span. The leaves may be either single or compound, but in any case their sizes fall most frequently within Raunkiaer's

---

[1] Many authors have compared the shape of savanna trees to that of an orchard apple or of a fruit tree in temperate regions.

mesophyllous class. Microphyllous leaves are less common, while tree species with smaller leaves are almost absent. But generally the large, compound leaves of certain trees have numerous fairly small leaflets. Most species have flat leaves with entire margins, or if compound, with entire leaflets. The *Pinus* species of some Central American and West Indian savannas, however, constitute a conspicuous exception. In these features of leaf size and form, woody savanna species differ sharply from xeromorphic species characteristic of arid climates.

But the most noteworthy feature of the leaves of savanna trees is the scleromorphism that results from several characteristics, such as the superabundance of mechanical tissues and the deposition of silica. The anatomy of some of these leaves has been discussed by several authors (Morretes and Ferri, 1959; Beiguelman, 1962; Morretes, 1966, 1969; Mérida and Medina, 1967). Many species have leathery leaves that in some cases, such as in *Curatella americana* and *Palicourea rigida*, reach a consistency as stiff as pasteboard. A quantification of this scleromorphism is provided by the leaf area/dry weight ratio (Table 5.2). Other anatomical features frequently found in the leaves of savanna trees are: thick cuticle and cuticular layers; stomata

in the bottom of deep depressions; and great development of hypodermis and colorless parenchyma. These morphological traits are often accompanied by a high total ash content and slow stomatal reactions.

Leaves of savanna trees live for about one year, hence most species appear as evergreen or as brevideciduous. In these cases, leaf fall proceeds simultaneously with the development of a new leaf crop, in such a way that the total green biomass on the plant decreases during that period, but the trees never remain entirely leafless. Truly deciduous species — that is, those remaining leafless for several months — are less frequent and they often represent pioneering forest species colonizing certain savannas. This seems to be the case with *Cochlospermum vitifolium*, *Genipa americana* and *Godmania macrocarpa*, and other deciduous trees in certain areas of the Venezuelan llanos, as well as with several tall forest trees in the Ivory Coast savannas (Lamotte, 1978). The Brazilian cerrados, with a woody flora richer than any other tropical savanna area, have a larger number of deciduous species, but nevertheless they are far less numerous than the brevideciduous and evergreen species (Rizzini, 1965). Most woody species change their

TABLE 5.2

Area/weight ratios of mature leaves (in $cm^2 \ g^{-1}$ dry weight) for several species of trees occurring either in the tropical deciduous forest or in the savannas of the Venezuelan llanos

Species	Area/weight ratio		
	Montes and Medina (1975)	Mérida and Medina (1967)	Sarmiento (unpubl.)
Deciduous forest tree			
Luehea candida	$317 + 39$		
Deciduous forest and savannas			
Godmania macrocarpa	113 ± 15		
Genipa americana	107 ± 18		
Savanna species			
Curatella americana	69 ± 6	96	86
Byrsonima crassifolia	57 ± 7	84	81
Byrsonima coccolobaefolia			86
Bowdichia virgilioides		116	92
Casearia sylvestris		144	125
Palicourea rigida			76
Roupala complicata			62

leaves during the dry season. Towards the first weeks of the rainy season, apical meristems close further activity, and the plants enter then into a period without morphogenetic activities that extends throughout the whole rainy season.

Another characteristic shared by many savanna trees is the development of thick, hard, often corky barks — a trait of obvious value for the protection of the cambium against burning injury. Ferri (1962) illustrates several striking examples of these barks in cerrado trees, such as *Erythroxylon suberosum* and *Connarus suberosus*. Hopkins (1962) measured bark thickness in thirteen species of a Nigerian savanna, finding a mean of 15.2 mm and a maximum of 31.6 mm in *Cussonia kirkii*.

Concerning the behavior of woody species in response to fire damage or any other mechanical injury, one of the most general responses is sprouting from stumps, or suckering from rootstocks or from deep lateral roots. In this way, the shoots are reconstituted more or less rapidly after a traumatic destruction, this process leading to the coppice appearance of many tree populations. Lawson et al. (1968) illustrated various examples of this behaviour in the Guinean savanna. These responses to injury by sprouting and suckering are almost universal, too, in all Neotropical savanna ecosystems.

The tortuous and irregular branching patterns of savanna trees, as well as their open crowns, seem to be more a consequence of certain endogenous rhythms of meristematic activity than a phenetic response to external agents, whether drought or fire. In fact, many species may develop these features even under irrigation and fire protection, thus suggesting that this growth pattern derives from the death of apical meristems after each period of leaf formation and a subsequent development of adventitious buds during the next growth season. In some species like *Curatella americana* and *Bowdichia virgilioides*, under a regime of annual burning, the whole annual shoot growth dies back and the next season growth starts from adventitious buds that develop from the vascular cambium of older branches. In this way, a tree behaves merely as a mechanical structure supporting a crop of annual branches. In other species, or even in these same species under different environmental conditions, some annual branches develop profusely, giving rise to adventitious buds that will originate the next crop of leaves and branches, and

so develop the permanent aerial structures of the tree. Apparently, in many tree species the vascular cambium assumes the role of apical meristems as the main replacement tissue, since it will be responsible for any subsequent aerial development, either vegetative or reproductive; this is clearly shown in *C. americana* and *B. virgilioides*. Furthermore, numerous woody species are able to maintain themselves almost indefinitely as hemixyles — that is with annual shoots and woody underground structures — and even to complete the reproductive process in this growth form. Obviously, this is a behavior of high adaptive value in ecosystems subjected to seasonal drought and recurrent fires.

Many authors referring to African and American savannas have remarked that woody species which elsewhere are tree-like occur in savannas as shrubs with several stems sprouting from a xylopodium or from rootstocks, and that they are able to regenerate from these structures following the destruction of aerial parts. All these facts are in accordance with the statement above about the crucial role played by adventitious buds arising from the vascular cambium of stems and roots in the regeneration of the aerial shoots after a traumatic destruction.

But besides maintaining a tree population in a coppice state by sprouting and suckering, fire injury has also been thought of as responsible for the gnarled tree habit characteristic of woody savanna species. However, this does not seem to be the case, since in typical savannas the apical meristems of most trees cease their annual cycle of activity according to an endogenous rhythm, and any further growth has to depend on the development of new meristems in the form of adventitious buds arising from the cambium. This type of growth, with seasonally active apical meristems and the replacement of shoots through adventitious budding may explain the tortuous branching pattern of these trees.

The response of trees to fire damage during their active phase of leaf and flower production is the immediate development of a new crop of leaves and flowers through the fire-induced formation of numerous adventitious buds in the cambium of old branches uninjured by fire. Even when the whole aerial structures have been damaged, as is frequently the case with small individuals and low suckers, renewal is ensured by sprouting from underground rootstocks or lateral thick roots. But if fire occurs during the wet season, when the trees are morphogenetically inactive, a new leaf crop is seldom produced. Many individuals die, and the surviving trees will not form a new leaf crop until the following dry season.

Another major consequence of fire and drought on the woody species is their maintenance as half-shrubs or hemixyles. In effect, many individuals may be maintained in this subshrubby stage for many years, so that in some species the tree form becomes almost exceptional, as occurs with *Casearia sylvestris*, *Byrsonima verbascifolia* and *Andira humilis*, for instance.

An interesting experiment has been reported by Labouriau et al. (1964) concerning the behavior of *Caryocar brasiliense*, a typical tree species of the Brazilian cerrados. This tree was cultivated from seeds; some young plants were kept under normal field conditions, others were irrigated during their first two dry seasons. In most saplings under the normal water regime, the shoot died back during the dry season and new growth started during the following rainy season from buds located at the base of the stem. Repetition of this behavior in successive years led to a geophytic habit, the plants gradually acquiring a large underground woody structure. By contrast, those plants which were irrigated during their first two dry seasons continued their development. Their branches remained alive throughout the year, sprouting again each wet season to form after five years a tree ten times higher than the subshrubby unirrigated individuals. This experiment clearly shows the influence of the dry season in the acquisition and persistence of the half-woody habit in what would otherwise be a potential tree. Some species under favorable conditions — such as when irrigated — maintain their branches which will give rise to further growth during the next cycle, acquiring then, more or less rapidly, the tree habit. The above experiment also indicates that fire reinforces the action of drought, but that its action is not crucial for acquiring either the subshrubby or the gnarled habit.

Another polemic point clarified by Laboriau et al. (1964) concerns the type of multiplication of cerrado trees. Some authors have supported the idea that seed establishment was extremely unlikely in savanna trees, and hence that most species relied on some form of vegetative propagation to maintain their populations. Through an extensive survey of the cerrado region, Labouriau et al. found 50 species of trees germinating under natural conditions, while 32 additional species were found as young plants probably originating through seed germination. These results suggest that germination may not be a major filtering process during the life cycle of a savanna tree, probably the fate of the young seedlings may be more important.

Besides trees, palms constitute another woody form of frequent and widespread occurrence in tropical savannas. There are two growth forms among savanna palms: the normal, monocaulous, tree-like rosette; and the acaulescent, dwarf palm habit. The first type is represented for instance by species of *Acrocomia*, *Butia* and *Syagrus* in the Neotropical seasonal savannas, by *Copernicia tectorum* in the hyperseasonal savannas of Venezuela, and by *Mauritia flexuosa* in *esteros* and other waterlogged communities throughout tropical South America. In Africa, *Borassus aethiopum* is a tall palm of wide geographical range, mostly occurring in wet soils. Several acaulescent palms of genera such as *Attalea*, *Acanthococos* and *Diplothemium* are common in the Brazilian cerrados (Fig. 5.1). They are a part of the herb layer of these seasonal savannas, and must really be considered as geophytes, since even their hard, large leaves dry out during the dry season. Another architectural type related to palms is the low, few-branched rosette of some Velloziaceae; they occur mostly in rocky habitats, but also become frequent in some areas of cerrados.

Woody vines and epiphytes are rather exceptional in savannas, among this latter form we may include a few bromeliads and cacti in South America. The aphyllous-succulent form is also not characteristic of savannas, the columnar cactus *Cereus jamacaru* is one of the few exceptions. Some bromeliads of the genera *Bromelia*, *Dyckia* and *Ananas* represent the succulent rosette form in the Neotropical area, while the same growth form in Africa is exemplified by the aloes. Succulent trees also appear to be exceptional, some Bombacaceae may be among the few examples of this life form.

In hyperseasonal savannas, a different woody form may predominate, represented by spiny legumes with deciduous, compound leaves, such as the species of *Acacia* in alluvial communities

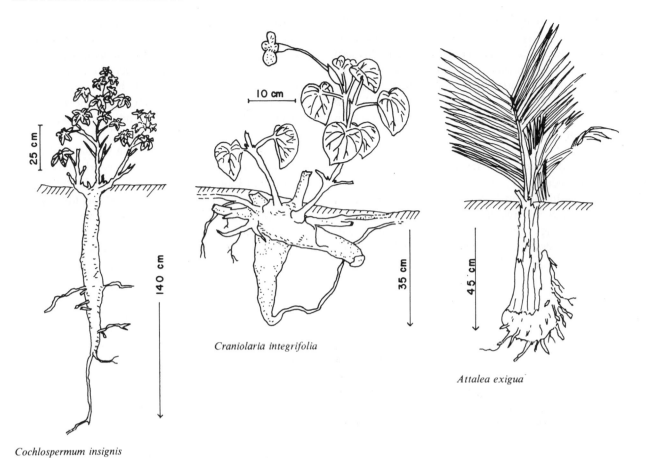

Cochlospermum insignis

Fig. 5.1. Some growth forms characteristic of the cerrado savannas. *Cochlospermum insignis* has a deep and prominent xylopodium; *Craniolaria integrifolia* has a branched, more superficial underground structure; the dwarf palm *Attalea exigua* shows a deep, corm-like organ. (After Rawitscher and Rachid, 1946; and Rachid, 1947.)

throughout tropical Africa, and species of *Acacia*, *Mimosa* and *Prosopis* in America. In many cases, these species may be considered rather as shrubs than as trees. When compared with the trees characteristic of well-drained soils, the differences sharply point out the enormous ecological dissimilarities between these two types of savanna ecosystems.

A last point, but by no means the least important, refers to the extensive nature of root systems in savanna trees, which apparently enable them to exploit the water and mineral resources of a great volume of soil. There are many references on the remarkable development of vertical roots in deep soils, particularly when the water table remains deep all year and there are no physical barriers to root penetration. According to Rawitscher et al. (1943) cerrado trees can develop enormous root

systems when growing in deep soils. They report the example of *Andira humilis*, a "subterranean tree" whose roots have been found to a depth of 18 m. Foldats and Rutkiss (1969) showed how the most widespread trees in the Venezuelan savannas have thick superficial roots that may reach twenty or more meters in length. Likewise, Van Donselaar-Ten Bokkel Huinink (1966) reported a similar type of root development in many of the trees of the Surinam savannas. Lebrun (1947), Lawson et al. (1968), and Menaut (1971) reported similarly on the vertical and horizontal root development of tree species in various African savannas.

The general picture emerging from these observations is that many trees have a double-purpose root system, with a main tap root able to reach great depths under favorable circumstances, and a very extensive crown of thick lateral roots emerging at various depths from the main root. In shallow soils only this latter system may develop.

In several species, like the widespread African species *Parinari curatellifolia* and *Daniellia oliveri*, root suckers develop after mechanical destruction of aerial parts; other species, like *Burkea africana* and many other savanna trees, have a remarkable capacity to regenerate by repeated sprouting of coppice shoots from the underground stump that gradually develops after successive shoot regenerations.

A final characteristic to notice in tree species is their capacity for rapid root elongation from the young seedling stages, a feature that obviously will greatly increase the chances of survival when the upper soil layers progressively dry out.

HALF-WOODY LIFE FORMS

Among the species without perennial aboveground structures, we have further distinguished those forms having woody underground organs from the perennial herbs having only fleshy rhizomes, bulbs or other non-lignified perennating organs. Plants with woody underground organs, but with all shoots annual, constitute one of the most characteristic life forms in American as well as in African savannas, either considering the number of species having this growth habit, or taking into account that they seem to be almost peculiar to these ecosystems. This life form has been variously called subshrubby, half-shrubby, suffrutescent, geofrutescent, geoxyle, hemixyle, half-woody, etc.

Two cases of different ecological and evolutionary significance have to be distinguished within this growth form: the permanent geoxyles, that will conserve this growth habit under any circumstance; and the traumatic geoxyles, in which this habit results from external injury, but which will revert to their normal tree habit when circumstances are favorable.

Seasonal savannas, particularly those ecosystems occurring on deep and well-drained soils, may show a rich diversity of subshrubs; but this life form appears more rarely or is completely absent in hyperseasonal savannas or waterlogged sites. To attain an evolutionary interpretation of this plant strategy, it may be important to point out that often the subshrub species of the savanna belong to genera mostly composed of forest trees or less commonly of forest woody vines. This is the case of *Anacardium pumilum*, *Andira humilis*, *Cochlospermum insignis*, *Jacaranda decurrens* and many other vicariant species that have acquired the suffrutescent growth habit. Other subshrubs belong to genera whose species include both herbs and subshrubs, like the genera *Cissus*, *Eriosema*, *Galactia*, *Indigofera*, *Tephrosia* and many others.

Warming (1892), in his unsurpassed study of the cerrado at Lagoa Santa, already called attention to this peculiar growth form, remarking how it was difficult to decide whether these species should be considered as perennial herbs, as subshrubs, or as shrubs. Rawitscher et al. (1943) and Rawitscher and Rachid (1946) described large woody, underground systems as quite frequent among cerrado species, analyzing the case of the extensive underground systems of *Anacardium pumilum* and *Andira humilis* that led them to the concept of "subterranean trees"; likewise they illustrated the conspicuous xylopodia of *Cochlosperum insignis* and *Craniolaria integrifolia* (Fig. 5.1) as typical of many cerrado subshrubs. All these species, as mentioned above, are vicariants of forest trees.

Rizzini and Heringer (1966) differentiated, among the cerrado species with this growth form, trees maintained in this suffrutescent habit on the one hand, and true subshrubs, either with xylopodia or with an extensive woody system, on the other. This latter case, which they called **diffuse underground systems**, may be either of caulinar or of radical nature, but only careful anatomical observation may distinguish between them. This functional convergence between organs of different nature is striking, since both types have the same key function in regeneration and clonal growth. Rizzini and Heringer listed among subshrubs with an extensive underground system of radical nature, the species *Anemopaegma arvense*, *Peschiera affinis* and *Coccoloba cereifera*, while *Annona pygmaea*, *Esenbeckia pumila*, *Chrysophyllum soboliferum* and others have caulinar underground structures.

In the flora of the Surinam savannas, the hemixyles are the best represented group. Van Donselaar-Ten Bokkel Huinink (1966) reported that, among 23 species of subshrubs, 19 have a conspicuous xylopodium, while 16 species of perennial "herbs" also have a woody stem base. The author proposed to name these forms *Geophyta geopodiosa.* In the seasonal savannas of the Venezuelan llanos, the geoxyles with large woody xylopodia are well represented, among them being, for instance, several species of *Galactia*, *Indigofera*, *Pavonia* and *Ichtyothere.* Extensive underground systems are less common than in the cerrados, some examples being species of *Psidium*, *Clitoria* and *Tephrosia.* On the other hand, species like *Byrsonima verbascifolia* and *Casearia sylvestris* which under some circumstances are low, gnarled trees, appear in this region almost exclusively as geoxyles.

Most African savannas also have a rich flora of half-shrubs. Lebrun (1947) gives several examples from the Lake Edward plain, such as *Vigna friesiorum* and *Cissus mildbraedii*, both with conspicuous xylopodia. In these savannas the suffrutescent species form 35% of the flora. Duvigneaud (1949) distinguished "steppes" from savannas in the lower Congo. His concept of "steppe" is equivalent to the type we have called savanna grassland, while his savannas comprise both hyperseasonal ecosystems and degraded forests. Duvigneaud considered that the two growth forms most characteristic of the "steppes" were perennial bunch grasses, and shrubs with xylopodia, this latter form frequently represented by vicariant species of rain-forest trees; on the other hand, the subshrub form is almost completely lacking in his savannas. In further papers he considered rhizomatous subshrubs with annual shoots that dry out completely during the rainless season as the ecological group characteristic of the Zambezian "steppes" and "steppic savannas" (Duvigneaud, 1955, 1958). Thus this growth form is characteristic of grasslands on the high plateaus covered by the Kalahari sands. Malaisse (1975) in his study of the Upper Shaba (Zaire) vegetation, followed the nomenclature of Duvigneaud for the plant formations, indicating, as a distinctive feature of the "dry steppes" on the sandy plateaus, the abundance of geofrutices, among which he listed *Syzygium guineense*, *Parinari capense* and *Eugenia malangensis.* In a similar way, the "tree steppes" are rich in subshrubby species, while the "humid steppes", with a high water table, only show one species with this growth habit: *Syzygium guineense.*

Schnell (1976–77) in his monograph on tropical African vegetation indicated that dwarf-woody species constitute one of the peculiar growth forms of the plateau savannas in the whole Congo area. Referring to a paper by Makany (1970), he listed as the more frequent geophytic subshrubs *Parinari pumila*, *Anisophyllea poggei* and *Landolphia thollonii*, all three vicariants of rain-forest trees.

In West Africa, on the other hand, there are relatively few species with a well-developed woody underground system in comparison with the Zambezian savannas, except traumatic forms of trees. In the Lamto savannas, César (1971) remarked on the abundance of suffrutescent forms in the herb layer, listing among them some species of *Cissus*, *Galactia*, *Rhynchosia* and *Eriosema.* However, these species do not seem to reach the huge underground development of other geoxyles. As these species are difficult to classify according to the system of Raunkiaer, César proposed for them the term **geochamaephytes**. In the same area of the Ivory Coast, Menaut (1971) discussed the case of the commonest tree components of the ecosystem which are able to maintain themselves for a long time in a subshrubby form, and even to complete their sexual reproduction in this state. Menaut considered that both drought and fire could be responsible for the subshrubby form of these otherwise normal trees, since only in some favorable years may the annual aerial growth develop further to form perennial above-ground structures.

After reviewing the variety of woody and half-woody growth forms occurring in most savannas, one could hypothesize an entire sequence of evolutive changes leading from species of "normal" trees to temporary hemixyles by the replacement of apical meristems by the vascular cambium as the main organogenetic tissue, and by the development of the capability to sprout from the stem base or from roots after a fire or drought injury. These temporary hemixyles possibly acquired later the capacity to flower in this stage, thus opening the way for a later evolutive change to permanent hemixyles; thus the original trees have been transformed by successive adaptive changes to geoxyle well suited to cope with the savanna environment, but unable to recover their ancestral tree habit. Obviously, this model only applies to those subshrubs which are vicariant of forest trees. It is interesting to notice that in the oldest savanna areas, such as the cerrados, which have evolved for a long time in close contact with tropical forests, this type of hemixyles predominates over the suffrutescent genera, while in younger ecosystems, such as the Orinoco llanos or the derived savannas of the Guinean zone, the reserve situation is true: almost all subshrubby species belong to genera without obvious ancestral tree forms.

THE HERBACEOUS PERENNIAL LIFE FORMS

To this group, devoid of woody structures of any kind, belong the dominant species in the herb layer of all types of tropical savannas, those that contribute most to the structure and primary production of the ecosystems and that control the possibilities and frequency of fires. Within this rich and diversified group, we will first discuss the major morphoecological features of the dominant growth form in any tropical savanna: the tussock grasses and sedges.

The tussock form is represented in the savannas by species that have perennating underground structures relatively close to ground level, while all aerial parts are entirely seasonal. The perennial organs are rather thin rhizomes that survive more than one year, characterized morphologically by their short internodes and scaly leaves. Through various branching patterns, the rhizomes give rise to the typical tuft or bunch of tussock species. The compactness of the tussock will depend on the length of the internodes and on the branching pattern of rhizomes. Tillers may be exclusively formed by the rhizome's apical meristem, or by the development of buds in the axils of the scales. Thus, a great variation of architectural models may be found within the same basic type of growth habit, such as: tussocks with or without aerial culms; culms that may or may not branch; rhizomes forming short internodes only, or changing periodically from short to long internodes; monopodial or sympodial branching; plants with or without runners or stolons; and so forth.

The architecture of two common Neotropical tussock grasses may illustrate the variability of developmental patterns within this life form. *Leptocoryphium lanatum* shows a modular structure where the basic modular unit is an upright shoot or tiller formed by a much reduced underground rhizome with short internodes and nodes with normal leaves. At a particular period of the annual cycle, the apical meristem of the rhizome changes to a reproductive stage, producing the single aerial

culm of each module: an inflorescence axis bearing reduced leaves. After that, some axillary buds start to develop and reproduce the module in a sympodial fashion. This proliferation of axillary buds on short vertical rhizomes gives rise to the compact corm-like underground structure of this plant (Fig. 5.2). Each module has definite growth and lives only one year, the perennation of the tussock is ensured by sympodial growth through several additional modules formed by the axillary buds of rhizomes.

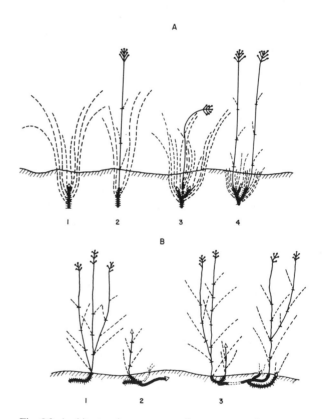

Fig. 5.2. Architectural patterns of two tussock grasses. A. *Leptocoryphium lanatum*: 1, the vertical rhizome with its apical bud, short internodes and leafy nodes; 2, the apical bud has changed to the reproductive phase forming the inflorescence, sole aerial axis of the plant; 3, after decay of the reproductive structures, axillary buds on the rhizome give rise to lateral branches; 4, the old rhizome and the aerial tiller it produced have decayed and new rhizomes repeat the growth process. B. *Trachypogon plumosus*: 1, a young tussock formed by a horizontal rhizome with short internodes that gives rise to a branched aerial tiller with long internodes and terminal inflorescences; 2, after the old tiller dies back, axillary buds on the rhizome form various lateral branches, some with short, others with long internodes; 3, by fragmentation of the old rhizome a loose tussock is composed of more or less independent tillers.

A different architectural design is shown by another widespread savanna grass: *Trachypogon plumosus*. This species has horizontally spreading rhizomes bearing scaly leaves, that may form either short or long internodes. Apical meristems give rise to the aerial tillers, which consist of long, branching culms, bearing normal leaves. When these meristems change to a reproductive stage, the inflorescences are formed and the development of a given tiller comes to the end. Buds in the axils of scale leaves give rise to branches of the original rhizome, thus originating a compact underground structure in which the branched rhizomes produce further rhizomes through the development of lateral buds and produce tillers from their apical meristems (Fig. 5.2).

These two examples illustrate the possibilities of obtaining different modular patterns through rhythmical functioning of meristems or through functional differentiation between apical and axillary buds, but in any case it is worthwhile to emphasize that the disentangling of these architectural designs remains yet at its very beginning.

Most tussock grasses and sedges have hard, scleromorphic leaves, either flat or revolute, glabrous or densely pubescent, with abundance of silica bodies that may be utilized to recognize the species. In fact, a series of contributions have built a catalogue for the identification of soil phytoliths in the cerrados for use as an auxiliary character in grass taxonomy (Sendulsky and Labouriau, 1966; Campos and Labouriau, 1969; Teixeira da Silva and Laboriau, 1971; Figueiredo and Handro, 1971).

Another morphological feature characteristic of many perennial savanna grasses is the protection of the apical bud from fire and dessication by a thick tunic formed by the old leaf sheaths; Rachid (1956) described this structure in several cerrado grasses, such as *Aristida pallens*, *Tristachya leiostachya*, and *Paspalum carinatum*. Tunicate species having their buds well protected at ground level are able to regrow rapidly after fire, the buds not only remain totally unharmed but revert to an active phase.

A majority of the species of tussock grasses belongs to the tribe Andropogoneae and are C_4 plants, but C_3 species are also well represented in tropical savannas, particularly on ill-drained, wetter habitats. Concerning tussock height, most species fall into the class Beard (1953) named as "tall bunches" (50–220 cm in vegetative stage), but some Paleotropical grasses are "high grasses" (120–240 cm tall); this is the case, for instance, in *Hyparrhenia* species, as well as in several species of *Pennisetum*, *Panicum* and other genera. In tropical American savannas, on the other hand, native high grasses are rather exceptional, only a few species of *Paspalum* and *Panicum* in hyperseasonal savannas reaching these heights. But one may recall here that some Old World bunch grasses, such as *Hyparrhenia rufa* and *Panicum maximum*, have become widely established in tropical America, colonizing modified habitats produced by forest clearing; nowadays, these alien grasses extend even over areas of disturbed savannas.

Bunch grasses and sedges may have their renewal buds immediately above the ground, in which case they have to be considered as hemicryptophytes; these buds stay well protected against desiccation and burning by a thick cover of old leaf sheaths. Other species are geophytes, the buds lie in the soil at depths down to ten or more centimeters.

In respect of allocation of energy and resources, the two major features that characterize tussock grasses are: first, a relatively weak reproductive effort during each annual cycle, since only some tillers in each bunch reach the flowering stage; and, second, a biomass allocation to the underground organs leading to a high ratio of below-ground to above-ground biomass Table 5.3 gives some data from dominant bunch grasses of the seasonal Venezuelan savannas.

Besides the tussock growth form, other architectural patterns found among the herbaceous savanna species include perennial grasses with isolated tillers. César (1971) remarked that in the Lamto savannas only two grasses belong to this type: *Imperata cylindrica*, with long rhizomes, and *Schyzachirium platyphyllum*, which extends by aerial stolons. In Neotropical savannas, particularly in hyperseasonal ecosystems, rhizomatous and stoloniferous grasses and sedges are not uncommon; sometimes they become the dominants of the herb layer. This is the case, for instance, with *Imperata brasiliensis*, *Paspalum chaffanjohnii* and *Leersia hexandra*, as well as several sedges of the genera *Cyperus*, *Kyllinga*, *Rhynchospora* and others. These species spread more rapidly than the slow-moving bunch grasses, thus becoming efficient colonizers on disturbed habitats; their extensive

TABLE 5.3

Some data concerning biomass allocation, reproductive effort and growth rates of five dominant species of perennial tussock grasses in Venezuelan savannas (H/E=ratio of below-ground to above-ground biomass, six months after fire; R/E=ratio of maximum reproductive biomass to maximum above-ground biomass)

Species	H/E	R/E	Average growth rates (mg dry weight per plant per day)		
			30 days after fire	132 days after fire	178 days after fire
Trachypogon vestitus	0.9	0.10	60	166	303
Axonopus canescens	1.1	0.05	27	63	98
Leptocoryphium lanatum	4.7	0.14	223	133	237
Sporobolus cubensis	1.9	0.11	690	172	161
Elyonurus adustus	1.8	0.08	367	272	284

Growth rates are minimum estimates assuming no decomposition, and are averages over the whole period considered. (from Sarmiento, unpublished data).

rhizomes give rise to new tillers, more or less apart from one another, that may in turn become centers for further spreading during the following season.

Finally, with regard to the growth forms of other perennial herbs, it may be stated that the most common life form of these species in tropical savannas is the bulbous geophyte, represented by numerous species of Amaryllidaceae, Iridaceae, Liliaceae, Orchidaceae and some other families. As will be shown later, these species have only a short active period, and remain most of the year as underground latent structures. Less frequent are rhizomatous geophytes and hemicryptophytes with aerial perennating structures at ground level, such as some orchids of the genus *Cyrtopodium*.

THE ANNUAL PLANTS

The representation of annuals in the savannas is highly variable, although their contribution to the biomass of the herb layer may be considered as generally inconspicuous. In the cerrados of the Lagoa Santa area in central Brazil, Warming (1892) remarked on the scarcity of annuals, since, of a total of about 730 species, the annuals hardly amounted to 30, forming thus less than 5% of this flora. Almost every author dealing with the cerrado

vegetation has noted the under-representation of annuals.

In the Venezuelan llanos, the importance of therophytes varies according to the habitat. They become more common in the driest sites, such as sand dunes and lateritic outcrops, and on the other hand, they colonize the unstable habitats created by seasonal fluctuations in swamps and other water-logged sites; likewise, they are well represented among the flora colonizing disturbed habitats. Ramia (1974) lists, for the whole area of the Venezuelan llanos, 55 species of annual grasses as against 145 perennial species in this family. In the llanos, the more widespread annual grasses belong to the genera *Andropogon*, *Aristida*, *Diectomis*, *Eragrostis* and *Gymnopogon*, while some frequent annual sedges belong to the genera *Cyperus*, *Fimbristylis* and *Scleria*. Among other families, some genera with a good representation of annual species in the savannas are *Borreria*, *Cassia*, *Euphorbia*, *Heliotropium*, *Hyptis*, *Polygala* and *Stylosanthes*.

In Lamto, César (1971) found only one annual grass. *Sorghastrum bipennatum*, while the proportion of therophytes varied among the different savanna communities in that area from less than 5% in the *Loudetia* grass savanna to more than 30% in the Andropogoneae shrub savanna. In the following section we will further discuss the behavior of annuals during the different seasons.

PHENOLOGY

In spite of the unquestionable interest that know-ledge of annual cycles and seasonal rhythms of plant species represent for a deeper understanding of the behavior of the whole ecosystem, as well as to provide a sound basis for any type of management practice, systematical phenological observations on tropical savanna ecosystems have been till now astonishingly scarce.

In this field, as in many others related to tropical ecology, Warming (1892) was the pioneer. Almost seventy years later, this approach to Neotropical savannas was followed by Van Donselaar-Ten Bokkel Huinink (1966) in Surinam and by Monasterio (1968) in Venezuela. Hoock (1971) in his study of the savanna vegetation of French Guiana, presented some relevant data on the phenological cycles of the most frequent species. More recently, Coutinho (1976) reported the chan-ges in phenological behavior of some cerrado species as a response to fire; Monasterio and Sarmiento (1976) presented a view of the main phenological strategies of plant species in the Venezuelan seasonal savannas; Ramia (1977, 1978) followed plant behavior along topographic gra-dients in the southern Venezuelan savannas; and Sarmiento (1978) discussed the general picture of plant rhythms related to other major ecological aspects of Neotropical savannas. In Africa, several early papers dealing with vegetation types and classification included phenological observations (Lebrun, 1947; Duvigneaud, 1949; Sillans, 1958; Koechlin, 1961), as do those of Morat (1973) for Madagascar, Perera (1969) for Sri Lanka (Ceylon), and Moore (1973) for the Australian grasslands. The first quantitative data on seasonal growth were presented by Hopkins (1970), who analyzed the growth patterns of several species in the derived savannas of Nigeria. More recently, the phenology of the Lamto savanna in the Ivory Coast has been considered by César (1971) in respect of herbs and by Menaut (1971) for woody species.

The basic phenological concepts that will be employed hereafter have been discussed in our previous paper (Monasterio and Sarmiento, 1976), together with certain methodological problems in-volved in sampling and measuring phenophases in natural populations; we refer to that paper for these two basic points. Our approach to the phenological behavior of a certain species in a given ecosystem is to consider this as simply a partial component of its global evolutionary strategy (Monasterio, 1982). This means that the adaptive mechanisms of any population may operate on every response relating the individuals, in each of their developmental phases and life-cycle stages, to the outside impulses and constraints of the biological and physical environment; hence their results may be expressed in an interplay of features and processes, including growth form, architecture, life cycle, annual and circadian rhythms, reproductive strategies, demo-graphic structure, and so on, each response consti-tuting a partial element of the global strategy of that population in that ecosystem.

The species that show similar patterns in their phenodynamics may be assembled into pheno-logical groups, each one representing a major phenological strategy which allows its species to cope successfully with the cyclic fluctuations of the environment, and of the ecosystem as a whole, through the perception of some periodic signals and impulses coming from the outside. We consider in the following pages the main strategies as they have been described in the various tropical savannas, leaving for a final section a general discussion on the significance of these rhythms to the functioning of the whole ecosystems. Hopkins (1968) em-phasized that no two savanna species show exactly the same annual rhythm; but, although this obser-vation may be correct, our groups seem to disclose roughly similar phenological patterns within spe-cies that otherwise may diverge from each other in many important aspects.

The phenological groups were established by taking into account certain crucial rhythmic activi-ties of the species (Table 5.4 and Fig. 5.3). A first major division sets apart species with assimilation all year from species with seasonal carbon gain, which consequently have a rest period devoid of any above-ground activity. Within this last group we differentiate the perennials, with latent under-ground structures, from the annuals which are present only as a seed bank during a certain period of the year. In turn, both groups are further subdivided — according to the reproductive pheno-dynamics — into species having precocious flower-ing (at the beginning of the rainy season); delayed flowering (from the middle to the last part of the rainy season); tardy flowering (during the dry

TABLE 5.4

The major phenological groups of savanna species and the features utilized for their characterization; the species are those after which the groups in question are named

				Perennials	Annuals
Carbon assimilation all year	Growth continuous or nearly so	flowering	precocious	A1. *Leptocoryphium lanatum*	
			delayed	A2. *Trachypogon plumosus*	
			tardy	A3. *Cassia moschata*	
			continuous	A4. *Evolvulus sericeus*	
			opportunistic	A5. *Imperata brasiliensis*	
	Growth seasonal	flowering	precocious	A6. *Jacaranda decurrens*	
			delayed	A7. *Piliostigma thonningii*	
			tardy	A8. *Curatella americana*	
Carbon assimilation seasonal (resting phase)		flowering	precocious	B1. *Curculigo scorzonaerifolia*	B2. *Spilanthes barinensis*
			delayed	B3. *Bulbostylis junciformis*	B4. *Aristida capillacea*
			tardy	B5. *Cochlospermum vitifolium*	B6. *Egletes florida*
			opportunistic		B7. *Phyllanthus sublanatus*

season); and opportunists, able to bloom in any period if suitable conditions occur.

Among the species with continuous carbon assimilation, all of which obviously are perennials, we distinguish a group that has continuous or nearly continuous growth from another group with strictly seasonal growth. Both groups are further subdivided on the basis of their reproductive dynamics, thus giving groups with precocious, delayed, tardy, continuous and opportunistic flowering behavior. The terms "precocious", "delayed" and "tardy" imply that one takes as the starting point for each annual cycle the onset of rains, but this is pure convention, since it might be argued that the "tardy" flowering species are in fact the most precocious, and vice versa.

By application of the above criteria, one may obtain phenological groups that combine species with comparable behavior in respect of some of their chief dynamic features, such as the duration of life cycle (perennials vs annuals), the seasonality of the assimilatory structures, the rhythm of meristematic activities (apical meristems of the shoot), and the timing of flowering. Within each group the species may differ in other important characteristics of their phenodynamics, but the groups are homogeneous enough to give us a glimpse of the

wide possibilities of phenological expression provided by the savanna ecosystems to their component species. To each phenological group we give the name of a typical savanna species showing that phenological behavior; we will briefly consider now the major characteristics of each of these groups.

PRINCIPAL PHENOLOGICAL GROUPS IN THE SAVANNA FLORA

A. Species with carbon assimilation all year

A.1. Species with continuous growth and precocious flowering. The *Leptocoryphium lanatum* group

The phenodynamics of these species may be summarized as follows (Fig. 5.4). A phase of vigorous shoot growth starts with the onset of rains after the long dry season, or even before that if the savanna has burnt during the last weeks of drought. Leafing and blooming, as well as tillering in the grasses, proceed simultaneously, or in some species inflorescence elongation precedes leafing. The growth rates during this early period of flush activity reach the highest annual values, they are in fact high by any standard of comparison (Table 5.3).

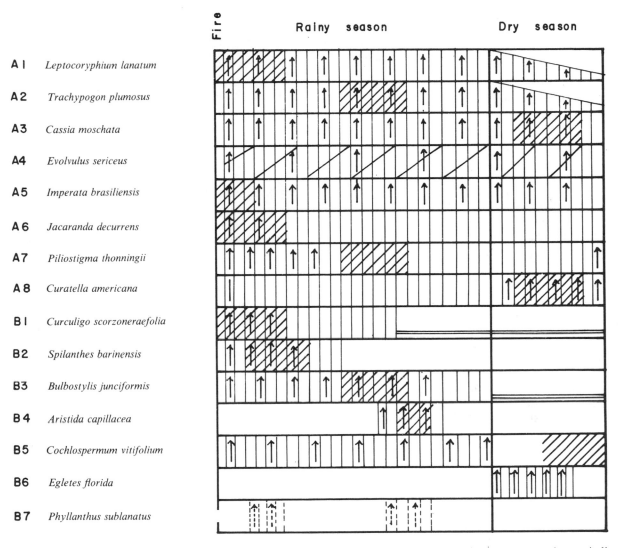

Fig. 5.3. Diagrammatic representation of the phenodynamics of the different phenological groups in the savannas. Arrows indicate leafing; vertical hatching refers to the presence of active green biomass; diagonal hatching indicates the flowering periods; horizontal hatching represents the persistence of perennating below-ground structures; white areas indicate the period when the species remain as a seed bank in the soil.

In a few weeks all the reproductive phenophases are over, including seed dispersal, so that during the rest of the rainy season the plants remain vegetative but actively growing. When drought becomes severe, the half-woody species belonging to this phenological group shed part of their leaves, while the grasses and sedges become increasingly yellowish, and end the season with a standing aboveground biomass almost totally formed of dead shoots. However, even during this unfavorable period, the plants continue to grow, producing new tillers and leaves, though at much slower rates and

with low efficiency, since this late growth does not proceed beyond the youngest stages and most of it dries out. This is what we consider as a semi-resting stage, since all plant activities sharply decrease, though the capability of these species to continue their growth is shown, even if the severe environmental stress hinders any further development of the newly formed organs.

In Neotropical savannas, this phenological group is represented by certain dominant tussock grasses such as *Leptocoryphium lanatum* and *Sporobolus cubensis*, by the common sedge *Bulbostylis*

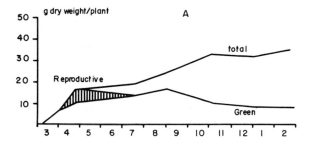

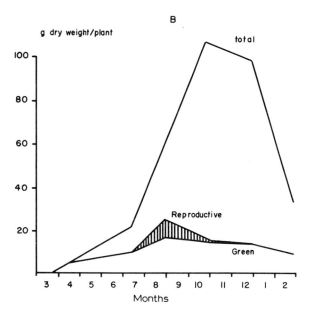

Fig. 5.4. Phenodynamics of two perennial grasses in the savannas of the western Venezuelan llanos when annual fire occurs at the beginning of March. A. *Leptocoryphium lanatum*. B. *Trachypogon vestitus*. The lines represent the development of the green, reproductive and total above-ground biomass during an annual cycle.

paradoxa, and by many half-shrubs such as *Clitoria guianensis*, *Stylosanthes capitata*, etc. In the Guinean savannas this behavior is shown by *Brachiaria brachylopha* and some half-woody species (César, 1971). César applied to these plants the name of long-cycled, early flowering species.

In the coastal savannas of the Guianas, under an ever-wet climate, *Leptocoryphium lanatum*, as well as many other species, flowers immediately after burning, during a short period of reduced rainfall in September (Hoock, 1971). The flowering process in this species appears thus as fire-induced, but in each area it blooms just once a year, differing thus

from the group of opportunistic species we consider later.

A.2. Species with continuous growth and delayed flowering. The *Trachypogon plumosus* group

These species start their vegetative growth once the rainy season is under way, or even before the first rains if the savannas have burnt late in the dry season. Their annual cycle resembles that of the preceding group of precocious evergrowing species, but with two main differences: first, the vegetative development starts smoothly with the rainy season, the highest growth rates only being attained several months afterward; and, second, sexual reproduction occurs towards the middle of the rainy season (Fig. 5.4).

A further distinction can be made among the species of this group according to their seasonal growth rates and the precise timing of flowering; some species are early-growing plants, developing rapidly from the beginning of the rainy season; while others are more slowly developing species, which start a phase of rapid shoot growth a couple of months later, to bloom towards the end of the rainy season. On this basis, César (1971) separated these species into two groups, one with immediate or early emergence and the other with late emergence; but we consider it is more a matter of continuous variation, from certain species emerging as soon as the rains come, to other species whose development is retarded till the last weeks of the rainy period.

After seed dispersal, these plants generally show further vegetative growth, but sooner or later they enter into a declining semi-resting phase that extends throughout the dry season; during this phase new tillers are formed but the mortality rate of the aerial biomass surpasses the assimilatory rates thus leading to an accumulation of standing straw at the end of the dry season.

This type of annual rhythm is shown by many dominant grasses and sedges, such as the common Neotropical grasses *Trachypogon plumosus*, *Axonopus canescens*, *Andropogon semiberbis*, and many others; the African grasses *Hyparrhenia diplandra*, *Andropogon schirensis* and *Loudetia simplex* behave in a similar way. Many sedges have the same phenological pattern, as for instance *Bulbosylis capillaris* in American savannas and *B. aphyllanthoides* in the Guinean communities.

A.3. Species with continuous growth and tardy flowering. The *Cassia moschata* group

This is a rather uncommon type of phenological behavior shown by some savanna trees. These species have continuous growth all the year, but they flower during the dry season; or in some cases they may flower twice a year, once during the dry season, and again some months afterwards. Monasterio (1968) reported this type of phenodynamics in several species of evergreen trees growing both in the gallery forests and the savannas of the central Venezuelan llanos, such as *Cassia moschata*, *Vochysia venezuelana* and *Copaifera officinalis*. In *Cassia moschata*, for instance (Fig. 5.3), leaf production and leaf fall occur simultaneously during the whole year, while flowering occurs early in the dry season, followed by a second peak during the change from the dry to the wet season. In other species, such as *Copaifera officinalis*, leaf production is rather aberrant, with small flushes at several periods of the year, and a single peak of flower production during the dry season. The same pattern, either with one or with two flowering periods, is found in several trees of the cerrado region.

A.4. Evergrowing and everflowering species. The *Evolvulus sericeus* group

This group includes a few perennial subshrubs and some rhizomatous herbs that have continuous growth and flowering all year. They occur in seasonal savannas with deep soils, where their extensive root systems reach water resources probably not available to other herbs and grasses. Some of these appear as completely arhythmical, with continuous leaf and flower production, while others, even if they grow and flower for a quite extended period, decrease their activities during the dry season.

This peculiar phenodynamics may surprise one in a highly seasonal ecosystem as most tropical savannas certainly are; however, the continuous development of the species may be understood on the basis of their life form and architectural design. In fact, from a more or less conspicuous underground xylopodium, successive crops of seasonal branches emerge and develop for a certain time, giving rise to new leaves and to flowers from axillary buds. When each shoot dries out after attaining a certain size and age, new branches appear, thus giving the plant its evergrowing and everflowering characteristic.

Some widespread half-woody species of Neotropical savannas may exemplify this behavior, such as *Evolvulus sericeus* and *Pavonia speciosa*, while the same phenodynamics is shown by some perennial herbs in the Lamto savannas (César, 1971), such as *Aneilema setiferum* and *Afromomum latifolium*.

A.5. Continuously growing, opportunistic species. The *Imperata brasiliensis* group

Some perennial grasses and many subshrubs apparently show the same phenological behavior of Group A.1 — that is, they grow continuously all the year and bloom at the beginning of the rainy season or even before, if a late fire burns the savanna. However, they have an important distinctive feature compared with those species, since these plants are able to bloom after any fire at whatever season. Therefore, we consider these species as opportunistic strategists that may profit from any favorable circumstance (a reduced competition in this case) to reproduce by rapid induction of flowering.

Coutinho (1976) thoroughly discussed this kind of behavior in the cerrados, giving a long list of species able to bloom after any fire, and another list with species that only bloom if the fire occurs during the dry season; these species would bloom at that time anyway, even without having been burnt. *Imperata brasiliensis* and *Elyonurus adustus*, two quite common savanna grasses, exemplify the first group of entirely pyrophilous (fire-tolerant) species blooming after any fire. Though not yet precisely reported, it seems quite possible that this same response may be found among the African savanna species. Meguro (1969) analyzed the flowering behavior of *Imperata brasiliensis* in southern Brazil under various experimental conditions, including fire, leaf-cutting and application of gibberellic acid. This grass does not bloom unless it is burned; a plant unburned during one growth season or more will flower after any fire during the next year, but once it blooms, a new flower induction by fire can only be obtained a year later. Mechanical removal of all leaves also induced flowering, but the response was much less intense than that induced by fire. The application of gibberellic acid enhanced flowering in those plants whose leaves were removed, but had no effects either on intact or on burned individuals. Meguro concluded that at least two different processes are involved in the induction of flowering in this species, one triggered by any kind of leaf removal, the other promoted by some other effect of fire, perhaps the temperature shock. In any case the behavior of *Imperata brasiliensis* exemplifies the phenological strategy of opportunistic perennials able to bloom after fire when they have remained in a vegetative phase for a certain minimum period.

A.6. Species with seasonal growth and precocious flowering. The *Jacaranda decurrens* group

This group consists of the evergreen species capable of photosynthetic assimilation all year, but showing seasonal growth and precocious flowering. Their active period of growth and flowering starts with the onset of rains or after the consumption of above-ground biomass by fire if burning has occurred late in the dry season. In a period ranging from a few weeks to a couple of months, the leaves are fully developed and all the reproductive phases accomplished, in such a way that during the remainder of the wet season the plants are green but without any further growth, the meristems seeming to degenerate or to die. During the dry season, the leaves may become partly senescent but nevertheless the plants remain photosynthetically active, though with a reduced green area.

This phenological behavior is shown by many half-woody species in Neotropical savannas, such as *Jacaranda decurrens*, *Cochlospermum regium* and *Psidium salutare*. They differ from group A.1 because leaf production is restricted to a short period of the year, and from group B.1 because the species in this latter group, as will be seen later, have a rest period of underground life.

A.7. Species with seasonal growth and delayed flowering. The *Piliostigma thonningii* group

Apparently a few woody evergreen species have seasonal growth and delayed flowering, thus uncoupling the processes of leaf formation and flower initiation. As an example, one may consider the phenodynamics of *Piliostigma thonningii*, a common low tree in the Lamto savannas, as reported by Menaut (1971). The period of active growth begins before the first rains, when the buds start to open. A late fire may destroy this early growth, but in any case leaf formation proceeds rapidly after fire and continues for two or three months. The first flowers open in June, four to five months after the start of

leafing, the flowering process and the ripening of fruits continuing for several months. The old leaves fall gradually as the new ones are formed at the start of a new growth cycle.

This phenodynamics is not very common among savanna woody species, since most evergreens with seasonal growth bloom at about the same time that they produce their new leaf crop, either during the dry season or immediately after the onset of rains. Only some deciduous species have the same reproductive rhythm as this group, flowering several months after leafing (Group B.2).

In Neotropical savannas *Roupala complicata* and *Kielmeyera coriacea* may be mentioned among the common trees having this behavior.

A.8. Evergreen trees with seasonal growth and tardy flowering. The *Curatella americana* group

To this phenological group belong the evergreen woody species that produce their leaves and flowers during the dry season (Fig. 5.5). It is in this apparently unfavorable period that these species change their leaves through the almost simultaneous progress of the two opposite processes of leaf fall and new leafing. Consequently, the individuals never remain leafless, there is nothing comparable to a rest phase, though the green area diminishes during these phenophases of leaf renewal. Due to this fact, the species in this group can be considered either as evergreens or brevideciduous, since they remain with a smaller leaf area during a short period of time.

Foldats and Rutkiss (1975) gave quantitative data on the phenology of *Curatella americana* in the Venezuelan llanos; we take it as a typical example of this type of behavior and will consider therefore its phenodynamics in more detail.

The reproductive phenophases start during the dry season concurrently with sprouting to produce new shoots. After several weeks, the flowering process is completed, but fructification and seed dispersal may continue for a short time during the rainy season. Some species sprout leaves before flowers; in others, as in *Curatella*, both processes go on simultaneously, while a third group blooms before leafing. Once a new leaf crop has been formed, the shoots remain unchanged for several months without any further growth; apical meristems cease functioning irreversibly, thus ending all morphogenetic activity for the rest of the rainy

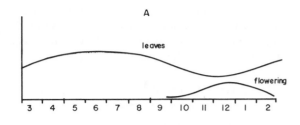

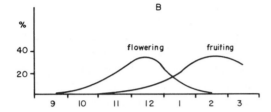

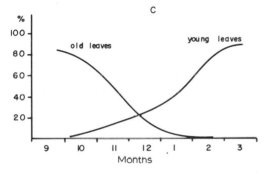

Fig. 5.5. Phenodynamics of *Curatella americana* in the Venezuelan llanos. A. A semi-quantitative representation of the annual cycle of leaf and reproductive biomass. B. Flowering and fruiting. C. Leaf production and leaf fall. In B and C the vertical axis indicates the percentage of the population at a given phenophase. (After Foldats and Rutkiss, 1975.)

season. Towards the end of this season, some leaves begin to yellow, and soon the entire foliage enters into a period of senescence preceding leaf fall. As we already remarked, a fire during the dry season does not hinder the normal cycle of phenological events; most old leaves simply fall and a flush of new leaves and flowers appears after a few days.

A peculiar feature apparent in many of these species, already noticed by Warming (1892) among the Lagoa Santa woody flora, is that in some cases, a second period of leafing and blooming occurs during the wet season, but this is much more irregular and occasional than the main active phase during the dry season.

In the cerrados as well as in the llanos, a fairly important proportion of the total tree flora shows this kind of phenodynamics. Some representative

examples besides *Curatella americana* are *Bowdichia virgilioides*, *Casearia sylvestris*, *Caryocar brasiliense*, *Salvertia convallariodora*, various species of *Byrsonima*, and many others. In the Lamto savannas, according to Menaut's observations, a few of the common trees, like *Crossopterix febrifuga* and *Bridelia ferruginea*, show this behavior, but in these savannas the woody species seem to show a greater variation in phenological patterns than does the Neotropical woody savanna flora.

B. Species with seasonal carbon assimilation

B.1. Perennial with precocious flowering. The *Curculigo scorzoneraefolia* group

The species with this phenological pattern start their development with the rainy season (Fig. 5.6); they give rise simultaneously to shoots and inflorescences, accomplishing all phases of sexual reproduction in a few weeks; thereafter the leaves persist for a while but soon they enter into a declining phase, to disappear entirely from the ground either towards the middle or at the end of the rainy season; only the perennating underground structures remain alive, together with the seed bank, during a long resting phase of dormancy.

This behavior characterizes many geophytes either in seasonal or in hyperseasonal savannas, particularly sedges and bulbous species of Amaryllidaceae, Liliaceae, Iridaceae and other monocotyledon families. *Curculigo scorzoneraefolia* in Neotropical savannas and *Curculigo pilosa* in the African savannas are good examples of this group. Some subshrubs show this same phenodynamics, such as *Vernonia guineensis* and *Rhynchosia sublanata* in Africa. *Ruellia geminiflora* and *Desmodium pachyrrhizum* in America.

As these species disappear from the ground during the dry season and their perennating organs remain more or less deeply buried in the soil, fire cannot have any direct effect on their phenology, therefore their cycle has necessarily to be synchronized with other environmental signals.

B.2. Annuals with precocious flowering. The *Spilanthes barinensis* group

Apparently, a few annuals show essentially the same phenological rhythms as the preceding group of perennials, with precocious growth and flowering, but of course during the rest phase only the

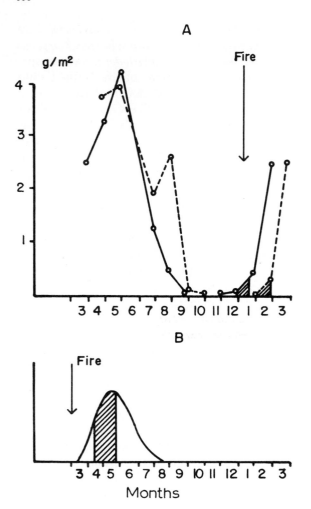

Fig. 5.6. Phenodynamics of two perennial species with seasonal carbon assimilation and precocious flowering. A. *Curculigo pilosa* as it behaves in two different savanna communities at Lamto, Ivory Coast (after César, 1971). B. *Curculigo scorzoneraefolia* in the savannas of the western Venezuelan llanos. In A the biomass data were obtained by random sampling; in B the curve only represents a subjective estimate of the development of the biomass; in both cases diagonal hatching indicates the flowering period.

seed bank persists in the soil. Their entire annual cycle occurs during the first months of the rainy season, they germinate after the first rains and bloom and disperse the seeds within a couple of months.

Most of these species have a more or less weedy ecological behavior. *Brachiaria plantaginea*, a widespread grass in wet savannas, constitutes a typical example, as well as the small Neotropical composite *Spilanthes barinensis*. Other species in the group are weeds of such genera as *Amaranthus* and

Croton, that differ from the two forementioned species by having a longer period of blooming. We could not find any example of African species belonging to this phenological group.

B.3. Perennials with delayed flowering. The *Bulbostylis junciformis* group

In this group are included all perennial species which have a definite rest phase, and which flower several months after their emergence, in the case of herbs and subshrubs, or several months after leafing, in the case of deciduous trees.

The active phase starts with the rains (Fig. 5.7), they produce the annual crop of leaves, and towards the middle of the wet season their reproductive phenophases begin. When the dry season becomes severe, the deciduous trees shed their leaves, while the whole aerial parts of herbs and subshrubs die back, only the perennating underground organs remaining alive.

In this phenological group are included many sedges, some other geophytic monocotyledons, a variety of subshrubs and a few deciduous trees. *Bulbostylis junciformis* and the bulbous *Cypella linearis* (Iridaceae) may typify this phenodynamics within the Neotropical herbaceous flora, while in the Guinean savannas the best examples are two widespread sedges, *Cyperus schweinfurthianus* and *Scleria canaliculato-triquetra* (César, 1971). Many half-shrubs also behave in this way, like *Tephrosia bracteolata* and *T. elegans* in Lamto and *Zornia*

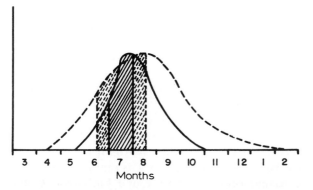

Fig. 5.7. Phenodynamics of *Bulbostylis junciformis*, a perennial sedge with seasonal assimilation and delayed flowering. The curves give a subjective estimate of the development of the green biomass and indicate the flowering period. The continuous curve corresponds to a savanna normally burnt in March, while the discontinuous lines indicate the behavior of this species when protected from fire.

reticulata and *Aeschynomene brasiliana* in Venezuela. Examples of deciduous trees flowering several months after leafing are *Genipa americana* in the Neotropics and *Terminalia glaucescens* in Africa.

B.4. Annuals with delayed flowering. The *Aristida capillacea* group

The life cycle of these species lasts between two and seven months, entirely within the rainy season, except in a few species whose reproductive activities extend to the first part of the dry period. From one to six months after the onset of rains, they germinate, rapidly develop the vegetative structures, and enter into the sexual reproductive phases. Even before completing seed dispersal, many of these annuals die back completely.

In order to obtain phenologically homogeneous groups it is necessary to distinguish between two sets of annuals with delayed flowering: the "long-cycled" species, whose life cycle spans three or more months; and the "ephemerals" that accomplish all their active phenophases in an exceptionally short time.

All long-cycle annuals have in common that they start their development at the middle, even toward the end of the rainy season (Fig. 5.8). This fact has been remarked both in African savannas (César, 1971) and in American communities (Monasterio, 1968). This means that germination takes place at least two months after the onset of rains; in some cases four months elapse between the start of the rainy season and the development of these annuals. Another distinctive functional trait shared by most annuals is their extended period of blooming, since the production of new leaves and new flowers proceed during almost their entire active cycle. As was noted in the section on life forms, the number of annual species varies greatly among the various savanna ecosystems, but in any case a large part of the annual flora shows this kind of phenodynamics. *Hyptis suaveolens*, a widespread weedy forb in northern South American savannas may typify this behavior as may do most annuals in Paleotropical savannas.

The ephemerals constitute a highly interesting group of species within the floras of the savannas, if not by their number or their importance in vegetation, at least by the very fact that this strategy occurs in savanna ecosystems. Their activities start

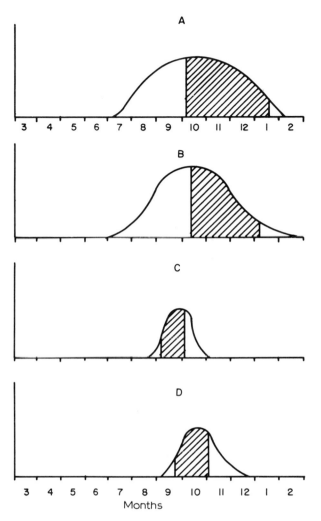

Fig. 5.8. Phenodynamics of four species of annuals in the savannas of the Venezuelan llanos. It is a subjective estimate of the development of the green and flowering biomass. A. *Cassia desvauxii*. B. *Hyptis suaveolens*. A and B represent two long-cycled annuals with delayed flowering. C. *Aristida capillacea*. D. *Gymnopogon foliosus*, C and D. are two ephemeral grasses.

very late (Fig. 5.8) in the last half of the rainy season; after germination they have an extremely rapid development, with immediate blooming, since they show practically no exclusively vegetative phase of growth, but begin to bloom from the very early stages of shoot development. Their decay is as fast as their previous development; they rapidly disappear from the ground, to remain just as a seed bank in or on the soil for a large part of the year.

Considering a whole local population, an ephemeral species may complete its active life cycle in about two months, but if one follows individuals, they may be found to pass through all developmental phases in no more than four or six weeks.

In Venezuelan savannas, some characteristic ephemerals are the grasses *Aristida capillacea*, *Digitaria fragilis*, *Gymnopogon foliosus* and *Microchloa indica* (this latter species also occurs in Zambezian "steppes"); as well as many forbs like *Borreria ocimoides*, *Polycarpaea corymbosa* and several species of *Polygala*. Outside tropical America, this phenological pattern has not been reported, though the occurrence of the same or of very similar species makes its existence quite probable.

B.5. Tardy flowering perennials. The *Cochlospermum vitifolium* group

This group includes all the deciduous woody species that flower during the dry season. This behavior is much more frequent among the canopy trees of the tropical deciduous forest than among savanna species, where, as already pointed out, most trees are evergreen or brevideciduous. In fact, among the deciduous trees growing in savannas, there are many species that also occur in the neighbouring forests, and they could be considered as pioneer forest species colonizing some unstable savannas. This seems to be the case, for instance, with most trees invading the derived savannas in Nigeria (Hopkins, 1962).

After the vegetative rest period when they remain leafless for a part or the whole of the dry season, the trees generally start to grow a few days before the onset of rains (Fig. 5.9), developing their new leaf crop in a couple of weeks to complete this process of leaf renewal at the very beginning of the rainy season. Other species may have their leaf flush after the onset of rains instead.

Once the new leaf crop has been produced, the phenodynamics of these species may show some slight variations among different species, some of them, like *Spondias mombim*, cease all vegetative development, while others, like *Cochlospermum vitifolium*, continue to produce new leaves throughout the wet season. In any case, shortly after the rains cease, the abscission process starts and more or less suddenly the whole foliage is shed. Flowering occurs during the leafless period in the dry season; this is thus a resting phase for vegetative activities only, and not for reproduction, except for the short period of complete rest after leaf fall and before blooming.

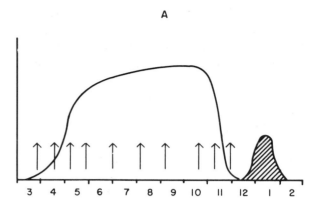

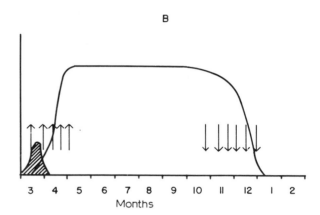

Fig. 5.9. Phenological behavior of two deciduous trees in the savannas of the Venezuelan llanos. A. *Cochlospermum vitifolium*. B. *Spondias mombim*. Upward arrows indicate leafing; downward arrows refer to leaf fall; diagonal hatching refers to flowering and the continuous curves indicate the annual cycle of leaf biomass. Notice that *C. vitifolium* shows continuous leafing throughout the rainy season, while *S. mombim* produces its foliage in a limited period during the change from the dry to the rainy season.

This phenological strategy of deciduous trees flowering during the leafless stage is more common among the rich arboreal flora of the Brazilian cerrados than in any other savanna area; in this region, this strategy is represented by many Bombacaceae, such as several species of *Bombax*, as well as by common species of other families, such as *Cochlospermum insignis*, *Terminalia argentea*, *Sterculia striata*, and many others. According to Hopkins (1968), *Annona senegalensis* and *Butyrospermum parkii* show the same behavior in the Nigerian savannas.

B.6. Tardy flowering annuals. The *Egletes florida* group

In most savannas, almost all annuals belong to a "wet season annual flora", which accomplish the whole active cycle during the favorable season of the year, when there are fewer probabilities of water shortage for these shallow-rooted species. But in *esteros* and similar seasonally waterlogged habitats, a "dry season annual flora" may occur, since its species may take advantage of the short period left betweeen the drainage of surface water and the exhaustion of water reserves in the upper soil layers.

These species germinate as soon as soil humidity becomes favorable, after the excess water has been drained off — that is, some time after the last rains (Fig. 5.3); they rapidly develop to complete their active phenophases, either vegetative or reproductive, before the soil becomes too dry; seed dispersal, however, may continue until the end of the rainless season.

Among the annuals behaving this way, a few species may be listed which occur in seasonally waterlogged areas of the Venezuelan llanos (Ramia, 1977, 1978), such as *Egletes florida*, *Trichospira verticillata* and *Heliotropium filiforme*.

B.7. Opportunistic annuals. The *Phyllanthus sublanatus* group

The developmental strategy of opportunistic annuals that may profit from any short favorable period to become active seems to be peculiar to a few species living under highly unreliable environments, such as extreme deserts. It is striking, therefore, that the same strategy may be found in tropical savannas under entirely different humidity conditions. César (1971) noted that *Phyllanthus sublanatus*, a small annual occurring in the Lamto savannas, shows this opportunistic behavior. This plant accomplishes its cycle in a few weeks whenever humidity conditions become favorable. We do not know yet of any similar case in other savannas.

A somewhat comparable phenological strategy is displayed by some annuals that can grow and reproduce at any time of the year because almost always they can find suitable environmental conditions. They occur, for instance, in habitats where, due to a permanently high water table, water shortage never becomes acute, but where the ground does not become waterlogged either. This is the case in wet savannas in northern Surinam, where Van Donselaar-Ten Bokkel Huinink (1966) reported that three species of annuals become active whenever conditions are favorable in any season of the year. These plants, which accomplish their cycle in a few months, could hardly be considered as annuals since it is obvious that they do not have any annual rhythmicity, therefore it seems better to consider them simply as short-cycled plants.

PHENOLOGICAL STRATEGIES AND THE DYNAMICS OF SAVANNA ECOSYSTEMS

Considering the whole spectrum of annual rhythms found in the savanna flora, the first fact to emphasize is the wide range of phenological strategies apparent among the species of these tropical ecosystems. In spite of the sharp seasonality of the vegetation that seems to reflect an acute water shortage during the long rainless season, every period in the year appears to be favorable at least to the accomplishment of certain phenophases in one or another group of plant species. Thus all perennial herbs, as well as a majority of the annuals, grow and flower during the wet season; but there are some species that are able to flower during the dry season, as occurs with the dry season annuals and the everflowering perennials.

In contrast with herbs, a majority of woody and half-woody species renew their leaves and bloom during the dry season, giving thus a clear indication that there are then certain water resources remaining available to these deep-rooted plants.

Among the species showing active growth during the wet season, it is quite interesting to notice that they behave as if a certain temporal division of the niche exists (Fig. 5.10), since some species start to grow with the first rains or soon after a late fire, entering immediately into their reproductive phenophases (precocious species); other populations instead emerge only gradually, develop their shoots slowly, and enter into their reproductive phases towards the middle of the rainy season or even during its last weeks. The annuals too show a differentiation between precocious and delayed species, with a small group that is able to develop even during the rainless period.

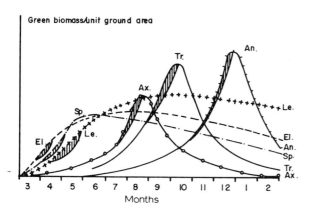

Fig. 5.10. The temporal division of the niche in the herb layer of a seasonal savanna in the Venezuelan llanos. This savanna is normally burnt in March. The annual cycle of the six dominant grasses is indicated by the development of the green biomass and by the flowering periods (vertical hatching). Notice the temporal displacement of flowering periods and the non overlapping of the periods of maximum growth-rates in each species. *El* = *Elyonurus adustus*; *Sp* = *Sporobolus cubensis*; *Le* = *Leptocoryphium lanatum*; *AX* = *Axonopus canescens*; *Tr* = *Trachypogon vestitus*; *An* = *Andropogon semiberbis*.

If the phenodynamics of each group is now considered against the background of seasonal changes at the whole ecosystem level, one may realize that, through their rapid response to fire or to drastic changes in soil humidity, the early growing species, either annuals or perennials, develop when the ground is practically devoid of any plant cover, since it has either been consumed by fire or, in unburnt savannas, it is mainly formed of dead leaves. Under these conditions they do not face any competitive pressure from other species, either for light, water or nutrients. Besides that, the less common species, either wind- or insect-pollinated, have the best conditions for effective pollination shortly after a fire, since their visibility and accessibility become maximum at this moment.

Some of these early growing and precocious flowering species, such as the annuals and the perennials with a resting phase, will disappear from the ground surface when the dominant grasses and other species with late emergence and delayed flowering reach their maximum ground cover, while the precocious perennial grasses will slow down their growth rates as the late-emergence grasses increase theirs.

The annuals with late emergence and delayed flowering have a strategy that at first sight does not

seem to give them many chances to compete with the perennial grasses and subshrubs. In effect, when these annuals germinate and start their fast developmental cycle, most savannas already have a closed herbaceous canopy. Under these circumstances, one can understand why these annuals have adopted one of two following divergent strategies. On the one hand, they tend to occupy the less favorable sites, where the herbaceous matrix becomes discontinuous, leaving open places where annuals may establish more easily. On the other hand, shade-tolerant plants may survive under a herb canopy that seldom is closed enough to prevent the filtering of a certain amount of light; in this case, these annuals prosper in a microenvironment that, if it is poorer in radiant energy, seems to compensate by being more humid and cooler.

If the phenodynamics of woody species is now examined, it can be seen that a majority of them belong to the *Curatella americana* group (A.8) with continuous photosynthetic assimilation, seasonal growth and tardy flowering (during the dry season). At this time, an overwhelming proportion of the herb layer has already changed to a standing dead cover, annuals have disappeared, and perennials with a semi-resting phase are reduced to their annual minimum. Under these circumstances, active trees do not find any competition from the herbaceous plants. Furthermore, the occurrence of natural fires greatly increases during this season, both because of climatic drought and because there is a continuous blanket of standing straw; if a fire does occur, it will consume only old decaying leaves, or, at the worst, a part of the new leaf crop, but it will not do irreparable harm to the trees. Besides these advantages, a change of foliage during the relatively less favorable season makes it possible to display a fully developed photosynthetic apparatus functioning at full rates when more favorable conditions arrive. This group also shows a phenodynamics perfectly coupled with the major environmental changes, providing that water availability during the dry season make possible the maintenance of a certain green area. If through the development of an extensive and deep root system these trees are able to exploit the water held in deeper soil layers out of reach of herbs, then there is an obvious advantage in changing leaves and reproducing during the period of relative water shortage and of greater fire risks, leaving to the rainy season the function of active assimilation permitting a storage of energy to support the dry season's activities.

The deciduous trees, instead, must be able to reconstitute their foliage as soon as possible when the wet season arrives, in order to take advantage of its favorable conditions to re-establish a positive carbon balance after the long period of zero assimilation; moreover, they neglect the growth possibilities offered during the rainless season, as if this period were totally unfavorable for any assimilation, being therefore at a net disadvantage with respect to evergreens which, through a strategy of profiting from deep water resources, may maintain a positive carbon balance throughout the year.

The phenological strategy of continuously growing perennials that flower tardily in the rainy season (the *Trachypogon plumosus* group, A.2) appears as the most successful, since a majority of the dominant grasses and sedges in different savanna ecosystems all over the tropical area belong to this group. This good ecological performance may be easily understood on the basis of the perfect coupling between the phenorhythms of these species and rainfall seasonality. In effect, during the period of water shortage, their aerial biomass rapidly decays, while as the rainy season progresses they gradually develop their shoots and reproductive structures, to reach the maximum growth rates during the reproductive phenophases that occur in the safest period of the year in respect of water availability. Moreover, these perennials effectively occupy the ground during most of the annual cycle, holding the space and depending much less than annuals on seed establishment. When perennial grasses with this developmental pattern reach their full size, they overtop all other species in the herb layer, being then the most competitive for light. The semi-resting phase during the dry season means that they retain the capability of growth; although this late growth does not contribute greatly to a better carbon balance, it represents a plastic response to adverse conditions, giving the possibility of finer adjustments to the particular humidity conditions actually met with each year. In this sense, they surpass species with a complete resting phase, since these are absolutely unable to respond in such a plastic way. The semi-resting period of perennial grasses and sedges implies that during the less

extreme years a greater green area could be maintained for a longer time, enhancing thus the possibilities of early growth when water becomes again available.

Another successful strategy in most savannas has been that of the *Leptocoryphium lanatum* group (A.1) — that of continuously growing perennials with precocious flowering. These species, through a rapid early growth at the beginning of the rainy season, may occupy a niche left temporarily void by the dominant grasses, that have slower emergence and delayed flowering. These two strategies may also be seen as successful adaptations to recurrent fires, since, when the savannas are burnt late in the dry season, the flames just consume the annual shoots, mostly dry at that moment, leaving intact and unharmed the perennial underground structures.

While the hydroperiodicity of the savanna environment seems to be the major external compulsion regulating plant development and rhythmicity, in many cases the direct signals triggering the physiological mechanisms responsible for the successive phenophases seem to be different from soil or atmospheric humidity. This fact is most obvious in woody species sprouting before the onset of rains, when the soil and the atmosphere still remain quite dry. This is the case of a majority of woody species, as already discussed; it is necessary then to consider other external signals of high constancy from year to year, that could be received by these plants. There are two such rhythmical environmental impulses, namely photo- and thermoperiodicity. Flowering of trees during the dry season starts after the shortest days and in a period of maximum daily fluctuations of temperature as well as of minimum night temperatures. Quite the opposite situation occurs in species blooming during the peak of the rainy season after the longest days and during the narrowest daily temperature fluctuations. Monasterio and Sarmiento (1976) discussed thoroughly the possible relations between rhythmicity and environmental impulses in the seasonal savannas and semi-deciduous forests of the Venezuelan llanos, we refer to that paper for further details concerning this point.

A last point worthy of further consideration relates to the influence of fire on the phenological behavior of savanna species. One may take for

granted that fire has always been a natural ecological factor inducing different adaptive responses
from the savanna populations. Most savanna species must have had a long evolutionary interplay
with recurrent fires. Natural burnings have more
probabilities of occurring during the last part of the
dry season, precisely in the period when nowadays
ranchers set fire to the savannas, for different
practical purposes.

Our main argument concerning fire action is that
a late burning (that is, a fire during the last weeks of
the dry season or at the very beginning of the rainy
season) does not alter drastically the normal course
of phenological events in the savannas. In fact, its
direct action on annuals will be nil, since the seed
banks may lie well protected in the soil. Herbaceous
perennials with a resting phase have already disappeared from the ground, and the semi-resting
species have at that time a great proportion of dead
biomass. Half-woody species and low trees may
lose a part or even the whole of their leaves and
annual branches, but in any case these are short-
lived structures that were passing through decaying
phases. Adult trees, taller than a certain minimum
threshold height normally attained by the flames,
remain fairly unharmed. The most obvious consequences of a late fire will be then: first, the maintenance of tree species in a half-shrubby habit;
second, slight damage to young trees by partial or
complete loss of their new leaf crop; and, third, a
triggering effect on perennial herbs, grasses, half-
shrubs and trees, promoting immediate leafing
and/or flowering. These two effects could hardly be
considered as deleterious.

When the savannas do not burn, these species
show essentially the same rhythmicity as when
burned, but the different species appear as less
synchronized with each other, and in some cases
their development is slower and less intense. A late
fire acts then as a synchronizing agent among the
species having similar phenological responses.

Furthermore, there are particular cases where
savanna species show some kind of fire-
dependency. Coutinho (1977) reported that fire
promotes fruit dehiscence in some subshrubs of the
cerrados, such as *Anemopaegma arvense* and
Jacaranda decurrens; afterwards the seeds will find
favorable conditions for anemochoric dispersal. But
the main dependence is shown by savanna species
that do not flower unless fire or other destructive

agents have eliminated the entire above-ground
biomass. Coutinho (1976) found that 150 species of
half-shrubs and perennial herbs have fire-induced
flowering, some of them blooming after fire whenever in the year it may occur (opportunistic strategists), other species responding only to a dry season
burning. Only in this last case it seems possible to
evoke a dependency on photoperiodicity.

Finally, Coutinho (1976) reported that in some
half-shrubs with a xylopodium, dormant buds were
structurally changed by the action of fire, to reproductive buds resulting in reproductive organs,
which he called a "pyro-morphogenetic effect".
One may see, then, how a number of characteristic
species belonging to the savanna flora have evolved
phenological patterns not only adequate to cope
with frequent fires, but that even take advantage of
burning to improve their performance in the ecosystem.

To conclude this chapter on life forms and
phenology of savanna species, we want to summarize our main arguments as follows. Apparently,
during a long evolutionary interaction between
plant populations and the environmental constraints characteristic of savanna ecosystems, certain forms and rhythms were originated as successful responses to the selective pressures which operate nowadays, as if natural selection had been
acting with similar selective pressures for a rather
long time. Most populations have optimized their
fitness through architectural and phenological
adaptations that produced plant forms, developmental patterns and annual rhythms leading to
good performance under the prevailing set of
outside challenges.

We may advance the hypothesis that hydroperiodicity on the one hand and recurrent fires
during the dry season on the other, have been the
principal environmental factors selectively filtering out inadequate responses, and allowing the
plant populations that evolved in a certain direction
a better reproductive efficiency. There are also
some reasons to suppose that the interplay between
plant populations with different growth forms —
annuals, perennial tussocks, half-shrubs, trees —
and with different phenodynamics — continuous
assimilation or rest phase; seasonal or continuous
growth; early or late emergence; precocious, delayed or tardy flowering — have led to a spatial and
temporal division of the niche, as suggested by the

division of soil water resources between herbaceous and woody species, or the sequential development of the various phenological groups during each annual cycle, as if they were replacing each other in the use of some critical resource.

The major morphological and developmental strategies of successful plant species have been either to shorten the life cycle and to synchronize its active phenophases with the most favorable season, thus escaping the filtering action of drought and fire, as annuals and resting perennials have done; or to resist these stresses through, first, a precise synchronization of the developmental processes with favorable periods, and, second, a type of resource allocation among the various plant structures favoring resistance and survival under these stresses, as most perennials have done. Sometimes, these adaptations involved original solutions, such as the substitution of the shoot apical meristems by the vascular cambium as the main replacement tissue; or the exceptional degree of subterranization of the biomass and of certain key functions attained by half-woody plants, where even the function of producing and restoring vegetative and reproductive organs relies on underground meristems.

Other types of phenological behavior seem rather to be a result of selective pressures no longer operative under the conditions where these species live nowadays in the savannas; this may be the case, for instance, for the ephemerals and the opportunistic strategists, since it may be reasonable to suppose that these strategies have evolved as responses to stronger environmental stresses.

To accept or reject these hypotheses much more observational and experimental work is needed; but in any case we think they may be useful as research guidelines, suggesting some profitable ways to get further insight into these very interesting populations and ecosystems.

REFERENCES

Aristeguieta, L., 1966. Flórula de la Estación Biológica de los Llanos. *Bol. Soc. Venez. Cienc. Nat.*, 110: 228–307.

Ataroff, M., 1975. *Estudios ecológico-poblacionales en dos especies de árboles de las sabanas de los Llanos*. Thesis, Facultad de Ciencias, Mérida, 51 pp.

Beard, J.S., 1953. The savanna vegetation of northern tropical America. *Ecol. Monogr.*, 23: 149–215.

Beiguelman, B., 1962. Contribuiçào para o estudo anatômico de plantas do Cerrado. *Rev. Biol.*, (*Lisboa*), 3: 97–123.

Braun-Blanquet, J., 1932. *Plant Sociology*. McGraw-Hill, New York, N.Y., 439 pp.

Campos, A.C. and Labouriau, L.G., 1969. Corpos silicosos de gramineas dos Cerrados. II. *Pesqui. Agropecu. Bras.*, 4: 143–151.

César, J., 1971. *Etude quantitative de la strate herbacée de la savane de Lamto* (*Moyenne Côte d'Ivoire*). Thesis, Faculté des Sciences de Paris, Paris, 95 pp.

Clements, F.E., 1920. *Plant Indicators*. Carnegie Inst. Wash. Publ; No. 290: 453 pp.

Coutinho, L.M., 1976. *Contribuição ao conhecimento do papel ecologico das queimadas na floração de especies do Cerrado*. Thesis, Universidade de São Paulo, São Paulo, 173 pp.

Coutinho, L.M., 1977. Aspectos ecológicos do fogo no cerrado. II. As queimadas e a dispersão de sementes em algumas espécies anemocoricas do estrato herbácco-subarbustivo. *Bol. Bot. Univ. São Paulo*, 5: 57–64.

Du Rietz, G.E., 1931. Life-forms of terrestrial flowering plants. *Acta Phytogeogr. Suec.*, 3: 1–95.

Duvigneaud, P., 1949. Les savanes du Bas-Congo. Essai de phytosociologie topographique. *Lejeunia*, 10: 1–192.

Duvigneaud, P., 1955. Etudes écologiques de la végétation en Afrique Tropicale. In: *Les Divisions Ecologiques du Monde*. C.N.R.S., Paris, pp. 131–148.

Duvigneaud, P., 1958. La végétation du Katanga et de ses sols metallifères. *Bull. Soc. R. Bot. Belg.*, 90: 127–286.

Esau, K., 1953. *Plant Anatomy*. Wiley and Sons, New York, N.Y., 735 pp.

Ferri, M.G., 1962. Problems of water relations of some Brazilian vegetation types, with special consideration of the concepts of xeromorphy and xerophytism. In: *Plant/Water Relationships in Arid and Semi-Arid Conditions, Proceedings of the Madrid Symposium*. UNESCO, Paris, pp. 191–197.

Figueireido, R.C.L. and Handro, W., 1971. Corpos silicosos de Gramíneas dos Cerrados. V. In: M.G. Ferri (Editor) *III Simpósio sôbre o Cerrado*. University of São Paulo, São Paulo, pp. 215–230.

Foldats, E. and Rutkiss, E., 1969. Suelo y agua como factores determinantes en la selección de algunas especies de árboles que en forma aislada acompañan nuestros pastizales. *Bol. Soc. Venez. Cienc. Nat.*, 115–116: 9–30.

Foldats, E. and Rutkiss, E., 1975. Ecological studies of chaparro (*Curatella americana* L.) and manteco (*Byrsonima crassifolia* H.B.K.) in Venezuela. *J. Biogeogr.*, 2: 159–178.

Hallé, F. and Oldeman, R.A.A., 1970. *Essai sur l'architecture et la dynamique de croissance des arbres tropicaux*. Masson, Paris, 178 pp.

Harper, J.L., 1967. A Darwinian approach to plant ecology. *J. Ecol.*, 55: 247–270.

Harper, J.L., 1977. *Population Biology of Plants*. Academic Press, London, 892 pp.

Hoock, J., 1971. Les savanes guyanaises: Kourou. *Mém. ORSTOM*, 44: 251 pp.

Hopkins, B., 1962. Vegetation of the Olokemeji Forest Reserve, Nigeria. I. General features of the Reserve and the research sites. *J. Ecol.*, 50: 559–598.

Hopkins, B., 1968. Vegetation of the Olokemeji Forest Reserve, Nigeria. V. The vegetation of the savanna site with special reference to its seasonal changes. *J. Ecol.*, 56: 97–115.

Hopkins, B., 1970. Vegetation of the Olokemeji Forest Reserve, Nigeria. VII. The plants of the savanna site with special reference to their seasonal growth. *J. Ecol.*, 58: 795–825.

Jeannoda-Robinson, V., 1977. *Contribution a l'étude de l'architecture des herbes*. Thesis, Université des Sciences et Techniques du Languedoc, Montpellier, 76 pp.

Koechlin, J., 1961. La végétation des savanes dans le Sud de la République du Congo Brazzaville. *Mém. ORSTOM*, No. 1: 310 pp.

Labouriau, L.G., Marques Valio, I.F. and Heringer, E.P., 1964. Sôbre o sistema reproductivo de plantas dos Cerrados. *An. Acad. Bras. Ciênc.*, 36: 449–464.

Lamotte, M., 1978. La savanne préforestière de Lamto, Côte d'Ivoire. In: F. Bourlière and M. Lamotte (Editors), *Problèmes d'ecologie: structure et fonctionnement des écosystemes terrestres*. Masson, Paris, pp. 231–311.

Lawson, G.W., Jenik, J. and Armstrong-Mensah, K.O., 1968. A study of a vegetation catena in Guinea savanna at Mole Game Reserve (Ghana). *J. Ecol.*, 56: 505–522.

Lebrun, J., 1947. *La végétation de la plaine alluviale au Sud du Lac Edouard*. Inst. Parcs Natl. Congo Belge, Bruxelles, 800 pp.

Malaisse, F., 1975, *Carte de la végétation du bassin de la Luanza*. Cercle Hydrobiol., Brussels, 42 pp.

Meguro, M., 1969. Fatores que regulan a floração en *Imperata brasiliensis* Trin. (Gramineae). *Bol. Fac. Fil. Ciênc. Letr. Univ. São Paulo Bot.*, 24: 105–125.

Menaut, J.C., 1971. *Etude de quelques peuplements ligneux d'une savane Guinéenne de Côte d'Ivoire*. Thesis, Faculté des Sciences, Paris, 141 pp.

Mérida, T. and Medina, E., 1967. Anatomía y composición foliar de árboles de las sabanas de *Trachypogon* en Venezuela. *Bol. Soc. Venez. Cienc. Nat.*, 111: 46–55.

Monasterio, M., 1968. *Observations sur les rhythmes annuels de la savane tropicale des "llanos" du Vénézuéla*. Thèse, Université de Montpellier, 108 pp.

Monasterio, M., 1982. Etudes écologiques dans la haute montagne tropicale. Les paramos du Vénézuéla. In press.

Monasterio, M. and Sarmiento, G., 1976. Phenological strategies of plant species in the tropical savanna and the semideciduous forest of the Venezuelan llanos. *J. Biogeogr.*, 3: 325–356.

Montes, R. and Medina, E., 1975. Seasonal changes in nutrient content of leaves of savanna trees with different ecological behavior. *Geó-Eco-Trop.*, 1: 295–307.

Moore. R.M. (Editor), 1973. *Australian Grasslands*. Australian National University Press, Canberra, A.C.T., 455 pp.

Morat, P., 1973. Les savanes du Sud-Ouest de Madagascar. *Mém. ORSTOM*, No. 68: 235 pp.

Morretes, B.L. 1966. Contribuição ao estudo da anatomia das folhas de plantas do cerrado. II. *Bol. Fac. Fil. Ciênc. Letr. Univ. São Paulo, Bot.*, 22: 209–244.

Morretes, B.L., 1969. Contribuição ao estudo da anatomia das folhas de plantas do cerrado. III. *Bol. Fac. Fil. Ciênc. Letr. Univ. São Paulo, Bot.*, 24: 1–32.

Morretes, B.L. and Ferri, M.G., 1959. Contribuição ao estudo da anatomia das folhas de plantas de cerrado. *Bol. Fac. Fil. Ciênc. Letr. Univ. São Paulo, Bot.*, 16: 7–70.

Oldeman, R.A.A., 1974. L'architecture de la fôret guyanaise. *Mém. ORSTOM*, No. 73: 204 pp.

Perera, N.P., 1969. The ecological status of the savanna of Ceylon. I. The upland savanna. *Trop. Ecol.*, 10: 207–221.

Rachid, M., 1947. Transpiração e sistemas subterrâneos da vegetação de verão dos campos cerrados de Emas. *Bol. Fac. Fil. Ciênc. Letr. Univ. São Paulo, Bot.*, 5: 5–135.

Rachid, M., 1956. Alguns dispositivos para proteção de plantas contra a seca e o fogo. *Bol. Fac. Fil. Ciênc. Letr. Univ. São Paulo, Bot.*, 13: 38–68.

Ramia, M., 1974. *Plantas de las Sabanas Llaneras*. Monte Avila, Caracas, 287 pp.

Ramia, M., 1977. Observaciones fenológicas en las sabanas del Medio Apure. *Acta Bot. Venez.*, 12: 171–206.

Ramia, M., 1978. Observaciones fenológicas en las sabanas del Alto Apure. *Bol. Soc. Venez. Cienc. Nat.*, 135: 149–198.

Raunkiaer, C., 1934. *The Life Form of Plants and Statistical Plant Geography*. Clarendon Press, Oxford, 632 pp.

Rawitscher, F., Ferri, M.G. and Rachid, M., 1943. Profundidade dos solos e vegetação em campos cerrados do Brasil meridional. *An Acad. Bras. Ciênc.*, 15: 267–294.

Rawitscher, F. and Rachid, M., 1946. Troncos subterrâneos de plantas brasileiras. *An. Acad. Bras. Ciênc.*, 18: 261–280.

Rizzini, C.T., 1965. Experimental studies on seedling development of Cerrado woody plants. *Ann. Mo. Bot. Gard.*, 52: 410–426.

Rizzini, C.T. and Heringer, E.P., 1962. Studies on the underground organs of trees and shrubs from some Southern Brazilian savannas. *An. Acad. Bras. Ciênc.*, 34: 235–247.

Rizzini, C.T. and Heringer, E.P., 1966. Estudo sobre os sistemas subterrâneos difusos de plantas campestres. *An. Acad. Bras. Ciênc.*, 38 (Supl.): 85–112.

Sarmiento, G., 1978. *Estructura y funcionamiento de sabanas neotropicales*. Universidad de Los Andes, Mérida, 367 pp.

Sarmiento, G. and Monasterio, M., 1971. Ecologia de las sabanas de América tropical. I. Análisis macroecológico de los Llanos de Calabozo, Venezuela. *Cuad. Geogr.*, 4: 1–126.

Sarmiento, G. and Monasterio, M., 1975. A critical consideration of the environmental conditions associated with the occurrence of savanna ecosystems in tropical America. In: F.B. Golley and E. Medina (Editors), *Tropical Ecological Systems*. Springer-Verlag, Heidelberg, pp. 223–250.

Schnell, R., 1971. *Introduction à la phytogéographie des pays tropicaux*. Gauthier-Villars, Paris, 951 pp. (2 vols.).

Schnell, R., 1976–77. *Introduction à la phtogéographie des pays tropicaux*, 3 et 4. La flore et la végétation de l'Afrique tropicale. Gauthier-Villars, Paris, 459; 378 pp.

Sendulsky, T. and Labouriau, L.G., 1966. Corpos silicosos de gramineas dos Cerrados. I. *An. Acad. Bras. Ciênc.*, 38 (Supl.): 159–186.

Sillans, R., 1958. *Les savanes de l'Afrique Centrale française*. Lechevalier, Paris, 423 pp.

Teixeira da Silva, S. and Labouriau, L.G., 1971. Corpos silicosos de gramineas dos Cerrados. III. *Pesqui. Agropecu. Bras.*, 6: 71–78.

Van Donselaar-Ten Bokkel Huinink,W.A., 1966. Structure, root systems and periodicity of savanna plants and vegetations in Northern Surinam. *Wentia*, 17: 1–162.

Warming, E., 1892. *Lagoa Santa*. University of São Paulo, São Paulo, 386 pp. (Portuguese edition 1973).

Warming, E., 1909. *Ecology of Plants*. Clarendon Press, Oxford, 492 pp.

Chapter 6

THE VEGETATION OF AFRICAN SAVANNAS

JEAN-CLAUDE MENAUT

INTRODUCTION

The large number of papers dealing with the classification of the savannas testifies to the unflagging interest tropical ecologists take in having a precise definition of the vegetation category usually called "savanna" and on its distribution. Discussions about the exact meaning of the word savanna are not only of academic interest. When dealing with an ecological problem, people describing the vegetation far too often use terms such as "grassland", "orchard savanna", etc., without defining them. Furthermore, such words are sometimes given a different meaning by different people. Therefore, it is difficult for the reader to have any precise idea of the kind of vegetation described, and any comparison is made difficult, if not impossible.

Most of the classifications in use are based on different criteria (floristic, physiognomic, climatic or edaphic) which emphasize one particular factor more or less arbitrarily chosen; others are very broad in scope and group together a variety of vegetation categories. None can be considered entirely valid or sufficiently detailed. Furthermore, modifications or additions to the existing classifications are often proposed, most of them dealing with local or transitional vegetation types. Moreover, existing terminologies are very different from each other and make any comparison difficult (Schnell, 1971–1977). On a continental scale, however, the Yangambi classification (Anonymous, 1956), or that of UNESCO (1969, 1973) offer a sound basis for further work. At the regional level, useful proposals have been made by Boughey (1957a,b), Monod (1957, 1963) and Greenway (1973) among others.

In this chapter the approach adopted has been very pragmatic and an attempt has been made to describe the various vegetation types generally considered as "savannas". The designations chosen are either very general (e.g. savanna), or descriptive (e.g. woodland) or vernacular (e.g. *miombo*); they must be considered as convenient labels, and no more. They are discussed and described as precisely as possible, to help the reader to make useful comparisons, particularly between different continents.

The word savanna is used here in the traditional way. Following the proposals of the Pretoria Symposium (Huntley and Walker, 1979), I use the term in a broad sense to embrace a continuum of physiognomic types, of which the most important is the wooded grassland. It is furthermore proposed that the concept of a "savanna biome type" be adopted for wooded grasslands of the tropics and subtropics, including biotically related seral and catenary series. Such a definition limits the savannas to tropical latitudes and includes a variety of vegetation types. It includes pure grasslands (e.g. the sand dunes of the Sahel or the seasonally waterlogged *dambos*) as well as woody formations with a discontinuous grass cover or even no grass cover at all (e.g. the miombo woodlands or the dry evergreen thickets). Despite their obvious differences, all these plant formations have a number of ecological characteristics in common; the savanna biome differs from the rain-forest biome as well as from the desert biome or the mediterranean biome by a number of climatic, floristic, faunistic and land-use features.

However, it is also obvious that the distinction between these adjacent biomes is not easy to make at the regional scale. The rain forest penetrates deeply into the savanna biome, often as gallery forests along the river banks, and sometimes as forest remnants on the plateaux; it is the forest–savanna mosaic. Nevertheless the two formations are usually separated by a sharp boundary due to the very contrasting floristic and physiognomic features of the two biomes. In the driest parts of the tropics and subtropics, a savanna–desert mosaic also occurs, as a result of different soil conditions. In this case no sharp boundary between the two biomes can be drawn. The arid savanna thus merges into desert, dwarf shrubs and desert grass communities becoming progressively dominant. Moreover, considerable changes can take place from year to year, depending on the climatic conditions of the year and also on local land-use practice.

In this chapter, to represent the limits of the African savanna biome I have made use of the AETFAT/UNESCO map of African vegetation (Keay, 1959b), a simplified version of which is shown in Fig. 6.1. The savanna belt encloses the West African and Congo forest blocks; it is limited, both in the north and in the south, by desert and semi-desert formations. Within the savanna belt itself patches of mountain vegetation and permanent wetlands can be found, depending on physiographic conditions; they are not discussed in this chapter, being dealt with in other volumes of this series.

THE AFRICAN SAVANNA FLORA

As compared with those of other continents, the flora of African and Madagascan savannas is relatively well known (Koechlin, 1963). A great number of floristic studies, local as well as regional, have been published. Furthermore, modern floras are available for the whole continent, including the *Flora of West Tropical Africa* (Hutchinson and Dalziel, 1927–1936), the *Flora of Tropical East Africa* (Turrill and Milne-Redhead, 1952–1959), the *Flora Zambesiaca* (Brenan et al., since 1960) and the *Flora of Southern Africa* (Dyer et al., since 1963). In addition, taxonomic revisions and floristic papers are frequently published, and the publications of the Association for the Taxonomic Study of the Flora of Tropical Africa (AETFAT) help the specialists to keep abreast of the recent literature in this field.

Numerous gaps remain however. There is a great need for books better adapted to field work in certain domains, such as range management, giving more details on the life history and ecology of living plants. Some excellent examples of publications of this kind are Aubréville (1950–1959) on West African trees, Edwards and Bogdan (1951) on Kenya grasses, Chippindall (1955) on southern African grasses, Keay et al. (1960–1964) on Nigerian trees, Jacques-Félix (1962) on tropical African grasses, Bosser (1969) on Malagasy grasses, and Rose Innes (1977) on Ghana grasses. Books of this kind are badly needed for many areas, and more emphasis should be directed towards acquiring accurate information on the distribution, biology and ecology of the various species involved.

Reviews of the plant geography of Africa have recently been published by Schnell (1971–1977) Knapp (1973) and Brenan (1978); they have been of a great help to me in compiling this chapter.

Floristic relationships of the African savanna flora

The affinities of the xeric African floras with those of Asia and even America are obvious. Whereas pantropical species are rare (ruderals excepted), many genera are represented on the three continents, and many more families. This situation results not only from passive transport or range extension of individual species, but also from very old pantropical land connexions, prior to the differentiation of endemic groups. The floristic relationships are particularly marked between Africa and southern Asia. Some xeric species can even be found from the westernmost part of Africa to the arid areas of northern India; such broad ranges can be explained by the recentness of the land connexions between the two continents. Conversely, the floristic affinities between Africa and America can be traced back to Cretaceous times. Needless to say, this is not the case for some plants, such as *Calotropis procera* or *Hyparrhenia rufa*, recently introduced by man (Aubréville, 1949). Floral affinities between Africa and Australia are much scarcer but nonetheless obvious — for instance, for *Acacia* spp., *Adansonia* spp. and *Cochlospermum* spp. Many savanna plants with a very large tropical range (ruderals and marsh species excepted) are trees or shrubs (e.g. *Acacia* spp., *Cochlospermum* spp., *Jatropha* spp., *Prosopis* spp., *Ziziphus* spp.).

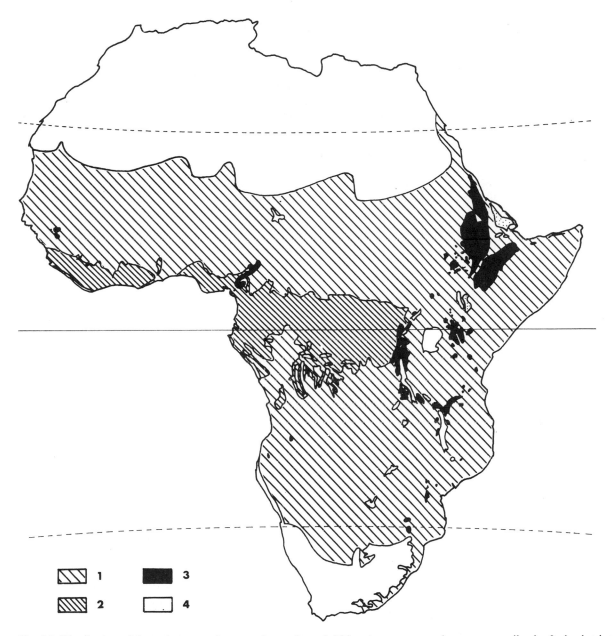

Fig. 6.1. Distribution of the major vegetation types in continental Africa. *1*=savannas and savanna woodlands; *2*=lowland moist forest; *3*=montane communities; *4*=deserts and mediterranean-type vegetation. (Simplified after Keay, 1959b.)

There is also a high proportion of herbaceous therophytes (Lebrun, 1947).

Characteristics of the African savanna flora

Following Engler (1910), many botanists agree to include the tropical parts of the African and Asian continents in a single Palaeotropic floristic kingdom (Good, 1974) or realm (Walter and Straka, 1970). However, the distinctive character of the flora of both continental, sub-Saharan Africa, and Madagascar is obvious (Emberger, 1968), most Malagasy plants being closely related to those of the continent.

The distinctive character of the tropical African floristic realm is made apparent by a certain degree of floristic isolation rather than by the existence of endemic groups (fifteen families, most of them restricted to forest habitats, seven of which are endemic to Madagascar). Such isolation is not only obvious in the case of forest floras (Aubréville, 1955), but also in that of the savanna floras. Long ago, Chevalier (1933) remarked that the floristic characteristics which set the tropics of Africa and South America apart are rather of a negative nature; many plant groups which are so characteristic of the Neotropics are entirely lacking in Africa. Indeed, only one species of bromeliad (*Pitcairnia feliciana*) is found in Africa and restricted to the Fouta Djallon Mountains of Guinea, and a single epiphytic or orophytic cactus, *Rhipsalis cassutha*, is widespread in the continent. *Ananas comosus* and *Opuntia* spp. formerly introduced by man are now becoming naturalized. It is worth noticing however that the introduced *Opuntia* spp, which so easily turn wild in North and South Africa and in coastal areas, do not easily become naturalized in savanna areas. Whereas the absence of some families which appeared since the breaking up of the former southern continent is easy to understand, that of primitive families now widespread in America and Asia is harder to account for. For instance, the Dipterocarpaceae, so numerous in the Asian tropics, are only represented in Africa by two genera (Monotoideae).

Since Christ (1892) it has been the usual practice to oppose the floras of the drier to those of the more humid parts of Africa. As pointed out by Schnell (1971) the savanna floras differ from those of the rain forests both by their structure (e.g. the prominent role of herbaceous species in savannas) and their biology (e.g. a rest period during the dry season). However, these differences are often not so clear-cut. Species belonging to a single genus can replace each other when the environment becomes dry (e.g. *Khaya* spp.). Sometimes their habit is very different in the two biomes (e.g. *Parinari* spp.); for instance, trees with a tall woody stem are replaced by shrubs with a gnarled trunk. However, other species maintain the same habit in the two different environments (e.g. *Lophira* spp.). It cannot therefore be excluded that some savanna species originated in a rain-forest habitat. Nevertheless this cannot be the case either for most herbaceous species, or for a number of woody genera which are never found in rain-forest environments. Quite likely, the present-day African savanna floras are made up of two components, one originating in the

humid tropics and the other, by far the more important, made up of true xerophytic species. This latter component includes both primitive African xerophytes (e.g. *Encephalartos* spp.) and intercontinental "migrants" (e.g. *Acacia* spp., *Prosopis* spp., *Ziziphus* spp.).

The flora of the African savannas therefore represents an unquestionable entity, at the geographical, floristic, ecological and physiognomic levels (Aubréville, 1949a). This speaks in favour of the great age of both Afro-tropical grasslands and dry forests. It could even be said that, in some cases at least, these plant formations antedated the rain forests themselves. This view has often been doubted (Bews, 1927; Lönnberg, 1929; Lebrun, 1947), but it is now held by a number of botanists (Monod, 1947, 1957; Aubréville, 1949a) and zoologists (Chapin, 1932; Dekeyser and Villiers, 1954).

Geographical distribution of savanna floras

The savannas of the dry parts of tropical Africa as a whole make up a distinct phytogeographic unit, in spite of the barrier provided by extended rain forest areas in the western and central parts of the continent. A number of species can be found throughout the savanna belt, from West to East and South Africa, whereas others display a bipolar distribution, being found north and south of the rain-forest blocks but not in East Africa. Needless to say, other patterns of distribution also exist, some species ranging from Senegal to Somalia, and others from Somalia to South Africa.

Many species of savanna plants are distributed all around the western and west-central rain-forest areas. They belong to a number of different families, such as *Actiniopteris radiata* (Adiantaceae), *Aristida hordeacea* (Poaceae), *Combretum aculeatum* (Combretaceae), *Grewia villosa* (Tiliaceae), *Parinari curatellifolia* (Rosaceae), *Piliostigma thonningii* (Caesalpiniaceae), *Acacia albida* (Mimosaceae) and *Diospyros mespiliformis* (Ebenaceae). The history of their crescent-shaped distribution around the forested core of the continent can be debated, but their present-day distribution can be easily explained by the geographical continuity of their distribution area.

Such is not the case for the true "diastemic taxa" discontinuously distributed north and south of the central forest block and missing on the eastern side of the continent (Monod, 1971). This pattern of distribution is obviously facilitated by the latitudinal zonation of the biomes north and south of the Equator in tropical latitudes (Troll, 1968; Aubréville, 1969), and by the physiographic and climatic changes which took place in East Africa in the recent geological past (since the start of the Quaternary). It is quite likely that the eastern savanna corridor has been interrupted a number of times, some local extinctions occurring on such occasions. The similarity of some components of the western African savanna flora with that of the Namib–Kalahari region has also led some authors to hypothesize a similar western corridor along the Atlantic coast (Lebrun, 1962; Monod, 1971).

The floristic homology of northern and southern African savannas, although discussed many times (White, 1971), still remains to be studied more closely at the species level. Monod (1971), has found, at the generic level, that 63% of the plant genera of the Sahelian zone, north of the Equator, were also found in the corresponding area of southern Africa. The similarity between the two areas still subsists once pantropical, Palaeotropical and aquatic species are excluded from the comparison. Engler (1921) and De Winter (1971) also emphasized the similarity of the northern and southern xeric floras of Africa, but they were mostly concerned with extra-tropical species. Aubréville (1949b) and Griffith (1961), on the other hand, have also clarified the floristic and physiognomic homologies between the open forests north and south of the Equator (i.e., between *Isoberlinia*, *Monotes* and *Uapaca* woodlands on the one side, and *Brachystegia*, *Isoberlinia* and *Jubernardia* woodlands on the other).

Floristic richness of dry tropical Africa

Most data on species richness apply to territories of varied area. In his comprehensive review of the floristic richness of the various parts of Africa, Lebrun (1960) quite rightly pointed out that to draw meaningful ecological and biogeographical conclusions it is mandatory to compare areas of a similar size. Following Cailleux (1953), a reference area of 10 000 km^2 is considered by him to be adequate, that is, to include all the species present in the district studied. Their grand total represents the "areal richness" (*richesse aréale*) of the region under study. The average values for the major vegetation zones of the African continent are given in Table 6.1.

However, too much emphasis should not be put on these figures. Some of the most intensively studied areas are very heterogeneous, floristically speaking, and some parts of the African continent have been far less adequately surveyed than others. However, the floristic richness of the African savannas stands out clearly: their average areal richness (*c.* 1750 species) is not much lower than that of the rain forests (*c.* 2020 species), contrary to what is the rule in the Neotropics. It is also worth noticing that kinds of savannas which are more similar physiognomically also have very comparable areal richness. As shown in Fig. 6.2, there is a great similarity between the distribution of the areas of comparable floristic richness and that of the major physiognomic categories of African vegetation. The greater areal richness of the forest–savanna mosaic, south of the Zaïre forest block, must also be emphasized, particularly that of the eastern part of the Zambesian domain. It therefore can be concluded that floristic richness is, in Africa, much greater south of the Equator than north of it, Madagascar being by far the richest area of all.

TABLE 6.1

Average areal richness (i.e. number of species per 10 000 km^2) of the various chorological territories

Region	Richness
Guineo-Congolese region, peripheral domain	
Northern district	1440
Southern district	1680
Sudano-Zambezian region	
Sahelian and Sudanian domains	1060
Zambezian domain	2590
Eastern transition zone	
Sahelian type	1270
Sudano-Zambezian type	2330
Kalahari domain	1020
Madagascar	5410

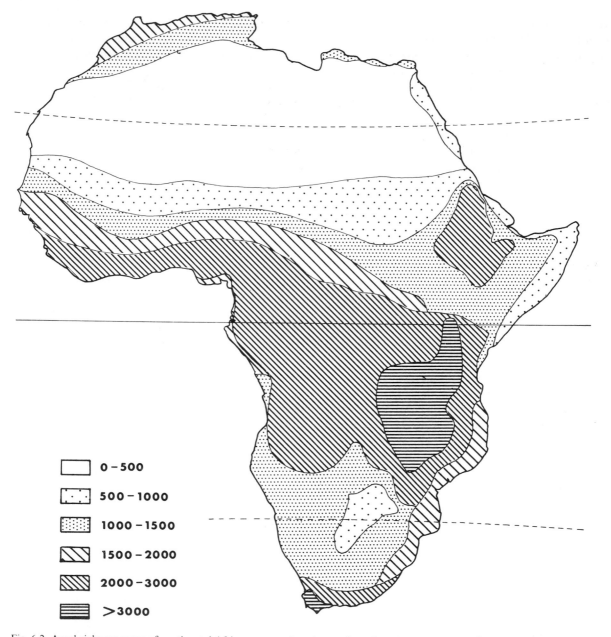

Fig. 6.2. Areal richness zones of continental Africa, expressed as the number of species per 10 000 km² (redrawn from Lebrun, 1960).

ADAPTIVE CHARACTERISTICS OF THE AFRICAN SAVANNA VEGETATION

The plants of the tropical African savannas differ from those of the rain forests by a number of morphological and functional characteristics which not only condition the physiognomy of the vegetation, but also have a direct bearing upon its dynamics and production.

Phenology

The phenology of savanna plants has already been discussed elsewhere in this book (Ch. 5). I shall limit my comments here to some ecological implications of the phenological cycle of the grass and woody components of the savanna vegetation in Africa.

Savanna trees are mostly composed of deciduous

species. The alternation of leaf and leafless conditions which gives the characteristic seasonal physiognomy of the wooded communities seems directly related to the regular occurrence of dry periods. No doubt leaf fall is one form of "adaptation" to drought, and its duration depends on the water conditions of the environment. It is well known that a number of species keep their leaves longer in a watered plain than on a parched slope. Thus, most evergreen formations within the savanna biome occur on sites with adequate moisture available during the dry season; however, some evergreen or semi-evergreen thickets occur on droughty sites with a poor concretionary soil. Trees may thus have two ways to survive a dry period (Walter, 1973): they may develop xeromorphic leaves that survive the dry season, or mesomorphic leaves falling during the dry season. According to Walter, the dominance of deciduous species comes from their better competitive ability. Although their leaves are only productive for a short period, less material is required for their formation and they are therefore more efficient. Thus deciduous species grow more rapidly, are better "producers" and are more successful in competitive situations.

A number of hypotheses have been advanced to explain the opening of buds. It is a common feature that most savanna species (both herbaceous and woody) begin to grow at the end, or even at the height, of the dry season. Following Aubréville (1949b), some authors still consider that the increase of the atmospheric vapour pressure occurring before the onset of the rainy season could be the triggering factor. Others suggest the possible role of the seasonal changes in photoperiod; Njoku (1958, 1963, 1964) showed that the onset of dormancy in some species is influenced by slight changes in day length. Many authors now feel that the trigger could be a rise in temperature (Lebrun and Gilbert, 1954; Walter, 1971). The maximum temperature and the largest diurnal ranges are reached before the rainy season occurs; the uniform mean temperatures within the tropics do not prevent plants from reacting to very small differences in temperature. Unfortunately, the supporting evidence is very meagre. In areas where the climate is very seasonal, leaf fall may occur very irregularly. An extreme case is that of *Acacia albida*, which is covered with fresh leaves in the dry season and becomes leafless during the rains (Lebrun, 1968).

Reproductive strategies

Floral biology and sexual reproduction

It is generally assumed that sun-loving plants have larger and more brightly coloured flowers than shade-loving species. Spectacular bursts of flowering often occur in savanna trees (Leguminosae), shrubs (e.g. *Cochlospermum*) and herbaceous plants (e.g. *Curculigo*). This is not the rule, however. Most savanna plants have small, discrete flowers. With very few exceptions, mostly in the southern part of the continent, the African savanna landscapes are not brightly dotted with flowers.

Though a number of woody species burst into bloom during the dry season, when the tree is leafless — which makes their flowering even more spectacular — most of them flower after the onset of the rains. Among herbaceous plants, gregarious flowering, and reiterated episodes of gregarious flowering in species sharing the same ecological preferences, are well known. The triggering role of a number of possible synchronizers (changes in photoperiod, ambient temperature or nutritive status; bush fires) has been advocated, but none of them alone can account for the flowering periodicity of all savanna species. The post-fire flowering of some geophytes excepted, very few savanna plants flower during the dry season. During the rains most species have their own preferential flowering period, thus giving evidence of the diversity of conditions inducing reproduction in savanna plants (Jaeger and Adam, 1966). However, some life forms sometimes flower all at the same time. In the Guinean savannas of the Ivory Coast, for example, all hemicryptophytes burst into bloom at the end of their vegetative cycle, towards the end of the rainy season (Menaut and César, 1979).

Seed dispersal by wind is very important since the environment becomes more open. Seeds adapted to dispersal by wind (e.g. having "wings") are especially numerous in savannas (e.g. among Combretaceae). In some genera (e.g. *Pterocarpus*) the seed wings are small in forest species and much larger in their savanna counterparts. Dispersal of seeds through external transportation by animals (by clinging to their fur or feathers) is particularly widespread in Sahelian savannas, where many seeds bear hooks or adhesive hairs (e.g., *Cenchrus biflorus*, *Guiera senegalensis*).

Vegetative reproduction

The presence of woody plants whose aerial parts are made of semi-woody stems springing from plagiotropic underground shoots has often been noticed in savannas. These aerial parts may constitute dense clumps, similar to grass tussocks or to bundles of suckers sprouting from a root stock after injury; sometimes they can even spread over an area of some dozens of square metres (Lebrun, 1947). This has often been considered as a case of vegetative reproduction, though it is difficult to define what one means by a reproducing individual in such a situation. Janzen (1975) did not consider this phenomenon as vegetative reproduction, even when the connexion with the "parent individual" is severed; according to him such plants are "hardly doing more than producing a diffuse crown". This is the case for *Piliostigma thonningii*: lateral shoots spread under the ground surface around the taproot, from which fine roots diverge. No adventive roots are seen on the underground shoots, which are apparently unable to separate from the mother plant (Menaut, 1971). However, Makany (1976) observed true vegetative reproduction taking place in *Landolphia tholloni*, *Ochna arenaria* and *Parinari pumila*. Their underground stems send out adventive roots, which eventually rot away here and there, and thus give rise to new individual plants. Vegetative reproduction does not apparently play an essential rôle in the multiplication of savanna trees, however.

Morphology of the species

Above-ground parts

The habit of savanna trees is very different from that of rain-forest species. Whatever the stage of development of their bole and crown, they never reach a very large size. They are low-branched and may ramify from their base. Crowns are generally very wide, and much more developed than the bole. Tree habits are very varied and individuals whose shape converges towards that of forest trees may be found as well as others which are typical shrubs by their mode of branching, but whose size reaches that of a tree. The shape can also vary a great deal within individuals belonging to the same species. Some trees, such as *Parkia bicolor*, *Parinari excelsa* or *Ceiba pentandra*, which are usually found in forest environments, are stunted and low-branched

in savannas. On the other hand, typical savanna trees, such as the baobab *Adansonia digitata*, whose trunk is typically broad and short, can have a bole 20 m high in some dry forests. It can happen that individuals of one species occur with different growth forms or life forms, even when growing in apparently similar ecological conditions. This is for instance the case of *Nauclea latifolia* which can look like either a sarmentose shrub or a tree, and that of *Piliostigma thonningii* which can either occur as a phanerophyte up to 6 m high or as a hemicryptophyte with a huge root system.

The thickness of the bark of savanna trees, so often noticed, has been interpreted as affording protection against repeated bush fires. This is likely in some cases, such as *Butyrospermum paradoxum* and *Cussonia barteri*. But the lack of a thick bark does not prevent numerous other species from thriving in regularly burnt savannas.

The occurrence of spines and thorns on savanna trees and shrubs (e.g., *Acacia* spp., *Balanites aegyptiaca*, *Euphorbia* spp.) is also frequent, particularly in the "thornbush" of the Sahel or in the south of Madagascar. It may afford some protection against browsing by some large mammals, but does not prevent foliage browsing by many others, such as the giraffe (*Giraffa camelopardalis*) or the black rhinoceros (*Diceros bicornis*). It is also one way of reducing water loss through evaporation, especially in areas such as Madagascar where native browsing animals were entirely lacking. In any case, spiny species are mostly found in some vegetation facies of the most xeric savannas; but even there, they seldom represent 10% of the overall population of trees and shrubs (Cornet and Poupon, 1977).

The leaves of savanna woody plants are, generally speaking, of a moderate or small size, and the percentage of microphyllous species increases with the dryness of the climate. The fan-palms (*Borassus aethiopum* and *Hyphaene thebaica*) excepted, plants with large leaves are never found in African savannas. Hard leaves, with a thick cuticle, are often found, but a large number of species have soft leaves, even in the most arid areas; in this case, however, leaves are shed at the beginning of the dry season. Long ago, Schimper (1903) noted that the savanna flora was mostly made of tropophytic species seasonally losing their green "hygrophilous" parts, whereas the more xeric floras of the desert and mediterranean areas included a larger percentage of plants with true xeromorphic perennial organs. At the time when the savanna woody plants shed their leaves, they very often also shed parts of the new shoots of the year, which partly accounts for their general slowness of growth.

Xeromorphism manifests itself through a wide range of morphological and physiological adaptations to reduce water loss (through thickened epidermis, sunken and protected stomata, small leaves sometimes reduced to scales, etc.) and to enhance water storage (the cactus-like *Euphorbia* spp., the water-storing trunk of *Adansonia* spp. or *Adenium obesum*). However, in adapting to the same environmental stress, savanna plants display an amazing variety of "strategies", and this fact must be strongly emphasized to prevent undue generalizations.

A number of growth and life forms well represented in tropical rain forests seldom occur in African savannas. Such is the case for parasitic plants. Epiphytes are very scarce: a few ferns, orchids and liverworts. Lianescent species are much more numerous: thick woody lianas, even, can be found in dry woodlands. Sarmentose species or vines (e.g., *Lonchocarpus cyanescens*, *Acacia pennata*, *Strophantus sarmentosus*) often occur. Xeromorphic and often succulent lianas, with very small leaves or no leaves at all (e.g., *Cissus quadrangularis*, *Seyrigia gracilis*) are found in the more xeric areas. Creepers (e.g. Cucurbitaceae) are also common.

Below-ground parts

Contrary to what is the rule for rain-forest species, savanna plants are renowned for their well-developed root systems, penetrating deeply into the soil.

Herbaceous plants, mostly perennials, always have an extensive root system, often forming a close mat of rootlets in the upper layers of the soil. Most of the roots (c. 80%) are located within the upper 30 cm of soil. The shortage of available water during the drier part of the year results in the exploitation of a large volume of soil by roots, in order to ensure an adequate water supply. It is only in the drier areas where annuals may predominate that the root system is less extensive.

A well-developed root system allows the plants to survive both drought and fire. At the end of the vegetation cycle, before the onset of the dry season,

the annual aerial parts wither; water and nutrient reserves have, by that time, been stored in roots through absorption and translocation from the shoots. A root system with such large reserves allows a rapid growth of the tops, which elongate rapidly before any important photosynthetic activity can take place. At that time of the year, as a result of the translocation of nutrients from roots to shoots, the root system reaches its lowest biomass values. The absorptive rootlets do not become active before the onset of the rains. Thus, the root system of herbaceous savanna plants is very sensitive to soil water conditions, and its biomass can be reduced at any time of the year, when a dry spell occurs. Underground organs of these plants react to a drought period in the same manner as the above-ground parts, by shedding their most delicate parts: their reaction is even more rapid (César, 1971).

The low woody plants behave much in the same way as herbaceous perennials. Their root systems are often disproportionately developed as compared with their aerial parts, which may be annual or partly perennial (White, 1976). Young seedlings develop a rapidly growing root system which increases their chances of survival. The annual withering of the above-ground organs exhausts a plant whose root system is not extensive enough. Large root systems allow rapid coppicing after defoliation (through drought, fire, browse or injury). Some species may even survive as hemicryptophytes, with a root system as old and well-developed as those of mature shrubs. The root/shoot ratio of most phanerophytes, initially high in their first developmental stages, decreases with increasing age and becomes less than one in mature shrubs and trees, even in the Sahel zone (Menaut, 1971; Poupon, 1979).

Most savanna shrubs and trees are reputed to have a downward growing tap-root, from which a number of lateral horizontal roots arise (Hopkins, 1962). A few data, sometimes mere chance observations, have given rise to the assumption that such shrubs and trees must have very deep roots to enable them to reach the subterranean water table in order to survive. Walter (1973) has shown that this is not the case, in most instances. In arid areas, root systems tend to flatten out, in order to provide the best opportunity to absorb water from the upper soil layers after relatively light rains. The situation is the same in most savanna plants. Most species have a very shallow, more or less horizontal, root system, and their tap-roots, when present, do not enter deeply into the soil. The most recent studies carried out in West Africa have shown that almost all the root biomass is found in the upper 50 cm of the soil; tap-roots, when present, rarely enter deeply into the soil (Menaut, 1971; Poupon, 1979). In a sandy South African savanna, root biomass is similarly concentrated (M.C. Rutherford, pers. comm; 1979) but tap-roots of some tree species do penetrate deeply into the soil although lateral roots usually predominate over tap-roots, especially in post-juvenile trees (Rutherford, 1980). Hopkins (1962) has also noted that, between 10 and 20 cm below the ground surface, the savanna soil is occupied by a network of long, twisting roots of trees and shrubs. This apparently contradicts Walter's hypothesis, which tends to explain the coexistence of woody and herbaceous plants in savannas by a partitioning of different soil layers, the roots of trees and shrubs "exploiting" the water resources of the deeper soil layers, and those of the herbaceous plants the upper soil layers. More precise studies on a number of study sites are needed in order to resolve this contradiction.

Life forms

The Raunkiaer (1934) classification of life forms has been widely used by African botanists, mostly because it combines plant ecology with plant physiognomy. However, Aubréville (1963) found it misleading for tropical botanists, because the limiting factor in low latitudes is seasonal water shortage and not low temperatures, as is the case in temperate countries. Nevertheless most of the authors working in African savannas have emphasized the protective role of the mature grass layer, which is in many ways the tropical counterpart of the winter snow cover; Lebrun (1947, 1964, 1966) thought that Raunkiaer's classification pretty well accounted for the roles played by water and temperature gradients within the various layers of savanna vegetation.

The variability of the life form "typical" of a given species is very characteristic of savanna plants. This has often been noticed when environmental conditions happen to change. Some species of *Cissus* can be classified as lianescent phanero-phytes in forest habitats and as geophytes in savannas (Tchoumé, 1966). The same is true for many woody plants from the Sahelo-Sudanian and Zambezian floristic realms (Koechlin, 1961; Lebrun, 1947; Gillet, 1968). The removal of certain ecological constraints, such as bush fires or browsing, apparently allows some species to resume their original life form more or less quickly. In ordinary savanna conditions, the different life forms often grade into each other and their classification is made difficult. Differences in interpretation are frequent: hemicryptophytes may include "pluriseasonal therophytes" (Sillans, 1958), and may also be classified as chamaephytes (Hopkins, 1962) or geophytes (Werger, 1978).

However, the "biological spectrum" — that is, the respective percentage of species with a given life form in a given environment — does not give a good idea of the physiognomy of the vegetation or of the ecological conditions prevailing in the area studied. It is much better to use spectra based upon the frequency of individuals (Hopkins, 1962), the biovolumes (Descoings, 1973) or the biomasses (César, 1971), as far as the grass layer is concerned. The example given by Hopkins (1962) is very striking. In a single savanna site, the original method gives a well-balanced spectrum for the grass layer (i.e. hemicryptophytes 33%; geophytes 29% and therophytes 38%); expressed as percentages of plants, the spectrum is very different (hemicryptophytes 71%; geophytes 14% and therophytes 15%), and highlights the life form which plays a major role in the physiognomy of the vegetation as well as in the productivity of the grass layer. Following Jacques-Félix (1962), Descoings (1975) has demonstrated the usefulness of accurately describing the structure of a savanna grass layer by biomorphological types (a combination of biological and morphological types), in which the modalities of ramification are taken into account. This method has been successfully used to distinguish the various facies of a *Trachypogon thollonii* savanna in the Congo (Descoings, 1976).

Many biological spectra based upon the percentage of species present are available in the literature, and it is tempting to compare those established in various vegetation zones and floristic realms. This should allow one to identify the role of some of the major environmental parameters. Unfortunately, spectra from specific sites belonging to a single region are in most cases very different from each other and from the spectrum of the whole region, local variations in soil composition and woody cover playing a major rôle in these differences. Divergences in interpretation between various authors make such comparisons even more difficult, if not misleading. The study by one author of a number of different sites along a gradient crossing adjacent vegetation zones would be necessary to allow meaningful comparisons. This has not yet been done. Table 6.2, comparing the biological spectra of a number of African shrub and tree-savannas (grasslands and woodlands excluded), exemplifies the difficulties mentioned; it nevertheless allows some conclusions to be drawn. The proportion of phanerophytes does not change much along the humidity gradient, although it diminishes slightly in the driest savannas. This is probably due to the regulating action of bush fires. Humid savannas are especially rich in geophytes and hemicryptophytes; these two life forms provide a good protection against the fierce fires fed by an abundant grass fuel. Chamaephytes seem to thrive in mesic savannas, and therophytes are largely dominant in arid savannas. In the latter case, severe environmental conditions are only withstood by opposite life forms, short-living "annuals" and woody perennials.

The relationships between annuals and perennials on the one hand, and between herbaceous and woody plants on the other, are very important features of the tropical African savannas and need to be discussed in some detail.

The balance between annuals and perennials

As shown in Table 6.2, the hemicryptophytes, which are dominant in humid and mesic savannas, give way to therophytes in arid savannas. When discussing the competitive ability of these two life forms, Walter (1971) stated that "the competitively weak annuals only persist as zonal vegetation in areas where climatic conditions are so unfavourable that other life forms cannot grow at all". He also demonstrated how ephemeral plants make an efficient use of the water available after rainfall. However, some perennial grasses (e.g. *Aristida sieberana*, *Panicum turgidum*) grow over sand dunes in West Africa, in places where the Sahel grades into the Sahara Desert. Therefore, perennials may well occur under conditions of very low rainfall, and can even become dominant in specific habitats. In such cases the edaphic factors (texture and water content of the soil) are far more important than climatic factors.

Most authors now agree to consider annuals as characteristic of more or less recent secondary formations, following overstocking by cattle, at

TABLE 6.2

Percentages of the various life forms in the savanna species of the various African botanical realms

Domain	Phanerophytes	Chamaephytes	Hemicryptophytes	Geophytes	Thero phytes
Guinean savannas (humid savannas)					
Ivory Coast (César, 1971)	28	–	42	6	24
Congo (Makany, 1976)	14	25	26	16	19
Nigeria (Hopkins, 1962)	30	–	23	21	25
Sudano-Zambesian savannas (mesic savannas)					
Central African Republic (Sillans, 1958)	13	24	8	11	40
Rwanda (Troupin, 1966)	29	30	12	7	20
Zaire (Lebrun, 1947)	38	44	9	4	5
Sahelian savannas (arid savannas)					
Southern Mauretania (Boudet and Duverger, 1961)	24	7	6	2	61
Northern Senegal (Cornet and Poupon, 1977)	19	2	2	2	75
Northern Chad (Gillet, 1961)	14	2	12	5	67

Decreasing humidity gradient ↓

least in mesic and arid savannas inhabited by pastoral tribes. Overgrazing tends to eradicate perennial grasses, which disappear entirely if overstocking continues for too long; then annuals become dominant. Bush fires could also contribute to eliminate perennials, by favouring annuals whose seeds ripen before burning takes place and survive on the ground (Guillaumet and Koechlin, 1971; Walter, 1971). When burning is repeated, or is started very late in the dry season, its influence is very much the same as that of overgrazing. In Africa, however, savannas are generally burned only once a year, before the regrowth of perennials. In such conditions fires are not destructive for geophytes (e.g. *Imperata cylindrica* with its large rhizomes) or for hemicryptophytes whose buds are protected by the dense and tight bases of tillers and leaf sheaths (Sillans, 1958; César, 1971). In humid and mesic savannas, nearly all the herbaceous biomass is made of perennials, and there are very few annuals, except on lithosols where they become dominant (e.g. *Loudetia togoensis*). Annual herbs often congregate on old termite mounds and in woody groves. They are represented by many species belonging to various families (Poaceae, Asteraceae, Fabaceae, Rubiaceae). They probably take advantage of the protection against fire afforded by woody areas, but microclimatic conditions could also be involved. In open grasslands at Lamto, for instance, annuals have a delayed life cycle and can only grow when the cover afforded by hemicryptophytes is dense enough (Menaut and César, 1979). According to Walter (1971), shading exerts a favourable effect, because evaporation is somewhat reduced and delicate plants are protected from sunshine. Competition between annuals and perennials also involves some competition at the rhizosphere level. Hemicryptophytes have long and dense root systems which impede the development of the thin spreading roots of annuals. Indeed, few annuals can grow between tussocks of perennials. It is only when the density of the latter diminishes, through competition with woody plants or unfavourable environmental conditions, that annuals can occur in large numbers. Sillans (1958) considered that hemicryptophytes make up the stable, basic component of the savanna grass layer and that therophytes only constitute a fleeting component, "filling the gaps" when the opportunity arises.

The relationships between herbaceous and woody plants

Contrary to what happens in other biomes, where herbaceous and woody plants are usually antagonistic, most savannas are characterized by the simultaneous occurrence of both forms.

The relationships between the grass and tree layers (that is, their "dependence", as the phytosociologists call it) has been discussed by many botanists working in open formations like tropical savannas. Finding that a given association based upon the analysis of the herbaceous layer alone can exist with or without a tree layer, or conversely that a single tree synusia can coexist with different herbaceous associations, some authors have concluded that the two synusiae were mutually independent. This is the case for Lebrun (1955) in his study of the Rwanda savannas. Moreover, he thought that the woody synusia was of secondary importance and could not be used to characterize their various types. In so doing he agreed with the view held by some plant ecologists who consider that woody plants are far less sensitive to environmental changes than herbaceous species.

However, Troupin (1966), working in the same Rwanda savannas, reached a different conclusion. He considered that, at least in certain circumstances, the tree synusia characterizes a plant community better than the grass synusia. Furthermore, the ecological characteristics and dynamics of the woody species would be better indicators of the dynamics of the plant community as a whole. Troupin nevertheless concluded that the two synusiae are mutually independent in most open savannas, that is, in those where the tree cover does not exceed 40 to 50%.

César and Menaut (1974), on the other hand, found that the floristic composition, structure and peak biomass of the grass layer of some Guinean savannas of the Ivory Coast were modified when the tree density increased. Furthermore, such a modification occurred from the very beginning of the appearance of a woody layer. Bouxin (1974) as well as Kennard and Walker (1973) have also found that the distribution and productivity of *Panicum maximum* were quickly affected by a scanty shrub cover. Troupin (1966) concluded that it was dangerous to generalize, the interrelationship of the two synusiae depending first of all on the environmental situation as a whole.

Walter (1971, 1973) held a different view of this problem. He considered that herbaceous and woody plants are strong competitors, and that their relative proportions depend on the water budget of the soil. He provided a thorough analysis of the relationship between perennial grasses and woody species, based on their water economy and root systems. The leaves of the grasses wither at the onset of the dry season and the aerial system does not lose any more water from transpiration. The plants become dormant, absorb almost no water, and live on what is stored in their underground parts. Grasses can accommodate themselves to a short wet period with low rainfall because they only need water during their growing season when shoots and leaves transpire actively. The available water is taken up from a rather small volume of upper soil by a fine and intensive root system. On the other hand, woody plants require a higher and more regularly distributed rainfall. They need to absorb some water even during the dry season to balance the continuous water loss through transpiration of their aerial parts, despite their better regulation of water losses. Thus, trees need to develop an extensive and deep root system.

On this basis, Walter tried to explain the relative distribution of herbaceous and woody plants in savannas. On stony soils with a low water-holding capacity, where the water is not uniformly distributed and often percolates to great depths, woody plants dominate over grasses. On soils of fine texture, absorbing and storing all the water received from precipitation, the "competitive equilibrium" between grasses and woody plants depends on the amount of rainfall. When the latter is low (less than 250 mm yr^{-1}), woody plants are missing and only pure grasslands are found. Under a regime of higher rainfall (250–500 mm yr^{-1}) grasses are still dominant under scattered woody plants. Beyond 500 mm yr^{-1}, there are more and more woody plants as the rainfall increases. As Walter (1971, 1973) put it, "the controlling element in all these cases is the grasses. The woody plants are merely tolerated ... Only when the rainfall reaches a level where the crowns of the trees link up to form a canopy whose shade prevents the development of grasses, is the competitive relationship reversed. Woody plants are now the dominant competitors ... and the grasses are obliged to adapt themselves". Walter also explained bush encroachment after overgrazing by the surplus of water made available to woody species when grasses are eaten up.

Such a balance between herbaceous and woody plants may well operate in those parts of southern Africa where Walter (1939) worked, but it is not so apparent in most other African savannas. Of course, the results of this "competitive equilibrium" are obvious in the most extreme situations: woody plants thrive either on rocky hills or in well-drained valleys, and are nearly absent on the sand dunes of the more arid regions. Questions arise when various admixtures of woody and herbaceous plants contribute to distinguish different savanna types. Furthermore, the diversity of soil categories and rainfall patterns make the interpretation of observational data even more difficult. This problem is still hotly debated and any generalisation would be premature at this stage. As emphasized by Sicot (1978) and Grouzis (1979), a thorough year-long analysis of the water available for both plant categories, considering both their above- and below-ground living biomass, will be necessary before one can expect to solve this problem.

As early as 1939, Michelmore attempted to identify the environmental factors limiting the development of the tree layer in his analysis of the East and South African grasslands. He showed that natural lowland grasslands were determined by soil conditions, either shallow dry soils or, more often, by an alternation of wet and dry conditions. Various adaptive strategies allow grasses to withstand more extreme situations than shrubs and trees which cannot cope with either prolonged drought or persistent waterlogging, which prevent root respiration. Working in the Sudan, Morison et al. (1948) proposed an interesting interpretation of the ways by which the distribution of herbaceous and woody plants are influenced by water availability along a catena (Fig. 6.3). In this case, burning is of secondary importance; it only accelerates a natural trend. Obviously, Fig. 6.3 is an oversimplified diagram, the trees being far from missing at lower levels. Morison et al., nevertheless, mentioned that the *Hyparrhenia* grassland gives way to *Acacia* savannas when poorly drained black clay soils replace loamy soils. Indeed, river beds are often colonized by a number of woody species, but flooding here is generally of short duration, and running water is apparently less harmful to tree growth than standing water. Only palm trees

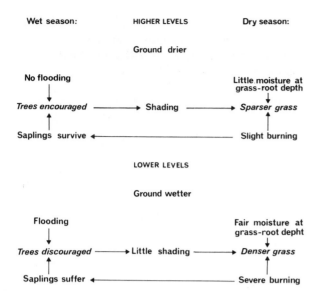

Fig. 6.3. Water conditions and fire as factors of the distribution of herbaceous and woody plants along a catena in the Sudan (after Morison et al., 1948).

(*Borassus aethiopum, Hyphaene thebaica, Phoenix reclinata*) and a few *Acacia* species can withstand flooding in most African savannas.

Poor soils have often been reported as responsible for the occurrence of tropical grasslands, especially in rain-forest areas. Such might be the case in the *esobe* grasslands of the Zaïre Basin (Robyns, 1936), but a poor soil could also be responsible for the maintenance of such grasslands, without having played a causal role in their origin (Germain, 1965). This might be the case for the small "enclosed savannas" of the lower Ivory Coast and other rain-forest areas. Their very poor lateritic soils do not allow reafforestation to take place quickly. The treeless "steppes" of the Congo (Makany, 1976) and of southern Zaïre (Duvignaud, 1952) might also have an edaphic origin, at least in part.

The impact of burning and overstocking is discussed elsewhere in this volume. However, it is worth mentioning some extreme cases where human pressure has been so strong and widespread that very large areas have now become almost treeless. Such is the case for some Andropogoneae and Paniceae (e.g. *Pennisetum purpureum*) grasslands, where the huge amount of grass fuel burnt each year excludes the presence of any woody species (Jacques-Félix, 1956). In the same way, the

Imperata cylindrica grasslands widespread in the rain-forest biome are maintained by repeated burning, and possibly also by the active exclusion of competitive species by the grass itself. Further north in the Sahel, recent severe drought episodes have greatly increased tree mortality and reduced fruit production (Poupon, 1977, 1979; Poupon and Bille, 1974). To this factor were added a high browsing pressure and the felling of trees by nomadic herdsmen to provide more fodder for their livestock. This explains why large areas of the West African Sahel now have a greatly reduced woody cover.

THE MAJOR CATEGORIES OF SAVANNA VEGETATION

Many classification systems for the African vegetation have been proposed, and a number of vegetation maps have been published. None of them are completely satisfying. Discussing world vegetation maps, Lieth and Van der Maarel (1976) even went so far as to conclude that "most of them are scientifically of restricted value. In many cases the underlying classification system is inconsistent and the description of the vegetation types discriminated is so curtailed that the boundaries are hardly justifiable for the connoisseur". Without being so pessimistic, it is quite true that most of these classificatory schemes have been hotly disputed. However, some of them are now considered by most African botanists as a good approximation to the actual situation.

As pointed out by Schnell (1977), the phytogeography of tropical Africa must take into account both the diversity of the physiognomic categories of vegetation determined by environmental factors, and its floristic structure which is also influenced by the past history and palaeogeography of the region. This view is now shared by most of those who wish to promote more comprehensive definitions of phytogeographical units. It now seems more and more necessary to take into account floristic, ecological and physiognomic features in order to achieve realistic classification schemes. Unfortunately these features are difficult to integrate into a single system, and even more difficult to represent on a single map. Most of those already published are based either on floristic structure, or on physiognomic types, a number of ecological parameters being used to characterize each unit.

Phytochorological units (Fig. 6.4)

A comprehensive review of the literature on the phytochorology of tropical Africa has been published by Monod (1957). Since then, it has been brought up to date by Troupin (1966), White (1971) and Brenan (1978), not to mention some regional studies such as those by Volk (1966), Jacques-Félix (1970) or Wickens (1976).

A number of biogeographers, such as Eig (1931) and Monod (1938), have emphasized the phytogeographic homologies between Africa and Asia, and postulate the existence of three major phytogeo-

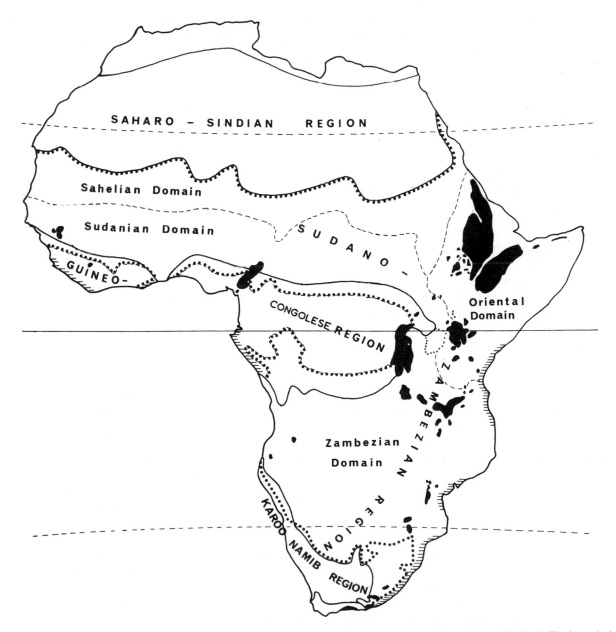

Fig. 6.4. The phytochorological units of tropical Africa. Montane and Afro-alpine communities are figured in black. The boundaries of the savanna biome are figured by heavy dotted line. (Adapted from Monod, 1957; Troupin, 1966; White, 1971; and Wickens, 1976).

graphic regions running more or less parallel to the Equator: the Saharo-Sindian region, mostly covered by steppes; the Sudano-Deccanian region, where most savannas and dry woodlands are found; and the Congo-Indian region, where rain forests are concentrated. As late as 1970, Jacques-Félix considered it necessary to distinguish the Sudanian region from its southern homologue, the Zambian region. In so doing, he agreed with earlier classifications dividing up the whole of tropical Africa into four (Engler, 1910), five (Robyns, 1948) or six units (Chevalier and Emberger, 1937). All of them agree in separating the Guinean, the Sudanian and the Ethiopian regions, but diverge somewhat in the number of units into which the eastern and southern parts of the continent should be divided.

Unfortunately, these classifications do not give enough importance to the basic contrast existing between the dry and humid tropics. Also they do not emphasize the relative symmetry of the various vegetation types on each side of the Equator. These two points have been particularly stressed by Lebrun (1947), who considered that tropical Africa could be divided up into two regions: (a) the Guinean region, characterized by a forest landscape and a rain-forest climax; and (b) the Sudano-Zambezian region, which surrounds it, and is dominated by grassland communities, originating from a dry forest climax. Each of these two regions is itself divided up into a number of domains.

This viewpoint was shared by Monod (1957), as well as Troupin (1966) and White (1965, 1971). Their conclusions were very similar, though the limits of some of the domains varied somewhat according to the authors' particular interest and personal field experience. Most of the differences relate to the arid zones and mountain areas. It is worth noticing, at this point, that White refrained from drawing precise contour lines for the various domains (Fig. 6.4). He also considered the Sahel "as an impoverished western extension of the floristically rich oriental domain" (part of the Sudano-Zambezian region).

At first sight it is tempting to look for conformities between the phytochorological regions (based upon floristic criteria) and the vegetation zones (based upon physiognomic criteria). There are indeed some remarkable coincidences. The Saharo-Sindian region can be equated with deserts and steppes, the Guineo-Congolese region with rain

forests, and it would be tempting to associate the Sudano-Zambezian region with savannas. Obviously, this cannot be done. A notable fraction of the savanna biome belongs to the Guineo-Congolese region, whereas the Sudano-Zambezian region is itself very heterogeneous.

Vegetation types

No mention will be made here of the general schemes of classification proposed at the world scale, but difficult to use in the African context. Descoings (1976) has already discussed the applicability of the classification proposed by Fosberg (1968) and UNESCO (1969) to herbaceous communities.

So far, the Yangambi classification (Anonymous, 1956) still remains the most satisfactory, despite its shortcomings. This view is supported by the many constructive comments it has stimulated (e.g. Trochain, 1957; Boughey, 1957a,b; Monod, 1963; Aubréville, 1965; Guillaumet and Koechlin, 1971). It is still widely in use, and has become the basis for the nomenclature of the vegetation map of Africa (Keay, 1959b), by far the most useful at this scale, pending a revised edition by White (Fig. 6.5).

However, despite its indisputable value for African biogeographers (Aubréville, 1969), this classification still remains somewhat ambiguous. It was originally conceived as a purely physiognomic classification. Unfortunately a number of other criteria have also been retained at various levels: ecological (e.g. herb swamp), physiological (e.g. deciduous forest), dynamic (e.g. secondary formation), floristic (e.g. bamboo forest), and physiographic (e.g. montane grassland). Nevertheless, the actual existence of the major physiognomic categories retained is beyond question. It will therefore be adopted in the present review (Table 6.3).

One of the major criticisms raised against the Yangambi classification is that it mostly makes use of the woody communities, only incidentally using the herbaceous communities, to identify the major vegetation categories. As rightly pointed out by Boughey (1957b), "African vegetation usually includes one or more woody plants in its upper layers. To a forester ... the whole plant cover of the continent is therefore envisaged as falling within a forestry classification". Insofar as the interdependence of the tree and grass layers can be questioned, it seems justified in our case to give

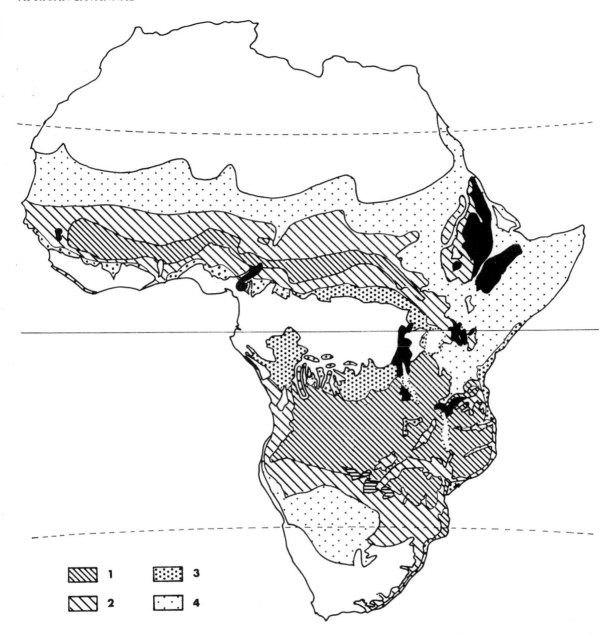

Fig. 6.5. The main savanna vegetation types of continental Africa: *1* = woodlands; *2* = tree and shrub savannas; *3* = forest and savanna mosaic; *4* = tree and shrub "steppes". Montane communities are figured in black. The vegetation zones not covered in this volume are left in white. (Simplified from Keay, 1959b.)

more importance to the herbaceous layer. Furthermore, Descoings (1973, 1976, 1978) has emphasized the imprecision of the distinction made between savannas, steppes and grasslands. This has led him to propose another way to classify herbaceous communities on the basis of their fine structural analysis (Descoings, 1975). Unfortunately such a system can only be used by a specialist and does not

allow a rapid comparison of communities. The view taken by Langdale-Brown (1959–1960) in his study of Uganda vegetation might represent a useful compromise. The Yangambi categories are adopted, but within each of them the individual communities are detailed on a phytosociological basis and grouped together according to their successional relationships.

TABLE 6.3

The vegetation categories of the African savanna biome, according to the Yangambi classification (English and French equivalents)

	English	French
1.	Closed forest formations	Formations forestières fermées
1.1.	Forest climax formations	Formations forestières climatiques
1.1.1.	Forests at low and medium altitudes	Forêts de basse et moyenne altitude
1.1.1.2.	Dry deciduous forest	Forêt dense sèche
1.1.1.3.	Thicket	Fourré
2.	Mixed forest–grassland formations and grassland formations	Formations mixtes forestières et graminéennes et formations graminéennes
2.1.	Woodland	Forêt claire
2.2.	Savanna	Savane
2.2.1.	Grass savanna	Savane herbeuse
2.2.2.	Shrub savanna	Savane arbustive
2.2.3.	Tree savanna	Savane arborée
2.2.4.	Savanna woodland	Savane boisée
2.3.	Steppe	Steppe
2.3.1.	Tree and/or shrub steppe	Steppe arborée et/ou arbustive
2.3.4.	Grass and/or herb steppe	Steppe herbacée et/ou graminéenne
2.4.	Grassland	Prairie

Rattray (1960) also provided a very useful map of the grass cover of Africa. He noted that "it is not possible to rely entirely on vegetation maps based on physiognomy as an accurate guide to the distribution of the main grass species". First of all, because of the lack of interdependence of grass and tree layers in a given zone, and also because "the climatic range of the grass constituents of the woody vegetation types does not necessarily coincide with that of trees". He also noted that "the grasses are extremely useful as indicators of soil conditions". Thus, he based his map on the distribution of the main grass species, but felt it necessary to add indications on the general vegetation physiognomy (based upon Yangambi terminology) and on the successional changes of the major grass associations. Having been conceived in a range-management perspective, such a small-scale map (1:10 000 000) is of limited value for the plant ecologist. The indicator value of the grass species selected by Rattray is only meaningful on a much larger scale, and the information given for the physiognomy of some communities is sometimes confusing — the rain forest being called woodland, for instance. However, this work constitutes a most interesting attempt to define herbaceous communities better.

The work of Phillips (1965) must also be mentioned. It was planned to provide a tentative compilation of the distribution of the major climatic regions based on potential and actual vegetation, on rainfall, humidity and temperature. The role of fire is also given much attention throughout the volume. This compilation is very useful and includes a lot of new information. Unfortunately the scale and the complexity of the accompanying map (1:20 000 000) make it difficult to use. It is definitely impossible to condense the large amount of information given into a map at a very small scale. It would also be unfair not to mention the work of Shantz and Marbutt (1923), who classified savannas on the basis of the physiognomy of the grass cover they support: high grass–low tree savanna, *Acacia*–tall grass savanna, *Acacia*–desert grass savanna. It was a pioneer effort which laid the ground for further work.

Closed formations

These are made up of trees and shrubs whose contiguous crowns generally prevent the development of a grass cover.

Dry deciduous forests

To include such forests in the savanna biome might appear surprising, and has in fact, been hotly debated. The physiognomy of these forest formations is very comparable to that of rain forests. However, most of the trees completely lose their leaves for some weeks at least, and the undergrowth is very often deciduous. These tropophilous forests grow in dry areas, within wide climatic extremes, and their constituent species have almost nothing in common with those inhabiting moist forests. Furthermore, they are frequently burned despite the sparsity of the grass cover, if any. Their degradation frequently leads to a woodland or a wooded savanna. They can therefore be considered as the last stage in the development of savannas. Their susceptibility to burning and clearing is great, and this fragility explains why they now only occur in Africa as isolated relicts.

The dry forests of the savanna biome are not to be confused with the moist semi-deciduous forests bordering the rain forest itself, and sometimes called "dry forests", as in Nigeria (see Clayton, 1958).

Evergreen facies. Some evergreen (or semi-evergreen) forests, are also included within the dry deciduous forests. Therefore, they are more appropriately called "dense dry forests". Such formations are difficult to classify, as they may include numerous "Guinean" species in spite of their geographical location within the Sudano–Zambezian region (Schmitz, 1971). These dense dry forests are considered as climax vegetation turned into woodland through regressive succession. They are well represented in Zaïre, where they have been studied by a number of botanists, and recently by Freson et al. (1974). They are not related to a given kind of soil but are found in areas where the annual rainfall averages 1300 mm falling for four consecutive months. The species richness of these *Entandophragma delevoyi* and *Diospyros hoyleana* forests is around one hundred species of spermatophytes, plus a number of mosses and liverworts.

In southern Central Africa, dry evergreen forests generally occur on soils with adequate moisture available throughout the dry season (*c.* seven months). They are often three-storeyed: a closed canopy of evergreen or semi-evergreen species, up to 25 m, a loose understorey reaching 10 to 15 m, and a dense layer of shrubs and scramblers below 5 m. Climbers occur in the three layers. The ground often remains bare but is sometimes covered with sparse rain-forest grasses. Good regeneration enables the maintenance of the forest, when burning does not occur. Three types of dry evergreen forests have been described by Fanshawe (1961). *Parinari–Syzygium* forests occur in a few relict patches, either in pure mature communities of *Parinari*, or at different stages of the maturation process (e.g. forming an understorey under an *Erythrophleum* canopy) or of the degradation of the forest (i.e. its development towards deciduous forest). Henkel (1931) considers them as climactic; they probably covered most of the country in the past, before being transformed by man into woodlands and savannas. A dense, low evergreen forest dominated by *Cryptosepalum pseudotaxus*, whose crowns are often tied together with such vines as *Combretum microphyllum* or *Uvaria angolensis*, occurs on the northern Kalahari sands. As for the *Marquesia* dry evergreen forests, they occur on deep sandy humic loams. Their dynamic relationships with deciduous woodlands and savannas have been studied by Lawton (1978), who emphasized the major role played by *Uapaca* spp. in the reconstitution of the dry evergreen forest from an initial deciduous woodland.

Dry evergreen forests have also been reported in the Sudanian zone. Now reduced to a few scattered remnants, they are restricted to rocky sites where fire does not occur. The dominant layer (15–20 m) is usually composed of a single species (*Gilletiodendron glandulosum* or *Guibourtia copallifera*), the understorey is easily penetrable (cover 40–50%) and the herbaceous layer, where grasses are practically lacking, is sparse (cover 1–5%). The flora is very different from that of the surrounding savannas, although few species are true rain-forest species. Jaeger (1956) has shown that the dominant species are well adapted, biologically speaking, to the constraints of the Sudanian climate. They have quite likely been eliminated from surrounding savannas by repeated burning. Schnell (1971) has described some dense dry *Uapaca togoensis* forests, either with a very open understorey where true savanna species are lacking or with a denser one where a number of savanna grasses are present in the herbaceous layer. This could confirm Lawton's (1978) observations in Zambia, on the role played by *Uapaca* spp. in the evolution of savannas and woodlands.

Deciduous facies. Dry deciduous forests are often considered as a regressive stage of the dry evergreen forests. In the western part of the continent, they are represented only by a few scattered remnants. They are generally composed of savanna species, including a number of sun-loving grasses (e.g. *Albizia–Anogeissus* forests, or *Parkia–Pterocarpus* forests). In the eastern part of the Central African Republic, Chevalier (1951) described dry deciduous forests of *Anogeissus* and *Isoberlinia* trees, with a herbaceous layer, which might be a relict of the primitive vegetation cover. This view is supported by the presence of *Encephalartos*. Some of these dry deciduous forests could also be considered as impoverished tropophilous extensions of the rain forest itself. For instance, the Bandia woodland in Senegal is made up of species characteristic of semi-deciduous humid forests (e.g. *Antiaris africana* and *Morus mesozygia*) and of typical savanna trees such as *Adansonia digitata* and *Parkia biglobosa*. The difficulty of guessing the origin of such dry forest relicts in savanna landscapes has been well emphasized by Schnell (1971). When environmental conditions become favourable, for instance when burning is prevented for a number of consecutive years, an "open" savanna becomes more and more "closed", through the increase in number and size of savanna trees and shrubs. Later on, it will be very difficult to differentiate such a closed savanna patch from a "relict" forest. Only the presence of some woody lianas and, above all, that of understorey species lacking in the surrounding savanna could be in favour of the latter interpretation.

In southern Central Africa, *Baikiaea plurijuga* forests are spread over areas large enough to appear on the AETFAT map. They are found in hot and dry regions (rainfall *c.* 650 mm) and on deep sandy soils with a high water-holding capacity (Cole, 1963). They have a continuous canopy over a fairly dense understorey of *Combretum*, *Acacia* and *Commiphora* spp. Being very sensitive to burning, they are only able to exist in areas where fires are not too fierce. They are considered as intermediate, from a climatic point of view, between *Brachystegia* deciduous woodlands and *Acacia* shrub "steppes". Sometimes they are also considered as woodlands rather than dense forests; in fact, they often thin out

at their southern edge to grade into open woodlands.

Thickets

A thicket is defined as a shrubby vegetation, evergreen or deciduous, usually more or less impenetrable by man, often in clumps, with a grass layer absent or discontinuous (Fig. 6.6). Thickets are often difficult to distinguish from low forests and may be looked upon as extreme types of woodland or bushland. They can be found in a great diversity of environments, close to the rain forest as well as near the desert edge.

Most thickets seem to be second-growth formations, developing in areas more or less protected against burning. *Dichrostachys* thickets, for instance, are found throughout tropical Africa on alluvial or other fertile soils, in areas with varied ranges of rainfall (Rattray, 1961). They frequently occur in coastal savannas (e.g. *Strychnos* thickets in Angola, or *Chrysobalanus* thickets in Ghana). They can also be composed of bamboos (e.g. *Oxytenanthera abyssinica*) and may be found throughout the more humid savannas of West Africa, or within various dry deciduous woodlands in East Africa.

In southern Central Africa, thickets generally form belts around dry evergreen forests, but can also be found within more open woodlands. They are considered as regressive stages, and may indeed represent seral stages which can again revert to dry deciduous forest communities when burning and clearing are prevented for long enough (Boughey, 1964). They are not very sensitive to fire, since the grass cover is lacking or very sparse (often made up of shade-loving panicoid annuals), and since coppicing is dense and rapid.

In East Africa, the best known area of thornless deciduous thicket is the *Itigi* type which occupies a large area in central Tanzania. It is composed of highly branched coppicing shrubs (e.g. *Baphia*, *Bussea* and *Pseudoprosopis* species) growing up to 5 m with interlaced crowns forming a thick and

Fig. 6.6. A thicket diagrammatic profile.

often continuous canopy. They occur under the same climatic conditions as deciduous woodlands but require deep soils with a high acidity and a light texture (consolidated swamp floor deposit) (Milne, 1947). Semi-evergreen thickets with *Euphorbia candelabrum* and Guinean savanna species occur in different parts of East Africa; they are well developed in the northwestern coastal zone of Lake Victoria (Langdale-Brown, 1959–1960).

In arid regions, thorny thickets (with *Acacia* spp.) are often found in natural depressions where the soil is heavier and has a better water-holding capacity. In the Karamoja region, *Acacia* thickets also occur as a result of overgrazing, as well as *Euphorbia–Sansevieria* succulent shrublands which sometimes attain thicket density. *Acacia mellifera* thickets are found further north, in the eastern part of the central Sudan, on clayey soils. Patches of *Acacia* spp. or thornless thickets are very frequent throughout the Sahel zone, from Somalia to Senegal. They may cover large areas, especially on gravelly soils, or on lateritic crusts, where bushes of *Pterocarpus lucens* and *Combretum* spp. often reach thicket density and become nearly impenetrable. Further south, in the Sudanian zone, dense thickets of *Acacia ataxacantha*, often mixed with *Combretum micranthum*, occur. They can grow as well on poor stony soils as on the more humid soils of shallow depressions.

Open formations

These are characterized by a herbaceous layer which varies according to the importance of the tree layer and the dryness of the habitat. These open formations are typically burnt each year during the dry season, and repeated burning has exerted a strong selective pressure in favour of fire-resistant species. The density of trees is very variable and depends on both soil conditions and the extent of human interference. A continuum of more or less open formations, ranging from the open forest on the one side to the grassland savanna and even the herbaceous "steppe" on the other, can therefore be found, even on a relatively small area. There is indeed no basic difference between an open forest (i.e. a woodland, according to the Yangambi classification) and a more or less densely wooded savanna. The borderline between these two vegetation types can sometimes be very imprecise. However,

some differences are noticeable: for instance, the trees of an open forest have a genuine tree habit, with straight boles, whereas the trees and shrubs of a wooded savanna are gnarled and stunted (Aubréville, 1965). Furthermore, the floristic composition of these two formations is often quite different.

Many attempts have been made to correlate the kind of tree cover and its density with some "climatic index". Walter (1971), for instance, distinguished the following sequence of vegetation zones in relation to decreasing summer rainfall in South-West Africa (Namibia): woodland → savanna → grassland → desert. He documented this sequence by a number of data on the competition between herbaceous and woody plants. Unfortunately, such relationships between plant formations on the one hand, and the amount of annual rainfall, or length of the dry season on the other, vary from place to place, even when one does not take into account the edaphic factors which induce local variations (Smith, 1949). Under present conditions, one may notice that the densest woody formations are generally situated in the moderately dry regions of the savanna biome; the density of trees and shrubs decreases towards both the less humid and the more humid areas. The disruptive effect of human activities (burning, clearing and overgrazing) is apparently more important in the latter situations.

Most of the African savanna biome is covered by open formations, but the importance of the closed ones must not be underrated. The various types of closed formations often found along a successional gradient help us to better understand the dynamics of the savanna formations as a whole, especially in relation to climatic and anthropic factors.

Woodlands

The upper stratum of woodlands is composed of deciduous, small- or medium-sized trees (8–20 m tall). For an observer, the feeling of openness is due not only to the fact that the tree crowns do not always touch each other, but also to their relative thinness (trees being often flat-topped), and to the lightness of the foliage (often made of compound leaves with small leaflets). Woodlands generally have no woody understorey; the herbaceous layer, sometimes sparse, is often made up of tall grasses, more or less mixed with forbs and suffrutescent vegetation (Fig. 6.7).

Fig. 6.7. A woodland diagrammatic profile.

Woodland flora. The African woodland flora has been studied, among others, by Duvigneaud (1959) and White (1965). It includes: (1) some rain-forest genera (e.g., *Afzelia*, *Brachystegia*, *Burkea*, *Daniellia*, *Erythrophleum*, *Parinari*, *Uapaca*); (2) some genera related to those of the rain forest (e.g. *Isoberlinia*, *Pseudoberlinia*); and (3) a few genera never found in the Guinean forest (e.g., *Anogeissus*, *Butyrospermum*, *Cochlospermum*).

The overwhelming importance of legumes is worth noticing. But another characteristic feature of these open forests is the dominance of a small number of woody species, which sometimes grow in almost pure stands (Aubréville, 1959). These few dominant and gregarious species have often led researchers to consider woodlands as secondary formations, these formations being generally less heterogeneous, floristically speaking than primary ones. Such a criterion is not absolute, however. Aubréville (1959) considered woodlands as a primitive vegetation type, which has been secondarily opened up and made more homogeneous by prolonged burning and clearing.

Woodland ecology. Throughout its range, this community experiences a prolonged seasonal drought of between four and seven months, during which not a single drop of rain may fall for four or five months in some areas (Eyre, 1968). This climatic constraint is obviously responsible for a number of characteristic features of the African woodlands.

Most species are deciduous, the duration of the leafless conditions during the dry season varying with the moisture conditions of the environment. However, it is noteworthy that defoliation does not occur simultaneously for all species and generally does not last for more than a few weeks, sometimes not more than a few days, whereas the drought lasts for many months. The foliage, commonly of a varnished appearance, develops before the onset of the rains, as the flowers can also do. Buds are often

well protected against desiccation by stipules or hairs. The stems are often thick barked. The trees are adapted to high temperatures and strong insolation. There are no epiphytes, and thick-stemmed woody lianas are scarce.

During the dry season, the deep litter of tree leaves, as well as the large amount of dry grass, provide abundant fuel for bush fires. These take place frequently and destroy the understorey and most saplings. Burning is certainly the major factor determining the distribution and spacing of trees.

There are various levels of adaptation to drought in African woodlands. *Daniellia* woodlands (Fig. 6.8) are rather sensitive to it, whereas *Acacia* woodlands are much more resistant. However, even the latter can evolve towards *Acacia* "steppes" when the climate becomes drier.

Distribution. The determinants of woodland distribution have been particularly well studied by Aubréville (1959). African woodlands occur in regions with a long dry season (up to seven months), receiving 650 to 1500 mm rainfall each year. On the whole, the open forests of the Sudanian domain, although growing under more humid conditions than those of the Zambezian domain, suffer from a higher saturation deficit. This deficit is attributable both to the higher ambient temperatures caused by lower altitudes and also to the effects of the dry "harmattan" winds blowing from the Sahara Desert. Sudanian woodlands are therefore more open, less tall and less species-rich than the Zambezian woodlands. Open forests are often found on bare rocky or sandy soils, strongly leached and far less fertile than those of the surrounding savannas. According to Aubréville, this all points to their primitive origin, burning only being responsible for their subsequent floristic impoverishment. Other authors hold quite the opposite view, and consider woodlands as secondary formations. Such is obviously the case for the *Anogeissus* groves developing on abandoned village sites (Sobey, 1978) and for some Guinean woodlands dominated by *Anogeissus* and *Terminalia*, growing on old farmlands (Jones, 1963).

African woodlands have a homologous distribution on each side of the Equator, but the woodland belt is much broader in southern Africa than north of the Equator. In both situations, these open forests never reach the coasts of the Atlantic

Fig. 6.8. A *Daniellia* savanna woodland, northern Ivory Coast (photo J.C. Menaut).

Ocean, the Sudanian woodlands not extending beyond the western Rift Valley. They are therefore restricted to continental climate areas, very arid during the dry season. Some woodlands, however, are also found close to the Indian Ocean coast, at least south of the Equator. African woodlands are never in direct contact with rain forests; they are always isolated from them by a savanna belt, or by treeless grasslands. This situation has led many botanists to think that these herbaceous formations are the result of repeated burning and clearing. Aubréville (1959) considered that these anthropic activities have not destroyed the rain forest, but have only prevented its regeneration after a dry climatic episode.

Sudanian woodlands. Three main categories have been recognized:

(1) The *Isoberlinia* woodland is the most widespread and occurs on plateau soils. It is often replaced by *Monotes* woodlands on eroded slopes, and further down by *Terminalia* woodlands on poorly drained clayey soils.

(2) The legume woodland is often hardly distinguishable from the savanna woodland made up of the same characteristic species.

(3) The *Boswellia* woodland takes the form of open stands located on rocky sites, in areas with a Sahelo-Sudanian climate.

The floristic structure and ecology of the Sudanian woodlands have been studied in a number of countries: the Ivory Coast (Miège, 1955; Adjanohoun and Aké Assi, 1967), Nigeria (Keay, 1952, 1959b) and the Central African Republic (Sillans, 1958). These open forests are often degraded and turn into savanna formations characterized by *Daniellia oliveri* and *Lophira lanceolata* in the south, and *Butyrospermum paradoxum* and *Parkia biglobosa* in the north (Ivory Coast and Upper Volta, for instance).

Woodlands of the Oriental domain. Only a few relict *Albizia* and *Strychnos* woodlands still exist in eastern Zaïre, often reduced to a few degraded tree clumps.

The Uganda woodlands have been carefully

studied by Langdale Brown (1959–1960). They might be considered as fire pseudo-climax formations, some of them existing in formerly cultivated areas. Such might be the case for the *Chlorophora excelsa–Pennisetum purpureum* association, rich in Guinean forest and savanna species.

Acacia woodlands occur all over East Africa. They often thrive on valley floors, whereas broadleaf deciduous woodlands (often with Combretaceae) occur on the more freely drained slopes (even with small altitudinal differences) or on rocky soils (Bunting and Lea, 1962).

In the southern part of the Oriental domain, *Brachystegia* woodlands (*miombo*) occur on well-drained plateau soils. They are not very numerous in Kenya, but nearly cover two thirds of Tanzania (*Brachystegia, Afzelia, Julbernardia* spp.). They are often replaced by woodlands of similar structure dominated by other leguminous trees, such as *Milletia, Dalbergia* and *Lonchocarpus* spp., in the extreme south eastern part of this domain (Polhill, 1968).

Zambezian woodlands. Three main categories can be recognized:

(1) the *Brachystegia* woodlands which extend over most of the plateaux of southern Central Africa above an altitude of 1000 m.

(2) the *Baikiaea* woodlands, also found in the same region, sometimes very similar to the preceding category, but associated with the Kalahari sands.

(3) the *Colophospermum* woodlands (*mopane*), more xerophilous, which are found south of the above two categories, at lower altitudes.

These three major kinds of Zambezian woodlands are themselves divided up into a number of distinct communities. Rattray (1961) has provided a very useful classification of these woodlands, and Cole (1963) has described the functioning and dynamics of these communities in relation to the geomorphology of the area. Although only dealing with the vegetation of Zambia and Zimbabwe (Rhodesia), these papers provide a useful background for the study of the woodlands of southern Central Africa as a whole.

Brachystegia–Julbernardia woodlands all have a rather similar physiognomy, in spite of both the large number of species involved and the local variations imposed by edaphic conditions. The dominant species have a typical tree habit, trees being 10 to 20 m tall on the average. The understorey is generally loose and the grass cover, mostly made of Andropogonae, is more or less sparse. These woodlands occur in areas whose rainfall ranges from 600 to 1500 mm yr^{-1}.

Brachystegia spiciformis woodlands thrive on the main watershed plateau, above an altitude of 1200 m, and on relatively well-drained lateritic soils. These woodlands cover large areas, but the local topographical features tend to split them into a number of small forest blocks.

Brachystegia boehmi woodlands develop mostly on escarpments, below the plateau level, on stony or gravelly soils, or in areas which are waterlogged during wet spells. They grow under warmer and drier conditions than the preceding categories, and merge with *Colophospermum* woodlands at low altitudes.

Julbernardia globiflora most often grows in mixed stands with *Brachystegia* spp., but can also be found in almost pure stands when environmental conditions become unfavourable for *Brachystegia* spp.

Uapaca kirkiana is found on well-drained, rocky soils, when rainfall is higher. This type of woodland may progressively turn into forest, and could have formed part of the primitive vegetation, according to Henkel (1931).

Baikiaea woodlands are limited to Kalahari sands, in areas with a rather low rainfall (500–750 mm). A number of species characteristic of *Brachystegia* woodlands also occur in this kind of open forest. Some of them, such as *Guibourtia coleosperma* or *Pterocarpus angolensis*, may even become dominant. *Baikiaea* woodlands have a fairly dense understorey. On Kalahari sands, the westernmost *Brachystegia* woodlands are often intermingled with semi-deciduous formations characterized by *Marquesia macroura*.

Colophospermum mopane woodlands are markedly xerophytic. They occupy an area parallel to that of *Brachystegia* woodlands, but further south, running from Angola to Moçambique. They also occur within the *Brachystegia* zone, at lower altitudes, especially along valleys with deep black clay soils having a high pH (e.g. the Zambezi Valley). They generally have a sparse understorey, and the grass cover is dominated by Paniceae.

Many ecological and even eco-physiological studies have been carried out in the miombo wood-

Fig. 6.9. Seasonal aspects of a miombo woodland in southern Zaire: during the rainy season (left) and during the dry season (right) (photo G. Goffinet).

lands (Fig. 6.9). The well-balanced hydration level of tissues (hydrature) of some miombo species, and their resistance both to heat ($+48°C$ for leaves and buds) and frost ($-4°C$ for stems), have been shown by Ernst (1971) and Ernst and Walker (1973). The geographical distribution of miombo vegetation has been found to be correlated with the temperature tolerance of the trees.

The reaction of miombo vegetation to fire and drought has also been carefully studied. Miombo woodland is resistant to fire, when burning takes place early during the dry season; it does not prevent the regeneration of trees (Trapnell, 1959) and the miombo can even be considered as a fire-maintained type of vegetation. However, fierce late fires are much more harmful and destroy the canopy woodland, reducing it to coppice, and even leading to what is locally known as *chipya* open vegetation. In this regressive stage, only fire-resistant woody species survive, such as *Diplorhynchus condylocarpon*, *Hymenocardia acida* and *Syzygium guineense*.

The observations made by Schmitz (1950) in Zaïre, and White (1968) in Zambia show that protection against fire greatly modifies the structure of the miombo woodland. The herbaceous stratum disappears and the woody understorey develops, leading to a dense dry forest. It is noteworthy that Trapnell (1959) found that the canopy remained unaltered and still characteristic of a miombo woodland, after being protected from fire for twenty consecutive years. It took almost thirty years for the miombo dominants to die and their saplings to be suppressed by the vigorous evergreen thicket

growth (White, 1968). The above observations are supported by experiments carried out in the Ivory Coast and Nigeria. After eighteen years of fire protection, a woody savanna vegetation has thickened but still successfully prevents forest establishment (Menaut, 1977). Keay (1959a) did not notice any obvious effect of fire protection in *Isoberlinia* woodlands; he only observed a little thickening of the canopy and understorey and a slight loosening of the grass cover. Thus he stated that the richness of the herbaceous flora and herbivorous fauna indicates that even in the primitive state these woodlands had a light and broken canopy with plenty of tall grasses.

The natural regeneration of the major miombo species has also been studied. Strang (1966) showed that almost all of the young tree growth was either from root suckers or coppice shoots; the failure of seedlings seems likely to be caused by their inability to withstand the stresses imposed by the rapid fluctuations between cool, moist conditions and very hot, dry ones. Due to its high timber value, the growth habit of *Pterocarpus angolensis* has been carefully studied. The surviving seedlings develop an extensive root system in order to successfully compete for water, whereas the shoots (whose length increases every year) always die back in the dry season (Boaler, 1966). It takes an average of seven years before they are able to pass into the sapling stage (Groome, 1955); during this period, reserves are stored in the root system until the plant can grow shoots strong enough to withstand fire (Groome et al., 1957). Sapling growth is closely correlated with humidity and available soil moisture (Jeffers and Boaler, 1966). Such reactions to fire and drought are quite characteristic of all dry plant communities, and have often been described in savannas (Lebrun, 1947; Sillans, 1958; César and Menaut, 1974).

Contrary to *Pterocarpus angolensis*, the "Rhodesian teak" *Baikiaea plurijuga* grows easily from seedlings. It can also grow from shoots, but less successfully (Martin, 1941). As is usually the case, the young sapling survival rate is very low after burning.

Savannas

Savannas were defined at Yangambi as follows: "formations of grasses at least 80 cm high, forming a continuous layer dominating a lower stratum, usually burnt annually. The leaves of grasses are flat, basal and cauline. Woody plants are usually present."

The word "savanna" groups together a number of facies, whose variety is the most characteristic feature. However, the very existence of a number of transition stages between woodlands and grasslands does not prevent the savanna from representing a well-individualized entity, quite distinct from the other plant communities which can be found within the savanna biome, at least in its typical aspect — the only one to which the following descriptions adequately apply (Monod, 1963). Most features of the wooded savannas have already been discussed (pp. 114–122) and emphasis will be put on grass savanna whose ecology is specially interesting.

Savanna woodlands (Fig. 6.10). These communities only differ from typical woodlands in the floristic composition of the dominant layer and the low-branched and contorted stem of most woody species. Furthermore, trees and shrubs generally form a light canopy. They do not fundamentally differ from tree and shrub savannas either, only being less

degraded. The density and size of trees are higher, and the woody plants more often have a definite tree habit. Savanna woodlands never occur on very large areas within the tree and shrub savannas.

Tree and shrub savannas. Here the woody plants are scattered over a continuous herbaceous layer. These more or less open formations have a density of woody species intermediate between those of the woodlands and the grasslands (Fig. 6.11).

These communities are rather heterogeneous: the many facies differ from each other mostly by the nature, physiognomy, size, density and distribution patterns of their woody components. One can, for instance, distinguish palm savannas, thorn-scrub savannas, orchard savannas, park savannas, grove savannas, etc. Such physiognomic descriptions are meaningful but cannot be too strictly formalized. How should one set a limit between tree savannas and shrub savannas, for example? Typically, a tree has a distinct bole, whose height generally exceeds 7 to 8 m, whereas a shrub has a short, more or less contorted stem, low-branched and not exceeding 7 to 8 m (Aubréville, 1963). In this way it has been possible to individualize *Butyrospermum paradoxum* and *Parkia biglobosa* tree savannas in West Africa, and a *Cussonia angolensis* tree savanna in Zaïre. In the same way *Hymenocardia acida* and *Piliostigma thonningi* shrub savannas, widespread throughout the whole Sudanian domain, have been identified. In point of fact, such a clear-cut distinction between trees and shrubs is somewhat unrealistic. Very frequently woody plants with a typical tree trunk do not reach a size large enough to be included in the tree layer, and conversely low-branched and contorted shrubs may reach typical

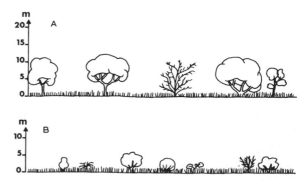

Fig. 6.11. Diagrammatic profiles of a tree savanna (A) and a shrub savanna (B).

Fig. 6.10. A savanna woodland diagrammatic profile.

tree height. In such cases, it is necessary to consider other criteria, such as the position of the reserve buds which are responsible for the mature habit of the individual plant (Champagnat, 1947). Furthermore, it seldom happens that a shrub savanna does not include at least a few dominant trees over the shrub layer.

In most tree and shrub savannas woody plants grow in clumps, surrounded by purely grassy areas. The origin of such a distribution pattern is sometimes easy to understand, as in the case of trees or shrubs growing on old termite mounds (termitaria peppering). In many other cases, its origin is far less clear, as for example that of the "tiger-skin formations" or "brousse tigrée" of the Sahel, which are ascribed to the joint effect of differences in soil composition, run-off and overgrazing. In regularly burnt areas the clumped distribution of woody plants is assumed to be due, for the most part, to the competition taking place between woody and herbaceous plants, which is intensified by burning. Within the grassy areas, the growth of woody seedlings is impeded by shading above ground, and by the dense "felt" of grass roots underground. Moreover, very few seedlings or saplings can withstand the heat of fierce fires. The particular breeding biology of the woody species also has to be taken into consideration.

Grass savannas. Trees and shrubs are normally missing in this category of savanna. However, grass savannas may be punctuated by woody plants isolated or in clumps. In this case, trees and shrubs mostly occur on termite mounds or in the fissures of an outcropping lateritic crust. At first sight, the physiognomy is very much like that of a tree or shrub savanna, but the herbaceous component of the grass savanna is a distinct ecological entity. A good example of this situation is provided by the palm savannas. The grass cover is only dominated by palm trees (*Borassus aethiopum*) (Fig. 6.12), widely scattered and randomly distributed, though their root systems form an astonishingly regular network throughout the whole community. The herbaceous layer remains very much the same over large areas (César, 1974). In such palm savannas other woody species are eliminated by seasonal waterlogging and burning, both conditions easily withstood by *Borassus* palms. It is no wonder, then, that the grass component of such palm savannas has been con-

sidered by a number of authors as a distinct entity, one of the many facies of the grass savannas.

Grass savannas may be edaphically conditioned ("natural"), occurring in strictly defined situations. They can also be the result of human activities ("secondary"), taking over very large areas formerly covered by woody formations, mostly humid forests.

Secondary grass savannas. Most of these are found around rain-forest areas, and can be considered to be the last regressive stage of the forest as a result of human activities. However, it is not at all impossible that grass savannas could, in certain circumstances, originate directly from wooded savannas, even when they are located within the forest itself.

In West Africa, for example, a number of small grassy glades isolated in the heart of the tall closed forest are believed to be relicts of savanna vegetation which penetrated the present forest zone during a major climatic shift, and were later left behind when the savanna retracted under a reversed climatic swing (Rose Innes, 1977). These "enclosed" grass savannas may occur on poor sandy soils, which were shown to have been formerly covered with grasses (more or less mixed with woody vegetation) before the establishment of the rain forest in surrounding areas with better soils (Leneuf and Aubert, 1956). Such savannas may also occur on very shallow soils covering an impervious hard pan, itself the result of past forest clearings in areas with a highly seasonal climate. All over the forest zone and around it, the prevailing humid conditions are very propitious to the growth of grass, and lead to the production of huge amounts of grass fuel (Adegbola, 1964). Despite the wetness of the climate, the open character of these savannas facilitates the drying of the standing grass crop and promotes fierce fires. Such severe burning prevents the establishment of pioneer forest species, as well as the regeneration of the savanna woody species themselves.

The grass savannas of the rain-forest areas sometimes form narrow strips close to the forest edge (e.g. the *Loudetia simplex* savannas). They can also spread over very large areas, as is the case for the tall and coarse grass communities dominated by *Pennisetum purpureum* or *Imperata cylindrica*. These species occur on lands lying fallow after cultivation. *Imperata cylindrica* can grow spon-

Fig. 6.12. The palm savanna of the International Biological Programme study area at Lamto, Ivory Coast (photo Y. Gillon).

taneously in clearings within drier types of forest; it can also succeed *Pennisetum purpureum* after a second cultivation period. In the latter case, *Imperata* growth goes together with an intense activity of termites, which quite likely plays an important role in the evolution of plant cover (Letouzey, 1968). *Imperata* savannas can ultimately return back to forest, after cultivation and the establishment of forest seedlings; they can also be invaded by Andropogoneae (e.g. *Hyparrhenia rufa*) and shrubs (e.g. *Annona senegalensis, Bridelia ferruginea*). It has often been reported that cultivation close to the forest edge was promoting forest recruitment (Spichiger, 1977).

Coastal grass savannas are fire-proclimax communities which may derive either from forests (Jacques-Félix, 1949) or from former wooded savannas (Adjanohoun, 1962). They occur on soils which are at least seasonally wet, the properties of which condition the dominant species (e.g., *Setaria anceps, Schizachyrium schweinfurthii, Vetiveria fulvibarbis*). Frequently, recurring fires inhibit the development of woody species which have to take refuge on weathered termite mounds (Lansbury et al., 1965).

In the western part of the Zaïre Basin, grass savannas occur on leached sands. *Loudetia demeusei* savannas normally contain woody species, but also often form wide expanses of purely grassy areas. *Loudetia simplex* and *Monocymbium ceresiiforme* grasslands, with semishrubs (low suffrutescent species), are also found on white Kalahari sands (Duvigneaud, 1949). They are often called steppes. Their flora, of southern origin (Zambesian region), shows that they are quite likely very old communities, although they are often considered as one of the regressive stages of the rain forest.

Within the Congo forest block, numerous grassy clearings dominated by *Hyparrhenia diplandra* are also found in various topographical situations. They were considered by Robyns (1936) as edaphic grasslands, but other botanists (De Wildeman, 1912; Germain, 1965) have been in favour of their anthropic origin. In any case, they are certainly fire-maintained communities, as fire protection leads to the settlement of pioneer woody species.

Secondary grass savannas are widespread throughout East and Central Africa (Lind and Morrison, 1974). They generally occur at higher elevations than in the western part of the continent, but are not to be confused with montane climax grasslands. They originate from evergreen forests or woodlands through tree felling and recurrent burning. Upland grasslands look very uniform, physiognomically speaking, but they are composed of a number of mixed herbaceous species, none of which is obviously dominant. None of them is restricted to such communities either, which suggests that the upland grasslands are a secondary type of grassland which has "borrowed" species from other habitats (Vesey-Fitzgerald, 1963). Conversely, when grasslands occur on steep slopes, sudden variations take place. In western Uganda, for instance, tall grass types dominated by *Hyparrhenia cymbaria* are found on the deep soils of the lower slopes and level summits, while poor grass types dominated by *Cymbopogon afronardus* occur on the steeper slopes with thin soils (Lang Brown and Harrop, 1962). Extensive areas of the higher and more humid regions of East Africa are covered

with short *Pennisetum clandestinum* grasslands. Somewhat drier conditions favour short *Cynodon* grasslands and, at lower elevations, *Themeda triandra* becomes dominant, when fires occur regularly and the grazing pressure is not too high (Pratt and Gwynne, 1977).

In southern Central Africa, humid forests seem to have turned into short grasslands dominated by *Themeda triandra* on fertile red soils, and by *Loudetia simplex* on poor waterlogged soils. Tall communities of *Hyparrhenia* spp., often associated with the plateaux of the main watershed, are considered as a fire subclimax to woodlands.

Large mammals, especially elephants (*Loxodonta africana*), have been reported to induce the transformation of wooded savannas and even woodlands to grasslands, especially in national parks where game densities are extremely high. Pure grasslands (e.g. *Sporobolus* communities close to river banks) are sparse and of small size (Lock, 1972; Schmidt, 1975). In most cases, low scrub and coppice shoots sprouting annually from buried stocks still occur inside the grass layer; however, such woody individuals are completely overwhelmed by grasses and do not provide any cover. Thus, if the disappearance of trees can be ascribed to elephants, the maintenance of the grass cover can be attributed to both grazing and burning, whose combination alone prevents any tree regeneration from taking place (Buechner and Dawkins, 1961). Exclusion of large mammals and prevention of fire induce the regeneration of a woody vegetation, often quite different from the former one (Lock, 1977; Harrington and Ross, 1974).

Grass savannas occurring over impervious hard pans are considered as natural, edaphically conditioned, communities. However, when one takes into consideration the origin of these lateritic crusts, it is logical to consider such grasslands as the ultimate stage, quite irreversible on the scale of the human life-span, in the degradation of former tree savannas through sustained human activities. The herbaceous vegetation of these natural grasslands is rather scanty, but their species richness is greater than that of the large periforest savannas. One finds, for instance, some tussock-forming (caespitose) perennials, such as *Loudetia simplex* or *Ctenium newtonii*, and a good number of annuals, sometimes of small size and short life-span. A number of hydrophytes can even grow there, when a thin water film covers the hard pan during the rains. During the dry season, plants suffer from the extreme drought, and burning prevents regeneration of the woody plants. These grass savannas can therefore be considered as the joint result of both a poor soil substrate and regular burning; if fires were prevented, they might be progressively colonized by a more luxuriant vegetation (Schnell, 1971).

Edaphic grass savannas. Alternating wet and dry environmental conditions favour the occurrence of primary ("natural") grasslands. Different categories may be recognized according to physiography and soil texture (Michelmore, 1939). Small valley grasslands occur at the upper reaches of the drainage line; soils may vary from sand to clay, but they always have a good colloidal content. The vegetation is usually a medium-dense, mixed herb mat, mainly composed of perennial bunch grasses providing a rather uniform appearance and height (Vesey-Fitzgerald, 1963). Riverine grasslands generally occur on more sandy soils, often deep and freely drained, but they vary in relation to the type of river flow. Lower parts of the river system are occupied by flood-plain grasslands, often concentrically zoned in relation to a central sump area; the substrate generally consists of deep clayey soils, dark, heavy, badly drained and exposed to great extremes of temperature, wetness and dryness.

In each type, the hydroseral zonation is determined by the height and duration of flooding. Differences in elevation between the edge of the formation and its centre are generally slight, but sharp steps mark the junction between perennially wet grasslands and seasonally dry grasslands, and between the latter and the woodland zone (Cole, 1963). The sump areas are covered by an aquatic meadow with *Vossia cuspidata* and *Oryza barthii* giving place to stands of *Echinochloa* spp., more adapted to alternating wet and dry conditions. The seasonally dry part is covered by a rich variety of grasses, sedges and forbs. The upper parts often bear tall bunch species, such as *Hyparrhenia rufa* and *Loudetia simplex* providing a rather uniform cover. This zone is often punctuated by termite mounds over which woody species (e.g. *Acacia* spp., *Combretum* spp.) grow. Regular late burning may even reach the *Echinochloa* stands and is responsible for the occurrence of a secondary fire climax ecotone between the flooded grasslands and

the surrounding woodland or savanna. This general scheme, characteristic of the very numerous valley grasslands of Central Africa is also valid, broadly speaking, for all other African grasslands.

However, some local variations are worthy of note. Extensive meadows of *Loudetia kagerensis* occur around the shores of Lake Victoria, on shallow, leached, very poor sandy soils. This short and delicate grass provides a sparse cover and is very different from the tall tufted *Loudetia* spp. found in other grasslands and savannas. The valley grasslands of the southwestern Sudan are dominated by *Hyparrhenia* spp. on alluvial drained loams (Morison et al., 1948). In the eastern part of the central Sudan, annually flooded grey clay soils are covered by *Brachiaria obtusiflora* grasslands; few other species occur, as *B. obtusiflora* is virtually the only species which can tolerate extremely wet conditions, although it is by no means restricted to wet sites (Bunting and Lea, 1962). The classical hydroseral succession (*Oryza* → *Echinochloa* → *Hyparrhenia*) is also found in West Africa, in the inner Niger Delta, and around Lake Chad. The mud flats of alkaline lakes do not usually bear any vegetation, but they are bordered by a sparse carpet of *Cynodon dactylon* or *Sporobolus robustus*.

In the foregoing pages the terms "grassland" and "savanna" have often been used in an indiscriminate way. In so doing I have not followed the Yangambi classification which restricted the use of the term "grassland" to the high montane grassland, the herb swamp (both excluded from the present volume) and aquatic grasslands such as the *Echinochloa pyramidalis* community. According to this narrow definition, only the most humid stage of the hydroseral succession described can be called a grassland. It does not appear very realistic to do so for a seral succession whose functional cohesion looked so obvious. On the other hand, it is also advisable not to overextend the meaning of the term "grassland". Too many people currently tend to apply it indiscriminately to all kinds of savannas, and speak for instance, of "wooded grasslands". This has the disadvantage of underestimating the role of the tree and shrub layers in the ecology of tropical savannas.

'Steppes'

A steppe is considered as an open herbaceous vegetation, sometimes also including woody plants.

Fig. 6.13. A diagrammatic "steppe" profile.

Steppes are usually not burned. Perennial grasses are widely spaced, and usually less than 80 cm high; their leaves are narrow, rolled or folded, mainly basal. Annual plants are very often abundant between the perennials (Fig. 6.13).

In such a vegetation, emphasis is thus laid upon the lower size of the grasses and the xeromorphy of their leaves which, for the most part, originate close to the soil surface. These diagnostic features are so questionable that the very use of the term "steppe" has been strongly criticized. The coining of the new term "pseudo-steppe" by Trochain (1951), to distinguish tropical and temperate steppes, did not clarify the situation.

Descoings (1973) has shown that the diagnostic criteria proposed at Yangambi did not withstand a careful structural analysis. This is also the position taken by Volk (1966) who showed, in a most pertinent way, that a single community could in turn be classified as a savanna or a steppe depending upon the season. Various authors also use very different terms to name similar formations. For instance, what is called "steppe à épineux" (thorny steppe) in West Africa is named *"Acacia* savanna" in East Africa, and "thornbush savanna" in southern Africa. The lack of burning is not a conclusive criterion either; many so-called steppes burn quite well when fires are started at the right time of the year. The "steppe" vegetation is not at all resistant to burning, but the traditional land-use techniques of the herdsmen forbid large-scale bush fires whose outcome are often tragic in xeric environments.

Most authors have, so far, used the term "steppe" to designate what is now better known as the African Sahel (Fig. 6.14), a zone which is clearly different, floristically as well as ecologically, from the Sudanian savannas. Such confusion between a vegetation category and a biogeographical realm is unfortunate. Furthermore, the steppe appearance (by analogy with the temperate steppes) of the vegetation is only obvious on the very borders of the Sahara Desert. This is what Maire (1933, 1940) used to call the *Acacia raddiana* and *Panicum turgidum* "desert savanna"!

Further south, within the Guinean zone which can be climatically wet but edaphically xeric, some

Fig. 6.14. The International Biological Programme study area of Fété Olé, in the Sahel zone of northern Senegal (photo F. Bourlière).

almost pure grass communities have also been called steppes. This has been the case in Zaïre (Duvigneaud, 1952) as well as in Congo (Koechlin, 1961), for example. Nevertheless, all these formations are regularly burnt, are poor in annuals and display a number of variants closely related to typical savannas.

Thus, the tropical "steppes" appear to be merely one stage in the degradation of the savannas, one way for them to adapt to more xeric conditions; this does not imply a basic functional difference between two plant communities. The characteristic features of the "steppes" correspond to those of the most xeric variants of the savanna biome.

THE SAVANNAS OF MADAGASCAR

The Malagasy "mini-continent" (c. 600 000 km^2) is very closely related, floristically speaking, to the African continent. However, its flora is unique in many ways, and the vegetation of the island is also very diverse due to the extreme variety of physiographic and climatic conditions.

The plant environment

The vegetation of Madagascar is influenced by two major categories of factors (Perrier de la Bathie, 1921). First, an important role is played by the physiography and the wind regime of the island; this combination accounts for the contrast between the windward and leeward floras. Second, human impact is responsible for the large-scale destruction of the native, mostly woody, vegetation and its substitution by impoverished communities, either woody (*savoka*) or grassy. Such secondary formations are generally made up of species widely distributed on the island.

The major categories of vegetation

The windward vegetation occurs in the eastern and central areas of the island, where the humid trade winds meet with high mountain ranges. These high-rainfall areas are (or were originally) covered with climatic moist forests, little influenced by soil conditions. Conversely, the leeward side of the island gets very little rain, and the aridity of the environment increases towards the western and southern coasts. This area is covered by edaphic

xeric communities, mostly conditioned by the water status of the sandy or limestone soils. Unfortunately, undisturbed native woody communities only persist presently in the southern part of the island; those of the western part are presently very degraded, and in large part replaced by secondary grass communities. This is also the case for the whole central part of the island, now only covered by pastures, though originally belonging to the windward humid and forested domain.

Savanna-like communities therefore now cover most of the island. Following Humbert (1955) the following phytogeographical units can be identified:

Eastern Malagasy region

Eastern and central domain

Low and medium-altitude district. The former rain forest, after repeated clearings, has been transformed first into an ericoïd bush (*Helichrysum*, *Philippia*, *Pteridium*), and second into *Aristida* grasslands after further clearing and burning. Western slopes district. The native forests (*Uapaca bojeri* and Sarcolaenaceae forests) already had a definite sclerophyllous character here owing to the lower rainfall; only a few patches of them persist presently and most have been turned into *Aristida* grasslands, regularly burnt and including very few woody species.

Western Malagasy region

Western domain

The native forests are dense, deciduous, and definitely xerophilous (*Commiphora*, *Dalbergia*, *Hildegardia* forests). Their structure varies according to soil conditions. The most peculiar are those growing on limestone soils, with their leafless and succulent plants, and trees with a swollen water-storing trunk. Very little of those forests presently remain; most of them have been replaced by *Hyparrhenia* and *Heteropogon* savannas, with a more or less dense shrub layer.

Southern domain

The southwestern part of the island is the most arid of all. Here the western deciduous forest progressively turns into dense thickets of Didieraceae and *Euphorbia*. Such a community, the most peculiar of the island, is comparatively little degraded, the succulent plants being resistant to fire.

With the exception of the southern thickets, most of the savanna communities of Madagascar are therefore secondary. The native woody communities, very diversified and well adapted to local environmental conditions, are all very sensitive to human interferences (Koechlin et al., 1974). However, man has moved to Madagascar only recently (*c*. 1700 B.P.) and his impact has possibly been magnified by the poor quality of the soils, which do not allow forest regeneration. It is also quite possible that a dry climatic episode took place immediately prior to human colonization, thus worsening the consequences of human interference. A recent increase in the dryness of the climate has, for instance, made the eastern forests even more vulnerable than before, according to Bourgeat (1972). Another factor to take into consideration is the lack of native Malagasy species which can play the pioneer role in secondary formations. Contrary to the case on the African continent, there is no native "secondary flora" in Madagascar; native species seem unable to compete successfully with prolific alien invaders, such as African savanna grasses. Furthermore, immigrant hemicryptophytes are fire-resistant, whereas native species are for the most part very sensitive to repeated burning.

Distinctive features of the Malagasy flora

As a result of its varied environmental conditions and its prolonged geographical isolation, Madagascar has an unusually rich flora, with a high degree of specific endemism, often exceeding 80% in the case of groups belonging to the native vegetation of the island (Koechlin, 1972). The species richness of the Malagasy flora, and its unusual characteristics, has motivated a large number of taxonomic studies, mostly under the influence of Perrier de la Bathie (1936) and Humbert (1959). Out of a grand total of *c*. 8500 species of vascular plants, about 1800 are restricted to the western part of the island, the domain of dry forests and savannas. Although far poorer in species than the eastern domain it nevertheless has a very high degree of endemism: 38% at the generic level and 89% at the specific level. The most peculiar endemic family is that of the Didieraceae, but other more

widespread families are also represented here by a large majority of endemic species. Endemism is particularly high in woody species, which by themselves amount to more than 80% of the spermatophytes of the whole island; this speaks for the great antiquity of the Malagasy dry flora. Conversely, most of the recently introduced plants are herbaceous annuals, growing in disturbed communities and almost never entering native woody formations.

Among the endemic species, a large number are considered as palaeoendemics no longer displaying any obvious evolutionary tendency, contrary to what happens in neoendemics. Some of these neoendemics are presently limited to very small areas, a single valley for instance, despite their high abilities for dispersal, a fact which points to their recent origin. Most neoendemics are supposed to have originated from taxa belonging to the ancient flora of the island, but many plants recently introduced by man (e.g. *Pavonia*, *Sida*, some Cyperaceae) have also already given rise to a number of varieties or even species (Koechlin, 1972). The large number of neoendemics has sometimes been interpreted as proof that Madagascar had been colonized by diaspores brought by wind and sea water. This interpretation has been questioned by Humbert (1959) who, on the contrary, emphasized the importance of the progressive geological isolation of the island. Although wind and water transportation of diaspores certainly played a role, most of the basic components of the Malagasy flora are of an old pantropical origin, a relict of the old Gondwana vegetation. The past land connexions of Madagascar also explain the presence of some species related to tropical Asian taxa on the eastern coast of the island. Madagascar started parting from Africa during the Permian era, but final and complete separation only took place during the Pliocene. The relationships between African and Malagasy floras have therefore been long-lasting, but the variety of the island environments, and their geographical isolation have promoted high speciation rates.

Vegetation types (Fig. 6.15)

The richness of the dry flora and the variety of the ecological conditions in Madagascar account for the small size of the different formations of the

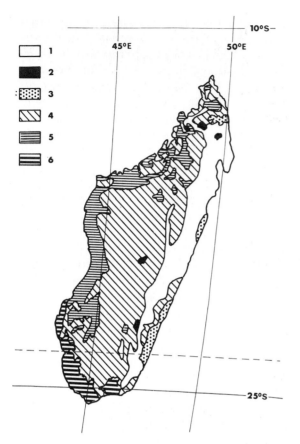

Fig. 6.15. The vegetation types of Madagascar: *1* = moist forests; *2* = montane communities; *3* = forest and savanna mosaic; *4* = open shrub and grass savannas; *5* = dry deciduous forests; *6* = thickets. (Redrawn from Humbert, 1955; and Koechlin et al., 1974.)

island, contrary to what is the rule in continental Africa where a single vegetation type can be spread over very large areas. Furthermore, due to the fragility of the primitive vegetation and soils, most climax plant communities now occur as isolated remnants within the secondary savanna formations.

Fortunately, some localized habitats (e.g. rocky substrates), naturally protected against burning and clearing, are still occupied by the original dry flora, whose particular morphological features account for the characteristic physiognomy of the Malagasy dry communities. Indeed they look quite different from their African counterparts. There are many trees with swollen, water-storing trunks ("bottle trees", such as many species of *Adansonia*, *Adenia*, *Cyphostemma* and *Pachypodium*), succu-

lent plants (*Aloe* spp., Crassulaceae, Asteraceae), and cactus-like plants (*Alluaudia*, *Didierea* spp.), contrary to what occurs on the continent. There are also many more representatives of small-leaved, leafless and spiny plants, as well as species tolerating complete dessication (Xerophyta, *Seleginella*, ferns). Although many vegetation categories of the Yangambi classification apply well to the Malagasy situation (Guillaumet and Koechlin, 1971), a number of native plant formations do not adequately fit them. This is especially the case for the xeric formations of the south.

Dry deciduous forests

These are still well represented in Madagascar where they constitute the climax stage in the western domain. Being apparently very fragile, they have generally turned into secondary savannas in most of the former forested range, except along the western coast. These forests belong to the *Commiphora*, *Dalbergia* and *Hildegardia* series and their local structure is strongly influenced by the edaphic conditions. They are characterized by a large number of deciduous woody species, many lianas and the quasi-absence of herbaceous plants, some geophytes excepted.

Forests on lateritic clays. These are found on soils derived from metamorphic rocks, often deep and rich in organic matter. The closed tree canopy reaches *c.* 15 m high with a few emergent trees 20 m tall. There is no obvious dominant species, legumes (e.g. *Dalbergia*) being relatively numerous. The open understorey is also rich in species.

Forests on sandy soils. These forests are much more widespread than the above-mentioned category. They occur on poor soils, where water conditions are extreme. The tree layer is low (*c.* 10 m) and the aridity of the climate entails an increased number of succulent and leafless plants with water-storing stems.

Forests on limestone soils. The tree layer of these forests is uneven and plants become even better adapted to xeric conditions. The karst topography honeycombed by tunnels and openings give rise to many micro-environments whose flora is very distinct. *Adansonia* and *Dalbergia* forests are found on the plateaux, whereas xerophytic shrublands, often dominated by Euphorbiaceae, replace them in less favourable conditions. The most peculiar representatives of the Malagasy flora are found in this milieu. As a rule, they are deciduous: the aerial parts can be shed seasonally, in part (e.g. *Begonia* spp., *Dorstenia cuspidata*, *Euphorbia enterophora*), or totally (e.g. Dioscoreaceae, Orchidaceae, Araceae, Taccaceae). Leaves can become fleshy or may be totally absent (*Euphorbia* spp., Asclepiadaceae, Orchidaceae) and water-storing stems and trunks are frequent (*Adansonia* spp., *Adenia* spp., *Cyphostemma* spp., *Pachypodium* spp.). There are even some evergreen sclerophyllous plants (*Diospyros* spp., *Polycardia* spp.).

Thickets

Most of the plant communities of the southern domain of Madagascar are usually included in this category. However, Koechlin et al. (1974) very rightly pointed out that true thickets only occur along the coast, most of the inland communities being dry forests representing the most arid stage of those mentioned above. The more xeric species are found in this community, which is often dominated by Didieraceae and *Euphorbia* spp. The number of endemic genera and species is greater here than anywhere else on the island.

In such thickets tree species are scarce, being mostly represented by succulent or leafless *Euphorbia* spp. Other dominant species (*Alluaudia* spp. and *Didierea* spp.; Fig. 6.16) often display a remarkable convergence of form and habit with American Cactaceae. The shrub layer is very dense, thorny and difficult to penetrate. The grass layer is discontinuous and sparse. It is made of some Poaceae and Velloziaceae, some crassulescent species (e.g. *Aloe* spp., *Kalanchoe* spp., *Senecio* spp., *Xerosicyos* spp.) and of reviviscent *Selaginella* spp.

Fortunately, these thickets grow on very poor soils where traditional agriculture is almost impossible; they cannot be browsed by domestic stock, and they do not burn easily.

Other kinds of thickets also occur in Madagascar: the heath formations ("ericoid bush") of the mountain areas which are outside the scope of this volume, and the secondary *savoka*.

Woodlands

These woody formations, which cover the western slopes of the central domain, are usually called

Fig. 6.16. An *Alluaudia* thicket near Fort Dauphin, southern Madagascar (photo C.M. Hladik).

sclerophyllous forests. They represent the most xeric facies of the evergreen rain forests of the eastern domain. Therefore they should not be included in this volume. However, these climax formations are presently so degraded that a number of their structural and ecological characteristics relate them more to the savanna than to the rain-forest biome.

The physiognomy of these woodlands is very uniform. They are made of small, contorted, low-branched trees and shrubs, which have a thick furrowed bark and long-lasting, tough and often dull leaves. The upper layer (*c.* 10 m tall) is dominated by *Uapaca bojeri* and various Sarcolaenaceae (another endemic family). The understorey is made of ericoid bushes belonging to Ericaceae (*Philippia* spp., *Vaccinium* spp.) or to Asteraceae (*Helichrysum* spp.). The canopy being quite open, the herbaceous layer is made of a number of grassy and suffrutescent species. Most of these plants are definitely heliophilous, some dominant species such as *Uapaca bojeri* only regenerating in full sunshine.

This community is very sensitive to burning, many understorey species being rich in essential oils; however, some trees (e.g. *Asteropeia densiflora, Sarcolaena oblongifolia, Uapaca bojeri*) are fire-resistant. The end result is a rapid development of a secondary grass layer (*Isalus* spp., *Loudetia* spp., *Schizachyrium* spp., *Trachypogon* spp.) and the formation of a typical woodland.

Savannas

The herbaceous communities which presently cover extensive areas of Madagascar obviously have a recent origin, and must therefore be considered as secondary savannas. Most of the participating species are not native; the flora as a whole is poor and the vegetation physiognomy very monotonous.

However, two main categories can be distinguished, each corresponding to the two major biogeographical subdivisions of the island.

Eastern region. In favourable climatic conditions (rainfall over 1000 mm yr^{-1}) the maintenance of secondary grasslands in these formerly forested areas is mostly due to poor soils. These are ferralitic soils, poor in mineral nutrients and virtually lacking in organic matter. Their surface is indurated and almost impervious. When this hard crust disappears, they are strongly leached and eroded, devoid of any vegetation.

The herbaceous layer of the eastern secondary savannas is discontinuous, and made up of xerophytic hemicryptophytes (with basal, narrow and plicately folded leaves), whose above-ground life cycle is very short. According to Koechlin (1972), such a steppe-like community is very similar to those already described in the Congo (Trochain and Koechlin, 1958). When the soil is less degraded, and richer in organic matter, the herbaceous layer becomes denser and includes typical savanna grasses (*Heteropogon contortus, Hyparrhenia rufa* and *Imperata cylindrica*).

The herbaceous flora is very poor, dominated by *Aristida rufescens* and *A. similis*, with a few other grasses and some forbs, mostly geophytes. Woody plants are virtually lacking, reduced to a few left-over representatives of the past forest. These east-

ern savannas obviously represent the very last regressive stage of the former eastern forests; the extremely poor soils seem to make any regeneration quite impossible.

Western region. There are also some steppe-like communities in the southern domain, but they have little in common with those which have just been described. Dominated by *Cenchrus ciliaris*, they are much more similar to some Saharo-Sahelian formations. They result from the destruction of the native thickets, through clearing and intensive grazing; they never occupy very large areas, however.

The major part of the western domain is now covered by savannas, with a dense and continuous layer of grasses (Andropogoneae and Arundinelleae), dotted with trees and shrubs. The grass cover, mostly dominated by *Heteropogon contortus*, is relatively low (c. 1 m) on poor and eroded soils (with *Aristida rufescens* and *Loudetia stipoides*), but reaches a greater height on better, more humid soils (with *Hyparrhenia* spp., *Hyperthelia dissoluta, Imperata cylindrica* and *Panicum maximum*). Therophytes are numerous, mostly within the grass layer itself (e.g. *Brachiaria* spp., *Bulbostylis* spp., *Zornia* spp.) or under the shade of shrubs (e.g. *Oplismenus* spp., *Setaria* spp.). Chamaephytes and geophytes are scarce. Trees and shrubs are small (c. 10 m tall) and scattered; they are either native (e.g. *Acridocarpus excelsus, Dicoma incana, Stereospermum variabile*) or introduced (e.g. *Sclerocarya caffra, Ziziphus mauritiana*). The woody plants are in general rather sensitive to burning; their aerial parts are often partly destroyed, but they survive thanks to their high sprouting ability correlated with their highly developed root system. Palm trees naturally resist fires, and are more numerous there than in continental African savannas (*Borassus madagascariensis, Hyphaene shatan, Medemia nobilis*).

On the whole, and contrary to what is the rule on the continent, the Malagasy savanna species appear to be poorly adapted to the savanna environment. The fire tolerance of the woody species is low, and their phenology (leafing and flowering periods) rather irregular. Added to their low species diversity and the scarcity of endemics, this all points to the recent and anthropic origin of the savanna biome in Madagascar.

145

REFERENCES

Adegbola, A.A., 1964. Forage crop research and development in Nigeria. *Niger. Agric. J.*, 1: 34–39.

Adjanohoun, E., 1962. Etude phytosociologique des savanes de basse Côte d'Ivoire (savanes lagunaires). *Vegetatio*, 11: 1–38.

Adjanohoun, E. and Aké Assi, L., 1967. Inventaire floristique des forêts claires sub-soudanaises et soudanaises en Côte d'Ivoire septentrionale. *Ann. Univ. Abidjan, Sér. Sci.*, 3: 89–148.

AETFAT, since 1953. *Bulletins périodiques.* Conservatoire et Jardin Botaniques, Genève.

Anonymous, 1956. Phytogeography, Yangambi. *C.S.A./C.C.T.A. Publ.*, 22: 35 pp.

Aubréville, A., 1949a. *Contribution à la paléohistoire des forêts de l'Afrique tropicale.* Larose, Paris, 99 pp.

Aubréville, A., 1949b. *Climats, forêts et désertification de l'Afrique tropicale.* Larose, Paris, 351 pp.

Aubréville, A., 1950. *Flore forestière soudano-guinéenne.* Soc. Ed. Geogr. Mar. Coll., Paris, 523 pp.

Aubréville, A., 1955. La disjonction africaine dans la flore forestière tropicale. *C.R. Soc. Biogéogr.*, 278, 280: 42–49.

Aubréville, A., 1959. Définitions physionomiques, structurales et écologiques des forêts claires en Afrique. *C.S.A./C.C.T.A. Publ.*, 52: 81–87.

Aubréville, A., 1960. Forêts claires. Extension géographique. *C.S.A./C.C.T.A. Publ.*, 52: 89–92.

Aubréville, A., 1963. Classification des formes biologiques des plantes vasculaires en milieu tropical. *Adansonia*, 3: 221–226.

Aubréville, A., 1965. Principes d'une systématique des formations végétales tropicales. *Adansonia*, 5: 153–196.

Aubréville, A., 1969. Essais sur la distribution et l'histoire des Angiospermes tropicales dans le monde. *Adansonia*, 2: 189–247.

Bews, J.W., 1927. *Studies in the Ecological Evolution of the Angiosperms.* London, 134 pp.

Boaler, S.B., 1966. The ecology of *Pterocarpus angolensis* D.C. in Tanzania. *Min. Overseas Devel., Overseas Res. Publ.*, No. 12: 128 pp.

Bosser, J., 1969. Graminées des pâturages et des cultures à Madagascar. *Mém. ORSTOM*, No. 35: 440 pp.

Boudet, G. and Duverger, E., 1961. *Etude des pâturages naturels sahéliens. Le Hodh (Mauritanie).* Vigot, Paris, 160 pp.

Boughey, A.S., 1957a. The physiognomic delimitation of West African vegetation types. *J.W. Afr. Sci. Assoc.*, 3: 148–165.

Boughey, A.S., 1957b. The vegetation types of the Federation. *Proc. Trans. Rhod. Sci. Assoc.*, 45: 73–91.

Boughey, A.S., 1964. Deciduous thicket communities in Northern Rhodesia. *Adansonia*, 4: 239–261.

Bourgeat, F., 1972. Contribution à l'étude des sols sur socle ancien à Madagascar. Types de différenciation et interprétation chronologique au cours du Quaternaire. *Mém. ORSTOM*, 57: 338 pp.

Bourlière, F., 1978. La savane sahélienne de Fété Olé, Sénégal. In: M. Lamotte and F. Bourlière (Editors), *Structure et fonctionnement des écosystèmes terrestres.* Masson, Paris, pp. 187–229.

Bouxin, G., 1974. Distribution des espèces dans la strate herbacée au sud du parc national de l'Akagera (Rwanda, Afrique Centrale). *Oecol. Plant.*, 9: 315–332.

Brenan, J.P. et al. (Editors), since 1960. *Flora Zambesiaca.* Crown Agents for the Overseas Governments and Administrations, London.

Brenan, J.P., 1978. Some aspects of the phytogeography of tropical Africa. *Ann. Mo. Bot. Gard.*, 65: 437–478.

Buechner, M.K. and Dawkins, M.C., 1961. Vegetation change induced by elephants and fire in Murchison Falls National Park, Uganda. *Ecology*, 42: 752–766.

Bunting, A.M. and Lea, J.D., 1962. The soils and vegetation of the Fung, East Central Sudan. *J. Ecol.*, 50: 529–558.

Cailleux, A., 1953. *Biogéographie mondiale.* P.U.F., Paris, 128 pp.

César, J., 1971. *Etude quantitative de la strate herbacée de la savane de Lamto.* Thesis, University of Paris, Paris, 95 pp.

César, J., 1974. *L'enracinement du palmier ronier.* Typescript, 15 pp.

César, J. and Menaut, J.C., 1974. Le peuplement végétal. *Bull. Liaison Chercheurs Lamto*, numéro spécial 2: 161 pp.

Chapin, J.P., 1932. The birds of the Belgian Congo, Part I. *Bull. Am. Mus. Nat. Hist.*, 45: 756 pp.

Champagnat, P., 1947. Les principes généraux de la ramification des végétaux ligneux. *Rev. Hortic., N. S.*, 30: 335–341.

Chevalier, A., 1933. Analogies et dissemblances entre les flores tropicales de l'Ancien et du Nouveau Monde. In: *C.R. Congr. Int. Géogr., Paris*, 2: 839–850.

Chevalier, A., 1951. Sur l'existence d'une forêt vierge sèche sur de grandes étendues aux confins des bassins de l'Oubangui, du Chari et du Haut-Nil (Bahr-el-Ghazal). *Rev. Bot. Appl.*, 31: 135–136.

Chevalier, A. and Emberger, L., 1937. Les régions botaniques terrestres. In: *Encyclopédie Française*, 5. Masson, Paris, pp. 1–7.

Chippindall, L.K.A., 1955. *The Grasses and Pastures of South-West Africa.* Central News Agency, Johannesburg, 527 pp.

Christ, H., 1892. La flore dite "ancienne africaine". *Arch. Sci. Phys. Nat. Genève*, 28: 1–48.

Clayton, W.D., 1958. Secondary vegetation and the transition to savanna near Ibadan, Nigeria. *J. Ecol.*, 46: 217–238.

Cole, M., 1963. Vegetation and geomorphology in Northern Rhodesia: an aspect of the distribution of the savanna of Central Africa. *Geogr. J.*, 129: 290–310.

Collins, N.M., 1977. Vegetation and litter production in southern Guinea Savanna, Nigeria. *Oecologia*, 28: 163–175.

Cornet, A. and Poupon, H., 1977. *Description des facteurs du milieu et de la végétation dans cinq parcelles situées le long d'un gradient climatique en zone sahélienne au Sénégal.* Rapport ORSTOM, Dakar, 30 pp.

Dekeyser, P.C. and Villiers, A., 1954. Essai sur le peuplement zoologique terrestre de l'ouest africain. *Bull. IFAN*, A 16: 957–970.

Descoings, B., 1973. Les formations herbeuses africaines et les définitions de Yangambi considérées sous l'angle de la structure de la végétation. *Adansonia, Ser. 2*, 13: 391–421.

Descoings, B., 1975. Les types morphologiques et biomorphologiques des espèces graminoïdes dans les formations herbeuses tropicales. *Nat. Monspel., Ser. Bot.*, 25: 23–35.

Descoings, B., 1976. *Approche des formations herbeuses tropicales par la structure de la végétation.* Thesis, University of Montpellier, Montpellier, 221 pp.

Descoings, B., 1978. Les formations herbeuses dans la classification phytogéographique de Yangambi. *Adansonia, Ser. 2,* 18: 243–256.

De Wildeman, E., 1912. Documents pour l'étude de la géobotanique congolaise. *Bull. Soc. R. Bot. Belg.,* 51: 5–406.

De Winter, B., 1971. Floristic relationships between the northern and southern arid areas in Africa. *Mitt. Bot. Staatssmml. München,* 10: 427–437.

Duvigneaud, P., 1949. Les savanes du Bas-Congo. Essai de phytosociologie topographique. *Lejeunia,* 10: 192 pp.

Duvigneaud, P., 1952. La flore et la végétation du Congo méridional. *Lejeunia,* 16: 95–124.

Duvigneaud, P., 1959. Forêts claires: Composition floristique, classification, affinités et dynamisme des peuplements. *C.S.A./C.C.T.A. Publ.,* 52: 115–121.

Dyer, R.A. et al., (Editors), since 1963. *Flora of Southern Africa.* Pretoria.

Edwards, D.C. and Bogdan, A.V., 1951. *Important Grassland Plants of Kenya.* Pitman and Sons, London.

Eig, A., 1931. Les éléments et les groupes phytogéographiques auxilliaires dans la flore palestinienne. *Feddes Repert. Sp. Nov. Regn. Veg., Beih.,* 63.

Emberger, L., 1968. *Les plantes fossiles dans leurs rapports avec les végétaux vivants.* Masson, Paris, 2e ed., 758 pp.

Engler, A., 1910. Die Pflanzenwelt Afrikas, insbesondere seiner tropischen Gebiete. In: A. Engler and O. Drude (Editors), *Die Vegetation der Erde,* IX, 1(1,2). Engelmann, Leipzig, pp. 479–1029.

Engler, A., 1921. Die Pflanzenwelt Afrikas, insbesondere seiner tropischen Gebiete. In: A. Engler and O. Drude (Editors), *Die Vegetation der Erde,* IX, 3(2). Engelmann, Leipzig, 878 pp.

Ernst, W., 1971. Zur Ökologie der Miombo Wälder. *Flora,* 160: 317–331.

Ernst, W. and Walker, B., 1973. Studies on the hydrature of trees in the Miombo woodland in south Central Africa. *J. Ecol.,* 61: 667–673.

Eyre, J.R., 1968. *Vegetation and Soils.* Arnold, London, 328 pp.

Fanshawe, D.B., 1961. Evergreen forest relics in Northern Rhodesia. *Kirkia,* 1: 20–24.

Fosberg, F., 1968. A classification of vegetation for several purposes. In: G.F. Peterken (Editor), *Guide to the Check Sheet for IBP Areas.* Blackwell, Oxford, pp. 73–120.

Freson, R., 1973. Contribution à l'étude de l'écosystème forêt claire (Miombo). Note 13: Aperçu de la biomasse et de la productivité de la strate herbacée du Miombo de la Luiswishi. *Ann. Univ. Abidjan,* E 6: 265–277.

Freson, R., Goffinet, G. and Malaisse, F., 1974. Ecological effects of the regressive succession Muhulu–Miombo–Savanna in Upper Shaba (Zaïre). In: *Proc. First Int. Congr. Ecol,* PUDOC, Wageningen, pp. 365–371.

Germain, R., 1965. Les biotopes alluvionnaires herbeux et les savanes intercalaires du Congo équatorial. *Acad. R. Sci. Outre-Mer, Cl. Sci. Nat. Med.,* 15: 399 pp.

Gillet, H., 1961. Pâturages sahéliens. Le ranch de l'Ouadi Rimé. *J. Agric. Trop. Bot. Appl.,* 8: 557–667.

Gillet, H., 1968. Le peuplement végétal du massif de l'Ennedi (Tchad). *Mém. Mus. Nat. Hist. Nat., Sér. B,* 17: 206 pp.

Good, R., 1974. *The Geography of Flowering Plants.* Longman, London, 4th ed., 557 pp.

Greenway, P.J., 1973. A classification of the vegetation of East Africa. *Kirkia,* 9: 1–68.

Griffith, A.L., 1961. Les forêts claires sèches d'Afrique au Sud du Sahara. *Unasylva,* 15: 10–21.

Groome, J.S., 1955. Muninge (*Pterocarpus angolensis* D.C.) in the western province of Tanganyika. *E. Afr. Agric. J.,* 21.

Groome, J.S., Less, H.M.N. and Wigg, L.T., 1957. A summary of information on *Pterocarpus angolensis. For. Abstr.,* 18: 3–8; 153–162.

Grouzis, M., 1979. *Structure, composition floristique et dynamique de la production de matière sèche de formations végétales sahéliennes (Mare d'Oursi, Haute-Volta).* Rapport ORSTOM, Ouagadougou, 59 pp.

Grunow, J.O. and Bosch, O.J.M., 1978. Above ground annual dry matter dynamics of the grass layer in a tree savanna ecosystem. In: *Proc. First Int. Rangeland Congress, Denver, Colo.,* pp. 1–19.

Guillaumet, J.L. and Koechlin, J., 1971. Contribution à la définition des types de végétation dans les régions tropicales (exemple de Madagascar). *Candollea,* 26: 263–277.

Harrington, G.N. and Ross, I.C., 1974. The savanna ecology of Kidepo valley National Park. Part I. The effects of burning and browsing on the vegetation. *E. Afr. Wildl. J.,* 12: 93–106.

Henkel, J.S., 1931. Types of vegetation in Southern Rhodesia. *Proc. Trans. Rhod. Sci. Assoc.,* 30: 1–24.

Hopkins, B., 1962. Vegetation of the Olokemeji Forest Reserve, Nigeria. I. General features of the reserve and the research sites. *J. Ecol.,* 50: 559–598.

Humbert, H., 1955. Les territoires phytogéographiques de Madagascar. *Ann. Biol., 3e Sér.,* 31: 439–448.

Humbert, H., 1959. Origines présumées et affinités de la flore de Madagascar. *Mém. Inst. Sci. Madagascar, Sér. B,* 9: 149–187.

Huntley, B.J. and Morris, T.W., 1978. Savanna ecosystem project. *S. Afr. Natl. Sci. Progr., Rep.* No. 29: 52 pp.

Huntley, B. and Walker, B.H. (Editors), in press. *Dynamic Changes in Savanna Ecosystems.* Proceedings of the Pretoria Symposium.

Hutchinson, J. and Dalziel, J.M., 1927–1936. *Flora of West Tropical Africa.* Second edition revised by R.W.J. Keay and edited by F.N. Hepper (1954–1972). Crown Agents, London.

Jacques-Félix, H., 1949. A propos des savanes côtières de l'Ouest Africain. *Bull. Agric. Congo, Belge,* 40: 732–733.

Jacques-Félix, H., 1956. Ecologie des herbages en Afrique intertropicale. *Agron. Trop.,* 11: 217–233.

Jacques-Félix, H., 1962. *Les graminées d'Afrique tropicale,* IRAT, Paris, 345 pp.

Jacques-Félix, H., 1970. Contribution à l'étude des Umbelliflorae du Cameroun. *Adansonia,* 10: 35–94.

Jaeger, P., 1956. Contribution à l'étude des forêts reliques du Soudan occidental. *Bull. IFAN, Ser. A,* 18: 993–1053.

Jaeger, P. and Adam, J.G., 1966. Sur le cycle annuel de la végétation en prairie d'altitude des monts Loma (Sierra Leone). *C.R. Acad. Sci. Paris,* D 263: 1724–1727.

Janzen, D.H., 1975. *Ecology of Plants in the Tropics.* Arnold, London, 60 pp.

Jeffers, J.N.R. and Boaler, S.B., 1966. Ecology of a miombo site, Lupa North Forest Reserve, Tanzania. I. Weather and plant growth. *J. Ecol.*, 54: 447–463.

Jones, E.W., 1963. The Cece Forest Reserve, Northern Nigeria. *J. Ecol.*, 51: 461–466.

Keay, R.W.J., 1952. *Isoberlinia* woodlands in Nigeria and their flora. *Lejeunia*, 16: 17–26.

Keay, R.W.J., 1959a. Derived savanna — derived from what? *Bull. IFAN, Ser. A*, 21: 28–39.

Keay, R.W.J., 1959b. *Vegetation Map of Africa.* Oxford University Press, Oxford, 24 pp.

Keay, R.W.J., Onochie, C.F.A. and Stanfield, P.F., 1960–1964. *Nigerian Trees.* Government Printer, Lagos.

Kelly, R.D. and Walker, B.H., 1976. The effects of different forms of land use on the ecology of a semi arid region in South-Eastern Rhodesia. *J. Ecol.*, 64: 553–576.

Kennard, D.G. and Walker, B.H., 1973. Relationships between tree canopy cover and *Panicum maximum* in the vicinity of Fort Victoria. *Rhod. J. Agric.*, 11: 145–153.

Knapp, R., 1973. *Die Vegetation von Afrika.* Fischer-Verlag, Stuttgart, 626 pp.

Koechlin, J., 1961. La végétation des savanes du Sud de la République du Congo. *Mem. Inst. Et. Centrafr.*, 10: 310 pp.

Koechlin, J., 1963. In: *Enquête sur les ressources naturelles du continent africain.* UNESCO, Paris, 448 pp.

Koechlin, J., 1972. Flora and vegetation of Madagascar. In: R. Battistini and G. Richard-Vindard (Editors), *Biogeography and Ecology in Madagascar.* Junk, The Hague, pp. 145 190.

Koechlin, J., Guillaumet, J. L. and Morat, P., 1974. *Flore et végétation de Madagascar.* Cramer, Vaduz, 650 pp.

Lang Brown, J.R. and Harrop, J.F., 1962. The ecology and soils of the Kibale grasslands, Uganda. *E. Afr. Agric. For. J.*, 27: 264–273.

Langdale-Brown, I., 1959–1960. The vegetation of Uganda. *Dep. Agric. Mem. Res. Div., Ser. 2, Vegetation*, 1–5: 154, 90, 106, 111, 45 pp.

Lansbury, T.J., Rose Innes, R. and Mabey, G.L., 1965. Studies on Ghana grasslands: yield and composition on the Accra plains. *Trop. Agric.*, 42: 1–18.

Lawton, R.M., 1978. A study of the dynamic ecology of Zambian vegetation. *J. Ecol.*, 66: 175–198.

Lebrun, J., 1947. *La végétation de la plaine alluviale au Sud du lac Edouard.* Publ. Inst. Parcs Natl. Congo Belge, Brussels, 800 pp.

Lebrun, J., 1955. *Esquisse du Parc National de la Kagera.* Inst. Parcs Natl. Congo Belge, Brussels, 2, 89 pp.

Lebrun, J., 1960. Sur la richesse de la flore de divers territoires africains. *Acad. R. Sci. Outre-Mer, Bull. Séances*, 6: 669–690.

Lebrun, J., 1962. Le couloir littoral atlantique, voie de pénétration de la flore sèche en Afrique guinéenne. *Acad. R. Sci. Outre-Mer, Bull. Séances*, 8: 719–735.

Lebrun, J., 1964. A propos des formes biologiques des végétaux en régions tropicales. *Acad. R. Sci. Outre-Mer, Bull. Séances*, 10: 926–937.

Lebrun, J., 1966. Les formes biologiques dans les végétations tropicales. *Mem. Soc. Bot. Fr.*, pp. 164–175.

Lebrun, J., 1968. A propos du rythme végétatif de l'*Acacia albida* Del. *Collect. Bot.*, 7: 625–636.

Lebrun, J. and Gilbert, G., 1954. Une classification écologique des forêts du Congo. *Publ. INEAC, Ser. Sci.*, 63: 89 pp.

Legrand, P., 1979. *Biomasse racinaire de la strate herbacée de formations sahéliennes.* Rapport ORSTOM, Ouagadougou, 28 pp.

Leneuf, N. and Aubert, G., 1956. Sur l'origine des savanes de la basse Côte d'Ivoire. *C.R. Acad. Sci. Paris*, 243: 859–860.

Letouzey, R., 1968. *Etude phytogeógraphique du Cameroun.* Lechevalier, Paris, 513 pp.

Lieth, H. and Van der Maarel, E., 1976. Classifying and mapping the world's vegetation. *Vegetatio*, 32: 73–74.

Lind, E.M. and Morrison, M.E.S., 1974. *East African Vegetation.* Longman, London, 257 pp.

Lock, J.M., 1972. The effects of Hippopotamus grazing on grasslands. *J. Ecol.*, 60: 445–467.

Lock, J.M., 1977. Preliminary results from fire and elephant exclusion plots in Kabalega National Park, Uganda. *E. Afr. Wildl. J.*, 15: 229–232.

Lönnberg, E., 1929. The development and distribution of the African fauna in connection with and depending upon climatic changes. *Ark. Zool.*, 21 A.N. 4: 1–33.

Maire, R., 1933. Etudes sur la flore et la végétation du Sahara central. *Mém. Soc. Hist. Nat. Afr. Nord*, 1933: 1–272.

Maire, R., 1940. Etudes sur la flore et la végétation du Sahara central. *Mém Soc. Hist. Nat. Afr. Nord*, 1940: 273–433.

Makany, L., 1976. Végétation des plateaux Teke (Congo). *Trav. Univ. Brazzaville*, 1: 301 pp.

Malaisse, F. and Strand, M.A., 1973. A preliminary miombo forest seasonal model. In: *Modeling Forest Ecosystems, I.B.P., Woodland Workshop, Oak Ridge, Tenn.*, pp. 291–295.

Malaisse, F., Malaisse-Mousset, M. and Balaimu, J., 1970. Contribution à l'étude de l'écosystème forêt dense sèche (Muhulu). 1. Phénologie de la défoliation. *Trav. Serv. Sylvicult. Pisc. Univ. Off. Congo*, 9: 11 pp.

Malaisse, F., Fréson, R., Goffinet, G. and Malaisse-Mousset, M., 1975. Litter fall and litter breakdown in Miombo. In: F.B. Golley and E. Medina (Editors). *Tropical Ecological Systems: Trends in Terrestrial and Aquatic Research.* Springer-Verlag, Berlin, pp. 137–152.

Martin, J.D., 1941. *Report on Forestry in Barotseland* (Baikiaea *Forest*). Government Printer, Lusaka.

Menaut, J.C., 1971. *Etude de quelques peuplements ligneux d'une savane guinéenne de Côte d'Ivoire.* Thesis, University of Paris, Paris, 141 p.

Menaut, J.C., 1977. Evolution of plots protected from fire since 13 years in a Guinea savanna of Ivory Coast. In: *Proc. 4th Int. Symp. Trop. Ecol., Panama*, 10 pp.

Menaut, J.C. and César, J., 1979. Structure and primary productivity of Lamto savannas (Ivory Coast). *Ecology*, 60: 1197–1210.

Michelmore, A.P.G., 1939. Observations on tropical African grasslands. *Ecology*, 27: 282–312.

Miège, J., 1955. Les savanes et forêts claires de Côte d'Ivoire. *Etud. Eburn.*, 4: 62–81.

Milne, G., 1947. Areal reconnaissance journey through parts of the Tanganyika Territory. *J. Ecol.*, 35: 192–265.

Monod, Th., 1938. Notes botaniques sur le Sahara occidental, ses confins sahéliens et remarques générales. *Mém. Soc. Bio-géogr.*, 6: 351–406.

Monod, Th., 1947. Notes biogéographiques sur l'Afrique de l'Ouest. *Port. Acta Biol.*, 2: 208–285.

Monod, T., 1957. Les grandes divisions chorologiques de l'Afrique. *C.S.A./C.C.T.A. Publ.*, 24: 147 pp.

Monod, T., 1963. Après Yangambi (1956): Notes de phytogéo-graphie africaine. *Bull. IFAN*, A 24: 594–655.

Monod, Th., 1971. Remarques sur les symétries floristiques des zones sèches Nord et Sud en Afrique. *Mitt. Bot. Staatssamml. München*, 10: 375–423.

Morison, C.G.T., Hoyle, A.C. and Hope-Simpson, J.F., 1948. Tropical soil vegetation catenas and mosaics: a study in the south western part of the Anglo-Egyptian Sudan. *J. Ecol.*, 36: 1–84.

Njoku, E., 1958. The photoperiodic response of some Nigerian plants. *J. W. Afr. Sci. Assoc.*, 4: 99–111.

Njoku, E., 1963. Seasonal periodicity in the growth and develop-ment of some forest trees in Nigeria. I. Observations on mature trees. *J. Ecol.*, 51: 617–624.

Njoku, E., 1964. Seasonal periodicity in the growth and develop-ment of some forest trees in Nigeria. II. Observations on seedlings. *J. Ecol.*, 52: 19–26.

Ohiagu, C.E. and Wood, T.G., 1979. Grass production and decomposition in Southern Guinea savanna, Nigeria. *Oeco-logia*, 40: 155–165.

Penning de Vries, F.W.T., 1978. *Résultats et perspectives du projet "Production primaire du sahel", esquisse mi-chemin.* Bamako, mimeographed report.

Perrier de la Bathie, H., 1921. La végétation malgache. *Ann. Mus. Colon. Marseille, 3e Sér.*, 9: 268 pp.

Perrier de la Bathie, H., 1936. *Biogéographie des plantes de Madagascar.* Soc. Ed. Géogr. Mar. Colon. Paris, 156 pp.

Phillips, J., 1965. Fire — as master and servant: its influence in the bioclimatic regions of Trans-Saharan Africa. In: *Proc. 4th Annual Tall Timbers Fire Ecology Conference*, 4: 7–109.

Polhill, R.M., 1968. Tanzania. *Acta Phytogeogr. Suec.*, 54: 166–178.

Poupon, M., 1977. Evolution d'un peuplement d'*Acacia senegal* dans une savane sahélienne au Sénégal de 1972 à 1976. *Cah. ORSTOM, Sér. Biol.* 12: 283–291.

Poupon, H., 1979. *Structure et dynamique de la strate ligneuse d'une steppe sahélienne au nord du Sénégal.* Thesis, University of Paris, Paris, 317 pp.

Poupon, M. and Bille, J.C., 1974. Recherches écologiques sur une savane sahélienne du Ferlo septentrional, Sénégal: Influence de la sécheresse sur la strate ligneuse. *Terre Vie*, 28: 49–75.

Pratt, D.J. and Gwynne, M.D., 1977. *Rangeland Management in East Africa.* Hodder and Stoughton, London, 310 pp.

Rattray, J.M., 1960. *The grass cover of Africa. FAO Agric. Stud. FAO Mem.*, No. 49: 168 pp.

Rattray, J.M., 1961. Vegetation types of Southern Rhodesia. *Kirkia*, 2: 68–93.

Raunkiaer, C., 1934. *The Life Forms of Plants and Statistical Geography.* Clarendon Press, Oxford, 632 pp.

Robyns, W., 1936. Contribution à l'étude des formations herbeuses du district forestier central du Congo Belge. *Mem. Inst. R. Colon. Belge, Sect. Sci. Nat. Med.*, 5(1).

Robyns, W., 1947. *Flore des Spermatophytes du Parc National*

Albert, I, Inst. Nat. Parcs Nat. Congo Belge, Brussels, 745 pp.

Robyns, W., 1948. *Flore des Spermatophytes du Parc National Albert, II,* Inst. Natl. Parcs Natl. Congo Belge, Brussels, 627 pp.

Rose Innes, R., 1977. *A Manual of Ghana Grasses.* Ministry of Overseas Development, London, 261 pp.

Rutherford, M.C., 1978. Primary production ecology in Southern Africa. In: M.J.A. Wenger (Editor), *Biogeography and Ecology of Southern Africa.* Junk, The Hague, pp. 621–659.

Rutherford, M.C., 1980. Field identification of roots of woody plants of the savanna ecosystem study area, Nylsvley. *Bothalia*, 13: 171–184.

Schimper, A.F.W., 1903. *Plant Geography upon a Physiological Basis.* Translation W.R. Fisher, Oxford, 839 pp.

Schmidt, W., 1975. Plant communities on permanent plots of the Serengeti plains. *Vegetatio*, 30: 133–145.

Schmitz, A., 1950. Principaux types de végétation forestière dans le Haut-Katanga. *C.R. Trav. Congr. Sci., Elizabethville*, 30 pp.

Schmitz, A., 1971. La végétation de la plaine de Lubumbashi (Haut-Katanga). *Publ. INEAC, Sér. Sci.*, 113: 388 pp.

Schnell, R., 1971. *Introduction à la phytogéographie des pays tropicaux, 1. Les flores — Les structures.* Gauthier-Villars, Paris, pp. 1–499.

Schnell, R., 1973. *Introduction à la phytogéographie des pays tropicaux, 2. Les milieux — les groupements végétaux.* Gauthier-Villars, Paris, pp. 500–951.

Schnell, R., 1977. *Introduction à la phytogéographie des pays tropicaux, 3, 4. La flore et la végétation de l'Afrique tropicale.* Gauthier-Villars, Paris, Vol. 3, 459 pp.; Vol. 4, 378 pp.

Shantz, M.L. and Marbut, C.F., 1923. The vegetation and soils of Africa. *Am. Geogr. Soc., Res. Ser.*, No. 13: 263 pp.

Sicot, A.M., 1978. *Cycle de l'eau et bilan hydrique dans les écosystèmes types du bassin versant de la Mare d'Oursi (Haute-Volta).* Rapport ORSTOM, Ouagadougou, 204 pp.

Sillans, R., 1958. *Les savanes de l'Afrique Centrale.* Lechevalier, Paris, 423 pp.

Smith, J., 1949. Distribution of tree species in the Sudan in relation to rainfall and soil texture. *Bull. Min. Agric., Khartoum*, 4.

Sobey, D.G., 1978. *Anogeissus* groves on abandoned village sites in the Mole National Park, Ghana. *Biotropica*, 10: 87–99.

Spichiger, R., 1977. Contribution à l'étude du contact entre flores sèche et humide sur les lisières des formations forestières humides semi-décidues du V baoulé et de son extension nord-ouest. *Bull. Liaison Chercheurs Lamto*, numéro spécial 1,261 pp.

Strang, R.M., 1966. The spread and establishment of *Brachystegia spiciformis* Benth. and *Julbernardia globiflora* (Benth.) Troupin in the Rhodesian highveld. *Com. For. Rev.*, 45: 253–256.

Strugnell, R.G. and Pigott, C.D., 1978. Biomass, shoot pro-duction and grazing of two grasslands in the Rwenzori National Park, Uganda. *J. Ecol.*, 66: 73–96.

Tchoumé, F.M., 1966. Morphologie et variation des formes

biologiques de quelques *Cissus* (Vitaceae) africains. *Mém. Soc. Bot. Fr.*, pp. 133–139.

Trapnell, C.G., 1959. Ecological results of woodland burning experiments in Northern Rhodesia. *J. Ecol.*, 47: 129–168.

Trochain, J.L., 1951. Nomenclature et classification des types de végétation en Afrique noire française (2ᵉ note). *Bull. Inst. Etud. Centrafr.*, 2: 9–18.

Trochain, J.L., 1957. Accord interafricain sur la définition des types de végétation de l'Afrique Tropicale. *Bull. Inst. Etud. Centrafr.*, 13–14: 55–93.

Trochain, J.L., 1980. *Ecologie de la zone intertropicale non désertique.* Université Paul Sabatier, Toulouse, 468 pp.

Trochain, J.L. and Koechlin, J., 1958. Les pâturages naturels du sud de l'A.E.F. *Bull. Inst. Etud. Centrafr.*, 15–16: 59–83.

Troll, C., 1968. The cordilleras of the tropical americas; aspects of climatic, phytogeographical and agrarian ecology. *Colloq. Geogr., Bonn*, 9: 15–56.

Troupin, G., 1966. Etude phytocénologique du Parc National de l'Akagera et du Rwanda oriental. *Inst. Natl. Rech. Sci., Butare*, No. 2: 293 pp.

Turrill, W.B. et al. (Editors), since 1952. *Flora of Tropical East Africa.* Crown Agents for the Overseas Governments and Administrations, London.

UNESCO, 1969. *A Framework for a Classification of World Vegetation.* UNESCO, Paris, 26 pp.

UNESCO, 1973. International classification and mapping of vegetation. *Ecol. Conserv.* 6: 93 pp.

Vesey-FitzGerald, D.F., 1963. Central African grasslands. *J. Ecol.*, 51: 243–273.

Volk, O.M., 1966. Die Florengebiete von Südwestafrika. *J. South West Afr. Sci. Soc.*, 20: 25–58.

Walter, H., 1939. Grassland, savanna und Busch der ariden Teile Afrikas in ihrer ökologischen Bedingtheit. *Jahrb. Wiss. Bot.*, 87: 750–860.

Walter, H., 1971. *Ecology of Tropical and Subtropical Vegetation.* Oliver and Boyd, Edinburgh, 539 pp.

Walter, H., 1973. *Vegetation of the Earth in Relation to Climate and the Eco-physiological Conditions.* Springer-Verlag, Berlin, 237 pp.

Walter, H. and Straka, H., 1970. *Arealkunde. Floristisch-historische Geobotanik. Einführung in die Phytologie, 3, Part 2,* Fischer-Verlag, Stuttgart.

Werger, M.J.A., 1978. Vegetation structure in the southern Kalahari. *J. Ecol.*, 66: 933–941.

White, F., 1965. The savanna woodlands of the Zambezian and Sudanian domains. *Webbia*, 19: 651–679.

White, F., 1968. Zambia. *Acta Phytogeogr. Suec.*, 54: 208–215.

White, F., 1971. The taxonomic and ecological basis of chorology. *Mitt. Bot. Staatsamml., München*, 10: 91–112.

White, F., 1976. The underground forests of Africa: a preliminary review. *Garden's Bull., Singapore*, 29: 57–71.

Wickens, G.E., 1976. The flora of Jebel Marra (Sudan Republic) and its geographical affinities. *Kew Bull. Add. Ser.*, 5: 1–368.

Chapter 7

INDIAN SAVANNAS

R. MISRA

INTRODUCTION

India is a vast country of savanna landscapes. Land-use practices under increasing human pressure (present average density $c.$ 200 km^{-2}) have replaced through millennia the ancient tropical subhumid and dry deciduous forests by savanna to the extent that hardly 10% of the land is covered by either relict forest or patches of tree plantation, and these are found in difficult precipitous hilly terrain only. An equal amount of the so-called forest lands are degraded to savannas. The farmlands (50% of the total area of the country) simulate savanna through the graminaceous crops (sugarcane, rice or maize in summer, and wheat, barley, legumes, etc., in winter) broken by planted fruit trees (*Aegle marmelos*, *Mangifera indica*, *Psidium guavava*, *Tamarindus indica*, *Zizyphus jujuba*, etc.), fuel trees (*Acacia arabica* and other *Acacia* species, *Casuarina equisetifolia* and others), shade trees (*Ficus religiosa* and other *Ficus* species; *Melia azadirachta*, *Madhuca indica* and others) and timber trees (*Albizia lebbeck*, *Dalbergia sissoo* and others). These tree species are raised around nucleate villages either in house compounds or in groves. Since the holdings are small and the villages are too near each other the country looks from an aeroplane like a vast savanna landscape.

The Indian subcontinent is entirely tropical except for the high altitude mountains — the Himalayas and the Nilgiris — which have temperate and arctic climates. The Indo-Gangetic Plain lying in the geographically subtropical region of northern India remains tropical in effect, on account of the high Himalayas shielding it from the northern cold wind and holding the monsoon and the southern warm winds within the country. Hence

less than 6% of the area of the country lying at altitudes higher than 1000 m remains extratropical.

Another remarkable phenomenon of Indian savanna vegetation is its physiognomic and floristic homogeneity throughout the country, on account of the human and biotic pressure which follows in intensity the humidity gradient. For instance, human and associated cattle populations increase in northern India from $c.$ 70 men per km^2 in the arid west to more than 700 in the humid east along a gradient of annual rainfall from 200 to more than 2000 mm.

PHYSIOGRAPHY AND CLIMATE

The Himalayas in the north and the Nilgiris in the southwest orient the winds and especially the monsoon in such a way as to produce forest vegetation all over the country by supplying enough moisture. Moreover, the rainfall induces strong seasonality in the climate. Peninsular India is based upon the Deccan plateau consisting of ancient Gondwanaland rocks with leached rocks and outcrops of Cretaceous basalt in central India. The hills are mostly flat-topped with the maximum elevation hardly rising above 1000 m save the Nilgiris (2500 m). This region is contained between lat. 8°N and the Tropic of Cancer over the Vindhyan Hills, and is drained by rivers flowing eastward into the Bay of Bengal and westward into the Arabian Sea.

In contrast to the Deccan plateau the north Indian plain and the tropical foothills of the outer Himalayas represent recent land (Pleistocene) created by the alluvial sediments of the bottom of the ancient Tethys Sea, made possible by silt brought down by the Indus, the Ganga (Ganges) and the Brahmaputra river systems cutting and eroding the still rising mountain ranges of the Himalayas. The Ganga Plain alluvium is shown at places to be more than 1650 m deep, consisting of alternating layers of sand, clay and stone pebbles.

The climate of India exhibits strong seasonality on account of the monsoon activity. The southwest monsoon of the Indian Ocean is active from June to September. It brings rain all over the country in varying intensity. The ocean wind unloads its moisture on the way as it travels inland and is cooled at higher altitudes when forced to rise against high mountains. Thus, the heaviest rain falls along the Nilgiris in the coastal plain of Kerala, in the northeastern part of India, and in the foothills of the eastern and central Himalayas. The rest of the country receives less rain in proportion to the distance from these centres, the minimum falling in the Thar Desert of Rajasthan (c. 200 mm). From October to December the wind direction is reversed from northwestern India to the southeast. This retreating southeast monsoon provides sporadic showers during otherwise dry months. But while turning in the Bay of Bengal towards the Tamil Nadu coast it has collected enough moisture to precipitate heavily in that area.

The monsoon pours rain into Kerala and northeastern India during the first week of June. It takes from fifteen days to one month to reach the extreme northwestern part of the country. Further, it tails off in September earlier in the northwest and continues till late in October or even till December in the northeastern part and the southern Kerala coast. Thus, the rainfall in these regions, besides being heavier, extends over a longer period, and so increases the span of the humid rainy season. The semi-arid regions of India experience a rainy season of three months, as against the seven months of the humid region. The monsoon accounts for more than 90% of the annual rainfall.

The relative humidity is high (50–90%) during the rainy season with the smallest diurnal fluctuation in temperature, which ranges between 24 and 36°C with an average of 29 to 30°C both at Varanasi in the north (lat. 25°20′ N) and Madras in the south (lat. 13°04′N). The fluctuation is much less and the average temperature is slightly lower in more humid regions of the country. The temperature similarity between the northern and the sou-

thern parts of the country during the dry summer months of April and May is maintained by the hot westerly wind known as *loo* which sweeps northern and central India. The mean temperature in May preceding the rainy season rises to 37.5°C with a mean maximum of 44.5°C. The Deccan plateau becomes comparatively cooler on account of breezes from the Indian Ocean.

The most contrasting temperature variation between the north and the south of the country is found in the cooler (winter) season extending from December to February. For instance, at Varanasi the mean maximum temperature for the season is from 25° to 31°C and the mean minimum from 11.7° to 18.1°C with a mean of 22.4°C. January is the coldest month and the absolute minimum temperature may go down to 6°C at Varanasi and 2 or 3°C in the semi-arid Rajasthan. Frost may occur in pockets. Madras temperatures in January, on the other hand, fluctuate between a mean minimum of 19.5° and a mean maximum of 29.6°C. Thus, it will be seen that temperature is at no time a critical factor in plant growth, but availability of moisture is so during the months of April to June. While trees and shrubs may build up tissue moisture storage to draw on at this time, the savanna grasses succumb to the moisture stress. The landscape at this time is characterized by dried-up grasses all over the country, except those on low-lying lands where the soil stores moisture.

K.C. Misra (1979) has recognized four ecoclimatic zones in tropical India. These are humid, moist subhumid, dry subhumid and semi-arid, with annual precipitation values of 1381, 1190, 942 and 492 mm respectively for the areas of grasslands studied under the International Biological Programme (I.B.P.). The average annual temperature for the semi-arid sites is 33°C against c 25° in the rest. The moisture indices calculated on Thornthwaite's (1948) formula for the four zones range from −22 to −33.

Literature on grassland in India has been based on analysis of the ground cover of the savannas, there being no exclusive grassland in India. The tree and the shrub elements on the Indian MAB (UNESCO Man and the Biosphere Programme) site have been only recently worked out (cf. K.P. Singh and Misra, 1978) along with the ground cover. However, data on trees are separately available which may help in computing tree savanna structure and function in the grassland study areas provided the distribution pattern by girth class is available.

CHARACTERISTICS OF SAVANNA VEGETATION

Overgrazing of savannas leads to their destruction through desertification and aridization. Bald hills and eroded land can be found everywhere in India. Eckholm (1976) recorded 43% of the Indian subcontinent as undergoing such erosion. Much of the savanna land has changed from "mesic" to "xeric" during the past four centuries. Moghul emperors used to hunt rhinoceros all over the Indo-Gangetic Plain where today one finds either dry deciduous thorn scrub or degraded grassland.

As one proceeds from the east to the west in northern India, say from Manipur to Rajasthan, the annual rainfall decreases from 3000 mm to 300 mm or less with the number of months of the dry season increasing from 3 to 9. The forests accordingly phase out in the east–west sequence of wet evergreen, moist mixed, moist deciduous, dry deciduous and dry evergreen types. The diversity, height and density of the trees also fall along this gradient, but the grass undergrowth follows a reverse trend.

The country has a very long human history. The early civilizations go back to the Harappa and Mohanjodaro periods ranging from 4000 to 2000 years B.C. Archaeological evidence including cave paintings indicates that as man navigated and moved along rivers fishing and hunting, and his number increased with the introduction of agriculture and animal husbandry, he cleared forest land for cultivation, grazing and settlement, wood for fire and timber, and burnt forest for trapping and eating wild animals. Thus, the once universal dense forests gave place to savanna formation, more easily and earlier in the drier western part of the country.

Degradation and destruction of the forests brought about savannization as the more vulnerable tree and shrub species succumbed to grazing of seedlings, lopping, felling, fire and other abuses by man. The thorny bushes and the hardy grasses, however, withstand them and very often provide a live hedge to some tree seedlings which ultimately characterize the savanna. It is evident that the trees, shrubs and grasses of the savanna are plastic to the climatic and biotic challenges. A mix of these species is found in the lower storeys of all the forests so that the floristic similarity among the savannas of India contrasts with the different types of forests from the humid evergreen to the semi-arid deciduous ones.

Soils of savanna

The savanna soils usually develop through calcification. In depressions and under canal irrigation salinity is most often found. Carbonates and sulphates of potassium and sodium preponderate over other salts. Raychaudhary (1966) has distinguished alluvial, black, red, lateritic, mountain and hill, arid and desert and saline and alkaline soils in India, and savanna occurs on all of them. The pH of the soil mostly varies from 6.5 to 9.0. It has a tendency to increase with depth. The texture of the soil is variable. The soils are generally poor in organic matter (0.09–4.7%), nitrogen (0.001–0.126%) and phosphorus (0.001–0.275%).

Life forms

The biological spectra of many grasslands have been worked out. Since the number of phanerophytes in savannas is limited to from 3 to 10% of the total number of species, it would be interesting to examine the biological spectra of the herbaceous and graminoid species of grazing lands in India as given by Yadava and Singh (1977). It has been observed that the percentages of cryptophytes and therophytes are higher and that of the hemicryptophytes is lower as compared to Raunkiaer's normal spectrum. This kind of spectrum reflects the high intensity of grazing in the Indian savannas. The therophytes are able to survive through seeds while the cryptophytes withstand grazing because of the hidden, subsurface position of their perennating buds.

Savanna types of India

Since all the tropical grasslands of India are savannas the literature on the former should be read by adding the tree and shrub component for each region in order to have a complete picture. A

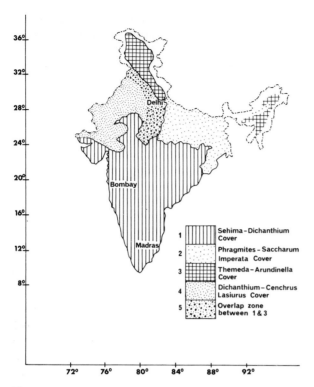

Fig. 7.1. Map of grass covers of India (after Dabadghao and Shankarnarayan, 1973)

broad classification of the Indian grasslands has been given by Dabadghao and Shankarnarayan (1973) as depicted in Fig. 7.1.

The *Sehima–Dichanthium* type (dry subhumid zone) covers peninsular India. It generally goes with the open thorn forests of southern India. The thorny bushes of the savanna are *Acacia catechu, Mimosa rubicaulis, Zizyphus* sp. and sometimes fleshy *Euphorbia*, along with low trees of *Anogeissus latifolia, Soymida febrifuga* and other deciduous species. It covers the central Indian plateau, the Chota Nagpur plateau and the Aravalli Range. The topography is undulating to hilly. Annual rainfall varies from 300 mm (Kutch) to 6350 mm (western Ghats) and the soil colour varies from pale-brown through brown to dark-grey. The floristic list includes 24 perennial grasses and 129 other herbaceous species of which 56 are legumes. The *Sehima* community is more prevalent on gravel soils as in the States of Gujarat, Maharashtra, Madhya Pradesh, Andhra Pradesh, Tamil Nadu and Karnataka. The cover of *Sehima* may be 87%. The *Dichanthium* community flourishes on level soil and may cover 80% of the ground.

The *Dichanthium–Cenchrus–Lasiurus* type (semi-arid zone) extends to the northern portion of Gujarat, Rajasthan (excluding Aravallis), western Uttar Pradesh, Delhi State and Punjab. The topography is broken by hill spurs and sand dunes. The annual rainfall ranges between 200 and about 700 mm on the eastern boundary. Eleven important perennial species of grass and 45 other herbaceous species (including 19 of Leguminosae) are listed. To this list may be added scattered shrubby growth of *Acacia senegal, Calotropis gigantea, Cassia auriculata, Prosopis spicigera, Salvadora oleoides* and *Zizyphus nummularia* which make the savanna look like scrub, from a distance.

The *Phragmites–Saccharum–Imperata* type (moist subhumid zone) covers the Ganga alluvial plain in northern India. The topography is level, low-lying and ill-drained. The rainfall may range from 1000 to 2000 mm. The cover consists of 19 principal grass species and 56 other herbaceous species including 16 legumes. *Bothriochloa pertusa, Cynodon dactylon* and *Dichanthium annulatum* are found in transition zones. The tree and shrub species commonly associated with the grasslands are: *Acacia arabica, Anogeissus latifolia, Butea monosperma, Phoenix sylvestris* and *Zizyphus nummularia* (in the palm savannas near the Sunderbans).

The *Themeda–Arundinella* grass cover extends to the humid montane regions and moist subhumid areas of Assam, Manipur, west Bengal, Uttar Pradesh, Punjab, Himachal Pradesh and Jammu and Kashmir. The savanna type is derived from the humid forests on account of shifting cultivation and sheep grazing. The tree and the bush elements are varied and numerous as given by Champion and Seth (1968) for the forest types.

The more common species of the savanna are listed in Appendix I (pp. 163–164).

Periodicity of savanna species

It will be seen that the Indian savannas are mostly derived from dry-deciduous forests and to some extent from subhumid deciduous forests which are estimated to have covered originally about 90% of the land area of India. The species constituting the savanna vegetation exhibit population fluxes due to succession, but are usually stabilized by continuing external forces of grazing, trampling, burning, lopping, browsing, shifting cultivation and timber removal. It is obvious that according to the time and intensity of the biotic and human pressure there would appear several stages of savanna particularly in regard to grassland/tree/shrub biomass ratios.

The majority of the grass and herb species flower, fruit and produce mature seeds during the rainy season. The rainy-season annuals and shoots of many perennial species dry up in October, and may remain as standing dead biomass. Decomposition of this matter may start even in this condition whenever there is a shower. The standing dead material is transformed into litter as it becomes weak and shatters by wind. Some winter-season annuals come up in October when there is a flush of tillering of the perennial grasses. These produce another crop of seeds. Singh (1967) grouped the species at Varanasi as: (1) Completing life cycle in one season, e.g. *Alloteropsis cimicina*, *Fimbristylis schoenoides* and *Lindernia nummulariaefolia*; (2) completing life cycle in one season but recurring in the next season also, e.g. *Eragrostis tenella*, *Panicum humile* and *Paspalum royleanum*; (3) germinating, flowering and fruiting in one season and completing seed maturation in the next season, e.g. *Trichodesma indicum* and *Volutarella divaricata*; (4) having two or more flowering and fruiting flushes in a year, e.g. *Dichanthium annulatum* and *Desmodium triflorum*; and (5) with one flowering and fruiting flush each year, although the period of flowering

may cover two seasons, e.g. *Evolvulus alsinoides* and *Indigofera ennaephylla*. It may be noted that overgrazing leads either to elimination of the perennial species when the annuals build up their populations or to an annual behaviour of the perennial species. For instance *Dichanthium annulatum* and *Heteropogon contortus* may persist in the field through the seed crops produced at the end of the rainy season. These germinate each year in July.

J.S. Singh (1967) has described the phenology of the *Dichanthium* community at Varanasi in some detail. Large number of seedlings of both perennials and annuals come up each year in early July. Simultaneously fresh sprouts emerge from rhizomes and rootstocks perennating in the soil. Active vegetative growth follows, so that by the middle of July the ground is filled up with seedlings and sprouts which compete for space during growth. Cattle which were languishing for want of forage till June are let loose for grazing, but the rate of growth of the vegetation exceeds its removal. Most of the species start flowering in the month of August and fruiting is common in September. It may be noted that many grass seeds show staggered germination due to different concentrations of chemical inhibitors which leach out gradually. By September the maximum number of grass tillers are produced (up to 10 000 m^{-2}; see Table 7.1). However, the density of different species fluctuates during the season.

Many grasses such as *Bothriochloa*, *Dichanthium*

TABLE 7.1

Seasonal variations in total plant density (shoots m^{-2}) in four tropical grasslands in India

Month	1 Kurukshetra (1970–71)	2 Varanasi (1964–65)	3 Ujjain (1971–72)	4 Ratlam (1971–72)
May	931	–	–	–
June	1490	4024	142	1213
July	1552	8214	344	2506
August	2041	9874	572	3934
September	2143	11 055	822	4700
October	1408	9312	862	1271
November	830	1997	376	1112
December	471	2857	104	983
January	564	3642	367	460
February	780	2196	555	89
March	817	3242	419	55
April	950	2623	267	44
May	1217	1963	255	21

and *Saccharum* grow in tussocks. In such cases tillering without much growth in length may continue till March or April. Nevertheless, most of the ground vegetation, especially on the uplands, dries up in October and November leaving the young growth hidden within the brown and dry tussock. Very often graziers set fire to the dead standing mass, which is too rough to be grazed. Active tillering with depauperate shoots follows such fires, so that protein-rich young shoots become available for grazing.

When the forests degrade into savanna the taller trees disappear first. The second and the third storeys consist of low trees and shrubs, many of which participate in savanna formation. These appear to be preadapted to xeric conditions. But the majority of them are deciduous. The leaves fall in the spring season (March–April), but young leaves come out early in some species and late in others. Their active growth period is again the rainy season, though smaller flushes of growth with diminutive foliage may continue in the dry seasons of winter and summer. The seeds in all cases germinate in July with the onset of rains. But most of the seedlings are eaten up by goat, sheep and cattle so that the population of trees and shrubs is reduced in the savanna.

Canopy architecture

Yadava and Singh (1977) have studied the community architecture for the grasslands of Kurukshetra (moist to dry subhumid) and given quantitative features of the canopy such as density, diversity, vertical projection, layer structure, leaf area and pigment concentration. Their conclusions are that the density of the community as a whole is highly correlated with the aerial biomass though not for individual species, but the community is not so dense as to be saturated. Similar results have been reported by Shankarnarayan et al. (1969) and by Mall et al. (1973a) with regard to height, basal area and biomass. Billore (1973) has shown a positive correlation between leaf area index and chlorophyll content for *Sehima* grassland near Ujjain (dry subhumid). Billore and Mall (1975) further showed that chlorophyll content is more highly correlated with production. Chlorophyll content varied from 0.56 mg m^{-2} to 1369.79 mg m^{-2} in different stands and in different months.

Kumar and Joshi (1972) recorded lower values (0.11–1.08 g m^{-2}) for grasslands at Pilani (semi-arid).

Yadava and Singh (1977) stated that "In a strongly seasonal climate with the growing conditions varying markedly across the year, the growth behaviour of principal species is adjusted so as to result in year-long community production". J.S. Singh (1968) has grouped various important perennials of the dry subhumid grassland of Varanasi as: (1) rainy season peak growth, with a sudden decline in winter (e.g. *Bothriochloa pertusa*), or a gradual decline (e.g. *Indigofera ennaephylla*); (2) continued growth from rainy to winter season and steady decline thereafter (e.g. *Desmodium triflorum*); and (3) two peak growing periods, one in the rainy and the other at the winter–summer interphase (e.g. *Volvulopsis nummularia*), and those with maximum growth in summer (e.g. *Convolvulus pluricaulis*).

In the case of grasses the biomass is more concentrated at the base. In some forbs and shrubs the biomass is maximum at mid-height. In fact the canopies of individual species are adjusted in the community to constitute a multilayered architecture for maximizing production under the stresses of climate and grazing. G. Misra and Singh (1978) have further shown that all the tropical savanna grasses in India are C_4 species, so that they are equipped for better performance under stresses of moisture, light intensity and carbon dioxide concentration.

J.S. Singh and Misra (1969) have discussed the relationship between diversity, dominance, stability and net production in the Varanasi grassland. They have reported an inverse relationship between dominance and diversity, net production and dominance, and stability and net production in the grassland, which is in continuous flux under the stresses of seasonal grazing and community successions and degradations.

K.P. Singh and Misra (1979) have reported the total biomass of a *Zizyphus* shrub savanna stand as 8891 kg ha^{-1}, the contributions being 29% by the shrub, 58.9% by the herbs and 11.9% by the litter. Further, the underground biomass constitutes about 33% of the total biomass. The shrub leaf area index is found to be 0.059 to 0.067 against 3.63 to 4.12 for the herb layer, both attaining peak values in September or October.

PRODUCTION AND NUTRIENT CYCLING

Productivity studies

Some of the recent data generated through the MAB 1 (cf. K.P. Singh and Misra, 1978) project will be discussed first as a bench-mark, for comparing with the I.B.P. data published elsewhere. The MAB site is the Chandraprabha sanctuary at Varanasi on the Vindhyan upland, where grazing has reduced the dry deciduous mixed forest to scrub and savanna in parts of the sanctuary during the recent past, so that a gradient of degeneration stages can be examined. Fig. 7.2 shows the structure of the forest during foliage and naked phases. Fig. 7.3 is that of a contiguous shrub savanna stand protected from grazing for two years, and Fig. 7.4 is that of a nearby savanna open to grazing throughout the year.

Table 7.2 gives the physical and chemical properties of the soils under different stands of the sanctuary, which lie on undulating rocky terrain with thin soil. The soil conditions appear to be uniform in their characters.

There happen to lie nearby two villages with 21 and 36 households of 99 and 276 persons, respectively. A few families hold 0.5 to 3 ha of cultivated land each, where paddy, barley and wheat are grown. The cattle population of cows (420), bullocks (83), buffaloes (62), goats and sheep (337) are equivalent to 700 cow units, assuming 2.5 goats or sheep as equivalent to 1 cow unit for herbage consumption. However, besides the local pressure during the year migratory people and cattle arrive and camp around these villages. In one locality five groups of graziers comprising a total of about 250 persons and 2750 cattle (75% of which were cows) were noted during the rainy season of

Fig. 7.2. A. Typical dry deciduous mixed forest of *Anogeissus latifolia* and *Diospyros melanoxylon* with full canopy in September, 1976. B. Same as in A after leaf fall in April 1977.

Fig. 7.3. Savanna stand (protected for two years), August 1976. *Heteropogon contortus* and *Bothriochloa pertusa* tussocks in the foreground and *Zizyphus jujuba* shrub in the background.

Fig. 7.4. Savanna open to grazing, August 1976. *Desmostachya bipinnata* and *Zizyphus jujuba* (small tree).

TABLE 7.2

Physical and chemical properties of the surface soils under savanna and forest in Chandraprabha sanctuary

Properties	Savanna	Forest
Colour (Muenchels' Chart)	reddish brown (2.5YR 5/4)	reddish brown (5YR 4/4)
Texture	sandy loam	sandy loam
Mechanical composition (%)		
clay	12.8–15.1	13.3–16.5
silt	27.9–30.1	26.7–30.2
sand	52.0–56.0	50.4–54.8
Bulk density (g cm^{-3})	1.34–1.42	1.23–1.37
Water-holding capacity (%)	40–46	43–49
Field capacity (%)	27–32	29–34
pH	6.6–7.2	6.6–7.2
Organic matter (%)	1.6–1.8	1.6–2.1
Total nitrogen (%)	0.05–0.07	0.06–0.09

TABLE 7.3

Vegetation composition of savanna stands protected for two years and grazed (F=frequency; D=density, number m^{-2}; BA=basal area, cm^2 m^{-2}; IVI=importance value index)

Species	Protected savanna				Grazed savanna			
	F	D	BA	IVI	F	D	BA	IVI
Rainy season								
Heteropogon contortus	54	137.5	125.09	52.09	12	0.8	0.006	1.36
Bothriochloa pertusa	74	153.4	110.10	51.41	36	10.7	4.38	9.24
Desmostachys bipinnata	–	–	–	–	92	110.8	153.36	133.34
Minor species		746.5	134.33	196.50		201.7	11.48	156.06
Total		1037.6	369.52	300		324.1	169.226	300
Winter season								
Heteropogon contortus	46	69.2	62.61	58.21	16	1.7	0.26	3.07
Bothriochloa pertusa	50	75.0	52.44	53.98	46	5.7	11.60	10.40
Desmostachya bipinnata	–	–	–	–	88	61.5	883.5	142.81
Minor species		339.0	53.76	187.81		93.9	137.21	143.72
Total		483.3	168.81	300		163.0	932.57	300
Summer season								
Heteropogon contortus	16	24.7	47.98	47.15	6	0.1	0.008	2.29
Bothriochloa pertusa	48	32.0	31.74	63.63	30	4.9	8.71	23.02
Desmostachya bipinnata	–	–	–	–	80	28.9	259.45	179.01
Minor species	–	146.2	22.03	189.22		15.1	7.87	95.68
Total		202.9	101.75	300		49.1	276.038	300

1977. Thus between July and January the human population is doubled and the cattle numbers increase about fourfold. After the return of the graziers, another inflow of people (about 400) is observed during March and April for plucking young leaves of the tree *Diospyros melanoxylon*, which are used for manufacturing country cigarettes as a village industry. The average grazing pressure as calculated from the forestry records is 0.36 cow unit per hectare.

The vegetation of the two savanna stands investigated in the area (cf. Figs. 7.3 and 7.4) consists of *Zizyphus jujuba* (shrub to small tree 3–4 m) and herbs (0.9 m). The protected stand is dominated by *Heteropogon controtus* and *Bothriochloa pertusa*, and the grazed one by *Desmostachya bipinnata*. Table 7.3 gives phytosociological descriptions of the two stands.

Table 7.4 gives the standing crop biomass in the savanna stands for different fractions. The leaf area index is minimum in the month of May and rises from 0.012 to the maximum of 3.92 in the month of September. The maximum amount of chlorophyll and carotenoid pigments in different life forms of

TABLE 7.4

Standing crop biomass (kg ha^{-1}) in savanna stands

Components	Protected stand		Grazed stand	
	weight	%	weight	%
Shrubs				
leaf	151.9	5.9	133.1	6.8
branch	1127.8	43.7	701.4	35.8
bole	565.8	22.0	498.0	25.4
root	732.8	28.4	628.6	32.0
Total	2578.4	100	1961.1	100
Herbs				
above-ground	3464.4	71.6	2556.9	64.2
below-ground	1373.1	28.4	1423.1	35.8
total	4837.5	100	3980.0	100
Litter				
shrub	41.6	2.8	32.9	4.0
herb	1434.0	97.2	777.0	96.0
total	1475.6	100	809.9	100
Stand total	8891.5		6751.0	
Stand live	5309.9	71.6	3889.4	65.5
above-ground	5309.9	71.6	3889.4	65.5
below-ground	2105.9	28.4	2051.7	34.5
total	7415.8	100	5941.1	100

the two stands (August–October) is set out in Table 7.5, from which it is seen that the carotenoids constitute one-fourth of the total pigments and that chlorophyll *b* is about one-third. The total chlorophyll content is roughly 20.6 kg ha^{-1} in the protected and 13.4 kg ha^{-1} in the grazed savannas, with the herbs contributing more than 95% of the total.

A compartment model of the biomass and energy dynamics for shrub and herb layers in the protected savanna stand is given in Fig. 7.5. The stand annually elaborates 13 183 kg ha^{-1} (or 58 905 × 10^3 kcal) total net production. The gross production would amount to 1.33 times this figure, as respiration loss by the grass is shown by Dwivedi (1970) to be one-third of the net production. On this basis the savanna shows energy utilization of 1.2% per annum. The grazed stand seems to run on two-thirds of this energy.

Of the total litter energy, 88% is dissipated during the year. This loss amounts to 52.3% of the total net production. Total energy dissipation is 43% of the total energy stored during the year. The net increment of biomass is 6278 kg ha^{-1} yr^{-1}.

It is but natural that corresponding figures of annual production, water use efficiency, and turnover generated earlier for savanna or grasslands through I.B.P. activities differ widely, depending on the soil moisture conditions of the sites situated with

varying depth of soil on varying topography even though with comparable annual rainfall.

The turnover of underground plant material seems to be quite rapid. Thus, J.S. Singh and Yadava (1973) have reported an annual rate of 97% for the dry subhumid grassland at Kurukshetra, and calculations based on the data of Kumar and Joshi (1972) indicate an average annual turnover rate of 93% for the semi-arid grassland at Pilani.

J.S. Singh and Joshi (1979) have presented an analysis of Indian grassland data. They concluded that, contrary to expectation, there is no significant relationship between above-ground net production and annual rainfall. They also calculated water use efficiency of the grasslands as the ratio of annual above-ground net production (g m^{-2}) to annual rainfall (mm). This ratio according to them ranges between 0.1 and 4.0 for the sites in India, and declines with higher rainfall. The Vananasi MAB savanna site yields a ratio of 4.0.

Discussing compartment transfers, J.S. Singh and Joshi (1979) have shown that net accumulation of dry matter in the underground parts is twice as much as in the shoots in the semi-arid grassland of Pilani, but the reverse is true for Kurukshetra (subhumid) grassland. In the moist subhumid grassland of Sagar the below- and above-ground parts share annual dry matter accumulation equally. The transfer from the standing dead to the litter

TABLE 7.5

Maximum amount of chlorophyll and carotenoid pigments in different life forms (mg g^{-1} dry wt), and in stands (g ha^{-1}) of savanna (p = protected; g = grazed)

Life form and stand	Chlorophyll *a*	Chlorophyll *b*	Total chlorophyll	Carotenoids
Life forms				
shrub S$_p$	3.08	1.17	4.25	1.21
S$_g$	3.10	1.18	4.28	1.17
herb S$_p$	2.47	1.39	3.86	1.13
S$_g$	2.06	1.35	3.41	1.05
Stand S$_p$				
shrub	467.8	177.7	645.5	183.8
herb	12 809.6	7208.6	20 018.3	5860.2
total	13 277.5	7386.4	20 663.9	6044.0
Stand S$_g$				
shrub	412.6	157.0	569.6	155.7
herb	7783.5	5100.8	12 884.3	3967.3
total	8196.1	5257.9	13 454.0	4123.0

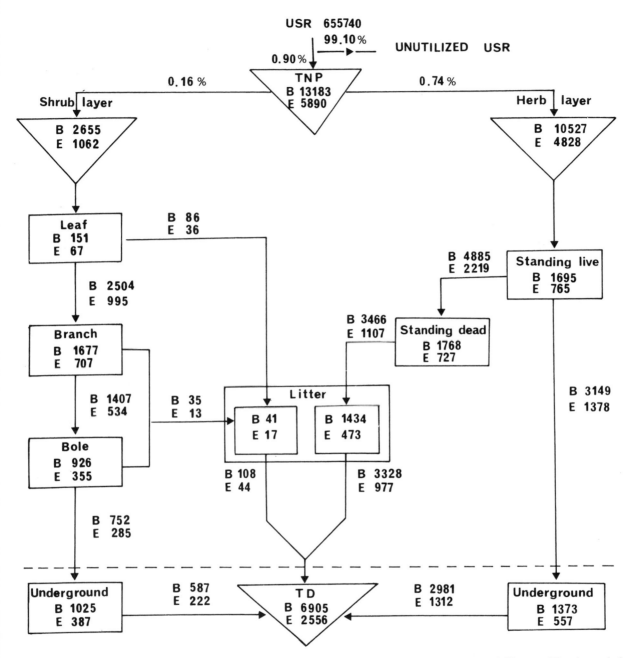

Fig. 7.5. Compartment model showing distribution of biomass and energy in the protected savanna stand. Biomass (B) units are in kg ha^{-1} for compartments, and in kg ha^{-1} yr^{-1} for flows between compartments. Corresponding values for energy (E) are in 10^4 kcal ha^{-1} yr^{-1}. Flow of usable solar radiation (USR) shows percents of energy captured and unused. TNP represents total net production and TD represents total disappearance in stand.

compartment in both semi-arid and moist sub-humid grasslands is faster than, or at par with, the accumulation of standing dead material; the reverse is true for the subhumid grassland. They further observed that, in comparatively dry grasslands, from 65 to 97% of the annual production is channelled underground, while the corresponding figure is only 26 to 47% in regions of higher rainfall.

J.S. Singh and Misra (1969) have shown that grazing creates relatively open canopies, thus making the invasion of annuals and other aliens possible and increasing plant diversity. Only coarse and unpalatable grasses remain, and the new arrivals are usually thorny species.

Besides grazing, harvesting of the biomass close to the soil surface is a common practice in India. J.S. Singh (1968) reported that harvesting of grasslands near the villages removed from 87 to 89% of the biomass. Cattle grazing usually removes 15 to 51% of the herbage in the same fields. However, it must be noted that harvesting is practised in the dry seasons only, and grazing efficiency is higher during the rainy season.

Nutrient cycling

The uptake, transfer and release of nutrients have been studied in many of the Indian grasslands. Most studies are, however, limited to nitrogen, phosphorus and potassium. Table 7.6 gives data for the standing quantities and cycling of these nutrients for forest and savanna stands of the MAB site. It will be seen that 19.7 to 20.5% of the total nitrogen in the forest system occurs in plant material, whereas in savanna the corresponding figure is 8%. For phosphorus the respective proportions are 29.3 to 33.3% and 5 to 7.4%; for potassium, they are 55.0 to 56.6% and 28.6 to 35.0% respectively for forest and savanna. The rest of the nutrients are held in the soil. It will further be seen that nutrient uptake in savanna is consistently greater than that in the forest, and that the savanna releases much greater amount of nutrients from the vegetation and the litter.

The uptake of nitrogen by the savanna is 107.9 to 207.8 kg ha^{-1} yr^{-1}. A corresponding figure reported by J.S. Singh et al. (1979) for savannas of

TABLE 7.6

Standing quantities and cycling of three nutrients in different stands (p = protected; g = grazed)

Stand forest (F) and savanna (S)	Standing quantities (kg ha^{-1})			Cycling (kg ha^{-1} yr^{-1})		
	plant material	soil (to 30 cm)	total	uptake	release	retention[1]
Nitrogen						
F$_p$	729.7	2970	3699	154.3	91.9	62.4
F$_g$	737.9	2843	3580	160.3	88.4	71.9
S$_p$	96.9	2387	2484	207.8	179.0	28.8
S$_g$	79.4	2384	2463	107.9	96.8	11.1
Phosphorus						
F$_p$	59.6	119	179	12.1	7.0	5.0
F$_g$	55.2	138	188	11.4	6.8	4.6
S$_p$	9.2	114	123	20.4	17.9	2.5
S$_g$	7.2	133	140	10.6	9.5	1.1
Potassium						
F$_p$	482.6	373	855	91.1	41.7	44.4
F$_g$	454.6	380	835	89.0	44.7	44.4
S$_p$	96.3	160	246	179.5	156.4	23.1
S$_g$	68.9	160	223	85.9	74.4	11.5

[1] Retention is the change in standing quantity; uptake minus retention is release. Release includes leaching losses.

the dry subhumid zone is 2.56 kg ha^{-1} yr^{-1}, which appears to be an under-estimate. The rate of uptake is maximum in the rainy season (56–71% of the annual total). However, the rate of release back to the soil is maximum during the winter (47–64% of the annual transfer). The savanna has, thus, a faster nutrient cycling and the turnover rate, as compared to the original forest stands on the MAB site, is double for nitrogen, almost three times as great for phosphorus, and nearly seven times as great for potassium.

It will be further seen from Table 7.7 that for unit energy storage the nutrients are taken up in the order of nitrogen > potassium > phosphorus. The energy requirement for nitrogen uptake is ten times higher than that of phosphorus. The forest stands pump in lower amounts of all the three nutrients for each unit of energy stored in the plant system. J.S. Singh et al. (1979) have observed that the savannas of dry regions appear to need less nitrogen to support the same magnitude of energy flow than the grasslands of humid regions.

No nutrient picture of the ecosystem would be complete without taking into consideration the input and output through rain water. Table 7.8

gives figures for such losses and gains for the MAB savanna site. Data on calcium and sodium are also added to those on nitrogen, phosphorus and potassium of the stand. The nutrient balance indicates a net gain of all the listed nutrients save calcium, which is lost to the extent of about 3 g m^{-2} yr^{-1} through run-off.

CONCLUSION

It is apparent from a review of Indian literature on savannas that much more work is needed to understand their structural and dynamic relationships under the abiotic and biotic stresses in relation to the total resilience of the system. Soil erosion following loss of vegetative cover may bring about total rupture, just as complete protection may lead to forest development. Relationships with moisture, nutrients and photosynthesis are the basic directions along which further work may be pursued.

TABLE 7.7

Nutrient uptake and energy flow relationships in different stands [values represent the ratio of nutrient uptake (mg) to energy stored (kcal. 10^3)] (p = protected; g = grazed)

Stand	Nitrogen	Phosphorus	Potassium
F_p	2900	228	1710
F_g	3100	222	1720
S_p	3520	346	3040
S_g	3050	300	2430

TABLE 7.8

Amounts of nutrients (mg m^{-2}), annually added through rain and lost through run-off, in savanna

Nutrients	Rain input	Run-off loss	Stand gain or loss
Ca	1513	4503	−2990
Na	2818	1815	+1003
K	3980	1801	+2179
NO$_3$-N	918	336	+582
Org-N	1418	554	+864
PO$_4$-P	421	99	+322
Total P	921	301	+620

APPENDIX 1: LIST OF SPECIES RECORDED FROM THE MAB RESEARCH SAVANNA STANDS IN THE CHANDRAPRABHA SANCTUARY

Species	Family
TREES	
Acacia catechu	Fabaceae
Aegle marmelos	Rutaceae
Anogeissus latifolia	Combretaceae
Bassia latifolia	Sapotaceae
Bauhinia tomentosa	Fabaceae
Bombax malabaricum	Malvaceae
Boswellia serrata	Burseraceae
Bridelia retusa	Euphorbiaceae
Buchanania lanzan	Anacardiaceae
Cassia fistula	Fabaceae
Diospyros melanoxylon	Ebenaceae
Eriolaena hockeriana	Sterculiaceae
Flacourtia ramontchi	Bixineae
Gardenia turgida	Rubiaceae
Grewia tiliaefolia	Tiliaceae
Lagerstroemia parviflora	Lythraceae
Lannea grandis	Anacardiaceae
Ougeinia dalbergioides	Fabaceae
Phyllanthus amblica	Euphorbiaceae
Pterocarpus marsupium	Fabaceae
Saccopetalum tomentosum	Annonaceae

APPENDIX I (continued)

Species	Family
Schleichera trijuga	Sapindaceae
Soymida febrifuga	Meliaceae
Stereospermum suaveolens	Bignoniaceae
Tectona grandis	Verbenaceae

SHRUBS

Holarrhena antidysenterica	Apocyanaceae
Mimosa himalayana	Fabaceae
Nyctanthes arbor-tristis	Oleaceae
Zizyphus glaberrima	Rhamnaceae
Zizyphus jujuba	Rhamnaceae

HERBS

Achyranthus aspera	Amarantaceae
Alysicarpus monilifer	Fabaceae
Aristida adscensionis	Poaceae
Bonnaya brachiata	Scrophulariaceae
Borreria hispida	Rubiaceae
Bothriochloa pertusa	Poaceae
Cassia tora	Fabaceae
Chloris barbata	Poaceae
Chloris incompleta	Poaceae
Convolvulus pluricaulis	Convolvulaceae
Corchorus acutangulus	Tiliaceae
Cyperus aristatus	Cyperaceae
Dactyloctenium aegypticum	Poaceae
Desmodium triflorum	Fabaceae
Digitaria bifasciculata	Poaceae
Digitaria sanguinalis	Poaceae
Eragrostis tenella	Poaceae
Eragrostis uniloides	Poaceae
Euphorbia hirta	Euphorbiaceae
Evolvulus alsinoides	Convolvulaceae
Heteropogon contortus	Poaceae
Kyllinga triceps	Cyperaceae
Oplismenus burmannii	Poaceae
Paspalidium flavidum	Poaceae
Paspalum scorbiculatum	Poaceae
Peristrophe bicalyculata	Acanthaceae
Rungia repens	Acanthaceae
Setaria glauca	Poaceae
Sida acuta	Malvaceae
Tridax procumbens	Asteraceae
Triumfetta neglecta	Tiliaceae
Vandellia crustacea	Scrophulariaceae
Vernonia cinerea	Asteraceae
Zornia diphylla	Fabaceae

REFERENCES

Ambasht, R.S., Maurya, A.N. and Singh, U.N., 1972. Primary production and turnover in certain protected grasslands of Varanasi, India. In: P.M. Golley and F.B. Golley (Editors), Papers from a Symposium on Tropical Ecology with an Emphasis on Organic Production. Institute of Ecology, University of Georgia, Athens, Ga., pp. 43–50.

Billore, S.K., 1973. Net Primary Production and Energetics of a Grassland Ecosystem at Ratlam, India. Thesis, Vikram University, Ujjain, 401 pp.

Billore, S.K. and Mall, L.P., 1975. Chlorophyll content as an ecological index of dry matter production. J. Indian Bot. Soc., 54: 75–77.

Chakravarty, A.K., 1968. Role of some perennial grasses of western Rajasthan in development of grasslands of the arid zone of India. In: R. Misra and B. Gopal (Editors), Proc. Symp. Recent Adv. Trop. Ecol. II. International Society for Tropical Ecology, Varanasi, pp. 620–630.

Champion, H.G. and Seth, S.K., 1968. The Forest Types of India: A Revised Survey. Manager of Publications, New Delhi, 404 pp.

Chaudhary, V.B., 1967. Seasonal Variation in Standing Crop and Energetics of Dichanthium annulatum Grassland at Varanasi. Thesis, Banaras Hindu University, Varanasi, 89 pp.

Chaudhary, V.B., 1972. Seasonal variation in standing crop and net aboveground production in Dichanthium annulatum grassland at Varanasi. In: P.M. Golley and F.B. Golley (Editors), Papers from a Symposium on Tropical Ecology with an Emphasis on Organic Production. Institute of Ecology, University of Georgia, Athens, Ga., pp. 51–57.

Coupland, R.T., 1979. Conclusion. In: R.T. Coupland (Editor), Grassland Ecosystems of the World. Analyses of Grasslands and their Uses. Cambridge University Press, Cambridge, pp. 335–355.

Dabadghao, P.M. and Shankarnarayan, K.A., 1973. The Grass Cover of India. Indian Council of Agricultural Research, New Delhi, 713 pp.

Dwivedi, R.S., 1970. A Comparative Study of Energetics and Cycling of Phosphorus Uptake ^{32}P in Wheat (T. aestivum L.) and Marvel Grass (D. annulatum (Forst). Stapf.) Thesis, Banaras Hindu University, Varanasi.

Eckholm, E.P., 1976. Losing Ground. Norton and Co., New York, N.Y., 223 pp.

Gill, J.S., 1975. Herbage Dynamics and Seasonality of Primary Productivity at Pilani, Rajasthan. Thesis, Birla Institute of Technology and Science, Pilani, 291 pp.

Gillon, Y. and Gillon, D., 1973. Recherches écologiques sur une savanna sahélienne du Ferlo Septentrional, Sénégal: Données quantitatives sur les Arthropodes. Terre Vie, 27: 297–323.

Gupta, R.K., Saxena, S.K. and Sharma, S.K., 1972. Aboveground productivity of grassland at Jodhpur, India. In: P.M. Golley and F.B. Golley (Editors), Papers from a Symposium on Tropical Ecology with an Emphasis on Organic Production. Institute of Ecology, University of Georgia, Athens, Ga., pp. 75–93.

Gupta, S.K., 1973. Influence of stage and height of cutting on photosynthetic and non-photosynthetic net primary production of Dichanthium annulatum. Trop. Ecol., 14: 173–181.

Hills, T.L. and Randall, R.E. (Editors), 1968. The Ecology of the Forest/Savanna Boundary. Proc. of the I.G.U. Humid Tropics Comm. Symp. Venezuela, 1964. Department of Geography, McGill University, Montreal, Ont., 129 pp.

Jain, H.K., 1971. *Production Studies in Some Grasslands of Sagar*. Thesis, Sagar University, Sagar, 186 pp.

Kumar, A., 1971. *Structure and Net Primary Community Production in the Terrestrial Herbaceous Vegetation at Pilani (Rajasthan) with Special Reference to Grasses*. Thesis, Birla Institute of Technology and Sciences, Pilani, 215 pp.

Kumar, A. and Joshi, M.C., 1972. The effect of grazing on the structure and productivity of vegetation near Pilani, Rajasthan, India. *J. Ecol.*, 60: 665–675.

Mall, L.P. and Billore, S.K., 1974a. Production relations with density and basal area in a grassland community. *Geobios*, 1: 84–86.

Mall, L.P. and Billore, S.K., 1974b. Seasonal variation in leaf area index of a grassland community. *Proc. Indian Natl. Sci. Acad.*, 40B: 430–432.

Mall, L.P. and Billore, S.K., 1974c. Dry matter structure and its dynamics in *Sehima* grassland community. I. Dry matter structure. *Trop. Ecol.*, 14: 105–118.

Mall, L.P. and Billore, S.K., 1974d. An indirect estimation of litter disappearance in grassland study. *Sci. Cult.*, 43: 506–507.

Mall, L.P., Billore, S.K., and Misra, C.M., 1973a. A study of the community chlorophyll content with reference to height and dry weight. *Trop. Ecol.*, 19: 81–83.

Mall, L.P., Misra, C.M. and Billore, S.K., 1973b. *Primary Productivity of Grassland Ecosystems at Ujjain and Ratlam Districts of Madhya Pradesh*. Progress Report, Indian National Science Academy Project. School of Studies in Botany, Vikram University, Ujjain, 30 pp.

Mall, L.P., Billore, S.K. and Misra, C.M., 1974. Relation of ecological efficiency with photosynthetic structure in some herbaceous stands. *Trop. Ecol.*, 15: 39–42.

Misra, C.M., 1973. *Primary Productivity of a Grassland Ecosystem at Ujjain*. Thesis, Vikram University, Ujjain, 172 pp.

Misra, C.M. and Mall, L.P., 1974. Energy storage and transfer in a tropical grassland community. *Trop. Ecol.*, 15: 22–27.

Misra, C.M. and Mall, L.P., 1975a. Architecture and standing biomass of a grassland community. *Ann. Arid Zone*, 141: 119–123.

Misra, C.M. and Mall, L.P., 1975b. Production and compartmental transfer of dry matter in a tropical grassland community. *Proc. Indian Natl. Sci. Acad.*, 41B: 452–456.

Misra, C.M. and Mall, L.P., 1975c. Photosynthetic structure and standing biomass of a grassland community. *Ann. Arid Zone*, 16: 76–80.

Misra, G. and Singh, K.P., 1978. Some aspects of physiological ecology of C_3 and C_4 grasses. In: J.S. Singh and B. Gopal (Editors), *Glimpses of Ecology*. Int. Sci. Publ., Jaipur. pp. 201–206.

Misra, K.C., 1979. Introduction. In: R.T. Coupland (Editor), *Grassland Ecosystems of the world: Analyses of Grasslands and Their Uses*. Cambridge University Press, Cambridge, pp. 187–196.

Misra, R., 1959. The status of the plant community in the upper Gangetic plain. *J. Indian Bot. Soc.*, 38: 1–7.

Misra, R., 1972. A comparative study of net primary productivity of dry deciduous forest and grassland of Varanasi, India. In: P.M. Golley and F.B. Golley (Editors), *Papers from a Symposium on Tropical Ecology with an Emphasis on Organic Production*. Institute of Ecology, University of Georgia, Athens, Ga., pp. 279–293.

Naik, M.L., 1973. *Ecological Studies on Some Grasslands of Ambikapur*. Thesis, Sagar University, Sagar, 180 pp.

Pandey, A.N., 1971. *Effect of Burning on* Dichanthium annulatum *Grassland*. Thesis, Banaras Hindu University, Varanasi, 120 pp.

Pandey, A.N., 1974. Short-term effect of burning on the aboveground production of *Dichanthium annulatum* grassland stands. *Trop. Ecol.*, 15: 152–153.

Pandey, H.N. and Kothari, A., 1972. Studies on the root distribution pattern and phosphate uptake rate of cropland and grassland species. In: V. Puri, Y.S. Murty, P.K. Gupta and D. Banerji (Editors), *Biology of Land Plants*. Sarita Prakashan, Meerut, pp. 36–43.

Pandeya, S.C., 1964a. Ecology of grasslands of Sagar, Madhya Pradesh IIA. Composition of the fenced grassland associations. *J. Indian Bot. Soc.*, 43: 557–560.

Pandeya, S.C., 1964b. Ecology of grasslands of Sagar, Madhya Pradesh. IIB. Composition of the association open to grazing or occupying special habitat. *J. Indian Bot. Soc.*, 43: 606–639.

Pandeya, S.C., 1977. *The Environment and* Cenchrus *Grazing Lands in Western India*. Department of Biosciences, Saurashtra University, Rajkot, 451 pp. (mimeograph).

Pandeya, S.C., Mankad, N.R. and Jain, H.K., 1974. Potentialities of net primary production of arid and semi-arid grazing-lands of India. In: *Proc. XII International Grassland Congress*, pp. 136–170.

Ramam, S.S., 1959. *Soil–Root Relationships in Grassland Communities at Varanasi*. Thesis, Banaras Hindu University, Varanasi, 154 pp.

Raychaudhary, S.P., 1966. *Land and Soil*. National Book Trust, New Delhi, 171 pp.

Sant, H.R., 1962. *A Study of the Reproductive Capacity of Some Grasses and Forbs of the Grounds of Banaras Hindu University*. Thesis, Banaras Hindu University, Varanasi, 255 pp.

Sant, H.R., 1964. Grazing susceptibility, number of grasses and forbs from the grazing ground of Varanasi. *Indian J. Range Manage.*, 17: 337–339.

Satyanarayan, Y., 1958b. Indigenous species in the stabilization of sand dunes of Rajasthan desert. *J. Soil Water Conserv. India*, 7; 47–51.

Shankar, V., Shankarnarayan, K.A. and Rai, P., 1973. Primary productivity, energetics and nutrient cycling in *Sehima–Heteropogon* grassland I. Seasonal variation in composition, standing crop and net production. *Trop. Ecol.*, 14: 238–251.

Shankar, V., Shankarnarayan, K.A. and Rai, P., 1975. Vertical distribution of aboveground and below-ground biomass in six communities of *Sehima–Dichanthium* cover. *Indian Biol.*, 7: 61–66.

Shankarnarayan, K.A., Shreenath, P.R. and Dabadghao, P.M., 1969. Studies on the height:weight relationship of six important range grasses of *Sehima–Dichanthium* zone. *Ann. Arid Zone*, 8: 61–65.

Shankarnarayan, K.A., Dabadghao, P.M., Rai, P., Upadhyaya, V.S. and Maurya, R.K., 1973. Fertilizers for the grasslands of *Heteropogon contortus* Beauv. in Jhansi, Uttar Pradesh. *Indian J. Agric. Sci.*, 43: 562–566.

Shankarnarayan, K.A., Dabadghao, P.M., Upadhyaya, V.S. and Rai, P., 1974. Studies on utilization of spear grass pasture I: Monthly variation in yield, chemical composition and performance of sheep. *Indian J. Anim. Sci.*, 44.

Sims, P.L. and Singh, J.S., 1971. Herbage dynamics and net primary production in certain ungrazed and grazed grass-lands in North America. In: N.R. French (Editor), *Preliminary Analysis of Structure and Function in Grasslands*, Range Science Department, Science Series No. 10, Colorado State University, Fort Collins, Colo., pp. 59–174.

Singh, A.K., 1972. *Structure and Primary Net Production and Mineral Contents of Two Grassland Communities of Chakia Hills, Varanasi.* Thesis, Banaras Hindu University, Varanasi, 154 pp.

Singh, J.S., 1967. *Seasonal Variation in Composition, Plant Biomass and Net Community Productivity in the Grasslands at Varanasi.* Thesis, Banaras Hindu University, Varanasi, 319 pp.

Singh, J.S., 1968. Net aboveground community productivity in the grasslands at Varanasi. In: R. Misra and B. Gopal (Editors), *Proc. Symp. Rec. Adv. Trop. Ecol. II.* International Society for Tropical Ecology, Varanasi, pp. 531–654.

Singh, J.S., 1969a. Influences of biotic disturbance on the preponderance and interspecific association of two com-mon forbs in the grasslands at Varanasi, India. *Trop. Ecol.*, 10: 59–71.

Singh, J.S., 1969b. Growth of *Eleusine indica* L. Gaertn. under reduced light intensities. *Proc. Indian Natl. Sci. Acad.*, 35B: 153–160.

Singh, J.S., 1970. Influence of the period of exposures to alternating low and high temperature in *Eleusine indica.* *Sci. Cult.*, 36: 472–473.

Singh, J.S., 1973. A compartment model of herbage dynamics for Indian Tropical Grasslands. *Oikos*, 24: 367–372.

Singh, J.S., 1976. Structure and function of tropical grassland vegetation of India. *Pol. Ecol. Stud.*, 2(2): 17–34.

Singh, J.S. and Joshi, M.C., 1979. Primary production. In: R.T. Coupland (Editor), *Grassland Ecosystems of the World. Analysis of Grasslands and their Uses.* Cambridge University Press, Cambridge, pp. 197–218.

Singh, J.S. and Misra, R., 1968. Efficiency of energy capture by the grassland vegetation at Varanasi. *Curr. Sci.*, 37: 636–637.

Singh, J.S. and Misra, R., 1969. Diversity, dominance, stability and net production in the grasslands at Varanasi, India. *Can. J. Bot.*, 47: 425–427.

Singh, J.S. and Yadava, P.S., 1973. Seasonal variation in composition, plant biomass, and net primary productivity of a tropical grassland at Kurukshetra, India. *Ecol. Monogr.*, 44: 351–375.

Singh, J.S., Lavenroth, W.K. and Steinhorst, R.K., 1975. Review and assessment of various techniques for estimating net aerial primary production in grasslands from harvest data. *Bot. Rev.*, 41: 181–232.

Singh, J.S., Singh, K.P., and Yadava, P.S., 1979. Static, com-partment models of energy flow and nutrient cycling. In: R.T. Coupland (Editor), *Grassland Ecosystems of the World. Analysis of Grasslands and Their Uses.* Cambridge University Press, Cambridge, pp. 231–240.

Singh, K.P. and Misra, R. (Editors), 1978. *Structure and Functioning of Natural, Modified and Sylvicultural Ecosystems of Eastern Uttar Pradesh.* Technical Report MAB Research Project, Banaras Hindu University, Varanasi, 161 pp. (mimeograph).

Thornthwaite, C.W., 1948. An approach towards a rational classification of climate. *Geogr. Rev.*, 38(1): 55–94.

Varshney, C.K., 1972. Productivity of Delhi grassland. In: P.M. Golley and F.B. Golley (Editors) *Papers from a Symposium on Tropical Ecology with an Emphasis on Organic Production.* Institute of Ecology, University of Georgia, Athens, Ga., pp. 27–42.

Vuattoux, R., 1970. Observation on the evaluation of the woody and shrubby strata of the Savanna of Lamto (Ivory Coast). *Ann. Univ. Abidjan, Sér. E, Ecol.*, 3: 285–315.

Vyas, L.N., Garg, R.K. and Agrawal, S.K., 1972. Net above-ground production in the monsoon vegetation at Udaipur. In: P.M. Golley and F.B. Golley (Editors), *Papers from a Symposium on Tropical Ecology with an Emphasis on Organic Production.* Institute of Ecology, University of Georgia, Athens, Ga., pp. 95–99.

Whyte, R.O., 1964. Grassland and fodder resources of India. *ICAR Sci. Monogr.*, No. 22: 553 pp.

Whyte, R.O., 1974. Grasses and grasslands. In: *Natural Resources of Humid Tropical Asia.* UNESCO, Paris, pp. 239–262.

Yadava, P.S. and Singh, J.S., 1977. Grassland vegetation. In: R. Misra, B. Gopal, K.P. Singh and J.S. Singh (Editors), *Progress in Ecology, 2.* Today and Tomorrow, New Delhi, p. 182.

Chapter 8

THE TRANSITION FROM OPEN FOREST TO SAVANNA IN CONTINENTAL SOUTHEAST ASIA

F. BLASCO

INTRODUCTION

The prevalence, side by side or in mosaic, of open forests and dense evergreen rain forests is a characteristic feature of continental Southeast Asia. Shrub savannas and grasslands are restricted to very small areas. It is no exaggeration to say that these two communities are an exception at low elevations in this region.

The most common, most extended and most characteristic vegetation type in this part of the tropics is the deciduous dipterocarp forest, often referred to as dry dipterocarp forest, corresponding to the *forêt claire* of French-speaking botanists — so called because it looks like an open forest in which the ground is covered with grass. "Open forest" is used in this chapter as a convenient general term. This peculiar community is most widespread in Cambodia, Laos, Thailand and Vietnam, below an elevation of 900 m. At higher altitudes practically all dipterocarps disappear.

Though everyone agrees that deciduous dipterocarp open forests are strongly influenced by human interference and annual burning, their status and origin still pose controversial questions. Their study requires careful investigations of environmental conditions and of the flora involved, as well as an understanding of the dynamics of woody species

DEFINITION, EXTENT, AND DISTRIBUTION

With the exception of relatively small areas in the highlands of northern Thailand and east of the Mekong River, no detailed vegetation mapping has ever been carried out in this part of Southeast Asia.

Taking into account personal experience for Cambodia (Legris and Blasco, 1972) and other data provided by several small-scale maps (Republic of Vietnam, National Geographic Service, 1969; U.S. Agency for International Development, 1978), the present distribution of deciduous dipterocarp open forests in Southeast Asia has been shown in Fig. 8.1. These forests fit the definition given by Legris and Blasco in 1972: "A plant community generally found at low elevation, having only one discontinuous tree layer, often ranging between 10 and 20 m high. Most of the trees have a diameter less than 40 cm, are deciduous and have small boles, with crowns rarely touching each other. Three species of Dipterocarpaceae are characteristic: *Dipterocarpus tuberculatus, Pentacme suavis* and *Shorea obtusa*. A ground herbaceous layer is always present, though very uneven in height and density" (Figs. 8.2 and 8.3).

An attempt was made by the UNESCO workshop (1973) to quantify the density of trees: "A woodland is an open stand of trees composed of trees at least 5 m tall with crowns not usually touching but with a coverage of at least 40 per cent. A herbaceous synusia may be present ... the boundary of 40 per cent coverage is convenient because it can be estimated with ease during the field-work: when the coverage of the trees is 40 per cent, the distance between two tree crowns equals the mean radius of a tree crown."

More information regarding the structure of some of these plant communities in northern Thailand is given in Table 8.1.

Deciduous dipterocarp forests cover a very large area. In Cambodia alone, the U.S. Agency for International Development recently quoted a figure of about 5 300 000 ha covered by these open forests

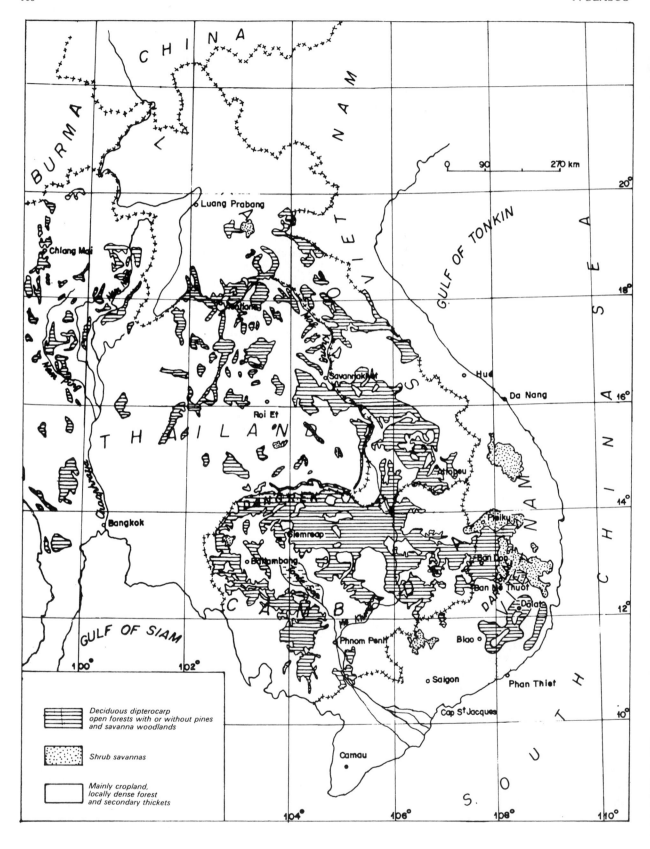

Fig. 8.2. Cambodian open dipterocarp forest near Stung Treng, eastern Mekong; end of the rainy season. See text, stages C and B. (Photo J. Fontanel.)

Fig. 8.3. Deciduous dipterocarp forest in **Thailand, Korat Plateau, Maha Salakhram** Province; end of the rainy season, about eight days after burning. See text, stage B. (Photo J. Boulbet.)

Fig. 8.1. Open forests and savannas in Southeast Asia.

TABLE 8.1

Relations between soils and vegetation: some examples from northern Thailand (data from **BIOTROP**, 1976, 1977)

Locality, altitude (m)	Vegetation type	Soil type and mother rock	Root penetration (cm)	Iron compounds	Soil texture S (surface) D (deep horizon)	pH[1]	E (erosion) D (drainage)	S (slope) F (fire) G (grazing)	Dominant trees	Average stand height (m)	Number of species in sample area (2000 m²)	Stand density (trees 0.1 ha⁻¹)
Ban Hued (Lampang), 370	open deciduous forest	red-yellow podzolic soils from sandstone	29	lateritic nodules, 10% boulders	S: sandy clay → D: loam	6–5.8	E: severe D: good	S: 15% F+G	*Pentacme suavis* *Shorea obtusa* *Odina wodier*	18	19	36
Tung Kwian (Lampang), 400	deciduous dipterocarp forest (moderate density)	lithosol on sandstone + quartzite + conglomerate	30	10% boulders	S: sandy → D: sandy clay	c. 5.8	E: severe D: rapid	S: 17% F	*Dipterocarpus obtusifolius* *Shorea obtusa* *Pentacme suavis*	20	22	40
Foot hill Doi Inthanon near Mae Klang, 585	deciduous dipterocarp forest (moderate density)	lithosol on sandstone + quartzite	52	abundant gravel	S: sandy → D: clay	5.6–5.2	E: slight D: poor	S: 20% F	*Dipterocarpus tuberculatus* dominant	14	15	40
Huey Bong (Amphoe Hod), Chiang Mai, 815	open dipterocarp forest with pines	lithosol on sandstone + quartz	50	boulders + lateritic nodules	S: sandy loam → D: sandy clay	c. 6	E: moderate D: rapid	S: 12% F + slight G	*Pinus merkusiana* and *Dipterocarpus tuberculatus*	26	8	40
Huey Bong (Amphoe Hod), Chiang Mai, 835	open dipterocarp forest with pines	reddish brown laterite from sandstone	39	few boulders + lateritic nodules	S: sandy loam → D: sandy clay	c. 5.8	E: moderate D: rapid	knolly terrain on a ridge; S: 6% F+G	*Pinus merkusiana* and *Dipterocarpus obtusifolius*	26	10	38
Mae Sariang (Khun Yuam), 640	dipterocarp forest with pines	red-yellow podzolic soils from sandstone + quartzite	54	no lateritic nodules	S: sandy D: clay	c. 5.8	E: slight D: moderate	S: 12% annual ground fire	*Pinus merkusiana* *Dipterocarpus tuberculatus* codominant (30–40 m)	29	9	43
Doi Suthep (Amphoe Muang) Chiang Mai, 630	deciduous dipterocarp forest (moderate density)	red-yellow podzolic soils derived from sandstone	35	lateritic nodules	S: sandy clay D: clay	c. 5.8	E: severe D: moderate	S: 30% annual fires no grazing	*Dipterocarpus obtusifolius* *Dipterocarpus tuberculatus*	17	12	24

[1] Soil:water, 1:1.

— that is, about 40% of the total forest area in the country. In Vietnam they also occupy more than 5 million hectares, mainly between Ban Me Thuot and Pleiku, where they make up the eastern extension of the Cambodian woodlands. In Thailand, their area, which was estimated by Ogawa et al. at about 45% of the total forest area in 1967, is receding rapidly, due to a tremendous demographic pressure. The Korat Plateau, in eastern Thailand, has a fairly high population density with nearly 90 inhabitants km^{-2}. Today the acreage of open forests in Thailand can be estimated at about 100 000 km^{-2} — that is, 33% of the forested area (Ogino, 1976). In Laos, it is extremely difficult to estimate the extent of deciduous dipterocarp open forests. As shown in Fig. 8.1, they occur mainly in southern Laos, between Attopeu and Savannakhet (Vidal, 1960b).

The above figures include some open forests in northern Thailand and in Cambodia which contain a pine, *Pinus merkusiana* (Cooling and Gaussen, 1970). This pine and the dipterocarps commonly found in the open forests of Southeast Asia progressively disappear towards the west. They are unknown in peninsular India. Hence, considering their main floristic components, these communities should be regarded as typical of Southeast Asia.

THE NATURAL ENVIRONMENT

The presence of deciduous dipterocarp open forests certainly results from the particular ecological conditions found in Southeast Asia — chiefly the bioclimatic conditions and soil properties.

Bioclimatic features (Fig. 8.4)

As shown in a bioclimatic map of tropical Asia (Gaussen et al., 1967), the variety of climates is very great, ranging from ever-wet to dry ones. The comparison of the distribution of vegetation types with that of the major climatic areas, already leads to the following remarks.

Under humid or perhumid climates (mean annual rainfall exceeding 2000 mm, dry season nil or short, not exceeding four months) open forests are almost unknown. This is the case in Malaysia, Sumatra and Kalimantan. In these areas the degradation of the rain forest leads to several re-

gressive stages, the last one often being a shrub savanna maintained by burning.

Under dry climatic conditions (rainfall < 1000 mm; six dry months or more), and contrary to what happens elsewhere in the tropics, savannas are almost completely lacking in Asia. In peninsular India (see Ch. 7), in central Burma and in the southeastern coastal region of Vietnam, stages of degradation include several secondary communities, but savannas are an exception.

Asiatic open forests seem to be restricted to semi-dry, or semi-humid warm areas (rainfall between 1000 and 2000 mm; four to seven dry months) particularly in Cambodia, Laos, Thailand and Vietnam. Therefore, as regards temperature conditions, it appears that the deciduous dipterocarp forests are communities of warm areas, where the mean temperature of the coldest month (January or February) is higher than 15°C, the absolute minimum temperature never falls below 8°C, and the mean annual thermic amplitude is either moderate or low. A minimum temperature of 4°C has, however, been recorded at Phnom Penh, and 7°C at Chiang Mai in northern Thailand. On the whole, these climatic conditions are found only in plains and on low plateaux, below 1000 m elevation and at latitudes barely exceeding 20°N. Outside these boundaries, most of the dipterocarps forming the canopy of Asiatic open forests disappear. Moreover, these communities are rare wherever the mean annual rainfall is less than 1000 mm or greater than 2500 mm. Their geographical distribution is always linked with that of monsoon climates, which are seasonally dry from December to March or from November to April. The annual number of rainy days ranges from 70 to 120 and rain falls from May to October (Fig. 8.4).

Therefore it appears that the climatic conditions suitable for open forests in Asia, can be summarized as follows:

(1) mean temperature of coldest month > 15°C
(2) absolute minimum temperature   > 8°C
(3) mean annual thermic amplitude   < 10°C
(4) mean annual rainfall between 1000 and 2000 mm
(5) mean annual number of dry months ranging from four to seven
(6) mean monthly saturation deficit during the dry season ranging from 7 to 12 mm

This apparently satisfactory climatic approach

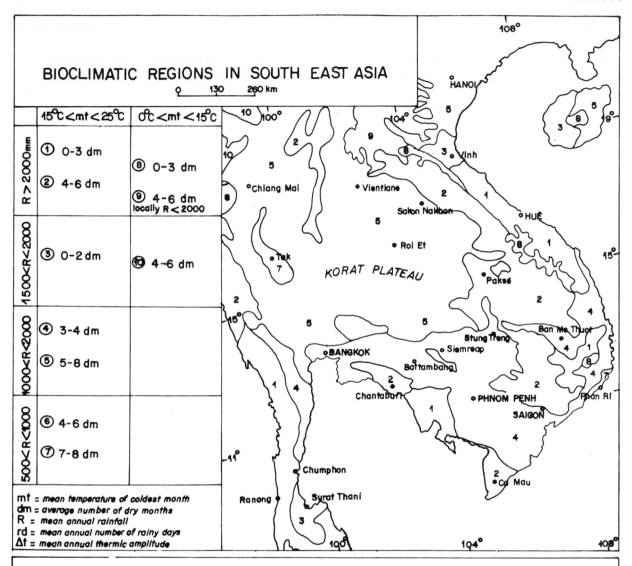

BIOCLIMATIC DATA FROM SOME DECIDUOUS DIPTEROCARP FOREST AREAS					
Station	Rmm	mt°C	Δt°C	dm	rd
Chiang Mai alt.320m _ Thai	1250	21°4	7°3	6	108
Roi Et alt.140m _ Thailand	1415	22°	6°	5-6	86
Vientiane alt.165m _ Laos	1730	20°5	6°	5-6	108
Siemreap alt.20m. Cambodia	1430	24°7	4°7	4-5	110
Stung Treng alt.50m. Cambodia	1832	24°3	5°7	4-5	115
Ban Me Thuot alt.540m _ Viet Nam	1845	22°	4°5	4-5	?

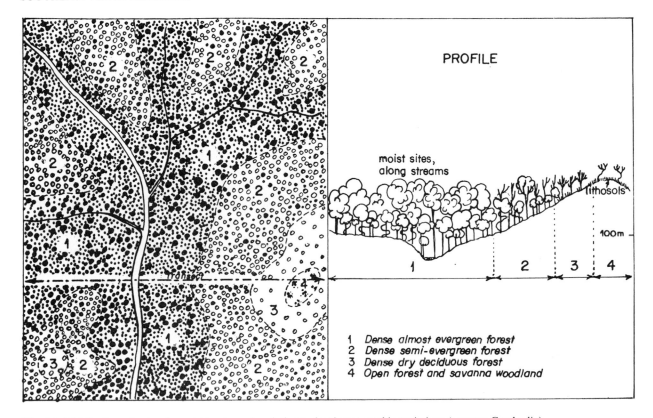

Fig. 8.5. Distribution of natural vegetation types in relation to local topographic variations (western Cambodia).

does not explain, however, why a mosaic of vegetation types, ranging from dense rain forests to deciduous open forests, often coexist in the seasonally dry climates of Cambodia, Vietnam and Thailand (Fig. 8.5). One can nevertheless assume, first, that the present climatic conditions suitable for open forests are the same as those required by most climax dense deciduous and semi-deciduous dipterocarp forests throughout tropical Asia; and, second, that the presence of open forests is generally correlated with some specific soil conditions.

The soils of open forests (Table 8.1)

Within the above-mentioned climatic zone, soil is a major factor controlling the distribution of deciduous dipterocarp open forests and maintaining their apparent stability. Depending on the parent material and local topographic circumstances, a great variety of soils can be observed. However, despite their diversity, these soil types are usually related to one of two main groups: acid lithosols or skeletal soils, or red and yellow podzolic soils and grey podzolic soils. Crocker (1963) for Cambodia, Moormann (1961) for Vietnam and Pendleton and Sarot Montrakun (1960) for Thailand, have carried out comprehensive soil surveys, and published detailed soil maps.

Acid lithosols are predominantly poor shallow eroded soils, found on hilly areas in northern Thailand, southern Cambodia and east of the Mekong. Depending on the site and the local topography, their depth varies. In general, they are stony soils, no more than 50 cm deep. On steep slopes, erosion is an important factor for these weakly developed soils. Often, they do not show distinct profiles. Unaltered geological material lies near the surface or breaks through in patches. In northern Thailand, their pH varies from 4.5 near the surface to 6.5 in the deeper parts. In moun-

Fig. 8.4. Bioclimatic regions in Southeast Asia, and bioclimatic data from some deciduous dipterocarp forest areas.

tainous Cambodian localities (Mondolkiri Province) some lithosols, derived from basaltic rocks (basic lithosols), are chemically richer, but they lack depth as they are too young. Therefore, all kinds of lithosols seem to be suitable for the establishment of open forests in Southeast Asia.

Grey podzolic soils and red and yellow podzolic soils are, by contrast with the previous group, soils of almost flat landscapes. They cover the major part of Southeast Asia. According to Dudal et al. (1974) they occupy more than half of the total area of Vietnam. In Thailand they cover the greater part of the Korat Plateau. A leached horizon and a clay accumulation horizon are more or less clearly differentiated, and iron compounds such as laterite and iron concretions are commonly found. Moreover, under open forests, the presence of hard or hardenable laterite in the subsoil, at various depths, is often noticed. It is not known how closely these features are related to the low chemical fertility of these old soils. However, according to Pendleton and Sarot Montrakun (1960), these iron compounds lock up in insoluble form phosphorus and other important elements, even when present only in traces, so that they cannot be absorbed by plant roots and are unavailable for plants use. A more important deficiency is their low water-holding capacity; such soils become extremely dry as soon as the rains stop. Complementary data are summarized in Table 8.1; they were obtained recently by Thai foresters working in northern Thailand (BIOTROP Research Projects, 1976, 1977).

Even if both acid lithosols and podzolic soils belong to very distinct groups, they exhibit interesting fundamental similarities:

(a) An acidity which decreases from a pH of 4.2 in some top soils to one of 6.5 near the parent material.

(b) Surface horizons which often belong to the "sandy loam" textural class, whereas the amount of clay often exceeds 40% in deeper horizons.

(c) A low water-holding capacity, a deficiency in nutrients and therefore a low fertility.

(d) Nearly everywhere the sites are disturbed by biotic or human interferences and burnt almost yearly.

MAIN TYPES OF DECIDUOUS DIPTEROCARP FOREST AND DERIVED SAVANNAS

A number of more or less open forests have been studied in Southeast Asia with regard to their stand density (number of trees per hectare), floristic composition and physical structure, and their origin has been discussed (Rollet, 1962, 1972; Legris and Blasco, 1972). All authors agree that they represent a physiognomic transition between dense deciduous dipterocarp forests on the one hand and shrub savannas on the other. In other words, they represent a gradient of regressive stages between the dense and almost closed forests, in which trees are the dominant component, and treeless grasslands.

Practically all forms of disturbed vegetation types that can be expected under the bioclimatic conditions already described, are to be found in every country of Southeast Asia. However, at the risk of over-simplification, it seems sufficient to focus attention on the five main regressive stages leading to herbaceous vegetation.

This series of five stages (A to E) is shown diagrammatically in Fig. 8.6. Both extremes are found only in small areas, whereas the intermediate stages are widespread. In this section an attempt will be made to describe their relationships and their evolution.

Stage A: Dense, semi-deciduous and deciduous forests

These are more or less closed communities usually possessing two strata of trees and a shrubby undergrowth. Grass cover, though very irregular and discontinuous, is always present, with grasses such as *Heteropogon triticeus* and *Arundinella setosa*, among others. Some dwarf bamboos are also common, particularly *Dendrocalamus strictus* and *Oxytenanthera albociliata*.

The upper stratum, quite variable in height (20–40 m high) includes some large evergreen trees, but most of them are deciduous. All the species of Dipterocarpaceae found in stages B and C can also be found there, as well as several legumes (*Dalbergia* spp., *Pterocarpus pedatus*, *Sindora cochinchinensis*, *Xylia xylocarpa*), and species of *Lagerstroemia* (Lythraceae), and of *Terminalia* and *Anogeissus* (Combretaceae), which are common in

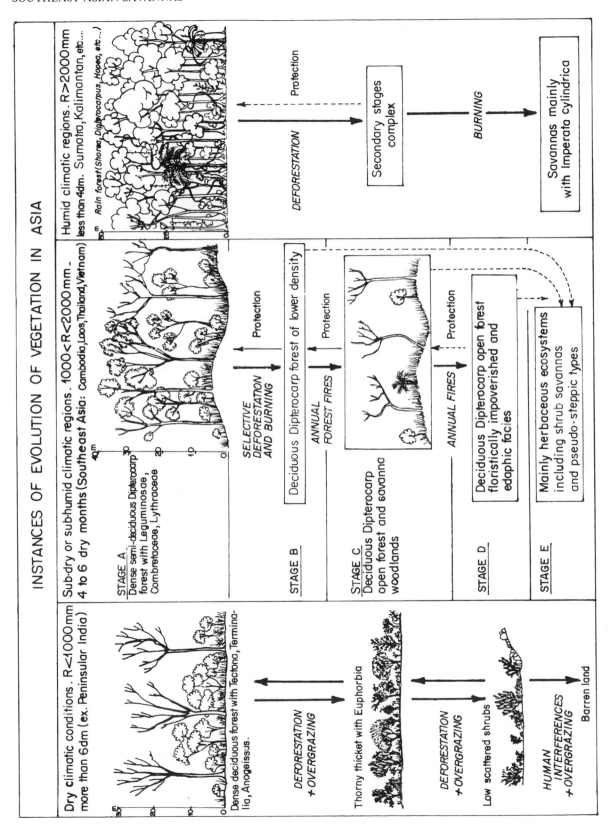

Fig. 8.6. Examples of evolution of vegetation in Asia.

the dense deciduous forests of Vietnam and Cambodia.

Some of the presently dense deciduous forests may not be climax communities, but derived by degradation either from moister forest types such as the "semi-evergreen forests" found in the area, or from forests analogous to the "mixed deciduous forests" found in western and northern Thailand, often bearing teak (*Tectona grandis*).

Both "semi-evergreen" and Thai "mixed deciduous" forests contain a variety of major commercial timber trees (*Anisoptera* sp., *Dipterocarpus intricatus*, *D. costatus*, *Lagerstroemia* sp., *Tetrameles* sp., and teak in Thailand) which have been selectively felled or periodically destroyed by forest fires for centuries.

It is noteworthy that conspicuous genera such as *Adina*, *Dipterocarpus*, *Irvingia*, *Lagerstroemia*, *Shorea*, *Terminalia*, and *Vitex*, provide species not only to stages A and B, but to stage C as well.

Nowadays, dense deciduous forests are restricted to sparsely populated areas such as the north of Cambodia, up to the Dangrek range (Boulbet, 1979), and some remote localities of the foothill zone of the Batambang Province. They never constitute large forest areas but always small patches intermingled with "open forest".

Stage B: Deciduous forests with a moderate tree density

These deciduous forests with a moderate tree density and thick woody undergrowth also exist in every country of Southeast Asia.

As a result of their physiognomy and floristic composition they can be considered as regressive stages, transitional between dense and open forest types. Grazing, burning, and human activities affect all the strata, particularly the ground vegetation and the top canopy.

In eastern Thailand, on the Korat Plateau near Sakon Nakhon (Phan Phu Phan hills), all the grasses and palatable species of the ground cover have been overgrazed by domestic animals and the risk of fire is obviously minimized. Nevertheless, occasional burning can be observed. Unpalatable shrubs, young shoots and some climbers invade the undergrowth which becomes denser and denser, whereas emerging trees are selectively felled.

In northern Thailand, near Chiang Mai (Huey

Bong, Amphoe Hod), some of these anthropic forest types are characterized by the presence of pines (*Pinus merkusiana*), *Dipterocarpus tuberculatus*, and *D. obtusifolius*. These trees are the main components of the uppermost forest layer (25–30 m in height). Generally these dry dipterocarp forests with pines are found at an elevation of from 500 to 900 m, where the flora is enriched with oaks (*Quercus kerrii*) and some orophytes such as *Aporosa villosa*, *Melanorrhoea usitata*, and *Phoenix humilis*. *Allopteropsis semialata* is one of the commonest grasses in these communities; it disappears as soon as the shade becomes too dense. *Brachiaria mutica* and *Heteropogon triticeus* are also grasses characteristic of this stage.

In Cambodia, Vietnam and probably also in Laos, similar deciduous forests, with or without pines, exist. The distinctive floristic feature is the mixture of species; some are more abundant in stages C and D (*Dipterocarpus intricatus*, *D. tuberculatus*, *Pentacme suavis*, *Shorea obtusa*), whereas others are more frequently found in stage A (*Anisoptera*, *Calamus*, *Dillenia*, *Irvingia*, *Lagerstroemia*, *Parinari*, *Xylia*). Stages A and B always have close floristic affinities; their differences are mostly physiognomic.

East of the Mekong River (Rollet, 1972) forest inventories over 46 ha have identified 82 tree species of 10 cm in diameter (DBH) or more. About 90% of the trees are small (10–30 cm). In northern Thailand trees are slightly larger; 40% of them have a small diameter (10–30 cm), and another 40% have a diameter varying between 30 and 50 cm (BIOTROP Research Project, 1976).

Stage C: Typical deciduous dipterocarp open forests

These open forests are found in every Southeast Asian country from Burma to Vietnam, where they are by far the most common and most extensive "natural" communities. This stage represents a simplification of the physiognomy and structure of the preceding stages, together with a floristic impoverishment accompanied by a deterioration of soil properties.

Generally trees form a light canopy and the stand is devoid of an intermediate storey. Some palms (*Corypha lecomtei*, *Phoenix humilis*) and *Cycas siamensis* are numerous or scattered amid an almost continuous grass layer dominated by a dwarf bamboo, *Arundinaria falcata*, reaching a maximum height of about 1 m. This grass cover, which is burnt every year, includes a variety of grasses (*Eulalia cumingii*, *E. tristachya*, *Themeda triandra* group) and numerous suffrutescent fire-resistant Fabaceae belonging mostly to genera *Crotalaria*, *Desmodium* and *Indigofera*.

The top canopy, very irregular in density and size, reaches 25 m; straight boles are exceptional. Its flora has great affinities with the previous stages, with scattered species of *Buchanania*, *Irvingia*, *Lagerstroemia*, *Sindora* and *Xylia*. However, five species of dipterocarps (*Dipterocarpus intricatus*, *D. obtusifolius*, *D. tuberculatus*, *Pentacme suavis* and *Shorea obtusa*) are dominant, together with *Terminalia alata* in the typical deciduous dipterocarp open forests.

Nowadays, in most places this community coincides with lithosols and red and yellow podzolic soils of very low potential. There is no clear evidence of a progressive dynamism of trees leading to more closed ecosystems (stages B and A). Finally, stage C appears to be in a state of dynamic equilibrium resulting from the combined effects of natural environmental factors and annual burning.

Stage D: Floristically impoverished dipterocarp open forests

This is probably the poorest woody stage to be found anywhere in Southeast Asia. It includes several floristic **facies** entirely dominated by one or two woody species. Thus, according to the dominant tree, several associations can be described, for instance:

(a) On steep eroded slopes, almost pure stands of *Shorea obtusa* and *Pentacme suavis* are found in northern Thailand and Vietnam.

(b) On flat areas with deep sandy soils near the surface, but poorly drained, two distinct facies exist east of the Mekong River, either with *Dipterocarpus obtusifolius* or with *D. tuberculatus*.

Generally the grass cover is also discontinuous, but a ground layer of grasses and herbaceous plants is always present (*Echinochloa colona*, *Sorghum nitidum*, *Themeda triandra*).

No clear explanation has yet been given on the origin of these very open communities. This has been a subject of much discussion. However, two unquestionable facts can be stated. First, the woody species of stage D are often found in denser forest communities but, having a wide ecological range, they survive the fires and the extreme adverse edaphic conditions. Second, some pure stands are also due to the gregariousness of some constituting species.

Stage E: Grass savannas

On the whole, "grass savannas", without trees, are seldom found in Southeast Asia (see Fig. 8.1). The only region where they cover appreciable areas is in Vietnam, on the Darlac Plateau, particularly near Pleiku and Ban Me Thuot, at elevations from 300 to 700 m.

Schmid (1958, 1974), who has studied the grasses and grasslands of this region has related their distribution to edaphic conditions. Two main types of grass savannas are to be distinguished:

(1) Climax grasslands, found in narrow low alluvial plains, are determined by an excessive and permanent soil moisture. Some hygrophilous grasses are characteristic of these boggy soils, periodically or almost continually flooded. Among them are *Echinochloa stagnina*, *Phragmites karka* and *Saccharum narenga*. Usually trees are unable to thrive in such highly hydromorphic soils.

(2) Secondary savannas are generally shrub savannas, wherein a number of shrubs, sub-shrubs, ferns and forbs are mixed with grasses. The size and density of the grassy components are extremely variable.

In dense tall grasslands, sometimes reaching a height of more than 1.5 m when in flower, small trees are present although they are very scattered. Floristically and physiognomically, these grasslands are nothing but very open woodlands in which the main grasses are *Arundinella setosa*, *Imperata cylindrica*, *Sorghum serratum*, *Themeda* (*gigantea* group). Among shrubs, subshrubs, and small trees, *Albizzia procera*, *Careya* sp., *Clerodendrum serratum*, *Dillenia ovata*, *Emblica officinalis*, *Grewia* sp., *Ziziphus rugosa* are the commonest.

In short discontinuous grasslands, with or without scattered trees (the *prairie-steppe* of French-speaking authors), *Aristida cumingiana*, *Arundinella setosa*, *Schizachyrium brevifolium* and *Themeda* sp. are common, whereas stunted *Dillenia ovata*, *Pentacme suavis* and *Shorea obtusa*, constitute a very scanty woody cover.

In hilly country most of these secondary herbaceous vegetation types are considered to be derived from several forest types due to shifting cultivation, known as "ray" in Indochina. The exact mechanisms of the successions are not yet known in which *Imperata cylindrica* and *Eupatorium odoratum* (an introduced composite), along with bracken (*Pteridium aquilinum*), successfully compete with woody species, with the help of annual fires. Nevertheless, on exposed areas, where erosion is extremely severe, the degradation leads to the poorest discontinuous grasslands referred to earlier.

FLORISTIC AFFINITIES AND ORIGIN OF THE OPEN FORESTS

The paucity of woody species in the open dipterocarp forests is well known. In northern Thailand the number of trees in each sample area (of 2000 m^2) ranges from 7 to 26, depending on the locality studied (Table 8.1). In eastern Cambodia, Rollet (1972) found 82 woody species in 46 hectares. In southern Laos their number was 135 in 73 hectares according to Ly Van Hoi (1952). It is clear that species richness increases very gradually with the size of the sample area.

These figures bring out the disparity between the paucity of species in open forests and the species richness of the evergreen rain forest, which ranges from about 150 per hectare to about 200 in 2 hectares (Van Steenis, 1950, in Whitmore, 1975). Obviously, the total number of vascular plants, including grasses, forbs and other suffrutescent dicotyledons is much higher, giving the open dipterocarp forest a greater floristic diversity.

General distribution of grasses

Very little is known on the biogeography of grasses in this part of the world. However, even a broad approach to their geographical distribution at the species level yields instructive information.

The majority of grasses presently found in open forests of Southeast Asia are found throughout tropical Asia, including India, and generally on poor eroded soils. Some have a still wider geographical distribution, reaching Australia (*Alloteropsis semialata*, *Heteropogon triticeus*, *Schizachyrium brevifolium*), or are found all over the tropical world (such as *Heteropogon contortus*, *Imperata cylindrica*, *Sehima nervosum*). Many of these species are highly polymorphic. Such is the case for *Arundinella setosa*, *Kerriochloa siamensis*, *Themeda gigantea* and *Themeda triandra*, among others. Each of the major species has developed several ecotypes adapted to a great diversity of local environmental conditions.

As a general rule, it seems that the grass community remains virtually the same in the different types of deciduous, open woodlands and savannas of Southeast Asia. Moreover, most of these grasses are also found in forest communities, as exemplified by *Allopteropsis semialata*, *Eulalia tristachya*, *Heteropogon triticeus*, *Kerriochloa siamensis* and *Sehima nervosum*.

Floristic affinities of the woody elements

The hitherto commonly held view is that the most characteristic trees of the open forests are three species of Dipterocarpaceae: *Dipterocarpus tuberculatus*, *Pentacme suavis* and *Shorea obtusa*. Two other dipterocarps found in these ecosystems (*D. intricatus* and *D. obtusifolius*) are also trees of dense forests.

In my opinion none of the so-called trees of the dipterocarp open forest is either restricted to that plant community or characteristic of it. Biogeographical and taxonomic arguments support this viewpoint.

Biogeographical arguments

A complete floristic inventory of Southeast Asian woodlands is not within the scope of this chapter. Table 8.2, however, summarizes the distribution and probable origin of the commonest tree species found in the woodlands of Southeast Asia.

Dipterocarpus tuberculatus, frequently found in the open forests, is also a species of mixed evergreen forests in Thailand (Smitinand, 1963), and it is present in similar moist communities in Bangla Desh (Chittagong area).

Pentacme suavis and *Shorea obtusa* are important components of Thai mixed evergreen and dry evergreen forests, though more frequently found in

TABLE 8.2

Probable origin of the main woody species found in Asiatic woodlands

Species found in (and probably originating from) dense semi-deciduous forests	Species found in (and probably originating from) dense moister forest types
Buchanania lanzan	Anisoptera cochinchinensis
Dillenia pentagyna	Dalbergia sp.
Dipterocarpus obtusifolius	Dipterocarpus intricatus
Dipterocarpus tuberculatus	Eriodendron anfarctuosum
Elaeodendron glaucum	Irvingia malayana
Emblica officinalis	Lagerstroemia sp.
Melanorrhea usitata	Pahudia cochinchinensis
Memecylon edule	Parinari anamense
Pentacme suavis	Pterocarpus pedatus
Shorea obtusa	Shorea cochinchinensis
Terminalia alata	Sindora cochinchinensis
Terminalia chebula	Syzygium sp.
Terminalia corticosa	Xylia xylocarpa

open forests and savanna woodlands. *Pentacme suavis* is a well-known species of northern Malaysia, where it plays an important role in dense seasonal forests on limestone hills.

Terminalia alata, one of the most constantly stunted species of the open woodlands, is one of the commonest and most widely distributed of Indian trees. It is found all over peninsular India, from the moist deciduous dense forests to the savanna woodlands. In India it is known as *Terminalia tomentosa*, and often found in teak forests (*Tectona grandis*).

Hence, data provided by plant geography do not afford any evidence to indicate that there is a flora proper to open forests and savanna woodlands in Southeast Asia.

To some extent, the dominance of dipterocarps in the vegetation of continental Southeast Asia is somewhat deceptive. Among important tree families, Annonaceae (with more than 60 species), Lauraceae (35 species), Meliaceae, Moraceae, Myrtaceae, Sapindaceae and Verbenaceae have more representatives than Dipterocarpaceae, which are represented by only 24 species in Thailand and Vietnam, according to Smitinand (1963) and Schmid (1974), and by 17 species in Laos. These figures are extremely low, compared to the neighbouring Malesian flora in which more than 500 species have been recorded.

The present-day flora of open woodlands in Southeast Asia is apparently the outcome of three major factors. First, a dominant influence of the Asiatic dense deciduous forest flora (*Buchanania, Careya, Emblica, Terminalia*); Second, an obvious influence of the Asiatic dense rain forest flora (most of the dipterocarps, *Irvingia, Pahudia, Sindora, Syzygium*). Third, as is the rule for all secondary plant communities, biotic factors which have favoured the invasion of alien plants (such as *Eupatorium*) and of pantropical species (many grasses). These factors have also considerably helped the eastward migration of xeromorphic plants such as *Capparis, Randia* and *Ziziphus*.

Taxonomic arguments and vicariation

It is well known that the Malaysian floristic region is the dispersion centre of dipterocarps. Practically all species are presently found only in rain forests. However, during the course of palaeoclimatic oscillations this family has given birth to closely related genera and species, which were able to thrive under drier climatic conditions, either in Africa (genus *Monotes*) or in Asia (genus *Pentacme*).

Regarding the taxonomic position of open-forest trees, many problems remain to be solved. Species are often split into several subspecies or ecotypes having a distinct geographical distribution. *Pentacme suavis* A.DC., for instance, has two varieties in Thailand: *P. suavis* var. *siamensis* and var. *tomentosa*. *Dipterocarpus obtusifolius*, which has a wide ecological distribution (evergreen forests, teak-bearing mixed deciduous forests) is extremely polymorphic. This leads to many taxonomic problems. For instance, *D. obtusifolius* Teysm. var. *costatus* Smit. has also been described as a separate species *D. costatus* Gaertnf. Similarly, *Shorea obtusa* Wall. var. *kohchangensis* Heim and *Shorea thorelii* Laness. are synonyms. *Terminalia alata* Heyne ex Roth. (= *T. tomentosa* W. and A.) is known under three varieties: *coriacea, crenulata* and *typica* (see Gamble, 1916). Theoretically, each of these tree varieties has its own range and ecological requirements. Most of the woodland species seem to come from the dense forest.

This vicariation also concerns the genus *Pentacme*, which is very closely related to the genus *Shorea*. As a matter of fact, it was only in 1870 that Kurz transferred it from *Shorea* Roxb. to *Pentacme*

A.DC. (Smitinand, 1958, p. 63). The main floristic difference between the two genera is very slight, based mainly on the stamens (shape of the connectives). A detailed taxonomic revision of all these plants, and also of the Lythraceae (*Lagerstroemia* group), is badly needed.

CONCLUSION

The majority of Southeast Asian "open forests" and savanna woodlands should be considered as sub-climax communities, stabilized by the influence of regular burning. Their highly simplified floristic structure results from drastic selection; only fire-resistant species capable of regeneration by vegetative mechanisms are able to survive in such habitats.

As a matter of fact, the dynamics of the open-forest vegetation remains to be studied in Southeast Asia. Natural seedling establishment, growth and response to light have not yet been adequately studied. However, it can be assumed that most of the tree seedlings die from exposure to poor soil conditions, are killed by grass fires, or cannot compete with the herbaceous cover which rapidly invades the gaps.

In Thailand, the increasing density of domestic animals accelerates the regressive evolution of these plant communities. In Cambodia, Vietnam and southern Laos, open forests are the best remaining habitat for larger wild ungulates, which include such rare species as the koupreys (*Bos sauveli*) (in Cambodia), bantengs (*Bos banteng*) and gaurs (*Bos gaurus*). However, their density is too low to affect the stability of plant communities.

The effect of shifting cultivation varies considerably from one place to another. It is strongly felt in Thailand and in the *hauts plateaux* of Vietnam, but is negligible in the driest parts of Cambodian lowlands. The result is the same everywhere: a destruction of dense forests, accompanied by an increasing frequency of ground fires. This leads to open forests in which only trees with a wide ecological range, allowing them to survive adverse edaphic conditions and recurrent fires, are able to survive.

REFERENCES

Aubréville, A., 1957. Au pays des eaux et des forêts. Impressions du Cambodge forestier. *Bois For. Trop.*, 52: 49–56.

Bejaud, M., 1934. *La forêt cambodgienne*. 190 pp. (mimeograph).

BIOTROP Research Projects, 1976, 1977. *Quantitative Studies of the Seasonal Tropical Forest Vegetation in Thailand*. Faculty of Forestry, Kasetsart University of Bangkok, Bangkok, Annual Report No. 1 (1976): 240 pp.; Annual Report No. 2 (1977): 369 pp.

Boulbet, J., 1979. Le Phnom Koulen et sa région. *Bull. E.F.E.O.*, 12: 1–136.

Boulbet, J., in press. *The Korat Plateau*. Ecole française d'Extrême-Orient, in press.

Castagnol, E.M., 1950. Problèmes de sol et l'utilisation des terres en Indochine. *Arch. Inst. Rech. Agron. Indochine*, 7: 61 pp.

Champsoloix, R., 1959. A propos de la forêt claire du Sud-Est asiatique. *Bois For. Trop.*, 64: 3–11.

Consigny, J., 1928. *Tournée de Pursat à Chhuk*. Mimeograph.

Cooling, E.N.G. and Gaussen H., 1970. In Indo-China *Pinus merkusiana* sp. nov. not *P. merkusii* Jungh & De Vries. *Trav. Lab. For. Toulouse*, VIII(1), art. VII: 8 pp.

Crocker, C.D., 1963. *Carte générale des sols du Cambodge à 1/1.000.000*. Royaume du Cambodge, Secrétariat d'Etat à l'Agriculture, Phnom Penh.

C.S.A., 1956. *C.S.A. Specialists' Meeting on Phytogeography. Yangambi 28th July–8th August, 1956. C.S.A./C.C.T.A. Publ., 22*: 35 pp.

Delvert, J., 1961. *Le paysan cambodgien*. Mouton et Cie, The Hague, Paris, 740 pp.

Dudal, R., Moormann, F. and Riquier, J., 1974. Soils of humid tropical Asia. In: *Natural Resources of Humid Tropical Asia. Nat. Resour. Res.*, 12: 159–178.

FAO/UNESCO, 1975. *Soils Map of the world. 1:5,000,000, 1, Legend*. FAO, Rome, 62 pp.

Fontanel, J., 1967. *Ratanakiri. Etude du milieu natural d'une région frontière du Cambodge*. Thesis, University of Grenoble, Grenoble, 373 pp.

Gamble, J.S., 1916/1935. *Flora of the Presidency of Madras*. Botanical Survey of India, Calcutta, 1389 pp. (3 vols).

Gaussen, H., Legris, P. and Blasco, F., 1967. Bioclimats du Sud-Est asiatique. *Trav. Sec. Sci. Tech. Inst. Fr. Pondichéry*, 3(4): 115 pp.

Khiou-Bon-Thonn, 1955. *Le climat du Cambodge*. Royaume du Cambodge, Ministère des Travaux Publics, Service Météorologique, Pnom Penh, 240 pp.

Küchler, A.W. and Sawyer, J.O., 1967. A study of the vegetation near Chiengmai: Thailand. *Trans. Kansas Acad. Sci.*, 70: 281–348.

Lecomte, O., 1969. *Flore du Laos du Cambodge et du Vietnam. No. 10; Combretaceae*. Muséum national d'Histoire naturelle, Paris, 119 pp.

Lecomte, H. and Gagnepain, F., 1907–1951. *Flore générale de l'Indochine*. Masson, Paris.

Legris, P. and Blasco, F., 1972. Carte internationale du Tapis végétal, "Cambodge", 1/1.000.000; notice. *Trav. Sec. Sci. Tech. Inst. Fr. Pondichéry, H. S.*, No. 11: 240 pp.

Ly Van Hoi, 1952. *Contribution à l'étude des forêts claires du Sud*

Laos. Centre de Recherches scientifiques et techniques, Saigon, 100 pp.

Martin, M., 1971. *Introduction à l'ethnobotanique du Cambodge.* Editions du Centre National de la Recherche Scientifique, Paris, 257 pp.

Maurand, P., 1938. *L'Indochine forestière.* Inst. Rech. Agron. For. Indochine, Hanoi, 150 pp.

Maurand, P., 1943. *L'Indochine forestière.* Inst. Rech. Agron. For. Indochine, Hanoi, 254 pp.

Maurand, P., 1965. Les forêts du Vietnam. *Bull. Trimest. Assoc. France-Vietnam*, 1965: 15–28.

Moormann, F.R., 1961. *The soils of the Republic of Vietnam.* Republic of Vietnam, Ministry of Agriculture, Saigon, 66 pp.

Ogawa, H., Koda, K. and Kira, T., 1967. A preliminary survey on the vegetation of Thailand. *Nature Life South East Asia*, 1: 21–157.

Ogino, K., 1976. Human influences on the occurrence of deciduous forest vegetation in Thailand. *Mem. College Agric. Kyoto Univ.*, 108: 55–74.

Pendleton, R.L., 1941. Laterite and its structural uses in Thailand and Cambodia. *Geogr. Rev.*, 31: 177–202.

Pendleton, R.L. and Sarot Montrakun, 1960. The soils of Thailand. In: *Proc. 9th Pac. Sci. Congr.*, 18: 12–32 (1 soil map).

Republic of Vietnam, 1969. *Vegetation map, 1/1,000,000.* National Geographic Service, Dalat.

Robbins, R.G. and Smitinand, T., 1966. A botanical ascent of Doi Inthanond. *Nat. Hist. Bull. Siam Soc.*, 21: 205–227.

Rollet. B., 1952. *Etude sur les forêts claires du Sud Indochinois.* Recherches forestières, Saigon, 250 pp.

Rollet, B., 1953. Notes sur les forêts claires du Sud de l'Indochine. *Bois For. Trop.*, 31: 3–13.

Rollet, B., 1962. *Inventaire forestier de l'Est Mékong.* Rapport FAO, No. 1500, Rome 184 pp.

Rollet, B., 1972. La végétation du Cambodge. *Bois For. Trop.*, 145: 23–38; 146: 3–20.

Samapuddhi, K., 1957. *The Forest of Thailand and Forestry Programs.* Royal Forest Department, Ministry of Agriculture, Bangkok, 70 pp.

Schmid, M., 1956. Note sur les formations végétales des hauts plateaux du centre Vietnam et des régions limitrophes. In: *Humid Tropics Research; Proceedings of the Kandy Symposium.* UNESCO, Paris, Recherches sur la zone tropicale humide, 1958, pp. 183–194.

Schmid, M., 1958. Flore agrostologique de l'Indochine. *Agron. Trop.*, 13: 1–320.

Schmid, M., 1960. La végétation naturelle et la mise en valeur du territoire du Vietnam. In: *Causerie sur le développement des ressources naturelles au Vietnam.* Secrétariat d'Etat à l'Agriculture, Saigon, 8: 39–56.

Schmid, M., 1963. Contribution à la connaissance des sols du Vietnam: le massif sud-Annamitique et les régions limitrophes. *Cah. ORSTOM, Sér. Pédol.*, No. 2: 15–72.

Schmid, M., 1964. Aperçu sur la végétation occupant les alluvions récentes de la partie méridionale de l'Indochine. In: UNESCO, *Les problèmes scientifiques des deltas de la zone tropicale et leurs implications. Actes du colloque de Dacca*, pp. 235–241.

Schmid, M., 1974. Végétation du Vietnam. Le massif sud-Annamitique et les régions limitrophes. *Mém. ORSTOM*, No. 74: 1–243.

Schmid, M., Godard, D. and De la Souchère, P., 1952. Les sols et la végétation au Darlac et sur le Plateau des Trois Frontières. *Arch. Rech. Agron. Cambodge, Laos, Vietnam*, 8: 1–107.

Schnell, R., 1962. Remarques préliminaires sur quelques problèmes phytogéographiques sur Sud-Est Asiatique. *Rev. Gén. Bot.*, 69: 301–365.

Smitinand, T., 1958. Identification keys to the Dipterocarpaceae of Thailand. *Nat. Hist. Bull. Siam Soc.*, 19: 57–83.

Smitinand, T., 1960. Shifting cultivation in Thailand. In: UNESCO, *Symposium on the Impact of Man on Humid Tropics Vegetation, Goroka*, p. 215

Smitinand, T., 1962. *Types of Forests of Thailand.* Ministry of Agriculture, Royal Forest Department, Bangkok, 12 pp.

Smitinand, T., 1963. Studies on the flora of Thailand Dipterocarpaceae and Lythraceae. *Dan. Bot. Ark.*, 23: 47–56.

UNESCO, 1973. International classification and mapping of vegetation. *Ecol. Conserv.*, 6: 92 pp.

U.S. Agency for International Development, 1978. *Land Use Map.* Committee for Coordination of Investigations of the Lower Mekong Basin, Bangkok. Land use map.

Vidal, J., 1960a. La végétation du Laos. *Trav. Lab. For. Toulouse*, 5, Section I, Vol. 1–2, 570 pp.

Vidal, J., 1960b. Les forêts du Laos. *Bois For. Trop.*, 70: 5–21.

White, R.O., 1974. Grasses and grasslands. In: UNESCO, *Natural Resources of Humid Tropical Asia. Nat. Resour. Res.*, 12: 239–262.

Whitmore, T.C., 1975. *Tropical Rain Forests of the Far East.* Clarendon Press, Oxford, 282 pp.

Zimmerman, C.C., 1937. Some phases of land utilization in Siam. *Geogr. Rev.*, 27: 383–393.

Chapter 9

TROPICAL SAVANNAS OF AUSTRALIA AND THE SOUTHWEST PACIFIC

A.N. GILLISON

INTRODUCTION

Definition

The term "savanna" conjures up a number of structural and floristic complexes that reflect different usage between continents and regions. In this treatment I will not explore these various uses as there will be unavoidable conflict. It is generally considered however that graminoids are central to the definition (Beard, 1953; Ramia, 1968; Bourlière and Hadley, 1970; Gillison, 1970a; Gillison and Walker, 1981; Walker and Gillison, 1982) and that savanna (*per se*) may be a variable complex of graminoids, with or without shrubs and trees. In Australia, the term "savanna" has been loosely applied to mean a grassy formation with a park-like appearance, usually with widely spaced trees with rounded crowns (Wood, 1950; Williams, 1955). This is similar to the African usage, but because much of Australia has formations with tree spacing varying from almost pure grassland to forest conditions, the term "woodland" has tended to replace savanna.

At a global level, savanna formations have been examined most intensively in South America and Africa. More recently, some attention has been given to the extent of savanna complexes in Australia (Walker and Gillison, 1982) and Papua New Guinea (Paijmans, 1976) and to a lesser extent the Pacific Islands. Some unique savannas particularly in Papua New Guinea require specific definition, in particular the high-altitude tree-fern savannas of Paijmans and Löffler (1974) and bamboo savannas. The latter are included if only because they have a significant "graminoid" component. For this chapter the definitions proposed by Walker and Gillison (1982) have been extended and modified.

The following definitions may be compared with the structural categories recognized by Specht (1981) for Australia as shown in Table 9.1. "Cover" refers to foliage projective cover of Specht (1981).

Grass savanna and grassland are interchangeable terms and refer to graminoid dominated formations where woody plants may be present as widely spaced individuals (up to 0.2% cover).

Bamboo savanna is a unique woody graminoid formation which may occur either as a local formation with complete canopy cover or as a significant understorey component dominated by other woody plants with <70% cover, or in grassland as widely spaced clumps with >0.2% cover.

Woodland savanna is a formation where single-stemmed woody plants over 3 m tall occur with >0.2% cover and where there is a >2% graminoid cover. Where trees <3 m tall occur with >0.2% cover the formation is very low woodland savanna. In some areas, trees may be replaced by tree ferns, pandans, palms or cycads.

Shrub savanna is a formation where multistemmed rather than single-stemmed woody plants occur with >0.2% cover and where there is a >2% graminoid cover.

As 'grasslands" *sensu stricto* are to be dealt with elsewhere in this series, only those savanna types with a significant woody component (>2% cover) will be dealt with in this chapter.

TABLE 9.1

Structural woody plant formations in Australia (from Specht, 1981)

Life form of tallest stratum		Foliage projective cover of tallest stratum				
		100–70% (4)[2]	70–50% (3+)[2]	50–30% (3−)[2]	30–10% (2)[2]	<10% (1)[2]
Trees[1] >30 m	(T)[2]	tall closed-forest	tall forest	(tall open-forest)[3]	(tall woodland)[3]	–
Trees 10–30 m	(M)	closed-forest	forest	open-forest	woodland	open-woodland
Trees <10 m	(L)	low closed-forest	low-forest	low open-forest	low woodland	low open-woodland
Shrubs[1] >2 m	(S)	closed-scrub	scrub	open-scrub	tall shrubland	tall open-shrubland
Shrubs 0.25–2 m						
sclerophyllous	(E)	closed-heathland	heathland	open-heathland	shrubland	open-shrubland
non-sclerophyllous	(C)	–	–	low shrubland	low shrubland	low open-shrubland
Shrubs <0.25 m						
sclerophyllous	(D)	–	–	–	dwarf open-heathland (fell-field)	dwarf open-heathland (fell-field)
non-sclerophyllous	(W)	–	–	–	dwarf shrubland	dwarf open-shrubland

[1] A tree is defined as a woody plant usually with a single stem; a shrub is a woody plant usually with many stems arising at or near the base.

[2] Symbols and numbers given in parentheses may be used to describe the formation, e.g. tall closed-forest = T4, hummock grassland = H2.

[3] Senescent forms of tall forest.

Study area

I have chosen to examine only the Australian biogeographic region. For present purposes, although this conflicts with the systems of Dasmann (1973) and Udvardy (1975), the region includes Papua New Guinea and the surrounding islands, the Timor Islands, and those islands of Micronesia and the southwest Pacific Basin that show attenuated but distinct phytogeographic relationships with this group of which western Samoa forms the eastern boundary (Fig. 9.1).

Within this region, savannas as I have defined them are barely significant in the smaller islands, and it is for this reason that the review deals mostly with the savannas of Australia and Papua New Guinea.

Use of the term "tropical"

In biogeographic terms, tropical has at best only a broad meaning at a global level (cf. Oliver, 1979). Within the tropical zone, there are numerous associations of biota at high altitudes that exhibit strong affinities with associations at "temperate" or higher latitudes and lower altitudes. Therefore, although the term tropical is applied in a geographic sense, I

have chosen a more specific environmental framework within this context that reflects ecological change along recognizable bioclimatic gradients. In this way, biotic associations with similar environments in both tropical and temperate regions may be more meaningfully related (see also Goulissachvili, 1963).

Historical influences

Within the region under study, there are broad divergences in savanna types that reflect differences in landscape history between continental Australia and the island groups. The characteristic features of the northern Australian savannas are ancient denuded or weathered landscapes of relatively low relief, usually with low nutrient soils, which have probably been fired by man since his advent on the continent some 32 000 years ago (Barbetti and Allen, 1972). In more recent times these lands have been grazed by domestic animals and more frequently fired. Remnant, closed "monsoon" forests in fire-shadow refugia suggest that much of the present northern Australian savanna may have been previously forested (Specht, 1958a, b; Kershaw, 1975, 1978; Stocker and Mott, 1981). Fluctuating climatic cycles during the Quaternary

Fig. 9.1. Map of Australia and the southwest Pacific. Region under study is indicated by dotted lines.

combined with the comparatively recent arrival of man have probably caused the retreat of these closed seasonal woody types to their present refugia. The structural and floristic content of these forests show close affinity with some of the seasonally dry "monsoon' forests of southern India, Malaysia (see also Whitmore, 1975) and South Africa. This is suggestive of a possible widespread forest type that may have a Gondwanian origin, where the subsequent breakup of the supercontinent left remnants of these forests that are for the most part highly drought-resistant but very susceptible to fire. Certain woody floristic elements (e.g. Combretaceae, Ebenaceae, Fabaceae, Sterculiaceae) have remained from these forests to occupy the woodland savannas in these areas. It is also likely that the occurrence of these elements in the woodland savannas of Papua New Guinea may reflect a former Miocene land connexion across the Sahul shelf with northern Australia (cf. Nix and Kalma, 1972).

Evidence for former land connexions with the island chains of the southwest Pacific is not clear. The extent of the Fijian Plate, the relatively recent emergence of the New Hebridean chain and the deep trenches separating New Caledonia from its neighbours make it difficult to speculate upon links between the similar floristic entities that presently occupy savanna landscapes in these areas. The phytogeography within this region (Van Balgooy, 1971; Gillison, 1975a) and the floristic makeup of some of the coastal savanna types (Gillison, 1970a; Heyligers, 1972a, b) suggest that many of the woody savanna species may be littoral in origin.

The major exceptions to this are *Eucalyptus* and to a lesser extent *Acacia* and the Proteaceae (*Banksia, Grevillea, Stenocarpus*) that are primarily Australian (cf. Johnson and Briggs, 1975). These have outliers in Papua New Guinea where in many cases their abundance can be related to the increasing activity of man in promoting inland savanna formations or, in some cases, to ultramafic outcrops. In Vanuatu (the New Hebrides) however, the only acacias (*A. spirorbis* and *A. simplicifolia*)

186

A.N. GILLISON

commonly occupy littorine and riparian habitats. There seems little doubt that the savannas of Papua and northern Australia represent parts of what was a connected floristic region. Good (1960, 1963) has suggested that the lowland Papuan flora is derived from that of Australia and that the remainder of the island represents a floristic disjunct. This may be so at present, but palynological and other evidence in northern Australia (Kershaw, 1978) suggests that there was a united floristic region of which there are extant floristic groups in Papua New Guinea. The development of dryland savanna in the Trans-Fly region of Papua New Guinea is almost certainly due in large part to hunting activities by man who has used fire as a tool (Brass, 1956; Bulmer, 1968; Gillison, 1972). In this respect the more recent historical influences on the savannas of the Papuan Trans-Fly and those of northern Australia are probably similar.

The combination of high seasonality and a similar fire regime has produced savannas of like morphology in both areas. Of the seven *Eucalyptus* species in Papua New Guinea, six are associated with savannas. Savanna genesis in forests of the dry seasonal southeast of Papua New Guinea is continuing. The increase in urban and suburban population in the Port Moresby region in recent years has been accompanied by a change in nearby seasonal closed forest to open grassland and thence to stabilized woodland savanna (Fig. 9.2) that is now subject to annual fires.

The genesis and maintenance of many "tropical" Australian savannas and those in the southwest Pacific are for the most part directly related to fires — largely man-made, although lightning plays a significant, albeit undocumented, role. The Australian continental savanna development is somewhat unique in that non-agricultural aboriginal man simply fired the landscape and was not involved in slash-and-burn subsistence agriculture.

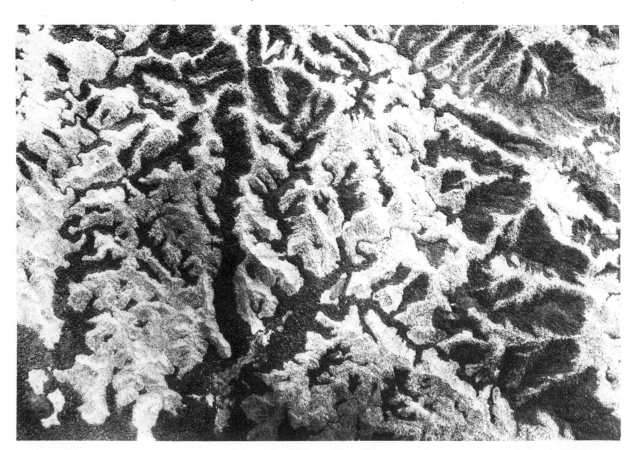

Fig. 9.2. Interdigitating megatherm seasonal eucalypt woodland savanna and semi-deciduous dry seasonal forest in southeast Papua (Province 3).

In the southwest Pacific islands, subsistence agriculture, accompanied by decreased rotation cycles and increased population pressures has also contributed to savanna development. Throughout the region, most authors agree that fire is a dominant factor in the development of grassland (savanna) formations (Lane-Poole, 1925a, b; Archbold and Rand, 1935; Brass, 1938, 1956, 1964; McAdam, 1954; Bateman, 1955; Havel, 1960; Perry, 1960; Robbins, 1960; Van Royen, 1963a, b; Reiner and Robbins, 1964; Taylor, 1964a, b; Fitzpatrick et al., 1966; Heyligers, 1965, 1967, 1972a; Paijmans, 1967, 1969; Gillison, 1969, 1970a, b, 1972; Coode and Stevens, 1972; and Paijmans and Löffler, 1974). In New Zealand, Cumberland (1963) has suggested that about 2.8×10^6 ha of forest were converted to grassland by the hunting activities of the Maori, and in Malaysia Dilmy (1960) estimated that 17×10^6 ha of grassland had been derived from man-made fire. Whether savannas are "natural" or "derived" is largely an academic point as, while there is no doubt that much has been "derived" from forested habitats by man in northern Australia and Papua New Guinea, man has nevertheless been an integral and therefore "natural" part of the savanna ecosystem for probably at least 30 000 years. The influence of "natural" phenomena such as frost, cold air ponding and fires caused by lightning strikes is evident in many areas of Papua New Guinea (Van Royen, 1963a; Brass, 1964; Wade and McVean, 1969; Gillison, 1972; Brown and Powell, 1974; Smith, 1977; Walker, 1981) although savanna genesis by these means is difficult to document. The incidence of lightning strike may be more significant than previously supposed; in Darwin, in northwestern Australia, within-horizon lightning strikes have been recorded in excess of 1000 in one month (R.J. Thistlethwaite, pers. comm.). In Papua New Guinea, lightning strikes have been recorded in coconut and cocoa plantations (Shaw, 1968). In some of the most remote areas of highland Papua New Guinea many such grassy pockets are frequented by man and are fired for hunting purposes so that the issue of primary influence again becomes clouded.

Periodic inundation is rarely documented as an environmental influence and its importance in this instance depends on the definition of what is a savanna. In the present chapter, the vast tracts of seasonal dry and wetland savanna complexes in Papua New Guinea and parts of northern Australia are greatly influenced by either periodic or episodic flooding, or else by a permanently high water table. The savannas of the seasonal Vunapa floodplain in southeastern Papua New Guinea reflect such a drainage mosaic. Fluctuating water levels in many cases do not allow progression to a "stable" savanna type but rather to a mixed vegetation type that is in a continual state of flux.

Classification of savannas

Within the region outlined in Fig. 9.1, I have classified savannas firstly within a bioclimatic context and secondly by the use of discrete structural and physiognomic criteria. The woodland savanna areas in Australia and Papua New Guinea to be discussed here are shown in Fig. 9.3.

Bioclimate

There are two reasons for using bioclimate: firstly, terms such as "tropical" or "perhumid" or "montane" have no easily defined quantitative functional relationship with vegetation and therefore cannot be used quantitatively to compare similar plant/environment situations at a biogeographic level; secondly, the establishment of a functional expression that employs physical aspects of plant and environment is more likely to provide a uniform and rational basis for comparing and describing plant and animal associations at a range of biogeographic scales, both within and between continents. For present purposes, climatically based systems such as those of Köppen, Thornthwaite or Holdridge are either too circular or else too rigid in their expression of plant/ environment criteria, and in these systems the important element of seasonality tends to be subjugated. I have chosen bioclimatic criteria because the criteria used by Dasmann (1973) or Udvardy (1975) for their divisions have little functional basis for purposes of describing vegetation in terms of the physical environment.

For these reasons, I have adopted the more explicit gradients of plant growth response mentioned by Nix (Ch. 3) as a function of several physical environmental measures [see also Walker and Gillison (1982) for details]. By interpolating known vegetation boundaries and plant life-form distributions within this array of gradients, it is

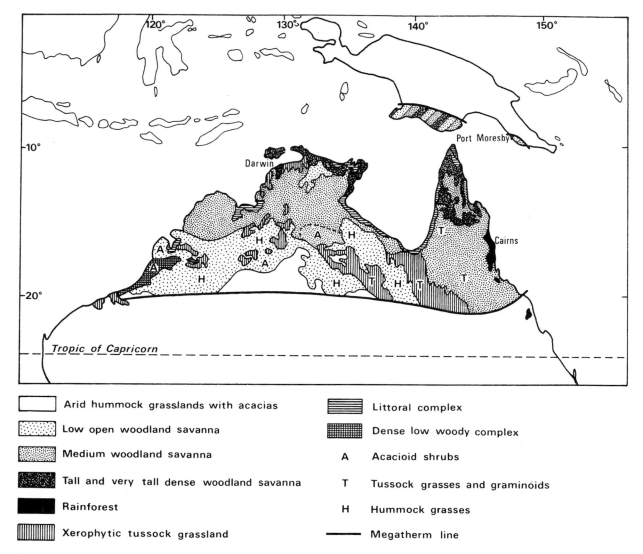

Fig. 9.3. Areas of woodland savanna in Australia and Papua New Guinea described in this chapter (grassland terms after Moore, 1970).

possible to arrive at patterns that provide a means of comparing vegetation on a geographic as well as a climatic basis. Thus, the tropical "montane" savannas of Papua New Guinea can be compared with the "temperate" savannas of southeastern Australia at much lower altitudes, with which they share considerable structural and to some extent floristic affinities. On this basis, a tentative bioclimatic map has already been produced for Australian woodland savannas by Gillison and Walker (1981) and Walker and Gillison (1982), and this is now provisionally extended to include the island of New Guinea (Fig. 9.4). For much of the area outside Australia the definition and re-

liability of boundaries is limited by lack of data. The map is divided into three major Regions that represent the megatherm, mesotherm and microtherm plateaus defined by Nix, with overlap zones. Regions are divided into Sub-Regions and Provinces and further defined by whether or not they exhibit strong seasonality (i.e. >60% coefficient of variation of soil moisture availability during a year — from a range of coefficients computed by Nix for weekly mean values). Thus, Province No. 1 is coded as AIS where A = megatherm, I = the Sub-Region (i.e. the lowest gradient level of the Region), and S = seasonal. The system together with the gradient range for each Province is outlined in Table 9.2.

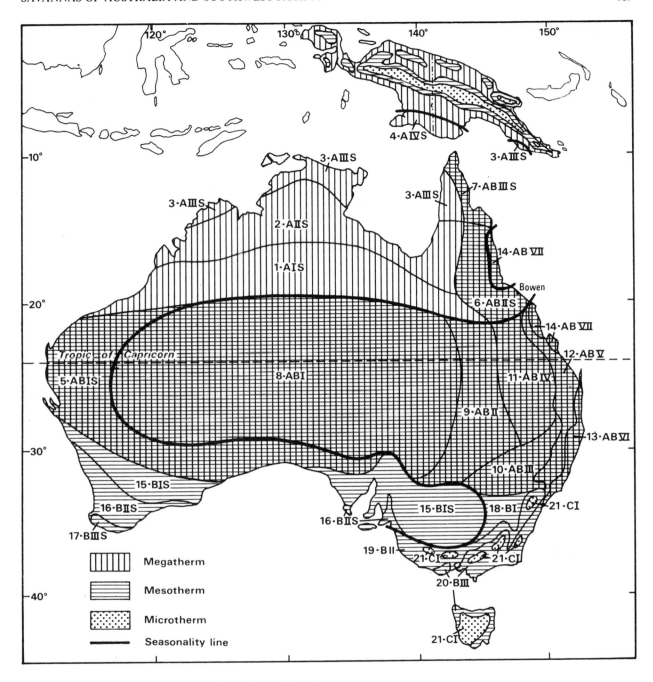

Fig. 9.4. Provisional bioclimatic regions of Australia and Papua New Guinea.

In using a bioclimatic framework of this kind the aim is to relate the distribution of major woody vegetation formations to primary environmental criteria which in the present case I believe to be climatic. At a secondary level there are geomorphic, edaphic and land-use influences that are often difficult to relate to the broader aspects of climate.

These influences are discussed in some of the examples presented in this chapter.

Structure and physiognomy

Throughout the region, woody savannas are rarely homogeneous in space, and this is particularly evident in megatherm southeast Papua

TABLE 9.2

Provincial bioclimatic Regions and Provinces for Australia and Papua New Guinea

Province No.	Code	Bioclimatic Region	Bioclimatic Sub-region	Seasonality regime[1] (cv%)	Thermal optima[2] (°C)	Gradient range[3]
1	AIS	megatherm	I	60–100	28/35	0.1–0.2
2	AIIS		II	100–113	28/35	0.2–0.4
3	AIIIS		III	81–90	28/35	0.4–0.5
4	AIVS		IV	>90	28/35	0.5–0.7
5	ABIS	megatherm–	I	51–60	28/19	0–0.1/0–0.15
6	ABIIS	mesotherm	II	50–104	28/19	0.15–0.4/0.15–0.3
7	ABIIIS		III	90–100	28/19	0.4–0.55/0.15–0.4
8	ABI	megatherm–	I	30–60	28/19	0.1/0–0.15
9	ABII	mesotherm	II	30–60	28/19	0.15–0.2/0–0.3
10	ABIII		III	30–60	28/19	0.1–0.15/0.2–0.3
11	ABIV		IV	28–50	28/19	0.2–0.25/0.2–0.3
12	ABV		V	20–40	28/19	0.25–0.4/0.3–0.45
13	ABVI		VI	9–30	28/19	0.15–0.45/0.45–0.75
14	ABVII		VII	5–60	28/19	0.5–0.7/0.45–0.75
15	BIS	mesotherm	I	80–100	19	0.15–0.3
16	BIIS		II	70–97	19	0.3–0.45
17	BIIIS		III	50–70	19	0.45–0.5
18	BI	mesotherm	I	85–60	19	0.15–0.3
19	BII		II	43–60	19	0.3–0.45
20	BIII		III	20–48	19	0.45–0.5
21	CI	microtherm (Aust.)	I	14–30	12	0.3–0.5 (approximate only for PNG)

[1]Seasonality as indicated by coefficient of variation (per cent) of weekly mean values for water availability throughout the year. A province is seasonal (S) if cv% $\geqslant 60$.

[2]Thermal optima for plant growth set at megatherm (28° and 35°C for C_3 and C_4 plants), mesotherm (19°C) and microtherm (12°C).

[3]Scale of 0–1 growth index for plant growth within a specified thermal optimum, taking into account known radiation and moisture levels (for further details see Nix (Ch. 3), and Walker and Gillison, 1982).

New Guinea where many savannas are still being formed. It is recognized therefore that environmental continua create difficulties in establishing a simplistic classification with arbitrarily defined physiognomic and structural classes. For this reason the combined ordinative classificatory system of Gillison and Walker (1981) is used, where the known range of structural data (height and cover of dominant stratum) is plotted in a "structuregram" (Fig. 9.5). This allows the reader to appreciate at a glance the structural variation one can expect within an otherwise physiognomically defined woody savanna type, such as a "box" or "tree-fern" savanna.

The term "structure" is used to define spatial dimensions whereas "physiognomy" refers to finer-scale morphology. This is consistent with Webb's (1959) recommendations and is in general usage within Australia, although there is conflict with the terminology of Fosberg (1973) for classification at a global level.

For convenience, the height classes of Gillison and Walker (1981) and the foliage projective cover classes of Specht (1981) are applied with the modification that "woodland" and "open forest" are combined (see Table 9.1 and Fig. 9.5). The use of floristic names in savanna classifications becomes limiting as geographic range is extended. For example, Gilbert River box (*Eucalyptus microneura*) and McArthur River box (*E. tectifica*) occur in different parts of northern Australia, but are physiognomically similar and often occupy similar

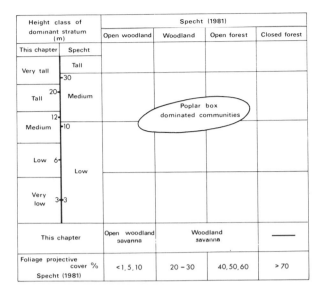

Fig. 9.5. Structuregram, showing the method used to indicate the structural (height/cover) range for any given bioclimatic Province or physiognomic vegetation type. (See examples in Appendix 1.)

Adaptations of savanna woodland plants

As with savanna plants in other parts of the world, there is a strong association between morphological character and the incidence of fire and environmental extremes such as seasonal drought and frost.

This is evident in many of the xeromorphic and sclerophyllous features to be found in both megatherm and microtherm extremes, and is apparent in bark and leaf characteristics and in the development and positioning of perennating organs.

Bark characters

In Australia, in the widespread eucalypt-dominant megatherm savanna woodlands, bark types are highly variable but indicate adaptation to radiation from both the sun and frequent fires. Some are highly reflective "gum-barked" (*Eucalyptus alba*, *E. papuana*) or else dark and deeply fissured as in the ironbarks (*E. crebra* and *E. jensenii*), or scaly-barked as in the bloodwoods (*E. polycarpa*, *E. dichromophloia*). Other fibrous-barked species such as Darwin stringybark (*E. tetrodonta*) and boxes (*E. microneura* and *E. tectifica*) are common, and in some cases are fibrous-barked only on the lower half of the trunk which may indicate adaptation to grass fires (*E. microneura*, *E. microtheca*, *E. miniata*, *E. tereticornis*). Bark characters are a useful diagnostic for most eucalypts, but are difficult to describe in quantitative terms. The term "box"-barked or "box" refers to a eucalypt that has a fibrous bark which is so named because of its supposed resemblance to the European box (*Buxus sempervirens*). Distinctions between different fibrous-barked eucalypts are best indicated visually (see Fig. 9.6). For additional bark descriptions, the reader is referred to Hall et al. (1970). Many other genera in these savannas also exhibit rough-barked features (*Atalaya*, *Casuarina*, *Erythrophleum*, *Petalostigma*) or a high bark albedo (some *Eugenia* spp.). The paperbarks (*Melaleuca* spp.) are most common in seasonally inundated areas which may also suffer periodic fires. This bark type consists of many laminations of thin papery bark which appears to protect the tree adequately against all but the most severe fires, and at the same time may provide resistance to water loss. Occasionally, rather bizarre fire-resistant adaptations occur as in

environments. Hence the use of the physiognomic term "box" rather than the floristic epithets in such cases is more ecologically meaningful, and it is partly for this reason that terms such as "iron-bark", "paperbark", "pandan" and so on are widely used in Australian literature and in the field.

The use of the terms "tree" and "shrub" present some difficulties within the Australian context. The European concept of a shrub is a woody plant that branches at or near the ground and which is "of smaller structure than a tree" (Carpenter, 1938). In parts of southern mesotherm Australia, multistemmed woody plants in excess of 20 m height are frequent and cannot be regarded easily as "trees". Conversely, in megatherm regions single-stemmed woody plants <1 m tall are also common and cannot be regarded simply as "shrubs". There is a conspicuous trend in the development of multistemmed (sympodial) woody growth in microtherm and very seasonal mesotherm areas, and single-stemmed (monopodial) growth in megatherm areas (except in very oligotrophic conditions), although monopodial growth is common in the mesotherm moist sclerophyll forests of eastern and south-western Australia. Leaf-fall (deciduousness) is an additional descriptor where the dominant stratum is periodically completely leafless.

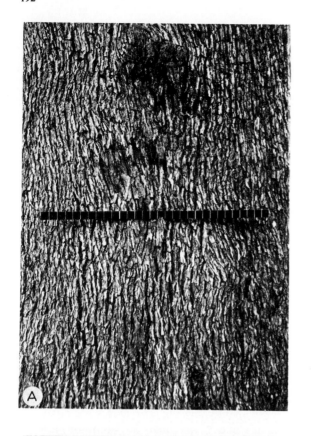

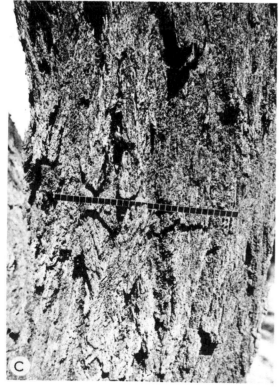

the banana-like growth habit of *Ensete calospermum* which is locally common with *Albizia procera* in savannas of the Markham rift valley of Papua New Guinea. This 4 m tall herbaceous species develops a swollen, water-filled bole that seems to resist all but the most intense grass fires.

Further information about fire-adaptation in the Australian region is provided by Gill (1981a, b) and Walker (1981).

Leaf characters

For the most part, leaves tend to be evergreen sclerophylls. In the megatherm savannas, most eucalypts have isobilateral pendulous leaves so that thermal effects from zenithal solar radiation are reduced. Like those of many other genera, they are leathery and high in volatile oils. Leaves tend to have thickened or waxy cuticles and in many cases are microphyllous or nanophyllous (*Desmodium*, *Petalostigma*) although large mesophylls also exist (*Barringtonia*, *Gardenia*). Highly reflective megaphylls are widespread in savannas with *Pandanus*, or palms (*Livistona*) and cycads (frond-megaphylls). Quite often leaves *sensu stricto* are replaced by leaf-like organs as in the phyllodineous species of *Acacia*, or phylloclades as in *Casuarina* and *Gymnostoma*, or else, if they are multi-costate, resemble phyllodes as in *Melaleuca* and some *Grevillea* spp. Green stems or aphyllous branchlets are locally common (*Apophyllum*, *Capparis*, *Sarcostemma*).

In the microtherm region of Papua New Guinea, highly reflective microphylls and nanophylls are common in *Rhododendron* and *Vaccinium*, with frond-megaphylls in the tree ferns *Cyathea* and *Dicksonia* where pinnae tend to be rolled, often with a dense indumentum.

Completely deciduous savanna plants rarely occur outside the megatherm seasonal environment. They usually indicate seasonally wet and dry cracking clays although they may be evident on sandy coastal soils (e.g. remnant dune systems) or on limestone karst. Typically deciduous genera are *Bauhinia*, *Bombax*, *Cochlospermum*, *Sterculia* and *Terminalia*. Although *Eucalyptus* is usually regarded as an evergreen genus, in the megatherm seasonal savannas species such as *E. alba*, *E. clavigera*, *E. confertiflora* and *E. papuana* are sometimes deciduous (cf. Story, 1969), but these are the exception rather than the rule.

Perennating organs

The position and kind of perennating organs usually indicate a recurring environmental condition (*sensu* Raunkiaer, 1934). In seasonal megatherm areas, with high fire frequency, shoots from latent epicormic buds on the upper stems of many myrtaceous species within the genera *Eucalyptus*, *Eugenia*, *Melaleuca*, *Tristania* and *Xanthostemon* allow recovery after fire or following defoliation from severe water stress or wind.

Several eucalypts have developed below-ground adaptations, as in the root-suckering *E. tetrodonta* and the rhizomatous species *E. jacobsiana*, *E. porrecta* and *E. ptychocarpa*. Other woody plant species associated with these eucalypts also exhibit similar characteristics (Lacey, 1974; Lacey and Whelan, 1976). In the same areas, geophytic expression is very marked; Araceae (*Amorphophallus*), Dioscoreaceae (*Dioscorea*) and Vitidaceae (*Ampelocissus*) develop extensive underground organs that are sometimes used as a famine resource by the indigenous population particularly in parts of Papua New Guinea.

Dispersal mechanisms

In common with other savannas around the world, the Australian woodland savannas contain species with seed well adapted to high longevity, fire resistance, and both wind and animal dispersal. In Myrtaceae of the savanna, fire-resistant woody capsules with small, hard, often wind-dispersed seeds are found in *Eucalyptus*, *Melaleuca*, *Tristania* and *Xanthostemon*, whereas *Eugenia* and *Syzygium* produce succulent fruit dispersed by birds, bats and ground-dwelling animals. Winged woody fruit are common in the Combretaceae (*Combretum*, *Terminalia*) and amongst the legumes, mem-

---

Fig. 9.6. Four common bark types in Australian eucalypt woodland savannas. A. "Box" bark, so called because of its supposed resemblance to the European box (*Buxus sempervirens*). B. "Stringybark", because of its conspicuous linear fibres. C. "Ironbark", because of its extreme hardness. D. "Gum" bark, usually smooth and light-coloured, sometimes with a peeling outer bark as in this photograph. The scales are in centimetres.

branous pods (*Albizia*) and brightly coloured hard seeds (*Abrus*, *Adenanthera*) assist in dispersal by wind and birds respectively. Apocynaceous genera (*Alstonia*, *Marsdenia*, *Parsonsia*) in the more humid seasonal savannas, produce masses of winged seeds that are often wind-dispersed following fire. The large woody fruit of *Owenia acidula*, the emu apple, of Australian savannas is said to be effectively dispersed by emus (*Dromaius novaehollandiae*), where passage through the gut of the bird enhances germinability.

The dominant grass species in many savannas have wind-dispersed seed, for instance *Imperata*, and even those with heavy seed such as *Rottboellia* may have seed dispersed in strong thermal convections associated with fire. Many are awned and may be thus dispersed by animals or else lodge locally in cracks in the ground at the onset of the dry season. Seed of the bunch spear grass, *Heteropogon contortus*, is capable of actively burying itself by spiral movements of the awn under moist conditions.

Evolution of megatherm eucalypt savannas

The paucity of experimental taxonomic information in "tropical" eucalypts limits speculation about floristic aspects of their evolution. The Australian mainland is also relatively stable in the sense that the distribution of eucalypt savannas corresponds for the most part with a pattern of recurrent fires. Where fires are excluded, as in refugia such as volcanic cones (Fig. 9.7)) rocky gorges or water courses, there is usually closed seasonal non-eucalypt forest or vine thicket. Without experimental evidence and where there is little observable change in the vegetation, it is difficult in such circumstances to deliberate on the genesis of the eucalypt-dominated woodland savanna.

In the seasonal southwest and in particular the southeastern coastal plains of Papua New Guinea, the situation is more dynamic. In the plains and foothills surrounding the Port Moresby area it is possible to observe the change in seasonal forest vegetation that has taken place with the rapid increase in the urban and suburban population over the past 20 to 30 years. Here, subsistence gardening followed by fire has paved the way for the initial development of tall to intermediate height savanna grassland. Aerial and ground examination has revealed a mosaic of these savannas and woodland savannas of mixed ages to the west of Port Moresby, from which it is possible to infer a general successional pathway from seasonal semi-deciduous vine forest to eucalypt savanna woodland. The development is portrayed in Fig. 9.8 and is as follows:

Stage 1. Semi-deciduous vine forest (cf. Webb, 1959; Heyligers, 1972a) on interfluves commonly with Anacardiaceae (*Dracontomelon*, *Mangifera*,

Fig. 9.7. Because of fire, semi-deciduous "monsoon" forests have retreated to fire refugia. This photograph shows such a forest in a fire refuge formed by a volcanic cone on a weathered basalt flow *c.* 40 km southeast of Mount Surprise, north Queensland.

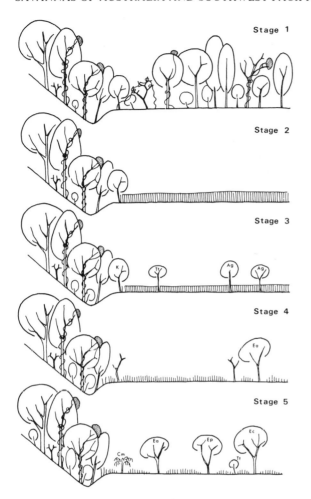

Fig. 9.8. Five stages suggested for the development of eucalypt woodland savannas from original closed "monsoon" semi-deciduous vine forest in a megatherm seasonal bioclimate.

Pleiogynium, Rhus, Spondias), Bombacaceae (*Bombax, Salmalia*), Burseraceae (*Canarium*), Combretaceae (*Combretum, Terminalia*), Dipterocarpaceae (*Anisoptera*), Fabaceae (*Albizia, Adenanthera, Pterocarpus*). Hernandiaceae (*Gyrocarpus*), Proteaceae (*Finschia, Helicia*) and Sterculiaceae (*Firmiana, Sterculia, Pterocymbium*).

Stage 2. Subsistence gardening followed by periodic burning, leading to tall to intermediate height grassland savanna.

Stage 3. Invasion of short-lived, scattered low trees such as *Antidesma ghaesembilla, Desmodium umbellatum, Kleinhovia hospita, Timonius timon.*

Stage 4. Increase in fire frequency with some elimination of low trees and gradual increase in short grasses. First appearance of eucalypts.

Stage 5. Domination of interfluves by eucalypts (*Eucalyptus alba, E. confertiflora, E. papuana*) and scattered woody understorey genera such as *Atylosia, Cycas, Desmodium, Timonius* and *Moghania*. Sharply defined edges with forest in fluvial "fire-shadow" zones.

According to Paijmans (1975), although mature eucalypts are largely fire-resistant, very frequent and fierce fires prevent regeneration and such fires cause a degradation to grassland. My own observations are that many of the eucalypt woodlands in the Port Moresby hinterland, for example, are fired at least annually. Although regeneration may be restricted by fire, overall survival is probably sufficient for maintenance. In many instances, seedlings regenerate under the parent tree (e.g. *Eucalyptus alba*) where there is usually a marked reduction in fuel.

This inferred succession/evolutionary pattern will vary with seasonal differences in soil-water storage, soil type and fire frequency; for instance, on the deeper wetter soils grasses will be taller, and there will also be non-eucalypt trees such as *Adenanthera pavonina, Albizia procera,* and *Deplanchea tetraphylla*. At higher altitudes *E. tereticornis* replaces the other eucalypt species.

I would contend that this general sequence also probably took place in northern Australia and was hastened by the advent of aboriginal man. Pollen analyses from the Atherton Tableland in northeast Queensland (Kershaw, 1975, 1978) suggest there may also have been cyclic development of eucalypt savannas and rain forest as a result of climatic arid and mesic cycles (see also Nix and Kalma, 1972; Bowler, 1976; Bowler et al., 1976). At present it appears that, whereas the northernmost Australian woodland savannas are relatively stable under current climate and land use, the savannas in megatherm seasonal areas of southeastern and to some extent southwestern Papua New Guinea are undergoing considerable change. These changes will probably proceed to a point where eucalypt-dominated savannas will eventually resemble those in northeastern megatherm seasonal Australia.

The savanna/forest boundary

In recent years there has been considerable focus on the dynamic aspects of savanna/forest boundaries (Aubréville, 1966, 1968; Eden, 1968;

Fosberg, 1968; Hills and Randall, 1968; Miège, 1968; and Taylor, 1968). Although such research is important for increasing our knowledge of savanna development, and hence management, very few studies have taken place within the Australian region. In Papua New Guinea, Gillison (1970a, b) examined a number of such boundary situations within a wide climatic range where it was found that next to climate, the main determinants of savanna development were topography and fire frequency. The position of a savanna boundary was, in most cases, fixed by micro-relief (drainage) or macro-relief (fire-shadows where vegetation in depressions escaped fire because of convectional systems (cf. Fig. 9.7). In all the boundaries investigated, both floristic and structural information suggested there was a distinct niche defined by the physical environment, occupied by a highly specialized group of plants characterized by a well-developed capacity for establishment by diverse vegetative as well as sexual means. This was in addition to a wide range of dispersal mechanisms and a close association with recognizable faunal groups, particularly birds and rodents, and, among the invertebrates, ants and butterflies.

There were also marked indications that, compared with those species in the adjoining savanna or forest, there was a greater diversity of plants able to fix atmospheric nitrogen by one means or another [Cycadaceae, Casuarinaceae, Eleagnaceae, Fabaceae, Myoporaceae, Rubiaceae (*Psychotria*) and Ulmaceae (*Trema*)]. A similarly high development of plants rich in phytochemical compounds, particularly alkaloids and volatile oils, suggests further that this assemblage of plants is an edge

community in its own right, rather than an overlap community or ecotone as commonly thought. The structural development of these edge situations in three different climatic aspects is displayed in Fig. 9.9. Fig. 9.9a shows a transition between a mesotherm short *Themeda* grassland to a *Castanopsis*-dominated oak forest; Fig. 9.9b shows a typical drainage effect associated with seasonally inundated swamps in a megatherm high-rainfall (>3000 mm yr^{-1}) area, and Fig. 9.9c depicts a typical sharply defined forest edge with a eucalypt woodland savanna on an adjoining interfluve in a megatherm dry seasonal habitat.

Although "dipterocarp savannas" (see Blasco, Ch. 8) are absent in Australia and the southwestern Pacific, some dipterocarp species are common in savanna-edge communities of Papua New Guinea where dipterocarps occur in the forest. In some megatherm grassland savannas in the Markham Valley for example, the dominant edge community species is juvenile *Anisoptera polyandra*, which suggests there may be an opportunistic element for establishment of this species in a savanna edge situation. Dipterocarpaceae and Proteaceae together with Myrtaceae and Fabaceae are also common along the boundaries of lowland megatherm seasonal savannas in southeast Papua New Guinea (see Fig. 9.9c).

It is difficult to generalize about soil characteristics across savanna-edge situations, but data from Papua New Guinea indicate that, in non-seasonal savannas with >2500 mm rain per annum, the savanna communities, and in particular those of the savanna edge, possess more fertile soils than the adjoining forest. In the megatherm seasonal low-

---

Fig. 9.9. Savanna/forest transitions in Papua New Guinea showing: a, mesotherm savanna (*S*) with transition (*T*) to *Castanopsis* oak forest (*F*); b, megatherm wet (>3000 mm yr^{-1}) savanna with transition to floodplain rain forest; and c, megatherm dry seasonal (<3000 mm yr^{-1}) eucalypt savanna on leached interfluves with transition to "monsoon" semi-deciduous vine forest with Dipterocarpaceae (from Gillison, 1970a). Legend to a: *C* = *Castanopsis acuminatissima*; *Ci* = *Cissus*; *Cy* = *Cyathea*; *D* = *Decaspermum*; *Eu* = *Euphorbia*; *F* = *Ficus dammaropsis*; *Fi* = *Ficus* sp.; *H* = *Hydriastele*; *I* = *Ilex ledermanii*; *M* = *Macaranga*; *O* = *Omalanthus populneus*; *S* = Sapindaceae; *Sy* = *Symplocos*; *V* = *Vaccinium albicans*; *W* = *Wendlandia paniculata*; *X* = *Miscanthus floridulus* and *Dicranopteris linearis*. Savanna is mostly *Themeda australis*.
Legend to b: *A* = *Asplenium*; *Ar* = *Arenga*; *C* = *Caryota*; *Cli* = *Cissus lineata*; *Ca* = *Calamus*; *F* = *Ficus*; *Fr* = *Fagraea racemosa*; *Gl* = *Gnetum latifolium*; *Hi* = *Horsfieldia irya*; *I* = *Ilex*; *L* = *Litsea*; *Med* = *Medinilla*; *Ms* = *Mussaenda scratchleyi*; *Nm* = *Nepenthes mirabilis*; *P* = *Pandanus*; *Pd* = *Pandanus* (dead); *Pf* = *Pittosporum ferrugineum*. Tall grasses are *Miscanthus* and *Rottboellia*, short grasses are *Imperata*, *Ischaemum*.
Legend to c: *Am* = *Amorphophallus*; *Ap* = *Anisoptera polyandra*; *Ba* = *Barringtonia*; *Bh* = *Buchanania heterophylla*; *Ca* = *Canarium australianum*; *Cr* = *Croton*; *Cs* = *Colona scabra*; *De* = *Derris*; *Dy* = *Dysoxylon*; *Ea* = *Eucalyptus alba*; *Er* = *Erythrina* cf. *variegata*; *Ga* = *Garuga floribunda*; *Go* = *Gonocaryum*; *Ja* = *Jagera*; *My* = *Myristica*; *St* = *Sterculia*; *Tt* = *Timonius timon*. Savanna is *Imperata* and *Themeda novoguineensis*.

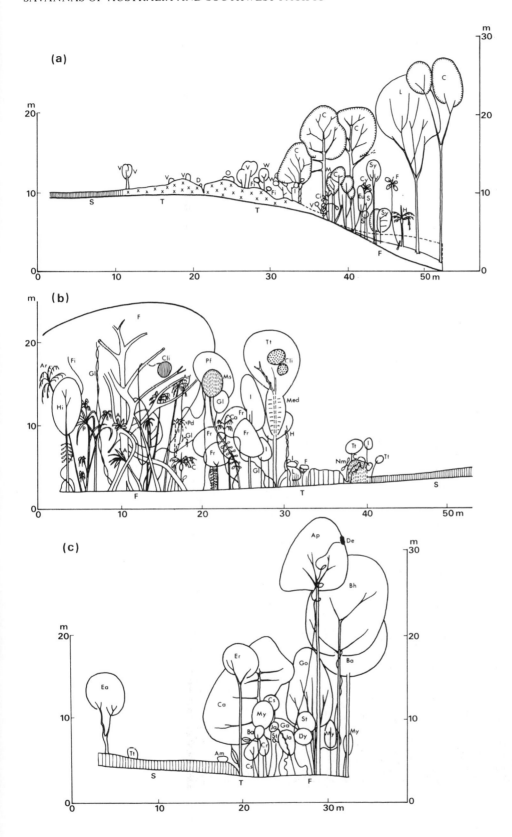

land savannas (<2500 mm rainfall yr^{-1}) dominated by eucalypts the trend is reversed except that
in some cases the edge community is richest in
nutrients. Relative changes in organic carbon and
total nitrogen in the soil profile from savanna to
forest are displayed in Fig. 9.10. Table 9.3 indicates
the changes in other soil properties associated with
a typical savanna-edge situation in a megatherm
seasonal site in Province 3 near Port Moresby,
Papua New Guinea (see Fig. 9.9). In many highland
areas of Papua New Guinea use is made of the fact
that the edge community is more fertile, and for this
reason subsistence gardens are commonly established there (see also Walker, 1966).

The physiology and life form of these edge
communities suggest that they are more resistant to
water stress than many of the plant species in the
adjoining savanna or forest. This characteristic
favours the expansion of the community at the end
of a drought, to the disadvantage of many savanna
species. Such expansion is usually short-lived as,
apart from the geophytes, most edge community
plants are highly susceptible to fire.

The "sharpness" of the edge community or
savanna "boundary" will therefore usually be influenced by fire frequency, although in microtherm
savannas frost may also play an important role (cf.
Brown and Powell, 1974). In the prolonged absence
of fire (and/or frost), it is likely that most savannas
will ultimately return to a forest cover. In areas with
reduced fire frequency, savanna definition sometimes becomes poor due to the suppression of
graminoids. In the present day, such re-invasion is
increasingly rare but it is likely that the megatherm
seasonal savannas within the region were once
forested with elements characteristic of present-day
savanna-edge communities and those from fire
refugia, such as the semi-deciduous vine thicket of
Webb (1959), more widely known in Indo-Malaysia
as "monsoon" thicket or forest.

TROPICAL AUSTRALIAN SAVANNAS

Megatherm seasonal

Province No. 1-AIS
In this Province the most common woodland
savanna type is composed of characteristically
stunted, low umbrageous and often single-

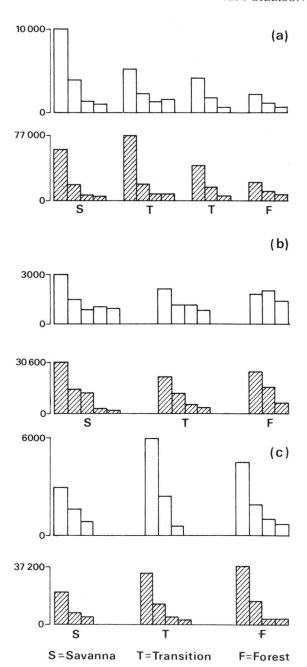

Fig. 9.10. Relative changes in soil organic carbon and soil total
nitrogen across profiles in Fig. 9.9 (a, b, c). Each bar represents a
separate soil horizon; A horizon on the left. Hachured bars
indicate organic carbon (p.p.m.) and open bars indicate total
nitrogen (p.p.m.). (From Gillison, 1970a.)

stemmed trees, widely spaced with a foliage projective cover (f.p.c.) of less than 10%, usually with a
low understorey of drought-resistant grasses such
as spinifex (*Triodia* and *Plectrachne* spp.), with or

TABLE 9.3

Changes in soil properties across a megatherm savanna/forest transition in Papua New Guinea (from Gillison, 1970a)

Community:	Savanna			Transition				Forest			
Horizon:	A	B	C	A	B₁	B₂	C	A	B₁	B₂	C
Depth (cm)	0–10	10–20	20–76	0–15	15–30	30–66	66–80	0–15	15–30	30–69	69–76
Bulk density	1.34	1.41	1.12	1.16	1.25	1.47	1.15	0.95	1.14	1.08	1.30
Sand (%)	60	57	42	85	77	57	55	85	70	50	55
Silt (%)	20	13	23	10	13	8	30	8	15	23	33
Clay (%)	20	30	35	5	10	35	15	7	15	27	12
Munsell Colour (moist)	10 YR 3/1	10 YR 3/3	10 YR 5/4	10 YR 2/1	10 YR 2/2	10 YR 2/2	10 YR 4/2	10 YR 2/2	10 YR 3/2	10 YR 4/2	10 YR 4/3
Organic carbon[1] (%)	1.93	0.71	0.44	3.17	1.30	0.40	0.20	3.72	1.31	0.33	0.15
Nitrogen (%)	0.30	0.16	0.10	0.60	0.24	0.06	—	0.46	0.19	0.11	0.09
C/N ratio	6:1	4:1	4:1	5:1	5:1	7:1	—	8:1	7:1	3:1	2:1
pH	6.5	6.5	6.9	6.7	6.5	7.1	7.3	6.9	6.9	7.4	7.7
Specific conductivity (mS cm⁻¹)	0.0275	0.0164	0.0162	0.0377	0.0156	0.0197	0.0188	0.0537	0.0268	0.0216	0.0188
Calcium (p.p.m.)	10 572	5868	7079	13 060	12 158	17 732	25 573	20 704	19 562	28 092	40 480
Magnesium (p.p.m.)	10 731	8463	13 466	5182	15 000	28 862	37 478	20 404	25 974	34 843	40 977
Phosphorus (p.p.m.)	677	777	622	995	854	665	688	756	796	714	844
Potassium (p.p.m.)	1987	909	1381	1658	1610	1661	3307	1718	1623	1742	1656
Iron (p.p.m.)	19 078	16 033	15 345	14 096	18 840	18 455	17 990	17 010	20 779	17 770	15 562
Manganese (p.p.m.)	3974	3967	3315	3980	4026	1428	1252	3350	3214	1324	1060

[1] Walkley-Black figure.

The specific conductivity row value $mS\ cm^{-1}$.

Fig. 9.11. Termite communities are very common in megatherm Australia and south-west Papua New Guinea. This colony of moundbuilding termites is typical of many of the eucalypt savanna communities on non-cracking clays in Provinces 1, 2 and 3.

Fig. 9.12. Low open snappy-gum (*Eucalyptus brevifolia*) woodland savanna with hummock grasses *Triodia* and *Plectrachne* spp. on skeletal quartzite soils. Ord River area, Northern Territory, Province No. 1.

without acacioid shrubs (Fig. 9.11). It occurs commonly on skeletal soils on quartzite ridges where a common tree species is *Eucalyptus brevifolia* (snappy gum) (Fig. 9.12) that extends in patches of dissected sandstone and quartzite plateaux from the Ord River in the northwest through the Northern Territory to the Leichhardt River in western Queensland. Another species of similar form that occurs towards the higher rainfall end of the gradient in the northwest, is the ironbark *Eucalyptus jensenii* (see Fig. 9.13). This eucalypt also commonly occurs as a dominant in medium dense to dense woodland communities in the northern Kimberleys. Low open woodland also exists on colluvial soils of heavy texture and on deep sands, where it is commonly dominated by *Eucalyptus pruinosa* with a mid-height grassland understorey (Fig. 9.14). Most trees in low open woodlands in Province AIS belong to the genus *Eucalyptus* with microphyll-sized isobilateral leaves. *E. pruinosa* is an exception, as it is usually a mesophyll or notophyll. There appears to be a compensating factor operating here, as the leaves are highly pruinose (with a waxy bloom) and hence, because of a high albedo (cf. Mulroy, 1979), are probably

better insulated from a relatively harsh environment. In general, woody plants in the drier end of the megatherm region conform to this pattern — that is, with increasing size the leaf becomes more glaucous or pruinose or thickened. No published data are available to indicate other physiological or functional trends such as divergence or convergence in stomate characters or in metabolic pathways.

Other typical low open woodlands include coolibah (*Eucalyptus microtheca*) woodland in areas that suffer transient flooding; ghost gum (*E. papuana*) woodland also in areas of low relief with a mid to low grassland understorey of *Eragrostis* and *Chrysopogon* spp.

Acacia-dominated woodland savannas often merge into scattered low forest; the dominant species include mulga (*A. aneura*), gidgee (*A. georginae*), lancewood (*A. shirleyi*) and ironwood (*A. estrophiolata*). The trees are typically sympodial in the southern extremes, and in the northern limits of this gradient are often monopodial and associated with understorey shrubs of *Carissa lanceolata* and *Eremophila* spp., and low- to mid-height grasses (*Aristida, Chrysopogon, Eragrostis, Themeda ave-*

Fig. 9.13. Low open ironbark (*Eucalyptus jensenii*) woodland savanna in Province No. 1.

Fig. 9.14. Low open silver-leaf box (*Eucalyptus pruinosa*) woodland savanna on heavy clays. West of Katherine, Northern Territory. Occurs mainly in Provinces 1 and 2.

acea). These *Acacia* species are mostly microphyllous with thickened xeromorphic phyllodes and occur on a wide range of soil types from calcareous skeletal soils and sandstones to red clayey sands. In general, reproduction is by seed with very little vegetative regeneration.

Province No. 2-AIIS

This Province contains the most important pastoral lands of northern Australia and is undoubtedly the most variable in terms of environmental range and variation in savanna type. It is a transition between the drier limit of eucalypt-dominated woodlands in Province 1 (that is itself transitional to the *Acacia*-dominated arid interior of Australia), and the taller, denser eucalypt woodlands and forests to the north (Province 3). The higher-rainfall extremes to the east and west contain the best formed woodlands, with depauperate types in the drier intermediate zone. Although there are eucalypt species that extend throughout the Province, Perry and Lazarides (1964) have suggested that the Great Artesian Basin which extends

south from the Gulf of Carpentaria may have acted as a migration barrier during periods of submergence. They quote, as an example, most of the ironbarks (*Eucalyptus crebra*, *E. drepanophylla*, *E. melanophloia*, *E. shirleyi*, *E. whitei*) which do not occur in the west although similar habitats exist. The boxes present an interesting case of structural convergence; whereas *E. brownii*, *E. leptophleba* and *E. microneura* occur in the east, they are replaced in similar habitats in the west by *E. tectifica* and *E. pruinosa*. On the other hand, the bloodwoods (*E. dichromophloia*, *E. polycarpa*) and many of the gums (e.g. *E. alba*, *E. camaldulensis*, *E. grandiflora*, *E. papuana*) are generally ubiquitous.

The overall distribution of savannas is governed primarily by rainfall pattern and secondarily by edaphic factors. In the drier extremes and on the poorer sandy soils, low open woodland is characteristically dominated by *Eucalyptus brevifolia*, occasionally with *E. perfoliata* and *E. ·pruinosa*. Woody understorey components vary in floristic composition and structure but throughout this formation may be *Acacia* spp., *Carissa lanceolata*, *Cochlospermum fraseri* and *Grevillea striata* (see

Speck and Lazarides, 1964). Other associated species of this formation are the bloodwood *E. dichromophloia* and the deciduous *Bauhinia cunninghamii*. In the east, an understorey of *Grevillea dryandri* and *Petalostigma quadriloculare* with *Acacia* spp. often occurs on skeletal soil systems west of Burketown. In the drier extremes, for the most part the characteristic grass layer consists of the hummock grasses *Plectrachne pungens*, *Triodia irritans* and *T. inutilis*, together with *Chrysopogon* spp. on interfluves and drainage floors. The independence of upper and lower strata has been pointed out by Perry (1960) and reiterated by Speck and Lazarides (1964), and this appears to be a general case throughout most woody savanna systems in Australia. There is a tendency for understorey woody components to have a wider ecological amplitude than those of the upper stratum, and this is particularly evident through Province 2 with its extensive transitions.

In the drier, usually low nutrient, sandy soils and skeletal stony soils — usually on sandstones — may be found two significant woody elements that may have been more widespread at one time. These are the gymnosperm "Cypress pine" (*Callitris columellaris*; Cupressaceae) and the ancient baobab or boab (*Adansonia gregorii*). Although the former extends throughout the north and east of Australia it tends to be restricted to areas of relatively low fire frequency. It occurs in both Provinces 2 and 3 (*q.v.*) in association with both monsoon forest and woodland savanna (Fig. 9.15). The baobab is regarded by some authors to be something of a relict (Armstrong, 1977). Rather like *Callitris*, it occupies semi-relictual habitats such as fire shadows, and is found in its best development on protected sandstone scree slopes (Fig. 9.16) and on rugged limestone country. It is also capable of occupying channels and drainage floors, but in almost every case it is associated with a number of deciduous plant genera — *Bauhinia*, *Brachychiton*, *Cochlospermum*, *Gyrocarpus* and *Terminalia* (see Speck and Lazarides, 1964 for additional detail).

The higher-rainfall areas and heavier, deeper soils support well-formed woodlands of which the box communities are among the most important. In the west these tend to be dominated by *Eucalyptus tectifica*, with the grasses *Sehima nervosum* and

Fig. 9.15. A conifer/broadleaf woodland savanna in the Kimberleys, Western Australia, on low-nutrient deep sandy soils (Province 2). *Callitris* (Cupressaceae) with *Eucalyptus* and dried-off grasses (*Sehima*, *Sorghum* spp.) in Provinces 2 and 3.

Fig. 9.16. Baobab (*Adansonia gregorii*) with a spinifex hummock grass layer occupying sandstone scree slopes northeast of Wyndham, Western Australia (Province 2).

Sorghum spp. (Fig. 9.17) common on red earths of volcanic origin, and usually with few shrubs. On heavy cracking clays *Eucalyptus tectifica* woodlands are often accompanied by a mixture of woody species including *E. dichromophloia* and *E. grandifolia* and a grass layer of *Chrysopogon*, whereas *Plectrachne pungens* is more common on sand plains. On lateritic gravel pans *Eucalyptus tectifica* may be associated with *E. perfoliata* and occasionally the shrub *Calytrix*. *Eucalyptus argillacea*, a species closely resembling *E. tectifica*, has a slightly wider ecological range (Speck and Lazarides, 1964) and occurs typically in a grassy woodland often with a moderately dense *Acacia* shrub layer and with a grass layer dominated by *Chrysopogon*. The remaining box woodland is characterized by *Eucalyptus pruinosa* (silver leaf box), that varies in structure from woodland to low-woodland savanna on red and yellow earths and cracking clays. This savanna type extends from the west to the Great Artesian Basin in Queensland (Perry and Lazarides, 1964) and may be variously associated with non-eucalypt woody species, such

as the trees *Bauhinia cunninghamii*, *Terminalia canescens* and *T. oblongifolia*, and the shrubs *Atalaya hemiglauca*, *Carissa lanceolata* and *Capparis* sp., with the grasses *Aristida*, *Sehima* and *Triodia* spp. The most ubiquitous box woodland on flood plains is that dominated by *Eucalyptus microtheca* (coolibah), which may be associated with low trees of *Bauhinia* and *Terminalia* and a grass layer of *Chrysopogon*. In the east, the most common box woodlands are dominated either by Gilbert River (or Georgetown) box (*E. microneura*), or Reid River box (*E. brownii*). Both occur on heavy, typically solodic soils, and may be either pure stands or else associated with the trees *Alphitonia excelsa*, *Bauhinia cunninghamii*, *Dolichandrone heterophylla*, *Erythrophleum chlorostachys*, *Petalostigma banksii*, *Terminalia aridicola* and *T. ferdinandiana*, with *Acacia* spp. *Eucalyptus brownii* woodlands represent an attenuation of the poplar box (*E. populnea*) woodlands of Province II and some shrub genera common to both occur in the Einasleigh uplands, such as *Alphitonia*, *Carissa*, *Dodonaea*, *Eremophila*, *Grevillea* and *Petalostigma*.

Fig. 9.17. McArthur River box (*Eucalyptus tectifica*) woodland on red earths of volcanic origin in the western Kimberleys, Western Australia, on transition between Province 2 and 3.

Paperbark (*Melaleuca*) woodlands are a significant feature throughout Province 2. Paperbarks occur in a variety of habitats ranging from quartzite plateaux in the west Kimberleys to the extensive outwash plains of the Gulf of Carpentaria. There are a number of species, notably *M. argentea, M. cajuputi, M. leucadendron, M. nervosa, M. quinquenervia* and *M. viridiflora*. Of these, *M. argentea* and *M. nervosa* are found in poorly drained solodic soils, and sometimes occur on outcropping sites but then usually over perched (wet season) water tables. *M. leucadendron* commonly occupies fringing swamps and waterways, but by far the most common are *M. quinquenervia* and *M. viridiflora*, in woodland savannas in the seasonally inundated solodic plains that drain into the Gulf of Carpentaria. The paperbarks range over a number of soil types from red and yellow earths, brown soils of light texture, to red and yellow podzolics (Perry and Lazarides, 1964). Associated species are *M. acacioides* and *M. symphyocarpa*, with low trees of deciduous *Bauhinia cunninghamii* and *Terminalia platyptera, Eucalyptus microneura, E. polycarpa* and *E. prui-*

nosa, together with *Celastrus cunninghamii, Dolichandrone heterophylla, Excoecaria parviflora, Gardenia vilhelmii*, and *Petalostigma banksii*. The most common grass species are *Aristida ingrata* and other *Aristida* spp. and *Chrysopogon fallax*.

Province No. 3-AIIIS

Medium to very tall woodland savannas and open forests are typical of this Province. With an increase in available water and higher temperatures, plant growth is more rapid than in Provinces 1 and 2 and leaf size increases markedly with canopy height in most cases. Associations of woollybutt and stringybark (*Eucalyptus miniata* and *E. tetrodonta*; Fig. 9.18) attain their best growth in this Province, the best formed stands of which occur on deep sands in Melville Island in the extreme west. This is a dense woodland with monopodial trees up to 25 m tall and a scattered understorey of trees and tall shrubs (*Acacia microcarpa, Alphitonia excelsa, Planchonia careya*), cycads (*Cycas media*) (Fig. 9.19) and often extensive palm communities of *Livistona humilis*. In the Mitchell Plateau, *L.*

Fig. 9.18. A stand of woollybutt (*Eucalyptus miniata*) woodland savanna showing typical monopodial form found in megatherm woodlands on deep soils. Near Darwin, Northern Territory, Province 3.

Fig. 9.19. Eucalypt woodlands east of Darwin, with *Cycas media* showing foliar regeneration following fire. Province 3.

eastonii (Fig. 9.20), occupies the transitional zone on basalt soils, between *Eucalyptus tectifica* woodlands towards the drier east and the rain forest of the western part of the Mitchell Plateau. According to Hnatiuk (1977) reproduction of the palm may be stimulated by annual burning.

Firing is commonplace and it is easy to speculate on a former cover of closed semi-deciduous vine forest that retreated in the face of fire to leave remnants of relatively fire-resistant cycads and palms that themselves became gradually occluded by eucalypts with the eventual establishment of a regular fire regime. Near the mainland coast, there are extensive mosaics of bloodwood woodlands (*Eucalyptus nesophila* and *E. porrecta*), while on Cape York the woodlands are mostly open forest dominated by stringybark and bloodwood (*E. tetrodonta*, *E.* aff. *polycarpa*) up to 25 m tall, with a lower tree layer of *Acacia rothii*, *Alphitonia excelsa*, *Grevillea glauca* and *Planchonia careya*. In the shrub layer *Cycas media* and *Xanthorrhoea johnsonii* are common with *Coelospermum reticulatum* and *Morinda reticulata* (see Pedley and Isbell, 1971). In woodlands of 10–50% foliage projective cover, graminoids are common, including the mid-height grasses *Heteropogon triticeus*, *Schizachryum fragile*, *Panicum mindanaense* and *Eriachne stipacea*. Across northern Australia, particularly in woollybutt woodlands, the grasses *Sorghum plumosum*, and *Chrysopogon fallax* are almost constant associates.

The bloodwood and paperbark woodlands in Province 2 are much more dense than those in Province 1 and are commonly dominated by *Eucalyptus dichromophloia* and *Melaleuca viridiflora*. *Erythrophleum chlorostachys* is generally ubiquitous throughout most open forests and in some of the denser woodlands. In the Adelaide and Alligator Rivers regions of the northwest, Story (1969) has recorded deciduous eucalypt woodlands on skeletal soils. Those species that were regarded as deciduous in such sites were *E. alba*, *E. confertiflora*, *E. dichromophloia*, *E. ferruginea*, *E. foelscheana*, *E. grandiflora*, *E. latifolia* and *E. papuana*. It is likely that in no other area of Australia is there such a preponderance of deciduous eucalypts, and this is almost certainly a reflexion of the high seasonality of the northwest. The most common,

Fig. 9.20. Eucalypt/palm (*Livistona eastonii*) woodland in northwestern Australia (Mitchell Plateau) occupying a transition zone between open woodland savanna and "monsoon" forest. Province No. 3.

Fig. 9.21. Deciduous woodland savanna. Dry-season aspect of *Bauhinia cunninghamii* with *Chrysopogon*, *Sehima* and *Sorghum* spp. grass layer. This type is common on seasonally inundated cracking clays in Provinces Nos. 2 and 3.

Fig. 9.22. *Pandanus spiralis* is common within seasonal tracts of eucalypt woodland savanna on heavy coastal soils that may be seasonally inundated. Near Adelaide River, Province 3.

completely deciduous non-eucalypt communities are those on seasonally inundated cracking clays, occupied by *Bauhinia cunninghamii* and *Terminalia* spp. (*T. aridicola*, *T. ferdinandianum* and *T. oblongifolia*). The grass understorey in these communities usually consists of mid-height *Chrysopogon*, *Sehima*, and *Sorghum* spp. that desiccate in the dry season (Fig. 9.21).

A notable feature of Province 3 is the extent of *Pandanus* spp. (mostly *P. spiralis*) in woodland and forest margins. It is most common on the coast and near floodplains (Fig. 9.22), but the genus occurs also òn sandstone outcrops in Arnhem Land where woody cover is otherwise very sparse. Paperbark woodlands of *Melaleuca symphyocarpa* (Fig. 9.23) also occur in the upper parts of floodplains.

Megatherm/mesotherm seasonal

Province No. 6-ABIIS

Much of this Province lies in a rain-shadow area west of the Great Dividing Range. Both topography and parent rock types are highly variable and

these changes are commonly reflected in the vegetation. It is therefore difficult to select a representative example although the box woodlands are most extensive and of generally similar form. The type is structurally simple with scattered umbrageous *Eucalyptus* species (*E. argillacea*, *E. leptophleba*, *E. microneura*, *E. moluccana* and *E. normantonensis*. Box woodlands with *E. brownii* are locally common. Low tree and shrub strata tend to be absent although *Grevillea glauca*, *G. parallela* and *Xanthorrhoea* are sometimes conspicuous. Mid-height to tall grasses are widespread, particularly *Heteropogon contortus*, *H. triticeus*, *Imperata cylindrica* and *Themeda australis*, with *Aristida*, *Eriachne*, *Schizachryum* and sedges. The more open up-slope and ridge communities are often ironbarks — *Eucalyptus drepanophylla* and *E. melanophloia*, with *E. shirleyi* on lithosols and on flooded country. Tall open woodland savanna with blue gum (*E. tereticornis*) and poplar gum (*E. alba*) (sometimes deciduous) are locally common (Fig. 9.24). The low to medium paperbark woodland savannas (*sensu* Story, 1970) with *Melaleuca* spp. so characteristic of the Mitchell–Normanby River

Fig. 9.23. Paperbark (*Melaleuca symphyocarpa*) woodland savanna with *Pandanus spiralis* on seasonally inundated clays. Province No. 3.

Fig. 9.24. *Eucalyptus alba* tall woodland savanna in northern Queensland with heavily grazed spear grass (*Heteropogon contortus*) on solodic soils. Province No. 6. (Photo: C.S. Christian, C.S.I.R.O.)

plains extend into this Province, where they become more dense (up to 40% f.p.c.). The gymnosperm *Callitris columellaris* occasionally forms semi-dense stands in sandy soils, often with ironbarks and bloodwoods (see Fig. 9.15). This is the only genus of the Cupressaceae in Australia, and its ecology is little understood, although it is known that regeneration from seed is dependent on the right combinations of fire frequency and rain (see also Stocker and Mott, 1981). It is an important component of megatherm seasonal savannas and often occurs in association with *Eucalyptus melanophloia* and *E. similis* in the east. As in northwestern Australia, woodland savannas on basalt appear to be derived from former closed semi-deciduous vine forest, of which *Callitris* is also a part. It is the occurrence of *C. columellaris* that has probably led Schmithüsen (1976) to map parts of the northwest as "cold-deciduous mixed conferous and broad-leafed forest".

In areas where fire has been excluded on basaltic soils southwest of Mount Garnet, woodland dominated by *Eucalyptus melanophloia* is being re-invaded by forest elements, notably *Acacia*, *Alphitonia* and *Flindersia*.

Mixed savanna types with *Eucalyptus clavigera*, *E. papuana* and *E. similis* as dominants occur throughout the Province in stands of up to 40% f.p.c. Understorey components such as *Cycas kennedyana*, *Grevillea pteridifolia*, *Petalostigma quadriloculare* and *Xanthorrhoea johnsonii* form scattered synusiae with little apparent relationship to environmental change (see also comment by Story, 1970). On deep sandy soils near the east coast, woodlands increase in canopy cover (*Eucalyptus alba*, *E. brownii*, *E. cambageana*, *E. polycarpa*, *E. tessellaris*) with more frequent monopodial lower tree layers (*Petalostigma pubescens* and *Melaleuca* spp. with *Alphitonia excelsa*, *Planchonia careya*, *Pandanus whitei*). Mid-height to tall grasses (*Alloteropsis semi-alata*, *Dichanthium sericeum*, *Elyonurus citreus*, *Heteropogon contortus*) are common on the better soils. On granite sands in the Bowen region, open forests of *Casuarina luehmannii* occur locally with deciduous *Cochlospermum* and *Terminalia*.

Province No. 7-ABIIIS

The low open forest and woodlands of Cape York have been described by Pedley and Isbell (1971). Apart from the occurrence of open forest similar to that in Province 3 (AIIIS) there are several formations unique to this area, notably low open forest dominated by *Eucalyptus dichromophloia* and *E. phoenicea* with an understratum of low trees and tall shrubs (*Acacia aulacocarpa*, *Grevillea glauca*, *Petalostigma banksii*) and mixed low open forest with *Callitris columellaris*, *Eucalyptus tetrodonta*, *Grevillea glauca*, *G. parallela*, *Tristania suaveolens*, and the shrubs *Acacia calyculata* and *Sinoga lysicephala* (some of which extends into southwestern Papua New Guinea). This type occurs on deep sandy yellow red earths and on deep sandy bleached grey earths (Pedley and Isbell, 1971). Other than this, the low open paperbark woodlands described in Province 3 are evident here as low dense woodland savanna on gleyed coastal podzolic soils and bleached grey earths.

Province No. 11-ABIV

Although this Province is not, strictly speaking, a "tropical" Province in the geographic sense, it does transgress the Tropic of Capricorn in its northern extremity and there are elements within it that undoubtedly occur well into the tropics. It is of particular biogeographic interest as many aspects are uniquely Australian and it represents a number of ecological tension zones.

For the most part, simple woodland savannas of mixed floristic types characterize this Province. The northern portion forms part of a distinct overlap with some of the megathermal seasonal elements, as indicated by the presence of the deciduous *Terminalia oblongata* and by some of the box woodlands with *Eucalyptus brownii* and *E. normantonensis*. The last is taxonomically close to poplar box (*E. populnea*), which is widespread throughout the southern part of the Province. Of the mixed woodlands, those dominated by ironbark (*E. melanophloia*), poplar box (*E. populnea*; Fig. 9.25), and

Fig. 9.25. Poplar box (*Eucalyptus populnea*) woodland with understorey of "false sandalwood" *Eremophila mitchellii*. Grass layer has been removed by grazing sheep. Southern area of Province No. 11.

brigalow (*Acacia harpophylla*; Fig. 9.26) are easily the most important. The structuregram for this Province indicates the range of the first two and the dotted lines indicate the extensive structural range of brigalow (Appendix I). Both brigalow and poplar box intergrade with one another in the northern part of the Province. The layering within woodlands forms no constant pattern, although there are some predictable relationships on certain soil types. For the Nogoa–Belyando area, Pedley (1967) has described "argillicolous" and "arenicolous" midstoreys; the former occurs mainly on heavier-textured soils ranging from cracking clays through texture-contrast and red and yellow earths, to shallow rocky soils. Typical associates are *Eremophila mitchellii* in the lower layer and *Carissa ovata* in the shrub layer. Tree wilga (*Geijera parviflora*), supplejack (*Ventilago viminalis*) and *Albizia basaltica* are common as low trees, *Acacia excelsa*, *Bauhinia carronii*, *Cassia brewsteri*, *Flindersia dissosperma*, and *Heterodendrum oleifolium* are frequent. Apart from *Eucalyptus populnea*, other associated eucalypts are *E. cambageana*, *E. melanophloia*, *E. orgadophila* and *E. thozetiana*.

"Arenicolous" midstoreys consist of low trees, often including *Callitris columellaris* with numerous *Acacia* species (*A.* sp. aff. *burrowii*, *A. cunninghamii*, *A. sparsiflora*, *A.* sp. aff. *cunninghamii* and others), *Alphitonia excelsa*, *Casuarina luehmannii*, *Eremophila mitchellii*, and *Petalostigma pubescens*.

The substrate varies from uniform sandy soils to relatively sandy texture contrast soils and red and yellow earths (Pedley, 1967). Vegetation on the sandier soils tends to be fired more frequently and this may reflect the high incidence of phyllodineous *Acacia* species where germination is often fire-stimulated.

The brigalow (*Acacia harpophylla*) formation is best developed within this Province (Fig. 9.26). This is a unique vegetation type and, like mulga (*A. aneura*) ranges over a wide variety of sites and exhibits considerable variation from closed to open community structure. Brigalow lands are of considerable agronomic importance and in recent years many thousands of hectares have been cleared for grazing and cultivation. Because of its capacity for vigorous vegetative regeneration following distur-

Fig. 9.26. Brigalow (*Acacia harpophylla*) "soft-wood scrub", a tall, relatively open wooded formation with a dense understorey of woody plants that may be removed by fire to produce a savanna with a grass layer dominated by *Acacia*. Provinces Nos. 11 and 12.

bance, brigalow now poses a major woody-weed problem. Coaldrake (1970) has estimated a former area of *c.* 4.9×10^6 ha for brigalow on soils of moderate to high fertility. The species is a woody plant, either single- or multi-stemmed, up to 25 m tall, with phyllodes capable of withstanding water deficits beyond 6.8 MPa (Tunstall and Connor, 1975), and whose margins are vertically orientated. Apart from its capacity to withstand droughting, it is capable of occupying the more solodic soils and cracking clays, particularly in areas where "gilgai" formations are common. Of all the Australian woodlands perhaps no other type occupies such a wide transition between major vegetation types. It is a common associate of "soft wood" scrubs — closed vegetation types regarded as semi-evergreen and semi-deciduous vine thicket (rain-forest types) by Webb (1959). Story (1967) has maintained that the "soft wood" shrub elements are not capable of simple definition due to their complexity, and Johnson (1964) divided brigalow into eleven dif-

ferent communities based essentially on floristic associations.

In the southern part of the Province, brigalow occurs in more open savanna communities, often in association with *Casuarina cristata* on the more solodic soil types, and in the Balonne–Maranoa area is often associated with bendee (*Acacia catenulata*) on low scarps (cf. Pedley, 1974). In the north, brigalow woodland savanna is often formed locally as a result of fire removing the woody understorey, which then becomes replaced by graminoids. Other artificial savannas are also formed by man where brigalow trees are left in pastures — for instance, of buffel grass (*Cenchrus ciliaris*). A significant associate of brigalow throughout this Province is the conspicuous "bottle tree" (*Brachychiton rupestris*) so called because of its resemblance to a bottle shape (Fig. 9.27), that is itself a common component of soft wood scrubs. Additional information on this vegetation type has been recorded by Speck (1968) for the

Fig. 9.27. A "bottle tree" (*Brachychiton rupestris*) remnant of semi-deciduous brigalow scrub, now in a derived savanna. Such trees are often left for cattle fodder in drought, as the edible soft inner parts of the stem have a high water content. (Photo: N. Speck, C.S.I.R.O.)

Dawson–Fitzroy area and by Story (1967) for the Isaac–Comet area.

Province No. 12-ABV

This Province includes the sub-coastal woodlands of central eastern Australia with a small extension into the tropics. Within it are a number of associations of considerable floristic and structural range. There are two basic structural overlaps — the eucalypt woodlands of medium height and density, and the low woodlands of the coastal fringe dominated by paperbarks (*Melaleuca* spp.). Edaphic differences play a major role here in differentiating between types. Many woodland savannas of the upper slopes and poorer soils are dominated by ironbarks (*Eucalyptus drepanophylla, E. crebra*), foot slopes are commonly dominated by woodland or "open forest" with boxes (*E. moluccana* subsp. *queenslandica*; Fig. 9.28) and bloodwoods (*E. dichromophloia, E. polycarpa*), while flats and alluvial plains on heavier soils support tall open woodland of *E. alba* and *E. tereticornis*. There has been considerable clearing of this region in recent years and much of the woodland on better

soils is now either under grazing or cultivation. Understorey components are variable with *Acacia* spp. and *Melaleuca* spp. common in the denser woodlands. Woodland savannas on the more uniform texture-contrast and heavier soil types usually support tall grassland of *Heteropogon contortus* and *Themeda australis*, and paperbark woodlands with *Melaleuca leucadendron*, *M. nervosa* and *M. viridiflora*, which occupy lands of poor drainage, are locally common. In these last woodland types (cf. Fig. 9.29), graminoid cover is composed of a high proportion of Cyperaceae, and groves of *Pandanus* are frequent.

A minor, but nevertheless significant, woodland savanna on leached aeolian coastal sands of very low nutrient status is the "wallum" that extends from Taree in New South Wales to Broad Sound in Queensland. In its northern extremities it is typically a eucalypt woodland, with *Eucalyptus exerta* and *E. signata*, usually associated with *Melaleuca quinquenervia* and the understorey shrubs *Banksia oblongifolia*, *B. robur* and *B. serrata*, (Fig. 9.30), and a dense sedge layer. A floristically rich ground layer is characterized by Cyperaceae, Droseraceae,

Fig. 9.28. Grey box (*Eucalyptus moluccana* subsp. *queenslandica*) tall woodland on poorly drained solodic soils, associated with *E. alba*. Grass layer has been heavily grazed by cattle. Province No. 12.

Fig. 9.29. Paperbark (*Melaleuca quinquenervia*) woodland savanna with understorey of *Banksia robur* and mixture of grasses and sedges. On seasonally inundated lands in sub-coastal Queensland, Province No. 14.

Fig. 9.30. Open eucalypt, paperbark woodland savanna on low-nutrient siliceous sands in central coastal Queensland. Dominant species are *Eucalyptus signata* (scribbly gum) and *Melaleuca quinquenervia* with a sedge (*Ptilanthelium*) layer. Seasonally inundated. Province No. 12.

Fabaceae, Juncaceae, Liliaceae, Myrtaceae, Restionaceae, Rutaceae, Proteaceae and Xanthorrhoeaceae (see also Coaldrake, 1961, 1975).

Province 14-ABVIII

As indicated in Fig. 9.4, this Province is restricted to pockets on the east coast of Queensland. The mesotherm/megatherm (AB) overlaps indicate the range of bioclimate that is the most mesic expression of tropical woodland savanna on the Australian continent. Although the area is restricted, there are a number of distinct woodland savanna types that mostly reflect different drainage patterns. In areas of semi-permanent standing water on either siliceous sands or heavy clays are woodlands of *Melaleuca quinquenervia* with *Banksia robur* in the understorey (Fig. 9.29). In such associations sedges are most common, particularly *Eleocharis*, *Gahnia*, and *Lepidosperma*. Such communities are of little pastoral value.

One of the most common communities on slightly better drained sites is the poplar gum–paperbark (*Eucalyptus alba–Melaleuca virid-*

iflora) woodland savanna, which is variously associated with the bloodwoods *E. intermedia* and *E. polycarpa*, and often has the understorey woody species *Pandanus whitei*, *Petalostigma quadriloculare* and *Planchonia careya*, with a mid-height grass layer of *Heteropogon contortus*, *Imperata cylindrica* and *Themeda australis* (Fig. 9.31). Such communities occur on coastal to sub-coastal plains with a fluctuating water table and are heavily used for pasture, or in more recent years have been drained and converted to sugar-cane growing. In the higher-rainfall areas, particularly below foot slopes, are to be found "brush-box" stands of *Tristania suaveolens* (Fig. 9.32) , sometimes with *Melaleuca viridiflora*, and mid-height to tall *Coelorhachis*, *Heteropogon* and *Imperata*. Such stands probably represent a disclimax that would return to forest and later rain forest, if fire was excluded.

On foothills, woodland savannas of ironbarks (*Eucalyptus drepanophylla*), stringybarks (*E. acmenioides*), and bloodwoods (usually *E. intermedia*) are found with *Cycas media* and/or

Fig. 9.31. Coastal woodland savanna of poplar gum (*Eucalyptus alba*) and paperbark (*Melaleuca viridiflora*) on poorly drained heavy clays with microrelief drainage. Mid-height grasses are commonly bunch spear grass (*Heteropogon contortus*) and blady grass (*Imperata cylindrica*) with some kangaroo grass (*Themeda australis*). Province No. 14.

Fig. 9.32. Seral brush box (*Tristania suaveolens*) dense woodland savanna on poorly drained coastal soils. Understorey has mid-height grasses *Heteropogon contortus* and *Imperata cylindrica*. Without fire this woodland would probably return to non-eucalypt rain forest. Province No. 14.

Xanthorrhoea johnsonii, and grasses *Heteropogon*, *Imperata* and *Themeda*. Other grass genera including *Bothriochloa*, *Coelorachis*, *Dichanthium*, *Digitaria*, *Ophiuros*, *Paspalum*, and *Sorghum* are common, and in parts of the Mackay and Cairns areas the introduced molasses grass (*Melinis minutiflora*) is actively invading areas dominated by *Imperata cylindrica*.

North of Mackay, a drier bioclimate exists in the Bowen–Townsville area, and a convergence is evident here from Fig. 9.4. This drier area has been described by Perry (1953), and on granite outcrops remnant "monsoon" thicket is evident with deciduous *Brachychiton*, *Cochlospermum* and *Pleiogynium* trees.

In the northern extremities of this Province are sub-coastal woodland savanna hillsides with communities that are structurally and to some extent floristically similar to moist megatherm areas of moderate seasonality in New Guinea. Such areas have mid-height to tall grasses, such as *Coelorhachis rottboellioides*, *Imperata cylindrica* and *Ophiuros exaltatus*, with an upper storey of

Tristania suaveolens and *Eucalyptus intermedia* that usually contains *Albizia procera*. Introduced grasses, especially *Panicum maximum* and *Melinis minutiflora*, are common together with exotic legumes, *Centrosema*, *Glycine* and *Phaseolus* spp. Often such savannas surround sugar-cane farms and are either deliberately or accidentally fired, usually annually; this has caused a retreat of the rain forest that would otherwise occupy such sites.

PAPUA NEW GUINEA

The island of New Guinea presents a bewildering complex of savanna types. Whereas the environment of the Australian continent has produced relatively stable vegetation, much of Papua New Guinea is in a state of flux controlled as it is by changing water tables in extensive lowlands and modified extensively by man-made fire. As mentioned previously, many of the savannas are considered to be in a successional stage. These aspects create difficulties in vegetation classification.

Similar bioclimates exist nevertheless in both Australia and Papua New Guinea, and this is reflected in a number of common floristic and structural types particularly in the megatherm Region.

Paijmans (1975) has mapped the savannas of Papua New Guinea and has described four main types: eucalypt savanna, *Melaleuca* savanna, mixed savanna and tree-fern savanna. Megatherm woodland savanna covers approximately 10 300 km².

Megatherm seasonal

Province No. 3-AIIIS

This Province occurs mainly in the dry seasonal sub-coastal region surrounding Port Moresby. It is a mixture of savanna types that is currently undergoing considerable change due to the increasing frequency of subsistence gardening and man-made fire. Drainage, as well as fire, is a critical determinant of savanna pattern. Poorly drained, seasonally inundated areas support a mosaic of medium paperbark (*Melaleuca*), savanna woodland and open swamps (Fig. 9.33). The most stable savanna occurs on well-drained interfluves (see Fig. 9.9c) and on low, coastal hills (Fig. 9.34), often with texture-contrast soils, and is typically a *Eucalyptus*-dominated medium woodland savanna with mixtures of *E. alba*, *E. confertiflora*, *E. papuana*, and with a mid-height graminoid layer variously dominated by *Themeda australis* and *T. novoguineensis*, with *Arundinella*, *Capillipedium*, *Heteropogon*, *Imperata* and *Sehima*. The sedges *Schoenus* and *Scleria* are also common, with shrubs *Atylosia*, *Desmodium*, *Moghania*, and other Fabaceae. Different associations of eucalypts with other woody species have been described in detail by Heyligers (1965, 1972a), who has pointed out that many of the environmental factors which may account for the distribution of the eucalypt species have not yet been established. The non-eucalypt savannas are dominated by the low tree species *Antidesma ghaesembilla*, *Desmodium umbellatum*, *Glochidion* and *Timonius timon*, usually with a tall grass layer of *Ophiuros tongcalingii* on deeper, heavier soils. Other common non-eucalypt woodland savannas are dominated by *Albizia procera* and *Acacia* spp., again on heavier soils. The palm-like *Cycas media* is often a common element of savannas with or without eucalypts. It seems certain that many of the associations are part of a complex successional mosaic influenced by increasing land use. There are close relationships with Province 3 savannas in Australia, and there are similar associations with fire-sensitive, semideciduous monsoon thicket that now exists in fire refugia in many localities in the Port Moresby hinterland (Fig. 9.35) (see also Heyligers, 1972a).

Surrounding the floodplains of Province 3 and transitional to the forests and eucalypt woodland savanna, a characteristic mixed savanna often develops that is recognizable by the palm *Livistona brassii* (Fig. 9.36) and many pandans and tall sedges. This palm-dominated transition has similarities with other transitional savannas in northern Australia, both in Province 3 where *L. eastonii* (Fig. 9.20) and *L. humilis* are frequent understorey elements in the more mesic woodlands of the northwest, and in Province 14 where *L. australis* commonly occupies transitions between coastal savannas and rain forest.

Province No. 4-AIVS

By far the most extensive and most complex woodland savannas occur in this Province, which ranges from the southwestern part of Papua New Guinea (the Fly–Digul River systems) to subcoastal areas to the east of Port Moresby (see Figs. 9.3 and 9.4). The vegetation has been described in various detail by Brass (1938), Van Royen (1963b) and Paijmans (1971, 1976), and much of the floristic information recorded here is from Paijmans (1971).

A prominent feature of the savannas of southwestern Papua is the extensive area subjected to seasonal inundation usually from November to May. This has a dramatic effect on the distribution pattern of the savannas and suggests that any classification should take into account a range of savanna types including swamp woodland savanna. The complex scale of fire and flood maintains a changing vegetation which is difficult to map and creates savannas of variable structure and floristics which intergrade with permanent open swamps at one extreme and dense mesophyll forest on extensive levee systems at the other.

The paperbark genus *Melaleuca* is dominant in many savanna formations in this Province possibly because of the periodic wetting and drying of the soils. In the more permanently wet swamps, medium-height swamp woodland savanna is domi-

Fig. 9.33. Typical meander pattern on the coastal plains of Papua New Guinea, produces a changing mosaic of different savanna types. *Melaleuca* and *Pandanus* are common tree species. Province No. 3. (Photo: P. Heyligers, C.S.I.R.O.)

Fig. 9.34. Oblique view of eucalypt woodland savanna near Port Moresby, Papua New Guinea. Dominant species are *Eucalyptus alba*, *E. confertiflora* and *E. papuana*, with *Themeda australis* grass layer. Province No. 3. (Photo: B. Ruxton, C.S.I.R.O.)

Fig. 9.35. Fired hillsides cause a retreat of some deciduous forest to gully refugia. This grassland may eventually become eucalypt woodland savanna. Port Moresby hinterland, Papua New Guinea. Province No. 3. (Photo: P. Heyligers, C.S.I.R.O.)

Fig. 9.36. Scattered palm (*Livistona brassii*) and *Pandanus* savanna typical of transition zones between swamp and either forest or eucalypt woodland savanna in Provinces Nos. 3 and 4, Papua New Guinea. (Photo: P. Heyligers, C.S.I.R.O.)

nated by *M. cajuputi* and *M. leucadendron*. Associated tree species include *Barringtonia tetraptera*, *Erythrina fusca*, *Mitragyne speciosa*, *Nauclea orientalis*, *Neonauclea* sp., and *Pandanus*. There is a general lack of shrubs in these wetter sites but there is commonly a dense tall graminoid understorey of *Phragmites karka* and often *Leersia hexandra* and *Oryza* spp. with sedges and the climbing fern *Stenochlaena*. Occasionally swamps dry out and fire sweeps through, temporarily removing the graminoid layer. *Pandanus* spp. also dominate some of the more open formations of the coastal plains, again in poorly drained and periodically wetted sites (Fig. 9.37).

Less permanently inundated paperbark savanna of medium height is commonly dominated by *Melaleuca symphyocarpa* and *M. viridiflora*, often associated with *Acacia leptocarpa*, *Dillenia alata* and *Tristania suaveolens* (Fig. 9.38). Here shrubs are more common, particularly *Sinoga lysicephala* together with low trees of *Glochidion* and *Pandanus* spp. There is usually a mid-height graminoid layer variously composed of the grasses *Arundinella nepalensis*, *Elyonurus citreus*, *Eriachne triseta*,

Germainia capitata and *Ischaemum barbatum* [according to Henty (1969), both *Elyonurus* and *Eriachne* displace *Themeda* in the southern part of the island]. The sedges *Schoenus* and sometimes *Scleria* also occur. Where fire is common, this graminoid layer may be largely removed and the regenerating sward grazed by wallabies (*Macropus agilis*) and the Timor deer (*Cervus timorensis*) (see Downes, 1969) so that a park-like appearance is maintained.

Tall woodland savanna with physiognomically mixed tree species is found on drier areas of slightly higher relief (Fig. 9.39). Typical low species are *Acacia* spp., *Banksia dentata* and *Grevillea glauca*, overtopped by *Eucalyptus brassiana*, *E. polycarpa*, *Melaleuca symphyocarpa*, *Tristania suaveolens*, *Xanthostemon brassii* and *X. crenulatus* with occasional *Deplanchea tetraphylla*, *Dillenia alata* and *Parinarium nonda*. A patchy shrub layer up to 3 m tall is common, with *Acacia simsii*, *Melastoma* and *Rhodamnia*. Low woody species of *Alstonia*, *Glochidion* and *Sinoga* may be present together with *Pandanus* and clump palms (?*Arenga* sp.). The mid-height graminoid layer is richer in sedges

Fig. 9.37. *Pandanus* savanna with grass and sedge layer, in poorly drained coastal plains in the Morehead area of southwestern Papua New Guinea. Province 4. (Photo: K. Paijmans, C.S.I.R.O.)

(*Rhynchospora, Scleria* and *Schoenus*) together with the rhizomatous *Arundinella nepalensis* and *Imperata cylindrica, Pseudopogonatherum irritans* and *Setaria surgens*. This tall savanna grades into a very tall type on drier sites where inundation is rare. Much the same tree species occur, with some broader-leaved species such as *Acacia mangium* being more common. Tall undershrubs such as *Cordyline* and *Leea* are also found in this type, which tends to be more dense and to carry more epiphytes and lianes. In both savanna types epiphytic *Dendrobium canaliculatum* is ubiquitous, together with other *Dendrobium* species and the unusual epiphytic "ant plant" *Myrmecodia*, together with the hemi-epiphytic asclepiads *Dischidia* and *Hoya*. The existence of epiphytes in these savannas is a feature in common with those in Australian Provinces 3 and 7, particularly in the Cape York area of northern Queensland. Most of the soils in Province 4 are low in nutrients and generally acidic; such conditions are usually indicated by the presence of the low shrub *Melastoma malabathricum* and the pitcher plant *Nepenthes mirabilis*.

On the lowland megatherm mid-height to tall grassland savanna plains of the Ramu and Markham valley systems of northern Papua New Guinea, occasional transitional woody trees occur (*Glochidion, Kleinhovia, Macaranga, Mallotus*) with the more permanent *Albizia procera* and *Antidesma ghaesembilla*. It is in this area that up until twenty years ago *Cycas media* formed extensive cycad savannas (Fig. 9.40). Since that time, pasture improvement for cattle grazing has seen the removal of many thousands of these cycads, as the foliage is considered toxic if eaten by stock. Today only a few remnants are visible in the Erap and Ramu River systems. There is an interesting convergence of plant form here with the microtherm tree-fern savannas (see next section).

Mesotherm and microtherm savannas

Exact bioclimatic equivalents for the Australian mesotherm and microtherm Provinces are difficult to identify in Papua New Guinea. Although one can calculate approximate temperature equivalents for shifts in altitude relative to latitude, changes towards the Equator in the solar radiation received,

Fig. 9.38. Open woodland savanna on the low Oriomo Plateau of southwestern Papua New Guinea. Trees are mostly *Acacia leptocarpa*, with shrubby *Sinoga lysicephala* and a mixed low grass and sedge layer. Province 4. (Photo: R. Pullen, C.S.I.R.O.)

Fig. 9.39. Tall, dense myrtaceous woodland savanna in southwestern Papua New Guinea in the Trans-Fly area. Characteristic dominant tree species are *Melaleuca symphyocarpa*, *Tristania suaveolens* and *Xanthostemon*. Province 4. (Photo: R. Pullen, C.S.I.R.O.)

Fig. 9.40. Cycad savanna in the megatherm Ramu Valley of Papua New Guinea. *Cycas media*, with the mid-height grass layer mainly of *Imperata cylindrica*. These savannas were once widespread in the Ramu and Markham Valleys and are now changing under grazing pressure. (Photo: R. Hoogland, C.S.I.R.O.)

and the considerable variation in topography and landform between Australia and the island of New Guinea, lead to significant climatic differences for plant growth that are very evident at higher altitudes. Not the least of these is the *Massenerhebung* effect associated with a change in climate with altitude as one approaches the coast, particularly along mountain chains (Richards, 1961; Robbins, 1964).

For microtherm elements there are sufficient climatic similarities between the high mountains (>3000 m above sea level) of Papua New Guinea and those in the Australian microtherm Province (approximately >1500 m) to allow useful comparisons. As with Australia, there are difficulties in locating a boundary between microtherm and mesotherm, and there is considerable overlap. In the absence of sufficient data on meteorology and plant growth response for Papua New Guinea, Table 9.4 outlines what I tentatively suggest as a broad relationship between altitude and bioclimate.

Mesotherm woodland savannas are not common in Papua New Guinea. The majority of mesotherm areas are in high-rainfall environments (>2500 mm yr^{-1}). The vegetation is either "Kunai" grassland, or else closed forest as in the densely populated Central Highlands, with only fast-growing, transitional softwood "nomad" species occurring between these extremes (cf. Van Steenis, 1958). Such "pure" grassland savannas are maintained by fire and associated gardening practices, usually in areas where wood is a precious fuel. In drier upland areas, such as the Northern and Central administrative Provinces, megatherm/mesotherm seasonal overlaps include areas of eucalypt woodland savanna with variously mixed species assemblages of *Eucalyptus alba*, *E. confertiflora*, *E. papuana* and *E. tereticornis* (the last occurring throughout the entire altitudinal range of eucalypt woodland).

TABLE 9.4

Approximate relationship between altitude and bioclimate for the island of New Guinea

Bioclimate	Altitude (m)
Megatherm	0–1000
Mesotherm	1000–3000
Microtherm	>2500

Structure ranges from medium open woodland to tall open forest, usually with an understorey of low trees including *Albizia procera*, *Antidesma ghaesembilla*, *Desmodium umbellatum* and *Timonius timon*, often with *Cycas media* and the epiphytes *Dischidia*, *Hoya* and *Usnea*. This vegetation type occupies much of the hill country inland from Port Moresby and in the Northern District inland from Oro Bay and the Pongani coast (cf. Fig. 9.34). In the upper limits (450 m above sea level), *E. tereticornis* often occurs in pure stands with understorey trees *Banksia dentata*, *Casuarina* (*Gymnostoma*) *papuana* and *Grevillea glauca*. These upland savannas usually occur on leached, shallow ridge soils or else deeply weathered acidic laterites — often with the shrubs *Melastoma malabathricum* and *Eurya* and the pitcher-plant *Nepenthes*. The common midheight grassland species is *Themeda australis,* with various sedges such as *Machaerina* and *Rhynchospora* spp. This type is maintained by fire, both man-made and caused by lightning, particularly on the southern mid-slopes of the Owen Stanley Range such as Efogi, Mount Brown and Mount Obree, and has been described by Lane-Poole (1925a, b), Paijmans (1967, 1976) and Gillison (1970a).

Microtherm savannas are mostly above the altitudinal limit of cultivation (*c.* 2500 m) and are a mixture of open montane or subalpine tussock grassland, ericaceous shrub savanna, tree-fern savannas and dense bamboo-lands.

Ericaceous shrub savannas (Fig. 9.41) typify transitional areas between closed (often gymnospermous/fagaceous) forest and open grassland. Such transitions may be relatively stable climatic/edaphic tension zones or else represent a successional phase following a reduction in fire and/or frost frequency in otherwise open grassland. In general, subalpine areas on high mountain massifs (up to *c.* 3600 m) with least access by the human population tend to have the most widespread shrub savannas — for instance, parts of the Albert Edward, Saruwaged and Victoria Plateaux. The effect of fire and frost in the maintenance of high mountain savannas has been discussed by a number of authors (Robbins, 1964, 1970; Brass, 1964; Robbins and Pullen, 1965; Street, 1966; Gillison, 1969, 1970b, 1972; Kalkman and Vink, 1970; Coode and Stevens, 1972; Smith, 1975, 1977). Ericaceous shrub savanna is characteristically

Fig. 9.41. Ericaceous shrub savanna with tree-fern savanna in background. Typical shrubs are *Rhododendron* and *Vaccinium* with tussocks of *Danthonia*. Microtherm Papua New Guinea. (Photo: J. Saunders, C.S.I.R.O.)

dominated by the shrub and shrub-like genera *Agapetes*, *Dimorphanthera* (e.g. *D. collinsii*, *D. cornuta*, *D. microphylla*), *Diplycosia rupicola* (scandent), *Rhododendron* (numerous spp.), *Vaccinium* (*V. cruentum*, *V. keysseri*) and the smaller, woody *Gaultheria mundula* (all Ericaceae or Vacciniaceae). Other woody species are commonly *Anaphalis*, *Coprosma*, *Detzneria*, *Eurya*, *Haloragis*, *Olearia*, *Styphelia* and *Xanthomyrtus* with occasional *Daphniphyllum*, *Saurauia* and *Symplocos*. Many of these genera reach their optimum growth along forest edges and in such situations often become scandent, particularly if the forest advances due to a lowering of fire frequency. When isolated in the more open grassland they are more rounded, erect and dense "shrubs" although some genera such as *Xanthomyrtus* may become small prostrate woody "reservist" plants in the herb layer when fire frequency is high — more than once in two years (Gillison, 1969, 1970b).

The associated herb and graminoid layer is usually extremely rich in species. Although species assemblages change considerably with altitude (e.g. between 2500 and 3600 m), there are a number of forbs with a wide ecological amplitude such as *Acaena*, *Astelia*, *Drapetes*, *Epilobium*, *Euphrasia*, *Gentiana*, *Lactuca*, *Myosotis*, *Oreomyrrhis*, *Plantago*, *Potentilla*, *Swertia*, *Trachymene* and *Viola*. Cushion graminoids such as the grass *Monostachya* and the sedge *Oreobolus* are common in the wetter sites. *Gleichenia vulcanica* and *Lycopodium cernuum* are also found in these savannas, these genera having perhaps the widest ecological amplitude of any savanna species in the southwest Pacific region.

The bamboolands occupy a rather specialized niche in the microtherm/mesotherm overlap. The two main genera are *Bambusa* and *Nastus*. They are both widespread throughout Papua New Guinea, and in areas of human habitation are widely used for construction purposes and in some cases for food. Both genera, but particularly *Nastus*, form a dense, almost impenetrable barrier on steep mountain slopes, in many cases as a dominant, single-layered community. The ecology of this vegetation type is poorly known, but it is almost certain that in many cases establishment and subsequent maintenance is due largely to manmade fire. Once established, it is difficult to eradicate. In these two respects, the bamboo communities exhibit typical aspects of savanna behaviour. From the structural point of view they are more an anomalous savanna type although in many areas in the central cordillera where mesotherm forest has been disturbed *Nastus* often establishes as an understorey stratum which may then either carry fire into the forest to reduce the cover of forest species further, or else, in the absence of fire, may become suppressed with the increasing closure of tree canopies.

Tree-fern savannas (Fig. 9.42) have been mentioned as such by Whyte (1968) and subsequently described more fully by Paijmans and Löffler (1974) and Paijmans (1976). The formation was not included by Burtt-Davy (1938) or by Fosberg (1961, 1967, 1973) in their vegetation classification system, and according to Paijmans and Löffler (1974) tree-fern savanna may well be unique to the island of New Guinea. It is a formation typically dominated by the tree-fern genera *Cyathea* and *Dicksonia*. The former genus is most common in the open savanna (e.g. *Cyathea atrox*), while *Dicksonia* tends to occur mainly along forest edges. Although tree ferns are generally regarded as forest dwellers, in the present case they appear to be ecological analogues of some megatherm palm species (e.g. *Livistona brassii*) that are capable of tolerating extremes of closed forest and exposure to fire. In the case of the microtherm tree ferns it is probably their tough vascular construction combined with a dense indumentum of scales and hairs that helps them to tolerate both fire and frost and water stress due to high solar radiation. The understorey may be composed of tussock grasses such as *Agrostis, Danthonia, Deschampsia, Deyeuxia, Hierochloe* and *Poa* and a mixture of forbs, with or without the shrub genera mentioned above under shrub savan-

Fig. 9.42. Microtherm tree-fern savanna in highland Papua New Guinea. *Cyathea* spp. dominate this formation. (Photo: R. Hoogland, C.S.I.R.O.)

nas. In some cases the tree ferns may support a rich epiphytic community variously composed of ericaceous genera and pteridophytes; and, like shrub savanna, these communities tend to occupy the transition zone between forest and open grassland, most commonly on foot slopes. On open valley floors the tree ferns are usually restricted to stream banks.

The formation is widespread throughout the high mountain massifs of New Guinea, particularly on Mount Albert Edward and Mount Dickson.

THE SOLOMON ISLANDS

These islands consist of a double elongated chain, extending about 1050 km if one includes Bougainville which is politically part of Papua New Guinea. The first known emergence of any present land area of the Solomons appears to have been at the end of the Oligocene. Broadly speaking, sediments deposited after this generally contain some ultrabasic material; and on Guadalcanal coralline limestone is locally common, as are calcarenites of similar age on all the other large islands. As well as present-day vulcanism in the New Georgia group, during the Pleistocene and Recent periods there has been a general emergence of all the larger islands with evident elevated coral-reef terraces (Thompson and Hackman, 1969). This emergent pattern is similar to that of Vanuatu (New Hebrides), as is the recurring pattern of hurricanes; however the Solomon Islands are richer floristically.

Climate data have been scarce and difficult to interpret (cf. Brookfield, 1969) but it is evident there is the typical southwest Pacific pattern of savanna development in rain-shadow areas, which presumably reflects relatively high seasonality and lower mean annual rainfall.

The savannas, or "grasslands and heaths", have been described by Whitmore (1969), who regarded them as probably anthropogenous. According to Whitmore, the most extensive of these lies along part of the north coast of Guadalcanal on Pleistocene sediments in the rain shadow of the high Kavo Range, where there is a series of grass-covered raised beaches with deeply dissected forested gullies (rather similar, it seems, to western Malekula). The Guadalcanal Plain to the east contains tracts of almost pure *Themeda australis* with small areas of *Imperata cylindrica*, and *Saccharum spontaneum* and *Phragmites karka* in the wetter locations. Whitmore also described the occurrence of *Pennisetum polystachyum* in disturbed areas; it would seem, from its expansion in other island areas in the southwest Pacific, that this species may well, in time, dominate the savannas of Guadalcanal.

In these savannas, scattered tree species (*Alstonia scholaris, Colona scabra, Commersonia bartramia, Desmodium umbellatum, Hibiscus tiliaceus, Kleinhovia hospita, Leucaena leucocephala, Macaranaga aleuritoides, M. tanarius, Premna corymbosa, Timonius timson*) and the shrubs *Melastoma polyanthum* and *Morinda citrifolia* recorded by Whitmore, indicate a close relationship with the coastal savannas of Papua New Guinea. In his account, Whitmore also described pandan savannas. With the exception of *Sararanga* (an endemic Solomon Islands genus of the Pandanaceae), these are reminiscent of those described by Paijmans (1971) in the Trans-Fly Plain (cf. Fig. 9.37). Apart from some grasslands in the Florida Islands, there are few other documented records of savannas in the Solomon Island region, although Whitmore (1969) described abundant scrambling bamboo (*Nastus productus*) in below-summit zones of Mounts Balbi, Kolombangara and Popomanesu.

NEW CALEDONIA

The megatherm woodland savannas of New Caledonia are somewhat unique within the southwest Pacific. The most common type is the *niaouli* woodland savanna dominated by the paperbark tree *niaouli* (*Melaleuca quinquenervia*). As a general savanna form it is widespread on New Caledonia, particularly where there is a history of recent disturbance. *Niaouli* savanna has been described by a number of authors (Guillaumin, 1952; Virot, 1956; Jaffré, 1974; Jaffré and Latham, 1974; Holloway, 1979). It is extremely variable both structurally and floristically and ranges from a dwarf form on extremely poor soils on peaks and mountain ridges to a tall woodland on deeper basalt and alluvial soils on lower coastal slopes. Guillaumin (1952) recognized niaouli savanna as one of the three major vegetation types on New

Caledonia, the other two being maquis (a low myrtaceous microphyll scrub/sedge formation) and forest. In the wetter east, the most common savanna form in coastal and sub-coastal areas is a niaouli-dominated community with ferns, shrubs and sedges. On the lower slopes of Mount Koghis is niaouli-dominated *Baeckea*/sedge maquis; in intermediate upland areas, on schists and gneiss, *Baeckea parvula* is common with *Gleichenia* and *Pteridium*, while this tends to change towards an association with *B. virgata* near habitation (Holloway, 1979). Also associated with these savannas on schist hills is an impoverished form of *Baeckia*/sedge maquis with *Dracophyllum*, *Scaevola* and *Spathoglottis*, while other burned and degraded areas may have the *Gleichenia* and *Lycopodium cernuum* association that is so common in other parts of the southwest Pacific.

On the coastal slopes of the drier western massifs lie extensive niaouli savannas with an *Imperata* grass understorey, occasionally with the introduced woody *Leucaena leucocephala* and *Psidium guayava*. This formation is best developed on deep black soils overlying basalt and, according to Holloway (1979), is sometimes intermingled with *Acacia farnesiana* and *A. spirorbis* grassland associates, the former being best developed near Anse Longue. Other associates of niaouli savanna in the west are *Gymnostoma* and *Pandanus* with *Baeckea ericoides* maquis and the sedges *Baumea* and *Costularia*. This type may also have the ferns *Gleichenia* and *Pteridium*.

Holloway (1979) has raised the question of what was the original vegetation that dominated the peaks and ridges currently occupied by niaouli savanna. Guillaumin (1952) and others have suggested that niaouli is a species essentially of a secondary vegetation type. Holloway has also pointed out that counterparts for niaouli savannas are lacking in other parts of Melanesia, the most comparable form being perhaps the *Eucalyptus* woodlands of Australia. He has also compared niaouli lands with the dry *talasiga* of Fiji (see p. 232).

Although Holloway discounted any significant parallel between the New Caledonian and New Guinean *Melaleuca* savannas, there are many areas where considerable similarities can be observed, particularly in the Trans-Fly and Trans-Oriomo areas of southwestern Papua New Guinea where, as

in northern Australia, *Melaleuca* species dominate much of the landscape, in some cases with "maquis"-like *Acacia* associates and shrub and sedge layers. The major ecological feature of paperbark woodland environments in both Australia and Papua New Guinea is that although some *Melaleuca* species (e.g., *M. argentea*, *M. nervosa*) may occupy ridge situations, the most extensive formations are dominated by other *Melaleuca* species (e.g., *M. quinquenervia*, *M. viridiflora*) in poorly drained, often seasonally inundated land, whereas the niaouli on New Caledonia covers extensive leached uplands. The explanation for these phenomena is not clear. While it appears certain that the spread of niaouli savanna is associated with human activity, such as firing, subsistence gardening and nickel mining, the rate of spread into a wide range of habitats in New Caledonia may be due to the lack of available competitive species (e.g. *Eucalyptus* spp.).

VANUATU (NEW HEBRIDES)

The Vanuatuan Islands lie northeast and east of New Caledonia and the Loyalty Islands and southeast of the Solomon Islands with Fiji to the east. The eleven main islands range from about 13° to about 21°S latitude and, until achieving independence recently as the new political entity of Vanuatu, were administered as the New Hebrides by Britain and France jointly.

Most of the oldest rocks in Vanuatu probably date from the Early Miocene, although some limestones in Mauro Island are possibly Eocene. Uplift of nearly all the present land surface of the islands is post-Pliocene and most of it is post-Pleistocene (Lee, 1975; Mallick, 1975). There is active vulcanism, and signs of uplift due to recent tectonism are evident in many areas particularly in the raised marine terraces of Malekula (Mallicolo). The recent geological history of the islands combined with active tectonic movement and frequent hurricanes has created a relatively unstable environment for a floristically poor flora (<400 angiosperm taxa). This instability of climate, land surface and low floristic richness have facilitated the expansion of savanna species and associated exotics, particularly *Leucaena*, in seasonal areas where fire frequency is more than biennial.

The annual climate varies from hot, very humid (c. 23–32°C with c. 4000 mm mean annual rainfall) with little seasonality in the north, to warm humid (c. 17–30°C; 2200 mm mean annual rainfall) with marked seasonality in the south. All these factors are related to differences in vegetation type and distribution, so that vegetation in the north has affinities with the islands of New Guinea and the Solomon Islands, while in the islands to the south of Efate there are more distinct affinities with Fiji, New Zealand and New Caledonia (cf. Van Balgooy, 1971; Gillison, 1975a, b; Schmid, 1975).

The savanna areas within Vanuatu are essentially maritime megatherm. There are insufficient data to determine seasonality levels with regard to available soil moisture, although it is likely that the southern islands of Tana (Tanna) and Aneityum (Anatom) are the most seasonal, together with the west coast savanna of Malekula and Espiritu Santo (Marina) to the north. In nearly all cases, savannas are distributed within rain shadows of the western coasts. In the south, Schmid (1975) has described open Acacia spirorbis forest on deep calcareous soils and on ferrallitic soils developed on basalts. The upper stratum is monospecific A. spirorbis with occasional undershrubs of Croton, Halfordia, Symplocos and Xylosma. Associated graminoids are the grasses Chrysopogon and Oplismenus, the sedge Gahnia and the ferns Dicranopteris and Pteridium.

On Aneityum, Schmid (1975) has also described a savanna dominated by Miscanthus floridulus, up to 2 m tall that has colonized areas previously occupied by forest particularly in the north and west, whereas to the east there is a gradual increase in shrubs and ferns, with small trees of Acacia, Commersonia and Glochidion or various "maquis" elements (cf. New Caledonia).

On Tana in the west between 100 and 300 m Schmid (1975) recorded extensive Miscanthus savanna, Miscanthus being replaced towards the central plateau by other grass species including Apluda, Imperata and Paspalum, associated with woody Asteraceae and Malvaceae. The same pattern of Miscanthus savanna occurs on Eromanga (Erromango), but on the plateaux at about 200 m the grass Chrysopogon aciculatus becomes dominant on the more compact soils.

The mountainous interior of Espiritu Santo has developed incipient areas of savanna that are a mixture of mid- to tall-grass savannas with various soft-wooded trees such as Alphitonia, Omalanthus, Schizomeria and Trema, usually associated with garden areas, although more frequent firing in recent years has caused a spread into forested hillsides (Fig. 9.43). The southwestern rain-shadowed coastal hills have already been largely converted to grassland.

The west coast of Malekula is likewise "savannized" with extensive tracts dominated by mixed mid-height to tall Imperata, Themeda and Miscanthus, associated with the trees Cordia dichotoma, Glochidion, and Hibiscus tiliaceus (Fig. 9.44). Leucaena leucocephala has replaced large areas of forest and savanna, as a dense layer 3 to 6 m tall. This replacement has occurred mostly on raised marine terraces on black cracking soils (Gillison, 1975a) and is likely to expand further with the continuing fire regime practised by the people of that area.

FIJI

Of the more than five hundred islands that form the Fijian Group, the two most prominent are the volcanic islands Viti Levu (c. 10 400 km²) and Vanua Levu (c. 6000 km²) where most of the population of 500 000 resides. Cultivation is mostly restricted to the river valleys and coastal areas, where considerable disturbance to the landscape has taken place. Although the climate is mild there are distinct seasons, and on the larger islands there are marked differences in climate and vegetation according to whether the aspect is windward (wet — between 3000 and 3600 mm annual rainfall) or leeward (dry — between 1650 and 2300 mm). It is in the latter areas where savannas are mostly developed up to about 450 m altitude (Fig. 9.45).

Although there is limited documentation for Fijian savannas, unstable, rather than stable, megatherm disclimax savannas appear to be the rule. As with many other southwest Pacific islands, most of the original vegetation is considered to have been closed forest of one kind or another (Twyford and Wright, 1965; Parham, 1972; Smith, 1979). Twyford and Wright (1965, in collaboration with Parham and Loweth) considered that the only exception to this would have been small areas of reed-swamp surrounding open water, and wind-

Fig. 9.43. Rugged interior of Espiritu Santo, near Mount Tabwemasana, Vanuatu (New Hebrides), showing incipient savanna development due to man-made fire.

Fig. 9.44. Recently fired coastal savanna in western Malekula, Vanuatu (New Hebrides). Dominant trees are the soft woods *Cordia*, *Hibiscus tiliaceus* and *Glochidion*, with mid-height to tall grass genera *Imperata*, *Themeda* and *Miscanthus*.

Fig. 9.45. Nadi foothills, Fiji. Firing and overgrazing have produced this savanna complex of the grass *Pennisetum polystachyon* with *Pteridium* (bracken) and the she-oak, *Casuarina equisetifolia*. (Photo courtesy of the Fiji Pine Commission.)

shorn scrub on some of the more exposed mountain peaks. Parham (1972) estimated that grassland associations occupied 1 500 000 acres (607 040 ha) and "forest and bush" 2 300 000 acres (930 800 ha).

Twyford and Wright (1965, p. 79) have also reported that whereas most of the forest on the drier leeward sides of the island has been destroyed, there are sufficient remnants to suggest that, on the larger islands, the leeward and most leached stable soils once supported a form of xerophytic open forest, variously composed of low trees such as *Acacia richii, Alphitonia zizyphoides, Dodonaea viscosa, Fagraea gracilipes* and the parasitic sandalwood (*Santalum yasi*), possibly in association with *Casuarina equisetifolia* and *Pandanus* spp. The ground cover was dominated by the pteridophytes *Dicranopteris linearis, Lycopodium cernuum* and *Pteridium esculentum* (the last usually indicates a fire history in other parts of the southwest Pacific).

Since then, man-induced changes, mainly through subsistence gardening and associated fire, have transformed much of the forest to a mosaic of disclimax grass and fernland with some woody elements. If fire is removed there is a gradual return to forest.

In recent years, the introduction of woody species and many grass species has considerably modified many of the disclimax associations. Twyford and Wright (1965) reported that in 1918 Seemann collected nineteen grass species of which only five were indigenous. Since then, about 140 additional grass species have been introduced and have become established, in particular "mission grass" (*Pennisetum polystachyum*), "wire grass" (*Sporobolus indicus*) and "Guinea grass" (*Panicum maximum*). Wire grass dominates the Momi Hills, a large area between Lautoka and Mba on Viti Levu, and large areas in the Macuata and Mbua provinces of Vanua Levu. The introduced guava (*Psidium guayava*) is commonly associated with these grasslands as a stunted woody element, together with the local *Casuarina equisetifolia, Pan-*

danus tectorius and the creeping legume *Atylosia scarabaeoides.*

Introduced Myrtaceae (*Psidium* and *Rhodomyrtus*) tend to dominate certain disclimax savanna mosaics, particularly on poor soils, in other parts of the southwest Pacific (particularly *Rhodomyrtus* on Norfolk Island and *Psidium* in parts of coastal northeastern Queensland). In Fiji, the introduced mimosoid tree *Leucaena leucocephala* is also a significant component in the drier grasslands of Viti Levu and on Naviti Island in the Yasawas. In parts of Guam (Micronesia) and in western Malekula (Vanuatu), *L. leucocephala* has now completely displaced the grasslands, while in the floristically richer (? and more competitive) Papua New Guinea it is confined to subcoastal pockets. The leguminous (also mimosoid) saman or rain tree (*Pithecellobium samanea*) also occurs around settled coastal areas of Viti Levu, while in the dry, seasonal Rigo coast of southeastern Papua New Guinea the same species appears to be invading mid-height grasslands associated with gallery forest.

Perhaps the only present-day Fijian vegetation that approximates a "natural" woodland savanna is that of the "sunburnt" *talasiga* lands where the reed grass (*Miscanthus floridulus*), which replaced what was formerly closed forest, has itself been largely replaced by ferns (*Dicranopteris linearis* and the bracken *Pteridium esculentum*), and wire grass (*Sporobolus indicus*) and the tree *Casuarina equisetifolia*. In Viti Levu, *Pandanus odoratissimus* talasiga is associated with mission grass (*Pennisetum polystachyon*) and on Macuata, the trunked cycad (*Cycas rumphii* f. *seemannii*) is common. Parham (1972) has also recorded *C. rumphii* f. *seemannii* on Vanua Levu, and listed *Miscanthus floridulus* as one of the significant species of the dry zone as well as the grasses *Dichanthium caricosum, Heteropogon contortus, Panicum maximum, Paspalum orbiculare* and *Sporobolus elongatus* that have become naturalized over large areas. Parham also mentioned that *Pennisetum polystachyum* was, at that time, increasing and spreading from Mbua to Macuata. The disclimax nature of the *talasiga* lands is evidenced in the shrub and low tree succession which takes place if burning is stopped for several years (e.g., *Acacia richii, Alphitonia zizyphoides, Decaspermum fruticosum, Dodonaea viscosa, Hibbertia lucens, Leucopogon cymbulae, Morinda citrifolia,*

Mussaenda raiateensis and *Syzygium richii*; cf. Twyford and Wright, 1965; Parham, 1972). In this respect, the disclimax is floristically similar to some of the subcoastal grasslands of eastern Papua New Guinea.

The continual firing of these lands has accelerated erosion and it is likely that the present pattern of land use will continue to cause concern in land stabilization, although in recent years the establishment of pine plantations (*Pinus caribaea*) may ameliorate this condition.

MICRONESIA

The islands that make up Micronesia are distributed over about 7.8×10^6 km^2 of ocean and have a total land area of about 1800 km^2 (see Fig. 9.1). There are, at the time of writing, four major political entities of the U.S. Trust Territory of the Pacific Islands, namely the Commonwealth of the Northern Marianas, the Marshall Islands, the Palau District (western Caroline Islands) and the Federated States of Micronesia (Yap State, Truk State, Ponape State and Kosrae State) that form the western Caroline Islands. In the same geographic zone of the southern Mariana Islands lies the U.S. island territory of Guam. The total population of the Trust Territory as at September 1979 was 116 666 (U.S.T.T. Office of Planning and Statistics). This population, which (together with the activities of World War II) has already had a dramatic effect on the vegetation, is expected to double within the next twenty years.

As with most small tropical islands in the southwest Pacific, the bioclimate is essentially maritime megatherm. Most Micronesian islands are isothermal (average *c.* 26.7°C with a maximum annual deviation of less than 2°C). Annual rainfall records vary from 2230 mm in Saipan (northern Marianas) to 4870 mm in Kolonia, Ponape.

Whilst there are some early accounts by Safford (1905) and some by Glassman (1952) of savanna types, the vegetation of the Micronesian islands has been described in greatest detail by Fosberg (1960) who has recorded that on the volcanic (and metamorphic) islands are grasslands of variable type, with patchy occurrences of trees and shrubs. As with Fiji and Vanuatu, there are no stable woodland savannas as such but rather a patchwork of disclimax savannas with woody elements where degree of cover varies with fire frequency.

According to Fosberg (1960), the plants that are generally characteristic of savanna vegetation are "*Miscanthus floridulus, Heteropogon contortus, Dimeria* and various other small grasses; many sedges, principally species of *Fimbristylis, Scleria* and *Rhynchospora;* ferns and related plants, such as *Gleichenia linearis, Lycopodium cernuum, Cheilanthus tenuifolia, Lygodium scandens,* and *Blechnum orientale*; shrubs and herbs, such as *Geniostoma, Eurya, Melastoma malabathricum, Pandanus, Myrtella benningseniana, Glossogyne tenuifolia, Nepenthes mirabilis, Morinda pedunculata,* several species each of *Hedyotis, Euphorbia,* and *Phyllanthus,* and a number of others". To this list may be added *Casuarina equisetifolia, Pandanus tectorius, Tacca leontopetaloides* and more recently the grass *Pennisetum polystachyum,* which at present

appears to be invading the *Imperata*-dominated hill grasslands of Moen (Truk) (Fig. 9.46). In these mid-height Moen grasslands, associated species are the grasses *Ischaemum muticum* and the tall *Saccharum spontaneum,* and the geophytes *Curculigo* sp. and *Tacca leontopetaloides.* Associated transitional fast-growing soft-wood trees are *Cananga odorata* and *Glochidion* sp. The invasive pattern of exotic grass and woody species in Micronesia tends to follow a similar pattern to that in other island groups such as Fiji and Vanuatu. In all these areas *Pennisetum polystachyum* is an active invader and, as mentioned earlier, the tree *Leucaena leucocephala* has displaced a number of grassland "savanna" areas on Guam, although this succession was assisted by purposive planting by man in the period immediately after World War II.

The general pattern of succession from pure

Fig. 9.46. Recently derived mid-height to tall grassland on the island of Moen. (Truk Group, Micronesia): *Imperata cylindrica, Ischaemum muticum, Pennisetum polystachyon* and *Saccharum spontaneum,* with scattered trees of *Cananga odorata, Ficus* sp., and geophytes *Curculigo* and *Tacca* spp.

grassland or fernland to forest, is, as in other places, governed by fire. It is highly likely that the Micronesian islands were also once covered with forest before the advent of man. None of the Micronesian forests can be said to be particularly susceptible to fire in its natural condition (Fosberg, 1960). Fosberg also recorded that savanna on volcanic soils is often sharply differentiated from forests where it abuts soil of limestone origin. This intriguing phenomenon has not yet been explained, although as he indicated, the rapid regeneration of forest on limestone soils may limit the persistence of savanna. There are few grassland areas on any of the "low" (coralline) islands. According to Fosberg (1960), J. Bridge (pers. comm.) and Bridge and Golditch (1948) noted a high correlation between bauxite deposits and the fern-dominated phase of the savanna vegetation. Such a community is still evident on bauxite outcrops on the north of Babelthaup Island in the Palau district.

Towards the northern Marianas there is an increase of savannas dominated by *Miscanthus floridulus*, associated with *Dimeria* and *Heteropogon*, with scattered trees of *Acacia* sp., *Casuarina equisetifolia* and *Leucaena leucocephala*. On Yap Island, the savanna or *tedh* is locally dominated by *Pandanus tectorius*, the grasses *Heteropogon contortus* and *Ischaemum* spp. together with the geophyte *Tacca leontopetaloides*, while on the weathered volcanic soils of Gagil and Tomil on Yap there is a low shrubby savanna dominated by the fern *Dicranopteris* (*Gleichenia*) *linearis*, together with the grasses *Imperata* and *Ischaemum* spp., and the acid-indicating low shrubs *Eurya* and *Melastoma malabathricum*.

Fosberg (1960) has noted that the endemics "*Myrtella benningseniana*, *Ischaemum longisetum*, *Dimeria chloridiformis*, *Hedyotis tomentosa*, *H. korrensis*, *Geniostoma micranthum* and *Phyllanthus saffordii* appear to be confined to the savanna". While this might seem to contradict the view that savannas are only recently derived, Fosberg suggested that these floristic elements may have been present originally as rare plants on peaks and crests, landslides and ravine walls, from whence they invaded the savannas.

Wetland savannas or open swamp graminoid formations also exist on the larger islands. These rather patchy vegetation units have been described in considerable floristic detail by Stemmerman and Proby (1978). It is the wetland areas that have been most utilized for subsistence gardens, in particular for the swamp taros *Alocasia esculentum* and *Cyrtosperma chamissonis*. As yet, the Micronesian does not use the dry savanna landscape to any great degree. Some medicinal plants exist, and *Miscanthus* and the fern *Lygodium* have structural uses. Although there are attempts at Airai on Palau to establish forest plantations in these disclimax formations, there are considerable difficulties involved in excluding fire.

The anticipated increase in land use in the coming years will almost certainly create an expansion of the disclimax savanna vegetation. This is already evident on Moen in the Truk group, on Ponape (Fig. 9.47) and at Airai on Babelthuap (Palau), where on the last-named *Pandanus* savanna is increasing significantly year by year. Unless some way is found to utilize these savanna lands it is likely that their expansion in the near future will lead to a significant depletion in land resource.

TIMOR (NUSA TENGGARA)

Aside from the island of New Guinea and its nearby eastern islands, the only other islands known to contain naturally occurring eucalypt savannas lie in the Nusa Tenggara (Lesser Sunda) group of Timor, Adonara, Lomblen, Pantar, Alor and Wetar[1]. Of these volcanic islands, Timor is the largest, with a central cordillera that rises to 2963 m at Mount Tatamailau, while the adjacent islands although smaller, are of similar topography.

Throughout the group, the eucalypt communities are extremely varied in structure. Of the two eucalypts — *Eucalyptus alba* and *E. urophylla* — the former is considered to form the more open woodland savanna while the latter tends to be mainly forest, and then at higher altitudes. Martin and Cossalter (1975a, b, 1976) have described in detail a wide range of habitats throughout the islands from which eucalypt seed was collected for provenance

---

[1] Martin and Cossalter (1975a) have reported that the Forestry Service of the North Celebes (Sulawesi) Province has confirmed the presence there of natural stands of *Eucalyptus alba*.

Fig. 9.47. An open savanna on the island of Ponape, Micronesia, with scattered transitional woody trees (*Commersonia*, *Glochidion*, *Macaranga*, *Melochia* and *Morinda*) and ferns (*Gleichenia* and *Pteridium*) with short grasses, among them *Ischaemum* spp.

trials. Most of the following details have been extracted from their reports.

Throughout the environmental range of the eucalypts, the thermal regime varies from 27°C at sea level to about 20°C at 1800 m (and presumably less for outlying communities on mountain summits such as Mount Moutis at 2827 m). Annual rainfall ranges from 500 mm to about 2070 mm depending on aspect and topography, while seasonality varies, in some cases up to six months having less than 50 mm of rain. The effect of such seasonality is borne out on the island of Pentar where *Eucalyptus urophylla* becomes completely leafless from the onset of the dry season. This species has been recorded from an altitude of about 420 m to 2427 m on the top of Mount Moutis on Timor. Although *E. urophylla* communities tend to achieve their best form at altitudes >1200 m, individual trees taller than 50 m have been recorded at 890 m,

and on Timor the communities are best developed on the southern exposures, whereas *E. alba* tends to dominate the drier northern aspects. *E. urophylla* sometimes forms tall dense woodland savanna with understorey trees *Albizia procera*, *Lagerstroemia* and *Schleichera oleosa* and the grass *Imperata*. At 1400 m near Quetrato in the Ermera region, *Eucalyptus urophylla* forms a more open community (20–25 trees ha^{-1}) with a ground layer of ferns and grasses, while on Alor it occurs with an understorey of *Acacia* spp. *Cycas*, tree-ferns and a layer of *Imperata*, 1.5 to 2 m tall.

Communities dominated by *Eucalyptus alba* range from dense mid-height forests on deeper soils at low altitudes to open woodland savannas and, at extremes, a low scrub (0.5–1 m tall) on exposed ridges with shallow soils. *E. alba* usually occurs at lower altitudes than *E. urophylla*, but overlaps are known and Martin and Cossalter (1975a, b, 1976) have reported that, in such cases, *E. alba* may occur as an understorey element. Other than this, the strongly seasonal regime is emphasized by the association of *E. alba* with typically deciduous tree species, *Albizia procera*, *Canarium commune* and *Erythrina*, with the occasional *Bombax*. (Such a formation is similar to the *Eucalyptus alba* woodland savannas behind Port Moresby in Papua New Guinea.) With these may be the semi-deciduous *Cassia javanica*, the evergreen *Timonius sericeus* and the exotic *Psidium guayava*.

The origin of these savannas is not clear, but it would seem that man-made fire has played a major role in the establishment of *Eucalyptus alba* woodland savannas. On Flores Island, Martin and Cossalter suggested that *E. urophylla* may represent a secondary forest that has colonized rain forest areas following devastation from volcanic eruptions.

LAND USE

Up to the present day, the tropical savannas of the region subject to most commercial use are those of megatherm northern and northeastern Australia, where the major industry throughout is the raising and fattening of beef cattle and sheep, with some dairy cattle in the more mesic Provinces of the east (Table 9.5). In recent years, the development of crops such as cotton and sugar cane has been tested on irrigated lands in the Ord River area of the northwest, with limited success, some of the biggest problems arising from bird and insect pests. Of these crops, it seems sugar cane may have the greatest short-term potential. However, as Courtenay (1977) has pointed out, the Ord scheme is not likely to be a model for further settlement in the more remote parts of the north. In the eastern Provinces the economic potential of already established, energy-rich crops such as oil seeds, may lead to further conversion of woodland savannas. In western Queensland, mining of silver, lead and zinc is an important industry where the world's largest single lead mine is located (Mount Isa). Major deposits of bauxite are currently being mined at Weipa in Cape York, manganese at Groote Eylandt in the Gulf of Carpentaria, and coal in the Bowen Basin in central sub-coastal Queensland.

For the cattle industry, the majority of pastures are based on native grass species, with introduced grass species increasing towards the wetter extremes and on deeper soils. Until recently, in a few isolated instances, there has been very low capital

TABLE 9.5

Tropical Australia: numbers and average density per square kilometre of sheep, beef cattle and dairy cattle by regional groupings, 1970–1971 (from Courtenay, 1977)

Region (Statistical divisions in brackets)	Sheep		Beef cattle		Dairy cattle	
	number	average density	number	average density	number	average density
Central Queensland (Rockhampton, Central Western)	1 966 837	10.50	2 058 396	7.29	64 632	0.23
North Queensland (Mackay, Townsville)	792	0.01	913 625	7.81	13 324	0.11
Far North Queensland (Cairns, Peninsula)	716	0.00	405 895	2.04	34 092	0.17
Western Queensland (Far Western, North Western)	3 046 887	4.66	1 438 606	2.20	940	0.00
Northern Territory	7 000	0.01	1 166 000	0.88	–	–
Tropical Western Australia (Kimberley, Pilbara, North West)	1 846 079	1.74	757 772	0.72	324	0.00
Total	7 868 311	2.16	6 740 294	1.85	113 312	0.03

input in the cattle industry in the region, with low returns per unit area. Typical management practice involves usually haphazard annual firing of grasses so that new growth provides a patchwork of more attractive but not necessarily more nutritious fodder. Stock distribution is controlled mainly by the position of watering points, which assist in the location of stock during "mustering" or herding operations. Originally, European cattle breeds such as Herefords and Shorthorns (*Bos taurus*) made up most of the cattle numbers, but in recent years this has changed to the Zebu type (*B. indicus*) such as Brahman and Santa Gertrudis, that are better suited to the megatherm seasonal regime. In more recent years, in the wetter northwest, the presence of feral buffaloes (*Bubalis bubalis*) has created additional management problems, both as an additional grazing agent and as foulers of waterholes (see also Perry, 1960; Stocker and Mott, 1981). According to Shaw and Norman (1970), properties in the Northern Territory–Western Australia area are the largest, averaging about 390 000 ha; those in Queensland tend to be smaller and more intensively operated. The importance of these savannas is indicated by the fact that they support about one quarter of the Australian beef cattle population. Carrying capacity ranges from one beast per 2 ha in southern Queensland to about one beast per 65 ha in Cape York (Provinces 6 and 7) (Sutherland, 1961). Although sheep outnumber cattle, returns for sheep raised in temperate regions outweigh those of the tropics where there are difficulties in mating and lambing due to higher temperatures (Courtenay, 1977).

A characteristic feature of the northern xerophytic mid-grass grazing lands of Moore (1970) in the semi-arid zone of Provinces 2 and 11 is that during drought years edible foliage from native trees and shrubs is browsed by domestic stock. In more mesic areas in normal years, it is the native understorey of the "tropical tall grass" category of Moore (1970) that provides basic fodder. In Province 3 in the Northern Territory–Western Australia area this is characterized by both annual and perennial sorghum (*Sorghum plumosum*) and kangaroo grass (*Themeda australis*).

The most widespread grass of the megatherm grazing lands is the warm season perennial kangaroo grass. However, whereas the associated *Sorghum* spp. are typical of the northwest of

Province 3, the woodland savannas east of and including Province 6 are characterized more by bunch spear grass (*Heteropogon contortus*), although this species is increasing in the northwest. Apart from the open Mitchell grass (*Astrebla* spp.) plains on heavy-textured soils in the southeast corner of Provinces 1 and 2, there are woodlands to the northeast in the drainage system of the Gilbert and Mitchell Rivers that support a native pasture of rather poor quality (*Aristida* spp. and *Chrysopogon fallax*). In all of these lands, seasonal desiccation of pasture, coupled with fire and flood, create a difficult environment for the cattle industry.

In some of the better drained areas south of the "Gulf Country" of western Queensland are the *Dichanthium* (blue grass)–*Eulalia* (brown top) downs (*D. fecundum*, *D. tenuiculum* and *Eulalia fulva*) while further south the blue grass is dominated by *D. sericeum* (Shaw and Norman, 1970). Native legumes *Alysicarpus*, *Crotalaria* and *Glycine* are important components of these pastures.

The more mesic woodland savannas on soils of medium texture in Province 14 and in the northern limits of Provinces 11 and 12 are important grazing lands dominated by *H. contortus* associated with the grass genera *Aristida*, *Bothriochloa*, *Cymbopogon*, *Dichanthium* and *Themeda*. Important introduced perennial grasses are buffel grass (*Cenchrus ciliaris*), Rhodes grass (*Chloris* spp.) and green panic (*Panicum maximum*), Pangola grass (*Digitaria decumbens*) and *Urochloa mosambicensis*, together with the legumes siratro (*Phaseolus atropurpureus*), Hunter River lucerne (*Medicago sativa*), *Centrosema*, *Desmodium*, *Glycine* and *Stylosanthes* spp. The introduction of such legumes is generally considered to be the most economic way to improve pastures in northern Australia. Considerable improvement of pastures in the Ramu and Markham Valleys of Papua New Guinea indicate that the same may be true of island savannas which are likely to be grazed by livestock. In recent years, large tracts of woodland in the eastern Australian Provinces have been cleared to make way for pastures containing mixtures of these species. There has been only limited success in some of these areas — for instance, in the brigalow and poplar box belt, and in northern areas dominated by paperbarks — as the emergence of woody weeds (both juveniles of the tree species, and shrubs) to the exclusion of grass has created a major management problem. In some cases ringbarking (girdling) or poisoning of trees has stimulated grass response, but in areas where there is insufficient fuel, firing of these lands only promotes the growth of shrubs at the expense of pasture.

In Papua New Guinea some of the savannas near Port Moresby support a limited beaf cattle and dairy enterprise. The former is more extensive in the megatherm savannas of the Ramu and Markham Valleys where the growing of sugar cane also has potential with the exception of cattle-ranching in Guam and parts of the northern Marianas, and the establishment of pine plantations in Fiji. Savannas are little used in the remainder of the southwest Pacific, and in many cases indicate a resource depletion.

SUMMARY

A provisional bioclimatic framework based on plant growth indices has been used to describe the savannas of the region. Within this framework, the Australian megatherm seasonal eucalypt-dominated woodland savannas stand out as the major savanna biome, while on that continent acacioid and *Melaleuca*-dominated woodland savannas are the next most important ecological entities. Whereas the eucalypt woodland savannas extend from Australia to the Timor Islands and the island of New Guinea, it is the *Melaleuca* savannas, and to a lesser extent acacioid savannas, that radiate to New Guinea, New Caledonia and Vanuatu (New Hebrides). In sum, with the exception of the more remote islands of Micronesia and the high-altitude microtherm savannas of Papua New Guinea, the evidence suggests that, throughout the region discussed, there is a strong radiation of major savanna elements from the Australian continent. The distribution of megatherm savannas is markedly related to areas of high seasonality, and in the islands, this is reflected in rain shadows.

Some savanna genera have a wide ecological amplitude, such as *Gleichenia* and *Lycopodium* that range from sea level to 4000 m, but it is the presence of exotics such as *Leucaena leucocephala*, *Pennisetum polystachyum*, and *Psidium guayava* that poses a major ecological threat, particularly to some of the smaller islands where competition from indigenous species is low.

The origin of savanna formations is almost certainly associated with fire and to some extent frost in certain areas, modified in many instances by seasonally fluctuating, soil water tables. The variable influence of these environmental factors creates vegetation mosaics and continua that make difficult any arbitrary system of classification.

In northern Australia, woodland savannas are more or less stable disclimaxes that are widely used as pasture lands — it is in many of the savannas of the smaller and more distant islands that there is the greatest immediate potential for change. There, with rapidly increasing populations and associated land-use pressures, the outlook is for expanding savanna areas that under present methods of land use will lead to a diminishing resource.

APPENDIX I: STRUCTUREGRAMS SHOWING THE RANGE OF HEIGHT AND FOLIAGE PROJECTIVE COVER FOR WOODY SAVANNAS IN EACH BIOCLIMATIC PROVINCE

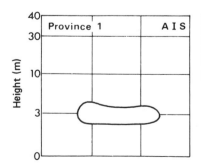
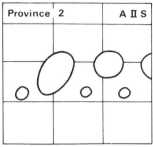
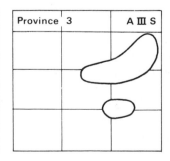

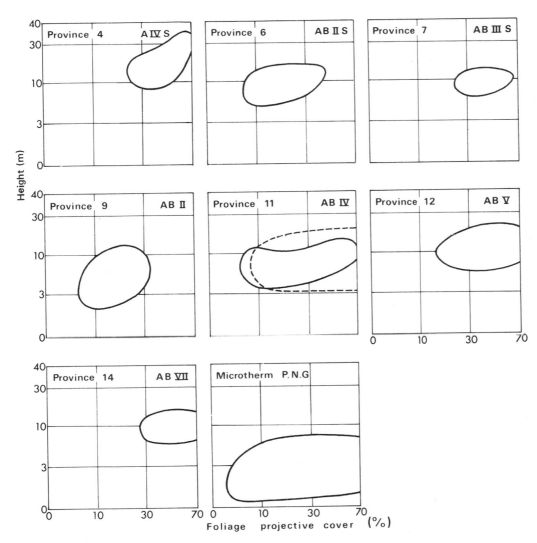

REFERENCES

Archbold, R. and Rand, A.L., 1935. Results of the Archbold Expeditions No. 7. Summary of the 1933–34 Papuan expedition. *Bull. Am. Mus. Nat. Hist.*, 68: 527–529.

Armstrong, P., 1977. Baobabs: remnant of Gondwanaland. *New Sci.*, 73: 212–213.

Aubréville, A., 1966. Les lisières forêt-savane des régions tropicales. *Adansonia*, VI: 175–187.

Aubréville, A., 1968. Les lisières forêt-savane des régions tropicales. In: T.L. Hills and R.E. Randall (Editors), *The Ecology of the Forest/Savanna Boundary. McGill Univ. Savanna Res. Ser.*, 13: 19–20.

Barbetti, M. and Allen, H., 1972. Prehistoric man at Lake Mungo, Australia, by 32 000 B.P. *Nature*, 240: 46–48.

Bateman, W., 1955. Forestry in the Northern Territory. *For. Timber Bur. Leaflet*, No. 72: 16 pp.

Beard, J.S., 1953. The savanna vegetation of Tropical America. *Ecol. Monogr.*, 23: 149–215.

Bourlière, F. and Hadley, M., 1970. The ecology of tropical savannas. *Annu. Rev. Ecol. Syst.*, 1: 125–152.

Bowler, J.M., 1976. Recent developments in reconstructing late Quaternary environments in Australis. In: R.L. Kirk and A.G. Thorne (Editors), *The Origin of the Australians.* Aust. Inst. Aboriginal Studies, Canberra, A.C.T., pp. 55–57.

Bowler, J.M., Hope, G.S., Jennings, J.N., Singh, G. and Walker, D., 1976. Late Quaternary climates of Australia and Papua New Guinea. *Quaternary Res.*, 6; 359–394.

Brass, L.J., 1938. Botanical results of the Archbold Expedition II. Notes on the vegetation of the Fly and Wassi Kussa Rivers, British New Guinea. *J. Arnold Arbor. Harv. Univ.*, 19: 174–193.

Brass, L.J., 1956. Results of the Archbold Expeditions No. 75. Summary of the fourth Archbold expedition to New Guinea (1953). *Bull. Am. Mus. Nat. Hist.*, 111: 83–152.

Brass, L.J., 1964. Results of the Archbold Expedition to New Guinea (1959). *Bull. Am. Mus. Nat. Hist.*, 127: 147–216.

Bridge, J. and Golditch, S.S., 1948. *Preliminary Report on the Bauxite Deposits of Babelthuap Island, Palau Group.*

Prepared by United States Geological Survey, Military Geology Section under direction of Office of the Engineer, General Headquarters Far East Command, 46 pp.

Brookfield, H.C., 1969. Some notes on the climate of the British Solomon Islands. *Philos. Trans. R. Soc. Lond.*, B 225: 207–210.

Brookfield, H.C. and Hart, D., 1966. *Rainfall in the Tropical Southwest Pacific*. A.N.U., Department of Geography, Canberra, A.C.T., Publ. G/3.

Brown, M. and Powell, J.M., 1974. Frost and drought in the highlands of Papua New Guinea. *J. Trop. Geogr.*, 38: 1–6.

Bulmer, R., 1968. The strategies of hunting in Papua New Guinea. *Oceania*, 38: 302–318.

Burtt-Davy, J., 1938. The classification of tropical woody vegetation types. *Imp. For. Inst. Pap.*, 13: 85 pp.

Calaby, J.H. and Gay, F.J., 1959. Aspects of the distribution and ecology of Australian Termites. In: A. Keast, R.L. Crocker and C.S. Christian (Editors), *Biogeography and Ecology in Australia*. W. Junk, The Hague, pp. 211–223.

Carpenter, J., 1938. *An Ecological Glossary*. Hafner, New York, N.Y., 319 pp. (1962 reprint).

Coaldrake, J.E., 1961. The ecosystem of the coastal lowlands ("Wallum") of southern Queensland. *CSIRO Aust. Bull.*, No. 283: 138 pp.

Coaldrake, J.E., 1970. The Brigalow. In: R.M. Moore (Editor), *Australian Grasslands* A.N.U., Canberra, A.C.T., pp. 123–140.

Coaldrake, J.E., 1975. The natural and social history of Cooloola. In: J. Kikkawa and H.A. Nix (Editors), *Managing Terrestrial Ecosystems. Proc. Ecol. Soc. Aust.*, 9: 307–335.

Coode, M.J.E. and Stevens, P.F., 1972. Notes on the flora of two Papuan Mountains. *Papua New Guinea Sci. Proc.*, 1971, 23: 18–25.

Courtenay, P.P., 1977. Tropical Australia. In: D.N. Jeans (Editor), *Australia a Geography*. Sydney University Press, Sydney, N.S.W., pp. 289–315.

Cumberland, K.B., 1963. Man's role in modifying island environments in the southwest Pacific: with special reference to New Zealand. In: F.R. Fosberg (Editor), *Man's Place in the Island Ecosystem*. B.P. Bishop Museum Press, Honolulu, Hawaii, pp. 187–205.

Dasmann, R.F., 1973. A system for defining and classifying natural regions for the purposes of conservation. *I.U.C.N. Occas. Pap.*, No. 7: 48 pp.

Dilmy, A., 1960. The effect of fire used by early man on the vegetation of the humid tropics. In: *Symposium on Impact of Man on Humid Tropics Vegetation*. T.P.N.G. Administration and Unesco Science Coop. Office for Southeast Asia, Goroka, pp. 119–122.

Downes, M.C., 1969. Deer in New Guinea, Part II. *Papua New Guinea Agric. J.*, 20: 95–99.

Eden, M.J., 1968. The norther Rupununi savannas of British Guiana: an ecoclimatic approach to a savanna environment. In: T.L. Hills and R.E. Randall (Editors), *The Ecology of the Forest/Savanna Boundary. McGill Univ. Savanna Res. Ser.*, No. 13: 50–54.

Fitzpatrick, E.A., Hart, D. and Brookfield, H.C., 1966. Rainfall seasonality in the tropical southwèst Pacific. *Erdkunde*, 20: 181–194.

Fosberg, F.R., 1960. The vegetation of Micronesia 1. General

description, the vegetations of the Mariana Islands, and a detailed consideration of the flora of Guam. *Bull. Am. Mus. Nat. Hist.*, 119: 1–70.

Fosberg, F.R., 1961. A classification of vegetation for general purposes. *Trop. Ecol.*, 2: 1–28.

Fosberg, F.R., 1967. A classification of vegetation for general purposes. In: G.F. Peterken (Compiler), *Guide to the Check Sheet for I.B.P. Areas*. I.B.P. Handbook 4, pp. 73–120.

Fosberg, F.R., 1968. Commentary. In: T.L. Hills and R.E. Randall (Editors), *The Ecology of the Forest/Savanna Boundary. McGill Univ. Savanna Res. Ser.*, No. 13: 29–30.

Fosberg, F.R., 1973. A working system for classification of world vegetation. *I.U.C.N. Occas. Pap.*, 5: 21 pp.

Gill, A.M., 1981a. Fire adaptive traits of vascular plants. In: H.A. Mooney, J.M. Bonnicksen, N.L. Christensen, J.E. Lotan and W.A. Reiners (Editors), *Fire Regimes and Ecosystem Properties*. U.S. Dep. Agric. Forest Science General Technical Report, Washington, D.C., pp. 208–230.

Gill, A.M., 1981b. Adaptive responses of Australian vascular plant species to fires. In: A.M. Gill, R.H. Groves and I.R. Noble (Editors), *Fire and the Australian Biota*. Australian Academy of Science, Canberra, A.C.T., pp. 243–271.

Gillison, A.N., 1969. Plant succession in an irregularly fired grassland area — Doma Peaks Region, Papua. *J. Ecol.*, 57: 415–428.

Gillison, A.N., 1970a. *Dynamics of Biotically Induced Grassland/Forest Transitions in Papua New Guinea*. Thesis, Australian National University, Canberra, A.C.T. (unpublished).

Gillison, A.N., 1970b. Structure and floristics of a montane grassland/forest transition, Doma Peaks Region, Papua. *Blumea*, 18: 71–86.

Gillison, A.N., 1972. The tractable grasslands of Papua New Guinea. In: J. Ward (Editor), *Change and Development in Rural Melanesia*. Fifth Waigani Seminar, UPNG and ANU, Port Moresby, pp. 161–172.

Gillison, A.N., 1975a. *Vegetation Types of the New Hebrides with Particular Reference to the Northern Islands*. Unpublished mimeo.

Gillison, A.N., 1975b. Phytogeography of the northern islands of the New Hebrides. *Philos. Trans. R. Soc. Lond.*, B 272: 385–390.

Gillison, A.N. and Walker, J., 1981. Woodlands of Australi... In: R.H. Groves (Editor), *Australian Vegetation*. Cambridge University Press, Cambridge, pp. 177–197.

Glassman, S., 1952. The flora of Ponape. *B.P. Bishop Mus. Bull.*, No. 209: 152 pp. (Repr. Kraus, New York, 5971).

Good, R., 1960. On the geographical relationships of the Angiosperm flora of New Guinea. *Bull. Br. Mus. Nat. Hist. Bot.*, 2: 205–226.

Good, R., 1963. On the biological and biophysical relationships between New Guinea and Australia. In: J.G. Gressitt (Editor), *Pacific Basin Biogeography*. Tenth Pac. Sci. Congr., B.P. Bishop Museum Press, Honolulu, Hawaii, pp. 301–308.

Goulissachvili, V.Z., 1963. Les savanes des pays tropicaux et subtropicaux. In: *Ecosystémes et Productivité Biologique*. Neuvième réunion technique, I.U.C.N., R.T. 9/II/4, pp. 2–8.

Guillaumin, A., 1952. Les Caractères de la végétation néocaledonienne. *C.R. Somm. Séanc. Soc. Biogéogr.*, 29: 82–86.

Hall, N., Johnston, R.D. and Chippendale, G.M., 1970. *Forest Trees of Australia*. Dep. National Development, Forestry and Timber Bureau, Canberra, A.C.T., 334 pp.

Havel, J.J., 1960. Factors influencing the establishment of ligneous vegetation in mid-mountain pyro and anthropogenic grasslands. In: *Symposium on Impact of Man on Humid Tropics Vegetation*. Administration of T.P.N.G. and Unesco Science Coop. Office for Southeast Asia, Goroka, pp. 119–122.

Henty, E.E., 1969. A manual of the grasses of New Guinea. *Papua New Guinea Dep. For. Bot. Bull*, No. 1: 215 pp.

Heyligers, P.C., 1965. Vegetation and ecology of the Port Moresby–Kairuku area. In: *Lands of the Port Moresby–Kairuku Area, Papua New Guinea. CSIRO Aust. Land Res. Ser.*, No. 14: 146–173.

Heyligers, P.C., 1967. Vegetation and ecology of Bougainville and Buka Islands. In: *Lands of the Bougainville and Buka Islands. CSIRO Aust. Land Res. Ser.*, No. 20: 121–145.

Heyligers, P.C., 1972a. Analysis of the plant geography of the semi-deciduous scrub and forest and the eucalypt savanna near Port Moresby. *Pac. Sci.*, 26: 229–241.

Heyligers, P.C., 1972b. Vegetation and ecology of the Aitape–Ambunti area. In: *Lands of the Aitape–Ambunti Area. Papua New Guinea. CSIRO Aust. Land Res. Ser.*, No. 30: 73–99.

Hills, T.L. and Randall, R.E. (Editors), 1968. *The Ecology of the Forest/Savanna Boundary. McGill Univ. Savanna Res. Ser.*, No. 13: 128 pp.

Hnatiuk, R.J., 1977. Population structure of *Livistonia eastonii* Gardn., Mitchell Plateau, Western Australia. *Aust. J. Ecol.*, 461–466.

Holloway, J., 1979. *A Survey of the Lepidoptera, Biogeography and Ecology of New Caledonia*. W. Junk, The Hague, 588 pp.

Jaffré, T., 1974. La végétation et la flore d'un massif les roches ultrabasiques de Nouvelle-Calédonie: le koniambo. *Candollea*, 29: 427–456.

Jaffré, T. and Latham, M., 1974. Contribution à l'étude des relations sol-végétation sur un massif des roches ultrabasiques de la côte ouest de la Nouvelle-Calédonie: le boulinda. *Adansonia, Ser. 2*, 14: 311–336.

Johnson, L.A.S. and Briggs, B.G., 1975. On the Proteaceae — the evolution and classification of a southern family. *Bot. J. Linn. Soc.*, 70: 83–182.

Johnson, R.W., 1964. *Ecology and Control of Brigalow in Queensland*. Queensland Dep. of Primary Industries, Brisbane, Qld., 72 pp.

Kalkman, C. and Vink, W., 1970. Botanical exploration in the Doma Peaks Region, New Guinea. *Blumea*, 18: 87–135.

Kershaw, A.P., 1975. Late Quaternary vegetation and climate in northeastern Australia. In: R.P. Suggate and M.M. Creswell (Editors), *Quaternary Studies*. Royal Society of New Zealand, Wellington, pp. 181–187.

Kershaw, A.P., 1978. Record of last interglacial–glacial cycle from north-eastern Queensland. *Nature*, 272: 159–161.

Lacey, C.J., 1974. Rhizomes in tropical eucalypts and their role in recovery from fire damage. *Aust. J. Bot.*, 22: 29–38.

Lacey, C.J. and Whelan, P.I., 1976. Observations on the ecological significance of vegetative reproduction in the Katherine–Darwin region of the Northern Territory. *Aust. For.*, 39: 131–139.

Lane-Poole, C.E., 1925a. *The Forest Resources of the Territories of Papua and New Guinea*. Government Printer, Melbourne, Vic., 209 pp.

Lane-Poole, C.E., 1925b. *Questionnaire by the British Empire Forestry Conference, 1923 regarding Papuan Forests*. Unpublished.

Lee, K.E., 1975. A discussion of the results of the 1971 Royal Society–Percy Sladen expedition to the New Hebrides: introductory remarks. *Philos. Trans. R. Soc. Lond.*, B 272: 269–276.

McAdam, J.B., 1954. *The Forests of the Territory of Papua and New Guinea*. Department of Forests, Port Moresby, 18 pp.

Mallick, D.I.J., 1975. Development of the New Hebrides archipelago. *Philos. Trans. R. Soc. Lond.*, B272: 277–285.

Martin, B. and Cossalter, C., 1975a. Les *Eucalyptus* des Îles de la Sonde. *Bois For. Trop.*, 163: 3–25.

Martin, B. and Cossalter, C., 1975b. Les *Eucalyptus* des Îles de la Sonde. *Bois For. Trop.*, 164: 3–14.

Martin, B. and Cossalter, C., 1976. Les *Eucalyptus* des Îles de la Sonde. *Bois For. Trop.*, 165: 3–20.

Miège, J., 1968. Relations savane forêt en base Côte d'Ivoire. In: T.L. Hills and R.E. Randall (Editors), *The Ecology of the Forest/Savanna Boundary. McGill Univ. Savanna Res. Ser.*, No. 13: 96–97.

Moore, R.M., 1970. Australian grasslands. In: R.M. Moore (Editor), *Australian Grasslands*. A.N.U. Press, Canberra, A.C.T., pp. 87–100.

Mulroy, T.W., 1979. Spectral properties of heavily glaucous and non-glaucous leaves of a succulent rosette plant. *Oecologia*, 38: 349–357.

Nix, H.A., 1974. Environmental control of breeding, post-breeding dispersal and migration of birds in the Australian region. In: *Proc. 16th Int. Ornithological Congress*. Australian Academy of Science, Canberra, A.C.T., pp. 272–305.

Nix, H.A. and Kalma, J.D., 1972. Climate as a dominant control in the biogeography of northern Australia and New Guinea. In: D. Walker (Editor), *Bridge and Barrier — The Natural and Cultural History of Torres Strait. Aust. Nat. Univ. Dep. Biogeogr. Geomorphol. Publ.*, BG/3: 61–91.

Oliver, J., 1979. A study of geographical imprecision: the tropics. *Aust. Geogr. Stud.*, 17: 3–17.

Paijmans, K., 1967. Vegetation of the Safia–Pongani area. In: *Lands of the Safia–Pongani Area, Papua New Guinea. CSIRO Aust. Land Res. Ser.*, No. 17: 142–167.

Paijmans, K., 1969. Vegetation of the Kerema–Vailala area. In: *Lands of the Kerema–Vailala Area, Papua New Guinea. CSIRO Aust. Land Res. Ser.*, No. 23: 95–116.

Paijmans, K., 1971. Vegetation, forest resources and ecology of the Morehead–Kiunga area. In: *Land Resources of the Morehead–Kiunga Area, Papua New Guinea. CSIRO Aust. Land Res. Ser.*, No. 29: 88–113.

Paijmans, K., 1975. *Explanatory Notes to the Vegetation Map of Papua New Guinea. CSIRO Aust. Land Res. Ser.*, No. 35: 25 pp.

Paijmans, K., 1976. Vegetation. In: K. Paijmans (Editor), *New Guinea Vegetation*. CSIRO and A.N.U. Press, Canberra, A.C.T., pp. 23–105.

Paijmans, K. and Löffler, E., 1974. High-altitude forest and grasslands of Mt. Albert Edward, New Guinea. *J. Trop. Geogr.*, 34: 58–64.

Parham, J., 1972. *Plants of the Fiji Islands.* Government Printer, Suva, revised edition, 462 pp.

Pedley, L., 1967. Vegetation of the Nogoa–Belyando area. In: *Lands of the Nogoa–Belyando Area, Queensland. CSIRO Aust. Land Res. Ser.*, No. 18: 138–169.

Pedley, L., 1974. Vegetation of the Balonne–Maranoa area. In: *Lands of the Balonne–Maranoa Area, Queensland. CSIRO Aust. Land Res. Ser.*, No. 34: pp. 180–203.

Pedley, L. and Isbell, R.F., 1971. Plant communities of Cape York Peninsula. *Proc. R. Soc. Qld.*, 82: 51–74.

Perry, R.A., 1953. The vegetation communities of the Townsville–Bowen region. In: *Survey of Townsville–Bowen region, 1950, CSIRO Land Res. Ser.*, No. 2: 44–54.

Perry, R.A., 1960. *Pasture Lands of the Northern Territory, Australia. CSIRO Aust. Land Res. Ser.*, No. 5: 55 pp.

Perry, R.A. and Lazarides, M., 1964. Vegetation of the Leichhardt–Gilbert area. In: *Lands of the Leichhardt–Gilbert area, Queensland. CSIRO Aust. Land Res. Ser.*, No. 11: 152–191.

Ramia, M., 1968. Classification and origin of savannas in the light of a world classification. In: T.L. Hills and R.E. Randall (Editors), *The Ecology of the Forest/Savanna Boundary. McGill Univ. Savanna Res. Ser.*, No. 13: 4–6.

Raunkiaer, C., 1934. *The Life Forms of Plants and Statistical Plant Geography.* Clarendon Press, Oxford, 632 pp.

Reiner, E.J. and Robbins, R.G., 1964. The middle Sepik plains, New Guinea. A physiographic study. *Geogr. Rev.*, 54: 20–44.

Richards, P.W., 1961. *The Tropical Rainforest.* Cambridge University Press, Cambridge, reprint of 1st ed., 450 pp.

Robbins, R.G., 1960. The anthropogenic grasslands of Papua New Guinea. In: *Symposium on Impact of Man on Humid Tropics Vegetation.* Administration of T.P.N.G. and Unesco Science Coop., Office for Southeast Asia, Goroka, pp. 313–329.

Robbins, R.G., 1964. The montane habitat in the Tropics. In: *The Ecology of Man in the Tropical Environment.* Proc. and Papers of the Ninth Technical Meeting of I.U.C.N. *I.U.C.N. Publ.*, N. S., No. 4: 163–171.

Robbins, R.G., 1970. Vegetation of the Goroka–Mount Hagen Area. In: *Lands of the Goroka–Mount Hagen Area, Papua New Guinea. CSIRO Aust. Land Res. Ser.*, No. 27: 104–118.

Robbins, R.G. and Pullen, R., 1965. Vegetation of the Wabag–Tari area. In: *Lands of the Wabag–Tari Area, Papua New Guinea. CSIRO Aust. Land Res. Ser.*, No. 15: 100–115.

Safford, W.F., 1905. The useful plants of Guam. *Conf. U.S. Natl. Herbariums*, 9: 1–416.

Schmid, M., 1975. La flore et la végétation de la partie méridionale de l'Archipel des Nouvelles Hébrides. *Philos. Trans. R. Soc. Lond.*, B 272: 329–342.

Schmithüsen, J., 1976. *Atlas zur Biogeographie.* Biographisches Institut, Mannheim, 80 pp.

Shaw, D.F., 1968. Lightning strike of *Cacao* and *Leucaena* in New Britain. *Papua New Guinea Agric. J.*, 3: 75–84.

Shaw, N.H. and Norman, J.T., 1970. Tropical and sub-tropical woodlands and grasslands. In: R.M. Moore (Editor), *Australian Grasslands.* A.N.U. Press, Canberra, A.C.T., pp. 112–122.

Smith, A.C., 1979. *Flora Vitiensis Nova, 1.* Pacific Tropical Botanical Garden, Lawai, Kauai, Hawaii, 495 pp.

Smith, J.M.B., 1975. Mountain grasslands of New Guinea. *J. Biogeogr.*, 2: 27–44.

Smith, J.M.B., 1977. Vegetation and microclimate of east- and west-facing slopes in the grasslands of Mt. Wilhelm, Papua New Guinea. *J. Ecol.*, 65: 39–53.

Specht, R.L., 1958a. The climate, geology, soils and plant ecology of the northern portion of Arnhem Land. In: R.L. Specht, and C.P. Mountford (Editors), *Records of the American–Australian Scientific Expedition to Arnhem Land, 3. Botany and Plant Ecology.* Melbourne University Press, Melbourne, Vic., pp. 333–314.

Specht, R.L., 1958b. Geographical relationships of the flora of Arnhem Land. In: R.L. Specht and C.P. Mountford (Editors), *Records of the American–Australian Expedition to Arnhem Land, 3. Botany and Plant Ecology.* Melbourne University Press, Melbourne, Vic., pp. 415–478.

Specht, R.L., 1981. The use of foliage projective cover. In: A.N. Gillison and D.J. Anderson (Editors), *Vegetation classification in Australia* CSIRO and A.N.U. Press, Canberra, A.C.T., pp. 10–21.

Speck, N.H., 1968. Vegetation of the Dawson–Fitzroy area. In: *Lands of the Dawson–Fitzroy area, Queensland. CSIRO Aust. Land Res. Ser.*, No. 21: 157–173.

Speck, N.H. and Lazarides, M., 1964. Vegetation and pastures of the west Kimberley area. In: *General report on lands of the west Kimberley area, W.A. CSIRO Aust. Land Res. Ser.*, No. 9: 140–174.

Stemmerman, L. and Proby, F., 1978. Inventory of wetland vegetation on the Caroline Islands, 1. Wetland vegetation types. VTN Pacific, Honolulu, Hawaii, 230 pp.

Stocker, G.C. and Mott, J.J., 1981. Fire in the tropical forests and woodlands of northern Australia. In: A.M. Gill, R.H. Groves and I.R. Noble (Editors), *Fire and the Australian Biota.* Australian Academy of Science, Canberra, A.C.T., pp. 425–439.

Story, R., 1967. Vegetation of the Isaac–Comet Area. In: *Lands of the Isaac–Comet area, Queensland. CSIRO Aust. Land Res. Ser.*, No. 19: 108–129.

Story, R., 1970. Vegetation of the Adelaide–Alligator area. In: *Lands of the Adelaide–Alligator Area, Northern Territory. CSIRO Aust. Land Res. Ser.*, No. 25: 114–130.

Story, R., 1970. Vegetation of the Mitchell–Normanby area. In: *Lands of the Mitchell–Normanby Area. CSIRO Aust. Land Res. Ser.*, No. 26: 75–88.

Street, J.M., 1966. Grasslands on the highland fringe in New Guinea. *Capricornia, Qld. Univ. Geogr. Soc.*, 3: 9–12.

Sutherland, D.N., 1961. The beef cattle industry in Queensland. In: *Introducing Queensland.* Government Printer, Brisbane, Qld. pp. 61–65.

Taylor, B.W., 1964a. Vegetation in the Wanigela–Cape Vogel area. In: *Lands of the Wanigela–Cape Vogel Area, Papua New Guinea. CSIRO Aust. Land Res. Ser.*, No. 12: pp. 69–83.

Taylor, B.W., 1964b. Vegetation of the Buna–Kokoda area. In: *Lands of the Buna–Kokoda Area, Papua New Guinea. CSIRO Aust. Land Res. Ser.*, No. 10: 89–98.

Taylor, B.W., 1968. Changes in the forest, savanna boundary in the per-humid lowland of New Guinea. In: T.L. Hills and R.E. Randall (Editors), *The Ecology of the Forest/Savanna Boundary. McGill Univ. Savanna Res. Ser.*, No. 13: 16–17.

Thompson, R.B. and Hackman, B.D., 1969. Some geological notes on areas visited by the Royal Society Expedition to the British Solomon Islands, 1965. *Philos. Trans. R. Soc. Lond.*, B 255: 189–202.

Tunstall, B.R. and Connor, D.J., 1975. Internal water balance of brigalow (*Acacia harpophylla* F. Muell.) under natural conditions. *Aust. J. Plant Physiol.*, 2: 489–499.

Troll, C., 1936. Termitensavennen. In: *Landekundliche Forschrift, Festrichrift für Norbert Krebs.* Englehorn, Stuttgart, pp. 275–312.

Twyford, I.T. and Wright, A.C.S., 1965. *The Soil Resources of the Fiji Islands, I.* Government of Fiji, Suva, pp. 77–86.

Udvardy, M., 1975. A classification of the biogeographical provinces of the world. *I.U.C.N. Occas. Pap.*, No. 8: 48 pp.

Van Balgooy, M.M.J., 1971. Plant geography of the Pacific. *Blumea*, (Suppl.) VI: 222 pp.

Van Royen, P., 1963a. *The vegetation of the island of New Guinea.* Department of Forests, Port Moresby, 70 pp. (unpublished mimeo)

Van Royen, P., 1963b. Notes on the vegetation of South New Guinea. *Sertulum Papuanum. Nova Guinea, Bot.*, 13: 195–241.

Van Steenis, C.G.G.J., 1958. Rejuvenation as a factor for judging the status of vegetation types: the biological nomad theory. In: *Study of Tropical Vegetation. Proc. Kandy Symposium UNESCO*, pp. 212–215.

Virot, R., 1956. La végétation Canaque. *Mem. Mus. Natl. Hist. Nat. N.S., B, Bot.*, 7: 1–398.

Wade, L.K., and McVean, D.N., 1969. Mt. Wilhelm Studies I. The alpine and sub-alpine vegetation. *Res. School Pac. Stud., Dep. Beogeogr. Geomorphol., Aust. Natl. Univ. Publ.*, B/G 1.

Walker, D., 1966. Vegetation of the Lake Ipea region, New Guinea highlands I. Forest, grassland and 'garden'. *J. Ecol.*, 54: 445–456.

Walker, J., 1981. Fuel dynamics in Australian vegetation. In: A.M. Gill, R.H. Groves and I.R. Noble (Editors), *Fire and the Australian Biota.* Australian Academy of Science, Canberra, A.C.T., pp. 101–127.

Walker, J. and Gillison, A.N., 1982. Australian savannas. In: B.J. Huntley and B.H. Walker (Editors), *Dynamics of Savanna Ecosystems.* Springer-Verlag, Heidelberg, in press.

Webb, L.J., 1959. A physiographic classification of Australian rain forests. *J. Ecol.*, 47: 551–570.

Whitmore, T.C., 1969. The vegetation of the Solomon Islands. *Philos. Trans. R. Soc. Lond.*, B 255: 259–270.

Whitmore, T.C., 1975. *Tropical Rain Forests of the Far East.* Clarendon, Oxford, 282 pp.

Whyte, R.O., 1968. *Grasslands of the Monsoon.* Faber and Faber, London, 325 pp.

Williams, R.J., 1955. Vegetation regions. In: *Atlas of Australian Resources.* Department of National Development. Canberra, A.C.T., 23 pp.

Wood, J.G., 1950. Vegetation of Australia. In: *The Australian Environment.* CSIRO, Melbourne, Vic., pp. 77–96.

Chapter 10

THE SAVANNAS OF TROPICAL AMERICA

G. SARMIENTO

THE MAIN ECOLOGICAL AND PHYSIOGNOMIC TYPES OF SAVANNAS IN TROPICAL AMERICA

Savanna formations constitute a substantial part of the vegetation cover of tropical America. The total area occupied by Neotropical savannas, considering natural communities alone, exceeds 2 million km². In some regions, such as the Brazilian *cerrado* or the Colombian and Venezuelan *llanos*, only narrow fringes of gallery forests bordering the streams interrupt the monotonous continuity of the savanna landscape. In other areas, such as Amazonia or Central Amèrica, the reverse situation occurs: savannas appear as more or less isolated patches of open vegetation amid a continuous cover of rain forests.

Neotropical savannas show a remarkable floristic similarity throughout their geographic range. In fact, though they may vary widely in composition from one community to another, a floristic list, if it belongs to a Neotropical savanna, may be easily recognized wherever it may come from. A common floristic stock unifies the savannas of tropical America from their northernmost areas in Cuba and southern Mexico to their austral limits in Paraguay and southeastern Brazil (Figs. 10.1, 10.25 and 10.26).

Present-day ecological and phytogeographic knowledge of American savannas is still uneven. In more thoroughly studied areas, such as the Venezuelan llanos, they begin to be understood at least in their more essential aspects, while in other regions they remain scarcely known beyond preliminary accounts of their environmental and structural features. The treatment here will therefore appear as geographically unbalanced, since it will rely heavily on the best known ecosystems.

Savanna types will be considered from a double viewpoint, one ecological, the other purely physiognomic or structural. From an ecological point of view, tropical savannas will be divided into four major categories according to the seasonality of the ecosystem (Sarmiento and Monasterio, 1975; Sarmiento, 1978). Seasonality represents one of the most essential features of a savanna ecosystem (see Ch. 1), whether the cyclic changes in the environmental impulses and constraints during the year are considered, or the biological rhythms of plant species and of the whole vegetation that accompany those external fluctuations.

The first ecological type is the semi-seasonal savanna. It occurs under weak seasonal alter-

Fig. 10.1. Major tropical savanna regions in South America.

nations exerted by a constantly or mostly wet climate. In this case one or two short dry seasons may represent the main rhythmic environmental strain. Under these circumstances, fires become natural events of much less frequent occurrence. Semi-seasonal savannas therefore change very little during the year. For instance, the semi-dormant phenophase of perennial grasses and sedges (see Ch. 5) is much less pronounced than in the other types. As a general rule, semi-seasonal savannas occur as scattered patches in regions with a continuous rain-forest cover.

A second type, and the most widespread in tropical America, is the seasonal savanna. Here, an extended rainless season increases the probabilities of dry season fires, and both factors, drought and burning, provide a neat rhythmicity in the functioning of the ecosystem. A good proportion of the savannas in the two major savanna regions of South America, the cerrado and the llanos, belongs to this seasonal type.

A third class of savanna is characterized by the alternation of two contrasting stresses during each annual cycle, one induced by drought and fire, the other by soil saturation. These hyperseasonal savannas experience a period of water shortage during the rainless season, and an extended period of water excess during all or part of the rainy season, when soils become waterlogged and asphyxiating. Changes in vegetation follow these environmental constraints. Hyperseasonal savannas may occur on poorly drained bottomlands in any climatic region, but they are particularly common on large, depressed regions with slow and ill-defined drainage, as the Gran Pantanal of Mato Grosso, the Bolivian llanos, and some areas of the Orinoco llanos.

In a fourth type of savanna ecosystem, the water excess period may last most of the year, while a period of acute water shortage either does not exist or is very brief. This is really a kind of seasonal swamp, and it is considered as a savanna only when grasses and sedges are the dominant plants. This type is called *estero*, using a common Spanish term widely applied to these ecosystems in several Latin American countries.

Another useful criterion in differentiating savanna types is based on the structural features of the vegetation. Various characteristics may be taken into account, referring either to the woody or to the herb component. A physiognomic system is followed here, which is simple enough to be useful, while permitting the differentiation of significant landscape types. Table 10.1 summarizes this physiognomic system of savanna classification. Four types are distinguished according to the importance attained by woody species. In the

TABLE 10.1

Main physiognomic types of tropical American savannas, with the corresponding Brazilian and Spanish-American names

Savanna type	Total cover of woody layers (%)	Average tree density (trees ha^{-1})	Brazilian name	Spanish-American name
Savannas				
Savanna grassland	–	–	*campo limpo*	*sabana pastizal*
Tree and/or shrub savanna	<2	500	*campo sujo*	*sabana abierta*
Wooded savanna	2–15	1000	*campo cerrado*	*sabana cerrada*
Savanna woodland	15–40	3000	*cerrado*	*sabana boscosa*
Woodlands or open forests				
Sclerophyllous woodland	>40	4000	*cerradão*	–
Savannas dotted with groves				
Savanna parkland	<40		–	*sabana parqueada*

savanna grassland, trees and shrubs, if they exist at all, have a dwarf form which does not exceed the height of the herb layer. A **tree** and/or **shrub savanna** has woody species scattered within a mostly continuous herb layer. Total cover of trees and shrubs is less than 2%. A **wooded savanna** has an open tree cover ranging between 2% and 15%, which corresponds to a total density of about 1000 trees ha.$^{-1}$ A **savanna woodland** has a tree cover above 15%, reaching cover values of 20% to 30% on the average, corresponding to a maximum density of about 3000 to 4000 trees ha^{-1}.

These four structural categories agree with the savanna formations differentiated by popular usage, corresponding for instance to the four types known as *campo limpo, campo sujo, campo cerrado,* and *cerrado,* recognized in the Brazilian literature which will be discussed later. A last type of woodland, where tree crowns touch each other but the herb layer persists, is known in Brazil as *cerradão,* and though it constitutes a useful reference to which the more open formations can be compared, I shall follow the usage of most authors who consider it more as a type of forest than of savanna.

As a last structural form, typical of some areas where savannas and tropical forests form intricate vegetation mosaics, a savanna of any one of the previous four types, may be dotted with small clumps or groves of trees and shrubs. This physiognomic unit will be called a **savanna parkland**. In fact it generally intergrades with true vegetation mosaics of savannas and rain forests.

With this two-dimensional system of savanna typification, the occurrence, habitat and composition of the main savanna types in tropical America will be discussed. The major savanna landscapes will be considered first, and subsequently the remaining regions where savannas only occur as minor components of the regional vegetation.

THE BRAZILIAN CERRADOS

The portuguese name of *cerrado,* or its plural *cerrados,* designates a natural region, a phytogeographic province and a series of plant formations with varying proportions of woody and herbaceous species. Though these floristically related plant formations prevail within the natural region, it seems convenient, in order to prevent misunderstandings, to review the three different meanings of the cerrado concept.

The natural region occupies an area of more than 1.8 million km^2 in the Brazilian Shield. Ab'Saber (1971) when considering the major morphoclimatic domains of that country, characterized the core area of cerrados in central Brazil as the realm of plateaux and high tablelands covered by savannas. The higher levels correspond to an old planation surface of Middle Tertiary age. The formerly continuous surface of these extensive tablelands has been deeply dissected and fragmented, now appearing as isolated high plains or *serras,* separated by wide interfluves. Gallery forests penetrate through these wide valleys, while the uplands are completely covered by savannas (Fig. 10.2).

The flat or gently rolling tablelands range in altitude from 1000 m to about 300 m above sea level. They descend southwards to the lowlands of the Mato Grosso Gran Pantanal, and northwards to the Amazon lowlands with their almost uninterrupted rain forests. To the east and south, the transition with the humid forest landscapes of the Atlantic region is gradual, while to the northeast a rather steep climatic gradient leads to the large depressions of the dry *caatinga* region (Fig. 10.3). The whole cerrado region has a tropical wet and dry climate (**Aw** type of Köppen, 1931) with intermediate rainfall between the wetter regions to the northwest and southeast and the drier northeastern areas. Annual rainfall is of the order of 1500 mm, with extreme values of 750 mm at the caatinga border and somewhat more than 2000 mm at the Amazonian border. The dry season lasts from three to five consecutive months, during the winter of the Southern Hemisphere.

The Cerrado Phytogeographic Province (Eiten, 1978) is characterized by its rich and special flora that neatly distinguishes these areas both from the neighboring forested Amazonian and Atlantic Brazilian Provinces and from the subtropical Chaco Province. This flora has, however, a quite close affinity to that of other Neotropical savannas that have not been considered by phytogeographers as parts of this Cerrado Province. Floristically, the Province extends not only through the main continuous core area of the cerrado in Central Brazil, but also to various cerrado outliers in the southeast (São Paulo) and the northeast (Rio Grande do Norte, Pernambuco).

Fig. 10.2. Landscape and vegetation patterns in the cerrado region. Seasonal savannas cover the extensive tablelands and their steep borders, while gallery forests occupy the deeply incised valleys.

The cerrado vegetation has been reviewed by Eiten (1972), and four symposia have been devoted to different aspects of its floristics, ecology, environmental conditions, etc. (Ferri, 1964, 1971, 1977; Labouriau, 1966). A comprehensive bibliography has also been published (EMBRAPA, 1976). In its wider sense, cerrado refers to several structural types of open vegetation, from dense woodlands to grasslands, most of which fit the definition of tropical savannas. Ecologically, the cerrados belong to the seasonal type of savanna; the hyperseasonal savannas and the esteros are rather restricted within this area, besides having a fairly different floristic composition. Brazilian literature refers to these two types of ecosystems as *campos*, not including them among the cerrado formations because of the floristic differences. Since the word campos has too broad a meaning, encompassing quite different vegetation types, the terms "hyperseasonal savannas" and "esteros" will be retained for these ecosystems of the cerrado region. The seasonal savannas, by far the more extensive, will be considered first, and then the other types,

Fig. 10.3. The core area of the cerrados in central Brazil and its neighbouring formations (after Ab'Saber, 1971).

restricted to small areas of particular habitat conditions.

Four structural types of cerrado vegetation have been recognized and designated by popular names (Eiten, 1972). They roughly correspond to the four previously defined physiognomic types. Cerradão refers to a woodland or open low forest, with a fairly continuous tree canopy. Crown cover averages 50%, and the ground layer of grasses, forbs and halfshrubs is inversely correlated with total crown cover of woody species. Cerrado, in its restricted structural sense, refers to a savanna woodland where the total woody cover is about 20%. Campo cerrado is a wooded savanna, where the scattered low trees have a total crown cover of about 3%, but the woody species still appear as a conspicuous part of the landscape (Fig. 10.4). Campo sujo ("dirty field") is applied to a tree and shrub savanna, with widely scattered woody species (Fig. 10.5). Finally, a pure or almost pure grassland is designated as campo limpo ("clean field"). These four physiognomic types appear more or less intermingled in almost every area of cerrado vegetation. They are useful for a preliminary, overall

characterization of the plant cover. One may note also that neither dense woodlands nor pure grasslands are separated from the mixed savanna communities, since all types intergrade with one another and constitute complex vegetation mosaics everywhere.

A quantitative analysis of cerrado vegetation reveals a continuous variation in physiognomy and species composition. Goodland (1971) analyzed 110 stands of cerrado vegetation, ranging from campo sujo to cerradão, in a region of Minas Gerais in central Brazil. His results show a continuous variation in all the sampled attributes along this physiognomic gradient. Thus, total basal area of trees varies continuously from 0.9 m^2 ha^{-1} in the most treeless grassland to 51.3 m^2 ha^{-1} in the cerradão; canopy cover ranges from 0 to 85%; ground cover from 30% in campo sujo to 2% in cerradão; tree density varies from 266 to 4925 trees ha^{-1}; number of tree species from 19 to 72; number of herb species from 79 to 21. The importance of most species, either trees, shrubs or grasses, also shows continuous variation along the physiognomic gradient. In a later paper (Goodland and

Fig. 10.4. A wooded savanna (campo cerrado) in the central region of the cerrado, near Brasilia.

Fig. 10.5. A tree and shrub savanna (campo sujo) in the cerrado near Brasilia.

Pollard, 1973), a significant correlation was established between the vegetation structural gradient and levels of phosphorus, nitrogen and potassium in the surface soil horizon.

The existence of a peculiar cerrado flora has already been noted. A few species occur over the whole area, but the most common situation is that of slight and gradual changes from one region to another. Widely separated areas may have only one-fourth of the species in common, but gradients of floristic similarity connect the two extremes. The precise phytosociology of cerrado communities still needs to be studied; it may be a difficult task since even now its flora and phytogeography are imperfectly known. Tables 10.2 and 10.3 give lists of the most common species of woody and herbaceous plants.

Every small area within the cerrado shows a rich and diversified flora. Heringer (1971) recorded more than 300 species in one hectare of protected cerrado near Brasilia. Sampling the area through twenty-five 20 × 20 m plots, he found that the number of species per plot varied from 52 to 117; these are high numbers even for tropical plant communities.

TABLE 10.2

Some of the most widespread trees in the cerrados

Agonandra brasiliensis	*Machaerium angustifolium*
Annona coriacea	*Magonia pubescens*
Antonia ovata	*Miconia argentea*
Aspidosperma dasycarpon	*Myrcia tomentosa*
Aspidosperma tomentosum	*Piptadenia peregrina*
Bombax gracilipes	*Piptocarpha rotundifolia*
Bowdichia virgilioides	*Platypodium elegans*
Byrsonima coccolobaefolia	*Pterodon pubescens*
Byrsonima crassifolia	*Qualea grandiflora*
Caryocar brasiliense	*Qualea multiflora*
Casearia sylvestris	*Roupala heterophylla*
Connarus suberosus	*Salvertia convallariodora*
Copaifera langdorffii	*Sclerolobium paniculatum*
Curatella americana	*Solanum lycocarpum*
Dalbergia violacea	*Strychnos pseudoquina*
Dimorphandra mollis	*Stryphnodendron*
Diospyros hispida	*barbadetimam*
Erythroxylon suberosum	*Symplocos lanceolata*
Eugenia dysenterica	*Tabebuia caraiba*
Hancornia speciosa	*Terminalia argentea*
Hirtella glandulosa	*Vernonia ferruginea*
Hymenaea stigonocarpa	*Vochysia elliptica*
Jacaranda brasiliana	*Xylopia aromatica*
Kielmeyera coriacea	*Zeyheria digitalis*

TABLE 10.3

Some of the commonest species in the ground layer of the cerrado savannas

SHRUBS

Anacardium humile	Jacaranda ulei
Andira humilis	Kielmeyera rosea
Byrsonima verbascifolia	Mimosa polycarpa
Cochlospermum regium	Palicourea rigida
Davilla eliptica	Peschiera affinis
Erythroxylon campestre	Vernonia thyrsoidea

HALF-SHRUBS OR HALF-WOODY SPECIES

Anemopaegma arvense	Jacaranda decurrens
Annona pygmaea	Kielmeyera neriifolia
Baccharis humilis	Lippia martiana
Borreria capitata	Mimosa gracilis
Clitoria guianensis	Parinari obtusifolia
Craniolaria integrifolia	Pfaffia jubata
Esenbeckia pumila	Stryphnodendron confertum
Evolvulus sericeus	Tibouchina gracilis
Gomphrena postrata	Vernonia elegans
Ichthyothere rufa	Viguiera robusta

GRASSES

Andropogon bicornis	Mesosetum loliiforme
Andropogon hirtiflorus	Panicum cayenense
Andropogon leucostachys	Panicum olyroides
Andropogon selloanus	Paspalum carinatum
Aristida capillacea	Paspalum gardnerianum
Aristida pallens	Paspalum pectinatum
Aristida tincta	Paspalum plicatulum
Axonopus aureus	Paspalum pulchellum
Axonopus capillaris	Sporobolus cubensis
Ctenium chapadense	Thrasya paspaloides
Diectomis fastigiata	Trachypogon canescens
Echinolaena inflexa	Trachypogon montufari
Elyonurus latiflorus	Trachypogon plumosus
Eragrostis maypurensis	Trachypogon vestitus
Leptocoryphium lanatum	Tristachya leiostachya

PALMS

Acanthococos emensis	Butia leiospatha
Acanthococos sericeus	Syagrus acaulis
Astrocaryum campestre	Syagrus campestris
Attalea exigua	

SEDGES

Bulbostylis paradoxa	Fimbristylis diphylla
Bulbostylis capillaris	Rhynchospora barbata
Cyperus flavus	Rhynchospora velutina
Dichronema ciliata	

To give a further picture of a cerrado community, the description given by Eiten (1975) of an undisturbed cerrado on the crest of the Serra do Roncador in Mato Grosso will be examined. Eiten refers to this vegetation as "a tree and shrub woodland" where woody plants, counting those of all heights, form an open cover of about 50%. There are a few scattered tall trees (from less than 7 to 12 or even 15 m). Lower trees and shrubs of all sizes fill in the rest of the space, but woody plants do not form definite layers. Among the most common trees, Eiten lists *Aspidosperma* spp., *Curatella americana*, *Davilla eliptica*, *Kielmeyera coriacea*, *Palicourea rigida*, *Qualea grandiflora*, *Salvertia convallariodora*, and various others. Besides these trees and shrubs, some low palms, 2 to 5 m tall, and a few species of acaulescent palms are usually present, giving a characteristic appearance to the vegetation. The ground layer appears slightly open; grasses are dominant, with intermixed sedges, perennial herbs, semi-shrubs, dwarf shrubs and vines. Species of *Andropogon*, *Axonopus*, *Ichnanthus* and *Paspalum* are the most common grasses. They reach 0.5 to 1.25 m tall when in flower. Open colonies of a low terrestrial *Bromelia* occur in a few areas. As may be gathered from this description, a typical cerrado (in this case a savanna woodland) is quite a rich formation, considering either life forms or number of species. The structural variation in respect of arrangements and importance of the different growth-forms is striking.

In the same area of the Serra do Roncador, Ratter et al. (1973) described two different communities of cerradão. The *Hirtella glandulosa* cerradão is a "low savanna woodland in which the taller trees vary from 5 to 12 m in height; there is no continuous canopy and the ground vegetation is very dense and difficult to walk through". Characteristic trees of this community, besides *Hirtella glandulosa*, are *Aspidosperma macrocarpon*, *Bowdichia virgilioides*, *Sclerolobium paniculatum*, *Xylopia sericea*, and others. Fig. 10.6 shows the structural profile of this community. The ground layer contains numerous cerrado species, among which are found many grasses, acaulescent palms and ground bromeliads. Another community in the same area is the *Magonia pubescens–Callisthene fasciculata* cerradão, in which the largest trees attain 15 m high and the ground layer is quite dense. As may be realized from these descriptions,

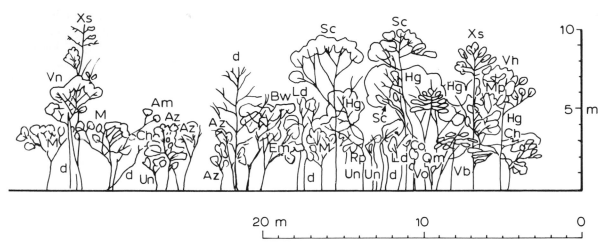

Fig. 10.6. A profile diagram (30 × 3m) of a *Hirtella glandulosa* cerradão in the Serra do Roncador, Mato Grosso (after Ratter et al., 1973). Key to the species: *Am* = *Aspidosperma macrocarpon*; (unknown) *Az* = *azeidinho*; *Bw* = *Bowdichia virgilioides*; *Ch* = *Chaetocarpus echinocarpus*; *Co* = *Connarus fulvus*; *Em* = *Emmotum nitens*; *Hg* = *Hirtella glandulosa*; *Ld* = *Lafoensia pacari*; *M* = *Miconia* sp.; *Mp* = *Maprounea guianensis*; *Rp* = *Roupala montana*; *Sc* = *Sclerolobium paniculatum*; *Qm* = *Qualea multiflora*; *Vb* = *Virola sebifera*; *Vh* = *Vochysia haenkeana*; *Xy* = *Xylopia amazonica*; *Xs* = *Xylopia sericea*; *Un* = unknown; *d* = dead.

even the densest types of woodlands have structural and floristic characteristics that relate them to the other savannas. Several authors, like Rizzini and Heringer (1962) and Rizzini (1963b), have considered the cerradão as the original forest type in the whole cerrado area, the other structural types being derived from it by burning and man's other activities. Though this is probably the case in some restricted, formerly forested areas, it seems difficult to accept this generalization as valid for the entire region.

Quite numerous cerrado outliers appear in the State of São Paulo as small islands within a landscape that was formerly forested. Borgonovi and Chiarini (1968) presented a map showing these southernmost intrusions of cerrado. They reach the border between the States of São Paulo and Parana, at about 24°S. Silberbauer-Gottsberger et al. (1977) analyzed the effects of the exceptional 1975 frost on these cerrados, concluding that frost seems to be one of the selective factors influencing the floristic composition of the cerrado at its southern limit, as many woody species were substantially damaged.

In northeastern Brazil, there also occur several small areas of disjunct cerrados among the prevailing caatinga vegetation (Valverde et al., 1962; Eiten, 1972). They occupy sandstone plateaux and tablelands. Their floristic and physiognomy are similar to those of typical cerrado in its central core

area. *Byrsonima cydoniaefolia*, *Curatella americana*, *Hancornia speciosa*, *Hirtella ciliata* and *Ouratea fieldingiana* are the commonest trees (Tavares, 1964).

The hyperseasonal savannas and esteros of the cerrado region

A belt of natural grassy campo occurs on the valley sides or plateau edges, in almost all areas of cerrados (Eiten, 1978). Generally, it is a narrow belt separating the upland cerrado from the gallery forest. In these areas the water table is near the surface for some or all of the year. Askew et al. (1970) noticed the existence of these grassy formations, either treeless or with occasional trees or palms, in northeastern Mato Grosso. Ratter (1971) and Ratter et al. (1973) considered two types of grasslands in that area: dry or hill grasslands, and moist valley campos. In fact, the dry grasslands that occur on shallow soils or on lateritic outcrops represent a type of cerrado: the campo limpo or savanna grassland. Their herbaceous and shrubby flora is quite similar to the ground flora of the other savanna formations already discussed. The valley-side campos may be considered either as hyperseasonal savannas or as esteros, depending upon both the amount of time they remain waterlogged during the rainy season and the degree of soil desiccation during the dry months.

Eiten (1975) described the valley-side campos in Serra do Roncador (Mato Grosso) as a pure grassland with a dominant grassy layer 0.5 to 1.25 m tall, intermixed with sedges, rushes, species of *Xyris* and a great number of herbs. Scattered *buriti* palms (*Mauritia vinifera*) appear in a few cases. Eiten noticed that the dominant grass species, as well as the rest of the flora, vary from one campo to another, though a few species are present in almost all campos. These formations are subjected to frequent fires, though less frequent than other types of savannas in the cerrado area.

Goldsmith (1974) presented a quantitative analysis of vegetation in one transect from cerrado to a gallery forest traversing the wet campo, in the Rio das Mortes area of Mato Grosso. Campos soils are hydromorphic, grey or black colored, with a high organic matter content and low base status. The mean water table depth is between 1 and 2 m. Grasses form the bulk of vegetation, with 29 species from 14 genera. There are also a large number of other monocotyledons, with nine genera of sedges and representatives from the Commelinaceae, Costaceae, Eriocaulaceae, Heliconiae, Iridaceae, Rapataceae and Xyridaceae. The total campo flora reaches about 200 species.

Other common types of hyperseasonal savanna and seasonal swamp form a well-defined mosaic with seasonal cerrados. Eiten (1975, 1978) described a vegetation pattern in Mato Grosso where cerrado trees and shrubs are clumped in round groves, 3 to 7 m in diameter, raised on earth mounds, 1 to 2 m high, which always contain one or two termite mounds. The groves form a regular pattern in a continuous matrix of hydromorphic grasslands, that occupy a perfectly flat area, flooded in the wet season, either treeless or dotted with buriti palms. The raised platforms seem to be artifacts built up by termites to keep their nests above the saturated soil. As will be discussed later, this same type of landscape occurs in the neighboring Pantanal region.

In summary, floristically rich seasonal savannas prevail in the cerrado region and its outliers. They form a continuum of structural types from grasslands to woodlands, probably correlated with site conditions. In any area, different structural types coexist side by side, giving rise to vegetation mosaics at different scales and of various complexity. Hyperseasonal savannas and esteros generally oc-

cur as treeless grasslands or sometimes as palm savannas with *Mauritia vinifera*. They occupy wet sites on valley sides or tableland margins throughout the area, though they become more important in Mato Grosso, towards its boundary with the Gran Pantanal.

THE COLOMBIAN–VENEZUELAN LLANOS AND RELATED SAVANNAS OF NORTHERN SOUTH AMERICA

The large sedimentary basin known as the llanos is a huge plain lying between the Andes and the Caribbean Cordilleras (to the west and north) and the broken landscapes bordering the Guiana Shield (to the east and south). In the popular sense, llanos means both a plain and an open landscape where savannas provide the most significant part of the plant cover. Though within this region there are several large areas covered by various types of tropical forests, and almost every watercourse is bordered by a fringe of gallery forest, most of the area is occupied by natural savannas. The llanos indeed constitute the major savanna region of northern South America.

The Guaviare River in central Colombia (Fig. 10.7) marks the limit between the nearly continuous Amazonian rain forest to the south and the savanna lands to the north. The Orinoco Delta constitutes a natural limit to the east, while the savanna landscape intermixes with rain forests on the hills limiting the plains a few kilometers south and east of the Orinoco. This big river is the major collector for the whole area, and its main right-hand tributaries, such as the Meta, Arauca and Apure, dissect and drain a large part of the llanos.

The regional climate is a typical tropical wet and dry climate, where annual rainfall increases from about 1000 mm in the eastern border to a maximum of 2200 mm at the Guaviare River on the southwestern margin. Correspondingly, the number of dry months decreases from five to six in the east to one or two in the southwest. Quaternary alluvial and aeolian sediments extend throughout the major part of the llanos. In the eastern portion, a tongue of Tertiary clays and shales extends southwards from the Cordillera de la Costa and approaches the Orinoco. This is the major area of tropical deciduous forest within the llanos.

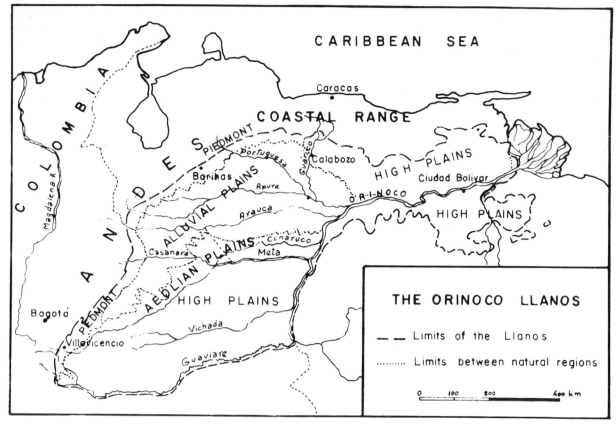

Fig. 10.7. The Orinoco llanos in Colombia and Venezuela. Four major regions have been distinguished: the piedmont, the high plains, the alluvial overflow plains and the aeolian plains.

On well-drained sites, the whole range of tropical soils appears, from Entisols on young alluvia surfaces, to Oxisols on the more ancient ones; but Alfisols and Ultisols are the two most widespread types. On badly drained terrains, Vertisols and Alfisols predominate.

Savannas are used as extensive rangelands for cattle raising, but modern agriculture penetrates a region extending from the best soils, near the piedmont, to the hinterlands. Some large irrigation programs already allow an intensive agriculture in lozalized areas, raising crops like rice, cotton, sorghum, corn and sesame.

Almost every ecological and physiognomic type of savanna occurs in the llanos, the dominant formations depending on topography and soil. Except in some quite inaccessible areas, such as parts of the Meta and Vichada Departments in Colombia, the physiognomy and main floristic types of savannas are fairly well known. An overall account of savanna vegetation in the Colombian

llanos may be found in FAO (1966) and Blyden-stein (1967). Tamayo (1964), Ramia (1967) and Medina and Sarmiento (1979) give a general picture of this vegetation in the Venezuelan llanos. Indeed, the Venezuelan llanos are perhaps one of the more thoroughly analyzed and best known areas of neotropical savannas.

Four main subregions may be distinguished within this natural region, differing from each other in age of parent materials, land forms and soil, and as a consequence in types of savanna formations and overall vegetation patterns. These four subregions — the piedmont, the high plains, the alluvial overflow plains and the aeolian plains — will be briefly considered in turn.

The piedmont savannas

The piedmont region is characterized by large alluvial fans and a system of alluvial terraces. Semi-deciduous tropical forests are widespread on these

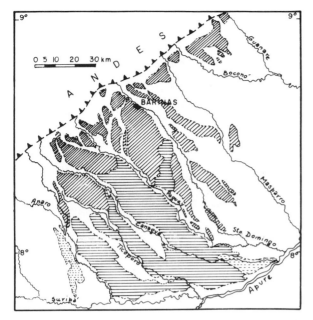

Fig. 10.8. Vegetation pattern in the piedmont region of the Venezuelan llanos. Oblique hatching indicates the areas with seasonal savannas; horizontal hatching shows the areas of hyperseasonal savannas; discontinuous horizontal hatching indicates the esteros, the white areas correspond to various types of rain forests. (After Sarmiento et al., 1971a.)

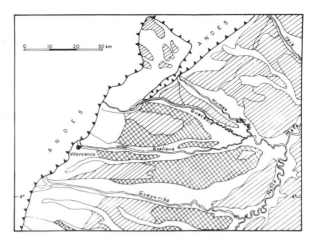

Fig. 10.9. Vegetation pattern in the piedmont region of the Colombian llanos. Oblique hatching indicates areas of seasonal savannas (savannas of *Melinis minutiflora*, *Trachypogon ligularis*, *Trachypogon vestitus*, *Trachypogon vestitus–Axonopus purpusii*; and savannas of *Paspalum pectinatum*); cross-hatching refers to areas with hyperseasonal savannas (*Leptocoryphium lanatum* savanna); white areas correspond to various types of rain forests. (After FAO, 1966.)

land forms, but savannas occupy a major portion of the landscape (Figs. 10.8 and 10.9). In the western Venezuelan llanos, the distribution, ecology and composition of the piedmont savannas were analyzed by Sarmiento et al. (1971a), Silva et al. (1971) and Monasterio et al. (1971).

A savanna woodland occupies the oldest Quaternary deposits, Q_{iv} according to Tricart and Millies-Lacroix (1962). Just a few species form the low and open tree layer, while the ground vegetation is dominated by hard tussock grasses and sedges and by halfshrubs (Table 10.4 and Fig. 10.10). The Q_{iii} terrace maintains wooded savannas intermingled with semideciduous forests. The more extensive Q_{ii} terrace was formerly covered by seasonal savannas that have now been replaced by croplands or grazing lands of introduced grasses, mainly the African *Hyparrhenia rufa*, covering large areas. The lowest and youngest terrace level (Q_i) is the domain of gallery forests, though on sandy soils seasonal savanna may occur with *Elyonurus tripsacoides* as the dominant grass. Hyperseasonal savannas, on the other hand, are restricted to a few localized bottom lands, in this piedmont region.

TABLE 10.4

Some of the commonest species in the seasonal savannas of the piedmont region of the Venezuelan llanos

TREES

Acrocomia sclerocarpa	*Cochlospermum vitifolium*
Bowdichia virgilioides	*Curatella americana*
Byrsonima coccolobaefolia	*Genipa americana*
Byrsonima crassifolia	*Roupala complicata*
Casearia sylvestris	*Xylopia aromatica*

SHRUBS AND HALF-SHRUBS

Clitoria guianensis	*Palicourea rigida*
Desmodium pachyrrhizum	*Pavonia speciosa*
Galactia jussieana	*Psidium guianense*
Ichthyothere terminalis	*Tephrosia adunca*

GRASSES AND SEDGES

Andropogon selloanus	*Leptocoryphium lanatum*
Andropogon semiberbis	*Panicum olyroides*
Axonopus canescens	*Paspalum gardnerianum*
Axonopus purpusii	*Paspalum plicatulum*
Bulbostylis capillaris	*Sporobolus cubensis*
Bulbostylis junciformis	*Thrasya petrosa*
Bulbostylis paradoxa	*Trachypogon montufari*
Dichronema ciliata	*Trachypogon plumosus*
Elyonurus adustus	*Trachypogon vestitus*

Fig. 10.10. A savanna woodland in the piedmont region of the Venezuelan llanos. *Byrsonima crassifolia*, *B. coccolobaefolia* and *Bowdichia virgilioides* are the main components of the tree layer, while the ground layer is dominated by *Andropogon semiberbis*, *Trachypogon plumosus* and various halfwoody species. (Photo from Vera, 1979.)

Silva and Sarmiento (1976a, b) analyzed the floristic composition and ecological relationships of these savannas as they occur on seven different soil series either on the Q_{iv}, Q_{iii} or Q_{ii} terraces. Each soil series was found to carry a characteristic community, but that the composition and importance of the species varied continuously, from the drier seasonal savannas on coarse soils developed on the Q_{iv} terrace, to the wetter hyperseasonal savannas on poorly drained sites of the Q_{ii} terrace. Only a few species were restricted to narrow parts of this gradient, while most of them occurred in more than one community (Fig. 10.11). This phytosociological continuum reflects environmental gradients related to soil-water conditions during the dry and wet seasons.

Sarmiento and Vera (1979a) followed the changes in soil water content during an annual cycle, in various sites along that gradient, confirming that soil-water availability was very different at the two extreme sites, though varying continuously along the gradient.

Sarmiento and Vera (1979b) measured the annual production of the ground layer by the harvesting method, in three seasonal and one hyperseasonal savanna, in this piedmont area of the western Venezuelan llanos. Seasonal formations attained a maximum standing crop of 522 to 604 g m^{-2}, without showing significant differences between communities and among different years. The hyperseasonal savanna reached a maximum of 705 g m^{-2}. As all these savannas were burned each year, the maximum above-ground biomass may be taken as a rough estimate of the aerial net primary production of the ground layer. Maximum below-ground biomass to a depth of 2 m ranged from 1148 to 1891 g m^{-2}, giving an estimated below-ground annual production of about 500 to 1300 g m^{-2}. In a savanna woodland on the Q_{iv} terrace, where total tree density was about 1000 trees ha^{-1} Vera (1979) determined an annual litter production of about 120 g m^{-2}. Sarmiento (1978) discussed some methodological and conceptual problems derived from these estimates of annual production,

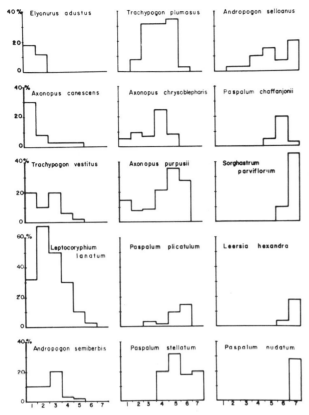

Fig. 10.11. The representation of different grasses in seven savanna communities along a humidity gradient in the piedmont region of the Venezuelan llanos. Number *1* corresponds to the driest savanna and number *7* to the most humid type. The vertical axis indicates total cover of the species in each of the seven ecosystems. Notice how most species show a wide range of occurrence, though with different cover in each community. Species like *Paspalum nudatum* restricted to a single community are rather the exception. (From Silva and Sarmiento, 1976a.)

concluding that the real production figures might be about 100% higher than these figures, if other variables like differential growth and mortality rates were taken into account.

The high plains

The high plains, or mesas, constitute a second subregional unit within the llanos. They occupy two distinct areas in the south and east, separated by a central tectonic depression (Fig. 10.7). These high plains appear as remnants of a former continuous tableland of Late Pliocene or Early Pleistocene age. The isolated mesas may show either a gently undulating surface or their surface may have been dissected giving rise to a hilly relief. A few relict mesas remain in some places, generally protected from erosion by hard lateritic layers. The parent material on the tablelands is rather coarse alluvium on which light-textured soils have evolved.

The savanna on the mesas in the eastern Venezuelan llanos was considered in a pioneer paper by Pittier (1942). He described this characteristic landscape of low plateaus, 200 to 300 m above sea level, deeply dissected by rivers flowing to the Orinoco. The savanna is a seasonal tree savanna (Fig. 10.12) dominated by *Trachypogon plumosus* or by *T. vestitus*, with *Andropogon selloanus*, *Axonopus canescens* and *Leptocoryphium lanatum* as subdominant grasses. The low, scattered trees belong almost exclusively to three species: *Bowdichia virgilioides*, *Byrsonima crassifolia* and *Curatella americana*. A wooded savanna occurs on the steep slopes of the incised valleys, with the forementioned species plus a few others, like *Byrsonima coccolobaefolia*, *Piptadenia peregrina* and *Roupala complicata*, while the valley bottoms are occupied either by *Mauritia flexuosa* palm swamps or by gallery forests.

The savanna on the mesas in the central Venezuelan llanos has been described by Blydenstein (1962) and Sarmiento and Monasterio (1969a, b, 1971). Here, isolated mesas remain as dissected remnants of a formerly continuous upland. As the denudation of these mesas left just a few meters or even less of coarse alluvium, underlying materials may appear at the surface, introducing a further diversity of habitats. As a result, these seasonal savannas are more varied than those in the eastern mesas, and they frequently form intricate patterns with forests and hyperseasonal savannas. On the mesas, seasonal savannas vary from tree savannas to savanna woodlands. Tree density and height, as well as floristic composition of the ground layer, depend on soil depth, occurrence of lateritic crusts and fire frequency (Sarmiento and Monasterio, 1969a, 1971). A savanna parkland prevails in areas where water shortage is less restrictive (Fig. 10.13). All these communities have a rather closed ground layer dominated by species of *Andropogon*, *Axonopus* and *Trachypogon*.

The wide valleys dissecting the mesas have a system of river terraces of different ages. Seasonal savannas, mostly treeless grasslands, occupy the

Fig. 10.12. A seasonal savanna with very scattered trees on the mesas of the eastern high plains in the Venezuelan llanos. *Curatella americana* is almost the only tree species over a ground layer dominated by species of *Andropogon*, *Axonopus* and **Trachypogon**.

Fig. 10.13. A savanna parkland on the high plains (mesas) of the central Venezuelan llanos (Calabozo Biological Station). The seasonal savanna is a *Trachypogon–Axonopus* community with scattered trees (*Bowdichia virgilioides*, *Byrsonima crassifolia*, *Curatella americana*). The groves are dominated by high trees like *Cassia moschata*, *Copaifera officinalis* and *Vochysia venezuelana*.

drier habitats, while hyperseasonal savannas occur on waterlogged sites, where they may either be grasslands dominated by *Leersia hexandra*, *Mesosetum loliiforme*, *Sorghastrum parviflorum* and several species of *Andropogon* and *Panicum*, or the palm *Copernicia tectorum* may be locally abundant as the single woody species (Fig. 10.14). Savanna parklands occur in some areas where the hyperseasonal savannas are dotted with groves of deciduous tropical forest (Fig. 10.15). On the younger terraces, a fairly continuous deciduous forest prevails, while the stream borders are mostly covered by gallery forests.

The high plains also occupy large extensions of the Colombian llanos, particularly between the Meta and Guaviare Rivers. They extend unchanged to southern Venezuela, and end at the Cinaruco

Fig. 10.14. A hyperseasonal savanna with palms (*Copernicia tectorum*) on a low alluvial terrace in a wide valley dissecting the mesas in the Venezuelan llanos.

River. The most common formation in this region is a seasonal tree savanna, with low trees mostly *Byrsonima crassifolia* and *Bowdichia virgilioides* and a ground layer dominated by species of *Bulbostylis*, *Leptocoriphium*, *Paspalum* and *Trachypogon*. Table 10.5 gives a list of the most frequent species in these savannas according to FAO (1966).

The Guiana Shield in southern Venezuela is bordered to the east and north by a fringe of seasonal savannas on uplands that represent a southern extension of the mesas. Williams (1942) and Tamayo (1964) give descriptions of these formations. They are mostly *Trachypogon* savannas with sparse trees, including the common species mentioned previously.

Primary production of the savannas on the mesas of the central Venezuelan llanos, in the Calabozo area, has been estimated in several papers (Blydenstein, 1962, 1963; San José and Medina, 1975, 1977; Medina et al., 1977). According to San José and Medina (1975), the *Trachypogon* savanna shows a peak of green biomass that ranges from 320 to 420 g m^{-2}, being higher when burned than when protected from fire. These figures were obtained during a year when rainfall was well above the mean, and the total standing crop of the ground layer reached a maximum of 730 g m^{-2}. In drier years, total standing crop scarcely surpasses 300 or 400 g m^{-2}, the figures are dependent on the precise date of burning. There are no data available on biomass or annual production of the woody layers. The *Andropogon* hyperseasonal savanna on the terraces reaches a peak of 653 g m^{-2} for the aboveground standing crop (San José and Medina, 1977). Apparently, this community is more productive than the seasonal savanna of the same region.

The alluvial overflow plains

The alluvial overflow plains occupy a vast depression in the central part of the llanos, between the piedmont and the high plains. These savannas have been considered by Ramia (1959, 1974b), FAO (1966), Blydenstein, (1967) and Sarmiento et al. (1971b). Vegetation varies along topographic catenas, which show differences of only 1 or 2 m between their highest and lowest points in this flat area. The upper part of the vegetation soil catena corresponds to natural levees or banks that border the streams, where sandy alluvium has been de-

Fig. 10.15. A hyperseasonal savanna parkland on alluvial soils of the Venezuelan llanos. Besides the palm *Copernicia tectorum*, the most frequent trees are *Annona jahnii, Platymiscium pinnatum, Pterocarpus podocarpus* and *Spondias mombim*.

posited. These sandy banks appear as narrow strips gently dipping away from the river to a wide flat area where silty alluvium predominates. The lowest part of the catena is formed by slowly draining decantation cuvettes, where clay particles have settled leading to the development of Vertisols. During the rainy season, only the levees remain unflooded, all the rest of the area being waterlogged for periods ranging from a few weeks to several consecutive months. The terrain is not flooded by overflowing streams but by the accumulation of rain water that drains quite slowly. Towards the end of the dry season the waters have receded, except from the permanent swamps.

Either gallery forests or seasonal savannas may occur on banks, according to the depth of the water table. On the higher banks tree savannas predominate, while the lower levees support a treeless grassland. In the first case, *Axonopus purpusii* and *Paspalum plicatulum* are the dominant species in the ground layer. In the treeless savannas the main grasses are *Sporobolus indicus* and *Imperata contracta*. Since cattle gather on these higher areas during the rainy season, most savannas are heavily

overgrazed, with a marked decrease in all palatable species and a corresponding increase in herbaceous and woody weeds.

Hyperseasonal savannas are the most widespread ecosystems in this low region, occupying the wide silty extensions between successive banks. These are mostly treeless savannas (Fig. 10.16), or less frequently palm savannas with *Copernicia tectorum* as the single woody species. The ground flora is rather rich, with many species of annual and perennial grasses and sedges (Table 10.6). The two main savanna communities occurring on this landform produced by slack waters are dominated respectively by species of *Andropogon* and *Mesosetum*.

The bottom lands support a vegetation either of esteros or of permanent swamps. Both types of ecosystems remain under 1 m or more of water during the rainy season, while they slowly dry out as the dry season approaches. The esteros are probably the best grazing lands of the whole llanos, since they can support a heavy carrying capacity during the critical dry months of the rainless season. Though these treeless savannas are rather

TABLE 10.5

Some of the commonest species in the savannas of the high plains region of the Colombian llanos (after FAO, 1966)

WOODY AND HALF-WOODY SPECIES

Bowdichia virgilioides	*Jacaranda lasiogyne*
Byrsonima crassifolia	*Lantana moritziana*
Byrsonima verbascifolia	*Palicourea rigida*
Cassia flexuosa	*Pavonia speciosa*
Clitoria guianensis	*Psidium eugenii*
Curatella americana	*Psidium guianense*
Ichthyothere terminalis	*Xylopia aromatica*

GRASSES AND SEDGES

Andropogon selloanus	*Panicum versicolor*
Andropogon hypogynus	*Paspalum carinatum*
Andropogon virgatus	*Paspalum minus*
Aristida capillacea	*Paspalum multicaule*
Axonopus canescens	*Paspalum pectinatum*
Axonopus pulcher	*Rhynchospora barbata*
Axonopus purpusii	*Rhynchospora globosa*
Bulbostylis junciformis	*Scleria hirtella*
Dichronema ciliata	*Setaria geniculata*
Eriochrysis holcoides	*Sorghastrum parviflorum*
Leptocoryphium lanatum	*Thrasya petrosa*
Mesosetum rottboellioides	*Trachypogon ligularis*
Panicum rudgei	*Trachypogon plumosus*
Panicum stenodes	*Trachypogon vestitus*

TABLE 10.6

Some of the commonest species in the savannas of the alluvial overflow plains in the Colombian llanos (from FAO, 1966)

WOODY AND HALF-WOODY SPECIES

Buettneria jaculifolia	*Melochia villosa*
Caraipa llanorum	*Mauritia minor*
Ipomoea crassicaulis	*Rhynchanthera grandiflora*
Ludwigia lithospermifolia	

GRASSES AND SEDGES

Andropogon bicornis	*Panicum laxum*
Andropogon hypogynus	*Panicum parvifolium*
Andropogon selloanus	*Panicum stenodes*
Andropogon virgatus	*Panicum versicolor*
Axonopus purpusii	*Paratheria prostrata*
Cyperus haspan	*Paspalum pulchellum*
Eriochrysis holcoides	*Rhynchospora barbata*
Eriochrysis cayennensis	*Rhynchospora globosa*
Leersia hexandra	*Scleria hirtella*
Leptocoryphium lanatum	*Setaria geniculata*
Mesosetum chaseae	*Sorghastrum parviflorum*
Mesosetum rottboellioides	*Trachypogon ligularis*

Fig. 10.16. An *Andropogon–Sorghastrum* hyperseasonal savanna grassland during the dry season in the alluvial overflow plains of the Venezuelan llanos. A gallery forest appears at the background.

poor in species, the two commonest grasses, *Leersia hexandra* and *Hymenachne amplexicaulis*, are both highly palatable. Swamps appear as more or less pure consociations of one or another hydrophilous species, like *Ipomoea crassicaulis*, *Ludwigia lithospermifolia*, *Thalia geniculata* or one of several species of sedges.

The areas annually flooded by the overflow of rivers and streams support another type of estero, a tall grassland dominated by *Paspalum fasciculatum*, a tussock grass that reaches 2.5 m high (Figure 10.17). This is a widespread formation in the lowest parts of the llanos, such as those near the Apure River and in the lowlands bordering the eastern mesas and merging into the Orinoco Delta.

González Jiménez (1979) measured above-ground production of the herb layer in the three main types of savannas along the topographic catena. During the year of data collection, the seasonal savanna on the sandy bank reached a maximum green biomass of 425 g m^{-2} and a maximum above-ground standing crop of almost 700 g m^{-2}. The hyperseasonal savanna in the loamy basin reached maximum figures of 550 and 800 g m^{-2} respectively for green and total biomass, while the estero had a peak green biomass of 900 g m^{-2}, almost equivalent to the peak standing crop (916 g m^{-2}). As these figures were obtained in savannas under their normal regime of annual burning, they represent rough estimates of the net aerial production of the ground layer. The figures may vary somewhat according to burning or fire protection, and also according to the precise time of burning.

The aeolian plains

An aeolian landscape extends as a continuous belt from the upper Meta River in central Colombia to the Cinaruco River in southern Venezuela (Fig. 9.7). It continues eastwards forming sparse reduced patches within the eastern high plains. This aeolian landscape represents the remnant of a former arid morphogenesis that took place during the Würm glacial period according to Tricart (1974). Its characteristic land forms are extensive

Fig. 10.17. An estero during the rainy season on the alluvial overflow plains of the Venezuelan llanos. *Paspalum fasciculatum* is the dominant grass. In the background a gallery forest.

dune fields superimposed on larger areas of loess-like material that was partly covered by younger alluvia.

A dry seasonal savanna occurs on the dunes, with a few trees or entirely treeless. *Byrsonima crassifolia* is almost the single tree species over a quite open ground layer, dominated by *Trachypogon ligularis* and *Paspalum carinatum* (Fig. 10.18). A *Mesosetum* hyperseasonal savanna occupies the depressions between the dunes and the extensive silty plains. In some areas, *Caraipa llanorum* appears as a characteristic tree, occurring only on this type of habitat and forming a wooded savanna. Along rivulets or on slowly draining lowlands, a grass and sedge estero occurs, with the *moriche* palm (*Mauritia minor*) as its only tree. These *morichales* may be fairly good grazing lands since the grasses remain green throughout the dry season.

Other savannas related to the llanos

Several savanna patches closely related to the seasonal savannas of the piedmont region of the llanos occur in various areas of northern Colombia and Venezuela. In some cases, such as along the middle Magdalena Valley, they occupy high terraces with coarse soils. In other areas they cover extensive alluvial fans, such as in the eastern piedmont of the Perijá Cordillera or in the north-western piedmont of the Venezuelan Andes, in the Maracaibo Lake basin, where tree and wooded savannas occur in old terraces of Early Pleistocene age. All these seasonal savannas have some species of *Trachypogon* as the leading species of the ground layer, while the trees belong to the three or four commonest tree species in the llanos.

Hyperseasonal savannas and esteros are widespread in the swampy depression of the lower Magdalena in northern Colombia, intermingled with permanent swamps and other types of hygrophyllous formations. Small patches of hyperseasonal savannas have also been described in Trinidad (Beard, 1953; Richardson, 1963) where they occur on an old poorly drained terrace of Late Pleistocene age, as small savanna islands encircled by a seasonal swamp forest.

Fig. 10.18. The aeolian plains in the region between the Cinaruco and the Meta rivers. In the foreground a *Mesosetum* hyperseasonal savanna grassland; in the background a dune covered by a seasonal savanna woodland. *Byrsonima crassifolia* is the only tree species over a ground layer dominated by *Trachypogon ligularis* and *Paspalum carinatum*.

Seasonal savannas extend widely over the low mountains of the Caribbean cordillera that border the llanos along their northern edge. They cover the ridge crests and mid-slopes, while semi-deciduous forests occupy creeks and valleys. These mountain savannas, as some other isolated patches that occur along the eastern Andean slopes, seem to represent seral stages stabilized as fire disclimax long ago (Vareschi, 1969; Sarmiento and Monasterio, 1969b). They are mostly tree savannas, sometimes with relict clumps of forest trees. The widespread species of seasonal savannas also occur here, and even the ground layer is dominated by widespread grasses and sedges, mostly species of *Andropogon*, *Axonopus*, *Elyonurus*, *Paspalum*, *Trachypogon*, together with some naturalized species like *Heteropogon contortus* and *Melinis minutiflora*.

Denevan and Chrostowski (1970) and Scott (1977) studied one isolated savanna area on an uplifted plateau, 1000 to 1200 m above sea level, in the eastern chains of the Peruvian Andes. This area, known as the Gran Pajonal, supports several associations that range from pure grasslands to tree and shrub savannas. *Byrsonima crassifolia* is the most conspicuous woody species, together with some shrubby Melastomataceae and Rubiaceae. *Andropogon lanatus*, *A. leucostachys*, and *Leptocoryphium lanatum* are the three dominant grasses. The introduced *Melinis minutiflora* may also be locally abundant. Scott (1977) emphasized the crucial influence of the local Amerindians in the origin and maintenance of these savannas.

As may be realized, the savannas on mountain slopes and other highlands that occur in various parts of tropical South America, are plant communities closely related to the upland formations of the llanos, in spite of the great differences between the habitats on which the two types of savannas occur. This similarity may reflect the fact that both types are fairly young ecosystems, the one colonizing relatively recent alluvial or aeolian materials, the other a product of human influences from pre-Spanish times.

SAVANNAS OF THE GUIANA PLATEAUX AND THEIR SOUTHERN BORDERS

In the Guiana region of southeastern Venezuela, there is a large area where the sandstones of the Precambrian Roraima Formation form landscapes of tablelands at different altitudes, with a relief of large horizontal stairs. They appear as a complex arrangement of flat-topped plateau levels bounded by very steeply sloping scarps and divided by broad and swampy river flats. The highest levels, with altitudes from 2000 to 3000 m, appear as solitary plateaux called *tepuys*[1]. The Auyan Tepuy, Roraima, Duida, Jaua, Cerro de la Neblina, are some of the major tepuys. Lower tablelands, from 800 to 1200 m above sea level, connect the higher plateaux. They extend as much smaller and isolated areas to Guyana and Surinam. The Venezuelan part of these lower-level plateaux is known as the Gran Sabana (Grand Savanna) and is almost entirely a savanna land. To the south, these highlands descend and penetrate the Brazilian and Guyanan territories where they form another extensive savanna land known as the Rio Branco-Rupununi savannas (Fig. 10.19). This savanna

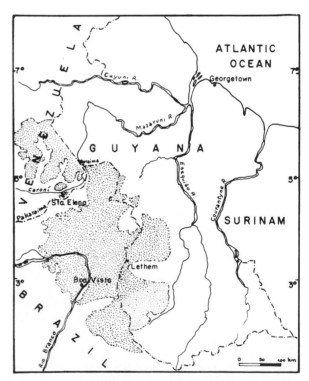

Fig. 10.19. The Rio Branco–Rupununi region in Brazil and Guyana. The Gran Sabana area in southeastern Venezuela is also shown.

---

[1]Tepuy is an Amerindian name used for a high tableland in the Guianas.

country, with altitudes from 300 to 100 m above sea level, extends to the Mucajai River, where the Amazonian forests begin rather abruptly. Climatic conditions on the higher tepuys are practically unknown, though they seem to be areas of heavy rainfall. The lower tablelands and the adjacent low-altitude regions have an annual rainfall of 1500 to 2000 mm, with three to five dry months.

Myers (1936) traveled through this area of the Venezuelan/Brazilian/Guyanan border and he gave one of the first accounts of its vegetation. He described several types of savanna formations. A *Trachypogon/Curatella* tree savanna was recognized as the most extensive type, apparently quite homogeneous floristically throughout the area. Myers also referred to the occurrence of moister flats with treeless sedge savannas, and seasonal swamps where the palm *Mauritia flexuosa* forms open stands. Savannas were recorded on the Roraima and Pakaraima highlands up to more than 2000 m above sea level.

Beard (1953) also presented a first-hand description of this area and its savannas. He recognized two principal types of site under savanna, which he termed the alluvial and the residual, repeated either on the valleys or on the uplands. The alluvial sites are treeless and dotted with termite mounds. The residual sites, with soils developed *in situ* or from ancient alluvia, frequently have a stony surface layer of ironstones and quartz pebbles. Here the savannas may have sparse low trees. Beard indicated that *Trachypogon plumosus* was the dominant grass in all savannas on well-drained soils. In swampy depressions, that as Beard remarked are similar to the esteros of the llanos, a different grass cover occurs, with isolated groves of *Mauritia flexuosa*.

Takeuchi (1960) gave a short description of the savannas in the Roraima Territory of Brazil, as Tamayo (1961) did for the Gran Sabana area. Steyermark (1967) presented floristic information for the Auyan Tepuy savannas and related vegetation formations. Van Donselaar (1968, 1969) extended his phytosociological and phytogeographic analysis of northern Surinam savannas to the restricted savanna patches in the Surinam/Brazil border, comparing both areas floristically and phytosociologically. But the most comprehensive ecological analysis of savanna lands in this area is that of a research team from McGill University that worked on the Rupununi savanna of the former British Guiana (Guyana). Several papers on plant communities (Goodland, 1964), ecology (Eden, 1964; Hills, 1969), geomorphologic evolution of the landscape (Sinha, 1968) and other aspects were produced. The Radam Brazil Project (Projeto RADAMBRASIL, 1973–78) inventoried geology, soils and vegetation formations in the whole area of the Rio Branco savannas. Finally, I had direct acquaintance with this area during a field trip in 1976, whose results still remain unpublished.

Main savanna types

Summarizing all the forementioned information, the following picture of the main savanna formations in this area emerges:

(1) Herbaceous swamps on the tepuys, above an altitude of 1500 m. Though they scarcely could be considered as savannas, since grasses do not play any significant role in this formation, this ecosystem may be taken as a useful reference point for comparison with the savannas. The same type of formation appears on sandy soils developed on the Roraima Sandstones in all known tepuys. In the Auyan Tepuy, for instance, Steyermark (1967) described several communities dominated by sedges (*Cephalocarpus*, *Everardia*, *Lagenocarpus*), Xyridaceae (*Abolboda*, *Xyris*), Eriocaulaceae (*Eriocaulon*, *Paepalanthus*, *Syngonanthus*), Bromeliaceae (*Brocchinia*, *Cottendorfia*), Rapateaceae, Velloziaceae, Orchidaceae, together with many species of shrubs.

(2) Savannas and seasonal swamps on white sands, at levels around 900–1200 m. This is the main level of the tablelands of the Gran Sabana. On white sands derived from the Roraima sandstones, an intermediate formation between the high tepuys and the savannas on the lower levels occurs (Fig. 10.20). The wetter communities are dominated by the families and genera of monocotyledons mentioned above, but with a greater representation of grasses. The drier areas maintain a treeless grass and sedge savanna (Fig. 10.21), with species of *Axonopus*, *Panicum*, *Rhynchospora*, *Scleria*, *Trachypogon*, etc. (Steyermark, 1967). Although the phenological changes of these species and communities have not been followed, one is apparently dealing with a dry type of semi-seasonal savanna, since the rainless season is rather short,

Fig. 10.20. The landscape of high tablelands in the Gran Sabana region. A savanna grassland on white sands appears in the foreground.

Fig. 10.21. A white sand savanna in the Gran Sabana. Notice the high proportion of bare sand among the open cover of sedges and grasses.

but soils permanently maintain conditions of sub-optimal humidity due to their coarse textures.

(3) Seasonal savannas on the lower tablelands, on alluvial fans and upper river terraces, at altitudes between 100 and 500 m. These are mostly open tree savannas with *Antonia ovata*, *Bowdichia virgilioides*, *Byrsonima coccolobaefolia*, *B. crassifolia*, *Curatella americana*, *Genipa americana*, *Plumeria inodora*, *Roupala complicata* and *Salvertia convallariodora* as the woody components. The ground layer is dominated by *Trachypogon plumosus*, with *Aristida setifolia*, *A. tincta*, *Axonopus canescens*, *Bulbostylis paradoxa*, *Echinolaena inflexa*, *Leptocoryphium lanatum* and *Thrasya paspaloides* as codominant species. The main half-woody species are *Byrsonima verbascifolia*, *Casearia sylvestris*, *Palicourea rigida* and *Psidium* spp. A similar type of seasonal savanna occurs on dunelands in southern Rupununi (Sinha, 1968).

(4) Savanna woodlands and woody savannas in the piedmont zone of the Kanuku Mountains and on the lower slopes of the Pakaraima Mountains. Here, the woody species are practically the same as in the tree savannas, though they have a much

greater density. In many cases, *Curatella americana* seems to be the most important tree.

(5) Hyperseasonal savannas on river flats. These are either pure grasslands, or areas where small trees occur on termite mounds. The drier communities are dominated by *Andropogon selloanus*, *Panicum laxum*, *Paspalum pulchellum* and *Sporobolus cubensis* and some other grasses and sedges. In the wetter sites, the most frequent species are *Andropogon bicornis*, *Paspalum densum*, *P. millegrana*, *P. pulchellum*, etc.

(6) Morichales and seasonal swamps. These are open stands of *Mauritia flexuosa* palms (Fig. 10.22) with a sedge and grass undergrowth in the swampier parts of river flats.

THE COASTAL SAVANNÁS IN THE GUIANAS

In the coastal region of the three Guianas (Guyana, Surinam and French Guiana) there is a belt of lowland savanna country quite distinct from the llanos and from the interior of Guiana. It extends further south across the Amapa Territory

Fig. 10.22. A stand of *Mauritia flexuosa* palms (morichal) on a river flat between the high tablelands of the Gran Sabana region.

of Brazil to the mouths of the Amazonas, thus establishing a link with the Amazon savannas which will be considered later.

The coastal savannas occupy a narrow and discontinuous belt on Plio-Pleistocene alluvial deposits, mainly coarse sands, located between the littoral swamps and mangroves and the continuous rain forest that covers most of the hinterland in these countries. The coastal climate is tropical and constantly wet, with annual rainfall from 2000 to 3800 mm, and two to four months with less than 50 mm rainfall.

The savanna formations of this coastal belt are relatively well-known. The first descriptions were made by Benoist (1925), a comprehensive study was carried out by Dutch workers in Surinam (Lanjouw, 1936; Heyligers, 1963: Van Donselaar, 1965), and a detailed analysis of French Guiana savannas was performed by Hoock (1971). Benoist recognized three main formations: dry savannas, with scattered trees (*Byrsonima crassifolia* and *Curatella americana*) over a grass layer dominated by *Aristida*, *Axonopus* and *Trachypogon*; intermediate savannas, dominated by sedges (*Lagenocarpus*, *Rhynchospora*, *Scleria*) with *Mauritia flexuosa* patches; and wet savannas or herbaceous swamps. Lanjouw (1936) differentiated three savanna formations in northern Surinam: a flat-watershed type of dry grass savanna with shrubs; a type of wet, sedge savanna on impermeable clays; and a type on leached soils and impermeable iron pans, where patches of bushes and small trees occur interspersed with open areas.

Heyligers (1963) analyzed the vegetation and soil of a white sand savanna in northern Surinam. These white sands form a low plateau consisting of coarse sands 30 m above sea level. Low types of forest, which Heyligers called "savanna forest" and "savanna wood", develop on this parent material. Besides these formations, Heyligers referred to a "savanna scrub" and "thicket" as more open and lower woody formations. Neither of these four types could really be considered as a savanna, but they are related to a last formation of true savanna in which small patches of bushes and woods appear intermingled with open stretches of grasses, sedges and herbs. *Trachypogon plumosus* is the dominant grass in this formation, while in wetter habitats the sedges *Lagenocarpus tremulus* and *Rhynchospora tenuis* become more important. Heyligers main-

tained that the *Trachypogon* savanna owed its origin to the destruction of the savanna woods and the prevention of its regeneration by repeated burning.

Van Donselaar (1965) established a formal classification of the Surinam savannas according to the Braun-Blanquet phytosociological system. All savannas in the area were grouped in a single class, which he named after the two most frequent and widespread species: *Trachypogon plumosus* and *Leptocoryphium lanatum*. Within this class, Van Donselaar distinguished three Orders, eight Alliances and 27 Associations. These phytosociological units, defined on the basis of their characteristic and differential species, reflect differences in soil moisture and type of parent material.

On very dry to moist sites a *Trachypogon–Axonopus* savanna is found. On white sand, this type appears as a grassland dotted with bushes dominated by *Terstroemia punctata* and *Matayba opaca*. The *Curatella–Trachypogon* savanna, so widespread throughout northern South America, occurs on red sand and loamy sand. On red sandy loam, this tree savanna becomes enriched with some additional tree species, while sedges become more abundant in the ground layer. Wet to very wet sites maintain savannas grouped in a second Order, named after the grass *Paspalum pulchellum*. On white sand, sedges together with *Xyris* and Eriocaulaceae prevail, while a treeless community characterized by the sedge *Bulbostylis lanata*, sometimes with bush islands of *Licania incana*, *Tetracera asperula* and *Tibouchina aspera* occurs on red loamy sand and sandy loam. An *Imperata brasiliensis–Mesosetum cayennense* grassland appears on sandy loam and heavier soils, sometimes with sparse trees (*Byrsonima crassifolia* and *Roupala montana*). Finally, on very wet soils, the Associations of the *Panicum stenodes* Order occur either as pure grass and sedge savannas or with *Mauritia flexuosa* groves along the rivulets.

Hoock (1971) recognized seven savanna types and biotypes in the savanna belt of French Guiana. Each type is characterized by the presence of certain ecological groups of species. On yellow sands, he distinguished one forest and two savanna types: a tall shrub savanna and a tall grass savanna. The shrub savanna, characterized by a mesophilous group of species, is an open tree savanna with *Curatella americana* as the most frequent sparse tree and *Axonopus purpusii*, *Leptocoryphium la-*

natum and *Trachypogon plumosus* as the dominant grasses. The grass savanna has a rather similar ground layer but no trees or shrubs. Two savanna communities occur on colluvial gray sands with podzols, a low savanna with nanophanerophytes and a low herbaceous savanna. In the first type, *Byrsonima verbascifolia* appears as a characteristic half-woody species scattered in a *Trachypogon plumosus–Leptocoryphium lanatum* grass layer. An ecological group of hygrophytes becomes important in the second type, on wetter sites, with species like the grasses *Echinolaena inflexa* and *Panicum stenodes* and the sedges *Rhynchospora globosa* and *R. graminea*. Two communities occur on podzolized white sands: a low shrub savanna with scattered trees (*Byrsonima crassifolia*) over a ground layer of sedges, *Xyris* and other herbs; and a shrub formation of the type Heyligers (1963) called "scrub savanna" with *Clusia fockeana*, *Hirtella strigulosa*, *Tetracera aspersa* and other common white sand shrubs. Finally, an hygrophilous herbaceous savanna occurs on hydromorphic soils, of the type called here hyperseasonal savanna, with tall species of *Andropogon* and *Paspalum* as its dominant grasses.

To summarize all the data heretofore presented on the coastal savannas of the Guianas, it may be emphasized that several savanna types occur in this narrow belt that extends from Guyana to Amapa (Brazil). They may be distinguished either on a phytosociological basis or major site features. A double distinction becomes evident, one based on type of parent material, the other on moisture conditions. White sand communities form a very special type of savanna that, as will be seen later, is ecologically and floristically related to the Amazon caatinga forests and their derived savannas occurring on the same type of parent material. By contrast, savanna formations on red soils with a more or less advanced latosolization, are more akin to the widespread types of savanna formation occurring in other areas of northern South America, particularly in the llanos and the interior Guiana highlands. Both types, savannas on white sand and on red soils, may be further subdivided either according to structural features forming a continuum from treeless savannas to savanna woodlands, or on the basis of a seasonality gradient. In the ever-wet climates of the Guiana savanna belt, these formations are semi-seasonal savannas, on well-drained sites, or hyperseasonal savannas and esteros on poorly drained areas.

THE SAVANNAS IN THE AMAZONIAN REGION

Though the Amazonian lowlands are really the domain of the continuous tropical rain forest, open vegetation, either savannas or swamps, is by no means absent from this huge natural region. These vegetation types appear either as narrow fringes along the flood plains or as scattered patches within a nearly continuous forest cover. As the knowledge of the region advances, a picture emerges of a landscape dotted with numerous natural clearings occupied by different types of savannas and swamps.

After the pioneer work of Bouillene (1926) and the first overall picture given by Ducke and Black (1953), several authors dealt with the Amazonian open vegetation as it occurs in various parts of this area. Braun and Ramos (1959) described the campos in the region of the Madeira, Purus and Ituxi Rivers in southwestern Amazonia (Fig. 10.23). In this zone, savannas are a part of topographic catenas that extend from cerrado-like savannas on the narrow interfluves, pass through the wet campos and end in the rain forest on the alluvial soils.

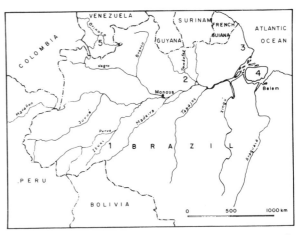

Fig. 10.23. Location of some savanna areas (campos) within the Amazonian region: *1* = the Puciari–Humaitá campos (after Braun and Ramos, 1959); *2* = the Aribamba campos (after Egler, 1960); *3* = the Amapá cerrados and campos (after Azevedo, 1967); *4* = the campos and swamps of the Marajo Island (after Meggers and Evans, 1957); *5* = the upper Orinoco savannas in southern Venezuela (after Eden, 1974).

The upper surface seems to be the remnant of an ancient tableland of Tertiary age. *Hancornia speciosa* is the characteristic savanna tree on well-drained latosols, where it occurs, together with *Curatella americana*, scattered in a grass carpet of *Aristida capillacea, Elyonurus* sp., *Leptocoryphium lanatum, Panicum rudgei* and other herbs and sedges.

Egler (1960) gave one of the first detailed accounts of the vegetation mosaics in the Lower Amazon region, near the mouth of the Trombetas River (Fig. 9.23). Here, both cerrado-like savannas and wet campos occur, besides gallery forests and a type of evergreen dry forest. The seasonal savannas occupy relict surfaces of an old tableland that has been preserved by hard lateritic crusts. The ground, paved with lateritic pebbles, has an open cover of grasses and sedges, with species like *Bulbostylis conifera, B. junciformis, Leptocoryphium lanatum, Paspalum carinatum, P. gardnerianum, Trachypogon* sp., etc. The isolated trees are mostly *Curatella americana, Qualea grandiflora* and *Salvertia convallariodora*. Seasonally waterlogged areas, on level sandstones, show a mosaic of open savannas (campos) and swamps. Besides grasses and sedges, these communities have many representatives of the families Rapataceae, Xyridaceae, Eriocaulaceae, Bromeliaceae and Orchidaceae. Floristic relationships may thus be established with the savannas on sandstones and those on the white sands of the interior Guiana region.

Azevedo (1967) also found a gradient of savanna formations in the Amapa Territory, in the northeastern part of the Amazon Basin (Fig. 9.23). Three main types of open vegetation occur there: cerrados, campos and seasonal swamps. The cerrado savannas have tree species like *Byrsonima crassifolia, Curatella americana, Palicourea rigida* and *Salvertia convallariodora*. The campos are treeless grasslands (hyperseasonal savannas) that, besides grasses and sedges, have many representatives of the monocotyledon families already mentioned. Seasonal swamps on flood plains (*campos de varzea*) are dominated by tall grasses and sedges, particularly by species of *Panicum* and *Paspalum*. The same three formations occur in Marajo Island nearby (Meggers and Evans, 1957).

In the upper Xingu area of Mato Grosso, Setzer (1967) reported that cerrados cover most of the interfluves, flat tablelands of about 400 m in eleva-

tion; while the flood plains, 100 m lower, are the domain of rain forests.

A mosaic of seasonal and hyperseasonal savannas has also been described in the Venezuelan Amazonas Territory. Eden (1974) reported that wooded savannas with *Curatella americana* and *Byrsonima crassifolia*, over a ground layer of *Axonopus, Leptocoryphium* and *Trachypogon*, appear as isolated patches on well-drained sites in several parts of the upper Orinoco. Waterlogged savannas dominated by species of *Andropogon, Paspalum* and *Rhynchospora* occur in poorly drained areas. Sometimes a few palms (*Mauritia flexuosa*) appear as the single woody component of these grasslands.

As may be realized from all these descriptions, similar vegetation gradients and mosaics appear in widely separated areas within the Amazonian region. Seasonal savannas occupy well-draines sites, while hyperseasonal savannas and esteros occur on poorly drained areas. Eiten (1978) believes that most of these non-forest ecosystems have to be excluded from the cerrado concept, but it must be remembered that Eiten's delimitation of the cerrado is essentially floristic. According to Eiten, only the drained savannas and the campos on well-drained soils or on soils with lateritic pebbles, contain a flora having affinities with the cerrado.

In a recent unpublished presentation covering the savannas in the Venezuelan Amazonas Territory, Huber (1979) considered three types of savannas: a llanos-like type, on hilly land with latosols; an Amazonian type on flat areas with white sands and podzolic soils; and flooded savannas on the poorly drained bottom lands. The llanos-like communities occur in small patches spread throughout the northern part of that Territory. They are tree savannas ecologically and floristically related to the savannas of the neighboring llanos. Their most common trees are *Bowdichia virgilioides, Curatella americana* and *Xylopia aromatica*, together with *Platycarpum orinocense*, a species restricted to this type of savanna in the Venezuelan Amazonia. Due to its highly permeable soil, the white sand savanna almost never remains waterlogged for long. According to Huber, this formation must be quite ancient, since in its still poorly known flora he found several new species of tropical African and Asian affinity. There are also open formations called *sabanetas*, transitional to the caatinga forest, dominated by the monocotyledon genera characteristic of white sands in the Guiana region. They remain flooded under half a meter of water over long periods.

To summarize the information presented in this short review, the same ecological and structural types already known from the llanos or from the Guianas do occur as isolated patches throughout the Amazonian region, with the possible exception of the upper Amazon in Peru and Brazil, where savannas have not yet been reported. Some new types, as yet unreported for other Neotropical regions, have also been described. The following savanna formations occur in the Amazonian region:

(1) Seasonal savannas, most of them with a structure of tree savannas. Apparently, wooded savannas or woodlands do not exist in this area. The tree savannas are quite similar to the seasonal formations occurring in the llanos, with perhaps a few additional species. The ground flora is quite similar both to that of the seasonal savannas in the llanos and to the herbaceous flora of the cerrados. The occurrence of seasonal savannas under an ever-wet climate may be explained in the following two ways. First, even in the Middle and Lower Amazonia, there are two or three months with reduced precipitations which simulate a short dry season. Second, these savannas might also represent relicts from a former extension of open formations throughout the Amazonian region. It seems logical to suppose that these relict savanna patches will tend to remain on the sites least suitable for recolonization by rain forests.

(2) Hyperseasonal savannas on periodically waterlogged lowlands. They seem to be equivalent to the wet campos of the cerrado region.

(3) Seasonal swamps, particularly of the varzea (flood-plain) type, dominated by tall grasses and sedges. They are closely related to the esteros of the llanos.

(4) Savannas on pure white sand with podzolized soils. This unit refers to a vegetation series occurring on this very special kind of substratum, that extends from a low sclerophyllous forest (Amazonian caatinga) to scrub and grass formations. Ecologically, they could be considered as semi-seasonal savannas. They extend throughout the whole area of white sand in the Amazonian and the Guianan regions without major floristic or structural changes.

These various savanna formations scattered throughout the Amazonian region constitute a link in the chain of savanna ecosystems which connects the Brazilian cerrados with the llanos and the interior Guiana savannas on one side, and with the coastal savanna region on the other side. Many authors have been reluctant to consider them as cerrados, and from a purely floristic viewpoint they are not; but they surely represent relicts from more widespread grassland formations connecting the main savanna areas of South America during earlier climatic phases (Eden, 1974; Sarmiento and Monasterio, 1975; Brown and Ab'Saber, 1979).

THE GRAN PANTANAL, THE LLANOS DE MOJOS AND THE CONNECTIONS WITH THE CHACO FORMATIONS

In the very center of the South American continent, where Brazil, Bolivia and Paraguay meet, there is a large lowland of more than 100 000 km², known as the Gran Pantanal ("great swampland"). This northernmost part of the Paraguay–Plata Basin is a tectonic depression where the Quaternary climatic fluctuations produced an intricate mosaic of actual and relictual landforms, either of alluvial or aeolian origin.

The main consequence of these changes is that the drainage network has not yet reached a state of equilibrium with the present-day climate, causing most of the area to be waterlogged during the rainy season.

The Gran Pantanal vegetation is a mosaic of forests, cerrados and campos (Valverde, 1972; Eiten, 1975). The cerrado savannas dominate the interfluves with sandy soils and a deep water table. Campos de varzea (esteros) occupy the flood plains, either as pure grasslands of tall, coarse grasses and sedges, or intermingled with palmlands of *Copernicia alba*. This is a typical formation of the Chaco lowlands that reaches its northern limit in this area. Wet campos occur on somewhat higher sites, with woody species restricted to termite mounds. Here two species of Bignoniaceae, *Tecoma* (*Tabebuia*) *caraiba* and *T. aurea*, constitute the only trees on a campo ground layer. A mosaic of cerrados, cerradãos and forest islands of Chacoan affinities occurs on higher ground.

Eiten (1975) noticed that a totally intergrading series may be formed in the pantanal from a typical cerrado to the typical pantanal landscape of extensive flat lowlands, periodically inundated or badly drained, with a grass cover and large cerrado groves on high platforms. This series, correlated with a gradient of drainage conditions, includes, besides the two extreme types, two intermediate vegetation patterns, the one previously discussed (see p.253) with small cerrado groves on raised platforms with termitaria, the other where the cerrado trees show a definite tendency to clump in circular groves while the rest of the landscape is left treeless.

Two different floristic stocks meet in the Gran Pantanal, forming mosaics of plant formations, some of them of definite cerrado affinity, others closely related to the Chaco vegetation. Upland types may be more or less clearly differentiated according to their relationships with the cerrado or the Chaco, as is the case with campos and palmlands. Hyperseasonal savannas and esteros show instead a more gradual transition. These formations, that dominate the landscape throughout the eastern border of the Paraguayan and Argentinian Chaco, have species in common with the corresponding communities in tropical America. Morello and Adamoli (1974) analyzed the distribution and the ecological characteristics of these esteros (also called *pajonales*) in the eastern Chaco. The esteros are consociations of tall and coarse species of *Paspalum* or *Panicum*, while *Sorghastrum agrostoides* is the dominant grass in hyperseasonal formations. Frequently there is a sparse cover of *Copernicia alba*.

In northern Bolivia, there is another large plain mostly covered by seasonally flooded savannas. This area, known as the Llanos de Mojos, extends between the Andes and the Brazilian Shield, being bounded to the south by the Chiquitos highlands. To the north, the imprecisely known border with the continuous Amazonian forest seems to be located near the Madre de Dios River (Fig. 10.24). According to Denevan (1966), 50% of this 180 000 km² area is grassy savanna, about 30% tree or palm savanna, and the remaining 20% is under forest.

Seasonal and hyperseasonal savannas and esteros seem to occur in the area, depending on the length of the flooding period. Savanna woodlands cover the ground rarely attained by flooding, with *Curatella americana* and *Tabebuia suberosa* as the

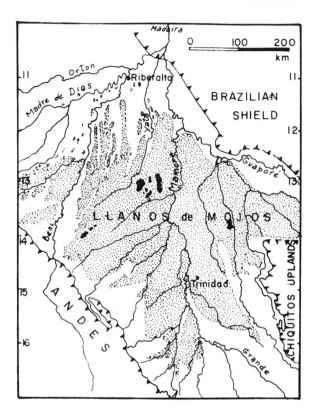

Fig. 10.24. The Llanos de Mojos in northern Bolivia. They occupy a large depression between the Andes, the western border of the Brazilian Shield and the Chiquitos Uplands (after Denevan, 1966).

two commonest trees (Denevan, 1966). This savanna type is popularly known as *arboleda*, while the name *chaparral* refers to a dense woodland with species of *Acacia*, *Cassia* and *Mimosa*. In some areas of arboledas, trees only grow on termite mounds as is the case in areas of Pantanal. Where flooding lasts for five to ten months, pure grasslands occur with species of *Leersia*, *Panicum* and *Paspalum*. *Sporobolus–Axonopus–Andropogon–Trichachne* grasslands or palm savannas with *Copernicia alba* and *Acrocomia totai* cover higher ground, with a shorter flooding period. Esteros with tall grasses and sedges occur on the ever-wet bottom lands.

Southeast of Mojos, along the continental divide between the Amazon and the Plata Basins, there is another poorly known region with various patterns of savannas and forests. Here, too, the Amazonian and Chacoan floras meet, resulting in a mosaic of formations of one type or another. According to

Hueck (1972), palm savannas with *Acrocomia totai* and *Attalea princeps* are common on dry soils near Santa Cruz, in eastern Bolivia. *Mauritia vinifera* appears instead as the characteristic palm of flooded savannas on the northern border of the Sierra de Velazco.

THE SAVANNAS OF SOUTHERN MEXICO, CENTRAL AMERICA AND THE CARIBBEAN ISLANDS

Neotropical savannas cover 20 000 km² in the States of Tabasco, Chiapas, Veracruz, Campeche, Oaxaca and Guerrero in southern Mexico (Flores Mata et al., 1971). They occur on coastal plains both of the Gulf of Mexico and the Pacific Ocean (Fig. 10.25), as well as in interior valleys, where a strong human influence makes their ecological interpretation difficult (Miranda, 1952). This same type of interior savanna extends to Honduras (Johannessen, 1963).

Puig (1972) analyzed the savannas of Tabasco, in the coastal plain of the Gulf, where he recognized four main types:

(1) An open tree savanna on upper topographic positions with ill-drained latosols. *Curatella americana*, *Byrsonima crassifolia* and the palm *Acrocomia mexicana* form the woody layer, while the commonest grasses and sedges are *Dichronema ciliata*, *Digitaria leucites*, *Paspalum pectinatum*, *P. plicatulum*, *P. punctatum*, *Trachypogon angustifolius*, etc.

(2) A denser wooded savanna, which is really a mosaic of areas of tree savanna and scrub patches.

Fig. 10.25. The major savanna regions of southern Mexico and Central America.

To the three trees mentioned above, Puig adds species of *Bauhinia*, *Calliandra*, *Calyptranthus*, *Clidemia*, *Cochlospermum*, etc. The soils are apparently similar to the previous type, but as overgrazing has prevented fires, an invasion of woody species has been induced.

(3) *Tasistales* or palm swamps, with the palm *Paurotis* (*Acoelorraphe*) *wrightii* over a grass and sedge cover of *Andropogon*, *Paspalum*, etc.

(4) *Encinares* or open woodlands of *Quercus oleoides*, with forest trees (*Vochysia*, *Terminalia*) and savanna trees (*Byrsonima*, *Xylopia*). The herb layer is poorly developed, with some grasses, sedges and legumes.

Gómez Pompa (1973), in his analysis of the vegetation of Veracruz (Mexico), recognized a special type of savanna woodland strongly related to lowland oak forest. Its tree species are *Acrocomia mexicana*, *Byrsonima crassifolia*, *Coccoloba barbadensis*, *Crescentia cujete* and *Curatella americana*. The upland savannas have many species in common with this type of savanna woodland, but they have a close ground cover of *Andropogon*, *Bulbostylis*, *Cassia*, *Dichronema*, *Paspalum*, *Rhynchospora*, etc. Finally, *Paurotis wrightii* forms pure stands of palm savannas on wet soils corresponding to the drier parts of swamps.

It is clear, then, that three types of savanna formations occur as more or less large patches in the coastal regions of tropical Mexico: a seasonal savanna on latosols with just a few tree species, apparently quite similar to other seasonal savannas of northern South America and Central America; a hyperseasonal savanna characterized by the palm *Paurotis wrightii*; and a third type not previously found elsewhere — a woodland formation apparently related to the tropical oak forests, though it is not yet clear if it should be considered as a secondary seral stage of that forest formation.

In British Honduras (Belize), Charter (1941) reported the occurrence of a series of savanna formations related to soil evolution on river terraces and on the coastal plain. At a certain stage of soil differentiation, when an impervious clay develops under a sandy layer, a *Quercus* savanna appears, later changing to a *Curatella–Byrsonima* community, and finally to a pine woodland (*Pinus caribaea*). On level terraces flooded for long periods, the vegetation is a sedge (hyperseasonal) savanna, dotted with clumps of the palm *Paurotis wrightii* and isolated trees of *Crescentia cujete* and *Cameraria belizensis*. There also are pine savannas in hilly areas on soils similar to the pine savanna soils of the plains. Charter compared these pine woodlands to those of the southeastern United States and the Bahamas.

Stretches of seasonal savannas are scattered through the Peten Province of Guatemala, either with *Pinus caribaea, Curatella americana* or *Byrsonima crassifolia* (Lundell, 1937). On the Caribbean coast of Honduras and Nicaragua, an extensive area of deeply weathered quartz gravel and sand supports a savanna vegetation. Parsons (1955) described the environment and vegetation of this Miskito savanna. It is a park-like savanna where pines are widely spaced, intermingled with a sparse tree cover of *Byrsonima crassifolia, Calliandra houstoniana, Curatella americana, Miconia* spp., and less commonly *Crescentia, Mimosa* and *Quercus*. The ground layer is rich in sedges (*Rhynchospora, Bulbostylis*, etc.) with *Trachypogon* sp. as the most important grass, followed by species of *Andropogon, Aristida, Leptocoryphium* and *Paspalum*. The highest layer, open to very open, up to 25 m high, has *Pinus caribaea* as its single species. On poorly drained soils Parsons reports the occurrence of palm groves (*Paurotis*).

Taylor (1963) analyzed these coastal pine savannas of Nicaragua, favoring the hypothesis that they represent secondary formations of the tropical rain forest, since they occur in an area of exceptionally high rainfall (2600–3500 mm, with three dry months). He pointed out that even in the area of deciduous tropical forest there are secondary savannas with *Crescentia alata* and several species of *Acacia, Haematoxylon, Pithecellobium*, etc., over a grass cover of *Aristida* and *Bouteloua*.

Seasonal savannas of the *Curatella* type have also been reported on upland soils of coastal Costa Rica and Panamá. Porter (1973) listed *Anacardium occidentale, Byrsonima crassifolia, Curatella americana, Psidium guajava, Xylopia aromatica* and *Xylopia frutescens* as the dominant trees in the Panamian savanna. Among the species of the ground layer he cites *Andropogon angustatus, Aristida jorullensis, Rhynchospora armerioides*, etc.

Thus, in several parts of Central America from Guatemala to Panama (Fig. 10.25), there are the same types of savannas already found in Mexico and northern South America. However a quite interesting new formation appears: the pine savanna, either on lowlands or highlands, where *Pinus caribaea* form a high tree layer. These tall trees are absolutely unknown in all other types of Neotropical savanna formations. Whether these pine savannas are secondary to rain forests or whether they constitute original types determined by soil evolution under particular conditions of climate and parent material, is not yet clear. A similar type of pine savanna occurs in Cuba, including the Isla de Pinos, and Hispaniola.

In Hispaniola, one of the Greater Antilles, there are savannas on the plains and plateaux (Beard, 1953). A treeless savanna covers an extensive area of the central plain with poorly drained soils. A tree savanna, either with pines (*P. occidentalis*) or with the typical orchard trees (*Byrsonima, Curatella*) covers a broken surface, remnant of a dissected plateau in the same area.

In several of the Lesser Antilles, there are reduced patches of savanna, such as the Gran Sabana of Dominica, with *Byrsonima* and *Sporobolus* (Beard, 1953; Howard, 1973). But it is in Cuba, the largest of the Antilles, that tropical savannas reach regional importance. Though there is no agreement among authors about the extent of natural savannas as opposed to man-induced open formations (Bennet, 1928; Waibel, 1943; Seifriz, 1943; Beard, 1953; Borhidi and Herrera, 1977), between 10 and 20% of the island seems to be covered by original or by semi-natural savannas (Fig. 10.26).

Seasonal savannas occur on three different types of habitats: shallow soils, often with an ironstone hardpan; quartz sands; and red soils developed on serpentine. An open formation with palms (*Sabal parviflora*) and other low trees (*Acacia, Caesalpinia, Pisonia*) occurs on shallow soils, with lateritic pebbles or hardpan (locally called *mocarrero*). The ground layer includes several species of *Andropogon, Panicum, Paspalum, Rhynchospora, Scleria* and *Setaria* (Borhidi and Herrera, 1977). Siliceous sandy soils, in western Cuba and the Isla de Pinos are covered by open woodlands of *Pinus tropicalis*, or sometimes *P. caribaea*, with scattered palms, such as *Colpothrinax wrightii, Copernicia curtissii* and *Paurotis wrightii*. The herb layer includes species of *Andropogon, Bulbostylis, Cyperus, Fimbristylis* and *Paspalum*, as well as many typical neotropical savanna species like *Leptocoryphium*

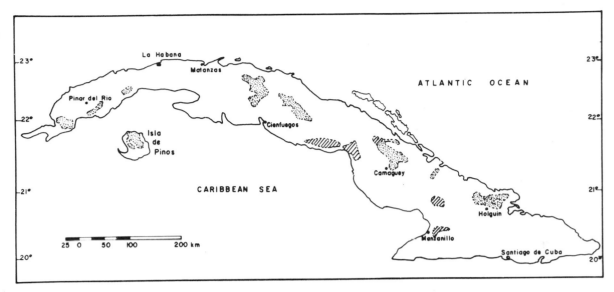

Fig. 10.26. The distribution of natural savannas in Cuba. Dotted areas correspond to seasonal savannas, hatched areas to hyperseasonal savannas. (Adapted from Borhidi and Herrera, 1977.)

lanatum, *Mesosetum loliiforme* and *Trachypogon filifolius*. Serpentine soils typically maintain a tree savanna that has as a distinctive physiognomic feature the abundance of low palms, like *Coccothrinax miraguama*, *C. pseudorigida*, *Copernicia macroglossa*, *C. pauciflora*, *C. ramosissima*, *C. yarey*, etc. (Borhidi and Herrera, 1977). Other trees are *Byrsonima crassifolia*, *Curatella americana*, *Rondeletia correifolia*, *Tabebuia lepidophylla*, *T. lepidota*, etc. The ground layer has several species of *Andropogon* and *Aristida*, besides *Imperata brasiliensis*, *Leptocoryphium lanatum* and other widespread Neotropical savanna grasses and sedges.

On the highlands of the Sierra de Nipe in eastern Cuba, at 400 to 600 m above sea level, Carabia (1945) reported open pine woodlands on limonitic soils derived from serpentine. According to him, grasses and sedges represent the true dominant species in these *pinares*. *Pinus cubensis*, a tree up to 25 m high, is the single tree species, with a density of about 200 trees ha^{-1}. A similar community occurs in the same area, but without pines. Carabia suggested that these pineless savannas are fire-induced.

Hyperseasonal savannas occupy hydromorphic soils on the flood plains of several of the main rivers of the island. Several palms of the genus *Copernicia* and various shrubby legumes of the genera *Acacia*, *Caesalpinia* and *Cassia* form the open woody layer, while the ground vegetation includes species of *Andropogon*, *Aristida*, *Bulbostylis* and *Rhynchospora*. Seasonal swamps have two distinctive palms: *Paurotis wrightii* and *Sabal parviflora*. In the grass and sedge layer, up to 2 m high, the more frequent species are *Cladium jamaicense*, *Hymenachne amplexicaulis*, *Panicum virgatum* and several species of *Cyperus*, *Eleocharis*, *Fimbristylis* and *Rhynchospora*.

In summary, in the Caribbean area, particularly in the two major islands Cuba and Hispaniola, seasonal and hyperseasonal savannas, as well as seasonal swamps, occupy substantial areas. Some of these savanna formations are floristically quite similar to continental savannas, either considering the tree or the ground flora. But it is in this insular region where pine savannas, also occurring in Central America, reach their greatest diversification, with various lowland or medium-altitude types occurring on different soil types, either silicious or derived from serpentine. Another point to note about Cuban savannas is the high degree of endemism shown by this flora. This fact appears clearly in palms, where several genera reach a remarkable specific diversification within these savanna formations.

THE FLORA OF THE NEOTROPICAL SAVANNAS: ITS SPECIFICITY AND AFFINITIES

The savanna formations in tropical America, as every series of plant formations spreading over an extensive geographical range, include a complex mixture of floristic elements of various provenance, age and affinity. This fact follows from the continuous paleogeographical and palcoecological changes that have taken place in the two continents of North and South America and their related islands. Since the Middle Mesozoic, when the Atlantic Ocean began to spread, pushing apart the continental masses hitherto united in a single supercontinent, the South American and to a lesser degree the Central American flora began to evolve more or less isolated from each other and from other floristic realms. In the Late Cenozoic their evolution culminated under the impacts of the cataclysmic changes induced by the still active Andean orogeny, the periodic climatic oscillations of the Glacial Ages, and their associated fluctuations in sea level. These events influenced decisively the present-day distribution of plant formations and their corresponding floras and faunas. [See Simpson Vuilleumier (1971), Van der Hammen (1974), and Brown and Ab'Saber (1979) for a general consideration of the paleogeographical evolution of South America and its ecologic consequences.]

The two major vegetation types of tropical lowlands in the Americas, rain forests and savannas, seem to have suffered successive expansions and contractions in their areas during the alternating dry and wet climatic phases. During the Last Glacial a dramatic constriction of rain forests took place, with a parallel expansion of savannas and to a lesser degree of thorn forests as well. [Ab'Saber (1977) presented a paleogeographic map with the natural domains of South America between 13 000 and 18 000 years ago.]

These frequent displacements of whole biomes along continental distances provided ample opportunities for speciation in isolation in forest and savanna refuges, during the drier and more humid climatic phases respectively. Furthermore, these disruptions of the vegetation equilibrium also facilitated wide interchanges between the floristic stocks of the various interacting and interconnecting plant formations. In this way one may suppose that a vast genetic flow took place, along with the diversification of characteristic taxa in each ecosystem at various moments of these cycles.

In spite of the resulting inherent heterogeneity in age, origin and evolutionary history of the component species, a characteristic savanna flora evolved in tropical America. In fact, the overall floristic picture provided by these formations is that of a rich flora in which at least half of the species appear to be largely restricted to savanna ecosystems, while the other half consists of more ubiquitous components. A first distinction can thus be made at the species level between a floristic stock mostly exclusive to Neotropical savannas, and the remaining alien species representing a more or less opportunistic, occasional and often more recent component of this flora.

One further distinction has to be made within the Neotropical savanna flora between the woody and half-woody elements on the one side and the herbaceous species on the other side, since the two groups seem to represent rather divergent evolutionary situations. Four different groups must therefore be considered separately: a cohort of woody species exclusive to these savannas; a numerous group of trees closely related to rain-forest species; a herbaceous and subshrubby flora peculiar to savannas; and an accesory non-woody flora without much ecological specificity arriving in these ecosystems from different sources.

In order to go more deeply into the analysis of the affinities of the Neotropical savanna flora, attention must now be shifted to the main patterns of distribution of plant genera within and between savanna formations. Genera are a more adequate analytical tool for providing a wider evolutionary perspective of floristic changes and relationships.

Among the woody species peculiar to Neotropical savannas are a number of taxa, phylogenetically more or less isolated, that do not have non-savanna species in the same genus. Table 10.7 lists some of these species. This list is rather limited, but several of the species are quite abundant and frequent savanna trees. All of them occur in the cerrados, though several extend to other formations. *Cochlospermum* has two Neotropical species, one *C. regium* is a half-woody plant exclusive to the cerrado, the other *C. vitifolium* is a tree occurring in savannas of northern South America and Central America and in seral stages of tropical deciduous

TABLE 10.7

Woody species belonging to the characteristic floristic element of Neotropical savannas (isolated species, sometimes monotypic)

Species	Family
Antonia ovata	Loganiaceae
Austroplenckia populnea	Celastraceae
Bowdichia virgilioides	Fabaceae
Bowdichia major	Fabaceae
Cochlospermum regium	Cochlospermaceae
Cochlospermum vitifolium	Cochlospermaceae
Curatella americana	Dilleniaceae
Diptychandra glabra	Caesalpiniaceae
D. aurantiaca	Caesalpiniaceae
Hancornia speciosa	Apocynaceae
Harpalyce brasiliana	Fabaceae
Magonia pubescens	Sapindaceae
Magonia glabrata	Sapindaceae
Pamphilia aurea	Styracaceae
Pterodon pubescens	Fabaceae
Pterodon polygalaeflorus	Fabaceae
Riedeliella graciliflora	Fabaceae
Salvertia convallariodora	Vochysiaceae
Spiranthera odoratissima	Rutaceae

forests as well. *Bowdichia virgilioides*, *Curatella americana*, *Hancornia speciosa*, *Magonia pubescens* and *Salvertia convallariodora* occur to a varying extent outside the cerrado region. It may be postulated that their taxonomic isolation reflects an evolution within these ecosystems which has continued long enough to allow a differentiation at the generic level, while the lack of diversification may suggest an advanced specialization. They probably represent a floristic paleoelement well adapted to the savanna environment.

There is another small group of genera of woody plants in which, even though each genus has several rain-forest species, many of their species belong exclusively to the savannas. Among the genera in this group may be mentioned *Anacardium*, *Bombax*, *Byrsonima*, *Kielmeyera*, *Stryphnodendron* and *Sweetia*. Several species in these genera are quite common savanna trees, like *Byrsonima crassifolia* and *B. coccolobaefolia* throughout the South American savannas, or *Kielmeyera coriacea* and *Stryphnodendron barbadetimam* in the cerrados. The genus *Bombax* has even reached a high degree of diversification in the cerrado, with more than ten species.

These two groups of genera of woody plants together constitute the characteristic floristic element of Neotropical savannas, either by their exclusive occurrence or by their greater evolutive radiation in these ecosystems. As may be seen in Chapter 5 on growth forms, a hypothetical chain of architectural and functional changes may have derived from ancestral forest trees some half-woody or even subterranean tree forms peculiar to savannas, such as *Anacardium humile* or *Byrsonima verbascifolia*.

A different situation occurs in many woody genera that have diversified mostly in the Amazonian or the Atlantic rain forests where they have many species. But in these genera, one or two species, quite closely related morphologically and taxonomically to some of the rain-forest representatives, occur exclusively in the cerrados. These savanna vicariants of rain-forest species constitute an important component of the rich woody flora of the Brazilian cerrados (Table 10.8). A comparable situation does not occur elsewhere. Several genera of forest trees, like *Hymenaea*, *Copaifera* or *Vochysia*, are common to both savanna and forest, but their representatives in the two ecosystems are not closely related.

Apparently the Brazilian rain forests have been the ecosystems where all these genera accomplished their greatest evolutionary radiation. The very fact of the existence of closely related species in the neighboring savannas suggests a long interplay between both types of ecosystems, as well as a relatively more recent speciation of the representatives in the savannas. One may speculate that the savanna vicariant species probably originated during the glacial periods of rain-forest contractions when savannas spread over large areas of the tropical American lowlands. The occurrence in some cases of a diversification below the species level suggests that this process is still actively proceeding (Table 10.8).

Another situation is represented by many genera of wide distribution in tropical areas which have representatives in several plant formations, including a few savanna species (Table 10.9). These species may be considered part of a less specific floristic stock able to adapt to many different ecological situations, including those of tropical savannas.

To summarize this discussion about the phyto-

TABLE 10.8

Some examples of pairs of closely related species (vicariants), one growing in
the Brazilian rain forests, the other in the cerrados (from Heringer et al., 1977)

Tropical rain forests (Amazonian and Atlantic)	Cerradão and cerrado
Aegiphila arborescens Vahl	*A. lhotzkyana* Cham.
Agonandra silvatica Ducke and	
Agonandra brasiliensis Miers f. *silvestre*	*A. brasiliensis* Miers f. *campestre*
Andira retusa H.B.K.	*A. humilis* Mart.
Aspidosperma duckei Huber and	
Aspidosperma pallidiflorum M. Arg.	*A. tomentosum* Mart.
Brosimum discolor Schott.	*B. gaudichaudii* Trec.
Callisthene dryadum A. Duarte	*C. fasciculata* (Spr.) Mart.
Caryocar villosum (Subl.) Pers.	*C. brasiliense* Camb.
Cenostigma tocantinum Ducke	*C. gardnerianum* Tul.
Connarus cymosus Planch.	*C. suberosus* Planch.
Copaifera lucens Dwyer	*C. langdorffii* Desf.
Copaifera trapezifolia Hayne	*C. oblongifolia* Mart.
Dalbergia nigra Fr. All.	*D. violacea* (Vog.) Malme
Dalbergia foliolosa Benth.	*D. spruceana* Benth.
Dimorphandra parviflora Benth.	*D. mollis* Benth.
Dioclea megacarpa Rolfe	*D. erecta* Hoehne
Diospyros hispida DC	*D. hispida* var. *camporum* Warm.
Emmotum glabrum Benth.	*E. nitens* (Benth) Miers
Enterolobium contortisiliquum (Vell.)	
Morong	*E. gummiferum* (Mart.) Macbr.
Erythrina verna Vell	*E. mulungu* Mart.
Ferdinandusa speciosa Pohl	*F. elliptica* Pohl
Hymenaea altissima Ducke and	
Hymenaea stilbocarpa Mart.	*H. stigonocarpa* Mart.
Kielmeyera excelsa Camb.	*K. petiolaris* (Spr.) Mart.
Lafoensia glyptocarpa Koehne	*L. densiflora* Pohl
Machaerium villosum Vog.	*M. opacum* Vog.
Maprounea guianensis Aubl.	*M. brasiliensis* St. Hil.
Mimosa obovata Benth.	*M. laticifera* Rizz. & Matt.
Peschiera affinis (M. Arg.) Miers	*P. affinis* var. *campestris* Rizz.
Piptadenia peregrina (L.) Benth.	*P. falcata* Benth.
Plathymenia foliolosa Benth.	*P. reticulata* Benth.
Psittacanthus decipiens Eichl.	*P. robustus* Mart.
Qualea jundiahy Warm.	*Q. multiflora* Mart.
Sclerolobium rugosum Mart.	*S. aureum* (Tul.) Benth.
Stryphnodendron polyphyllum Benth.	*S. barbatimam* (Veu.) Mart.
Swartzia macrostachya Benth.	*S. grazielana* Rizz.
Tabebuia chrysotricha (Mart.) Standl.	*T. ochracea* (Cham.) Standl.
Terminalia hylobates Eichl.	*T. argentea* Mart & Zucc.
Tragia amoena M. Arg.	*T. lagoensis* M. Arg.
Vochysia tucanorum (Spr.) Mart.	*V. thyrsoidea* Pohl.
Zeyhera tuberculosa (Vell.) Bur.	*Z. digitalis* (Vell.) Hoehne

geographical significance of the woody flora in the
American savannas, one may first point out that the
overall pattern of speciation in this flora suggests a
persistent contact between rain forests and savannas, with a heavy interchange of floristic stocks
from the former to the latter over a rather extended
period. In this way the Amazonian and the Atlantic
rain forests appear as the main sources nourishing
primarily the cerrados and through them the other
American savannas. But evidently this has not been

TABLE 10.9

Woody genera with most of their species in rain forests but with a few savanna species

Agonandra (Opiliaceae)	*Ilex* (Aquifoliaceae)
Andira (Fabaceae)	*Jacaranda* (Bignoniaceae)
Annona (Annonaceae)	*Laplacea* (Theaceae)
Aspidosperma (Apocynaceae)	*Licania* (Rosaceae)
Astronium (Anacardiaceae)	*Lonchocarpus* (Fabaceae)
Bauhinia (Fabaceae)	*Matayba* (Sapindaceae)
Blepharocalyx (Myrtaceae)	*Ouratea* (Ochnaceae)
Brosimum (Moraceae)	*Palicourea* (Rubiaceae)
Calliandra (Fabaceae)	*Platycarpum* (Rubiaceae)
Caraipa (Hypericaceae)	*Pinus* (Pinaceae)
Caryocar (Caryocaraceae)	*Qualea* (Vochysiaceae)
Chaunochiton (Olacaceae)	*Quercus* (Fagaccac)
Clusia (Hypericaceae)	*Roupala* (Proteaceae)
Diospyros (Ebenaceae)	*Symplocos* (Symplocaceae)
Esenbeckia (Rutaceae)	*Vochysia* (Vochysiaceae)
Ficus (Moraceae)	*Xylopia* (Annonaceae)
Hirtella (Rosaceae)	

TABLE 10.10

Some characteristic herbaceous species in Neotropical savannas

Axonopus (Poaceae)	*Lagenocarpus* (Cyperaceae)
Brasilia (Asteraceae)	*Leptocoryphium* (Poaceae)
Bulbostylis (Cyperaceae)	*Mesosetum* (Poaceae)
Curculigo (Amaryllidaceae)	*Orthopappus* (Asteraceae)
Chaetium (Poaceae)	*Polycarpaea* (Caryophyllaceae)
Ctenium (Poaceae)	*Rhynchospora* (Cyperaceae)
Diectomis (Poaceae)	*Thrasya* (Poaceae)
Echinolaena (Poaceae)	*Trachypogon* (Poaceae)
Gymnopogon (Poaceae)	*Tristachya* (Poaceae)
Hoehnephyton (Asteraceae)	*Torresea* (Fabaceae)
Kyllinga (Cyperaceae)	

the only floristic source of the savanna flora; other tropical, subtropical and even temperate formations have also contributed, as seems to be indicated by the distribution of numerous genera listed in Table 10.9. Moreover, in the case of the northernmost formations occurring in Mexico, Central America and the Caribbean, some boreal genera have contributed a few species to the peculiar flora of the savannas, as for instance, some large temperate genera like *Quercus* and *Pinus*.

The case appears somewhat different with the herbaceous and subshrubby savanna flora. A few of these genera are more or less restricted to tropical American savannas or to tropical savannas in general. This is the case with certain small genera, either monotypic or with few species, like *Brasilia*, *Orthopappus* and *Hoehnephyton* among the Asteraceae, *Diectomis* in the grasses and *Torresea* in the legumes. Other genera show a wider diversification in Neotropical savannas, like the grasses *Axonopus*, *Mesosetum* and *Trachypogon*, or the sedges *Bulbostylis*, *Lagenocarpus* and *Rhynchospora* (Table 10.10).

However, these two groups of herbaceous genera concentrated in tropical savannas only include a minority of the herbaceous savanna flora. Most of the remaining species, composing by far the bulk of this flora, belong to widespread and non-specialized taxa whose species occur in almost every type of ecosystem, though with a concentration in seral communities, transitory niches and disturbed habitats. These genera undoubtedly represent more opportunistic evolutionary strategies; thanks to this behavior they have been able to occupy the widest geographical ranges and to show the richest diversification. Table 10.11 lists some of these taxa, all of them with 100 or more species, only a few of which are peculiar to Neotropical savannas. Many of these species appear among the codominants in the herb layer of every Neotropical savanna. Nevertheless, it is a rather difficult task to identify with any acceptable degree of certainty the origin and affinities of this more opportunistic floristic stock. Because of the lack of adequate factual data, this point remains for the moment highly speculative.

If one's purpose were to elucidate the center of origin and diversification of the herbaceous savanna flora as a means of reconstructing its history, a more suitable research object could be certain monotypic elements restricted to, but of widespread occurrence, in these formations — like *Leptocoryphium lanatum* — whose evolutive steps might be followed through the morphological, functional and biochemical variability of its populations among the different ecosystems where it occurs. The same may be said of some other small grass genera, such as *Ctenium*, *Echinolaena*, *Thrasya* and *Trachypogon*.

The palms constitute a particular case within the savanna flora since a few genera, like *Coccothrinax*, *Colpothrinax*, *Copernicia* and *Mauritia*, have attained a significant diversification within savanna ecosystems as characteristic elements of hyperseasonal savannas and esteros (Table 10.12). The

TABLE 10.11

Some large plant genera having some representatives in Neotropical savannas (total number
of species and distribution after Willis, 1973)

Species	Family	Total number of species	Distribution
Aeschynomene	Fabaceae	150	tropical and subtropical
Andropogon	Poaceae	113	tropical and subtropical
Aristolochia	Aristolochiaceae	350	tropical and subtropical
Aster	Asteraceae	500	cosmopolitan
Baccharis	Asteraceae	400	American
Borreria	Rubiaceae	150	warm
Cassia	Fabaceae	600	tropical and warm temperate
Crotalaria	Fabaceae	550	tropical and subtropical
Croton	Euphorbiaceae	750	tropical and subtropical
Cuphea	Lythraceae	250	American
Desmodium	Fabaceae	450	tropical and subtropical
Eriosema	Fabaceae	140	tropical and subtropical
Eupatorium	Asteraceae	1200	cosmopolitan
Euphorbia	Euphorbiaceae	2000	cosmopolitan
Evolvulus	Convolvulaceae	100	tropical and subtropical
Galactia	Fabaceae	140	tropical and subtropical
Gomphrena	Amaranthaceae	100	American
Hyptis	Lamiaceae	400	warm American
Indigofera	Fabaceae	700	warm
Lantana	Verbenaceae	150	tropical, American, African
Lippia	Verbenaceae	220	tropical, American, African
Mimosa	Mimosaceae	500	tropical and subtropical
Pavonia	Malvaceae	200	tropical and subtropical
Phaseolus	Fabaceae	240	tropical and subtropical
Polygala	Polygalaceae	600	cosmopolitan
Rhynchosia	Fabaceae	300	tropical and subtropical
Salvia	Lamiaceae	700	tropical and temperate
Senecio	Asteraceae	3000	cosmopolitan
Sida	Malvaceae	200	warm
Stachytarpheta	Verbenaceae	100	American
Stevia	Asteraceae	150	tropical and subtropical American
Tibouchina	Melastomataceae	200	tropical American
Vernonia	Asteraceae	1000	cosmopolitan
Viguiera	Asteraceae	150	warm American

TABLE 10.12

Palm genera with representatives in Neotropical savanna
formations

Acanthococos	*Copernicia*
Acrocomia	*Diplothemium*
Astrocaryum	*Mauritia*
Attalea	*Orbignya*
Bactris	*Paurotis*
Butia	*Sabal*
Coccothrinax	*Syagrus*
Colpothrinax	

genus *Copernicia*, for instance, has several South
American species extensively occurring in tropical
savannas and extending to dry forests and subtropical grasslands. Thus *C. alba* occurs in the
subtropical grasslands of the Gran Chaco; *C.
cerifera* in the dry formations of northeastern Brazil
(caatingas); *C. tectorum* and *C. sanctae-martae* in
northern South American savannas. But this genus
has attained a most remarkable speciation within
the island of Cuba, where about ten species occur in
its savannas. Moreover the Cuban savannas and
the cerrados are the two Neotropical areas where

savanna palms show a particularly rich diversification, often with quite peculiar growth forms like the acaulescent dwarf palms (see Ch.5). On the other hand, palms appear among the primeval floristic elements of Neotropical savannas. Van der Hammen (1972) reported *Mauritia* pollen in the Paleocene of northern South America as an indication of the early occurrence of wet savanna-like formations in this area. Other palms, such as species of *Astrocaryum*, *Syagrus* and some other genera, appear instead as more recent additions to the savanna flora from rain-forest stocks, where these genera have a large number of species.

Another interesting point to notice concerning the relationships of the Neotropical savanna flora is its almost total lack of affinity with some nearby dry floras. In the cerrado, for instance, it is surprising that, even though this formation borders the dry caatinga for thousands of kilometers, all along this border the two types remain quite distinct, either floristically or in vegetation structure. Though certain transitional communities exist (Eiten, 1972), they seem to be rather impoverished types of each formation than ecotones where cerrado and caatinga elements mix with each other. Similarly, a savanna thorn forest contact occurs in Venezuela between the llanos and the semi-arid and arid formations of the Caribbean coast (Sarmiento, 1976), but the distinction between them always remains clear-cut. Floristically the number of common species is almost nil, while the genera common to both are large genera of the type referred to previously as unspecialized or ubiquitous, like *Cassia*, *Croton*, *Gomphrena*, *Mimosa* and *Sida*. Quite exceptionally an element characteristic of the dry floras enters into the savannas, such as one species of *Plumeria* in the Rio Branco–Rupununi savannas or *Prosopis juliflora*, a ubiquitous tree in areas with seasonal water shortage, that may occasionally enter the savannas along the savanna/thornscrub border in northern South America. Similarly a few floristic elements common to savannas and arid formations are really rain-forest genera that not only have a few savanna representatives, but also extend to dry vegetation types. Among them are *Aspidosperma*, *Tabebuia*, and a few other woody genera. Typical elements of arid formations, like the Cactaceae and Agavaceae, are almost unrepresented in savannas. Thus, the only savanna cactus, *Cereus jacamaru*, does not have close relationships with species of dry formations, but with species of tropical deciduous and subtropical forests.

A particular case worth separate consideration concerns communities occurring on pure white sand, either in the Amazonian region (*campinas* or Amazonian caatingas) or in the Guianas. These white sand formations — bush, open woodlands and savannas — have a peculiar woody flora, in which, besides a few common savanna species that sometimes may occur, there is a new floristic element, unrelated to the savanna flora of other formations. This characteristic component shows the closest affinity with the flora of the Guiana highlands (Maguire, 1971), mainly with species typical of the scrub, low forest, and herbaceous swamps that characterize the high sandstone plateaux of the Roraima Formation (tepuys). Steyermark (1966, 1967) analyzed these isolated highland formations from both floristic and phyto-geographical viewpoints. He stressed not only the high degree of endemism within this flora, but also the relationships with the Amazonian lowlands, particularly with the open formations referred to previously.

A whole set of peculiar plant families, particularly among the monocotyledons, characterizes the tepuys as well as the open lowland communities. Among the herbaceous flora may be mentioned species in the families Bromeliaceae, Cyperaceae, Eriocaulaceae, Marantaceae, Orchidaceae, Rapataceae, Thurniaceae, and Xyridaceae. The most characteristic genera among the woody flora are: *Antonia*, *Clusia*, *Cupania*, *Humiria*, *Matayba*, *Rapanea*, *Ternstroemia*, etc. The grass family, on the other hand, is conspicuously underrepresented, even in the herbaceous formations, and for this reason it is hard to apply to most of them the name of savannas, the denomination of campinas being preferable (Lisboa, 1975).

Huber (1979) stressed the floristic peculiarities of the Amazonian-type savannas in the Venezuelan Amazonas Territory, where several highly interesting floristic connections with Paleotropical areas are becoming evident with the advancing knowledge of their flora, suggesting the considerable antiquity of this particular kind of savanna.

Outside the white sand areas, the plant communities exhibiting the greatest number of Guianan elements undoubtedly are certain types of esteros and morichales occurring throughout tropical America on permanently wet, sandy soils (Aristiguieta, 1969; Egler, 1960). When discussing the origin and evolution of the savanna flora in Surinam, Van Donselaar (1965) suggested that most of its species originated from an ancestral savanna stock living on the high plateaus of the Roraima Formation and migrating to and evolving in the lowlands as the plateaus were lifted, broken and isolated from one another. The small southern Surinam savannas have a strong floristic bond with those of Northern Surinam, and also with those of central Amazonia (Van Donselaar, 1968).

The flora of the savannas on siliceous sands in the Caribbean and Central America (eastern Cuba, Isla de Pinos, Belize, Honduras, etc.), though it also appears quite rich in exclusive species, has a definite boreal affinity as indicated not only by the pines but also by numerous shrubs and subshrubs (Seifriz, 1943; Carabia, 1945). In Cuba these savannas also have certain remarkable endemics like the dwarf cycads of the genus *Zamia*.

A final point to mention refers to the genera shared in common by American and African savannas. Disregarding the man-induced recent acquisitions of the Neotropical savannas, there are several typical genera common to both areas. Outstanding examples are the grass genera *Ctenium*, *Sorghastrum*, *Trachypogon* and *Tristachya*, as well as certain other genera in various families. These continental disjunctions at the generic level sharply point out the antiquity of savanna-like formations. Some kinds of proto-savannas must certainly have been in existence before the splitting apart of South America and Africa, since the alternative hypothesis of an independent parallel evolution with a later adaptation to savannas in both sets of species seem much harder to maintain. In any case, it is interesting to notice that these disjunctions only occur in herbaceous genera, the woody savanna component having evolved independently in each continent. A corollary is that the primeval savannas were mostly herbaceous or with a poor woody flora.

FLORISTIC DIVERSITY IN NEOTROPICAL SAVANNAS

The seasonal savannas of the cerrado region have one of the richest savanna floras in the world. This is particularly true of the woody flora. Already in the last century, Warming (1892) gave a list of more than 700 cerrado species for the area of Lagoa Santa in Minas Gerais alone. Eiten (1963) analyzed the cerrado flora in a restricted region in eastern São Paulo, at 22°S — that is, near the southern border of this formation. In an area of about 50 km^2 he recorded 237 species, including several weeds but excluding many of the common woody species because they were not in bloom when the area was visited. Rizzini (1963b) gave the first comprehensive floristic list for the whole cerrado region, which included 537 species of trees and shrubs alone. In this flora Rizzini differentiated between forest species (227), cerrado elements (226) and species from the campos (84), thus making distinctions between the proper and the alien flora as well as between the woody species of the more closed-canopy types of cerrados and those species exclusive of the open campos. Goodland (1970) gave a list of 600 species and 336 genera in the flora of an area of about 15 000 km^2 in western Minas Gerais, in the core area of the cerrados. This list includes 73 species of grasses. Rizzini (1971) enlarged his list of cerrado woody plants to 653, and in the latest inventory of the flora in the whole cerrado region (Heringer et al., 1977) a total of 774 woody species are recorded, with a complementary list of 127 herbaceous and subshrubby species, 108 grasses and 54 orchids — that is, an overall figure of 1063 species (Table 10.13), 718 of which pertain to the floristic element specific to savannas.

As may be realized from these figures, the woody flora of the cerrados, referring exclusively to the seasonal savannas in this area, appears to be impressively diversified. According to Heringer et al. (1977), of the 774 woody species, 429 compose the proper floristic stock of the savannas, 300 species belong to forest formations and the remaining 45 to other vegetation types. This figure of 429 woody savanna species is not approached by any other savanna flora in the world. On the other hand, the herbaceous flora, including the grasses, is not so rich as the woody one, with something more than 300 listed species of which about one third are grasses. These figures are much lower than those of temperate South American grasslands (*pampas*) where the number of grass species is well above 400, according to the floras of Burkart (1969) and Cabrera (1970).

TABLE 10.13

Floristic richness of various Neotropical savanna formations

Formation	Area (km^2)	Number of trees and shrubs	Number of subshrubs, half-shrubs, herbs, vines, etc.	Number of grass species	Total number of species
Cerrado in northwestern São Paulo (Eiten, 1963)	50	45	175	17	237
Cerrado in western Minas Gerais (Goodland, 1970)	15 000	~200	~330	73	~600
Whole cerrado region (Heringer et al., 1977)	2 000 000	429 (774)[1]	181	108	718 (1063)[1]
Rio Branco savannas (Rodríguez, 1971)	40 000	40	87	9	136
Rupununi savannas (Goodland, 1966)	12 000	~50	291	90	431
Northern Surinam savannas (Van Donselaar, 1965)	~3000	15	213	44	272 (445)[2]
Central Venezuelan llanos (Aristeguieta, 1969)	3	69 (16)[3]	175	44	288
Venezuelan llanos (Ramia, 1974a)	250 000	43	312	200	555
Colombian llanos (FAO, 1966)	150 000	44	174	88	306

[1] Total flora including other plant formations.
[2] Total flora including bushes.
[3] Number of savanna trees excluding groves.

Similarly, the flora of the wet campos, either within the cerrado region or in neighboring areas of Amazonia, appears as relatively much less diversified. Eiten (1963) recorded 108 species in this community type in northwestern São Paulo, while Andrade Lima (1959) for the savannas near Monte Alegre and Egler (1960) in his floristic inventory of the Ariramba campos, both areas in lower Amazonia, each record about 300 species, including 26 grasses, but these inventories correspond to several formations occurring in those small areas: wet campos, esteros, cerrados and campinas. Rodríguez (1971) gave a first list, certainly far from comprehensive, of the flora of the Rio Branco savannas in the Roraima Territory of northern Brazil. His list includes a total of 136 species: 87 herbs and subshrubs, 9 grasses and 40 woody species. Incidentally, it is interesting to note how the woody flora of these savannas, though not at all comparable in richness to the cerrado woody flora, seems to be at least equally or perhaps even more varied than the tree and shrub flora of the Orinoco llanos.

Goodland (1966) presented a more complete inventory of the flora in the Rupununi savannas of Guyana, an extension of the Rio Branco savannas in the same country. There are a total of 398 herbaceous and subshrubby species, including more than 90 grasses, together with 33 trees — that is, a global figure of 431 species. Thus, this rather well-known formation has a rich herbaceous flora, but the number of woody species remains much lower than those recorded for the cerrados.

In the small savanna area of northern Surinam (coastal savannas), Van Donselaar (1965) records a list of 272 species, including 44 grasses and 15 trees and shrubs. These savannas, though diversified into several community types are floristically rather poor, particularly in respect of woody elements, but if the shrub communities on white sand were also taken into account, the total number of species increases to 445, with a significantly larger number of trees (Table 10.13). The small southern Surinam savannas include 314 species collected so far (Van Donselaar, 1968).

In the seasonal savannas of the central Venezuelan llanos, Aristiguieta (1966) reported 54 tree species, 15 shrubs, 35 vines, 44 grasses and 140 herbs and subshrubs, making a total of 228 species, for the 250 ha of the Los Llanos Biological Station. This is a long list for such a small area. However, most trees, shrubs and vines belong to the groves (*matas*) scattered within the savanna in the characteristic pattern called "savanna parkland". There are only 22 tree species in the actual savanna of this biological station.

Most other areas within the Venezuelan llanos are still poorer in woody species. In his preliminary account of the flora of the whole Venezuelan llanos, Ramia (1974b) listed 555 species, including in this figure any type of savanna as well as many widespread weeds. The grass family is the largest, with 200 species; trees and high shrubs including palms, total 43 species. In the Colombian llanos (FAO, 1966), a total of 88 grasses, 30 sedges, 144 herbs and subshrubs, and 44 trees, palms and shrubs are listed, for a grand total of 306 species in this savanna flora. But this was not a complete floristic inventory, which explains the lower number of species in comparison with the nearby Venezuelan llanos.

The floristic knowledge of savanna formations in Central America, Mexico and the Caribbean Islands still remains too fragmentary to make possible comparisons with South American formations. The only generalization possible is that, on the basis of actual knowledge, Cuban savannas seem to be richer than the other savannas in the area. This fact may be due to a greater ecological diversification on several different types of parent

material including some unusual substrata such as silicious sands and serpentine rocks which are responsible for an endemic flora with remarkably high endemism (Seifriz, 1943; Carabia, 1945).

The various formations of hyperseasonal savannas and esteros are much more difficult to compare with each other, because of their more patchy distribution at a large ecological scale and their more continuous area at a small, continental scale. In fact, the fragmentation of these formations on the landscape renders dubious a comparison between vegetation types divided into patches of every possible size. On the other hand, the distribution of the habitats of these formations along rivers and bottom lands leads to greater floristic continuity without significant gaps that might induce great differences in composition. If only one particular type of community is considered, as for example the morichales in the llanos (Aristiguieta, 1969), a list of 193 species is obtained including 49 grasses and 21 woody species. These figures compare well with those reported for the wet Brazilian campos (Eiten, 1963).

To sum up all this information on floristic richness of different Neotropical savanna vegetations, it seems that the seasonal savannas have reached their highest floristic diversification in the cerrado area. This floristic richness heavily depends on the number of woody species, while the herbaceous and subshrubby flora does not seem to be more diversified in this formation than in other American areas. This conclusion could be biased by an imperfect knowledge of the herbaceous element, but though this factor may somewhat alter the previous figures, even in well-known families like the grasses the total number of species in the cerrados is scarcely one half that of the grass flora of the Venezuelan llanos. This latter region of northern South America harbors the richest herbaceous savanna flora, while on the other hand its list of woody species is surprisingly reduced. The Rupununi savannas in Guyana, located between the two areas mentioned, maintain an intermediate position concerning floristic diversity, since their herbaceous and woody floras show figures between those of the cerrados and the llanos, particularly taking into account the small area occupied by this Guyanan formation.

A possible explanation of these contrasting gradients of impoverishment in woody species and enrichment in herbaceous elements as one passes from the cerrado to the llanos, may rely on the more continuous interplay between savannas and rain forests in the Brazilian area, with a fluctuating record of replacements and displacements all along their evolutionary and paleogeographical history. The Orinoco seasonal savannas, on the other hand, besides being geologically younger, have been almost encircled by drier lowland formations, like tropical deciduous and thorn forests, or by middle-altitude montane rain forests that have a totally different floristic stock adapted to a wholly different set of environmental conditions.

SUMMARY AND CONCLUSIONS

A wide variety of savanna ecosystems exists in tropical America. These ecosystems may be classified into several ecological and physiognomic types, or an ordination may be made along two main axes of variation: one structural, the other ecological. The structural gradient shows a continuous variation between two extreme types: a treeless grassland and a closed woodland that looks very much like a low, sclerophyllous forest formation. The other axis relates to seasonality gradients. It has semi-seasonal savannas as one extreme, where ever-wet climates determine a feeble seasonal water stress. The gradient continues with seasonal savannas, under climates or conditions leading to a rather long period of water shortage; and then with formations subjected to alternate periods of water deficiency and waterlogging during each annual cycle (hyperseasonal savannas). Finally, there are those ecosystems that remain under conditions of excessive soil water during a major part of the year (esteros), approaching the situation of permanently waterlogged swamps.

The occurrence of each type depends therefore both on climate and on topographic situation, but the parent material may also be important, particularly when it influences soil drainage, as is the case with coarse white sands. The physiognomic types depend more on soil fertility, depth of water table, occurrence and depth of hard plinthite, fire frequency, etc.

In each region the same savanna types appear on similar sites, generally disposed along equivalent environmental gradients or topographic catenas.

All Neotropical savannas also share a common floristic stock. Many species occur in similar formations of neighboring areas, while a few extend practically throughout tropical America, occurring in related savanna types anywhere in the Neotropics. This is the case, for instance, for trees like *Bowdichia virgilioides*, *Byrsonima crassifolia* and *Curatella americana*, and also for grasses and sedges like *Leptocoryphium lanatum*, *Trachypogon plumosus* and several species of *Andropogon*, *Axonopus*, *Bulbostylis*, *Paspalum*, *Rhynchospora*, etc. But even considering this floristic similarity that homogenizes Neotropical savannas, two particular areas show a much more diversified savanna flora. These are the cerrados and, perhaps to a lesser degree, the Cuban savannas, specifically the formations on serpentine and on siliceous soils.

To go further in the characterization of savanna communities following one or another of the widely used phytosociological systems is difficult due to the lack of adequate floristic knowledge and to the intrinsic weakness of phytosociological methodologies. In general, one may notice that phytosociological classifications, like those already discussed (Van Donselaar, 1965; Hoock, 1971), seem to be successful at a regional level. But extrapolation to regions further away from the areas of origin becomes rather problematical.

REFERENCES

Ab'Saber, A.N., 1971. A organização natural das paisagens inter e subtropicais brasileiras. In: M.G. Ferri (Editor), *III Simposio sôbre o Cerrado*. University of São Paulo, São Paulo, pp. 1–14.

Ab'Saber, A.N., 1977. Espaços ocupados pela expansão dos climas secos na América do Sul, por ocasiao dos periodos glaciais quaternários. *Univ. São Paulo, Inst. Geogr., Paleoclimas*, 3: 19 pp.

Andrade Lima, D. de., 1959. Viagem a os campos de Monte Alegre. *Bol. Técn. Inst. Agron. Norte (Belem)*, 36: 99–149.

Aristiguieta, L., 1966. Flórula de la Estación Biológica de Los Llanos. *Bol. Soc. Venez. Cienc. Nat.*, 110: 228–307.

Aristiguieta, L., 1969. Consideraciones sobre la flora de los morichales llaneros al Norte del Orinoco. *Acta Bot. Venez.*, 3: 3–22.

Askew, G.P., Moffatt, D.J., Montgomery, R.F. and Searl, P.L., 1970. Interrelationships of soil and vegetation in the savanna–forest boundary zone of north-eastern Mato Grosso. *Geogr. J.*, 136: 370–376.

Azevedo, L.G., 1967. Tipos eco-fisionómicos de vegetação do Territorio Federal de Amapá. *Rev. Bras. Geogr.*, 29: 25–51.

Beard, J.S., 1953. The savanna vegetation of northern tropical America. *Ecol. Monogr.*, 23: 149–215.

Bennet, H.H., 1928. Some geographic aspects of Cuban soils. *Geogr. Rev.*, 18: 62–82.

Benoist, R., 1925. La végétation de la Guyane Francaise II. Les savanes. *Bull. Soc. Bot. Fr., Sér. 5*, 72: 1066–1076.

Blydenstein, J., 1962. La sabana de *Trachypogon* del alto llano. *Bol. Soc. Venez. Cienc. Nat.*, 102: 139–206.

Blydenstein, J., 1963. Cambios en la vegetación después de protección contra el fuego. *Bol. Soc. Venez. Cienc. Nat.*, 103: 223–238.

Blydenstein, J., 1967. Tropical savanna vegetation of the Llanos of Colombia. *Ecology*, 48: 1–15.

Borgonovi, M. and Chiarini, J.V., 1968. Cobertura vegetal do Estado de São Paulo. *Rev. Bras. Geogr.*, 30: 39–50.

Borhidi, A. and Herrera, R.A., 1977. Génesis, características y clasificación de los ecosistemas de sabana de Cuba. *Cienc. Biol.*, 1: 115–130.

Bouillene, R., 1926. Savanes équatoriales de l'Amerique du Sud. *Bull. Soc. R. Bot. Blg.*, 58: 217–223.

Braun, E.H.G. and Ramos, J.R.A., 1959. Estudo agroecológico dos campos Puciari–Humaitá, Estado do Amazonas e Territorio Federal de Rondonia. *Rev. Bras. Geogr.*, 21: 443–497.

Brown, Jr., K.S. and Ab'Saber, A.N., 1979. Ice-Age forest refuges and evolution in the neotropics: correlation of paleoclimatological, geomorphological and pedological data with modern biological endemism. *Univ. São Paulo, Inst. Geogr., Paleoclimas*, 5: 30 pp.

Burkart, A., 1969. *Flora Ilustrada de Entre Rios II. Gramineas.* Instituto Nacional de Tecnología Agropecuaria, Buenos Aires, 551 pp.

Cabrera, A., 1970. *Flora de la Provincia de Buenos Aires. Gramineas.* Instituto Nacional de Tecnología Agropecuara, Buenos Aires, 624 pp.

Carabia, J.P., 1945. The vegetation of Sierra de Nipe, Cuba. *Ecol. Monogr.*, 15: 321–341.

Charter, C.G., 1941. *Reconnaissance Survey of the Soils of British Honduras.* Government Printer, Trinidad.

Denevan, W.M., 1966. The aboriginal cultural geography of the Llanos de Mojos of Bolivia. *Ibero-Americana*, 48: 185 pp.

Denevan, W.M. and Chrostowski, M.S., 1970. The biogeography of a savanna landscape. The Gran Pajonal of Eastern Peru. *McGill Univ. Savanna Res. Ser.*, 16: 1–87.

Ducke, A. and Black, G.A., 1953. Phytogeographical notes on the Brazilian Amazon. *An. Acad. Bras. Cienc.*, 25: 1–46.

Eden, M.J., 1964. The savanna ecosystem — northern Rupununi, British Guiana. *McGill Univ. Savanna Res. Ser.*, 1: 1–216.

Eden, M.J., 1974. Paleoclimatic influences and the development of savanna in southern Venezuela. *J. Biogeogr.*, 1: 95–109.

Egler, W.A., 1960. Contribuiçoes ao conhecimento dos campos da Amazonia. I. Os Campos do Ariramba. *Bol. Mus. Para. E. Goeldi, Bot.*, 4: 1–36.

Eiten, G., 1963. Habitat flora of Fazenda Campininha, São Paulo, Brazil. In: M.G. Ferri (Editor), *Simpósio sôbre o Cerrado.* University of São Paulo, São Paulo, pp. 181–231.

Eiten, G., 1972. The cerrado vegetation of Brazil. *Bot. Rev.*, 38: 201–341.

Eiten, G., 1975. The vegetation of the Serra do Roncador. *Biotrópica*, 7: 112–135.

Eiten, G., 1978. Delimitation of the cerrado concept. *Vegetatio*, 36: 169–178.

EMBRAPA. 1976. *Cerrado: Bibliographia Analítica.* Empresa Brasileira de Pesquisa Agropecuaria, Centro de Pesquisa Agropecuaria dos Cerrados, Brasilia, 361 pp.

FAO, 1966. *Reconocimiento Edafológico de los Llanos Orientales, Colombia. III. La Vegetación Natural y la Ganadería en los Llanos Orientales.* FAO, Rome, 233 pp.

Ferri, M.G. (Editor), 1964. *Simpósio sôbre o Cerrado.* University of São Paulo, São Paulo, 423 pp.

Ferri, M.G. (Editor), 1971. *III Simpósio sôbre o Cerrado.* University of São Paulo, São Paulo, 239 pp.

Ferri, M.G. (Editor), 1977. *IV Simpósio sôbre o Cerrado. Bases para Utilizaçao Agropecuária.* University of São Paulo, São Paulo, 405 pp.

Flores Mata, G., Jiménez López, J., Madrigal Sánchez, X., Moncayo Ruiz, F. and Takaki Takaki, F., 1971. *Tipos de Vegetación de la República Mexicana.* Susecretaría de Planeación, Dirección de Agrología, México, 59 pp.

Goldsmith, F.H., 1974. Multivariate analysis of tropical grassland communities in Mato Grosso, Brazil. *J. Biogeogr.*, 1: 111–122.

Gómez Pompa, A., 1973. Ecology of the vegetation of Veracruz. In: A. Graham (Editor), *Vegetation and Vegetational History of Northern Latin America.* Elsevier, Amsterdam, pp. 73–148.

González Jiménez, E., 1979. Primary and secondary productivity in flooded savannas. In: *Tropical Grazing Land Ecosystems. Nat. Resour. Res.*, 14: 620–625.

Goodland, R., 1964. *The Phytosociological Study of the Northern Rupununi Savanna, British Guiana.* Thesis, McGill University, Montreal, Que., 156 pp.

Goodland, R., 1966. South American savannas. Comparative studies Llanos and Guyana. *McGill Univ. Savanna Res. Ser.*, 5: 52 pp.

Goodland, R., 1970. Plants of the cerrado vegetation of Brazil. *Phytologia*, 20: 57–78.

Goodland, R., 1971. A physiognomic analysis of the cerrado vegetation of Central Brazil. *J. Ecol.*, 59: 411–419.

Goodland, R. and Pollard, R., 1973. The Brazilian cerrado vegetation: a fertility gradient. *J. Ecol.*, 61: 219–224.

Heringer, E.P., 1971. Propogaçao e sucessao de especies arboreas do cerrado em funçao do fogo, do capim, da capina e de aldrin. In: M.G. Ferri (Editor), *III Simpósio sôbre o Cerrado.* University of São Paulo, São Paulo, pp. 167–179.

Heringer, E.P., Barroso, G.M., Rizzo, J.A. and Rizzini, C.T., 1977. A flora do Cerrado. In: M.G. Ferri (Editor), *IV Simpósio sôbre o Cerrado.* University of São Paulo, São Paulo, pp. 211–232.

Heyligers, P.C., 1963. Vegetation and soil of a white-sand savanna in Suriname. In: *The Vegetation of Suriname, 3.* Van Eederfonds, Amsterdam, 148 pp.

Hills, T.L., 1969. The savanna landscapes of the Amazon basin. *McGill Univ. Savanna Res. Ser.*, 14: 1–34.

Hoock, J., 1971. Les savanes guyanaises: Kourou. *Mém. ORSTOM*, No. 44: 1–251.

Howard, R.A., 1973. The vegetation of the Antilles. In: A.

Graham (Editor), *Vegetation and Vegetational History of Northern Latin America*, Elsevier, Amsterdam, pp. 1–38.

Huber, O., 1979. The ecological and phytogeographical significance of the actual savanna vegetation in the Amazon Territory of Venezuela. In: *Fifth Int. Symp. Assoc. Trop. Biol., Macuto.*

Hueck, K., 1972. *As Florestas da America do Sul.* University of Brasilia, Brasilia, 466 pp.

Johannessen, C.L., 1963. Savannas of interior Honduras. *Ibero-Americana*, 46: 173 pp.

Köppen, W., 1931. *Grundriss der Klimakunde.* W. de Gruyter, Berlin, 2nd ed., 478 pp.

Labouriau, L.G., (Editor), 1966. Segundo Simpósio sôbre o Cerrado. *An. Acad. Bras. Cienc.*, 38 (Supl.): 346 pp.

Lanjouw, J., 1936. Studies on the vegetation of the Suriname savannahs and swamps. *Ned. Kruinkd. Arch.*, 46: 823–851.

Lisboa, P.L., 1975. Estuados sôbre a vegetaçao das Campinas Amazónicas. II. Observacoes gerais e revisão bibliográfica sôbre as campinas amazónicas de areia branca. *Acta Amazónica*, 5: 211–223.

Lundell, C.L., 1937. The vegetation of Peten. *Carnegie Inst. Wash. Publ.*, 478: 244 pp.

Maguire, B., 1971. On the flora of the Guayana highland. In: W. Stern (Editor), *Adaptive Aspects of Insular Evolution*, Washington State University Press, Pullman, Wash., pp. 63–78.

Medina, E. and Sarmiento, G., 1979. Ecophysiological studies in the *Trachypogon* savanna. In: *Tropical Grazing Land Ecosystems. Nat. Resour. Res.*, 14: 612–619.

Medina, E., Mendoza, A. and· Montes, R., 1977. Balance nutricional y producción de materia orgánica en las sabanas de *Trachypogon* de Calabozo, Venezuela. *Bol. Soc. Venez. Cienc. Nat.*, 134: 101–120.

Meggers, B.J. and Evans, C., 1957. *Archeological Investigations at the Mouth of the Amazon.* Bull. Bureau Am. Ethnol., Washington, D.C., No. 167, 664 pp.

Miranda, F., 1952. *La Vegetación de Chiapas.* Gobierno del Estado, Tuxtla Gutiérrez, 334 pp.

Monasterio, M., Sarmiento, G. and Silva, J., 1971. Reconocimiento ecológico de los Llanos Occidentales. III. El Sur del Estado Barinas. *Acta Cient. Venez.*, 22: 153–169.

Morello, J. and Adamoli, J., 1974. *Las Grandes Unidades de Vegetación y Ambiente del Chaco Argentino. Segunda Parte: Vegetación y Ambiente de la Provincia del Chaco.* I.N.T.A., Serie Fitogeográfica 13, Buenos Aires, 130 pp.

Myers, J.G., 1936. Savannah and forest vegetation of the interior Guiana plateau. *J. Ecol.*, 24: 162–184.

Parsons, J.J., 1955. The Miskito pine savanna of Nicaragua and Honduras. *Ann. Assoc. Am. Geogr.*, 45: 36–63.

Pittier, H., 1942. *La Mesa de Guanipa. Ensayo de Fitogeografia.* Asociación para la Protección de la Naturaleza, Carácas, 46 pp.

Porter, D.M., 1973. The vegetation of Panama: a review. In: A. Graham (Editor), *Vegetation and Vegetational History of Northern Latin America.* Elsevier, Amsterdam, pp. 167–201.

Puig, H., 1972. La sabana de Huimanguillo, Tabasco, México. In: *Mem. I Congr. Latinoam. Bot., México*, pp. 389–411.

RADAM BRASIL, 1973–78. *Levantamento dos Recursos Naturais (Geología, Geomorfologia, Pedologia, Vegetaçao,*

Uso Potencial da Terra), 1–18. Departamento de Pesquisas de Recursos Minerais, Rio de Janeiro.

Ramia, M., 1959. *Las Sabanas de Apure.* Ministerio de Agricultura y Cria, Carácas, 134 pp.

Ramia, M., 1967. Tipos de sabanas en los Llanos de Venezuela. *Bol. Soc. Venez. Cienc. Nat.*, 112: 264–288.

Ramia, M., 1974a. Estudio ecológico del Módulo Experimental de Mantecal (Alto Apure). *Bol. Soc. Venez. Cienc. Nat.*, 128–129: 117–142.

Ramia, M., 1974b. *Plantas de las Sabanas Llaneras.* Monte Avila Editores, Caracas, 287 pp.

Ratter, J.A., 1971. Some notes on two types of cerradao occurring in northeastern Mato Grosso. In: M.G. Ferri (Editor), *III Simpósio sôbre o Cerrado.* University of São Paulo, São Paulo, pp. 100–102.

Ratter, J.A., Richards, P.W., Argent, G. and Gifford, D.R., 1973. Observations on the vegetation of northeastern Mato Grosso. I. The woody vegetation types of the Xavantina–Cachimbo Expedition area. *Philos. Trans. R. Soc. Lond.,* B. 266: 449–492.

Richardson, W.D., 1963. Observations on the vegetation and ecology of the Aripa savannas, Trinidad. *J. Ecol.*, 51: 295–313.

Rizzini, C.T., 1963a. Nota prévia sôbre a divisão fitogeográfica do Brasil. *Rev. Bras. Geogr.*, 25: 3–64.

Rizzini, C.T., 1963b. A flora do cerrado. Analise floristica das savanas centrais. In: M.G. Ferri (Editor), *Simpósio sôbre o Cerrado.* University of São Paulo, São Paulo, pp. 125–177.

Rizzini, C.T., 1971. Arvores e arbustos do cerrado. *Rodriguesia*, 38: 63–77.

Rizzini, C.T. and Heringer, E.P., 1962. *Preliminares acerca das formaçoes vegetais e do reflorestamento no Brasil Central.* Serviço de Informação, M. da Agricultura, Rio de Janeiro, 79 pp.

Rodríguez, W.A., 1971. Plantas dos campos do Rio Branco (Territorio de Roraima). In: M.G. Ferri (Editor), *III Simpósio sôbre o Cerrado.* University of São Paulo, São Paulo, pp. 180–183.

San José, J.J. and Medina, E., 1975. Effect of fire on organic matter production and water balance in a tropical savanna. In: F.B. Golley and E. Medina (Editors), *Tropical Ecological Systems.* Springer-Verlag, Berlin, pp. 251–264.

San José, J.J. and Medina, E., 1977. Producción de materia orgánica en la sabana de *Trachypogon*, Calabozo, Venezuela. *Bol. Soc. Venez. Cienc. Nat.*, 134: 75–100.

Sarmiento, G., 1976. Evolution of arid vegetation in tropical America. In: D.W. Goodall (Editor), *Evolution of Desert Biota.* Texas University Press, Austin, Texas, pp. 65–99.

Sarmiento, G., 1978. *Estructura y Funcionamiento de Sabanas Neotropicales.* Universidad de Los Andes, Mérida, 367 pp.

Sarmiento, G. and Monasterio, M., 1969a. Studies on the savanna vegetation of the Venezuelan Llanos. I. The use of association-analysis. *J. Ecol.*, 57: 579–598.

Sarmiento, G. and Monasterio, M., 1969b. Corte ecológico del Estado Guárico. *Bol. Soc. Venez. Cienc. Nat.*, 115–116: 83–106.

Sarmiento, G. and Monasterio, M., 1971. Ecología de las sabanas de América tropical. I. Análisis macroecológico de los Llanos de Calabozo, Venezuela. *Cuad. Geogr.*, 4: 126 pp.

Sarmiento, G. and Monasterio, M., 1975. A critical consideration of the environmental conditions associated with the occurrence of savanna ecosystems in tropical America. In: F.B. Golley and E. Medina (Editors), *Tropical Ecological Systems*. Springer-Verlag, Berlin, pp. 223–250.

Sarmiento, G. and Vera, M., 1979a. Composición, estructura, biomasa y producción de diferentes sabanas en los Llanos de Venezuela. *Bol. Soc. Venez. Cienc. Nat.*, 136: 5–41.

Sarmiento, G. and Vera, M., 1979b. La marcha anual del agua en el suelo en sabanas y bosques tropicales en los Llanos de Venezuela. *Agron. Trop.*, 27: 629–649.

Sarmiento, G., Monasterio, M. and Silva, J., 1971a. Reconocimiento ecológico de los Llanos Occidentales. I. Las unidades ecológicas regionales. *Acta Cient. Venez.*, 22: 52–61.

Sarmiento, G., Monasterio, M. and Silva, J., 1971b. Reconocimiento ecológico de los Llanos Occidentales. IV. El oeste del Estado Apure. *Acta Cient. Venez.*, 22: 170–180.

Scott, G.A.J., 1977. The role of fire in the creation and maintenance of savanna in the Montaña of Peru. *J. Biogeogr.*, 4: 143–167.

Seifriz, W., 1943. Plant life of Cuba. *Ecol. Monogr.*, 13: 375–426.

Setzer, J., 1967. Imposibilidade do uso racional do solo no Alto Xingu, Mato Grosso. *Rev. Bras. Geogr.*, 29: 102–109.

Silberbauer-Gottsberger, I., Morawetz, W. and Gottsberger, G., 1977. Frost damage of cerrado plants in Botucatu, Brazil, as related to the geographical distribution of the species. *Biotropica*, 9: 253–261.

Silva, J. and Sarmiento, G., 1976a. La composición de las sabanas en Barinas en relación con las unidades edáficas. *Acta Cient. Venez.*, 27: 68–78.

Silva, J. and Sarmiento, G., 1976b. Influencia de factores edáficos en la diferenciación de las sabanas. Análisis de componentes principales y su interpretación ecológica. *Acta Cient. Venez.*, 27: 141–147.

Silva, J., Monasterio, M. and Sarmiento, G., 1971. Reconocimiento ecológico de los Llanos Occidentales. II. El Norte del Estado Barinas. *Acta Cient. Venez.*, 22: 60–71.

Simpson Vuilleumier, B., 1971. Pleistocene changes in the fauna and flora of South America. *Science*, 173: 771–780.

Sinha, N.K.P., 1968. Geomorphic evolution of the northern Rupununi Basin, Guyana. *McGill Univ. Savanna Res. Ser.*, 11: 131 pp.

Steyermark, J., 1966. Contribuciones a la flora de Venezuela. 5. *Acta Bot. Venez.*, 1: 9–256.

Steyermark, J., 1967. Flora del Auyan-tepui. *Acta Bot. Venez.*, 2: 5–370.

Takeuchi, M., 1960. The structure of the Amazonian vegetation. I. Savanna in northern Amazon. *J. Fac. Sci. Univ. Tokio*, 7: 523–533.

Tamayo, F., 1961. Exploraciones botánicas en el Estado Bolívar. *Bol. Soc. Venez. Cienc. Nat.*, 98–99: 25–180.

Tamayo, F., 1964. *Ensayo de Clasificación de Sabanas de Venezuela*. Escuela de Geografía, Facultad de Humanidades y Educación, U.C.V., Caracas, 63 pp.

Tavares, S., 1964. Contribuiçao para o estudo da cobertura vegetal dos tabuleiros do Nordeste. *Sudene Bol. Recut. Nat.*, 2: 13–25.

Taylor, B.W., 1963. An outline of the vegetation of Nicaragua. *J. Ecol.*, 51: 27–54.

Tricart, J. and Millies-Lacroix, A., 1962. Les terrasses quaternaires des Andes vénézuéliennes. *Bull. Soc. Géol. Fr. Sér*, 74: 201–218.

Tricart, J., 1974. Existence de periodes seches au quaternaire en Amazonie et dans les regions voisines. *Rev. Géomorphol. Dyn.*, 23: 145–158.

Valverde, O., 1972. Fundamentos geográficos do planejamento do Município de Corumbá. *Rev. Bras. Geogr.*, 34: 49–144.

Valverde, O., Mesquita, M.G. and Scheisivar, L., 1962. Geografia económica do nordeste potiguar. *Rev. Bras. Geogr.*, 24: 3–42.

Van der Hammen, T., 1972. Historia de la vegetación y el medio ambiente del Norte Sudamericano. In: *Mem. I Congr. Latinoam. Bot. México*, pp. 119–134.

Van der Hammen, T., 1974. The Pleistocene changes of vegetation and climate in tropical South America. *J. Biogeogr.*, 1: 3–26.

Van Donselaar, J., 1965. An ecological and phytogeographic study of northern Surinam savannas, *Wentia*, 14: 1–163.

Van Donselaar, J., 1968. Phytogeographic notes on the savanna flora of southern Surinam (South America). *Acta Bot. Neerl.*, 17: 393–404.

Van Donselaar, J., 1969. Observations on savanna vegetation-types in the Guianas. *Vegetatio*, 16–17: 271–312.

Vareschi, V., 1969. Las sabanas del Valle de Caracas. *Acta Bot. Venez.*, 4: 427–522.

Vera, M., 1979. *Producción de hojarasca y retorno de nutrientes al suelo en una sabana arbolada*. Facultad de Ciencias. Universidad de Los Andes, Mérida, 86 pp.

Waibel, L., 1943. Place names as an aid in the reconstruction of the original vegetation of Cuba. *Geogr. Rev.*, 33: 376–396.

Warming, E., 1892. *Lagoa Santa*. University of São Paulo. São Paulo, 386 pp. (Portuguese edition, 1973.)

Williams, L., 1942. *Exploraciones Botánicas en la Guayana Venezolana*. Servicio Botánico, Ministerio de Agricultura y Cría, Carácas, 468 pp.

Willis, J.C., 1973. *A Dictionary of the Flowering Plants and Ferns*. Cambridge University Press, Cambridge, 7th ed., 1245 pp.

Chapter 11

THE INVERTEBRATES OF THE GRASS LAYER

YVES GILLON

INTRODUCTION

The most abundant invertebrates of the savanna grass layer are the arthropods. Among them, spiders and insects are most prominent, though ticks may become very numerous wherever their mammalian hosts are abundant. The status of phytophagous mites (Acari) has not yet been elucidated in tropical savannas.

Occasionally, when atmospheric moisture is high enough, land gastropods may constitute a rather important part of the invertebrate community of the grass layer. For example, the density of snails may reach seven to eight individuals per square metre on the Laguna Verde study site, Mexico (Halffter and Reyes Castillo, 1975), and up to ten individuals per square metre in some natural grasslands of Mount Nimba, Guinea (Lamotte et al., 1962).

THE DOMINANT GROUPS OF SAVANNA ARTHROPODS

The relative importance of the various taxonomic categories of arthropods is very different in open savannas and in tropical forest habitats, although comparison is made difficult by the lack of any reliable sampling method for the forest canopy. Furthermore, some arthropod groups play a major trophic role in savanna communities and are represented there by very characteristic taxa (Pollet, 1972; D. Gillon and Gillon, 1974; Lachaise, 1974).

The relative numbers and biomasses of the major arthropod groups in the grass layer of two West African study sites are represented in Fig. 11.1 (see also Tables 11.1 and 11.2 for Lamto).

Among primary consumers, the short-horned grasshoppers (Acrididae) rank first in the two areas studied; at the beginning of the dry season they represent 49.7% of the arthropod biomass at Fété Olé (Senegal), and 41.8% at Lamto (Ivory Coast). A very similar percentage (50%) has been found by Morello (1970) in periodically burnt pastures of the Argentinian *chaco*. Other primary consumers well represented in West African savannas are jumping plant lice (Psyllidae), white flies (Aleurodidae), plant lice (Aphididae), leafhoppers (Jassidae), and plant bugs (Coreidae, Lygaeidae and Pentatomidae) among Hemiptera, Elateridae among Coleoptera, Chloropidae among Diptera, Satyridae and Pieridae among butterflies, as well as some noctuid moths whose larvae (army worms) feed upon grass.

By far, the most numerous invertebrate secondary consumers in the two African sites studied are the spiders. At Lamto, they make up from 15 to 29% of the arthropod biomass according to the season, and 4 to 12% at Fété Olé. In both areas, four families are particularly well represented in the spider community, but in different proportions: the Thomisidae, Salticidae, Drassidae and Lycosidae (Blandin, 1974; D. Gillon and Gillon, 1974). Together with the ants (see Ch. 24), their trophic impact upon other invertebrates must be considerable. In the savannas of the Argentinian chaco, the mygale spiders are also considered by Morello (1970) as one of the most important invertebrate predators. Other predators and parasites well represented in West African savannas are the Mantodea among orthopteroid insects, the Reduviidae among Hemiptera, the Diopsidae, Chloropidae, Bombyliidae, Dolichopodidae and Tachinidae among Diptera, the Meloidae among

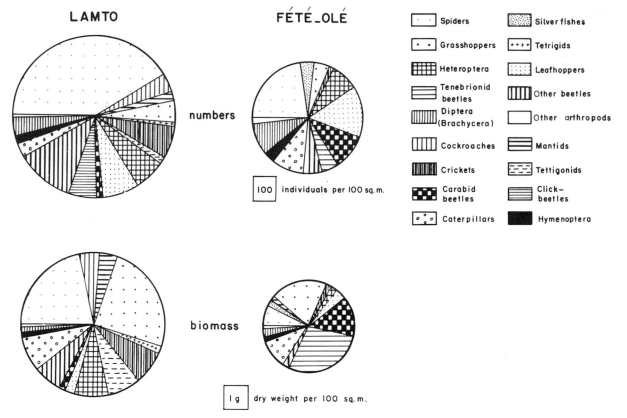

LAMTO   FÉTÉ_OLÉ

numbers

Spiders   Silver fishes

Grasshoppers   Tetrigids

Heteroptera   Leafhoppers

Tenebrionid
beetles   Other beetles

Diptera
(Brachycera)   Other arthropods

Cockroaches   Mantids

Crickets   Tettigonids

Carabid
beetles   Click–
beetles

Caterpillars   Hymenoptera

100   individuals per 100 sq.m.

biomass

1 g   dry weight per 100 sq.m.

Fig. 11.1 Numbers and biomasses of the major groups of grass layer arthropods, in two contrasted West African savannas, at two different times of the year, the rainy season (left) and the dry season (right). After Gillon and Gillon (1974).

Coleoptera, and the Sphecoidea among Hymenoptera.

Saprophages are numerous among savanna arthropods. The most common are obviously the termites (see Ch. 23), but other groups are also well represented, such as the Calliphoridae and Drosophilidae among Diptera. The silverfishes (Thysanura), cockroaches (Blattidae) and crickets (Gryllidae) are far less numerous in savannas than in forest habitats. The same is true for Diplopoda (Chilognatha and Polydesmoidea).

Despite many common characteristics, there are however some significant differences between the arthropod communities of the driest and the more humid savannas. Some of these differences are shown on Fig. 11.1 which compares the relative importance of the major taxonomic groups at different seasons at Fété Olé and Lamto. The most obvious is the predominance of tenebrionid beetles (Tenebrionidae) in the Sahelian savanna. Nineteen species have been found there, representing from 19

to 58% of the total arthropod biomass depending on the season; two species were even particularly abundant during the dry season (D. Gillon and Gillon, 1974). Other groups well represented in xeric African savannas are the Solifugae and scorpions among the Arachnida, and the Embioptera among the insects.

The seed-eating insect community is also probably different in dry and humid tropics (D. Gillon and Gillon, 1974). The few studies carried out so far on seed beetles (Bruchidae) all point out that these seed predators are extremely host-specific (Center and Johnson, 1974; Janzen, 1977).

The savanna/forest boundary sometimes has also a peculiar insect fauna. Pollet (1972), for example, has found in the Ivory Coast that Lygaeidae, Chrysomelidae, Languridae, Pipunculidae, Tenthredinidae and Chalcididae were particularly numerous at the forest/savanna ecotone. Drosophilidae are also very numerous there, more than 100 species having been collected

TABLE 11.1

Densities (ind m^{-2}) of the various groups of Arthropoda in the grass layer of the Lamto savanna (Ivory Coast) during a yearly cycle (D. Gillon and Gillon, 1974)

	Jan.	Febr.	March	April	May	June	July	Aug.	Sept.	Oct.	Nov.	Dec.
Spiders	22.50	13.57	8.78	13.30	10.60	13.07	19.05	19.97	13.42	15.80	13.32	17.50
Opiliones							0.10	0.12	0.10	0.05	0.12	0.12
Scorpions			0.09		0.02	0.05	0.02			0.05	0.07	0.02
Blattaria	1.55	0.37	0.54	0.37	1.12	0.97	1.65	3.47	1.65	1.05	2.05	2.02
Mantodea	0.35	0.57	0.27	0.20	0.37	0.65	0.60	0.87	0.40	0.52	0.40	1.32
Acridids	0.17	1.95	2.36	2.47	1.90	2.07	1.87	1.67	1.47	1.20	0.90	1.30
Tetrigids	0.05	0.05	0.06	0.05	0.45	0.45	0.80	1.10	0.45	0.37	0.50	0.42
Tridactylids						0.05	0.12	0.25	0.87	0.15	0.05	0.10
Gryllidae	1.85	0.40	0.40	0.77	0.60	2.37	4.42	3.50	4.15	2.75	3.37	1.75
Tettigonids		0.02	0.17	0.07	0.25	0.27	0.52	0.90	1.95	1.82	1.10	0.12
Pentatomids	0.32	0.30	0.60	0.85	0.45	0.27	0.65	0.80	0.45	0.50	1.20	0.72
Coreids		0.05	0.50	0.30	0.35	0.12	0.22	0.22	0.10	0.17	0.27	0.50
Lygaeids	0.25	0.47	0.69	2.25	1.70	0.70	0.45	0.55	0.70	0.17	0.22	0.05
Reduvids	0.17	0.22	0.39	0.60	0.55	0.67	0.92	1.47	0.40	0.70	0.72	0.72
Homoptera	0.85	0.62	2.66	2.02	1.75	2.65	3.17	5.15	3.70	4.45	2.42	1.67
Carabid beetles	0.25	0.05	0.17	0.25	0.27	0.57	0.92	0.42	0.72	0.47	0.50	0.30
Tenebrionids	0.17	0.12	0.10	0.02	0.17	0.15	0.12	0.10	0.05	0.10	0.07	0.30
Langurids	0.12	0.10	0.15	0.05		0.22	0.35	0.15		0.45	0.60	0.10
Elaterids	2.12	4.07	0.45	1.45	1.50	2.75	2.20	3.40	0.95	1.20	1.82	3.37
Coleoptera (others)	0.82	0.50	1.06	1.02	18.52	2.95	2.57	2.57	4.50	7.12	3.15	2.47
Lepidoptera	1.00	1.12	1.00	1.05	0.32	0.60	0.35	0.67	0.50	2.40	1.72	1.37
Hymenoptera	0.22	0.27	0.65	0.32	0.35	0.57	0.40	0.45	0.57	0.82	0.62	0.22
Diptera	3.27	0.30	0.95	0.55	0.40	1.25	0.67	1.65	2.80	2.92	1.17	1.07
Others	0.02	0.02	0.17	0.14	0.17	0.02	0.30	0.12		1.25	0.35	0.07
Total	36.05	25.14	22.21	28.10	41.81	33.44	42.44	49.57	39.90	46.48	36.71	37.60

at Lamto by Lachaise (1974, 1979). Some species, such as *Glossina palpalis*, can also be found more abundantly at the forest edge, whereas other tse-tse flies are restricted either to savannas (*G. morsitans*) or to woodlands and gallery forests (*G. tachinoides*).

MORPHOLOGICAL CHARACTERISTICS OF SAVANNA ARTHROPODS

Size of individuals

Tropical insects are usually famous for their large size, as well as for their often unusual morphology. However, such a widely held notion needs to be supported by quantitative data. Species richness being extremely great in the tropics, and collecting being generally selective, it might well happen that large and spectacular insects are not actually more numerous here than small and dull ones, but merely more conspicuous and hence more often collected. Unfortunately, a comparison of the size distribution of adult arthropods in various temperate and tropical communities has never been done.

However, the comparison of the live weights of all the arthropods sampled in the same way, by the same observers, on two 25-m^2 plots, in two very different West African savannas (Fété Olé and Lamto) has disclosed significant size differences between individuals in the two samples. The live weight of an average arthropod was 6.7 mg at Fété Olé, as compared with 10.2 mg at Lamto (D. Gillon and Gillon, 1974). Two explanations come to mind to account for such a size difference. First, the duration of the rainy season during which single-brooded (univoltine) species achieve their post-embryonic development is much shorter at Fété Olé than at Lamto, hence the growth period is limited. Second, the physical structure of the grass layer itself is very different in the two savannas, the grass being shorter with thinner blades in the north; a

TABLE 11.2

Biomasses (mg dry wt. m^{-2}) of the various groups of Arthropoda in the grass layer of the Lamto savanna during a yearly cycle (D. Gillon and Gillon, 1974)

	Jan.	Febr.	March	April	May	June	July	Aug.	Sept.	Oct.	Nov.	Dec.
Spiders	100.37	45.12	42.50	76.25	39.85	59.60	59.10	79.58	39.48	72.98	84.17	78.90
Opiliones							0.47	0.40	0.61	0.37	0.60	0.60
Scorpions			1.37		0.35	0.27	0.91			0.60	1.42	0.25
Blattaria	10.25	0.62	5.22	2.37	4.58	13.17	11.01	23.25	16.36	10.15	34.83	30.50
Mantodea	3.06	5.12	18.56	8.87	6.52	15.47	10.77	9.62	3.51	41.21	4.36	13.37
Acridids	11.25	51.37	86.25	155.12	72.93	48.80	53.86	38.63	54.75	61.73	72.05	161.00
Tetrigids	0.44	0.62	0.69	0.50	5.38	7.90	7.00	15.41	4.56	3.48	4.86	4.75
Tridactylids						0.09	0.15	0.51	1.40	0.42	0.06	0.18
Gryllidae	7.12	15.75	9.47	55.00	11.38	19.16	23.87	24.71	23.48	16.41	51.73	19.69
Tettigonids		1.25	10.56	2.81	9.71	13.08	5.96	2.76	9.72	80.35	129.36	2.77
Pentatomids	2.87	4.56	6.75	11.68	9.02	5.71	19.98	17.85	8.33	6.43	20.07	9.12
Coreids		0.87	5.25	5.62	3.07	2.10	2.70	1.80	0.38	1.30	2.51	3.87
Lygaeids	0.62	1.31	1.06	5.75	3.70	1.26	1.10	2.32	0.90	0.50	0.51	0.50
Reduvids	1.87	3.00	3.39	9.50	2.62	9.78	8.40	11.66	4.32	8.77	24.06	5.62
Homoptera	0.75	2.00	6.34	7.00	3.38	8.40	9.02	10.46	5.28	10.50	9.81	6.25
Carabid beetles	2.00	0.25	3.81	8.87	5.76	7.58	12.43	4.63	15.86	15.43	11.18	5.65
Coleoptera (others)	6.00	6.75	9.06	8.65	48.32	15.20	18.28	12.58	16.30	25.80	15.15	16.87
Lepidoptera	29.31	14.81	24.47	16.90	2.50	8.57	21.92	16.07	4.63	32.73	95.13	19.87
Hymenoptera	1.50	1.37	4.65	3.12	2.22	1.62	1.02	1.23	0.73	4.66	3.85	0.87
Diptera	7.62	0.87	3.19	1.62	1.75	3.70	1.86	2.68	6.36	7.77	5.37	4.31
Others	0.12	1.25	0.68	0.93	2.01	0.21	1.01	1.46		2.00	0.85	
Total	185.15	156.89	243.27	380.56	235.05	241.67	270.82	277.61	216.96	403.59	571.93	384.94

large size would not therefore be adaptive for most arthropods of the grass layer in the Sahel. Obviously, these two explanations are not mutually exclusive.

It has also been noticed by Lachaise (1974) that the adults of related species in two families of Diptera (Anthomiidae and Calliphoridae) were smaller in savannas than in rain forests, the latitude being the same for both study sites.

Within a given taxonomic group, the average individual weight varies during the annual cycle, according to the phenology and voltinism (number of generations per year) of the species concerned. It can remain the same throughout the year, as in the case of spiders (Fig. 11.2), or can display minimum values during the rains as is the case for crickets or acridids. The seasonal variations of the average body weight of the arthropod community as a whole are rather small; they depend essentially on the weight of the univoltine species, whose individual weight is generally high and whose seasonal changes in abundance are clear-cut.

The lightest individual body weights were observed in our studies at the end of the rainy season:

5.3 mg in September in northern Senegal and 5.4 mg in September in the southern Ivory Coast (D. Gillon and Gillon, 1974). Average body weights are rather meaningless at the beginning of the rainy season, when most arthropods hatch and a large percentage of immature stages is found in any population.

External morphology

The colour and external appearance of arthropods are deeply influenced by the habitat of their adult stages and their feeding habits. Any elongated, green or straw-coloured acridid has a good chance of belonging to the grass layer community, although most arthropods living in the grass layer are not grass-mimics. Heteroptera, for instance, never mimic grass blades. The best grass-mimics are found among the larger species, such as phasmids of the genus *Gratidia*, which are even more thread-like than most other stick insects. On the other hand, the smallest insect species, such as jassids, tend to resemble seeds rather than grass.

The adaptive function of the cryptic elongated

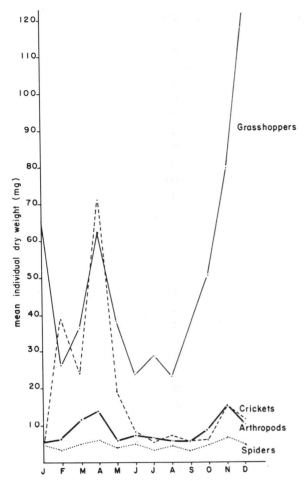

Fig. 11.2. Monthly variations of the average dry body weight of spiders, crickets and acridids, compared to that of the average body weight of all sympatric arthropods of the grass layer, in the Guinean savanna of Lamto, Ivory Coast.

body form of savanna acridids is enhanced by special concealing attitudes. The very mobile species which settle on the grass (such as members of the genera *Acrida*, *Chromotruxalis* and *Truxalis* in West Africa) have an elongated body and long, thin hindlegs which are kept well away from the body axis (Fig. 11.3). On the other hand, the more sedentary species which spend a long time on the same grass support (such as *Cannula*, *Leptacris* and *Mesopsis*) have shorter and stronger hindlegs, kept close to their body when at rest (Fig. 11.4). Among African acridids, all species with elongated bodies are graminivorous, whereas forbivorous species never display such a body-build; the same situation occurs in the llanos of Venezuela, and apparently in other tropical grasslands as well.

Similar concealing habits are also found among savanna mantids. In Africa, those belonging to genus *Pyrgomantis* (Fig. 11.5) resemble *Cannula*, whereas *Leptocola* are more similar to phasmids. Even the egg-case (ootheca) of *Pyrgomantis pallida* is elongated and cryptically coloured; the female crouches on the ootheca which is fixed on a grass blade during the entire "incubation" period. Both the adult and the egg-case are straw-coloured, and have the same elongated appearance and a similar size (Fig. 11.4). When praying-mantis species are grass- or stick-mimics, with elongated forelegs, their defensive posture is quite different from the usual "frightening" behaviour of Mantinae; they promptly protract their forelegs so that their resemblance to a grass blade or a stick is intensified (Edmunds, 1972, 1976).

Plant mimicry can also be enhanced by an increased pilosity of the cuticle in some savanna insects. A good example is *Bocagella acutipennis hirsuta*, an acridid living on the composite *Vernonia guineensis*, a common forb in the Guinean savanna of Ivory Coast. The numerous hairs on its thorax and legs, together with its very slow movements, make it almost unnoticeable on the flower head of the plant. Another example is *Anablepia granulata* which feeds exclusively on *Brachiaria* grasses.

Among heterometabolous insects, selective pressures are not only exerted upon adults but also, and probably to a much greater extent, upon the smaller juvenile stages. It is worth noting, in this connexion, that grass- and twig-mimicry are much scarcer among holometabolous insects. In African savannas at least, twig- and leaf-mimicry is found among Lepidoptera but their caterpillars generally feed more upon woody plants than on grass.

Batesian mimicry is also common in savanna butterflies, the model in this case being unpalatable sympatric butterflies. The most extensively studied example is indeed that of the African swallowtail, *Papilio dardanus*, whose females are highly polymorphic. Thirty-one different morphs have been identified so far (Clarke and Sheppard, 1960; Ford, 1971; Owen, 1971; Edmunds, 1974). Throughout much of tropical Africa the females of *Papilio dardanus* are excellent mimics of various species of Danaidae, and to a lesser extent of *Bematistes* spp. (Acraeidae). Where the models or potential models are abundant, as in lowland forest areas, most females of *Papilio dardanus* are good mimics, but

Fig. 11.3. *Chromotruxalis liberta*, a cryptic acridid of the Lamto savanna, Ivory Coast. (Photo by Y. Gillon.)

where the models are rare, as in some dry savannas, non-mimetic and highly variable female forms occur (Ford, 1971).

Ant-mimicry is also common in tropical savannas. In West Africa, young stages of a number of tettigonids (*Eurycorypha*), mantids (*Gonypetella*), coreids [*Euthetus leucostictus*, *Mirperus jaculus* (Fig. 11.6) and *Riptortus dentipes*], and mirids (*Diocoris agelastus*, *Formicopsella* spp., *Leaina belua*, *Myombea bathycephala* and *Systellonotidea triangulifer*), very closely resemble ants, at least for a time. The well known spiders belonging to the genus *Myrmarachne* remain perfect ant-mimics throughout their life-span (Fig. 11.7). In Ghana, each species of *Myrmarachne* is positively associated with a different species of ant, and there is mutual exclusion between the three species of dominant ants. Furthermore, the early instars of *Myrmarachne* mimic different, smaller, species of ants from the adults (Edmunds, 1978). The larvae

of some assassin bugs of the subfamily Piratinae, camouflage themselves by carrying the bodies of the ants they have killed on their backs.

Concealing coloration and mimicry are scarce among ground-living savanna arthropods. Among crickets, only the bodies of those belonging to genus *Eucyrtus* are somewhat elongated. Some insects however may camouflage themselves by using sand grains; such is the case in Africa for the tetrigid *Pantelia horrenda*, the tenebrionid beetle *Vietta senegalensis* and the pentatomid *Thoria gillonae*.

Colour change to match a particular background

A cryptic animal cannot properly harmonize with its surroundings without an adequate overall colour resemblance to its background: hence, the predominance of greens and browns among the colours of insects living in the savanna grass layer.

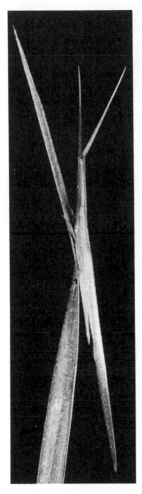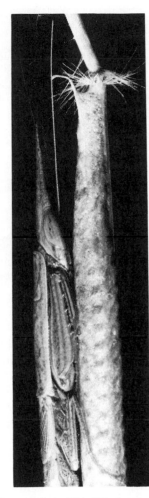

Fig. 11.4. *Cannula gracilis*, another cryptic acridid of the Lamto savanna (Ivory Coast), generally found closely applied against grass blades. (Photo by Y. Gillon.)

Fig. 11.5. Resting attitude of a female *Pyrgomantis pallida* on its egg case fixed upon a stem of *Loudetia simplex*. (Photo by Y. Gillon.)

Some tettigonids do bear some purple body markings, but these purple marks also bear a striking resemblance to the purple areas of anthocyanin pigment that occur on very young, or very old, leaves in the habitats in which these grasshoppers live.

The tettigonid *Homorocoryphus nitidulus*, for example, has six distinct colour forms in Uganda: green, brown, green with purple stripes, brown with purple stripes, green with a purple head, and brown with a purple head. The frequency of the six forms in a large random sample from the derived savanna around Kampala (Owen, 1976) is shown in Table

11.3. Green was about twice as frequent as brown, and together these two forms accounted for nearly 97% of the sample. In the habitat studied, green was the predominant colour, brown came next, and purple (on green or brown leaves) was the rarest. Hence the three conspicuous colours in the habitat of this grasshopper were also the same colours that occurred in the grasshopper themselves. The percentages would probably be different in populations living in other savanna areas, and possibly at different times of the year. A wholly purple form has indeed been found in the small patches of savanna enclosed within the lowland rain forest of the lower Ivory Coast (Bellier et al., 1969).

The overall colour of the savanna grass layer also drastically changes with the seasons, and species have to adapt their colour so that they can harmonize with two or even three different backgrounds. This is achieved through environmentally induced polymorphism (Edmunds, 1974), a phenomenon quite common among orthopteroid insects living in the grass layer. Indeed, when two generations follow one another during the same year, the green form predominates during the rainy season and the brown form during the dry season. If a polyvoltine savanna species is monomorphic, then it is generally straw-coloured.

In the acridid *Acrida turrita*, Ergene (1950) found that green larvae placed on a yellow background changed to yellow at the next moult. However, the quality of light reflected from the background is but one of the factors determining whether a grasshopper changes colour or not. Dryness has been shown to induce a green coloration in *Acrida bicolor* and *Schistocerca vaga* (Okay, 1956; Rowell and Cannis, 1971), and Owen (1976) has noted that in *Homorocoryphus nitidulus* there was a statistical association of green with females and brown with males.

Ground-living insects are very seldom green-coloured, regardless of the number of generations per year.

After grass fires, most savanna insects, the acridids particularly, turn partially or totally black (Burtt, 1951). A wholly black female mantid *Danuria buccholzi* has even been found at Lamto, at the very beginning of the dry season, before the outbreak of the grass fires; being wingless, this mantid could not have come from very far away.

Although young and adult acridids turn black

Fig. 11.6. An ant-mimic, the young of the coreid bug *Mirperus jaculus*, on *Cyperus zollingeri*, a savanna sedge, Lamto, Ivory Coast. (Photo by Y. Gillon.)

very quickly — in less than two days during the dry season — they take a much longer time to revert to their original colour (Y. Gillon, 1971). It would be interesting to know whether the same ability exists in species whose young are found only during the rainy season. If not, this would tend to demonstrate that the selection pressure exerted by grass fires has indeed a very long history.

Whereas the gregarious phase individuals of the migratory locusts are not cryptically coloured, the solitary phase individuals keep the same cryptic colours and patterns as the solitary locust species.

Except for some beetles (Chrysomelidae, Meloidae), very few savanna insects display warning colorations. Those which do so, such as some pyrgomorphids (*Phymateus* spp., *Zonocerus* spp.) and *Staurocleis magnifica* when adult, feed exclusively on forbs.

Fig. 11.7. Another ant-mimic, the spider *Myrmarachne* sp. on a blade of *Imperata cylindrica*, Lamto, Ivory Coast. (Photo by Y. Gillon.)

TABLE 11.3

Relative frequency of colour forms of *Homorocoryphus nitidulus* at Kampala, Uganda (after Owen, 1976)

Colour form	Number examined	Percentage
Green	6682	63.34
Brown	3521	33.38
Green with purple stripes	305	2.89
Green with purple head	34	0.32
Brown with purple head	5	0.05
Brown with purple stripes	2	0.02

The brightly coloured yellow or red hind-wings of some savanna acridids (*Chromotruxalis liberta*, *Gastrimagus africanus*, *Ornithacris* spp.) are always concealed when at rest. The overall coloration of those species of crickets (such as *Xenogryllus eneopteroides*) which climb grass stems and blades is more brownish than those which remain at ground level which are darker.

ADAPTATIONS TO THE SEASONAL WATER SHORTAGE

Most of the savanna insects, especially those inhabiting the more xeric grasslands, are well adapted to the seasonal water shortage — by means of a diapause during which development is arrested. Diapause can occur at various stages of the life history of a species — the egg, the pupa, or even the adult before the breeding season; more rarely at the immature stages. However, some mosquito larvae can stand prolonged dehydratation, and larvae of the chironomid *Polypedilum vanderplanki* can survive ten years of cryptobiosis (Hinton, 1960).

Most of these resistant stages survive the seasonal drought underground. Soil is loose enough during the rains to let most insects dig into it, and it becomes hard enough during the dry season to protect them from many predators. It also affords shelter against overheating during grass fires, as shown in Chapter 30. The humidity of the habitat, which in turn depends on the characteristics of the soil, has been shown to be one of the major environmental factors influencing the distribution of tropical acridids (Duranton and Lecoq, 1980).

Another way for many insects to escape the seasonal drought is to move temporarily to habitats where living conditions are less severe, especially along the forest/savanna boundary. In the derived savanna in the Ivory Coast, for example, plataspidids and langurids undertake massive seasonal movements, whereas acridids, pentatomids and membracids are far less mobile, and carabid beetles and whiteflies (Aleurodidae) do not move at all (Pollet, 1972).

SEASONAL VARIATIONS IN ABUNDANCE

It is difficult to appreciate globally and accurately the seasonal changes in the numbers of an invertebrate community, even in the savanna grass layer, since the reliability of the available collecting techniques differ greatly from one taxonomic category to another. It is, therefore, only possible to compare population densities within one category of organisms (or possibly categories of closely related organisms) at different times of the year. Even in this case, it must be kept in mind that the various developmental stages of many invertebrates cannot be properly sampled quantitatively by the same technique used for the adults. Hence, a fair knowledge of the phenological cycle of at least the most abundant species present within the community is necessary.

Overall variations of the invertebrate community

Some sampling techniques are, however, more reliable than others for quantitative inter-seasonal comparisons. Such is the case for methods involving the careful hand collection of all the arthropods present in a study plot of fixed size, enclosed whenever possible in a mobile cage (a "biocoenometer", Y. Gillon and Gillon, 1965). In three African grasslands, such a method has shown that the invertebrate fauna of the grass layer is definitely more abundant during the rainy season than during the dry one.

In the derived savanna of Lamto, Ivory Coast, the overall arthropod biomass increases with the standing crop of grass (Y. Gillon and Gillon, 1967). However, whereas the average grass biomass reaches 1 kg m^{-2} (fresh weight) for at least six months during the year, that of the arthropods of the grass layer seldom attains 1 g m^{-2}. The lower biomass values are found during the dry season, in January and February, whereas peak values are reached in October and November, at the end of the long rains (Tables 11.1 and 11.2).

However, the changes in population density do not follow the same pattern in all taxonomic categories. Four different patterns can be distinguished: (1) an increase in abundance ending with a peak during the rainy season itself; (2) an increase in abundance ending with a peak at the very beginning of the next dry season; (3) an increase continuing until the start of the annual grass fires; and (4) a multimodal pattern, with two or more peaks in numbers, or even no definite change in abundance at all.

Most insects of the grass layer follow the first pattern: cockroaches, acridids, tetrigids, lygeids, reduvids, and adult carabid beetles living above ground level [those living in the upper parts of the soil are more abundant during the dry season, according to Lecordier and Girard (1973)]. The long-horned grasshoppers (Tettigonidae) and the caterpillars of moths and butterflies follow the second pattern, whereas the Homoptera (mostly represented by jassids) follow the third one. Mantids have two peaks of abundance, as well as rutelid beetles (Girard and Lecordier, 1979). Variations in the abundance of spiders do not conform to any regular pattern.

It is worth remembering at this stage that an increase in arthropod numbers does not necessarily imply an increase in biomass, because the older the individuals become the less numerous they are. However, within a given species, the increase in body weight is generally more rapid than the decrease in numbers (Y. Gillon, 1973). Furthermore, since the larger arthropod species have a single breeding season per year, it follows that the overall maximum arthropod biomass is reached at the end of the rainy season, and not earlier. It is at that time of the year that the amount of available food is the greatest, at least for secondary consumers. The situation is somewhat different for primary consumers, because the increase in grass biomass does not go hand in hand with its nutritional value. Quite on the contrary, it is at the beginning of the rains, when fresh grass appears, that its nutrient content is the highest — precisely at the time when most acridids hatch from the egg.

Most seasonal variations in numbers or biomass imply changes in the demographic structure of the populations concerned. Every season can thus be characterized by the predominance of a particular age class among univoltine species.

In the drier Sahelian savanna of Fété Olé, northern Senegal, the overall arthropod community is four times more numerous during the short rainy season than during the long dry season (Fig. 11.8), although certain species can be more abundant during the drier part of the year (Y. Gillon and Gillon, 1973). The arthropod community of the mountain grasslands of Mount Nimba, Guinea, is also more numerous during the wettest part of the year, although large orthopteroid insects are more conspicuous during the dry season (Lamotte, 1947). Though using a much cruder technique of sampling, Dingle and Khamale (1972) also found an increased insect biomass during the long rains in the *Themeda triandra* savanna of the Athi Plains, Kenya (Table 11.4).

Seasonal variations in numbers of univoltine species

The development of arthropods, which breed only once a year, is either spread over many months, or is interrupted by a long quiescent phase during the dry season.

Such species do not maximize their natality. Their intrinsic rate of increase r is indeed influenced more by the time necessary to reach their reproductive phase than by the rate at which females produce off-spring (Lewontin, 1965). This limitation of natality is balanced by a better adaptation of each phase of the life-cycle to the prevailing environmental conditions. This does not imply that all univoltine species become adult at the same time of the year, as a number of different strategies can be adopted by sympatric species.

For example, among the nineteen species of thomisid spiders whose life-cycle has been studied by Blandin (1972) in the Guinean savanna of Lamto, Ivory Coast, eleven were univoltine. Among them, three different phenologic cycles can be distinguished: (1) in some of them, males become adult earlier than females (protandry), and females outlive males; (2) in other species, adults of both sexes disappear during the dry season, from December to February; and (3) in still other species, some breed at the beginning of the rains, from April to June (*Runcinia sjoestedti*, *Stiphropus niger* and *Thomisops lesserti*, for instance), whereas others do

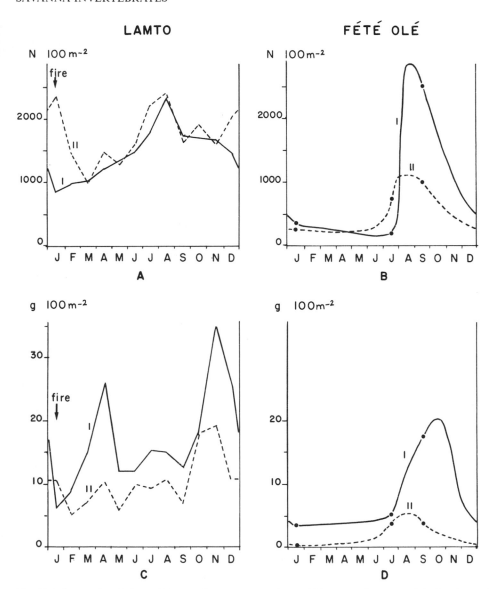

Fig. 11.8. Seasonal variations of the numbers (upper graphs) and biomasses (lower graphs) of arthropods in two African savannas, the Guinean savanna of Lamto, Ivory Coast (the two graphs on the left side of the figure) and the Sahelian savanna of Fété Olé, Senegal (on the right side of the figure). Primary consumers (*I*) and secondary consumers (*II*) are figured separately. The same sampling techniques were used in the two cases. Biomasses are expressed in dry weight per 100 m². (After D. Gillon and Gillon, 1974.)

so at the end of the rainy season (*Firmicus haywoodae*, *Proxysticus egenus*; Fig. 11.9).

Species belonging to a single genus can display different phenologic cycles: adults of *Tibellus seriepunctatus* are found from April to June, while those of *T. demangei* are found from August to November. Univoltine and multivoltine spiders can also be found within a same genus: *Thanatus dorsilineatus* has a single breeding season, whereas *T. lamottei* and *T. pinnetus* are bivoltine.

Protracted life-cycles can result from a slow post-embryonic development. *Runcinia sjoestedti* and *Tibellus seriepunctatus* hatch in September and October, and reach the adult stage six to seven months later; these spiders spend the dry season as juveniles. Conversely, *Diaea puncta* hatches in January, at the peak of the dry season, whereas the last adult individuals have disappeared months earlier; in this case, it is the embryo which develops slowly or undergoes an embryonic diapause. On the

TABLE 11.4

Seasonal fluctuations in insect numbers and biomass in sweep samples from the
Athi Plains, Kenya (after Dingle and Khamale, 1972)

		Number of species	Number of individuals	Standing crop biomass (mg dry wt.)
October	1969	87	204	1161
November	1969	117	561	1160
January	1970	127	1002	901
March	1970	102	791	1315
May	1970	214	1133	4162
July	1970	121	994	1432

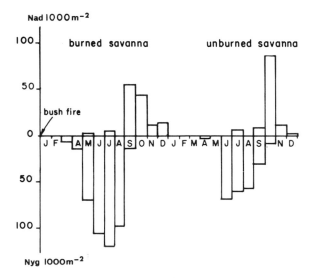

Fig. 11.9. Seasonal variations in numbers of the spider *Firmicus haywoodae* (Thomisidae) in the Lamto savanna. The number of adult individuals (*Nad*) are indicated by the top histograms, and the immatures (*Nyg*) by the bottom histograms. (Data from Blandin, 1972.)

same percentage as that found by Y. Gillon (1974) at Lamto, much further south, where the average annual rainfall is over 1200 mm and the short dry season is far less severe. Among the 23 species whose life-cycle has been studied at Maradi, Niger, by Launois and Launois-Luong (pers. comm., 1979), twelve were univoltine, four bivoltine and seven trivoltine.

The ovaries of the acridid females undergoing an imaginal diapause during the dry season remain at a very early stage of development (pre-vitellogenesis) during the drought, and sexual maturity coincides with the onset of the first rains. The triggering of sexual maturation can be very rapid, as in *Acridoderes strennus* at Saria, where it takes place as soon as the atmospheric moisture begins to increase, in April. The first eggs are laid in June (*Acorypha clara*, *Tylotropidius gracilipes*) and the adults are found from early September (*Catantops axillaris*) to early December (*Rhabdoplea munda*), with a peak in October when the grass cover reaches its maximum standing crop (Lecoq, 1978a).

Incidentally, *Rhabdoplea munda* raises a very intriguing problem. At Saria, all individuals belong to the macropterous form, only one micropterous individual having ever been found by Lecoq (pers. comm., 1979). Further south, in the Lamto Guinean savanna, the situation is different, only half of the individuals belonging to the dry-season generation are macropterous, the others being micropterous. In the lowland rain forest of West Africa, most acridid species are also micropterous.

When diapause occurs at the egg stage, it takes place before blastokinesis (Dempster, 1963). Whereas the eggs of *Zonocerus variegatus* begin to hatch in May, at Saria, those of *Orthochtha grossa* do not do so before August, at a time when most

other hand, the post-embryonic development of *Diaea puncta* is twice as fast as that of the preceding species. So far, no adult diapause stage is known for these Lamto spiders.

The invertebrate group having the most clearly understood life cycle in tropical savannas is the acridids. Their univoltine species either enter diapause at the embryonic stage or as adults. Such univoltine acridids are always very numerous in the savanna communities studied (Phipps, 1968). Out of a total of 76 species studied by Lecoq (1978a) in the Saria area, Upper Volta, where the annual rainfall is only 836 mm with a single long dry season between the middle of October and the end of May, 50 species (66%) are univoltine. This is the sympatric species have already reached their adult stage. Female *Orthochtha grossa* do not lay their eggs before October, at which time most other sympatric acridids become adult before entering diapause during the dry season (Lecoq, 1978a).

In the Lamto area, the eggs of univoltine acridid species begin to hatch in early March [*Acorypha johnstoni*, *Chloroxyrrhepes virescens* (Fig. 11.10) and *Dnopherula bifoveolata*]. Of the three short-horned grasshoppers just mentioned, the first two undergo an embryonic diapause and the last experiences an adult diapause. Other univoltine species continue to hatch until the end of May (*Bocagella acutipennis*, *Dictyophorus griseus*,

Machaeridia bilineata and *Tanita parva*, all undergoing adult diapause). Young *Eucoptacra anguliflava* hatching at that time of the year can come from eggs laid either before or after the dry season, adults being found throughout the year.

Two Lamto species, however, undergo diapause during the rainy season, which incidentally implies that it cannot be considered the most favourable period for all species in this taxonomic group. *Petamella prosternalis* eggs are laid at the onset of the rains, but do not hatch before August or September (Fig. 11.11), thus implying a rainy season embryonic diapause, as defined by Phipps (1968). On the other hand, *Gastrimargus africanus* enters diapause at the adult stage.

If the hatching time of all univoltine species studied in different parts of West Africa are compared, it becomes apparent that there are always some species which hatch before or after the main hatching period of most of the others (Fig. 11.12). The hatching of half of the sympatric species is

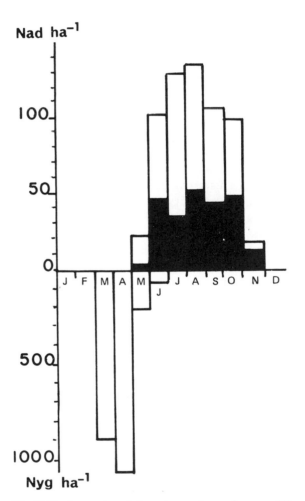

Fig. 11.10. Seasonal variations in numbers of young (*Nyg*, bottom histograms) and adult (*Nad*, top histograms) of *Chloroxyrrhepes virescens* (Acrididae, Tropidopolinae) in the Lamto savanna; an example of embryonic diapause and early hatching. Females in black. (After Y. Gillon, 1974.)

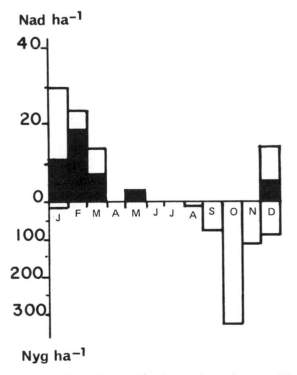

Fig. 11.11. Seasonal variations in numbers of young (*Nyg*, bottom histograms) and adult (*Nad*, top histograms) of *Petamella prosternalis* (Acrididae, Tropidopolinae) in the Lamto savanna; an example of adult diapause with late hatching. Females in black. (After Y. Gillon, 1974.)

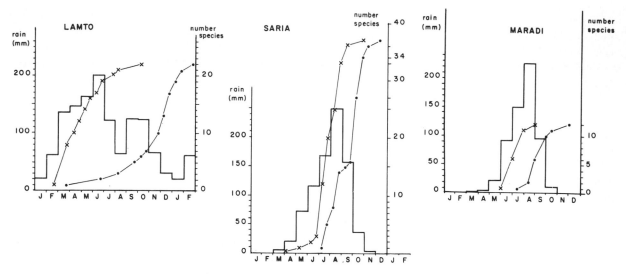

Fig. 11.12. Fifteen days summation of the number of acridid species breeding once a year, at the time of hatching (X) and of complete development (●), in three different West African savanna study sites, Lamto (Ivory Coast), Saria (Upper Volta) and Maradi (Niger). The histograms represent the average monthly rainfall. (Based on data by Gillon, Lecoq and Launois Luong.)

completed at different times of the year in different localities, but always at the end of the second month of the rainy season in the three study sites concerned. According to Lecoq (pers. comm., 1979), the bend of the hatching graph separates the species with an embryonic diapause, and those undergoing an early imaginal moult, from those with a late imaginal moult. In all three cases, the latest hatchings occur long after the beginning of the emergence of the early hatching species.

Seasonal variations in numbers of multivoltine acridid species

The case of the acridid *Rhabdoplea munda*, previously described, shows that the number of breeding generations per year for a given species can vary in different parts of its range. The same thing happens in multivoltine species living in different environments. *Oedalus senegalensis*, for example, can breed once, twice or three times annually, not only in different localities, but also in the same area, depending upon the yearly environmental conditions (Batten, 1969; Launois, 1978b). In this case the number of broods per year is higher in more humid conditions. The reverse can also occur, however; *Eucoptacra anguliflava* and *Gastrimargus africanus* have only one generation per year at Lamto, Ivory Coast, against two and three re-

spectively in the drier savannas of Upper Volta. The population turnover of these two acridids is apparently slower in what one might consider to be a more favourable environment. Working with crickets in Japan, Masaki (1978) has shown how complex the determinism of latitudinal variation on the annual number of generations is, involving as it does a subtle interplay of genetic and ecological variables.

Large size acridids are generally those whose life-cycles are the longest, about a year, involving an embryonic diapause. A large locust which has several generations annually implies an unusual ability to synthesize living tissues and to metabolize a considerable amount of food. Such is indeed the case for *Locusta migratoria* and *Schistocerca gregaria* which devastate natural and cultivated vegetation when they feed.

The most important synchronizer of the major stages of the life-cycle, hatching particularly, is the occurrence of the first rains at the end of the dry season, as established by Louveaux (1972) in *Locusta migratoria* in the Malagasy grasslands.

Seasonal variations in numbers of savanna pentatomids

Univoltine species are rare among the small-size pentatomids of the grass layer in West African wet savannas. At Lamto, only one single brooded species (*Deroplax nigropunctata*) out of a total of fifteen, has been found by D. Gillon (1974). Mating takes place in October, followed by the death of the males, and laying occurs in November and December (Fig. 11.13). Young stages of *Deroplax nigropunctata* develop during the dry season, and an imaginal diapause takes place during the rainy season. All the other sympatric species of pentatomids are multivoltine, with an inactivity period during the dry season. *Ennius morio* for instance, has two distinct breeding generations with population peaks in June and November, the beginning and the end of the rainy season (Fig. 11.14).

To summarize, the life-cycle of sympatric heterometabolous insects in a given savanna can be extremely variable, even within one univoltine species. However, definite population peaks can be observed in West African savannas, both at the

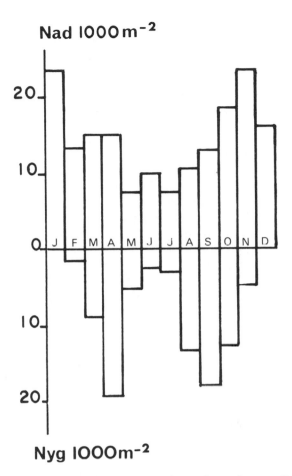

Fig. 11.14. Seasonal variations in numbers of young (*Nyg*, bottom histograms) and adults (*Nad*, top histograms) of *Ennius morio* (Pentatomidae, Pentatominae) in the Lamto savanna. (After D. Gillon, 1973.)

onset and at the end of the rainy season, as noticed long ago by Lamotte (1947).

Seasonal variations in numbers of holometabolous insects

The seasonal variations in abundance of holometabolous insects are poorly understood in tropical savannas, because the study of their population dynamics involves different sampling techniques for larvae, pupae and adult individuals. Nevertheless, the abrupt increase in the numbers of most butterflies and moths at the beginning of the rainy season implies a synchronizing role (direct or indirect) of the first rains, as in the case of heterometabolous insects. However, many species also

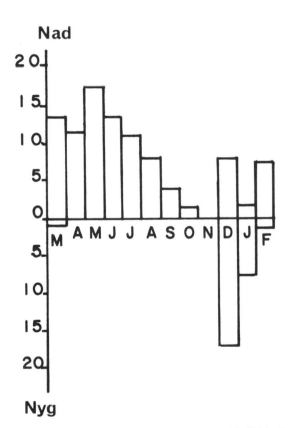

Fig. 11.13. Seasonal variations in numbers of individuals per 1000 m^{-2} of young (*Nyg*, bottom histograms) and adults (*Nad*, top histograms) of *Deroplax nigropunctata* (Pentatomidae, Scutellerinae) in the Lamto savanna. (After D. Gillon, 1973.)

have a second breeding generation at the end of the rainy season.

Adult holometabolous insects are often long-lived. Some carabid beetles can live for a year at Lamto (Lecordier et al., 1974), and tenebrionid beetles can live for several years (Y. Gillon and Gillon, 1974). It is therefore difficult to establish the number of breeding generations, and it becomes necessary to study both the seasonal variations in the larval populations, and the breeding behaviour of the adults. In the case of the rutelid beetle *Anomala curva* studied by Girard and Lecordier (1979) at Lamto, adults occur nine months out of twelve, but the duration of the larval stage is less than six months. Adults are more often seen flying around from February to April and from September to November. Two breeding generations are therefore likely, especially since all the other sympatric rutelid species are least often captured in light traps at the peak of both the rainy and dry season (Fig. 11.15). But the situation may be further complicated by differences between adjacent micro-habitats within the same savanna landscape. In pure Lamto grassland the density of adult *Anomala curva* exhibits two peaks in abundance (in February and in October) in the upper layer of the soil where they take shelter outside their periods of flying activity. But the populations sampled at the base of *Borassus* palms, whether in the middle of the savanna or near the forest edge, show a single peak in June or July (Fig. 11.16).

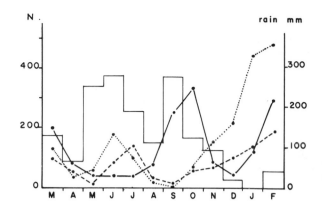

Fig. 11.16. Seasonal variations in numbers of rutelid beetles per 1000 m² in different micro-habitats of the Lamto savanna: in the open grassland (heavy line), at the bottom of *Borassus* palms (dashed line), and at the forest edge (dotted line). Histogram of the rains in the background. (After Girard and Lecordier, 1979.)

Generally speaking, polyvoltine species have a better reproductive potential than univoltine species, and some also benefit from the adaptive advantages of seasonal delays in development. In these species, diapause and quiescence are apparently not under strict genetic control, but are triggered by external synchronizers. Unfortunately very little is known about these timing devices in tropical latitudes; the seasonal variations of photoperiod are very limited, and the onset of the first rains alone is a highly unreliable predictor of the rainy season, at least in the drier (Sahelian) savannas.

To complicate the issue further, the breeding of many species under laboratory conditions has disclosed the frequency of polymorphism among savanna insects, as well as the frequent occurrence of an overlapping of generations.

Bivoltine butterfly species can be represented by different forms at different times of the year. The wet season form of the nymphalid *Precis octavia* is bright orange with black markings, while the dry season form is intricately patterned with dark markings and blue spots and has almost no orange. In this case, rainfall and humidity, rather than temperature, are considered as the main stimulus affecting the production of these seasonal forms (Owen, 1971, 1976). Some satyrid butterflies which have quite distinct wet- and dry-season forms in the Sudanian savanna, also have intermediate forms (Fig. 11.17) in the Guinean savanna of the Ivory

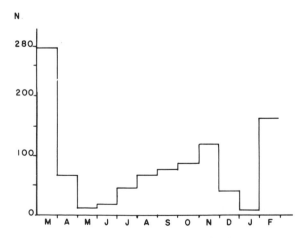

Fig. 11.15. Seasonal changes in above-ground activity of rutelid beetles in the Lamto savanna, based upon the number of individuals caught in light traps. (After Girard and Lecordier, 1979.)

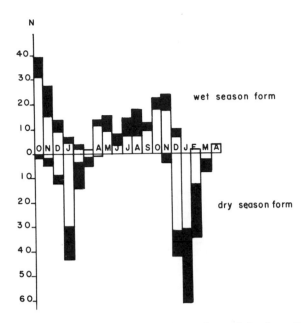

Fig. 11.17. Seasonal variations in numbers of the butterfly *Bicyclus milyas* (Satyridae) in the Lamto savanna. Females are indicated in black for both the rainy-season form (top histograms) and the dry-season form (bottom histograms). (Data from Condamin and Vuattoux, 1972.)

Coast, further south, where seasonal conditions are less contrasted (Condamin and Vuattoux, 1972). In many cases males outnumber females, the latter developing more slowly. In Guinean savannas where the dry season is short, dry-season forms appear much more quickly than the wet-season forms, as shown by *Bicyclus milyas* at Lamto (Fig. 11.16).

THE SEASONAL MIGRATIONS OF SAVANNA INSECTS

One of the major differences between the insect fauna of wet and dry savannas is the lower incidence of seasonal migrations in the former *versus* its frequent occurrence in the latter. In the Guinean and more humid savannas, often interspersed with gallery forests (the forest savanna mosaic), the seasonal variations of climate are far less marked and much more predictable than in the drier savannas. Furthermore, in the latter situation, the seasonal movements of the intertropical convergence zone carry with them a large number of adult insects adapted to long-range movements.

Acridid migrations

It is noteworthy that acridids from the dry savanna areas are far more mobile than those of the wet savannas. They are much more prone to be captured in light traps, especially because their activity peaks take place during the night. Clark (1969), for instance, has described the night movements of the Australian plague locust *Chortoicetes terminifera* in relation to storms. At Maradi, Niger, Launois (1978b) has also found a positive correlation between the number of *Oedaleus senegalensis* caught in light traps and temperature. The ability to move over long range is correlated with obvious morphological characteristics. Out of the 103 species of acridids studied by Launois (1978b) in the Sudanian and Sahelian savannas of West Africa, only a few species of *Chrotogonus* belonged to the micropterous form, whereas 10 out of the 101 species of the Lamto Guinean savanna had micropterous females, according to Y. Gillon (1973b). It has already been mentioned that almost all *Rhabdoplea munda* individuals are short-winged in the Guinean savanna of the Ivory Coast during the rainy season, whereas the long-winged form is the rule in the Sahel. This is not to say that seasonal movements do not occur among wet savanna acridids; they do, but far less frequently. In the Ivory Coast, *Ornithacris turbida* and *O. magnifica* probably follow the harmattan during the dry season, entering the Sudanian savannas of West Africa at that time; if this were not so, it would be difficult to understand why the young instars of these species are far less numerous than the adults in these areas (Y. Gillon, 1973b).

The seasonal changes in demographic structure of *Oedalus senegalensis* populations at Saria, Upper Volta, have enabled Lecoq (1978a) to establish arrival and departure schedules for this insect in the area (Fig. 11.18). The numbers of first-generation adult individuals falls sharply in June, before the maturation of ovaries in the females. This outward movement corresponds to the northward movement of the intertropical convergence zone. Four months later, adult females of the third generation suddenly appear in numbers, this inward movement coinciding with the southward movement of the intertropical convergence zone, at the end of the rains. Similar observations have been made on *Catantops haemorrhoidalis* in the same area

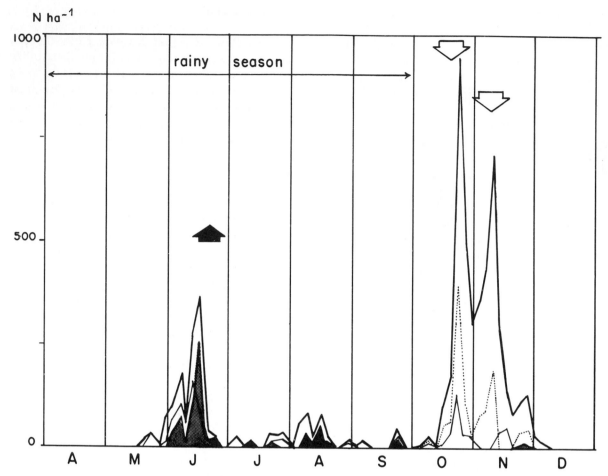

Fig. 11.18. Frequency of occurrence of *Oedalus senegalensis* at Saria, Upper Volta. Heavy line: overall population density; hachured: females with soft cuticle; thin line: non-breeding females with hard cuticle; dashed line: breeding females. The black vertical arrow indicates the departure of adults, and the two white arrows the arrival of other adults. (After Lecoq, 1978a.)

(Duranton et al., 1979). Therefore, the beginning and the end of the rains play as decisive a role in triggering mass population movements as they do in initiating breeding among insect communities of these savannas.

Direct observations by radar operating at ground level (Schaeffer, 1976; Reynolds and Riley, 1979a,b), or from aircraft (Rainey, 1975), have provided much information on the altitude, direction and importance of insect flights. It has even been possible to capture some flying individuals within the night swarms, using nets towed by slow-flying planes. Even species long considered as sedentary and solitary, such as *Aiolopus simulatrix* and *Catantops axillaris*, have been captured in this way (Schaeffer, 1976). The build-up of these

swarms takes place progressively, and probably increases the opportunity for mating (Rainey, 1976). This might particularly be the case when swarming occurs regularly in well-defined localities, such as large river valleys in Mali (Reynolds and Riley, 1979a).

Among true migratory locusts the crowding together of many young nymphs results in increased activity, which in turn is associated with the development of morphological characteristics of a gregarious phase. Such a gregarization process results in the well-known locust plagues. In savanna areas plague locusts are represented by several species of *Anacridium* and *Nomadacris*, which feed mostly on trees and shrubs, as well as by *Schistocerca americana* in the New World and

Schistocerca gregaria and *Locusta migratoria* in the Old World. Of these last two species, the former is most frequent in sub-desertic areas, and the latter in Sahelian savannas, reaching North Africa and even southern Europe on rare occasions. The gregarious phase of *Locusta migratoria* has its outbreak area in the flood plains of the middle Niger Valley, from where its swarms invade large regions.

Solitary phase individuals of this species, however, can also undertake large-scale movements, possibly enabling four different generations to follow one another in the course of a year. Eggs are laid during the dry season in the south of the outbreak area, as soon as the floods recede; they give rise to adults which move northward and breed there. The next generation lays its eggs at the onset of the rains, and the third generation moves further north during the short rains; it is the adults of the fourth generation which move back towards the south.

A similar succession of four generations in solitary *Locusta migratoria* has also been described in southwest Madagascar by Lecoq (1975). The largest individual flight range recorded among colour-marked solitary locusts is 300 km (Davey, 1956, 1959).

Air movements usually carry locusts towards low-pressure areas, where the occurrence of rainfall is the most likely, thus optimizing their reproductive success and the chances of survival of the first instars.

Migrations of the cotton stainers (*Dysdercus* spp.)

Long-range migratory movements are also well known among other savanna insects: the army-worms, caterpillars of the noctuid moth *Leucania unipuncta*, and the cotton stainers — pyrrhocorid bugs belonging to the genus *Dysdercus*, which feed upon the seeds of Malvales.

The long-range movements of cotton stainers had already been suspected by Golding (1928) in Nigeria, but a more comprehensive study of *Dysdercus voelkeri*, carried out in Ivory Coast by Duviard (1977), threw a new light on the problem.

Two kinds of large-scale seasonal movements were identified: (1) a dry-season invasion of Guinean savannas, the southernmost part of the species range, during which great damage is done to cotton fields; and (2) an early rainy-season migration in the Sudanian savannas, the northernmost part of the *D. voelkeri* range. This migration has not been given much attention in the past, since it causes far less damage to crops (Sarel Whitfield, 1933).

These seasonal movements are accompanied by a change in colour of the insects. Those moving south are yellow in colour; in Guinean savannas they give rise to a new generation whose adults are red. This generation in turn produces orange individuals at the beginning of the rains before disappearing for good from the Guinean savannas, at the same time that their staple food, the seed of Malvales, becomes unavailable. Most of the rainy season migrants in the Sudanian savannas are also orange-coloured, but their local progeny is yellow — the very colour of the bugs moving south later on with the intertropical convergence zone.

Duviard's interpretation of the seasonal movements and colour changes of *D. voelkeri* explains the observed facts better than the alleged short-range movements between cotton fields and wild host-plants postulated by Pearson (1958).

In moving seasonally between two adjacent savanna zones, *D. voelkeri* populations actually manage to maintain themselves in a very stable climatic environment: the dry season in the Guinean savannas, which is so frequently interspersed with showers, does not differ much, in fact, from the rainy season in the Sudanian savannas.

The "natural marking" of *D. voelkeri* generations might itself be due to seasonal changes in illumination, a more intense light turning red pterines into yellow ones. Vuillaume (1969) has indeed described the influence of solar radiation upon the synthesis of chromoproteins.

The possible role of a dietary change is far more questionable; the body colouring of the adults is achieved before they start feeding (Duviard, 1977). Moreover, the first meal of an adult female coincides with the end of its ability to undertake large-scale, and even small-scale movements, its flight muscles being more or less destroyed by histolysis (Edwards, 1969). Group living might also play a role at this stage to reinforce the effect of feeding (Gatehouse and Hall, 1976). Immigrant females cannot become sexually mature before beginning to feed, and mating also stimulates both muscular histolysis and ovarian maturation.

The circumstances of migrating flights

The flight periodicity, direction and range of migrating swarms of a few savanna insects are now adequately documented. Yet very little is known on the conditions of these long-range movements.

For an insect to benefit from an air current means that wind velocity must at least be the same as the velocity this insect reaches in active flight. Schaeffer (1976) indeed recorded that locust swarms tracked by radar had an average speed of 3 m s^{-1}, and that flight duration averaged three hours — figures well in line with the performances of insects flying in laboratory conditions.

To take advantage of the kinetic energy of a low-level jet stream, the migrating insect first has to reach a certain elevation, as the lower levels of air move very slowly or not at all. To do so, it must first cross the boundary layer — that is, the layer of air near the ground within which insects are able to control their movements relative to the ground because their flight speed exceeds wind speed (Taylor, 1974).

The directions followed by insects at take-off are not very meaningful; environmental conditions such as light intensity, air temperature, atmospheric moisture and barometric pressure are more significant. They play a definite role in the initiation of upwelling air currents (thermals). Some indications of the optimal conditions for night migration can be drawn from the results of light-trapping; more insects are generally caught in pitch-dark nights, although giant water bugs (Belostomatidae), *Diplonychus nepoides* particularly, prefer full moon nights (Duviard, 1974). It is possible that they are attracted by the reflexion of moonlight on the savanna pools (Cullen, 1969).

As mentioned earlier, large-scale migratory movements often take place at the turning points between the dry and rainy seasons. This is why it is best to conclude, at least in West Africa, that the four significant periods in the annual cycle are: first, the dry season; then, the tornado period corresponding to the advance of the intertropical convergence zone; third, the rainy season; and fourth, the retreat of the intertropical convergence zone. The "short dry season" occurring in the middle of the rains in subequatorial climates does not appear to have much phenological importance in Guinean savannas.

All locust migration circuits result in swarms moving in a direction which is most likely to take them to the area containing the most suitable habitats for that particular season. Some swarms, however, can be led by strong winds in a wrong direction to perish *en masse* in the sea. There are many instances of swarms of *Schistocerca gregaria* ending up in the Red Sea, and similar incidents have been recorded for other migratory species: *Chortoicetes terminifera* in Australia (Farrow, 1975) and *Zonocerus variegatus* in the Ivory Coast (Duviard, 1977).

CONCLUSIONS

In spite of the many differences found among the arthropod faunas of the various categories of savannas, some basic similarities do exist which confer a pattern common to all of them, at least in Africa.

First of all, the ubiquity of acridids, which together with caterpillars during the rainy season make up the major groups of primary consumers. It might well be that in some cases their trophic impact on the vegetation is more important than that of grazing mammals.

Among secondary consumers, the same dominant role is played by spiders and ants (see Ch. 24). In West African savannas four families of spiders represent 60 to 80% of the total number of spiders present, both at Lamto and at Fété Olé (D. Gillon and Gillon, 1974; Blandin, 1974).

Most of the major faunal differences between dry and wet savannas are due, at least in Africa, to other taxonomic groups: Solifuga, scorpions, Thysanura, Embioptera and tenebrionid beetles are true faunistic "markers" of the driest savannas, whereas Phalangida, cockroaches, tetrigids and tridactylids are always scarce, if not absent, in such habitats. The dichotomy between the more seasonal savannas and those closer to the forest edge is even more marked during the dry season; at that time of the year some groups, such as jassids or carabid beetles, and some life stages such as caterpillars, are no longer active though present during the rainy season.

The second common characteristic of all arthropod faunas in savanna ecosystems is the high prevalence, at the individual as well as at the population level, of adaptive mechanisms which help the species to avoid the most drastic conditions of the dry season. Such "escape strategies" take advantage of a wide range of morphological, physiological and behavioural adaptations which permit the animals either to spend the dry season in a more or less torpid stage, or to move away from the sun-parched grass to more congenial environments, some centimetres below the ground or hundred of kilometres away.

REFERENCES

Batten, A., 1969. The Senegalese grasshopper *Oedaleus senegalensis* Krauss. *J. Appl. Ecol.*, 6: 27–45.

Bellier, L., Gillon, D., Gillon, Y., Guillaumet, J.L. and Perraud, A., 1969. Recherches sur l'origine d'une savane incluse dans le bloc forestier du Bas Cavally (Côte d'Ivoire) par l'étude des sols et de la biocoenose. *Cah. ORSTOM, Sci. Biol.*, 10: 65–94.

Blandin, P., 1972. Recherches écologiques sur les araignées de la savane de Lamto (Côte d'Ivoire). Premières données sur les cycles des Thomisidae de la strate herbacée. *Ann. Univ. Abidjan*, E 5: 241–264.

Blandin, P., 1974. Les peuplements d'araignées de la savane de Lamto. *Bull. Liaison Chercheurs Lamto, Numéro Spéc.*, III: 107–135.

Burtt, E., 1951. The ability of adult grasshoppers to change colour on burnt ground. *Proc. R. Entomol. Soc. Lond.*, A 26: 45–48.

Center, T.D. and Johnson, C.D., 1974. Coevolution of some seed beetles (Coleoptera: Bruchidae) and their hosts. *Ecology*, 55: 1096–1103.

Clark, D.P., 1969. Night flights of the Australian plague locust, *Chortoicetes terminifera* Walk, in relation to storms. *Aust. J. Zool.*, 17: 329–352.

Clarke, C.A. and Scheppard, P.M., 1960. The evolution of mimicry in the butterfly *Papilio dardanus*. *Heredity*, 14: 163–173.

Condamin, M. and Vualtoux, R., 1972. Les Lépidoptères Satyridae des environs de Lamto (Côte d'Ivoire). *Ann. Univ. Abidjan*, E 5: 31–57.

Cullen, M.J., 1969. The biology of giant water bugs (Hemiptera: Belostomatidae) in Trinidad. *Proc. R. Entomol. Soc. Lond.*, A 44: 123–136.

Davey, J.T., 1956. A method of marking isolated adult locusts in large numbers as an aid to the study of their seasonal migrations. *Bull. Entomol. Res.*, 46: 797–802.

Davey, J.T., 1959. The African Migratory Locust (*Locusta migratoria migratorioïdes* R. & F.) (Orth.) in the central Niger Delta. II. The ecology of *Locusta* in the semi-arid lands and seasonal movements of populations. *Locusta*, No. 7: 180 pp.

Dempster, J.P., 1963. The population dynamics of grasshoppers and locusts. *Biol. Rev.*, 38: 490–529.

Dingle, H. and Khamale, C.P.M., 1972. Seasonal changes in insect abundance and biomass in an east african grassland, with reference to breeding and migrations in birds. *Ardea*, 59: 216–221.

Duranton, J.F. and Lecoq, M., 1980. Ecology of locusts and grasshoppers (Orthoptera, Acrididae) in Sudanese west Africa. I. Discriminant factors and ecological requirements of acridian species. *Acta Oecologica, Sér. A, Oecol. Gen.*, 1: 151–164.

Duranton, J.F., Launois, M., Launois-Luong, M.H. and Lecoq, M., 1979. Biologie et écologie de *Catantops haemorrhoidalis* en Afrique de l'ouest (Orthopt. Acrididae). *Ann. Soc. Entomol. Fr. (N.S.)*, 15(2): 319–343.

Duviard, D., 1970. Recherches écologiques dans la savane de Lamto (Côte d'Ivoire); l'entomocénose de *Vernonia guineensis* Benth. (Composées). *Terre Vie*, 24: 62–79.

Duviard, D., 1974. Flight activity of Belostomatidae in central Ivory Coast. *Oecologia*, 15: 321–328.

Duviard, D., 1977. *Ecologie de* Dysdercus voelkeri *en Afrique occidentale: migrations et colonisation des nouveaux habitats*. Thesis, Université de Paris-Sud, Paris, 168 pp.

Edmunds, M., 1972. Defensive behaviour of Ghanaian praying mantids. *Zool. J. Linn. Soc.*, 51: 1–32.

Edmunds, M., 1974. *Defence in Animals. A Survey of Anti-Predator Defences*. Longman, London, 357 pp.

Edmunds, M., 1976. The defensive behaviour of Ghanaian praying mantids with a discussion of territoriality. *Zool. J. Linn. Soc.*, 58: 1–37.

Edmunds, M., 1978. On the association between *Myrmarachne* spp. (Salticidae) and ants. *Bull. Br. Arachnol. Soc.*, 4: 149–160.

Edwards, F.J., 1969. Environmental control of flight muscle histolysis in the bug *Dysdercus intermedius*. *J. Insect Physiol.*, 15: 2013–2020.

Ergene, S., 1950. Untersuchungen über Farbanpassung und Farbwechsel bei *Acrida turrita*. *Z. Vergl. Physiol.*, 32: 530–551.

Farrow, R.A., 1975. Offshore migration and the collapse of outbreaks of the Australian Plague Locust (*Chortoicetes terminifera* Walk.) in South-east Australia. *Aust. J. Zool.*, 23: 569–595.

Ford, E.B., 1971. *Ecological Genetics*. Chapman and Hall, London, 3rd ed., 410 pp.

Gatehouse, A.G. and Hall, M.J.R., 1976. The effect of isolation on flight and on the preoviposition period in unmated *Dysdercus superstitiosus*. *Physiol. Entomol.*, 1: 15–19.

Gillon, D., 1974. Etude biologique des espèces d'Hémiptères pentatomides d'une savane préforestière de Côte d'Ivoire. *Ann. Univ. Abidjan*, E 7: 213–303.

Gillon, D. and Gillon, Y., 1974. Comparaison du peuplement d'invertébrés de deux milieux herbacés ouest africains: Sahel et savane préforestière. *Terre Vie*, 28: 429–474.

Gillon, Y., 1971. The effect of bush fire on the principal acridid species of an Ivory Coast Savanna. In: *Proc. Annu. Tall Timbers Fire Ecology Conference*, 1971, pp. 419–471.

Gillon, Y., 1972. Caractéristiques quantitatives du développement et de l'alimentation d'*Anablepia granulata* (Ramme, 1929) (Orthoptera: Gomphocerinae). *Ann. Univ. Abidjan*, E 5: 373–393.

Gillon, Y., 1973a. Bilan énergétique de la population d'*Orthochtha brachynecmis* Karsch, principale espèce acridienne de la savane de Lamto. *Ann. Univ. Abidjan*, E 6: 105–125.

Gillon, Y., 1973b. *Etude écologique quantitative d'un peuplement acridien en milieu herbacé tropical.* Thesis, Université de Paris VI, Paris, 322 pp.

Gillon, Y., 1974. Variations saisonnières de populations d'acridiens dans une savane préforestière de Côte d'Ivoire. *Acrida*, 3: 129–174.

Gillon, Y. and Gillon, D., 1965. Recherche d'une méthode quantitative d'analyse du peuplement d'un milieu herbacé. *Terre Vie*, 19: 378–391.

Gillon, Y. and Gillon, D., 1967. Recherches écologiques dans la savane de Lamto (Côte d'Ivoire). Cycle annuel des effectifs et des biomasses de la strate herbacée. *Terre Vie*, 21: 262–277.

Gillon, Y. and Gillon, D., 1973. Recherches écologiques sur une savane sahélienne du Ferlo septentrional, Sénégal. Données quantitatives sur les Arthropodes. *Terre Vie*, 27: 297–323.

Gillon, Y. and Gillon, D., 1974. Recherches écologiques sur une savane sahélienne du Ferlo septentrional, Sénégal. Données quantitatives sur les Ténébrionides. *Terre Vie*, 28: 296–306.

Girard, Cl. and Lecordier, Ch., 1979. Les Rutélides de la savane de Lamto (Côte d'Ivoire): Diversité et structure de quelques peuplements (*Coleoptera*). *Ann. Soc. Entomol. Fr. (N.S.)*, 15: 171–178.

Golding, F.D., 1928. Notes on the bionomics of Cotton stainers (*Dysdercus*) in Nigeria. *Bull. Entomol. Res.*, 18: 319–334.

Halffter, G. and Reyes-Castillo, P., 1975. Analisis cuantitativo de la fauna de Arthropodes de Laguna Verde. *Fol. Entomol. Mex.*, 30: 1–32.

Hinton, H.F., 1960. Cryptobiosis in the larva of *Polypedilum vanderplanki* Hint. (Chironomidae). *J. Insect Physiol.*, 5: 286–300.

Janzen, D.H., 1977. The interaction of seed predators and seed chemistry. In: *Coll. Int. du CNRS*, 265: 415–428.

Lachaise, D., 1974. Les peuplements de Diptères de la savane de Lamto. *Bull. Liaison Chercheurs Lamto, Numéro Spéc.*, III: 59–105.

Lachaise, D., 1979. *Spéciation, coévolution et adaptation des populations de Drosophilides en Afrique tropicale.* Thesis, Université de Paris VI, Paris, 294 pp.

Lamotte, M., 1947. Recherches écologiques sur le cycle saisonnier d'une savane guinéenne. *Bull. Soc. Zool. Fr.*, 72: 88–90.

Lamotte, M., Aguesse, P. and Roy, R., 1962. Données quantitatives sur une biocenose ouest africaine: la prairie montagnarde du Nimba (Guinée). *Terre Vie*, 16: 351–370.

Launois, M., 1978a. *Manuel pratique d'identification des principaux Acridiens du Sahel.* Gerdat, Paris, 303 pp.

Launois, M., 1978b. *Modélisation écologique et simulation opérationnelle en acridologie. Application à* Oedaleus senegalensis (*Krauss, 1877*). Gerdat, Paris, 212 pp.

Lecoq, M., 1975. *Les déplacements par vol du criquet migrateur malgache en phase solitaire: leur importance sur la dynamique des populations et la grégarisation.* Thesis, Université de Paris XI, Paris, 272 pp.

Lecoq, M., 1978a. Biologie et dynamique d'un peuplement acridien de zone soudanienne en Afrique de l'Ouest (Orthoptera, Acrididae). *Ann. Soc. Entomol. Fr. (N.S.)*, 14: 603–681.

Lecoq, M., 1978b. Les déplacements par vol à grande distance

chez les Acridiens des zones sahélienne et soudanienne en Afrique de l'ouest. *C.R. Acad. Sci. Paris*, D 286: 419–422.

Lecordier, Ch. and Girard, C., 1973. Peuplements hypogés de carabiques (Col.) dans la savane de Lamto (Côte d'Ivoire). *Bull. I.F.A.N.*, 35(A): 361–392.

Lecordier, Ch., Gillon, D. and Gillon, Y., 1974. Les carabiques (Col.) de Lamto (Côte d'Ivoire). Etude d'un cycle saisonnier. *Ann. Univ. Abidjan*, E 7: 533–585.

Lewontin, R.C., 1965. Selection for colonizing ability. In: H.G. Baker and G.L. Stebbins (Editors), *The Genetics of Colonizing Species.* Academic Press, New York, N.Y., pp. 202–210.

Louveaux, A., 1972. Action des facteurs climatiques sur le développement de *Locusta migratoria capito* Sauss. à Madagascar sur le plateau de l'Horombe. *Ann. Zool. Ecol. Anim.*, Fascicule hors série, pp. 189–224.

Masaki, S., 1978. Seasonal and latitudinal adaptations in the life cycles of crickets. In: H. Dingle (Editor), *Evolution of Insect Migration and Diapause.* Springer-Verlag, Berlin, pp. 72–100.

Morello, J., 1970. Ecologia del Chaco. *Bol. Soc. Arg. Bot.*, 11(Supl.): 161–174.

Okay, S., 1956. The effect of temperature and humidity on the formation of green pigment in *Acrida bicolor* (Thunb). *Arch. Int. Physiol. Biochem.*, 64: 80–91.

Owen, D.F., 1971. *Tropical Butterflies. The Ecology and Behaviour of Butterflies in the Tropics, with Special Reference to African Species.* Clarendon Press, Oxford, 214 pp.

Owen, D.F., 1976. *Animal Ecology in Tropical Africa.* London, Longman, 132 pp.

Pearson, E.O., 1958. *The Insect Pests of Cotton in Tropical Africa.* Commonwealth Institute of Entomology, London, 355 pp.

Phipps, J., 1968. The ecological distribution and life cycles of some tropical African grasshoppers (Acrididae). *Bull. Entomol. Soc. Nigeria*, 1: 71–97.

Pollet, A., 1972. Contribution à l'étude du peuplement d'insectes d'une lisière entre forêt galerie et savane éburnéennes. I. Données générales sur les phénomènes. *Ann. Univ. Abidjan*, E 5: 395–473.

Pollet, A., 1974. Contribution à l'étude des peuplements d'insectes d'une lisière entre savane et forêt galerie éburnéennes. II. Données écologiques sur les principales espèces constitutives de quelques grands groupes taxonomiques. *Ann. Univ. Abidjan*, E 7: 315–357.

Rainey, R.C., 1969. Effects of atmospheric conditions on insect movements. *Q. J. R. Meteorol. Soc.*, 95: 424–434.

Rainey, R.C., 1975. New prospects for the use of aircraft in the control of flying insects and in the development of semi-arid regions. In: *5th Int. Agric. Aviation Congr., Kenilworth*, pp. 229–233.

Rainey, R.C., 1976. Flight behaviour and features of the atmospheric environment. *Symp. R. Entomol. Soc. Lond.*, 7: 75–112.

Reynolds, D.R. and Riley, J.R., 1979a. Radar observations of concentrations of insects above a river in Mali, West Africa. *Ecol. Entomol.*, 4: 161–174.

Reynolds, D.R. and Riley, J.R., 1979b. Radar based studies of the migration flight of grasshoppers in the Middle Niger area of Mali. *Proc. R. Soc. Lond.*, B 204: 67–82.

Rowell, C.H.F. and Cannis, R.L., 1971. Environmental factors affecting the green brown polymorphism in the Cyrtacanthacridine grasshopper *Schistocerca vaga* (Scudder). *Acrida*, 1: 69–77.

Sarel Whitfield, F.G., 1933. The bionomics and control of *Dysdercus* (Hemiptera) in the Sudan. *Bull. Entomol. Res.*, 24: 301–313.

Schaeffer, G.W.S., 1976. Radar observations on insect flight. *Symp. R. Entomol. Soc. Lond.*, 7: 157–197.

Taylor, L.R., 1974. Insect migration, flight periodicity and the boundary layer. *J. Anim. Ecol.*, 43: 225–238.

Vuillaume, M., 1969. *Les pigments des invertébrés. Biochimie et biologie des colorations.* Masson, Paris, 184 pp.

Chapter 12

AMPHIBIANS IN SAVANNA ECOSYSTEMS

MAXIME LAMOTTE

INTRODUCTION

Of the three existing orders of Amphibia, only the Anura are widely distributed in tropical savannas, Urodela and Apoda being scarcely, if at all, present in this environment. However, toads, frogs and tree frogs having, like any other typical amphibian, to live a double life — the first part in (or close to) water and the second part on land — cannot permanently live very far from water. Their distribution in the savanna landscape will therefore be closely linked with the presence of bodies of water, whether permanent or seasonal, where reproduction can take place.

The close dependence of amphibians on water is not limited to the need to find suitable reproductive sites. Being very sensitive to desiccation, they generally need a permanently damp atmosphere during the terrestrial stage of their life-cycle. To resist the highly seasonal conditions of most savannas, amphibians have to adopt one or more of the following strategies: (1) to live only in those parts of the savannas, like swamps or marshy areas, where humid conditions prevail throughout the year; (2) be active only during that part of the year where the atmospheric humidity is the highest, that is, the rainy season, and retreat during the dry season to live in some place where conditions are sufficiently moist to allow them to survive; (3) be active only at night when moisture is higher than during the day; or (4) live permanently underground in moist and loose soil, like the African Hemisinae.

The humidity constraint, much more imperative for amphibians than for any other class of terrestrial vertebrates, helps to explain much about their distribution, seasonal changes in density, life-cycle and activity rhythms in savanna areas.

ADAPTIVE CHARACTERISTICS

Pattern and morphology of savanna amphibians

A great many Anura living in tropical savannas share a number of colour and morphological characteristics which may be considered as adaptive to a predominantly grassland environment.

Many small climbing frogs spending most of their adult life in the grass layer display distinctive cryptic or disruptive patterns, as well as characteristic rest attitudes, whose concealing function can hardly be questioned (Fig. 12.1). In species like *Hyperolius nasutus* and *H. lamottei* which live among narrow and long grass blades, the longitudinal dark dorsolateral stripes are also very thin. Species living among broader leaves can either exhibit a cryptic coloration, being more or less uniformly green or tan, or adopt a mottled pattern. Others, like *Kassina senegalensis* or *K. cassinoides* have a disruptive pattern made of broader, sometimes broken, longitudinal dark stripes. *Phrynomerus microps* wears a reddish spot shaped like an arrow head on its back.

More strictly terrestrial Anura can also exhibit dark longitudinal stripes; this is the case of various species of *Ptychadena* and *Phrynobatrachus*, elongated frogs which edge their way through dense grass. On the other hand, toads of the genus *Bufo* which pursue their prey in more open places, often on reddish-brown lateritic soil, have a mottled cryptic pattern of dark and light dots.

Metatarsal cornified tubercles used for digging are found in many burrowing forms; they may become much enlarged in species inhabiting the more xeric areas (Fig. 12.2).

Spatulate toes and glandular discs are adaptations for climbing common to the anurans living

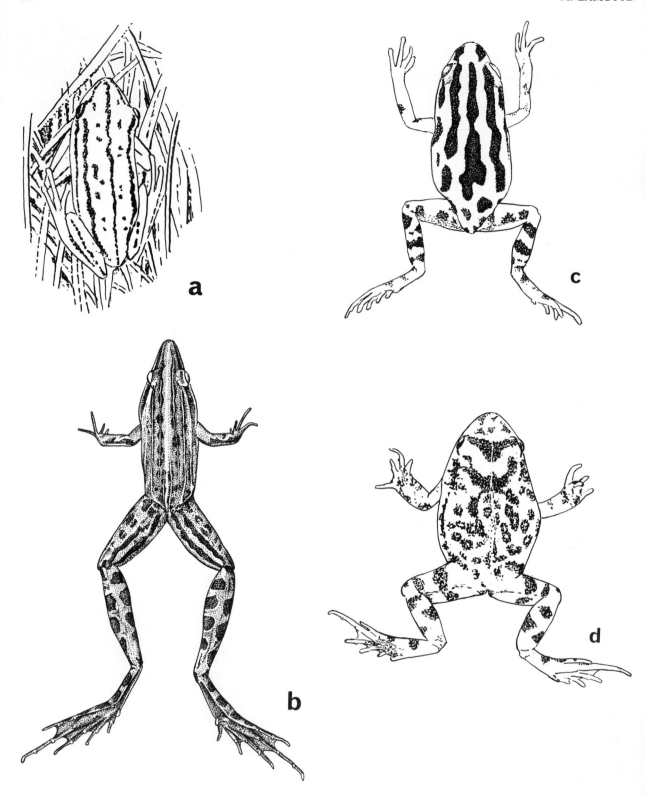

Fig. 12.1. Characteristic patterns of some West African savanna amphibians. a. *Hyperolius lamottei*. b. *Ptychadena tournieri*. c. *Kassina senegalensis*. d. *Tomoptera delalandii*.

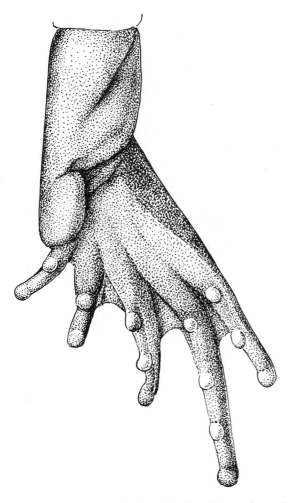

Fig. 12.2. Metatarsal cornified tubercles of *Leptopelis viridis*.

in long grass as well as in trees. They are particularly well developed on the fingers and toes of African tree frogs such as *Hyperolius*, *Kassina* and *Leptopelis*.

Circadian activity rhythms

As already mentioned, very few amphibians can withstand the climatic conditions which prevail in the open air during the day in tropical savannas. Consequently most species are active by night, and spend the day time in some shady shelter. By taking advantage of the relative dampness of the night to feed, a number of anurans can live in the dry savannas. Frequently their resistance to desiccation is also made easier by some adaptations of their integument: thickening of the epidermis, and in-

creased number of mucous glands and "warts". An increased secretion of mucus helps to keep the skin moist and to regulate body temperature through evaporative cooling of the surface. This is well exemplified by some African toads like *Bufo pentoni*, *B. xeros*, and some species of *Tomoptera* which even enter sub-desertic areas in Africa. Other dry savanna species are confined to the areas surrounding the seasonal ponds which are formed during the rains, and start to breed as soon as any major rains occur. Such is the case for *Pyxicephalus adspersus* which ranges from Senegal to southern Africa throughout the savanna belt.

Very few savanna anurans can manage to be active during the day, except by requiring special conditions. This is the case of *Phrynobatrachus francisci*, which frequently enters water, however, to prevent desiccation. In the mountain grasslands of Mount Nimba (Ivory Coast/Guinea/Liberia), the small viviparous toad *Nectophrynoides occidentalis* is also active in daylight during the rains, taking advantage of the quasi-permanent drizzle. During the dry season it lives underground (Lamotte, 1959).

Seasonal activity rhythms and breeding cycle

Free water and a damp atmosphere being an absolute necessity for the breeding of anurans, it is no wonder that the reproduction of tropical savanna frogs and toads is attuned to the rain schedule. This is especially important for those species which breed in temporary bodies of water, although anurans laying their eggs in permanent water bodies are also influenced by the seasonality of rainfall. The number and duration of the wet and dry seasons in a given locality varies greatly according to latitude and other geographic and physiographic factors as shown in Chapter 3. In some areas there are two wet and two dry seasons, of unequal duration, during the same calendar year; in other areas a single and long dry season can follow the rains. The further one proceeds from the rainforest areas, the more generally unpredictable becomes the rainfall, in time, duration and amount.

Very little is so far known on the ways in which most savanna amphibians spend the driest period of the year. The most interesting observations have been made in Australia. According to Tyler (1976) several species of *Cyclorana* and *Neobatrachus*, and one *Limnodynastes* species, all inhabitants of dry areas, secrete a special outer skin layer which separates from the remainder in one piece so that it forms a cocoon. Lee and Mercer (1967) found that this cocoon surrounded the frog and was complete across the eyes and cloaca, sometimes sealing the mouth as well. In two frogs that had been underground for seven and ten months respectively, the cocoon was dry and flexible, and attached only at the head and the extremities of the limbs. The cocoon was shown to reduce water loss. Other burrowing species, like *Cyclorana platycephalus* of Central Australia, store water in their subcutaneous lymph sacs, becoming surrounded by "loose, floppy bags of water" (Tyler, 1976). Some African species, like *Bufo pentoni*, are also capable of spending many months underground between their short breeding periods. It is possible, however, that they do not remain inactive during the whole dry season, and that they continue to feed upon termites, as the southern African Hemisinae (Wager, 1965) and some Australian frogs (Frazer, 1973). In the more xeric savannas, a number of anurans also dig underground chambers in which they lay (and brood) their eggs, sometimes kilometres away from free water. Such is the case for *Breviceps adspersus* in southern Africa, and of some species of *Arthroleptis* in West Africa (Lamotte and Lescure, 1977; see also Loveridge, 1975).

For those species of anurans living in the dry savannas which continue to depend on free water to lay their eggs, a first condition to ensure successful reproduction is the ability to breed as soon as any major rain occurs. The eggs must then hatch quickly and the tadpoles must develop extremely fast. This is what happens in the population of *Bufo pentoni* living in northern Senegal (Forge and Barbault, 1977). Females lay their eggs, up to 10 000 per clutch, 24 h after the first rains, when the temporary ponds are filled, and the embryonic development is very rapid, most tadpoles hatching about 30 h after egg laying. Later in the rainy season, up to four successive breeding episodes have been observed in the same pond, each laying of eggs following heavy rains. The Australian frog *Cyclorana platycephalus* also has a very rapid development; in a mere sixteen to eighteen days from egg laying the young emerge from the water as baby frogs. Good reviews of the biology of xeric amphibians have been given by Main (1968) and Mayhew (1968).

Wherever the rainy season lasts for more than three months, the triggering of spawning by the first rains is less rigid. There is even some evidence that the reactivity to rainfall changes during the course of the year, being higher at the beginning of the breeding season than later (Balinsky, 1969). Most savanna anurans tend to prefer temporary pools of standing water for breeding. Such water bodies can be mere seasonal pools in shallow depressions, or small ponds left in temporary river beds. The first are mostly available during the rains themselves, while the latter remain available for some time after the beginning of the next dry season. If both kinds occur in the same area, breeding opportunities can be greatly increased.

Whereas large species spawning in temporary rain pools like *Bufo pentoni*, *Tomoptera delalandi* and *Pyxicephalus adspersus*, cannot generally have more than one generation per year, small frogs and toads whose tadpoles develop quickly can have two generations yearly. Such is the case in Africa for the genus *Phrynobatrachus*. The frogs of the genus *Arthrolepis*, whose development is direct and takes place in moist soil, may also have two generations per year (Fig. 12.3). Multiple breeding is also likely in some larger species living in the palm savanna of the Ivory Coast, like *Ptychadena maccarthyensis* (Barbault and Trefaut Rodrigues, 1978).

The nature and distribution of water bodies, whether uniformly distributed or clumped, have a decisive influence on the abundance and distribution of savanna amphibians. These water bodies serve not only as breeding sites, but also as refuge areas where adults can concentrate and "aestivate"

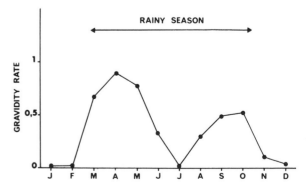

Fig. 12.3. Seasonal variations of the average gravidity rate in adult female *Arthroleptis poecilonotus* (after Barbault and Trefaut Rodrigues, 1979a).

during the dry season. Indeed, it must never be forgotten that amphibian eggs can never serve as a "resistant stage"; only adults can withstand the harsh conditions of the dry season. They do it in many different ways. Out of the breeding season *Kassina senegalensis* lives in burrows of mole rats in South Africa (Balinsky, 1969); and *Hyperolius nitidulus* and other Lamto tree frogs are found in the apical buds of tall *Borassus* palms (Vuattoux, 1968).

At least in the Transvaal highveld there is no significant dependence of anuran egg-laying either on the level of ambient temperature, or on its changes (Balinsky, 1969).

As in other ectothermic vertebrates, clutch size and yearly egg production are strongly influenced by the size of the mother: the larger the female, the greater the number of eggs laid (Fig. 12.4).

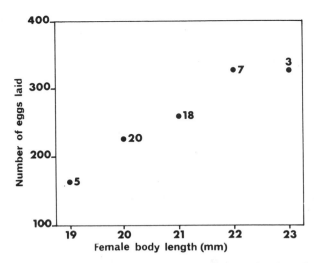

Fig. 12.4. Variation of the size of the egg clutch as a function of size of the mother in *Phrynobatrachus accraensis*. The number against each point indicates the number of clutches studied. (After Barbault and Trefaut Rodrigues, 1979b.)

AMPHIBIAN POPULATIONS AND COMMUNITIES

The number of amphibian species is never very high in tropical savannas. In Africa, for example, 24 species of anurans are known to exist in the Guinean savanna at Lamto (Lamotte, 1967; Barbault, 1974b), 18 species in the Nylsvley Nature Reserve in Transvaal (Jacobsen, 1977), 12 species in the Transvaal highveld near Blairgowrie (Balinsky, 1969), and only 8 in the Fété Olé Sahelian savanna, in northern Senegal (Bourlière, 1978).

In some situations the number of individuals can, however, be much more impressive. At Lamto, for instance, it can reach several thousand individuals per hectare in shallow depressions during the rains, whereas it remains very much lower on the plateaus (Table 12.1 and Fig. 12.5). The dry-season populations are always much reduced. In a given area, population density depends very much on the average duration of water in pools, which varies from year to year, and can range from one to ten months at Lamto (Fig. 12.6). In the Transvaal highveld, Balinsky (1969) has shown that if the waterholes last less than twenty days they do not give any species an opportunity to metamorphose, and any tadpoles that develop in such temporary water bodies inevitably die when they dry up. Pools lasting up to 45 days enable some species with a

TABLE 12.1

Numbers (individuals ha^{-1}) and biomasses (g ha^{-1} live weight) of anurans in the Lamto savannas, Ivory Coast (after Barbault, 1972)

	Average numbers		Biomass
	minimum	maximum	
Regularly burnt savanna			
Plateau	18	104	75–308
Depression	10–30	2000–3000	
Unburnt savanna			
Plateau	40	179	30–150
Depression	100–300	2000–3000	

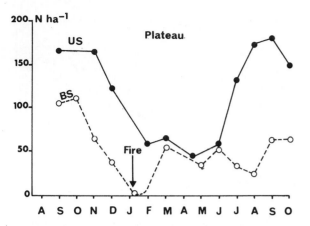

Fig. 12.5. The seasonal variations of the numbers of amphibians per hectare in a plateau savanna at Lamto (Ivory Coast) between September 1964 and October 1965. Seasonal changes in unburnt (*US*) and burnt (*BS*) parcels of the same savanna have been shown separately. (After Barbault, 1974a.)

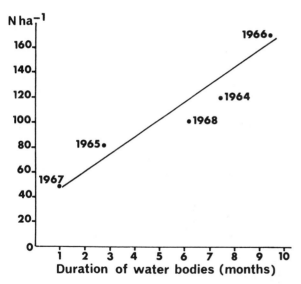

Fig. 12.6. Correlation between the maximum population density of amphibians per hectare each year, and the average duration of water in pools, at Lamto, Ivory Coast (after Barbault, 1972).

short larval life to metamorphose. Bodies of water lasting up to 90 days are necessary for other species to complete their development, and two *Rana* species can only metamorphose in permanent or very long lasting bodies of water.

The wettest lowland savannas where swampy areas are numerous are consequently those which are richest in anurans, both in number of species and in number of individuals. In West Africa, for

instance, they harbour numerous species of *Afrixalus*, *Hyperolius* and *Ptychadena*.

Seasonal flooding also occurs frequently on the vast lateritic plateaux and helps to maintain some anuran populations far from any permanent water. In Guinea, for example, representatives of the genus *Ptychadena* (*P. tournieri* and *P. maccarthyensis*) and *Phrynobatrachus* are found at ground level, whereas *Hyperolius lamottei*, *H. weidholzi*, and *Afrixalus fulvovittatus* feed among the grass, and *Leptopelis viridis* and *Kassina senegalensis* ambush their prey in the trees and shrubs. During the dry season, these anurans disappear underground or move to the nearest permanent swamp.

In high-grass savannas on more permeable soils, seasonal pools are far less numerous despite the higher rainfall — from 1300 to 1500 mm. Anurans are consequently less abundant. In West Africa such savannas are inhabited by a few tree frogs (*Leptopelis* sp.), some ground-living frogs (*Arthroleptis* sp., *Phrynobatrachus* sp.), some toads (*Bufo* sp.) and some large and more mobile species such as *Ptychadena maccarthyensis* and *P. oxyrhynchus*. Here again, the populations are not uniformly distributed, but are concentrated around the longer-lasting water bodies, some species probably moving seasonally into gallery forests and woodland areas.

In the driest savannas, species richness as well as population density decrease, anurans becoming more and more dependent upon the number and duration of the seasonal bodies of water. In the Sahelian savannas they may even fail to breed for one or several (?) consecutive years in case of prolonged drought.

In the mountain grasslands the situation varies from one mountain to the other. On Mount Nimba the viviparous *Nectophrynoides occidentalis* can reach a density of more than 7000 individuals per hectare during the rains, despite the lack of any rain pool, whereas species with aquatic tadpoles (*Ptychadena tournieri* and *Hyperolius lamottei*) are always scarce. *Arthroleptis crusculum*, a frog which lays its eggs in the soil and has a direct development, is not at all common either. Lower on the slopes *Astylosternus occidentalis*, a species whose tadpoles live in the mountain streams, is also found.

Further east, the mountain grasslands of Mount Cameroun (Cameroon) do not harbour any viviparous anuran, but only oviparous *Wolterstorffina parvipalmata*, very similar in appearance to *Nectophrynoides*. Further west, in the mountain grasslands of Simandou and Mount Loma, anurans (mostly *Arthroleptis* sp.) are also scarce.

Population structure and dynamics

The study of population structure and dynamics of savanna amphibians has hardly begun. Barbault and Trefaut Rodrigues (1978, 1979a, b) have studied a number of anurans from the Guinean savanna of Lamto, Ivory Coast. *Ptychadena oxyrhynchus* breeds from February to May. Average clutch size is 3476 eggs per clutch. Metamorphosis takes place three to four weeks after egg laying. Immature frogs outnumber adults from June to September, and sexual maturity is reached eight to nine months after metamorphosis when females reach a body length of 59 mm. In December and January the whole population is adult. Sex ratio among adult frogs is more or less equal. *Ptychadena maccarthyensis* breeds from March to October. The average clutch size is 1333 eggs per clutch. Sexual maturity is also reached less than nine months after metamorphosis, when females reach a body length of 43 mm. Immature frogs predominate in the population from July to January. Sex ratio at the frog stage is also more or less equal. *Arthroleptis poecilonotus*, another terrestrial ranid, lays a smaller number of large eggs twice or three times a year (average clutch size: 10–30 eggs) within the soil and the development of the young frog takes place entirely within the egg. Hatching occurs 15 to 20 days after laying. This species breeds twice during the year at Lamto (Fig. 12.7), and hatchlings born during the first breeding season (March to June) reach their sexual maturity during the second breeding season (August to December). The average life-expectancy of this frog is low, between three and four months. The small viviparous toad *Nectophrynoides occidentalis*, from the Nimba mountain grasslands, has an even lower fertility — no more than six to nine young per female per year, but it can live as long as three or four years.

Mortality rate at the tadpole stage is very high for all species breeding in water. It has never been estimated in any savanna species. This high rate is undoubtedly due mainly to predation, but data on the fate of a number of batches of *Pyxicephalus adspersus* tadpoles, which live gregariously (Table

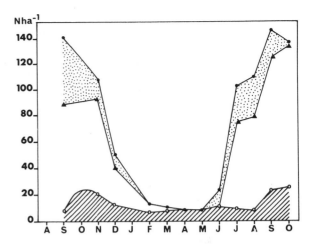

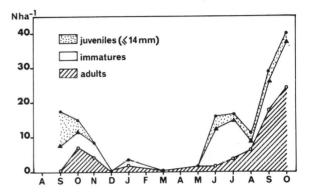

Fig. 12.7. Seasonal variations of the demographic structure of two populations of *Arthroleptis poecilonotus* living in unburnt (top graph) and regularly burnt (bottom graph) plots of a plateau savanna at Lamto, Ivory Coast (after Barbault and Trefaut Rodrigues, 1979a).

12.2), also show the importance of another mortality factor: the premature drying-up of seasonal pools (Balinsky, 1969).

The most important predators of eggs and tadpoles are fish, other anurans and aquatic insects, mainly giant water bugs (Belostomatidae) and water beetles (Dytiscidae) in Africa. As soon as the surviving young frogs and toads emerge from water they fall prey to a number of predators, ranging from large carabid beetles to snakes and birds. At Lamto seven species of snakes (*Causus rhombeatus, Chlorophis heterodermus, C. heterolepidotus, Crotaphopeltis hotamboiea, Elapsoidea guntheri, Natriceteres olivacea,* and *Philothamnus semi-variegatus*) feed exclusively on amphibians, and four others also take a large toll of them (Barbault,

TABLE 12.2

The fate of batches of *Pyxicephalus adspersus* eggs in a study site in Transvaal highveld (Balinsky, 1969)

Year	Total of batches followed	Dried up as eggs	Dried up as tadpoles	Died as result of pollution in drying pools	Metamorphosed	Probably metamorphosed
1950–51	1	0	0	1	0	0
1951–52	3	1	1	1	0	0
1952–53	7	1	1	4	1	0
1953–54	7	0	4	1	1	1
1954–55	8	4	0	0	4	0
All years	26	6	6	7	6	1

1974b). About thirty species of birds, ranging from storks to birds of prey, also eat frogs and even toads, as also do a few mammalian carnivores.

The survivorship curve of a population of the viviparous toad *Nectophrynoides occidentalis* from the Nimba mountain grasslands (Guinea) is shown in Fig. 12.8.

Diet and trophic relationships

Whereas most tadpoles feed upon algae, diatoms and bacteria, toads and frogs are secondary consumers, feeding mainly on various kinds of invertebrates. The kind of food eaten by any species depends upon a number of variables.

(1) **Availability of prey.** Most anurans feed on a wide spectrum of prey, but a few species specialize on certain food items. Ants and termites are a favourite food for some frogs and toads. In the Sahelian savanna of Fété Olé (Senegal), *Bufo pentoni* and *Tomoptera delalandii* are very fond of termites, three or four toads being consistently observed on each *Macrotermes* mound at times of swarming (Lepage, pers. comm., 1979). *Bufo regularis* is an ant specialist throughout the African savannas; ants constitute 93% of its prey at Lamto (Barbault, 1974c), 88% in Senegal (Lescure, 1971) and 87% in the Upemba National Park, Zaire (Inger and Marx, 1961). At Lamto, *Phrynomerus microps* is an exclusive ant-eater, and *Hemisus marmoratus* a termite and ant specialist (Barbault, 1974b). Most of the savanna anurans, however, can better be considered as "generalists".

(2) **Body size of the predator** is a second factor in prey selection. Small species like *Phrynobatrachus*

accraensis and *P. gutturosus* feed mostly on very small insects and spiders, whereas larger frogs like *Ptychadena maccarthyensis* and *P. oxyrhynchus* prey upon invertebrates 1 to 3 cm long. Other anurans, small snakes and even rodents are only

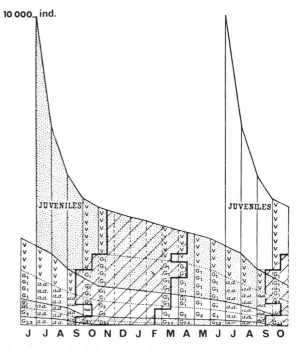

Fig. 12.8. Survivorship curve of a female cohort of the viviparous toad *Nectophrynoides occidentalis* from the Nimba mountain grasslands. Of the young females (V) born in June 20% become pregnant (G) the following September, and about 40% in November; the other members of the same cohort will not be inseminated before September of the following year. Females having already given birth during the preceding year (aa) can be found from June to August. The striped area corresponds to the age groups which "hibernate" underground. (After Lamotte, 1959.)

taken by the largest species. Barbault (1974b) has shown at Lamto that the average length of the ingested prey increases with the body length of the anuran predators (Fig. 12.9).

(3) **Food preferences.** Given a similar availability of potential prey, some species display definite food preferences, as shown in Fig. 12.10. However, this does not necessarily mean a specific choice for a given food item; it might as well reflect differences in the distribution of potential prey within the various vegetation layers in which sympatric (but not syntopic) anurans hunt for food. Ground-living toads and frogs do not meet the same kind of prey, in similar proportions, at ground level, as do tree frogs in high grass, shrubs or trees. Species active by day and those which feed during the night do not prey upon the same kinds of invertebrates, either.

The position of amphibian communities within the savanna ecosystem

To establish the energy budget of a population, one first needs to establish the energy budget of an individual at the different stages of its life-cycle. Once its structure and dynamics in field conditions are known, one can extrapolate to the population as a whole. If this can be done for all species of a given trophic category within the community, one

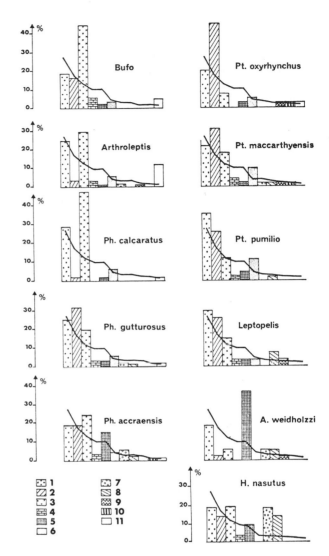

Fig. 12.10. Frequency histograms of the major categories of prey (ants and termites excluded) ingested by eleven species of Lamto anurans. The curve indicates the abundance of the same food items in the arthropod community during the same period. Legend: *1* = spiders; *2* = Orthoptera; *3* = Coleoptera; *4* = Heteroptera; *5* = Homoptera; *6* = Blattaria; *7* = Diptera; *8* = Hymenoptera (ants excepted); *9* = caterpillars; *10* = Mantodea; *11* = Myriapoda. Abbreviations as in Fig. 12.9. (After Barbault, 1974c.)

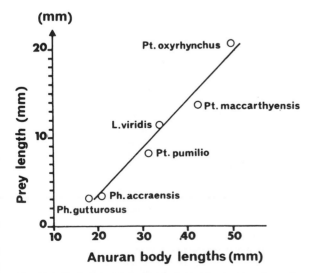

Fig. 12.9. Variations of the average prey size (ants and termites excluded) as a function of the average body length of some Lamto anurans (after Barbault, 1974b): *L* = *Leptopelis; Ph* = *Phrynobatrachus; Pt* = *Ptychadaena.*

can then evaluate the role the trophic group plays within the energy flow of the whole ecosystem.

Needless to say, such a study has not yet been carried out at any study site. However, a tentative estimate has been made in the case of the Lamto savanna.

As a basis for the individual energy budget of savanna amphibians, the data obtained on the small viviparous toad from the Nimba mountain grasslands, *Nectophrynoides occidentalis*, have been used (Lamotte, 1972). During its three years average life-span, of which five to six months each year are spent in fasting and inactivity, the following ratios between ingested energy (*I*), assimilated energy (*A*) and production (*P*) have been estimated on the basis of repeated laboratory measurements:

	P/I	P/A	A/I
First year of life	0.310	0.364	0.852
Second year of life	0.168	0.214	0.788
Third year of life	0.100	0.120	0.831

Taking into account the five to six months of inactivity which occur each year, the following average ratios have been calculated:

$$P/I = 0.117 \quad P/A = 0.143 \quad A/I = 0.816$$

On this basis, one may adopt an average P/I ratio of 0.30 for short-lived savanna amphibians with an accelerated growth rate and a short life-span, and a P/I ratio of 0.10 for large, long-lived species which have repeated periods of inactivity during their life-span. It is quite possible that the P/I ratio is even lower in some burrowing forms which spend a great deal of energy moving underground. For most amphibians, however, production represents between 0.25 and 0.17 of the energy ingested, the remainder being lost through respiration and "rejecta".

At the population level, an indirect estimate of production can be made on the basis of the turn-over of biomass ($P/\bar{B}$) in a given cohort. This annual rate of biomass turnover ranges from 2.5 to 3.0 for most amphibians, reaching 5.0 to 6.0 in very small species not surviving for more than six months, and being as low as 1.0 in the longest-lived frogs and toads.

Using the above figures, the role of an average amphibian community of 627 kJ ha^{-1} in the energy flow of the Lamto savanna ecosystem can be estimated as follows:

consumption 18 810 kJ ha^{-1}yr^{-1}
net production 1881 kJ ha^{-1}yr^{-1}
$P/\bar{B}$ 3/0
P/I 0.10

If these figures are compared with the average consumption of all animal consumers (5 434 000 kJ ha^{-1}yr^{-1}) and their net production (1 453 000 kJ ha^{-1}yr^{-1}), it becomes obvious that the role of amphibians in the Lamto ecosystem is a very small one, from the bioenergetics viewpoint at least. In more seasonal and less species-rich savannas it probably becomes almost negligible. However, the role of the tadpole populations in the bodies of water so widespread during the rains in many savannas remains to be studied.

REFERENCES

Balinsky, B.I., 1969. The reproductive ecology of amphibians of the Transvaal highveld. *Zool. Afr.*, 4: 37–93.

Barbault, R., 1972. Les peuplements d'Amphibiens des savanes de Lamto, Côte d'Ivoire. *Ann. Univ. Abidjan*, E 5: 59–142.

Barbault, R., 1974a. Les peuplements d'Amphibiens et de Reptiles de la savane de Lamto. *Bull. Liaison Cherch. Lamto, Numéro Spéc.*, pp. 2–37.

Barbault, R., 1974b. Observations écologiques dans la savane de Lamto, Côte d'Ivoire: Structure de l'herpétocénose. *Bull. Ecol.*, 5: 7–25.

Barbault, R., 1974c. Le régime alimentaire des Amphibiens de la savane de Lamto, Côte d'Ivoire. *Bull. I.F.A.N.*, A 36: 952–972.

Barbault, R., 1976. Structure et dynamique d'un peuplement d'Amphibiens en savane protégée du feu (Lamto, Côte d'Ivoire). *Terre Vie*, 30: 246–263.

Barbault, R. and Trefaut Rodrigues, M., 1978. Observations sur la reproduction et la dynamique des populations de quelques Anoures tropicaux. I. *Ptychadena maccarthyensis* et *Ptychadena oxyrhynchus*. *Terre Vie*, 32: 441–452.

Barbault, R. and Trefaut Rodrigues, M., 1979a. Observations sur la reproduction et la dynamique des populations de quelques Anoures tropicaux III. *Arthroleptis poecilonotus*. *Trop. Ecol.*, 20: 64–77.

Barbault, R. and Trefaut Rodrigues, M., 1979b. Observations sur la reproduction et la dynamique des populations de quelques Anoures tropicaux IV. *Phrynobatrachus accraensis*. *Bull. I.F.A.N.*, A41: 417–428.

Bourlière, F., 1978. La savane sahélienne de Fété Olé, Sénégal. In: M. Lamotte and F. Bourlière (Editors), *Structure et Fonctionnement des Ecosystèmes Terrestres*. Masson, Paris, pp. 187–229.

Forge, P. and Barbault, R., 1977. Ecologie de la reproduction et du développement larvaire d'un Amphibien déserticole *Bufo pentoni* Anderson 1893, au Sénégal. *Terre Vie*, 31: 117–125.

Frazer, J.F.D., 1973. *Amphibians*. Wykeham Publications, London and Winchester.

Inger, R.F. and Marx, H., 1961. The food of amphibians. *Explor. Parc Natl. Upemba, Bruxelles*, 64.

Jacobsen, N.H.G., 1977. An annotated checklist of the amphibians, reptiles and mammals of the Nylsvley Nature Reserve. *S. Afr. Natl. Sci. Progr., Rep.* 21: 1–65.

Lamotte, M., 1959. Observations écologiques sur les populations naturelles de *Nectophrynoides occidentalis* (Famille des Bufonidae). *Bull. Biol.*, 93: 355–413.

Lamotte, M., 1967. Les Batraciens de la région de Gpakobo, Côte d'Ivoire. *Bull. I.F.A.N.*, A 29: 218–294.

Lamotte, M. and Lescure, J., 1977. Tendances adaptatives à l'affranchissement du milieu aquatique chez les Amphibiens anoures. *Terre Vie*, 31: 225–311.

Lamotte, M. and Meyer, J.A., 1978. Utilisation des taux de renouvellement $P/\bar{B}$ dans l'analyse du fonctionnement énergétique des écosystèmes. *C.R. Acad. Sci. Paris*, D 286: 1387–1389.

Lee, A.K. and Mercer, E.H., 1967. Cocoon surrounded desert-adapted frogs. *Science*, 159: 87–88.

Lescure, J., 1971. L'alimentation de *Bufo regularis* Reuss et de la grenouille *Dicroglossus occipitalis* (Günther) au Sénégal. *Bull. I.F.A.N.*, A 33: 446–466.

Loveridge, J.P., 1976. Strategies of water conservation in southern African frogs. *Zool. Afr.*, 11: 319–333.

Main, A.R., 1968. Ecology, systematics and evolution of Australian frogs. *Adv. Ecol. Res.*, 5: 37–86.

Mayhew, W.W., 1968. Biology of desert amphibians and reptiles. In: G.W. Brown (Editor), *Desert Biology, 1*. Academic Press, New York, N.Y., pp. 195–356.

Tyler, M.J., 1976. *Frogs*. The Australian Nature Library; Collins, Sydney, N.S.W., 256 pp.

Vuattoux, R., 1968. Le peuplement du palmier rônier (*Borassus aethiopum*) d'une savane de Côte d'Ivoire. *Ann. Univ. Abidjan*, E 1: 1–138.

Wager, V.A., 1965. *The Frogs of South Africa*. Purnell, Cape Town.

Chapter 13

REPTILES IN SAVANNA ECOSYSTEMS

ROBERT BARBAULT

INTRODUCTION

Although it is possible to find some information on the life history and autoecology of a number of savanna reptiles in many taxonomic and faunistic papers, comprehensive studies of reptilian populations and their role within the tropical savanna communities are still lacking. This chapter has therefore been centered upon the results of my own work, carried out during the past decade in the Lamto savanna, Ivory coast (Barbault, 1971, 1974a, 1975a, 1977). The scanty data available in the literature will then be compared with my own results and some tentative hypotheses put forward; I hope they will stimulate further research.

Most of the information presently available relates to Squamata (that is, lizards and snakes). With the exception of some insular habitats such as the Aldabra Atoll and the Galapagos Islands, which cannot be considered as representative of the situation prevailing in continental savanna areas, populations of tropical tortoises have never been studied. As for crocodilians, they only enter the savanna biome along rivers and in marshy areas, and therefore play no role in savanna communities.

The only family of Squamata which is restricted to the dry tropics (savannas and deserts) of Africa and Asia is that of the Lacertidae. All other families can be found both in the dry and humid tropics, even though some genera are better represented in the savannas than in the tropical woodlands. Such is the case, for instance, with the following genera: *Mabuya* and *Agama* among the Sauria, and *Echis* and *Psammophis* among the Ophidia. At the species level, however, habitat preferences are much sharper, every species being restricted either to a savanna or a forest habitat. Strictly arboricolous species can however enter the savannas through the arboreal layer of the ecosystem.

Generally speaking, tropical savanna reptiles belong to one of the following categories: (a) cosmopolitan families such as the Colubridae and the Viperidae (the Australian region excepted, for the latter family); (b) pan-tropical ("tropicolitan") families such as the Gekkonidae and Scincidae among the lizards, and the Typhlopidae, Boidae and Elapidae among the snakes; (c) families restricted either to the Old World tropics (e.g. Agamidae and Chamaeleontidae) or to the New World tropics (e.g. Iguanidae).

POPULATION DENSITIES AND THEIR SEASONAL VARIATIONS

Within a given savanna ecosystem, reptilian population densities can vary greatly from one area to another. This is due mostly to three different categories of variables: topographic factors (shallow areas liable to seasonal flooding *versus* plateaux); the existence of a tree stratum and its importance; and finally the "fire regime" (whether or not bush fires occur regularly).

Topographic factors greatly influence the population density of snakes. At Lamto, they were three to four times more numerous in marshy bottoms than on the plateau. This was mostly due to a sharp increase of batrachophagous species feeding upon the dense population of Amphibia (Barbault, 1971, 1972). On the other hand, lizard population densities did not differ much between these two savanna environments.

The existence of a shrub and tree stratum influences only the density of lizards, not that of snakes. At Lamto lizards were twice as numerous in the tree savanna than in places where only a few trees were present (Fig. 13.1).

The importance of the "fire regime" of the savanna must not be underestimated either. In our study area, the average density of both snakes and lizards was twice as great in the unburnt plots of savanna as in those which were burnt yearly (Fig. 13.1; and Barbault, 1977). This higher density of reptiles is easy to understand. When grass is protected from fire, litter accumulates, the physical structure of the grass layer changes and so does the microclimate, favouring an increase of the invertebrate prey populations.

Seasonal variations of reptilian population densities apparently occur in all kinds of savannas. At Lamto, densities were lower during the dry season and increased during the rains (Fig. 13.1). Population build-up was basically due to natality, and decrease in mortality, but population movements between unburnt and burnt parts of the savanna also occur and have to be taken into consideration (Barbault, 1974b, 1977).

Since long-term studies are scarce, and there is no information on the seasonal population changes in most of the localities studied (Table 13.1), it is difficult to compare the existing data on reptilian densities and biomasses in different tropical savannas. However, the comparison of the data collected by Western (1974) in various xeric localities of northern Kenya (Table 13.2), where the annual rainfall was less than 300 mm, with my own data for Lamto, where the annual rainfall averages 1500 mm, is interesting. The density of lizards, censused by sight in August 1978 in an *Acacia tortilis* dry savanna of northern Kenya, reached 680 individuals per hectare — a figure ten times higher than that found in the "lizard-rich" savanna of the southern Ivory Coast. Lizards seem to be less affected by an increased aridity than other reptiles, which depend more upon such prey as amphibians and other animals poorly adapted to xeric conditions; they might even profit from a reduced snake competition.

THE SEASONAL REPRODUCTIVE CYCLE

The main features of the reproductive cycle of tropical lizards — and to a lesser extent of tropical

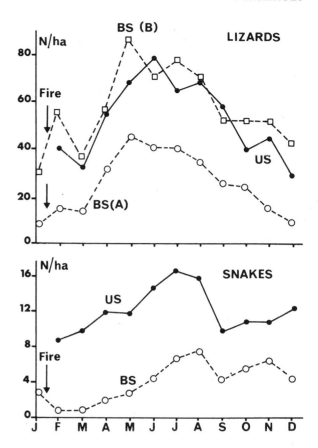

Fig. 13.1. Seasonal variations of the number of lizards (top graphs) and snakes (bottom graphs) in burnt (*BS*) and unburnt (*US*) plots of a plateau savanna at Lamto (Ivory Coast). The number of lizards being very different in an almost treeless plot [*BS*(A)] and a more densely wooded plot [*BS*(B)], they have been shown separately. The strictly arboricolous gekkos are not included in the figures. (After Barbault, 1977).

snakes — can be outlined on the basis of the numerous field data gathered during the past two decades (Fitch, 1970; Saint-Girons and Pfeffer, 1971; Barbault, 1975b, 1976a). Broadly speaking, most savanna species breed during the rains (Fitch, 1970, 1973; Licht and Gorman, 1970; Barbault, 1971, 1974a; Saint-Girons and Pfeffer, 1971; Gorman and Licht, 1974). The breeding period is most strictly limited in the driest areas where periods of reduced activity occur (Mayhew, 1968; Fitch, 1970; Saint-Girons and Pfeffer, 1972). Wherever the seasonal drought is not too severe, reproduction tends to spread over a longer period (Fig. 13.2; see also Sexton et al., 1971; Fitch, 1973; Barbault, 1974a,b), even extending over the whole year as in rain-forest species (Inger and Greenberg,

TABLE 13.1

Densities (N ha^{-1}) of some lizard and snake populations in tropical savannas

	Number	Country	Author
LIZARDS			
Agama agama	1–187	Kenya	Western (1974)
A. atra	51–155	South Africa	Burrage (1974)
Ameiva quadrilineata	17–45	Costa Rica	Hirth (1963)
Basiliscus vittatus	5–10	Costa Rica	Hirth (1963)
Calotes nemoricala	41–54	Andra Pradesh (India)	Subba Rao (1970)
Cordylus cordylus	4–288	South Africa	Burrage (1974)
Eremias spekii	23–600	Kenya	Western (1974)
Latastia longicauda	2–5	Kenya	Western (1974)
Lygodactylus picturatus	5–62	Kenya	Western (1974)
Mabuya brevicollis	7	Kenya	Western (1974)
M. buettneri	0–30	Ivory Coast	Barbault (1975a)
M. maculilabris	5–22	Ivory Coast	Barbault (1975a)
M. quinquetaeniata	2	Kenya	Western (1974)
Hemidactylus brookii	30	Kenya	Western (1974)
Panaspis nimbaensis	1–10	Ivory Coast	Barbault (1975a)
Riopa sundevalli	12	Kenya	Western (1974)
Sitana ponticeriana	22–30	Andra Pradesh	Subba Rao (1970)
Pachydactylus tuberculosus	14–26	Kenya	Western (1974)
Varanus exanthematicus	0.1–0.5	Kenya	Western (1974)
V. komodoensis	0.02	Indonesia	Darevskij and Malev (1965), in Darevskij and Terentev (1967)
V. niloticus	0.4–0.6	Kenya	Western (1974)
SNAKES			
Dasypeltis scabra	0.4–0.9	Ivory Coast	Barbault (1977)
Echis carinatus	0.3–2.5	Ivory Coast	Barbault (1977)
Oxybelis aeneus	0.28–0.35	Honduras	Henderson (1974)
Psammophis sibilans	0.5–1.8	Ivory Coast	Barbault (1977)

1966). This is exemplified by the study undertaken by Sexton et al. (1971) on three species of *Anolis* living in Panama: *A. auratus* which is found only in savannas, *A. tropidogaster* which lives at the forest/savanna boundary, and *A. limifrons*, a forest species. Whereas egg production stops completely during the dry season in the first two species, it proceeds throughout the year in *A. limifrons*.

The proximate and ultimate causes of such seasonal breeding patterns are difficult to disentangle. The correlation with the occurrence of the rains is not a simple one (Fitch, 1970; Licht and Gorman, 1970). For instance, Barbault (1974a, 1976a) has shown that three sympatric lizards living in Lamto (Ivory Coast) have different breeding patterns (Fig. 13.3). Furthermore, Vanzolini and Rebouças-Spieker (1976) have found that different populations of *Mabuya caissara* and *M. macro-*

rhyncha from the State of São Paulo (Brazil) can breed at different times, indicating very local and efficient control mechanisms on a spatial and ecological scale much narrower than accepted by current ecological thinking. The same investigators have even found intra-populational differences in breeding strategies. Many more long-term and comparative studies of reptilian population dynamics will have to be undertaken before such local control mechanisms can be identified.

LIFE-HISTORY STRATEGIES

Tropical lizards being much better known than any other reptilian categories, discussion will be limited to the suborder Sauria.

TABLE 13.2

Density distribution of lizards (numbers per hectare) in some representative habitats of semi-arid northern Kenya (after Western, 1974)

Representative habitats:	*Acacia* *tortilis*	*Cordia* *gharaf*	*Salvadora* *persica*	Outcrops	*Acacia* *nubica*	Lava desert
Agama rupelli						4.0
A. agama	3.8			187.0	0.8	1.3
Latastia longicaudata		6.8	1.8		5.4	
Eremias spekii	600.0	188.0	22.7		30.0	
Lygodactylus picturatus	9.6		62.0			
Pachydactylus tuberculosus	24.0	14.0	26.0			
Hemidactylus brookii	32.0		29.0			
Varanus exanthematicus	0.1		0.2	0.5		
V. niloticus	0.6	0.4				
Riopa sundevalli			12.0			
Mabuya quinquetaeniata			2.0			
M. brevicollis				7.0		
Holodactylus sp	10.0		14.0			
Total N ha^{-1}	680.1	209.2	169.7	194.5	36.2	5.3
kg ha^{-1}	2.49	0.96	1.2	3.99	0.15	0.19

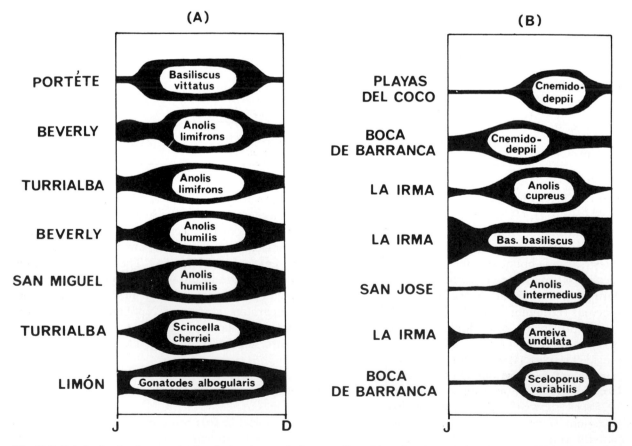

Fig. 13.2. Relative levels of egg production during the year in fourteen Costa Rican lizard populations in areas of high rainfall of the Caribbean slope (A), and in areas having a severe dry season (B). (After Fitch, 1973.)

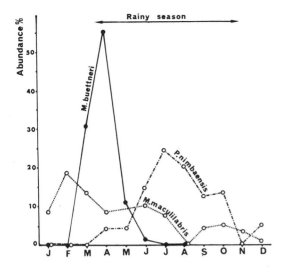

Fig. 13.3. Monthly emergence of hatchlings (0–1 month old) for three sympatric species of lizards from the Lamto savanna. The figures are the emergences during the month in question as percentages of the annual total. *M = Mabuya*; *P = Panaspis*. (After Barbault, 1976a.)

Growth and sexual maturity

The growth rate of tropical juvenile lizards is very similar to that of many temperate species (Tinkle, 1967). It averages 0.2 to 0.4 mm day^{-1} (Hirth, 1963; Harris, 1964; Alcala, 1966; Blanc, 1970; Fitch, 1973; Barbault, 1974a,c; Cissé and Karns, 1977). The smaller species usually have

smaller growth rates than do the larger ones (Fig. 13.4).

The age at which sexual maturity occurs is one important factor which greatly influences both the dynamics of a population and its demographic strategy (Cole, 1954; Tinkle, 1969; Tinkle et al., 1970). In most cases sexual maturity is reached between four and eight months of age (Sexton et al., 1963; Alcala, 1966; Telford, 1971; Fitch, 1973; Barbault, 1974a,c). The larger species mature later. *Iguana iguana* and *Ctenosaura similis* do not breed before two or even three years of age (Fitch, 1973). This is also probably the case for the Varanidae.

Clutch size

The number of eggs per clutch varies greatly from species to species in tropical as well as in temperate lizards; it ranges from a single egg to dozens (Fitch, 1970). Tinkle et al. (1970) maintained that the statistically significant relationship between the average clutch size, C, and the average female body length, L (in mm):

$$C = 0.109 \, L - 1.371$$

which they found among temperate lizards does not hold true for tropical species. However, their tropical sample (sixteen species) was very heterogeneous

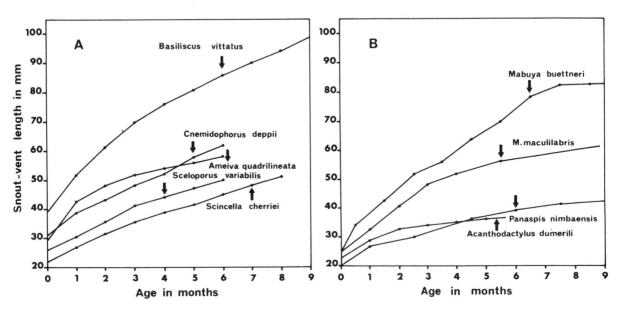

Fig. 13.4. Growth curves for some female lizards from various open habitats in Costa Rica (A), Ivory Coast and Senegal (B) (after Fitch, 1973: Barbault, 1974b; and Cissé and Karns, 1977). Arrows indicate the time at which adult size levels are reached.

TABLE 13.3

Linear correlation between average clutch size (P) and average female body length (L) in tropical savanna lizards (after Barbault, 1975b) and temperate lizards (after Tinkle et al., 1970)

	Number of species	Average clutch size P	Average female body length (mm) L	Equation of the regression line	r
Tropical Scincidae	15	5.82	67.80	$P=0.113\,L-1.841$	0.787**
Tropical Agamidae	13	8.45	90.00	$P=0.118\,L-2.170$	0.623*
Temperate lizards	53	5.42	62.30	$P=0.109\,L-1.371$	0.602**

* Significant at the 0.05 level; ** significant at the 0.01 level.

and included lizards belonging to different families and living in very different environments. Within two single taxonomic groups (tropical Agamidae and Scincidae), Barbault (1975b) has found a good correlation between clutch size and female length only among savanna species and not among rain-forest species (Table 13.3, Fig. 13.5). It can therefore be concluded that under highly seasonal climates — whether temperate or tropical — female lizards which breed seasonally lay the maximum number of eggs compatible with their adult body weight. The slope of the regression line however depends not only on the morphology of the species considered, but also on the different selective pressures acting upon them (Andrews and Rand, 1974;

Vitt and Congdon, 1978). These pressures influence the average reproductive effort (Fig. 13.6), and, for a given clutch volume, the balance between two alternative strategies: a small number of large eggs and a large number of small ones. Unfortunately the present lack of data prevents any further discussion on this point.

Number of clutches per year

The scanty data available on the yearly egg production of tropical lizards are summarized in Table 13.4. The fecundity of tropical species appears to be similar, if not slightly superior, to that of comparable temperate species (Barbault, 1975b).

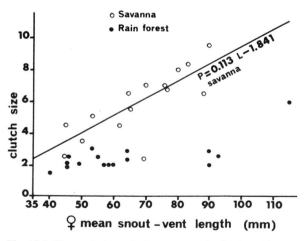

Fig. 13.5. The variation of the average clutch size (P) as a function of the average body length from snout to anus (L) in female Scincidae living in savannas (white circles) and rain forests (dark circles). (After Barbault, 1975b.)

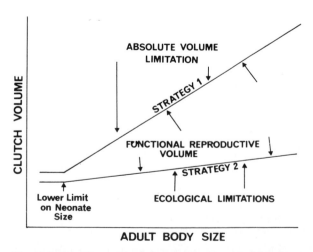

Fig. 13.6. Graphic model relating clutch volumes of lizards to adult body sizes. There are two alternative strategies: some species, utilizing a "sit and wait" strategy (*1*), exhibit high clutch volumes, whereas others, utilizing a widely foraging strategy, exhibit low clutch volumes (*2*). (After Vitt and Congdon, 1978.)

TABLE 13.4

Average annual egg production in tropical lizards

	Number of clutches per year	Number of eggs per year	Country	Authors
Savanna species				
Agama agama	3–4	18–24	Ivory Coast	Barbault (in press)
	2–3	11–16	Nigeria	Harris (1964)
Mabuya buettneri	1.5	13	Ivory Coast	Barbault (1974a)
M. maculilabris	5.5	30	Ivory Coast	Barbault (1974a)
Panaspis nimbaensis	5	13	Ivory Coast	Barbault (1974a)
Forest species				
Anolis cupreus	18	18	Costa Rica	Fitch (1973)
A. intermedius	10–12	10–12	Costa Rica	Fitch (1973)
A. limifrons	18	18	Barro Colorado I,	Andrews and Rand (1974)
Anolis tropidolepis	12	12	Costa Rica	Fitch (1973)
Basiliscus basiliscus	5–7	50–70	Costa Rica	Van Devender (1975)
Chamaeleo lateralis	4	55	Madagascar	Blanc (1970)
C. pardalis	1	19	Réunion	Bourgat (1969)
Cnemidophorus deppii	6	17	Costa Rica	Fitch (1973)
Ctenosaura similis	1	20–34	Costa Rica	Fitch (1973)
Corythophanes percarinatus[1,2]	1	7	Guatemala	McCoy (1968)
Dasia smaragdina	1	2	Philippines	Alcala (1966)
Draco volans	1	7	Philippines	Alcala (1966)
Iguana iguana	1	35	Panama	Rand (1968)
Mabuya blandingi	7–8	19–22	Ivory Coast	Barbault (1974c)
M. multifasciata[2]	1	7	Philippines	Alcala (1966)
Panaspis kitsoni	5–8	10–16	Ivory Coast	Barbault (1974c)
Sceloporus malachitus[1,2]	1	6	Costa Rica	Marion and Sexton (1971)

[1] Mountain habitat; [2] ovoviviparous species.

There is no obvious difference between rain-forest and savanna lizards, but the latter apparently compensate their smaller number of clutches per year by a higher clutch size.

Reproductive strategies of savanna *versus* forest species

Long ago Rollinat (1934) suggested that the breeding cycle of temperate zone reptiles was correlated with their yearly cycle of fat deposition (Saint-Girons, 1957; Tinkle, 1962; Saint-Girons and Kramer, 1963; Hahn and Tinkle, 1965; Hoddenbach, 1966). Such a correlation has also been found in a number of tropical species living in highly seasonal environments (Marshall and Hook, 1960; Chapman and Chapman, 1964; Marion and Sexton, 1971; Licht and Gorman, 1970). It is generally agreed that these fat reserves are used for vitellogenesis. During their investigations in the rain forests of Borneo, Inger and Greenberg (1966) found small adipose bodies in polyoestrous lizards and concluded that the energy ingested was immediately invested in reproduction — a process which makes the production of large clutches of eggs most unlikely. To produce as many eggs as their savanna counterparts, these forest lizards with negligible fat reserves require more time. They space out their clutches and produce a smaller number of eggs each time. Wherever climatic conditions and the abundance of food do not change much during the year, as is apparently the case in most rain forests, selection pressure should favour those species capable of spreading their egg production evenly throughout the year. This is the best way for them to ensure an efficient use of the available resources and to decrease intraspecific competition. Quite on the contrary, the species living in highly seasonal environments where resources are scarce during the "lean season", have to produce as many eggs as possible at the right time of the year — that is, when the maximum food is available. The laying period can vary between species, but the ultimate goal is to have the young born at a time when their survival and growth will be optimized.

Survival rate

Very little is known about the mortality rates and life expectancies of tropical lizards in field conditions. Only a few species of small size have been studied so far. Barbault (1974a, 1976a) stresses the short life expectancy at maturity for the three species of Scincidae he studied in the Lamto savanna (Ivory Coast). On the average it reaches 2.2, 7.1 and 9.2 months respectively in *Mabuya buettneri*, *Panaspis nimbaensis* and *Mabuya maculilabris*

(Fig. 13.7). Females slightly outlive the males. The study of a population of *Acanthodactylus dumerili* from Senegal by Cissé and Karns (1977) has shown that the recapture rates of marked individuals were very low, implying a "substantial annual turnover". This is very similar to the situation found by Fitch (1973) in a number of forest lizards from Central America.

Discussion

The few species of small-size savanna lizards which have been adequately studied to date can all be considered as *r* strategists: all of them mature early, produce a large number of eggs and have a short life expectancy. The close correlation between an *r* demographic strategy and life in a harsh unpredictable climate, advanced by some ecologists (MacArthur and Wilson, 1967; Pianka, 1970), is therefore far from being firmly established. Although seasonal, the savanna climate is in most places quite predictable and allows a rapid growth, an early sexual maturity, and the production of a number of clutches in succession throughout the rest of the year. Such a high rate of reproduction is moreover balanced by a strong predation pressure (Barbault, 1976b).

However, a high predation rate does not obligatorily lead to the adoption of an *r* strategy. As pointed out by Maiorana (1976), another means of coping with an increased predation pressure is to hide from predators and escape mortality. Such a "submergent behavior strategy" implies a slowing down of the food-searching activities, which in turn might lead to demographic characteristics reminiscent of a *K* strategy: slow growth, delayed sexual maturity, protracted reproduction and long lifespan. It is quite possible that this strategy, differing from the true *K* strategy which implies a strong competition pressure, is adopted by lizards living in the litter, or the canopy, of the rain forest, or by those which are active by night (Gekkonidae). The extension of the reproductive effort over a large number of small clutches might be part of this intermediate strategy, backed up by the sustained food production of rain-forest environments. For the time being, however, so many ecological and biological variables interact to determine the adoption of a given demographic strategy by a given species that it is hard to single out the most important one (Barbault and Blandin, 1980).

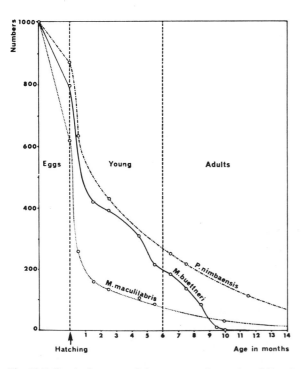

Fig. 13.7. Survival curves of three sympatric species of lizards from the Lamto savanna (after Barbault, 1975a).

THE PLACE OF REPTILIAN COMMUNITIES WITHIN SAVANNA ECOSYSTEMS

Species diversity

Concluding a comparative study of the species diversity (H' index of Shannon) in the reptilian communities of various Ivory Coast savannas, and considering the conclusions of other workers (Pianka, 1967; Arnold, 1972; Scott, 1976), Barbault (1976c) has proposed the following hypotheses:

(1) Species diversity in savanna lizard communities is probably correlated with the physical diversity of their habitat.

(2) Species diversity in savanna snake communities tends to increase with the physical diversity of the habitat and with the duration of the rainy season.

(3) This last relationship is quite probably an indirect one, based on the increased diversity of the prey species, which in turn may be attributed to variations in the structural diversity of the habitat (for lizards, birds and possibly small mammals) or to the duration of the dry season (for Amphibia, see Fig. 13.8).

Grouping snakes from various latitudes into feeding guilds (anuran-eaters, lizard-eaters, mammal-eaters, etc.), Arnold (1972) has also found that much of the variance in snake species densities was related to prey species densities (Fig. 13.9).

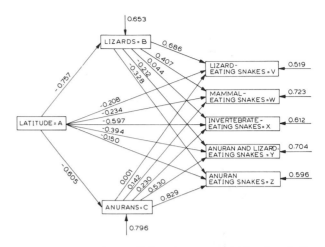

Fig. 13.9. Correlation coefficients between variables along indicated single paths. All variables except latitude are species densities. (After Arnold, 1972.)

Trophic relationships and community structure

The broad relationships found by Arnold (1972) between the abundance of snake species and that of their prey species, imply close interactions between the structure and dynamics of the communities concerned. During his study of the Lamto savanna (Ivory Coast), Barbault (1971, 1974d, 1977) indeed found a close relationship of the distribution of the various species of snakes, their population densities, and even their activity and abundance

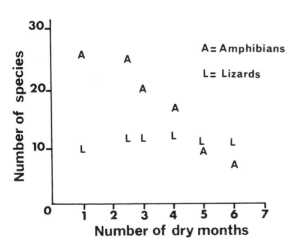

Fig. 13.8. Number of species of litter amphibians and lizards in Costa Rican and Panamian samples from areas with differing lengths of the dry season (after Scott, 1976).

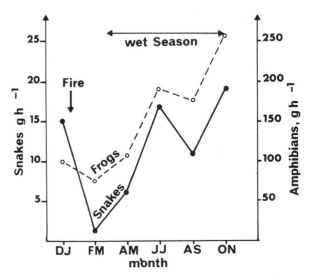

Fig. 13.10. Simultaneous variations of the biomasses of amphibians and frog-eating snakes in Lamto (after Barbault, 1971).

cycles, with the distribution, abundance and dynamics of their prey population (Figs. 13.10 and 13.11). This relationship is supported by the following observations: the seasonal variations of the biomass of frog-eating snakes and of their prey always take place synchronously; the species richness and population density of these frog-eaters increase in the wet shallow depressions which are rich in amphibians; a prolonged drought period brings about a decline of the peak yearly density of snakes (from 1.5 ind. ha^{-1} in 1966 to 0.5 ind. ha^{-1} at mid-1968), as well as a decrease in amphibian numbers.

So far, no study has been carried out in tropical savannas on the means by which resource partitioning is achieved within each reptilian guild.

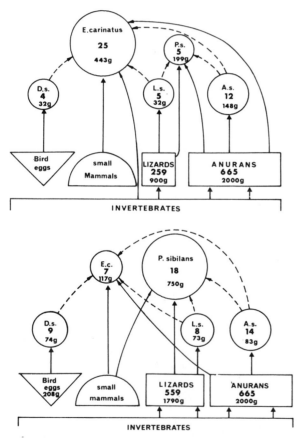

Fig. 13.11. Trophic relationships within reptilian communities at Lamto. Ivory Coast. The top figure refers to a grass-savanna community, and the bottom figure to a tree-savanna community. The number of individuals and the fresh weights are average values for 10 ha. *E.c* = *Echis carinatus*; *P.s.* = *Psammophis sibilans*; *D.s.* = *Dasypeltis scabra*; *L.s.* = other lizard-eating snakes; *A.s.* = other frog-eating snakes. (After Barbault, 1977.)

The role played by reptiles in savanna ecosystems

Within savanna ecosystems most reptiles are secondary consumers group, only tortoises being primary consumers.

Contrary to what happens on some islands where they are now protected (Galapagos, Aldabra), land tortoises are never present in large numbers in continental savannas. This might be due either to competition with wild or domestic grazing mammals, or to a heavy predation on their eggs and young. Their trophic impact on the vegetation is therefore almost negligible.

Most of the savanna lizards prey upon invertebrates. At Lamto, at least, they consume but a small part of the available biomass (Fig. 13.12) and cannot therefore regulate their numbers (Barbault, 1975c). The situation is probably different in the semi-arid zone where the lizard standing crop may be much more important (Western, 1974).

As a result of their high population turnover — the $P/\bar{B}$ ratio approximates 3 in a number of small species studied by Barbault (1974d) — tropical lizards can make up an important food source for many vertebrate predators, either occasionally or customarily: in Africa, for instance, a number of snake and bird species are true lizard specialists (Barbault, 1971, 1974d, 1977; Thiollay, 1974).

Savanna snakes are basically eaters of small vertebrates, but their trophic impact is certainly slight as compared to that of many other predators.

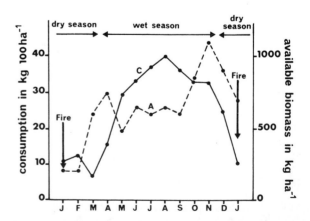

Fig. 13.12. Seasonal variations of the arthropod biomass (*A* in kg) and of the food consumption of Scincidae (*C* in kg) in an unburnt and sparsely wooded plot of savanna at Lamto, Ivory Coast. Ants and termites are *not* included in the total arthropodan biomass of the grass layer. (After Barbault, 1975c.)

They might however help to maintain diversity within the reptilian community itself (Barbault, 1974d, 1977).

Janzen's hypothesis

In a recent paper Janzen (1976) has advanced the theory that a large biomass of large wild herbivores in a habitat should severely depress the biomass of reptiles in that habitat. Consequently the following situations should occur:

(1) Specialized predators on reptiles should be able to turn to carrion from kills of large herbivores during times when their usual prey is absent, etc. The availability of this food should mean that reptile specialists will exercise a more omnipresent predator pressure where large herbivorous mammals are abundant.

(2) Predators that feed frequently on large herbivore kills and the carrion from them may take reptiles as incidental food items. The larger the biomass of herbivores in the habitat, the larger should be the biomass of such carnivores, and the greater should be their incidental depressant impact on the reptile biomass.

(3) Through intensive grazing, browsing, and trampling, large herbivores should greatly reduce the cover available for reptiles, the small vertebrate prey of snakes, and the insects available to reptiles.

Attractive as it is, this theory is only supported so far by some incidental observations and the scanty evidence of travelling naturalists — for example, the density of reptiles appears to be low in a wide variety of African habitats where "game" is plentiful. It needs to be given serious consideration.

REFERENCES

Alcala, A.C., 1966. *Populations of Three Tropical Lizards on Negros Islands, Philippines*. Thesis, Stanford University, Stanford, Calif., 259 pp.

Andrews, R.M. and Rand, A.S., 1974. Reproductive effort in anoline lizards. *Ecology*, 55: 1317–1327.

Arnold, S.J., 1972. Species densities of predators and their prey. *Am. Nat.*, 106: 220–236.

Barbault, R., 1971. Les peuplements d'Ophidiens des savanes de Lamto (Côte d'Ivoire). *Ann. Univ. Abidjan*, E 4: 133–194.

Barbault, R., 1972. Les peuplements d'Amphibiens des savanes de Lamto (Côte d'Ivoire). *Ann. Univ. Abidjan*, E 5: 61–142.

Barbault, R., 1974a. Structure et dynamique d'un peuplement de Lézards: les Scincidés de la savane de Lamto (Côte d'Ivoire). *Terre Vie*, 28: 352–428.

Barbault, R., 1974b. Structure et dynamique des populations naturelles du Lézard *Mabuya buettneri* dans la savane de Lamto (Côte d'Ivoire). *Terre Vie*, 28: 272–295.

Barbault, R., 1974c. Ecologie comparée des Lézards *Mabuya blandingi* (Hallowell) et *Panaspis kitsoni* (Boulenger) dans les forêts de Lamto (Côte d'Ivoire). *Terre Vie*, 28: 272–295.

Barbault, R., 1974d. Observations écologiques dans la savane de Lamto (Côte d'Ivoire: structure trophique de l'herpétocénose. *Bull. Ecol.*, 5: 7–25.

Barbault, R., 1975a. Les peuplements de Lézards des savanes de Lamto (Côte d'Ivoire). *Ann. Univ. Abidjan*, E 8: 147–221.

Barbault, R., 1975b. Observations écologiques sur la reproduction des Lézards tropicaux: les stratégies de ponte en forêt et en savane. *Bull. Soc. Zool. Fr.*, 100: 153–168.

Barbault, R., 1975c. Place des Lézards dans la biocénose de Lamto: relations trophiques; production et consommation des populations naturelles. *Bull. I.F.A.N.*, A 37: 467–514.

Barbault, R., 1976a. Population dynamics and reproductive patterns of three African skinks. *Copeia*, 1976: 483–490.

Barbault, R., 1976b. Contribution à la théorie des stratégies démographiques: recherches sur leur déterminisme écologique chez les Lézards. *Bull. Soc. Zool. Fr.*, 101: 671–693.

Barbault, R., 1976c. Notes sur la composition et la diversité spécifiques d'une herpétocénose tropicale (Bouaké, Côte d'Ivoire). *Bull. I.F.A.N.*, A 38: 445–456.

Barbault, R., 1977. Structure et dynamique d'une herpétocénose de savane (Lamto, Côte d'Ivoire). *Geó-Eco-Trop.*, 4: 309–334.

Barbault, R. and Blandin, P., 1980. La notion de stratégie adaptative: sur quelques aspects écophysiologiques, démographiques et synécologiques. In: R. Barbault, P. Blandin and J.A. Meyer (Editors), *Recherches d'écologie théorique: les stratégies adaptatives*. Maloine, Paris, pp. 1–27.

Blanc, F., 1970. Le cycle reproducteur chez la femelle de *Chamaeleo lateralis* Gray, 1931. *Ann. Univ. Madagascar (Sci.)*, 7: 345–358.

Bourgat, R., 1969. Recherches écologiques et biologiques sur le *Chamaeleo pardalis* Cuvier 1829 de l'île de la Réunion et de Madagascar. Thesis, University of Languedoc, Montpellier, 211 pp.

Bourn, D., 1976. The giant tortoise population of Aldabra (Cryptodira: Testudinidae). *Zool. Afr.*, 11: 275–284.

Burrage, B.R., 1974. Population structure in *Agama atra* and *Cordylus cordylus cordylus* in the vicinity of De Kelders, C.P. *Ann. S. Afr. Mus.*, 66: 1–23.

Chapman, B.M. and Chapman, R.F., 1964. Observations of biology of lizard *Agama agama* in Ghana. *Proc. Zool. Soc. Lond.*, 143: 121–132.

Cissé, M. and Karns, D.R., 1977. Aspects of the ecology of *Acanthodactylus dumerili* Milne-Edwards (Sauria: Lacertidae) in Senegal. *Bull. I.F.A.N.*, A 39: 190–218.

Cole, L.C., 1954. The population consequences of life history phenomena. *Q. Rev. Biol.*, 29: 103–137.

Darevskij, I.S. and Terentev, P.V., 1967. Estimation of energy flow through amphibians and reptile populations. In: K. Petrusewicz (Editor), *Secondary Productivity of Terrestrial Ecosystems*. Polish Academy of Sciences, Warsaw, pp. 181–195.

Fitch, H.S., 1970. Reproductive cycles of lizards and snakes. *Univ. Kansas Mus. Nat. Hist., Misc. Publ.*, 52: 1–247.

Fitch, H.S., 1973. Population structure and survivorship in some Costa Rican lizards. *Occas. Pap. Mus. Nat. Hist. Univ. Kansas*, 18: 1–41.

Gorman, G.C. and Licht, P., 1974. Seasonality in ovarian cycles among tropical *Anolis* lizards. *Ecology*, 55: 360–369.

Hahn, W.E. and Tinkle, D.W., 1965. Fat cycling and experimental evidence for its adaptive significance to ovarian follicle development in the lizard *Uta stansburiana*. *J. Exp. Zool.*, 158: 79–86.

Harris, V.A., 1964. *The life of the Rainbow Lizard*, Hutchinson and Co., London, 174 pp.

Henderson, R.W., 1974. Aspects of ecology of the neotropical vine snake *Oxybelis aeneus* (Wagler). *Herpetologica*, 30: 19–24.

Hirth, H.F., 1963. The ecology of two lizards on a tropical beach. *Ecol. Monogr.*, 33: 83–112.

Hoddenbach, G.A., 1966. Reproduction in Western Texas *Cnemidophorus sexlineatus* (Sauria: Teiidae). *Copeia*, 1966: 110–113.

Inger, R.F. and Greenberg, B., 1966. Annual reproduction patterns of lizards from Borneo rain-forest. *Ecology*, 47: 1007–1021.

Janzen, D.H., 1976. The depression of reptile biomass by large herbivores. *Am. Nat.*, 110: 371–400.

Licht, P. and Gorman, G.C., 1970. Reproductive and fat cycles in Caribbean *Anolis* lizards. *Univ. Calif. Publ. Zool.*, 95: 1–52.

MacArthur, R.H. and Wilson, E.O., 1967. *The Theory of Island Biogeography*. Princeton University Press, Princeton, N.J., 203 pp.

MacFarland, C.G., Villa, J. and Toro, B., 1974. The Galapagos giant tortoises (*Geochelone elephantopus*). Part I: Status of the surviving populations. *Biol. Conserv.*, 6: 118–133.

Maiorana, V.C., 1976. Predation, submergent behavior, and tropical diversity. *Evol. Theory*, 1: 157–177.

Marion, K.R. and Sexton, O.J., 1971. The reproductive cycle of the lizard *Sceloporus malachitus* in Costa Rica. *Copeia*, 1971: 517–526.

Marshall, A.J. and Hook, R., 1960. The breeding biology of equatorial vertebrates: reproduction of the lizard *Agama agama lionotus*. *Proc. Zool. Soc. Lond.*, 134: 197–205.

Mayhew, W.W., 1968. Biology of desert amphibians and reptiles. In: G.W. Brown (Editor), *Desert Biology*, Academic Press, New York, N.Y., pp. 195–356.

McCoy, C.J., 1968. Reproductive cycle and viviparity in Guatemalan *Corytophanes percarinatus* (Reptilia: Iguanidae). *Herpetologica*, 24: 175–178.

Pianka, E.R., 1967. On lizards species diversity: North American flatland deserts. *Ecology*, 48: 333–351.

Pianka, E.R., 1970. On r- and K- selection. *Am. Nat.*, 100: 592–597.

Rand, A.S., 1968. A nesting aggregation of iguanas. *Copeia*, 1968: 552–561.

Rollinat, R., 1934. *La vie des Reptiles de la France Centrale*. Delagrave, Paris, 340 pp.

Saint-Girons, H., 1957. Le cycle sexuel chez *Vipera aspis* dans l'ouest de la France. *Bull. Biol.*, 91: 284–350.

Saint-Girons, H. and Kramer, E., 1969. Le cycle sexuel chez *Vipera berus* en montagne. *Rev. Suisse, Zool.*, 70: 191–221.

Saint-Girons, H. and Pfeffer, P., 1971. Le cycle sexuel des Serpents du Cambodge. *Ann. Sci. Nat.*, 12° Sér., 13: 543–572.

Saint-Girons, H. and Pfeffer, P., 1972. Notes sur l'écologie des Serpents du Cambodge. *Zool. Meded.*, 47: 5–87.

Scott, N.J., Jr., 1976. The abundance and diversity of the herpetofaunas of tropical forest litter. *Biotropica*, 8: 41–58.

Sexton, O.J., Heatwole, H. and Meseth, E., 1963. Seasonal population changes in the lizard *Anolis limifrons* in Panama. *Am. Midl. Nat.*, 69: 482–491.

Sexton, O.J., Ortleb, E.P., Hathaway, L.M., Ballinger, R.E. and Licht, P., 1971. Reproductive cycles of three species of anoline lizards from the isotherm of Panama. *Ecology*, 52: 201–215.

Subba Rao, M.V., 1970. Studies on the biology of two selected lizards of Tirupati. *Br. J. Herpetol.*, 4: 151–154.

Telford, S.R., Jr, 1971. Reproductive patterns and relative abundance of two microteiid lizard species in Panama. *Copeia*, 1971: 670–675.

Thiollay, J.M., 1973. Place des Oiseaux dans les chaînes trophiques d'une zone préforestière en Côte d'Ivoire. *Alauda*, 41: 273–300.

Tinkle, D.W., 1962. Reproductive potential and cycles in female *Crotalus atrox* from northwestern Texas. *Copeia*, 1962: 306–313.

Tinkle, D.W., 1967. The life and demography of the Side-Hotched lizard, *Uta stansburiana*. *Misc. Publ. Mus. Zool. Univ. Mich.*, 132: 1–180.

Tinkle, D.W., 1969. The concept of reproduction effort and its relation to the evolution of life histories of lizards. *Am. Nat.*, 103: 501–516.

Tinkle, D.W., Wilbur, H.M. and Tilley, S.G., 1970. Evolutionary strategies in lizard reproduction. *Evolution*, 24: 55–74.

Van Devender, R.W., 1975. *The Comparative Demography of Two Local Populations of the Tropical Lizard* Basiliscus basiliscus. Thesis, University of Michigan, Ann Arbor, Mich.

Vanzolini, P.E. and Rebouças-Spieker, R., 1976. Distribution and differentiation of animals along the coast and on continental islands of the state of São Paulo, Brasil. 3. Reproductive differences between and within *Mabuya caissara* and *M. macrorhyncha* (Sauria, Scincidae). *Pap. Avulsos Zool. S. Paulo*, 29: 95–109.

Vitt, L.J. and Congdon, J.D., 1978. Body shape, reproductive effort and relative clutch mass in lizards: resolution of a paradox. *Am. Nat.*, 112: 595–608.

Western, D., 1974. The distribution, density and biomass density of lizards in a semi-arid environment of northern Kenya. *E. Afr. Wildl. J.*, 12: 29–62.

Chapter 14

BIRDS IN SAVANNA ECOSYSTEMS

C.H. FRY

DESCRIPTION

Bird life is generally abundant and diverse even in less promising-looking tropical savanna habitats; in relatively luxuriant mesic woodland savanna it is so rich and varied that in limited space it is hard to represent it. At one locality in the grassy *Isoberlinia* woodlands of Nigeria, an area less than 10 000 km² in extent, over 375 bird species have been recorded of which about 250 breed there (Fry, 1973). In a much more arid savanna in the Sahel zone of Senegal, about 120 species have been recorded in a quadrat only 25 ha in extent and 67 were breeding residents (Morel and Morel, 1978). Only tropical rain forest has a greater abundance of bird species, but for bird morphological diversity even rain forest is exceeded by savannas, because savanna habitats vary so much, from lush grassy woodlands to treeless steppe.

Throughout the vast areas of tropical savannas of Africa and Madagascar and the lesser areas in South America, southern Asia, Malaysia and northern Australia, there live nearly 1500 bird species representing some 100 families. They include ostriches and other spectacular ground-living birds, scores of scavenging and predatory hawks, numerous songbirds and songbird-like non-passerines, and many species of essentially aquatic families that have adapted to plains and even to arboreal environments. Despite the profusion some taxa can be identified as contributing importantly to savanna avifaunas, particularly when considered regionally.

Africa

About 700 species in 57 families occur in the lowland savannas of tropical Africa. The impor-

tance of a family in the ecosystem may be evaluated by the absolute number of species it contributes; but a further consideration is the proportion of its savanna to its non-savanna species, since a family which is (for instance) confined to savanna will probably have its species more obviously adapted to that biome than are the species in a forest family only marginally represented in savanna. Table 14.1 identifies the twelve most numerous families in African savannas (each with over 20 species, in bold-type face), as follows: The largest is the Ploceidae (weavers) with 61 species, graminicolous and arboreal birds which feed mainly on standing and fallen grain. Only some 20 species occur outside Africa; but in Africa as many species are found in other habitats (mainly forest) as in savanna. Whether African savannas were the original habitat of the family cannot be inferred safely; clearly, however, weavers have a long evolutionary history, with several small genera and at least one large one, the bishops and whydahs (*Euplectes*) confined to grasslands. The Estrildidae (weaverfinches, sometimes treated as confamilial with the Ploceidae), a Palaeotropical family, have numerous African savanna species too, and are mostly small granivorous birds feeding on or near the ground. To judge from the distribution of the 75 world species of Alaudidae (larks), the family arose in the savannas and deserts of Africa and Asia. In Africa the 51 species are exclusively ground-dwelling birds of grassy and desert biomes, eating mainly seeds (20 of them are restricted to temperate-zone southern Africa or to desert, and 31 are found in tropical grasslands). Another large family of feeders upon fallen grain is the Phasianidae (francolins). Of insectivorous birds the Sylviidae (warblers) have much the greatest number of species in savanna,

TABLE 14.1

The species composition of savanna avifaunas

Neotropics			Africa			Australia		
Family	Neotropical region	tropical savannas	Family	Africa south of 20°N	tropical savannas	Family	Australia (except Tasmania)	tropical savannas
Rheidae	2**	2	Struthionidae	1**	1	*Dromaiidae*	1**	1
Tinamidae	45	12	Ardeidae	20	2			
Ciconiidae	3	2	Ciconiidae	7	2			
Anhimidae	3**	3	Threskiornithidae	8	1			
Cathartidae	6	4	Aegypiidae	9**	9			
Aquilidae	67	**21**	Aquilidae	43*	**33**	Aquilidae	17	9
Falconidae	23	10	Falconidae	12*	11	Falconidae	6**	6
Cracidae	40	10	*Sagittariidae*	1**	1			
Phasianidae	30	3	Phasianidae	41	**22**	Phasianidae	3**	3
			Numididae	7	3	Megapodiidae	3	1
			Turnicidae	3**	3	Turnicidae	6	4
Rallidae	51	4	Rallidae	20	1	*Pedionomidae*	1**	1
Cariamidae	2	1	Otididae	15*	11	Otididae	1**	1
Charadriidae	15	2	Charadriidae	17	7	Charadriidae	8	2
Burhinidae	2	1	Burhinidae	3	2	Burhinidae	2	1
			Glareolidae	10	6	Glareolidae	1**	1
			Pteroclidae	8**	8			
Columbidae	67	16	Columbidae	32	17	Columbidae	18*	14
Psittacidae	139	**25**	Psittacidae	16	11	Psittacidae	53	17
Nyctibiidae	5	2	*Musophagidae*	22	11			
Cuculidae	31	11	Cuculidae	22	14	Cuculidae	11	6
Strigidae	39	13	Strigidae	28	10	Strigidae	8	5
Caprimulgidae	37	11	Caprimulgidae	17	12	Caprimulgidae	3	1
Apodidae	28	3	Apodidae	22	7	Apodidae	5	2
Trochilidae	328	**24**	*Coliidae*	6**	6	Podargidae	3	1
Trogonidae	24	3	Trogonidae	3	1	*Aegothelidae*	1**	1
Momotidae	8	3	Alcedinidae	16	6	Alcedinidae	10	5
Todidae	5	2	Meropidae	18	12	Meropidae	1**	1
Galbulidae	15	2	Coraciidae	7*	6	Coraciidae	1**	1
Bucconidae	32	7	*Phoeniculidae*	7	5			
			Upupidae	1**	1			
Ramphastidae	41	3	Bucerotidae	22	11			
Furnariidae	217	**27**	Capitonidae	42	**22**			
Formicariidae	227	16	Indicatoridae	15	6			
Picidae	108	18	Picidae	26	13			
Alaudidae	1**	1	Alaudidae	51	**31**	Alaudidae	1**	1
Hirundinidae	25	7	Hirundinidae	31	16	Hirundinidae	4**	4
Motacillidae	8	2	Motacillidae	24*	18	Motacillidae	1**	1
Vireonidae	36	4	Campephagidae	10	2	Campephagidae	7	4
Icteridae	89	**25**	Pycnonotidae	50	8	*Grallinidae*	1**	1
Parulidae	78	7	*Prionopidae*	9	5	Artamidae	6*	5
			Laniidae	56	**27**	*Cracticidae*	9	4
Turdidae	56	4	Turdidae	94	**31**	*Ptilonorhynchidae*	9	3
Dendrocolaptidae	48	8	Timaliidae	33	13	Timaliidae	18	8
Sylviidae	14	2	Sylviidae	148	**56**	Sylviidae	9	6
Rhinocryptidae	28	2	Muscicapidae	67	**26**	Muscicapidae	35	11
Cotingidae	88	10	Paridae	17	11	*Maluridae*	23	12
Pipridae	52	4	Sittidae	1**	1	Sittidae	5	3
Tyrannidae	353	**72**	Nectariniidae	70	**23**	*Acanthizidae*	39	14
Phytotomidae	3	1	Zosteropidae	4	2	*Pachycephalidae*	19	7

TABLE 14.1 (*continued*)

Neotropics			Africa			Australia		
Family	Neotropical region	tropical savannas	Family	Africa south of 20°N	tropical savannas	Family	Australia (except Tasmania)	tropical savannas
Fringillidae	272	**65**	Fringillidae	37	15	*Ephthianuridae*	5*	4
Troglodytidae	68	7	Ploceidae	121	**61**	*Climacteridae*	6	1
Mimidae	25	5	Estrildidae	67	**36**	Estrildidae	18*	13
Coerebidae	37	5	Sturnidae	45	**23**	Dicaeidae	8	4
Tersinidae	1**	1	Dicruridae	3	2	*Meliphagidae*	66	**32**
Thraupidae	205	**23**	Oriolidae	6	2	Oriolidae	4	2
Corvidae	38	5	Corvidae	9	5	Corvidae	4*	3
Totals	3165	521		1500	708		460	227

African families are in systematic sequence. Neotropical and Australian ones have been rearranged somewhat so that families shared in common with Africa are on the same line. Introduced species and non-breeding visitors excluded.

Italicized are families endemic (or virtually so) to: Neotropical region; Africa; Australasian region. Numbers in bold type are savanna representations exceeding twenty species; * 75–99%; ** 100%, of regional species are in savanna.

Principal literature sources consulted in compiling this Table were: for the Neotropics: Sick (1965, 1966), Schauensee (1966), Haffer (1967, 1974), Howell (1969), Fry (1970a), Short (1975), Schauensee and Phelps (1978); for Africa: Benson and Irwin (1966), Moreau (1966), Hall and Moreau (1970), Snow (1978), Fry (1980a); and for Australia: Slater (1970, 1975), Macdonald (1973), Hall (1974), and Storr (1977).

but more species inhabit African non-savanna biomes and there is an even greater number of warbler species outside than within that continent. Among several genera of essentially graminicolous African warblers, *Cisticola* is pre-eminent; there are 40 world species and only one is not African. All are associated with grasses, in savannas and marshes, except for one or two evolving as rock-, shrub- and desert-dwelling species. The Sylviidae grade ecologically with another large family of insectivores, the Muscicapidae (flycatchers). The Laniidae (shrikes) prey on larger, grounded arthropods; and the Turdidae (thrushes) and Sturnidae (starlings) are large families in savanna of species with mixed diets. Three other large families are the Aquilidae (eagles, kites, sparrow-hawks) with 33 savanna species, the Nectariniidae (sunbirds, insectivores and nectarivores) with 23, and the Capitonidae (barbets, mainly berry eaters) with 22.

Eight small families are confined in Africa to savanna (double asterisked in Table 14.1; Fig. 14.1), notable ones being the Sagittariidae (the secretary bird, a snake predator that forages partly by walking) and Coliidae (mousebirds, small frugivores), both being endemic to Africa and the latter constituting the only order of birds endemic to that continent; the Struthionidae (ostriches); Aegypiidae (vultures); and the Pteroclidae (sandgrouse). All presumably originated either in African savanna or in Afro-Asiatic steppe, but three endemic African families perhaps arose in other habitats and invaded savanna only subsequently: the Numididae (Guinea fowl), Phoeniculidae (wood-hoopoes) and Prionopidae (helmetshrikes). Last, I would identify as characteristic elements in Africa's savanna landscapes the Falconidae (falcons), the Otididae (bustards, large cursorial omnivores which have almost certainly undergone their main evolutionary radiation in African savannas: Snow, 1978), the coraciiform families and in particular the Bucerotidae (hornbills), and the common and ubiquitous Columbidae (doves).

Madagascar

Except for the highland biomes clothing the mountains of Madagascar's long axis and for the remaining eastern lowland rain forests, practically all of the island is tropical deciduous-woodland savanna. The avifauna is distantly related with that of Africa, but lacks over twenty African families

Fig. 14.1. Representative birds of African savannas. Facing left: a pigeon (*Columba guinea*), bustard (*Otis kori*), plover (*Vanellus coronatus*), button-quail (*Turnix sylvatica*), sandgrouse (*Pterocles exustus*), ostrich (*Struthio camelus*), secretary bird (*Sagittarius serpentarius*), coucal (*Centropus senegalensis*) and crow (*Corvus albus*). Facing right (at twice scale of birds facing left): a shrike (*Lanius collaris*), viduine weaver (*Vidua orientalis*), grass-warbler (*Cisticola ruficeps*), pipet (*Anthus similis*), lark (*Mirafra africana*), weaver (*Ploceus cucullatus*), bishop weaver (*Euplectes afer*), wheatear (*Oenanthe pileata*) and finch-lark (*Eremopterix nigriceps*).

including many with important savanna representation (bustards, colies, wood-hoopoes, barbets, woodpeckers, shrikes, etc.) (Dorst, 1972). Ground birds, which feature so prominently in African savannas, are represented by only ten genera, including five in the endemic families Brachypteraciidae (ground-rollers) and Mesoenatidae (mesites). The largest is *Coua* (centropine cuckoos), with ten species, mostly cursorial savanna insectivores. *The* plains birds of Madagascar, exterminated in historical times, were the elephant-birds (Aepyornithidae). Nine species in three fossil gen-

era are known, and they were massively built, cursorial, flightless birds resembling ostriches and probably leading similar lives as grassland omnivores. Other endemic families that occur marginally in wooded savannas are the arboreal Leptosomatidae (cuckoo-roller) and Vangidae (vangashrikes).

Asia

India's savanna avifauna is like that of Africa, but shorn of its most characteristic elements.

Several African families are not represented in the xerophilous woodlands of the Indian subcontinent, where the only additional families are the Eurylaemidae (broadbills, arboreal frugivores), Hemiprocnidae and Artamidae (crested-swifts and wood-swallows, both aerial insectivores) and are of little moment. Further east, such savannas as occur in Indo-China, Sumatra, Borneo and Sulawesi contain representatives of several more families, but none of them are of particular importance and there is no well-defined savanna bird fauna there at all.

Australia

Woodland savannas are found in southern New Guinea, but they are small in area and their bird fauna is related with that of the far more extensive savannas of tropical Australia. New Guinea has a very rich bird fauna, and some families represented in savanna there do not occur in Australia. In the absence of any characteristic savanna element in the New Guinea avifauna I shall, however, consider that island no further.

The arid plains, dry woodlands, dissected sandstone savannas, mulga and saltbush country of tropical latitudes in Australia contain a bird fauna of about 230 species representing 44 families (Table 14.1). Seven contribute ten or more species, namely the doves, Psittacidae (parrots), Maluridae (blue-wrens), Acanthizidae (Australian warblers), flycatchers, weaverfinches, and Meliphagidae (honey-eaters), but only the last is of the same magnitude as the dozen largest African savanna families. Thirty-two meliphagids inhabit savanna, where they are almost exclusively arboreal, feeding on insects, fruit and nectar. Parrots, for which Australia has been a major centre of radiation, are essentially arboreal, exploiting nectar (lorikeets), nuts and seeds; but many species habitually feed on the ground, competing with doves as gleaners of grain. At least ten other birds, of the families Megapodidae (megapodes), Phasianidae, Turnicidae (bustard-quails), Pedionomidae (plains wanderer) and Otididae, belong to the same guild of ground-foraging grassland granivores. The numerous granivorous songbirds in African savanna have as their broad equivalents in Australia only the weaverfinches; but a number of other songbirds have evolved there as ground-foragers of seeds and invertebrates, notably the

magpie lark *Grallina* (Grallinidae), quail-thrushes (*Cinclosoma* spp., Orthonychidae) and other babblers (Timaliidae), songlarks (*Cinclorhamphus* spp., Sylviidae), scrubwrens (*Calamanthus*, etc., Acanthizidae) and chats (Ephthianuridae). The grassbirds *Megalurus* and *Eremiornis* (Sylviidae) and especially the fairy-wren *Malurus* and grass-wren *Amytornis* (Maluridae) occupy the same niches in Australian savanna as do *Cisticolas* in African. Several of these families are endemic to Australia or Australasia, and a further endemic of significance is the plains-dwelling emu (Dromaiidae).

Neotropics

The greater part of the Neotropical lowlands is covered with forest, but savanna vegetation varying from short grass sward to grassy xerophilous woodlands, desert steppe and thorny caatinga is widespread. Such savannas are found in western Cuba, in fragments (mostly elevated) from Jalisco to Mexico Provinces and Chiapas, Mexico, from southern Guatemala to Lake Nicaragua, in a large area of northern Columbia and western Venezuela, around and east of the Sierra Pacaraima of Brazil and Guyana, and in a vast zone from the caatinga of Ceara across the southern half of Brazil to the chaco of eastern Bolivia and northern Paraguay (Haffer, 1967). The habitat requirements of some birds in the last region are poorly documented, and so its savanna avifauna cannot be enumerated precisely. However, the real number of Neotropical savanna species is probably little greater than the 521 shown in Table 14.1. While that amount is only three-quarters of the African savanna total (some function, perhaps, of respective extents of savannas), the Neotropical region has over twice as many landbird species in all habitats — mainly rain forest, mountains and islands — as Africa, so that the savanna component is a substantially smaller proportion of the Neotropical than of the African avifauna. Emphasizing that point is the fact that, of fifty families contributing to the Neotropical savanna avifauna, only four have all of their species in savanna, comparable figures for Africa and Australia being eight and thirteen.

By far the most important contribution to the Neotropical savanna avifauna, in terms of species abundance, is made by the Tyrannidae (tyrant-flycatchers), a huge and widespread New World family, and the Fringillidae (finches). The latter are the only passerine granivores and as such they occupy exclusively the same place in the Neotropical savanna ecosystem as the weavers and weaverfinches in Africa. The Tyrannidae are nearly all insectivorous "flycatchers" and "shrikes" and only a few are ground-feeders or otherwise associated with grassy vegetation. Six more families each contribute over 20 species to the savanna: the Aquilidae (as in Africa), parrots, Trochilidae (hummingbirds), Furnariidae (spinetails, etc.), Icteridae (caciques, etc.) and Thraupidae (tanagers). Characteristic large ground birds of the open savanna eating mainly vegetable matter are the rheas (Rheidae) and screamers (Anhimidae), and one eating mainly animal matter is the seriema *Cariama* (Cariamidae). All three families are endemics. The seed-eating "partridge" niche is filled mainly by another endemic family, the tinamous (Tinamidae) and, as in savannas elsewhere, doves are a prominent component of the Neotropical granivorous fauna. Apart from the finches, rather few songbirds are obviously adapted as graminicoles, herbicoles, or as exploiters of food on or near the ground. They include several genera of furnariids, an extremely diversified Neotropical family: the miner *Geositta*, earthcreeper *Upocerthia*, lark-like brushrunner *Coryphistera*, and the spinetails *Synallaxis*, *Certhiaxis*, etc.

The Neotropical savanna avifauna is far more arboreal than those of Africa or Australia. Woodpeckers (Picidae), woodcreepers (Dendrocolaptidae), foliage-gleaners (Furnariidae) and numerous other birds wholly dependent on woody vegetation abound particularly in forest but also in savanna woodlands. The parrots are mostly arboreal frugivores, not ground-foraging granivores; the ubiquitous hummingbirds depend on nectariferous flowers and are thus quite incidental to grassland; and families like the Trogonidae (trogons), Galbulidae (jacamars), Ramphastidae (toucans), Cotingidae (cotingas) and Pipridae (manakins) belong to forest in tropical America and only have savanna representation by virtue of a few species occurring in *cerradão*, the densely wooded ecotone between forest and open *cerrado* woodland (Fry, 1970a).

An example of a pure-grass savanna avifauna

Some 225 bird species occur in the Serengeti savanna, lying along the Kenya/Tanzania border, but only where the grassy plains are dotted with trees and thickets. In the treeless region of 10 000 km² extent in the southeast of the Serengeti–Mara, less than 50 birds occur regularly (A.R.E. Sinclair, pers. comm., 1978). They comprise: a small vulture (numbered *1* in Table 14.2), quails (*2, 3, 10*), a large bustard (*4*), lapwings (*5, 6*), coursers (*7, 8*), a tern (*9*), sandgrouse (*11, 12*), a large pigeon (*13*) and a small dove (*14*), a coucal (*15*), an owl (*16*), twelve larks and pipits (*17–23, 24–28*), a wheatear (*29*), six *Cisticola* grass-warblers (*30–35*), a shrike (*36*), a crow (*37*), eight *Euplectes* and other weaverbirds (*38–45*), and weaverfinches (*46–48*). All except some of the grass-warblers feed on the ground (Table 14.3), mainly on seeds and insects searched for by perambulation. Most of the birds (especially those with nidifugous young: *2–8* and *10–12*) are markedly cursorial, and are ground-nesters.

Other African grasslands have similar avifaunas of from 30 to 70 species (Winterbottom, 1972) with much the same generic composition as the above example.

AVIFAUNAL COMPARISONS

Intracontinental comparisons have been discussed at length, for Africa by Moreau (1966), and in inter-related papers by Fry (1980a) and Short (1980). Their principal value lies in facilitating the analysis of inter-biome variation in species densities, biomasses and community structures. Whilst the primary faunistic distinction in tropical lowland is between rain forest and "nonforest" (or savanna) biomes, and whilst many bird species range widely over several biomes, each main savanna biome has numerous endemic species. Thus, over one-third of all African passerine species is restricted to a single biome: 7.5% in savanna woodland, 12% in *Acacia* grassland, and 15% in lowland rain forest (Hall and Moreau, 1970). In Australia, at least that proportion is similarly restricted (Keast, 1974). The formative patterns of habitat specialization and of speciation have been discussed, with examples from tropical Australian

TABLE 14.2

A representative pure-grass savanna avifauna (Serengeti-Mara, Tanzania)

Species	Weight[1] (g)	Species	Weight[1] (g)
1 *Neophron percnopterus*	(1500)	25 *A. caffer*	(15)
2 *Coturnix coturnix*	90	26 *A. cervinus*	18
3 *C. delegorguei*	75–85	27 *A. novaeseelandiae*	23
4 *Otis kori*	12000	28 *A. lineiventris*	(18)
5 *Vanellus coronatus*	160–200	29 *Oenanthe pileata*	20
6 *V. melanopterus*	170	30 *Cisticola galactotes*	16
7 *Cursorius temminckii*	65–70	31 *C. aridula*	7.5
8 *C. africanus*	(115)	32 *C. brunnescens*	(7)
9 *Sterna leucoptera*	40–80	33 *C. chiniana*	13–18
10 *Turnix sylvatica*	42	34 *C. natalensis*	14–23
11 *Pterocles exustus*	170–200	35 *C. angusticauda*	(7)
12 *P. gutturalis*	290–340	36 *Lanius collaris*	37–41
13 *Columba guinea*	325	37 *Corvus capensis*	700
14 *Oena capensis*	35–40	38 *Ploceus rubiginosus*	(35)
15 *Centropus toulou*	95–125	39 *P. spekei*	(35)
16 *Asio capensis*	300–340	40 *Quelea erythrops*	15–20
17 *Mirafra africana*	(30)	41 *Euplectes orix*	15
18 *M. africanoides*	25	42 *E. gierowii*	(20)
19 *M. rufocinnamomea*	24	43 *E. axillaris*	19–25
20 *Eremopteryx leucopareia*	(12)	44 *E. macrourus*	19–25
21 *E. signata*	(12)	45 *E. albonotatus*	16–23
22 *Calandrella cinerea*	22	46 *Ortygospiza atricollis*	11
23 *C. rufescens*	(22)	47 *Vidua macroura*	13
24 *Anthus vaalensis*	(24)	48 *V. hypocherina*	12

[1] Weights in parentheses are estimated.

TABLE 14.3

Foraging modes and principal foods of species listed in Table 14.2

Feed on ground			Feed in grass
Cursorial	ambulatory (hopping)	search in flight or from elevated perch	
2–3 seeds, insects	31–32 insects	1 carrion	30, 33–35 insects
4, 37 omnivorous	38–48 seeds	9 insects	
5–8, 10, 17–23 insects, seeds			
11–14 seeds		16 vertebrates	
24–29 insects		15, 36 invertebrates	

Species 11–14 have limited progression on ground and fly readily; 33 inhabits bushes (where available) rather than grass.

mulga, grassland, *gibber* desert and *Triodia* sandhill desert, by Ford (1974), Keast (1974) and Harrison (1975). The species multiplication that results from these processes gives rise to a degree of species packing which is a function of a biome's vegetational complexity and climatic stability. Tropical grass sward uninterrupted by any woody vegetation is structurally simple and, as shown, carries rather few bird species; wooded savannas accommodate far more (see p.337). Simple enumeration of the passerine species mapped by Hall and Moreau (1970) in each principal biome in a longitudinal transect between 20° and 25°E through Africa gives the following values: sub-Saharan steppe 46, northern dry savanna woodlands 111, northern moist savanna woodlands 119, lowland forest 186, southern moist savanna woodlands 293, southern dry savanna woodlands 143, Kalahari/Karroo steppe 106, and Cape *macchia* 103. (The disparity between the northern and the southern moist savanna woodland values arises partly from the inadequate ornithological exploration between 20° and 25°E in the north, and partly from the complexity of the forest/savanna interface in the south which, at the map scale of 1:40 000 000, has led to my scoring many forest species in savanna.)

Data on bird densities in African savanna biomes are given by Winterbottom (1978) and they vary from 6.2 birds ha^{-1} in *Brachystegia* woodland to 14.9 in *mopane* woodland and up to 17.6 in grassland. Further density and biomass data from West Africa are given below (pp.352–353).

Intercontinental comparisons of savanna bird community structure have recently been published (Fry, 1980a; Short, 1980). Comparing the *Isoberlinia* woodlands of the northern tropics of Africa with the cerrado in Mato Grosso State, Brazil (which bear a remarkable physiognomic resemblance to each other), Fry (1980a) found that the bird faunas are systematically distant, common families contributing half of the *Isoberlinia* and half of the cerrado avifaunas and only three genera being shared (*Glaucidium, Caprimulgus, Turdus*). However, arising perhaps from similar vegetal structures of these two woodlands, the avifaunas are much alike in their proportions of raptors, ground-birds, other non-passerines, and passerines, and of seven food-foraging categories (Table 14.4). The only notable difference in proportions of feeding categories (cerrado having more insectivorous and fewer granivorous birds than has *Isoberlinia* woodland) reflects the radiation in the Neotropics of the essentially insectivorous Tyrannidae and in Africa of the essentially granivorous Ploceidae and Estrildidae.

Ecological analogues

In the above description some reference has been made to supposed ecological equivalence of unrelated birds in each of the three main savanna regions. It is tempting to pursue the matter in some detail, because some birds can be found that not only appear to be ecological analogues of each other in distant savannas, but also converge morphologically. However, discussion of ecological analogues must involve their taxonomic appraisal, which is beyond the scope of this review; moreover, for most tropical birds insufficient is known about

TABLE 14.4

Percentage composition by feeding categories of avifaunas of (A) cerrado in Mato Grosso, Brazil and (B) *Isoberlinia* savanna in north tropical Africa (adapted from Fry, 1980a)

	Raptors		Ground-birds		Other non-passerines		Passerines		Totals	
	A	B	A	B	A	B	A	B	A	B
Vertebrates, carrion	8.4	13.5							8.4	13.5
Insects taken on ground	2.8	2.1	0.7		2.8	4.3	5.7	4.9	12.0	11.3
on trees				1.1	4.2	7.6	27.5	22.2	31.7	30.9
in air					7.0	4.9	9.9	4.3	16.9	9.2
Fruit				0.6	6.4	5.9	4.2	3.2	10.6	9.7
Seeds			2.8	4.3	1.4	3.8	11.3	14.6	15.5	22.7
Nectar					4.2		0.7	2.7	4.9	2.7
Totals cerrado	11.2		3.5		26.0		59.3		100.0	
Isoberlinia		16.2		5.4		26.5		51.9		100.0

their biology to substantiate any opinion that elements in allopatric avifaunas fill exactly "the same" place in their respective ecosystems. Such claims have often been made, generally on the basis of the supposed ecological analogues being about the same size and proportions and in particular having similarly shaped beaks. The mouthparts of birds are undoubtedly closely correlated with (hence, indicative of) diet; and in the absence of detailed knowledge of food and foraging biology we may yet be fairly sure that, for instance, the many representatives of diverse families of passeriform birds in American, African, Indian, Malagasy, New Guinea and Australian savannas, weighing 30 g or more and having a robust hook-tipped beak, prey on larger arthropods, to which extent they will be in part ecologically analogous. Nectarivory provides an example of the polyphyletic evolution of a demonstrably adaptive structure, the brush-tipped tongue. That organ is possessed by the following unrelated taxa, all having nectarivorous representatives in wooded savannas: in America, the Trochilidae and the honeycreepers (Thraupidae); in Africa, the Nectariniidae and *Promerops* (perhaps a starling, Sturnidae: Sibley and Ahlquist, 1974); in Madagascar, *Foudia* (Ploceidae); in India, a babbler (Timaliidae); and in Australia, *Trichoglossus* (Psittacidae), *Ephthianurus* (Ephthianuridae), Artamidae, Dicaeidae and Meliphagidae (Parker, 1973; Staub, 1973; Schodde and Mason, 1975).

Notwithstanding the pitfalls, the best examples of apparent ecological analogues may be given. Each principal savanna region has a huge, flightless, cursorial, grazing omnivore (Davies, 1976): the ostrich *Struthio camelus* in Africa (and Arabia), the emu *Dromaius novaehollandiae* in Australia, the rheas *Rhea americana* and *Pterocnemia pennata* in South America, and, formerly, the elephant-birds in Madagascar. All are usually classified in different orders. The American vultures (Cathartidae) have most of their species inhabiting open savannas, where they are soaring scavengers like the Old World vultures (Aegypiidae); in Africa, however, the vulture species are more diverse and on the whole larger than American ones, in adaptation to the much greater abundance and diversity of mammal carcases on the African plains (Houston, 1978). The Seriema *Cariama cristata* of Brazilian savanna strongly resembles some African bustards (another

gruiform family) in appearance and life style. Neotropical jacamars (Galbulidae, usually placed in the Piciformes) are markedly convergent with Palaeotropical bee-eaters (Meropidae, Coraciiformes), each family being adapted to the specialized habit of preying on venomous bees and wasps. Two woodland savanna representatives, *Galbula ruficauda* and *Merops bulocki*, provide one of the best examples of ecological analogues among all birds, having anatomical, plumage, life-history and behavioural resemblances adaptive to almost exactly the same diets (Fry, 1970b). At a lower taxonomic level, different tribes of arboreal woodpeckers have given rise independently to ground-foraging forms: *Colaptes* (Colaptini) in America, *Geocolaptes* and *Campethera* (Campetherini) in Africa, and *Picus* (Picini) in Asia (Short, 1971). Tropical examples are *Colaptes campestris* in Brazilian grassland, *Campethera bennettii* throughout the wooded savannas to the south of the Congo rain forest, *Picus xanthopygaeus* in Sri Lanka and *P. canus* in Sumatra open woodland. Since the evolution of terrestriality among lineages of arboreal birds is such a feature of the savanna biome, the adaptations of these woodpeckers to ground-foraging may be given in detail. They tend to evolve: a walking from a hopping gait; ground-nesting; a narrow, curved beak used for probing; dull plumage with "flash" colour patterns and reduced sexual recognition marks; greater sociality; and far-carrying voices. The adaptive nature of these properties is discussed by Short (1971). Diverse genera of the American family Icteridae resemble certain African birds (Lack, 1968), and further investigation may show that the parallels are profound. The "best" ecological analogue in tropical grasslands are the icterid *Agelaius* with the African ploceid *Euplectes*, and the icterid *Sturnella* (including *Pezites*) with the African *Macronyx* (pipits, Motacillidae). *Agelaius* and *Euplectes* are sexually dimorphic black and red or yellow ground-foraging seed- and insect-eaters which breed semi-colonially and polygamously in rank grass and marshes; *Sturnella* and *Macronyx* inhabit short grass sward, are monomorphic in cryptic browns above and yellow or bright pink below with a black pectoral chevron, and are omnivorous, solitary ground-nesters delivering territorial songs in flight.

TROPHIC CONSIDERATIONS

A good deal more is known about the ecology of savanna birds in Africa than of those elsewhere, and African data will predominate in the remainder of this review.

Aside from scavengers, predators of vertebrates, and some frugivores and large omnivores, land birds in tropical thinly wooded grassland may be characterized broadly as either "granivores" (i.e., eaters of seeds, some small fruits and other vegetable matter) or "insectivores" (insects, spiders, some other invertebrates). Only a minority are exclusively seed-eaters or insect-eaters; most birds exploit both food sources in varying part. In passing, one may note that birds almost entirely fail to utilize that food of which savanna has the most to offer, namely grass; some ratites, notably rheas, feed to an extent by grazing, but as a Class birds have not begun to compete with mammals as grazers.

Grain

A few small granivores take seed *in situ*, but the large majority forage on the ground for fallen seed. In the arid Sahelian savanna of Africa, the principal granivores belong to four families, the Columbidae, Pteroclidae, Ploceidae and Estrildidae; but many other birds compete with them for the grain harvest — finch-larks (*Eremopteryx* spp.), viduines (Ploceidae: Viduinae), finches, francolins and larger birds like guinea fowl and bustards (the last being omnivorous). In the Senegalese Sahel the quantity of rainfall influences the productivity of savanna grasslands, in proportion to the duration of availability to plants of the water table. In a good year with 110 days of available ground water, grass productivity is as high as 1300 kg ha^{-1} dry weight (Bille, 1974). The seed production of that grass will then be as much as 30 kg ha^{-1} dry weight, of which about one-third is consumed by animals, the remainder being ultimately decomposed (Bille and Poupon. 1974). Birds take a substantial proportion of that third, up to 4.3 kg ha^{-1} (Morel and Morel, 1972a). Seed production is confined to the short wet season, at the Sahel latitudes less than three months. Thereafter grain is readily available on the soil surface, the ground being in general too hard-baked for seeds to sink in. The initial depletion of the

harvest must be great, since many granivores (all of the ploceids, for instance) breed during the rains and their populations are at the annual peak immediately afterwards (Morel and Morel, 1970). The enormously abundant ploceid weaver *Quelea quelea* may be given as an example. It breeds in the Sahel zone in immense concentrations, and in a huge breeding colony a sample hectare of 12 400 nests was found to consume 1845 kg of seeds in eighteen days [and also, for use in nest-building, 186 kg dry weight of grass, the standing biomass of which was 700 kg ha^{-1} (G. Morel, 1968a)]. During the ensuing dry season the store of grain is gradually consumed, any remaining at the end being lost for graminivorous birds practically overnight by germination when the first rains fall in the next wet season. That is a time of potentially great hardship for granivores, and *Q. quelea* adapts to the situation by migrating southward to cross the northward-advancing rain front, about June, to enter a region well south of it where rain has been falling for long enough for grass to have grown and its fruit to be maturing (Ward, 1965a, 1971).

Granivorous birds have, in general, rather unspecialized diets (G. Morel, 1968a), and take any seed within limits imposed by the bird's size, with consequently poor resource partitioning (Rubenstein et al., 1977; but see Cody, 1968). Congeners may reduce interspecific competition, however, by partitioning the available seed resources between them, at least in times of food abundance. The diets of seven species of doves (*Oena, Turtur* and five *Streptopelia* species), sympatric in Senegal, have been shown by Morel and Morel (1972b) and M.Y. Morel (1980) to comprise a spectrum of grass seeds, each dove having a different spectral window: *Panicum* in *Oena capensis*, *Panicum* and *Brachiaria* in *Streptopelia vinacea*, *Sorghum* in *S. senegalensis*, *Colocynthis*, *Sorghum* and other cultivated crops in *S. decipiens*, and so on. Even if their diets are unspecialized, the foraging techniques of granivorous birds may be decidedly specialized. For instance, the small estrildid *Vidua* has a fast "kick-jump" technique used in foraging on dry soil, which probably serves to expose the minute seeds of the grass *Eleusine coracana* and to allow them to be eaten before they can be seized by the commensally foraging estrildid *Lonchura*; and the canary *Serinus* feeds on the out-of-reach fruiting head of the herb *Tridax procumbens* by bending the stem to the ground and treading on it (Fry, 1975). Granivorous passerines have stout, conical beaks which are usually regarded as adaptive to diet (Bock, 1964), but their stoutness may also resist wear, as in *Quelea* (Ward, 1965a). A further anatomical specialization of many granivores, including *Quelea*, apart from the muscular gizzard, is the crop, a distensible food reservoir (Ward, 1978).

Insects

Insectivorous birds in grassland are either graminicoles, feeding mainly on prey gleaned in the herb layer, or are ground-foragers by swooping (rollers, shrikes, many flycatchers) or walking [pipits (*Anthus*), wheatears (*Oenanthe*), coursers (*Cursorius*)]. Graminicoles and swoopers generally nest above ground, and cursorial ground-feeding birds on the ground. Where savanna grass is interspersed with woody growth, arboreal insectivores (creepers, gleaners, etc.) constitute a further category. As with seed-eaters, many features of insectivorous birds are evidently adaptive to diet. To name some, flycatchers (or at least tyrant-flycatchers) have snap-closing jaw ligaments helping them rapidly to seize prey in flight; bee-eaters have an innate bee-devenoming behaviour which improves with experience (Fry, 1969); and oxpeckers (*Buphagus*), aberrant starlings, have woodpecker-like adaptations of feet and tail enabling them to exploit large African savanna mammals[1] for their ectoparasites (Attwell, 1966). Insects constitute a dispersed food resource, and exploiting it efficiently requires behavioural specializations among avian predators, such as territoriality and searching by leaf-gleaning. Above the herb layer, some insectivores do not actively seek out their prey but adopt a sit-and-wait stratagem (certain flycatchers, shrikes, kingfishers, etc.). Seasonal scarcity of insect food imposes upon many insectivorous birds the necessity to migrate, and of the huge numbers of lands birds of Palaearctic origin that winter in African savannas, nearly all are insectivorous. Sixty of the Nigerian species which Elgood et al. (1973) found to be intra-tropical migrants are mainly or exclusively insectivorous (but surprisingly they found no evidence for migration in many insectivorous species inhabiting the highly seasonal Sudanian savannas). Often, however, insects are super-abundant, and they are then exploited by a variety of avian opportunists. The best example is the swarming emergence, after rain, of the winged reproductives of ants and, particularly, termites. Thiollay (1970b) recorded over 150 bird species feeding at such swarms, including raptors, frugivorous hornbills and barbets, and granivorous weavers; between them they consumed up to 30% of an emerging swarm. A further indication of the impact of termites in the economy of birds is the determination by Lepage (1972) that in the Senegalese Sahel the standing crop biomass of *Macrotermes bellicosus*, one of eight termite species feeding entirely upon grass material, averaged 4.75 to 5 kg ha^{-1}, and that where particularly abundant that species can consume half the grass produced. Winged termites are lipid-rich, and have 1.5 times the calorific value of seeds, dry weight for weight (Ward, 1965a); they are defenceless, easily caught and ingested, so their popularity with birds is readily understood. Another occasion when granivorous birds may hunt insects, is when breeding: many provide their newly hatched young with insects in adaptation to the presumed requirement of fast-growing nestlings for specific amino-acids [Ward (1965b) for *Quelea quelea*; Morel and Morel (1973) for *Passer luteus*].

Mixed-species foraging flocks

As with finches in the temperate zone, granivores, mainly ploceids and estrildids, form mixed-species feeding flocks throughout the tropical savannas. In some cases it appears to be largely an epiphenomenal event — the concentration of non-territorial, unaggressive species at a newly found patch of abundant food — rather than a purposeful aggregation of species for, say, enhanced protection from predators. In some cases the mixing of species appears less haphazard, more predictable; as examples, in Africa the estrildids *Lagonosticta* spp. and

---

[1] Avian feeding symbioses in relation to large animals are rather common in tropical savannas. In Africa yellow wagtails (*Motacilla flava*) walk among the feet of grazing cattle, catching the insects they disturb (Wood, 1976), and in South America their close ecological analogue is the cattle tyrant (*Machetornis rixosus*). The cattle egrets (*Ardeola ibis*), a dry-land heron, have a similar habit and take larger insects (Siegfried, 1971, 1972; Dinsmore, 1973). Carmine bee-eaters *Merops nubicus*, again in Africa, sometimes ride goats, antelopes, ostriches and bustards to prey on the insects that they flush.

Uraeginthus spp. are seen together as often as separately, while mixed flocks of *Ploceus* weavers, of *Streptopelia* doves, and of ground-feeding *Tockus* hornbills are commonplace. An important adaptive advantage in the habit has been proposed by Rubenstein et al. (1977) for Costa Rican finches. Studying mixed foraging flocks of the grassquit (*Tiaris*) and two seedeaters (*Sporophila* spp., the Neotropical ecological analogue of the Palaeotropical weaverfinch genus *Lonchura*), they found that a bird foraged for longer in company than on its own, which probably promotes feeding efficiency by way of social learning. All of these flocking species are prone to roost gregariously, and in Africa the weavers *Ploceus cucullatus* and *P. nigerrimus*, for instance, may even breed in mixed colonies, which assemblages Ward and Zahavi (1973) argued might function as "information centres" for the efficient exploitation of patchy food sources. Ward (1972) has also argued that the massing of sandgrouse of up to four species at drinking pools, at regular times of day, has the same function.

An important tropical phenomenon is mixed-species feeding by arboreal insectivorous birds, mainly in forests and mesic woodlands but to some extent also in thinly wooded and *Acacia* savannas. Again, it is thought that joining a heterogeneous foraging flock of bark- and foliage-gleaning insectivores, flycatchers, even nectarivores and occasional granivores, serves to optimize a bird's hunting for food which is probably patchily distributed (Brosset, 1974; Croxall, 1976; Greig-Smith, 1978). Studying mixed-species flocks in woodland savanna in Ghana, Greig-Smith (1978) found that over fifty species are involved (usually about ten per flock), but only two, a warbler and a tit, at all regularly. All stages of the annual cycle were represented, and each main species habitually flocked monospecifically. Probably a similar adaptive value will be found for such associations of ground-feeding savanna insectivores as that of starlings (*Spreo* spp.) with buffalo-weavers (*Dinemellia* and *Bubalornis* spp.) (personal observations).

THE ANNUAL CYCLE

The concensus of current opinion is that bird populations in the tropics, far from enjoying a cornucopia of year-round food abundance, are limited by periods (however short) of environmental adversity, generally food scarcity (see, for instance Ward, 1965a) as rigorously as are boreal species. At least in the best studied tropical birds, like *Quelea quelea*, the timing of events in the annual cycle can be seen as adaptive to resisting such adversity and to maximizing reproductivity.

Reproduction, moult and migration are the crucial energetic events in the avian year. For birds of tropical savannas, moult is at least as important as it is for high-latitude species, being necessitated by the abrasive nature of grass and thorny vegetation (whether solar radiation has any deleterious effect on plumage, other than bleaching, is not known); but migration, although much evident within the tropics, is considerably less energetic or hazardous than for boreal and temperate-zone land birds. A further energetic consideration is that, because of high ambient temperatures, tropical birds expend much less energy in homoiothermy than do others. Moreau (1972) calculated that the maintenance needs of Palaearctic birds wintering in Africa average one-third less than on their breeding grounds, as a result of the greater warmth and of shorter daily periods of activity. Using a variety of techniques Thiollay (1976) estimated the daily consumption of food by insectivorous birds in Ivory Coast savanna–forest mosaic, and found that on annual average their daily rations were only half of those of birds in high latitudes.

Breeding seasons (Fig. 14.2)

In his pioneer review, Moreau (1950) emphasized the number of land birds that breed during the rainy season (which for the Sudan, whence much of his data came, coincides with the Northern Hemisphere summer). However, in a series of papers the Morels have modified that view (Morel and Morel, 1962, 1970; G. Morel, 1968b; M.Y. Morel, 1973a) and shown that, in the Senegalese Sahel, ground-nesters breed in all months, peaking about March to September, and also that tree- and grass-nesting granivores breed in all months, peaking about August to February (the wet season there being July to October). Any one species has a limited breeding season. For instance, the ploceids *Ploceus*, *Quelea*, *Euplectes* and *Passer* nest exclusively in the rains, breeding in three, four or at most five

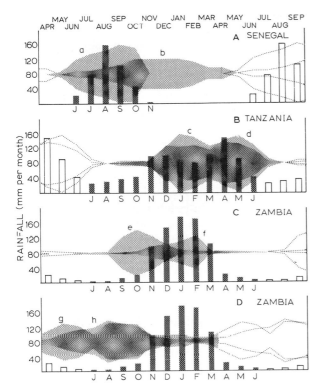

Fig. 14.2. Breeding seasons of some African savanna birds in relation to mean annual rainfall. Bars show monthly rainfall, with one 12-month cycle heavy-shaded, Senegal, in the northern tropics, and Zambia, in the southern, each experience one rainy season (six months apart from each other); equatorial Tanzania has two annual peaks of rainfall. A. Richard Toll, Senegal: a = weavers; b = weaverfinches. B. Serengeti, Tanzania: c = above-ground-nesting insectivores; d = ground-nesting insectivores, each related significantly to different rainfall peaks. C. Zambia: e = flycatchers; f = Cisticola warblers. D. Zambia and adjacent countries of similar latitude: g = raptors; h = waders (Charadriiformes). (Redrawn from Morel and Morel, 1970; Maclean, 1976b; Sinclair, 1978.)

months, according to species; estrildids start breeding within the rains and continue through the dry season to February or even April, so that some species have a breeding season of almost nine months; and most species of doves breed in all months. A marked degree of temporal partitioning in breeding is often shown by sympatric congeners (Beals, 1970; Fry, 1972). The nesting season of one individual bird is of course much shorter than that of its species, and it seldom reproduces twice during the year (Thiollay, 1970a). Insectivorous bird species are far more restricted to the wet season for reproduction (Bourlière and Hadley, 1970). Raptors (Aquilidae, Falconidae, Sagittariidae,

Strigidae) and scavengers (Aegypiidae) lay their eggs in the dry season, and the young fledge early in the rains (L.H. Brown, 1970), a rule well illustrated by Pomeroy (1978) for the marabou *Leptoptilos crumeniferus*, a scavenging stork. Some of the foregoing remarks do not apply in equatorial East Africa, where the rainfall is rather unpredictable in timing and amount. Even there, however, in the instance of large vultures with necessarily long individual nesting periods of up to $5\frac{1}{2}$ months, breeding is evidently timed so that the young fledge when feeding conditions for them are likely to be at their annual best (Houston, 1976), which, ornithologists believe, is broadly the case with all birds everywhere.

Opportunistic breeding in response to episodic rainfall is much less a character of African and South American than of Australian savannas (Serventy and Marshall, 1957; Immelmann and Immelmann, 1967), because on the whole rain is more regular and predictable in the first two continents than in the last. In the Kalahari *sandveld*, an African steppe at 26°S experiencing highly irregular rainfall, Maclean (1970) found that predators and many sedentary and migrant insectivores breed seasonally (or in the case of a courser, all year) regardless of rainfall; but that other birds, mainly small granivores, are nomadic and breed after rainfall at any time of the year. The exemplar of nomadism in southern African steppe is the gregarious starling *Creatophora cinerea*, whose movements and breeding are dictated by those of the locusts that they mainly feed on.

After several months of drought, the first rains generate almost immediately a great increase in the abundance of flying insects, a combination of swarming termites and ants with an airborne fauna of insects carried on the violent converging winds that have produced the rain (Schaefer, 1976; Sinclair, 1978). Moreover, where the rainy season is regular, a flush of herbaceous and tree foliage precedes its onset by several weeks, which results in increasing populations of locally bred phytophagous insects. For insectivorous birds this burgeoning insect fauna — or even the advent of rain itself — evidently provides the proximate environmental stimulus for endogenous events precipitating sexual behaviour. On the Serengeti plains, Sinclair (1978) found a lag of one or two months between the increase of invertebrate prey and the hatching of insectivorous birds' clutches (with smaller ground-nesting species breeding earlier than larger ones, and, amongst above-ground nesters, the aerial feeders significantly later than canopy and ground feeders). He suggested that the lag might reflect the birds having to wait for nest-concealing plant growth, or for insect larvae to feed to their young. Elsewhere in Africa, insectivores start breeding with or even prior to the onset of the rains — in *Acacia* savanna (Moreau, 1950), *Brachystegia* woodland (Reynolds, 1968; Vernon, 1978), and among graminicoles only 100 km away from Serengeti (Beesley, 1973). The same applies to frugivorous and omnivorous hornbills (Kemp, 1976). Further correlations have been drawn by Maclean (1976b); for instance, in any one savanna grassland-nesting *Cisticola* spp breed slightly later than woodland-nesting flycatchers, and granivores that feed their young with seeds nest a little later than those that feed their young with insects.

Throughout the seasonal tropics, the timing of bird breeding appears to be controlled by environmental factors associated with the availability of water, mediated by nutrition (Dingle and Khamala, 1972) and physiological condition (Jones and Ward, 1976), and not, as among temperate-zone birds, by photo- or thermoperiodicity (Ruwet, 1964). Environmental influences regulate breeding and migration likewise in Australia, as critically demonstrated by Nix (1976).

Moult

A further contrast with temperate-zone birds is provided by moult regimes in tropical savannas. In temperate birds moult and reproduction are exclusive, synchronized and of relatively short duration. But among tropical birds prenuptial moult may exceed seven weeks' duration in one individual (Davidson, 1978) and, owing to poor synchronization, a local population of a species often takes three to six months to moult (Fry, 1970c; Hanmer, 1978) and in some cases up to nine months. Moreover, in numerous species breeding commences long before moulting has finished (Foster, 1975), the energetic interpretation of which remains obscure.

Seasonal movements

Migrations are a widespread and important phenomenon in tropical savannas. They are best understood probably in West Africa, because the single rainy season and the simple latitudinal arrangement of vegetation zones there impose a single annual north–south oscillation on most migrants, whether of African or Palaearctic birds. Elgood et al. (1973) showed that at least 120 species, or 28%, of Nigerian savanna-breeding birds are migratory. As a rule they move "with the sun" and the rain-bearing intertropical convergence of airmasses that the sun generates — that is, northward in March and April and southward in September and October — and most savanna migrants cross one, two or three of the latitudinally arranged vegetation-zone boundaries. Fifty-four grassland species of diurnal raptors have been studied in the Ivory Coast by Thiollay (1977, 1978): 16 are more or less sedentary, 14 partial or erratic migrants, and 24 long-distance migrants (African and Palaearctic). The timing of their northward "spring" migrations is closely correlated with the onset of the rains, and the abundance of raptors at any one latitude and season is correlated with that of their prey, particularly grasshoppers (Thiollay 1978). Granivores migrating "across" rather than "with" the advancing rain front are an exception to the rule in West Africa (Ward, 1971). A pause for breeding may divide the "spring" migration into two distinct movements, as with the kingfisher *Halcyon leucocephala* in Nigeria; and in the climatically more complex East Africa some species, like the sunbird *Nectarinia erythroceria*, may have four migrations relating two breeding seasons and two "wintering" localities each year.

The influence of the oscillating inter-tropical front on the movements of birds, mediated by vegetation growth and food availability, is clearly seen also in trans-equatorial migrants like the African dry-land stork *Ciconia abdimii*, which breeds in the north-tropical savannas in the wet season and migrates from 3000 to 5000 km to winter in south-tropical savannas also in the wet season (Snow, 1978). The nightjars *Macrodipteryx vexillarius* and *Caprimulgus rufigena* migrate likewise (breeding, however, in the southern tropics: Snow 1978), and several other savanna species may do so too (Elgood et al., 1973).

Migrations of a less spectacular nature character-ize a great many savanna birds, and they vary from predictable migrations of entire populations to partial migration, irregular nomadism, and local foraging. The birds most sensitive to rain- and groundwater-induced changes in food supply are the ground-feeding insectivores (Wood, 1979), like the native wheatear *Oenanthe* and the wintering wagtail *Motacilla flava*. Bush fires are responsible for a considerable amount of local movement. Although frequent and widespread in savanna (Hopkins, 1974), their occurrence is essentially patchy. While it lasts, each fire produces an abun-dance of flushed food for insectivorous and preda-tory birds (Thiollay, 1971b), many of which are clearly attracted from afar. After a burn, the immediate loss and subsequent accelerated re-growth of leaf cover affects food-finding by such birds and also by granivores, causing nomadism and patchy distribution in numerous species (Greig-Smith, 1980). Like localized foraging move-ments, such nomadism is essentially random, but on a larger scale nomadism is far from a random process (even if an unpredictable one): among Kalahari Desert birds Liversidge (1980) has shown that it serves to keep a bird in a constant environment although its terrain is ever changing.

More than 180 Palaearctic species spend the northern winter months (September to April) in Africa; nearly all of the land-bird species winter in savanna, not forest, and they are more abundant in the northern than the southern tropics of Africa (Moreau, 1972). Moreover, within the northern savannas much the greatest diversity of Palaearctic species is in the dry sub-Sahelian zone, rather than the moister *Isoberlinia* woodlands further south (Elgood et al., 1966). This great influx coincides in the arid northern savannas with the season of unrelieved drought, when the environment appears to become progressively more hostile, and it seems paradoxical that only a small proportion of Palae-arctic immigrants winter in southern tropical sa-vannas, where the annual oscillation of the inter-tropical convergence ensures plenty of rain during the northern temperate winter months. These facts beg many ecological questions (G. Morel, 1973) which have not yet been satisfactorily researched.

Properties of intra- and inter-continental mi-grations, such as route fidelity (Elgood et al., 1973) are much alike, although the journey is far less

hazardous for savanna birds than for inter-continental migrants. Pre-migratory fattening, which is universal in inter-continental migrants, has so far been reported also in an African savanna oriole, a bee-eater, an estrildid, two *Euplectes* species (Jones and Ward, 1977, and references therein) and *Quelea quelea* (Ward and Jones, 1977). *Quelea* usually fattens itself on flying termites, and so its fattening is associated with the onset of rain. Ward and Jones (1977) have also found that the differing amounts of fat laid down by each of three races correlates with the anticipated distance to be flown: from 1.5 g fat for 300 km in the first race to 4 g fat for 1200 km in the third.

SOCIAL ORGANIZATION

In recent years it has emerged that the social organization of an animal species is in large part a function of its habitat, and it so happens that a pioneering study (Crook, 1964, 1965) involved tropical savanna birds. A vast amount has yet to be learnt about the selective forces involved in shaping a specific mating system and social pattern, but comparative studies of primates, antelopes, carni-vores and certain birds are giving considerable insight. Crook (1964) studied the socio-ecology of 90 ploceine weaverbirds and found that savanna species mostly forage gregariously and breed colo-nially. That is in sharp contrast with forest species, and relates with the facts that forest weavers eat insects, which are territorially defendable, and savanna ones exploit seeds, which are not. A further contrast with forest relatives is that savanna weavers tend to be polygynous, to have a con-spicuous male courtship display, either static (in *Ploceus*) or aerial (in *Euplectes*), and to be sexually dimorphic. Since most other savanna granivores are monogamous (Lack, 1968), this set of trends is unlikely to be a correlate of diet. Instead, polygamy and its associated attributes probably stem from the differing abilities of males to monopolize good nest sites within the colony: it would better profit a female to mate polygynously with a successful male than monogamously with an unsuccessful one. The nests of some weaverbirds are thorny or otherwise protected, but *Euplectes* species, all of which in-habit marshy or savanna grassland, have to build nests of grass. They are thus vulnerable to predation, which is probably why *Euplectes* species are much less aggregated in nesting colonies than are *Ploceus*. There is a strong tendency among birds for polygyny to be associated with grassy habitats; thirteen out of fourteen polygynous North American passerines breed in marshes or prairies, which Verner and Willson (1966) accounted for in terms of inter-male variation in competing for good nest sites and of feeding-potential variation between territories. So far as they are known, all sixteen *Euplectes* species are polygynous, and sexual selection has produced extreme dimorphism and complex courtship in the five species in which the male is long-tailed, culminating with *E. jacksoni* of the East African montane grassland, a terrestrial bird with lekking display[1]. There is more than a passing resemblance between these long-tailed *Euplectes* weavers and some of the whydahs, particularly *Vidua orientalis*. Viduine whydahs are also graminivorous ploceids, but being brood-parasites (on estrildine weaverfinches), their sexual dimorphism and adorned males must have evolved under selection pressures different from those of *Euplectes*.

Communal breeding

Communal, or co-operative, breeding in birds involves aiding behaviour by non-breeding individuals usually of reproductive age at the nest of a breeding pair. It is a widespread tropical or rather hot-climate phenomenon, and a majority of the 120 or so communal breeding species live in savanna habitats. During the 1970s it excited much interest because the evolution of aid-giving behaviour, if altruistic, is inexplicable by Darwinian theory. However, in the few cases where field researches have so far produced indications, aid-giving behaviour proves to be selfish, not altruistic — that is, a stratagem for gain in fitness by the helper (Ligon and Ligon, 1978; Emlen and Demong, 1980; Woolfenden, 1980). Researches have revealed several socio-ecological correlates of communal breeding, suggesting why the habit may have evolved, but understanding is yet poor and a great many questions remain to be answered (J.L. Brown, 1974, 1978; Emlen, 1978; Dyer and Fry, 1980).

DEMOGRAPHIC ATTRIBUTES

Density and biomass

The total number of bird species and the avian biomass fluctuate seasonally at any one savanna locality, the difference between the least and greatest values within the annual cycle being related with the degree of seasonality imposed by the rainfall and water-table regime. In the arid thorn-bush (Sahel) savanna of Fété Olé (Senegal), where studies by a team of biologists have resulted in the best understanding of ecological considerations available for any savanna (Bourlière, 1978), the number of species per monthly census on a 25-ha plot varied from 15 to 46, the number of individual birds per 10 ha from 18 to 88 (monthly means), and avian biomass from 0.11 to 0.51 kg ha^{-1} (Morel and Morel, 1972a). Seasonal variations, the product of immigration and emigration by Palaearctic as well as by African birds and of breeding, varied significantly from low values in the dry season (May) of the species diversity (26), density (3 birds ha^{-1}), and biomass (0.20 kg ha^{-1}) to peaks in the wet season (August–September) of the species diversity (35), density (7 birds ha^{-1}), and biomass (0.41 kg ha^{-1}) (Morel and Morel, 1978). Rainfall, or rather ground-water level, is the key factor which ultimately determines bird abundance, mediating its effect through vegetation growth and arthropod production. In a normal rainfall year, an annual average of 6.3 birds ha^{-1} occurred, but in the drought year of 1972–73 there were only 2.9, with biomass and species diversity reduced concomitantly (Morel and Morel, 1974). A dry Sudanian savanna in northern Cameroun was slightly richer, and in a normal rainfall year the lowest monthly biomass was 0.52 kg ha^{-1} (December) and the highest 2.20 kg ha^{-1} (July) (Greling, 1972). A region of forest savanna mosaic at a lower latitude, in the Ivory Coast, had a comparable avian biomass but with less seasonal variation, the lowest monthly value being 0.87 kg ha^{-1} (December) and the highest 1.41 kg ha^{-1} (July) (Thiollay, 1970a, 1971a). A part of the fluctuation in densities and biomasses which the above authors have enumerated in great

---

[1]The aggregation of males of a polygynous species in a small conventional display-ground where visiting females are inseminated.

detail stems from the monthly difference in the balance between natality and mortality of sedentary populations, but a greater part results from the seasonal movements of populations. Those of native species are important in tropical savannas but, particularly in the northern tropics of Africa, the annual influx of inter-continental migrants can be more so. In one equatorial habitat, migrating Palaearctic passerines comprised 50% of the entire non-granivorous bird biomass (Britton, 1974).

Productivity and survival

In "aseasonal" tropical rain forests birds have small clutches, commonly of two eggs, and breed only once a year, so that their annual productivity (and mortality) is much lower than among temperate-zone species. Savanna avifaunas are intermediate in these respects, on the whole closer to equatorial forest than to high-latitude species. Much interspecific variation is found at a single savanna locality; in the Senegalese Sahel savanna at Richard Toll the lark *Eremopteryx leucotis* lays one clutch of one egg a year, but the weaverfinch *Lagonosticta senegala* has up to four clutches of 3 or 4 eggs (M.Y. Morel, 1964, 1973b). The lark, being common, must enjoy excellent survival, but owing largely to parasitism by a viduine weaver only 28% of the weaverfinch's eggs result in fledged young, and its annual adult survival is only 30%. But adaptive reasons for interspecific variation are not always evident, as with three woodland savanna bee-eaters (*Merops* spp.) at Zaria, Nigeria, having specific clutch sizes of 2.5, 3 and 4.5 (Dyer and Fry, 1980).

In *Acacia* grassland the fecundity of some species is further increased by young birds breeding precociously, at about five months in sandgrouse, doves and estrildids (M.Y. Morel, 1973a). Clearly such species must have high annual mortality and low life expectancy at fledging. By contrast, numerous other land birds of large and small species, particularly in mesic savannas and rain forest, survive to great age, and the inference is that the longevity of some individuals is part of good survivorship of populations as a whole (Fry, 1980b). Their mean adult annual survival may be as high as 90%. Implications for the population dynamics of such birds have hardly yet been appreciated; good survivorship may be pre-requisite for the evolution of communal breeding; and that a disproportionate amount of fitness may be vested in a few long-lived individuals is a crucial consideration for the management of threatened species (Koeman et al., 1978).

MORPHO-PHYSIOLOGICAL ADAPTATIONS

Locomotory adaptations

Aside from the ambulatory and cursorial locomotion characteristic of ground-dwelling savanna birds and self-evidently adaptive to their habitat (epitomized perhaps by the secretary bird, *Sagittarius serpentarius*), the most fertile area for research is the morphometric comparison of closely allied species. Their adaptive attributes may thus be demonstrated, as has already been done for several large Nearctic and Palaearctic genera including the grass-dwelling warblers *Locustella* and *Acrocephalus* (Leisler, 1977, 1980). Such studies have not yet been undertaken for tropical savanna birds; *Cisticola*, *Ploceus* and *Euplectes*, would be rewarding subjects.

Protective adaptations

Predation pressure on tropical birds is undoubtedly very high and particularly affects nesting. Building impregnable hard-mud nests, and nesting in close association with stinging wasps (Smith, 1980) afford some protection; much more widespread, however, are small, cryptic nests, large impenetrable thorn nests, and the use of natural or excavated tree- and ground-holes. Many of the larger grassland and steppe birds are obliged by the lack of alternative sites to nest on the ground, when eggs and the incubating sex(es) are normally cryptically coloured and patterned.

Water and thermoregulatory adaptations

Obtaining sufficient water is evidently not difficult for birds of wooded savannas, because the trees are normally in contact with ground water even after prolonged drought. It is probably more difficult for birds in treeless grassland, particularly granivores, but they can always move to more mesic regions. Even flightless ratites obtain sufficient water from their food to permit them to use it for evaporative cooling, although ostriches conserve body water by employing convective cooling too. The budgerigar (*Melopsittacus undulatus*), the only non-passerine granivore that can survive without drinking water, has physiological and behavioural means of conserving water comparable with those of the ostrich (Weathers and Schlenbaechler, 1976). Water problems are far more acute for nestlings; doves carry water in the crop and give it to their young by regurgitation, and sandgrouse carry water to their young by soaking the specialized belly feathers (Thomas and Robin, 1977). Preventing overheating is important for grounddwelling savanna birds, and crucially so for steppe and desert birds. They variously exhibit behavioural, ecological and physiological adaptations, but these are more the province of desert than of savanna ornithology and so the reader is referred to reviews by Maclean (1974, 1976a).

REFERENCES

Attwell, R.I.G., 1966. Oxpeckers and their associations with mammals in Zambia. *Puku*, 4: 17–48.

Beals, E.B., 1970. Birds of *Euphorbia–Acacia* woodland in Ethiopia: habitat and seasonal changes. *J. Anim. Ecol.*, 39: 277–297.

Beesley, J.S.S. 1973. The breeding seasons of birds in the Arusha National Park, Tanzania. *Bull. Br. Ornithol. Club*, 93: 10–20.

Benson, C.W. and Irwin, M.P.S., 1966. The *Brachystegia* avifauna. *Ostrich, Suppl.*, 6: 297–321.

Bille, J.C., 1974. Recherches écologiques sur une savane sahélienne du Ferlo septentrional, Sénégal: 1972, année sèche au Sahel. *Terre Vie*, 28: 3–20.

Bille, J.C. and Poupon, H., 1974. Recherches écologiques sur une savane sahélienne du Ferlo septentrional, Sénégal: régénération de la strate herbacée. *Terre Vie*, 28: 21–48.

Bock, W.J., 1964. Bill shape as a generic character in the cardinals. *Auk*, 76: 50–61.

Bourlière, F., 1978. La savane sahélienne de Fété Olé, Sénégal. In: M. Lamotte and F. Bourlière (Editors), *Structure et Fonctionnement des Ecosystèmes Terrestres*. Masson, Paris, pp. 187–229.

Bourlière, F. and Hadley, M., 1970. The ecology of tropical savannas. *Annu. Rev. Ecol. Syst.*, 1: 125–152.

Britton, P.L., 1974. Relative biomass of Ethiopian and palaearctic passerines in west Kenya habitats. *Bull. Brit. Ornithol. Club*, 94: 108–113.

Brosset, A., 1974. Étude d'une niche écologique complexe en forêt équatoriale. In: P. Pesson (Editor), *Écologie forestière*. Gauthier-Villars, Paris, pp. 335–341.

Brown, J.L., 1974. Alternate routes to sociality in jays with a theory for the evolution of altruism and communal breeding. *Am. Zool.*, 14: 63–80.

Brown, J.L., 1978. Avian communal breeding systems. *Annu. Rev. Ecol. Syst.*, 9: 123–155.

Brown, L.H., 1970. *African Birds of Prey*. Collins, London, 320 pp.

Cody, M.L., 1968. On the methods of resource division in grassland bird communities. *Am. Nat.*, 102: 107–147.

Crook, J.H., 1964. The evolution of social organisation and visual communication in the weaverbirds (Ploceinae). *Behaviour, Suppl.*, 10: 1–178.

Crook, J.H., 1965. The adaptive significance of avian social organisation. *Symp. Zool. Soc. Lond.*, 14: 181–218.

Croxall, J.P., 1976. The composition and behaviour of some mixed-species bird flocks in Sarawak. *Ibis*, 118: 333–346.

Davidson, N.C., 1978. Weight, prenuptial moult and feeding of bishop birds in northern guinea savanna in Ghana. *Bull. Nigerian Ornithol. Soc.*, 14: 54–65.

Davies, S.J.J.F., 1976. The natural history of the Emu in comparison with that of other ratites. In: *Proc. 16th Int. Ornithol., Congr.*, pp. 109–120.

Dingle, H. and Khamala, C.P.M., 1972. Seasonal changes in insect abundance and biomass in an East African grassland with reference to breeding and migration in birds. *Ardea*, 59: 216–221.

Dinsmore, J.J., 1973. Foraging success of Cattle Egrets, *Bubulcus ibis*. *Am. Midl. Nat.*, 89: 242–246.

Dorst, J., 1972. Evolution and affinities of the birds of Madagascar. In: G. Richard-Vindard and R. Batistini (Editors), *Biogeography and Ecology of Madagascar*. *Monogr. Biol.*, 30: 615–627.

Dyer, M. and Fry, C.H., 1980. The origin and role of helpers in bee-eaters. In: *Proc. 17th Int. Ornithol. Congr., Berlin 1978*, pp. 852–858.

Elgood, J.H., Fry, C.H. and Dowsett, R.J., 1973. African migrants in Nigeria. *Ibis*, 115: 1–45; 375–411.

Elgood, J.H., Sharland, R.E. and Ward, P., 1966. Palaearctic migrants in Nigeria. *Ibis*, 108: 84–116.

Emlen, S.T., 1978. The evolution of co-operative breeding in birds. In: J.R. Krebs and N.B. Davies (Editors), *Behavioural Ecology, an Evolutionary Approach*. Blackwell, Oxford, pp. 245–281.

Emlen, S.T. and Demong, N.J., 1980. Bee-eaters: an alternative route to co-operative breeding? In: *Proc. 17th Int. Ornithol. Congr., Berlin 1978*, pp. 895–901.

Ford, J., 1974. Speciation in Australian birds adapted to arid habitats. *Emu*, 74: 161–168.

Foster, M.S., 1975. The overlap of molting and breeding in some tropical birds. *Condor*, 77: 304–314.

Fry, C.H., 1969. The recognition and treatment of venomous and non-venomous insects by small bee-eaters. *Ibis*, 111: 23–29.

Fry, C.H., 1970a. Ecological distribution of birds in northeastern Mato Grosso State, Brazil. *An. Acad. Brasil. Ciênc.*, 42: 275–318.

Fry, C.H., 1970b. Convergence between jacamars and bee-eaters. *Ibis*, 112: 257–259.

Fry, C.H., 1970c. Migration, moult and weights of birds in northern Guinea savanna in Nigeria and Ghana. *Ostrich, Suppl.*, 8: 239–263.

Fry, C.H., 1972. The biology of African bee-eaters. *The Living Bird*, 11: 75–112.

Fry, C.H., 1973. Avian indicators of increasing environmental aridity at Zaria. *Savanna*, 2: 126–128.

Fry, C.H., 1975. Viduine foraging behaviour and *Lonchura* commensalism. Grey canary feeding technique. *Bull. Nigerian Ornithol. Soc.*, 11: 41–42.

Fry, C.H., 1980a. An analysis of the avifauna of African northern tropical woodlands. *Ostrich. Suppl.*, 14: 77–87. (Proc. 4th Pan-Afr. Ornithol. Congr., 1976).

Fry, C.H., 1980b. Survival and longevity among tropical landbirds. *Ostrich, Suppl.*, 14: 333–343. (Proc. 4th Pan-Afr. Ornithol. Congr., 1976).

Greig-Smith, P.W., 1978. The formation, structure and function of mixed-species insectivorous bird flocks in west African savanna woodland. *Ibis*, 120: 284–297.

Greig-Smith, P.W., 1980. Ranging behaviour, density and diversity of birds in savanna and riverine forest habitats in Ghana. *Ibis*, 122: 109–116.

Greling, C. De, 1972. Sur les migrations et mouvements migratoires de l'avifaune éthiopienne, d'après les fluctuations saisonnières des densités de peuplement en savane soudanienne au nord Cameroun. *Oiseau R.F.O.*, 42: 1–27.

Haffer, J., 1967. Zoogeographical notes on the "nonforest" lowland bird faunas of northwestern South America. *Hornero*, 10: 315–333.

Haffer, J., 1974. Avian speciation in tropical South America. *Nuttall Ornithol. Club, Publ.*, 14: 390 (Cambridge, Mass.).

Hall, B.P. (Editor), 1974. *Birds of the Harold Hall Australian expeditions 1962–70*. British Museum of Natural History, London, 300 pp.

Hall, B.P. and Moreau, R.E., 1970. *An Atlas of Speciation in African Passerine Birds*. British Museum of Natural History, London, 423 pp.

Hanmer, D.B., 1978. Measurements and moult of five species of bulbul from Moçambique and Malawi. *Ostrich*, 49: 116–131.

Harrison, C.J.O., 1975. Speciation in arid habitats: some further comments. *Emu*, 75: 157.

Hopkins, B., 1974. *Forest and Savanna*. Heinemann, London, 2nd ed., 207 pp.

Houston, D.C., 1976. Breeding of the White-backed and Rüppell's Griffon Vultures, *Gyps africanus* and *G. rueppellii*. *Ibis*, 118: 14–40.

Houston, D.C., 1978. The effect of food quality on breeding strategy in Griffon Vultures (*Gyps* spp.). *J. Zool., Lond.*, 186: 175–184.

Howell, T.R., 1969. Avian distribution in Central America. *Auk*, 86: 293–326.

Immelmann, K. and Immelmann, G., 1967. Verhaltensökologische Studien an afrikanischen und australischen Estrildiden. *Zool. Jahrb. Syst.*, 94: 609–686.

Jones, P.J. and Ward, P., 1976. The level of reserve protein as the proximate factor controlling the timing of breeding and clutch-size in the Red-billed Quelea *Quelea quelea*. *Ibis*, 118: 547–574.

Jones, P.J. and Ward, P., 1977. Evidence of pre-migratory fattening in three tropical granivorous birds. *Ibis*, 119: 201–203.

Keast, J.A., 1974. Avian speciation in Africa and Australia: some comparisons. *Emu* 74: 261–269.

Kemp, A.C., 1976. Factors affecting the onset of breeding in African hornbills. In: *Proc. 16th Int. Ornithol. Congr.*, pp. 248–257.

Koeman, J.H., Den Boer, W.M.J., Feith, A.F., De Jongh, H.H. and Spliethoff, P.C., 1978. Three years' observation on side effects of helicopter applications of insecticides used to exterminate *Glossina* species in Nigeria. *Environ. Pollut.*, 31–59.

Lack, D., 1968. *Ecological Adaptations for Breeding in Birds*. Methuen, London, 409 pp.

Leisler, B., 1977. Die ökologische Bedeutung der Lokomotion Mitteleuropäischer Schwirle (*Locustella*). *Egretta*, 20: 1–25.

Leisler, B., 1980. Morphologie und Habitatnutzung europäischer *Acrocephalus*-Arten. In: *Proc. 17th Int. Ornithol. Congr., Berlin 1978*, pp. 1031–1037.

Lepage, M., 1972. Recherches écologiques sur une savane sahélienne du Ferlo septentrional, Sénégal: données préliminaires sur l'écologie des termites. *Terre Vie*, 26: 383–409.

Liversidge, R., 1980. Seasonal changes in the use of avian habitat in southern Africa. In: *Proc. 17th Int. Ornithol. Congr., Berlin 1978*, pp. 1019–1024.

Ligon, J.D. and Ligon, S.H., 1978. Communal breeding in green woodhoopoes as a case for reciprocity. *Nature*, 276: 496–498.

Macdonald, J.D., 1973. *Birds of Australia*. A.H. and A.W. Reed, Sydney, N.S.W., 307 pp.

Maclean, G.L., 1970. The breeding seasons of birds in the southwestern Kalahari. *Ostrich, Suppl.*, 8: 179–192.

Maclean, G.L., 1974. Arid-zone adaptations in southern African birds. *Cimbebasia*, A, 3: 163–176.

Maclean, G.L., 1976a. Adaptations of sandgrouse for life in arid lands. In: *Proc. 16th Int. Ornithol. Congr.*, pp. 504–516.

Maclean, G.L., 1976b. Factors governing breeding of African birds in non-arid habitats. In: *Proc. 16th Int. Ornithol. Congr.*, pp. 258–271.

Moreau, R.E., 1950. The breeding seasons of African birds — 1. Land birds. *Ibis*, 92: 223–267.

Moreau, R.E., 1966. *The Bird Faunas of Africa and its Islands*. Academic Press, London, 424 pp.

Moreau, R.E., 1972. *The Palaearctic–African Bird Migration Systems*. Academic Press, London, 384 pp.

Morel, G., 1968a. L'impact écologique de *Quelea quelea* (L.) sur les savanes sahéliennes: raisons du pullulement de ce plocéidé. *Terre Vie*, 22: 69–98.

Morel, G., 1968b. Contribution à la synécologie des oiseaux du sahel sénégalais. *Mém ORSTOM*, No. 29: 179 pp.

Morel, G., 1973. The Sahel zone as an environment for Palaearctic migrants. *Ibis*, 115: 413–417.

Morel, M.Y., 1964. Natalité et mortalité dans une population naturelle d'un passereau tropical, le *Lagonosticta senegala*. *Terre Vie*, 18: 436–451.

Morel, M.Y., 1973a. Cycles annuels de quelques oiseaux granivores des savanes africaines semi-arides. *Bull. I.F.A.N., Sér. A*, 35: 180–185.

Morel, M.Y., 1973b. Contribution à l'étude dynamique de la population de *Lagonosticta senegala* (Estrildides) à Richard-Toll (Sénégal). Interrelation avec le parasite *Hypochera hypochera chalybeata* (Müller) (Viduinés). *Mém. Mus. Natl. Hist. Nat., Sér A, Zool.*, 78: 1–156.

Morel, M.Y., 1980. The coexistence of seven species of doves in a semi-arid tropical savanna of northern Senegal. In: *Proc. 4th Pan-Afr. Ornithol. Congr.*, 1976, pp. 283–290.

Morel, G. and Morel, M.Y., 1962. La reproduction des oiseaux dans une region semi-aride: la vallée du Sénégal. *Alauda*, 30: 161–203; 241–269.

Morel, G. and Morel, M.Y., 1970. Adaptations écologiques de la reproduction chez les oiseaux granivores de la savane sahélienne. *Ostrich, Suppl.*, 8: 323–331.

Morel, G. and Morel, M.Y., 1972a. Recherches écologiques sur une savane sahélienne du Ferlo septentrional, Sénégal: l'avifaune et son cycle annuel. *Terre Vie*, 26: 410–439.

Morel, G. and Morel, M.Y., 1972b. Étude comparative du régime alimentaire de cinq espèces de tourourelles dans une savane semi-aride du Sénégal. Premiers resultats. In: J. Dinowski and S.C. Kendeigh (Editors), *Proc. General Meeting of Working Group on Granivorous Birds, IBP, IT Section. The Hague, 1970.*

Morel, G. and Morel, M.Y., 1973. Premières observations sur la reproduction du Moineau Doré, *Passer luteus* (Licht.) en zone semi-aride de l'ouest africain. *Oiseau R.F.O.*, 43: 97–118.

Morel, G. and Morel, M.Y., 1974. Recherches écologiques sur une savane sahélienne du Ferlo septentrional, Sénégal: influence de la sécheresse de l'année 1972–1973 sur l'avifaune. *Terre Vie*, 28: 95–123.

Morel, G. and Morel, M.Y., 1978. Recherches écologiques sur une savane sahélienne du Ferlo septentrional, Sénégal. Étude d'une communauté avienne. *Cah. l'ORSTOM, Ser. Biol. Anim*, 13: 3–34.

Nix, H.A., 1976. Environmental control of breeding, post-breeding dispersal and migration of birds in the Australian Region. In: *Proc. 16th Int. Ornithol. Congr.*, pp. 272–305.

Parker, S.A., 1973. The tongues of *Ephthianura* and *Ashbyia*. *Emu*, 73: 23–25.

Pomeroy, D.E., 1978. Seasonality of Marabou Storks *Leptoptilos crumeniferus* in eastern Africa. *Ibis*, 120: 313–321.

Reynolds, J.F., 1968. Notes on the birds observed in the vicinity of Tabora, Tanzania, with special reference to breeding data. *J.E. Afr. Nat. Hist. Soc.*, 27: 117–139.

Rubenstein, D., Barnett, R., Ridgely, R. and Klopfer, P., 1977. Adaptive advantages of mixed species feeding flocks among seed-eating finches in Costa Rica. *Ibis*, 119: 10–21.

Ruwet, J.C., 1964. La périodicité de la reproduction chez les oiseaux du Katanga. *Gerfaut*, 54: 84–110.

Schaefer, G.W., 1976. Radar observations of insect flight. In: R.C. Rainey (Editor), *Insect Flight, Symp. R. Entomol. Soc.*, 7. J. Wiley and Sons, New York, N.Y., pp. 157–197.

Schauensee, R.M. De, 1966. *The Species of Birds of South America with their Distribution*. Acad. Natl. Sci., Livingston, Philadelphia, Penn., 576 pp.

Schauensee, R.M. De and Phelps, W.H., 1978. *A Guide to the Birds of Venezuela*. Princeton University Press, Princeton, N.J., 424 pp.

Schodde, R. and Mason, I.J., 1975. Occurrence, nesting and affinities of the White-throated Grass-Wren *Amytornis woodwardi* and the White-lined Honeyeater *Meliphaga albilineata. Emu*, 75: 12–18.

Serventy, D.L. and Marshall, A.J., 1957. Breeding periodicity in Western Australian birds: with an account of unseasonal nestings in 1953 and 1955. *Emu*, 57: 99–126.

Short, L.L., 1971. The evolution of terrestrial woodpeckers. *Am. Mus. Novit.*, 2467: 1–23.

Short, L.L., 1975. A zoogeographical analysis of the South American Chaco avifauna. *Bull. Am. Mus. Nat. Hist.*, 154(3): 165–352.

Short, L.L., 1980. Chaco woodland birds of South America, some African comparisons. In: *Proc. 4th Pan-Afr. Ornithol. Congr.*, 1976, pp. 147–158.

Sibley, C.G. and Ahlquist, J.E., 1974. The relationships of the African sugarbirds (*Promerops*). *Ostrich*, 45: 22–30.

Sick, H., 1965. A fauna do cerrado. *Arqu. Zool. São Paulo*, 12: 71–93.

Sick, H., 1966. As aves do cerrado como fauna arboricola. *An. Acad. Brasil. Ciênc.*, 38: 355–363.

Siegfried, W.R., 1971. The food of the cattle egret. *J. Appl. Biol.*, 8: 447–468.

Siegfried, W.R., 1972. Aspects of the feeding ecology of cattle egrets (*Ardeola ibis*) in South Africa. *J. Anim. Ecol.*, 41: 71–78.

Sinclair, A.R.E., 1978. Factors affecting the food supply and breeding season of resident birds and movements of Palaearctic migrants in a tropical African savanna. *Ibis*, 120: 480–497.

Slater, P., 1970. *A Field Guide to Australian Birds. Non-passerines*. Rigby, Adelaide, S.A., 428 pp.

Slater, P., 1975. *A Field Guide to Australian Birds. Passerines*. Scottish Academic Press, Edinburgh, 309 pp.

Smith, N.G., 1980. Some evolutionary, ecological, and behavioural correlates of communal nesting by birds with wasps and bees. In: *Proc. 17th Int. Ornithol. Congr., Berlin, 1978*, pp. 1199–1205.

Snow, D.W. (Editor), 1978. *An Atlas of Speciation in African Non-passerine Birds*. British Museum of National History, London, 390 pp.

Staub, F., 1973. Birds of Rodriguez Island. *Proc. R. Soc. Arts Sci. Mauritius*, 4(1): 17–59.

Storr, G.M., 1977. Birds of the Northern Territory. *W. Aust. Mus. Spec. Publ.*, 7: 1–130.

Thiollay, J.M., 1970a. Recherches écologiques dans une savane de Lamto (Côte d'Ivoire): le peuplement avien. Essai d'étude quantitative. *Terre Vie*, 24: 108–144.

Thiollay, J.M., 1970b. L'exploitation par les oiseaux des essaimages de fourmis et termites dans une zone de contact savane-forêt en Côte-d'Ivoire. *Alauda*, 38: 255–273.

Thiollay, J.M., 1971a. L'avifaune de la région de Lamto (Moyenne Côte-d'Ivoire). *Ann. Univ. Abidjan*, E 4(1): 5–130.

Thiollay, J.M., 1971b. L'exploitation des feux de brousse par les oiseaux en Afrique occidentale. *Alauda*, 39: 54–72.

Thiollay, J.M., 1976. Besoins alimentaires quantitatifs de quelques oiseaux tropicaux. *Terre Vie*, 30: 229–245.

Thiollay, J.M., 1977. Distribution saisonnière des repaces diurnes en Afrique Occidentale. *Oiseau R.F.O.*, 47: 253–294.

Thiollay, J.M., 1978. Les migrations de rapaces en Afrique occidentale: adaptations écologiques aux fluctuations saisonnières de production des écosystèmes. *Terre Vie*, 32: 89–133.

Thomas, D.H. and Robin, A.P., 1977. Comparative studies of thermoregulatory and osmoregulatory behaviour and physiology of five species of sandgrouse (Aves: Pterocliidae) in Morocco. *J. Zool., Lond.*, 183: 229–249.

Verner, J. and Willson, M.F., 1966. The influence of habitats on mating systems of North American passerine birds. *Ecology*, 47: 143–147.

Vernon, C.J., 1978. Breeding seasons of birds in deciduous woodland at Zimbabwe, Rhodesia, from 1970 to 1974. *Ostrich* 49: 102–115.

Ward, P., 1965a. Feeding ecology of the Black-faced Dioch *Quelea quelea* in Nigeria. *Ibis*, 107: 173–214.

Ward, P., 1965b. The breeding biology of the Black-faced Dioch *Quelea quelea* in Nigeria. *Ibis*, 107: 326–349.

Ward, P., 1971. The migration patterns of *Quelea quelea* in Africa. *Ibis*, 113: 275–297.

Ward, P., 1972. The functional significance of mass drinking fights by sandgrouse: Pteroclididae. *Ibis*, 114: 533–536.

Ward, P., 1978. The role of the crop among Red-billed Queleas *Quelea quelea*. *Ibis*, 120: 333–337.

Ward, P. and Jones, P.J., 1977. Pre-migratory fattening in three races of the Red-billed quelea *Quelea quelea* (Aves: Ploceidae), an intra-tropical migrant. *J. Zool., Lond.* 181: 43–56.

Ward, P. and Zahavi, A., 1973. The importance of certain assemblages of birds as "information-centres" for food-finding. *Ibis*, 115: 517–534.

Weathers, W.W. and Schlenbaechler, D.C., 1976. Regulation of body temperature in the Budgerygah, *Melopsittacus undulatus*. *Aust. J. Zool.*, 24: 39–47.

Winterbottom, J.M., 1972. The ecological distribution of birds in southern Africa. *Monogr. Percy FitzPatrick Inst. Afr. Ornithol.*, 1: 1–82.

Winterbottom, J.M., 1978. Birds. In: M.J.A. Werger, (Editor), *Biogeography and Ecology of Southern Africa. Monogr. Biol.*, 31: 949–979.

Wood, J.B., 1976. *The Biology of Yellow Wagtails* Motacilla flava *L. Overwintering in Nigeria*. Thesis, University of Aberdeen, Aberdeen, 327 pp.

Wood, J.B., 1979. Changes in numbers of overwintering Yellow Wagtails and their food supplies in a West African savanna. *Ibis*, 121: 228–231.

Woolfenden, G.E., 1980. The selfish behavior of avian altruists. In: *Proc. 17th Int. Ornithol. Congr., Berlin, 1978*, pp. 886–889.

Chapter 15

THE SAVANNA MAMMALS: INTRODUCTION

F. BOURLIÈRE

In the mind of most people, the word savanna is intimately associated with the image of huge herds of game animals quietly grazing in endless grassy plains under a scorching African sun. Such a stereotype is partly justified in the sense that tropical savannas and savanna woodlands indeed represent an optimal habitat for grazing mammals, wild and domestic, large and small. However, a huge standing crop biomass of ungulates is (or better was) only characteristic of some African savannas, particularly those of the eastern and southern parts of this continent. Here, the great plains harboured very large numbers of a variety of animals until the first decades of the present century — a state of affairs which persists nowadays only in a few national parks and game reserves. By contrast, the savannas of the western part of the African continent have for a long time, been much poorer in game animals.

The East African situation cannot be generalized to other tropical continents either (Table 15.1). Tropical American grasslands and savanna woodlands appear in most cases almost devoid of large wild herbivorous mammals. As for tropical Asia, human population pressure has, since many years, almost wiped out the herds of wild cervids, bovids, rhinoceroses and elephants which once roamed through the virgin native grasslands. Only in a few protected areas, mostly in savanna woodlands and flood plains, do appreciable numbers of large herbivorous mammals still persist. In primeval Australia, the biomass of large marsupials quite likely never approached that of the wild ungulates in East African plains.

Once the wild herbivores were eradicated, or thinned out, domestic stock soon took their place. From times immemorial, pastoral tribes have driven their herds in Old World grasslands. More recently, modern ranching techniques have been used in attempting to develop and maintain stable biomasses of domestic ungulates on natural or artificial tropical pastures, without damaging the environment.

No wonder then, that the study of the wild ungulates and of their mammalian competitors and predators, has been a favourite topic for savanna ecologists. The study of the ecology and behaviour of the larger species has also been facilitated by the openness of the savanna landscape, as well as by the protection that large and spectacular mammals have enjoyed in some world-famous national parks. The following five chapters review some of the most significant advances made during the past few decades, with a particular emphasis on the adaptive strategies of the various taxonomic groups involved. More details, particularly in the behavioural field, can be found in a number of excellent books (Geist and Walther, 1974; Moss, 1975; Leuthold, 1977; Delany and Happold, 1979).

TABLE 15.1

Ungulate biomass in some national parks and other protected areas

Location	Habitat type	Number of ungulate species	Live biomass ($t\ km^{-2}$)	Domestic stock ($t\ km^{-2}$)	Reference
Africa					
Tarangire Game Reserve, Tanzania	open *Acacia* savanna	14	1.1	–	Lamprey (1964)
Kafue National Park, Zambia	tree savanna	19	1.3	–	Dowsett (1966)
East Tsavo National Park, Kenya	open *Commiphora Acacia* woodland	13	4.4	–	Leuthold and Leuthold (1976)
Nairobi National Park, Kenya	open savanna	17	5.7	–	Foster and Coe (1968)
Serengeti National Park, Tanzania	open and tree savannas	20	8.2	–	Sinclair and Norton-Griffiths (1979)
Ruwenzori National Park, Uganda	open savanna and thickets	11	12.0	–	Petrides and Swank (1966)
Ruwenzori National Park, Uganda	same habitat, overgrazed	11	27.8–31.5	–	Petrides and Swank (1965)
Virunga National Park, Zaire	open savanna and thickets, overgrazed	11	23.6–24.8	–	Bourlière (1965)
South Asia					
Gir Forest, Gujarat, India	dry deciduous woodland and tree savanna	6	0.4	6.2	Berwick (1974)
Wilpattu National Park, Sri Lanka	open forest and scrub	7	0.7	?	Eisenberg and Lockhart (1972)
Kanha National Park, Madhya Pradesh, India	open *Shorea robusta* forest and grass meadows	10	0.9-1.2	2.9–3.0	Schaller (1967)
Karnali–Bardia National Park, Terai, Nepal	open *Shorea robusta* forest and grass flood plain	6	2.8–3.1	47.9	Dinerstein (1980)
Kaziranga Wildlife Sanctuary, Assam, India	grass flood plain	9	3.8	?	Spillett, in Seidensticker (1976)
Chitawan National Park, Terai, Nepal	tall grass and riverine forest	6	18.5	41.8	Seidensticker (1976)
South America					
Estacion Biologica de los Llanos, Masaguaral, Venezuela	mosaic of savanna types	2	0.3[a]	–	Eisenberg et al. (1979)

[a] To which a biomass of 0.3 t km^{-2} must be added, representing the population of capybaras (*Hydrochoerus capybara*), the large rodent species which is the ecological counterpart of a small ungulate.

REFERENCES

Berwick, S., 1974. *The Community of Wild Ruminants in the Gir Forest Ecosystem.* Thesis, Yale University, New Haven, Conn., 226 pp.

Bourlière, F., 1965. Densities and biomasses of some ungulate populations in Eastern Congo and Rwanda, with notes on population structure and lion/ungulates ratios. *Zool. Afr.,* 1: 199–207.

Delany, M.J. and Happold, D.C.D., 1979. *Ecology of African Mammals.* Longman, London, 434 pp.

Dinerstein, E., 1980. An ecological survey of the Royal Karnali —Bardia wildlife reserve, Nepal. Part III. Ungulate populations. *Biol. Conserv.,* 18: 5–38.

Dowsett, R.J., 1966. Wet season game populations and biomass in the Ngoma area of the Kafue National Park. *Puku,* 4: 135–146.

Eisenberg, J.F. and Lockhart, M., 1972. An ecological reconnaissance of Wilpattu National Park, Ceylon. *Smithson. Contrib. Zool.,* 101: 1–118.

Eisenberg, J.F., O'Connell, M.A. and August, P.V., 1979. Density, productivity, and distribution of mammals in two Venezuelan habitats. In: J.F. Eisenberg (Editor), *Vertebrate Ecology in the Northern Neotropics.* Smithsonian Institution Press, Washington, D.C., pp. 187–207.

Foster, J.B. and Coe, M.J., 1968. The biomass of game animals in Nairobi National Park. *J. Zool. Lond.,* 155: 413–425.

Geist, V. and Walther, F. (Editors), 1974. The behaviour of ungulates and its relation to management. *I.U.C.N. Publ., N.S.,* 24: 1–940.

Lamprey, H.F., 1964. Estimation of the large mammal densities, biomass and energy exchange in the Tarangire Game Reserve and the Masai steppe in Tanganyika. *E. Afr. Wildl. J.,* 2: 1–46.

Leuthold, W., 1977. *African Ungulates. A Comparative Review of Their Ethology and Behavioral Ecology.* Springer-Verlag, Berlin, 307 pp.

Moss, C., 1975. *Portraits in the Wild. Behavior Studies of East African Mammals.* Houghton Mifflin Co., Boston, Mass., 363 pp.

Petrides, G.A. and Swank, W.G., 1965. Population densities and the range-carrying capacity for large mammals in Queen Elizabeth National Park, Uganda. *Zool. Afr.,* 1: 209–225.

Schaller, G.B., 1967. *The Deer and the Tiger. A Study of Wildlife in India.* University of Chicago Press, Chicago, Ill., and London, 370 pp.

Seidensticker, J., 1976. Ungulate populations in Chitawan valley, Nepal. *Biol. Conserv.,* 10: 183–210.

Sinclair, A.R.E. and Norton-Griffiths, M., 1979. *Serengeti. Dynamics of an Ecosystem.* University of Chicago Press, Chicago, Ill., and London, 389 pp.

Chapter 16

RODENTS AND LAGOMORPHS

D.C.D. HAPPOLD

INTRODUCTION

Rodents and lagomorphs form an important, but usually unnoticed, component of the mammalian fauna of the tropical savannas of Africa, South America and Australia. The majority of species are small, nocturnal and rarely seen. They are found in most savanna habitats from seasonally arid savannas with low annual rainfall (300 mm) to moist savannas (1500 mm) bordering rain forest, and from sea level to high alpine grasslands. The large variety of microenvironments in the savanna supports a great diversity of species, each adapted to a specialized niche and way of life. They utilize terrestrial, subterranean and arboreal habitats, consume many types of insect and plant foods, and play a significant part in the savanna ecosystem. Despite their importance, information on tropical savanna rodents and lagomorphs is very patchy; some species are relatively well known, others are known only by name; the ecology of some African species is comparatively well studied, whereas there are practically no ecological studies on South American and Australian species. Because of these limitations, the majority of this chapter discusses the ecology of African rodents and lagomorphs. It is assumed that the various characteristics of tropical savannas which regulate rodent populations, and determine the way of life of each species, are similar in all tropical savannas, wherever they may be.

PATTERNS OF DISTRIBUTION

The wealth of families, genera and species of rodents in tropical savannas is a result of many factors. The diversity of forms follows the general pattern of increased diversity in the tropics compared with temperate regions (Pianka, 1978) which is due, particularly, to competition (Dobzhansky, 1950), the long evolutionary time of relatively stable climate and vegetation, and the spatial heterogeneity of habitats. The three major continental savanna regions show some similarity at the family level, but practically no similarity at the genus or species level. Some "old" cosmopolitan families occur in all tropical savannas, but others of more recent evolutionary origin are more restricted in distribution. The majority of African species[1] belong to the families Muridae and Cricetidae. Sciuridae (squirrels) are the principal arboreal rodents, and the Muscardinidae (dormice), Thryonomyidae (cane rats), Rhizomyidae (bamboo rats), Hystricidae (porcupines), Pedetidae (springhares) and Bathyergidae (mole rats) are each represented by a few species (Delany and Happold, 1979). The Lagomorpha are mainly represented by several species of the genus *Lepus* which tend to replace each other geographically. The South American rodent fauna is dominated by the family Cricetidae, which has evolved to produce many species which occupy similar niches to the Muridae of Africa (Keast, 1972). The "pastoral" species (Hershkovitz, 1962, 1969) include the phyllotine, signodont and akodont groups, with the genera *Calomys*, *Zygodontomys*, *Phyllotis*, *Akodon*, *Baiomys* and *Eligmodontia*. Most of these cricetines are vole-like, although the last genus is similar to the Old World gerbils. There are several genera of

---

[1] Nomenclature of African mammals follows Meester and Setzer (1971–77).

arboreal Sciuridae, as in Africa (Hershkovitz, 1969). Caviomorph rodents comprise an important group of large rodents (which are not considered here — see Chapter 18) and appear to be the ecological equivalents of the African Thryonomyidae, Hystricidae and Cephalophini (Artiodactyla, Bovidae). The fossorial Ctenomyidae live in a similar way to the African Bathyergidae. There is one indigenous lagomorph, *Sylvilagus brasiliensis*, except in parts of the savannas of Venezuela where it has been displaced by the North American *S. floridianus* (Hershkovitz, 1969).

In comparison, the Australian fauna is rather sparse. Rodents are represented only by the Muridae and include *Rattus* and *Melomys*, several genera of the indigenous subfamily Conilurinae (*Conilurus*, *Mesembriomys*, *Pseudomys*, *Xeromys*, *Zyzomys*), and the aquatic *Hydromys*. The arboreal equivalents of the Sciuridae are the marsupial Phalangeridae and Petauridae, but there are no ecological equivalents of the fossorial Bathyergidae and Ctenomyidae. Similarly, there are no Lagomorpha, as the introduced *Oryctolagus*, common in temperate and arid Australia, has not penetrated the tropical savannas (Myers, 1971).

Even though there are similarities at higher taxonomic levels, savanna areas in the same continent show differences at the genus and species level. Differences may be in species composition, or the representation of a particular species in the total rodent fauna. The fauna of any region is due to several historical and ecological characteristics, which include:

(1) The past and present distribution of the savanna in relation to rain forest and arid regions, which has broadly determined the geographical range of each species. This is especially noticeable in Africa due to successive periods of extreme aridity followed by high rainfall.

(2) The average annual rainfall, and the pattern of annual rainfall, which, in turn, influences the vegetation and the annual primary production.

(3) Increasing altitude which generally results in lower mean annual temperatures, lower minimum temperatures and reduced productivity.

(4) Certain biotic factors such as predation and competition.

(5) Last, and perhaps most importantly, the heterogeneity of the habitat. Features such as soil characteristics (texture, porosity, salt concentrations, moisture), topography, floral composition and physiogonomy, result in a mosaic of habitats each producing a slightly different microenvironment.

The number of species, and the abundance of each species, vary according to the locality (Table 16.1). Habitats which are adjacent may differ due to slight, though important, differences in microhabitat, and consequently savannas within a particular country may show great diversity in species composition (Table 16.2).

Selection of a particular microenvironment within its geographical range is an important feature of the ecology of any rodent. As the majority of species are small and have restricted home ranges, the fine details of the microenvironment are of crucial importance. For many species, the precise distribution is quite well known and is illustrated diagramatically in Fig. 16.1. Depending on the geographical locality, similar ecological niches are occupied by different sets of species (Table 16.1, Fig. 16.1). Replacement of one species by another along a geographical gradient is also a common feature of savanna rodents. These trends may be illustrated by two examples. In West Africa, the gradual shortening of the wet season from the Equator northwards (and the associated changes in vegetation) are correlated with changes in the rodent fauna (Table 16.3). Similarly, increasing altitude results in the gradual replacement of lowland savanna species by those better adapted to montane grassland conditions and faunal diversity decreases with altitude (Delany and Happold, 1979).

The reliability of estimates of species composition may often be in doubt. Although a gross estimate is possible, differential reaction of species to traps, efficiency of different trap designs, season of the year, number of trapping days and the size of the trapped area all affect the estimate (Neal and Cock, 1969; Happold, 1975; Bellier, 1967). In the Ivory Coast, for example, smaller areas contained fewer species than larger areas, and the percentage abundance of a species changed according to the size of the survey area (Bellier, 1967). A large area is more likely to contain more microhabitats and therefore a greater number of species, with perhaps a different composition, compared to smaller areas.

LIFE FORMS

Exploitation of a particular savanna habitat partly depends on the morphological and physiological characteristics of the species concerned. The majority of small terrestrial savanna rodents are unspecialized murid-like animals whose size, shape, fur texture, colour and patterning (although very varied) appear to bear little relationship to habitat. The only exceptions to this are the montane species (e.g. *Otomys*, *Phyllotis*) which have longer, denser fur than non-montane species. Most are surface dwellers, living in surface nests or in burrows during the day. Some species have elongated limbs and can move rapidly (Lagomorpha, some cricetines) or are bipedal (*Pedetes*). Rocky inselbergs or *kopjes* are often inhabited by scrambling species, some of which have spinous hairs (*Acomys*, *Zyzomys*). Moist and aquatic habitats are utilized by mesic and aquatic species, even though many of these have no specialized swimming ability (*Dasymys*, *Oryzomys palustris*, *Otomys irroratus*, *Pelomys*, *Rattus sordidus colletti*). It is more likely that these species need the cover provided by the mesic environment, and are less physiologically adapted to shortages of water than other savanna species. The only rodents which are truly aquatic, possessing webbed digits, are the Australian *Hydromys* and the South American *Holochilus* and *Myocastor*.

The arboreal habitat is less utilized by rodents than might be expected considering the abundance of trees in some savannas although it is used by a number of other non-rodent mammals (Africa: primates, fruit bats; Australia: possums, gliders, fruit bats; South America: primates, sloths, fruit bats). The following families and genera are the principal arboreal savanna rodents:

Africa	Australia	South America
Arboreal		
Sciuridae Muscardinidae	*	Sciuridae
Semi-arboreal		
Hylomyscus *Arvicanthis***	*Conilurus* *Mesembriomys*	?

*Ecological niche occupied by marsupials.
**In exceptional circumstances (see p. 392).

Most of the arboreal species have specialized tails and limbs. Indirectly trees and shrubs are important also to terrestrial rodents as they determine the local microclimates, affect the growth and composition of the herb layer, and provide fruits and seeds which are eaten by many terrestrial species.

Fossorial species show particular specializations to their subterranean habitat, and most are incapable of surviving outside their burrows systems. They are expert diggers, and each species digs and burrows in its own specialized way using (singly or in combination) the teeth, forefeet and hindfeet. Burrowing may be accomplished by a single individual or by several working together. Fossorial species are not usually sympatric (unlike surface-dwelling species), which suggests that there is probably only one subterranean niche which can be exploited (Pearson, 1959). The geographical distribution of each species is determined mainly by soil type, soil temperature, and the ability of the species to regulate its temperature in the burrow (see also p. 385).

THE SAVANNA AS A HABITAT FOR RODENTS

During the annual cycle, the savanna varies from a relatively lush habitat with abundant cover, moisture and food during and immediately after the wet season, to one of extreme severity during the dry season when cover is minimal, food may be scarce and the diurnal temperature range is large. The lengths of each of these contrasting situations is related to latitude and rainfall. Dry savannas (300–900 mm of rain per annum) experience the longest periods of severe conditions (8–10 months per annum) whereas moist savannas may experience similar conditions for three or four months each year. The seasonal cycle is relatively predictable and constant in moist savannas, but less predictable and subject to greater variation in dry savannas adjacent to the arid zones.

Habitat diversity

The heterogeneity of habitats in the savanna provides many microenvironments and microhabitats for rodents. Savanna characteristics which change in space and time include plant profiles, plant species diversity, soil characteristics, presence

TABLE 16.1

Rodent populations in tropical savannas

Locality	Vegetation	Number of species	Rodent species contributing more than 10%	Density (ind. ha^{-1})	Biomass (g ha^{-1})	Annual rainfall (mm)	Reference
Feté Olé, Senegal	dry savanna	–	*Taterillus pygargus*	0–6	0–216	200	Poulet (1972)
Dabou, Ivory Coast	savanna, with palm trees at low altitude	9	*Dasymys incomtus* 19% *Lemniscomys striatus* 19% *Uranomys ruddi* 32% *Lophuromys sikapusi* 32%	–	–	–	Gautun et al. (1969)
Foro, Ivory Coast	*Panicum* wooded savanna	14[a]	*Lemniscomys striatus* 47% *Praomys daltoni* 24% *Tatera* sp. 10% *Mus* (*Leggada*) sp. 10%	–	–	1150	Gautun (1975)
Lamto, Ivory Coast	unburnt savanna	9	*Uranomys ruddi* 15% *Dasymys incomtus* 18% *Mylomys dybowskyi* 15%	8–21	520–1645	1400	Bellier (1967)
Olokemeji, Nigeria	derived savanna	11	*Mus minutoides* 56% *Praomys daltoni* 32%	4 –50 (dry) (wet)	70–540	1400	Anadu (1973)
Blukwa, N.E. Zaire	wooded savanna	10	*Lemniscomys striatus* 18% *Otomys tropicalis* 37% *Lophuromys flavopunctatus* 15%	63	3795	–	Misonne (1963)
Lwiro, Zaire	elephant grass/ bush	12	*Oenomys hypoxanthus* 27% *Mus minutoides* 17% *Dendromus mesomelas* 13%	361	16 458	1700	Dieterlen (1967a)
Lwiro, Zaire	grass bush	11	*Dasymys incomtus* 14% *Mus minutoides* 13% *Mus triton* 11% *Lophuromys flavopunctatus* 10% *Dendromus mesomelas* 16% *Otomys irroratus* 20%	236	11 152	1700	Dieterlen (1967a)
Kivu, Zaire	*Pennisetum*, and secondary growth	14	*Lophuromys aquilus* 45% *Oenomys hypoxanthus* 22%	–	–	–	Rahm (1967)
Crater area, Ruwenzori National Park, Uganda	grassland	10	*Lemniscomys striatus* 40% *Lophuromys sikapusi* 18% *Praomys natalensis* 25% *Mus triton* 12%	17 –63[b] (dry) (wet)	672–2221[c]	–	Cheeseman (1975)
South Turkana, Kenya	dry savanna, mixed habitats	9	*Arvicanthis niloticus* *Acomys cahirinus* *Tatera robusta*	–[d]	169	166	Coe (1972)

Location	Habitat	No.	Species				Reference
Ngorongoro Crater, Tanzania	moist savanna	6 genera	—		386[e]	—	Foster (in De Vos, 1969)
Serengeti, Tanzania	grassland	5	*Arvicanthis niloticus* 86%		—	—	Misonne and Verschuren (1966)
	savanna	5	*Otomys tropicalis* *Praomys natalensis*	19 –67 (dry) (wet)	940 –1870 (dry) (wet)	~900	Sinclair (1975)
Ethiopia	afro-alpine grasslands (3720 m)	5	*Arvicanthis niloticus* 77% *Stenocephalemys griseicauda* 10%	65–250	4900–16 500	1200	Müller (1977)
Matope, Malawi	*Acacia* savanna	10	*Praomys natalensis* 41% *Saccostomus campestris* 19% *Tatera leucogaster* 17% *Aethomys chrysophilus* 13%		—	92	Hanney (1965)
Zomba, Malawi	high-altitude grasslands	6	*Lophuromys flavopunctatus* 78% *Mus triton* 12%		—	213	Hanney (1965)
Checcayani, Peru	altiplano grasslands (*ichu*) (4000 m)	5	*Akodon boliviensis* 38% *Calomys sorella* 50%	16	371	—	Dorst (1971)
	altiplano, rocky valley	5	*Phyllotis pictus* 11% *Phyllotis osilae* 43% *Calomys sorella* 50%	28	960	—	Dorst (1971)
Tola, Peru	*Festuca* grass— *Lepidophyllum* bush (4500 m)	6	*Eligmodontia typus* 63% *Auliscomys sublimis* 16% (*Ctenomys opimus* common but not included in % abundance)	3–4	352[f]	—	Pearson and Ralph (1978)
Venezuela	llanos of *Hyparrhenia* (*banco*)	2	*Sigmodon alstoni* 90% *Zygodontomys microtinus* 10%		—	1550	Soriano (1977)
Masaguaral, Venezuela	llanos, with palms and matas	5	*Rhipidomys venezuelae* 29% *Oryzomys bicolor* 10% *Zygodontomys brevicauda* 53%	7	425	1500	Eisenberg et al. (1979)
Humpty Doo, N.T., Australia	floodplains	5	*Rattus sordidus coletti*	0 –140 (wet) (dry)		—	Redhead (1979)
Mt. Erimbari, New Guinea	*Imperata* grassland	3	*Rattus exulans* 97%	19	900	2250	Dwyer (1978b)

[a] Includes very small numbers of three "forest" species.
[b] Live trapping, varies according to season.
[c] Removal trapping, varies according to season.
[d] Varies according to precise habitat.
[e] Includes one insectivore.
[f] 55 g ha^{-1} if *Ctenomys* omitted.

TABLE 16.2

Trapping results in several grassland habitats in Uganda using five sorts of live and breakback traps
(Delany, 1964a)

Species	Habitat				
	dense grassland[1]	scrub grassland[1]	elephant-grass[2]	sedge swamp[3]	savanna[4]
Trap nights:	395	376	178	307	980
Total rodents:	62	38	6	62	51
% capture:	16	10	4	20	5
Mylomys dybowskyi	5	–	–	–	
Arvicanthus niloticus	1	10	–	–	1
Lemniscomys striatus	29	7	–	–	1
Praomys natalensis	19	17	3	–	5
Mus minutoides	2	2	–	–	3
Mus triton	2	–	–	10	–
Lophuromys sikapusi	2	–	2	–	–
Tatera valida	2	–	1	–	–
Graphiurus murinus	–	1	–	–	–
Dasymys incomtus	–	–	–	1	–
Mus bufo	–	–	–	6	–
Mus sp.	–	1	–	3	–
Lophuromys flavopunctatus	–	–	–	22	–
Lophuromys woosnami	–	–	–	5	5
Dendromus mesomelas	–	–	–	5	–
Delanymys brooksi	–	–	–	1	–
Otomys denti	–	–	–	2	–
Otomys tropicalis	–	–	–	7	–
Thamnomys dolichurus	–	–	–	–	2
Lemniscomys barbarus	–	–	–	–	6
Acthomys kaiseri	–	–	–	–	7
Praomys fumatus	–	–	–	–	2
Acomys cahirinus	–	–	–	–	5
Acomys subspinosus	–	–	–	–	5
Saccostomus campestris	–	–	–	–	8
Steatomys parvus	–	–	–	–	1
Tatera nigricauda	–	–	–	–	4
Taterillus emini	–	–	–	–	1

[1] Ruwenzori National Park; [2] Kichwamba; [3] Echuya, at high altitude; [4] Moroto, savanna scrub.

of rocks and termitaria, and local topography (Fig. 16.1). These, in turn, determine the availability, variety and nutritive value of food for herbivorous and gravinorous species. In general, habitats showing a high degree of heterogeneity have more species and a larger biomass of rodents than less heterogeneous habitats. This pattern, when considered in conjunction with variations in plant productivity, explains the changes in rodent diversity and numbers from heterogeneous high-productivity habitats associated with high-rainfall areas, to less heterogeneous low-productivity habitats associated with low-rainfall areas.

Fig. 16.1. Diagram of the microdistribution of tropical savanna rodents. a. Sahelian savanna, Senegal (Poulet, 1972). b. Guinean savanna, Nigeria (Happold, unpubl.). c. High-altitude grasslands, Kivu, Zaire (Dieterlen, 1967a). d. Low-altitude grasslands, Zaire (Misonne, 1963). e. Grasslands, Northern Territory, Australia (J.H. Calaby, pers. comm., 1979).

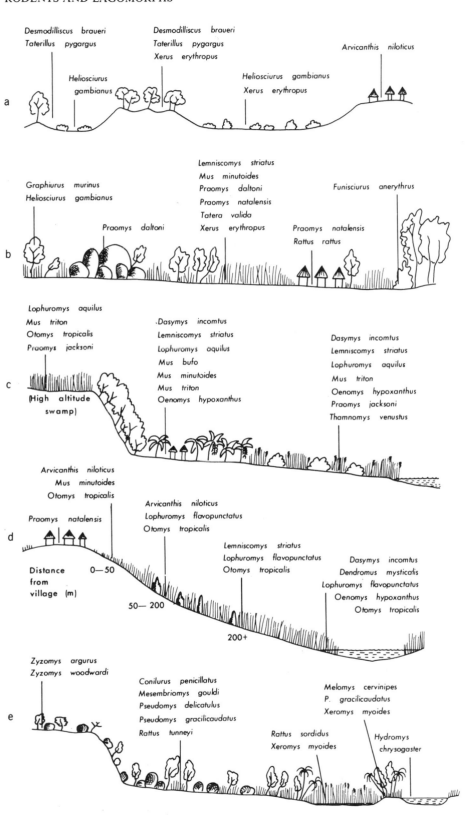

a

Desmodilliscus braueri
Taterillus pygargus

Heliosciurus gambianus

Desmodilliscus braueri
Taterillus pygargus
Xerus erythropus

Heliosciurus gambianus
Xerus erythropus

Arvicanthis niloticus

b

Graphiurus murinus
Heliosciurus gambianus

Praomys daltoni

Lemniscomys striatus
Mus minutoides
Praomys daltoni
Praomys natalensis
Tatera valida
Xerus erythropus

Praomys natalensis
Rattus rattus

Funisciurus anerythrus

c

Lophuromys aquilus
Mus triton
Otomys tropicalis
Praomys jacksoni

(High altitude swamp)

Dasymys incomtus
Lemniscomys striatus
Lophuromys aquilus
Mus bufo
Mus minutoides
Mus triton
Oenomys hypoxanthus

Dasymys incomtus
Lemniscomys striatus
Lophuromys aquilus
Mus triton
Oenomys hypoxanthus
Praomys jacksoni
Thamnomys venustus

d

Arvicanthis niloticus
Mus minutoides
Otomys tropicalis

Praomys natalensis

Distance from village (m)

0—50

50—200

200+

Arvicanthis niloticus
Lophuromys flavopunctatus
Otomys tropicalis

Lemniscomys striatus
Lophuromys flavopunctatus
Otomys tropicalis

Dasymys incomtus
Dendromus mysticalis
Lophuromys flavopunctatus
Oenomys hypoxanthus
Otomys tropicalis

e

Zyzomys argurus
Zyzomys woodwardi

Conilurus penicillatus
Mesembriomys gouldi
Pseudomys delicatulus
Pseudomys gracilicaudatus
Rattus tunneyi

Rattus sordidus
Xeromys myoides

Melomys cervinipes
P. gracilicaudatus
Xeromys myoides

Hydromys chrysogaster

TABLE 16.3

The percentage occurrence of *Tatera valida*, *Praomys daltoni* and *Mus minutoides* in the grasslands of derived savanna, southern Guinean savanna and northern Guinean savanna of Nigeria (Happold, 1975)

Locality	Latitude (N)	Species			Number in sample
		Tatera valida	*Praomys daltoni*	*Mus minutoides*	
Borgu	09°52'	100	0	0	16
Wawa	09°54'	41	59	0	22
Kishi	09°15'	7	93	0	27
Igbetti	08°42'	14	71	14	7
Ibgo-Ora	07°27'	8	81	11	37
Oke-Iho	08°00'	4	55	41	22
Olokemeji	07°26'	0	10	90	10

Rainfall

The annual rainfall is probably one of the most important characteristics of the savanna biome, as it determines and regulates many of the characteristics which are important to rodents. The annual rainfall determines, to a large extent, the annual primary production of the trees, grasses and herbs (Phillipson, 1975; Coe et al, 1976). In addition, the pattern of rainfall and the number of rainy days determines the efficiency of the rainfall. The most important criteria are the amount of water that is available for plant growth and the length of time and constancy of this water. A single rain storm of 20 mm is less "efficient" in these terms than several rain storms totalling the same precipitation during the course of a week. This has been elegantly shown in Senegal (Bille and Poupon, 1974) by the relationship between rainfall, number of rainy days, and annual plant biomass. This affects the food supply for rodents, and the number of individuals which can be supported.

Rainfall may also affect rodents in the opposite way. Local flooding, and the seasonal flooding of riverine plains, temporarily eliminates some habitats for rodent occupation (p. 390).

Primary productivity

The annual primary production is an important parameter as a measure of the potential food availability for herbivorous and granivorous rodents. Generally, increases in primary production support an increasing biomass of primary consumers. Regions of lower rainfall normally have a short growing season, whereas higher rainfall regions may have an eight-month growing season. Recent studies have shown that the quality and nutritive value of savanna grasses is related to their stage of growth (Dougall, 1963; Jarman, 1974; Sinclair, 1975). Old and dead grasses are nutritionally of little value, so that figures of primary production alone may not give a true indication of *available* food (Sinclair, 1975). Rodents eat a variety of leaves, stems, roots, seeds and fruits, and therefore the primary production and quality of food is likely to be an important regulator of rodent populations. This is true also for rain-forest and arid zone tropical rodents (Rahm, 1970; Poulet, 1972; Happold, 1974).

Burning of the savanna vegetation destroys much of the annual primary production of grasses and herbs, and over the course of years determines the structure and composition of grasses, herbs and trees. Annual burning affects rodents mostly by reduction in food availability and cover (see p. 388).

Soil

Many species of small rodents dig burrows where they hide during the day and rear their young. The characteristics of the soil are of vital importance to their burrowing activities. Savanna soils vary from coarse grade sands and laterites to dense loams and clays. The soil moisture content depends on soil type, topography and the season of the year, and these features determine the ease with which rodents may burrow. The varied architectures of rodent burrows are species-specific (Fig. 16.2), but their construction also depends on the properties of the soil. Soils which are too dry and hard, and those which are too wet and flooded, are avoided. Soil structure and moisture are particularly important for fossorial species; in these species, burrowing activity is most intense during the late wet season and early dry season, and almost non-existent when the soil is hard at the end of the dry season and when the soil is flooded at the height of the rains (Genelly, 1965). One important characteristic of burrows is the relative constancy of the burrow temperature, and the lack of diurnal fluctuations compared with the soil surface and the air above the soil.

The microdistribution of several species in the same locality may be due to differing soil preferences. In Senegal, for example, *Taterillus pygargus* and *T. gracilis* prefer sandy well-drained soils, *Tatera robusta* prefer a clay or loam soil which does not drain so easily, *Tatera valida* prefer an intermediate type of soil, and *Praomys natalensis* occur in many soil types except pure sand (Hubert et al., 1977). Similarly, the position of the burrows of *Cricetomys gambianus* is related to soil characteristics (Ajayi and Tewe, 1978).

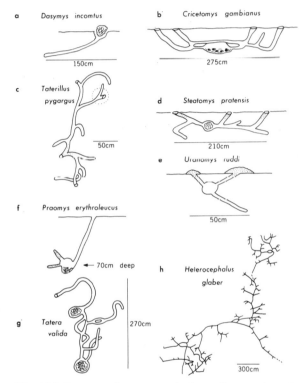

Fig. 16.2. Burrows of savanna rodents. a. *Dasymys incomtus:* surface nest, single shallow burrow often leading to stream (after Hanney, 1965). b. *Cricetomys gambianus:* subterranean nest, often used as food store, several entrances (after Hanney, 1965). c. *Taterillus pygargus:* subterranean burrows, maybe enlarged area for nest, several entrances (often plugged with sand) and burrow shafts ending near soil surface, spill heap near entrance (after Hubert, 1977). d. *Steatomys pratensis:* subterranean nest, several entrances often plugged when not in use (after Hanney, 1965). e. *Uranomys ruddi:* subterranean nest, shallow burrows sometimes used as food stores, spill heaps on surface (after Bellier, 1968). f. *Praomys erythroleucus:* subterranean nest, single entrance, small burrow shafts from nest (after Hubert, 1977). g. *Tatera valida:* subterranean nest, or nests, one entrance with several burrow shafts ending near soil surface (after Verheyen and Verschuren, 1966). h. *Heterocephalus glaber:* very large complex burrow system used for permanent residence and foraging, many spill heaps on surface, no permanent entrance holes (after Jarvis and Sale, 1971).

Insects

Several species of rodents are partly insectivorous, and the availability of insects determines where and when such species may survive. The biomass of insects in the savanna is lowest at the end of the dry season, increases during the wet season, and declines towards the beginning of the dry season (see Ch. 11 and Gillon and Gillon, 1967, 1973). The biomass of insects on an annual basis is the highest in moist savanna and lowest in dry savanna (Gillon and Gillon, 1973), but may vary greatly from year to year (Lamotte, 1975). Generally habitats of low rainfall support insects for fewer months of the year than higher rainfall areas. Most insectivorous rodent species tend to live in higher rainfall areas, although insects may form a significant part of the diet in some species of dry savanna (e.g. *Tatera valida*) when grasses and seeds are scarce (D.C.D. Happold, unpubl.).

All the characteristics of the savanna biome form a series of constraints which singly, or synergistically, determine the numbers, structure and habits of savanna rodents. The importance of any one of these characteristics depends on the species of rodent, where it lives, and on the time of the year. The ways in which these characteristics are related to the ecology of rodents is most obvious in relation to food and feeding habits, reproduction, population structure and numbers and biomass. These will be considered in the following pages.

FOOD

The annual cycles of plant growth, maturation and decay determine the food available for rodents and lagomorphs. Grasses and herbaceous plants are edible and nutritious for only part of their growing period (Bell, 1971; Sinclair, 1975), usually during the early and middle part of the wet season. The chemical composition of plants and their crude protein content changes with age (Dougall, 1963; Hedin, 1967). Seeds and fruits are not available until plants are mature, and by this time green forage may be less abundant. These limitations might suggest that particular foods are limiting at certain times during the wet season but the different life cycles of savanna plants result in different times of germination, growth and the production of seeds and fruits (Hopkins, 1965, 1968; Bille and Poupon, 1974; Poupon and Bille, 1974). Generalized feeders probably have a continual supply of food in normal years. Specialized feeders are likely to be more restricted in the quantity and availability of their food, and the habitats where they can live.

The number of plant species in the herbaceous layer in moist savanna increases at the beginning of the wet season, and then declines as the wet season progresses. Hopkins (1968) found that the average number of plant species in 25 quadrats (each 0.25 m^2) in the moist savanna of Nigeria was from 10 to 12 in the dry season after burning, increased to 25 to 30 during the rains, and then declined to 18 to 20 at the end of the rains. In the Ivory Coast, the number of plant species in 200 m^2 of *Hyparrhenia* savanna was between 45 and 55 and the plant species diversity was highest in savanna which had not been burnt (Roland, 1967). These figures suggest that savanna rodents have a wide selection of plants as potential food, particularly in unburnt savanna.

The length of the growing season, and the plant biomass produced is dependent on the efficiency of the rains (Bille and Poupon, 1974), and therefore rodents in moist savanna have a longer period of adequate food compared with rodents in dry savanna. This has important consequences for regulation of the reproductive cycle and the popu-

lation structures of rodents living in these two extreme situations.

The most adverse period of the year is during the dry season when green forage is scarce or absent, and dead forage is nutritionally poor. Seeds, fruits and insects may be available but become scarcer as the length of the dry season and the rate of exploitation increases. Fire during the dry season reduces or eliminates many forms of food over large areas. Some food may remain in unburnt patches, but these can only be used by rodents already resident in such areas and by those which are capable of dispersing over burnt savanna lacking protective cover.

New growth of good quality forage begins within a few days of the first rain storms in dry savannas. In moist savanna, burning stimulates the sprouting of some plant species within a few weeks of burning and before the rain begins, and in this situation, rodents have access to a new supply of forage during the dry season.

The availability of insects for insectivorous species fluctuates during the course of the year (p. 371), and so these species, like the herbivore ones, are likely to experience periods of abundance and shortage of food.

In summary, the food resources of savanna rodents and lagomorphs fluctuate in abundance, quality and variety during the year. Food of adequate nutritional value is available for more months of the year in moist savannas than in dry savannas.

Food partitioning

Rodents and lagomorphs are selective feeders, and each species in the same savanna eats different foods, or different proportions of the same food. Small rodents in the Ruwenzori National Park in Uganda are either herbivores, omnivores or insectivores (Delany, 1964b; Cheeseman, 1975) (Table 16.4). Of all the potential foods that are available, each species selects those components which it requires and which it is capable of masticating and digesting. A second feature which emerges from this and other studies is that rodents are capable of utilizing a wide spectrum of savanna foods, thereby reducing possible competition for a particular resource. Detailed information on which plant species are selected by herbivorous rodents

TABLE 16.4

The ecological separation of rodents in the crater area of Ruwenzori National Park, Uganda (Cheeseman, 1975)

Species	Feeding habits	Major habitat preference	Activity pattern
Aethomys kaiseri	omnivorous (mainly herbivorous)	bush	mainly nocturnal
Arvicanthis niloticus	herbivorous	*Themeda*	nocturnal and diurnal
Lemniscomys striatus	omnivorous (mainly herbivorous)	all types	crespuscular
Lophuromys sikapusi	insectivorous	*Imperata* and *Cymbopogon*	crepuscular
Mus minutoides	omnivorous (mainly insectivorous)	*Imperata* and *Cymbopogon*	strictly nocturnal
Mylomys dybowskyi	herbivorous	*Imperata*	crepuscular
Otomys irroratus	herbivorous	*Cymbopogon*	?
Praomys natalensis	omnivorous (mainly herbivorous)	all types	strictly nocturnal
Zelotomys hildegardeae	insectivorous	*Imperata* and *Cymbopogon*	strictly nocturnal

(analogous to studies on large herbivores) is sparse, but it is likely that each species feeds on different plant species, or on different combinations of a certain number of species as occurs in two sympatric species of *Lepus* in Kenya (Table 16.5).

Most rodents eat a variety of general food types. Analysis of stomach contents of five species in alpine grasslands of Peru showed that three species were mainly herbivores but also ate some insects, and two mainly insectivorous species also consumed some herbs and seeds (Table 16.6; see also Dorst, 1971, 1972). In the dry savanna of northern Kenya, a more detailed analysis showed that most species ate a variety of grasses, herbs, seeds and insects, but the percentage composition of each food category varied significantly between the species (Table 16.7).

The ability of a species to change its food requirements to whatever is available means a larger and more constant food source. In Kenya, *Arvicanthis niloticus* living in grasslands adjacent to crops, fed on a combination of cereals (maize and wheat), seeds, grass, other herbs and insects. At the beginning of the wet season, they fed mostly on new grasses, but during the middle and end of the wet season the diet was predominantly seeds and cereals, and grass formed a small proportion. Dry season foods were mainly cereals and dicotyledonous weeds (Taylor and Green, 1976). Insects were eaten throughout the year in small quantities. Other species in the same habitat (*Praomys natalensis, Rhabdomys pumilio*) also showed seasonal shifts in food consumption, although *Otomys* tended to feed on grasses at all times of the year.

The diet eaten by each species is also determined by foraging behaviour. During the dry season in Kenya, *Praomys natalensis* forages for seeds and cereals in open areas devoid of cover, whereas other species which need to remain under cover are unable to utilize these food resources (Taylor and Green, 1976). The nutritional quality and availability of the food, and the foraging patterns, affect many aspects of the life history of a species, especially survival and reproduction. The importance of these aspects of rodent ecology has been poorly studied compared with the wealth of information on diet and feeding in large herbivores, yet the relationships between food, nutrition, and other aspects of rodent ecology are essential for understanding how rodents live in the savanna ecosystem.

TABLE 16.5

Food selection in three species of Lagomorpha in Kenya, expressed as mean proportion (%) of different plant items in faeces (from Stewart, 1971a,b)

Plant species	Lepus capensis[1]	Lepus crawshayi[1]	Pronolagus crassicaudatus[2]
Aristida sp.	0.4	0.9	–
Chloris, Microchloa	1.1	–	–
Cynodon dactylon[3]	1.7	3.8	–
Digitaria sp.	8.6	18.8	–
Eragrostis sp.	5.3	4.6	–
Eragrostis/Harpachne	1.7	1.8	–
Hyparrhenia sp.	0.2	10.7	0.8
Themeda triandra[3]	3.6	0.8	1.3
Heteropogon	–	–	0.6
Ischaemum	–	–	33.5
Dicotyledon epidermis	6.3	1.4	0.3
Unidentified grass sheath and epidermis[4]	37.2	29.4	19.6
Unclassified stem fibres	26.5	21.1	42.8

[1] Mount Margaret, Rift Valley; [2] Ngong Hills; [3] the commonest two species of grasses; [4] probably mainly grasses.

TABLE 16.6

Food selection in five species of alpine savanna rodents in Peru during the wet season, January–March (+ + = principal food; + = occasional food; – = recorded) (data from Dorst, 1971)

Species	Food type		
	vegetable material	insect material	seeds
Phyllotis osilae	+ +	+	+
Phyllotis pictus	+ +	–	–
Chinchillula sahamae	+ +	+	–
Akodon boliviensis[1]	+	+ +	–
Calomys sorella	–	+ +	–

[1] Two other rare species, A. jelskii and A. amoenus, have similar food requirements.

Food storage and fat reserves

The effects of reduced food availability during the dry season may be minimized by the storage of food during the wet season for future use, and by increasing the fat reserves of the body which may be utilized when additional energy is required.

Hoarding or storage of food is less common in savanna rodents than might be expected, and is only found in some granivorous species. Grasses, herbs and insects, which are the main food of so many species, cannot be stored, and most species apparently do not have the behavioural characteristics which enable food storage. It seems as if a better strategy for species survival is that numbers and food utilization are reduced during the dry season. Food storage occurs in Xerus (ground squirrels), Cricetomys (pouched rats) (Ewer, 1965, 1967) and several species of cricetine rodents living in dry savanna. Some fossorial rodents (Cryptomys hottentotus, Tachyoryctes splendens) also store roots and tubers which are eaten when the soil is too dry or too wet to make new foraging burrows.

Many workers on African rodents have noticed that abdominal, subcutaneous and testicular fat deposits vary during the course of the year. In Kenya, the fat deposits of Arvicanthis niloticus and Praomys natalensis increased during the early part of the dry season when the animals were feeding mostly on seeds. These reserves were utilized when grasses were growing at the beginning of the wet season and before reproduction commenced, and fat deposits remained low throughout the breeding season (Taylor and Green, 1976). In Uganda, the liver fat content of Lemniscomys striatus (and Praomys natalensis to a lesser extent) increased during the early part of the rains but decreased during the months of reproduction, especially in females (Field, 1975). In these two species, there

TABLE 16.7

Food selection in six genera of small rodents from south Turkana, Kenya, expressed as %
occurrence in stomach contents (from Coe, 1972)

Genera	n	Grass	Leaves/ flowers	Coarse plants	Seeds	Soft fruits	Insects
Arvicanthis	22	83	–	17	–	–	–
Tatera	16	–	13	21	26	26	14
Taterillus *Gerbillus*	27	–	7	24	25	42	2
Acomys	17	–	–	32	42	18	7
Xerus	3	–	30	–	53	11	5

appears to be a very sensitive relationship between rainfall and fat deposition: liver-fat content increased (more so in females than in males) in response to increasing cumulative rainfall. Fat content remained high during the rains, and decreased when the rains ceased. Fluctuations in fat content occur very rapidly, which suggests that slight changes in the nutritional status of the food easily affects the deposition or utilization of fat reserves. Field (1975) has emphasized the important relationship between food quality, foraging behaviour and body-fat levels. In Uganda, *Lemniscomys striatus* is diurnally active and feeds on low-protein grasses in the dry season, and at this time fat levels are low. During the wet season, it feeds on high-protein insects and termites and fat deposits increase. In contrast, *Praomys natalensis* is mostly nocturnal, feeds on insects at all times, and probably does not experience a protein deficiency at any time of the year. From this limited information, it appears that rainfall, through food availability, affects body-fat levels, and these in turn affect survival and reproductive activity. It is likely that there are considerable species differences in this respect. These studies indicate the fine adjustment between environment, food, and life histories, and how slight variations in the environment and the responses of the rodents to these variations may have considerable effects on the ecology of each species.

REPRODUCTION

Many studies on tropical savanna rodents have shown that reproductive activity is related to rainfall. The precise relationship between reproduction and rainfall is uncertain, but it is likely that increased moisture, abundant and nutritious food, good cover, renewal of depleted fat reserves and a more favourable microclimate initiate reproductive activity. There are considerable variations in the responses of rodents to these changes, as illustrated in the following examples, and consequently the beginning, duration and end of the reproductive season is extremely varied.

In eastern Zaire, rainfall occurs in all months, but there is a "dry season" with less than 60 mm of rain per month from June to August. All species of rodents (except *Otomys tropicalis*) breed throughout the year but with a reduction in reproductive activity at the end of the "dry season" and the beginning of the wet season (Dieterlen, 1967b) (Fig. 16.3). In Uganda, where there are two "dry seasons" each year, *Arvicanthis niloticus* breeds throughout the year, but *Lemniscomys striatus*, *Praomys natalensis* and *Mus triton* breed only towards the end of the wet season and during the early dry season, this pattern being repeated twice each year (Delany and Neal, 1969) (Fig. 16.4). In Guyana where there are also two periods of reduced rainfall each year, *Holochilus sciureus* breeds throughout the year (Twigg, 1965); the fecundity rate varies monthly in adult females from 30% to 70% but there is no precise relationship to rainfall.

In the moist savannas of southern Nigeria, there is one dry season of three or four months. There are two peaks of reproductive activity in *Praomys daltoni*; the first lasts for three months at the end of the rains and the beginning of the dry season (Anadu, 1973) (Fig. 16.5). The five-month interval between the reproductive peaks occurs during the wet season and includes two months when there is a reduction in rainfall. *Mus minutoides* in the same

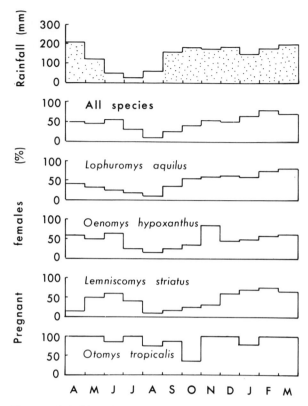

Fig. 16.3. Rainfall and the percentage of adult females visibly pregnant in savannas of Kivu, Zaire (Dieterlen, 1967b). Dotted areas indicate months with more than 50 mm rain.

locality shows a similar, but not identical, pattern. It is difficult to understand why both species cease reproduction in the middle of what appears to be a favourable time of the year. This situation is not unique, as it also occurs in *Praomys daltoni* in the Ivory Coast (Gautun, 1975). In contrast, *Lemniscomys striatus* does not breed until the third month of the wet season and then continues, to a greater or lesser extent, throughout the rains and into the dry season (Fig. 16.6). In both localities, *Tatera valida* exhibits an even more restricted period of reproductive activity.

In dry savannas, as in Senegal, where the dry season lasts for seven or eight months, two or three months elapse after the rains begin before females become reproductively active (Hubert, 1977). Depending on the rainfall pattern, and the primary production, breeding extends for from three to six months after the end of the rains, a much longer dry-season reproductive period than for rodents in the higher rainfall areas. Initiation and length of the

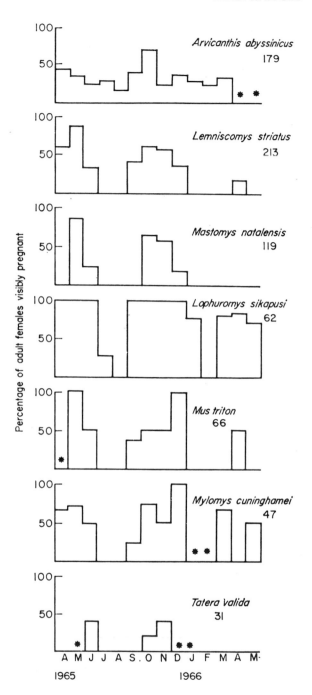

Fig. 16.4. The percentage of adult females visibly pregnant in Ruwenzori National Park, Uganda. Asterisks indicate months when no adult females were sampled. The numbers under the specific names indicate sample size. (From Delany and Neal, 1969.)

breeding season shows species-specific differences as in other localities (Fig. 16.7).

Most species in the savannas of northern

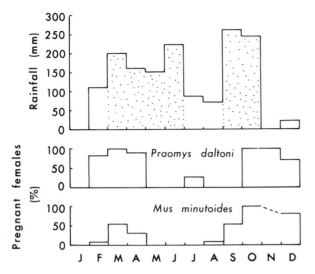

Fig. 16.5. Monthly rainfall and the percentage of adult females visibly pregnant in derived savanna of southern Nigeria (after Anadu, 1973). Dotted areas indicate months with more than 110 mm rain.

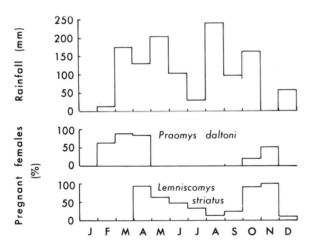

Fig. 16.6. Monthly rainfall and the percentage of adult females visibly pregnant in savanna in the Ivory Coast (after Gautun, 1975).

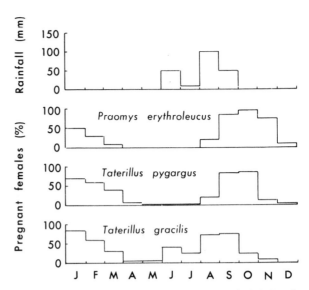

Fig. 16.7. Monthly rainfall and the percentage of adult females visibly pregnant in wooded savanna of Senegal in 1972 (after Hubert, 1977).

Australia are reproductively active in January, one to two months after the beginning of the wet season. Breeding continues for five to six months, extending into the early dry season. All species which have been investigated show peaks of reproductive activity, and there may be a few individuals in particularly favourable microenvironments which are still in reproductive condition in the middle of the dry season. The rock-dwelling *Zyzomys* spp. are exceptional as they appear to

breed throughout the year (J.H. Calaby, pers. comm., 1979).

In all localities, the percentage of pregnant females in the population (a convenient measure of reproductive activity) fluctuates each month. For many species, all eligible females are pregnant at the beginning of the reproductive season, although in species which show continuous breeding (as in Zaire and Uganda) the percentage at any one time is lower than in species with seasonal reproduction. Two strategies appear to have evolved which are, in part, related to environmental conditions: continuous breeding with a smaller percentage of adult females breeding at one time, and seasonal breeding with a much higher percentage of adult females breeding at one time. When considered with variations in litter frequency and number of young per litter, variations in the annual production of young are extremely large.

A possible exception to the general pattern of reproduction is *Arvicanthis niloticus*, which lives in the afro-alpine grasslands (3700 m) of Ethiopia. In this environment, the damp cold climate of the wet season appears to inhibit reproductive activity. The majority of the young are born in the first three months of the eight-month dry season and the young remain with their mother for longer than young of lowland species (Müller, 1977).

In Kenya, the hare *Lepus capensis* probably

breeds throughout the year, with seasonal variations in fecundity. Litter sizes increase during the "short rains" (Flux, 1969), although numbers tend to remain similar throughout the year (Eltringham and Flux, 1971). Hares do not seem to show fluctuations in fecundity (or population numbers) to the same extent as the small rodents.

These examples indicate how closely the reproductive cycle in most species is related to rainfall. Quite small variations in rainfall can initiate a sequence of events which promote or inhibit reproductive activity. The effects of the slight reduction of rainfall during the wet season in West Africa (Figs. 16.5 and 16.6) lead to a dramatic fall in reproductive activity, as does the reduced rainfall in Uganda in January and February (Fig. 16.4).

The same relationship occurs in species which have a very large geographical range, and encounter many different rainfall regimes. The versatility of species such as *Praomys natalensis*[1] enables each population to adapt to the local conditions so that the species as a whole exhibits a great variety of reproductive patterns (Fig. 16.8).

Fossorial species seem to be exceptions to the general rule. In Zaire, the largest number of *Tachyoryctes splendens* breed during the dry season although a low level of fecundity is maintained throughout the year (Rahm, 1971). This pattern is probably related to the lower chances of flooding in the burrows during the dry season.

Rodents appear to be extremely sensitive to rainfall patterns and this enables them to finely regulate the periods of reproduction in relation to abnormally long or short wet seasons, and to maximize their reproductive capacity with the minimum wastage of energy. Initiation of reproduction, although correlated with rainfall, is not easy to understand. Variations in day length are not usually considered important at tropical latitudes, although reduction in day length (perhaps in conjunction with other environmental factors) appears to inhibit reproductive activity in some West African species (Anadu, 1973; Gautun, 1975). Abundance and quality of food, presence of oestrogenic substances in the vegetation and water availability, appear to be necessary even though some species are able to begin reproduction while environmental conditions are harsh. All except fossorial and alpine species reduce reproductive activity when rainfall is low or absent. The different

patterns of reproductive activity are a measure of the species-specific responses to the environmental conditions, and although some general trends are recognizable, each species is capable of modifying its reproduction in relation to the changing environmental conditions of a particular time and place.

Besides the length of the reproductive season, savanna rodents can vary their production in several ways, some of which are species-specific and others which are related to environmental conditions. These include: (1) litter size, and variations in litter size in space and time; (2) number of females breeding at one time; (3) number of litters per adult female per breeding season; and (4) age and season at which the young become reproductively active. Average litter sizes in savanna rodents range from 3 to 12 according to species (Table 16.8). Sympatric species show considerable variation, too, suggesting that litter size is partly related to phylogeny. However, within a single species, litter size may change in time and space. Litter sizes tend to be larger in the wet season (when presumably the chances of survival are greater), and smaller when reproduction continues into a less favourable time of the year (Table 16.9). Populations of the same species in different localities which span a wide range of environmental conditions also have varying litter sizes (Table 16.10); this may suggest that localities where litter sizes are largest are closest to the optimum habitat for the species. Litter sizes in hares, *Lepus capensis*, are smaller at higher altitudes than at lower ones (Flux, 1969), a feature which may also occur in rodents.

The number of females breeding at one time is a measure of the synchrony between breeding activity and the environment. In many species a high percentage of adult females breed when conditions are "correct" after a period of anoestrus (Figs. 16.5 and 16.6). The reproductive rate may remain high, or it may fluctuate to produce peaks and troughs of reproductive activity. A high reproductive rate implies that some adult females are producing

---

[1] The *Praomys natalensis* group is taxonomically confusing and probably includes several closely related species; recent work suggests that *P. natalensis* occurs mainly in southern and eastern Africa, and *P. huberti* and *P. erythroleucus* are the West African representatives of the group.

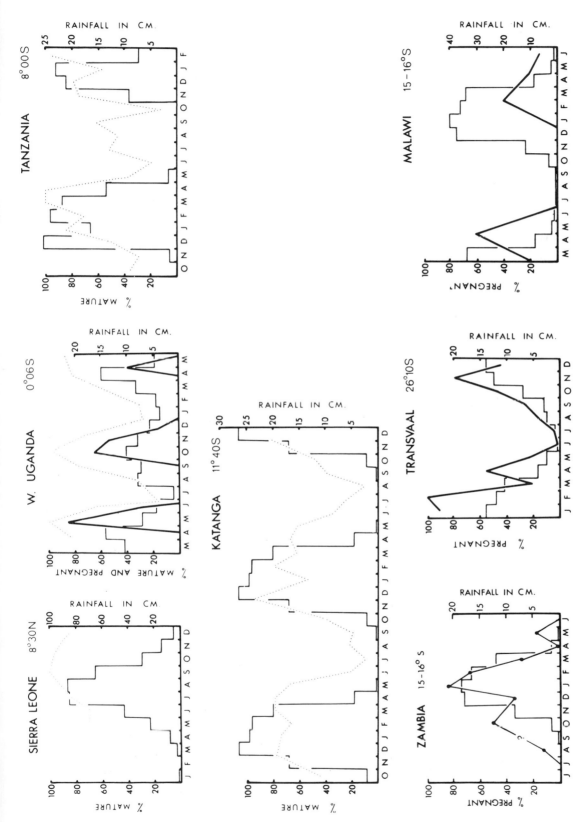

Figure 16.8. Reproductive cycles of *Praomys notalensis* in different parts of Africa. Adult females as % of adult females are indicated by solid lines. Histograms represent mean monthly rainfall. [From Neal, 1977a, after Brambell and Davis, 1941 (Sierra Leone); Neal, 1977a (Uganda); Chapman et al., 1959 (Tanzania); Pirlot, 1954 (Katanga, Zaire); Sheppe, 1972 (Zambia); Hanney, 1965 (Malawi) and Coetzee, 1965 (Transvaal, South Africa).]

TABLE 16.8

Litter size in free-living tropical savanna rodents

Species	n	Litter size		Locality	Reference
		mean	range		
Acomys cahirinus	–	3.2	2–5	Malawi	Hanney (1965)
Aethomys chrysophilus	37	3.1	1–5	S. Africa	Brooks (1972)
Aethomys namaquensis	4	3	–	S. Africa	Meester and Hallett (1970)
Arvicanthis niloticus	44	4.5	3–10	Uganda	Neal (1967)
Lemniscomys striatus	58	4.8	2–8	Uganda	Neal (1967)
Lophuromys sikapusi	46	3.2	1–5	Uganda	Neal (1967)
Mus minutoides	13	3.4	2–6	Nigeria	Anadu (1973)
Mus triton	14	4.4	1–6	Uganda	Neal (1967)
Mylomys dybowskyi	21	4.3	1–6	Uganda	Neal (1967)
Otomys irroratus	13	2.8	2–4	S. Africa	Davis (1972)
Praomys daltoni	18	5.6	3–10	Nigeria	Anadu (1973)
Praomys natalensis	41	12.1	6–19	Uganda	Neal (1967)
Rhabdomys pumilio	–	4.5	2–7	Malawi	Hanney (1965)
Tatera brantsi	51	2.9	–	S. Africa	Meester and Hallett (1970)
Taterillus pygargus	–	4	–	Senegal	Poulet (1972)

TABLE 16.9

Variations in average litter size at different times of the year; wet seasons are indicated by italics (– = no data) (from Coetzee, 1965; Dieterlen, 1967b)

Species and locality	Jan.	Febr.	March	April	May	June	July	Aug.	Sept.	Oct.	Nov.	Dec.
Praomys natalensis (southern Africa)	*12*	–	*12*	*11*	11	6	2	6	*5*	*7*	*9*	*10*
Oenomys hypoxanthus (Kivu, Zaire)	*2.5*	*2.3*	*2.7*	*2.1*	*2.5*	2.5	2.8	2.2	*1.9*	*1.9*	*2.8*	*2.5*

TABLE 16.10

Variation in average litter size of *Praomys natalensis* from different localities in Africa (after Coetzee, 1965)

Locality	Average litter size
Sierra Leone	11.8
Uganda	12.1
Tanzania (Rukwa Valley)	11.2
Malawi	11.0
South Africa (Transvaal)	9.5

litters in quick succession, and hence contributing many young to the population. Species which produce large litters (e.g. *Praomys natalensis*) tend to have short reproductive periods under normal conditions; species with small litters (e.g. *Lophuromys*, *Otomys*) tend to have prolonged reproductive seasons. There are many intermediate conditions between these extremes. The number of young produced is necessarily a compromise between the number that can be supported in the niche utilized by the species, and the necessity to produce as many young as possible to ensure survival of the species and the maximum exploitation of resources. Thus, savanna rodents have evolved numerous strategies to accomplish these ends. Why some species are so common and numerous compared with others is uncertain, but the "success" of any species is a reflexion of its reproductive strategies, the size of its niche, and whether it is a "generalist" or a "specialist".

POPULATION SIZE

An important corollary of seasonal breeding is fluctuations in population numbers. Populations of all species studied so far show considerable variations in numbers during the course of the year. Before the breeding season, populations tend to be small and to consist mostly (or exclusively) of adult individuals. Population numbers rise two or three months after the beginning of the breeding season when young are weaned and become independent. The length of the breeding season, changes in fecundity rate and litter size all affect the rate at which populations increase and the final maximum population size. As each species in the population usually exhibits differences in all these characteristics, there is no uniformity in the timing and magnitude of population increase in each species. Population numbers usually decline as breeding ceases (either during or at some time after the wet season), and when environmental conditions deteriorate (e.g., dry season, after burning, during food shortages).

In southern Nigeria, the numbers of *Mus minutoides* increase rapidly during the dry season and early wet season, but gradually decline as the wet season proceeds. *Praomys daltoni*, in the same locality, has more stable population numbers although peaks of population numbers occur between February and April and in November and December (Anadu, 1973) (Fig. 16.9). Four species in moist savanna of Uganda illustrate asynchronous fluctuations in numbers (Cheeseman, 1975). *Lemniscomys striatus* numbers increase towards the end of each wet season and the beginning of the dry season, as do those of *Lophuromys sikapusi* (Fig. 16.10). Because of so many variables, it is difficult to make generalizations on population

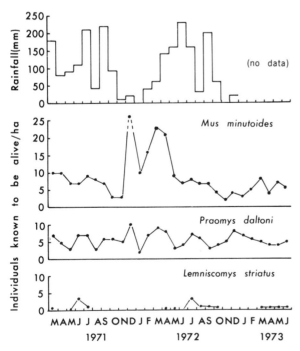

Fig. 16.9. Monthly rainfall and fluctuations in population numbers of *Mus minutoides*, *Praomys daltoni* and *Lemniscomys striatus* in derived savanna of Nigeria (Olokemeji) from March 1971 to June 1973. Burning of savanna occurs in December or early January each year. (After Anadu, 1973.)

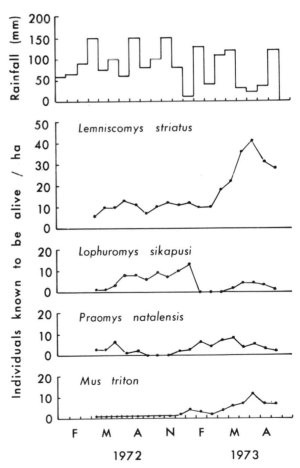

Fig. 16.10. Monthly rainfall and fluctuations in population numbers of *Lemniscomys striatus*, *Lophuromys sikapusi*, *Praomys natalensis* and *Mus triton* in savanna of Ruwenzori National Park, Uganda from January 1972 to October 1973 (after Cheeseman, 1975).

fluctuations. However, the following remarks are generally applicable to most populations:

(1) The total number of individuals in the population fluctuates during the course of the year, and is usually related to the length and timing of the wet and dry seasons.

(2) The percentage abundance of each species in the total population is varied, and fluctuates during the year.

(3) Each species shows differential patterns of population numbers, and therefore its numerical contribution to the total population changes each month.

(4) The timing of population peaks and troughs is a reflexion of species-specific reproductive characteristics.

(5) Although there are general patterns of population fluctuations, the numbers in the same month of different years are not necessarily the same. Similarly, population peaks and numbers on an annual basis may be dissimilar [compare, for example, June 1972 with June 1973 in Ruwenzori (Fig. 16.10), and December 1971 with December 1972 at Olokemeji, Nigeria (Fig. 16.9)].

The different numbers each month indicate monthly changes in the addition and loss of individuals, and these changes are related to seasonal shifts in environmental conditions and reproductive physiology. It is possible for rodent numbers to increase considerably above "normal" when conditions are abnormally favourable, and then to be considerably reduced during abnormally harsh conditions. The relationship between environment and population numbers allows considerable flexibility in relation to environment change.

Although information on population structure of savanna rodents is sparse, two studies illustrate patterns which are probably typical of many localities. The ratio of young:adult, or the abundance of individuals in different age-classes, is related to reproductive strategies and survival as described above for population numbers. In the dry savanna of Senegal, *Taterillus pygargus* breeds for up to eight months of the year except during the end of the dry season and the early wet season (Hubert, 1977). The percentage of adult females breeding during the breeding season is low, litter size is usually small (2–4 young per litter), and therefore the production of young is small and spread out over many months. The percentage of mature

animals decreases during the dry season due to the production of young and mortality of old individuals (Fig. 16.11), and then increases throughout the short wet season as young become adult. The more or less continuous production of young, even though in small numbers, ensures that in most months at least some of the population are juveniles. Two phases of population structure are evident: November to May when fecundity is high and the proportion of juveniles increases, and June to October when fecundity is lower and when young born prior to this period become adult causing a fall in juvenile numbers. In this species, characterized by continuous but low fecundity, a population composed entirely of adults or entirely of young does not occur.

A more accurate assessment of population structure is to divide the population into "age-classes". In moist savanna of Uganda, *Lemniscomys striatus* and *Praomys natalensis* breed in May and June and from October to December (Neal, 1977a,b). Consequently juveniles form the largest proportion of the population from June to August and from December to February (Fig. 16.12). Each of these cohorts may be clearly followed through the months following birth; the result is that the population is composed either mostly of young, or mostly of adults, depending on the time of year. There is naturally some overlap of each cohort, but the pattern is quite distinct from that described above for *Taterillus pygargus*. A similar situation exists in populations of *Mus minutoides* and *Praomys daltoni* in moist savanna of Nigeria; both species alternate between periods of high and low

Fig. 16.11. Monthly rainfall, population structure of *Taterillus pygargus* and the percentage of adult females visibly pregnant in Sahel savanna of Senegal in 1970. Population structure (histogram) is expressed as the percentage of adult individuals in the sampled population. (After Poulet, 1972.)

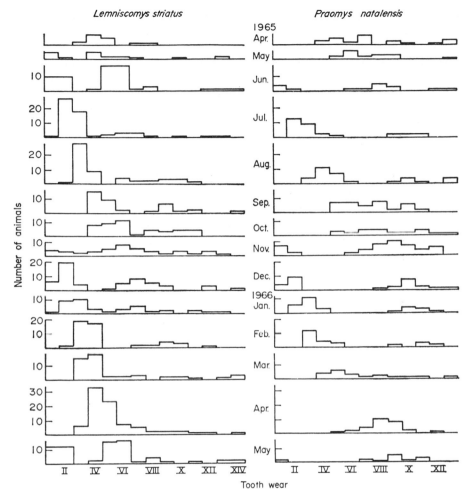

Fig. 16.12. Population structure of *Lemniscomys striatus* and *Praomys natalensis* in Ruwenzori National Park (Uganda) from April 1965 to May 1966. Tooth wear classes (*II–XIV*) indicate the relative age of individuals in the population. (From Delany, 1972; after Neal, 1967.)

reproductive activity, and both show marked differences in population structure at different seasons of the year (Anadu, 1973).

BIOMASS

The total standing crop, or biomass, and the seasonal variations in biomass are useful ecological characteristics as they give an indication of the probable resource requirements of the population. Biomass is not related in a constant way to numbers. For example, a biomass of 40 g may comprise eight *Mus minutoides* (5 g each) or one *Praomys daltoni*. Similarly a large number of young during the breeding season may have the same biomass as a few adults during the non-breeding season. Nevertheless, biomass does provide a rough guide to numbers and vice versa, and changes in numbers are reflected by changes in biomass.

Numerous studies indicate that habitats vary in their biomass of rodents (Table 16.1). Dry savannas and alpine grasslands support a lower biomass than moist savannas. Most of the available figures refer to small terrestrial murids and cricetids; the larger rodents, arboreal and fossorial species and lagomorphs are often omitted and therefore direct comparisons are not always valid. In general, the maximum biomass that can be supported at the most favourable time of the year ranges from about

200 g ha^{-1} in dry savanna to about 2000 g ha^{-1} in moist savanna. Higher biomasses have been reported in exceptional situations — for instance 6760 g ha^{-1} in colonies of fossorial *Ctenomys* in Peru (Pearson, 1959), up to 10 000 g ha^{-1} in lush swampy grasslands of Zaire (Verschuren, 1966), and 16 000 g ha^{-1} in high-rainfall savanna of Kivu (Dieterlen, 1967a).

The biomass fluctuates during the course of the year. During the early part of the breeding season, the biomass increases but not in synchrony with the number of individuals. Maximum biomass is usually attained towards the end of the breeding season as young attain adult weight. After the breeding season, biomass declines (as do population numbers) due to mortality and emigration associated with the dry season. In general, biomass in moist savanna fluctuates by a factor of three or four between the highest and lowest levels. In dry savanna, where environmental extremes are greater, biomass (and numbers) may fluctuate to an even greater extent. Due to the very rapid response of rodents to changing climatic conditions, the rapid changes in biomass (often in as little as two months) are a good indication of changing environmental conditions and resource availability. There is only a short "time lag", very much shorter than for large mammals.

The biomass of large mammals is generally proportional to the annual primary production (Coe et al., 1976). This relationship may be used to predict the potential biomass of large mammals, and appears to be a more sensitive measure than vegetation structure and composition, and average annual rainfall. Data for small mammals indicate a similar trend, although the extremely high biomass values are not fully explainable by this analysis. More information is required to decide whether this relationship is as useful for predicting biomass of small mammals as it is for large mammals.

Plagues

A combination of especially good environmental conditions, and the high reproductive potential of some species, may lead to plagues or ratadas. The prerequisite environmental conditions are a high primary production during the previous wet season so that there is abundant food during the following dry season, a short and less stressful dry season

allowing good rodent survival, and high primary production during the wet season of the plague. Species which can take advantage of these conditions produce many young quickly and in rapid succession, mature rapidly, and can tolerate very high population numbers without inhibition of the reproductive cycle. The variability of the environment in tropical savannas occasionally results in all these conditions simultaneously. Plagues have been recorded in a number of species including *Praomys natalensis* (Chapman et al., 1959; Coetzee, 1975), *Arvicanthis niloticus* (Poulet and Poupon, 1978), *Rhabdomys pumilio* (Taylor, 1968) and *Tatera brantsi* (Davis, 1964) in Africa; *Holochilus sciureus* (Twigg, 1965), *Phyllotis* spp., *Akodon* spp., *Oryzomys nigripes* and *Calomys laucha* (Hershkovitz, 1962) in South America; and *Rattus sordidus villosissimus* (Carstairs, 1976; Newsome and Corbett, 1975) in Australia. During a plague, the high population numbers over-exploit available food resources, fecundity tends to fall, and predation probably increases. Numbers gradually return to the average for the habitat. Additional food sources, such as fields of crops and stored cereals, are conducive to plague conditions.

UTILIZATION OF SPACE

Most rodents, due to their small size and limited mobility, utilize only a restricted area around their domicile. Trapping results show that most small terrestrial murids and cricetids have home ranges of from 100 to 300 m^2 (Table 16.11). For some species, the home ranges of males and females differ, but in no consistent way. Species living in dry savanna (e.g. *Taterillus pygargus*) have larger home ranges than species of moist savanna, probably due to the wider dispersion of resources in drier regions. Slight differences in habitat structure and diversity also cause variations in the home range, as in *Phyllotis darwini* in South American alpine grasslands (Pearson and Ralph, 1978).

Similarly, the area over which a rodent forages depends on the season of the year. In the wet season when resources are more abundant and less dispersed, movements of most species (except *Arvicanthis niloticus*, see Table 16.11) are smaller than during the dry season when resources are dispersed. In the wet season, therefore, there are more indi-

TABLE 16.11

Home ranges of tropical savanna rodents

Locality	Species	Home range (m²)		Reference
		males	females	
Ethiopia (afro-alpine grassland)	*Arvicanthis niloticus*	1400 (dry season)	600	Müller (1977)
		2750 (wet season)	950	
Uganda (tall grass savanna)	*Lemniscomys striatus*	261	256	Cheeseman (1975)
Uganda (farmland)	*Lophuromys flavopunctatus*	293	293	Delany and Kasiimeruhanga (1970)
Uganda (tall grass savanna)	*Lophuromys sikapusi*	280	264	Cheeseman (1975)
	Mus triton	123	109	
	Praomys natalensis	251	198	
Senegal (wooded savanna)	*Tatera robusta*	1500	1400	Hubert (1977)
	Tatera valida	800	600	
	Taterillus gracilis	750	700	
Senegal (Sahel savanna)	*Taterillus pygargus*	1100	300	Poulet (1972)

viduals (due to population increase), each using a smaller area of savanna than during the dry season. In *Lemniscomys striatus* in Uganda, home range size was shown to be inversely related to population size. In this, and other species to a lesser extent, home range decreased as population numbers increased (Cheeseman, 1975), but whether this was due to interactions between individuals or to the more abundant food resources is uncertain. Prior to the breeding season, males tend to increase their home ranges.

Home range size is also related, in part, to food requirements; herbivores tend to have smaller movements than insectivorous species (Cheeseman, 1975) (Table 16.12).

TABLE 16.12

Home range of Uganda rodents in relation to feeding habits (Cheeseman, 1975); the average distance between captures is a measure of home range size

Species	Food habitat	Average distance between captures (m)
Mylomys dybowskyi	exclusively herbivore	28.5
Lemniscomys striatus	mostly herbivore	22.3
Praomys natalensis	mostly herbivore	27.4
Mus triton	mostly insectivorous	34.2
Lophuromys sikapusi	purely insectivorous	35.3
Zelotomys hildegardeae	purely insectivorous	35.2

Burrowing species

The distribution of fossorial species is determined by the soil type (Genelly, 1965; Jarvis and Sale, 1971). The home ranges of many fossorial species are restricted to their burrow system; for species which feed exclusively on subterranean roots and bulbs (*Heterocephalus*, *Tachyoryctes*), the foraging area is extended as required and over-utilized parts of the home range are abandoned (Jarvis and Sale, 1971; Rahm, 1971). Large complex burrows, with many short feeding burrows, extending over areas of up to one hectare, are characteristic of some species (Fig. 16.2). In *Cryptomys hottentotus*, most home range extensions occur during the dry season after the rains, when the soil is neither too wet, nor too dry, for digging. Movements in the burrow are usually at night, and the extensions to the home range are indicated by the number of new spill heaps; *Tachyoryctes splendens*, for example, produces one or two spill heaps each day (Rahm, 1971) (Fig. 16.13).

Other species of fossorial rodents, such as *Ctenomys opimus* and *Ctenomys peruanus* of alpine grasslands of South America, forage in their burrows and on the surface (Pearson, 1959). The former makes new burrow entrances surrounded by spill heaps, and then forages up to one metre from the burrow entrance; the latter forages over a wider area around the entrance. New entrances and spill

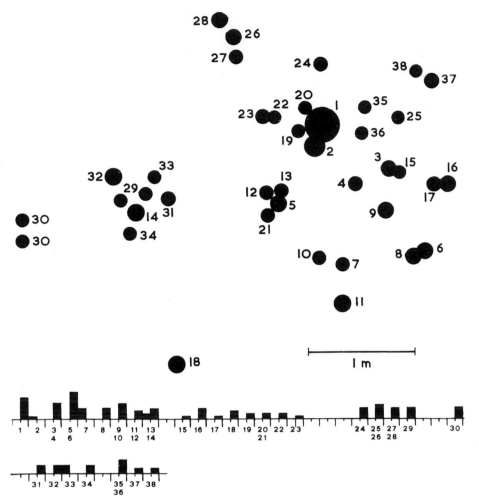

Fig. 16.13. The pattern of spill heaps produced during 37 days in January–December by *Tachyoryctes splendens* in Kivu, Zaire. The chronological sequence of spill heap production is shown by the numbers *1* to *38*, and by the calendar of events. (From Rahm, 1971.)

heaps are made when new foraging areas are required.

Activity patterns

Most savanna rodents are nocturnal, though a few are diurnal and crepuscular as well. Nocturnal activity as well as abundant cover allow protection from predators and from the climatic extremes of the day. Squirrels are mostly diurnal, lagomorphs are mostly nocturnal and/or crepuscular, and some murids (e.g., *Arvicanthis*, *Lemniscomys*, *Lophuromys*) may be crepuscular. Thus, the majority of savanna rodents are active during the hours when light intensity and temperature are at their lowest values.

Nocturnal rodents show peaks of activity, and the activity patterns of several sympatric species are specifically distinct (Fig. 16.14; see also Cheeseman, 1975, 1977). Thus, at any particular time of the night some species are active and others less so, and this situation changes during the course of the night. This is a method of allocating time and resources so that runways, patches of cover and feeding areas are utilized all night, and not over-utilized at one particular time of the night. It presumably spreads the chances of predation over the whole population provided that predator activity patterns are also spread throughout the night. Rodent activity patterns usually show one or more peaks of activity each night, but they may change with changing environmental conditions: night-

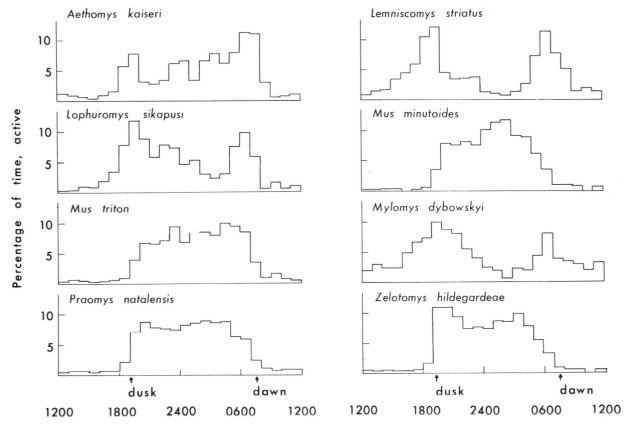

Fig. 16.14. The activity patterns of eight species of rodents in the tall grasslands in Ruwenzori National Park, Uganda (from Delany and Happold, 1979; after Cheeseman, 1975).

time rainfall and nights of full moon result in reduced activity (as indicated by reduced success in trapping) and insectivorous species may show different patterns of activity depending on the abundance of a preferred prey. Crepuscular species show reduced activity when the grass cover is reduced after burning. Temperature also affects activity patterns: *Arvicanthis niloticus* in afro-alpine grasslands, where nocturnal temperatures are very low (Müller, 1977), is diurnal whereas low-altitude *Arvicanthis niloticus* tends to be nocturnal. On Mount Kenya, rodents are mostly diurnal, due to severe frosts at night (Coe, 1969).

The association of different activity patterns with varied food requirements and microdistribution (Table 16.3) results in reduction of competition for resources between sympatric species. This principle is probably applicable for all rodent populations; it enables several species of similar size to occur sympatrically (a very noticeable feature of savanna rodents), and is likely to be of particular importance when population numbers are high.

EFFECT OF MAJOR SEASONAL CHANGES

During the course of the year, savanna fires and flooding of swamps and river floodplains result in major changes of the environment which have profound effects on rodents. Similarly, long periods of drought extending over two or more years and of much longer duration than the normal dry season result in particularly harsh conditions. These changes affect survival, reproduction and population characteristics of rodent populations.

Fire

It is assumed by many authors (e.g., Daubenmire, 1968; Hills, 1974; West, 1965) that fire has been a regular feature of tropical savannas, and it is reasonable to suppose that fire may have been an important evolutionary factor in determining the characteristics of savanna rodents. Most tropical savannas are burnt to a greater or lesser extent every two or three years (Hills, 1974). The extent of savanna burning varies greatly depending on the locality and the year. In Ruwenzori National Park (Uganda), from 1 to 32% was burnt in any one year, and during a three-year period 55% of the area of the park was burnt at least once (Eltringham, 1978); but in some non-reserved areas of Africa, the extent of burning is much greater. Fire itself appears to have little effect on rodents, and mortality from fire is probably rare. Although temperatures in the grass layer may reach 900°C (Hopkins, 1965), the temperatures in burrows a few centimetres below the soil surface rarely increase to a lethal limit because the passage of the fire is rapid. Larger

species of non-burrowing rodents and lagomorphs are able to move away from fires. Of far greater importance are the reduction of cover, increased chances of predation, higher radiation and soil temperatures before the regrowth of the vegetation, and the elimination of many food resources. Early fires soon after the end of the rains do not burn all the savanna: low-lying areas, thickets, and areas where the moisture content of the grass prevents burning, may be left intact and form a mosaic of refuges where food and cover are available. Sprouting of new shoots by many savanna grass species in some areas within two to three weeks of the fire provides a new source of high-quality food.

Fires affect rodent populations to the greatest extent by reduction (or elimination) of food supplies and cover, and the different floristic structure of the vegetation produced by regular burning (Fig. 16.15). On an annual basis unburnt areas

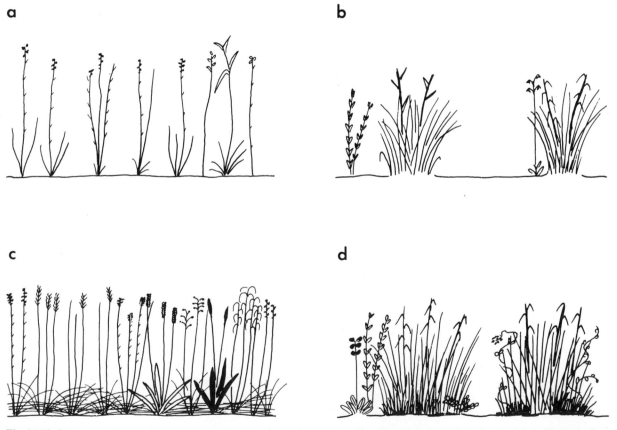

Fig. 16.15. Schematic diagram of the grass-savanna profile in burnt and unburnt savanna grasslands in the Ivory Coast. a. Burnt *Loudetia* grasslands. b. Burnt *Andropogon* grasslands. c. Unburnt *Loudetia* grasslands. d. Unburnt *Andropogon* grasslands. The differences in plant density, amount of litter, and number of plant species between burnt and unburnt savanna are particularly noticeable. (After Roland, 1967.)

TABLE 16.13

The numbers of rodents in 900 m² of burnt and unburnt savanna in Ivory Coast (after Bellier, 1967)

Genera		Burnt savanna	Unburnt savanna
Dasymys		2	35
Lemniscomys		7	12
Lophuromys		1	13
Mus		2	2
Praomys		4	27
Steatomys		2	1
Tatera		0	11
Uranomys		3	28
Not identified		0	6
	Total	21	133
Biomass (including shrews) (g ha⁻¹)		1250	11 445

TABLE 16.14

The number of rodents per ha before and after savanna burning (after Bellier, 1967; Anadu, 1973; Cheeseman, 1975)

Locality	Before burning	After burning	Number of days after burning before post-burn sampling
Lamto, Ivory Coast	10–16	12	30
Ruwenzori National Park, Uganda	32	20	few days
Olokemeji, Nigeria	74	54	2
	33	24	2

(more litter, better cover, more food at certain times of the year) have larger populations of rodents than regularly burnt areas (Table 16.13, Fig. 16.16). Insectivores and omnivores, which forage for insects in the litter layer (*Dasymys*, *Lophuromys*, *Mylomys*) show a considerable preference for non-burnt areas.

Most populations are at their highest towards the end of the rains and in the early dry season before the grasses are burnt. Surveys immediately before a fire and within a few days or weeks afterwards show a reduction in population numbers after the fire (Table 16.14); this reduction is partly due to emigration, but also to mortality. The reduction is not as great as might be expected. Some species show a much greater reduction in numbers than others (Table 16.15), which suggests that these species either show a higher mortality or higher emigration than "fire-adapted" species. Even though the numbers may show only a small reduction, the individuals forming the population may change; for example, in moist savanna in Nigeria, twenty new *Mus minutoides* entered the population within two or three weeks after the fire, but none of these individuals were found immediately after it. However, despite the "loss" of individuals, the population declined by only a few animals due to immigration of more new individuals (Anadu, 1973). Savanna rodents appear to show considerable mobility at this time of the year, making it difficult to distinguish the various components which result in decline of population numbers.

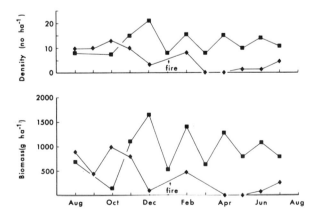

Fig. 16.16. Monthly fluctuations in the density and biomass of small rodents in unburnt and regularly burnt savannas in the Ivory Coast: ◆ = burnt savanna; ■ = unburnt savanna (Bellier, 1967).

TABLE 16.15

The numbers of rodents before and after savanna burning in Uganda as catches per thousand trap nights (Neal, 1970)

Species	Before burning (28 June–3 July)	After burning (4–8 July)
Arvicanthis niloticus	5	0
Lemniscomys striatus	103	44
Lophuromys sikapusi	21	0
Praomys natalensis	49	67
Mus minutoides	3	0
Mus triton	9	7
Mylomys dybowskyi	21	0
Otomys irroratus	3	7
Tatera valida	21	15
Zelotomys hildegardeae	0	0
Total	245	140

For several weeks or months after a fire, a burnt area is devoid of cover and food supplies are reduced. The length of time before growth of new vegetation and a renewal of food supplies depends on the grass species, the climatic regime, and rainfall. Moist savannas with a shorter dry season recover from the effects of burning more rapidly than dry savannas, and this is reflected in the responses of the rodent populations. The responses of each species to these environmental conditions is indicated by the rate and timing of increase in numbers. Burning appears to improve the food resources and other habitat requirements for some species while depleting them for others; omnivores flourish best after a burn due to their exploitation of many food sources (Cheeseman, 1975). Some species respond rapidly (e.g., *Lemniscomys striatus, Mus minutoides, Praomys daltoni, Praomys natalensis*), within two or three months, while others respond more slowly (Fig. 16.17; see also Neal, 1970; Anadu, 1973). Patterns of burnt and unburnt vegetation, and the frequency of burning, are critical to the composition, size and recovery of rodent populations. Not enough information is available on rodents in burnt dry savannas where six to eight months elapse between burning and the commencement of the wet season, to show if they follow the same responses as rodents in moist savanna. Certainly some rodent populations may be called "fire-adapted" in respect of species and numbers, as is vegetation subject to frequent burning.

Floods

Floodplains, swamps and low-lying areas with high rainfall alternate between periods of good cover and high primary production which favour large populations, and periods of flooding which usually result in heavy mortality. During the dry season, *Dasymys incomtus, Mus minutoides, Pellomys fallax* and *Praomys natalensis* live on the floodplains of the Kafue River, Zambia (Sheppe, 1972). These species show considerable mobility during their colonization of the 10 to 40 km wide floodplain as the floods recede. Their distributions are related to habitat preferences; at first, rodents occur throughout the floodplain, but as the grasses dry out and are trampled by large mammals they are restricted to low-lying areas and the damp creek

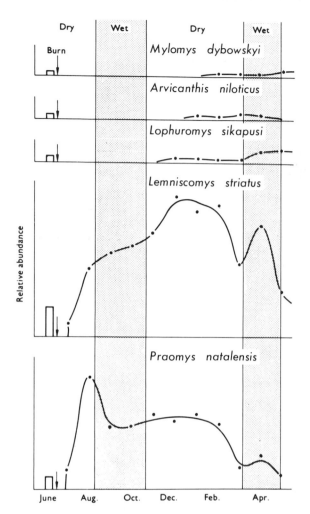

Fig. 16.17. The recolonization of savanna grassland by rodents in Ruwenzori National Park, Uganda after burning in later June (from Delany, 1972; after Neal, 1970).

beds. Reproduction begins during the dispersal and colonization of the floodplains, so populations are composed of many young individuals. Quick-reproducing species, such as *Praomys natalensis*, are able to colonize large areas of floodplain; the slower breeders and those with restricted habitat preferences (e.g. *Pellomys fallax*) spread more slowly and over smaller areas. Many species continue to breed throughout the dry season and into the wet season before flooding occurs. Local rains cause flooding in some regions, which then become uninhabitable. Widespread flooding throughout the plains occurs towards the end of the rains and in the early dry season. Refugia of high ground are extremely important for the survival of the rodents;

these are either "islands" in the floodplain, or the floodplain banks. Mortality is high, and distribution is restricted. The height of the floods determines the number of "island refugia" or levees and therefore the numbers of rodents which survive and can later recolonize the floodplain when the water subsides.

The rodents of the floodplain show seasonal cycles of fat deposition: decrease during the dry season due to reproduction and lactation, and increase during the rains and flooding. Fat reserves appear to be closely correlated with food and water availability. Total population numbers also fluctuate: they are low after the floods, increase during the early dry season, decline at the end of the dry season, increase locally during the rains and decline abruptly during flooding, remaining low until the floods recede.

A similar pattern of fluctuations in population numbers, reproductive seasons and colonization occurs in *Rattus sordidus colletti* living on the floodplains of the Adelaide River in northern Australia (Redhead, 1979) (Fig. 16.18). This species has a high reproductive rate, and colonizes the floodplains from the surrounding woodlands and levees along the river bank. The length of the breeding season is related to the rainfall pattern on the floodplains before flooding occurs. If the rains are short, the length of the breeding season is restricted and population numbers are low; if the rains are prolonged, or unseasonal rain occurs in the dry season, further breeding may occur. The floodplains are suitable for breeding only between the ends of the floods and the drying of the swamp vegetation. The high fecundity of this species is probably related to this normally short breeding season. As on the Kafue floodplain, the height of the flood affects survival. The environmental conditions which regulate rodent populations in both these habitats are different to those described previously. Here, the annual cycle of flooding, position of refugia, dispersal ability, and the extent and timing of breeding determine the size and distribution of populations. Only certain species can take advantage of these specialized conditions, but these show considerable ecological adaptability enabling them to take advantage of the seasonal and annual variations of their environment.

Species which live in low-lying marshy areas, subject to regular flooding, move to the edges of the

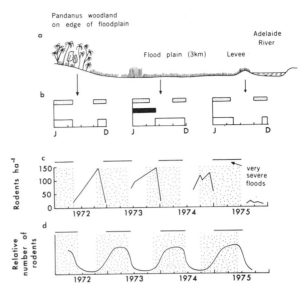

Fig. 16.18. The utilization of the floodplains of the Adelaide River in northern Australia by *Rattus sordidus colletti* (drawn from data in Redhead 1979). a. Diagram of the floodplain habitat. b. Utilization of the plain edge (left), floodplain (centre), and levee (right) by *R. sordidus* in relation to the wet season and floods. Dotted area = wet season; black area = floods; open boxes = time of utilization of habitat by *Rattus*. c. Population numbers in the floodplain from 1972 to 1975. Very severe floods in 1974 to 1975 reduced the survival of *Rattus* to a much lower level than in former years, resulting in very small population numbers in the floodplain during the dry season of 1975. d. Diagram of utilization of the woodland plain edge by *R. sordidus*.

marsh during seasonal flooding. In Venezuela, *Sigmodon alstoni* prefers the low-lying *banco* savanna (*Hyparrhenia*) during the dry season, but moves into the non-flooded *bajio* savanna (*Andropogon*) during the wet season (Soriano, 1977). Other species, like the African *Otomys irroratus*, remain in marshy areas even when flooded (Davis, 1972).

Flash floods, due to extremely heavy deluges may cause temporary flooding, as in the *Themeda–Pennisetum* grasslands of Serengeti (Misonne and Verschuren, 1966). When this occurs, the ground is covered by water, burrows are flooded and rodents seek refuge, if possible, on higher ground or on termitaria. Mortality may be as high as 25% due to drowning and while exposed to cold night temperatures when the fur is wet. Although these conditions may occur periodically in discrete areas, their effects are geographically limited, and the resilience of many populations probably enables a quick return to normal numbers.

Drought

Although the length and severity of the dry season is extremely variable, especially in dry savannas, prolonged drought with below-average rainfall for several successive years occurs occasionally in savannas. Conditions then tend to resemble those of arid zones, as during the prolonged drought in the Sahel region of Africa from 1970 to 1973 when the amount and efficiency of the rains decreased, primary production was extremely low, and there was only a limited amount of food for rodents. *Taterillus pygargus* responded by reduced recruitment and lower population numbers, and therefore a lower level of exploitation of the limited food resources. Population numbers remained low, even after the brief "wet season" during the drought. As *T. pygargus* tends to be an *r*-strategist, it was capable of returning to its original population numbers very quickly after the first good rains which marked the end of the drought (Fig. 16.19; see also Poulet, 1974). In contrast, *Desmodilliscus braueri* which is normally uncommon became more abundant as the numbers of *T. pygargus* decreased (Bourlière, 1978). Generally, species of drier regions are capable of a rapid

increase in numbers when conditions are especially favourable.

The importance of aestivation for savanna rodents is uncertain. It seems, however, that the commonest strategy under adverse conditions is reduction of population numbers and less resource utilization, rather than aestivation. Individuals that survive do not aestivate as a matter of course, and perhaps maintain metabolic rates at, or close to, normal levels. One notable exception is *Steatomys* spp. which aestivate during the dry season, often in groups of up to twenty individuals, but if aestivation is a successful strategy it is surprising that this species is so rare in most habitats.

EFFECTS OF HABITAT CHANGE

The species composition and numbers of rodents change in a quickly reversible and predictable manner in relation to seasonal cycles, but also as a result of sudden or progressive alterations to the habitat. The two most obvious habitat changes are those associated with dense populations of large mammals, and with man-made agricultural systems.

Effects of large animals

Under certain conditions (e.g. in reserved areas), high population numbers of large herbivorous mammals result in the reduction of grass cover and trees, excessive trampling of the soil, and gradual alterations of the flora (Buechner and Dawkins, 1961; Laws, 1968, 1970). Under these conditions the rodent population tends to be smaller and of different composition to that in similar savanna where large mammals are scarce or absent (Delany, 1964b; Cheeseman, 1975) (Table 16.16). The rodents are mainly affected by the reduced vegetation cover and consequent reduction in food. Herbivorous species are affected to the greatest extent; omnivores and insectivores are less affected as they, and their insect prey, find refuge amongst shrubs which form dense bushes under these conditions. Burning of the savanna when large mammals are numerous increases the rate of habitat alteration and the consequent effect on rodents. In contrast, lagomorphs tend to be commoner in areas where there is heavy grazing by large mammals, but

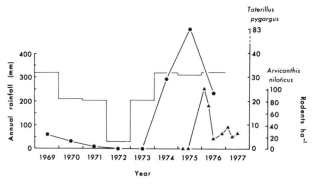

Fig. 16.19. Mean population numbers of *Taterillus pygargus* (●) and *Arvicanthis niloticus* (▲) in the Sahel savanna of Senegal from 1969 to 1977. Histogram indicates annual rainfall. Note the decline of *T. pygargus* during the drought years of 1970–73 and the rapid increase in numbers after the drought. The unusual invasion of the savanna by *Arvicanthis niloticus* in 1975–77 was temporary as this species is not adapted to the normal condition of the Sahel. (From Poulet, 1972, 1978; Poulet and Poupon, 1978; Bourlière, 1978.)

TABLE 16.16

The effects of large herbivorous mammals on rodents in Uganda (Delany, 1964b)

	Locality		
	crater area	Mweya peninsula (before burning)	scrub area
Large-mammal abundance	+	+ + +	+ +
Trap nights	1053	497	518
Trap success (%)	13	7	3
Number of small rodent species	9	4	5
Number of insectivorous or omnivorous species	3	2	4
Number of individual rodents	139	35	17

not over-exploitation (Eltringham and Flux, 1971; J.E.C. Flux, pers. comm., 1979). In Australian savannas, feral buffalo and pigs play a similar role to the large herbivores in Africa (J.H. Calaby, pers. comm., 1978), as do feral pigs in the New Guinea highland savannas (Dwyer, 1978a). In South America, there are no equivalent large grazing mammals (except domestic cattle) and small rodents are less affected by this type of habitat alteration.

Effects of agriculture and human habitation

Agriculture results in many alterations to the natural habitat: removal of grass and herb cover, reduction of tree cover, soil disturbance, and temporary loss of food supplies. As crops grow, a new habitat develops which is often rich in food resources, especially if the farm is a mosaic of plant species as in most tropical savanna farmlands. Rodents show two reactions to agricultural systems: some species flourish and are often more numerous than in undisturbed savanna, while others are unable to adapt and become rare or absent. Species which adapt well to farmlands are *Arvicanthis niloticus, Praomys natalensis, Rhabdomys pumilio, Tatera* spp. and *Taterillus* spp. and *Uranomys ruddi* in Africa. *Melomys littoralis* and *Rattus sordidus* in Australia (McDougall, 1944; Redhead, 1971), and *Holochilus sciureus* (Twigg, 1965), *Calomys laucha, Phyllotis darwini* (Hershkovitz, 1962) and *Zygodontomys lasiurus* (Karimi et al., 1976) in South America.

In developed farmlands of cereals and tubers (perhaps only two or three months after planting in high rainfall areas), rodent populations are often higher than in undisturbed savanna. For example, in Uganda, ten species were found in a mixed coffee–cassava–grassland farm, at a density of about 16.75 individuals ha^{-1} (Delany and Kansiimeruanga, 1970), which is much higher than in adjoining grasslands and most other savanna habitats (cf. Table 16.1). Although quantitative data are sparse, the situation is probably similar in most tropical savanna farmlands. In slower-growing palm plantations, species numbers are reduced at first, and do not increase for several years (Fig. 16.20), and the rodent fauna is still changing after four to seven years (Bellier, 1965).

Irrigated farmlands, where plant growth and cover are more luxuriant than in adjoining non-irrigated farmlands, often have larger rodent populations due to higher fecundity rates and prolonged breeding seasons. Irrigated channels and adjacent fields along the Senegal River supported an average

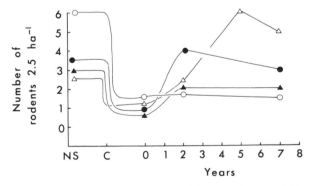

Fig. 16.20. The population numbers of *Dasymys incomtus* (●), *Lemniscomys striatus* (▲), *Lophuromys sikapusi* (△), and *Uranomys ruddi* (○) in natural savanna, during the establishment of a palm plantation, and during the first eight years of the plantation. *NS* = Natural savanna; *C* = clearing of savanna. (Bellier, 1965.)

of 175 *Praomys natalensis* ha^{-1} ($=4700$ g ha^{-1}) and 205 *Arvicanthis niloticus* ha^{-1} ($=21\ 500$ g ha^{-1}), although the densities and biomasses along the irrigation channels themselves were even higher (Poulet, in press). The two species showed different rates of increase and population dynamics, yet both species eventually attained similar densities. The larger size and biomass of *A. niloticus* resulted in this species being more of a pest than *Praomys natalensis*. Similar situations occur in other regions of Africa, indicating the potential problem that rodents may cause to irrigated crops.

Rodents colonize farmlands from adjoining savanna, and therefore the speed of colonization and the numbers of individuals depends on the grass and shrub cover surrounding the farm. The extent of the savanna farm boundary, and the shape of the farm, influence the numbers and species of rodents in the farm (Taylor, 1968; Redhead, 1971; Taylor and Green, 1976). Rodents tend to be more numerous on the edges of farmland (and consequently do more damage) than in the central part of the farm well removed from the edge. The foraging pattern of each species, and whether they make domiciles in the farm or brief incursions from the surrounding savanna, determine rodent numbers and species composition. Dense weed cover among crops tends to result in higher populations than where weeds are removed. The good supply of food and cover in farmlands initiates and prolongs reproductive activity, and an increasing number of young individuals supplement the numbers due to immigration. The combination of good crop growth, good weed cover and adequate water (rainfall or irrigation) leads to very high population numbers or plagues (p. 384). Species which invade farmlands tend to be opportunistic species characterized by large litter sizes, high fecundity and rapid population turnover. Soil type, harvesting, and cultivation techniques are also relevant to the stability and fluctuations of rodent populations in savanna farmlands and may be used to control rodent numbers (Taylor, 1968; Green and Taylor, 1975; Taylor and Green, 1976).

Human activities alter rodent populations and composition in other ways. Areas around villages where grass is cut and which may be farmed, tend to have a fauna differing from that in areas further away (Misonne, 1963) (see Fig. 16.1). Some species colonize food stores (*Praomys natalensis*,

Arvicanthis niloticus and the introduced *Rattus rattus*); in these conditions populations are often high due to more or less continuous reproduction and recruitment.

There is a continuum of habitats between natural savanna, farmlands and houses. Each has a slightly different climatic regime and food availability, and a different set of environmental variables. There are some striking differences in the ecological characteristics of species which can exploit all three habitats. Populations of *Rattus exulans* in New Guinea montane savanna showed lower average weights, higher rates of injuries, lower litter sizes, shorter breeding seasons, and greater fluctuations in numbers, than populations in nearby farmlands and houses (Dwyer, 1978a,b). These differences affected population size and structure, and rate of dispersal. It is probable that savanna individuals are more disposed to disperse than farmland and house individuals, due to the greater fluctuations of their environment; their aggressiveness (as indicated by the high injury rate) allows settling in farmlands where resources are abundant although space is limited (Dwyer, 1978b).

EFFECTS OF RODENTS ON THE SAVANNA ECOSYSTEM

In African savanna ecosystems, the biomass of rodents is generally less than that of the larger grazing and browsing herbivores. Comparative data are scarce, and there are considerable differences between habitats. In protected areas, large mammals may be numerous and consequently small mammals are relatively rare, and areas of high primary production where there are few large mammals may have a large biomass of small mammals. Dry savannas support a lower biomass of both large and small mammals, and in some situations the biomass of small mammals (mostly rodents) may exceed that of large ones. In South American savannas where there are no large indigenous non-rodent herbivores, rodents form the bulk of the mammalian biomass. Despite their smaller biomass, rodents and lagomorphs probably exploit and utilize savanna resources to a greater extent than is normally appreciated (Delany and Happold, 1979). Large and small mammalian herbivores, and insects, are the main exploiters of

TABLE 16.17

Rodent population parameters and food utilization in tall grass savanna of Serengeti, Tanzania (from Sinclair, 1975)

Month	Mean wt of individual (g)	Biomass (g ha^{-1})	Density (ind. ha^{-1})	Food utilization (kg ha^{-1} month^{-1})
February	27.9	1970	67	8.57
June	37.4	1160	31	4.65
October	48.9	940	19	3.13

savanna vegetation, and the importance of each of these depends on their relative numbers and biomass at different seasons of the year. Consequently the importance of rodents varies considerably between habitats.

Analysis of grass production, rodent populations and rodent rates of ingestion in the tall grasslands of the Serengeti has shown that small mammal biomass, density, and food utilization vary seasonally, as does the grass primary production (Sinclair, 1975) (Table 16.17). Large mammals are abundant in this habitat, and consequently the effects of small mammals are relatively smaller than in many other localities. In most months, grasses are not limiting although available food may be scarce. Insects are important herbivores and probably eat young grass that could be eaten by rodents, but the insects themselves provide food for some rodent species. The five species of rodents consume about 69 kg dry weight of grass ha^{-1} yr^{-1}, monthly consumption ranging from 3.3 to 8.8 kg ha^{-1} dry weight (Sinclair, 1975). This is only about 1.2% of the annual primary production. However, as much of the primary production is destroyed by

fire and eaten by termites, the proportion of the edible production utilized by rodents is 4.2% of the total herbivore (mammals, insects) take-off, and nearly 6% of the mammalian take-off. Short-grass habitats have fewer small mammals and lesser rates of utilization, and inselbergs (kopjes) have more small mammals and higher rates of utilization (Table 16.18). When the low average annual biomass of rodents (1320 g ha^{-1}) in the Serengeti is compared with many other moist savanna habitats (Table 16.1), it is likely that rodents and lagomorphs exploit some savanna ecosystems to a considerable extent.

Habitat modification by rodents and lagomorphs is generally not as obvious or as marked as that caused by large herbivores. Grass, seeds, and tubers are eaten and destroyed, but some seeds and fruits are disseminated by rodents, especially by species which hoard and store food. Fossorial species move soil and allow aeration and soil mixing, as do other burrowing species to a lesser extent. The rapid rates of growth in tropical savannas result in rapid regeneration of any vegetation destroyed by rodents. It is likely that the influence of rodents and

TABLE 16.18

The annual grass production and herbivore utilization (kg ha^{-1} yr^{-1}) in the tall grass, short grass and kopjes habitats of the Serengeti, Tanzania (Sinclair, 1975)

	Tall grass		Short grass		Kopjes	
	kg ha^{-1}	% of production	kg ha^{-1}	% of production	kg ha^{-1}	% of production
Annual grass production	5978	100.0	4703	100.0	5978*	100.0
Small mammal utilization	69	1.2	4	0.1	259	4.3
Ungulate utilization	1122	18.8	1597	34	122	2.0
Grasshopper utilization	456	7.6	194	4.1	484	8.1

*Kopjes had the same type of grasslands as the surrounding tall grasslands, and their production was not measured separately. The main difference between tall grasslands and kopjes is the lower herbivore utilization (consumption) on the kopjes because of the fewer number of ungulates (A.R.E. Sinclair, in litt., 1981).

lagomorphs on tropical savanna grasslands is not as great as on temperate grasslands (De Vos, 1969).

Plagues of small rodents are well known for their over-exploitation of resources. An outbreak of *Praomys natalensis* in Tanzania in 1955 which denuded the vegetation around cereal stores and villages (Chapman et al., 1959) is typical of the habitat exploitation of a temporarily abundant rodent. Similarly, large numbers of *Arvicanthis niloticus* in Senegal dispersed into the drier parts of the Sahel savanna in 1975 and 1976 from their normal mesic and cultivated habitat (Fig. 16.19); under these unusual conditions, the rodents climbed trees and damaged 90% of the shoots of *Commiphora africana* and 80% of the shoots of *Acacia senegal* (Poulet and Poupon, 1978). Both these examples indicate that some species of rodents temporarily over-exploit their habitat and can alter the habitat as drastically as can the large herbivores.

Rodents and lagomorphs form part of the extensive food webs of the tropical savannas, and provide food for many species of terrestrial and aerial predators, such as small carnivores, large reptiles, hawks and owls. Practically nothing is known about the predation pressure on these small mammals, or the effects of predation on population numbers. Analysis of owl pellets suggests that these birds prey selectively on certain species, and that the activity patterns and microenvironment of

rodents may influence predation pressure (Davis, 1959; Laurie, 1971; Demeter, 1978). It seems likely that the small predators found in large numbers in tropical savannas exploit rodents and lagomorphs to a considerable extent, but there is no information on the effects of this predation.

POSTSCRIPT

Rodents in tropical savannas have evolved many life forms and many strategies which are related to the predictability and seasonality of the environment. These range from species which tend to be *K*-strategists in moist savannas to those which are more akin to *r*-strategists in dry savannas. The large diversity of species suggests the existence of a wide range of ecological niches occupied by rodents although some, for instance arboreal frugivores, are barely utilized. It may be that the generalist nature of some species has precluded the evolution of more species, as Pearson and Ralph (1978) have shown for tropical South American ecosystems where habitats with a high plant diversity contain many individuals of a few mammalian species, and fewer individuals of many avian species. This concept seems to apply equally to African savannas. Rainfall is undoubtedly the most important ecological factor for savanna rodents, as it regulates the timing and amount of plant growth

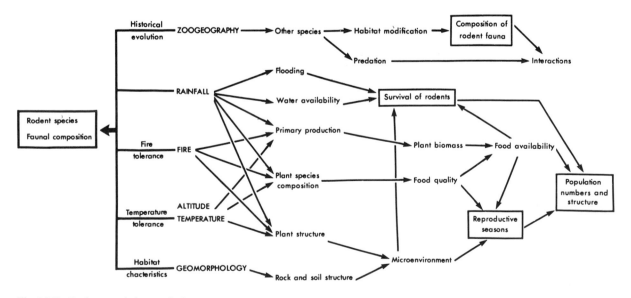

Fig. 16.21. Rodents and the tropical savanna ecosystem.

which, themselves, influence many parameters (Fig. 16.21) of rodent populations. Despite their generally small size, the biomass of rodents is not insignificant because of their large numbers. As primary consumers, they utilize a variety of plant products, and they provide food for a (probably large) variety of secondary consumers. Thus, although they form a "usually unnoticed component of the mammalian fauna" (p. 363), rodents and lagomorphs are a diverse and important group of species and are an integral part of the tropical savanna ecosystem.

ACKNOWLEDGEMENTS

I am grateful to Dr P.A. Anadu, Dr J.H. Calaby, Dr C.L. Cheeseman, Professor M.J. Delany, Dr J. Flux, Dr B. Hubert, Dr B.R. Neal, Professor O.P. Pearson, Mr T. Redhead and Mr P. Soriano who have allowed me to use figures from their publications or have sent me unpublished information.

Paul Parey (Publishers) kindly granted permission to reprint Fig. 16.8, originally published in *Zeitschrift fur Säugetierkunde*, 42: 228 (1977).

REFERENCES

Ajayi, S.S. and Tewe, O.O., 1978. Distribution of burrows of the African giant rat *Cricetomys gambianus* in relation to soil characteristics. *E. Afr. Wildl. J.*, 16: 105–112.

Anadu, P.A., 1973. *The Ecology and Breeding Biology of Small Rodents in the Derived Savanna Zone of South-Western Nigeria*. Thesis, University of Ibadan, Ibadan, 215 pp.

Bell, R.H.V., 1971. A grazing ecosystem in the Serengeti. *Sci. Am.*, 225(1): 86–93.

Bellier, L., 1965. Évolution du peuplement des rongeurs dans les plantations industrielles de palmier à huile. *Oleagineux*, 20: 573–576.

Bellier, L., 1967. Recherches écologiques dans la savane de Lamto (Côte d'Ivoire): densités et biomasses des petits mammifères. *Terre Vie*, 21: 319–329.

Bellier, L., 1968. Contribution a l'étude d'*Uranomys ruddi* Dollman. *Mammalia*, 32: 419–446.

Bille, J.C. and Poupon, H., 1974. Recherches écologiques sur une savane sahélienne du Ferlo septentrional, Sénégal: la régénération de la strate herbacée. *Terre Vie*, 28: 21–48.

Bourlière, F., 1978. La savane sahélienne de Fété Olé, Sénégal. In: *Problèmes d'écologie: structure et fonctionnement des écosystèmes terrestres*. Masson, Paris, pp. 187–229.

Brambell, F.W.R. and Davis, D.H.S., 1941. Reproduction in the multimammate rat (*Mastomys erythroleucus*) of Sierra Leone. *Proc. Zool. Soc. Lond.*, 111B: 1–11.

Brooks, P.M., 1972. Post-natal development of the African bush rat, *Aethomys chrysophilus*. *Zool. Afr.*, 7: 85–102.

Buechner, H.K. and Dawkins, H.C., 1961. Vegetation change induced by elephants and fire in Murchison Falls National Park, Uganda. *Ecology*, 42: 752–766.

Carstairs, J.L., 1976. Population dynamics and movements of *Rattus villosissimus* (Waite) during the 1966–69 plague at Brunette Downs, N.T. *Aust. Wildl. Res.*, 3: 1–9.

Chapman, B.M., Chapman, R.F. and Robertson, I.A.D., 1959. The growth and breeding of the multimammate rat, *Rattus* (*Mastomys*) *natalensis* (Smith) in Tanganyika territory. *Proc. Zool. Soc. Lond.*, 133: 1–9.

Cheeseman, C.L., 1975. *The Population Ecology of Small Rodents in Grassland of Rwenzori National Park, Uganda*. Thesis, University of Southampton, Southampton.

Cheeseman, C.L., 1977. Activity patterns of rodents in the Renzori National Park, Uganda. *E. Afr. Wildl. J.*, 15: 281–287.

Coe, M.J., 1969. Microclimate and animal life in the equatorial mountains. *Zool. Afr.*, 4: 101–128.

Coe, M.J., 1972. The South Turkana Expedition. IX. Ecological studies of the small mammals of South Turkana. *Geogr. J.*, 138: 316–338.

Coe, M.J., Cummings, D.H. and Phillipson, J., 1976. Biomass and production of large African herbivores in relation to rainfall and primary production. *Oecologia*, 22: 341–354.

Coetzee, C.G., 1965. The breeding season of the multimammate mouse *Praomys* (*Mastomys*) *natalensis* (A. Smith) in the Transvaal highveld. *Zool. Afr.*, 1: 29–39.

Daubenmire, R., 1968. Ecology of fire in grassland. *Adv. Ecol. Res.*, 5: 209–266.

Davis, D.H.S., 1959. The Barn Owl's contribution to ecology and palaeoecology. *Ostrich, Suppl.*, 3: 144–153.

Davis, D.H.S., 1964. Ecology of wild rodent plague. In: D.H.S. Davis (Editor), *Ecological Studies in Southern Africa*. W. Junk, The Hague, pp. 301–314.

Davis, R.M., 1972. Behaviour of the vlei rat *Otomys irroratus* (Brants 1827). *Zool. Afr.*, 7: 119–140.

Delany, M.J., 1964a. A study of the ecology and breeding of small mammals in Uganda. *Proc. Zool. Soc. Lond.*, 142: 347–370.

Delany, M.J., 1964b. An ecological study of the small mammals in the Queen Elizabeth Park, Uganda. *Rev. Zool. Bot. Afr.*, 70: 129–147.

Delany, M.J., 1972. The ecology of small rodents in tropical Africa. *Mammal Rev.*, 2: 1–42.

Delany, M.J. and Happold, D.C.D., 1979. *Ecology of African Mammals*. Longman, London, 434 pp.

Delany, M.J. and Kansiimeruhanga, W.D.K., 1970. Observations on the ecology of rodents from a small arable plot near Kampala, Uganda. *Rev. Zool. Bot. Afr.*, 81: 418–425.

Delany, M.J. and Neal, B.R., 1969. Breeding seasons in rodents in Uganda. *J. Reprod. Fertil., Suppl.*, 6: 229–235.

Demeter, A., 1978. Food of a barn owl *Tyto alba* in Nigeria. *Bull. Niger. Ornithol. Soc.*, 14(45): 9–13.

De Vos, A., 1969. Ecological conditions affecting the production of wild herbivorous mammals on grasslands. *Adv. Ecol. Res.*, 6: 137–183.

Dieterlen, F., 1967a. Ökologische Populationstudien an Muriden des Kivugebietes (Congo). I. *Zool. Jahrb. Abt. Syst. Ökol. Geogr. Tiere*, 94: 369–426.

Dieterlen, F., 1967b. Jahreszeiten und Fortpflanzungsperioden bei den Muriden des Kivusee-Gebietes (Congo). I. Ein Beitrag zum Problem der Populationsdynamik in den Tropen. *Z. Säugetierkd.*, 32: 1–44.

Dobzhansky, T., 1950. Evolution in the tropics. *Am. Sci.*, 38: 209–221.

Dorst, J., 1971. Nouvelles recherches sur l'écologie des rongeurs des haut plateaux Péruviens. *Mammalia*, 35: 515–547.

Dorst, J., 1972. Morphologie de l'estomac et régime alimentaire de quelques rongeurs des hautes Andes du Perou, *Mammalia*, 36: 647–656.

Dougall, H.W., 1963. Average chemical composition of Kenya grasses, legumes and browse. *E. Afr. Wildl. J.*, 1: 120.

Dwyer, P.D., 1978a. Rats, pigs and men: disturbance and diversity in the New Guinea Highlands. *Aust. J. Ecol.*, 3: 213–232.

Dwyer, P.D., 1978b. A study of *Rattus exulans* (Peale) (Rodentia: Muridae) in the New Guinea Highlands. *Aust. Wildl. Res.*, 5: 221–248.

Eisenberg, J.F., O'Connell, M.A. and August, P.V., 1979. Density, productivity and distribution of mammals in two Venezuelan habitats, In: J. F. Eisenberg (Editor), *Vertebrate Ecology in the Northern Neotropics.* Smithsonian Institution Press, Washington, D.C., pp. 187–207.

Eltringham, S.K., 1978. The frequency and extent of uncontrolled grass fires in the Rwenzori National Park, Uganda. *E. Afr. Wildl. J.*, 14: 215–222.

Eltringham, S.K. and Flux, J.E.C., 1971. Night counts of hares and other animals in East Africa. *E. Afr. Wildl. J.*, 9: 67–72.

Ewer, R.F., 1965. Food burying in the African ground squirrel, *Xerus erythropus. Z. Tierpsychol.*, 22: 321–327.

Ewer, R.F., 1967. The behaviour of the African Giant Rat (*Cricetomys gambianus* Waterhouse). *Z. Tierpsychol.*, 24: 6–79.

Field, A.C., 1975. Seasonal changes in reproduction, diet and body composition of two equatorial rodents. *E. Afr. Wildl. J.*, 13: 221–236.

Flux, J.E.C., 1969. Current work on the reproduction of the African hare, *Lepus capensis* L. in Kenya. *J. Reprod. Fertil., Suppl.*, 6: 225–227.

Gautun, J.-C., 1975. Périodicité de la reproduction de quelques rongeurs d'une savane préforestière du centre de la Cote d'Ivoire. *Terre Vie*, 29: 265–287.

Gautun, J.-C., Bellier, L. and Heim de Balsac, H., 1969. Liste préliminaire des rongeurs de la savanne de Dabou (Cote d'Ivoire). *J. W. Afr. Sci. Assoc.*, 14: 219–223.

Genelly, R.E., 1965. Ecology of the common mole-rat (*Cryptomys hottentotus*) in Rhodesia. *J. Mammal.*, 46: 647–665.

Gillon, Y. and Gillon, D., 1967. Recherches écologiques dans la savane de Lamto (Cote d'Ivoire): cycle annuel des effectifs et des biomasses d'Arthropodes de la strate herbacee. *Terre Vie*, 21: 262–277.

Gillon, Y. and Gillon, D., 1973. Recherches écologiques sur une savane sahélienne du Ferlo septentrional, Sénégal: données quantitatives sur les Arthropodes. *Terre Vie*, 27: 297–323.

Green, M.G. and Taylor, K.D., 1975. Preliminary experiments in habitat alteration as a means of controlling field rodents in Kenya. In: L. Hansson and B. Nilsson (Editors), *Bio-

control of Rodents.* Ecol. Bull. 19. Swedish Nat. Sci. Council, Stockholm, pp. 175–186.

Hanney, P., 1965. The Muridae of Malawi (Africa: Nyasaland). *J. Zool. Lond.*, 146: 577–633.

Happold, D.C.D., 1974. The small rodents of the forest-savanna-farmland association near Ibadan, Nigeria, with observations on reproduction biology. *Rev. Zool. Afr.*, 88: 814–836.

Happold, D.C.D., 1975. The effects of climate and vegetation on the distribution of small rodents in western Nigeria. *Z. Säugetierkd.*, 40: 221–242.

Hedin, L., 1967. Recherches écologiques dans la savane de Lamto (Cote d'Ivoire): la valeur fourragère de la savane. *Terre Vie*, 21: 249–261.

Hershkovitz, P., 1962. Evolution of neotropical cricetine rodents (Muridae) with special reference to the phyllotine group. *Fieldiana, Zool.*, 46: 1–524.

Hershkovitz, P., 1969. The evolution of mammals on southern continents. 6. The recent mammals of the neotropical region: a zoogeographic and ecological review. *Q. Rev. Biol.*, 44: 1–70.

Hills, T.L., 1974. The savanna biome: a case study of human impact on biotic communities. In: I.R. Manners and M.W. Mikesell (Editors), *Perspectives on Environment. Assoc. Am. Geogr., Publ.* 13: 342–373.

Hopkins, B., 1965. Observations on savanna burning in the Olokomeji Forest Reserve, Nigeria. *J. Appl. Ecol.*, 2: 367–381.

Hopkins, B., 1968. Vegetation of the Olokomeji Forest Reserve, Nigeria. V. The vegetation of the savanna site with special reference to its seasonal changes. *J. Ecol.*, 56: 97–115.

Hubert, B., 1977. Ecologie des populations de rongeurs de Bandia (Sénégal), en zone sahélo-soudanienne. *Terre Vie*, 31: 33–100.

Hubert, B., Leprun, J.-C. and Poulet, A., 1977. Importance écologique des facteurs édaphiques dans la repartition spatiale de quelques rongeurs au Sénégal. *Mammalia*, 41: 35–59.

Jarman, P.J., 1974. The social organisation of antelope in relation to their ecology. *Behaviour*, 48: 215–267.

Jarvis, J.U.M. and Sale, J.B., 1971. Burrowing and burrow patterns of East African mole rats *Tachyoryctes, Heliophopius* and *Heterocephalus. J. Zool. (Lond.)*, 163: 451–479.

Karimi, Y., De Almeida, C.R. and Petter, F., 1976. Note sur les rongeurs du nord-est du Brésil. *Mammalia*, 40: 257–266.

Keast, A., 1972. Comparisons of contemporary mammal faunas of southern continents. In: A. Keast, F.C. Erk and B. Glass (Editors), *Evolution, Mammals and Southern Continents.* State University of New York, Albany, N.Y., pp. 433–501.

Lamotte, M., 1975. The structure and function of a tropical savanna ecosystem. In: F. Golley and E. Medina (Editors), *Tropical Ecological Systems.* Springer-Verlag, Berlin, pp. 179–222.

Laurie, W.A., 1971. The food of the barn owl in the Serengeti National Park, Tanzania. *J. E. Afr. Nat. Hist. Soc. Nat. Mus.*, 28: 1–4.

Laws, R.M., 1968. Interactions between elephant and hippopotamus populations and their environment. *E. Afr. Agric. For. J.*, 33 (Spec. Iss.): 140–148.

Laws, R.M., 1970. Elephants as agents of habitat and landscape change in East Africa. *Oikos*, 21: 1–15.

McDougall, W.A., 1944. An investigation of the rat pest problem in Queensland cane fields. 2. Species and general habits. *Qld. J. Agric. Sci.*, 1: 48–78.

Meester, J. and Hallett, A.F., 1970. Notes on post-natal development of certain southern African Muridae and Cricetidae. *J. Mammal.*, 51: 703–711.

Meester, J. and Setzer, H.W. (Editors), 1971–77. *The Mammals of Africa: an Identification Manual.* Smithsonian Institution, Washington, D.C.

Misonne, X., 1963. Les rongeurs du Ruwenzori et des régions voisines. *Explor. Parc Natl. Albert, 2nd Ser.*, 14: 1–164.

Misonne, X. and Verschuren, J., 1966. Les rongeurs et lagomorphes ·de la région du Parc National du Serengeti (Tanzanie). *Mammalia*, 30: 517–537.

Müller, J.P., 1977. Populationökologie von *Arvicanthis abyssinicus* in der grassteppe des Semien Mountains National Park (Äthiopien). *Z. Säugetierkd.*, 42: 145–172.

Myers, K., 1971. The Rabbit in Australia. In: P.J. den Boer and G.R. Gradwell (Editors), *Dynamics of Populations. Proceedings of the Advanced Study Institute on "Dynamics of Numbers in Populations", Oosterbeck, 1970*, pp. 478–506.

Neal, B.R., 1967. *The Ecology of Small Rodents in the Grassland Community of the Queen Elizabeth National Park, Uganda.* Thesis, University of Southampton, Southampton.

Neal, B.R., 1970. The habitat distribution and activity of a rodent population in Western Uganda, with particular reference to the effects of burning. *Rev. Zool. Bot. Afr.*, 81: 29–50.

Neal, B.R., 1977a. Reproduction of the multimammate rat, *Praomys (Mastomys) natalensis* (Smith) in Uganda. *Z. Säugetierkd.*, 42: 221–231.

Neal, B.R., 1977b. Reproduction of the punctated grass-mouse, *Lemniscomys striatus*, in the Ruwenzori National Park, Uganda (Rodentia: Muridae). *Zool. Afr.*, 12: 419–428.

Neal, B.R. and Cock, A.G., 1969. An analysis of the selection of small African mammals by two break-back traps. *J. Zool., Lond.*, 158: 335–340.

Newsome, A. and Corbett, L.K., 1975. Outbreaks of rodents in semi-arid and arid Australia: causes, preventions, and evolutionary considerations. In: I. Prakash and P.K. Ghosh (Editors), *Rodents in Desert Environments.* W. Junk, The Hague, pp. 117–153.

Pearson, O.P., 1959. Biology of the subterranean rodents, *Ctenomys*, in Peru. *Mem. Mus. Hist. Nat. Javier Prado*, 9: 1–55.

Pearson, O.P. and Ralph, C.P., 1978. The diversity and abundance of vertebrates along an altitudinal gradient in Peru. *Mem. Mus. Hist. Nat. Javier Prado*, 18: 1–97.

Phillipson, J., 1975. Rainfall, primary production and 'carrying capacity' of Tsavo National Park (East), Kenya. *E. Afr. Wildl. J.*, 13: 171–201.

Pianka, E.R., 1978. *Evolutionary Ecology.* Harper and Row, New York, N.Y., 397 pp.

Pirlot, P.L., 1954. Pourcentages de jeunes et périodes de reproduction chez quelques rongeurs du Congo belge. *Ann. Mus. Congo Tervuren Zool.*, 1: 41–46.

Poulet, A.R., 1972. Caractéristiques spatiales de *Taterillus pygargus* dans le Sahel sénégalais. *Mammalia*, 36: 579–606.

Poulet, A.R., 1974. Recherches écologiques sur une savanne sahélienne du Ferlo septentional, Sénégal: quelques effects de la sécheresse sur le peuplement mammalien. *Terre Vie*, 28: 124–130.

Poulet, A., 1978. Evolution of the rodent population of a dry bush savanna in the Senegalese sahel from 1969 to 1977. *Bull. Carnegie Mus.*, 6: 113–117.

Poulet, A., in press. The 1975–1976 outbreak of rodents in an irrigated farmland of northern Senegal. In: *Proceedings of the Symposium on Small Mammals, Problems and Control, Laguna, Philippines, 1977.*

Poulet, A.R. and Poupon, H., 1978. L'invasion d'*Arvicanthis niloticus* dans le Sahel Sénégalais en 1975–1976 et ses conséquences pour la strate ligneuse. *Terre Vie*, 32: 161–193.

Poupon, H. and Bille, J.C., 1974. Recherches écologiques sur une savane sahélienne du Ferlo septentrional, Sénégal: influence de la sécheresse de l'année 1972–1973 sur la strate ligneuse. *Terre Vie*, 28: 49–75.

Rahm, U., 1967. Les muridés des environs du Lac Kivu et des régions voisines (Afrique Centrale) et leur écologie. *Rev. Suisse Zool.*, 74: 439–519.

Rahm, U., 1970. Note sur la reproduction des Sciuridés et Muridés dans la forêt équatoriale au Congo. *Rev. Suisse Zool.*, 77: 635–646.

Rahm, U., 1971. Oekologie und biologie von *Tachyoryctes ruandae* (Rodentia, Rhizomyidae). *Rev. Suisse Zool.*, 78: 623–638.

Redhead, T.D., 1971. Dynamics of a sparse population of the rat *Melomys littoralis* Lönnberg (Muridae) in sugarcane and natural vegetation. In: *Proc. 14th Conf. Int. Soc. Sugarcane Tech., La.*

Redhead, T.D., 1979. On the demography of *Rattus sordidus colletti* in monsoonal Australia. *Aust. J. Ecol.*, 4: 115–136.

Roland, J.-C., 1967. Recherches écologiques dans la savane de Lamto (Cote d'Ivoire): données preliminaires sur le cycle annuel de la végétation herbacée. *Terre Vie*, 21: 228–329.

Sheppe, W., 1972. The annual cycle of small mammal populations on a Zambian floodplain. *J. Mammal.*, 53: 445–460.

Sinclair, A.R.E., 1975. Resource limitations in tropical grasslands. *J. Anim. Ecol.*, 44: 497–520.

Soriano, P.J., 1977. *Caracterizacion y variaciones estacionales en communinades de pequeños mamiferos de los llanos occidentales de Venezuela.* Licenciado en Biologia thesis, Universidad de los Andes, Merida, 41 pp.

Stewart, D.R.M., 1971a. Diet of *Lepus capensis* and *L. crawshayi*. *E. Afr. Wildl. J.*, 9: 161–162.

Stewart, D.R.M., 1971b. Food preferences of *Pronolagus*. *E. Afr. Wildl. J.*, 9: 163.

Taylor, K.D., 1968. An outbreak of rats in agricultural areas in Kenya in 1962. *E. Afr. Agric. For. J.*, 34: 66–77.

Taylor, K.D. and Green, M.G., 1976. The influence of rainfall on diet and reproduction in four African rodent species. *J. Zool. Lond.* 180: 367–390.

Twigg, G.I., 1965. Studies on *Holochilus sciureus berbicensis*, a cricetine rodent from the coastal region of British Guiana. *Proc. Zool. Soc. Lond.*, 145: 263–283.

Verheyen, W. and Verschuren, J., 1966. *Rongeurs et Lagomorphes.* Exploration du Parc National de la Garamba, No. 50. Institut des Parcs Nationaux du Congo, Brussels, pp. 1–71.

Verschuren, J., 1966. Densités de population et biomasses des rongeurs en fonction des biotopes. *Ann. Mus. R. Afr. Centr., Ser. 8°, Sci. Zool.*, 144: 169–179.

West, O., 1965. *Fire in Vegetation and its Use in Pasture Management with Special Reference to Tropical and Subtropical Africa.* Commonwealth Agricultural Bureaux, Farnham Royal, 53 pp.

Chapter 17

THE ADAPTATIONS OF AFRICAN UNGULATES AND THEIR EFFECTS ON COMMUNITY FUNCTION

A.R.E. SINCLAIR

INTRODUCTION

The great diversity of African ungulates in the savanna regions is one of the most noticeable features of African ecosystems. Indeed there are as many species in the family Bovidae (buffalo, antelope, numbering about 78) as there are in the most diverse rodent family Muridae (79, according to Bigalke, 1972a). Whereas one might expect diversity in rodents which are fast-reproducing (so having many generations) and habitat-specific (so limiting gene flow), such radiation for large mammals is unusual and peculiar to Africa. Two questions relate to this radiation: firstly, how did this come about? Secondly, how do the many species coexist at the present time? Although reference is sometimes made to the diversity of habitats in savanna (Leuthold, 1977), this diversity is unsubstantiated, for there is no evidence that the savanna vegetation is any more complex or heterogeneous in Africa (south of the Sahara) than in other tropical continents. Savanna is rather more grassland than forest, and grassland is structurally less complex than forest. Hence savanna vegetation may even be more uniform at a gross level in Africa than elsewhere.

The African savanna vegetation is described in Chapter 6 (pp. 109–149), so I will make only brief reference here to those aspects of direct relevance to the herbivores. For these purposes I consider a range of habitats from semi-arid scrub that borders the three desert regions, through mesic open *Acacia* woodland, with a crown cover of 30% or less, and *Brachystegia*, *Colophospermum* woodland in Central Africa, with a nearly closed canopy (80–90% crown cover), to moist long grasslands on flood-plains along major rivers and in areas border-ing the lowland forest of Zaire and West Africa. Thus, there is a climatic gradient within the savanna from arid to humid.

The ungulates include three orders: the Proboscidea, 1 species (elephant); Perissodactyla, 6 species (rhinoceros and zebras); and the Artiodactyla. The latter order comprises five families: Suidae, 3 species (pigs); Hippopotamidae, 2 species (hippopotamus); Tragulidae, 1 species (water chevrotain); Giraffidae, 2 species (giraffes); and the Bovidae which are by far the most important group.

Although not ignoring the other groups I will concentrate on the bovids, for, being so similar, they provide the most interesting cases of adaptation. It seems that there have been several adaptive radiations in this taxon. The bovids can be divided into groups (subfamilies or tribes) whose species are similar in ecology and behaviour. However, parallel or convergent adaptations of species in different tribes is also a feature of their evolution. The taxonomy of the Bovidae is somewhat uncertain, but I will follow Ansell (1971) in most cases (Latin names are given in the Appendix). The species distributions within Africa have been documented by Dorst and Dandelot (1972).

The main bovid subgroups are as follows: the Tragelaphini (eland, kudu, etc.), which are large species inhabiting thickets and which are closely related to the buffalo (Bovini); the Cephalophini (duikers), which are all small, forest or scrub animals; the medium-sized Reduncini, which are humid savanna or swamp species like reedbuck, waterbuck and lechwe; the large Hippotragini of dry woodland and arid environments, including roan, sable and oryx; the large Alcelaphini which are mesic plains grazers such as topi, hartebeest and wildebeest; the Antelopini or gazelles which are small arid-adapted running types; and the Neotragini, a collection of very small forest- and thicket-hiding species.

In general, species within each subgroup above have some adaptations to the habitats listed common to the group. However, there are, in addition, some species such as the impala which are not closely related to other groups, and therefore they are placed on their own.

RADIATION OF THE AFRICAN UNGULATES

The palaeoclimate of Africa

The radiation of the ungulates can be considered in the light of the fluctuating environments and change in the vegetation of Africa during the Pliocene–Pleistocene, for it was in this period that the major radiation of bovids took place (Langer, 1974).

In the early Eocene the perissodactyls and proboscideans predominated, and this continued into the Miocene; there were many more species in these groups during that period than at the present day. Langer (1974) suggested that it was the evolution of the advanced ruminant stomach in the Oligocene that gave the bovids an advantage at this time: the efficient digestion of cellulose from coarse plant material such as grass enabled the bovids to replace the horses and elephants. While the pigs, hippopotami, and chevrotains also have divided stomachs to various lesser degrees, they are now clearly less numerous than the bovids. Langer (1974) proposed that at least pigs and hippopotami developed their digestive system independently from the bovids. Thus, by the Pliocene the bovids had developed an efficient digestive system. However, it was the characteristics of their environment that allowed them to diversify to the extent that they have in the Ethiopian region.

The vegetation of Africa today (see Fig. 6.1) is characterized by a central lowland forest region with an arm extending along the coast of West Africa. In concentric bands surrounding the forest are zones of moist savanna, drier *Acacia* savanna, and finally semi-arid areas abutting on three deserts — the Sahara in the north, the Somali desert in the north-east, and the Namib in the south-west.

There is now considerable evidence to suggest that at least since the Pliocene the climate has fluctuated to such an extent that the mean temperature changed several times from 2°C above to perhaps 5°C below the present-day conditions (Moreau, 1966; Van Zinderen Bakker and Coetzee, 1972). Moreau (1966) suggested that a lowering of mean temperature by 1°C would lower the present montane forest limit by 200 m to 1300 m. However, there is dispute in the literature over whether lowering the temperature decreases or increases aridity, but one can assume that there have been several advances and retreats of the forests and deserts in Africa (Livingstone, 1975).

During wetter periods, the lowland forest would have expanded eastwards to join with the Ethiopian highland forests and reach the coast at the Red Sea. The forest would also have reached the East African coast via the plateau and chain of mountains in northern Tanzania and further south (Moreau, 1966). Under these conditions the savanna areas would have been restricted to three separated areas — North Africa, the Somali horn, and south-western Africa, equivalent to the desert areas of today (Fig. 17.1).

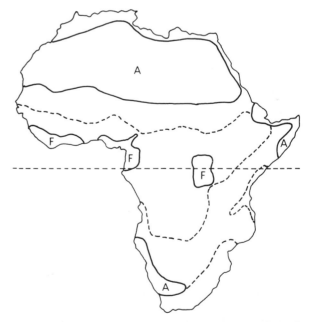

Fig. 17.1. The position of the three arid refugia (*A*) and forest refugia (*F*) during extremes of climate. Under present conditions, the arid areas, can be linked by a drought corridor of three consecutive months with less than 10 mm rainfall. (After Leuthold, 1977.)

At the same time the great swamplands in the heart of Africa, stretching from the upper Nile in Sudan, through the western Rift Valley lakes, Lake Victoria, and south to the Kafue–Zambesi River drainage, would have been far more extensive (Britton, 1978). Though Lake Tanganyika is steep-sided with little opportunity for papyrus swamps, the gaps between swamps would have been far narrower in more humid times.

In drier periods, the semi-arid savanna would have expanded to form a continuous belt around the forest. Even today one can draw a "drought corridor" (Fig. 17.1) delimiting areas with three consecutive months having less than 10 mm rainfall per month.

Perhaps most importantly, the intermediate climatic conditions supporting mesic savanna would have occurred during both expansion and contraction of the forests and deserts. This means that mesic savanna in any given locality would have occurred twice as often as, but for shorter periods of time than, the two extreme vegetation types.

Morton (1972) has likened the process of adaptive radiation resulting from these fluctuations in vegetation types to an "evolutionary pump" driven by climate. Every time the forests expanded and contracted, animals adapted to savanna experienced alternate increase and relaxation of selection pressures as their own habitats contracted and expanded. New types of animals could evolve to fill these new expanded habitats. Then during contraction the populations would be fragmented and isolated. In isolation they would diverge to an extent depending on the length of time they were isolated. As conditions returned to normal and habitats expanded again, the various isolated populations would have met and perhaps through reinforcement diverged further into true species if they were not already sufficiently different.

This situation suggests the hypothesis that ungulates adapted to mesic habitats would show a more continuous range through Africa divided into many contiguous (parapatric) ranges of closely related species or subspecies, resulting from the frequent fragmentation of their habitats. In contrast species adapted to deserts should be fewer in number, reflecting the fewer occasions for fragmentation, but more distinct from each other and widely separated, reflecting the longer periods of isolation. In the following section I review the distributions of

ungulates; in general the pattern of distribution supports the above hypothesis.

UNGULATE DISTRIBUTION PATTERNS

The forest

The differentiation between forest and savanna has probably been in existence since the Miocene (Moreau, 1952), and as a result a distinct fauna has evolved within the forest. One group of bovids, the duikers, have radiated in this environment, with species endemic to particular areas. There is some evidence to suggest that, in the periodic retreats of the forest, there remained three refugia (Fig. 17.1): the far west from Liberia to Ghana — being the most isolated due to the Dahomey gap, present even today through a peculiarity of climate; the high forests of Cameroon and Gabon; and the northeastern Congo drainage (Bigalke, 1972a). The large forest tragelaphine, the bongo, and the giant forest hog show just this tripartite disjunct distribution. The pygmy hippopotamus occurred in the two western of these forest refugia. The three forest races of buffalo also conform with this pattern (Sinclair, 1977a); it appears that some differentiation occurred in isolation, and subsequent expansion from these centres has allowed a complex series of clines to develop within the forest form.

The small and highly sedentary duikers, with faster life cycles, would have differentiated faster than the slower-breeding buffalo. One species, Abbott's duiker, is found isolated on East African mountains, possibly stranded there after the last forest retreat. The grey duiker is the only one to leave the forest, and it now inhabits thickets in broad-leaved savanna, showing a broad concentric distribution from west to southern Africa.

It is possible that the tragelaphines were originally forest browsers that have since expanded into savanna. Bushbuck remain forest dwellers, but they can use the narrow riverine forests, or even thickets in savanna, and are ubiquitous in Africa. Greater kudu show a similar distribution, but are no longer associated with forest. They now use shrub and thicket habitat in broad-leaved woodland. Sitatunga have become specialized to living in forest swamps. and show convergence in some behavioural traits with the unrelated lechwe.

The swamps and moist savanna

Bovids adapted to wet savanna are found in the Reduncini. In general, these specialized species exhibit a fragmented northern and southern distribution. Those showing the most extreme adaptation to wetlands are the lechwe, living in non-forest swamps. There is one species in the "sudd" on the Nile, and a group of species in Zambia and Botswana, 3000 km to the south (Fig. 17.2A). Lechwe are highly sedentary and are now showing divergence in the various isolated southern swamps. Less extreme wetland redundancies are the kob species adapted to wet grasslands. Again there is a northern form, the kob occurring from West Africa to Uganda, and a southern form, the puku in Zambia.

One reedbuck (*Redunca fulvorufula*) is adapted to montane grasslands. The same species occurs in two isolated areas: the mountains of Ethiopia and Kenya in the north, and in South Africa to the south. This pattern is similar to that of the two tragelaphine species, the mountain nyala in Ethiopia, and nyala in lowland forests from Moçambique south to Natal. Presumably these are relicts from a time when montane forest and grassland covered most of eastern Africa (Moreau, 1966).

The reduncines of mesic grasslands close to permanent water — the other reedbucks and the waterbucks — also have northern and southern species but their distributions are essentially contiguous, suggesting an expansion of range subsequent to allopatric divergence. In the case of the waterbucks there are a few small areas where the two forms meet and hybrids are produced — one area is the Nairobi National Park (Kiley-Worthington, 1965)). This suggests that allopatric separation was short and the evolution of reproductive isolation incomplete.

The deserts and semi-arid scrub

There are several groups of bovid species which are adapted to arid conditions and which have widely separated distributions. Examples are oryx (Fig. 17.2A) with one species in each arid refuge — gemsbuck in the Namib, beisa oryx in the northeast, and scimitar-horned oryx in the Sahara. The gazelles show a similar pattern with springbuck in

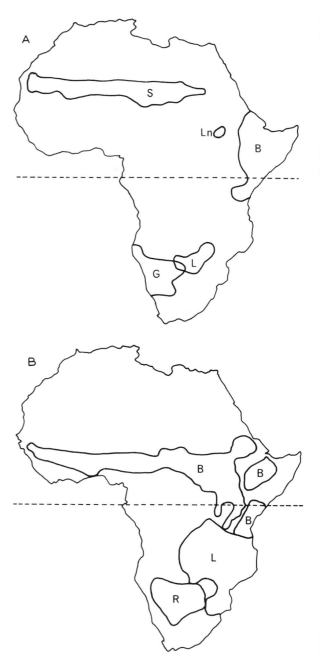

Fig. 17.2. A. The distribution of three species of oryx (*S* = scimitar-horned oryx; *B* = beisa oryx; *G* = gemsbuck) and two species of lechwe (*Ln* = Nile lechwe; *L* = lechwe). Note the wide separation of ranges in these desert- and swamp-adapted species. B. The distribution of hartebeest species showing the contiguous ranges forming a concentric band (*B* = bubal hartebeest, several races; *L* = Lichtenstein's hartebeest; *R* = red hartebeest).

the south, a group of species (e.g. Grant's gazelle, Speke's gazelle) in the northeast, and another group (e.g. red-fronted gazelle, dorcas gazelle) in the north. Several of these species differ in size and have overlapping range and habitats (e.g. Grant's gazelle, Thomson's gazelle). The greater number of species in the north and northeast suggests that isolation and rejoining of populations was more frequent there, in contrast to the more isolated south.

The northeast arid region (which now includes the Somali horn, northern and eastern Kenya, extending into eastern Tanzania) seems to have been particularly vulnerable to repeated isolation because of the narrow neck along the Red Sea, and the highland plateau in the south on the Tanzania/Kenya border which would have supported forest in more humid times. Populations of more mesic bovid groups appear to have been isolated there, and in isolation became adapted to arid conditions, some types showing remarkable evolutionary convergence. Thus Hunter's antelope is an arid-adapted topi, much lighter coloured than other *Damaliscus* species, and restricted to northeastern Kenya. It appears to have adapted to a habitat more appropriate for a hartebeest (*Alcelaphus*) which perhaps by accident did not occur there when Hunter's antelope first became isolated. Lesser kudu are also restricted to this northeastern region and may be only recently derived from the more ubiquitous greater kudu. Subsequent overlap with the latter species may have selected for size divergence in the lesser kudu. It would be interesting to see if "character displacement" in greater kudu has occurred — that is, whether this species is larger or shows interactive segregation (changes its habits) in areas overlapping with lesser kudu, compared to other parts of its range.

The vegetation of northeastern Africa is characterized by scrub with little grass, so that there is a prevalence of browse and a paucity of grazing opportunities. In this situation there has been convergence, in some forms, for a long neck to reach browse and the habit of standing up on the hind legs. This can be seen in the dibatag, which, although placed with the gazelles, is probably an arid-adapted reedbuck (Dorst and Dandelot, 1972) trapped in that area during a period of forest encroachment. Overlapping in range and habitat with dibatag is the gerenuk, but the latter is larger. Gerenuk are also placed with the gazelles, but I suspect they are really a desert-scrub form of impala. The range of the gerenuk and impala are almost mutually exclusive. Gerenuk are characterized by their long necks and they are well known for standing on their hind legs to reach leaves on bushes (see, for instance, Leuthold 1977). Dama gazelles show similar convergence in the Sahara (Dorst and Dandelot, 1972).

The mesic savanna

Savanna with intermediate rainfall (500–1000 mm per annum) supports a large number of ungulate species, the mesic-adapted types, and the majority of them show a continuous distribution from north to south (in contrast to the separated ranges of swamp and arid species), either as a single polytypic species or a series of congeneric parapatric species replacing one another. Only a few species show a north–south separation: the white rhinoceros, for example, is found in northern Uganda and South Africa, though it formerly extended north to Zambia.

The very large ungulates with catholic feeding habits, such as elephant, buffalo and hippopotamus, are found as one species in all habitats from forest to dry savanna over Africa. The only divergence seen in elephant and buffalo is in the smaller forest races. Clearly the changes in climate would not have limited these species to any real extent. The same applies to the warthog and bushpig, which eat short green grass, roots and tubers in forest glades and throughout the savanna.

Some ungulate species have a nearly complete concentric range in Africa — the black rhinoceros, greater kudu, roan antelope, grey duiker, oribi and klipspringer are good examples. They feed on widespread shrubs or are grazers (roan, oribi) in broad-leaved (*Combretum, Terminalia*) woodlands. Indeed this vegetation is the common element for all these species, and its significance in allowing their continuous undifferentiated distribution through Africa has so far not been stressed. It suggests that this vegetation has been sufficiently widespread throughout the climatic changes for these ungulates not to have been isolated. Alternatively, but from fossil evidence less probably, they have only recently expanded their range through Africa.

Giraffes form a nearly concentric range of adjacent subspecies. Perhaps the best examples of continuous bands subdivided into adjacent congeneric species are found in the alcelaphine grazers. *Alcelaphus* is found as various races (nine types) of the bubal hartebeest across the whole of the Sahel zone from West Africa down to Tanzania (Fig. 17.2B). It is replaced in southern Tanzania by Lichtenstein's hartebeest, which in turn is replaced by the red hartebeest in southern Africa. *Damaliscus* shows a similar series with topi in the north, tsessebe in southern Africa, and bontebuck and blesbuck in the far south. There is a gap in southern Tanzania and Zambia, however. Other sets of contiguous species are seen in the reedbucks and waterbucks, as already mentioned.

Finally there is a group of species whose ranges occur only in the southern half of the concentric savanna band. Sable and impala are two of these, reaching from South Africa to Kenya. The two wildebeest species show a similar distribution. The three zebra species form a series from the southwest to the northeast, the southern and northern forms (mountain zebra, Grevy's zebra) being desert-adapted, while the intermediate Burchell's zebra has a very similar distribution to that of the wildebeest. Zebra and wildebeest are the typical plains grazers, forming mixed herds and exhibiting large-scale movements in response to unpredictable climates. I suggest they have co-evolved over a long period of time, their nomadic habits enabling them to move when climates changed, thus precluding fragmentation and divergence of populations. This is, therefore, the opposite extreme to the sedentary species specialized to specific habitats such as swamps or deserts, where divergence and speciation have been prevalent. But it remains a mystery why these migrant species have not moved into West Africa.

In summary, the evidence from the present distribution of the ungulates supports the hypothesis that radiation occurred as a result of repeated allopatric separation of populations through climatic changes with subsequent divergence. Those species adapted to extreme climates — very moist (forest, swamps) or very dry (deserts, semi-arid scrub) — have had fewer opportunities for subsequent mixing because of sedentary habits in the former and the great distances between the populations in the latter. Thus, fewer races (or species) are formed in these types compared to the ungulates using intermediate grassland habitats, which would have experienced frequent isolation

but for shorter lengths of time before rejoining. In these forms many races (or species) are found, producing a continuous distribution. A few of these intermediate forms are nomadic enough to be able to move with the ebb and flow of the climate, thus preventing differentiation. Another group is adapted to broad-leaved (e.g. *Combretum*) woodland, the one vegetation type which may have been always present despite the climate, so also preventing species differentiation.

With each contraction of range, competition between non-ruminants and ruminants would have favoured the latter, allowing them to dominate the fauna. I emphasize that it was the changing conditions, the periodic swing from moist to dry and back, which provided the conditions for ruminants to compete. This hypothesis of dynamic environments is the converse to arguments which suggest that the tropics are stable over long periods of time, and as such allow speciation to progress.

The holarctic region also experienced climatic fluctuations, but one extreme (the Ice Ages) was so severe that it simply eradicated all species in those regions, so countering radiation.

ADAPTATION AND CO-EXISTENCE UNDER PRESENT CONDITIONS

The above discussion on the radiation of ungulates emphasizes that many of the species replace each other geographically. Nevertheless any one area still supports species that overlap both geographically and ecologically. The Serengeti area, for example, contains 28 species of ungulates. This leads to the second question: how do they co-exist in the face of this overlap?

I shall first describe the adaptations of the various species before considering how these may result in co-existence.

Climatic adaptation

For most of African savanna, the major climatic influences on ungulates are high solar radiation, high ambient temperatures and restricted water supplies. In higher-rainfall savanna the last factor is less important, though the first two remain. In arid areas all are important and their effects on animals are interrelated. Adaptations to reduce these effects can be divided into physiological and behavioural.

Physiological adaptations to avoid high heat loads involve sweating and panting (Robertshaw and Taylor, 1969). Where water is plentiful, large animals such as buffalo, eland and waterbuck use sweating for evaporative cooling (Taylor, 1968a; Taylor et al., 1969). The buffalo, for example, was found to keep body temperatures between 37.4°C and 39.3°C when water was freely available, and did not allow body temperature to increase beyond 40°C even when water was restricted. Under these conditions of restricted water, buffalo were unable to reduce evaporative water loss to any great extent (Taylor, 1970a, b). Waterbuck showed a similar inability; at 40°C ambient temperature, waterbuck lost 12% of their body weight in 12 h compared to 2% for desert-adapted oryx (Taylor et al., 1969). Both buffalo and waterbuck are typical species of mesic and humid savanna, and can reduce heat stress by sweating but cannot tolerate water restriction as well.

Whereas large animals can afford to lose water by sweating, smaller animals such as gazelles cannot, and they employ panting instead (Taylor, 1968a; Maloiy, 1973; Ghobrial, 1974). Panting is also employed in species that live in arid habitats (oryx) or shadeless plains with intense solar radiation and where water is restricted (wildebeest, hartebeest: Taylor et al., 1969; Finch, 1972). Wildebeest in particular have evolved enlarged surface areas in the nasal passages and paranasal sinuses to promote cooling by air flow during panting. They maintain a large anatomical dead space in the lungs, and panting is shallow to prevent excessive loss of carbon dioxide. (Fossil wildebeest from East Africa 2 million years ago, shown to me by J.M. Harris in Nairobi, did not have the enlarged nasal passages, and resembled present-day hartebeest in this respect. The jaw shape was similar to modern wildebeest, indicating that feeding differences evolved before these climatic adaptations.) Panting, therefore, reduces heat stress without losing as much water as in sweating.

Coat colour and structure are important in reducing heat loads from solar radiation. Finch (1972) found that the lighter coat of hartebeest reflected 42% of short-wave solar radiation, compared to only 22% in the darker eland. The dense fur of hartebeest allowed less penetration of radiation compared to the sparse coat of eland. But, in both species, Finch found that re-radiation of long-wave thermal radiation was greater than that absorbed, and this accounted for 75% of total heat loss. The rest was by evaporative heat loss — sweating in eland, panting in hartebeest.

Other physiological adaptations to arid conditions include tolerance to high body temperature. Some gazelles can allow body temperature to rise before panting is necessary: 43°C in Thomson's gazelle and even 46°C in Grant's gazelle. Taylor (1972) showed that this strategy allows Grant's gazelle to exist in more arid environments than Thomson's gazelle does. The dorcas gazelle, however, apparently does not let body temperature rise, and employs other adaptations (Ghobrial, 1974). Large species such as eland and oryx can absorb heat with a slower rise in body temperature compared to small species.

Water conservation in arid environments include restriction of urine output, concentrating the urine, and resorbing water from the faeces. Maloiy (1973) showed that dikdik have the lowest water content in faeces and highest urine concentration of any species measured. Oryx was next in these respects, followed by hartebeest, impala and eland. In general, water requirements were lowest in oryx, Grant's and Thomson's gazelles, intermediate in wildebeest and hartebeest, and high in eland and buffalo (Taylor, 1968a; MacFarlane and Howard, 1972).

Behavioural adaptations can supplement physiological adaptations. Despite the eland's inability to conserve water, it nevertheless survives in arid environments by behavioural compensation (Taylor, 1968a). Both eland and impala reduce heat loads by feeding and moving at night while remaining in shade during the day (Jarman and Jarman, 1973). Where there is no shade on open plains, topi stand facing away from the sun to expose the smallest surface area (Jarman, 1977). The narrow thermal neutral zone of eland results in their metabolic rate increasing during low ambient temperatures at night, so that metabolic water is produced, while the heat produced is dissipated in the cool night air (Taylor, 1969). In addition, the feeding behaviour of eland and impala results in selection for the most succulent food, which is normally browse from shrubs in the dry season (Taylor, 1969; Jarman, 1973).

Selection for browse is a common feature of all

species not restricted to the vicinity of surface water (Bigalke, 1972b; Western, 1975). Obligate grazers, therefore, are less well adapted to arid areas. Taylor (1968b) has suggested that oryx and Grant's gazelle select for hygroscopic shrubs (*Disperma* sp.) in northern Kenya. These shrubs contain only 1% free water in the day, but absorb water from the air at night when the relative humidity is higher, so that they contain up to 43% free water. By feeding at night these animals increase their water intake. Taylor (1968b) calculated that this may be sufficient for their needs. However, King et al. (1975) found that oryx in areas where *Disperma* was absent could obtain sufficient water from the vegetation only when it was green. Under dry conditions oryx changed from day to night feeding, and Root (1972) has found them digging for succulent tubers. Similarly, dorcas gazelle obtained sufficient water from non-hygroscopic *Acacia* spp. when they were green, but in the dry season they had to migrate to wetter areas (Ghobrial, 1974). In fact, migration to water sources in the dry season (Fig. 17.3), is the common adaptation shown by grazers (Western, 1975; Pennycuick, 1975).

In summary, behavioural adaptations to heat stress and lack of water include night feeding, selection for succulent foods, and migration.

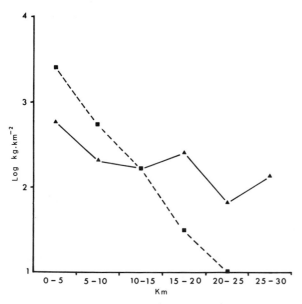

Fig. 17.3. Biomass densities at different distances from water during the dry season, Amboseli, Kenya. Water-independent browsers (triangles) have access to a greater area than water-dependent grazers (squares). (From Western 1975.)

Food acquisition

Movements

As Western (1975) has noted, independence from water requires ungulates to become browsers (not all browsers, however, are independent of water, especially greater kudu, bushbuck and duikers in forest). Mixed feeders (impala, Thomson's gazelle, elephant) and grazers are required to move to dry-season water supplies. The difference in water requirements between these groups dictates their life-history strategy with respect to how they obtain food and when they reproduce.

Small browsers in dry areas (e.g. dikdik, steinbuck) can remain in small territories year-round because their food requirements are small (Jarman, 1974). Larger species which form herds (Grant's gazelle, oryx, eland) with high food requirements become nomadic. This is because browse food is dispersed in the form of scattered bushes. The animals consume it quickly and have to move on; eland in particular are known to move large distances (Blankenship and Field, 1972), and in the Serengeti they perform movements similar to those of the migrating wildebeest and zebra. Perhaps because browse is restricted in the dry season, some species show convergence in developing long necks (dama gazelle, dibatag, gerenuk). These species, as well as dorcas gazelle and springbuck, have adopted the habit of standing on their hind legs to reach leaves otherwise out of reach (Bigalke, 1972b; Dorst and Dandelot, 1972).

Seasonal movements by species dependent on water are dictated by the spatial distribution of rainfall. If an area is uniformly dry then animals will converge on permanent water supplies such as swamps and rivers, as described for Amboseli (Kenya; Western, 1975), Tarangire (Tanzania; Lamprey, 1964) and the Zambesi Valley (Jarman, 1972). In the wet season the ungulates disperse from these centres. Predictable water supplies are characteristic of eastern Africa. However in the semi-arid areas of southern Zimbabwe, Botswana, and the Namib and Kalahari, water is found in hard-pan depressions, only some of which are filled sufficiently to persist well into the dry season each year, depending on accidents of rainfall. In this situation water supplies are unpredictable both in time and space. Ungulate concentrations shift from year to year, and the distinction between wet-season and dry-season range is less clear than in East Africa (B.H. Walker, per. comm., 1979).

Some movements are determined by the distribution of rainfall in the dry season, rather than by the location of permanent water. This is seen in the Serengeti migrations (Pennycuick, 1975; Maddock, 1979), in which wildebeest, zebra and Thomson's gazelle move towards higher-rainfall areas in the dry season and lower-rainfall areas in the wet season. The general area of high rainfall is predictable, so the overall pattern of migration is also predictable. But the exact location of rainstorms, and hence green food, is unpredictable within this area, and the animals are nomadic on a small scale of 10 to 50 km. Nomadic movements towards rain probably explain the migrations of springbuck (Bigalke, 1972b) and dama gazelle (Dorst and Dandelot, 1972), in the same way as they do for Thomson's gazelle.

Dependence on water constrains animals to finding food within a certain distance from surface water (10–15 km in Amboseli; Western, 1975). Pennycuick (1979) has suggested that the running ability of ungulates is an adaptation to increase the foraging distance from water. He argued that large size also increases this distance, so that larger species such as wildebeest have an advantage over smaller ones such as gazelle.

In general, because most water-dependent species are constricted in their dry-season range, one would expect that food resources are limiting populations at that season; where measurements have been made this appears to be the case — for instance, impala and other species in South Africa (Hirst, 1969), and elephant, buffalo, wildebeest and topi in East Africa (Malpas, 1977; Sinclair, 1975, 1977a, 1979b; Duncan, 1975). The unusually low rainfall of 1970–71 in Tsavo (Kenya) provided a disturbance which demonstrated that lack of food for calves and their mothers within reach of water, not lack of water itself, caused the mortality of elephants (Cornfield, 1973; Phillipson, 1975) — it was a famine rather than a drought (Fig. 17.4).

Reproduction

Since food supply is potentially, and in some cases actually, limiting populations, one might expect that reproduction has evolved to produce young at the time of greatest food abundance, which in most cases is in the wet season. Births

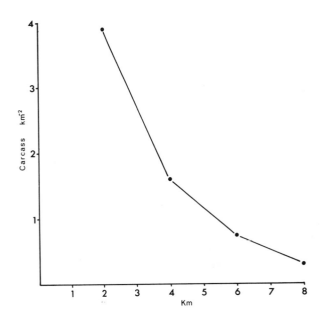

Fig. 17.4. Carcass density of elephants at different distances from water. They died in the Tsavo drought 1970–71. Note that most died close to water where little food was available. (From Corfield, 1973.)

should occur then because lactation imposes a higher nutritional demand on the mother than any other stage in the reproductive cycle (Sadleir, 1969; Sinclair, 1977a). Most studies are content to show a correlation between births and rainfall (or food) as evidence for adaptation towards alleviating nutritional stress. Better evidence, however, comes from observations of newborn survival under abnormal conditions: for example, in late 1960 and early 1961 the rains failed completely in Nairobi National Park, and the entire wildebeest calf crop of early 1961 died (Gosling, 1969). A similar situation occurred there in early 1974 (Hillman and Hillman, 1977). Therefore, failure of rain caused low survival, which wildebeest normally avoid by giving birth in the rains.

Seasonality of births should be more pronounced in water-dependent species, than in those that are independent of water and can range further afield. Circumstantial evidence for this hypothesis comes from observations on the proportions of juveniles in populations of Tsavo ungulates (Leuthold and Leuthold, 1975). Water-dependent grazers (zebra, hartebeest, warthog) showed a higher seasonality of young in the population than water-independent browsers (giraffe, lesser kudu). Small samples of other species produced inconclusive results, but oryx, though arid-adapted, is largely a grazer and also showed seasonal peaks of young.

There is a considerable literature showing wet-season peaks of young in the population; these include elephant (Hanks, 1969, 1972; Laws, 1969a; Smuts, 1975), white rhinoceros (Owen-Smith, 1972), Burchell's and Grevy's zebra (Klingel, 1969, 1972), hippopotamus (Laws and Clough, 1966; Fairall, 1968), warthog (Fairall, 1968; Clough, 1969; Leuthold and Leuthold, 1975; Bourlière et al., 1976), buffalo (Grimsdell, 1973; Sinclair, 1977a), eland (Skinner and Van Zyl, 1969; Skinner et al., 1974), giraffe (Hall-Martin et al., 1975), greater kudu, bushbuck, nyala (Wilson, 1965; Fairall, 1968; Underwood, 1978), waterbuck, kob in some areas (Leuthold, 1966; Fairall, 1968; Spinage, 1969; Herbert, 1972), sable, gemsbuck (Fairall, 1968; Skinner et al., 1974; Sekulic, 1978), wildebeest, hartebeest and topi (Fairall, 1968; Gosling, 1969; Estes, 1976; Sinclair, 1977b; Duncan, 1975; Skinner et al., 1973, 1974), impala (Fairall, 1968; Anderson, 1972; Jarman and Jarman, 1973), gazelles (Robinette and Archer, 1971; Bigalke, 1972b; Skinner et al., 1974; Bradley, 1977), steinbuck and grey duiker (Fairall, 1968).

In many cases information on the occurrence of young has been used as evidence for seasonal breeding, and that such seasonality is an adaptation to promote increased survival of young. But the data on the occurrence of young could be equally as well interpreted as reflecting survival success from a uniform birth distribution throughout the year: those born in the wet season survive better than those born in the dry season, resulting in a peak of survival in the wet season. Indeed many of the species noted above do produce young throughout the year. This alternative hypothesis is rarely considered by the authors, and hence data on the occurrence of young as evidence for seasonal births must be viewed with caution.

Some of the studies have used the frequency distribution of foetus weights as evidence for birth peaks (e.g., Laws and Clough, 1966; Laws, 1969a; Grimsdell, 1973; Sinclair, 1977a) which largely overcomes the above objection. But there remains a second problem, for foetus measurements do not distinguish between alternative strategies: (1) species which conceive at the same time each year and not at other times even if conditions are suitable; (2)

opportunistic species which conceive whenever conditions allow; and (3) species which conceive uniformly through the year except during the rare bad conditions of a famine: the end of the famine would create synchronized conceptions, and the resulting breeding peak could be maintained in the population for several years before it damped out. An example may have occurred in East African elephants (Laws, 1969a). Most field studies extend over only one to three years, so would not detect this damping.

Species using alternative (1) above have an inflexible strategy as an adaptation to predictable changes in climate (often seen in north-temperate regions). These species have evolved a response to proximate cues (J.R. Baker, 1938) which trigger conception and result in births during optimum conditions — these conditions being the ultimate evolutionary cause of seasonal breeding. If the proximate cue is some predictable factor such as photoperiod then one can conclude that the species is following strategy (1). In African ungulates only one species, the wildebeest, has been shown to adopt this strategy (Sinclair, 1977b). This uses a combination of lunar and solar photoperiod to synchronize conceptions near the Equator, but probably uses solar photoperiod alone at higher latitudes (Spinage, 1973). Wildebeest not only give birth in the rains, but they produce 80% of their calves within three weeks; most other species have a birth peak covering two or three months. Clearly it is not just nutrition which is dictating the synchrony in wildebeest.

Spinage (1973) has suggested that solar photoperiod may also trigger conceptions in impala, hartebeest, topi and warthog, and hence these species would follow strategy (1). He based this on the evidence that these species populations north of the Equator had birth peaks six months out of phase with populations south of the Equator. However, rainfall is also correlated with the position of the sun in the sky (and hence photoperiod), and these species may be responding to nutritional cues instead, and hence are adopting strategy (2). Good evidence for species using alternative (2) comes from comparisons of populations in areas differing in rain seasons but not in photoperiod; buffalo for example show two birth peaks six months apart in Uganda, but only one in the Serengeti (Grimsdell, 1973, Sinclair, 1977a). Photoperiod could not have been used under these circumstances, and nutrition was most likely both the proximate and ultimate cause.

Even better evidence that nutrition is the proximate and ultimate cause of birth seasons comes from studies on lechwe (Child and Von Richter, 1969; Lent, 1969; Sheppe and Osborne, 1971, Sayer and Van Lavieren, 1975; Schuster, 1976). Lechwe live on riverine floodplains that are seasonally flooded. At the height of the floods animals are compressed into less-preferred surrounding woodlands while their preferred habitats are inaccessible (Fig. 17.5). Optimum food conditions occur when water is at the lowest level exposing the greatest area of floodplain, and it is then that the peak of births occur. In Zambia this peak is in the dry season (August–October) because floods recede two or three months after the rains cease. However, in the Okavango swamp of Botswana, the peak of births is in December and January, which is mid-rains. This is because floods recede nearly nine months after the previous rains that fell much further upstream, producing a longer time lag than in Zambia. Hence the birth season is unrelated to photoperiod and local rainfall. But the timing is consistent with that of other ungulates in being related to nutrition.

Species which follow alternative (3) — uniform distribution of births through the year — may be those such as elephant and rhinoceros whose calving interval spans several years. Perhaps because lactation covers several seasons, birth in any one wet or dry season may have only marginal effects on calf survival (Laws and Parker, 1968) and adaptations for seasonal births have not evolved. This seems to me a weak argument, particularly in the light of strong birth seasonality by elephants in southern Africa (Hanks, 1969; Smuts, 1975). Similarly, the argument that in areas of erratic rainfall it pays ungulates to be aseasonal (Leuthold and Leuthold, 1975) can be countered by the argument that, despite high variation, there is still a higher probability of rain in some time periods than in others, and it would pay animals to adapt birth times accordingly. Because of the difficulty mentioned earlier that birth peaks may or may not be temporary effects following a climatic disturbance, combined with a lack of sufficient data in most

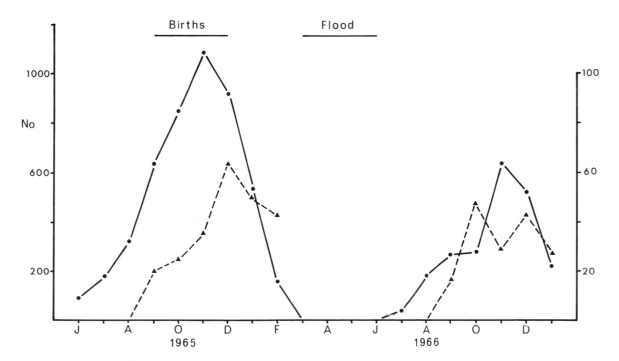

Fig. 17.5. The numbers of lechwe (circles, left axis) increase on the Chobe River floodplain (Botswana) when waters recede, exposing the greatest area of high-quality food. The figures for calves per 100 females (triangles, right axis) show that births occur at this time. (From Child and Von Richter, 1969.)

studies, it is difficult to suggest any species that definitely follows strategy (3). Perhaps water-independent browsers may find seasonal effects less severe, and so could spread births over a longer period; but even these species (giraffe, lesser kudu) appear to find food less abundant in the dry season because of low shoot production, compared to the rainy season (Leuthold, 1971; Leuthold and Leuthold, 1972; Hall-Martin, 1974; Hall-Martin and Basson, 1975).

So far I have suggested that seasonal variation in nutrition acts as a selection pressure primarily on the timing of births, and conception times are merely a consequence of this. Experimental evidence on domestic ungulates supports this view. For example, A.A. Baker (1969) showed that occurrence of oestrus in cattle was more sensitive to female body condition before the previous birth, than to her nutritional intake immediately prior to oestrus. This arose because females in good condition at birth (i.e. with high fat reserves) were not reduced by lactation to such a low body condition that oestrus was prevented. Therefore climatic events dictating nutritional intake during lactation have less effect on the mother than her starting body condition. The experimental work is supported by comparative studies; most African ungulates give birth during good nutritional periods, but because of different gestation lengths many species conceive in the dry season when nutrition is poor.

If the body condition of the mother giving birth is so important, then selection may be acting on birth seasons not so much to allow good nutrition *during* lactation but more to provide high nutrition *prior* to lactation to allow the mother to build up body condition (Fig. 17.6). That is to say, the period

before birth may be the critical season dictating the subsequent calf's survival. This hypothesis is supported by a comparison of large and small species. Large species such as buffalo or giraffe give birth in the latter half of the rains or even in the early dry season (Hall-Martin et al., 1975; Sinclair, 1977a) because the period needed to build up body condition is relatively long. Small species (gazelle, warthog) or those such as topi and eland that feed on young shoots high in protein can pick up body condition quickly and so give birth in the early wet season (Skinner and Van Zyl, 1969; Du Plessis, 1972; Bigalke, 1972b; Skinner et al., 1974; Leuthold and Leuthold, 1975).

In conclusion, female body condition appears to be the primary determinant for timing of reproduction, not only in tropical ungulates, but in temperate ungulates (D.C. Houston, per. comm., 1978) and in birds also (Perrins, 1970; Sinclair, 1978).

Interspecific competition

Where food supply has been studied, it appears to be limiting populations in the dry season, as mentioned earlier. If several species are eating this same food in the dry season then inter-specific competition is taking place, and thus competition should act as a selection pressure for adaptations to reduce it and allow the observed coexistence.

Food selection

There are a number of morphological differences which allow species to eat different food. Several adaptations are associated with browsing or grazing. Some browsers (giraffe, tragelaphines) probably evolved in forest and retained this feeding style in savanna, allowing some species (eland, lesser kudu) to become arid-adapted. Other browsers (dibatag, gerenuk) may have evolved from grazers as an adaptation to arid conditions. Browsing bovids have long narrow jaws suitable for selecting individual leaves or shoots. Grazers have wide jaws and incisors suitable for less selective grass mowing. The stomach of browsing ruminants has a relatively small rumen with many papillae, a large reticulum and small omasum. The absorptive surface is large, allowing uptake of nutrients from the rapid digestion of soft dicotyledonous food. Grazers, with their coarse food, have converse

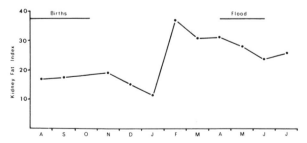

Fig. 17.6 The kidney fat index of lechwe on the Kafue River (Zambia) shows that body condition reaches a peak before births, and a low point during lactation after births. (From Sayer and Van Lavieren, 1975.)

stomach structure. The large rumen permits maximum delay for fermentation of fibrous food. Papillae are lacking in parts of the rumen due to the abrasive type of food, and the mucosal surface is less because digestion is slow (Hofman, 1968; Hofmann and Stewart, 1972). Some grazers (kob, waterbuck) eating soft aquatic monocotyledons show intermediate stomach morphology.

Within the browsers or grazers there is overlap in terms of plant species eaten. For example, Field and Ross (1976) found that half the tree species eaten by giraffe in Kidepo National Park, Uganda, were also eaten by elephant. In Ruwenzori Park, Uganda, Field (1972) found that the grass *Sporobolus pyramidalis* was the most frequent in the food of buffalo, waterbuck and warthog. The grass *Themeda triandra* was eaten by these species and also by kob, and in the Serengeti it was eaten frequently by most of the larger grazing ungulates (personal observation). Whether a grass species is eaten depends on the presence or absence of more acceptable species, so that in some areas a grass is eaten while in others it is ignored (Field et al., 1973). The congeneric lechwe, puku and waterbuck on the Chobe River floodplains of Botswana all ate the succulent aquatic grasses when they could reach them, but even during the floods when animals were restricted to the upper levels there was considerable overlap in grass species eaten (Child and Von Richter, 1969). Similar overlap has been noted amongst East African ungulates (Talbot and Talbot, 1962; Lamprey, 1963). Jarman (1971) recorded that overlap amongst browsers in the Zambesi Valley was highest in the wet season. In the dry season overlap declined initially as each species concentrated on a small range of food staples which differed from those of other species and acted as a "food refuge". Perhaps most importantly, overlap increased again in the late dry season when, by inference, resources became so scarce that animals were forced to eat less attractive alternatives — an indication that interspecific competition was promoting separation of food preferences.

Food preference is indicated if a greater proportion of a component is recorded in the diet than in the accessible vegetation. Detailed studies of diet have shown that ungulates prefer certain species of plants as well as certain parts of those plants (Gwynne and Bell, 1968; Sinclair and Gwynne,

1972; Duncan, 1975; Field, 1975; Kreulen, 1975; Jarman and Sinclair, 1979). Thus, buffalo prefer green grass leaves longer than about 10 cm, wildebeest prefer leaves shorter than this, and topi prefer leaves in tall grass. The common feature of the preferred components is that they have higher protein content relative to the rest of the accessible vegetation (Duncan, 1975; Sinclair, 1977a). Grass leaves have higher protein than stems, for example. Von Richter and Osterberg (1977) reported that the grass species eaten by lechwe, puku and waterbuck were those with the highest protein. Ungulates that eat a mixture of grass and browse — springbuck, impala (Fig. 17.7), eland and elephant, for example — concentrate on grass in the wet season and switch to browse in the dry season (Jarman, 1971, 1974; Field, 1971; Bigalke, 1972b; Rodgers, 1976; Leuthold, 1977). Crude protein content of dicotyledon leaves, even in the dry season, is considerably higher ($>10\%$) than that of grass leaves (*c.* 4%). Despite the presence in mature dicotyledon leaves of phenolic compounds that inhabit digestion, browse probably provides more protein than grass. By selecting for these components, therefore, the animals can increase their protein intake.

Preliminary studies on amino-acid requirements of various ungulates show that all species have the same basic needs (Crawford et al., 1968). It was concluded that all species are trying to obtain the same resource, protein, but do so in different ways.

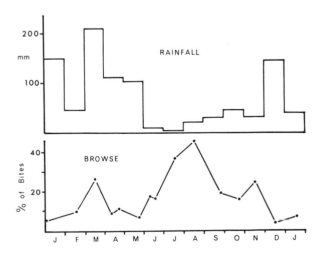

Fig. 17.7. The amount of browse in the diet of impala increases during the dry season when the quality of grass food decreases. (From Jarman and Sinclair, 1979, with permission of Chicago University Press.)

Field studies show that there is considerable overlap in these feeding methods.

Habitat selection

Because of food limitation in the dry season and overlap in food preference, interspecific competition would select for adaptations in animals to obtain food from different habitats. Use of the same habitats at different times of day is sometimes stated to be an adaptation to avoid competition. But this argument applies only to interference competition (i.e. behavioural exclusion of one species by another) and not to exploitation competition (i.e. direct use of a resource). Interference amongst African ungulates has not been studied in any detail, but it appears to be relatively unimportant; indeed one of the noticeable features of these mammals is the frequent association of many species. Only in extreme cases, such as the invasion of topi habitats by the thousands of Serengeti wildebeest, is interference observed through the retreat of topi (Duncan, 1975). Exploitation competition is, therefore, the dominant process. Since food is not renewable in the dry season, spatial separation may be an adaptation to counteract competition.

A few species have adapted to special habitats and therefore avoid overlap with others — for example, mountain nyala in the heath zone of Ethiopia, sitatunga with their enlarged splayed hooves for walking in forest swamps, and klipspringer that stand on the tips of their hooves to facilitate running on rock outcrops. However, most species use as their habitat a part of the vegetation continuum that extends from moist riverine forest to dry treeless plains.

Pienaar (1974) gives a qualitative account of the habitats used by South African species. Lamprey (1963), in a now classic study, used foot transects to quantify vegetation use by ungulates in Tarangire. He showed that each species preferred particular habitats but that there was overlap with other species. This overlap was most pronounced amongst the plains grazers, wildebeest and zebra. Lamprey also drew attention to vegetation edges or ecotones as preferred habitats for impala — thicket was used for protection, shade and dry-season browse, while open grassland was used for grazing. Sable in southern Africa is also an "edge species" living between *Brachystegia* woodland and valley grassland (Pienaar, 1974).

Ferrar and Walker (1974) used a more quantitative approach to describe the habitat preferences of the herbivores in Kyle National Park, Zimbabwe (Rhodesia). Using multivariate analysis they isolated three major habitat parameters that together described the distribution of the species. These were the gradients: (1) rocky to flatland, (2) short sparse herbs to tall dense herbs, and (3) open grassland to thicket or forest patches. They found certain species clearly distinct — klipspringer on rocks, bushbuck in thicket, and tsessebe on flat grassland. There was also a group of mixed feeders of different sizes (rhinoceros, eland, zebra, sable and impala) which showed no clear separation. Although each species demonstrated selectivity for vegetation types at a high level of significance, there was a considerable degree of overlap in selection for the three main vegetation categories (Fig. 17.8).

Lamprey (1963) noted in the *Acacia* woodlands of Tarangire (Tanzania) the change in habitat preferences as species moved towards the river in the dry season. Jarman (1972) recorded similar

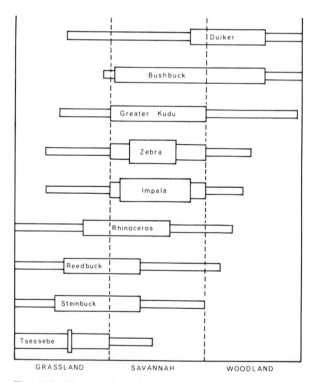

Fig. 17.8. The degree of habitat use by ungulates in Kyle National Park, Zimbabwe (Rhodesia). Width reflects the degree of preference. Note the large overlap between species (From Ferrar and Walker, 1974.)

seasonal habitat shifts in the mopane woodlands (*Colophospermum mopane*) of the Zambesi Valley. He pointed out that vegetation growth slows down in the dry season first on the upper parts of the valley slopes (termed the *catena*) and last on the flood plain adjacent to the river. The upper part of the catena usually has sandy porous soil while the lower has silt, therefore holding water longer. Grasses with shallow roots dried out before the bushes and trees stopped growth. Consequently, grazers (waterbuck, warthog, zebra) moved onto the floodplain before the mixed feeders (impala, elephant), while the browsers (eland, greater kudu, rhinoceros) arrived only late in the season. Nevertheless, overlap in habitats increased to a maximum by the late dry season, when food was potentially in short supply.

To conclude, the restricted food supply in the dry season, combined with the significant overlap in food and habitat preferences, provides the necessary conditions for interspecific competition to occur. The critical experimental evidence, namely the removal of one species resulting in an increase in another's population, is rare. Eltringham (1974) reported one case where hippopotamus removal resulted in an increase in buffalo population. Conversely, increasing wildebeest numbers may have slowed the increase of a buffalo population in the Serengeti (Sinclair, 1977a). Despite the lack of

critical observations, I suspect interspecific competition has been a dominant selection pressure affecting body size, food and habitat preferences.

Grazing succession, coexistence and facilitation

The succession of species moving down the catena as the dry season progressed was first described by Vesey-Fitzgerald (1960, 1965) for the Rukwa Valley floodplains in southern Tanzania. Large species (elephant, buffalo) moved down from the woodlands into the tall swamp vegetation first. Their grazing and trampling broke down the coarse swamp grasses and provided a shorter sward for the following zebra and topi. Without elephant, the smaller topi were unable to penetrate the dried-out swamp vegetation and were then restricted to a much smaller band of short grass at the edge of the swamp. Vesey-Fitzgerald suggested that elephant were providing a habitat for other species, and hence were "facilitating" rather than competing with them.

Bell (1970, 1971) used Vestey-Fitzgerald's hypothesis to explain the movement of grazers down the catena in the Serengeti: as short grasses on ridge tops dried out, the large species (buffalo, zebra) moved downhill first (Fig. 17.9) followed by progressively smaller species (topi, wildebeest and Thomson's gazelle). These populations were all

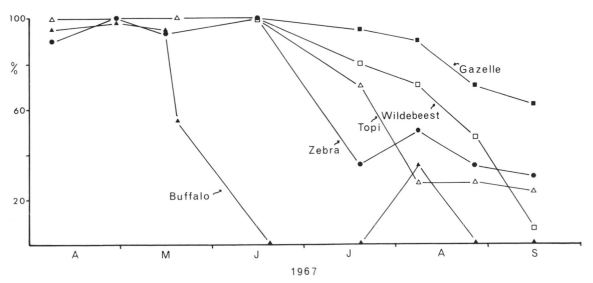

Fig. 17.9. Percent of the population of different species using short-grass areas on ridge tops (upper catena) in the Serengeti. The larger species leave earlier than the smaller. (From Bell, 1970.)

resident in a small area and did not form part of the main migratory populations. Bell suggested that the less selective large species removed much of the coarse stems, thereby exposing the more nutritive grass leaves and herbs at the base of the plants for the smaller grazers. Therefore, the smaller grazers depended upon the larger species for their food.

The grazing succession of plains ungulates has been recorded in other areas (e.g. Amboseli, D. Western, pers. comm., 1978) and appears to be a general phenomenon. However, the concept of facilitation — that is, one species depending upon another — is not a necessary condition to explain the succession. Such a sequence of movements can be adequately explained simply on the basis of size, metabolic rate and fermentation rate (Janis, 1976; Jarman and Sinclair, 1979).

Large species need to eat larger quantities of food to meet their requirements. But requirements increase with size curvilinearly (Fig. 17.10) rather than linearly because the specific metabolic rate declines with size. Since there is a finite amount of time in a day to find food, larger species must meet their high requirements by being unselective and eating abundant poor-quality grass (buffalo), grass stems (zebra) or even woody material (elephant).

Although smaller ruminants require smaller absolute amounts of food, the food needs to be digested at a fast rate. Because ruminants are dependent on micro-organisms for their nutrients, the rate of food made available for digestion is dependent on both the fermentation rate, and the time that food can stay in the rumen to be fermented. This residence time is in turn found to be directly proportional to the size of the stomach, which is related to the size of the animal. A small stomach (with short food residence time) can be compensated for by very fast fermentation.

Fermentation rate is dependent on the concentration of nitrogen in the food (Balch and Campling, 1965). Although ruminants can save urea nitrogen by recycling it through the saliva and gut wall, the important variable is the food protein content. Therefore a small animal, with high metabolic rate, must eat high-protein food to maintain a high fermentation rate, and because of this one finds a correlation between fermentation rate and metabolic rate (Moir, 1965). Fig. 17.10 shows a family of lines for different food protein levels indicating the rate at which food becomes available

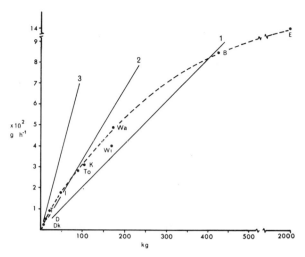

Fig. 17.10. The required rate of food intake (broken line) increases curvilinearly with body size in ungulates. For a given food quality, the rate at which it is fermented and made available for digestion increases linearly with body size (solid lines): *1* = the slowest rate from low-quality food is sufficient for large grazers; *2* = intermediate rate is sufficient for medium-sized grazers but insufficient for mixed feeders; *3* = the highest rate sufficient for the smallest animals is achieved only by selective browsing. *B* = buffalo; *E* = elephant; *D* = grey duiker; *Dk* = dikdik; *I* = impala; *K* = hartebeest; *T* = Thomson's gazelle; *To* = topi; *Wa* = waterbuck; *Wi* = wildebeest. (Data from Sinclair, 1977a; Malpas, 1977; Hoppe, 1977.)

for absorption by the gut after fermentation. For a given food quality, the rate at which food is made available is shown to increase linearly with body size. In fact it is more likely to curve upwards. It can be seen that small ruminants must eat high-protein food to match their high metabolic rate. Fig. 17.11

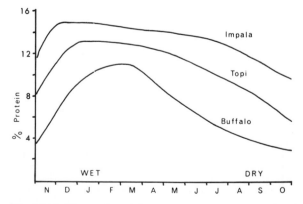

Fig. 17.11. The quality of food in the diet is highest in small animals, reflecting their greater food selection. (From Jarman and Sinclair, 1979, with permission of Chicago University Press.)

shows that feeding selection for protein does indeed increase in smaller species as an adaptation to meet these physiological requirements.

The inverse relationship between food protein quality and body size has important consequences for the co-existence of species. Large species with high food requirements can move down the catena first to eat the coarse grasses, but they cannot remain until it is all eaten, for by then their intake rate would be too low; they have to move on to other ungrazed areas, leaving scattered remnants in the form of leaves and herbs. Smaller animals can support themselves on these higher-quality remnants. This means that the larger species obtain most of the food, and in general their populations have higher numbers than those of smaller species; but larger species cannot completely exclude the smaller ones.

Conversely the smaller species cannot eat much of the food of the larger species because of the relationship shown in Fig. 17.10, and so cannot exclude larger species. Therefore ungulate species of different sizes can coexist by eating the same food but at different degrees of food dispersion. Such an explanation does not require one species to facilitate another by making food more available.

However, facilitation may still occur in some cases, and species may be adapted to take advantage of it. Vesey-Fitzgerald's example in the Rukwa Valley is a likely case though not proven. Bell (1971) predicted for the Serengeti case that a decrease in zebra numbers would cause a decline in wildebeest and gazelle. Alternatively wildebeest would not increase without zebra numbers increasing. Recent evidence from perturbations in the Serengeti (Sinclair, 1979b) shows a five-fold increase in wildebeest numbers, while those of zebra have remained unchanged. Therefore facilitation is unlikely to be important in this case.

However, for the wildebeest–Thomson's gazelle interaction, McNaughton (1976) provided evidence that facilitation was occurring. By grazing down mature grass, wildebeest stimulate the plant to produce new high-protein shoots, which do not appear in the ungrazed grass. Gazelles concentrate in areas previously grazed by wildebeest as an adaptation to make use of this new food supply. This is one reason why Serengeti gazelles can maintain populations (700 000; Bradley, 1977) larger than would be expected for a small (16 kg) ruminant.

In summary, coexistence of ungulate species using the same habitats and eating the same food supply can occur if the species are of different sizes. Larger species obtain most of the food and have the largest populations. At the same time their activities may provide food that would not otherwise be available for smaller species, and the latter have adapted their behaviour to make use of this extra food.

Social organization and sexual selection

Interspecific competition appears to have selected for species sufficiently different in body size to allow co-existence. Jarman (1974) was the first to outline the relationship between body size, type of food eaten and the typical size of social group (see also Estes, 1974; and Leuthold, 1977). Jarman pointed out that high-quality food such as dicotyledon shoots, buds and flowers come as widely spaced, discrete units that are eaten whole. An animal feeding on these highly scattered items increases the spacing between them, so that another animal following behind would find difficulty in obtaining enough food. Hence small animals live solitarily or in pairs at the most.

At the other extreme, low-quality food occurs as abundant bunches of grass. An animal rarely consumes a whole grass bunch, merely taking a bite out of it before moving on. Hence spacing between grass bunches does not alter much until many animals have grazed the sward. Therefore larger species can occur in larger social groups.

Recognizing the essential continuum from selective browsers to unselective grazers, Jarman (1974), for simplicity, outlined the following social categories:

(A) Small (3–20 kg) solitary or pair-forming species feeding as selective browsers on flowers, twig tips, fruits, seed pods and young shoots. Habitats are thickets with a variety of herbs. Includes duikers, dikdiks, klipspringer, steinbuck and suni antelope. One male and female defend a territory. Sexes similar in size.

(B) Small to medium-size (20–100 kg) species that are either entirely grazers or browsers, but in both cases very selective for plant parts such as grass shoots or young dicotyledon leaves. Group size is usually from 3 to 6, with one male and several females. Territorial or permanent home range.

Includes oribi, reedbucks, gerenuk, bushbuck and lesser kudu. Greater sexual dimorphism than A.

(C) Medium-size (50–150 kg female) species that are often mixed feeders, with diet changing seasonally from grass in the rains to browse in the dry season. Group size is variable (6–60) with one territorial male and groups of females and young that wander through the territories. The females have a larger home range, including a dry-season area and a wider wet-season area. Includes gazelles, waterbuck, kob, lechwe, impala, greater kudu and possibly sable. Extremely sexually dimorphic.

(D) Medium to large (100–250 kg) species that are grazers selecting for grass leaf. Female groups range from six to several hundred, at times aggregating to thousands. Home range is large as in C, sometimes separating wet- and dry-season ranges with long-distance migrations. Single territorial males, with less extreme sexual dimorphism. Includes wildebeest, hartebeest, topi, and I suggest Grevy's zebra also.

(E) Large (>200 kg) species that are unselective grazers and browsers of low-quality food. Group size of females (twenty to several hundred) shows aggregations in the wet season. Movements are seasonal within a large home range. Adult males are non-territorial and form a dominance hierarchy. They move with the females. Includes buffalo, eland, elephant, Burchell's zebra, giraffe, and possibly oryx and roan antelope. However, Estes and Estes (1974) have suggested that oryx may be territorial, so fitting more with category D. Sexual dimorphism slight.

These categories show that increasing body size allows a more abundant food supply, and this in turn permits animals to form larger groups (as a protection against predators, see pp. 419–420 below). The size of group determines how the male obtains his mates. Males of very small species can keep females in their territories year-round, and this may be the only way of finding females in oestrus. In larger species, the larger female group cannot remain within one territory and so it wanders through many of them. In this situation males must compete for space to breed, and territories do not necessarily provide year-round food supplies. Indeed, territoriality disappears in severe dry seasons for species in groups C and D (Jarman, 1974; Leuthold, 1977; Jarman and Jarman, 1979).

Intraspecific competition between males for a mating territory has led, through sexual selection, to larger males with elaborate weapons. Such selection has not operated on the females, which have remained at a smaller size because of other selection pressures. Hence in group C particularly there appears the extreme sexual dimorphism.

Inter-male competition for mates is most severe when opportunities for mating are limited in space or time. Under these circumstances territories are reduced in size and become clustered in small areas callled leks. In kob, females are attracted to the congregation of males on the lekking grounds, and they prefer the males in the centre (Leuthold, 1966; Buechner and Roth, 1974; Floody and Arnold, 1975). Although the centre males obtain the highest frequency of matings, they hold a territory for one or two days only. Males on the periphery of the lek obtain fewer matings but hold their territories for up to a year. Longer time in the outer territory may compensate for lower mating frequency, especially if there are some females in oestrus at all seasons, as appears in the case in kob.

Space to set up territory appears to be the limiting factor in kob. Lechwe also develop leks and in these species time may be as limiting as space (Schuster, 1976), for rutting occurs when flood waters are rising and inundating the areas suitable for territories. There would, therefore, be a premium on obtaining many matings in a short time.

From the description of De Vos and Dowsett (1966) puku may set up leks, but this needs confirmation. In the Serengeti migratory wildebeest, males set up small territories (30 m diameter) in the path of the moving female herds. These territories last only a few hours before the males move on and set up again (personal observation). In fact, I consider these wildebeest are setting up a series of temporary leks similar to the lechwe. Again, the wildebeest are limited by time and space for mating opportunities. This strategy is in contrast to that of wildebeest resident in Ngorongoro Crater: these have much larger permanent territories (Estes, 1969).

Males of the large species (group E) have not adopted territorial behaviour. Instead, they have developed a dominance hierarchy, and so can remain with the females throughout the time when oestrus is likely to occur. A dominance hierarchy seems to occur only in species where groups have fixed membership so that individuals can recognize each other. Fixed membership appears to be necessary where young animals have to learn from parents the location of restricted food or water sources within a large home range (Sinclair, 1977a), and the dominance hierarchy may be simply a consequence of this.

In summary, body size, and food and habitat requirements, dictate the size of social group typical of a species, and hence the degree of inter-male competition for mates. This competition is the selection pressure leading to sexual dimorphism.

Anti-predator adaptations

Hamilton (1971) has argued that the risk from predation can be reduced if individuals simply stayed close to others by forming herds. Therefore, animals should form herds wherever possible, and Jarman (1974) has suggested that predation pressure determines the lower size limit for such groups. Small species in group A cannot form groups, so they adopt a strategy of hiding: they are cryptically coloured, they hide in thickets and are nocturnal or crepuscular, and they "freeze" when alarmed. The presence of other individuals is a disadvantage because the others might attract predators. Hence predation acts in the same direction as interspecific competition by selecting for small size and a solitary existence.

Species in group B "freeze" if they see the predator first, but they run once the predator sees them and try to hide again (e.g. reedbuck). The strategy of those in C is determined by their habitat: those in thicket (kudu) "freeze" and run if detected, while those on open plains (gazelles) move away to a safe distance and then keep the predator in sight. Impala have a special "star burst" behaviour if ambushed by a predator: with their exceptional jumping ability, the group bounds outwards in all directions producing a "confusion effect" that distracts the predator. Sitatunga and lechwe show convergence by running into water and submerging. Puku and waterbuck also run into water but do not submerge (De Vos and Dowsett, 1966; Lent, 1969; Jungius, 1971; Bigalke, 1972a).

The plains species of D have little cover so make no attempt to hide. They run when chased, making use of the anonymity of the herd but they do not try to run faster than the chasing predators (hyaena, wild dog), probably because these predators have a long-distance running capability superior to the ungulates (C.R. Taylor, pers. comm., 1979). On the other hand, sprinting ability to out-distance ambush predators (lion, cheetah, leopard) seems to be all-important (Elliot et al., 1977).

Some of the large species in group E, like zebra and eland, use similar anti-predator strategies to those in D, but others like buffalo are very slow and would rarely outrun a lion. Instead buffalo use group defence behaviour to protect each other, making use of horns and large size (Sinclair, 1977a). Elephant adopt a similar strategy (Douglas-Hamilton, 1972). The fixed membership of herds in these species probably results in a higher degree of genetic relatedness than in the population at large; and this could result, through kin selection, in the evolution of group defence. It is noteworthy that group defence has not been definitely confirmed in species of groups A to D.

Protection of newborn young is in most species accomplished by the juveniles hiding in cover away from adults, termed "lying out" behaviour (Estes, 1974; Jarman, 1974; Leuthold, 1977). At least in hartebeest, the calf walks away from the mother and lies down shortly after birth (Gosling, 1969). Exceptions to this pattern are seen in the migrating species of groups D and E, such as wildebeest, zebra and eland and those with fixed-membership herds (buffalo, elephant). These species have young which follow the mother and never "lie out".

Particularly in wildebeest, the young are born on open plains devoid of cover. The calves are extremely precocious, taking only 5 to 10 min to walk (compared to 30 min in hartebeest; Gosling, 1969), and after 24 h they can run fast. These appear to be anti-predator adaptations. Kruuk (1972) and Schaller (1972) have documented that hyaenas, lions and wild dog all concentrate their attention on wildebeest calves during the birth season, which suggests that wildebeest experience unusually severe predation pressure, and have evolved special behaviour to minimize it. One such behaviour pattern appears to be the extremely synchronized birth period. By producing most of the young at the same time of year, the risk of predation is minimized for each calf (Estes, 1969; Kruuk, 1972). Another adaptation is the time of day when calves are born: most births occur between 07:30 and 10:30 h (Sinclair, 1977b), as soon as possible after the morning hunting period of hyaenas, so as to allow the maximum time before the evening hunting period (Fig. 17.12).

In conclusion, Bertram (1979) has pointed out that most ungulates are predated by both chasing and stalking predators. Therefore, the prey cannot specialize by being both a good long-distance runner and a fast sprinter, and so they remain vulnerable to both types of predator. Jarman (1974) noted that smaller ungulates are exposed to a wider array of predator species than larger prey. Therefore, predation would select for prey species to become larger, and such a trend would only be counteracted by interspecific competition.

ADAPTATION AND COMMUNITY FUNCTION

The adaptations that I have reviewed so far result from a variety of selection pressures that are often opposing each other. Furthermore the intensity of these selection pressures varies from place to place, so that in some areas seasonal extremes of climate are most important, whereas in others predation is more significant. The particular set of selection pressures and their intensity determines the types of species that can exist in an area, and hence de-

termines how the community functions. By comparing different areas, or by observing the effects of human disturbances (perturbations), one can gain insight on what regulates the community.

As a first example, the Serengeti ecosystem is dominated by the migration strategy of the wildebeest (Sinclair, 1979a), which is an adaptation to spatial changes in food supply. A consequence of this is that predators (lion, hyaena) cannot regulate the wildebeest population, and predator populations are themselves regulated by the abundance of alternative resident prey (topi, hartebeest, warthog) which are eaten when wildebeest move away (Hanby and Bygot, 1979). The wildebeest are regulated by their food supply (i.e. through intraspecific competition), and this in turn increases interspecific competition with other herbivores, particularly buffalo. It has been possible to draw these conclusions because of the perturbation resulting from the removal in 1962 of the exotic disease rinderpest. This had originally imposed heavy juvenile mortality on wildebeest and buffalo, but, after the removal of the disease, wildebeest numbers increased from about 250 000 to 1.3 million in 1978, and buffalo have shown a similar rate of increase.

If the wildebeest did not have to migrate but remained in the same area year-round, then they would be more available to predators, which in turn could take a higher proportion of the prey population and might even regulate it. This hypothesis can be tested with evidence from the Ngorongoro Crater (immediately adjacent to the Serengeti) where the same prey species, wildebeest, zebra and Thomson's gazelle, are resident because grass growth occurs throughout the year. There, lion and hyaena together kill some 14% of the wildebeest each year compared to only 1% in the Serengeti (Elliot and Cowan, 1978). Furthermore, the wildebeest population has not increased despite the removal of rinderpest in 1962, but has remained constant (between 10 000 and 15 000) from 1958 to 1978. This suggests that the predators are regulating their prey, and the increased wildebeest calf survival after rinderpest could have been counteracted by increased predation. The main point is that the different patterns of rainfall in the two areas, and the adaptations to them shown by wildebeest, determined whether their populations were regulated by food or predators.

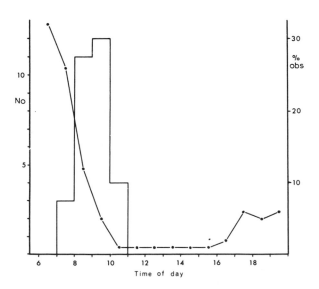

Fig. 17.12. The distribution of wildebeest births through the day (histogram). Solid line shows the proportion of all observations of hunting hyaenas recorded by Kruuk (1972) plotted against time of day. (From Sinclair, 1977a, with permission of Chicago University Press.)

Wildebeest provide a second example from the Kruger Park, South Africa, showing that there may be two stable states for a community, regulated by either predators or food (Smuts, 1978; Walker and Noy-Meir, 1979). In this area wildebeest are resident. An increase in rainfall created long grass which the wildebeest do not like, and the population broke up into many small scattered groups making them more vulnerable to predators. The population was perturbed artificially through a reduction cull by management considering, perhaps mistakenly, that there were too many animals. When culling stopped, the wildebeest population continued to decline as a result of predation. These authors interpreted this as a change in stability state: culling reduced the population below a threshold, where the increased density-dependent effect of predators took over. Future monitoring is needed to confirm this.

Perturbations have also thrown some light on the adaptations of elephant and their effect on vegetation. It has been suggested (Laws, 1969a, b, 1970) that the only regulatory response in elephant populations is through changes in reproductive rate, which is too slow to prevent elephants destroying the tree cover and turning it into grassland. However, it is highly unlikely that reproductive change is the only regulatory response, and evidence by Corfield (1973) has shown that mortality of juveniles and adult females is a much faster density-dependent response. Law's (1969b) prediction that Tsavo Park, Kenya, would be turned into grassland is thus in question.

A common explanation throughout Africa for vegetation changes caused by elephant populations is that the latter have been unnaturally compressed into refuges due to expanding cultivation. Whereas this may be so for some areas, there is no evidence to support this for Tsavo, and a more likely explanation is that the reduction in tree cover is a return to a more natural balance between elephants and vegetation; that previous to 1947 (when the park was designated) had been unnatural: Laws (1969b) reported that for several decades before 1947 heavy mortality by humans was imposed on elephants, and therefore one must suppose an unnaturally dense growth of trees had been able to grow in the area. When poaching mortality was prevented in 1947, the elephant population appears to have responded to their unusually abundant food by increasing in population. The tree population may only be responding as expected for a herbivore–plant interaction and should reach a balance in due course (Caughley, 1976a, b) provided culling does not upset it.

These few examples show that perturbations of ecosystems can tell one a great deal about the adaptations of organisms, the selection pressures involved and how the community is regulated. By way of conclusion, perturbations also show that it is unwise to assume animals are not adapted — an assumption often used to justify interference in the name of management of natural areas.

ACKNOWLEDGEMENT

I thank G.G.E. Scudder and B.H. Walker for helpful comment on earlier drafts.

APPENDIX I: SCIENTIFIC AND COMMON NAMES OF SPECIES MENTIONED IN THIS CHAPTER

ORDER
Family
  Tribe
   Species
	Common name

CARNIVORA	
Canidae	
Lycaon pictus	wild dog
Hyaenidae	
Crocuta crocuta	hyaena
Felidae	
Acinonyx jubatus	cheetah
Panthera leo	lion
P. pardus	leopard
PROBOSCIDEA	
Elephantidae	
Loxodonta africana	elephant
PERISSODACTYLA	
Rhinocerotidae	
Ceratotherium simum	white rhinoceros
Diceros bicornis	black rhinoceros
Equidae	
Equus burchelli	Burchell's zebra
E. grevyi	Grevy's zebra
E. zebra	mountain zebra

APPENDIX I *(continued)*

ORDER	Common name
Family	
Tribe	
Species	

ARTIODACTYLA

Suidae
Hylochoerus meinertzhageni	giant forest hog
Phacochoerus aethiopicus	warthog
Potamochoerus porcus	bushpig

Hippopotamidae
| *Choeropsis liberiensis* | pygmy hippopotamus |
| *Hippopotamus amphibius* | hippopotamus |

Tragulidae
| *Hyemoschus aquaticus* | water chevrotain |

Giraffidae
| *Giraffa camelopardalis* | giraffe |

Bovidae
Bovini
| *Syncerus caffer* | buffalo |

Tragelaphini
Taurotragus oryx	eland
Tragelaphus angasi	nyala
T. buxtoni	mountain nyala
T. euryceros	bongo
T. imberbis	lesser kudu
T. scriptus	bushbuck
T. spekei	sitatunga
T. strepsiceros	greater kudu

Reduncini
Kobus ellipsiprymnus	waterbuck
K. defassa	defassa waterbuck
K. kob	kob
K. leche	lechwe
K. megaceros	Nile lechwe
K. vardoni	puku
Redunca arundinum	southern reedbuck
R. fulvorufula	mountain reedbuck
R. redunca	bohor reedbuck

Hippotragini
Hippotragus equinus	roan
H. niger	sable
Oryx dammah	scimitar horned oryx
O. gazella	gemsbuck, Beisa oryx

Alcelaphini
Alcelaphus buselaphus	hartebeest (bubal)
A. caama	red hartebeest
A. lichtensteini	Lichtenstein's hartebeest
Connochaetes gnou	black wildebeest

Alcelaphini *(continued)*
C. taurinus	wildebeest
Damaliscus dorcas	blesbuck, bontebuck
D. hunteri	Hunter's antelope
D. korrigum	topi
D. lunatus	tsessebe

Aepycerotini
| *Aepyceros melampus* | impala |

Antilopini
Antidorcas marsupialis	springbuck
Gazella dama	dama gazelle
G. dorcas	dorcas gazelle
G. granti	Grant's gazelle
G. rufifrons	red-fronted gazelle
G. spekei	Speke's gazelle
G. thomsoni	Thomson's gazelle
Litocranius walleri	gerenuk

Ammodorcadini
| *Ammodorcas clarkei* | dibatag |

Cephalophini
| *Cephalophus spadix* | Abbott's duiker |
| *Sylvicapra grimmia* | grey duiker |

Neotragini
Madoqua kirki	dikdik (Kirk's)
Nesotragus moschatus	suni antelope
Oreotragus oreotragus	klipspringer
Ourebia ourebi	oribi
Raphicerus campestris	steinbuck

REFERENCES

Anderson, J.L., 1972. Seasonal changes in the social organization and distribution of the impala in Hluhluwe Game Reserve, Zululand. *J. S. Afr. Wildl. Manage. Assoc.*, 2: 16–20.

Ansell, W.F.H., 1971. Part 14, Perissodactyla and Part 15, Artiodactyla. In: J. Meester and H.W. Setzer (Editors), *Identification Manual for African Mammals*. Smithsonian Institution, Washington, D.C., 14, 1–16; 15, 1–84.

Baker, A.A., 1969. Post partum anoestrus in cattle. *Aust. Vet. J.*, 45: 180–183.

Baker, J.R., 1938. Evolution of breeding seasons. In: G.R. de Beer (Editor), *Evolution: Essays on Aspects of Evolutionary Biology Presented to Professor E.S. Goodrich*. Clarendon Press, Oxford.

Balch, C.C. and Campling, R.C., 1965. Rate of passage of digesta through the ruminant digestive tract. In: R.W. Dougherty (Editor), *Physiology of Digestion in the Ruminant*. Butterworths, Washington, D.C., pp. 108–123.

Bell, R.H.V., 1970. The use of the herb layer by grazing ungulates in the Serengeti. In: A. Watson (Editor), *Animal*

Populations in Relation to Their Food Resources. Blackwell, Oxford, pp. 111–123.

Bell, R.H.V., 1971. A grazing system in the Serengeti. *Sci. Am.*, 224: 86–93.

Bertram, B.C.R., 1979. Serengeti predators and their social systems. In: A.R.E. Sinclair and M. Norton-Griffiths (Editors), *Serengeti: Dynamics of An Ecosystem*. Chicago University Press, Chicago, Ill., pp. 221–248.

Bigalke, R.C., 1972a. The contemporary mammal fauna of Africa. In: A. Keast, F.C. Erk and B. Glass (Editors), *Evolution, Mammals and Southern Continents*. New York State University Press, Albany, N.Y., pp. 141–194.

Bigalke, R.C., 1972b. Observations on the behaviour and feeding habits of the Springbok, *Antidorcas marsupialis*. *Zool. Afr.*, 7: 333–359.

Blankenship, L.H. and Field, C.R., 1972. Factors affecting the distribution of wild ungulates on a ranch in Kenya. *Zool. Afr.*, 7: 281–302.

Bourlière, F., Morel, G. and Galat, G., 1976. Les grands mammifères de la basse vallée du Sénégal et leurs saisons de reproduction. *Mammalia*, 40: 401–412.

Bradley, R.M., 1977. *Aspects of the Ecology of the Thomson's Gazelle in the Serengeti National Park, Tanzania*. Thesis, Texas A & M University, College Station, Texas.

Britton, P.L., 1978. Seasonality, density and diversity of birds on a papyrus swamp in western Kenya. *Ibis*, 120: 450–466.

Buechner, H.K. and Roth, H.D., 1974. The lek system in Uganda kob antelope. *Am. Zool.*, 14: 145–162.

Caughley, G., 1976a. Wildlife management and the dynamics of ungulate populations. In: T.H. Coaker (Editor), *Applied Biology, 1*. Academic Press, New York, N.Y., pp. 183–246.

Caughley, G., 1976b. Plant–herbivore systems. In: R.M. May (Editor), *Theoretical Ecology*. W.B. Saunders, Philadephia, Penn., pp. 94–113.

Child, G. and Von Richter, W., 1969. Observations on ecology and behaviour of lechwe, puku, and waterbuck along the Chobe River, Botswana. *Z. Säugetierkd.*, 34: 275–295.

Clough, G., 1969. Some preliminary observations on reproduction in warthog, *Phacochoerus aethiopicus* Pallas. *J. Reprod. Fertil., Supply.*, 6: 323–337.

Corfield, T.F., 1973. Elephant mortality in Tsavo National Park, Kenya. *E. Afr. Wildl. J.*, 11: 339–368.

Crawford, M.A., Patterson, J.M. and Yardley, L., 1968. Nitrogen utilization by the Cape buffalo (*Syncerus caffer*) and other large mammals. In: M.A. Crawford (Editor), *Comparative Nutrition of Wild Animals. Symp. Zool. Soc., Lond.* 21: 367–379.

De Vos, A. and Dowsett, R.J., 1966. The behaviour and population structure of three species of the genus *Kobus*. *Mammalia*, 30: 30–55.

Dorst, J. and Dandelot, P., 1972. *A Field Guide to the Larger Mammals of Africa*. Collins, London, 2nd ed., 287 pp.

Douglas-Hamilton, I., 1972. *On the Ecology and Behaviour of the African Elephant: the Elephants of Lake Manyara*. Thesis, Oxford University, Oxford.

Duncan, P., 1975. *Topi and Their Food Supply*. Thesis, Nairobi University, Nairobi, 283 pp.

Du Plessis, S.S., 1972. Ecology of blesbok with special reference to productivity. *Wildl. Monogr.*, No. 30: 1–70.

Elliott, J.P. and Cowan, I. McT., 1978. Territoriality, density,

and prey of the lion in Ngorongoro Crater, Tanzania. *Can. J. Zool.*, 56: 1726–1734.

Elliott, J.P., Cowan, I.McT. and Holling, C.S., 1977. Prey capture of the African lion. *Can. J. Zool.*, 55: 1811–1828.

Eltringham, S.K., 1974. Changes in the large mammal community of Mweya peninsula, Rwenzori National Park, Uganda, following removal of hippopotamus. *J. Appl. Ecol.*, 11: 855–866.

Estes, R.D., 1969. Territorial behaviour of the wildebeest (*Connochaetes taurinus* Burchell, 1823). *Z. Tierpsychol.*, 26: 284–370.

Estes, R.D., 1974. Social organization of the African Bovidae. In V. Geist and F. Walther (Editors), *The Behaviour of Ungulates and its Relation to Management. I.U.C.N. Publ., N.S.*, 24: 166–205.

Estes, R.D., 1976. The significance of breeding synchrony in the wildebeest. *E. Afr. Wildl. J.*, 14: 135–152.

Estes, R.D. and Estes, R.K., 1974. The biology and conservation of the giant sable antelope, *Hippotragus niger variani* Thomas, 1916. *Proc. Acad. Nat. Sci. Philadelphia*, 126: 73–104.

Fairall, N., 1968. The reproductive seasons of some mammals in the Kruger National Park. *Zool. Afr.*, 3: 189–210.

Ferrar, A.A. and Walker, B.H., 1974. An analysis of herbivore/habitat relationships in Kyle National Park, Rhodesia. *J. S. Afr. Manage. Assoc.*, 4: 137–147.

Field, C.R., 1971. Elephant ecology in the Queen Elizabeth National Park, Uganda. *E. Afr. Wildl. J.*, 9: 99–123.

Field, C.R., 1972. The food habits of wild ungulates in Uganda by analysis of stomach contents. *E. Afr. Wildl. J.*, 10: 17–42.

Field, C.R., 1975. Climate and the food habits of ungulates on Galana Ranch. *E. Afr. Wildl. J.*, 13: 203–220.

Field, C.R. and Ross, I.C., 1976. The savanna ecology of Kidepo Valley National Park. II. Feeding ecology of elephant and giraffe. *E. Afr. Wildl. J.*, 14: 1–16.

Field, C.R., Harrington, G.N. and Pratchett, D., 1973. A comparison of grazing preferences of buffalo (*Syncerus caffer*) and Ankole cattle (*Bos indicus*) on three different pastures. *E. Afr. Wildl. J.*, 11: 19–30.

Finch, V.A., 1972. Energy exchanges with the environment of two East African antelopes, the eland and the hartebeest. In: G.M.O. Maloiy (Editor), *Comparative Physiology of Desert Animals. Symp. Zool. Soc., Lond.*, 31: 315–326.

Floody, O.R. and Arnold, A.P., 1975. Uganda kob (*Andenota kob thomasi*): Territoriality and the spatial distribution of sexual and agonistic behaviours at a territorial ground. *Z. Tierpsychol.*, 37: 192–212.

Ghobrial, L.I., 1974. Water relations and requirement of the Dorcas gazelle in the Sudan. *Mammalia*, 38: 88–107.

Gosling, L.M., 1969. Parturition and related behaviour in Coke's hartebeest, *Alcelaphus buselaphus cokei* Gunther. *J. Reprod. Fertil., Suppl.*, 6: 265–286.

Grimsdell, J.J.R., 1973. Reproduction in the African buffalo, *Syncerus caffer*, in Uganda. *J. Reprod. Fertil., Suppl.*, 19: 301–316.

Gwynne, M.D. and Bell, R.H.V., 1968. Selection of grazing components by grazing ungulates in the Serengeti National Park. *Nature, Lond.*, 220: 390–393.

Hall-Martin, A.J., Skinner, J.D. and Van Dyk, J.M., 1975. giraffe as determined by analysis of stomach contents. *J. S.*

Afr. Wildl. Manage. Assoc., 4: 191–202.

Hall-Martin, A.J. and Basson, W.D., 1975. Seasonal chemical composition of the diet of Transvaal lowveld giraffe. *J. S. Afr. Wildl. Manage. Assoc.*, 5: 19–21.

Hall-Martin, A.J., Skinner, J.D. and Van Dyk, J.M., 1975. Reproduction in the giraffe in relation to some environmental factors. *E. Afr. Wildl. J.*, 13: 237–248.

Hamilton, W.D., 1971. Geometry for the selfish herd. *J. Theor. Biol.*, 31: 295–311.

Hanby, J.P. and Bygott, J.D., 1979. Population changes in lions and other predators. In: A.R.E. Sinclair and M. Norton-Griffiths (Editors), *Serengeti: Dynamics of an Ecosystem.* Chicago University Press, Chicago, Ill., pp. 249–265.

Hanks, J., 1969. Seasonal breeding of the African elephant in Zambia. *E. Afr. Wildl. J.*, 7: 167.

Hanks, J., 1972. Reproduction of the elephant (*Loxodonta africana*) in the Luangwa Valley, Zambia. *J. Reprod. Fertil.*, 30: 13–26.

Herbert, H.J., 1972. The population dynamics of the waterbuck *Kobus ellipsiprymnus* (Ogilby, 1883) in the Saabi-Sand Wildtuin. *Mamm. Depicta, Beih. Z. Säugetierkd.*, 6: 1–63.

Hillman, J.C. and Hillman, A.K.K., 1977. Mortality of wildlife in Nairobi National Park, during the drought of 1973–1974. *E. Afr. Wildl. J.*, 15: 1–18.

Hirst, S.M., 1969. Predation as a regulating factor of wild ungulate populations in a Transvaal lowveld nature reserve. *Zool. Afr.*, 4: 199–231.

Hofmann, R.R., 1968. Comparisons of the rumen and omasum structure in East African game ruminants in relation to their feeding habits. In: M.A. Crawford (Editor), *Comparative Nutrition of Wild Animals. Symp. Zool. Soc., Lond.*, 21: 179–194.

Hofmann, R.R. and Stewart, D.R.M., 1972. Grazer or browser: a classification based on stomach structure and feeding habits of East African ruminants. *Mammalia*, 36: 226–240.

Hoppe, P.P., 1977. Comparison of voluntary food and water consumption and digestion in Kirk's dikdik and suni. *E. Afr. Wildl. J.*, 15: 41–48.

Janis, C., 1976. The evolutionary strategy of the Equidae and the origins of rumen and cecal digestion. *Evolution*, 30: 757–774.

Jarman, M.V. and Jarman, P.J., 1973. Daily activity of impala. *E. Afr. Wildl. J.*, 11: 75–92.

Jarman, P.J., 1971. Diets of large mammals in the woodlands around Lake Kariba, Rhodesia. *Oecologia (Berl.)*, 8: 157–178.

Jarman, P.J., 1972. Seasonal distribution of large mammal populations in the unflooded middle Zambezi valley. *J. Appl. Ecol.*, 9: 283–299.

Jarman, P.J., 1973. The free water intake of impala in relation to the water content of their food. *E. Afr. Agric. For. J.*, 38: 343–351.

Jarman, P.J., 1974. The social organization of antelope in relation to their ecology. *Behaviour*, 48: 215–266.

Jarman, P.J., 1977. Behaviour of topi in a shadeless environment. *Zool. Afr.*, 12: 101–111.

Jarman, P.J. and Jarman, M.V., 1973. Social behaviour, population structure and reproductive potential in impala. *E. Afr. Wildl. J.*, 11: 329–338.

Jarman, P.J. and Jarman, M.V., 1979. The dynamics of ungulate

social organization. In: A.R.E. Sinclair and M. Norton-Griffiths (Editors), *Serengeti: Dynamics of an Ecosystem.* Chicago University Press, Chicago, Ill., pp. 185–220.

Jarman, P.J. and Sinclair, A.R.E., 1979. Feeding strategy and the pattern of resource partitioning in ungulates. In: A.R.E. Sinclair and M. Norton-Griffiths (Editors), *Serengeti: Dynamics of an Ecosystem.* Chicago University Press, Chicago, Ill., pp. 130–163.

Jungius, H., 1971. The biology and behaviour of the reedbuck (*Redunca arundinum* Boddaert 1785) in the Kruger National Park. *Mamm. Depicta. Beih. Z. Säugetierkd.*, 5: 1–106.

Kiley-Worthington, M., 1965. The waterbuck (*Kobus defassa* Ruppell 1835 and *K. ellipsiprymnus* Ogilby 1833) in East Africa: Spatial distribution. *Mammalia*, 29: 177–204.

King, J.M., Kingaby, G.P., Colvin, J.G. and Heath, B.R., 1975. Seasonal variations in water turnover by oryx and eland on the Galana Game Ranch Research Project. *E. Afr. Wildl. J.*, 13: 287–296.

Klingel, H., 1969. The social organization and population ecology of the plains zebra (*Equus quagga*). *Zool. Afr.*, 4: 249–263.

Klingel, H., 1972. Social behaviour of African Equidae. *Zool. Afr.*, 7: 175–185.

Kreulen, D.A., 1975. Wildebeest habitat selection on the Serengeti plains, in relation to calcium and lactation: a preliminary report. *E. Afr. Wildl. J.*, 13: 297–304.

Kruuk, H., 1972. *The Spotted Hyena.* Chicago University Press, Chicago, Ill., 335 pp.

Lamprey, H.F., 1963. Ecological separation of the large mammal species in the Tarangire Game Reserve, Tanganyika. *E. Afr. Wildl. J.*, 1: 63–92.

Lamprey, H.F., 1964. Estimation of the large mammal densities, biomass and energy exchange in the Tarangire Game Reserve and the Masai steppe in Tanganyika. *E. Afr. Wildl. J.*, 2: 1–46.

Langer, P., 1974. Stomach evolution in the Artiodactyla. *Mammalia*, 38: 295–314.

Laws, R.M., 1969a. Aspects of reproduction in the African elephant, *Loxodonta africana. J. Reprod. Fertil., Suppl.* 6: 193–217.

Laws, R.M., 1969b. The Tsavo Research Project. *J. Reprod. Fertil., Suppl.*, 6: 495–531.

Laws, R.M., 1970. Elephants as agents of habitat and landscape change in East Africa. *Oikos*, 21: 1–15.

Laws, R.M. and Clough, G., 1966. Observations on reproduction in the hippopotamus, *Hippopotamus amphibius* (Linn.). In: I.W. Rowlands (Editor), *Comparative Biology of Reproduction in Mammals. Symp. Zool. Soc., Lond.*, 15: 117–140.

Laws, R.M. and Parker, I.S.C., 1968. Recent studies on elephant populations in East Africa. In: M.A. Crawford (Editor), *Comparative Nutrition of Wild Animals. Symp. Zool. Soc., Lond.*, 21: 319–359.

Lent, P.C., 1969. A preliminary study of the Okavango lechwe (*Kobus leche leche* Gray). *E. Afr. Wildl. J.*, 7: 147–157.

Leuthold, B.M. and Leuthold, W., 1972. Food habits of giraffe in Tsavo National Park, Kenya. *E. Afr. Wildl. J.*, 10: 129–141.

Leuthold, W., 1966. Variations in territorial behaviour of

Uganda kob, *Adenota kob thomasi* (Neumann 1896). *Behaviour*, 27: 214–257.

Leuthold, W., 1971. Studies on the food habits of lesser kudu in Tsavo National Park, Kenya. *E. Afr. Wildl. J.*, 9: 35–45.

Leuthold, W., 1977. *African Ungulates*. Springer-Verlag, Berlin, 307 pp.

Leuthold, W. and Leuthold, B.M., 1975. Temporal patterns of reproduction in ungulates of Tsavo East National Park, Kenya. *E. Afr. Wildl. J.*, 13: 159–169.

Livingstone, D.A., 1975. Late Quaternary climatic changes in Africa. *Annu. Rev. Ecol. Syst.*, 6: 249–280.

MacFarlane, W.V. and Howard, B., 1972. Comparative water and energy economy of wild and domestic mammals. In: G.M.O. Maloiy (Editor), *Comparative Physiology of Desert Animals. Symp. Zool. Soc., Lond.*, 31: 261–296.

McNaughton, S.J., 1976. Serengeti migratory wildebeest: Facilitation of energy flow by grazing. *Science*, 191: 92–94.

Maddock, L., 1979. The 'migration' and grazing succession. In: A.R.E. Sinclair and M. Norton-Griffiths (Editors), *Serengeti: Dynamics of an Ecosystem*. Chicago University Press, Chicago, Ill., pp. 104–129.

Maloiy, G.M.O., 1973. The water metabolism of a small East African antelope: the dikdik. *Proc. R. Soc. Lond.*, Ser. B. 184: 167–178.

Malpas, R.C., 1977. Diet and the condition and growth of elephants in Uganda. *J. Appl. Ecol.*, 14: 489–504.

Moir, R.J., 1965. The comparative physiology of ruminant-like animals. In: R.W. Dougherty (Editor), *Physiology of Digestion in the Ruminant*. Butterworths, Washington, D.C., pp. 1–14.

Moreau, R.E., 1952. Africa since the Mesozoic: with particular reference to certain biological problems. *Proc. Zool. Soc., Lond.*, 121: 869–913.

Moreau, R.E., 1966. *The Bird Faunas of Africa and its Islands*. Academic Press, New York, N.Y., 424 pp.

Morton, J.K., 1972. Phytogeography of the West African mountains. In: D.H. Valentine (Editor), *Taxonomy, Phytogeography and Evolution*. Academic Press, New York, N.Y., pp. 221–236.

Owen-Smith, N., 1972. Territoriality: the example of the white rhinoceros. *Zool. Afr.*, 7: 273–280.

Pennycuick, C.J., 1979. Energy costs of locomotion, and the concept of "foraging radius". In: A.R.E. Sinclair and M. Norton-Griffiths (Editors), *Serengeti: Dynamics of an Ecosystem*. Chicago University Press, Chicago, Ill., pp. 164–184.

Pennycuick, L., 1975. Movements of the migratory wildebeest population in the Serengeti area between 1960 and 1973. *E. Afr. Wildl. J.*, 13: 65–87.

Perrins, C.M., 1970. The timing of birds' breeding seasons. *Ibis*, 112: 242–255.

Phillipson, J., 1975. Rainfall, primary production and 'carrying capacity' of Tsavo National Park (East), Kenya. *E. Afr. Wildl. J.*, 13: 171–202.

Pienaar, U. de V., 1974. Habitat-preference in South African antelope species and its significance in natural and artificial distribution patterns. *Koedoe*, 17: 185–195.

Robertshaw, D. and Taylor, C.R., 1969. A comparison of sweat gland activity in eight species of East African bovids. *J. Physiol.*, 203: 135–143.

Robinette, W.L. and Archer, A.L., 1971. Notes on aging criteria

and reproduction of Thomson's gazelle, *Gazella thomsonii*. *E. Afr. Wildl. J.*, 9: 83–98.

Rodgers, W.A., 1976. Seasonal diet preferences of impala from south east Tanzania. *E. Afr. Wildl. J.*, 14: 331–333.

Root, A., 1972. Fringe-eared oryx digging for tubers in Tsavo National Park (East). *E. Afr. Wildl. J.*, 10: 155–157.

Sadleir, R.M.F.S., 1969. *The Ecology of Reproduction in Wild and Domestic Mammals*. Methuen, London, 321 pp.

Sayer, J.A. and Van Lavieren, L.P., 1975. The ecology of the Kafue lechwe population of Zambia before the operation of hydro-electric dams on the Kafue River. *E. Afr. Wildl. J.*, 13: 9–37.

Schaller, G.B., 1972. *The Serengeti Lion*. Chicago University Press, Chicago, Ill., 480 pp.

Schuster, R.H., 1976. Lekking behaviour in Kafue lechwe. *Science*, 192: 1240–1242.

Sekulic, R., 1978. Seasonality of reproduction in the Sable antelope. *E. Afr. Wildl. J.*, 16: 177–182.

Sheppe, W. and Osborne, T., 1971. Patterns of use of a flood plain by Zambian mammals. *Ecol. Monogr.*, 41: 179–205.

Sinclair, A.R.E., 1975. The resource limitation of trophic levels in tropical grassland ecosystems. *J. Anim. Ecol.*, 44: 497–520.

Sinclair, A.R.E., 1977a. *The African Buffalo*. Chicago University Press, Chicago, Ill., 355 pp.

Sinclair, A.R.E., 1977b. Lunar cycle and timing of mating season in Serengeti wildebeest. *Nature*, 267: 832–833.

Sinclair, A.R.E., 1978. Factors affecting the food supply and breeding season of resident birds and movements of palaearctic migrants in a tropical African savannah. *Ibis*, 120: 480–497.

Sinclair, A.R.E., 1979a. Dynamics of the Serengeti ecosystem: process and pattern. In: A.R.E. Sinclair and M. Norton-Griffiths (Editors), *Serengeti: Dynamics of an Ecosystem*. Chicago University Press, Chicago, Ill., pp. 1–30.

Sinclair, A.R.E., 1979b. The eruption of the ruminants. In: A.R.E. Sinclair and M. Norton-Griffiths (Editors), *Serengeti: Dynamics of an Ecosystem*. Chicago University Press, Chicago, Ill., pp. 82–103.

Sinclair, A.R.E. and Gwynne, M.D., 1972. Food selection and competition in the East African buffalo (*Syncerus caffer* Sparrman). *E. Afr. Wildl. J.*, 10: 77–89.

Skinner, J.D. and Van Zyl, J.H.M., 1969. Reproductive performance of the common eland, *Taurotragus oryx* in two environments. *J. Reprod. Fertil., Suppl.*, 6: 319–322.

Skinner, J.D., Van Zyl, J.H.M. and Van Heerden, J.A.H., 1973. The effect of season on reproduction in the black wildebeest and red hartebeest in South Africa. *J. Reprod. Fertil., Suppl.*, 19: 101–110.

Skinner, J.D., Van Zyl, J.H.M. and Oates, L.G., 1974. The effect of season on the breeding cycle of plains antelope of the western Transvaal highveld. *J. S. Afr. Wildl. Manage. Assoc.*, 4: 15–23.

Smuts, G.L., 1975. Reproduction and population characteristics of elephant in the Kruger National Park. *J. S. Afr. Wildl. Manage. Assoc.*, 5: 1–10.

Smuts, G.L., 1978. Interrelations between predators, prey and their environment. *BioScience*, 28: 316–320.

Spinage, C.A., 1969. Reproduction in the Uganda defassa waterbuck, *Kobus defassa ugandae*, Neumann. *J. Reprod. Fertil.*, 18: 445–457.

Spinage, C.A., 1973. The role of photoperiodism in the seasonal breeding of tropical ungulates. *Mammal Rev.*, 3: 71–84.

Talbot, L.M. and Talbot, M.H., 1962. Food preferences of some East African wild ungulates. *E. Afr. Agric. For. J.*, 27: 131–138.

Taylor, C.R., 1968a. The minimum water requirements of some East African bovids. In: M.A. Crawford (Editor), *Comparative Nutrition of Wild Animals. Symp. Zool. Soc., Lond.*, 21: 195–206.

Taylor, C.R., 1968b. Hygroscopic food: a source of water for desert antelopes? *Nature*, 219: 181–182.

Taylor, C.R., 1969. Metabolism, respiratory changes, and water balance of an antelope, the eland. *Am. J. Physiol.*, 217: 317–320.

Taylor, C.R., 1970a. Strategies of temperature regulation: effect on evaporation in East African antelopes. *Am. J. Physiol.*, 219: 1131–1135.

Taylor, C.R., 1970b. Dehydration and heat: effects on temperature regulation of East African ungulates. *Am. J. Physiol.*, 219: 1136–1139.

Taylor, C.R., 1972. The desert gazelle: a paradox resolved. In G.M.O. Maloiy (Editor), *Comparative Physiology of Desert Animals. Symp. Zool. Soc., Lond.*, 31: 215–227.

Taylor, C.R., Robertshaw, D. and Hofmann, R., 1969. Thermal panting: a comparison of wildebeest and zebu cattle. *Am. J. Physiol.*, 217: 907–910.

Taylor, C.R., Spinage, C.A. and Lyman, C.P., 1969. Water relations of waterbuck, an East African antelope. *Am. J. Physiol.*, 217: 630–634.

Underwood, R., 1978. Aspects of kudu ecology at Loskop Dam Nature Reserve, Eastern Transvaal. *S. Afr. J. Wildl. Res.*, 8: 43–47.

Van Zinderen Bakker, E.M. and Coetzee, J.A., 1972. A reappraisal of late-Quaternary climatic evidence from tropical Africa. In: *Palaeoecology of Africa, 7*. A.A. Balkema, Cape Town, pp. 151–181.

Vesey-Fitzgerald, D.F., 1960. Grazing succession amongst East African game animals. *J. Mammal.*, 41: 161–170.

Vesey-Fitzgerald, D.F., 1965. The utilization of natural pastures by wild animals in the Rukwa Valley, Tanganyika. *E. Afr. Wildl. J.*, 3: 38–48.

Von Richter, W. and Osterberg, R., 1977. The nutritive values of some major food plants of lechwe, puku and waterbuck along the Chobe River, Botswana. *E. Afr. Wildl. J.*, 15: 91–98.

Walker, B.H. and Noy-Meir, I., 1979. Stability and resilience of savannah ecosystems. Paper presented at the *Workshop on Dynamic Changes in Savanna Ecosystems, Kruger National Park, May 1979*, in press.

Western, D., 1975. Water availability and its influence on the structure and dynamics of a savannah large mammal community. *E. Afr. Wildl. J.*, 13: 265–286.

Wilson, V.J., 1965. Observations on the greater kudu (*Tragelaphus strepsiceros* Pallas) from a tsetse control hunting scheme in Northern Rhodesia. *E. Afr. Wildl. J.*, 3: 27–37.

Chapter 18

UNGULATES AND LARGE RODENTS OF SOUTH AMERICA

JUHANI OJASTI

INTRODUCTION

Herbivore populations form an important functional group in the energy and nutrient dynamics, information flow and control in most ecosystems (Golley, 1973; Petrusewicz and Grodzinski, 1975; Wiener, 1975). By consuming and otherwise interacting with vegetation the herbivores reduce the standing crop of plants and, depending on the extent of this activity, they may stimulate or restrain the primary production, improve the plant cover for other consumers, or compete with them. Due to selective consumption of certain kinds and parts of plants, the herbivore impact is non-random and regulates the floristic composition and successional trends of the community. This results in the long run in coevolutionary relations between plants and animals. The herbivores also speed up the mineral recycling by fragmentation of plant matter and by releasing excretions, such as urine, with nutrients readily usable by the primary producers (Odum, 1971, p. 104).

The energy contained in the plant tissue consumed by herbivores supports all animal life, and part of it is stored as animal biomass, available for higher-order consumers, including man. The activity of herbivores results in increased structural, genetic and biochemical diversity and information content of ecosystems. The grazing food chain is especially important in grassland ecosystems where most of the plant biomass is both available and suitable as food. Wiegert and Evans (1967) concluded that stable grasslands tolerate consumption levels of 30 to 45% of the primary production without signs of deterioration. The primary consumer level of grasslands is composed of a wide array of animals, such as many kinds of insects and small rodents, but large grazing mammals are the most important in this context, especially in some savanna ecosystems of Africa (Bourlière and Hadley, 1970; Petersen and Casebeer, 1971; Chapter 17; this volume).

The purpose of this chapter is to review the ecology of the native grazing mammals of South American savannas and discuss their role within this ecosystem. The term savanna, in an ecological and geographical sense, is used as defined by Sarmiento in Chapter 9 of this volume. The account is based mainly on recent research[1] on the capybara (*Hydrochoerus hydrochaeris*) and the white-tailed deer (*Odocoileus virginianus*) in the Venezuelan *llanos*, but efforts were made to review the available information on large grazers of other savanna areas of the continent as well.

GRAZING MAMMALS OF SOUTH AMERICA

Most large herbivores of grasslands are ungulates. The recent South American ungulates comprise three species of tapirs (*Tapirus bairdii, T. pinchaque, T. terrestris*), three peccaries (*Catagonus wagneri, Dicotyles tajacu, Tayassu pecari*), four camelids (*Lama glama, L. guanicoe, L. pacos, Vicugna vicugna*) and eleven species of deer (*Blastocerus dichotomus, Hippocamelus antisensis, H. bisulcus, Mazama americana, M. chunyi, M. gouazoubira, M. rufina, Pudu mephistophiles, P. pudu, Odocoileus virginianus, Ozotoceros bezoarticus*) ac-

[1]Research supported by Fondo Nacional de Investigaciones Agropecuárias and Consejo Nacional de Investigaciones Cientificas y Tecnológicas, Caracas (Venezuela).

cording to Cabrera (1961) and Wetzel et al. (1975). Most of them, such as the tapirs, peccaries and the spike-antlered brocket deer (*Mazama* spp.) and *Pudu* are forest dwellers, even though some of them, in particular *Dicotyles tajacu* and *Mazama gouazoubira*, may live in savanna woodlands and feed on open areas to some extent (Allen, 1916; Avila Pires, 1959; Sick, 1965). Llamas and *Hippocamelus* live mostly in open country of the Andean altiplano, pampas or other non-tropical areas. Only the branch-antlered white-tailed deer (*Odocoileus virginianus*) pampas deer (*Ozotoceros*) and swamp deer (*Blastocerus*) can be regarded as components of the tropical savanna ecosystem.

The Neotropical fauna is known to be extremely diverse (Fittkau, 1969; Haffer, 1969; Müller, 1973), but only 21 species of ungulates live in tropical America versus 91 in Africa. Keast (1972), assuming that the number of species ought to be proportional to the area of the continent, pointed out that 55 kinds of ungulates should occur in South America, on African standards. The scarcity of the recent large herbivores in general and grassland ungulates in particular in South America contrasts also with the rich past faunas of the continent (Simpson, 1950; Patterson and Pascual, 1972).

Strong paleontological evidence indicates that South America was separated from other continents through the Tertiary up to the Late Pliocene (Raven and Axelrod, 1975; Webb 1978). The first South American ungulates appeared in the Late Cretaceous and attained their maximum diversity, similar to the present African ungulate fauna, in the Early Miocene, with endemic orders such as Astrapotheria, Condylartha, Litopterna and Notoungulata. Glyptodonts and later the giant sloths and caviomorph rodents increased the richness of large herbivores. In the late Miocene and Pliocene the autochthonous fauna decreased in diversity, presumably due to increasing aridity (Webb, 1978). Later, after gradual establishment of the land connection with North America in the Pliocene, seven orders of mammals, including Proboscidea, Perissodactyla and Artiodactyla, invaded the continent from the north. Some old endemics were replaced by the immigrants, but in general the new groups, such as mastodons, horses, tapirs, camelids, peccaries and deer contributed in diversifying the grazing mammal fauna. In the Late Pleistocene however, approximately 10 000 years

before the present, the last ancient ungulates and half of the northern immigrants disappeared (Martin, 1967; Webb, 1978). Prehistoric hunters and climatic changes have been suggested as prime causes of these world-wide extinctions (Martin and Wright, 1967). These affected selectively the large herbivores (Guilday, 1967), but why this reduction was heavier in South America than elsewhere remains uncertain. The poverty of the present ungulate fauna of the continent is attributed partly to the lack of bovids, the most successful group of grazers of tropical savannas, which never reached South America in the wild. Some authors (Dubost, 1968; Keast, 1972; Bourlière, 1973) have pointed out that caviomorph rodents occupy the medium- to large-herbivore niche in some South American ecosystems. The capybara is included among the large grazers of grasslands, and hence discussed in this paper, whereas other sizable tropical caviomorphs (*Agouti, Dasyprocta, Dinomys*) are restricted to forests.

Domestic livestock, including cattle, horses, goats, sheep and pigs have been introduced to the continent since early colonial times (Cabrera and Yepez, 1960). They have been very successful in South American savannas and are extremely important in ecosystem dynamics. However, they are beyond the scope of this chapter. Some exotic game ungulates have been introduced into the continent (Petrides, 1975), but none of them has established populations in tropical savannas.

DEER OF SOUTH AMERICAN SAVANNAS

The cervids in general are adapted to the niche for browsing herbivores in forests. In South America, however, they also extend to tropical savannas. even though they may be able to coexist with primary grazers on marginal habitats only, as pointed out by Guilday (1967).

The pampas deer (*Ozotoceros bezoarticus*) is restricted to dry open grasslands of central and southern Brazil, Paraguay, Uruguay and northern Argentina (Fig. 18.1B). It is a delicately built small deer (total length 135 cm, height at withers 70 cm approximately), reddish-brown above, pure white below, with thin three-pointed antlers (Burmeister, 1854; Goeldi, 1893; Miranda Ribeiro, 1919). Part of its range (Fig. 18.1C) is shared with the swamp

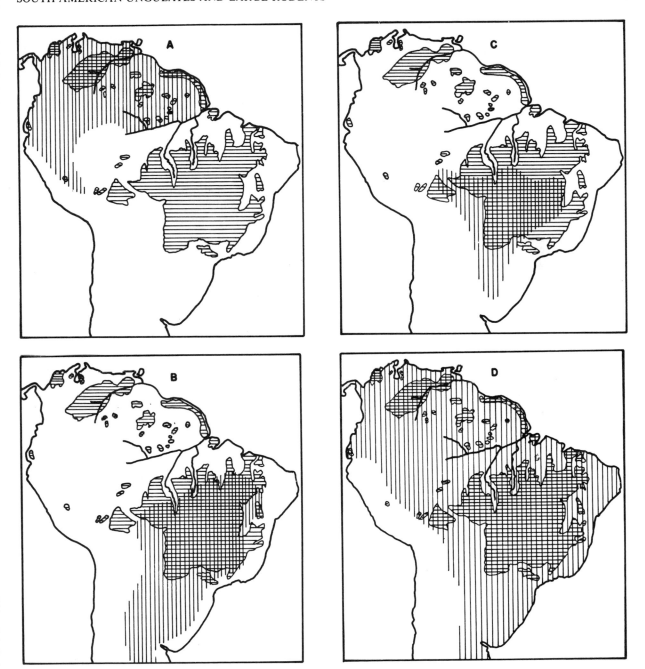

Fig. 18.1. Distribution of savannas (shaded by horizontal lines, after Sarmiento and Monasterio, 1975) and their large herbivores (vertical lines) in South America. A. The white-tailed deer (*Odocoileus virginianus*) according to Miranda Ribeiro (1919) and Brokx (1972a). B. Pampas deer (*Ozotoceros bezoarticus*), redrawn from De Carvalho (1973). C. Swamp deer (*Blastocerus dichotomus*) after Miranda Ribeiro (1919). D. Capybara (*Hydrochoerus hydrochaeris*), redrawn from Gonzalez Jimenez (1977).

deer (*Blastocerus dichotomus*), the largest South American cervid (total length 200 cm, height at withers 130 cm, weight up to 100 kg according to Coimbra-Filjo, [1972]). It lives on swamps, floodplains and other wet savannas. Its hoofs are wide and connected with a membrane as an adaptation to roaming on soft inundated ground. The antlers are usually double-forked and very heavy (Rengger, 1830; Kreig, 1941; Hofmann et al., 1976).

The deer of savannas north of Amazonas (Fig.

18.1A) is referred to the white-tailed deer (*Odocoileus virginianus*), unique among cervids in its wide distribution and ability to cope with different types of habitats. It lives both on the high Andean ranges from Venezuela to Peru (subspecies *goudoti*, *ustus* and *peruvianus*), and on lowlands including thorn woodlands, deciduous forests and savannas, but avoids closed evergreen forests. The population of the Venezuelan and Colombian llanos, Guyana, Surinam and northern limits of Brazil belong to *O. v. gymnotis*, and those of savanna islands (*campinas*) of the Amazonian forest in northern Brazil and in French Guiana are referred to *O. v. cariacou* (Avila Pires, 1958). The white-tailed deer of South America differs from its northern relatives in its smaller size [averages for bucks in Venezuela, according to Brokx (1972a): total length 144 cm, height at withers 82 cm, weight 50 kg], and by uncommon occurrence of the metatarsal gland (Hershkovitz 1958; Quay, 1971).

Very little is known about the biology of South American deer. The published reports deal mainly with systematics and nomenclature. Most information on the natural history of pampas and swamp deer can be traced to naturalists of the last century, particularly Rengger (1830). The appraisal of the relations between these animals and savanna ecosystems is therefore very tentative.

Food

Dietary adaptations form an important link between consumer populations and particular kinds of ecosystems. Brokx and Andressen (1970) report rumen contents of 69 white-tailed deer of the Venezuelan llanos. Leaves and twigs of *Mimosa* were the most important food item, followed by grasses and forbs, mainly *Caperonia* and *Desmodium*. Fruits of savanna trees, such as *Copernicia tectorum* and *Genipa caruto*, were also abundant. Osgood (1912) also pointed out the preference of these deer for fruit, especially *Pithecolobium*. The consumption of grasses increased and of fruits decreased from woodland savannas of Guárico to open plains of Apure (Brokx and Andressen, 1970). Escobar and Gonzalez Jimenez (1976), in a study of fecal material of various herbivores in Apure, concluded that dicotyledonous plants of gallery forest furnished the basic diet of white-tailed deer, while grasses accounted

only for 12%. Thus, this deer behaves as a browsing rather than a grazing herbivore in the llanos region, but is able to take advantage of grass and fruits as additional food items.

The pampas deer is probably the most specialized ungulate of South American savannas, but unfortunately no data have been found about its diet. The swamp deer feeds on grasses and swamp plants according to Rengger (1830), Goeldi (1893) and Krieg (1941). Nogueira Neto (1973), based on observations on confined animals, confirmed that this deer is definitively a grazer, whereas recent field studies of Hofmann et al. (1976) indicated that browsing is more important.

Behavior

In agreement with the general habit of ungulates of open habitats, the deer of South American savannas live in herds, but the group size is rather small. Rengger (1830) and Krieg (1941) gave five as maximum for swamp deer. Nogueira Neto (1973) concluded that it is only partially social, and Hofmann et al. (1976) reported solitary individuals only. The pampas deer is reported to live in pairs or groups up to ten (Rengger, 1830; Allen, 1916; Miranda Ribeiro, 1919). The white-tailed deer lives in family groups, and herds up to twenty are often observed in high-density areas on treeless plains of the llanos. Rengger (1830) pointed out that the male of swamp deer lives alone when growing new antlers and that solitary bucks are also common in pampas deer; but in general terms the available information suggests that males are more permanent in the herds of Neotropical deer than in the case of their northern allies.

The deer of South American savannas are probably sedentary. The only Neotropical ungulate which may be inclined to nomadic life is the white-lipped peccary (*Tayassu pecari*), a forest dweller. The white-tailed deer lives on rather small, permanent and overlapping home ranges (Leopold, 1959; Brokx, 1972a). Permanent water holes are important in seasonally dry habitats. The home ranges tend to be elongate, arranged side by side along rivers or around pools in such a way that each area has a drinking place at one end. The animals stay within their home ranges when partly flooded, but concentrate their activities in the higher portions of the area (Brokx, 1972a). The swamp deer uses permanent trails in tall grass, and shifts from the vicinity of ponds in the dry season to higher swamp islands during the flood (Hofmann et al., 1976).

The savanna deer avoid exposure to the heat load of noon by resting in shade. The swamp and pampas deer are regarded as crepuscular and nocturnal (Rengger, 1830; Burmeister, 1854; Goeldi, 1893; MacDonagh, 1941; Hofmann et al., 1976). The pampas deer is shy and much disturbed by human activities, but swamp deer may become quite tame and graze with domestic livestock by daytime when protected (Coimbra-Filjo, 1972). The white-tailed deer grazes in the morning and evening where protected, but becomes strictly nocturnal if harassed (Grimwood, 1969; Brokx, 1972a).

Seasonality

Alternating rainy and dry seasons form the fundamental periodicity of the savanna ecosystems. Seasonality of reproduction and related processes are therefore expected of the associated consumer populations. However, breeding, cast and growth of antlers takes place at any time of the year in the South American deer (Rengger, 1830; Krieg, 1941; Hershkovitz, 1958; Goss, 1963). Brokx (1972b) showed that does of the white-tailed deer are continuously polyestrous in the llanos region, but the main rut of adult bucks in the dry season (January to May) resulted in a peak of fawning late in the rainy season (August to November). A secondary maximum of births was evident in February and March, and attributed to sexual activity of young maturing bucks in the rainy season. According to Brokx (1972a), the bucks tend to rut earlier each year with advancing age, because the sexual and antler cycle is about eleven months (antlers worn seven months, grown four months). Does ovulate readily after parturition, and probably reproduce at intervals shorter than one year. Leopold (1959) and Brokx (1972a) reported differences in the main fawning season between regions. The apparent lack of synchronization in breeding biology is attributed therefore to accelerated reproduction and locally different selective forces. largely unknown at present.

Seasons of maximum reproduction are quoted for some populations of pampas and swamp deer

(Cabrera and Yepes, 1960), but no quantitative data exist on sexual and antler cycles. Twins are uncommon in South American deer, and unless the frequency of births is increased, as probably is the case in the white-tailed deer, the breeding potential of the tropical deer is lower than that of their northern allies.

The dry season is the critical period for deer in the llanos region. Most natural bodies of water are temporary. During the drought the wildlife rely mostly on large rivers, and drinking places arranged by the ranchers for cattle. The availability of water is probably important in determining the carrying capacity of a given area. Many deer habitats are overgrazed by cattle in the dry season, and both the quantity and the quality of most forages is low. Cover is also poor during that season, especially after fire, and the animals are more exposed to predators and hunters than during the rest of the year.

Abundance

The impact of a consumer population on the energy and nutrient dynamics of an ecosystem is proportional to the numbers present. Early data on Neotropical white-tailed deer reviewed by Brokx (1972a) indicate that they were formerly much more abundant than at present. In the same way, swamp and pampas deer were reported as common by naturalists of the last century whereas MacDonagh (1941), Coimbra-Filjo (1972), De Carvalho (1973) and Nogueira Neto (1973) have regarded them as uncommon or extinct in most areas.

Brokx (1972a) estimated the population densities of white-tailed deer on twenty ranches in the Venezuelan llanos. The levels were low, from 0.2 to 4 deer km^{-2}, in most areas, but estimates up to 12.5 were obtained for some ranches, and 30 to 50 deer km^{-2} for their most heavily stocked areas. Leopold (1959) quoted 12 to 15 deer km^{-2} for good deer ranges in Mexico. Recent estimates of Hofmann et al. (1976) for swamp deer on undisturbed habitats are from 0.5 to 0.7 deer km^{-2}. Brokx (1972a) suggested 4 to 8 deer km^{-2} (150 to 300 kg km^{-2}) as the optimum stocking rate for white-tailed deer in the llanos region. Assuming a net production rate of 30% (Leopold, 1959), the annual production would be 70 kg km^{-2} yr^{-1}, approximately. In most areas however, the population and production levels are very low and unlikely to be significant in the energy flow of savanna ecosystems.

In summary, much remains to be studied in the ecology of cervids of South America. The present review suggests that they fit quite well into the large herbivore niche of savanna ecosystems, but lack specific adaptations and seem unable to maintain levels high enough to be a major force in ecosystem dynamics. Their present role is further obscured by the impact of domestic livestock and diminished by the lack of management. They may be of Andean origin and secondarily adjusted to savanna environment as proposed by Guilday (1967). The proof that, as grazers they are only of secondary importance is the extraordinary success of introduced livestock, which invaded many South American grasslands before the settlers and often form feral populations.

THE CAPYBARA AS GRAZING MAMMAL

The capybara (*Hydrochoerus hydrochaeris*) is the last remnant of a family of large South American caviomorph rodents adapted to gramineous diet and wetland habitats (Kraglievich, 1930; Kraglievich and Parodi, 1940). It is the largest living rodent: total length 121 cm, height at withers 57 cm, weight 49 kg (averages of 100 adults from Apure, Venezuela; according to Ojasti, 1973). The maximum recorded weight is 91 kg for a specimen from southern Brazil (Mones, 1973). Widely distributed in tropical America east of the Andes (Fig. 18.1D), it was portrayed as an animal of forests and swamps along rivers by naturalists of the last century (Azara, 1802; Humboldt, 1820; Rengger, 1830; Tschudi, 1846; Darwin, 1853; Burmeister, 1854). In Venezuela it lives in many kinds of lowland habitat, such as forested riversides of Guayana, and swamps and mangrove forests of the Orinoco Delta and Lago de Maracaibo basin, but attains the highest population levels on hyperseasonal, largely treeless savannas of the llanos. These savannas include higher, unflooded and sometimes forested areas called *bancos*, wide flat plains known as *bajios* and more deeply inundated depressions, *esteros*. The distribution of large- and medium-sized mammals along this gradient is illustrated in Fig. 18.2. Most genera are closely

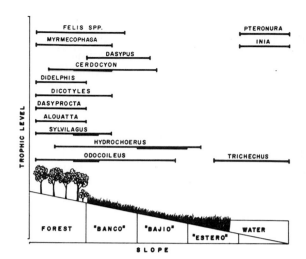

Fig. 18.2. The ranges of large to medium-sized mammals of the llanos region along the gradient from high to seasonally inundated habitats, arranged according to ascending trophic levels. The preferred habitat is indicated by the heavy line.

associated with forested habitats, whereas the white-tailed deer, crab-eating fox (*Cerdocyon thous*) and a small endemic armadillo (*Dasypus sabanicola*) live mostly on open areas, preferring bancos. The ecological range of the capybara extends from forest to estero with preference for bajios, which comprise 80% of this type of savanna (Ramia, 1967); it is thus the most important native herbivore in the region.

Adaptations

In order to be successful in the savanna environment, a herbivore needs to be adapted to utilize grass, the most abundant available food. The dentition of the capybara, including typical rodent incisors, enables it to cut the grass down to ground level, and to utilize very small plants. Its three ever-growing molars and one premolar on each side, composed of transverse ridges of dentine, bordered by enamel and separated from each other by layers of cement, and ordered in anteriorly convergent tooth rows (Fig. 18.3) assures a most efficient grinding (Ojasti, 1973). The stomach is simple, with a capacity of 2 l in adults; but the saccular cecum is very well developed, constituting three-fourths of the volume of the digestive tract (Garrod, 1876; Gonzalez Jimenez, 1978; Parra, 1978). It digests grass, including fiber, equally as well as sheep. Due

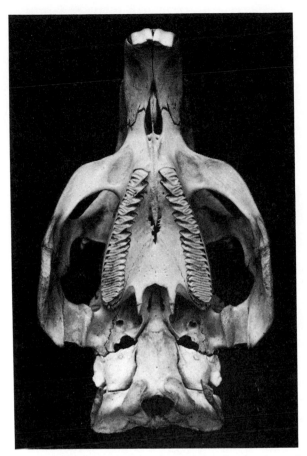

Fig. 18.3 Skull of adult capybara, ventral view showing the tooth rows.

to its higher endogenous fecal nitrogen, the ca-pybara is less efficient in retaining proteins from low-quality forage, but it digests more than sheep from diets rich in concentrates (Gonzalez Jimenez and Escobar, 1975). Its high digestive performance is attributed mainly to microbial fermentation in the caecum, but many details of the process are not yet known.

The capybara (Fig. 18.4) is a heavily built rodent with rather short limbs of cursorial type, provided with four digits on the forefeet, three on the hindfeet, with hoof-like nails. It is able to run fast when pursued, but for short distances only, 200 to 300 m. It also takes advantage of aquatic situations for protection and feeding. Its aquatic adaptations include partially webbed feet, an anal poach which covers external genitals, and the nostrils, eyes and ears located close to the dorsal plane of the head, which enables it to breathe, see and hear while

almost entirely submerged, Krumbiegel (1940) con-cluded that the adaptations of the capybara to aquatic life are moderate and restricted to external features. The size increase along the phylogenetic development of the capybara was associated with increasing complexity of tooth structure (Kraglie-vich and Parodi, 1940) and semiaquatic habits.

Individual strategies

The energy input of a herbivore population can be enhanced by appropriate selection of the plants and their parts consumed. Field observations, feed-ing trials (Ojasti, 1973, 1978) and identifications of plants consumed from fecal material (Escobar and Gonzalez Jimenez, 1976) indicate that capybara is a very selective grazer. The bulk of its diet consists of grasses of inundated savannas (*Hymenachne amplexicaulis, Leersia hexandra*) and wet bajios (*Panicum laxum*), all widely consumed by domestic livestock as well. Tiny dry-season annual grasses, such as *Paratheria prostrata* and *Reimarochloa acuta*, are important for capybara but not suitable for cattle, due to their small size. Small and medium-sized grasses of bancos (*Axonopus* sp., *Sporobolus indicus*) and sedges are also taken in the dry season when the dietary overlap with cattle is probably wider than in the rainy season (Gonzalez Jimenez, 1977). Aquatic weeds, such as *Eichhornia* are consumed in small amounts, probably as a complementary item when palatable grasses are scarce. Osgood (1912) reported the Panamian ca-pybara (*Hydrochoerus isthmius*) as feeding on sub-merged algae, but in the llanos region these rodents clip the grass to the water level and avoid eating below the surface (Ojasti, 1978). When food is plentiful the capybara eat only the topmost leaves of the preferred species, whereas during the drought they subsist on coarse bunchgrasses and other low-quality roughages.

Bartoli (1976) and Gonzalez (1977) have shown that the protein content and digestibility of impor-tant food plants of capybara (*Leersia hexandra, Panicum laxum*) decrease from the peak value early in the rainy season to a minimum in the dry season. Capybara responds to the seasonal variations of forage quality by accumulating subcutaneous fat during the rainy season, to be used as an energy source in the dry season (Ojasti, 1973). The fresh-weight caloric equivalent of adult capybara is

Fig. 18.4. Capybaras on natural habitat; an adult female with half-grown offspring.

known to vary from 1055 to 1583 cal g^{-1} (4415 to 6623 J g^{-1}), depending on the season and nutritional level (Ojasti, 1978).

The daily activities of the capybara are adjusted to the strong daily variations of temperature and radiation of open tropical habitats. Most grazing takes place in late afternoon and at night, but on cloudy days of the rainy season they are more active by daytime. During the hottest hours they rest under shade if available, or in shallow water (Ojasti, 1973; Azcárate y Bang, 1978). The daily bath is taken at noon and may last for hours. It may contribute to thermoregulation in treeless savanna habitats. The capybara is quite tame and confident by nature, but turns shy and nocturnal if persecuted.

Social behavior

In agreement with the general case of large grazers of savannas, the capybara lives normally in herds. The median size of 1535 herds recorded by Ojasti (1973) was 10.5. Solitary individuals made only 3.4% of the population. A normal herd is a closed unit composed of a dominant male, several adult females with their young and subordinate males as peripheral elements. Azcárate y Bang (1978), in his detailed field study of the behavior of this rodent, postulated that its behavioral pattern evolved largely as a response to its role as a grazer, and prey of large carnivores. The protection provided by the dominant male to others, especially to the young, is the main factor keeping the herd together. The alarm behavior is very efficient, and the preferred resting places are often located between water and cover so as to allow the herd to take to water or brush depending on the kind of predator. When pursued on forested areas or in tall grass, the animals disperse and hide themselves one by one, but on open savannas the herd sticks closely together. This behavior allows one to drive a capybara herd like cattle on open habitats, which is very important in their management and harvesting (Ojasti, 1973). On the other hand, fights for social rank in confinement often results in serious injuries or death (Cruz, 1974; Donaldson et al., 1975; Ojasti, 1978).

On areas suitable for year-long occupation, capybaras live on permanent home ranges about 10 ha in size (Ojasti, 1973; Azcárate y Bang, 1978), scent-marked with the secretion of a gland on the snout of males and of anal glands. In the dry season, however, the herds are forced to concentrate around the last remaining water holes, forming aggregations of hundreds of animals in high-density areas. At the onset of the rainy season many herds disperse again to the surrounding savannas, and live around temporary ponds until the next dry season. Thus, the capybara populations exhibit a scattered pattern from May to November and a crowded phase from December to April (see also Macdonald, 1981).

Demographic strategies

The capybara is able to breed throughout the year. However, the reproductive effort at the population level is coupled with the periodicity of the savanna ecosystem. The onset of the rains initiates a period of maximum sexual activity in May and June. After a gestation of five months (Zara, 1973; Gonzalez Jimenez, 1977), most litters are born late in the rainy season between September and November. The females may copulate again 20 to 25 days after delivery (Azcárate y Bang, 1978), but the fraction pregnant in the dry season is low, around 20% (Ojasti, 1973). More females breed in the dry season in areas where water and green pastures are plentiful, due to dikes or slow natural drying of floodplains. Two breeding cycles can be expected on optimal habitats, but one litter per year is the rule over most of the llanos. The capybara attains sexual maturity in eighteen months, and the average litter size is four (Ojasti, 1970a). The

general reproductive strategy of caviomorph rodents is a small litter of precocious young (Weir, 1974). The newborn capybara has complete dentition and pelage, and eats grass in a few days. The litter size however is larger than in most other caviomorphs. This enables the capybara populations to withstand heavy natural mortality and to grow fast when the conditions are favorable.

The mortality of new born in open habitats is high, often exceeding 50%, and represents a major factor regulating the population levels (Ojasti, 1973, 1978; Bone, 1977). Late dry season is the critical period for adults. Water and pastures become scarce and of low quality, the cover is poor, and the animals are crowded and exposed to predators, including man. Adults may lose weight, young animals grow slowly and the bulk of annual mortality of adults takes place from January to April. Accordingly, the demographic strategy of the capybara in the llanos region is to breed and grow in the rainy season and survive the drought.

Population levels

The present population levels of capybara depend largely on management. Its important natural predators, such as jaguar (*Felis onca*) and Orinoco crocodile (*Crocodylus intermedius*), are almost extinct. The habitat conditions are improved by water sources provided by ranchers for their livestock and by moderate grazing by cattle, which keeps the pasture low and stimulates palatable regrowth. As a result, on areas where capybaras are harvested on a limited scale, the present population levels are probably much higher than under unmodified conditions. On the other hand, over most of the area the capybaras are uncommon or locally extinct due to overhunting.

The maximum population density on large areas is 0.6 ind. ha^{-1}, but, when computed as a function of dry-season capybara habitat within a ranch, density estimates up to 2 ind. ha^{-1} (6000 kg km^{-2}) result (Ojasti, 1973; Cordero, 1977) on areas shared with cattle, horses and deer. Shortage of preferred grasses and increase of weeds and spiny shrubs is common in high-density areas along water courses. Stocking rates of 3 ind. ha^{-1} or more resulted in a gradual decrease of plant biomass, and inundated savanna becomes unsuitable for capybara in large enclosures, due to trampling and selective grazing.

The level of 1.6 ind. ha^{-1} was in a steady state with high plant biomass and secondary production [27 kg fresh weight ha^{-1} yr^{-1}, or 4.63 kcal m^{-2} yr^{-1} [18.98 k J m^{-2} yr^{-1})]. Consumption in this enclosure was 500 kg dry weight ha^{-1} yr^{-1} or 3.5% of the annual primary production (Ojasti, 1978). The capybara fits to the model of an efficient grazer (Noy-Meir, 1975), insensitive to food shortage until a critical threshold level, where forage becomes suddenly limiting and a population crash takes place. The condition of pastures is therefore an important guideline in the management of capybara, because the animal population is not able to adjust itself smoothly to the carrying capacity.

Management

Capybaras are hunted for cheap meat for local use, for hides which are generally exported (Carvalho, 1967; Grimwood, 1969), or as vermin to protect pastures and crops. In the Venezuelan llanos however, it is subject to market hunting for meat since colonial times (Codazzi, 1841). The dried, salted capybara meat is sold in city markets to be used for Lenten dishes, due to the tradition that the meat of this amphibious rodent may be consumed during the "meatless" period of Lent. Due to commercial cropping, subject to formal controls only, the populations decreased, and the hunting was closed in 1962 for five years. In 1968 a policy of controlled harvesting was established. Hunting licenses are granted to ranchers with an exploitable stock on their lands. The population size is estimated by game officers on each ranch where licenses are requested, and about one-third of the population is removed annually. The procedure allows a harvest of 20 000 to 40 000 animals per year, results in stable population levels, and is rewarding for the ranchers (Ojasti and Medina, 1972). The success of the controlled cropping has stimulated research on the rearing of capybaras in confinement and the industrial processing of its products (Piccinini et al., 1971; Nogueira Neto, 1973; Fuerbringer, 1974; Cruz, 1974; Gonzalez Jimenez et al., 1976). The capybara combines the rodent type of reproduction with the energetic advantages of large herbivores (Bourlière, 1975; Parra, 1978), which makes it very suitable for meat production in seasonally wet savannas.

PRESENT STATUS AND FUTURE OF NATIVE HERBIVORES

The animals discussed here are among the few big game mammals of South America and face a heavy hunting pressure. Suitable game laws now exist in most countries, but they are commonly ignored. Most hunting is done by peasants, and many rural communities rely on "bush meat" as their main source of animal protein (Mondolfi, 1965; Carvalho, 1967; Ojasti, 1970b; Castro et al., 1976). Hunting of all game is practised everywhere at any time by all possible means (Leopold, 1959; Grimwood, 1969; Avila Pires, 1975; Brokx, 1982). Law enforcement, if any, is restricted to control stations along roads for urban sport hunters. As a result, the game populations are greatly reduced. The pampas deer, widely destributed in the past, is now reduced to a few isolated relict populations (De Carvalho, 1973) and the swamp deer presents a very similar picture (Coimbra-Filjo, 1972; Avila Pires, 1975). Its total population was estimated as 800 (De Carvalho, quoted by Nogueira Neto, 1973). However, recent aerial surveys (Schaller and Vasconcelos, 1978) gave an estimate of 5000 to 6000 deer in Pantanal, Mato Grosso, Brazil. The most important game mammal of northern South America, the white-tailed deer, is also seriously overhunted and reduced in numbers, but, due to its adaptability, it has managed to persist over most of its range (Grimwood, 1969; Brokx, 1982). The capybara populations are also declining in most areas, but probably they are not endangered to the same extent as the endemic deer.

Hunting is considered the prime cause of decreased herbivore populations, but alterations of habitats may be important as well. The clearing of forests for agricultural and pastoral lands eradicates the forest-adapted species, whereas some others, such as the white-tailed deer, may take advantage of the changed habitats and secondary growth (Leopold, 1959). On the other hand, some grassland species may be otherwise disturbed by human activities, or hunted due to crop damage in agricultural districts. Foot and mouth disease, carried by the domestic animals, decimated pampas and swamp deer (Coimbra-Filjo, 1972). Savanna fire improves the pasture for deer (Cabrera and Yepez, 1960), but overgrazing by cattle is often quoted as a limiting factor for deer populations.

The future of native large mammals in South America has been considered in terms of measures to avoid extinctions, and options for long-term management. Because laws protecting wildlife cannot be enforced on a country-wide basis, national parks and other protected areas are suggested for preserving endangered species (Mondolfi, 1965; Carvalho, 1967; Coimbra-Filjo, 1972; Hofmann et al., 1976), and Nogueira Neto (1973) has proposed the rearing of indigenous vertebrates in confinement. However, he pointed out that only young pampas and swamp deer adjust to captivity, and seldom breed. Some enlightened landowners take care of native fauna on their properties, and, as stated by Avila Pires (1975), they may provide better protection than often understaffed official agencies. In summary, urgent efforts of protection and research are required to save the endangered deer from extinction.

The management of these herbivores beyond specific protected areas is more complex. Human populations grow fast in Latin America, new roads are built, more natural areas are cleared for cultivation, while some lands now in use are destroyed by erosion; pollution becomes a problem in some areas, and more city people can afford to hunt for sport. Leopold (1959) and Brokx (1982) point out the promising possibilities for management of white-tailed deer in tropical America. However, models of game management and administration in developed nations may in most South American countries be operative only on restricted areas and for urban sport hunters. Millions of peasant hunters who depend on wildlife as daily food outweigh the conservation efforts of governments and concerned citizens. This situation will probably continue until the game populations are exhausted or the economic and educational levels in rural South America are much improved.

REFERENCES

Allen, J.A., 1916. Mammals collected on the Roosevelt Brazilian expedition, with field notes of Leo E. Miller. *Bull. Am. Mus. Nat. Hist.*, 35: 559–610.

Avila Pires, F.D., 1958. Contribuição ao conhecimento dos cervideos Sul-Americanos (Ruminantia — Cervidae). *An. Acad. Bras. Cien.*, 30: 583–598.

Avila Pires, F.D., 1959. As formas Sul-Americanos do "veado virá". *An. Acad. Bras. Cienc.*, 31: 547–556.

Avila Pires, F.D., 1975. Cervidos Neotropicales: Estado actual y futuro. In: *Flora y fauna silvestre y su medio ambiente en el Continente Americano. Publ. Biol., Inst. Invest. Cientif., Univ. Auton. Nuevo Leon*, 1: 155–167.

Azara, F. de, 1802. *Apuntamientos para las historia natural de los quadrupedos del Paraguay y Rio de la Plata*. Imprenta Ibarra, Madrid, 328 pp.

Azcárate y Bang, T., 1978. *Sociobiologia del chigüire* (Hydrochoerus hydrochaeris). Dissertation, Universidad Complutense, Madrid, 154 pp.

Bartoli, D.J., 1976. *Valor nutrivito y contenido mineral en relación a parámetros ambientales y la fenología de los pastos tropicales*: Panicum laxum *Swartz y* Paspalum chaffanjonii *Maury*. Thesis, Univsidad Central de Venezuela, Caracas, 293 pp.

Bone, T.G., 1977. *Un modelo de simulación para la explotación comercial del chigüire* (Hydrochocrus hydrochaeris). Thesis, Universidad Central de Venezuela, Caracas, 86 pp.

Bourlière, F., 1973. The comparative ecology of rain forest mammals in Africa and tropical America: Some introductory remarks. In: B.J. Meggers, E.S. Ayensu and W.D. Duckworth (Editors), *Tropical Forest Ecosystems in Africa and Tropical America. A Comparative Review*. Smithsonian Institution Press, Washington, D.C., pp. 279–292.

Bourlière, F., 1975. Mammals, small and large: the ecological implications of size. In: F.B. Golley, K. Petrusewicz and L. Ryskowski (Editors), *Small Mammals, Their Productivity and Population Dynamics*. Cambridge University Press, Cambridge, pp. 1–8.

Bourlière, F. and Hadley, M., 1970. The ecology of tropical savannas. *Ann. Rev. Ecol. Syst.*, 1: 125–152.

Brokx, P.A., 1972a. *A Study of the Biology of the Venezuelan White-Tailed Deer* (Odocoileus virginianus gymnotis *Wiegmann 1833*) *with a Hypothesis on the Origin of South American Cervids*. Dissertation, University of Waterloo, Waterloo, Ont., 355 pp.

Brokx, P.A., 1972b. Ovarian composition and aspects of the reproductive physiology of Venezuelan white-tailed deer (*Odocoileus virginianus gymnotis*). *J. Mammal.*, 53: 760–773.

Brokx, P.A., 1982. White-tailed deer of South America. In: L.K. Halls (Editor), *Ecology and Management of the White-Tailed Deer*. Stackpole Company, Harrisburg, Penn., in press.

Brokx, P.A., and Andressen, F.M., 1970. Analisis estomacales del venado caramerudo de los Llanos Venezolanos. *Bol. Soc. Venez. Cienc. Nat.*, 28: 330–353.

Burmeister, H., 1854. *Systematische Übersicht der Thiere Brasiliens. I Teil, Säugethiere*. G. Reimer-Verlag, Berlin, 341 pp.

Cabrera, A., 1961. Catalogo de los mamiferos de America del Sur. Parte II. *Rev. Mus. Argent. Cienc. Nat. "Bernardino Rivadavia", Cienc. Zool.*, 4: 306–732.

Cabrera, A. and Yepes, J., 1960. *Mamiferos Sud Americanos. 11*. Ediar, Buenos Aires, 160 pp.

Carvalho, J.C., 1967. A conservação da natureza e recursos naturais na Amazonia Brasileira. In: H. Lent (Editor), *Atas do Simposio sobre a biota Amazonica*, 7. Conselho Nacional de Pesquisas, Rio de Janeiro, pp. 1–47.

Castro, N., Revilla, J. and Neville, M., 1976. Carne de monte como una fuente de proteinas en Iquitos, con referencia especial a monos. *Rev. For. Peru*, 5: 19–32.

Codazzi, A., 1841. *Resumen de la geografia de Venezuela*. H. Fournier and Co., Paris, 648 pp.

Coimbra-Filjo, A.F., 1972. *Mamiferos ameacados de extincão no Brasil*. Academia Brasileira de Ciencieas, Rio de Janeiro, 98 pp.

Cordero, G.A., 1977. *Estudio comparativo de poblaciones de chigüire* (Hydrochoerus hydrochaeris) *de sabana y bosque del Llano*. Thesis, Universidad Central de Venezuela, Caracas, 64 pp.

Cruz, C.A., 1974. Notas sobre el comportamiento del chigüiro en confinamiento. In: *Primer Seminario sobre Chigüiros y Babillas*. Inderena, Bogota, 1–45 pp.

Darwin, C., 1853. *Journal of Researches into the Natural History and Geology of the Countries Visited during the Voyage of H.M.S Beagle Round the World under Command of Capt. Fitzroy*. R.N. Harper and Brothers, New York, N.Y., 351 pp.

De Carvalho, C.T., 1973. *O veado campeiro — situação e distribuição* (Mammalia, Cervidae). Boletim Tecnico No. 7, Secretaria de Estado dos Negrocios da Agricultura, São Paulo, pp. 9–26.

Donaldson, S.L., Wirtz, T.B. and Hite, A.E., 1975. The social behavior of capybaras (*Hydrochoerus hydrochaeris*). *Int. Zoo Yearb.*, 15: 201–206.

Dubost, G., 1968. Les niches écologiques des forets tropicales Sud-Americaines et Africaines, sources de convergences remarquables entre Rongeurs et Artiodactyles. *Terre Vie*, 22: 3–28.

Escobar, A. and Gonzalez Jimenez, E., 1976. Estudio de la competancia alimenticia de los herbivoros mayores del Llano inundable con referencia especial al chigüire (*Hydrochoerus hydrochaeris*). *Agron. Trop. (Maracay)*, 26: 215–227.

Fittkau, E.J., 1969. The fauna of South America. In: E.J. Fittkau, J. Illies, H. Klinge, G.H. Schwabe and H. Sioli (Editors), *Biogeography and Ecology in South America*. W. Junk, The Hague, pp. 624–658.

Fuerbringer, J., 1974. El chigüiro, su cría y explotación racional. *Orient. Agropecu. (Bogotá)*, 99: 5–59.

Garrod, A.H., 1876. On the caecum coli of the capybara (*Hydrochaerus capybara*). *Proc. Zool. Soc. Lond.*, 1874: 20–23.

Goeldi, E.A., 1893. *Os mamiferos do Brasil*. Alves & Cia., Rio de Janeiro, 182 pp.

Golley, F.B., 1873. Impact of small mammals on primary production. In: J.A. Gessaman (Editor), *Ecological Energetics of Homeotherms*. Utah State University Press, Logan, Utah, 142–147 pp.

Gonzalez, J.A., 1977. *Valor nutritivo y contenido mineral en relación a parámetros ambientales y la fenologia de dos pastos tropicales*: Leersia hexandra (*Swartz*) *y* Axonopus purpusii (*Mez*) Chase. Thesis, Universidad Central de Venezuela, Caracas, 177 pp.

Gonzalez Jimenez, E., 1977. El capibara — una fuente indígena de carne de la America tropical. *Rev. Mundial Zootecn.*, 21: 24–30.

Gonzalez Jimenez, E., 1978. Digestive physiology and feeding of capybara (*Hydrochoerus hydrochaeris*). In: M. Rechcigl (Editor), *Handbook Series in Nutrition and Food, G, 1: Diets*

for Mammals. CRC Press, Cleveland, Ohio, pp. 163–177.

Gonzalez Jimenez, E. and Escobar, A., 1975. Digestibilidad comparada entre chigüires (*Hydrochoerus hydrochaeris*), conejos y ovinos, con raciones de diferentes proporciones de forraje y concentrados. *Agron. Trop. (Maracay)*, 25: 283–290.

Gonzalez Jimenez, E., Cerda, J., Bello, A. and Marin, P., 1976. *II Seminario sobre chiguires y babas*. Conicit, Maracay 55 pp. (mimeographed).

Goss, R.J., 1963. The deciduous nature of deer antlers. In: *Mechanisms of Hard Tissue Destruction*. Publication No. 75, American Association for Advancement of Science, Washington, D.C., pp. 339–369.

Grimwood, I.R., 1969. *Notes on the Distribution and Status of Some Peruvian Mammals 1968*. Special Publication No. 21, American Committee for International Wild Life Protection, New York, N.Y., 86 pp.

Guilday, J.E., 1967. Differential extinction during late Pleistocene and recent times. In: P.S. Martin and H.F. Wright (Editors), *Pleistocene Extinctions. The Search for a Cause*. Yale University Press, New Haven, Conn., 121–140 pp.

Haffer, J., 1969. Speciation in Amazonian forest birds. *Science*, 165: 131–137.

Hershkovitz, P., 1958. The metatarsal glands in white tailed deer and related forms of the Neotropical region. *Mammalia*, 22: 537–546.

Hofman, R.K., Ponce del Prado, C.F. and Otte, K.C., 1976. Registrato de dos nuevas especies de mamíferos para el Peru, *Odocoileus dichotomus* (Illiger-1811) y *Chrysocyon brachyurus* (Illiger-1811), con notas sobre su habitat. *Rev. For. Peru*, 5: 61–81.

Humboldt, A. de, 1820. *Voyages aux régions équinoxiales du Nouveau Continent fait en 1799, 1800, 1801, 1802, 1803 et 1804 par Al. de Humboldt et A. Bonpland. VI*. N. Maze, Paris, 318 pp.

Keast, A., 1972. The evolution of mammals on Southern Continents VII. Comparisons of the contemporary mammalian faunas of the Southern Continents. In: A. Keast, F.C. Erk and B. Glass (Editors), *Evolution, Mammals and Southern Continents*. State University of New York, Albany, N.Y., pp. 433–501.

Kraglievich, L., 1930. Los más grandes carpinchos actuales v fósiles de la subfamilia Hydrochoerinae. *An. Soc. Cientí. Argent.*, 110: 233–250; 340–358.

Kraglievich, L. and Parodi, L.J., 1940. Morfologia normal y morfogénesis de los molares de los carpinchos y caracteres filogenéticos de este grupo de roedores. In: *Obras completas y trabajos cientificos inéditos de Lukas Kraglievich*. Buenos Aires, 3: 439–455.

Krieg, H., 1941. Im Lande des Mähnenwolfes. Fortsetzung: Katzen, Kugelgürteltier, Sumpfhirsch. *Zool. Garten*, 13: 333–355.

Krumbiegel, I., 1940. Die Säugetiere der Südamerika-Expeditionen Prof. Dr. Kriegs. 6. Wasserschweine und Viscaciidae. *Zool. Anz.*, 132: 97–115.

Leopold, A.S., 1959. *Wildlife of Mexico. The Game Birds and Mammals*. California University Press, Berkeley, Calif., 568 pp.

MacDonagh, E.J., 1941. Etología del venado en el Tuyú. *Not. Mus. La Plata*, 5: 49–68.

Macdonald, W., 1981. Dwindling resources and the social behaviour of capybaras (*Hydrochoerus hydrochaeris*). *J. Zool., Lond.*, 194: 371–391.

Martin, P.S., 1967. Pleistocene overkill. In: P.S. Martin and H.F. Wright (Editors), *Pleistocene Extinctions. The Search for a Cause*. Yale University Press, New Haven, Conn., pp. 75–120.

Martin, P.S. and Wright, H.E., 1967. *Pleistocene Extinctions. The Search for a Cause*. Yale University Press, New Haven, Conn., 453 pp.

Miranda Ribeiro, A. de, 1919. Os veados do Brazil, secundo as collecões Rondon e de varios museus nacionales e extrangeiros. *Rev. Mus. Paulista*: 11: 1–99.

Mondolfi, E., 1965. Nuestra Fauna. *El Farol (Caracas)*. No. 214: 2–13.

Mones, A., 1973. Estudios sobre la familia Hydrochoeridae (Rodentia). I. Introducción e historia taxonomica. *Rev. Bras. Biol.*, 33: 277–283.

Müller, P., 1973. *The Dispersal Centers of Terrestrial Vertebrates in the Neotropical Realm*. W. Junk, The Hague, 244 pp.

Nogueira Neto, P., 1973. *A criação de animais indigenas vertebrados (peixes, anfibios, repeteis, aves, mamiferos)*. Tecnapis, São Paulo, 327 pp.

Noy-Meir, I., 1975. Stability of grazing systems: An application of predator–prey graphs. *J. Ecol.*, 63. 459–481.

Odum, E.P., 1971. *Fundamentals of Ecology*. W.B. Saunders, Philadelphia, Penn., 3rd ed., 574 pp.

Ojasti, J., 1970a. Datos sobre la reproducción del chigüire (*Hydrochoerus hydrochaeris*). *Acta Cientif. Venez.*, 21 (Supl. 1): 27 (abstract).

Ojasti, J., 1970b. La fauna silvestre produce. In: *La ciencia en Venezuela en 1970*. Universidad de Carabobo, Valencia, pp. 277–294.

Ojasti, J., 1973. *Estudio biológico del chigüire o capibara*. Fondo Nacional de Investigaciones Agropecuarias, Caracas, 275 pp.

Ojasti, J., 1978. *The Relation between Population and Production of the Capybara* (Hydrochoerus hydrochaeris). Dissertation, University of Georgia, Athens, Ga., 204 pp.

Ojasti, J. and Medina, P.G., 1972. The management of capybara in Venezuela. *Trans. N. Am. Wildl. Nat. Resour. Conf.*, 37: 268–277.

Osgood, W.H., 1912., Mammals from western Venezuela and eastern Colombia. *Field Mus. Nat. Hist., Zool. Ser.*, 10: 31–62.

Parra, R., 1978. Comparison of foregut and hindgut fermentation in herbivores. In: G.G. Montgomery (Editor), *The Ecology of Arboreal Folivores*. Smithsonian Institution Press, Washington, D.C., pp. 205–229.

Patterson, B. and Pascual, R., 1972. The fossil mammals of South America. In: A. Keast, F.C. Erk and B. Glass (Editors), *Evolution, Mammals and Southern Continents*. State University of New York, Albany, N.Y., pp. 247–309.

Petersen, J.C.B. and Casebeer, R.L., 1971. A bibliography relating to the ecology and energetics of East African large mammals. *E. Afr. Wildl. J.*, 9: 1–23.

Petrides, G.A., 1975. The importation of wild ungulates into Latin America, with remarks on their environmental effects. *Environ. Conserv.*, 2: 47–52.

Petrusewicz, K. and Grodzinski, W.L., 1975. The role of herbivore consumers in various ecosystems. In: *Productivity of World Ecosystems*. National Academy of Sciences, Washington, D.C., pp. 64–70.

Piccinini, R.S., Vale, W.G. and Wagner Gomes, F., 1971. *Criadouros artificiais de animais silvestres. I. Criadouro de capivaras*. Ministerio do Interior, Belem, 33 pp.

Quay, W.B., 1971. Geographic variation in the metatarsal "gland" of the white-tailed deer (*Odocoileus virginianus*). *J. Mammal.*, 52: 1–11.

Ramia, M., 1967. Tipos de sabanas en los Llanos de Venezuela. *Bol. Soc. Venez. Cienc. Nat.*, 28: 264–288.

Raven, P.H. and Axelrod, D.I., 1975. History of the flora and fauna of Latin America. *Am. Sci.*, 63: 420–429.

Rengger, J.R., 1830. *Naturgeschichte der Säugethiere von Paraguay*. Schweighauserschen Buchhandlung, Basel, 394 pp.

Sarmiento, G. and Monasterio, M., 1975. A critical consideration of the environmental conditions associated with the occurrence of savanna ecosystems in tropical America. In: F.B. Golley and E. Medina (Editors), *Tropical Ecological Systems. Trends in Terrestrial and Aquatic Research*. Springer-Verlag, Berlin, pp. 223–250.

Schaller, G.B. and Vasconcelos, J.M.C., 1978. A marsh deer census in Brazil. *Oryx*, 14: 345–351.

Sick, H., 1965. A fauna do Cerrado. *Arq. Zool.*, 12: 71–93.

Simpson, G.G., 1950. History of the fauna of Latin America. *Am. Sci.*, 38: 361–389.

Tschudi, J.J., 1846. *Untersuchungen über die Fauna Peruana*. St. Gallen, 262 pp.

Webb, S.D., 1978. A history of savanna vertebrates in the New World. Part II: South America and the great interchange. *Annu. Rev. Ecol. Syst.*, 9: 393–426.

Weir, B.J., 1974. Reproductive characteristics of hystricomorph rodents. In: I.W. Rowlands and B.J. Weir (Editors), *The Biology of Hystricomorph Rodents. Symp. Zool. Soc. Lond.*, 34: 265–301.

Wetzel, R.M., Dubos, R.E., Martin, R.L. and Myers, P., 1975. *Catagonus*, and "extinct" peccary alive in Paraguay. *Science*, 189: 379–381.

Wiegert, R.G. and Evans, F.C., 1967. Investigations of secondary productivity in grasslands. In: K. Petrusewicz (Editor), *Secondary Productivity of Terrestrial Ecosystems, II*. Polish Academy of Science, Warsaw, pp. 499–518.

Wiener, J.G., 1975. Nutrient cycles, nutrient limitation and vertebrate populations. *Biologist*, 57: 104–124.

Zara, J.L., 1973. Breeding and husbandry of the capybara *Hydrochoerus hydrochaeris*. *Int. Zoo Yearb.*, 13: 137–139.

Chapter 19

THE GRAZING AUSTRALIAN MARSUPIALS

A.E. NEWSOME

INTRODUCTION

Tropical Australia has one of the richest verte-brate faunas in the continent. Cape York with its rich mixture of rain forests and eucalypt savanna holds the richest assemblage. Arnhem Land, whose vegetation is basically savanna, is the next richest (J. Calaby, pers. comm., 1979) though its fauna is not yet well known. Settlement in the region remains sparse and there have been few scientific studies of any depth until recently. The American–Australian Scientific Expedition to Arnhem Land in 1948 (Specht, 1964) was the first of the modern surveys, followed by others beginning in 1961 (Frith and Calaby, 1974; J. Calaby, unpubl.). Scientific enquiry has intensified considerably in the past decade, and this chapter reports some of that progress.

Only one native species has been studied ecologically, the agile or sandy wallaby, *Macropus agilis* (Bolton, 1974; Bolton et al., in press); growth and reproduction have been studied in captivity also (Kirkpatrick and Johnson, 1969; Merchant, 1976; Dudzinski et al., 1978). Knowledge of other native species or groups is scattered. The taxonomy of the euros (*Macropus* spp.) and rock wallabies (*Petrogale* spp.) in particular have been studied through blood proteins and chromosomes (Richardson, 1970; Richardson and Sharman, 1976; G. Sharman, D. Briscoe and G. Maynes, pers. comm., in Poole, 1978), and there are some notes on the biology of the former group (Russell and Richardson, 1971). Some of the commoner small ground mammals (Dasyuridae, Rodentia) have been studied for their reproductive biology (J. Calaby and M. Taylor, unpubl.) and one rodent, *Rattus sordidus colletti*, for its ecology (Redhead,

1979, and unpubl.). The biology and ecology of the two large desert macropodids further inland, the euro or hill kangaroo (*M. robustus*) and the red or plains kangaroo (*M. rufus*) have been studied (summarized by Newsome, 1975). The introduced ruminants in the tropics, cattle (*Bos taurus* and *B. indicus*), sheep (*Ovis aries*) and water buffalo (*Bubalus bubalis*), are raised extensively for meat and wool. Their study has been intensifying (e.g., Andrews, 1976; Tulloch, 1968, 1969, 1970). European rabbits form high populations in parts of the subtropics (see p. 446) but extend north to around latitude 18°S. Though studied intensively in temperate Australia (see Myers, 1970), not even their distribution in the tropics is established (see Fig. 19.6).

This chapter has two major parts, one biogeographical and the other ecological, and uses knowledge of these species for comparisons. The Great Dividing Range in the east with its rain forest and wet sclerophyll forests is excluded. Most information exists for the Northern Territory, the central third of the tropical region, the area which I know best.

BIOGEOGRAPHY

Vegetation and climate

The term "savanna" is not used in modern classifications of Australian vegetation (Moore and Perry, 1970; Specht et al., 1974). Rather, vegetation is classified on the structure, height, cover, etc., of its major elements. One must go back to the early descriptions of Prescott (1931) to find the term, but it does not comply with the definition of savanna as used for this book. Instead, Prescott's savanna is extensive hummock grassland (*Astrebla* spp.) mapped as such by Moore and Perry (1970) and in this chapter (Fig. 19.1); in subtropical Queensland, however, there are large tracts of savanna as defined here (R. Specht, pers. comm., 1979).

North of these grasslands are woodlands extending to the coast (Fig. 19.1). Moore and Perry identified three kinds of woodland, which have been reduced to two in Fig. 19.1. Their tropical subhumid woodlands coincide with Prescott's savanna woodlands, the descriptive term retained here (Fig. 19.1). And their semi-arid shrub woodlands and semi-arid low woodlands further inland are termed woodland savannas here (Fig. 19.1); Prescott saw them as a mixture of savanna (i.e. grassland) and woodland savanna. South of the grasslands lie spinifex (*Triodia* spp.) deserts (dunes and plains), with extensive *Acacia* shrublands 10–15 m tall (Moore and Perry, 1970) surrounding the central Australian mountain ranges, and the range country near the northwest coast in Western Australia.

The dominant vegetation changes accordingly. In the savanna woodlands major trees are *Eucalyptus tetrodonta* and *E. miniata* (Moore and Perry, 1970) with an understorey basically of annual grasses (*Sorghum* spp.). In the woodland savanna, some other trees predominate in various places, e.g., the bean tree (*Bauhinia cunninghamii*), bloodwood (*Eucalyptus terminalis*), and, in Queensland, brigalow (*Acacia harpophylla*). These broad concentric bands of vegetation across northern Australia follow the patterns of rainfall. The monsoonal savanna woodlands (of latitude 12–14°S) have an annual rainfall of 1400 to 750 mm, the sub-monsoonal woodland savanna (14–22.5°S) has 750 to 500 mm, and the extensive grasslands further inland (18–22.5°), 500 to 375 mm. Rainfall in the deserts and *Acacia* shrubs (18–32°) is only an erratic 300 to 100 mm.

There is a distinct monsoonal season near the coast where the rains are heaviest, falling between November and March. During the rest of the year, water recedes to permanent lagoons, swamps and watercourses, pastures dry out and green herbage

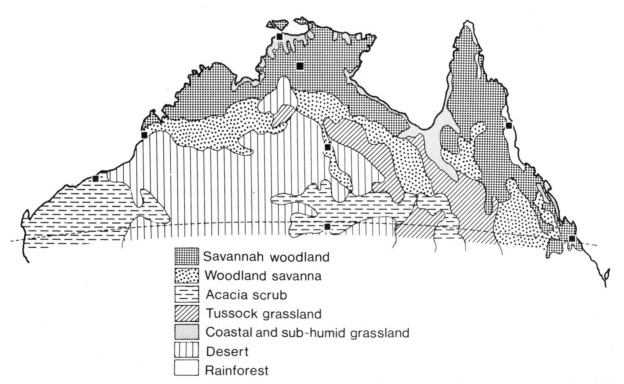

Savannah woodland
Woodland savanna
Acacia scrub
Tussock grassland
Coastal and sub-humid grassland
Desert
Rainforest

Fig. 19.1. Broad vegetation map of tropical Australia (after Moore and Perry, 1970). Major towns and cities shown here are subsequently named in Fig. 19.2.

becomes scarce. Extensive grass fires break out throughout the savanna woodlands, maintaining the understorey of grass with few shrubs (see Catling and Newsome, 1981). Further inland, in the sub-monsoonal woodland savannas and grassy plains, the rainfall lessens and becomes variable, a trend maintained into the desert. Mean maximum temperatures in summer (January) vary little throughout tropical Australia, being around 32°C at the northern coast and 34°C near the Tropic of Capricorn. Respective mean minima in winter are 20°C and 5°C (Bureau of Meteorology, 1956).

The mean number of months when rainfall causes pasture growth is more important to the biota than the actual rainfall. Average periods exceed five months in the savanna woodlands, range from three to five in woodland savanna, one to two in the grassland, and less than one in the semi-arid and arid regions (Nix, 1976). Nix has combined the essentially non-linear responses of plants to regimes of light, heat and moisture to form a linear scale from zero to one called a Growth Index. Growth Indices for the summer quarter (December–February) are mapped in Fig. 19.2 showing, for instance, that pastures reach 0.6 to 0.8 of their maximum potential growth in the savanna woodland, but only 0.1 to 0.2 in the drier regions. Isoclines of 0.6, 0.4 and 0.2 approximate the southern boundaries of the savanna woodlands, woodland savanna and grassland. Isoclines for the winter quarter (June–August) move a little fur-

ther south except as marked on Fig. 19.2; the greatest shift is for the value 0.1 in the desert.

Species, speciation and distribution

Kangaroos and wallabies

Kangaroos and wallabies, so rare in the tropical grasslands, are common and even abundant in species as well as numbers further north in the savanna woodlands, with gradations in between. A total of fourteen species with several subspecies are recorded. Their distributions including outlying records are shown in Figs. 19.3 to 19.5.

The similarity of the band pattern in these distribution maps to those for the vegetation and Growth Indices is striking. The grey kangaroo (*Macropus giganteus*), euros and wallaroos[1] (*M. robustus*), the antilopine kangaroo (*M. antilopinus*; Fig. 19.3), the agile wallaby (*M. agilis*; Fig. 19.4), and the rock wallabies (*Petrogale penicillata, P. brachyotis, P. concinna, P. burbidgei* and a new species; Fig. 19.5) inhabit regions included in the savanna woodland. The specific habitats of the euro, wallaroo and rock wallabies are mountain ranges and other rocky eminences. *Macropus robustus*, a euro found throughout the Australian

---

[1] These terms are almost synonymous but not really interchangeable. The term "wallaroo" is restricted to the large dark form along the Great Dividing Range and the east, and "euro", to other forms except the antilopine kangaroo.

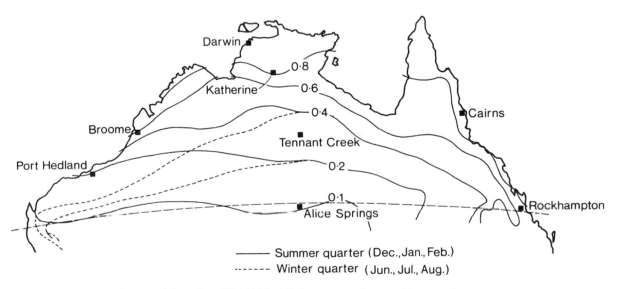

Fig. 19.2. Isoclines for Growth Indices (from Nix, 1976). All place names given on this map only.

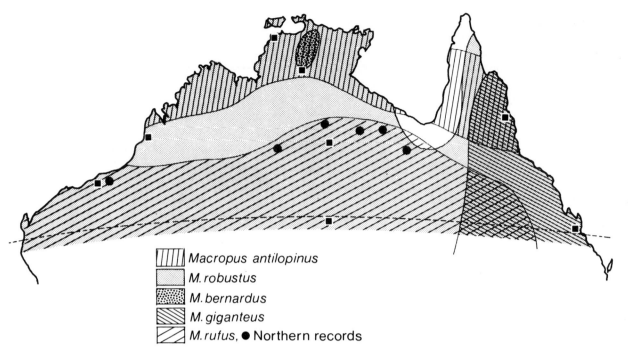

Fig. 19.3. Distribution of large kangaroos (after Richardson and Sharman, 1976; Poole, 1978; J. Calaby, pers. comm., 1979; and A. Newsome, unpubl.).

mainland, has its four subspecies represented in tropical Australia (Poole, 1978) (Fig. 19.3). The spectacled hare wallaby (*Lagorchestes conspicillatus*), and the northern nail-tailed wallaby (*Onychogalea unguifera*; Fig. 19.4) inhabit the woodland savanna, and the red kangaroo (*Macropus rufus*; Fig. 19.3), the arid and semi-arid *Acacia* shrublands and associated grasslands further inland. Distributions overlap. Indeed, some species penetrate well beyond their accepted prime habitats — for instance *M. agilis* and *Lagorchestes conspicillatus* (Fig. 19.4).

The astonishing and unexplained thing is the speciation within the monsoonal tropics. There are three species of euro varying from the gracile forest form, *Macropus antilopinus*, to the robust mountain forms. *M. bernardus* and *M. robustus* with four subspecies. There are also six species of rock wallaby (the five above plus *Petrogale rothschildi*) with nineteen subspecies (G. Sharman, D. Briscoe and G. Maynes, pers. comm., in Poole, 1978) (Fig. 19.5). Elsewhere in Australia, there is only one distinct species from these groups, *P. xanthopus*; *P. penicillata* has three subspecies there, and *Macropus robustus* two (Poole, 1978). There is

another species of grey kangaroo in temperate Australia, *M. fuliginosus* with three subspecies, and *M. g. giganteus* is also there (Poole, 1978). It is not easy to see why the tropics should be so rich taxonomically, and explanations fall back on the environmental heterogeneity which is a tautology.

A few species not typical of the tropical region extend into it along the eastern seaboard (Fig. 19.4). One of the most attractive of the wallabies, the whiptail or pretty face (*M. parryi*) follows the eastern slopes of the Great Dividing Range north well into the tropics to about latitude 15°S. The eastern nail-tail (*Onychogalea fraenata*) is now extinct over the greater part of its range, which once encompassed the woodlands of the western slopes of the Divide north from the Murray–Darling River basins around latitude 33°S. It has recently been rediscovered (G. Gordon, in Poole, 1978), locally only, in the extreme north of its range (P. Johnston, pers. comm., 1979). Then there is the northern rat kangaroo (*Bettongia tropica*) known from two isolated pockets in the east. A wallaby fitting the description of a rat kangaroo was clearly seen about 70 km southeast of Darwin on the sub-coastal plains of Arnhem Land late one evening in

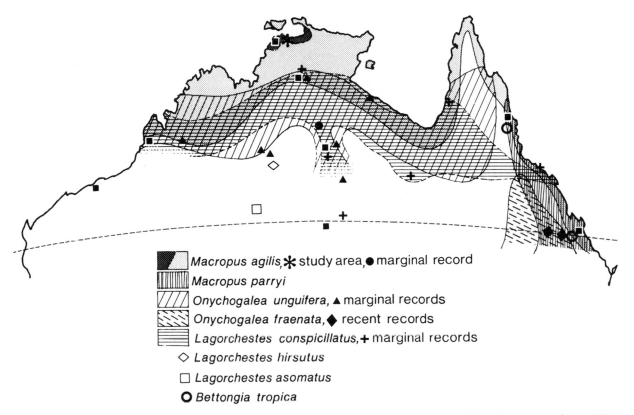

Fig. 19.4. Distributions of wallabies. The extinct *Onychogalea lunata* and *Bettongia lesueur* from arid regions are not shown. (From J. Calaby and P. Johnston, pers. comm., 1979; Bolton et al., in press; and Newsome, 1971, and unpubl.).

August, 1958 (A. Newsome, G. Letts and J. Fawcett, unpubl.).

A highlight of the state of knowledge is the recent, classical study of chromosomal morphology and of blood and tissue proteins in the tropical rock wallabies (*Petrogale* spp.) by G. Sharman, D. Briscoe and G. Maynes (pers. comm. in Poole, 1978). It reduces the described species from eight (Ride, 1970) to six, fuses two genera, *Petrogale* and *Peradorcas*, accepts a recently discovered species, *Petrogale burbidgei*, and establishes another species as yet undescribed (see Fig. 19.5). And an indication of the species' sympatry can be gleaned from B. Richardson's (pers. comm.) observation at Nourlangie Rock in Arnhem Land of euros and rock wallabies (*Macropus bernardus* and *Petrogale brachyotis*) on the flank of the Rock grazing down to the edge of the surrounding plain, where there were antilopine wallaroos (*Macropus antilopinus*) and agile wallabies (*M. agilis*). All drank at the same waterhole near the Rock.

Introduced herbivores

Some feral mammals demonstrate the same basic pattern of distribution as the macropodids — for instance, the water buffalo (*Bubalus bubalis*) in the monsoonal tropics, and the rabbit *Oryctolagus cuniculus* in the arid and semi-arid regions (Fig. 19.6). The latter species has limits close to the Gulf of Carpentaria at the edge of the Eyrean sub-zone (see p. 446). These limits are about 400 km further north than those proposed by Myers (1970). Warrens are found along watercourses and in limestone country in the mountain ranges, and similarly in pockets in the desert country wherever the soil is suitable, and even on the Barkly Table-lands (the hummock grasslands of Fig. 19.1). The outlying record from the southeast of the Gulf of Carpentaria (Ratcliffe and Calaby, 1958) could not be substantiated in 1963 (F. Sweeney, A. Hockey, in litt.) and is probably erroneous (J. Calaby, pers. comm., 1979). The outliers found recently near the northeast coast (Myers, 1970) may have been

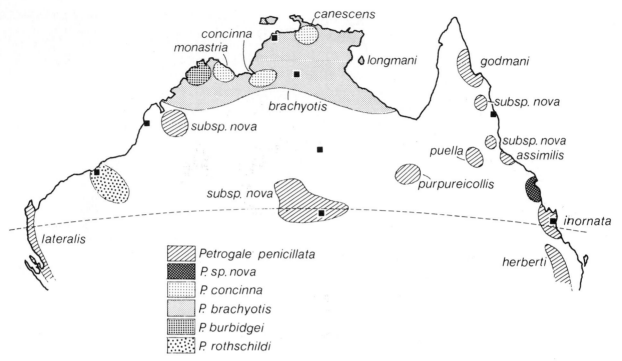

Fig. 19.5. Distributions of rock wallabies. Hatchings are full species, and names are sub-species of *Petrogale penicillata* and *P. brachyotis.* (From G. Sharman, D. Briscoe and G. Maynes, pers. comm., in Poole, 1978; and J. Calaby, pers. comm., 1979.)

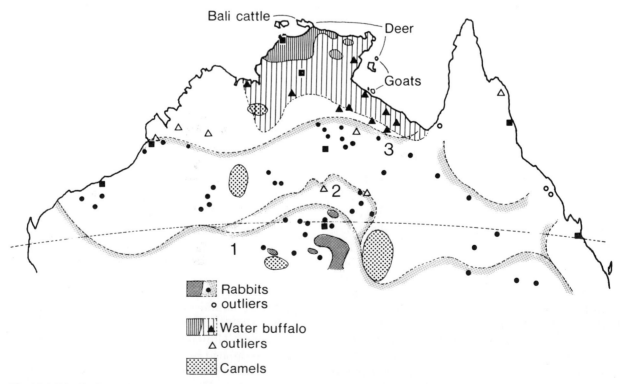

Fig. 19.6. Distributions of feral herbivores. Numbered lines represent concepts of the northern limits to distribution of rabbits: *1*, Myers (1970); *2*, Letts (1979); *3*, Newsome (unpubl.). Other species after Letts (1979) and Long (1972).

enabled through the gross clearing of the brigalow (*Acacia harpophylla*) savanna over the past decade or so for cattle grazing.

Buffaloes are most abundant on the sub-coastal plains of Arnhem Land (Fig. 19.6), but are scattered throughout the savanna woodlands to the southeast corner of the Gulf of Carpentaria (Letts, 1979). There are occasional records well outside this range, mostly of wandering individuals, or small groups. The most southerly record, well into semi-arid central Australia, came during the amazing run of wet years experienced there since 1967. Basically, the buffaloes and rabbits extend to the accepted limits of the Torresian and Eyrean sub-zones (see below).

Two other feral animals, banteng cattle (*Bos javanicus*) from Bali originally (Calaby, 1975), and the camel (*Camelus dromedarius*) need brief mention. The latter is thinly spread through the desert regions, having been cleared from the *Acacia* scrublands; its northern limits are similar to those for the rabbit. The banteng is restricted to Cobourg Peninsula, Arnhem Land, where it was liberated about 150 years ago (Letts, 1964). The reasons for so limited a distribution are unclear; but the animals are commonest near large plains of *Fimbristylis rara* which grows there (Chippendale, 1974) and nowhere else so well in the sub-coastal region. Buffalo, though released there also, are rare. Other releases were pigs (*Sus scrofa*), now widespread through the Torresian sub-zone especially coastally, and Timor ponies (*Equus caballus*), still restricted to Cobourg Peninsula like the bantengs.

Beef cattle, mostly *Bos taurus*, with some *B. indicus* and their crosses, are raised on all land except the desert, and the hummock grasslands in Queensland which is used for sheep, (*Ovis aries*; Fig. 19.7). Heat infertility is a problem in rams at these low latitudes (Entwistle, 1974), and predation from dingoes (*Canis familiaris dingo*) is another limiting factor. Their distribution essentially determines the distribution of sheep and cattle in the tropics, the area under sheep being surrounded by a fence proof against dingoes.

Faunal sub-zones and the Carpentaria Salient

Native mammals

Spencer (1896) first classified Australia into three basic biogeographic sub-zones; the Torresian (northern tropical); the Bassian (southern temperate); and the Eyrean (desert) regions. Modern knowledge and techniques have improved knowledge of the distributions of species and their analysis. For example, Kikkawa and Pearse (1969) numerically divided the distribution of land birds into faunal areas, provinces and sub-regions. They showed a

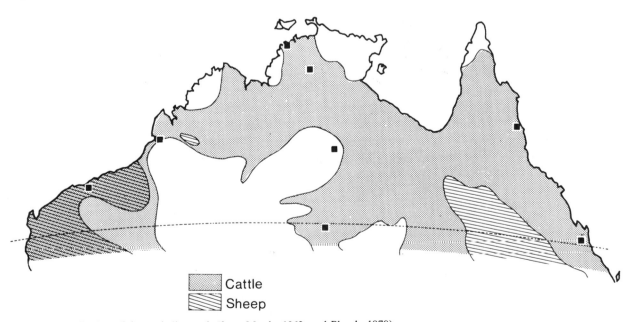

Cattle
Sheep

Fig. 19.7. Distribution of domestic livestock (from Moule, 1962; and Plumb, 1970).

very strong east–west division separating the area from the Great Dividing Range to the coast from the rest of Australia (Fig. 19.8). There was also quite a strong north–south division roughly along latitude 20°S. To the north of this line was a weaker north–south division, separating the Kimberleys, Arnhem Land and the Gulf of Carpentaria (northern coastal) from the rest (northern interior). This overall analysis of birds remained consistent, however, with Spencer's initial concept of Torresian, Bassian and Eyrean regions.

Similarly, Schodde and Calaby (1972) examined Australo-Papuan bird and mammal faunas across Torres Strait, and demonstrated examples of two enclosed biogeographical provinces in northeastern Australia, within the Torresian sub-zone of Spencer and the eastern division of Kikkawa and Pearse (Fig. 19.8). There were two small areas of lowland and hill biota near the tip of Cape York attributed to a biogeographical unit named by these authors the Irian Division, which is the commonest in New Guinea. The biota of an extensive strip of mountain rain forest (>1200 m altitude) further south was named the Tumbunan Division (the name indicates

the *ancestral* nature of the groups).

Macdonald (1969) introduced a third idea of a barrier to east–west movement across the Gulf of Carpentaria. A related concept is presented here, of a salient of Eyrean (arid) fauna at the Gulf which separates some elements of the Torresian fauna of the Kimberleys and Arnhem Land from that of Cape York (Fig. 19.8). Examples here come from the Macropodidae. *Macropus giganteus* is on Cape York but does not cross to the western side of the Gulf of Carpentaria (Fig. 19.3). *Petrogale brachyotis* is on the west of the salient but does not cross it in an easterly direction to Cape York (Fig. 19.5). *Macropus antilopinus* (Fig. 19.3) is on either side of the Gulf but has not been recorded in its southeastern corner. And the form of euro in the lower McArthur River area (Fig. 19.8), about 50 km from the coast in the central western part of the Gulf, is not a Torresian but the Eyrean form, *M. robustus erubescens* (Fig. 19.3) (J. Calaby, pers. comm., 1979). The identity of the latter form and also of *Petrogale brachyotis* was established on morphology, chromosomes and biochemistry. Also, the Eyrean *Macropus rufus* has been seen by

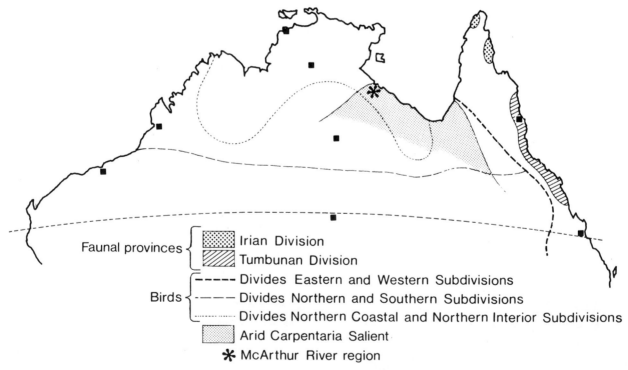

Faunal provinces { ▦ Irian Division
                   ▨ Tumbunan Division

Birds { ----- Divides Eastern and Western Subdivisions
        --- Divides Northern and Southern Subdivisions
        ········· Divides Northern Coastal and Northern Interior Subdivisions

▨ Arid Carpentaria Salient
✳ McArthur River region

Fig. 19.8. Major biogeographic boundaries for northern birds (after Kikkawa and Pearse, 1969), the enclosed faunal provinces in the northeast (from Schodde and Calaby, 1972), and the proposed Eyrean Salient at the Gulf of Carpentaria. All lie within the Torresian biogeographic sub-zone — an example of complexity in wildlife habitats.

J. Calaby (pers. comm., 1979) within 120 km of the coast (Fig. 19.3) — a group of four animals, two of them copulating.

Calaby's unpublished survey along the lower McArthur River strongly supports this concept. Of 31 species of mammal recorded, 22 were characteristic of the Torresian sub-zone and four were widely distributed in Australia (e.g., the water rat, *Hydromys chrysogaster*). *But* there were five that were characteristic of the Eyrean sub-zone. This intermixture of forms is therefore of great interest. For example, among the small marsupials, the Dasyuridae, there were the northern *Antechinus bilarni* and *Planigale ingrami*, but also the central *Antechinus macdonnellensis*; and among the rodents, the northern *Rattus tunneyi*, *Pseudomys gracilicaudatus*, *P. delicatulus*, and *Zyzomys*, but also the Eyrean *Rattus villosissimus*[1] and *Leggadina forresti*. The reptiles and amphibians also support the concept of an arid salient to the Gulf (J. Calaby, pers. comm., 1979). They form a rich fauna of around 90 species there, and are an admixture of forms from the tropical northwest (Arnhem Land and the Kimberleys) with a sizeable Eyrean element.

The evidence indicates, therefore, that there is a corridor for Eyrean fauna to reach the Gulf of Carpentaria, and that it may provide a barrier to movement east–west of Torresian fauna. Kikkawa and Pearse (1969) distinguished eastern and western groupings of birds; their grouping of species from the Kimberleys, Arnhem Land and the Gulf of Carpentaria ends at the southeastern corner of the Gulf, where Macdonald (1969) proposed a barrier also. These conclusions support the larger concept of a Carpentaria Salient. Its cause probably lies with climatic fluctuations in the short term from truly sub-monsoonal to truly semi-arid, superimposed upon particular land forms, such as heavy soils for tussock grasslands (Fig. 19.1), sands and rocky outcrops where the xeric spinifex (*Triodia*) grows.

Ecological Domains

As the foregoing section indicates, broad patterns of distribution obscure underlying diversity. Studies to elucidate that diversity in the vegetation have been under way for over twenty years, beginning with Christian's (1952, 1958) ideas of land systems. Much of the Northern Territory, for example, has been mapped, land units being recognized by an integration of geology, topography, hydrology, soils and vegetation [see Story (1960) for the region of study areas mentioned on pp. 442–447]. The degree of accuracy can be as fine as needed, but usually no smaller than units covering about 5 km^2. No similar maps exist for the fauna, which are more widespread and usually utilize related land units, or adjacent ones including ecotones — for instance, *Macropus rufus* in central Australia (Newsome, 1965a). Though knowledge of the fauna is still inadequate, I have combined it with the detailed map of pasture lands (Perry, 1960) in an attempt to delineate what I have termed the Ecological Domains of the Northern Territory (Fig. 19.9). Each Domain is described briefly in Table 19.1 where the distribution of the Macropodidae provides one faunistic proof.

It may be noted that the savanna woodlands of Fig. 19.1 contain parts of five Ecological Domains (Nos. 2–5, 7). The Macropodidae for the included Arnhem Land Escarpment (No. 2) may be compared with those of the Koolpinyah Woodlands (No. 3), the true savanna woodlands of that region. Other differences are ones of abundance — for instance, the agile wallaby, between Domains *1* to *10*, decreases in abundance inland with increasing aridity. It may be noted that certain species are concentrated at the transition from strictly Torresian to strictly Eyrean faunas — *Lagorchestes conspicillatus* and *Onychogalea unguifera*, for instance — though each may have outliers within those faunas.

ECOLOGY: REPRODUCTION IN A SEASONAL ENVIRONMENT

Continuous breeding: the agile wallaby

General comments

The agile wallaby is the most abundant macropodid in monsoonal Australia (Fig. 19.4), es-

---

[1] *Rattus villosissimus* is basically a form of the tussock grasslands (Fig. 19.1). It can form great irruptions during which it disperses very widely indeed (Newsome and Corbett, 1975). During 1968–69, it expanded into the northern woodlands, far to the south into the Simpson Desert and beyond, well into Queensland (L. Corbett, T. Redhead, unpubl.), and was recorded for the first time in Western Australia (Calaby, 1974).

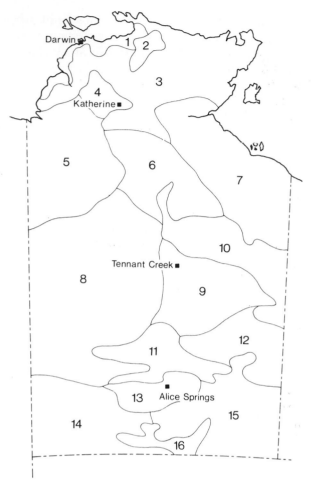

Fig. 19.9. Ecological Domains in the Northern Territory — a further example of complexity in wildlife habitats.

pecially around the river systems of sub-coastal Arnhem Land. It is a large wallaby which shelters in scrubs and woodlands flanking the watercourses and lagoons. In the late afternoon, the animals emerge to graze the river flats and glades, sometimes in great numbers. Like most macropodids, the species is dimorphic and males reach sizes around 30 kg, about half as big again as females. There is no evidence of social structure, the mothers and young having the only noticeable bond, as in many macropodids. Most data presented here are from Bolton (1974) and Bolton et al. (in press).

Reproduction

The breeding cycle. The agile wallaby matures at 12 to 14 months and breeds all year round in captivity (Merchant, 1976), having the typical macropodid breeding cycle. Pregnancy is short, around 29 days, and the length of oestrous cycle around 32 days. Following birth, the female returns to oestrus. If she is mated, the new embryo develops for a short while only, to the blastocyst stage (about 100 cells), and then remains dormant and unimplanted until the pouch-young is weaned between six and seven months of age, or is lost prematurely from the pouch. Pregnancy then continues normally, ensuring rapid replacement of young.

Anoestrus. Though females may breed continually in the wild, sexual maturity can be delayed in harsh seasons, and some adults become anoestrous (Bolton et al., in press). The sequence is the same as for the red kangaroo in the arid zone (Newsome, 1964a). The females fail to prepare reproductively for oestrus; no Graafian follicles develop to ripen in the alternate uterus and ovary, as is usual during pregnancy. Thus, there is no post-partum oestrus; yet the female remains reproductive to the extent that the young is suckled. To become fully reproductive again, these anoestrous, lactating females return to oestrus and mate, so regaining the normal condition during lactation (see above). Again the sequence is the same as for the red kangaroo (Newsome, 1964b).

The importance of nutrition for males. Though there was no evidence of loss of fertility as a result, average testis weight declined from 12 to 10 g from the rainy to the dry season. The loss was probably due to shrinkage of interstitium, which is thought to be influenced by nutrition (Leathem, 1970). There was a similar response in the red kangaroo to severe drought in central Australia, and also impairment of spermatogenesis due to high ambient temperatures (Newsome, 1973). As mentioned above, rams are rendered infertile by high temperatures in sub-monsoonal regions (see Entwistle, 1974, for a summary), but there were no signs of impaired spermatogenesis in the agile wallaby.

The importance of nutrition for females. There was marked contrast in breeding success between wallabies on irrigated, improved pastures receiving fertilizer on farmlands, and on native pastures. There was high (almost full) breeding and survival

TABLE 19.1

Ecological Domains of the Northern Territory and the distribution of the Macropodidae

Domain	Description
1. Sub-coastal plains	Swamp-sedgelands (*Eleocharis*, *Oryza*) drained and flooded by the major northern rivers. Monsoon forests and *Melaleuca* swamps minor. Domains 1, 2 and 3 form the great reservoir of western Torresian fauna; *Macropus agilis* abundant on the ecotone with Domain 3.
2. Arnhem Land Escarpment	Steep rugged spectacular cliffs and mountains rising from the northern plains. Monsoonal. Sole habitat for *Macropus bernardus* and *Petrogale concinna canescens*; *Macropus robustus alligatoris* and *Petrogale brachyotis* present.
3. Koolpinyah Woodlands	The common tropical savanna woodlands of *Eucalyptus tetrodonta* and *E. miniata* over annual grasses (*Sorghum* spp., *Heteropogon*), subject to annual fires. Monsoonal. *Macropus agilis* common especially along streams and adjacent to Domain 1; habitat of *M. antilopinus* also.
4. Tipperary Woodlands	Similar to Domain 3 but trees different (*Eucalyptus tectifola*, *E. latifolia*) and grasses perennial (*Themeda australis*, *Sorghum plumosum*). Monsoonal. *Macropus agilis*, *M. antilopinus*, *M. robustus*, *Petrogale brachyotis*, *Lagorchestes conspicillatus*, *Onychogalea unguifera* common to rare.
5. Victoria River	Broken rugged variable country; well-watered and drained. Sub-monsoonal to semi-arid. Mitchell Grass Downs (Domain 10) and Tanami Desert (Domain 8) in the south. Macropodidae as for Domain 4 but more uncommon, and including *Petrogale c. concinna*.
6. Muranji	Wooded (*Eucalyptus latifolia*, *Acacia shirleyi*) plus open grassy plains (*Aristida pruinosa*, *Astrebla pectinata*). Sub-monsoonal to semi-arid. Macropodids rare, *Macropus agilis*, *Lagorchestes conspicillatus*, *Onychogalea unguifera*.
7. Carpentaria	Similar to Domain 3, but sub-monsoonal to semi-arid. Macropodids uncommon to rare, as for Domain 5 but without *Petrogale concinna*, and *Macropus robustus* is the desert form, *erubescens*.
8. Tanami Desert	Sand-plain with some dunes mostly with soft spinifex (*Triodia pungens*); some salt lakes; rugged ranges in southeast with associated vegetation (*Acacia aneura* dominant). Arid to semi-arid. *Macropus rufus*, *M. robustus erubescens*, *Onychogalea unguifera*, *Lagorchestes hirsutus*, *Petrogale penicillata* all unusual to rare; *L. asomatus* known from one skull; *L. conspicillatus* probably.
9. Elkedra	Low rugged ranges separating the northern spinifex *Triodia pungens* from the southern *T. basedowii* on the sand-plains; mixed open association of *Acacia* and *Eucalyptus*. Semi-arid. *Macropus rufus*, *M. robustus erubescens*, *Petrogale penicillata* probably, *Onychogalea unguifera*, *Lagorchestes conspicillatus*.
10. Mitchell Grass Downs	Vast open grassy plains (*Astrebla pectinata* and *Iseilema* spp.); treeless except along streams and on small outliers of desert. Sub-monsoonal to semi-arid. Macropodids uncommon and patchy in wooded areas; *Lagorchestes conspicillatus*, *Onychogalea unguifera*; the Torresian *Macropus agilis* in the north, overlapping with the Eyrean *M. rufus*, both very rare.
11. Mulga Shrublands	Open shrublands (mostly *Acacia aneura*) and short grasses (*Aristida*, *Eragrostis*) and forbs. Semi-arid. *Macropus rufus* common to abundant on watercourses and any open plains. *Lagorchestes conspicillatus* rare, *Onychogalea lunata* and *Bettongia lesueur* now extinct.
12. Gidyea Shrublands	Vegetation and macropodids as for Domain 10, except dominant shrubs *Acacia georginae*.
13. MacDonnell Ranges	Rugged spectacular mountain ranges and gorges. Trees sparse (*Eucalyptus papuana*, *E. terminalis*, *Acacia aneura*) and mixed arid grasses; relict vegetation (*Livistona*, *Macrozamia*) in gorges. Semi-arid. *Macropus robustus erubescens* and *Petrogale penicillata* common in ranges, and *Macropus rufus* in some valleys. Extinctions as for Domain 11.
14. Amadeus Desert	Mostly spinifex (*Triodia basedowii*) sand-plains, but large dune fields with *Casuarina decaisneana* forming parklands in places; some *Acacia aneura*: some mountain ranges; extensive salt lakes. Arid. *Macropus rufus*, *M. robustus erubescens*, *Petrogale penicillata* uncommon to rare: *Lagorchestes hirsutus* perhaps extinct.
15. Simpson Desert	Tall extensive dune fields stabilized by *Triodia* spp.; swales broad; trees sparse to rare (*Acacia*). Arid. *Macropus rufus* rare: other macropodids uncertain.
16. Erldunda	Mosaic of open tree-less plains of shrubs, *Atriplex vesicaria* and *Kochia astrotricha*; *Triodia*, *Acacia* sand-plains and dunes. Arid. *Macropus rufus*, *M. robustus erubescens* rare: other macropodids uncertain.

of young on farmland but high losses in bushland (Table 19.2, Fig. 19.10). Monthly production of young ranged from 5.3 to 18.1 per 100 females on the native pastures (significance of difference: $P < 0.01$), but held relatively steady around 10.8 throughout the year on farmland. These results highlighted the importance of nutrition in a markedly seasonal (monsoonal) environment with poor soils. Pastures on the farmland were more nutritious than in the wild, and probably much more digestible (Table 19.3). Thus, given the chance of nutritious food, the wallaby was able virtually to realize its reproductive potential — the farmland constituted a natural experiment for the wallaby.

By contrast, breeding on native pastures was best in the dry season, and survival of young best from the dry season into the early rains (Fig. 19.10 and Table 19.2). The dry season can be as short as three months, but sometimes as long as seven months; on native pastures, adults and young lost considerable weight. Though this indicates a food shortage, it cannot have been critical, for breeding was best then. Green herbage continued to sprout around permanent water as it receded during the dry season, and wallabies congregated there. Such places, though patchily distributed, are common, and provide cool shade, water to drink, and green feed as well. In a very long dry period, though food shortage in many areas may border on the absolute at times, the problem is solved by the mobility of the wallabies. Also, grass fires are common, both natural and man-made, promoting green shoots that attract wallabies into the waterless savanna woodlands to feed.

The surprise was the loss of breeding and mortality of pouch-young in the rainy season, the time when pastures regenerate. The length of time be-

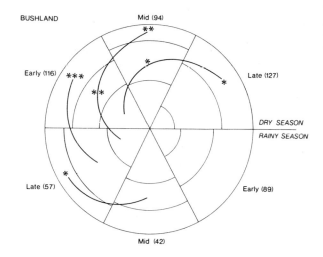

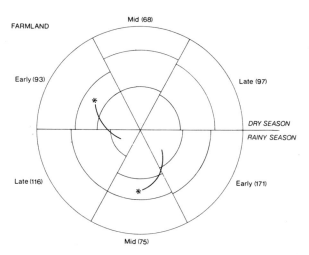

Fig. 19.10. Mortality in pouch-young of *Macropus agilis* on different pastures in Arnhem Land. Radial segments represent the frequencies of two-monthly cohorts. Total samples for the six sub-seasons are in brackets. Spiral lines join cohorts of significantly different frequencies: *$P < 0.05$; **$P < 0.01$; ***$P < 0.001$. (After Bolton et al., in press.)

TABLE 19.2

Breeding in adult female agile wallabies

	Bushland		Farmland	
Season:	rainy	dry	rainy	dry
Number of individuals observed	319	430	562	402
Normal breeding cycle (%)	82.4	96.0	93.7	98.6
Emerging from anoestrus (%)	2.6	1.4	4.7	0.5
Anoestrous (%)	15.0	2.6	1.6	0.9

From Bolton et al. (in press).

TABLE 19.3

Nutritional value of pastures grazed by agile wallabies

		Bushland		Farmland	
Season:		rainy	dry	rainy	dry
Protein (%)		3–6	1–3	8–10	4–8
Phosphorus (%)		0.03–0.05	0.02	0.09–0.14	0.04
Digestibility (%)		26.8	32.9[a]	50.3[b]	43.5[b]

[a] Values for native pastures from farmland area.
[b] Introduced legume, *Stylosanthes humilis*. Calculated from Wesley-Smith (pers. comm., 1972) and Bolton (1974).

tween the end of the dry season (i.e. the beginning of the rainy season) and flooding of the land was important, the two responses above being related inversely to this interval (Fig. 19.11). Though one cannot be sure, these striking relationships seem to be due to submergence of the usual dry-season feeding grounds, many of which are low-lying and the first to be flooded. Other factors cannot be ruled out. For example, breeding was worst following the longest — that is, the severest — dry season. Perhaps the wallabies had not had enough time to recover from the drought before the floods arrived. Certainly, during that rainy season, weight relative to body length fell sharply to its lowest point (Bolton et al., in press). Also in that season, 44% of females were anoestrous, the highest proportion recorded; in the preceding drought, there were 12% anoestrous, the highest figure for any dry season in the study.

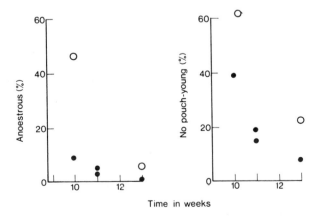

Fig. 19.11. Loss of fecundity and pouch-young in *Macropus agilis*, and the shortness of time between the end of the dry season and the onset of flooding. ○ = farmland; ● = bushland. (From Bolton et al., in press.)

The continuous pattern of breeding in the agile wallaby contrasts strongly with the pattern of four macropodids studied in winter-rainfall areas of Australia. The species are *Macropus eugenii*, *M. giganteus*, *M. rufogriseus* and *Setonix brachyurus*. Though they have the same reproductive cycle as other macropodids including *Macropus agilis*, the patterns of breeding are so arranged that young leave the pouch in spring and early summer when pastures are best (Tyndale-Biscoe, 1977). This represents a seasonal response to nutrition. The tammar (*M. eugenii*) is most remarkable. In southern Australia, with its winter rainfall and dry summer, the tammar has an obligate seasonal quiescence of about six months. During that time, females carry a quiescent blastocyst but no young in the pouch. At the summer solstice, however, pregnancy is renewed (Sadleir and Tyndale-Biscoe, 1977).

The other three species are not so strictly seasonal, but achieve the same timing of weaning by means of prolonged pouch-life or a seasonal anoestrus. Two other marsupials in southern Australia are known to time their breeding for a similar result: the possum, (*Trichosurus vulpecula*) and the marsupial mouse (*Antechinus stuartii*) (Tyndale-Biscoe, 1977). So it may not be phylogeny or a particular breeding cycle which forces continuous reproduction upon the agile wallaby. The adaptive forces probably lie in uncertainty as indicated above, through flooding of feeding grounds.

Being of medium size, even minor flooding may pose difficulties for wallabies, though some were seen wading in swamps feeding on emergent tips. An animal of small size, a native rat (*Rattus sordidus colletti*), is totally flooded out of such low-lying areas despite successful colonizations during the dry season; its populations survive only on high ground (Redhead, 1979). Wallabies are more mobile, and it is to the higher patches of savanna woodland and monsoon forests that they move to avoid the floods. Wallabies may, however, be caught in the swollen rivers and drowned. Some rescued by Bolton (1974) were totally exhausted, and their pouch-young had drowned also, a risk that must exist even when wallabies wade in shallower water. That the floods were involved in depressing breeding and survival of young is clear from Fig. 19.11 above.

It seems paradoxical that, in a seasonal environment, the agile wallaby should breed like the desert macropodids *M. rufus* and *M. robustus erubescens* (Newsome, 1975), that is, opportunistically. The uncertainty for red kangaroos in central Australia lies with the food supply (Newsome, 1966) caused by too little rain of uncertain arrival (Newsome, 1965b). The uncertainty for the agile wallaby, it is suggested, is not too little rain, but too much (150 cm) falling over too short a period (4–5 months).

Though there is a suggestion of seasonal breeding in results from two tropical kangaroos, *Macropus antilopinus* and *M. robustus alligatoris* (Russell and Richardson, 1971), the samples were small. Only 37 females were collected, and their twelve young were born from March to July. It is more likely that they breed opportunistically too.

Seasonal breeding; cattle and water buffalo, the introduced herbivores

In contrast, in the same study area, cattle and water buffalo, two introduced ruminants, breed seasonally (Tulloch, 1968: Andrews, 1976) (Fig. 19.12). Both species, however, are capable of breeding all the year round, like the wallaby. The reason for seasonality in the buffalo was not clear to Tulloch writing in 1968, but Andrews's (1976) detailed study of cattle presents a compelling argument.

In tropical Australia, cattle range over large properties, usually over 2000 km² in size and more. There are few fences, and bulls run with the herds all the time. In other words, the introduced cattle are subject to the natural environment, just as the feral buffalo and the agile wallaby. Fig. 19.13 shows the importance of rainfall *at any time of the*

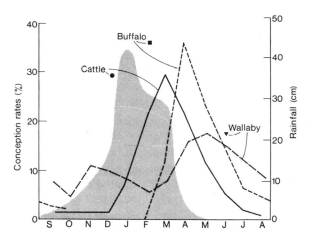

Fig. 19.12. Conception rates of *Babulus babulis*, *Bos taurus* and *Macropus agilis* in relation to the rainy season, Arnhem Land (after Tulloch, 1968; Andrews, 1976; Bolton et al., in press; respectively). Hatching represents rainfall; symbols represent timing of peaks for births.

year to reproduction in cattle in the monsoonal savanna woodlands, and also 700 km to the southeast in the tussock grasslands (Andrews, 1976).

Note that there *is* a breeding response to unseasonal rain, for instance in June. But Andrews (1976) very nicely demonstrated that cows that do not breed during the main rains pay a penalty. So great is the cost of breeding and lactating off-season that those cows may not conceive again till up to three years later (Fig. 19.14). There are, then, very strong adaptive reasons for breeding in the rainy season, which presumably apply to the buffalo as well. The long gestation periods — nine months for cattle and ten for buffalo — plus a lactation anoestrus allow these introduced ruminants to both calve and mate during the rainy season. Many antelopes and bovines in the African savanna do the same (Petersen and Casebeer, 1971; Sinclair, 1977; Skinner et al., 1977), presumably for nutritional reasons also.

By multiple regression analysis, Andrews (1976) was able to partition major environmental effects on conception rates in the tussock grasslands in cattle free of diseases (Table 19.4). Five factors accounted for 67.5% of the variability in conception rates ($P < 0.001$). In order of influence, they were lactation (negative), rainfall in the previous month (positive), ambient temperature (negative), the particular year of sampling (negative) and the body condition of the cattle (positive). As

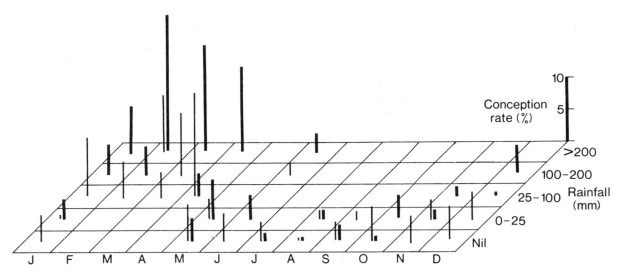

Fig. 19.13. Conception rates in *Bos taurus* in monsoonal savanna woodlands (heavy line) and tussock grasslands (light line) relative to rainfall (after Andrews, 1976).

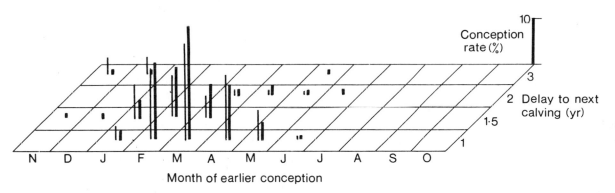

Fig. 19.14. Delays in returning to oestrus in *Bos taurus* depending on timing of earlier breeding, in monsoonal savanna woodlands (heavy line) and tussock grasslands (light line) (after Andrews, 1976).

TABLE 19.4

Factors influencing conception in *Bos taurus* on tussock grasslands (from Andrews, 1976)

	Regression slope	Regression slope in standard measure
Lactating or not	−0.234	−0.388
Rainfall in previous month (mm)	0.002	0.326
Temperature (°C)	−0.042	−0.254
Year of conception[1]	−0.125	−0.152
Body condition[2]	0.039	0.091
Age (yr)	−0.004	−0.020

$n > 5000$; $P < 0.001$; $R = 0.675$.

[1] 1967–1971 ordered as 1 to 6.

[2] Scored from 2 to 10 to represent a scale from emaciated to fat, measured around the butt of the tail.

rainfall (positive) and high ambient temperatures (negative) come at the same time of year, some measure of the environmental "tight-rope" walked by the European breeds of cattle (short-horns) can be seen. There was no similar influence of temperatures on cattle of Asiatic origin (Brahman/shorthorn crosses) further north in the savanna woodlands near Darwin (Andrews, 1976).

Uncertainty in a seasonal environment

Climate

Even though McAlpine (1976) described the rainfall of Arnhem Land as reliable monthly, seasonally, annually and geographically, there are no less than five moisture systems influencing the beginning, extent and yield of the rainy season. During a study of the wallaby from 1966 to 1970 (Bolton et al., in press), the dry season began as early as the 19th and as late as the 30th week and lasted from as little as 10 weeks to as much as 23 weeks. The rainy season as such began with more precision, between the 40th and 42nd weeks. The periods of submergence, however, lasted from 17 to 43 weeks.

One rainfall event deserves special mention, the cyclone. Usual between December and April (McAlpine, 1976) with a long-term average of 6.5 per decade (Lorrensz, 1977), cyclones come with about equal frequency in each month from December to March (Table 19.5). The torrential rains cause great floods, exacerbating those already in train, and submerging the plains inhabited by the wallabies. A small number of cyclones come off-season. Soon after cyclonic rains in May 1968 (Bolton et al., in press) the wallabies lost weight sharply, but recovered to have a good year. So, depending on timing, cyclones can worsen floods in the rainy season, or erase the dry season by keeping pastures green all year.

Diseases and parasites

Other negative influences on breeding in cattle are the parasitic diseases (Andrews, 1976; and pers. comm., 1980). One of these, brucellosis, increased in incidence from nil with no rain, to 8% with 126 mm, and 12 and 13% with 371 and 174 mm of rain, respectively. *Brucella abortus* does not influence conception rates but calf mortality, which was 61% in an infected herd but 16% in a nearby uninfected herd. Mortality was highest in the rainy season, 62%, compared with 32% in the dry season. The corresponding seasonal losses on the uninfected property were 14% and 18%.

Other organisms affecting reproduction in cattle in Andrews's studies were *Leptospira pomona*, *L. tarassovi*, *L. hardjo* and *Trichomonas fetus*. The incidence of antibodies for *Leptospira* near Darwin were highest in the rainy season (20.7 to 72.7%, depending on the parasite species) and lowest in the dry season (0 to 41.7%). Where the bulls were free of *Trichomonas fetus*, the pregnancy rate was 61% five months after mating, but only 26% in infected herds.

So, though cattle breed best in the rainy season, high ambient temperatures, disease, body condition and lactation are inimical at that time. One other factor not often considered but operating worst following the rainy season — floods — must be mentioned. Blood-sucking insects are an utter plague. The blood lost to mosquitoes alone has been measured in Queensland cattle at 166 ml per night, the equivalent to losses from cattle tick (*Boophilus microplus*) (Standfast and Dyce, 1968). In other words, a litre of blood must be replaced every six days. Irritation from these phlebotomous Diptera

TABLE 19.5

Incidence of cyclones per decade (1909–1975) (from Lorrensz, 1977)

Latitude	Dec.	Jan.	Feb.	Mar.	Apr.	May to Nov.	Total number
105–135°E	4.1	8.5	9.2	11.1	2.9	1.4 ⎱	454
135–165°E	4.1	11.7	10.8	10.8	5.3	3.5 ⎰	
Darwin area	1.7	0.9	1.1	1.8	–	–	44

is another problem (Colbo et al., 1977), interfering with feeding. Water buffalo cover themselves with a coat of mud from wallowing, probably to ward off the insects. This was a practice of Aboriginals also, besides using smoky fires.

The extent of these problems for wallabies (and other macropodids) is unknown. There was no evidence for suppressed breeding due to high ambient temperatures, and the little known about diseases indicates nothing remotely similar to the problems in cattle (D. Spratt, pers. comm., 1980). The insects are another matter; though never studied on the wallaby, they must be a nuisance at least. If they exsanguinate wallabies to the same extent as cattle and buffalo, then one cause of depressed breeding in the rainy season could be this parasitic nutritional drain. Biting midges (probably *Culicoides* spp.) are known to have blinded red kangaroos (*Macropus rufus*) in southwestern Queensland during great floods in the 1960s, so that, surrounded by plenty of food, they died of starvation, depopulating large areas (N. Wilson, pers. comm., 1961). At the time, horses and cattle were driven into frenzies, rushing wildly about and colliding with fences, trees, etc. (see Colbo et al., 1977).

CONCLUSIONS

It must be emphasized that, unlike the great African savannas, the savanna woodlands, woodland savannas, savannas and grasslands in tropical Australia are of low elevation (<300 m). The Arnhem Land Escarpment rises a little higher, to <700 m in places. These vegetative formations lie therefore mostly on a great plain sectioned by meandering rivers which flood extensively in the rainy season, forming great sheets of water subcoastally.

Such a description sounds bland; but the environment is variable even over short distances, and there is a rich fauna. Hence, recent land classifications have demonstrated the lack of blandness physically and botanically, and the few thorough faunal surveys have shown that animal distributions can be complex also. The Macropodidae, for instance, have speciated best in this region — to a remarkable degree compared with the rest of Australia. And the rock wallabies (*Petrogale*), the

best studied taxon, is now known to have a large array of sub-species as well. Thus, though it is customary to describe the tropical fauna as belonging to the Torresian sub-zone as if that embraced the one fauna, there are strong divisions between east and west, and less strong but evident between north and south in northwestern and northern Australia.

Of great interest is an admixture of Torresian and Eyrean (arid-zone) mammals at the Gulf of Carpentaria. This corridor for Eyrean forms has been named the Carpentarian Salient here, and helps explain the strong east–west division in the Torresian fauna. There are also species of wallaby, *Lagorchestes conspicillatus* and *Onychogalea unguifera*, characteristic of the transition between the Torresian and Eyrean sub-zones. At the next level of complexity, ecological domains within the Northern Territory have been mapped here using the Macropodidae as examples. There are sixteen of them. Six are within the Torresian sub-zone, and four or five straddle the transition to the Eyrean sub-zone, which has six exclusively. Several species of placental herbivores (e.g., rabbits, cattle, water buffalo) introduced to northern Australia have distributions which adhere to these patterns of faunal sub-zones and ecological domains.

The faunal ecology of the region is poorly known. Reproduction has been studied thoroughly in only one macropodid, *Macropus agilis*, and in the two ruminants, *Bos taurus* and *Bubalus bubalis*. The ecologies of the two arid-zone kangaroos, *Macropus rufus* and *M. robustus*, have also been studied and are drawn upon for comparisons. The most interesting comparison lies between the agile wallaby and the ruminants. Like their counterparts in the African savanna, the ruminants both calve and mate in the wet season. Breeding in the agile wallaby however is opportunistic like its desert counterparts, and is worst in the rainy season. Indeed, the wallaby's breeding most resembles that of a native rat, *Rattus sordidus colletti*, also studied in the region.

The patterns of reproduction in these species appear to depend on the following: (1) body size; (2) gestation length; and (3) alternation and severity of drought and flooding. Sheer size determines whether the monsoonal rains are a catastrophe, as for *Rattus sordidus colletti*, a seasonal check, as for *Macropus agilis*, or a stimulus for breeding, as for cattle and water buffalo. Size also confers mobility. Gestation length decides whether breeding is opportunistic, as for the rats and wallabies, or seasonal, as for the ruminants. Phylogeny is not the sole determinant of the breeding pattern, however, ᵤecause macropodids with similar basic reproductive cycles have evolved seasonal or pseudoseasonal breeding patterns in the mediterranean climate of southern Australia.

Seasonality in breeding implies dependability of the seasons. Table 19.6 shows two kinds of environmental dependability for different climates in Australia from north to south: the annual variability of the rainfall, and the ability of plants to grow throughout the year. It may be noted that the period from January to March is optimal for plant growth in northern Australia. Cattle and buffalo completely appreciate this, but the agile wallaby does so only partially (see Fig. 19.12). At a much lower growth potential in southern Australia, only a little better than in the desert at its maximum in spring and early summer in fact, macropodids wean their young from June to August.

Opportunist breeding is associated, then, with climatic patterns that produce the highest and lowest potential for growth, the monsoonal and desert environments respectively. This apparent anomaly, however, may be resolved by the high annual variability in rainfall there (Table 19.6). It is argued above that that variability, uncertainty in the midst of plenty, is responsible for the breeding pattern in the agile wallaby in northern Australia. This wallaby is distributed coastally well southwards into Queensland (Fig. 19.4) where the climate, though subtropical, is milder. Breeding there may more resemble that in the farmland in northern Australia, being regular throughout the year.

Other aspects of uncertainty in the tropical environment are best documented for the introduced cattle. High temperatures reduce conception rates, and rainfall promotes them. However, cows that breed off-season, in response to the unusual rains that sometimes come in the dry season, pay a high penalty and may not breed again for eighteen months to three years. The adaptive pressure to breed in the rainy season is therefore intense. Body condition also controls rates of conception, probably due to nutrition of the pastures, a major factor with the agile wallaby also. Disease, an unknown factor in the wallaby, markedly reduces conception rates and calving in cattle. Bloodsucking insects appear to place a considerable nutritional drain on animals, as well as being a severe irritation and interfering with feeding at times.

SUMMARY

The tropical savanna woodlands and woodland savannas of Australia are characterized by low coastal elevation, poor soils, dominance of eucalypts in the upperstorey and of tall rank grasses of low nutritional value in the understorey. There is, however, a considerable diversity of habitats due to vegetation, soils, elevation, dissected drainage, and, possibly, the proximity continentally of the arid region which projects into the tropical regions at

TABLE 19.6

Comparison of seasons continentally (from Nix, 1976)

Latitude (°S)	Climate (mean annual rainfall, mm)	Growth Indices				Months of optimal plant growth	Annual coefficient of variation in rainfall (%)
		DJF	MAM	JJA	SON		
12–14	tropical, monsoonal, summer rain (1500)	>0.8	0.6–0.8	–	0.1–0.2	JFM	75–100
22–24	arid, desert, rain uncertain (250)	0.1–0.2	<0.1	<0.1	<0.1	DJF	75–100
32–34	mediterranean; winter rain (500)	0.1–0.2	0.1–0.2	0.1–0.2	0.2–0.4	SON	50–75
40	temperate, mostly winter rain (1000)	0.2–0.4	0.2–0.4	<0.1	0.2–0.4	OND	50–75

the Gulf of Carpentaria. There is, in turn, a rich diversity of fauna.

The rainfall is high and the weather hot, torrid and humid for about half the year. Much of the land is flooded then, and the low-lying feeding grounds for herbivores are submerged, producing a relative shortage of food. Attendant nutritional problems, parasites and hazards reduce fecundity further. The dry season is the most favourable time for the native grazer, the agile wallaby, though breeding of introduced ruminants is concentrated on the rainy season as in the African savannas. Paradoxically for a seasonal environment, breeding in the agile wallaby is opportunistic as in the desert kangaroos. Annual rainfalls of the monsoonal and desert regions, though very different in quantity, have, however, similarly high variability.

The East African savanna contrasts markedly with that of Australia. It is considerably elevated, so reducing temperature and humidity but sustaining rainfall, and many soils are volcanic, ensuring high productivity. Lowland ecosystems in the tropics have received little attention world-wide, and there is no substantial body of data for large herbivores in such an ecosystem other than in Australia.

ACKNOWLEDGEMENTS

I wish to express my gratitude and thanks to many people: Dr B.L. Bolton and Mr J. Merchant for permission to use our joint data on the agile wallaby; Dr L.G. Andrews for permission to use his data on cattle; Dr J.H. Calaby for his thorough knowledge of the Australian fauna and flora so freely provided, plus unpublished data on the distributions of fauna; Prof G.B. Sharman and Drs D. Briscoe and G. Maynes for kind permission to use unpublished data on the taxonomy of *Petrogale*; Mr T. Redhead for his valuable insights into the ecological functioning of northern Australia; Mr R.N. Wesley-Smith for his data on the nutrition of pastures; Mr F. Knight for his excellent figures; all these people plus Mr C.S. Christian and Mr H. Nix, Dr T. Riney and Dr D. Happold for helpful ideas and discussions; and Dr D Goodall for pursuading me to contribute this chapter. Dr K. Johnson confirmed the identity of the southernmost record of *Macropus agilis*, and Dr P. Johnston provided the distribution records and recent localities of *Onychogalea fraenata*.

REFERENCES

Andrews, L.G., 1976. *Reproductive Performance of Beef Cattle in the Northern Territory*. Thesis, James Cook University, Townsville, Qld., 187 pp.

Bolton, B.L., 1974. *An Ecological Study of the Agile Wallaby on the Coastal Plains of the Northern Territory with a Comparison between Improved and Unimproved Areas*. Thesis, University of Queensland, St. Lucia, Qld., 121 pp.

Bolton, B.L., Newsome, A.E. and Merchant, J.C., in press. Reproduction in the agile wallaby, *Macropus agilis*, in tropical lowlands of north Australia.

Bureau of Meteorology, 1956. *Climatic Averages Australia*. Melbourne, Vic., 107 pp.

Calaby, J.H., 1974. *Rattus villosissimus* (Waite) — A new mammal record for Western Australia. *Rec. W. Aust. Mus.*, 3: 82–83.

Calaby, J.H., 1975. Introduction of Bali cattle to northern Australia. *Aust. Vet. J.*, 51: 108.

Catling, P.C. and Newsome, A.E., 1981. Responses of the Australian vertebrate fauna to fire: an evolutionary approach. In: M. Gill, R.H. Groves and I.R. Noble (Editors), *Fire and the Australian Biota*. Australian Academy of Sciences, Canberra, A.C.T., pp. 273–310.

Chippendale, G.M., 1974. Vegetation. In: H.J. Frith and J.H. Calaby (Editors), *Fauna Survey of the Port Essington District, Coburg Peninsula, Northern Territory of Australia. CSIRO Div. Wildl. Res. Tech. Pap.*, No. 28: pp. 21–47.

Christian, C.S., 1952. Regional land systems. *J. Aust. Inst. Agric. Sci.*, 18: 140–146.

Christian, C.S., 1958. The concept of land units and land systems. *Proc. 9th Pacif. Sci. Congr.*, 20: 74–81.

Colbo, M.H., Fallis, A.M. and Reye, E.J., 1977. The distribution and biology of *Austrosimulium pestilens*, a serious biting-fly pest following flooding. *Aust. Vet. J.*, 53: 135–138.

Dudzinski, M.L., Newsome, A.E. and Merchant, J.C., 1978. Growth rhythms in pouch-young of the agile wallaby in Australia. *Acta Theriol.*, 23: 401–411.

Entwistle, K.W., 1974. Reproduction in sheep and cattle in the Australian arid zones. In: A.D. Wilson (Editor), *Studies of the Australian Arid Zone*. CSIRO, Melbourne, Vic., pp. 85–97.

Frith, H.J. and Calaby, J.H. (Editors), 1974. Fauna survey of the Port Essington District, Cobourg Peninsula, Northern Territory, Australia. *CSIRO Div. Wildl. Res. Tech. Pap.*, No. 28: 208 pp.

Kikkawa, J. and Pearse, K., 1969. Geographical distribution of land birds in Australia — A numerical analysis. *Aust. J. Zool.*, 17: 821–840.

Kirkpatrick, T.H., 1970. The agile wallaby in Queensland. *Qld. Agric. J.*, 96: 169–170.

Kirkpatrick, T.H. and Johnson, A.D., 1969. Studies of Macropodidae in Queensland. 7. Age estimation and reproduction in the agile wallaby (*Wallabia agilis* (Gould)). *Qld. J. Agric. Anim. Sci.*, 26: 691–698.

Leathem, J.H., 1970. Nutrition. In: A.D. Johnson, W.R. Gomes and N.L. van Demark (Editors), *The Testis, III*. Academic Press, New York, N.Y., pp. 169.

Letts, G.A., 1964. Feral animals in the Northern Territory. *Aust. Vet. J.*, 40: 84–88.

Letts, G.A., (Editor), 1979. *Feral Animals in the Northern Territory. Report of the Board of Inquiry.* Government Printer, Darwin, N.T., 236 pp.

Long, J.L., 1972. Introduced birds and mammals in Western Australia. *Agric. Protection Board W.A. Tech. Ser.*, No. 1.

Lourensz, R.S., 1977. *Tropical Cyclones in the Australian Regions July 1909 to June 1975.* 111 pp.

McAlpine, J.R., 1976. Climates and water balances. Lands of the Alligator Rivers Area, Northern Territory. *CSIRO Land Res. Ser.*, No. 38: 35–49.

Macdonald, J.D., 1969. Notes on the taxonomy of *Neositta. Emu*, 69: 169–174.

Merchant, J.C., 1976. Breeding biology of the agile wallaby, *Macropus agilis* (Gould) (Marsupialia: Macropodidae) in captivity. *Aust. J. Zool.*, 13: 735–759.

Moore, R.M. and Perry, R.A., 1970. Vegetation. In: R.M. Moore (Editor), *Australian Grasslands.* Australian National University Press, Canberra, A.C.T., pp. 59–73.

Moule, G.B., 1962. The ecology of sheep in Australia. In: A. Barnard (Editor), *The Simple Fleece.* Melbourne University Press, Melbourne, Vic., pp. 82–105.

Myers, K., 1970. The rabbit in Australia. In: *Proc. Adv. Study Inst. Dynamics Numbers Popul., Oosterbeek, 1970,* pp. 478–506.

Newsome, A.E., 1964a. Anoestrus in the red kangaroo, *Megaleia rufa* (Desmarest). *Aust. J. Zool.*, 12: 9–17.

Newsome, A.E., 1964b. Oestrus in the lactating red kangaroo, *Megaleia rufa* (Desmarest). *Aust. J. Zool.*, 12: 315–321.

Newsome, A.E., 1965a. The abundance of red kangaroos, *Megaleia rufa* (Desmarest), in central Australia. *Aust. J. Zool.*, 13: 735–759.

Newsome, A.E., 1965b. Reproduction in natural populations of the red kangaroo, *Megaleia rufa* (Desmarest), in central Australia. *Aust. J. Zool.*, 13: 289–299.

Newsome, A.E., 1966. The influence of food in the red kangaroo in central Australia. *CSIRO Wildl. Res.*, 11: 187–196.

Newsome, A.E., 1971. Competition between wildlife and domestic livestock. *Aust. Vet. J.*, 47: 577–586.

Newsome, A.E., 1973. Cellular degeneration in the testis of red kangaroos during hot weather and drought in central Australia. *J. Reprod. Fertil., Suppl.*, 19: 191–201.

Newsome, A.E., 1975. An ecological comparison of the two arid-zone kangaroos of Australia, and their anomalous prosperity since the introduction of ruminant stock to their environment. *Q. Rev. Biol.*, 50: 389–324.

Newsome, A.E. and Corbett, L.K., 1975. Outbreaks of rodents in Australian deserts: origins, preventions and evolutionary considerations. In: I. Prakash and P. Ghosh (Editors), *Rodents in Desert Environments.* W. Junk, The Hague, pp. 117–153.

Nix, H.A., 1976. Environmental control of breeding, post-breeding dispersal and migration of birds in the Australian region. In: H.J. Frith and J.H. Calaby, *Proc. 16th Int. Ornithol. Congr.* Australian Academy of Science, Canberra, A.C.T., pp. 272–305.

Perry, R.A., 1960. Pasture lands of the Northern Territory, Australia. *CSIRO Land Res. Ser.*, No. 5: 55 pp.

Petersen, J.C.B. and Casebeer, R.L., 1971. A bibliography relating to the ecology and energetics of East African large

mammals. *E. Afr. Wildl. J.*, 9: 1–23.

Plumb, T.W. (Editor), 1970. *Atlas of Australian Resources, 2nd Ser. Livestock.* Geogr. Sec., Dep. National Mapping, Canberra, A.C.T.

Poole, W.E., 1978. The status of the Australian Macropodidae. In: *The Status of Endangered Australian Wildlife.* Royal Zoological Society, Adelaide, S.A., pp. 13–27.

Prescott, J.A., 1931. The soils of Australia in relation to vegetation and climate. *CSIRO Bull.*, 52: 82 pp.

Ratcliffe, F.N. and Calaby, J.H., 1958. Rabbit. In: *Australian Encyclopedia. 7.* Angus and Robertson, Sydney, pp. 340–347.

Redhead, T.D., 1979. On the demography of *Rattus sordidus colletti* in monsoonal Australia. *Aust. J. Ecol.*, 4: 115–136.

Richardson, B.J., 1970. *A morphological and biochemical study of two closely related marsupial genera,* Megaleia *(the red kangaroo) and* Osphranter *(the wallaroo).* Thesis, University of New South Wales, Kensington, N.S.W., 292 pp.

Richardson, B.J. and Sharman, G.B., 1976. Biochemical and morphological observations on the wallaroos (Macropodidae: Marsupialia) with a suggested new taxonomy. *J. Zool. Lond.* 179: 499–513.

Ride, W.D.L., 1970. *A Guide to the Native Mammals of Australia.* Oxford University Press, Melbourne, 249 pp.

Russell, E.M. and Richardson, B.J., 1971. Some observations on the breeding, age structure, dispersion and habitat of populations of *Macropus robustus* and *Macropus antilopinus* (Marsupialia). *J. Zool. Lond.*, 165: 131–142.

Sadleir, R.M.F.S. and Tyndale-Biscoe, C.H., 1977. Photoperiod and the termination of diapause in the marsupial, *Macropus eugenii. Biol. Reprod.*, 16: 605–608.

Schodde, R. and Calaby, J.H., 1972. The biogeography of the Australo-Papuan bird and mammal faunas in relation to Torres Strait. In: D. Walker (Editor), *Bridge and Barrier: The Natural and Cultural History of Torres Strait.* Australian National University Press, Canberra, A.C.T., pp. 257–300.

Sinclair, A.R.E., 1977. *The African Buffalo.* University of Chicago Press, Chicago, Ill., 355 pp.

Skinner, J.D., Nel, J.A.H. and Miller, R.P., 1977. Evolution of time of parturition and different litter sizes as an adaptation to changes in environmental conditions. In: J.H. Calaby and C.H. Tyndale-Biscoe (Editors), *Reproduction and Evolution.* Australian Academy of Science, Canberra. A.C.T., pp. 39–44.

Specht, R.L. (Editor), 1964. *Records of the American-Australian Scientific Expedition to Arnhem Land. 4. Zoology.* Melbourne University Press, Melbourne, Vic., 533 pp.

Specht, R.L., Roe, E.M. and Boughton, V.H., 1974. Conservation of major plant communities in Australia and Papua New Guinea. *Aust. J. Bot. Suppl. Ser.*, No. 7: 667 pp.

Spencer, B., 1896. *Report on the Work of the Horn Scientific Expedition to Central Australia. Pt. II. Zoology. Mammalia.* Melville, Mullen and Slade, Melbourne, Vic., 52 pp.

Standfast, H.A. and Dyce, A.L., 1968. Attacks on cattle by mosquitoes and biting midges. *Aust. Vet. J.*, 44: 585–586.

Story, R., 1960. Vegetation of the Adelaide–Alligator Rivers Area (incl. maps). *CSIRO Land Res. Ser.*, No. 25: 114–130.

Tulloch, D.G., 1968. Incidence of calving and birth weights of domesticated buffalo in the Northern Territory. *Proc. Aust. Soc. Anim. Prod.*, 7: 144–147.

Tulloch, D.G., 1969. Home range in the feral water buffalo, *Bubalus bubalis* Lydekker. *Aust. J. Zool.*, 17: 143–152.

Tulloch, D.G., 1970. Seasonal movements and distribution of the sexes in the water buffalo, *Bubalus bubalis*, in the Northern Territory. *Aust. J. Zool.*, 18: 399–414.

Tyndale-Biscoe, C.H., 1977. Environment and the control of breeding in kangaroos and wallabies. In: H. Messel and S.T. Butler (Editors), *Australian Animals and their Environment*. Shakespeare Head Press, Sydney, pp. 63–79.

Chapter 20

MAMMALS AS SECONDARY CONSUMERS IN SAVANNA ECOSYSTEMS

FRANÇOIS BOURLIÈRE

INTRODUCTION

In this chapter I shall consider all those mammals which feed mostly, or to a large extent, upon other animals, whatever may be their size, from tiny insects to large ungulates. Unfortunately our knowledge of the food and feeding habits of most tropical species is still very meagre, all the more so since many of these consumers have mixed diets, which also change seasonally. It is therefore very difficult to assign some mammals to any definite trophic category. Much of what is said in the following pages must consequently be considered as merely tentative, pending further research.

In addition to the large predators *sensu stricto*, the trophic category of "secondary consumers" includes mammals commonly considered as "insectivores", which prey upon many terrestrial arthropods and most kinds of small invertebrates. Among them, special attention must be given to those species which specialize upon ants and termites, a food source particularly abundant in most tropical savannas. The true predators, belonging to the Order Carnivora, feed upon a large variety of other vertebrates, some of them of large size. They catch them alive in most cases, but sometimes scavenge upon dead animals, even carrion. Sanguinivorous mammals are represented by New World bats; though few in species, the populations of these bats are sufficiently large to give rise to epidemiological problems for man and livestock. Animal prey never forms a large part of the diet of most omnivorous primates or rodents, but its dietary importance may nevertheless not be negligible.

THE SPECIES OF MAMMALIAN SECONDARY CONSUMERS

The number of species of secondary consumers in five tropical study sites is given in Table 20.1. It will at once be noted that both the total number of species, and their percentage within the entire mammalian fauna, is very similar in the two African communities where species richness is the greatest: the rain-forest site of Makokou (Gabon), and the Serengeti savannas (Tanzania). However, many rain-forest carnivores feed as much on fruit as on animal prey, as exemplified by the coati, *Nasua narica* (Kaufman, 1962) and the African palm civet, *Nandinia binotata* (Charles-Dominique, 1978).

In three other communities a striking difference becomes apparent between the two major tropical biomes: the percentage of secondary consumers is much greater in the savannas than in the rain forest, a situation very similar to that found in birds (see Ch. 14). The highest percentage of carnivores and "insectivores" is found in the impoverished Sahelian savanna of Fété Olé (Senegal).

So far, there are no comparable figures for the Asian or Australian tropics, but the percentages of secondary consumers given by Harrison (1962) for the terrestrial mammals of the lowland rain forests of the Malay Peninsula (24%) and of the Cape York Peninsula (28%), also illustrate the relative scarcity of mammalian secondary consumers in tropical rain forests.

Most bats, wherever they are found, feed upon flying insects; in the tropics a small number of bats prey upon other vertebrates. The smaller percentage of secondary consumers among the Barro Colorado (Panama) bats is, no doubt, due to the

TABLE 20.1

Number of species of secondary consumers in five tropical study sites.

	Forests		Savannas		
	Makokou (northeastern Gabon)	Barro Colorado (Panama)	Serengeti National Park (Tanzania)	Garamba National Park (Zaire)	Fété Olé (northern Senegal)
Total number of species of non-volant mammals	*92*	*70*	*104*	*52*	*27*
Secondary consumers					
Marsupialia	–	6	–	–	–
Insectivora	12	–	5	9	4
Primates	2	2	1	–	–
Edentata	–	3	–	–	–
Pholidota	3	–	1	1	–
Rodentia	2	–	2	1	–
Carnivora	12	5	27	8	9
Tubulidentata	1	–	1	1	1
Total	*32*	*16*	*37*	*20*	*14*
Percentage of non-volant mammals	34.7	22.8	35.5	38.4	51.8
Total number of species of flying mammals	*34*	*46*	*25*	*21*	*>5*
Insectivorous, carnivorous and sanguinivorous Chiroptera	27	27	19	18	>5
Percentage of flying mammals	79.4	58.7	76.0	85.7	100
References	1	2	3	4	5

References: 1, Emmons, Gautier-Hion and Dubost (pers. comm.); 2, Eisenberg and Thorington (1973), Wilson (1973), Bonaccorso (1975); 3, Heindrichs (1972), revised after J. Kingdon and J. Verschuren (pers. comm.); 4, J. Verschuren (pers. comm.); 5, Poulet (1972).

greater number of nectarivorous species in the Neotropics.

Taxonomic distribution

The mammalian secondary consumers found in the terrestrial ecosystems within the tropics belong to nine different orders. However, the greatest trophic impact is exerted by only five of these orders in the tropical savannas: the Chiroptera, which provide most of the aerial insectivores; the Carnivora, which are major terrestrial predators; and a few Edentata, Pholidota and Tubulidentata, which prey almost exclusively upon social insects.

Very few American marsupials are savanna dwellers. Most didelphids are arboreal, but there are a few exceptions like the short bare-tailed opossum *Monodelphis brevicaudata*, a forest species often found on the ground in the Venezuelan *llanos*, or some *Marmosa* species inhabiting Argentine *pampas*, Brazilian savannas, and Andean *paramos*; all of them are known to feed mostly upon insects. Likewise, a few species of marsupial mice (Dasyuridae) are found in northern Australia preying upon insects and even lizards. Such is the case for *Antechinus bellus*, *Planigale ingrami*, *P. subtilissima*, *P. tenuirostris* and *Sminthopsis larapinta*. The bandicoot *Macrotis lagotis* is also very fond of termites which can become its staple food.

Most of the tropical shrews (Soricidae) are also forest dwellers. A few, however, are clearly restricted to savannas. Such is the case for *Crocidura flavescens*, *C. luna*, *C. pasha*, *C. sericea* and *C. turba* in Africa. Other Insectivora of some importance as secondary consumers in dry African savannas are the elephant shrews (e.g. *Elephantulus rufescens*) and the hedgehog *Erinaceus albiventris*. All of them are mostly insectivorous, occasionally preying upon small vertebrates.

Among mammals, the aerial insectivore role is the monopoly of the Chiroptera. Most of the genera of bats living in tropical savannas specialize in feeding on insects on the wing, frugivorous and nectarivorous species being mostly restricted to forest areas. Bats also contribute a few other food specialists, such as blood feeders, fish eaters and predators on terrestrial vertebrates. The common vampire *Desmodus rotundus* is the most widespread and abundant of all the bats feeding upon the blood of other mammals and birds; it is found from Mexico to Argentina, and inhabits both arid and humid regions in the tropics and subtropics. Its population numbers probably increased following the introduction of cattle and horses by Europeans into Latin America. *Noctilio leporinus*, which, at least in Venezuela, lives in savannas and pastures as well as in marshes and forests, catches small surface-swimming fish with the long claws of its feet, but it also feeds upon moths, beetles, winged ants and cockroaches. *Vampyrum spectrum*, a carnivore, is found in the gallery forests of the llanos. The false vampires of the Old World (genera *Megaderma* in Asia and *Cardioderma* in Africa) are good examples of truly carnivorous bats. Their prey includes other bats and rodents, birds, small lizards and frogs, together with large insects; when eating vertebrates, these bats first suck the blood of their prey and then eat the flesh. *Megaderma lyra* inhabits the dry tree savannas of central and eastern peninsular India, as well as the humid tropics of western Malaysia. *Cardioderma cor* is found in the dry *Acacia* bush of Kenya, Somalia and northern Tanzania.

There are very few predominantly insectivorous species among savanna primates. In Africa, the prosimian *Galago senegalensis* is found in *Acacia* and *Combretum* "bush" and in *Isoberlinia* woodland. Insects and gums are the most important

items of its diet; in semi-arid areas, however, fruits of *Balanites aegyptiaca* may seasonally become its principal food. In South America, tufted-ear marmosets (*Callithrix jacchus*) are the only monkeys found in the gallery forests and tree savannas of eastern Brazil; they have a mixed diet of fruits, gums and insects.

Most of the New World Edentata are forest animals, though a few species may be found in more open environments. Such is the case for the giant ant-eater (*Myrmecophaga tridactyla*), which is entirely terrestrial and more often found in the palm savannas of the Venezuelan llanos than in moist evergreen forest. It feeds exclusively on a few species of ants, the colonies of which it seldom destroys completely. These ant-eaters must travel at night for long distances (up to 11 000 m) at a rapid pace in search of their food. Their home range is consequently very large (about 2500 ha). The golden ant-eater (*Tamandua tetradactyla*), although a forest animal in most of its range, can also enter the palm savanna and be sympatric with the giant ant-eater, as shown by Montgomery and Lubin (1977). On the same study site in Venezuela, golden ant-eaters feed on many species of ants and termites, visiting ground colonies more often than those located in trees. Their home range is about one-sixth of that of the giant ant-eaters. Both species utilize the same "cropping" strategy, which enables them to use a highly specialized, though abundant, food source on a sustained-yield basis. Some armadillos also live in open or tree savannas in South America. *Priodontes giganteus* and *Tolypeutes matacus* are more myrmecophagous than any of the other species. *Dasypus novemcinctus* and *Euphractus sexcinctus* are omnivore carnivores whose diet varies greatly with season and locality (J.F. Eisenberg, pers. comm.).

The Old World pangolins (Order Pholidota) are the ecological counterparts of the New World ant-eaters and armadillos, which also "crop" ant and termite colonies. Most species are arboreal and restricted to forest areas. The ground pangolin (*Manis temmincki*), however, is found in East and South African savannas and savanna woodlands. It feeds mainly on the juvenile stages of some ants and termites, avoiding those belonging to genera *Trinervitermes* and *Macrotermes* (Sweeney, 1956).

The only living representative of the Order Tubulidentata, the aardvark (*Orycteropus afer*) is widely distributed in the African savannas. Its distribution is undoubtedly determined by the abundance of the termites and ants on which it feeds, the same nests being repeatedly "cropped". The aardvark also eats large quantities of insect larvae, including those of the scarab beetles buried in dung pellets (Melton, 1976).

A few savanna rodents are predominantly or exclusively insectivorous. Such is the case in East Africa for *Lophuromys flavopunctulatus* and *Zelotomys hildegardeae*. The former eats mostly ants and termites, but also frogs and other small animals, carrion and vegetable matter. The latter is fond of myriapods and dung beetles.

The bulk of mammalian secondary consumers are nevertheless included in the various families of the Order Carnivora. African savannas harbour, by far, the largest number of species. Among canids, the hunting dog (*Lycaon pictus*) preys on small- to medium-sized animals (Van Lawick-Goodall and Van Lawick-Goodall, 1970; Schaller, 1972), whereas the Abyssinian wolf (*Canis simiensis*) specializes upon small rodents, particularly mole rats (Morris and Malcolm, 1977). The three species of jackals, the golden jackal (*Canis aureus*), the black-backed jackal (*C. mesomelas*) and the side-striped jackal (*C. adustus*) are much more opportunist in their choice of food (Lamprecht, 1978). All of them are capable of existing as independent predators if small game is abundant. Failing this, they may scavenge, eat small arthropods and even some vegetable matter such as the fruits of *Balanites aegyptiaca*. In Botswana (Smithers, 1971) and the Kalahari (Bothma, 1966), the bulk of the food in the stomachs of *Canis mesomelas* consists of insects, mostly locusts, beetles and termites. In the Namib desert, plant food, especially seeds and fruits of *Euclea pseudebenus*, seeds of *Salvadora persica* and grass are usually the predominant component of the scats of *Canis mesomelas* (Stuart, 1976). In the Serengeti Plains, *C. mesomelas* preys more upon arthropods during the rainy season and more upon mammals in the dry season, whereas *C. aureus* has the same diet throughout the year (Lamprecht, 1978). Very little is known of the diet of the two foxes, the sand fox (*Vulpes pallida*) and the Cape fox (*V. chama*) living in the more arid parts of the northern and southern savannas. Beetles and small rodents are the staple food of *V. chama* in Botswana (Smithers, 1971). The bat-eared fox

(*Otocyon megalotis*) appears to be mostly insectivorous, termites, ants and beetles being the food items most frequently recorded in Botswana as well as in the Serengeti Plains (Smithers, 1971; Ewer, 1973; Nel, 1978; Lamprecht, 1979).

Three species of mustelids are found in the African savannas: the striped polecat (*Ictonyx striata*), the striped weasel (*Poecilogale albinucha*) and the honey badger (*Mellivora capensis*). Not much is known of their habits; the first one preys upon small vertebrates and large invertebrates, the second on small mammals, whereas the third is known to be fond of scorpions, spiders, bee larvae and sweet fruits, and also feeds on snakes and carrion (Smithers, 1971; Kingdon, 1977).

Among the African viverrids, the two savanna species *Genetta genetta* and *G. tigrina* more or less share the same kind of food: small rodents, reptiles, amphibians and large invertebrates. But when found in the same locality, *G. tigrina* specializes on small rodents, while *G. genetta* also takes a variety of reptiles and invertebrates, particularly solpugids (Solifugae) and scorpions (Smithers, 1971). The African civet *Viverra civetta* is even more catholic in its taste; it feeds upon almost every kind of insect (including the stink grass-hopper *Zonocerus elegans*), myriapods, snails, crabs, and also on amphibians, reptiles (tortoises included), ground-living birds, rodents and carrion. It can also subsist entirely on vegetable food, from fruit to grass. The diet of most mongooses still remains to be studied. That of the banded mongoose (*Mungos mungo*) is, at least in the Ruwenzori National Park (Uganda), made up almost entirely of arthropods, mostly dung beetles and millipedes (Neal, 1970; Rood, 1975). Adult beetles were also the commonest item of diet found in the stomachs of four out of the eight species of Botswana mongooses studied by Smithers (1971). In South Africa, the yellow mongoose (*Cynictis penicillata*), though an opportunist feeder, mostly eats termites (Zumpt, 1968; Herzig-Straschill, 1977). *Herpestes sanguineus* and *H. ichneumon* are efficient killers of small vertebrate prey (Rautenbach and Nel, 1978).

The four species of African hyaenids differ greatly in their diet. In many places, the spotted hyaena (*Crocuta crocuta*) is almost exclusively a scavenger feeding on refuse. In the Serengeti National Park and Ngorongoro Crater, on the other hand, the same species kills most of its prey itself, specializing upon medium-sized ungulates (Kruuk, 1972). The striped hyaena (*Hyaena hyaena*) is omnivorous, scavenging a great deal and feeding also on insects, fruits and small vertebrates (Kruuk, 1976). The brown hyaena (*Hyaena brunnea*) feeds mostly on kills left by other predators and on wild melons and small rodents and hares (Owens and Owens, 1978); ostrich eggs and termites are also eaten in the Kalahari Desert (Mills, 1978). The aardwolf (*Proteles cristatus*) is exclusively insectivorous, and specializes on some species of surface-foraging termites, mostly *Trinervitermes* (Kruuk and Sands, 1972; Cooper and Skinner, 1978).

The African felids are true predators of any available warm-blooded prey. In the Serengeti National Park, the lion (*Panthera leo*) is known to have eaten eighteen kinds of mammals and four kinds of birds (Schaller, 1972). The leopard (*Panthera pardus*) has an even more varied diet, preying upon over thirty different species of mammals and birds in the same area (Bertram, 1978). On the other hand, the cheetah (*Acinonyx jubatus*) has a much narrower range of prey species — nine kinds of mammals in the Serengeti (Schaller, 1972) and nine in the Nairobi National Park (Eaton, 1974). Both the caracal (*Felis caracal*) and the serval (*F. serval*) have a preference for small rodents. Rats and mice also constitute the staple food of the wild cat (*F. libyca*) (Stuart, 1977), but Solifugae and Orthoptera were also important in the stomach contents studied by Smithers (1971).

Mammalian secondary consumers are far less numerous in other tropical savannas. In Indian grasslands, canids are represented by the golden jackal (*Canis aureus*) and the Indian fox (*Vulpes benghalensis*). Here their mixed diet is apparently very similar to that of their African relatives (Schaller, 1967; Johnsingh, 1978). The sloth bear (*Melursus ursinus*) is the only tropical bear occasionally entering open areas. Its diet is mostly made up of fruits, termites and honey. This bear has not been observed preying on mammals, and only rarely feeds on carrion (Laurie and Seidensticker, 1977). When found in Indian savannas, the honey badger (*Mellivora capensis*) has apparently the same food habits as in Africa. Viverrids are poorly represented in open areas in India. The small Indian civet (*Viverricula indica*) has a mixed diet of small terrestrial vertebrates, insects, fruits and roots. The common mongoose (*Herpestes ed-*

wardsi) and the small Indian mongoose (*H. auropunctatus*) also have very similar feeding habits. The diet of the striped hyaena in India does not differ much from the diet of the same species in Africa. The tiger (*Panthera tigris*) is basically a forest predator, only occasionally moving into grassy swamps in the Terai or in Kaziranga; its diet in such marginal areas still remains to be studied. The jungle cat (*Felis chaus*) readily lives in open country where it preys upon small rodents and lizards (Schaller, 1967).

Tropical American savannas are even poorer in mammalian carnivores and terrestrial insect-eaters. Canids are represented by the maned wolf (*Chrysocyon brachyurus*), the South American fox [*Dusicyon (Lycalopex) vetulus*] and the crab-eating fox (*Cerdocyon thous*). The first species feeds primarily on small vertebrates [especially wild guinea pigs (*Cavia* spp.)], invertebrates and fruit; the second apparently does not take any vegetable food, whereas the third species has a more catholic diet, being a predator of small mammals and an eater of invertebrates, fruits and eggs of iguana or turtles, as well as a scavenger (Langguth, 1975; Montgomery and Lubin, 1977). The pampas cat (*Lynchailurus colocolo*) and the jaguarondi (*Herpailurus yaguaroundi*) are the only South American felids entering llanos and tree savannas. The diet of the latter ranges from ground-nesting birds to deer. *Procyon cancrivorus* and *Grison vittatus* are reasonably abundant in the Venezuelan llanos.

Before the introduction of the dingo (*Canis familiaris dingo*) by man, the northern Australian savannas did not harbour any wild placental carnivore.

POPULATIONS OF MAMMALIAN SECONDARY CONSUMERS

Nowhere in the tropics has an estimate of the population density of the whole carnivorous and "insectivorous" mammalian fauna ever been made. The few data available relate to the numbers of large predators in five African reserves (Schaller, 1972). They are summarized in Table 20.2. Lions contribute the highest biomass in all areas except the Ngorongoro Crater, where spotted hyaenas occupy the top position. Three reserves out of five support about 1 kg of large predator per 100 kg of

TABLE 20.2

Density (ind. km^{-2}) and biomass (kg km^{-2}) of large predators in five African reserves (after Schaller, 1972)

	Ngorongoro (260 km²)	Manyara (91 km²)	Serengeti Unit (25 500 km²)	Nairobi (115 km²)	Kruger (19 084 km²)
Lion					
Density	0.27	0.38	0.08–0.09	0.22	0.06
Biomass	25.3	37.7	7.6–9.2	21.3	5.7
Leopard					
Density	0.22	0.11	0.03–0.04	0.09	0.03
Biomass	2.3	3.3	0.9–1.2	2.6	1.0
Cheetah					
Density	v.o.	v.o	0.01	0.13	0.01
Biomass			0.3–0.4	4.9	0.5
Spotted Hyaena					
Density	1.84	0.11	0.14	0.10	0.08
Biomass	68.1	4.0	5.1	3.8	2.9
Wild Dog					
Density	v.o		0.01		0.02
Biomass			0.1–0.2		0.3
Total large predators					
Density	2.33	0.60	0.27	0.54	4.91
Biomass	95.7	45.0	14.0–16.1	32.6	10.4
Total prey biomass	10 363	7785	4222	3052	1034
Kg prey per 1 kg of predator	108	174	262–301	94	100

v.o. = visitors only.

large mammalian prey. The ratio in the Serengeti National Park is much smaller, as 62% of the prey biomass consists of migratory ungulates which spend the rainy season in the plains, out of reach of most lions and leopards, which are mostly resident in the woodlands.

Resource partitioning in a community of mammalian secondary consumers has never been comprehensively studied. However, Rautenbach and Nel's (1978) paper gives an insight into the ways the available prey are shared by the 33 species of carnivores found in Transvaal: Fig. 20.1 summarizes their most important findings. In the majority of coexisting species, interspecific competition is avoided through different food sources, differences in prey size (correlated with the body size of the predators), different times of activity and differential use of habitat types.

PREDATOR–PREY RELATIONSHIPS

With the exception of a few food specialists — chiefly consumers of social insects — most carni-

vorous and "insectivorous" mammals are very adaptable in their feeding habits, as described in the preceding pages. However, within a rather broad spectrum of prey, some prey are definitely "preferred" to others in a given locality, and the factors governing prey selection need to be studied in order to understand the predator–prey relationships within the savanna community.

Unfortunately, such studies have seldom been carried out, and then only for large predators such as the African lion. The percentages of kills of the more important prey species recorded in six different study sites are shown in Table 20.3. A glance at these figures indicates that in each area two to five species contribute about three quarters of the food. However, the frequencies of different prey vary greatly from one area to the other. This phenomenon can be explained by a number of factors.

A first factor is prey availability, which can be very variable between different habitats within the same area. This is exemplified by the different percentages of Thomson's gazelles (*Gazella thom-*

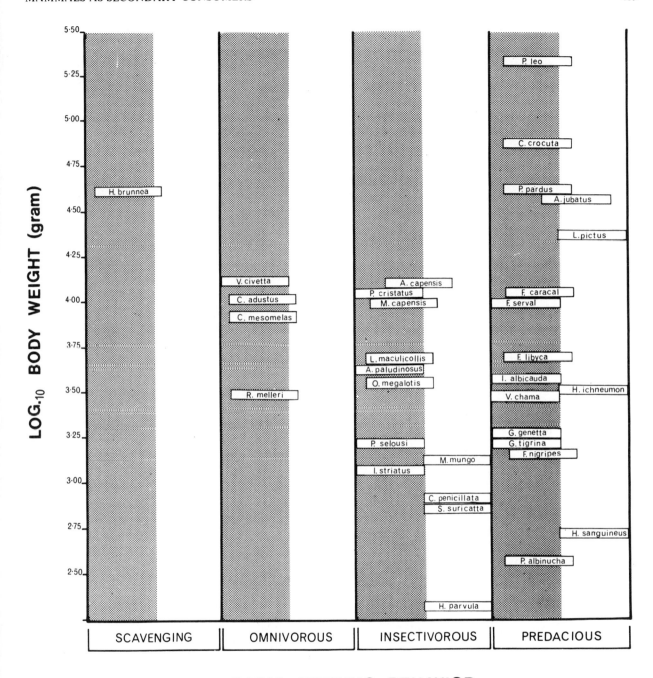

Fig. 20.1. Prey partitioning in Transvaal (South Africa) Carnivora. The four basic categories are presented as vertical columns, each subdivided by a stippled column denoting nocturnal activity and an unstippled column denoting day-light activity. Species are vertically spaced against the *x*-axis representing the log-value of the mean adult body weight in grams. (After Rautenbach and Nel, 1978.)

soni) killed in two adjacent parts of the Serengeti Ecological Unit, the plains and the Seronera area. Most of these gazelles spend the rains on the eastern short-grass plains and move westwards during the dry season, not entering the *Acacia* woodlands but remaining at their edge near Seronera.

The second factor is size of prey. It has long been recognized that large and medium-sized Carnivora

TABLE 20.3

Percentages of total lion kills constituted by the more important prey species in different African savannas

	Serengeti (plains)	Serengeti (Masai and Seronera prides)	Manyara National Park	Nairobi National Park	Kafue National Park	Kruger National Park
Percentage of kills						
wildebeest	56.7	22.0	2.0	25.0	6.1	23.6
zebra	28.9	15.8	16.0	14.6	7.3	15.8
Thomson's gazelle	7.5	50.0	–	0.9	–	–
buffalo	–	2.4	62.0	–	30.5	9.2
hartebeest	0.4	0.2	–	34.5	16.3	–
impala	–	0.2	11.0	0.9	2.0	19.7
warthog	–	2.2	–	6.9	9.5	1.9
Number of kills	280	552	100	116	410	12 313
Number of prey species recorded	9	17	7	11	19	38
Reference	1	1	1	2	3	4

References: 1, Schaller (1972); 2, Rudnai (1973); 3, Mitchell et al. (1965); 4, Pienaar (1969).

generally prey upon herbivores of about the same weight as themselves, avoiding animals that are much smaller than themselves. Only those predators like the lion that hunt in organized groups may succeed in overcoming animals much larger than themselves (Bourlière, 1963). This view has been supported by recent field work. Rudnai (1973), for instance, has found that 87.4% of the lion kills in the Nairobi National Park comprised adult mammals of medium size. The low percentage of small-size prey (11.2%) might be underestimated because of an obvious observation bias. However, the most likely explanation is that small mammals and birds are not hunted much by lions because the energy output in trying to subsist on such prey is not commensurate with the energy intake.

The third factor is "preference" for a certain type of prey. This becomes obvious when, in a given locality, one compares the frequency of species in kills and in the whole prey population. In Nairobi National Park, for instance, Rudnai (1973) has shown that wildebeest (*Connochaetus taurinus*) contributed 25.0% of the kills, although this species made up only 7.0% of the prey population. This was not the case for the syntopic kongoni (*Alcelaphus busephalus*), which contributed 34.5% of the kills and 38.5% of the prey population. This does not necessarily mean that lions prefer the taste of

wildebeest meat to that of the kongoni, and consequently make an effort to catch wildebeest. They may simply catch them more easily. Bertram (1978), however, has found what looked like specialization in different prides, and "pride traditions" cannot therefore be excluded. Malcolm and Van Lawick (1975) have also reported that some wild dog packs in the Serengeti specialize in hunting zebras rather than Thomson's gazelles, even when the two species form mixed herds.

IMPACT OF CARNIVORES UPON HERBIVORE POPULATIONS

Whether or not carnivore populations can control the size of populations of primary consumers has often been debated. Unfortunately long-term studies of game populations have seldom been undertaken in the tropics, and conclusive observations are scanty.

The best data, as far as savannas are concerned, are those from the Serengeti Research Institute, summarized in Tables 20.4 and 20.5. Table 20.4 illustrates an attempt to estimate the total extent of annual mortality among the most numerous large mammal species in the Serengeti area (Houston, 1974). It assumes stable numbers for all species,

TABLE 20.4

Estimate of total annual mortality of most numerous large ungulates in the Serengeti region (simplified after Houston, 1974)

Species	Population size	Number of young born per year	First year mortality (%)	Weight of first-years at death (kg)	Adult mortality (%)	Average adult weight (kg)	Combined weight of adults and first-years (kg)
Wildebeest	800 000	300 000	60	50	12	180	26 280 000
Thomson's gazelle	500 000	550 000	80	5	20	20	4 200 000
Zebra	250 000	100 000	70	70	12	230	11 800 000
Buffalo	70 000	20 000	50	80	6	600	3 320 000
Topi and kongoni	50 000	25 000	50	40	10	130	1 150 000
Impala	75 000	35 000	75	17	10	50	821 250
Total							47 280 000
migratory species							42 280 000
resident species							5 291 250

although it is probable that for some of them (wildebeest and buffalo) recruitment has exceeded mortality in recent years. In such conditions, first-year and adult ungulates dying each year from all causes in the Serengeti Ecological Unit represent a potential food supply of 47 571 250 kg for both predators and scavengers in the area.

By comparison the amount of food eaten by these predators has been estimated in Table 20.5 (where jackals are not included for lack of information). The total food intake of all the large Serengeti predators amounts to no more than 11 631 400 kg — that is about 25% of the total food available from the animals dying per year. Predation therefore cannot be considered as a major cause of mortality in this case, as about three quarters of the "dead biomass" dies from causes other than predation. It is this very considerable

food supply that various vultures (*Gyps* spp.) exploit (Houston, 1974). These vultures are readily able to follow large populations of migratory ungulates which vary in abundance and location with the seasons — a food supply which more sedentary mammalian scavengers cannot easily use. By contrast, predation has been a major factor limiting wildebeest populations in the Ngorongoro Crater, where this species is sedentary (Kruuk, 1972; Schaller, 1972).

Most large predators do not capture their favourite prey at random. Age and condition are important factors influencing the vulnerability of prey. Lions, for instance, take a greater proportion of very young animals from most prey species than would be expected on the basis of their relative numbers in the live population. Old animals are caught relatively often too. This does not mean that

TABLE 20.5

Estimate of food eaten by predators per year in the Serengeti region (after Houston, 1974)

Predator	Estimated population size	Weight of carcasses used per day per animal (including wastage) (kg)	Weight of food eaten per year per animal (kg)	Total weight of food taken by predator species per year (kg)
Hyaena	3000	3	1095	3 285 000
Lion	2400	6	2190	5 256 000
Cheetah	200?	6	2172	434 400
Wild dog	250	7.5	2740	685 000
Leopard	900	6?	2190	1 971 000
			Total	11 631 400

lions catch only young, old or weak victims. As pointed out by Bertram (1978) most of their prey are healthy subadult and adult animals. Moreover, adult males and females of a given species are frequently not equally represented among lion prey. In general more males are killed than females, up to twice as many in some species. This is probably due to the fact that adult male ungulates are more often territorial than females, and therefore more isolated and vulnerable. Schaller (1972) has shown that lions were generally more likely to be successful when hunting single prey individuals than when hunting prey which were in groups. As Bertram (1978) put it, several pairs of eyes, ears and nostrils are better than one at detecting the approach of a predator.

Prey selection and predation pressure have not yet been studied in savanna populations of small mammalian carnivores and "insectivores". Their trophic impact upon prey populations of small vertebrates and invertebrates is however likely to be negligible, compared to the pressure exerted by birds of prey and invertebrate predators, particularly spiders and ants.

Whatever the size of the prey and its reproductive potential, it must not be forgotten that predation constitutes only one of the several forces operating upon populations in an undisturbed ecosystem. Perhaps its most important function, at least in the tropical savannas, is to dampen the tendency of some herbivorous populations to increase beyond the carrying capacity of their range, thus preventing severe oscillations (Schaller, 1972).

EFFECT OF PREY CHARACTERISTICS ON THE ECOLOGY AND BEHAVIOUR OF PREDATORS

Whereas predators do not regulate the size of their prey populations, at least in most cases, there is a distinct possibility that prey abundance and behavioural characteristics do influence both the population size and the hunting strategies of their predators.

The first point can be illustrated by the recent situation in the Serengeti National Park. The increased dry-season rainfall which has taken place since 1971 in this area has improved the productivity of the grasslands, and this larger grass production in turn has led to an increase of both migrant and resident ungulate herbivores. In the same period, the number of resident lion prides has increased by 33%, the mean pride size has increased from fifteen to nineteen, and a threefold increase in the proportion of cubs surviving to adulthood has been noticed. Meanwhile the number of nomadic lions on the plains during the wet season has been substantially reduced. Spotted hyenas have also built up their numbers during the same time interval, and the cheetah population is now considered as flourishing (Hanby and Bygott, 1979). The only large predator whose numbers have diminished over the past ten years is the wild dog; its numbers were already low as compared to those of other large carnivores (Frame et al., 1979; Hanby and Bygott, 1979). In a situation such as this, it is most likely that it was the progressive increase in prey numbers which triggered that of their major predators.

The reverse situation has also been observed. The night counts carried out by Poulet (1974) during the drought spell from 1970 to 1972 in northern Senegal have shown that the numbers of *Canis aureus*, *Vulpes pallida*, *Felis lybica* and *Genetta genetta* declined in numbers at the same time that their mammalian prey became scarcer.

The second point — the relationships between the spatial distribution and the social structure of the prey on the one hand, and the hunting strategy and the social organization of the predator on the other hand — has also attracted a great deal of attention in the last few years (Bertram, 1978, 1979). First of all, it appears that the rule according to which small and scattered food resources are associated with a solitary existence, whereas large and clumped food sources call for a more social mode of life, generally holds true for mammalian predators as well as for mammalian herbivores (Sinclair, 1979). However, sociality takes very different forms, and presumably has different evolutionary origins, in the many species which live in long-lasting social groups. Bertram (1979) has convincingly pointed out that this diversity in social systems can be related to a number of different but interdependent factors.

Obviously group living and its frequent outcome, group hunting, has some definite advantages for a mammalian predator. As already mentioned (Bourlière, 1963; Kruuk, 1975), the group-hunting predators are able to kill prey larger relative to their own body size than the solitary predators. Even within a given species, the short-term advantage of hunting in groups can be demonstrated in comparing the success rate of hunting episodes carried out by a single versus several individuals. For instance, Kruuk (1972) reported a success rate of 15% for a single spotted hyaena, 23% for two hyaenas, and 31% for three and more hyaenas during hunts for wildebeests. Schaller (1972) also observed that lionesses had a success rate of about 15% when hunting alone for gazelles, zebras and wildebeests, and that this was approximately doubled when two or more lionesses hunted together. For the two species of Serengeti jackals, *Canis mesomelas* and *C. aureus*, Lamprecht (1978) also reported a success rate of 36.4% for individual hunts, and 84.6% for animals hunting in groups of two or three.

Furthermore, cooperation between members of a social group of carnivores is not limited to group hunting. Other kinds of what appears to be truly altruistic behaviour have recently been reported. Adult wild dogs feed other adults by regurgitating meat for them; adults can remain on guard with the pups while the rest of the group hunts, nonetheless receiving food when the hunters return (Schaller, 1972); an all-male pack was even observed successfully to rear puppies whose mother had died (Estes and Goddard, 1967). Adult lionesses rear their cubs communally (Schaller, 1972; Bertram, 1975). All brown hyaenas belonging to a single "clan" also raise their cubs together; they bring food to the den, including those without cubs, and suckle cubs other than their own (Owens and Owens, 1979). Bonded pairs of black-backed jackals (*Canis mesomelas*) hunt cooperatively and share food; parental care of cubs is shared approximately equally by both parents, and the male regurgitates to his mate during the period of lactation. Moreover, adult "helpers" are frequently observed in this species, and it has been established in some instances that these "helpers" were pups of the previous year's litter. Such helpers contribute food directly to the pups, and also regurgitate to the mother during the period of lactation (Moehlman, 1979). Male guarding has been reported in the banded mongoose (*Mungos mungo*; Rood, 1974), and "helpers" at the nest are also known in the dwarf mongoose (*Helogale parvula*; Rood, 1978). In the latter species, subordinate adult females and yearlings act not only as babysitters, but also collect insects near

the den site and carry them back to the young. Dwarf mongoose also are capable of showing considerable altruism in their care of a sick member of the pack, modifying their behaviour in such a way as to assist the invalid (Rasa, 1976).

So much altruism directed, in most if not all cases, towards close relatives has quite naturally been interpreted in terms of kin selection. It is therefore all the more interesting that, in two instances at least, cooperation between social carnivores did increase the number of young raised by altruistic individuals. Among black-backed jackals studied by Moehlman (1979), the presence of "helpers" correlated positively with pup survivorship. On average, each "helper" added 1.5 surviving pups to the litter. Similarly, cooperation also confers lifetime advantages on Serengeti lions. Bygott et al. (1979) have shown that, compared with singletons and pairs, male lions in groups of three or more produced more surviving offspring: the total number of cubs surviving to one year old was correlated with their father's group size.

To what extent altruistic behaviour such as the examples just mentioned can be considered as more profitable to mammalian predators than to vegetarians and "omnivorous" mammals is still a matter of debate. In spite of the fact that far fewer field studies of carnivores have been carried out than has been the case for ungulates and primates, more instances of cooperation and altruism are known among carnivores than among other mammalian orders. Close cooperation between troop members is indeed more important for the chase and capture of highly mobile mammalian prey than for the consumption of grass, leaves or fruits, or even the capture of invertebrates with limited powers of locomotion. As Sinclair (1979) put it, "ungulates unlike predators, do not face the problem of their food being larger than themselves and running away from them"; this applies to most primates as well. The adoption of a predator life style might well have been in itself a strong incentive for the development of cooperation and altruistic behaviour in mammals, all the more so since there is a continuous evolutionary race between the hunter and the hunted, leading to an endless refinement of both offensive and defensive strategies.

REFERENCES

Bertram, B.C.R., 1975. Social factors influencing reproduction in wild lions. *J. Zool., Lond.*, 177: 463–482.

Bertram, B.C.R., 1978. *Pride of Lions*. Scribners, New York, N.Y., 253 pp.

Bertram, B.C.R., 1979. Serengeti predators and their social systems. In: A.R.E. Sinclair and M. Norton-Griffiths (Editors), *Serengeti. Dynamics of an Ecosystem*. The University of Chicago Press, Chicago, Ill., pp. 221–248.

Bonaccorso, F., 1975. *Foraging and Reproductive Ecology in a Community of Bats in Panama*. Thesis, University of Florida, Gainesville, Fla., 122 pp.

Bothma, J. du P., 1966. Notes on the stomach contents of certain Carnivora from the Kalahari Gemsbok Park. *Koedoe*, 9: 37–39.

Bourlière, F., 1963. Specific feeding habits of African carnivores. *Afr. Wildl.*, 17(1): 21–27.

Bygott, J.D., Bertram, B.C.R. and Hanby, J.P., 1979. Male lions in large coalitions gain reproductive advantages. *Nature*, 282: 839–841.

Charles-Dominique, P., 1978. Ecologie et vie sociale de *Nandinia binotata* (Carnivores, Viverrides): comparaison avec les prosimiens sympatriques du Gabon. *Terre Vie*, 32: 477–528.

Cooper, R.L. and Skinner, J.D., 1978. Importance of termites in the diet of the aardwolf *Proteles cristatus* in South Africa. *S. Afr. J. Zool.*, 14: 5–8.

Eaton, R.L., 1974. *The Cheetah. The Biology, Ecology and Behavior of an Endangered Species*. Van Nostrand, New York, N.Y., 178 pp.

Eisenberg, J.F. and Thorington, R.W. Jr., 1973. A preliminary analysis of a neotropical mammal fauna. *Biotropica*, 5: 150–161.

Estes, R.D. and Goddard, J., 1967. Prey selection and hunting behaviour of the African wild dog. *J. Wildl. Manage.*, 31: 52–70.

Ewer, R.F., 1973. *The Carnivores*. Weidenfeld and Nicolson, London, 494 pp.

Frame, L.H., Malcolm, J.R., Frame, G.W. and Van Lawick, H., 1979. Social organization of African wild dogs (*Lycaon pictus*) on the Serengeti Plains, Tanzania 1967–1978. *Z. Tierpsychol.*, 50: 225–249.

Hanby, J.P. and Bygott, J.D., 1979. Population changes in lions and other predators. In: A.R.E. Sinclair and M. Norton-Griffiths (Editors), *Serengeti. Dynamics of an Ecosystem*. The University of Chicago Press, Chicago, Ill., pp. 249–262.

Harrison, J.L., 1962. The distribution of feeding habits among animals in a tropical rain forest. *J. Anim. Ecol.*, 31: 56–64.

Heindrichs, H., 1972. Beobachtungen und Untersuchungen zur Ökologie und Ethologie, insbesondere zur sozialen Organisation ostafrikanischer Säugetiere. *Z. Tierpsychol.*, 30: 146–189.

Herzig-Straschill, B., 1977. Notes on the feeding habits of the yellow mongoose *Cynictis penicillata*. *Zool. Afr.*, 12: 225–229.

Houston, D.C., 1974. The role of griffon vultures *Gyps africanus* and *Gyps ruppelii* as scavengers. *J. Zool., Lond.*, 172: 35–46.

Johnsingh, A.J.T., 1978. Some aspects of the ecology and behaviour of the Indian fox *Vulpes benghalensis* (Shaw). *J. Bombay Nat. Hist. Soc.*, 75: 397–405.

Kaufmann, J.H., 1962. Ecology and social behavior of the coati, *Nasua narica* on Barro Colorado island, Panama. *Univ. Calif. Publ. Zool.*, 60: 95–222.

Kingdon, J., 1977. *East African Mammals. Volume III. Part A. Carnivores*. Academic Press, London, 476 pp.

Kruuk, H., 1972. *The Spotted Hyena. A Study of Predation and Social Behavior*. University of Chicago Press, Chicago, Ill., 335 pp.

Kruuk, H., 1975. Functional aspects of social hunting by carnivores. In: G. Baerends, C. Beer and A. Manning (Editors), *Function and Evolution in Behaviour*. Oxford University Press, Oxford, pp. 119–141.

Kruuk, H., 1976. Feeding and social behaviour of the striped hyaena (*Hyaena vulgaris* Desmarest). *E. Afr. Wildl. J.*, 14: 91–111.

Kruuk, H. and Sands, W.A., 1972. The aardwolf (*Proteles cristatus* Sparrman, 1783) as predator of termites. *E. Afr. Wildl. J.*, 10: 211–227.

Lamprecht, J., 1978. On diet, foraging behaviour and interspecific food competition of jackals in the Serengeti National Park, East Africa. *Z. Säugetierkd.*, 43: 210–223.

Lamprecht, J., 1979. Field observations on the behaviour and social system of the bat-eared fox *Otocyon megalotis* Desmarest. *Z. Tierpsychol.*, 49: 260–284.

Langguth, A., 1975. Ecology and evolution in the South American Canids. In: M.W. Fox (Editor), *The Wild Canids*. Van Nostrand, New York, N.Y., pp. 192–206.

Laurie, A. and Seidensticker, J., 1977. Behavioural ecology of the sloth bear (*Melursus ursinus*). *J. Zool., Lond.*, 182: 187–204.

Malcolm, J.R. and Van Lawick, H., 1975. Notes on wild dogs (*Lycaon pictus*) hunting zebras. *Mammalia*, 39: 231–240.

Melton, D.A., 1976. The biology of aardvark (Tubulidentata, Orycteropidae). *Mammal Rev.*, 6: 75–88.

Mills, M.G.L., 1978. Foraging behaviour of the brown hyaena (*Hyaena brunnea* Thunberg 1820) in the southern Kalahari. *Z. Tierpsychol.*, 48: 113–141.

Mitchell, B., Shenton, J. and Uys, J., 1965. Predation on large mammals in the Kafue National Park, Zambia. *Zool. Afr.*, 1: 297–318.

Moehlman, P.D., 1979. Jackal helpers and pup survival. *Nature*, 277: 382–383.

Montgomery, G.G. and Lubin, Y.D., 1977. Prey influences on movements of neotropical anteaters. In: R.L. Phillips and C. Jonkel (Editors), *Proceedings of the 1975 Predator Symposium*. University of Montana, Missoula, Mont., pp. 103–131.

Morris, P.A. and Malcolm, J.R., 1977. The Simien fox in the Balé mountains. *Oryx*, 14: 151–160.

Neal, E., 1970. The banded mongoose *Mungos mungo* Gmelin. *E. Afr. Wildl. J.*, 8: 63–71.

Nel, J.A.J., 1978. Notes on the food and foraging behavior of the bat-eared fox *Otocyon megalotis*. *Bull. Carnegie Mus. Nat. Hist.*, 6: 132–137.

Owens, D.D. and Owens, M.J., 1979. Communal denning and clan associations in brown hyenas (*Hyena brunnea*,

Thunberg) in the Central Kalahari desert. *Afr. J. Ecol.*, 17: 35–44.

Owens, M.J. and Owens, D.D., 1978. Feeding ecology and its influence on social organization in brown hyaenas (*Hyaena brunnea*, Thunberg) of the Central Kalahari desert. *E. Afr. Wildl. J.*, 16: 113–135.

Pienaar, U. de, 1969. Predator–prey relations amongst the larger mammals of the Kruger National Park. *Koedoe*, 12: 108–176.

Poulet, A.R., 1972. Recherches écologiques sur une savane sahélienne du Ferlo septentrional, Sénégal: les Mammifères. *Terre Vie*, 26: 440–472.

Poulet, A.R., 1974. Recherches écologiques sur une savane sahélienne du Ferlo septentrional, Sénégal: quelques effets de la sécheresse sur le peuplement mammalien. *Terre Vie*, 28: 124–130.

Rasa, O.A.E., 1976. Invalid care in the dwarf mongoose (*Helogale undulata rufula*). *Z. Tierpsychol.*, 42: 337–342.

Rautenbach, I.L. and Nel, J.A.J., 1978. Coexistence in Transvaal Carnivora. *Bull. Carnegie Mus. Nat. Hist.*, 6: 138–145.

Rood, J.P., 1974. Banded mongoose males guard young. *Nature*, 248: 176–177.

Rood, J.P., 1975. Population dynamics and food habits of the banded mongoose. *E. Afr. Wildl. J.*, 13: 89–111.

Rood, J.P., 1978. Dwarf mongoose helpers at the den. *Z. Tierpsychol.*, 48: 277–287.

Rudnai, J.A., 1973. *The Social Life of the Lion.* Washington Square East Publishers, Wallingford, Penn., 122 pp.

Schaller, G.B., 1967. *The Deer and the Tiger. A Study of Wildlife in India.* University of Chicago Press, Chicago, Ill., 370 pp.

Schaller, G.B., 1972. *The Serengeti Lion. A Study of Predator–Prey Relations.* University of Chicago Press, Chicago, Ill., 480 pp.

Sinclair, A.R.E., 1979. Dynamics of the Serengeti ecosystem. Process and pattern. In: R.E. Sinclair and M. Norton-Griffiths (Editors), *Serengeti, Dynamics of an Ecosystem.* University of Chicago Press, Chicago, Ill., pp. 1–30.

Smithers, R.H.H., 1971. The mammals of Botswana. *Salisbury, Mus. Mem.*, 4: 1–340.

Stuart, C.T., 1976. Diet of the black-backed jackal *Canis mesomelas* in the Central Namib desert, South West Africa. *Zool. Afr.*, 11: 193–205.

Stuart, C.T., 1977. Analysis of *Felis lybica* and *Genetta genetta* scats from the central Namib desert, South West Africa. *Zool. Afr.*, 12: 239–241.

Sweeney, R.C.H., 1956. Some notes on the feeding habits of the ground pangolin *Smutsia temminckii. Ann. Mag. Nat. Hist.*, 12: 893–896.

Van Lawick-Goodall, H. and Van Lawick-Goodall, J., 1970. *Innocent Killers.* Collins, London, 222 pp.

Wilson, D.E., 1973. Bat faunas: a trophic comparison. *Syst. Zool.*, 22: 14–29.

Zumpt, I.F., 1968. The feeding habits of the yellow mongoose *Cynictis penicillata*, the Suricate *Suricata suricata*, and the Cape ground-squirrel *Xerus inauris. J. S. Afr. Vet. Med. Assoc.*, 39: 89–91.

Chapter 21

THE SOIL FAUNA OF TROPICAL SAVANNAS.
I. THE COMMUNITY STRUCTURE

PATRICK LAVELLE

INTRODUCTION

The fauna of the soil compartment of terrestrial ecosystems is rich in species and varied in function. Following Bachelier (1971), it can be considered that the two major characteristics which determine the place of a given soil animal within an edaphic community are its adult body size and its mode of respiration. Indeed, both of these characteristics emphasize the ways in which the organism is adapted to the two major features of the soil environment: the very limited space available for living animals within the soil, and the watery nature of this milieu.

Three main life-styles can be observed among soil animals. The smallest are definitely aquatic and spend their whole lifetime within the very small amount of water found in soil pores of various size. Such hydrobionts — soil protozoans, rotifers and nematodes — very seldom reach a body length of 0.2 mm; they constitute the soil microfauna.

The micro-arthropods and enchytraeid earthworms are a little larger in size. Although depending heavily on soil moisture, they have an aerial mode of life, and their size which ranges from 0.2 to 4 mm, enables them to move freely within the soil. They make up the soil mesofauna.

Burrowing animals constitute the soil macrofauna. Their size being larger than that of soil pores, they have to force their way actively through the soil. Some of them do it by using powerful fossorial fore legs; this is the case for mole crickets (Gryllotalpidae) and cicada larvae (Cicadidae). Other animals, such as earthworms, literally eat their way through the soil, whereas ants and termites build elaborate subterranean networks of galleries and nest chambers where they spend most of their lifetime.

The structure of a soil animal community is based upon the trophic relationships of its member species. The staple food source is the litter, made up both of leaves and dead wood on the soil surface, and the roots and soil organic matter within the soil itself. In most tropical savannas, the litter is scarce as it is regularly destroyed by grass fires (see Ch. 4 and Ch. 30); roots consequently represent the major food source for soil animals (Lavelle and Schaefer, 1974). They are eaten live or dead, or as amorphous humic material once decomposed. It is also possible, if not likely, that many consumers of dead organic material also feed actively upon living micro-organisms. Such trophic relationships between animals and micro-organisms within the soil being so far little studied, four major trophic categories of soil animals can provisionally be distinguished: the litter eaters, the root eaters, the soil eaters and the predators.

The structure of the animal communities of savanna soils is also greatly influenced by some environmental factors, even more so than in many other terrestrial ecosystems. Seasonality of climate acts strongly upon soil moisture and soil temperature, thus exerting a strong selective pressure upon soil organisms.

TAXONOMIC STRUCTURE OF SOIL ANIMAL COMMUNITIES IN TROPICAL SAVANNAS

Four study sites have been chosen to illustrate the taxonomic structure of soil animal communities in tropical savannas, as compared with those of the tropical rain forest and temperate latitudes. To make this inter-biome comparison even more valid, the sites studied were sampled in the same way, at the same time of the year and by the same observer in three cases out of four (Table 21.1).

The derived (southern Guinean) savanna of Lamto is located in the south of the Ivory Coast. The mean annual rainfall is 1275 mm and the dry season is of a short duration and not rigorous; the soil pF never exceeds 4.2 for more than a month, on the average (Athias et al., 1975). The Laguna Verde pastures, in the State of Vera Cruz, Mexico, were created a century ago at the expense of a mature rain forest. The mean annual rainfall reaches 1500 mm and some forest relics allow a com-

parison with the parent soil animal community (Lavelle et al., 1981).

The Foro Foro northern Guinean savanna is located about 250 km north of Lamto in the Ivory Coast. Although the annual rainfall averages 1150 mm, the climate is harsher than in the southern Guinean savanna and the soil remains at a pF of over 4.2 for at least two months every year (Lavelle, unpubl. results). The temperate grassland of Spiboke, shown here by comparison, is located in the south of Sweden; the mean annual rainfall is 520 mm and the average yearly soil temperature is 4.8°C (Persson and Lohm, 1977).

The microfauna of tropical savanna soils has seldom been studied, and quantitative studies are

TABLE 21.1

Numbers (N; ind. m^{-2}) and biomasses (B; $g\,m^{-2}$) of various soil animal groups in various tropical and temperate sites

Taxa	Lamto[1] (Ivory Coast) southern Guinean savannas		Foro Foro[2] (Ivory Coast) northern Guinean savannas		Laguna Verde[2] (Mexico) pastures		tropical forest		Spiboke[1] (Sweden) temperate grassland	
	N	B	N	B	N	B	N	B	N	B
MICROFAUNA										
Protozoa	32×10^6	0.21								
Nematoda	1.1×10^6	0.35								
MESOFAUNA										
Enchytreidae	700	0.30							23 800	3.40
Acari	17 500	0.18			25 000		49 300		112 000	0.52
Collembola	1 800	0.045			4400		7000		109 000	0.56
Other micro-arthropods	5100	0.47			3200		2400		1100	0.08
Total micro-arthropods	25 100	1.00			32 600		58 700		222 100	1.16
MACROFAUNA										
Earthworms	230	49.0	460	22.3	700	47.0	132	9.80	133	23.72
Chilopoda	7	0.037	44	0.22	14	0.10	106	0.76	0	0
Diplopoda	43	0.326	59	0.51	19	0.60	311	12.54	2	0.12
Arachnida	4	0.025	3.2	0.10	74	0.90	129	1.40	197	0.16
Coleoptera	230	0.285	28	1.30	320	45.0	122	2.40	1420	11.60
Diptera	16	0.028	0.5	0.05	60	0.10	79	0.60	5600	2.61
Hymenoptera	500	2.00	1400	2.10	570	1.09	1400	1.62	109	0.096
Isoptera	910	1.95	1200	2.80	2		500	1.07	0	0
Others	75	0.094	12	0.30	71	1.20	232	3.39	103	0.28
Total macrofauna	2015	53.75	3207	29.68	1830	96.00	3011	33.58	7564	38.59

[1]Annual average. [2]Wet season.

particularly scarce. The density of soil protozoans reaches 32×10^6 ind. m^{-2} at Lamto (Couteaux, 1976, 1978; Buitkamp, 1971); this is a low density, as compared with the figures published for some temperate sites, which range from 100 to 500×10^6 ind. m^{-2} (Bachelier, 1971). The situation is the same for soil nematodes. Their density in Lamto has been estimated at 1.1×10^6 ind. m^{-2} at Lamto (Malcevschi, 1978) and from 3 to 8×10^6 ind. m^{-2} in various moist Uganda savannas (Banage and Visser, 1967); in temperate grasslands the nematode density averages 10×10^6 ind. m^{-2} (Bachelier, 1971).

The mesofauna of tropical savanna soils is better known, and the results reached by the various investigators are generally in accordance. The enchytraeid earthworms, very numerous in cold and acid soils, are very scarce here; their density never exceeds 1000 ind. m^{-2}, whereas it averages 23 800 ind. m^{-2} at Spiboke, Sweden, and reaches 100 000 ind. m^{-2} in the arctic tundra and 750 000 ind. m^{-2} in the boreal forest (Swift ct al., 1979). The average density of micro-arthropods is of the order of 20 000 to 30 000 ind. m^{-2} (Salt, 1952, 1955; Covarrubias et al., 1964; Ryke and Loots, 1967; Belfield, 1971; Athias, 1974; Lavelle et al., 1981). Acari are the most numerous, whereas Collembola are far less abundant. Micro-arthropods are more numerous in tropical rain-forest soils; in the Laguna Verde (Mexico) relict forest patches they reach 58 700 ind. m^{-2}, and 72 700 ind. m^{-2} in an Amazonian forest (Beck, 1971). In Sweden, densities can reach 222 100 ind. m^{-2} at Spiboke.

Four taxonomic groups are dominant over the others in the macrofauna: earthworms, termites, ants and beetles, the latter being mostly represented by larvae.

In the tropical savannas and grasslands studied to date, the earthworm communities reach densities ranging from 234 to 700 ind. m^{-2}, with biomasses varying from 22.3 to 49.0 g fresh weight m^{-2}. Such values clearly exceed those found in nearby rain-forest areas (for instance 9.8 g m^{-2} in Laguna Verde relict forest patches, and 3.5 g m^{-2} in the Lamto gallery forests); they are of the same order of magnitude as the Spiboke biomass (23.7 g m^{-2}), but much lower than those of warmer temperate regions in Europe, where figures ranging from 56 to 287 g fresh weight m^{-2} have been reported (Edwards and Lofty, 1972). Earthworms are there-

fore the dominant taxonomic group in many tropical savanna soils.

The numbers of ants and termites found in savannas are very high, but their biomass is small, due to their diminutive size. At best, they reach a few grams per square metre; but the specific roles of these insects and their omnipresent activities make them very important components of the soil animal community. Furthermore, these arthropods are less dependent on environmental conditions than earthworms; therefore, their relative importance increases in the drier savannas where the earthworm community is far less numerous (see Ch. 22).

Termites are very numerous in many tropical grasslands, and their functions within the soil compartment of the savanna ecosystem are very diverse (see Ch. 23); however, their role in Laguna Verde grasslands was very limited as their populations were small. This might be a general feature of Neotropical grasslands, but needs to be supported by more data. The density of ants does not vary much between the various savanna categories, although it increases in shrub savannas where resources are more varied than in open grasslands (see Ch. 24).

Beetles are of secondary importance, as compared to termites and ants. In some situations, however, they can reach a sizable biomass. Such was the case in Laguna Verde grasslands, where the biomass of beetle larvae was of the same order of magnitude as that of earthworms. Myriapoda are not numerous in savanna soils, being more abundant in forest environments.

On the whole, the densities and biomasses of animal communities of the tropical savanna soils are lower than those of temperate grasslands, but greater (at least in biomass) than those of tropical rain forests. However, this relative poverty is more apparent than real. In tropical latitudes the high metabolic rate of soil invertebrates is much greater than that of a similar biomass of related species in temperate conditions, due to the high soil temperatures prevailing throughout the year.

DISTRIBUTION IN SPACE

One of the fundamental characteristics of the soil environment is its disposition in more or less distinct horizontal layers, termed soil horizons,

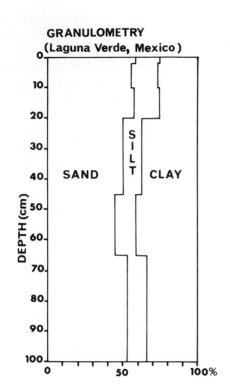

GRANULOMETRY
(Laguna Verde, Mexico)

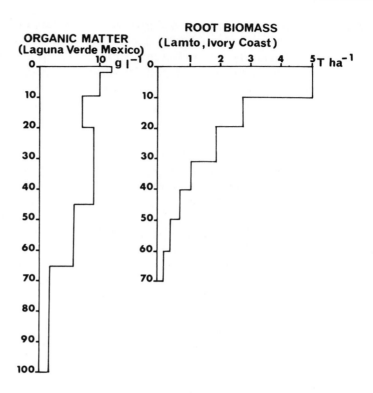

ORGANIC MATTER
(Laguna Verde Mexico)

ROOT BIOMASS
(Lamto, Ivory Coast)

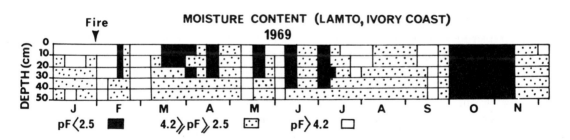

MOISTURE CONTENT (LAMTO, IVORY COAST)
1969

pF ⟨2.5 ■   4.2 ⟩ pF ⟩ 2.5 ⦂   pF ⟩ 4.2 ☐

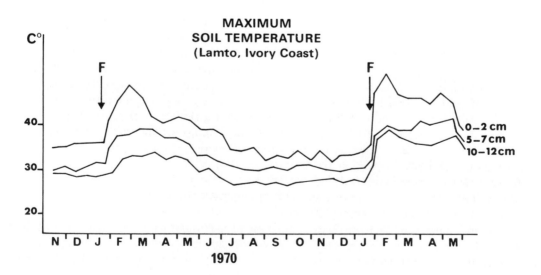

MAXIMUM
SOIL TEMPERATURE
(Lamto, Ivory Coast)

which differ in their physical, chemical and biological characteristics (Fig. 21.1). This also applies to savanna soils, and it is possible to identify species peculiar to the litter and others that are found mostly in deeper soil horizons (Table 21.2). The vertical layering of soil animals is the end result of the vertical gradients in temperature, organic matter content, texture and moisture existing within the soil.

Besides this vertical stratification, the distribution of soil animals in an horizontal plane is also influenced by the plant cover in savanna communities. For example, the soil arthropods are always more numerous under grass tussocks than under the bare areas that are found between clumps of grass (Athias, 1974; Malcevschi, 1978). This is shown in Table 21.3 for the Lamto savanna. The microrelief also plays a role; some earthworm populations apparently congregate under areas where rain water flows out most frequently (Fig. 21.2) (Lavelle, 1978). The presence of trees also influences the distribution of soil animals. The end result of all these factors is an horizontal mosaic of areas characterized by their microrelief and vegetation, each being in turn subject to vertical gradients of food availability and microclimatic

TABLE 21.2

Vertical distribution of various animal taxa in Laguna Verde (Mexico) pastures, and in near-by forest patches; the figures given are the percentages of the total population sample found in different soil layers

Taxa	Pastures				Taxa	Forest patches				
Depth (cm):	0–10	10–20	20–30	30–40		litter	0–10	10–20	20–30	30–40
Glossoscolecidae	91.9	6.8	1.2		Blattaria	63.3	13.3	23.3		
Carabidae (larvae)	91.6	8.4			Araneae	62.9	26.3	5.9	3.2	
Isopoda	90.9	9.1			Isopoda	54.6	34.3	11.1		
Elateridae (larvae)	88.2	7.9	1.3	2.6	Hemiptera	41.4	44.8	1.4		
Blattaria	86.2	13.8			Formicidae	29.2	54.9	15.9	9.5	7.4
Hemiptera	84.1	13.6	0.3		Iulidae	28.7	58.7	7.6	4.9	
Melolonthinae (larvae)	83.3	12.8	3.7		Chilopoda	21.9	54.4	18.5	4.2	1.0
Diptera (larvae)	83.0	13.2	3.8		Diptera (larvae)	18.5	43.1	30.8	7.7	
Araneae	82.2	11.1	6.8		Coleoptera (adults)	16.2	37.8	27.0	6.8	
Megascolecidae	81.3	11.4	3.2	0.2	Symphyla	9.0	74.6	14.9	1.5	
Coleoptera (adults)	76.2	17.8	1.5	4.4	Elateridae (larvae)	7.5	71.2	18.2	3.0	
Formicidae	72.0	15.3	9.4	3.3	Mermithidae	4.7	9.4	32.9	52.9	
Eumolpinae (larvae)	30.0	19.1	35.5	15.5	Chrysomelidae (larvae)	4.0	50.0	21.0	14.0	3.3
Mermithidae	25.7	31.3	37.4	8.7	Polydesmida	3.7	90.0	4.2	2.1	
Symphyla	17.5	60.8	14.8		Oligochaeta	0.1	70.2	22.4	6.2	
					Melolonthinae (larvae)		53.3	30.8	15.0	0.8

TABLE 21.3

The influence of the rhizosphere on the density of soil nematodes and soil micro-arthropods in the Lamto savannas (after Athias, 1974; and Malcevschi, 1978)

	Nematodes (ind. kg^{-1} of dry earth)	Micro-arthropods (ind. m^{-2}, yearly average)
Near the rhizosphere	8975 ± 1890	16 600
Far from the rhizosphere	2395 ± 135	36 900

Fig. 21.1. Variations with depth of some microclimatic and trophic soil parameters (after César and Menaut, 1974; Lavelle, 1978; Lavelle et al., 1981).

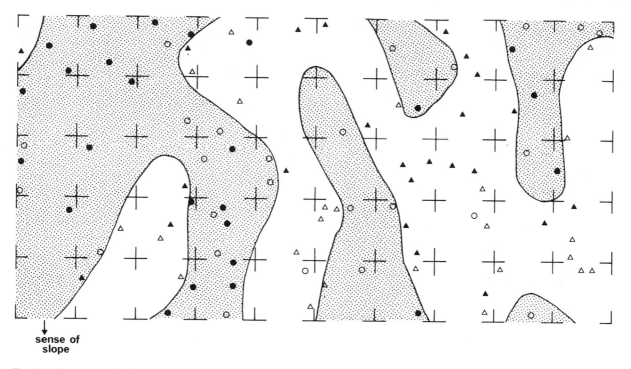

sense of slope

Fig. 21.2. Horizontal distribution pattern of eudrilid earthworm populations in a grass savanna at Lamto, Ivory Coast. Classification of the twelve monthly samples according to their density: ● = 1–3 months; ○ = 4–6 months; △ = 7–9 months; ▲ = 10–12 months. Dotted areas indicate zones of high concentration of worms (Lavelle, 1978). Seale: 15 mm = 10 m.

conditions. Such a pattern of distribution of soil animals is not at all restricted to savanna ecosystems, but it is quite probably more pronounced here than elsewhere, because of the marked seasonal variations in temperature and moisture gradients, as well as the clumped distribution of grass roots in the top soil.

THE TROPHIC STRUCTURE OF SOIL ANIMAL COMMUNITIES IN TROPICAL SAVANNAS

It is still far too early to attempt a description of the food web of the animal community in the soil of tropical savannas. Too little is known concerning the diet and feeding habits of the hundreds of participating species, mostly among the arthropods. All that can be done at present is to divide up the major taxonomic categories into a few broad trophic groups, in order to understand the ways in which the available resources are allocated (Table 21.4).

The soil eaters, mostly represented by earthworms and some humivorous termites, appear to be the most important group; these soil animals always represent more than 50% of the soil animal biomass. Root eaters, very scarce in the Lamto savanna (0.3% of the biomass), amount up to 40.3% in Laguna Verde grasslands; this is due to the abundance, possibly seasonal, of beetle larvae in this study site. Detritivores, mostly litter eaters, represent 17.6% of the soil animal biomass in annually burnt savannas at Lamto. This percentage increases with the importance of bush encroachment. After ten years of fire protection in Lamto it reaches 38.5%. The acompanying increase of predators, from 4.6 to 8.9%, can be explained by the fact that these animals mostly prey upon detritus feeders.

In the rain forest the dominant group is also constituted by the detritus eaters and their predators (65.9% at Laguna Verde), to the detriment of the soil eaters (30.6%) and the root eaters (3.8%).

Inasmuch as the limited number of data can allow one to draw any preliminary conclusion, the soil fauna of tropical savannas can be characterized by a predominance of soil-eating animals, which are always much more numerous than in temperate

TABLE 21.4

Biomass percentages of the four main trophic categories of soil animals at Lamto and Laguna Verde (after Athias et al., 1975; and Lavelle et al., 1981)

	Soil eaters	Root eaters	Detritus eaters	Predators
Lamto, burnt savanna	77.0	0.3	17.6	4.6
unburnt savanna	52.5	0.3	38.5	8.9
Laguna Verde, pastures	54.2	40.3	5.5	
forest patches	30.6	3.8	65.6	

grasslands. The increasing importance of detritus eaters and predators, combined with the progress of bush encroachment, also appears to be a characteristic feature of these communities.

ENVIRONMENTAL FACTORS INFLUENCING THE STRUCTURE OF ANIMAL COMMUNITIES IN SAVANNA SOILS

The soil environment is actually much more complex than it commonly appears. Climatic, edaphic and biotic parameters interact continuously, not to mention the interference of man. Only a multivariate analysis of the data collected at different times of the year on a single site, together with the comparison of different facies of a single savanna type, can help to identify the roles played by the various environmental factors.

Such an analysis has been carried out in the Lamto savanna by Lavelle and Meyer (1980). Eight factors have been identified which are responsible for 63.7% of the total variance for the thirteen variables studied. Most of these factors can be linked to environmental variables such as grass fires, soil moisture, soil temperature, microclimatic consequences of an increase in shrub cover and physical structure of the soil.

As discussed in Chapter 30, the effect of grass fires is manifold. The rise in temperature of the upper parts of the soil profile during the passage of the burning front not only immediately kills a number of organisms, but decreases soil moisture so that the pF rises above 4.2, a value which is lethal for many hygrophilous species. However, the most important consequence of grass fires for soil animals is the destruction of the litter which results in a drastic reduction of food resources for the soil animals. Repeated destruction of the litter in tropical savannas probably explains the predominance of soil eaters in this kind of environment, the only abundant and continuously available food sources being the roots and the soil organic matter.

The water regime of the soil obviously depends on the yearly amount of rainfall and its seasonal pattern of distribution; but it is also determined by the water-holding capacity of the soil, the local topography and the nature of the vegetation cover. As shown in Chapter 22, earthworms are most sensitive to changes in soil moisture, but this is also true of some other taxonomic groups such as the Collembola and Myriapoda. Conversely, the soil mites (Acari) apparently favour drier conditions (Athias, 1974; Athias et al., 1975) and social insects are little affected by this factor.

Changes in soil temperature have multivarious consequences. They influence the circadian rhythms of activity of many species (Lavelle, 1971; Lévieux, 1971) and the overall activity level of some populations (Lavelle, 1978), as well as their vertical movement within the soil (Belfield, 1971; Athias, 1974).

The nature of the plant cover is also important, as it affects both the microclimatic conditions of the soil below the vegetation layer, and the amount of food made available to soil organisms in the litter. Whereas the shade afforded by vegetation tends to reduce water loss at the soil level, it must be remembered that the evapotranspiration of shrubs exceeds that of grasses. This leads to a decrease in the water content of the soil at depths ranging from 20 to 40 cm, where the density of shrub roots is the highest. At Lamto, for instance, the increase in numbers of litter-eating animals corresponds to a decrease in soil-eating earthworms when shrub cover becomes more important in fire-protected savannas.

The chemical and physical structure of the soil is

also of great importance for soil animals. The amount of organic matter available in tropical savannas is always small; it does not exceed 1 or 2% in the topsoil, but its local variations can influence the local distribution of soil animals (Lavelle et al., 1979). The granulometric composition of the soil is often even more important, as it influences its water-holding capacity and its porosity — hence the ability of animals to penetrate into it.

The depth of the soil is also an important factor for soil animals which have to move in deeper horizons during the dry season. In some instances the shallowness of the soil, or the presence of a hard gravelly horizon, can be important limiting conditions for soil organisms.

REFERENCES

Athias, F., 1974. Les Microarthropodes du sol. In: *Analyse d'un écosystème tropical humide: la savane de Lamto (Côte d'Ivoire). V. Les organismes endogés. Bull. Liaison Chercheurs Lamto, Numéro Spéc.*, 5: 55–89.

Athias, F., Josens, G. and Lavelle, P., 1975. Traits généraux du peuplement endogé de la savane de Lamto (Côte d'Ivoire). In: *Fifth Int. Congr. Soil Zoology, Prague, 1973. Progress in Soil Zoology.* Academia, Prague, pp. 389–397.

Bachelier, G., 1971. La vie animale dans les sols. In: P. Pesson (Editor), *La vie dans les sols.* Gauthier-Villars, Paris, pp. 1–82.

Banage, W.B. and Visser, S.A., 1967. Microorganisms and Nematodes from a virgin bush site in Uganda. In: O. Graff and J.E. Satchell (Editors), *Progress in Soil Biology.* North-Holland, Amsterdam, pp. 93–101.

Beck, L., 1971. Bodenzoologische Gliederung und Charakterisierung des amazonischen Regenwaldes. *Amazoniana*, 3: 69–132.

Belfield, W., 1971. The effect of shade on the arthropod population and nitrate content of a west African soil. In: *Organismes du sol et production primaire. IVe Colloquium Pedobiologiae. Ann. Zool. Ecol. Anim.*, Volume hors série, pp. 557–567.

Buitkamp, U., 1979. Vergleichende Untersuchungen zur Temperaturadaptation von Bodenciliaten aus klimatische verschiedenen Regionen. *Pedobiologia*, 19: 221–236.

César, J. and Menaut, J.C., 1974. Le peuplement végétal des savanes de Lamto. In: *Analyse d'un écosystème tropical humide: la savane de Lamto (Côte d'Ivoire). Bull. Liaison Chercheurs Lamto, Numéro Spéc.*, 2: 1–161.

Couteaux, M.M., 1976. Etude quantitative des Thécamoebiens d'une savane à *Hyparrhenia* à Lamto (Côte d'Ivoire). *Protistologica*, 12: 563–570.

Couteaux, M.M., 1978. Etude quantitative des Thécamoebiens édaphiques dans une savane à *Loudetia* à Lamto (Côte d'Ivoire). *Rev. Ecol. Biol. Sol*, 15: 401–412.

Covarrubias, R., Rubio, F. and Di Castri, F., 1964. Observaciones ecologico-cuantitativas sobre la fauna edafica de zonas semi-aridas del norte de Chile. *Monogr. Ecol. Biogeogr. Chile, Bol. Prod. Anim.*, A 2: 1–109.

Edwards, C.A. and Lofty, J.R., 1972. *Biology of Earthworms.* Chapman and Hall, London, 283 pp.

Lavelle, P., 1971. Etude préliminaire de la nutrition d'un ver de terre africain *Millsonia anomala* (Acanthrodrilidae, Oligochètes). In: *Organismes du sol et production primaire. IVe Congrès International de Zoologie du Sol. Ann. Zool. Ecol. Anim.*, Volume hors série, pp. 133–145.

Lavelle, P., 1978. *Les vers de terre de la savane de Lamto (Côte d'Ivoire): peuplements, populations et fonctions dans l'écosystème.* Thesis, University of Paris VI, Paris, 301 pp.

Lavelle, P. and Meyer, J.A., 1980. Du tri des données à l'élaboration de modèles de simulation: exemple de l'étude écologique des vers de terre de la Savane de Lamto (Côte-d'Ivoire). In: J.P. da Fonseca (Editor), *Colloque Informatique et Zoologie.* Informatique et Biosphère, Paris, pp. 311–326.

Lavelle, P. and Schaefer, R., 1974. Les sources de nourriture des organismes endogés. In: *Analyse d'un écosystème tropical humide: la savane de Lamto (Côte d'Ivoire). Bull. Liaison Chercheurs Lamto, Numéro Spéc.*, 5: 27–37.

Lavelle, P., Maury, M.E. and Serrano, V., 1981. Estudio cuantitativo de la fauna del suelo en la region de Laguna Verde, Vera Cruz. Epoca de lluvias. In: P. Reyes Castillo (Editor), *Estudios Ecologicos en el Tropico Mexicano.* Instituto de Ecologia, Mexico, Publ. No. 6, pp. 65–100.

Lévieux, J., 1971. *Données écologiques et biologiques sur le peuplement en fourmis terricoles d'une savane préforestière de Côte d'Ivoire.* Thesis, University of Paris VI, Paris, 300 pp.

Malcevschi, S., 1978. Considérations quantitatives sur la nématofaune d'une savane herbeuse à Andropogonées de Lamto (Côte d'Ivoire). *Rev. Ecol. Biol. Sol.*, 15: 487–496.

Persson, T. and Lohm, U., 1977. Energetical significance of the annelids and arthropods in a Swedish grassland soil. *Ecol. Bull.*, 23: 1–211.

Ryke, P.A. and Loots, C.C., 1967. The composition of the microarthropod fauna in a South African soil. In: O. Graff and J.E. Satchell (Editors), *Progress in Soil Biology.* North-Holland, Amsterdam, pp. 538–546.

Salt, G., 1952. The arthropod population of pasture soil in some East African pastures. *Bull. Entomol. Res.*, 43: 203–320.

Salt, G., 1955. The arthropod population of soil under elephant grass in Uganda. *Bull. Entomol. Res.*, 46: 539–550.

Swift, M.J., Heal. O.W. and Anderson, J.M., 1979. *Decomposition in Terrestrial Ecosystems.* Blackwell, Oxford, 371 pp.

Chapter 22

THE SOIL FAUNA OF TROPICAL SAVANNAS.
II. THE EARTHWORMS

PATRICK LAVELLE

DISTRIBUTION

There are five large families of earthworms in the world, the Moniligastridae, Megascolecidae, Eudrilidae, Glossoscolecidae and Lumbricidae (Jamieson, 1971), but only four of them are represented in the tropics. The six genera of Moniligastridae are restricted to tropical Asia and eastern Africa; the four subfamilies and 124 genera of Megascolecidae are spread all over the tropics in the eastern and western hemispheres; the 44 genera of Eudrilidae are found in tropical Africa, and the 56 genera of Glossoscolecidae live in the Neotropics (Edwards and Lofty, 1972; Reynolds and Cook, 1976). The four sub-families and 26 genera of Lumbricidae are basically Palaearctic in distribution and are mostly limited to temperate and cold latitudes. They are sometimes found in warm temperate climates, such as those of the extreme south of Africa, Australia and New Zealand, though never entering tropical savannas (Lee, 1959; Reinecke and Ljungström, 1969).

The structure and morphology of the four tropical families of earthworms closely resemble those of the most primitive Oligochaeta. Except in the Moniligastridae and some Eudrilidae, the gizzard and the clitellum are located in the anterior part of the animal (segments 5–6 for the gizzard and 13–20 for the clitellum). These organs are placed much further back in Lumbricidae, which allows for the development of a strong musculature, very seldom found in tropical earthworms (Bouché, 1972).

ECOLOGICAL CATEGORIES

The obvious relationships existing between structure and life history of earthworms have led many authors to individualize a number of life-forms (Franz, 1950; Lee, 1959; Bouché, 1971, 1977). Among the Lumbricidae, Bouché thus recognized the following three major ecological categories:

(1) The litter-dwelling or **epigeic** earthworms live in the soil litter or other media rich in organic matter such as dung, or under bark. They are of small size and very active, coloured red in forest or formerly forested habitats and green in grasslands. Their burrowing musculature is reduced and they spend the harsh period(s) of the year as cocoons. Their life-span is short, their energy metabolism high and their population turnover rapid.

(2) The soil-dwelling or **hypogeic** earthworms live within the soil and feed upon its humic substances or sometimes on dead roots. Their size is moderate to large and they are not pigmented. Their burrowing musculature is well developed, and they can withstand unfavourable environmental conditions by becoming inactive, resuming an active life as soon as temperature and/or soil moisture again reach suitable levels. Their energy metabolism is low, and population turnover is slow.

(3) The **anecic** earthworms are for the most part large-sized animals which spend most of their lifetime within the soil, occasionally coming to its surface to feed on litter material. Their colour is usually dark brown, pigmentation being often limited to the dorsal, if not anterodorsal, part of the body in species which usually leave their terminal segments in their burrow. The burrowing musculature is well developed, and these earthworms become inactive when adverse conditions prevail. Their population dynamics appear to be variable from one species to another.

Hence, the adaptive strategies of the three above defined life-forms are very different. Litter-dwelling earthworms offset the heavy mortality rate due to the instability and insecurity of their environment by a high fecundity and a rapid growth rate, in turn made possible by an energy-rich food, the litter. On the other hand, the soil-dwelling earthworms being less exposed in their rather stable soil environment, can offset the disadvantage of a reduced fecundity and slow growth rate resulting from an energy-poor diet. Anecic earthworms combine the advantages of both categories.

These life-forms were defined by Bouché (1971) for Palaearctic lumbricids, but they apply to other families as well. However, true anecic earthworms have not yet been found in tropical habitats (Lavelle, 1978; Lavelle et al., 1981). The presence of a gizzard in the anterior part of the animal has apparently made the development of a strong musculature impossible. On the other hand, the soil-dwelling and soil-eating (**geophagous**) earthworms are very numerous in the tropics, and three ecological sub-categories can be distinguished — oligohumic, mesohumic, and polyhumic — depending upon the amount of humic substances present in their preferred habitat.

Oligohumic earthworms live in soils very poor in organic matter — that is, in the deeper portion of the soil profile. **Mesohumic** species are found between the soil surface and a depth of about 20 cm, where the soil is moderately rich in organic substances. On the other hand, **polyhumic** earthworms feed upon pockets of soil enriched with organic debris, such as small decomposing roots, or in the thin surface layer rich in organic matter; their small size may also lead them to ingest the smallest soil particles which are most often also the most energy-rich.

PROXIMATE AND ULTIMATE FACTORS INFLUENCING THE ACTIVITY AND DISTRIBUTION OF SAVANNA EARTHWORMS

A number of environmental parameters, both physical and biotic, play a key role in the activity cycle and distribution of tropical earthworms and help one to understand their community structure.

Soil moisture

This is quite certainly the most important of all environmental variables for earthworms in tropical soils. It obviously depends first of all on the amount of rainfall and its pattern of seasonal distribution. But other factors also have to be taken into consideration, among them the physical characteristics of the soil (its water-holding capacity), the local topography, the nature and density of the vegetation cover, and the regular occurrence of grass fires (Clément, 1980). As shown in Fig. 22.1, the water content of a savanna soil not only varies continuously throughout the year, but is very different in the various facies of the same savanna landscape. At Lamto (Ivory Coast), the soil of the study plot, located in a shallow depression in grass-derived savanna, is more humid than that of the shrub savanna plot on the plateau. The dryest soils of the Lamto area are those of the gallery forests, since the evaporation rate is the highest there. Generally speaking, grass savanna soils are more humid than those of the shrub and tree savannas for the same reason (Lavelle, 1978).

As a consequence of their cutaneous respiration, earthworms are extremely sensitive to drought. As soon as soil moisture reaches a given threshold, usually a little above the wilting point for plants (pF 4.2), earthworms empty their gut, coil up inside a little earth cavity plastered with mucus, and become inactive (Lavelle, 1971a). If the rains are delayed for too long, most of them will die, but their cocoons are usually far more resistant to drought than adult individuals, and will ensure the future of the population. The duration of the dry season is therefore of prime importance in determining earthworm distribution; they cannot live in the driest (Sahelian) savannas. Soil moisture also influences the activity of earthworms, as shown in Fig. 22.2.

Soil temperature

It is well established that under natural conditions the soil temperature of tropical savannas remains remarkably stable throughout the year. However, when any disturbance takes place which destroys most of the plant cover, whether it be by burning or mowing, the temperature conditions are totally changed and the upper layers of the soil can become hotter than the surrounding atmosphere. For instance, a few days after burning, when a thick layer of black ashes still remained on the ground, Athias (1974) recorded a temperature exceeding

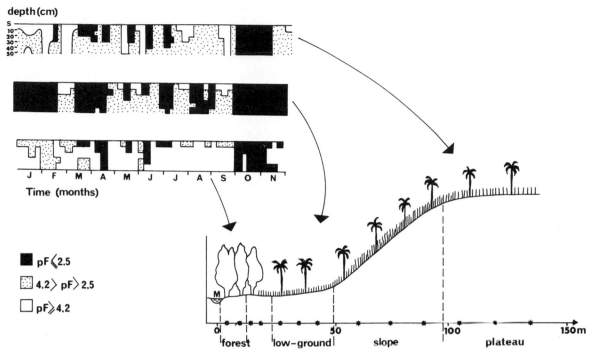

Fig. 22.1. Water regime of soils at different topographic levels in the same savanna landscape (Lamto, Ivory Coast, 1969; after Lavelle, 1978).

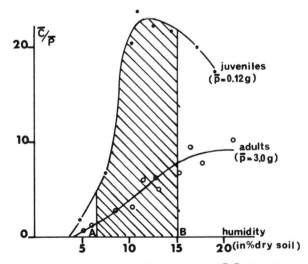

Fig. 22.2. Variations in the soil consumption ($\bar{C}/\bar{P}$) of *Millsonia anomala* (Megascolecidae) as a function of soil moisture, all other factors being equal (*A* and *B* are the limits of monthly mean values of soil moisture observed in the field and $\bar{p}$ is the average weight of the two age categories of worms) (after Lavelle et al., 1974).

50°C in the uppermost few centimetres of the soil at Lamto.

Soil temperature definitely influences the activity of earthworms. Even slight thermal variations can change the rate of ingestion of earth by *Millsonia anomala*, a geophagous species; this is particularly true in younger individuals (Fig. 22.3). The surface activity of these same earthworms is also affected by soil temperature. When soil surface temperature increases, earthworms have to move into the deeper parts of the soil, where nutrients exist in smaller quantities (Fig. 22.4).

Other factors

The granulometric and mineral composition of the soil can also modify earthworm distribution, mostly through their influence on soil water-holding capacity. The same applies to the plant cover, but only in part, as the level of earthworm populations will depend upon the amount of litter and roots produced by the vegetation. Once dead, this plant material will be consumed by the worms immediately, or after transformation by microorganisms. The amount and nutritional quality of the food thus made available deeply influences the growth rate of young earthworms, as exemplified by the sharp differences in growth rate of young

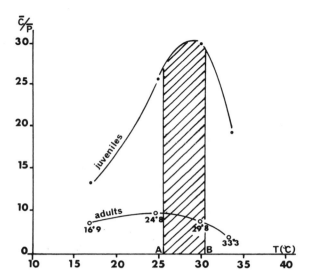

Fig. 22.3. Variations in the soil consumption ($\bar{C}/\bar{P}$) of *Millsonia anomala* as a function of soil temperature, all other factors being equal (*A* and *B*: see Fig. 22.2) (after Lavelle, 1975).

Millsonia anomala raised in earth samples taken at different depths in the soil (Fig. 22.5).

Among biotic factors, the influence of predation and parasitism is hard to estimate in field conditions. Bird predation on tropical earthworms seldom occurs in tropical savannas, as a result of the thickness of the grass cover and the subterranean habits of most of the worms. Other predators do exist, however: doryline ants, mole crickets, shrews and mongooses, not to mention predatory earthworms of the genus *Agastrodilus*. Some parasites are also known, including small ectoparasitic Enchytreidae. Nevertheless, predation pressure is probably quite small. Senapati (1980) found fungal infections in senescent and dead *Octochaetona surensis*.

NUMBERS AND BIOMASS OF SAVANNA EARTHWORMS

The available data are tabulated in Table 22.1. However, it is difficult to compare the figures obtained, as both the extraction methods and the areas sampled are not the same. The formaldehyde technique widely used in temperate habitats gives fairly satisfactory results as long as the worms are not aestivating (Lakhani and Satchell, 1970; Bouché, 1969). However, this method is far less suitable for tropical earthworms which, unlike

anecic lumbricids, do not live in extensive burrow systems with surface apertures. The efficiency of the method then becomes very low, which explains the values obtained by Block and Banage (1968) in Uganda grasslands, and Madge (1969) in Nigerian forests. Reinecke and Ljungström (1969), as well as Lavelle (1971a), have found this technique totally unreliable in other African savannas.

Hand sorting, sometimes combined with flotation, is therefore used by most authors. The method has the advantage that it can be adapted to extract earthworm cocoons and thus enables all stages of the populations to be estimated, but it is unfortunately very time-consuming. I have therefore resorted to a compromise, using hand sorting to collect earthworms in large samples, and flotation on small sub-samples to work out correction factors (Lavelle, 1978).

The size of the samples is also quite different in the various study sites mentioned in Table 22.1, ranging from 1.4 to 288 m². The depth of soil sampled (from 20 to 60 cm below soil surface) is also variable, as well as the duration of the studies (from 2 to 24 months). Comparisons must therefore be made with caution.

The average annual rainfall on all the sites studied exceeded 1000 mm. Three of them, Lamto and Foro Foro in Ivory Coast and Kabanyolo in Uganda, can be considered as undisturbed savannas.

At Lamto, the average annual rainfall is 1275 mm; the dry season is limited to two months, and is not very severe as showers occur from time to time. On the average, the numbers of earthworms range from 91 to 400 m^{-2}, depending upon both the year and the savanna *facies* considered. The corresponding biomass figures (wet weight) range from 134 to 544 kg ha^{-1}. The average values for the five *facies* studied and the four years concerned are 215 ind. m^{-2} and 325 kg ha^{-1}.

At Foro Foro, a tree savanna 250 km further north, the rainfall still reaches 1150 mm, but the dry season lasts longer and is harsher than that of Lamto. During the 1978 rains, the number of earthworms ranged from 460 to 582 ind. m^{-2}, and the wet biomass from 170 to 223 kg ha^{-1}.

The Kabanyolo "bush" near Kampala has an average rainfall of 1500 mm, but the formaldehyde technique allowed only 13 ind. m^{-2} to be extracted, representing a wet biomass of only 8.4 kg ha^{-1}.

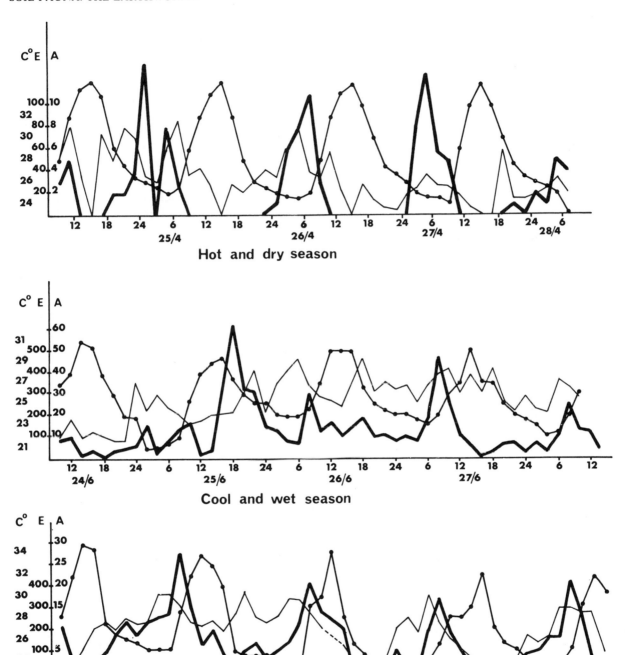

Fig. 22.4. Seasonal variations in the circadian production of surface casts by *Millsonia anomala* (*A*, heavy line) and eudrilids (*E*, normal line) on a 10-m² quadrat at Lamto. Air temperature (line with dots) in C°. (After Lavelle, 1978).

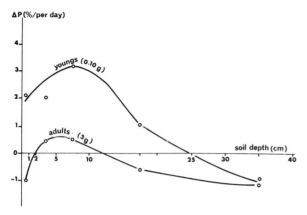

Fig. 22.5. Daily growth increment of *Millsonia anomala*, for young and adult individuals, in soils taken at different depths in the Lamto savanna (pF 2.5; $\bar{t}$ 25 6 C) (after Sow, 1979).

The four other sites are artificial pastures established in formerly forested areas. The highest earthworm densities were found in the Laguna Verde hill pastures (Vera Cruz, Mexico), where the annual rainfall averages 1800 mm. During the rainy season, densities approximate 1000 ind. m^{-2}, the wet biomass reaching 492 kg ha^{-1} (Lavelle et al., 1981). At a lower altitude, between 350 and 50 m above sea level, similar figures were obtained: 600 to 800 ind. m^{-2} and 358 to 555 kg ha^{-1}.

The Indian grassland studied by Dash and Patra (1977) was much poorer in earthworms: 64 to 800 ind. m^{-2} depending upon the season, with an average fresh biomass of 302 kg ha^{-1}. At Sambalpur, the mean annual earthworm densities ranged from 174 ind. m^{-2} in the grazed pasture to 247 ind. m^{-2} in the ungrazed plot. Biomasses were among the highest known so far: 410 kg ha^{-1} in the grazed pasture and 560 kg ha^{-1} in the ungrazed plot. The figures obtained by Breymeyer (1978) in a humid Panama grassland are much lower. This is probably due to the small size of her total sampling area (1.4 m^2) and to the shallowness of her soil samples (20 cm).

If data from the Uganda and Panama sites, where the methods used were inadequate, are excluded there is a steady increase in numbers and biomasses of earthworms with increasing rainfall (Fig. 22.6). The highest values are reached in a grassland grown in former rain-forest areas. It must nevertheless be noted that such figures are much smaller than those recorded in temperate grasslands (560 to 2870 kg ha^{-1}; Edwards and Lofty, 1972; Nowak, 1975), being closer to those found in temperate forests (370 to 680 kg ha^{-1}; same sources).

On the other hand, earthworms are more abundant in tropical savannas than in nearby tropical forests. No more than 130 ind. m^{-2} (98 kg ha^{-1}) were found in relict patches of rain forests in the Laguna Verde area, and 75 ind. m^{-2} (34 kg ha^{-1}) in the Lamto gallery forests. This is quite likely due to the relative dehydration of the forest soils at Lamto. During 1969, a dry year in this part of Ivory Coast (annual rainfall 960 mm), the soil moisture remained below the wilting point for plants (pF 4.2) for 25 to 90 days in the various savanna *facies*, as against 217 days in the gallery forest. This might help to explain why an increase in earthworm number and biomass can sometimes follow deforestation, as at Laguna Verde. This can last as long as the soil has not lost all of its nutrients, as could have been the case in Breymeyer's Panama savanna, if her findings are confirmed by a more satisfactory sampling procedure. It is also possible that all forest earthworms are not equally able to withstand such a drastic environmental change.

COMMUNITY STRUCTURE: RESOURCE PARTITIONING

Knowledge of the community structure of savanna earthworms is at present limited to five sites, two in Ivory Coast (Lamto and Foro Foro), one in Mexico (Laguna Verde), and two in India (Berhampur and Sambalpur).

Lamto

The Lamto earthworm community has the largest number of member species: four Eudrilidae and eleven Megascolecidae (Table 22.2). These fifteen species belong to three different trophic categories: arboreal detritivores (four species), litter detritivores (two species), and soil eaters (geophagous worms) (six species). The three remaining species have specialized diets: one, *Dichogaster* sp. is a termite commensal only found at the bottom of *Trinervitermes* mounds. The two others, *Agastrodrilus multivesiculatus* and *A. opisthogynus*, are quite probably predators of other earthworms, at least in part.

TABLE 22.1

Numbers and biomass of earthworms in various savanna study sites

Locality	Savanna category	Rainfall (mm)	Duration of study (months)	Surface sampled (m²)[1]	Depth (cm)	Extraction method[2]	Numbers (ind. m⁻²)	Biomass (g wet wt. m⁻²)	Reference
Panama	artificial pasture	1900	3	1.4	20	HS	12–15	0.15–0.36	Breymeyer (1978)
Laguna Verde, Vera Cruz (Mexico)	hill pasture (800 m)	1800	2	5+0.28	40	HS,W	948	49.2	Lavelle et al. (1981)
	lowland pastures (50–350 m)	1500	2	10+0.28	40	HS,W	620–787	35.8–55.5	
Kabanyolo (Uganda)	"bush"	1500	2	10.7	?	F	13	0.84	Block and Banage (1968)
Berhampur, Orissa (India)	*Cynodon dactylon* and *Hygrorhiza* sp. grasslands	1250	18	11.3	20	HS	64–800	30.2	Dash and Patra (1977)
Sambalpur, Orissa (India)	grazed irrigated pasture	1343	13	4.1	40	W	17.4	41.0	Senapati (1980)
	ungrazed irrigated pasture	1343	19	5.9	40	W	24.7	56.0	Senapati (1980)
Lamto (Ivory Coast)	grass savanna	1183	24	288+23.04	60	HS,W	188	38.0	Lavelle (1978)
	shrub savanna	1183	24	288+23.04	60	HS,W	287	48.6	Lavelle (1978)
	unburnt shrub savanna	1276	12	288+23.04	60	HS,W	400	35.9	Lavelle (1978)
Foro Foro (Ivory Coast)	shrub savanna	1150	2	20	60	HS,W	460–582	17.0–22.3	Lavelle (unpubl.)

[1]When two different extraction methods have been used on two different soil samples, two figures are indicated.

[2]Hand-sorting (HS), formalin (F) or soil-washing (W) techniques.

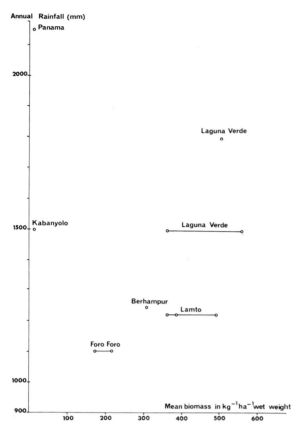

Fig. 22.6. Relationship between earthworm biomass and average annual rainfall in various tropical savanna study sites.

All arboreal detritivores live in "hanging soils"; they are forest species which have found micro-climatic and dietary conditions reminiscent of their original habitat in the small pockets of soil existing between the leaf bases of savanna palms. *Dichogaster bolaui* and *D. saliens* are pantropical in their distribution. *D. saliens* and *Chuniodrilus vuattouxi* can also be found at ground level, in rotten tree stumps. *Dichogaster baeri*, a forest species, can also marginally enter the savanna, in the shallow depressions close to gallery forests, and also in grasslands protected from burning for at least fifteen years — that is, already invaded by bush. All these worms are small and dark coloured, *Chuniodrilus vuattouxi* excepted. *Dichogaster baeri* is the only large arboreal earthworm, 15 cm long when adult.

The two litter detritivores, *Dichogaster agilis* and *Millsonia lamtoiana*, respectively brown- and green-coloured, live in the upper few centimetres of the soil and feed upon litter material. *Millsonia lamtoiana* is a large worm which comes out in the open by night, and also sometimes during the day after heavy rains, or at any time during the mating season; it feeds on large plant debris already partly decomposed and mixed with some soil. *Dichogaster agilis* is smaller, and remains at the soil litter interface, feeding on a mixture of soil and already decomposed plant material. Therefore, these two detritivores are not typical epigeic worms according to Bouché's criteria, being able to dig into the soil and ingest earth with the litter debris they feed on.

All the five remaining species are unpigmented soil-eating (geophagous) earthworms, very seldom found at the soil surface. Some live in the topsoil and produce surface casts. *Millsonia anomala* is the most numerous species in this trophic group; a mesohumic worm, it can reach a body length of 17 cm when adult.

The Eudrilidae are polyhumic worms which live at an average depth of 18 cm, though responsible for numerous casts. Being of small size they feed only on small and energy-rich soil particles, and on tiny organic debris such as decomposing rootlets, or the thin organic layer of the topsoil. The organic matter content of their casts is therefore higher than those of *Millsonia anomala* (Table 22.3).

Dichogaster terrae-nigrae and *Millsonia ghanensis* are both large oligohumic species reaching an adult body length of 70 and 30 cm respectively. They live in the deeper layers of the soil, where the organic matter content is the lowest.

The niches of the different members of the earthworm community are therefore well defined (Lavelle et al., 1980). Spatio-temporal overlap between niches has been estimated using the O_{jk} index (Pianka, 1974). The calculated values for each of the three variables selected (horizontal distribution, vertical distribution, and seasonal cycle of population density) are tabulated in Table 22.4. As the overlap ratios are high for seasonal cycles (0.82 to 0.95, with an average ratio of 0.90), niche separation is mostly achieved through occupation of different savanna facies (index ranging from 0.13 to 0.96; average 0.70) and preference for different soil layers (index ranging from 0.02 to 0.93; average 0.47).

The three variables considered being apparently independent, a resultant matrix can be established by multiplying the calculated values for every species. The results are shown in Table 22.5: the index ranges from 0.01 to 0.80, with an average

TABLE 22.2

Some characteristics of the Lamto savanna earthworms

Species	Family	Habitat	Vertical distribution (cm), range and average depth	Food habits	Maximum wet weight (g)
Chuniodrilus vuattouxi	Eudrilidae	"hanging soils" in palm trees	0 to +1800	detritivore	0.08
Dichogaster saliens	Megascolecidae	"hanging soils" in palm trees	0 to +1800	detritivore	0.15
Dichogaster bolaui	Megascolecidae	"hanging soils" in palm trees	0 to +1800	detritivore	0.18
Dichogaster baeri	Megascolecidae	"hanging soils" in palm trees	0 to +1800	detritivore	2.50
Dichogaster agilis	Megascolecidae	savanna soil	0 to −30(6)	detritivore	0.60
Millsonia lamtoiana	Megascolecidae	savanna soil	0 to −20(7)	detritivore	22.00
Millsonia anomala	Megascolecidae	savanna soil	0 to −50(8)	mesohumic soil eater	6.00
Chuniodrilus zielae	Eudrilidae	savanna soil	0 to −60(18)	polyhumic soil eater	0.20
Chuniodrilus palustris	Eudrilidae	savanna soil	0 to −60(18)	polyhumic soil eater	0.20
Stuhlmannia porifera	Eudrilidae	savanna soil	0 to −60(18)	polyhumic soil eater	0.25
Dichogaster terrae-nigrae	Megascolecidae	savanna soil	0 to −60(23)	oligohumic soil eater	28.00
Millsonia ghanensis	Megascolecidae	savanna soil	0 to −60(32)	oligohumic soil eater	16.00
Agastrodrilus opisthogynus	Megascolecidae	savanna soil	0 to −60(29)	oligohumic soil eater and possibly carnivorous	3.50
Agastrodrilus multivesiculatus	Megascolecidae	savanna soil	0 to − 60(29)	oligohumic soil eater and possibly carnivorous	4.50
Dichogaster sp.	Megascolecidae	base of termite mounds	?	polyhumic soil eater?	2.00

value of 0.35. Thus spatio-temporal niche separation is quite satisfactory, except for the following two groups of species.

(1) *Dichogaster agilis, Millsonia anomala* and *M. lamtoiana*, which live in the topsoil of, preferably, shrub savannas — whether annually burnt or not. In this case, niche separation is achieved by food partitioning: *Millsonia anomala* is a soil eater, whereas the other two species are detritus-feeders. The large *M. lamtoiana* feeds on almost intact litter material, and the small *Dichogaster agilis* eats tiny plant debris already decomposed.

(2) With regard to the second group of worms (*Agastrodrilus opisthogynus, Millsobia ghanensis* and the Eudrilidae), the species concerned are of different sizes and diets: *Millsonia ghanensis* lives in the deeper, energy-poor soil layers, while the Eudrilidae feed on energy-rich soil, and *Agastrodrilus opisthogynus* appears to be partially carnivorous.

Other earthworm communities

Only five species are found at Foro Foro: *Dichogaster agilis* and *Millsonia lamtoiana*, both detritivores; *Stuhlmannia porifera*, a polyhumic geophagous eudrilid; *Agastrodrilus dominicae* which is very close to *A. opisthogynus* in morphology and habits; and *Dichogaster terrae-nigrae*, a large soil eater which is represented here by a small-sized

TABLE 22.3

Granulometric composition and organic matter content of casts from Eudrilidae
and *Millsonia anomala* in a grass savanna of Lamto (Ivory Coast)

	Eudrilidae	*Millsonia anomala*	Control soil (0 to -40 cm)
Clay (<2 μm)	5.0	5.2	3.3–3.7
Fine silt (2–20 μm)	7.3	8.6	5.7–6.5
Coarse silt (20–50 μm)	12.0	11.5	7.5–8.5
Fine sand (50–200 μm)	40.0	27.0	25.5–29.0
Coarse sand (200–2000 μm)	32.0	46.5	50.5–56.5
C(%)	10.1	5.8	1.9–8.2
N(%)	0.76	0.57	0.17–0.57
C/N	13.3	10.2	10.8–15.3
Organic matter (%)	1.7	1.0	0.6–1.4

form, living closer to the soil surface (16 cm on the average, instead of 28 cm at Lamto).

In the Laguna Verde grasslands, only three species are found: *Diplocardia koebeli*, a polyhumic soil eater of small size (maximum wet weight: 0.30 g); *Ponthoscolex corethrurus*, a mesohumic geophagous species reaching a body weight of 1 g when adult; and *Dichogaster* sp., a small detritivore weighing no more than 0.40 g (Lavelle et al., 1979b).

The Indian pastures of Sambalpur harbour five species whose diets remain unknown. Adult weights range from 0.09 g (*Ocnerodrilus occidentalis*) to 1.22 g (*Lampito mauritii*), the three other species (*Drawidia willsii*, *D. calabi*, *Octochaetona surensis*) weighing 0.13 g, 0.54 g and 1.08 g, respectively.

The Berhampur community is composed of only two species: a detritivore *Lampito mauritii* (maximum body weight 0.60 g) and a small soil-eating member of the Ocnerodrilidae (M.C. Dash, pers. comm., 1979).

Comparisons

Although the number of communities studied is very small, a few tentative conclusions can be reached.

First, soil-eating (geophagous) worms are the most numerous in the four savanna communities studied. This is quite probably due to the suitable and well-buffered microclimatic conditions in tropical savanna soils, as well as to the large amount of organic matter provided, at all soil depths, by the rapid decomposition of litter and roots. The amount of nutrients thus made available to soil organisms is also increased by the lack of detritivorous earthworms, ascribable to the sharp seasonal changes in environmental conditions at the soil/litter interface.

Second, under similar climatic conditions, species richness is much greater in "natural" savannas than in man-made and man-maintained pastures.

Third, even in the more species-rich savannas, niche overlap is apparently very limited: size differences, time and space partitioning, as well as food partitioning, all contribute to reduce, if not avoid, interspecific competition.

Fourth, the adult size of a species seems to be closely related to the duration of the dry season. The populations of large species are apparently more sensitive to drought than those of small species. At Lamto, where the dry season is restricted to one or two months per year, most earthworms are of a large size, and their adult weight exceeds 1 g, even reaching 20 to 30 g for the largest species. At Sambalpur, where the annual rainfall is very similar to that of Lamto, but where the dry season extends over six months, only small earthworms are found, the largest adult individuals not exceeding 1.5 g in weight.

TABLE 22.4

Spatio-temporal niche overlap (O_{jk} index) in the earthworm community of Lamto savanna (Ivory Coast)

A. For vertical distribution

M.l.	0.93					
M.a.	0.91	0.96				
D.t.	0.11	0.44	0.49			
Eu.	0.38	0.48	0.62	0.57		
A.o.	0.08	0.10	0.27	0.33	0.91	
M.g.	0.02	0.05	0.24	0.29	0.77	0.88
	D.a.	M.l.	M.a.	D.t.	Eu.	A.o.

B. For horizontal distribution

M.l.	0.95						
M.a.	0.73	0.89					
D.t.	0.13	0.55	0.74				
C.z.	0.94	0.53	0.79	0.75			
S.p.	0.54	0.95	0.82	0.47	0.41		
A.o.	0.76	0.56	0.97	0.71	0.72	0.92	
M.g.	0.64	0.37	0.89	0.55	0.53	0.95	0.96
	D.a.	M.l.	M.a.	D.t.	C.z.	S.p.	A.o.

C. For seasonal cycle

M.l.	0.80					
M.a.	0.94	0.91				
D.t.	0.84	0.92	0.95			
Eu.	0.93	0.87	0.93	0.94		
A.o.	0.82	0.91	0.92	0.99	0.95	
M.g.	0.80	0.89	0.91	0.94	0.87	0.95
	D.a.	M.l.	M.a.	D.t.	Eu.	A.o.

Legend: D.a. = *Dichogaster agilis*; M.l. = *Millsonia lamtoiana*; M.a. = *Millsonia anomala*; D.t. = *D. terrae-nigrae*; Eu. = Eudrilidae; C.z. = *Chuniodrilus zielae*; S.p. = *Stuhlmannia porifera*; A.o. = *Agastrodrilus opisthogynus*; M.g. = *M. ghanensis*.

TABLE 22.5

Total niche overlap in the earthworm community of the Lamto savanna; same symbols as in Table 22.4

M.l.	0.71						
M.a.	0.62	0.65					
D.t.	0.01	0.21	0.34				
C.z.	0.33	0.40	0.46	0.40			
S.p.	0.19	0.11	0.47	0.25	0.41		
A.o.	0.04	0.05	0.24	0.23	0.62	0.79	
M.g.	0.01	0.02	0.18	0.15	0.36	0.64	0.80
	D.a.	M.l.	M.a.	D.t.	C.z.	S.p.	A.o.

SEASONAL VARIATIONS IN ACTIVITY AND NUMBERS

A marked seasonality of environmental conditions is one of the major characteristics of tropical savannas. During the dry season the soil moisture often drops to give a pF of 4.2 and even 4.7 at Lamto. Earthworms then become inactive, the inactivity threshold ranging from pF 4.2 to 3.0 according to the species (Lavelle, 1971a). However, the water content of the soil also varies with depth, and this explains why the seasonal cycles of different species can differ to a certain extent.

Activity cycles

In the derived savanna at Lamto a proportion of earthworms, variable according to the species concerned, become inactive from December to March (the "long" dry season). Then comes the rainy season during which worms are all very active before entering a second period of inactivity in August and September, during the "short" dry season. Activity levels are high again in October and November.

However, there are marked differences between the activity cycles of worms living in the various kinds of savanna, as shown in Fig. 22.7. In 1972, an average year at Lamto (yearly rainfall 1276 mm), the number of months per year during which over 50% of the earthworm population remained inactive ranged from three in old unburnt savannas with strong bush encroachment, and two in shrub savannas, to none in grass savannas. In the same savanna categories, the periods during which 10 to 50% of the worm population remained inactive were respectively of four, four and five months.

Seasonal changes in numbers and biomass

Obviously there are some relationships between seasonal variations in earthworm activity and their changes in number. The fact that every species has its own pattern of seasonal variation in numbers (Fig. 22.8) implies that environmental variables do not have the same importance for all sympatric species. Indeed, this can be established by calculating the relationship between population density and each of the eight environmental factors previously identified by multivariate analysis (Table 22.6).

Thus, population numbers of *Millsonia lamtoiana* and *Dichogaster agilis*, both detritivores, and of the polyhumic soil-eating Eudrilidae are negatively correlated with a factor corresponding to the trophic effects of grass fires. This factor alone accounts for 26 to 56% of the seasonal variance of their numbers. On the other hand, *Millsonia ghanensis* and *Agastrodrilus* spp., which live in the deeper soil layers, are affected very little by environmental factors; these are responsible for only 21 to 23% of the variance of their numbers.

In between these two species groups, the numbers of *Millsonia anomala*, a mesohumic earth feeder, and of *Dichogaster terrae-nigrae*, an oligohumic earth feeder, are influenced by other environmental factors. The abundance of the first species depends on the water regime of the soil related to the presence of a shrub cover; that of the second species is mostly influenced by drainage conditions of the soil (Lavelle, 1978).

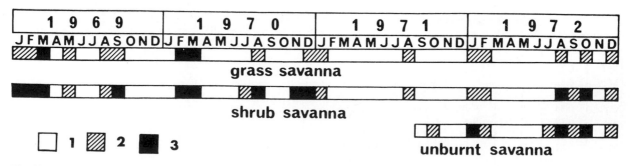

Fig. 22.7. Seasonal variations in earthworm activity in different savanna types at Lamto (after Lavelle, 1978). 1: all worms active; 2: 0 to 50% of quiescent individuals; 3: 50 to 100% of quiescent individuals.

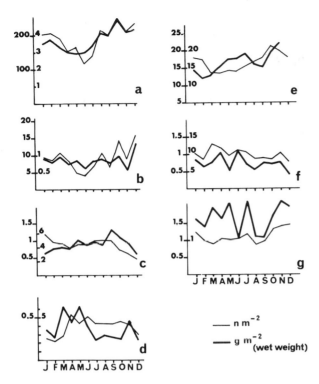

Fig. 22.8. Average seasonal changes in numbers and biomass of Lamto earthworms: a, Eudrilidae; b, *Dichogaster agilis*; c, *Millsonia ghanensis*; d, *M. lamtoiana*; e, *M. anomala*; f, *D. terrae-nigrae*; g, *Agastrodrilus opisthogynus* (after Lavelle, 1978).

Seasonal variations of the functional structure of the populations

Seasonal variations in activity and numbers also go together with changes in the functional structure defined by the weight distribution of the worms, the percentage of inactive individuals, the vertical distribution of populations in the soil and the production of cocoons. Therefore, stages of population decrease, increase and *status quo* alternate during the year, but in different ways (Fig. 22.9).

Lavelle and Meyer (1976) and Lavelle (1978) have made use of a technique of multivariate analysis of mixed data to identify seven different population stages in the Lamto earthworm community. Each of these stages differs from the others by the number and biomass of worms (axis I), the percentage of active individuals (axis II), and relative distribution in depth of the population (axis III). When monthly samples are taken, the seasonal changes in the structure and function of the populations can be followed throughout the year. All sympatric species are then found not to behave in the same way. In the shrub savanna, for instance, even two congeneric species, *Millsonia anomala* and *M. lamtoiana*, do not display the same patterns of change in functional structure.

TABLE 22.6

Correlations between the density of earthworm populations and the current values of the major environmental factors previously identified by multivariate analysis (Lavelle, 1978); values are given in percentage of the variance explained; the symbols + or − indicate whether the relationship is positive or negative

	Grass fires	Dry and hot spells	Cool spells	Thickness of vegetation cover	Soil drainage	Percentage of the total variance explained (%)
Millsonia lamtoiana	56 −	2 −	3 −	1 −	1 +	63
Eudrilidae	44 −		1 −	5 +	3 +	54
Dichogaster agilis	26 −					26
Millsonia anomala	2 −			17 −	3 +	22
Agastrodrilus spp.	2 −	7 −			7 +	15
Millsonia ghanensis	6 +	4 −	3 +		5 −	18
Dichogaster terrae-nigrae	1 −	5 −		3 +	48 +	58

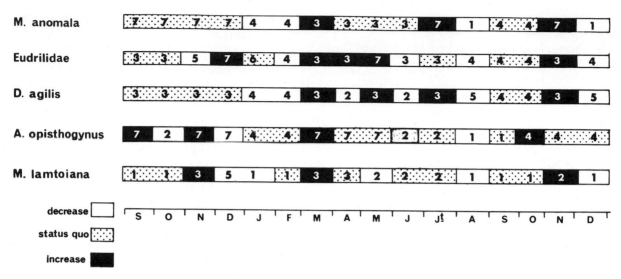

Fig. 22.9. Seasonal changes in the functional structure of some earthworm populations at Lamto, from September 1971 to December 1972. Stages are ranked from *1* to *7*. *A* stands for *Agastrodrilus*, *D* for *Dichogaster*, and *M* for *Millsonia*. (After Lavelle, 1978.)

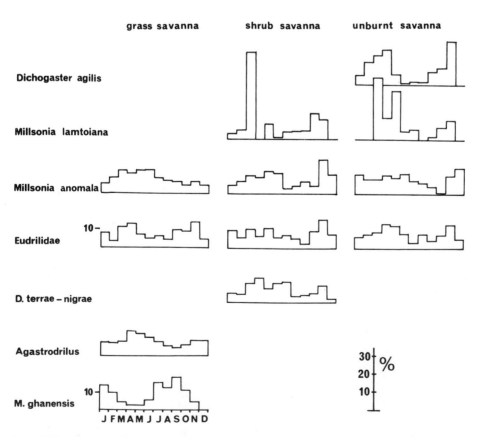

Fig. 22.10. Monthly variations in production of Lamto earthworms in three savanna types in 1972 (after Lavelle, 1978).

Seasonality in production

Seasonality in production, hence the trophic impact of the population, is closely linked to such seasonal changes. In the Lamto savanna, production is highly seasonal among detritivorous species living in the litter, such as *Millsonia lamtoiana* and *Dichogaster agilis*. Seasonality in production is also perceptible, however, even in species living deeper in the soil (Fig. 22.10). On the whole, the highest figures are reached during the rainy season, and the lowest during the dry season, or when there is a temporary excess of rain. This seasonality in production is also much more obvious in shrub savannas than in grass savannas, being closely correlated with the instability of the water regime of the soil in the former.

Seasonality in production is even more accentuated in savannas where the climate is much more seasonal than in Lamto. At Berhampur, for example, *Lampito mauritii* populations are only active for a few months each year (Dash and Patra, 1977).

POPULATION ECOLOGY

Little is known concerning the population structure and dynamics of savanna earthworms; so far, only a few species of the Lamto area have been adequately studied.

Demographic profiles

Demographic profiles can be established on the basis of the following three parameters: the dura-tion of the growth period (D, in months), the life expectancy of the hatchlings (E_v, in months), and the number of cocoons produced per individual per year (F). It has been shown (Lavelle, 1979a) that these three parameters vary together (Table 22.7). They can be combined in a demographic index:

$$D = 10^3 \, F/C \cdot E_v$$

which is directly related to another index, called the ecological index:

$$E = \log W_x \cdot \bar{p}$$

which summarizes the life-style of the species, on the basis of its maximum body weight (W_x) and the average depth at which they are living within the soil($\bar{p}$).

In the community studied, population turnover is all the more rapid when species are small, and live close to the soil surface; this is mediated by an accelerated growth, a short life expectancy and a high fecundity. The small size and high metabolic rate of the smaller earthworms allow them to grow quickly when enough food is available, as is the case in the litter. On the other hand, species living close to the soil surface are more exposed to predation and to the disruptive action of periodic grass fires and severe drought spells.

In a community like this, earthworm species can be ranked along an r–K gradient, according to the values of their D index, indicating their ability to increase their numbers, an ability which itself depends on their ecological characteristics assessed by their E index (Fig. 22.11).

TABLE 22.7

Some demographic characteristics of seven of the most numerous populations of Lamto earthworms (Lavelle, 1979)

	Duration of the growth period (in months)	Number of coccoons produced per adult, per year	Expectation of life at hatching (months)	Demographic index	Ecological index
	C	F	E_v	$D = 10^3 F/C \cdot E_v$	$\log W_x \cdot \bar{p}$
Chuniodrilus zielae	18	13.0	3.3	219	3.2
Dichogaster agilis	15	10.7	3.4	210	3.0
Millsonia anomala	20	6.2	6.2	50	32
Millsonia lamtoiana	24	3.1	7.5	17	210
Dichogaster terrae-nigrae	36	1.9	11.6	4.5	575
Agastrodrilus opisthogynus	24	1.3	11.1	4.9	87
Millsonia ghanensis	42	1.3	10.6	2.9	512

Outside Lamto, there is also some information on the demographic characteristics of two other species, *Lampito mauritii* in India (Dash and Patra, 1977; and pers. comm.) and *Ponthoscolex corethrurus* in Mexico (pers. obs.).

Lampito mauritii lives at an average depth of 12 cm, seldom reaching 20 cm. Its size is small, the maximum adult body weight not exceeding 0.60 to 0.70 g (wet weight). Annual fecundity is probably high, as the population density can change from 16 to 240 ind. m^{-2} in three months. Such a rapid increase in numbers implies that every adult produces at least fifteen young per year — that is, seven to 10 cocoons, as each cocoon gives rise to one or two young (M.C. Dash, pers. comm.). The average life expectancy at birth ranges from 3.5 to 4.0 months, and growth is very rapid. Sexual maturity occurs at one year of age. Nevertheless, the worms continue to grow afterward, the maximum body weight being reached between 12 and 15 months of age. The demographic index (*D*) is therefore of the order of 200 to 300, which means that this worm has the ability to build up its numbers very quickly. Its ecological index (*E*) is low, and the species would stay close to *Chuniodrilus zielae* and *Dichogaster agilis* on Fig. 22.11.

Ponthoscolex corethrurus, a mesohumic soil eater, is somewhat larger, reaching a maximum adult weight of 1.0 g. It lives in the topsoil, seldom burrowing deeper than 10 cm below the soil surface (5.9 cm deep on average). The growth is rapid, young worms increasing their body weight by 7.5% daily, during their first five weeks of life. Fecundity is high, at least three to four cocoons being produced per adult during the reproduction period, in field conditions, and five cocoons per adult during six weeks of laboratory rearing.

Similar data are unfortunately lacking for temperate species. However, the tropical savanna earthworms already studied are definitely more prolific than many European lumbricids, whose fecundity is low and growth rate slow.

Demographic strategies

The above-mentioned data have already shown that tropical earthworm species living in a single savanna can easily be ranked along a *r*–*K* continuum. At Lamto, the oligohumic soil eaters *Dichogaster terrae-nigrae* and *Millsonia ghanensis* are typical *K*-strategists. Both are large species,

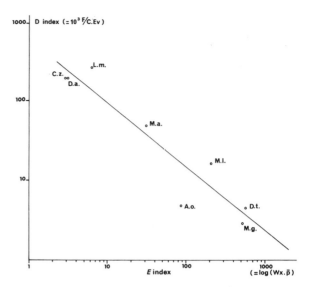

Fig. 22.11. Relationship between the ecological, $E = \log W_x \cdot \bar{p}$, and demographic indices, $D = 10^3 \, F/C \cdot E_v$, for eight species of tropical earthworms (W_x = maximum weight of the species; $\bar{p}$ = average depth at which the earthworm population is found; F = fecundity of adult individuals — that is, number of cocoons produced per year; C = total duration of the growth period; E_v = life expectancy of young at birth). Legend: *C.z.* – *Chuniodrilus zielae*; *D.a.* = *Dichogaster agilis*; *L.m.* = *Lampito mauritii*; *M.a.* = *Millsonia anomala*; *M.l.* = *M. lamtoiana*; *A.o.* = *Agastrodrilus opisthogynus*; *D.t.* = *D. terrae-nigrae*; *M.g.* = *M. ghanensis*. (After Dash and Patra, 1977; Lavelle, 1979a.)

growing slowly and having a low fecundity; both also live in the deeper soil layers, a rather stable environment, but one poor in organic matter. Their *D* index is less than 5.

Small arboreal detritivores, such as *Dichogaster bolaui* and *D. saliens*, the topsoil detritivore *D. agilis*, and the small polyhumic eudrilids are at the other end of the *r*–*K* gradient. Their size is small, their growth rate accelerated, and their fecundity high. They live in the energy-rich layers of the soil, but are at the same time more exposed to predators and climatic accidents. Their *D* index values are 200 or more. The Indian species *Lampito mauritii* also belongs to this category of *r*-strategists.

The large detritivore *Millsonia lamtoiana* (*D* = 17) and the smaller mesohumic soil eater *M. anomala* (*D* = 50) are in an intermediate position on the *r*–*K* continuum. However, *M. anomala* can temporarily adapt its demographic characteristics to adverse conditions; when rainfall is less than usual, its *D* index can reach 107 in shrub savannas.

In drier Sudanian savannas, such as at Foro Foro, the large *K*-strategists disappear first, as well as *Millsonia anomala*. Only *Dichogaster terrae-nigrae* remains, but it changes its diet and reduces its body length. Further north, the only earthworms left are *r*-strategists, small detritivores and polyhumic soil-eating eudrilids living in the topsoil. Humid grasslands growing in formerly forested areas are also colonized by *r*-strategists, such as the pantropical *Ponthoscolex corethrurus*. No *K*-strategists are found in such marginal habitats.

Population dynamics

Here again, the few data available are those of Lamto, where the population dynamics of a very common species *Millsonia anomala* has been comprehensively studied (Lavelle, 1978; Lavelle and Meyer, 1976, 1977).

Rate of growth. Field observations and laboratory experiments show that four major variables are able to influence the growth rate of *Millsonia anomala*: its body weight (i.e. its age), and the moisture, temperature and organic matter content of the soil.

The rate of growth is more rapid when the worm is young. Then, it decreases rapidly, up to sexual maturity, to diminish more slowly during the remainder of the life-span.

Soil moisture is very important to ensure optimal growth, especially for adult worms which are less resistant to drought than immature ones. The growth of adult *Millsonia anomala* is discontinued when the water content of the soil is lower than 10%, as against 8% for immature individuals. Maximum daily growth increments are reached at a soil moisture content little above 20% for adult, and 13% for young worms (pF 2.5 = 12%; pF 4.2 = 4% in such soils) (Fig. 22.2).

Soil temperature is also important. Maximum growth increments are obtained at 24.8°C for adult individuals and 29.8°C for young ones; growth stops below 19.0°C and above 35°C, for all ages.

All other factors being equal, it is the organic matter of the soil which is determinant. A larger growth increment is observed in the soil layers where organic matter is the most abundant — that is, in the first 10 cm of soil. However, the quality of organic material consumed is also important; optimal growth rates are reached at depths ranging from 5 to 10 cm, and not closer to the soil surface.

Mortality rate. Correlations between monthly mortality rates of *Millsonia anomala* and a number of population and environmental variables show that two of the environmental parameters exert a major influence on mortality. Dry and hot spells increase it significantly, whereas burning of the grass has the opposite effect. Population density does not play any role in the mortality of this species.

Natality rate. A similar approach has been used to determine the most important variables influencing natality. In this case the most highly correlated variables are not those which operate at the time of hatching, but three months earlier. Spells of dry and hot weather, here again, have an unfavourable effect on population dynamics, diminishing the number of cocoons produced by the adults; overcrowding has similar effects. On the other hand, the thickness of the vegetation cover, whether it be grass or shrubs, has a favourable effect on natality, probably by keeping the soil moist for a longer time.

A simulation model showing the ways by which the different population and environmental variables interact has been proposed by Lavelle and Meyer (1977).

The population dynamics of some other sympatric species of Lamto earthworms is, however, influenced by environmental factors in a way different from that for *Millsonia anomala*. As shown on Table 22.6, *M. lamtoiana*, *Dichogaster agilis* and the eudrilids which live in the litter or the top soil, are more affected by burning and less dependent on the thickness of the vegetation cover than *M. anomala*. In general, earthworm populations with a high turnover appear to be more susceptible to environmental influences than those with a slow turnover; susceptibility to population pressure is also greater in *K*-strategists.

THE EFFECTS OF EARTHWORMS ON SAVANNA SOILS

The mechanical role earthworms play within the soil has been known for a long time, but the details of this action are still very poorly understood, especially in the tropics. First of all, they bury and incorporate plant remains from the litter into the soil, assimilating and mineralizing part of the organic matter ingested. However, the unassimilated soil is also modified during its passage through the worm's gut, and these modifications have some impact on soil micro-organisms. Furthermore, the production of surface casts by a number of species improves drainage and aeration of the soil itself.

Energy transformation by savanna earthworms

The fate of the energy ingested during the year by a *Millsonia anomala* population in a Lamto grass savanna has been described by Lavelle (1977). Living in soils poor in organic matter (1% on the average), this species ingests huge amounts of soil yearly. For example, every young individual ingests 20 to 36 times its own body weight of dry soil every day, the ratio diminishing as the worm grows older, however. This means that an average *Millsonia anomala* can ingest from 3 to 5 kg of soil per year, and that a natural population of 215 000 ind. ha^{-1} ingests about 500 tons ha^{-1} yr^{-1} (wet weight). It has been estimated that the amount of soil which passes yearly through the entire earthworm community at Lamto approximates 850 to 1200 tons ha^{-1}.

The assimilation efficiency (A/I ratio) of soil-eating earthworms is low, from 7 to 10%. Most of the assimilated energy is used for respiration and mucus production, and less than one-tenth for growth and reproduction. Therefore, the ecological growth efficiency (P/I ratio) is extremely low, about 0.3%. Similar values are common among soil-eating worms in Lamto (*Dichogaster terrae-nigrae*, 0.9%; *Millsonia ghanensis*, 0.9%), as well as in Europe (*Allobophora rosea*, 0.2 to 0.6%; Bolton and Philippson, 1976). The ecological growth efficiency is, however, slightly better in larger species such as *Millsonia ghanensis* and *Dichogaster terrae-nigrae*, than in smaller ones such as *M. anomala*.

Litter-dwelling earthworms are a little more efficient. Litter consumption by the two Lamto species, *Dichogaster agilis* and *Millsonia lamtoiana* ranges from 180 to 980 kg ha^{-1} yr^{-1} depending upon the savanna categories concerned, and their production varies from 25.3 to 216 kg ha^{-1} yr^{-1}, respectively. The P/I ratio is also higher for the larger species (1.8%) than for the smaller one (0.4%).

Soil mixing and its consequences

Most of the energy assimilated by earthworms being used for muscular activity, it is easy to understand why the mechanical role of these animals is so important. In the 1000 t ha^{-1} of soil which on the average pass every year through the gut of Lamto earthworms, there are about 14.7 tons of organic matter, which correspond to one-third of the humus fraction of this soil (*c.* 44 t ha^{-1}). Less than 3% of this organic matter is mineralized, but what is left is no doubt in a condition very different from what it was previously: structural changes make it much more susceptible to the action of soil micro-organisms.

Wherever the earthworm community is less diversified than in Lamto, as in Sudanian savannas or in man-made tropical pastures, the mechanical action of the worms is less important. However, the few observations so far made at Foro Foro and Laguna Verde show that it is still far from negligible.

Soil mixing by earthworms changes its texture, which becomes fine grained in its upper layers, where their mechanical action is more important. The production of surface casts also prevents the packing down of the topsoil. The volume of material raised above soil surface (30–40 m^3 ha^{-1} yr^{-1} at Lamto) corresponds to an equal amount of air below ground level. In the long run, the accumulation of casts can lead to a uniform layer of fine-grained soil.

Interactions between earthworms and microorganisms

Some laboratory experiments carried out on *Millsonia anomala* casts by Lavelle et al. (1980) point to a possible competition between worms and soil micro-organisms for the exploitation of the more easily digestible and energy-rich substrates. Such a competition would not be apparent when food is in excess, but would appear when conditions are less favourable. However, earthworms can also profit from micro-organisms when they feed upon degradation products of substances, such as cellulose or hemicellulose, which they cannot digest themselves.

CONCLUSIONS

Tropical savannas can harbor a rich community of earthworms under favourable environmental conditions — that is, when rainfall ranges from 1000 to 1500 mm, and is more or less evenly spread over the whole year. In such cases, up to ten to twenty species can occur together, adequately sharing the food resources of the soil environment. Competition is avoided by food and space partitioning, and the variety of niches goes hand in hand with different demographic structures and strategies.

In man-made pastures located in formerly forested areas, species diversity is lower and much depends on the adaptive potentialities of the former forest species to take advantage of these new environments, as well as on occasion for pantropical species to invade these new habitats. In this perspective, the limited ability for dispersal of the earthworms is an obvious handicap. However, once established, the newcomers can play a very important ecological role, as the amount of humic material produced from litter and roots in moist tropical pastures can be important, and as the water regime of such grasslands is much more favourable for earthworms than that of a forest.

In drier savannas, the role of earthworms becomes much less important when rainfall decreases and becomes more and more seasonal and unpredictable. The long dry season of the Sahelian savannas makes permanent settlement by earthworms impossible.

REFERENCES

Athias, F., 1974. Les conditions microclimatiques dans le sol. In: *Analyse d'un écosystème tropical humide: la savane de Lamto (Côte d'Ivoire). V. Les organismes endogés. Bull. Liaison Chercheurs Lamto, Numéro Spéc.*, 1974: 5–27.

Block, W. and Banage, W.B., 1968. Population density and biomass of earthworms in some Uganda soils. *Rev. Ecol. Biol. Sol*, 5: 515–522.

Bolton, P.J. and Phillipson, J., 1976. Burrowing, feeding, egestion and energy budget of *Allobophora rosea* (Savigny) (Lumbricidae). *Oecologia*, 23: 225–245.

Bouché, M.B., 1969. Comparaison critique de méthodes d'évaluation des populations de Lombricidés. *Pedobiologia*, 9: 26–34.

Bouché, M.B., 1971. Relations entre les structures spatiales et fonctionnelles des écosystèmes illustrées par le rôle pédobiologique des Vers de terre. In: P. Pesson (Editor), *La Vie dans les Sols, aspects nouveaux, études expérimentales.* Gauthier-Villars, Paris, pp. 189–209.

Bouché, M.B., 1972. Lombriciens de France. Ecologie et Systématique. *Ann. Zool., Ecol. Anim.*, Numéro hors série, 1972, 671 pp.

Bouché, M.B., 1977. Stratégies lombriciennes. *Ecol. Bull.*, 23: 122–132.

Breymeyer, A., 1978. Analysis of the trophic structure of some grassland ecosystems. *Pol. Ecol. Stud.*, 4: 55–128.

Clément, D., 1980. *Modélisation des flux thermiques et hydriques dans le sol de la savane de Lamto (Côte-d'-Ivoire).* Thesis, Université de Paris, 126 pp.

Dash, M.C. and Patra, U.C., 1977. Density, biomass and energy budget of a tropical earthworm population from a grassland site in Orissa, India. *Rev. Ecol. Biol. Sol*, 14: 461–471.

Edwards, C.A. and Loftty, J.R., 1972. *Biology of Earthworms.* Chapman and Hall, London, 283 pp.

Franz, H., 1950. *Bodenzoologie als Grundlage der Bodenpflege.* Akademie Verlag, Berlin.

Jamieson, B.G.M., 1971. A review of the Megascolecoid earthworm genera (Oligochaeta) of Australia. I. Reclassification and checklist of the Megascolecoid genera of the World. *Proc. R. Soc. Qld.*, 82: 75–86.

Lakhani, K.H. and Satchell, J.E., 1970. Production by *Lumbricus terrestris* (L.). *J. Anim. Ecol.*, 39: 473–492.

Lavelle, P., 1971a. Etude préliminaire de la nutrition d'un Ver de terre africain *Millsonia anomala* (*Acanthodrilidae*, Oligochètes). *Ann. Zool., Ecol. Anim.*, Numéro hors série, pp.133–145.

Lavelle, P., 1971b. Recherches sur la démographie d'un Ver de terre d'Afrique: *Millsonia anomala* (*Acanthodrilidae*, Oligochètes). *Bull. Soc. Ecol.*, 2: 302–312.

Lavelle, P., 1975. Consommation annuelle de terre par une population naturelle de Vers de terre (*Millsonia anomala* Omodeo, *Acanthodrilidae*, Oligochètes) dans la savane de Lamto (Côte d'Ivoire). *Rev. Ecol. Biol. Sol*, 12: 11–24.

Lavelle, P., 1977. Bilan énergétique des populations naturelles du Ver de terre géophage *Millsonia anomala* (*Acanthodrilidae*, Oligochètes) dans la savane de Lamto (Côte d'Ivoire). In: *3e Coll. Ecol. Trop. Lumumbashi, 1975. GeoEcoTrop.*, 1: 149–157.

Lavelle, P., 1978. Les Vers de terre de la savane de Lamto (Côte d'Ivoire): peuplements, populations et fonctions dans l'écosystème. *Publ. Labor. Zool. ENS.*, 12: 301 pp.

Lavelle, P., 1979. Relations entre types écologiques et profils démographiques chez les Vers de terre de la savane de Lamto (Côte d'Ivoire). *Rev. Ecol. Biol. Sol*, 16: 85–101.

Lavelle, P. and Meyer, J.A., 1976. Les populations de *Millsonia anomala* (*Acanthodrilidae*, Oligochètes): structure, variations spatio-temporelles et production. Application d'une analyse multivariée (programme Constel). *Rev. Ecol. Biol. Sol*, 13: 561–577.

Lavelle, P. and Meyer, J.A., 1977. Modélisation et simulation de la dynamique, de la production et de la consommation des populations du Ver de terre géophage *Millsonia anomala* (Oligochètes, *Acanthodrilidae*) dans la savane de Lamto (Côte d'Ivoire). In: *Sixth Int. Congr. Soil Zoology, Upssala, 1976. Ecol. Bull.*, 25: 420–430.

Lavelle, P., Douhalei, N. and Sow, B., 1974. Influence de l'humidité du sol sur la consommation et la croissance de

Millsonia anomala (Oligochètes, *Acanthodrilidae*) dans la savane de Lamto (Côte d'Ivoire). *Ann. Univ. Abidjan*, E 7: 305–314.

Lavelle, P., Sow, B. and Schaeffer, 1979. The geophagous earthworms community in the Lamto savanna (Ivory Coast): Niche partitioning and utilization of soil nutritive resources. In: Proc. *Seventh Int. Congr. Soil Zoology, Syracuse, N.Y., 1979*, pp. 653–672.

Lavelle, P., Maury, M.E. and Serrano, V., 1981. Estudio cuantitativo de la fauna del suelo en la region de Laguna Verde (Vera Cruz, Mexico). Epoca de lluvias. In: P. Reyes Castillo (Editor), *Estudios Ecologicos en el Tropico Mexicano*. Instituto de Ecologia, Mexico, pp. 65–100.

Lee, K.E., 1959. The earthworm fauna of New Zealand. *N.Z. D.S.I.R. Bull.*, 130: 1–486.

Madge, D.S., 1969. Field and laboratory studies on the activities of two species of tropical earthworms. *Pedobiologia*, 9: 188–214.

Nowak, E., 1975. Population density of earthworms and some

elements of their production in several grassland environments. *Ekol. Pol.*, 23: 459–491.

Pianka, E.R., 1974. Niche overlap and diffuse competition. *Proc. U.S. Natl. Acad. Sci.*, 71: 2141–2145.

Reinecke, A.J. and Ljungström, P.O., 1969. An ecological study of the earthworms of the Mooi River in Potchefstroom, South Africa. *Pedobiologia*, 9: 106–111.

Reynolds, J.W. and Cook, D.G., 1976. *Nomenclatura Oligochaetologica*. University of New Brunswick, Fredericton, N.B., 217 pp.

Senapati, B.K., 1980. *Aspects of Ecophysiological Studies on Tropical Earthworms. Distribution, Population Dynamics, Production, Energetics, and Their Role in Decomposing Process.* Thesis, Sambalpur University, Sambalpur, 154 pp.

Sow, B., 1979. *Influence de la matière organique du sol sur la consommation et la croissance de trois espèces de Vers de terre géophages des savanes de Lamto (Côte d'Ivoire).* Thesis, University of Abidjan, Abidjan, 43 pp.

Chapter 23

THE SOIL FAUNA OF TROPICAL SAVANNAS.
III. THE TERMITES

GUY JOSENS

INTRODUCTION

Termites are obviously one of the dominant groups of tropical invertebrates. Their mounds are often a distinctive feature of many savanna landscapes in both hemispheres. These omnipresent insects can also be found in significant numbers at all levels, from deep in the soil to the tops of some trees. Their developmental biology, nutritional physiology and social behaviour have been actively studied during the past few decades, and our present state of knowledge has been summarized in a number of books (Krishna and Weesner, 1969, 1970; Lee and Wood, 1971; Wilson, 1971; Brian, 1978: Hermann, 1979, 1981).

However, the nature as well as the importance of their role in the functioning of savanna ecosystems remain poorly understood. This is largely due to the difficulty of sampling their natural populations, particularly for those species nesting below the soil surface. The study of their life-cycle in natural conditions also raises many methodological problems. The quantitative evaluation of their trophic impact, as well as of their production, still remains very tentative, being in many cases based on the extrapolation of laboratory observations.

Fortunately, a few comprehensive research projects, at the community and ecosystem levels, have recently been carried out or are currently in progress in Africa, Malaysia and Australia. These projects are beginning to yield valuable information on the structure and dynamics of termite communities and their function within the ecosystems. Coupled with similar programmes carried out in non-tropical habitats such as the Sonoran Desert of Arizona, these recent findings allow some meaningful comparisons.

The present contribution is not merely based upon a review of the existing literature; it also relies upon my personal fieldwork in the Ivory Coast, and the still unpublished results of a number of field workers who have kindly allowed me to quote some of their recent findings.[1]

As pointed out later in this chapter, the termite communities of different biogeographical realms differ greatly in their taxonomic structure, and correspondingly in their feeding adaptations. Therefore, the factors playing a key role in species distribution, and the adaptive strategies of the various taxonomic categories, will emerge more clearly from comparisons made between communities within a single region than from between communities characteristic of different continents. This is why the present discussion is mostly based on studies carried out in Africa.

THE SAVANNA TERMITE COMMUNITIES

Feeding habits in relation to geographical distribution

The food and feeding habits of termites, as well as the dynamics of their nutrition, have been reviewed by Noirot and Noirot-Timothée (1969), Sands (1969), Bouillon (1970), Lee and Wood (1971), Wood (1978) and La Fage and Nutting (1978). On the other hand, the geographical distribution of the Isoptera has been discussed by Krishna, Bouillon, Paulian, Harris, Roonwall, Gay and Calaby, Bess, Weesner and Araujo — all of them in 1970.

---

[1] I am particularly grateful in this respect to R. Buxton, M. Lepage and E.J. Roose, who provided me with unpublished material.

Feeding habits

The natural diets of termites basically range from living vegetation to sound, decaying or completely rotten dead plant tissues. A number of species have shifted to "humus" consumption, or even soil feeding, while the manure of large herbivorous mammals satisfies the appetites of many species. A peculiar diet is that of the subfamily Macrotermitinae, the so-called fungus-growing termites, which collect plant material in order to build up their "fungus combs". After an interval of five to eight weeks (Josens, 1971a, 1977), this material, impoverished in lignin and probably enriched with nitrogenous compounds (and possibly also vitamins) by the symbiotic fungi, is eaten by the termites. The origin of the comb material has been a matter of dispute for a long time, but general agreement has now been reached since Grassé (1978) has revised his previous position. The collected plant material is either ingested immediately, or carried into the nest, soaked with saliva and stored for some time prior to ingestion. After a short transit through the gut, the almost undigested material is deposited in the form of **mylospheres** [so called by Grassé (1978) and quite different from true faeces] onto the "fungus combs". Mouthparts are used to cement the mylospheres together. True faeces are deposited on the walls of nest chambers and galleries, as well as on the fungus combs.

Cannibalism has been reported as either an accidental or a normal feeding habit in several species. Among other hypothetical functions, cannibalism has been considered as adaptive, enabling species usually living on dead plant material poor in nitrogen to obtain this essential nutrient. Josens (in press) has estimated that cannibalism accounts for more than 20% of the energy requirements of *Trinervitermes geminatus* workers fed with dead *Imperata cylindrica* in laboratory conditions.

Intercontinental distribution

Termites have successfully colonized the warm latitudes of all continents. However, their function within the various ecosystems differs from one continent to the other, because their diets and feeding habits vary greatly between the various biogeographical realms.

The "lower termites" — that is, all families except the Termitidae — are most prominent in Palaearctic and Nearctic regions, as well as in Australia south of the Tropic of Capricorn. All of them are wood feeders.

By contrast, the "higher termites" — that is, the Termitidae — predominate in tropical latitudes. Among them, two major trophic categories display unexpected geographical distributions: (a) the fungus-growing termites, the Macrotermitinae, which are restricted to the Ethiopian and Oriental biogeographical realms; and (b) the true humus and soil feeders, very diversified in the most humid parts of the Ethiopian region; they are also found in the Oriental and Neotropical regions, but do not occur in Australia.

These differences in distribution can, in part, be explained by historical reasons. For instance, the Macrotermitinae originated in Africa during the Tertiary (Emerson, 1955) and therefore did not have the opportunity to enter either the Neotropics or the Australian region. However, one species of *Microtermes* successfully reached Madagascar, and another has been observed in the Moluccas. In addition to historical reasons, adaptive characteristics can be put forward to explain distribution patterns on a continental scale.

Continental and regional distribution

On the basis of the West African data then available, Wood (1976) found a relationship between the number of species of ground living termites and latitude — that is, the rainfall and primary production gradient. According to him, more relationships of this kind can be expected. For example, the number of species feeding upon soil and decomposing litter can be expected to grow with increasing annual rainfall, since a high humidity implies a higher rate of decomposition of organic matter (Noirot, 1958–1959). In the same way, the number of species of Macrotermitinae which prefer sound litter to make their fungus combs should decrease with increasing annual rainfall.

Table 23.1 summarizes the available information for Africa, and provides a comparison between eight savanna study sites and one rain-forest community in Cameroon. The reader wishing more comparative information on tropical forest termite communities is referred to a recent review by Wood (1979). However, it must be kept in mind that most of these studies have *not* been carried out in a single well-defined habitat or biotope, but rather in a

TABLE 23.1

Taxonomic composition and diet structure of termite communities from various African savannas, related to annual rainfall, and compared with a rain-forest termite community

Locality	Type of vegetation	Author	Latitude	Longitude	Annual rainfall (mm)	Percentage organic matter in top soil	Number of species eating				Number of species						Total	Total region[a]
							Grass and grass litter	fresh wood and leaves litter	decomposing litter	humus and soil	Kalotermitidae	Rhinotermitidae	Termitinae	Apicotermitinae	Macrotermitinae	Nasutitermitinae		
Fété Olé, Senegal	wooded steppe (small sand dunes) (reference area)	Lepage (1974)	16°N	15°W	375	3.6 −22	3	11	0	3	0	3	8	0	3	5	19	23
Tsavo east, Kenya	bush and wooded steppe	Buxton (1979)	3.5°S	39°E	400	3	8	17	0	3	4	1	9	0	8	3	25	28
Cap Vert, Senegal	woody sub-Guinean savanna ("Niaye") (biotope "4")	Roy-Noël (1974b)	15°N	17°W	550	1.7 −6.2	6	17	1	8	1	3	12	2	6	4	28[b]	37
Cap Vert, Senegal	woody savanna with cultivated plots (biotope "8")	Roy-Noël (1974b)	15°N	17°W	575	3	6	11	0	10	0	0	12	3	6	4	25[b]	37
Zaria, Nigeria	woodland to open grassland (cleared) (areas "I", "II" and "III")	Sands (1965b)	11°N	7.5°E	1170	1 −2.5	11	15	1	9	0	0	10	3	11	5	29[b]	
Mokwa, Nigeria	primary and secondary woodland	Wood et al. (1977)	9°N	5°E	1175	1.5 −2	10	17	1	9	0	1	9	3	11	7	31	
Lamto, Ivory Coast	grassy to woody derived savanna	Josens (1972)	6°N	5°W	1290	1 −2.5	10	18	3	13	2	0	10	4	9	10	36	54[c]
Youhouli, Ivory Coast	grass savanna	Bodot (1967)	5°N	4.5°W	2000	1.5	10	(0)	1	12	0	0	7	5	7	3	22	
Edea, Cameroon	equatorial rain forest	Collins (1977a)	3.5°N	10°E	>3000	0	0	8	4	31	0	1	25	9	5	3	43	

[a] Including species collected in adjacent biotopes.

[b] When the author mentioned the name of a genus followed by "spp.", I counted it as two species, which may still be an underestimate.

[c] Twenty-two species were found in a (non intensive) census of riparian forests and 42 species were caught at light traps, out of which 9 had never been found in the area.

cluster of adjacent habitats, including for example wet and dry facies of a single savanna landscape, or a gradient of human disturbance ranging from "climax" communities to heavily disturbed areas, such as croplands or clearings. Furthermore, the sampling techniques used were very often dissimilar and the "searching effort" very unequal. These biases being taken into consideration, some general trends nevertheless emerge, together with some interesting anomalies. It must be emphasized that these trends only apply, for the time being, to a large part of Africa, and are not necessarily valid for other tropical areas, pending further research (see also Fig. 23.1).

(1) The number of species within a given cluster of biotopes (column "Total" in Table 23.1) increases with increasing annual rainfall. When different habitats have been sampled in the same region, species richness is 1.2 to 1.5 times greater (column "Total region"). The main exception in Table 23.1 is the small number of species found by Bodot (1967). It is true that her sampling was rather extensive and that she might have missed some species for this reason; however, she did collect some very cryptic termites, though some rather conspicuous wood-eating genera (*Coptotermes*, *Fulleritermes*, *Microcerotermes*, *Nasutitermes* and *Pseudacanthotermes*) were lacking in her samples. Bodot herself explained their absence by the lack of a woody stratum and the occurrence of a very sandy soil in her study sites.

Another unexpected result is the high number of termite species which successfully inhabit dry regions such as Fété Olé in northern Senegal, and east Tsavo in Kenya. This reflects the ability of some termites to withstand particularly variable conditions (such is the case for Kalotermitidae; Noirot, 1958–1959) or, alternatively, to escape them. Some species even penetrate deep into the soil to obtain water from the water table; this is the way *Psammotermes* sp. (Grassé and Noirot, 1948) and *Macrotermes bellicosus* (Grassé and Noirot, 1961; Lepage, 1974) withstand the seasonal drought in West Africa.

(2) The number and percentage of species of humus feeders increase with increasing annual rainfall. This is reflected by the regular increase of Apicotermitinae with decreasing latitude, and the increasing percentage of humivorous Termitinae. The Nasutitermitinae are represented by only one

humivorous species at Mokwa, three at Lamto, one at Youhouli, and none at Edea. The major and most interesting anomaly in that overall trend is the high number of humus feeders found by Roy-Noël (1974b) in the Cap Vert Peninsula, near Dakar. However, this exception is only apparent: the climate in this part of Senegal is actually much moister than one might expect with an annual rainfall between 500 and 600 mm. From November to April (six out of the eight dry season months) an average of eighteen nights per month are characterized by abundant formation of dew, which moistens the soil down to a depth of about 1 cm. An amount of 300 g of dew per square metre (equivalent to a rainfall of 0.3 mm) is not a rare event (Masson, quoted by Roy-Noël, 1974a). This occult and evenly spread precipitation, though discrete, is nevertheless sufficient to stimulate decomposition processes and favour the existence of humivorous termites.

It must be remembered at this stage that the top soil, in areas with 1000 to 2000 mm rainfall, only contains 1 to 2.5% of organic matter, whereas that of drier areas can be much richer. Up to 22% organic matter has been found in the hollows between former dunes at Fété Olé, where organic material tends to accumulate from year to year. On the whole, dry soils constitute a very efficient barrier against colonization by humus-eating animals. The most important among these, the earthworms, are for the most part restricted to regions with an annual rainfall exceeding 1000 mm (see Ch. 22).

Among termites, those Apicotermitinae which possess soldiers appear to be even more sensitive to drought than the other subfamilies (only one species of *Allognathotermes* is present near Mokwa). Only genera which do not possess soldiers, such as *Adaiphrotermes*, *Aderitotermes*, *Astalotermes*, etc., are able to colonize regions with an annual rainfall as low as 550 mm, at least in areas such as the Cap Vert Peninsula, where the water input is predictable.

The humivorous Termitinae can enter dry regions, where the genus *Cubitermes* is frequently associated with lateritic crusts; despite their overall parched aspect, these lateritic crusts can hold some moist soil pockets and provide deep fissures where the microclimate is far milder than on the soil surface. This was the case in the sites studied by

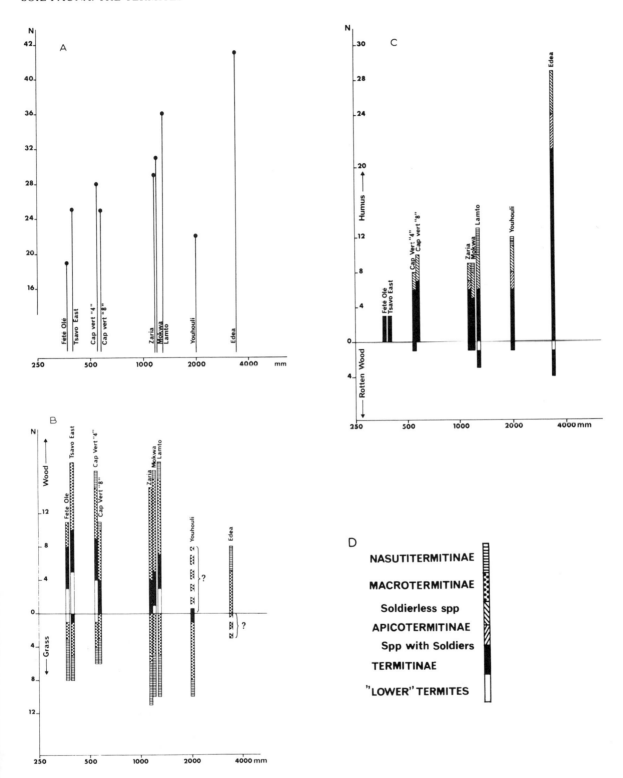

Fig. 23.1. Species richness (A), and number of termite species feeding on sound wood and leaf litter, and grass and grass litter (B), humus and rotten wood (C), as a function of the annual rainfall in nine African study sites. The question marks on B indicate that some species present in the community might feed on that particular food item, if it were present on the site.

Lepage and Roy-Noël in Senegal. In east Tsavo, *Cubitermes* occur on deep red soils (Buxton, 1979). Other humus feeders also occur in such dry areas: *Angulitermes*, *Promirotermes* and *Tuberculitermes*. Unfortunately, next to nothing is known of the structure and average depth of their subterranean nests.

(3) The species richness of the termites regularly feeding on decomposing wood follows the same humidity gradient as the humus feeders, with an easily understandable gap in the grass savannas studied by Bodot (1967). The species most consistently associated with rotten wood are *Termes* spp. (anywhere), *Amitermes evuncifer* [in most savannas, except the Cap Vert Peninsula where this species also eats sound and even living wood (Roy-Noël, 1974b)], and *Coptotermes intermedius* at Lamto (where it is also found in dry wood and decaying stems of *Borassus* palms). Of course, the borderline between feeding on sound and rotten wood is not always clear-cut, even for Kalotermitidae, the famous dry-wood termites. In a riparian forest at Lamto the author has found a colony of *Cryptotermes* sp., whose nest was built in a rotting fallen log. This colony probably started when the tree was still living, but it remained vigorous and even contained alates in this unusual situation.

(4) Dead wood and dead leaves are generally considered as the most primitive diet of the Isoptera. Termites only occasionally attack living plants. However, in areas under cultivation where litter is generally removed from the fields, the species diversity of the termite community tends to decrease and its biomass tends to increase, as a few species become severe pests for maize, yam, sugarcane and other tropical crops (Bigger, 1966; Wood et al., 1977).

From the data presented in Table 23.1 it is not possible to infer any gradient of species diversity for wood- and litter-feeding termites as a whole. However, gradients do exist for some restricted groups of xylophagous termites.

The "primitive" families — that is, Kalotermitidae and Rhinotermitidae — are better represented in drier climates. The complete absence of Kalotermitidae at Fété Olé, where alates were never caught by light traps, is more difficult to explain.

Wood-feeding Nasutitermitinae (*Fulleritermes*

and *Nasutitermes*) are represented by a larger number of species in the moister conditions, including the peculiar savanna landscape of the Cap Vert Peninsula. The species richness of the major group of wood consumers, the Macrotermitinae, is reduced both in the drier areas where wood production is low, and in the more humid regions where the decomposition of wood is more rapid.

Even at the local level there is a clear positive relationship between the number of xylophagous termites and increasing annual rainfall. Working in east Tsavo, Kenya, Buxton (1979) has compared the termite fauna of eight study sites, located within 50 km of each other, where the annual rainfall ranged from 200 to 500 mm. On each site, the termites were lured with thirty wood discs. On six out of the eight sites, the following relationship was found:

$$N = 0.016 \ R - 2.4$$

where N is the number of species attacking the baits, and R the annual rainfall in mm. Of course, such a technique only permits the sampling of soil-inhabiting wood-feeding termites.

(5) Grass-feeding termites include such specialists as five species of *Trinervitermes*, although *T. occidentalis* also feeds on dead wood (Sands, 1961a), as do some opportunists such as most of the Macrotermitinae which feed partly on grass. In the Youhouli savanna, grass probably constitutes their only source of food; elsewhere, the only important fungus-growing grass eaters are found among species belonging to the genera *Macrotermes* or *Odontotermes*, and, to a lesser extent, to *Pseudacanthotermes*.

The species diversity and population density of *Trinervitermes* spp. is reduced in two opposite situations: (a) In the driest savannas, because of their low primary productivity; in such a situation the termites must temporarily turn to a diet of wood when they are short of the staple food (Lepage, 1974; Buxton, 1979). (b) In the more humid situations, despite high grass production. The five species present at Lamto and Mokwa reached much lower densities on these sites than in the savannas close to Zaria, or in those studied in Upper Volta by Roose (1976), despite much lower grass production at the two latter sites.

This paradox might be explained by the fact that *Trinervitermes*, *T. geminatus* at least, clearly prefers dead to living grass (Sands, 1961a; Ohiagu and Wood, 1976). In the more humid grasslands, the rapid decomposition of dead organic material reduces the amount of dead grass available to the termites, and consequently reduces the carrying capacity of the habitat for species feeding on dead grass. Sands (1961a) has studied the coexistence and ecological segregation of five sympatric species of *Trinervitermes* in West Africa, and the problem has also been discussed by Bouillon (1970).

(6) From the previous five points, one might be led to conclude that the amount of annual rainfall (including such "occult precipitation" as dew), plus the amount and quality of available food, determine the structure of termite communities in tropical savannas. Since the quantity of food is broadly correlated with the amount of annual rainfall, the importance of food quality can only be demonstrated by unusual situations, such as those prevailing in the Cap Vert Peninsula or the grass savannas of the southern Ivory Coast.

This remains true as long as soil conditions are suitable for termites. Besides the Youhouli savannas, Bodot (1967) studied other grass savannas, under comparable climatic conditions, and found a very low species richness (only six to eight species per site). She attributed this to very poor soil conditions— up to 20% clay and less than 1% organic matter in the top soil. In such conditions humivorous termites are lacking, with one exception (one species of Apicotermitinae, not possessing soldiers). In this case, pedological factors are di-

rectly responsible for the lack of suitable food.

Roy-Noël (1974a) has also emphasized the role of soil factors to explain differences in species richness in some of her habitats in the Cap Vert Peninsula. A minimum amount of clay in the soil is necessary for mound-building species. In a site with almost purely sandy soil, with less than 1% organic matter, only six termite species were found, not a single humus feeder among them. No termites were ever found on halomorphic soils.

A very interesting study of the roles of various environmental factors on termite communities has been carried out by Pomeroy (1978) in Uganda. A very extensive survey of the large mounds, at least 1 m in height, of *Macrotermes bellicosus* and *M. subhyalinus* was made throughout the country. On plots of 4.0 ha, 638 counts of mound densities were made, and 14 environmental variables were recorded (4 climatic and 9 pedological, plus estimates of local human population density). Then a multivariate analysis was carried out. The best correlations were obtained from parabolic transformations, reflecting the existence of optimal and limiting values in most factors (Table 23.2). The density of large *Macrotermes* mounds is most strongly correlated with annual rainfall and mean minimal temperature. Clay content, at a depth of one metre, is also considered to be important, at least for *M. subhyalinus*, a species very selective in its use of soil particles. Pomeroy's results also suggest that *M. bellicosus* may distinctly influence the rate of litter disappearance, and subsequently

TABLE 23.2

Optimal and limiting values of some factors for *Macrotermes bellicosus* and *M. subhyalinus* (based on Pomeroy's Uganda survey, 1978)

	M. bellicosus			M. subhyalinus		
	lower limit	optimal value	upper limit	lower limit	optimal value	upper limit
Annual rainfall (mm)	700	1250	1900	300	1150	2000
Mean minimum temperature (°C)	12	18	>24	9	17	23
Mean maximum temperature (°C)	23	32?	⩾37	21	29	37
Clay in the soil (%)	minimum density at 30% (spurious result?)			–	~20	~60
pH	<5	5.6	7	<5	6	>7
C/N ratio	?	7	17	~5	12.5	~20

Fig. 23.2. A mound of *Macrotermes bellicosus* in the Comoe National Park, Ivory Coast. The standing height of the man in front is 180 cm.

the humus content of the soil. Previous work by Leprun and Roy-Noël (1976) in the Cap Vert Peninsula also pointed out that *Macrotermes bellicosus* was restricted to ferralitic ironstones and ferrugineous soils containing kaolinite clay, while *M. subhyalinus* lived in a wide range of soil types as long as they contained a minimum of 5 to 10% clay. The latter species uses cracking clay to build its nests.

TERMITE MOUNDS IN TROPICAL SAVANNAS: THE SMALL-MOUND PARADOX

One of the basic characteristics of the Isoptera is to build closed nests. Such nests may be mere networks of simple galleries in wood or soil, or elaborate constructions below and/or above ground level, if not in trees. The structure and

function of those nests has been reviewed by Noirot (1970, 1977), Bouillon (1970) and Lee and Wood (1971).

Termite mounds are a conspicuous and sometimes spectacular feature of many savanna landscapes (Fig. 23.2). Only a few genera are able to construct such sophisticated structures, however. In Africa, for example, most of the above-ground (epigeous) and/or arboreal nests are built by only seventeen of the hundred or so genera known to inhabit the continent (Table 23.3). Out of these seventeen genera, only one, *Trinervitermes*, is restricted to savanna ecosystems, the others being specific to rain-forest areas or found in both types of environments. Such is the case for *Astalotermes quietus* at Lamto, where the species is found both in savannas and riparian forests. However, this termite never builds its temporary pseudoecies (that is, fragile structures of soil on low shrubs connected to the subterranean nest by covered passageways) in savannas even if suitable props are available. Several other termite genera can also secondarily establish themselves in old nests built by other species, and eventually build their own epigeous nests, at times under peculiar circumstances. This

TABLE 23.3

List of African termite genera which build epigeous or arboreal nests[1]

Genus	Rain forests and riparian forests	Savannas and woodlands
Amitermes	×	×
Apilitermes	×	
Astalotermes	×	
Cephalotermes	×	
Cubitermes	×	×
Megagnathotermes	×	
Microcerotermes	×	×
Noditermes	×	
Procubitermes	×	×
Termes	×	
Thoracotermes	×	
Macrotermes	×	×
Odontotermes	×	×
Protermes	×	×
Pseudacanthotermes	×	×
Nasutitermes	×	×
Trinervitermes		×

[1] Some species within these genera only build epigeous nests in certain regions, and not in others.

was the case for an arboreal nest of *Schedorhino-termes lamanianus* found by the author in the gallery forest of the Bandama River, Ivory Coast. The species normally builds its nests in the soil or in logs. However, in this case, the original nest had been flooded during the annual overflow of the river and its inhabitants had taken refuge on a nearby tree, where they secondarily built a carton[1] nest.

The potential ability to erect mounds quite likely exists in many termite genera. On the basis of a very simple hypothesis, Deneubourg (1977) has proposed a mathematical model which satisfactorily accounts for the first stages of the building behaviour previously observed by Grassé (1939, 1959). It shows, among other things, how a regular structure can take shape without any need for the termite builders to measure the distance between the adjacent pillars they erect.

The adaptive value of epigeous mounds of termites in savanna environments is still a matter of dispute among termitologists. In the case of ant nests it is often claimed that one of the major functions of ant mounds is to increase the nest temperature by intercepting the largest possible amount of solar radiation (see Ch. 24). This might help to explain why ant mounds and ant arboreal nests are so scarce in savanna environments, where an excess of heat would be detrimental to the brood. Obviously, all termite mounds are not built to achieve the same aim, and small mound nests appear to be irrational in savanna environments for the following reasons:

(a) Small mounds are exposed to strong circadian variations in temperature. They become dangerously hot during the day (Table 23.4) and they lose their heat by infrared radiation during the night more quickly than the surrounding soil which is sheltered by the grass cover.

(b) Small mounds are exposed to desiccation during the dry season.

(c) Heavy showers often damage them during the rainy season, and this implies extra repair work.

(d) Being located partly below and partly above ground level, they have to face two categories of enemies instead of one.

Obviously the first two disadvantages do not exist in rain-forest environments. Temperature inside termite nests only fluctuates slightly in accordance with the surrounding air in tropical forest conditions. It can even be maintained slightly above the outside temperature, as is the case for *Cephalotermes rectangularis* (Lüscher, 1961; Noirot, 1970; Collins, 1977a). Lüscher (1961) never recorded a relative humidity within such nests lower than 96.2%. These two reasons might explain, at least in part, the abundance of small termite mounds and termite arboreal nests in tropical forest conditions.

In contrast, large mounds with compact walls, such as those of *Macrotermes*, do not suffer the

---

[1] The carton material that makes up all or part of the nests or mounds of many termites consists of excreta, often mixed with mineral soil particles, and occasionally with undigested, comminuted plant tissue (Lee and Wood, 1971).

TABLE 23.4

Mean minimum and mean maximum temperatures in a nest[1] of *Trinervitermes geminatus* at Lamto (after Josens, 1971b)

Location of the measurement	Wet season		Dry season	
	min °C	max °C	min °C	max °C
Near the top of the epigeous nest (inside)	21.4	51.3	20.1	54.0
Centre of the nest (at soil level)	26.1	36.0	29.2	36.1
Subterranean chamber at a depth of 40 cm	26.4	28.1	27.3	28.8
Air	21.4	31.9	22.4	35.7
Soil at a depth of 5 cm	25.0	32.4	25.8	34.2
Soil at a depth of 40 cm	26.9	27.5	28.0	29.7

[1] Height 37 cm and diameter at the base 86 cm; each measurement is a mean of observations over four days.

same handicaps as the small ones. To a large extent, their internal atmosphere becomes independent of external environmental conditions, as the huge mass of earth provides a thermal inertia comparable to that of the soil at a depth of 20 to 30 cm (Ruelle, 1964).

In another respect, above-ground nests provide some substantial advantages to social insects, even when these nests are small in size: (a) they facilitate the social activities of colony members; (b) they increase oxygen and carbon dioxide exchanges with the outside; (c) they reduce the competition for space within the ground compartment of the ecosystem: and (d) make easier the defense against subterranean enemies.

It is quite possible that many termites are basically able to build epigeous structures, and that they will do it provided that, first, the colony members are numerous enough when the termite colony is mature; second, the individuals tend to aggregate in dense clumps producing a large amount of carbon dioxide in a small enclosure; and, third, the surrounding soil or wood does not permit an adequate gas exchange.

This hypothesis obviously needs to be supported by experimental evidence. However, a number of field observations can be interpreted in this way. When nests of *Macrotermes bellicosus* are covered with polyvinyl domes, carbon dioxide concentration around the nests may be increased up to 2%, and new parts of the mounds which are subsequently built have very porous walls (Ruelle, 1964). A similar observation has been made by the author with *Trinervitermes geminatus* nests (Josens; unpubl.).

In savannas on light sandy soils, *Odontotermes* nests were all built entirely underground, whereas in the compacted soil of a football field "chimneys" protruding above-ground were built, presumably in order to facilitate gas exchanges with the outside (Ruelle, 1964).

Further support of the hypothesis just presented can also be found in the fact that the number of colony members in a subspherical arboreal nest of *Nasutitermes costalis* is almost proportional to the outer surface area of the nest (Wiegert and Coleman, 1970). The same applies to colonies of *Trinervitermes geminatus* (Josens, 1972). Following a heavy shower, it has also been noticed that damaged epigeous nests of the same species are

quickly repaired. However, some small holes are kept open and guarded by small soldiers for a couple of hours before being closed. This could be considered as a way to increase ventilation within the nest while its outer wall is still wet and therefore less permeable to gases (Josens, 1972).

TERMITE POPULATIONS

Population movements and changes in numbers

Termite species which build *small* mounds in savanna areas have to avoid the extreme temperatures reached at some times (of the day and/or the year) near the walls of their epigeous nests. For instance, *Trinervitermes geminatus* will never be found close to the overheated top of the mounds during the hottest part of the day. However, not all the population movements or daily fluctuations in numbers can be explained in this way: those observed by Sands (1965a) in the central core of the nests of this species could not clearly be related to changes in microclimatic conditions. On the other hand, seasonal changes in numbers are considered to be related to an hypothetical annual cycle of activity related to climatic conditions. More termites were observed within the epigeous nests from June to September — that is, during the rainy season when foraging activities are at their lowest. The highest percentages of workers within the same nests were also found at that time of the year.

Similar population movements have also been reported in other termite genera, such as *Amitermes* and *Cubitermes*, especially when the species live in areas where the dry season is very severe. At that time of the year the termites desert the above-ground part of their nests (Bouillon, 1970). Working in the more humid savannas of the southern Ivory Coast where it rains every month, Bodot (1969) did not notice any such seasonal population movements. However, the relative proportion of the various life-history stages, changes seasonally; the larvae, in particular, reach their lowest numbers at the end of the main "dry" season, when alates are being produced.

The huge mounds of *Macrotermes* have to be considered separately. The large amount of earth they are made of entails temperature inertia, and their clayey inner structures as well as the presence of fungus combs leads to a more or less constant humidity within the nest. Whether or not termites are able to regulate their nest temperature actively is not yet clear. However, it is now quite certain that they can go deep down into the soil to fetch water from the water table in order to keep the humidity of their nests high enough (Noirot, 1970). Some experiments also strongly suggest that they can actively regulate the ventilation of their nests (Ruelle, 1964).

True active temperature regulation has been reported in various species of Australian termites which have to face "cold" winters when they live south of the Tropic of Capricorn. *Nasutitermes exitiosus* and three species of *Coptotermes* maintain a high nest temperature by aggregating within the centre of their nests, thereby releasing their metabolic heat into a small space (Holdaway and Gay, 1948; Greaves, 1964). Subterranean species living in diffuse nests may experience a constant temperature throughout the year, at least in tropical latitudes. However, this does not apply to humidity, and it probably accounts for the vertical movements of hypogeous Macrotermitinae, such as *Ancistrotermes cavithorax*, *Microtermes toumodiensis* and *Pseudacanthotermes militaris* in Ivory Coast savannas. Colonies of these species move deeper into the soil and dig out new chambers for their fungus combs during the dry season, moving to shallower levels during the rainy season (Josens, 1972, 1977). In so doing they inhabit wetter, deeper horizons when the upper soil is unsuitably dry but allows high gaseous exchanges, and *vice versa*. Similar observations were made in Nigeria on *Microtermes* spp. These termites moved downwards during the dry season, especially in cultivated plots (Wood et al., 1977). Seasonal movements were less marked in woodland areas where food (roots in this case) was always available at some depth. However, even in a secondary woodland, the total weight of the fungus combs constructed by *Microtermes* spp. was reduced by about half at the end of the dry season.

Other termite genera, *Ancistrotermes*, *Macrotermes*, *Microcerotermes*, *Trinervitermes* and even humus feeders such as *Adaiphrotermes* and *Cubitermes*, also undertake downward movements during the dry season (Wood et al., 1977).

Termite swarms

The production and swarming of alates is the most striking expression of seasonal population fluctuations in termite colonies. This matter has been reviewed by Nutting (1969).

In tropical climates, most species produce their alates during, or at the end of the dry season, and swarming occurs at the beginning of the wet season. It generally takes place after a heavy rainfall or a day later, at sunset or at night.

From daily observations and light trappings at Lamto (Ivory Coast) during the whole year of 1968, it appeared that the alates were not yet ready to swarm in January: no alate took flight in spite of an unexpected amount of rainfall (35 mm, Fig. 23.3). Swarming only started at the end of February, at the time when the wet season normally begins, and a maximum of swarms was noticed in April despite a rainfall deficit of *c.* 100 mm. The release of alates of several species at one time, when the humidity of the air and that of the soil are suitable, probably has an adaptive value: it reduces predation rate by "saturating" the hunger of the predators.

There are however some exceptions to the general rule, such as *Pseudacanthotermes militaris* (Grassé and Noirot, 1951). At Lamto, swarms of this species may be observed on sunny afternoons over a long period of time, from July to October, which includes part of the main rainy season, the short dry season and the short wet season. Males and females pair in flight, high above ground level. According to some observations made in Ghana by Williams (1973) supernumerary males aggregate in all-male alate clouds around the top of tall trees and "serve to distract predators from the male–female pairs descending elsewhere".

In even more seasonal climates, alate swarming is strongly dependent upon rainfall. Such is the case at Fété Olé in northern Senegal (Fig. 23.3), where not a single alate was recorded taking flight during the long and severe dry season. Moreover, during the wet months there is a definite relationship between the number of swarming species and the amount of rainfall. Most species swarm early during the rains, but the five *Trinervitermes* species do so during the second half of the rainy season (Lepage, 1974).

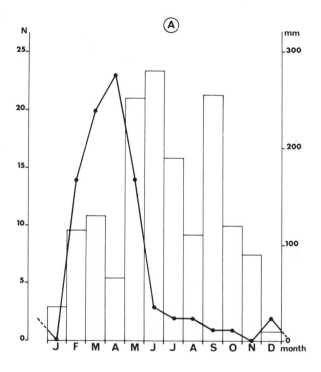

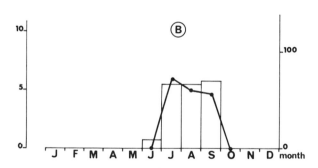

Fig. 23.3. Number of termite species (*N*) swarming every month of a same year (heavy line) in two different study sites: A, Lamto (1968); and B, Fété Olé (1970). The monthly rainfall pattern is shown (bar diagram) in the background. Data from Josens (1972) for Lamto and Lepage (1974) for Fété Olé.

Population densities

Termites with above-ground nests

The density of termite mounds in a given savanna is generally easy to calculate, but to obtain a reliable estimate of the population within each colony is a more difficult task. This matter has been discussed by Lee and Wood (1971), and Baroni-Urbani et al. (1978).

Table 23.5 summarizes the results obtained in some West African savannas. Obviously, most mounds are built here by termites belonging to the genus *Trinervitermes*, by far the dominant genus among those building above-ground nests in this part of Africa. The density of *Trinervitermes* mounds reaches its highest values in areas with a single long dry season, and an annual rainfall of about 1000 mm; this density decreases in both drier and moister climates, as previously mentioned. Since *T. geminatus* enlarges its nest by increasing the size of the existing mound, as well as by building new ones (Sands, 1961b), one wonders whether the average mound size is influenced by the outside environment or by inner social or environmental feedback alone. Despite the different ways used by the various authors to estimate the mean basal diameter of the mounds, some comparisons can be made. As shown in Table 23.6, the results obtained in Upper Volta, northern Nigeria and the Ivory Coast are very similar, but *Trinervitermes* mounds in northern Senegal and in the woodland facies of the Lamto savanna are distinctly larger. Therefore, the outside environment seems able to influence the size of *Trinervitermes* mounds, though still in an unexplained way. *T. geminatus* mounds were also measured at Mokwa (Nigeria), but the method used (Wood et al., 1977; Ohiagu, 1979) does not provide any opportunity of comparison with other measurements.

An estimate of the population inhabiting above-ground termite nests can be made by taking small core samples from the centre of the mounds with a weighted auger (Sands, 1965a), or by sorting the population of whole mounds, as well as that of the subsoil down to a depth of 0.5 to 1 m (Josens, 1972; Lepage, 1974; Wood et al., 1977; Ohiagu, 1979). A correlation of the sizes of the mounds with the populations which inhabit them can be calculated in this way. For instance, the regression for the Lamto population of *T. geminatus* is:

$$P = 1.57D - 11.3 \quad (r = 0.966)$$

where P is the number of thousands of termites within the mound; and D is the mean basal diameter of the mound in cm. At Mokwa, Wood et al. (1977) obtained a different relationship for each season. However, the correlation coefficient never exceeded 0.78, and Wood et al. found that a large

TABLE 23.5

Termite mound densities (numbers ha^{-1}) in various West African savannas

Location	Coordinates	Annual rainfall (mm)	Ami-termes	Cubi-termes	Macro-termes	Odonto-termes	Trinervi-termes	Total	Author
Fété Olé (Senegal) wooded steppe; average of 1969 and 1972 data	16°N 15°W	375			0.49		6.17	6.66	Lepage (1974)
Gonse (Upper Volta) savanna with *Butyrospermum parkii* and *Parkia biglobosa*	12°N 1°W	850					1300	1300	Roose (1976)
Saria (Upper Volta) four year old fallow	12°N 2°W	850					800	800	Roose (1976)
Zaria (Nigeria) undisturbed closed *Isoberlinia* woodland	15°N 7.5°E	1170			14.8	5.0	108.6	128	Sands (1965b)
Zaria (Nigeria) totally cleared grassland	15°N 7.5°E	1170					530.6	531	Sands (1965b)
Mokwa (Nigeria) mature woodland, average of 1973–74 data	9°N 5°E	1175			6.92		232.5	239	Wood et al. (1977)
Lamto (Ivory Coast) undisturbed *Loudetia* grassland with *Borassus* palm trees	6°N 5°W	1290	6		(+)		45	51	Josens (1972)
Lamto (Ivory Coast) undisturbed woodland with *Borassus* palm trees	6°N 5°W	1290	1		(−)		72	73	Josens (1972)
Youhouli (Ivory Coast) *Loudetia* grassland	5°N 4.5°W	2000	51.4	45.4	1.85		47.4	136	Bodot (1967)
Boubouri (Ivory Coast) *Anadelphia* grassland	5°N 4.5°W	2000	4.3				9.3	13.6	Bodot (1967)

TABLE 23.6

Mean basal diameter of *Trinervitermes geminatus* mounds in various savannas of West Africa

Location	Mean basal diameter (cm)	Author
Fété Olé (Senegal)	38	Lepage (1974)
Gonse (Upper Volta)	22.3	Roose (1976)
Zaria (N Nigeria)	25.5	Sands (1961b)
Lamto (Ivory Coast) all but one savanna facies	24.7	Josens (1972)
Lamto (Ivory Coast) dense woodland	50.7	Josens (1972)

part of the population (about one third) lived within the soil, outside of the mounds. Subsequently, Ohiagu (1979) raised this proportion to two thirds.

A better relationship was found between the volume of a mound and its termite population, in the case of *Macrotermes bellicosus*:

$$\log P = 0.564 \log V + 4.15 \quad (r = 0.951)$$

where P is the population (number of individuals); and V is the volume of the mound, calculated as a perfect cone in dm^3, or l (Wood et al., 1977).

This regression is valid up to a volume of 560 l — that is, a mound about 1.5 m high. The population apparently no longer increases in larger mounds, remaining somewhat constant, above half a million individuals.

Termites with subterranean nests

The hypogeous termites, and some subterranean parts of epigeous nests as well, are very difficult to sample quantitatively for the following reasons: (a) the termites live in aggregative structures; (b) their nest system may be very deep; (c) they can move quickly through their galleries and escape the sampling device; and (d) some of them are very small. The termite species spending their whole life within the soil have seldom been adequately studied, though they are much more numerous than the mound-building species.

Digging out pits of 1 m^2, and sorting the termites by hand in the field (Josens, 1972) is not a reliable method, since the volume of soil to handle is very large. The smallest individuals may easily be missed

during the sorting, and a large proportion (may be up to 75%) of most species may escape during digging. Furthermore, the depth of the pit is seldom sufficient. Wood et al. (1977) used a Jarrat hand-operated auger of 10 cm diameter, and took soil cores down to a depth of 2 m. The limited volume of soil thus obtained was then sorted by hand. The "inoculation" within a "virgin" soil of a known number of termites has enabled Wood et al. to check the reliability of this method. As could be expected, it depends on the species of termite concerned: about 90% of the active and light-coloured *Amitermes* and *Ancistrotermes* were recovered, but more than 20% of the slow-moving or dark-coloured *Microcerotermes* and *Trinervitermes* were missed. Wood et al. then tried the Salt and Hollick flotation method and reached in this way an efficiency of 95% in sorting *Microtermes*. Such a method, however, is very time-consuming and was used only as a check.

Wood et al. (1977) claim that their estimates of termite population densities are the most reliable to date, and they are probably right. Nevertheless, a source of error still persists: an unknown part of the termite population escaped from the sampled soil through their galleries while the auger was drilled into the soil (see note on p. 524).

Taking into account the disparity of reliability between the various population estimates published in the past, a comparison of these results is of limited value. The matter has recently been critically reviewed by Wood and Sands (1978).

Termites and soils

The influence of termite populations on soil dynamics has been a matter of dispute for a long time, and has been discussed at length by Lee and Wood (1971) (see also Ch. 4). For example, termites have been held responsible for the formation of lateritic crusts — a hypothesis which has now been almost abandoned. On the other hand, some species can bring up deep soil through the hard pans in order to build their nests, cover their runways and make their sheetings over their food sources.

More quantitative studies of the relationships between termites and soils are badly needed, however. The following two examples can illustrate this point. At Fété Olé, northern Sénégal, Lepage

(1974) attempted to quantify the yearly amount of soil moved in its various activities by *Macrotermes subhyalinus*, a mound-building species. With a density of 0.5 nest ha^{-1}, and an average annual growth of 1 m^3 per nest (i.e. about 1000 kg of soil ha^{-1}), the building of new nests, and the enlargement of old ones, bring up hardly more soil than the sheetings which cover about 475 m^{2}ha^{-1} and correspond to about 800 kg ha^{-1} of soil. Therefore, 0.5 nest ha^{-1} of *M. subhyalinus* causes the displacement of about 1800 kg of soil ha^{-1} yr^{-1}. At Gonse, Upper Volta, Roose (1976) attempted to estimate the influence on soil dynamics of a very high density of *Trinervitermes* mounds: 1300 living mounds ha^{-1} — a grand total of about 8100 kg of aboveground structures. The mean longevity of the mounds was estimated at seven years, the erosion of abandoned nests at 400 kg ha^{-1} yr^{-1}, and the erosion of living mounds at 800 kg ha^{-1} yr^{-1}, whereas the growth of living mounds amounted to 400 kg ha^{-1} yr^{-1}. The total amount of soil brought up by this species alone in Gonse was about 1200 kg ha^{-1} yr^{-1}. This species collects its food in the open air, without any sheeting. Therefore, the soil brought up by this species alone from a depth of 20 to 40 cm (bottom of layer A and top of layer B) amounts to a deposit of 1200 kg ha^{-1} yr^{-1}, of which 50 to 400 kg are washed away by sheet erosion (*érosion en nappe*) during heavy showers. Clay and fine silt are thus selectively removed and this generates an impoverished sandy top soil, which accumulates at a rate of 0.06 mm yr^{-1}. By way of compensation, the gallery network of the termites facilitates the penetration of rainwater into the soil and its access to the water table.

PRODUCTION AND CONSUMPTION

Production

Termites, especially the alates, are among the most generally appreciated prey of many secondary consumers in tropical ecosystems. One might therefore expect Isoptera populations to be *r*-strategists, but this is not at all the case for most species.

On the one hand, the production of alates varies considerably, from one species to another. It ranges from slightly over 1% of the population in *Trinervitermes geminatus* in the Ivory Coast to 28 to 43% in

Odontotermes obesus in India. The available data are summarized by Wood and Sands (1978). In many species the production of alates represents a high energetic expenditure, and even in the case of the "parsimonious" *T. geminatus*, this production accounts for 10.6% of the total annual production of the colony (Josens, 1982).

A large majority of alates perish, being devoured by a variety of predators before settling. However, the number of alates which succeed in entering the soil appears to be larger than previously supposed, at least in some cases. Darlington et al. (1977) have, for instance, reported that 2500 alate pairs ha^{-1} of *Hodotermes mossambicus* could settle successfully, though 52% of them died within the following eight days. In the population of *Macrotermes michaelseni* studied by M.G. Lepage (pers. comm.) near Kajiado (Kenya), seventeen swarms were observed during a two month period. Out of an estimated number of 100 000 alates produced per hectare, about 25% successfully reached soil level, out of which 20% succeeded in pairing. Ten hours later, 377 pairs ha^{-1} only had settled. Two weeks later, 96 of them were still living, and only 44 pairs remained two months later.

It therefore appears that massive swarming can, so to speak, momentarily "saturate" the potential predators, allowing a number of alates to settle without being harmed. As a matter of fact, in such a species as *M. michaelseni* whose colonies average a density of four nests ha^{-1} and have a mean expectation of life of ten years, 0.4 pair ha^{-1} yr^{-1} have to successfully settle and grow into a mature society to maintain the initial density.

On the other hand, the production of neuters in termite colonies is considered to be rather similar for most species. The P/B ratio is close to 3.0, reaching more than 5.0 only among Macrotermitinae (Wood and Sands, 1978). Once more, *Trinervitermes geminatus* appears as a thrifty species. My first P/B ratio for this species was estimated to be 1.0, a rather unrealistic figure, though the production of alates was not taken into account (Josens, 1973). Further considerations and laboratory measurements have now led me to conclude that P/B should be raised to 2.6 (Josens, 1982). Even in this case, *T. geminatus* appears to be a typical *K*-strategist which invests more than 14% of its annual energy expenditure in soldier production and defence of the colony. In contrast, the Macrotermitinae which have a high production of alates and neuters, appear to be *r*-strategists. This may be the consequence of a high predation pressure, as supposed by Wood and Sands (1978), or the outcome of the high nutritional quality of their symbiotically pre-digested food.

Caution must however be exerted in interpreting the values of the P/B ratio, as the B value is based on estimates of population density, data which in many studies cannot be considered as quite reliable.

Consumption

In "natural" savannas where the influence of man is limited to hunting, gathering and burning, almost all species of termites are litter or soil eaters. They therefore play the role of decomposers, although they do not actually accelerate the recycling of biotic elements. The part of the litter they consume — sometimes more than 30% of that available — is protected from the drastic mineralization carried out by the bushfires, and incorporated into their nests. When the colonies are diffuse, the termite faeces are deposited onto the walls of their galleries and thus become accessible to the soil microflora. In concentrated nests, the faeces — or part of them at least — can temporarily remain held up within the nests themselves. Moreover, the assimilation efficiency of the termites is very high, ranging from 54 to 93% (Wood, 1978), and the biotic elements are efficiently incorporated into the termite biomass. The interference of predators is then needed to recycle these elements within the ecosystem.

For the time being, there is no reliable technique which allows us to estimate the consumption of soil by humivorous termites. An attempt has been made by Hébrant (1970) on *Cubitermes exiguus*. Assuming a concentration of 0.11% assimilable organic matter within the soil (cellulose and derived products), and supposing an assimilation efficiency of only 15%, he found that a population of 510 nests ha^{-1}, representing a live biomass of 11.5 kg ha^{-1}, would consume about 115 000 kg soil ha^{-1} yr^{-1}.

The consumption of grass eaters such as *Trinervitermes geminatus*, which carry pieces of grass to their nests, has been measured in field conditions by Ohiagu and Wood (1976). To do so, they took into account the number of foraging holes in each nest, the harvesting rate observed at some of these holes, the average weight of the grass fragments carried by the termites, and an estimate of the amount of grass ingested by the termites while foraging. On this basis (see also Ohiagu, 1979), consumption was estimated to be 7.3 mg per gram of living termite per day, though Wood (1978) quoted a much higher value (27 mg g^{-1} day^{-1}) for the same species. On the other hand, laboratory measurements have shown that the daily consumption of food by *T. geminatus* doubles when alate nymphs are present within the nests (Josens, in press). On an annual basis, the consumption in this species approximates 10 mg of dry grass per gram of living termite per day.

Food consumption by other grass or litter eaters, especially in the case of lower termites, has been estimated mainly in laboratory conditions (Wood, 1978, table 4.1). Field estimates of litter consumption may be based on the rate of disappearance of the litter when termites are present and absent (Wood et al., 1977; Lepage, 1981). A variant of this method was used by Lepage (1974) in the form of line transects, particularly suitable in sparse vegetation. The feeding activity of the termites was measured as the length of soil sheetings covering litter intercepted by a line. This line transect had a minimum length of 2500 m, in order to avoid the effects of an aggregative distribution of the sheetings. From Lepage's measurements, it is possible to extrapolate to the surface covered by the sheetings of each species; and by comparison between quadrats with intact litter and the amount of litter remaining under the sheetings, it is possible to extrapolate the local litter consumption.

Baiting techniques with palatable food have been criticized by Wood (1978) as giving overestimated — though potential — consumption rates. However, the soot-marking technique (Josens, 1971a) cannot be considered as a "classical" baiting technique, although it uses a marked bait. Soot-marked wood is provided in small amounts in order to measure the rate of turnover of the fungus combs made by the Macrotermitinae. The soot-marked wood is not distributed evenly on the termite nests and, in this way, the fungus combs are subsequently often ornamented with a mosaic of marked and unmarked pellets or mylospheres. The amount of litter deposited onto the combs can be deduced from their speed of turnover and from their overall mass which is *independently* measured. Thus, the overall estimated consumption of the Lamto Macrotermitinae can probably be considered as reliable, though the individual consumption cannot be estimated because the population estimates were not reliable (see note on p. 524).

All the recent studies of food consumption in termites point out the clear-cut difference existing between the fungus-growing species and other litter feeders. Wood and Sands (1978) ascribed to the former an average consumption of 60 mg (dry weight) per gram of termite (fresh weight) per day, and to the latter a consumption of 30 mg g^{-1} day^{-1}. Of course, this is an oversimplification and more accurate figures will certainly become available in the future. However, such a difference between the Macrotermitinae and the other sub-families is quite probably not exaggerated; it might even be greater than presently suspected.

In his thesis, Buxton (1979) calculated the relationship between wood consumption by termites and annual rainfall:

$$C = 0.14 AR - 37.7$$

where C is the consumption of wood in g m^{-2}yr^{-1}; and AR is the annual rainfall in mm.

A similar relationship can be found between the food consumption of whole termite communities, soil-eating species excluded, and annual rainfall (Fig. 23.4). In the Sahelian savanna of Fété Olé (Senegal), about 10% of the above-ground primary production was consumed by termites, grass being the preferred food (Lepage, 1974). In the eastern part of the Tsavo National Park (Kenya), termites apparently fed mostly on wood, more than 90% of the woody litter having been consumed during the study period[1]. However, grass and leaf consumptions were not estimated by Buxton (1979). At Kajiado (Kenya) a single species, *Macrotermes michaelseni*, has been studied by Lepage (1981), using three different techniques producing quite consistent results. The consumption figures obtained in this way were very high, and it can be assumed that *M. michaelseni* populations represented a very large part of the termite community of the site concerned; Lepage's figures were therefore used in the regression calculation. For Mokwa (Nigeria) two sets of conflicting data are available. The report of Wood et al. (1977) suggested that

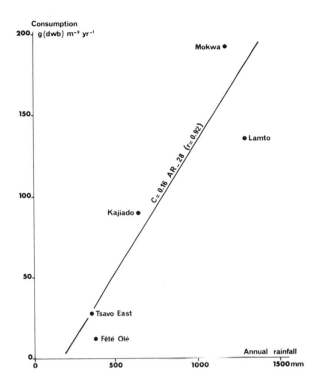

Fig. 23.4. Relationship between annual litter consumption by termite communities and annual rainfall in five African study sites.

wood was the major food source for the termites in this area, whereas Wood and Sands (1978) later reported a food consumption twice as high as previously recorded, including more than 50% of grass. This latter figure was recently corroborated by Ohiagu and Wood (1979), and it has been used for the regression calculation. Whereas leaf litter is almost neglected by termites in Mokwa, it is used at Lamto as much as dead wood, grass representing less than 30% of the total termite consumption (Josens, 1972).

However, the regression which has been calculated:

$$C = 0.16 AR - 28 \quad (r = 0.92)$$

where C is the litter consumption in g m^{-2}yr^{-1}, and AR is the annual rainfall in mm, should be approached with caution until more comprehensive data become available. Once this regression is

---

[1] This might be an overestimate due to the use of a baiting method.

better established, it might provide a useful tool for better understanding of the functioning of tropical ecosystems. Indeed, it is interesting to compare it with the relationship found by Collins (1977b) between litter production and rainfall:

$$LP = 0.59 AR - 112 \quad (r = 0.94)$$

where LP is the annual litter production in g m^{-2}yr^{-1}; and AR is the annual rainfall in mm.

On the basis of this comparison, the trophic impact of termite consumers would appear to be almost constant in relative value. As a matter of fact, consumption does increase slightly, from 27% of the litter production in moist savannas with 2000 mm annual rainfall to 31% in dry savannas with 300 mm annual rainfall, and to 34% in even drier areas with 400 mm annual rainfall.

CONCLUSIONS

Termite communities exhibit several gradients as one goes from the predesertiç steppes to the semi-deciduous tropical forests. Their species diversity, population density, production and consumption are deeply influenced by the climatic gradient and the related changes in primary production and speed of litter decomposition.

The most obvious role of termite communities in savanna environments is the consumption of litter. If one leaves aside the soil-eating species, the food consumption of which still remains completely unknown, it appears that termites consume from a quarter to a third of the litter produced, the fungus growing Macrotermitinae playing a predominant role.

Although a good deal of work on termite ecology has been carried out, mostly in Africa, under the impetus of the International Biological Programme, the gaps in knowledge still remain enormous. Among the most urgent points to be tackled, the following must be given top priority: (a) better estimates of the population size and structure of subterranean species at different stages of their yearly life cycle, which implies the design of new sampling techniques, or the improvement of old ones, in order to prevent too many individuals escaping from the soil sample during boring operations; (b) reliable quantitative estimates of the consumption of soil-eating species, as well as a better knowledge of the nature of their food; and (c) better quantitative evaluations of the role played by termites in soil dynamics, particularly by mound-building species, without overlooking the importance of soil sheeting in feeding areas.

REFERENCES

Araujo, R.L., 1970. Termites of the neotropical region. In: K. Krishna and F.M. Weesner (Editors), *Biology of Termites, 2.* Academic Press, New York, N.Y., pp. 527–576.

Baroni-Urbani, C., Josens, G. and Peakin, G.J., 1978. Empirical data and demographic parameters. In: M.V. Brian (Editor), *Production Ecology of Ants and Termites.* Cambridge University Press, Cambridge, pp. 5–44.

Bess, H.A., 1970. Termites of Hawaii and the Oceanic Islands. In: K. Krishna and F.M. Weesner (Editors), *Biology of Termites, 2.* Academic Press, New York, N.Y., pp. 449–476.

Bigger, M., 1966. The biology and control of termites damaging field crops in Tanganyika. *Bull. Entomol. Res.,* 56: 417–444.

Bodot, P., 1967. Etudes écologiques des termites des savanes de Basse Côte d'Ivoire. *Insectes Soc.,* 14: 229–258.

Bodot, P., 1969. Composition des colonies de termites: ses fluctuations au cours du temps. *Insectes Soc.,* 16: 39–54.

Bouillon, A., 1970. Termites of the Ethiopian region. In: K. Krishna and F.M. Weesner (Editors), *Biology of Termites, 2.* Academic Press, New York, N.Y., pp. 153–280.

Brian, M.V. (Editor), 1978. *Production Ecology of Ants and Termites.* The International Biological Program No. 13. Cambridge University Press, Cambridge, 409 pp.

Buxton, R., 1979. *The Role of Termites in the Ecology of Tsavo National Park, Kenya.* Thesis, University of Oxford, Oxford, 224 pp.

Collins, N.M., 1977a. Oxford expedition to the Edea Marienberg Forest Reserve, United Republic of Cameroon, 1973. *Bull. Oxford Univ. Explor. Club, N. S.,* 3: 5–15.

Collins, N.M., 1977b. Vegetation and litter production in Southern Guinea savanna, Nigeria. *Oecologia,* 28: 163–175.

Darlington, J.P.E.C., Sands, W.A. and Pomeroy, D.E., 1977. Distribution and post settlement survival in the field by reproductive pairs of *Hodotermes mossambicus* Hagen (Isoptera, Hodotermitidae). *Insectes Soc.,* 24: 353–358.

Deneubourg, J.L., 1977. Application de l'ordre par fluctuations à la description de certaines étapes de la construction du nid chez les Termites. *Insectes Soc.,* 24: 117–130.

Emerson, A.E., 1955. Geographical origins and dispersion of termite genera. *Fieldiana, Zool.,* 37: 465–521.

Gay, F.J. and Calaby, J.H., 1970. Termites from the Australian region. In: K. Krishna and F.M. Weesner (Editors), *Biology of Termites, 2.* Academic Press, New York, N.Y., pp. 393–448.

Grassé, P.-P., 1939. La reconstruction du nid et le travail collectif chez les termites supérieurs. *J. Psychol. Pathol. Gén.*, pp. 370–396.

Grassé, P.-P., 1959. La reconstruction du nid et les coordinations inter-individuelles chez *Bellicositermes natalensis* et *Cubitermes* sp. La théorie de la stigmergie, essai d'interprétation du comportement des termites constructeurs. *Insectes Soc.*, 6: 41–53.

Grassé, P.-P., 1978. Sur la véritable nature et le rôle des meules à champignons construites par les termites Macrotermitinae (Isoptera, Termitidae). *C.R. Acad. Sci. Paris, Sér. D*, 287: 1223–1226.

Grassé, P.-P. and Noirot, C., 1948. La "climatisation" de la termitière par ses habitants et le transport de l'eau. *C.R. Acad. Sci. Paris*, 227: 869–871.

Grassé, P.-P. and Noirot, C. 1951. Nouvelles recherches sur la biologie des Termites champignonnistes (Macrotermitinae). *Ann. Sci. Nat., Zool., 11° Sér.*, 13: 291–342.

Grassé, P.-P. and Noirot, C., 1961. Nouvelles recherches sur la systématique et l'éthologie des termites champignonnistes du genre *Bellicositermes* Emerson. *Insectes Soc.*, 8: 312–359.

Greaves, T., 1964. Temperature studies of termite colonies in living trees. *Aust. J. Zool.*, 12: 250–262.

Harris, W.V., 1970. Termites of the Palaearctic Region. In: K. Krishna and F.M. Weesner (Editors), *Biology of Termites, 2*. Academic Press, New York, N.Y., pp. 295–313.

Hébrant, F., 1970. *Etude du flux énergétique chez deux espèces du genre* Cubitermes *Wasmann (Isoptera, Termitinae), termites humivores des savanes tropicales de la région Ethiopienne*. Thesis, University of Louvain, Louvain, 227 pp.

Hermann, H.R., 1979. *Social Insects, 1*. Academic Press, New York, N.Y., 437 pp.

Hermann, H.R., 1981. *Social Insects, 2*. Academic Press, New York, N.Y., 491 pp.

Holdaway, F.G. and Gay, F.J., 1948. Temperature studies of the habitat of *Eutermes exitiosus* with special reference to the temperatures within the mound. *Aust. J. Sci. Res.*, B 1: 464–493.

Josens, G., 1971a. Le renouvellement des meules à champignons construites par quatre Macrotermitinae (Isoptères) des savanes de Lamto-Pacobo (Côte d'Ivoire). *C.R. Acad. Sci. Paris, Sér. D*, 272: 3329–3332.

Josens, G., 1971b. Variations thermiques dans les nids de *Trinervitermes geminatus* Wasmann, en relation avec le milieu extérieur dans la savane de Lamto (Côte d'Ivoire). *Insectes Soc.*, 18: 1–14.

Josens, G., 1972. *Etudes biologiques et écologiques des termites (Isoptera) de la savane de Lamto Pakobo (Côte d'Ivoire)*. Thesis, Free University of Brussels, Brussels, 262 pp.

Josens, G., 1973. Observations sur les bilans énergétiques dans deux populations de termites à Lamto (Côte d'Ivoire). *Ann. Soc. R. Zool. Belg.*, 103: 169–176.

Josens, G., 1977. Recherches sur la structure et le fonctionnement des nids hypogés de quatre espèces de Macrotermitinae (Termitidae) communes dans les savanes de Lamto (Côte d'Ivoire). *Mém. Cl. Sci. Acad. R. Belg.*, 42(5): 123 pp.

Josens, G., 1982. Le bilan energétique de *Trinervitermes geminatus* Wasmann (Termitidae Nasutitermitinae). I. **Mesures de** biomasses, d'équivalents énergétiques, de longévité **et de** production en laboratoire. *Insectes Soc.*, 29, in press.

Josens, G., in press. Le bilan énergétique de *Trinervitermes geminatus* Wasmann. II. Mesures de consommation en laboratoire. *Insectes Soc.*, 29, in press.

Krishna, K., 1970. Taxonomy, phylogeny and distribution of termites. In: K. Krishna and F.M. Weesner (Editors), *Biology of Termites, 2*. Academic Press, New York, N.Y., pp. 127–152.

Krishna, K. and Weesner, F.M. (Editors), 1969. *Biology of Termites, 1*. Academic Press, New York, N.Y., 598 pp.

Krishna, K. and Weesner, F.M. (Editors), 1970. *Biology of Termites, 2*. Academic Press, New York, N.Y., 643 pp.

La Fage, J.P. and Nutting, W.L., 1978. Nutrient dynamics of termites. In: M.V. Brian (Editor), *Production Ecology of Ants and Termites*. Cambridge University Press, Cambridge, pp. 165–232.

Lee, K.E. and Wood, T.G., 1971. *Termites and Soils*. Academic Press, New York, N.Y., 252 pp.

Lepage, M.G., 1974. *Les termites d'une savane sahélienne (Ferlo Septentrional, Sénégal): peuplement, populations, consommation, rôle dans l'écosystème*. Thesis, University of Dijon, Dijon, 344 pp.

Lepage, M.G., 1981. L'impact des populations récoltantes de *Macrotermes michaelsoni* (Isoptera; Macrotermitinae) dans un écosystème semi-aride (Kajiado, Kenya). II. La nourriture récoltée, comparaison avec les grands herbivores. *Insectes Soc.*, 23: 309–319.

Leprun, J.C. and Roy-Noël, J., 1976. Minéralogie des argiles et répartition des nids épigés de deux espèces du genre *Macrotermes* du Sénégal Occidental (Presqu'île du Cap Vert). *Insectes Soc.*, 23: 535–548.

Lüscher, M., 1961. Air-conditioned termite nests. *Sci. Am.*, 205: 138–145.

Masson, H., 1948. Condensations atmosphériques non enregistrables au pluviomètre. *Bull. I.F.A.N.*, 10: 1–181.

Noirot, C., 1958–1959. Remarques sur l'écologie des Termites. *Ann. Soc. R. Zool. Belg.*, 89: 151–159.

Noirot, C., 1970. The nests of termites. In: K. Krishna and F.M. Weesner (Editors), *Biology of Termites, 2*. Academic Press, New York, N.Y., pp. 73–125.

Noirot, C., 1977. Nest construction and phylogeny in termites. In: *Proc. 8th I.U.S.S.I. Int. Congr. Wageningen, 1977*, pp. 177–180.

Noirot, C. and Noirot-Timothée, C., 1969. The digestive system. In: K. Krishna and F.M. Weesner (Editors), *Biology of Termites, 1*. Academic Press, New York, N.Y., pp. 49–88.

Nutting, W.L., 1969. Flight and colony foundation. In: K. Krishna and F.M. Weesner (Editors), *Biology of Termites, 1*. Academic Press, New York, N.Y., pp. 233–282.

Ohiagu, C.E., 1979. Nest and soil populations of *Trinervitermes* spp. with particular reference to *T. geminatus* (Wasmann), (Isoptera), in Southern Guinea savanna near Mokwa, Nigeria. *Oecologia*, 40: 167–178.

Ohiagu, C.E. and Wood, T.G., 1976. A method for measuring rate of grass-harvesting by *Trinervitermes geminatus* Wasmann (Isoptera, Nasutitermitinae) and observation on its foraging behaviour in southern Guinea Savanna,

Nigeria. *J. Appl. Ecol.*, 13: 705–713.

Ohiagu, C.E. and Wood, T.G., 1979. Grass production and decomposition in Southern Guinea savanna, Nigeria. *Oecologia*, 40: 155–165.

Paulian, R., 1970. The termites of Madagascar. In: K. Krishna and F.M. Weesner (Editors), *Biology of Termites, 2.* Academic Press, New York, N.Y., pp. 281–294.

Pomeroy, D.E., 1978. The abundance of large termite mounds in Uganda in relation to their environment. *J. Appl. Ecol.*, 15: 51–63.

Roonwall, M.L., 1970. Termites of the oriental region. In: K. Krishna and F.M. Weesner (Editors), *Biology of Termites, 2.* Academic Press, New York, N.Y., pp. 315–391.

Roose, E.J., 1976. *Contribution à l'étude de l'influence de la mésofaune sur la pédogénèse actuelle en milieu tropical.* Rapport ORSTOM. Centre d'Adiopodoumé, Ivory Coast, 56 pp.

Roy-Noël, J., 1974a. Recherches sur l'écologie des Isoptères de la presqu'île du Cap Vert (Sénégal). Introduction et première partie: le milieu. *Bull. I.F.A.N.*, A 36: 291–371.

Roy-Noël, J., 1974b. Recherches sur l'écologie des Isoptères de la presqu'île du Cap Vert (Sénégal). Deuxième partie: les espèces et leur écologie. *Bull. I.F.A.N.*, A 36: 525–609.

Ruelle, J.E., 1964. L'architecture du nid de *Macrotermes natalensis* et son sens fonctionnel. In: A. Bouillon (Editor), *Etudes sur les termites africains.* Masson, Paris, pp. 325–351.

Sands, W.A., 1961a. Foraging behaviour and feeding habits of five species of *Trinervitermes* in West Africa. *Entomol. Exper. Appl.*, 4: 277–288.

Sands, W.A., 1961b. Nest structure and size distribution in the genus *Trinervitermes* (Isoptera, Termitidae, Nasutitermitinae), in West Africa. *Insectes Soc.*, 8: 177–186.

Sands, W.A., 1965a. Mound population movements and fluctuation in *Trinervitermes ebenerianus* Sjöstedt (Isoptera, Termitidae, Nasutitermitinae). *Insectes Soc.*, 12: 49–58.

Sands, W.A., 1965b. Termite distribution in man-modified habitats in West Africa, with special reference to species segregation in the genus *Trinervitermes* (Isoptera, Termitidae, Nasutitermitinae). *J. Anim. Ecol.*, 34: 557–571.

Sands, W.A., 1969. The association of termites and fungi. In: K. Krishna and F.M. Weesner (Editors), *Biology of Termites, 1.* Academic Press, New York, N.Y., pp. 495–524.

Weesner, F.M., 1970. Termites of the Nearctic region. In: K. Krishna and F.M. Weesner (Editors), *Biology of Termites, 2.* Academic Press, New York, N.Y., pp. 477–525.

Wiegert, R.G. and Coleman, D.C., 1970. Ecological significance of low oxygen consumption and high fat accumulation by *Nasutitermes costalis* (Isoptera, Termitidae). *Bioscience*, 20: 663–665.

Williams, R.M.C., 1973. Evaluation of field and laboratory methods for testing termite resistance of timber and building materials in Ghana, with relevant biological studies. *C.O.P.R., Trop. Pest Bull.*, 3: 64 pp.

Wilson, E.O., 1971. *The Insect Societies.* The Belknap Press of Harvard University Press, Cambridge, Mass., 548 pp.

Wood, T.G., 1976. The role of termites in decomposition processes. In: J.M. Anderson and A. Macfadyen (Editors), *The Role of Terrestrial and Aquatic Organisms in Decomposition Processes.* Blackwell, Oxford, pp. 145–168.

Wood, T.G., 1978. Food and feeding habits of termites. In: M.V. Brian (Editor), *Production Ecology of Ants and Termites.* Cambridge University Press, Cambridge, pp. 55–80.

Wood, T.G., 1979. The termites (Isoptera) fauna of Malesian and other tropical rain forests. In: A.G. Marshall (Editor), *Transactions of the Sixth Aberdeen–Hull Symposium on Malesian Ecology*, pp. 113–130.

Wood, T.G. and Sands, W.A., 1978. The role of termites in ecosystems. M.V. Brian (Editor), *Production Ecology of Ants and Termites.* Cambridge University Press, Cambridge, pp. 245–292.

Wood, T.G., Johnson, R.A., Ohiagu, C.E., Collins, N.M. and Longhurst, C., 1977. *Ecology and Importance of Termites in Crops and Pastures in Northern Nigeria.* Project Report 1973–76, C.O.P.R., London, 131 pp.

NOTE ADDED IN PROOF

Using a flotation method at Lamto in 1980–1981, G. Josens measured a density of *c.* 12 000 ind. m^{-2} in a dense woodland, a figure ten times higher than his previous estimate. Consequently, the consumption rates of the Lamto Macrotermitinae range from 50 to 100 mg dry wt. g^{-1} fresh wt. day^{-1}.

Chapter 24

THE SOIL FAUNA OF TROPICAL SAVANNAS.
IV. THE ANTS

JEAN LÉVIEUX

INTRODUCTION

Not much is known at present about the ecology of savanna ants, despite the very important role they obviously play in the functioning of the savanna ecosystems. To date, the most extensive observations have been made in Africa, and pending further research, care should be taken in extending the conclusions to other continents. Even the African savannas themselves must not be considered as forming a single ecological entity; there are indeed many fundamental differences between savannas where the average annual rainfall ranges from 800 to 1400 mm, and those that have an annual rainfall less than 800 mm.

THE SAVANNA ANT COMMUNITIES

There are few papers devoted to the study of tropical savanna ants and most of them deal with taxonomic problems. The habitat preferences of most species are poorly known, and very little attention has been given to the structure of ant communities.

Geographical distribution of ant faunas

A striking parallel exists, on the world scale, between the geographical distribution of mammals and that of ants (Emery, 1893: Von Ihering, 1894; Wheeler, 1910). Some of the conclusions drawn by Brown (1973) from his comparison of tropical American and African rain-forest ant faunas can also be extended to savannas.

The African savannas share many genera and a number of species with the oriental region — for instance *Camponotus*, *Brachyponera*, *Monomorium*, *Oecophylla*, *Odontomachus*, *Pachycondyla* (Ponerinae), *Polyrhachis* (Formicinae), *Tetramorium* (Myrmicinae). Other genera such as *Probolomyrmex* are equally represented in all tropical continents (Taylor, 1965b). Conversely, whereas a number of genera that probably radiated early in ant history are common to Africa and the Neotropics (*Anochetus*, *Cerapachys*, *Discothyrea*, *Leptogenys*, *Odontomachus*, *Pachycondyla*, *Platythyrea*, *Probolomyrmex*, *Sphintomyrmex* (Ponerinae), *Crematogaster*, *Leptothorax* (Myrmicinae), *Camponotus* (Formicinae), none of those that radiated later has species inhabiting both continents. The only exceptions are those species spread by human commerce, the so-called "tramp species" (Brown, 1972). A number of widely spread Neotropical genera share species with the Indo-Australian region, but not with tropical Africa (Kusnezov, 1963). Other ants, like the Tribe Basicerotini, are very heterogeneously distributed (Brown and Kempf, 1960; Taylor, 1968). As for Melanesia, it was probably populated by oriental taxa, via New Guinea (Wilson, 1959a, b; Greenslade, 1969).

There are, therefore, important differences between the African and Neotropical savanna ant faunas. Furthermore, a number of genera have a restricted distribution or very specialized habits, for example *Atta* and *Acromyrmex* in tropical America. The present distribution results, at least in part, from the evolution since the Miocene of a number of African and Asiatic species, which gave rise to the present dominant taxa (Emery, 1891; Wheeler, 1914; Brown, 1973). Recently discovered fossils, besides helping to promote a new phylogeny for the family (Brown, 1954; Wilson et al., 1967), help to understand how some highly adaptive genera of

525

myrmicines (*Crematogaster*, *Pheidole*) and formicines (*Camponotus*) were able to play a major role in the colonization of tropical savannas and the functioning of their ecosystems. Detailed information on the distribution, origin and evolution of ant faunas is provided by Borgmeier (1955), Wilson et al. (1956), Brown (1958), Wilson (1958a, b, c, 1959a, b, 1964), Kusnezov (1963), Wilson and Taylor (1964, 1967), Taylor (1965a) and Gotwald (1977).

Most modern taxonomists agree that many ant taxa originated in the tropical rain forests. However, the role of tree falls, clearings and other open areas so numerous in the forest cannot be disregarded. They might have played the role of centers of speciation and dispersion for a number of species. According to Wilson (1959c), for instance, savannas were at first marginal habitats in the Pacific islands, but their later extension provided new opportunities for expansion and speciation. It is likely that similar situations occurred even more frequently in large continental land masses such as Africa and South America.

General characteristics of savanna ant communities

Ants are *par excellence* tropical insects, and their species richness regularly decreases from tropical to cold latitudes. For example, Kusnezov (1963) enumerated 72 genera and 357 species in the Amazon Basin, against 260 species in the São Paulo area, 160 around Tucuman, 83 in the arid parts of Argentina and only 21 in Tierra del Fuego. Within a narrower latitudinal range, the various Australian environments harbor more than 200 species each (P.J.M. Greenslade, pers. comm.). Such large-scale comparisons can, however, be misleading, and more information is provided by a detailed examination of the ant communities of well-defined habitats.

The soil-dwelling ants

Savanna ants living in the soil layer belong to sub-families Ponerinae, Myrmicinae and Formicinae.

An intensive study of the ant community of the Guinean savanna of Lamto (Ivory Coast) was undertaken from 1963 to 1975. A grand total of *c.* 230 species has been found. More than 30 of them are very common (*Camponotus acvapimiensis*,

Acantholepis canescens and *Pheidole termitophila* particularly). Among those living within the clayey soil, 42% of the species belong to the Myrmicinae, 30% to the Ponerinae, and 15% to the Formicinae. This high percentage of ponerine ants is undoubtedly due to the proximity of the gallery forest (Lévieux, 1973). Ponerinae and Dorylinae are indeed two sub-families characteristic of forest habitats in Africa. At Lamto, within the gallery forest itself, the percentages are quite different: 60% Ponerinae (*Mesoponera caffraria* being the most numerous), 20% Myrmicinae, and 15% Formicinae. Still further south, in the N'Douci rain forest, the percentage of ponerines reaches 85%. Amblyoponinae, always scarce in all habitats, range from the semi-deciduous forest to the damp parts of undisturbed savannas.

Forest dorylines can occasionally enter savannas close to the forest edge (Schneirla, 1933, 1938, 1956, 1971; Leroux, 1975, 1977a, b). At Lamto, 57% of the nests of *Anomma nigricans* are found in the gallery forests, which account for only 20% of the study area. Population density there reaches 32 colonies km^{-2}, compared with 3 colonies km^{-2} in the adjacent savanna.

At least in Africa, dry savannas harbor a much poorer ant fauna than wet savannas: less than 90 species (Lévieux, unpubl.). Here again, Formicinae (*Acantholepis*, *Camponotus*, *Polyrhachis*) and Myrmicinae (*Messor*, *Myrmicaria*, *Pheidole*, *Tetramorium*, several genera of the tribe Dacetini) are the most numerous. Dorylines are represented by only a few species (*Dorylus* spp., *Rhogmus* spp), and only ponerines with a very wide distribution range are found there (*Brachyponera senaarensis*, *Leptogenys pavesii*, *Megaponera foetens*, *Paltothyreus tarsatus*).

Such a species richness corresponds closely to that of the low-altitude grassland fauna of the Tucuman area in Argentina (Kusnezov, 1963). In the more arid grasslands of America, about 30 ant species can be found. This is very close to what seems to be the rule in West African Sahelian savannas. Delye (1976) has identified 39 species around Niamey (Niger), out of which 50% were Myrmicinae, 33% Formicinae (among them five species of *Camponotus*), and only four Ponerinae and one Dorylinae.

The wet tropical savannas of northern Australia also harbor more species of ants (120) than the drier savannas further south. Here the fauna includes both arid-zone species (*Melophorus*, *Meranoplus*, *Tetramorium*) and ants characteristic of more humid climates (*Odontomachus*, *Opisthopsis*, some *Polyrhachis*). The overall community structure is nevertheless reminiscent of that of other arid zone areas, with a predominance of ground-living species and few inhabiting the soil itself or the arboreal layer (Greenslade and Mott, 1978; P.J.M. Greenslade, pers. comm.).

The relative abundance of the various species of ants in the Lamto savanna is very representative of a situation which is prevalent in many other tropical habitats. Five of the soil-inhabiting species are very common; for instance, *Componotus acvapimensis* can be found in 90% of the samples. Four species (*Acantholepsis canescens* and *Polyrachis viscosa* among them) are present in more than half of the samples, and five other species (*Crematogaster* sp., *Hypoponera coeca*, *Paratrechina weissi* and *Tetramorium* sp.) are found in 25 to 50% of the samples. Sixty or so species are occasional. Dry Ivory Coast savannas harbor a smaller number of species, but here again some of them are much more abundant than others. The most commonly found are *Myrmicaria eumenoides*, *Brachyponera senaarensis*, *Messor galla* and *Camponotus* of the *sericeus* group. A similar situation is encountered in the Mokwa savanna, Nigeria, and in some habitats of the Imatong Mountains, Sudan (Weber, 1943).

The population density of the most abundant Lamto species (*Camponotus acvapimensis*) is 350 adult nests ha^{-1}, which means about 2.2×10^6 workers ha^{-1}, i.e. 220 ind. m^{-2}. When all species are considered, the overall ant density probably approaches 20 millions ha^{-1}, that is about 2000 ind. m^{-2}. Such a value closely matches that found in an undisturbed rain forest in Panama: 15.5 millions ha^{-1} (Williams, 1941). The biomass of *C. acvapimensis* at Lamto averages 0.35 g m^{-2}, and the total standing crop for all ant species averages 0.75 g m^{-2} (both values in dry weight) (Lévieux, 1976c).

By hand-sorting of soil samples, P. Lavelle (pers. comm., 1979) has found an overall ant density of *c.* 1000 ind. m^{-2}, larvae included, in the Foro Foro *Lophira lanceolata* savanna, Ivory Coast. In a derived savanna close to Veracruz, Mexico, Lavelle et al. (1981) counted 150 to 580 ants m^{-2}, against 780 to 1400 ants m^{-2} in the forest soils of the same area. Although accounting for *c.* 25% by numbers

of the soil macrofauna found in this Mexican grassland, ants only accounted for 0.3 to 1.0% of their live biomass. In the forests of the same region, the live biomass of ants was higher, 1.6 to 3.0 g m^{-2} (5 to 8% of the biomass of all soil organisms). The Sahelian savanna of Fété Olé, Senegal, contains at least seventeen species of ants, the density of nests being estimated to 60 ha^{-1} (M. Lepage, pers. comm., 1979).

The density reported in a dry Australian savanna for a single *Iridomyrmex* species is only 37 ind. m^{-2} (Greenslade, 1976). Therefore, there is a marked reduction of the population density of ants in the more xeric tropical savannas, whereas the highest densities are reached in the humid tropics, in derived savannas as well as in lowland forests.

To these figures for density of sedentary species must be added those of the driver-ants which invade their habitat from time to time. Their numbers are impressive. At Lamto, the average population of an *Anomma nigricans* nest averages 3 to 6 million individuals, producing 2 to 3 million larvae per month. The live biomasses of such colonies range from 31 to 55 kg (9 to 16 kg dry weight). The average standing crop of these driver-ants in the derived savanna under study was estimated to reach 0.7 to 1.3 kg ha^{-1}, dry weight (J.M. Leroux, pers. comm., 1979). Population estimates for colonies of other dorylines are equally impressive: 5.0×10^5 ind. ha^{-1} for *Neivamyrex nigrescens*, 1.5 to 2.2 $\times 10^7$ for *Anomma wilverthi*, 1.0 to 5.0 $\times 10^5$ for *Eciton burchelli* (from data of Raigner and Van Boven, 1955; Schneirla, 1957; Rettenmeyer, 1963; and compilations of Brian, 1965, 1978).

The figures mentioned must also be compared with the few available for temperate latitudes (Letendre et al., 1971; Jensen, 1978). Dry weight biomasses of 0.06 g m^{-2} and 0.008 g m^{-2} respectively have been reported for *Lasius niger*, in some North American (Odum and Pontin, 1961) and British (Pickles, 1937) habitats. A figure of 0.08 g m^{-2} has been given for *Pogomyrex badius* in South Carolina by Golley and Gentry (1964), and Brian (1956) reported a dry biomass of 3.0 g m^{-2} for four sympatric species in England. Thus, in some temperate habitats at least, ants can reach local biomasses as high as those found in some tropical savannas. However, when very large areas are considered, the savannas usually harbor higher ant densities and biomasses, because of a greater homogeneity of environmental conditions — as long as they remain undisturbed.

The dominant role played by the more "advanced" groups of ants (that is, the Dolichoderinae and Myrmicinae) in the tropical savannas, is probably related to favourable living conditions specific to this ecosystem. Brown (1973) already suggested that radiant heat controls the distribution of ants. by providing the warmth necessary to allow them to forage efficiently, and by allowing their larvae to develop quickly. He noted that ants are exceedingly scarce in most tropical mountain forests, whether it be in Colombia or India, whereas they may be locally abundant at much higher altitudes on treeless slopes of the Andes or the Himalayas. This view is supported by my own field work in Ivory Coast. In humid savannas in the center of the country the density of ant nests per hectare is higher than in the shady rain forest of the south. It is no wonder then, that the combination of a high radiant energy imput with a wide spectrum of food resources creates optimal conditions for ants in the more humid African savannas, leading to an overall nest density of c. 3500 nests ha^{-1} (Lévieux, 1973).

The ants of the grass and tree layers

Though exploited by ants the grass and arboreal layers of the savannas are not generally densely populated.

Grass itself is seldom used by ants. A few species, however, can build their nests in hollow grass stems, as *Crematogaster impressa* does in the Ivory Coast (Delage Darchen, 1971).

Trees and shrubs are mostly colonized by Myrmicinae (*Cataulacas* spp., *Crematogaster heliophila*, *Leptothorax* spp., *Pheidole* spp., *Pseudomyrmex* spp., *Tetramorium* spp.) and Formicinae (*Camponotus* spp., *Oecophylla longinoda*, *Polyrhachis* spp.; Ledoux, 1949).

A number of ants of the soil layer (*Acantholepis* spp., *Camponotus* spp., *Pheidole* spp.) can occasionally penetrate the grass and arboreal layers, settling part of their brood there, while the mother queens remain in their underground nests. However, a complete colony of *Camponotus acvapimensis* was once found in a hollow branch of *Crossopteryx*, three metres above soil level.

Close to the forest/savanna boundary, mother queens of truly arboreal forest species can enter the savanna and establish new colonies in hollow trees. Such is the case for *Crematogaster gambiensis* at Lamto. These colonies are usually short-lived, and their density is very low, just a few per hectare.

Some species of trees can shelter many ants, however, especially those which have developed symbiotic relationships with them. Such is the case for many *Acacia* species, both in the Old and the New World. In East Africa, species of ants are associated with *Acacia drepanolobium*, and their density can reach 29.25×10^6 ind. ha^{-1} (Hocking, 1970).

Soil-inhabiting ants can also make use of the soil pockets held between the leaf bases of the apical bud of palms. Such pockets of "hanging soil" can be found in 97% of the *Borassus* palms in the Lamto savanna. Even large species, such as *Paltothyreus tarsatus*, can be found there. In these microhabitats, ants can account for 99.9% of the numbers and 97% of the biomass of the soil organisms; up to 34 000 individuals of *Crematogaster* sp. have been found in a single *Borassus* apical bud. The most frequently found species in Lamto palms were *Oecophylla longinoda* (which also sometimes builds its nests on the palm leaves), *Camponotus compressiscapus*, several species of *Pheidole* and *Crematogaster*, and *Anochetus africanus* (Vuattoux, 1968).

The myrmecofauna of gallery forests is very different from that of the surrounding savannas, and is rich in forest species.

THE ADAPTIVE CHARACTERISTICS OF SAVANNA ANTS

The partitioning of nest sites

The behavioural flexibility of ants has enabled them to make use of a large variety of microhabitats within one environment. In a tropical savanna ants can be found everywhere, from deep in the soil to the top of trees, as shown diagramatically in Fig. 24.1 and Table 24.1, both, based upon my own field work in Ivory Coast.

A definite stratification exists within the soil itself. The nests of the more primitive genera (*Amblyopone*, *Apomyrma*, *Hypoponera*) are found in the deeper layers of the soil. always below 20 cm; they consist of a network of narrow tunnels linking

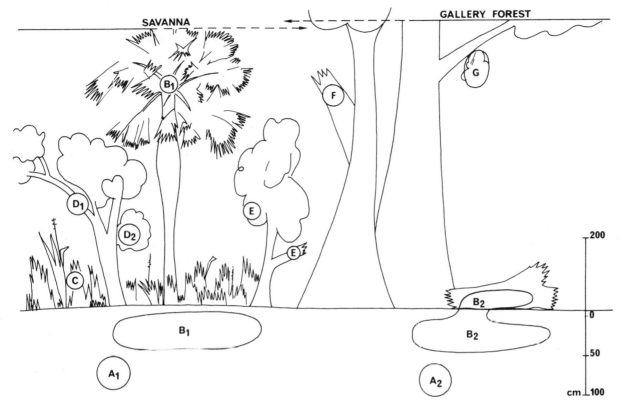

Fig. 24.1. The stratification of nest sites in the savanna gallery forest boundary in Ivory Coast. For an explanation of the symbols used, see Table 24.1.

together small chambers (Lévieux, 1976b). The nests of more "progressive" genera (*Camponotus, Myrmicaria, Paltothyreus, Pheidole, Polyrhachis, Tetramorium*) range from a depth of about 60 cm to the soil surface. Seen from above, nothing betrays their presence, except small earth craters around the nest entrances. Such is the case for the nest of *Camponotus maculatus* and *Polyrhachis schistacea* in wet savannas, and for those of *Camponotus* gr. *sericeus, Myrmicaria* spp. and *Polyrhachis* gr. *militaris* in dry savannas.

The nomadic ants (Dorylinae, *Leptogenys* spp., *Megaponera* spp.) never stay in the same place for long. Their frequent changes in nesting sites are sometimes (among Dorylinae) dependent upon the ovarian cycle of the queen (Schneirla, 1934, 1938), and do not result proximately from the depletion of the food resources around their temporary bivouacs, as earlier supposed (Müller, 1886; Vosseler, 1905). Whereas driver and army ants basically belong to the tropical forest myrmecofaunas, and enter humid savannas only close to the forest edge,

there are nonetheless some genuine nomadic ants in the world savannas, such as *Leptogenys* spp., *Megaponera faetens* and *Termitopone* spp. (Prell, 1911; Arnold, 1914; Schouteden, 1924; Collart, 1925, 1927; Wheeler, 1936; Wilson, 1958a, b; Lévieux, 1966; Sheppe, 1970; Longhurst and Howse, 1977; Longhurst et al., 1978).

The vertical stratification of nesting sites is even more apparent above the soil level. *Crematogaster* species are among the few ants which can establish their colonies in the grass and shrub layer, as already mentioned (Delage Darchen, 1974). On the other hand, many species can nest in trees. Some do so in the hollows of trees, either under the bark (*Melissotarsus* spp., *Pheidole* spp.) or in sound wood (*Cataucalus* spp., *Polyrhachis* spp.) (Delage Darchen, 1972). Some *Crematogaster* species build large carton[1] nests around branches. They are

---

[1] Entomologists have adopted the French word *carton* to signify a substance resulting when a paste made by masticating cellulose or lignin material dries.

TABLE 24.1

Vertical distribution of the nesting sites of Ivory Coast ants (see also Figs. 24.1 and 24.2)

GROUP A. Soil-dwelling species (and part of the ground-living ones)

Savanna (A₁)

Examples: *Amblyopone pluto, Amblypone* spp., *Apomyrma stygia, Asphinctopone silvestrii, Hypoponera* gr. *coeca, Megaponera foetens, Paltothyrens tarsatus, Plectroctena subterranea*

Gallery forest (A₂)

Examples: *Anochetus schoutedeni, Anochetus* spp., *Myopias silvestrii, Mystrium silvestrii, Psalidomyrmex foveolatus*

GROUP B. Ground-living species

Savanna (B₁)

Examples: *Acantholpis* gr. *arenaria, A. canescens, Aneleus* sp., *Camponotus acvapimensis, C. carbo, C. congolensis, C. maculatus, C. vestitus, Odontomachus troglodytes, Pachycondyla* sp., *Pheidole sculpturata, P. speculifera, Polyrhachis viscosa, P. schistacea, Tetramorium guineenese, T. sericeiventre, Xiphomyrmex* sp.

Gallery forest (B₂)

Examples: *Anochetus* sp., *Camponotus maculatus, Leptogenys conradti, Monomorium* sp., *Odontomachus assiniensis, Pachycondyla caffraria, Pheidole bucholzi*

GROUP C. Ants of the grass layer

Example: *Crematogaster impressa.*

GROUP D. Arboreal ants, tree savanna

Nests in hollow branches (D₁)

Examples: *Acantholepis* spp., *Crematogaster* spp., *Camponotus* gr. *sericeus*

Leaf carton nests (D₂)

Examples: *Oecophylla longinoda, Tetramorium aculeatum*

GROUP E. Arboreal ants, gallery forest

Nests in lower branches (E)

Examples: *Atopomyrmex mocquerysi, Camponotus solon, Melissotarsus titubans, Platythyrea conradti, P. modesta, Polyrhachis militaris*

Carton nests in lower branches

Examples: *Polyrhachis laboriosa, P. fissa*

Nests in upper branches (F)

Examples: *Cataulacus* sp., *Polyrhachis* sp.

Carton nests in upper branches (G)

Example: *Crematogaster* spp.

made of wood powder mixed with saliva or other secretions (*Crematogaster depressa* and *C. buchneri*), or of carton mixed with soil particles (*C. ledouxi*). Other ants make use of various fibres, silk included; such is the case of *Macromischoides aculeatus, Oecophylla smaragdina, O. longinoda, Polyrhachis gagates* and *P. laboriosa* (Santschi, 1909; Collart, 1932; Strickland, 1951a, b; Ledoux, 1958; Wilson, 1959d; Soulié, 1961; Bolton, 1973). Most of these truly arboreal ants are peculiar to gallery forests, but can sometimes be found in tree savannas. I have never found any elaborate arboreal ant nests in the West African dry savannas; this is probably correlated with the harsh climatic conditions that prevail there during the dry season.

Many tropical plants have hollow stems and thorns in which some ants preferentially live. This is the case for many *Acacia* species in dry savannas of both hemispheres, as well as for various species of *Barteria, Cuviera, Macaranga* and *Vitex* in Africa and Asia, and for some *Cordia* and *Triplaris* species in the Neotropics. Some of the ants which live in these plants are simply those which would move into any suitable cavity, regardless how it has been formed (*Crematogaster* spp., *Monomorium* spp., *Tetraponera* spp., *Cataucalus* spp.; (Wheeler, 1942). Others are obligatory associates of ant-plants, such as *Azteca muelleri, Pachysima aethiops, Pseudomyrmex ferruginea* and *Viticicola tessmani.* Although the presence of ants can sometimes attract ant-eating vertebrates such as woodpeckers (Escherich, 1911), and cause some harm to the plant, in most cases their presence is definitely beneficial; ants preserve "their" trees from attacks by herbivorous animals of various sorts, including coreids, membracids, chrysomelid and buprestid beetles, and even leafcutting ants (Brown, 1960; Janzen, 1966, 1967, 1969a, b).

Microclimatic adaptations

With such a broad range of nest sites, from deep in the soil to the tree canopy, tropical ants are exposed to a variety of microclimatic conditions to which they adapt successfully.

Savanna ants living within the soil have two ways of coping with adverse microclimatic conditions: by attuning their daily activity schedules to the circadian changes in temperature and air moisture, as discussed later, and by changing seasonally the depth at which they keep their brood.

In contrast to forest undergrowth (Cachan and Duval, 1963; Wilson, 1959a), fallen branches, dead tree trunks and even stones are scarce on the soil surface in savannas. Surface shelters providing buffered microclimatic conditions are usually lacking, and the only possibility left to the ants is to change the depth of their underground nests in order to benefit from more or less constant environmental conditions throughout the year. In a Guinean savanna in the Ivory Coast, for instance, the average soil temperature at a depth of -5 cm varies from season to season by $4°$ to $5°C$, as against a variation of only $2°C$ in a nearby gallery forest. At a depth of -25 cm, the seasonal variation of temperature is negligible ($\pm 1°C$), and the nest atmosphere is always saturated with water (Kullenberg, 1955; Weber, 1959; Lévieux, 1972b). The ants living within the soil can therefore escape the constraints of the outside local climate by simply moving their nests from a depth of -25 cm during the rainy season to a depth of -45 cm during the dry season; in this way inside nest temperature remains around $26°C$ and relative humidity is kept near saturation. In dry savannas, where the surface temperature of the soil can reach $40°$ to $50°C$ in the middle of the day during the seasonal drought, ant nests are located much deeper in the soil: -80 cm for *Myrmicaria* spp., -100 cm for *Messor* spp., and -230 cm for *Pogonomyrmex* spp. (Wray, 1938; Talbot, 1943; Tevis, 1958; Delye, 1968; Lévieux and Diomandé, 1978).

Another characteristic of tropical savanna ants is the absence of mound nests. This is easy to understand, since it has been conclusively shown that a major function of a mound is to raise the temperature of the nest by increasing the amount of sunlight it intercepts (Steiner, 1923, 1929; Hodgson, 1955; Tevis, 1958; Zahn, 1958; Scherba, 1959, 1962; Cloudsley-Thomson, 1962; Wilson, 1963; Sudd, 1967).

Very little is known about the microclimatic conditions within the nests of ants inhabiting the grass and tree layers. I have already mentioned the scarcity of formicids within the grass cover itself. This is all the more understandable since most of the grass is periodically destroyed by fire in the savannas. Although the burning front does not greatly increase the soil temperature (Pitot and Masson, 1951; Lévieux, 1973; see also Ch. 30), it completely destroys the grass plus the few *Cremato-*

gaster nests built within the hollow stems.

However, the more important effects of grass fires upon the savanna myrmecofauna are long term. In the Ivory Coast, the ant communities of annually burnt Guinea savannas are predominantly composed of "advanced" species (*Acantholepis* spp. *Camponotus* spp.), to the detriment of most ponerines, such as *Pachycondyla* spp. and *Hypoponera* spp. When the ant communities of two *Loudetia* savanna plots, one burnt annually and the other protected from fire, were compared over a number of years at Lamto, a dozen or so ponerine species (*Asphinctopone* spp., *Myopias* spp.) were found to disappear completely, others (*Paltothyreus* spp.) became scarcer, and the number of nests of the remaining species (such as *Mesoponera caffraria*) were reduced by half. Conversely, the population density of some typical savanna species increased by 50% in the regularly burnt plot (*Acantholepis canescens*, *Camponotus acvapimensis*). A few species (some *Crematogaster*, *Paratrechina weissi*, *Pheidole termitophila*) were apparently unharmed by burning. With such a change in faunal composition, microclimatic factors are certainly not the only ones to consider; the changes in food resources and competition are also important (Lévieux, 1972a, 1973).

Breeding strategies

The production of eggs by the queens has not yet been studied quantitatively in tropical savanna ants. The few figures available refer to some nomadic species only marginally entering the grasslands [*Eciton burchelli*: 2×10^5 eggs per month (Schneirla, 1971); *Anomma nigricans*: 3 to 4 $\times$ 10^6 eggs per month (Raigner and Van Boven, 1955)]. It is therefore impossible to make any comparison with the figures published for temperate latitudes (Golley and Gentry, 1964; Brian and Petal, 1972; Brian, 1978; Nielsen, 1978).

The periodicity of swarming is better understood (Table 24.2). In the Ivory Coast, mating flights are dependent upon the rain regime of the area. For some *Camponotus* species, the periods at which "alates" emerge from the nest even closely correspond to the occurrence of rain showers (Lévieux, 1973). For most species the production of the sexual brood takes place inside the nest in the middle of the dry season, at a time when the outside

TABLE 24.2

The swarming periods of sixteen sympatric species of ants of the Lamto derived savanna, Ivory Coast

Species	January	February	March	April	May	June	July	August	September
Amblyopone mutica									
Amblyopone pluto									
Centromyrmex sellaris									
Pachycondyla brunoi									
Pachycondyla caffraria									
Pachycondyla pachyderma									
Odontomachus troglodytes									
Paltothyreus tarsatus									
Pheidole megacephala									
Pheidole termitophila									
Tetramorium sp.									
Camponotus acvapimensis									
Camponotus congolensis									
Camponotus maculatus									
Paratrechina weissi									

conditions are the worst; on the other hand, swarming occurs when the first rains loosen the surface soil, helping the young queens to establish their new colonies. The whole breeding pattern is therefore highly adaptive. Some other species however (*Polyrhachis viscosa*, some *Tetramorium* spp.), swarm during the rainy season, in the absence of any obvious climatic variations but at a time when the climate closely resembles that prevailing inside the rain forest (daily temperature ranging from 25 to 28°C, relative humidity near saturation 15 to 18 h per day, reduced illumination).

Such a breeding pattern never occurs in the drier savannas, where all mating flights occur during the rainy season.

The partitioning of food resources

Savanna ants provide some remarkable examples of the many ways by which a number of sympatric species can share the available food resources, avoid interspecific competition and optimize their use of the environment (Wilson, 1963; Leston, 1970; Schoener, 1971; Carroll and Janzen, 1973).

The partitioning of the foraging space is made easier by the stratification of the nesting sites of different species. Taking into consideration the activity ranges of the workers of these species, it is possible to visualize the distribution of foraging activities within the habitat (Fig. 24.2).

Genera strictly limited to the soil layer (*Amblyopone*, *Apomyrma*, *Hypoponera*, *Mystrium*) never forage in the open, except possibly during the night in the rainy season; they exploit the soil resources to a depth of about 100 cm. Other genera are able to forage both within the upper soil layers, and in the open. Ants of the genus *Paltothyreus*, for instance, hunt in the soil at depths ranging from −40 to −60 cm during the dry season, and in the surface litter during the rains. Most of the terrestrial ants, however, only forage at ground level (*Anochetus*, *Pheidole*, *Tetramorium*), some of them also exploring the grass layer (*Camponotus*, *Polyrhachis*).

Different species of ponerine ants coexisting in the same soil avoid competition by using three different strategies: (1) every species exploits a limited soil volume, at depths ranging from a few centimetres to one metre; (2) diets are specialized, *Amblyopone* preying for instance upon chilopods

SAVANNA   GALLERY FOREST

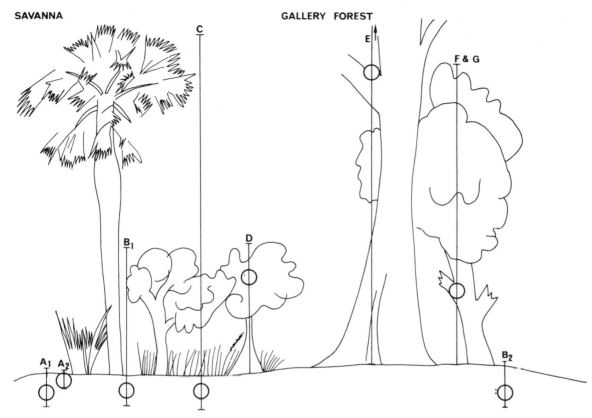

Fig. 24.2. The foraging zones of some ants, in the savanna gallery forest boundary in Ivory Coast. For an explanation of the symbols used, see Table 24.1.

(Wilson, 1971; Gotwald and Lévieux, 1972), *Plectroctena* upon diplopods, etc. (Table 24.3); and (3) the density of their colonies remains small. In this way, interspecific competition is minimized, if not entirely avoided. Ground-living ponerines, such as *Megaponera* spp. and some *Leptogenys*, also employ these strategies.

"Higher" ant sub-families, such as Formicinae, Pseudomyrmicinae and Myrmicinae, which live at the ground level or in the vegetation, are in most cases "generalists" in their feeding habits, eating a variety of animal and plant foodstuffs (Table 24.4; Gotwald, 1972; Lepage, 1974). In this case, foraging areas or foraging spheres are well separated in space, and furthermore there is a time separation between the feeding periods of sympatric species; this is all the more obvious when the ants belong to the same genus (Fig. 24.3). This strategy is adopted by most "higher" terrestrial and arboreal species (Lévieux, 1977).

Within the soil itself, where the density of nests is

TABLE 24.3

The staple food of sympatric specialized predators in the Lamto savanna

Species	Staple food
Soil-dwelling species	
Amblyopone pluto	
Amblyopone mutica	Chilopoda: Geophilomorpha
Apomyrma stygia	
Plectroctena subterranea	Diplopoda, Iulidae
Plectroctena lygaria	eggs of Diplopoda
Hypoponera gr. *coeca*	Collembola
Discothyrea oculata	eggs of Arthropoda
Centromyrmex sellaris	Isoptera
Ground-dwelling species	
Leptogenys conradti	
Leptogenys sp.	Isopoda, Oniscoidea
Megaponera foetens	Isoptera

always much higher than above ground, the risk of interspecific competition is greater than in the grass layer (Lévieux, 1972a). Indeed, many colonies of

TABLE 24.4

Some examples of food preferences among Ivory Coast savanna ants

ANIMAL FOOD	
Earthworms	*Paltothyreus*
Arachnida	
spiders	*Atopomyrmex, Camponotus, Platythyrea, Tetramorium*
Myriapoda	
Diplopoda	*Paltothyreus, Plectroctena, Psalidomyrmex*
Chilopoda	*Amblypone, Apomyrma, Camponotus*
Crustacea	
Isopoda, Oniscoidea	*Leptogenys*
Insecta	
Collembola	*Hypoponera*
Odonata, Zygoptera	*Camponotus*
Dermaptera	*Paltothyreus*
Dictyoptera	
Mantinae	*Camponotus, Platythyrea*
Orthopteroidea	*Camponotus, Platythyrea*
Isoptera	*Acantholepsis, Camponotus, Crematogaster, Megaponera, Paltothyreus, Pheidole, Platythyrea, Polyrhachis, Tetramorium*
Hemiptera	
Heteroptera	*Paltothyreus*
Homoptera	
Cicadidae	*Camponotus*
Ricaniidae	*Atopomyrmex, Camponotus, Platythyrea*
others	*Acantholepis, Camponotus, Crematogaster, Polyrhachis*
Lepidoptera	
Noctuidae	*Camponotus, Platythyrea*
Hesperidae	*Camponotus, Platythyrea*
Pyralidae	*Camponotus, Platythyrea*
Geometridae	*Camponotus*
others	*Paltothyreus*
Coleoptera	*Acantholepsis, Atopomyrmex, Camponotus, Paltothyreus, Platythyrea, Tetramorium*
Diptera	*Camponotus, Pheidole, Platythyrea, Tetramorium*
Hymenoptera	*Crematogaster, Camponotus, Paltothyreus, Pheidole, Platythyrea*
PLANT FOOD	
Leaves	*Crematogaster, Platythyrea*
Seeds	*Atopomyrmex, Pheidole, Platythyrea, Tetramorium*
Latex	*Atopomyrmex*
Sap and others	*Atopomyrmex, Crematogaster, Camponotus, Platythyrea*

soil ants actively control exclusive "foraging areas", especially those that feed upon seeds (Brian, 1965; Hölldobler, 1974, 1976).

However, neither the amount of animal prey nor plant food can be considered as a limiting factor for the ant population in wet tropical savannas. These populations even make very little use of the large amount of seeds produced by the grass cover. Some ants (*Pheidole, Tetramorium*) store them for a few weeks, but the very humid conditions within the soil prevent any long-term storage.

Many savanna ants also feed upon the honeydew

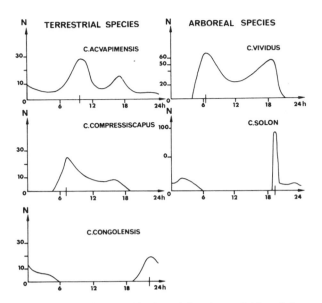

Fig. 24.3. The time partitioning of foraging activities of five sympatric species of *Camponotus* in derived savanna in the Ivory Coast. N = number of workers leaving the nest per minute.

of aphids, white flies and scale insects, as well as on the sap exuding from fresh plant wounds. This consumption is obviously seasonal, as the number of sap-sucking insects is influenced by the seasonal drought and the subsequent grass fires. During the dry season, however, some savanna ants are able to tend scale insects feeding upon grass roots within their nests. In the New World, several species of *Myrmecocystus* can seasonally store sugars in their gasters (Creighton and Crandall, 1954).

Whereas ants inhabiting derived and other wet savannas have no need to store food for long periods, those living in drier savannas must do so to withstand seasonal food shortage. This is particularly true for granivorous species (*Messor* spp., *Pheidole* spp., *Pogonomyrmex* spp. *Veromessor* spp.) whose importance within the ant communities increases with the aridity of the environment. Unfortunately the biology of these ants has only been studied in desert areas (Cole, 1934, 1937; Wheeler and Creighton, 1934; Creighton, 1953; McCluskey, 1963; Clark and Comanor, 1973a,b; Bernstein, 1974; Lavigne and Rogers, 1974; Rogers, 1974; Schumacher and Whitford, 1976; Whitford et al., 1976).

Members of the myrmicine tribe Attini (*Atta*, *Acromyrmex*, *Trachymyrmex*), restricted to the New World, are found in forest areas, though also invading derived savannas, cultivated land and deserts (Fowler and Robinson, 1977). They do not consume foliage for food, but use it as a culture medium for the fungi they eat. They have no counterparts in other continents. In Africa, for instance, invertebrate leaf eaters mostly consist of acridoids and caterpillars of moths and butterflies (more than 100 species at Lamto, according to R. Vuattoux, pers. comm., 1979), whereas fungus-growing is the privilege of Macrotermitinae.

Species richness in most tropical habitats has so far prevented any comprehensive study of the partitioning of food resources within ant communities. I have therefore attempted to identify the staple diet of the 113 species of ants living in the derived savanna of Lamto (Ivory Coast). Most of them (75%) can be considered as omnivores, though consuming a lot of seeds during the dry season. Twenty six (23%), however, are specialized carnivores, the forest ponerines entering the savanna from the nearby gallery forests accounting for this rather high percentage. Farther north, in the Sudanian savanna of Ferkéssédougou, only 11 (17%) out of the 63 sympatric species found are carnivores.

A comparison of the trophic preferences found within six ant communities ranging from a tropical rain forest to a temperate woodland, has been attempted in Table 24.5. This comparison is extremely tentative, since so little is known about the diet of so many species, and their seasonal variations. Some so-called omnivores can thus occasionally be considered as carnivores, granivores, or honeydew consumers. This explains why the same species was sometimes included in two different trophic categories (some totals thus exceeding 100% on Table 24.5). However, it remains clear that the omnivorous species are most numerous everywhere, the percentage of specialized carnivores decreasing regularly from the humid tropics to the temperate latitudes. The exclusive granivores (*Messor*, etc.) only play a major role in highly seasonal areas, with a long dry season which allows for long-term storage of seeds. In more humid habitats, some predators (*Pheidole*, *Tetramorium*) can become granivorous, even in rain forest (W.L. Brown, pers. comm., 1980).

Very little is known about the trophic impact of ants upon their basic food resources. Data exist only for predatory species. *Megaponera foetens*, for

TABLE 24.5

Trophic preferences within six different ant communities

Location	Latitude	References	Omnivorous species (%)	Carnivorous species (%)	Seed-harvesting species (%)		
					true harvesters	occasional harvesters	total
Rain forest (Tai, Ivory Coast)	4°N	Diomandé (1981)	66	33	0	0	0
Humid savanna (Lamto, Ivory Coast)	5°N	Lévieux (1971)	66	30	0	15	15
Dry savanna (Fété Olé, Senegal)	17°N	M. Lepage (pers. comm., 1979)	50	10	10	30	40
Desert (Béni-Abbès, Algeria)	30°N	Delye (1968)	66	15	25	15	40
Limestone plateau (Quercy, France)	44°N	Passera (1967)	85	5	5	10	15
Semi-deciduous temperate forest (Quebec, Canada)	46°N	Letendre et al. (1971)	85	20	0	15	15

instance kills 142 *Macrotermes* m^{-2} yr^{-1}, which represent a live weight of 1.42 g (Longhurst et al., 1979). The very large production of arthropods in many tropical savannas, during most of the year, allows such a heavy predatory toll to be taken. The many spiders present in the same environment could represent potential food competitors, but the two groups use very different techniques to catch their prey, and probably exploit different taxa, or different life stages of the same species.

CONCLUSIONS

There are too many gaps in our present knowledge of the ecology of savanna ants to allow any definite conclusions to be drawn. Indeed, one of the main objectives of the present review is to identify the fields in which research is most urgently needed.

One point, however, emerges quite clearly from the data at hand: the basic difference between wet and dry savannas. The number of formicid species, the taxonomic structure of their communities, the proportion of terrestrial and arboreal species and of granivorous and carnivorous taxa, the breeding patterns, are all different. Broadly speaking, the situation in the more humid savannas is consistently reminiscent of that in the tropical rain forest, whereas the situation in the drier savannas resembles that of the desert steppes. In this, as in many other ways, tropical savannas represent a

broad transition zone between these two contrasting habitats.

Among the many parameters that play an important role in the ecology of savanna ants, rainfall obviously ranks first. Not only does the yearly amount of rain play a fundamental role in influencing the production of their food sources, but the seasonal pattern of rainfall seems to regulate their breeding periodicity. Another important factor to consider is the behavioural adaptability of ants and their ability to reduce competition by an efficient partitioning of their foraging activities in time and space.

It must never be forgotten that savanna landscapes are very often true mosaics of somewhat different and intermingled habitats harboring unstable plant and animal communities. The impact of man is also almost always present, and its consequences are still very poorly understood, especially upon soil communities. To untangle better the action of the various environmental parameters, opportunity should never be missed to study the consequences of some large-scale "development" schemes which may amount to large-scale experiments (Greenslade, 1969, 1971; Room, 1971; Majer, 1972; Leston, 1973a;,b; Greenslade and Mott, 1978).

ACKNOWLEDGEMENT

The author is greatly indebted to Professor W.L.

Brown Jr., Cornell University, for his generous help and criticisms.

REFERENCES

Arnold, G., 1914. Nest changing migration of two species of ants. *Proc. Rhodes. Sci. Assoc.*, 12: 25–51.

Bernstein, R.A., 1974. Seasonal food abundance and foraging activity in some desert ants. *Am. Nat.*, 108: 490–498.

Bolton, B., 1973. The ant genus *Polyrhachis* F. Smith in the Ethiopian region (Hymenoptera Formicidae). *Bull. Br. Mus. Nat. Hist., Entomol.*, 28: 285–369.

Borgmeier, T., 1955. Die Wanderameisen der neotropischen Region. *Stud. Entomol.*, 3: 1–717.

Brian, M.V., 1956. The natural density of *Myrmica rubra* and associated ants in west Scotland. *Insectes Soc.*, 3: 474–487.

Brian, M.V., 1965. *Social Insect Populations.* Academic Press, London, 135 pp.

Brian, M.V., 1978. *Production Ecology of Ants and Termites.* International Biological Program, Synthesis series, 13. Cambridge University Press, Cambridge, 409 pp.

Brian, M.V. and Petal, J., 1972. Productivity investigations on social insects and their role in the ecosystems. *Ekol. Polsk.*, 20: 184 pp.

Brown, W.L., Jr., 1954. Remarks on the internal phylogeny and subfamily classification of the family Formicidae. *Insectes Soc.*, 1: 22–32.

Brown, W.L., Jr., 1958. A review of the ants of New Zealand. *Acta Hymenopterol.*, 1: 1–50.

Brown, W.L., Jr., 1960. Ants, acacias and browsing mammals. *Ecology*, 41: 587–592.

Brown, W.L., Jr., 1972. The geographical distribution of ants, past and present. In: *Proc. 14th Int. Congr. Entomol., Canberra A.C.T.*, p. 189.

Brown, W.L., Jr., 1973. A comparison of the hylean and Congo West African rain forest. In: B.J. Meggers, E.S. Ayensu and W.D. Duckworth (Editors), *Tropical Forest Ecosystems in Africa and South America: a Comparative Review.* Smithsonian Institution Press, Washington, D.C., pp. 161–185.

Brown, W.L., Jr. and Kempf, W.W., 1960. A world revision of the ant tribe Basicerotini. *Stud. Entomol., N.S.*, 3: 161–250.

Cachan, P. and Duval, J., 1963. Variations microclimatiques verticales et saisonnières de la forêt sempervirente de basse Côte d.Ivoire. *Ann. Fac. Sci. Univ. Dakar*, 8: 1–87.

Carroll, C.R. and Janzen, D.H., 1973. Ecology of foraging by ants. *Annu. Rev. Ecol. Syst.*, 4: 231–257.

Clark, W.H. and Comanor, P.L., 1973a. The use of the western harvester ant, *Pogonomyrmex occidentalis* (Cresson) seed stores by Heteromiid rodents. *Occas. Pap., Biol. Soc. Nev.*, 34: 1–6.

Clark, W.H. and Comanor, P.L., 1973b. A quantitative examination of spring foraging of *Veromessor pergandei* (Mayr) in northern Death Valley (California) (Hymenoptera Formicidae). *Am. Midl. Nat.*, 90: 467–474.

Cloudsley-Thompson, J.L., 1962. Microclimate and the distribution of terrestrial arthropods. *Annu. Rev. Entomol.*, 7: 199–222.

Cole, A.C., 1934. An ecological study of ants of the southern desert shrub region of the United States. *Ann. Entomol. Soc. Am.*, 27: 388–405.

Cole, A.C., 1937. An annotated list of the ants of Arizona (Hymenoptera Formicidae). *Entomol. News*, 48: 97–101.

Collart, A., 1925. Quelques observations sur les fourmis *Megaponera. Bull. Cercle Zool. Congo*, 2: 26–28.

Collart, A., 1927. Notes sur la biologie des fourmis congolaises. *Rev. Zool. Afr.*, 14: 293–353.

Collart, A., 1932. Une fourmi qui utilise la soie des Araignées (*Polyrhachis laboriosa* F. Smith). *Bull. Mus. R. Hist. Nat. Belg.*, 8: 1–14.

Creighton, W.S., 1953. New data on the habits of the ants of the genus *Veromessor. Am. Mus. Novit.*, 1612: 1–18.

Creighton, W.S. and Crandall, R.H., 1954. New data on the habits of *Myrmecosystus melliger* Forel. *Biol. Rev.*, 16: 2–6.

Delage Darchen, B., 1971. Contribution à l'étude écologique d'une savane de Côte d'Ivoire (Lamto): les fourmis des strates herbacées et arborées. *Biol. Gabon.*, 7: 461–496.

Delage Darchen, B., 1972. Une fourmi de Côte d'Ivoire, *Melissotarsus titubans* n. sp. (Hymenoptera Formicidae). *Insectes Soc.*, 19: 213–226.

Delage Darchen, B., 1974. Ecologie et biologie de *Crematogaster impressa* Emery, fourmi savanicole d'Afrique. *Insectes Soc.*, 21: 13–34.

Delye, G., 1968. *Recherche sur l'écologie, la physiologie et l'éthologie des fourmis du Sahara.* Thesis, Marseille University, Marseille, 155 pp.

Delye, G., 1976. Contribution à l'étude entomologique de la faune de la République du Niger II Les fourmis de la région de Niamey et du massif de l'Aïr. *Bull. I.F.A.N.*, A 38: 156–165.

Diomandé, T., 1981. *Etude du peuplement en fourmis terricoles des forêts ombrophiles et des zones anthropisées de la Côte d'Ivoire méridionale.* Thesis, Université d'Abidjan, 254 pp.

Emery, C., 1891. Le formiche dell'ambra siciliana nel museo mineralogico dell' Universita di Bologna. *Mem. R. Accad. Sci. Bologna*, 5: 567–591.

Emery, C., 1893. Beiträge zur Kenntnis der nordamerikanischen Ameisenfauna. *Zool. Jahrb., Abt. Syst.*, 7: 633–682.

Escherich, K., 1911. Zwei Beiträge zum Kapitel "Ameisen und Pflanzen" (vgl. auch Wassmann). *Biol. Centralbl.*, 31: 44–61.

Fowler, H.G. and Robinson, S.W., 1977. Foraging and grass selection by the grass eating ant *Acromyrmex landoli fracticornis* (Forel) (Hymenoptera Formicidae) in habitats of introduced forage grasses in Paraguay. *Bull. Entomol. Res.*, 67: 659–666.

Golley, F.B. and Gentry, J.B., 1964. Bioenergetics of the southern harvester ants, *Pogonomyrmex badius. Ecology*, 45: 217–225.

Gotwald, W.H., Jr., 1972. *Oecophylla longinoda*, an ant predator of *Anomma* driver ants (Hymenoptera Formicidae). *Psyche*, 79: 348–356.

Gotwald, W.H., Jr., 1977. The origins and dispersal of army ants of the subfamily Dorylinae. In: *Proc. 8th Int. Congr., Int. Union. Stud. Soc. Insects, Wageningen*, pp. 126–127.

Gotwald, W.H., Jr. and Lévieux, J., 1972. Taxonomy and biology of a new west African ant belonging to the genus *Amblyopone* (Hymenoptera Formicidae). *Ann. Entomol. Soc. Am.*, 65: 383–396.

Greenslade, P.J.M., 1969. Insect distribution patterns in the Solomon Islands. *Philos. Trans. R. Soc, Lond.*, 255: 271–284.

Greenslade, P.J.M. 1971. Interspecific competition and frequency changes among ants in Solomon Islands coconut plantations. *J. Appl. Ecol.*, 8: 323–352.

Greenslade, P.J.M., 1976. The meat ant *Iridomyrmex purpureus* (Hymenoptera Formicidae) as a dominant member of ant community. *J. Aust. Entomol. Soc.*, 15: 237–240.

Greenslade, P.J.M. and Mott, J.J., 1978. Ants (Hymenoptera Formicidae) of native and sown pastures in the Katherine area, N.T., Australia. In: *Proc. Sec. Aust. Conf. Grassland Invertebrate Ecology, Massey University, Palmerston North,* 15 pp.

Hocking , B., 1970. Insect associations with the swollen thorn acacias. *Trans. R. Entomol. Soc. Lond.*, 122: 211–255.

Hodgson, E.S., 1955. An ecological study of the behavior of the leaf cutting ant *Atta cephalotes. Ecology*, 36: 293–304.

Hölldobler, B., 1974. Home range orientation and territoriality in harvesting ants. *Proc. Natl. Acad. Sci. U.S.*, 71: 3274–3277.

Hölldobler, B., 1976. Recruitment behavior, home range orientation and territoriality in harvester ants, *Pogonomyrmex. Behav. Ecol. Sociobiol.*, 1: 3–44.

Janzen, D.H., 1966. Coevolution of mutualism between ants and acacias in Central America. *Evolution*, 20: 249–275.

Janzen, D.H., 1967. Interaction of the bull's horn acacia (*Acacia cornigera* L.) with an inhabitant (*Pseudomyrmex ferruginea* F. Smith) in eastern Mexico. *Kansas Univ. Sci. Bull.*, 47: 315–558.

Janzen, D.H., 1969a: Birds and the ant acacias interaction in Central America with notes on birds and her myrmecophytes. *Condor*, 71: 240–256.

Janzen, D.H., 1969b. Allelopathy by myrmecophytes: the ant *Azteca* as an allelopatic agent of *Cecropia. Ecology*, 50: 147–153.

Jensen, T.F., 1978. An energy budget for a field population of *Formica pratensis* Retz. *Nat. Jutl.*, 20: 203–226.

Kullenberg, B., 1955. Quelques observations microclimatiques en Côte d'Ivoire et Guinée française. *Bull. I.F.A.N.*, A 27: 127–132.

Kusnevoz, N., 1963. Zoogeografia de las hormigas en Sud America. *Act. Zool. Lilloana*, 19: 25–186.

Lavelle, P., Maury, M.E. and Serrano, V., 1981. Las comunidades animales en suelos de pastizales y selvas pertenecientes a la region de Laguna verde (Vera Cruz, Mexico) durante la temporada de Lluvas: estudia cuantativo preliminar de la composicion y estructura. In: R. Reyes Castillo (Editor), *Estudios Ecologicos en el Tropico Mexicano.* Instituto de Ecologia, Mexico, Publ. No. 6: 1–187.

Lavigne, R.J. and Rogers, L.E., 1974. An annotated bibliography of the harvester ant *Pogonomyrmex occidentalis* (Cresson) and *Pogonomyrmex owyheei* Cole. *Sci. Monogr.*, 26: 1–18.

Ledoux, A., 1949. Etude du comportement et de la biologie de la fourmi fileuse *Oecophylla longinoda* Latreille. *Ann. Sci. Nat., Zool.*, 12: 313–461.

Ledoux, A., 1958. La construction du nid de quelques fourmis arboricoles de France et d'Afrique tropicale. In: *Proc. 10th*

Int. Congr. Entomol., Montréal, Qué, 2: 521–528.

Lepage, M., 1974. *Les termites d'une savane sahélienne (Ferlo septentrional, Sénégal): Peuplements, Populations, consommation, rôle dans l'écosystème.* Thesis, Dijon University, Dijon, 344pp.

Leroux, J.M., 1975. *Recherches sur les nids et l'activité prédatrice des Dorylinae* Anomma nigricans *Illiger (Hym. Formicidae).* Diploma, University of Paris VI, Paris, 97 pp.

Leroux, J.M., 1977a. Densité des nids et observations sur les nids de *Dorylinae Anomma nigricans* Illiger (Hym. Formicidae) dans la région de Lamto (Côte d'ivoire). *Bull. Soc. Zool. Fr.*, 102: 51–62.

Leroux, J.M., 1977b. Formation et déroulement des raids de chasse d'*Anomma nigricans* Illiger (Hym. Formicidae) dans une savane de Côte d'Ivoire. *Bull. Soc. Zool. Fr.*, 102: 445–458.

Leston, D., 1970. Entomology of the cocoa farm. *Annu. Rev. Entomol.*, 15: 273–294.

Leston, D., 1973a. Ecological consequences of the tropical ant mosaïc. In: *Proc. 7th Int. Congr., Int. Union, Stud. Soc. Insects, London*, pp. 235–242.

Leston, D., 1973b. The ant mosaïc. Tropical tree crops and the limiting of pests and diseases. *Pans*, 3: 311–341.

Letendre, M., Francoeur, A., Beique, R. and Pilou, J.G., 1971. Inventaire des Fourmis de la station de Biologie de l'Université de Montréal. St. Hyppolyte, Québec (Hym. Formicidae). *Nat. Can.*, 98: 591–606.

Lévieux, J., 1966. Note préliminaire sur le comportement de chasse de *Megaponera foetens. Insectes Soc.*, 13: 117–126.

Lévieux, J., 1972a. Le rôle des formis dans les réseaux trophiques d'une savane préforestière de Côte d'Ivoire. *Ann. Univ. Abidjan, E 5*: 143–240.

Lévieux, J., 1972b. Le microclimat des nids et des zones de chasse de *Camponotus acvapimensis* Mayr. *Insectes Soc.*, 19: 63–79.

Lévieux, J., 1973. Etude du peuplement en fourmis terricoles d'une savane préforestière de Côte d'Ivoire. *Rev. Ecol. Biol. Sol*, 10: 379–428.

Lévieux, J., 1976a. Etude de la structure du nid de fourmis arboricoles d'Afrique tropicale. *Ann. Univ. Abidjan, C.* 12: 5–22.

Lévieux, J., 1976b. Etude de la structure du nid de quelques espèces terricoles de fourmis tropicales. *Ann. Univ. Abidjan, C* 12: 23–33.

Lévieux, J., 1976c. Densité et biomasses de *Camponotus acvapimensis* (Hymenoptera Formicidae) dans une savane de Côte d'Ivoire. *Terre Vie*, 30: 264–275.

Lévieux, J., 1977. La nutrition des fourmis tropicales V Eléments de synthèse. les modes d'exploitation de la biocoenose. *Insectes Soc.*, 24: 235–260.

Lévieux, J. and Diomande, T., 1978. La nutrition des fourmis granivores I Cycle d'activité et régime alimentaire de *Messor galla* Emery et de *Messor* (= *Cratomyrmex*) *regalis* Emery (Hymenoptera Formicidae). *Insectes Soc.*, 25: 127–140.

Longhurst, C. and Howse, P.E., 1977. Predatory behaviour of *Megaponera foetens* on termites in Nigeria. In: *Proc. 6th Int. Congr. Int. Union. Stud. Soc. Insects, Wageningen,* pp. 121–122.

Longhurst, C., Johnson, R.A. and Wood, T.G., 1978. Predation by *Megaponera foetens* (F.) (Hymenoptera Formicidae) on termites in the Nigerian Southern Guinea savanna. *Oecologia*, 32: 101–107.

Longhurst, C., Johnson, R.A. and Wood, T.G., 1979. Foraging recruitment and predation by *Decamorium uelense* (Santschi) (Formicidae Myrmicinae) on termites in Southern Guinea Savanna, Nigeria. *Oecologia*, 38: 83–91.

McCluskey, E.S., 1963. Rhythms and clocks in harvester and Argentine ants. *Physiol. Zool.*, 36: 273–292.

Majer, J.D., 1972. The ant mosaïc in Ghana cocoa farms. *Bull. Entomol. Res.*, 62: 151–160.

Müller, W., 1886. Beobachtungen an Wanderameisen (*Eciton hamatum* F.). *Kosmos*, 18: 81–93.

Nielsen, M.G., 1972. An attempt to estimate energy flow through a population of workers of *Lasius alienus* (Forst) (Hymenoptera Formicidae). *Nat. Jutl.*, 16: 99–107.

Nielsen, M.G., 1978. Production of sexuals in nests of the ant *Lasius alienus* (Först) (Hymenoptera Formicidae). *Nat. Jutl.*, 20: 251–254.

Odum, E.P. and Pontin, A.J., 1961. Population density of the underground ant *Lasius flavus* as determinated by tagging with P_{32}. *Ecology*, 42: 186–188.

Passera, L., 1967. Peuplement en fourmis terricoles du rebord méridional des Causses jurassiques du Quercy: la lande calcaire à buis. *Vie milieu*, C 18: 189–205.

Pickles, W., 1937. Populations, territories and biomasses of ants at Thornhill, Yorkshire in 1936. *J. Anim. Ecol.*, 6: 54–61.

Pitot, A. and Masson, H., 1951. Quelques données sur la température au cours des feux de brousse aux environs de Dakar. *Bull. I.F.A.N.*, A 13: 711–732.

Prell, H., 1911. Biologische Beobachtungen an Termiten und Ameisen. *Zool. Anz.*, 38: 243–253.

Raignier, A. and Van Boven, J., 1955. Etude taxonomique, biologique et biometrique des *Dorylus* du sous genre *Anomma* (Hymenoptera Formicidae, Dorylinae). *Ann. Mus. R. Congo Belge, Sci. Zool.*, 2: 1–359.

Rettenmeyer, C.W., 1963. Behavioral studies of army ants. *Kansas Univ. Sci. Bull.*, 44: 281–465.

Rogers, L.E., 1974. Foraging activity of the western harvester ants in the short grass plain ecosystems. *Environ. Entomol.*, 3: 420–424.

Room, P.M., 1971. The relative distribution of ants species in Ghana's cocoa farm. *J. Anim. Ecol.*, 40: 735–751.

Santschi, F., 1909. Formicidae nouveaux ou peu connus du Congo français. *Ann. Soc. Entomol. Fr.*, 78: 349–400.

Scherba, G., 1959. Moisture regulation in mound nests of the ant *Formica ulkei* Emery. *Am. Midl. Nat.*, 61: 499–508.

Scherba, G., 1962. Mound temperature of the ant *Formica ulkei* Emery. *Am. Midl. Nat.*, 67: 373–385.

Schneirla, T.C., 1933. Studies on army ants in Panama. *J. Comp. Psychol.*, 15: 267–299.

Schneirla, T.C., 1934. Raiding and other outstanding phenomena in the behavior of army ants. *Proc. Natl. Acad. Sci.*, 20: 316–321.

Schneirla, T.C., 1938. A theory of army ants behaviour based upon the analysis of activities in a representative species. *J. Comp. Psychol.*, 25: 51–90.

Schneirla, T.C., 1956. The army ants. *Smithsonian Inst. Publ. Rep.*, 4230: 379–406.

Schneirla, T.C., 1957. A comparison of species and genera in the ant subfamily Dorylinae with respect to functional patterns. *Insectes Soc.*, 4: 259–298.

Schneirla, T.C., 1958. The behavior and biology of certain nearctic army ants. Last part of the functional season, southeastern Arizona. *Insectes Soc.*, 5: 215–255.

Schneirla, T.C., 1971. *Army Ants, a Study in Social Organisation.* Edited by H.R. Topoff, W.H. Freeman and Co., San Francisco, Calif., 349 pp.

Schoener, T.W., 1971. Theory of feeding strategies. *Annu. Rev. Ecol. Syst.*, 2: 369–404.

Schouteden, H., 1924. Une fourmi qui razzie les termites, *Megaponera foetens. Rev. Zool. Afr.*, 12 (Suppl.): 36–38.

Schumacher, A. and Whitford, W.G., 1976. Spatial and temporal variation in Chuahuan desert ant faunas. *Southwest Nat.*, 21: 1–8.

Sheppe, W., 1970. Invertebrate predation of termites in the African savanna. *Insectes Soc.*, 17: 205–218.

Soulié, J., 1961. Les nids et les comportements nidificateurs des fourmis du genre *Crematogaster* d'Europe, d'Afrique du Nord et d.Asie du Sud-Est. *Insectes Soc.*, 8: 213–297.

Steiner, A., 1923. Über die Temperaturverhältnisse in den Nestern der *Formica rufa* var *rufo-pratensis* For. *Mitt. Naturforsch. Ges. Bern*, 3: 61–66.

Steiner, A., 1929. Temperaturuntersuchungen in Ameisennestern mit Erdkuppeln, im Nest von *Formica exsecta* Nyl. und in Nestern unter Steinen. *Z. Vergl. Physiol.*, 9: 1–66.

Strickland, A.H., 1951a. The entomology of the swollen shoot of cocoa I. The insect species involved with notes on their biology. *Bull. Entomol. Res.*, 41: 725–748.

Strickland, A.H., 1951b. The entomology of the swollen shoot of cocoa II The bionomics and ecology of the species involved. *Bull. Entomol. Res.*, 42: 65–103.

Sudd, J.H., 1967. *An Introduction to the Behaviour of Ants.* Edward Arnold, London, 200 pp.

Talbot, M., 1943. Population studies of the ant *Prenolepis imparis* Santschi. *Ecology*, 24: 31–44.

Taylor, R.W., 1965a. New Melanesian ants of the genera *Simopone* and *Amblyopone* (Hymenoptera Formicidae) of zoogeographic significance. *Brevoria*, 221: 1–11.

Taylor, R.W., 1965b. A monographic revision of the rare tropicopolitan ant genus *Probolomyrex* Mayr (Hymnoptera Formicidae). *Trans. Entomol. Soc. Lond.*, 117: 345–365.

Taylor, R.W., 1968. Notes on the Indo-Australian Basicerotine ants (Hymenoptera Formicidae). *Aust. J. Zool.*, 16: 333–348.

Tevis, L., 1958. Interrelations between the harvester ant *Veromessor pergandei* (Mayr) and some desert Ephemerals. *Ecology*, 39: 695–704.

Von Ihering, H., 1894. Die Ameisen von Rio Grande do Sul. *Berliner Entomol. Z.*, 39: 321–446.

Vosseler, J., 1905. Die ostrafrikanische Treiberameise (Siafu). *Pflanzer, Dar es Salam*, 19: 289–302.

Vuattoux, R., 1968. Le peuplement du palmier Rônier (*Borassus ethiopum*) d'une savane de Côte d'Ivoire. *Ann. Univ. Abidjan*, E 1: 1–138.

Weber, N.A., 1943. The ants of the Imatong Mountains, Anglo-Egyptian Sudan. *Bull. Mus. Comp. Zool.*, 93: 263–389.

Weber, N.A., 1959. Isothermal conditions in tropical soils. *Ecology*, 40: 153–154.

Wheeler, W.M., 1910. *Ants, Their Structure, Development and Behavior.* Columbia University press, New York, N.Y., 663 pp.

Wheeler, W.M., 1914. The ants of the baltic amber. *Schrif. Physik. Ökon. Ges. Königsberg*, 55: 1–142.

Wheeler, W.M., 1936. Ecological relations of Ponerine and other ants to termites. *Proc. Am. Acad. Arts Sci.*, 71: 159–243.

Wheeler, W.M., 1942. Studies of Neotropical ants and their plants. *Bull. Mus. Comp. Zool.*, 80: 195–211.

Wheeler, W.M. and Creighton, W.S., 1934. A study of the ant genera *Novomessor* and *Veromessor*. *Proc. Am. Acad. Arts Sci.*, 69: 341–387.

Whitford, W.G., Johnson, P. and Ramirez, J., 1976. Comparative ecology of the harvester ants: *Pogonomyrmex barbatus* (F. Smith) and *P. rugosus* (Emery). *Insectes Soc.*, 23: 117–131.

Williams, E.C., 1941. An ecological study of the floor fauna of the Panama rain forest. *Bull. Chicago Acad. Sci.*, 6: 63–124.

Wilson, E.O., 1958a. The beginning of nomadic and group-predatory behavior in the Ponerine ants. *Evolution*, 12: 24–31.

Wilson, E.O., 1958b. Studies on the ant fauna of Melanesia: I. The Tribe Leptogenyini. II. The Tribe Amblyoponini and Platythyreini. *Bull. Mus. Comp. Zool.*, 118: 101–153.

Wilson, E.O., 1958c. Studies on the ant fauna of Melanesia III *Rhytidoponera* in western Melanesia and the Moluccas IV The tribe Ponerini. *Bull. Mus. Comp. Zool.*, 119: 303–371.

Wilson, E.O., 1959a. Studies on the ant fauna of Melanesia V The tribe Ondontomachini. *Bull. Mus. Comp. Zool.*, 120: 483–510.

Wilson, E.O., 1959b. Studies on the ant fauna of Melanesia VI The tribe Cerapachyini. *Pac. Insects*, 1: 39–57.

Wilson, E.O., 1959c. Adaptive shift and dispersal in a tropical ant fauna. *Evolution*, 13: 122–144.

Wilson, E.O., 1959d. Some ecological characteristics of ants in New Guinea rain forest. *Ecology*, 40: 437–447.

Wilson, E.O., 1963. The social biology of ants. *Annu. Rev. Entomol.*, 8: 345–361.

Wilson, E.O., 1964. The true army ants of the Indo-Australian area. *Pac. Insects*, 6: 427–483.

Wilson, E.O., 1971. *The Insect Societies.* Harvard University Press, Cambridge, Mass., 548 pp.

Wilson, E.O. and Taylor, R.W., 1964. A fossil ant colony: new evidence of social antiquity. *Psyche*, 71: 93–103.

Wilson, E.O. and Taylor, R.W., 1967. The ants of Polynesia. *Pac. Insects Monogr.*, 14: 1–109.

Wilson, E.O., Eisner, T., Wheeler, G.C. and Wheeler, J., 1956. *Aneuretus simoni*, a major link in ant evolution. *Bull. Mus. Comp. Zool.*, 115: 81–105.

Wilson, E.O., Carpenter, E.M. and Brown, W.L., Jr., 1967. The first mesozoic ant with the description of a new subfamily. *Psyche*, 74: 1–19.

Wray, D.L., 1938. Notes on the southern harvester ant (*Pogonomyrmex badius* latr.) in North Carolina. *Ann. Entomol. Soc. Am.*, 31: 196–201.

Zahn, M., 1958. Temperatursinn, Wärmehaushalt und Bauweise der roten Waldameisen (*Formica rufa* L.). *Zool. Beitr.*, 3: 127–194.

Chapter 25

THE ROLE OF SOIL MICRO-ORGANISMS IN SAVANNA ECOSYSTEMS

PAUL KAISER

INTRODUCTION

Soil micro-organisms transform organic and mineral matter. The mineralization of organic substances into carbon dioxide, organic acids, amino acids and mineral nitrogen goes hand in hand with the synthesis of new organic compounds: microbial polysaccharides, proteins and humic substances.

Of course, microbial activity is influenced by environmental factors such as water availability, oxygen tension and temperature. Around the roots, the organic matter exuded greatly affects microbial activity, and this will in turn either stimulate or inhibit plant growth. This is particularly true for tropical grasslands.

Litter decomposition

The quantity of litter subject to microbial decomposition varies a great deal from one site to another (Bell, 1974). It also depends on the intensity of grazing (Rhoades et al., 1964). Though fire also drastically reduces, or even eliminates, surface litter, the quantity present before the fire is usually

restored within a few years (Dix and Butler, 1954, in Clark and Paul, 1970; Daubenmire, 1968).

Following decomposition by successive attacks of fungi and bacteria, the litter is ingested by the soil microfauna (Clark and Paul, 1970). Unfortunately these processes have seldom been studied in tropical savannas (Bell, 1974). In India, Sharma (1967), Rai (1968) and Khanna (1970) have investigated the succession of fungi growing on different types of herbaceous litter in grasslands. From their comparison of several ecosystems Rodin and Bazilevitch (1967) concluded that the rate of decomposition was rapid and that minerals were quickly returned to the soil; in some cases at least, there was apparently no humification.

Food resources for the soil organisms have been quantitatively studied at Lamto (Ivory Coast), and the results are summarized in Table 25.1. In such savannas the major part of the litter is composed of grass residues, but 15 to 25% of it consists of shrub leaves. These leaves play an important role just after the annual bush fires, at a time when the grass litter has not yet built up.

TABLE 25.1

Litter and root biomasses and annual decomposition of litter and roots in three savanna types at Lamto (Ivory Coast) in tons dry matter ha^{-1} (from Schaefer, 1974b)

	Maximum litter biomass	Average root biomass	Litter decomposition per year	Root decomposition per year	Total decomposition per year (roots + litter)	Stock of humic substances
Regularly burnt grass savanna	2	19	0.6	13	13.6	35.3
Regularly burnt shrub savanna	4–8	12	1.5–2.2	10	11.5–12.2	73.8
Unburnt shrub savanna	8	12	10.6–10.9	19	29.6–29.9	73.8

Root decomposition

Roots constitute the second major source of organic matter for soil micro-organisms.

In the *Loudetia* grass savanna of Lamto, and in an adjoining shrub savanna dominated by *Hyparrhenia*, the rates of root decomposition were respectively 68 and 52% per year (Lavelle and Schaeffer, 1974).

Nineteen types of grasslands have been studied in India by different authors, whose data have been summarized by Saksena (1955). The decomposition rates of roots ranged from 45 to 58% per year. These values are from two to four times greater than those calculated for temperate grasslands. But the decomposition of roots does not take place evenly throughout the year. According to Pandeya et al. (1975), 33% of the roots disappeared after the monsoon between October and November, 27% in December and January, and there was no root decomposition during the spring months.

Interactions between micro-organisms and soil fauna during litter decomposition

The roles of micro-organisms and soil animals in litter decomposition have never been ascertained in tropical savannas. Banage and Visser (1967) estimated the number of micro-organisms and nematodes in a shrub savanna in Uganda. The soil was covered with a thin litter which was continuously eaten by termites. About 67% of the nematodes fed usually or occasionally on micro-organisms, the others on roots. The numbers of

nematodes feeding upon micro-organisms ranged from 8 to 31 per gram of soil. In the Lamto savannas, Schaefer (1974a) has calculated that soil invertebrates assimilated about 20% of the total decomposed organic matter per year. Such a percentage is higher than those found in temperate ecosystems.

Humic substances synthesis

The microbial decomposition of plant organic matter is coupled with the synthesis of humic substances (Dommergues and Mangenot, 1970). Schaeffer (1974b) has analyzed the different categories of humic substances found in the Lamto savannas (Table 25.2). The savanna litter produced mostly fulvic acids, whereas forest litter led to the formation of brown humic acids. However, brown and very stable grey humic acids were also found in savanna soils.

THE MICROFLORA OF SAVANNA SOILS

The same "functional" categories of soil micro-organisms are found in savannas as elsewhere. In Paris, Pochon and Bacvarov (1973) made microbial counts on dried soil samples from the Lamto savanna. Table 25.3 summarizes their results. The small number of microbes is noticeable, together with the absence of *Azotobacter*; nitrifying bacteria are nevertheless present. Schaefer concluded, however, that the microflora is very active, although apparently not very numerous. This high biological

TABLE 25.2

The humic substances of three savanna soils (0–5 cm depth) at Lamto (Ivory Coast) (after Schaefer, 1974b)

	Soil type	Total organic C (%)	Extractable C (% of total org. C)	Fulvic acid C [FA] (p.p.m. C)	Brown humic acid [bHA] (p.p.m. C)	Grey humic acid [gHA] (p.p.m. C)	HA/FA	gHA/bHA	Exch. Mg^{2+} (%)
Loudetia grass savanna	Vertisol	2.4	28–33	845–1250	860–1350	5120	7.1	6.0	0.096–0.133
	Sandy hydromorphic soil (slope)	0.7–1.7	25	1030	620	2510	3.0	4.1	0.021
Hyparrhenia shrub savanna	ferruginous tropical soil (plateau)	1.0–1.1	24–30	550	550	800–1000	2.7–2.8	1.4–2.0	0.004–0.009

activity hinders the accumulation of organic matter and limits the development of the humic fraction. For the *Loudetia* savanna, Pochon and Bacvarov (1973) found a marked decrease in the populations of fungi and bacteria during the dry season. This *Loudetia* savanna was characterized by a very fast and frequent shift between aerobic and anaerobic phases, which was responsible for the large number of anaerobic nitrogen fixers in October. The number of bacteria found in the soil of the nearby gallery forest was similar to that found in the *Loudetia* savanna soil. Also at Lamto, Pochon and Bacvarov (1973) have examined the microbial population in a soil profile covered with *Hyparrhenia* (Table 25.4). The number of bacteria per gram of dry soil remained more or less the same in the various horizons examined, diminishing slightly with depth. This probably stems from the fact that the humic horizon is very deep.

In the Congo Republic, De Boissezon (1961) studied the physical, chemical and microbial characteristics of three savannas soils. Tables 25.5 and 25.6 summarize his data. The first soil was a red eroded soil developed over sandstone parent material and covered by an *Aristida* shrub savanna. Its biological activity (carbon dioxide output) was weak. The soil surface was partly covered by lichens, which apparently modify certain biochemical properties of the surface horizon (Table 25.7). The overall mineral richness (*Aspergillus* test) was nil. In the two other savanna soils studied carbon cycling proceeded normally, whereas nitrogen cycling was slow. In these three Congo savanna soils, the mineralization of organic nitrogen was weak or nil. The distribution of the organic matter to a relatively greater depth in the sandy soils of the Batéké Plateau was responsible for the still important biological activity at a depth of 45 cm in this

TABLE 25.3

Counts of the total microflora and of functional groups of bacteria in thousands per gram of dry soil, in the soils of the Lamto savanna and the adjacent gallery forest (after Pochon and Bacvarov, 1973)

	Total micro-flora	Anaerobic N-fixers	Anaerobic cellulolytic bacteria	Aerobic cellulolytic bacteria	Nitrifiers
October 1969					
Hyparrhenia savanna	8000	7	1.3	5.2	1.5
Loudetia savanna	90 000	23	4.5	7.0	3.2
Gallery forest	40 000	10	2.6	17.4	3.5
May 1970					
Hyparrhenia savanna	30 000	7	0.8	12.4	2.3
Loudetia savanna	20 000	4	2.0	16.4	4.7
Gallery forest	90 000	13	2.7	15.3	3.7

TABLE 25.4

Counts of bacteria (thousands per gram of dry soil) in the various horizons of a savanna soil profile at Lamto, Ivory Coast, February 1971 (after Pochon and Bacvarov, 1973)

Depth (cm)	Total micro-flora	Anaerobic N-fixers	Anaerobic cellulolytic bacteria	Aerobic cellulolytic bacteria	Nitrifying bacteria NO_2^-	NO_3^-
0–10	40 000	4.00	0.17	0.20	0.06	4.0
10–20	35 000	1.40	0.17	0.14	0.05	3.4
20–30	30 000	0.70	0.17	0.14	0.04	3.0
40 (gravel)	25 000	0.17	0.08	0.10	0.02	2.0
75	30 000	0.25	0.10	0.10	0.03	2.8

TABLE 25.5

The carbon cycle and tests for global biological activity in three savanna soils in the Congo Republic (after De Boissezon, 1961)

Soil type and vegetation	CO_2 evolution	Mineralization of organic carbon	Density of cellulolytic micro-organisms	Glucose index	Saccharase	Global mineral richness (*Aspergillus* test)
Eroded red soil *Aristida* shrub savanna	weak	moderate	moderate (fungi)	weak	weak	very weak
Plateau clayey soil *Hyparrhenia* shrub savanna, yearly burnt	moderate	moderate	high	high	high	high
Sandy soil shrub savanna with *Trachypogon thollonii*	high at the surface, weak deeper	moderate	high (fungi)	weak	high to very high	null

TABLE 25.6

The nitrogen cycle of three savanna soils of the Congo Republic (after De Boissezon, 1961)

Soil type and vegetation	Counts of aerobic nitrogen fixers	Potential ammonification (urea added)	Counts of nitrifiers	Mineralization of organic nitrogen
Eroded red soil, *Aristida* shrub savanna	very low: *Azotobacter* + *Beijerinckia*	low to moderate	very low	moderate
Plateau clayey soil, *Hyparrhenia* shrub savanna, yearly burnt	high: *Azotobacter* + *Beijerinckia*	low to moderate	low	very low
Sandy soil, shrub savanna, with *Trachypogon thollonii*	low	moderate	very low	very low

TABLE 25.7

Influence of lichens on the biological activity of an eroded red soil under *Aristida* savanna, Congo Republic (after De Boissezon, 1961)

Depth (cm)	Saccharase (mg reducing sugars per 100 g of soil)	CO_2 evolution (C–CO_2 mg kg^{-1} of soil in 7 days at 30°C)
0–5 (without lichens)	700	60
0–5 (with lichens)	900	327

profile (Table 25.8). De Boissezon (1961) has also studied the seasonal variations in biological activity for these soils. Carbon dioxide production remained low throughout the year in the red eroded soil under the *Aristida* shrub savanna. Biological activity was higher in the sandy soil of the Batéké Plateau under the *Trachypogon* shrub savanna, but it also did not vary seasonally. In the clay soil under

TABLE 25.8

The influence of depth on the biological activity of a plateau sandy soil, Batéké savanna, Congo Republic (after De Boissezon, 1961)

Depth (cm)	CO_2 evolution (C–CO_2 mg kg^{-1} of soil in 7 days at 30°C)	Rate of mineralization (C–CO_2/$C_{org.}$ × 100)
0–15	101	0.67
40–50	54.6	0.42

Hyparrhenia shrub savanna, carbon dioxide output was lowest at the end of the rains, whereas two peak periods were noted, the first in March and the second in September, at the end of the dry season. The rate of organic carbon mineralization did not change much during the year.

The effect of fire on soil microbial activity has been studied in three different sites, in the Lamto savanna in Ivory Coast (Schaefer, 1974a) and in two Malagasy grasslands by Dommergues (1954) and Moureaux (1959b). According to Schaefer (1974a), fires reduce the fungal and bacterial populations but promote the increase of actinomycete populations. This explains the slight increase in the total population of micro-organisms in the soil of the burnt savanna (Table 25.9). The dynamics of this change is not yet understood. However, Tiwari and Rai (1977), working in the Varanasi Botanical Garden (India), have shown that both the number of species of fungi and the density of their colonies were reduced immediately after a fire, a return to the pre-fire situation being achieved in two months. In a Malagasy grassland, on the other hand, Dommergues (1954) found that some months after a fire the number of cellulolytic bacteria was still barely half that before the fire. This decrease does not take place immediately after the fire, and is limited to the upper soil layer. It might therefore be due not to the direct action of fire itself, but rather to the disappearance of the litter. In addition

Moureaux (1959b), also working in Malagasy grasslands, pointed out that the solar heating of the soil surface following the burning of the grass layer is much greater than in an unburnt area. Here again, the decrease in numbers of cellulolytic bacteria might be due more to the disappearance of the grass cover than to the fire itself (Table 25.10).

Moureaux (1959a), studying the biological activity of the soil in tropical pastures set afire during the dry season, also found a transitory increase in biological activity after a fire (Table 25.11). This "beneficial" action of burning was, however, short-

TABLE 25.10

Surface temperatures (°C) of soil in burnt and unburnt grasslands in Madagascar; temperatures measured in November at 11 a.m. (after Moureaux, 1959a)

Depth (cm)	Bare soil under burnt savanna	Soil under unburnt savanna
surface	56	40
2.5	35	30
5	29.5	26.5

TABLE 25.11

Values of the glucose index[1] in soils under a burnt and an unburnt grassland, near Tananarive, Madagascar (after Moureaux, 1959a)

Burnt plots	Unburnt plots
31.5	18.8
25.7	18.8
46.6	34.5
54.2	37.2
67.7	47.1
62.0	53.8

[1] 20 g of soil are mixed with 0.1 g of glucose, humidified and incubated at 30°C for 24 h. Then the residual glucose is measured. Glucose index: I_g = consumed glucose mg/initial glucose mg × 100.

TABLE 25.9

Counts of micro-organisms (per gram of soil) in a savanna soil, before and after burning, Lamto, Ivory Coast (after Schaefer, 1974a)

	Fungi	Bacteria	Actinomycetes	Total
Unburnt savanna	103 000	945 200	1 901 800	2 950 000
Burnt savanna	30 600	758 400	2 598 600	3 387 000

lived; a notable loss of fertility was the rule in the regularly burnt parcels, as opposed to the unburnt ones. Working at Lamto, Darici (1978) found a higher rate of carbon mineralization in the soils of annually burnt plots of *Loudetia* savanna than in unburnt ones.

The influence of erosion and of a concomitant decrease in plant cover has been studied in Malagasy grasslands by Bosser et al. (1956). Four grassland soils were compared (Table 25.12) ranging from an area entirely covered with vegetation, where erosion was nil or very weak (PRa soil), to another heavily eroded area where vegetation did not cover more than 25% of the surface area (PRd soil). Intermediate situations occurred in sites PRb and PRc where vegetation covered respectively 90% and 30% of the surface area and where sheet erosion was taking place accordingly in increasing proportions. Table 25.12 shows that the biological activity of the soil decreases when erosion increases. This is likely to be due to the progressive disappearance of the humic horizon of the soil. Following repeated burning of such grassland soils in a hilly country, erosion proceeds rapidly and leads to a compact red soil in which organic matter has almost entirely disappeared, and whose biological activity is very weak or even nil.

Again in Madagascar, Moureaux (1962) compared the biological activity of: (a) a black forest soil, rich in humus, and located on a hill top; (b) a black grassland soil, also rich in humus and located on a hill top; (c) a ferrallitic *Aristida* grassland soil, poor in humus, from a hill slope; and (d) a bare ferrallitic soil, heavily eroded. The results of this

investigation are shown in Table 25.13: when the same kinds of soils are compared, the biological activity is always smaller in grassland than in forest soils. Furthermore, the biological activity is much reduced in heavily eroded ferrallitic soils which are poor in humus.

In Vietnam, Castagnol (1952), Nguyen Cong Vien (1953) and Castagnol and Nguyen Cong Vien (1951) also found that fungi, actinomycetes and bacteria were much more numerous in grassland soils than in soils devoid of vegetation. The dry season did not influence the populations of microorganisms, but temperature changes had a marked effect on them.

Seasonal variations in biological activity occur in tropical grassland soils. On the Malagasy plateaux Moureaux (1959c) compared two sites. In the first one, two periods of maximum biological activity

TABLE 25.13

Biological activity of four soils (see text) near Tananarive (after Moureaux, 1962)

	Soil A	Soil B	Soil C	Soil D
CO_2 (C–CO_2 mg kg^{-1} of soil in 7 days)	65.0	47.7	42.3	23.0
Glucose index[1]	95.4	58.2	42.6	11.1
Nitrification (p.p.m. NO_3 after 4 weeks at 30°C)	85.6	34.2	3.8	7.3
Global mineral richness[1]	1 132	1 700	289	548
Global index of fertility[1]	95.3	68.8	8.1	3.4

[1]These indices are described by Moureaux (1957b), "global mineral richness" being based on a test with *Aspergillus niger*.

TABLE 25.12

Carbon and nitrogen cycles in four grassland soils in Madagascar (after Bosser et al., 1956)

Soil	Nx	Na	N–NO_3			CO_2	I_g	Sacc	Humic acids (%)	C (%)	N (%)	C/N	pH
			imm	4w	min								
PRa	40	820	0	9.3	9.3	114	81.9	9.3	1.1	2.6	0.16	16.7	4.8
PRb	20	170	0.9	1.4	0.5	94	20.7	9.6	0.5	1.1	0.09	12.0	–
PRc	10	410	0.6	7.6	7.0	76	16.5	10.5	0.26	1.4	0.08	18.0	5.0
PRd	20	20	0.3	0.9	0.6	49	0	6,7	0.07	0.6	0.06	10.0	–

Nx = counts of *Nitrosomonas* colonies per gram of soil. Na = counts of nitrogen fixer colonies per gram of soil. N–NO_3: imm = initial amount of soil nitrate (p.p.m.). 4w = amount of soil nitrate (p.p.m.) after four weeks of incubation at 30°C at field humidity. min = mineralizable nitrogen (p.p.m.), by the method of Drouineau and Lefèvre (1949). CO_2 = C–CO_2 mg kg^{-1} from 20 g of soil in five days at 30°C and field humidity. I_g = glucose index, according to Moureaux (1957b). Sacc = saccharase: mg of reducing sugar per 100 g of soil (soil incubated with saccharose at 30°C for 24 h).

were found, one at the beginning of the rainy season, the other at the outset of the dry season; there were also two periods of minimum biological activity, at the end of the rains and towards the close of the dry season. The fall in biological activity of the soil which occurred at the end of the rains was due to the leaching of fertilizing elements caused by the rains. The soil of the second site studied was a heavily eroded lateritic soil on which some grass tussocks were still able to grow; between these tussocks lichens, which are known to stimulate the growth of *Azotobacter*, were found. The climate of the area was relatively dry. Nevertheless, the biological activity of the soil increased during the dry season. The heavy erosion taking place during the rains prevented any strong increase of biological activity during the warmer period of the year.

Pandeya et al. (1975) made monthly bacteria enumerations, over a year, in a grazed pasture soil located at Rajkot, India. Starting in September, the number of bacteria diminished in the superficial horizon of the soil, the lowest number being found during the winter; then the number of bacteria increased slightly and stabilized until the onset of the monsoon. During the rains the number of soil bacteria decreased again. The seasonal drop in numbers of soil bacteria in the superficial horizon was probably due to a decrease in the necessary nutrients. Pandeya et al. (1975) have indeed established that the quantity of litter did diminish at the beginning of the rains, because strong winds stripped the grass litter from the soil surface.

Fungi play an important role in all soils. Anderson and Domsch (1973) established that 78% of the mineralization of organic carbon in temperate-zone litter and cultivated soils was due to fungi and 22% was attributable to bacteria. The soil pH is often acidic in tropical soils and this also favours fungal activity at the expense of bacterial activity.

Rambelli and his colleagues (Rambelli, 1971; Rambelli and Bartoli, 1972; Rambelli et al., 1973) have studied the fungal flora of the savanna soils of the Lamto area, Ivory Coast. They have isolated more than sixty species, some of which are very rare, such as *Angulimaya, Chaetocerastoma, Gonytrichium* and *Periconia*. No precise relationships with soil types or vegetation could be demonstrated; on the other hand, a relative uniformity of the fungal densities throughout the year is suggested by the samplings made at the end of both dry and rainy seasons. Maggi et al. (1977) analyzed the microflora of *Loudetia simplex* roots in the Lamto savanna. Among the most abundant species were *Cephalosporium* sp., *Fusarium* sp., *Penicillium lilacinum*, *Spicaria griseola* and *Stemphylium* sp., *Aspergillus* and *Penicillium* are the genera represented by the greatest number of species. According to Rambelli (1971), the roots of *Loudetia simplex* are rich in mycorrhizae. These mycorrhizae are of the same type as those observed on the roots of forest species. This kind of association, which constitutes a true symbiosis, is very frequent. The presence of the fungus allows the root to explore a greater soil volume, leading to a better absorption of nutrients. This is especially important in the absorption of phosphorus, which is in far lower concentrations in tropical soils than in temperate ones.

Saksena (1955) was the first to make a quantitative study of the soil fungal flora; his observations were made in tropical pastures near the town of Sagar, Madhya Pradesh, India. The soil investigated was a black cotton soil, and the vegetation was dominated by grasses. Twenty eight species of fungi were isolated and the number of fungal spores decreased with depth, as follows:

Depth (cm)	Number of spores g^{-1}
0–15	70 520
15–30	26 160
30–45	9760

The great number of fungi found was most likely due to the large quantity of organic matter in the soil (4.1%) and to the large amount of nitrogen, calcium, potassium, phosphorus and iron present. The number of fungal spores was also a function of the moisture present. Certain species with thick cell walls were able to withstand long periods of drought; such was the case for *Absidia spinosa, Aspergillus niger, A. terreus, Hormodendrum cladosporioides* and *Penicillium nigricans*. Some fungi were only found in the deeper horizons, whereas others grew close to the soil surface.

In other studies performed at Varanasi, Dwivedi (1960, 1969), Mishra (1965, 1966), Sharma (1967) and Rai (1968) traced the succession of fungal populations in a pasture soil covered by Dichanthium, and on the decaying leaves of two meadow plants, Setaria glauca and Saccharum munja. Curvularia pallescens, Fusariella indica and Hormiscium gelatinosum were the most common cellulolytic species found. Mishra (1966) studied the seasonal variations of the soil fungal flora at Varanasi. The populations, which were at their lowest level during the dry and hot spring months, started to increase at the beginning of the rains. Then decomposition of organic matter began very rapidly; the fungi started growing quickly and sporulation occurred in July. During August and September some areas were flooded and the fungal population diminished in all soil horizons, increasing again in October when the water receded. From December to February fungi were again less numerous in the soil as water and organic matter became scarcer, reaching their lowest densities in March. Some species, such as Aspergillus niger, A. terreus, Chaetomium globosum, Poecilomyces fusisporus and Thielavia terricola, were present throughout the year. After having studied the seasonal distribution of soil fungi in four different kinds of tropical pastures, Mishra (1966) concluded that no general law concerning the distribution of fungal species in relation to the floristic structure of the pasture could be stated, though a few fungus species appeared to be restricted to some plant associations.

Roy and Dwivedi (1962) and Dwivedi (1966) studied the vertical distribution of fungi in six grassland plots at Varanasi. Aspergillus, Cladosporium and Trichoderma were the most frequently isolated. The Mucorales were scarce and confined to the upper soil horizons. Penicillium species were rare, except for P. funiculosum; all the plots studied contained Rhizopus nigricans and Mucor luteus. Saksenaea vasiformis was found in the middle and the lower horizons of two plots, whereas Choanephora cucurbitarum, Mucor circinelloides, M. oblongisporus and Syncephalastrum racemosum were scarce. Several genera of Actinomycetes were found. Some species of Fungi Imperfecti such as Fusarium nivale and Neocosmophora vasinfecta were common, while others such as Aspergillus variecolor, Chaetomium sp., Eremascopsis spinosa, Fusarium sp., Penicillium brefeldianum, P. spiculisporum, P. javanicum, Royella albida and Thielavia setosa were rare.

Eicker (1974) has also identified, to the genus level, the fungi found in an alkaline soil from a Transvaal (South Africa) grass savanna. Among the 53 genera identified, the most commonly found were Aspergillus, Chaetomium, Fusarium, Gliocladium, Mucor, Penicillium, Spicaria and Trichoderma.

SOIL PRODUCTION OF CARBON DIOXIDE

Measurement of the carbon dioxide output of the soil can be used to estimate the amount of organic carbon mineralized by soil micro-organisms. The most comprehensive study of carbon dioxide output from the soils of tropical savannas has been done at Lamto (Ivory Coast) by Villecourt (1973) and Schaefer (1974b).

In the various soil horizons of a Hyparrhenia savanna, the rate of carbon mineralization per day remained about the same throughout the soil profile; the output of carbon dioxide per unit of dry matter, however, was proportional to the amount of organic carbon in the sample (Table 25.14).

In a Loudetia savanna of the same region, the soil output of carbon dioxide was found to vary with the amount of water available in the soil, and continued well beyond the permanent wilting point of the vegetation. The observed differences in carbon dioxide output were largely influenced by the pattern of rainfall in previous months.

The circadian rhythm of carbon dioxide production by the Lamto soils was investigated in October 1969, in Hyparrhenia and Loudetia savannas, and in a nearby gallery forest. Whereas the highest daily temperature was reached around 3 p.m., the peak output of carbon dioxide took place around 6 p.m.; this delay corresponds to the time required for the carbon dioxide to diffuse from the soil into the atmosphere. Close to the grass tussocks this circadian change in carbon dioxide output is quite likely to correspond to the circadian change in metabolic activity of the roots; however, a similar rhythm has been found on bare soil. In the soils of gallery forest, both temperature and output of carbon dioxide remained relatively constant throughout the day.

The output of carbon dioxide per unit of dry matter at Lamto is a function of the amount of organic carbon available. The rate of carbon

TABLE 25.14

Carbon dioxide evolution measured in a soil profile under a *Hyparrhenia* savanna at Lamto, Ivory Coast, in October 1973 (after Schaefer, 1974b)

Depth (cm)	C (mg day^{-1} kg^{-1})	Rate of mineralization[1] per day	Organic C (%)
0–6	22.45	0.233	0.965
6–15	15.45	0.184	0.842
15–25	13.46	0.289	0.465
25–35	9.13	0.189	0.510
35–45	11.11	0.219	0.508

[1]Rate of mineralization: $C-CO_2/C_{org} \times 100$.

mineralization is indicative of the susceptibility to degradation of the energetic substrates. As shown in Table 25.15, the humic organic matter of the soil can be completely degraded and eliminated as carbon dioxide under a *Loudetia* savanna, whereas only part of the soil organic matter is degraded to carbon dioxide in the same soil, under an *Hyparrhenia* savanna. In the vertisol under *Loudetia* cover, only half of the soil organic matter is degraded into carbon dioxide because of the protective effect exerted by montmorillonite on the humic substances. The soil sampled close to the rhizosphere provides an insight into the metabolic activities taking place around the roots (Table 25.16). *In vitro* incubations of these soils show that the energetic substrates of the roots are more easily degradable under *Loudetia* than under *Hyparrhenia* cover. Although the biomass of roots in the *Loudetia* savanna is higher (2 kg m^{-2} versus 1.2 kg m^{-2}), the amount of accumulated carbon is lower.

Darici (1978) determined the rate of carbon mineralization of Lamto soils incubated in the laboratory under optimal conditions of humidity and temperature. The rates of mineralization during thirty days were as follows: vertisol, about 2.0%; burnt *Loudetia* savanna, about 4.5%; unburnt *Loudetia* savanna, 2.0 to 3.5%; *Hyparrhenia* savanna, 2.0 to 2.5%. With the exception of the soil under *Hyparrhenia* cover, the addition of ammonium nitrate decreased the rate of carbon dioxide output. In all cases gelatin induced an increase

TABLE 25.15

Annual balance of the CO_2 evolved and the soil humus in three type of savannas at Lamto, Ivory Coast (after Schaefer, 1974b)

Type of savanna	CO_2 output[1]	Organic matter in soil (t ha^{-1})	
		soil humus	roots
Loudetia (hydromorphic sandy soil)	32.53	31.57	23.74
Hyparrhenia (ferruginous soil)	55.32	70.47	15.32
Loudetia (Vertisol)	96.70	200.80	?

[1]Expressed as tons of decomposed organic matter ha^{-1} yr^{-1}.

TABLE 25.16

Carbon dioxide output of rhizosphere soils under *Loudetia* and Andropogoneae savannas at Lamto, Ivory Coast (after Schaefer, 1974b)

Type of savanna	Organic C (%)	C–CO$_2$ (mg day^{-1} kg^{-1}) soil	CO$_2$ evolved[1]	Rate of mineralization day	D 50 (days)[2]
Loudetia	0.703	16.34	3.13	0.232	215
Andropogoneae	1.324	26.23	2.36	0.198	262

[1] Expressed as tons of decomposed organic matter ha^{-1} yr^{-1}.
[2] Time required for a 50% decomposition of the soil organic matter.

in carbon dioxide output, this effect being particularly marked for the soil of the burnt *Loudetia* savanna. Examining the relationships between the carbon and nitrogen cycles in Lamto, Darici (1978) pointed out that the major limiting factor is the easily assimilable carbon and not the nitrogen. Indeed, the addition of mineral nitrogen does not increase mineralization, whereas the addition of urea and gelatin clearly stimulates the activity of micro-organisms.

Comparing various soils from the semi-arid zone of West Africa (Mauretania, Senegal, Mali and Guinea), Dommergues (1962) showed that the organic matter in savanna soils continued to be biodegraded during the long dry season. When various substances such as cellulose, starch or glucose were added to soil samples, the biological mineralization of carbon from them could still take place at humidity levels far lower than the wilting point.

In Senegal the carbon dioxide output of soils during the rainy season reached relatively high values between 35 and 50°C, and again towards 65°C. This increases the depletion of organic matter in soils exposed to insolation in recently burnt or overgrazed savannas (Moureaux, 1967). In ferralitic soils of the Malagasy plateaux, where the grass cover is often discontinuous, the *in vitro* measurement of carbon dioxide output also fell by about 40% when grass cover decreased from 100 to 30% (Moureaux, 1967). On these soils the traditional bush fires beginning at the end of the dry season stimulated the activity of the soil microflora at the beginning of the following rains. This could be due, at least in part, to changes in the physical structure of the superficial horizon and to the deposit of ashes which increases the soil pH and the

availability of bases. This stimulation of biological activity was, however, short-lived. Due to the effect of erosion, this activity rapidly became weaker than that of the unburnt plots in the same pastures. In the southwestern part of Madagascar, the carbon dioxide output of ferrugineous tropical soils was 50% higher under forest than under savanna cover (Moureaux, 1959c).

THE NITROGEN CYCLE

Mineralization of organic nitrogen

It is now well established that, under tropical conditions, primary productivity depends to a large extent on the mineralization of organic nitrogen, and that the stock of mineral nitrogen in the soil is small, particularly in grassland areas.

The investigations carried out by De Rham (1973) in the Lamto savannas corroborate this rule. *In situ*, a very weak accumulation of mineral nitrogen is found, as much NH$_4^+$ as NO$_3^-$, during the whole year cycle. The net *in situ* yearly production of mineral nitrogen was estimated at 2.0 kg ha^{-1} during 1963, and between 4.0 and 5.0 kg ha^{-1} during 1966 in the savanna, whereas it reached 30.0 kg ha^{-1} in an isolated grove of trees located in the middle of the same savanna, during 1965. Such figures are much lower than the 72.0 kg ha^{-1} found in the Lamto gallery forest, or the 135.0 kg ha^{-1} found in the rain forest of Yapo in lower Ivory Coast, on clay soil. Samples of savanna soils incubated in humid containers in the laboratory had a nitrogen production equal to 80.0 kg ha^{-1}, which clearly demonstrates the influence of humidity on nitrogen production.

Darici (1978) also studied nitrogen mineralization in Lamto soils incubated in the laboratory under optimal conditions of temperature and moisture. Great differences were found between the various kinds of savannas existing in the area. The smallest ammonia accumulation was found in burnt *Loudetia* savanna, followed by the soil of *Hyparrhenia* savanna, the unburnt *Loudetia* savanna and the vertisol. Nitrate accumulation was almost nil, even less than 10 p.p.m. The addition of ammonium nitrate to the soil samples induced alternating phases of immobilization and mineralization, ending in an imbalance. No denitrification took place. The addition of urea or gelatin to the samples brought about an intense ammonification, with phases of production and absorption of ammonia, evidencing a nitrogen deficit in the soil of the burnt *Loudetia* savanna. Contrary to De Rham's finding (1973), Darici did not notice any nitrification taking place after incorporation of these substrates. The rate of nitrogen mineralization was greater under unburnt *Loudetia* savanna than under that which was regularly burnt.

In his study of three different kinds of shrub savanna in the Congo Republic, De Boissezon (1961) found a mediocre or low rate of mineralization of organic nitrogen (Table 25.6). After the addition of urea ammonification levels remained low to mediocre.

Studies carried out by Dommergues (1962) in West Africa have shown that ammonification can take place at a moisture level far lower than the wilting point. Since nitrification bacteria are far less resistant to soil desiccation, ammonium nitrogen can accumulate during the dry season. This leads to an intense nitrification at the beginning of the rains and to a strong risk of leaching if the plant cover is reduced (Blondel, 1971a,b). Moreover, ammonification is not very sensitive to an excess of moisture, and its thermal optimum is 45°C (Moureaux, 1967).

Working in an *Acacia albida* tree savanna in Senegal, Jung (1967) found a low ammonifying activity under grass cover. This activity was greater under the trees themselves, but it diminished during the rains.

In the course of their studies of variously eroded pasture soils in Madagascar (Table 25.13), Bosser et al. (1956) found that nitrate nitrogen always remained low. The quantities of mineralizable nitrogen were higher in an eroded soil with a 30% grass cover than in soil with a 90% cover.

Nitrification

Numerous authors have noticed an inhibition of the nitrification processes and a low nitrate nitrogen content in the soils of temperate meadows. As a rule nitrification is also very weak in tropical savannas (Mills, 1953; Meiklejohn, 1955; Munro, 1966; De Rham, 1973).

Nitrification activity has been studied in Malagasy soils by Dommergues (1954), Bosser et al. (1956) and Moureaux (1957b, 1959b). It was first estimated on the basis of the number of nitrifying bacteria in soil samples, and then by the rate of nitrate nitrogen production after incubation of the samples. The nitrifying activity was found to be weak in the pasture soils of the plateaux, all somewhat eroded and subject to repeated bush fires. The situation was the same for ferrugineous tropical soils (*sables roux*) from the western part of the island (Moureaux, 1959c), whereas nitrification activity was very high in neighbouring forest soils.

Nitrification has also been studied in the Congo Republic by De Boissezon (1961) and Martin (1967). In addition to the techniques already mentioned they estimated nitrification in liquid medium (Kauffman and Boquel, 1951). Nitrifying activity was found to be very low in the various savannas studied, sometimes even nil in the clay soils of the plateaux along the Niari Valley. Clearing and traditional agricultural practices, however, improved the situation.

An absence of nitrification was also found in savanna soils close to Abidjan, Ivory Coast (Berlier et al., 1956). De Rham (1970) similarly found either an absence or a very low level of nitrification in the Lamto soils further north.

In Senegal, nitrification was found to be rather rapid in most ferrugineous tropical soils under savanna (Dommergues, 1959, 1960; Jung, 1967; Moureaux, 1967). It was often weaker in vertisols, but increased when their texture was lighter.

Denitrification

Denitrification takes place only in the absence of air, and in the presence of nitrate and easily assimilable carbon. Unfortunately, very little is known about the importance of this process in savanna soils at present. It might take place when the ground is saturated with water during the rains.

Denitrification activity has generally been estimated on the basis of the numbers of denitrifying bacteria in soil samples. De Boissezon (1961), for instance, found moderate and high densities of these bacteria in the three savannas he studied in the Congo Republic. However, this does not mean that these bacteria were actually very active *in situ*. More modern techniques based upon the reduction of nitrous oxide when in contact with the soil *in situ* (Garcia, 1974) should allow a better understanding of the role played by the denitrification process in the nitrogen cycle. The first assays, performed on Lamto savanna soils, show a strong *potential* denitrification activity in the sandy soils of the *Loudetia* savanna, a moderate one in the ferrugineous tropical soils of the plateaux, and a weak one in the vertisols. The addition of nitrate and organic carbon stimulates the activity in the three soil categories. This evidence indicates that a population of denitrifying bacteria does exist in these soils, but that its activity is limited by the availability of organic carbon.

Nitrogen fixation

The main groups of nitrogen-fixing bacteria of the temperate zone have also been found in tropical savanna soils: *Azotobacter, Clostridium,* Cyanobacteria and *Rhizobium.* In the ferralitic acid soils, however, *Azotobacter* is replaced by *Beijerinckia.* *Beijerinckia* (especially *B. indica*) is nearly always found in ferralitic soils (Derx, 1953; Becking, 1961; Dommergues, 1963; Hilger, 1964) and in the rhizosphere of several tropical grasses (Döbereiner, 1974). For example, *Beijerinckia* has been found in Senegal and Guinée (Dommergues, 1956, 1959; Champion et al., 1958), in the Ivory Coast (Kauffman and Toussaint, 1951), in the Congo Republic (De Boissezon, 1961), in Madagascar (Dommergues, 1952a) and Mauritius (Moureaux, 1957a), and in the north of Australia (Tchan, 1953).

Excluded from the acidic soils, *Azotobacter* (*A. chroococum* mainly) is found in most neutral or alkaline soils of tropical grasslands: in the Vertisols of Senegal (Dommergues, 1960), in western Madagascar (Dommergues, 1953; Moureaux, 1959c) and in Australia (Tchan, 1953, 1955). However, *Azotobacter* was not found in the ferrugineous tropical soils of Togo (Lamouroux, 1956). *Clostridium*, well adapted to a wide pH range, is

found in all soils, for instance in the Ivory Coast (Boquel et al., 1953; Berlier et al., 1956; Schaefer, 1974a) and in the pastures of the Malagasy plateaux (Dommergues, 1952a, b). The density of *Clostridium*, often exceeding 100 000 bacteria per gram of soil, is much greater than that of *Beijerinckia*, which suggests that the former plays a more important role in nitrogen fixation. Dommergues (1971) has also noted the nitrogen-fixing role played by a *Clostridium* species, in association with *Derxia*, within the rhizosphere of *Paspalum virgatum* in the ferralitic soils of Madagascar.

Other intensively studied nitrogen-fixing bacteria, *Azospirillum lipoferum* and *A. brasiliensis*, are also very common in tropical soils. More than half of the roots of grasses from Senegal, Gambia, Liberia, Nigeria and Brazil studied by Döbereiner et al. (1976) contained large numbers of these bacteria, while they were found in less than 10% of the samples from more temperate regions (southern Brazil, U.S.A., Kenya). *Azospirillum lipoferum* is more often found in alluvial soils than in eroded hilly country; it can be found sporadically in acidic soils.

The nitrogen-fixing Cyanobacteria are usually observed at the surface of tropical pastures whose grass cover is discontinuous, in continental Africa as well as in Madagascar (Moureaux, 1959a, b, 1965; Meyer and Laudelout, 1960; Schaefer, 1974a). Raud (1977) has also mentioned their presence in the Lamto savanna, Ivory Coast. They belong to the genera *Nostoc, Tolypothrix* and *Calothrix*, and are particularly abundant at the soil surface. At Lamto, Cyanobacteria were more abundant in savanna soils than in those of the nearby gallery forest. However, the contribution of nitrogen-fixing Cyanobacteria to the nitrogen cycle in tropical savannas is difficult to ascertain. Blondel (quoted by Boyer, 1970) noticed an increased nitrogen content in the soil under an algal crust in Senegal. Shtina (1960) has also pointed out that the activity of Cyanobacteria may increase in tropical savannas after bush fires, attributable both to their resistance to high temperatures and to the high pH of the ashes.

In the humid tropics, nitrogen fixation can also take place in the phyllosphere, where suitable conditions are found in the leaf sheaths of grasses. Water stays inside the grass sheath, and the plant itself exudes various nutritive substances, chiefly carbohydrates, which favour the multiplication of various micro-organisms. Ruinen (1971, 1974) has studied this phenomenon in the grass *Tripsacum laxum* in Ivory Coast; the concentration of carbohydrates in the exudate ranged from 400 to 10 000 p.p.m. according to the quantity of water available and the age of the leaf; the number of micro-organisms varied from 10^5 to 10^{10} ml^{-1} of liquid. Bessems (1973) obtained similar results on Surinam grasses. Numerous and varied nitrogen-fixing organisms were isolated on their leaves: *Beijerinckia*, *Azotobacter*, *Pseudomonas*, *Myconostoc*, *Spirillum* and Enterobacteriaceae. However, the amount of nitrogen fixed in the phyllosphere represents only a small percentage of that fixed in the rhizosphere.

The roots of most tropical grasses are able to fix an appreciable amount of nitrogen (Döbereiner, 1974; Dart and Day, 1975; Döbereiner and Day, 1975, 1976). The grass *Paspalum notatum* (at least its variety '*batataïs*') exhibits an important nitrogenase activity, its roots containing *Azotobacter paspali* and *Azospirillum lipoferum* (Döbereiner and Campello, 1971; Döbereiner, 1974, 1977). This grass is able to colonize large areas of Latosols in southeastern and central Brazil, and it can fix up to 100 kg nitrogen ha^{-1} yr^{-1} (Döbereiner et al., 1972). Nitrogen fixation by the *Paspalum notatum* nitrogen-fixing bacteria symbiosis has been confirmed by measuring the incorporation of $^{15}N_2$ (De Polli et al., 1977). The incidence of light is crucial to nitrogen fixation in this associative symbiosis: at 25% of the optimal light intensity, there were only 9000 *Azotobacter* cells per gram of rhizospheric soil, whereas at 100% of optimal light intensity the cell number reached 18 000 g^{-1} (Döbereiner and Campello, 1971). Nitrogen fixation in the entire plant also shows a fluctuation between day and night. If the duration of the dark period is increased, nitrogenase activity diminishes and eventually falls to zero; nitrogenase activity is restored when the plant is again illuminated (Döbereiner, 1974).

Azospirillum lipoferum can also be associated with the grasses *Panicum maximum* and *Digitaria decumbens* (Döbereiner et al., 1976; Day and Döbereiner, 1976). Once inoculated with this bacterium, *Panicum maximum* is able to produce more dry matter per unit of time than uninoculated control plants. Such inoculation can fix from 39 to 42 kg nitrogen ha^{-1} (Smith et al., 1976).

The daily rhythm of nitrogen fixation in the Lamto *Loudetia* savanna has been studied by Balandreau and Villemin (1973) and Balandreau (1975). Two peaks of nitrogen fixation were found, one during the day, and another during the night; the nocturnal peak was more important that the diurnal one. If the day maximum is related to the exudation of recently synthesized sugars by the plant, the night maximum might be associated with the exudation of the hydrolysis products of starch accumulated during the day. The amount of nitrogen fixed in the Lamto *Loudetia* savanna has been estimated at about 12 kg nitrogen ha^{-1} yr^{-1}, including the amount fixed by *Clostridium*. Nitrogen fixation in the Lamto *Hyparrhenia* savanna was smaller: 9 kg nitrogen ha^{-1} yr^{-1}. The dry season causes a decrease in the amount of nitrogen fixed in the savanna. However, nitrogen fixation compensates for loss due to grass fires and leaching. In any case, the values obtained at Lamto are far higher than those obtained in a Mediterranean *Brachypodium ramosum* meadow — 2 kg nitrogen ha^{-1} yr^{-1}. The energy input is indeed greater in tropical savannas; the increased solar radiation stimulates photosynthesis, which in turn increases root exudation. An increase in the nitrogen content of savanna soils (20 p.p.m. in four weeks) has also been reported in Nigeria (Odu and Vine, 1968).

The soil moisture, which determines oxygen tension, plays a great role in nitrogen fixation. The numbers of *Beijerinckia* and *Derxia* increase with moisture (Döbereiner, 1974), whereas the nitrogen-fixing bacteria and fungi reach their maximum numbers at the field capacity or at the wilting point.

The symbiosis of legumes with *Rhizobium* is a very effective way of fixing nitrogen. Inventories of nodule-forming *Rhizobium* species living in native tropical legumes have been carried out in Senegal (Jaubert, 1952), Ivory Coast (Berlier, 1958) and Zaire (Bonnier and Seeger, 1958). Nodulation occurs frequently in savanna legumes, whereas it is rarely found in forest legumes. However, during their investigations at Lamto, Balandreau and Villemin (1973) pointed out that, though nitrogen-fixing activity was high per gram of dry nodules, the contribution of nodule-forming *Rhizobium* to the nitrogen budget of the savanna as a whole was small. This was due to the low number of nodules per plant, and even more importantly to the scarcity of legumes within the savanna. For instance, *Vigna filicaulis* had an above-ground biomass of 180 g ha^{-1}, as compared to an overall grass biomass of 4230 kg ha^{--1} in the *Loudetia* savanna.

The *Rhizobium* symbionts of some native legumes from the Bamako region, Mali, have been studied by Sanogho et al. (1978). The isolated strains belong to the cow-pea group and show great variation in their ability to fix nitrogen. The nodules are formed during the rainy season. Nitrogenase activity has been found *in situ* in *Indigofera nummulariifolia*.

Numerous attempts have been made to introduce legumes inoculated with specific *Rhizobium* strains into tropical pastures. Some of these experiments have been successful and have led to an increased nitrogen fixation, but others have failed (Dommergues, 1967; Roche and Velly, 1961; Bonnier and Brakel, 1969; Aufeuvre, 1973; Norris, 1972; Date, 1969, 1975).

Termites also play an important nitrogen fixation role in tropical savannas; nitrogen fixation is as important in a termite nest as under a grass tussock (Schaefer, 1974b).

Enzymatic activity

The biological transformations which take place within the soil are catalyzed by extracellular enzymes (Kiss et al., 1975). Contrary to what happens in temperate countries, the enzyme content of grassland soils is always smaller than that of forest soils.

Bauzon et al. (1977) have studied the enzyme content of the Lamto savannas and of nearby gallery forests. The enzyme content of these savannas was low, even lower than that of other savannas studied by Mouraret (1965) in the Central African Republic. At Lamto, the limiting factors are quite likely the small amount of exchangeable cations and the small quantity of clay in the soils.

Generally speaking, the enzymatic activity of tropical soils is lower than that of temperate soils. However, this reduced activity is balanced by better temperature conditions over a longer period during the year. The end result might therefore be the same on a yearly basis.

CONCLUSIONS

So little is known about the microbiology of savanna soils that it would be dangerous to draw any definite conclusions from the scanty data available. A few points emerge, however.

First, the rate of decomposition of organic matter is high (two to four times greater than in temperate grasslands) and this restricts the amount of humus in the soil. A large part, sometimes even all organic matter produced during the year can be mineralized and turned into carbon dioxide. This mineralization is due to the combined action of fungi, actinomycetes and bacteria.

Second, nitrogen mineralization is almost always weaker in savanna soils than in those of temperate grasslands. The production of mineral nitrogen is also smaller than that of rain-forest soils. However, the small amount of easily assimilable carbon, not the nitrogen, is the limiting factor in tropical savanna soils. Nitrification is also weak, as is generally the case in grasslands. The major groups of nitrogen-fixing bacteria and algae are present in savanna soils, and tropical grasses are able to fix a noticeable amount of nitrogen through symbiosis with nitrogen-fixing bacteria in the rhizosphere. The amount of nitrogen fixed in this way is higher than in temperate grasslands. Fungi sometimes associate with grasses to form mycorrhizae.

The overall biological activity of savanna soils depends on numerous factors; the amount of plant cover and of organic carbon are the more important variables. Bare and eroded soils have much lower biological activity than those under dense grass cover. While in temperate conditions biological activity is greater in forest than in grassland soils, the reverse is true in tropical countries. The action of grass fires is both direct and indirect. However, many more investigations will have to be carried out before one can expect a proper understanding of the effects of fire on tropical savanna soils.

Seasonality of climate has an important influence on microbial activity in savanna soils. In some areas two peaks of activity have been observed, one at the beginning of the rains and the other at the beginning of the dry season. However, microbial activity still continues far beyond the wilting point of the plants. Circadian variations in carbon dioxide output and nitrogen fixation have also been described.

Ultimately, the evolution of tropical savanna soils depends on the balance reached between the synthesis of organic matter and its mineralization in the soil.

REFERENCES

Anderson, J.P.E. and Domsch, K.H., 1973. Quantification of bacterial and fungal contributions to soil respiration. *Arch. Mikrobiol.*, 93: 113–127.

Aufeuvre, M.A., 1973. Cytologie et évolution de *Rhizobium cowpea* dans le nodule d'arachide. *C.R. Acad. Sci. Paris, Sér. D*, 277: 921.

Balandreau, J., 1975. Mesure de l'activité nitrogénasique des micro-organismes fixateurs libres d'azote de la rhizosphère de quelques Graminées. *Rev. Ecol. Biol. Sol*, 12: 273–290.

Balandreau, J. and Villemin, G., 1973. Fixation biologique de l'azote moléculaire en savane de Lamto (basse Côte d'Ivoire). Résultats préliminaires. *Rev. Ecol. Biol. Sol*, 10: 25–33.

Balandreau, J., Millier, C. and Dommergues, Y., 1974. Diurnal variations of nitrogenase activity in the field. *Appl. Microbiol.*, 27: 662–665.

Banage, W.B. and Visser, S.A., 1967. Micro-organisms and nematodes from a virgin bush site in Uganda. In: O. Graff and J.E. Satchell (Editors), *Progress in Soil Biology*. North-Holland, Amsterdam, pp. 93–101.

Bauzon, D., Aubry, A.M., Van den Driessche, R. and Dommergues, Y., 1977. Contribution à la connaissance de la biologie des sols de la savane de Lamto, Côte d'Ivoire. *Rev. Ecol. Biol. Sol*, 14: 343–361.

Becking, J.H., 1961. Studies on nitrogen-fixing bacteria of the genus *Beijerinckia*. I. Geographical and ecological distribution in soils. II. Mineral nutrition and resistance to high levels of certain elements in relation to soil type. *Plant and Soil*, 14: 49–81; 297–322.

Bell, M.K., 1974. Decomposition of herbaceous litter. In: C.H. Dickinson and G.J.F. Pugh (Editors), *Biology of Plant Litter Decomposition*. Academic Press, London, pp. 37–67.

Berlier, Y., 1958. *La nodulation chez les légumineuses de basse Côte d'Ivoire*. ORSTOM, Paris, 39 pp.

Berlier, Y., Dabin, B. and Leneuf, N., 1956. Comparaison physique, chimique et microbiologique entre les sols de forêt et de savane sur les sables tertiaires de la basse Côte d'Ivoire. In: *6e Congr. Int. Science du Sol* (Paris), V, 81: 499–502.

Bessems, E.P.M., 1973. *Nitrogen Fixation in the Phyllosphere of Gramineae*. Thesis, Wageningem, Agric. Res. Rep. No. 786.

Blondel, D., 1971a. Contribution à l'étude du lessivage de l'azote du sol sableux (Dior) au Sénégal. *Agron. Trop.*, 26: 687–696.

Blondel, D., 1971b. Contribution à la connaissance de la dynamique de l'azote minéral au Sénégal. *Agron. Trop.*, 26: 1303–1333.

Bonnier, C. and Seeger, J., 1958. Symbiose *Rhizobium*-légumineuse en région équatoriale. Seconde communication. *Publ. INEAC, Sér. Sci.*, No. 76: 66 pp.

Bonnier, C. and Brakel, J., 1969. *Légumineuses–Rhizobium. Lutte biologique contre la faim*. Duculot, Gembloux, 148 pp.

Boquel, G., Kauffmann, J. and Toussaint, P., 1953. Recherche de l'influence du climat et de la végétation sur la flore microbienne des sols tropicaux. *Agron. Trop.*, 8: 476–481.

Bosser, J., Moureaux, C. and Pernet, R., 1956. Evolution biologique de deux sols à Madagascar. In: *6ème Congr. Int. Science du Sol, Paris*, III, 67: 399.

Boyer, J., 1970. *Essai de synthèse des connaissances acquises sur les facteurs de fertilité des sols en Afrique intertropicale francophone*. ORSTOM, Paris, 175 pp.

Castagnol, E.M., 1952. Contribution à l'étude des terres rouges basaltiques et dacitiques des Hauts-Plateaux de Sud de l'Indochine. *Arch. Rech. Agron. Cambodge, Laos, Vietnam*, 12: 123p.

Castagnol, E.M. and Nguyen Cong Vien, 1951. Etude de la flore microbienne des sols du Tonkin. *Arch. Rech. Agron. Cambodge, Laos, Vietnam*, 11: 54 pp.

Champion, F., Dugain, F., Maignien, R. and Dommergues, Y., 1958. Les sols de bananeraies et leur amélioration en Guinée. *Fruits*, 13: 415–462.

Clark, F.E. and Paul, E.A., 1970. The microflora of grasslands. *Adv. Agron.*, 22: 375–435.

Darici, C., 1978. *Effet du type d'argile sur quelques activités microbiennes dans divers sols tropicaux. Comparaison d'un sol à allophane, d'un vertisol à montmorillonite et d'un sol ferrugineux tropical à illite et kaolinite*. Thesis, Université Paris Sud-Centre d'Orsay, Orsay, 98 pp.

Dart, P.J. and Day, J.M., 1975. Non symbiotic nitrogen fixation in soil. In: N. Walker (Editor), *Soil Microbiology*. Butterworths, London, pp. 225–252.

Date, R.A., 1969. A decade of legume inoculant quality control in Australia. *J. Aust. Inst. Agric. Sci.*, 35: 27–37.

Date, R.A., 1975. Principles of *Rhizobium* strain selection. In: P.S. Nutman (Editor), *Symbiotic Nitrogen Fixation in Plants*. Cambridge University Press, Cambridge, pp. 137–150.

Daubenmire, R., 1968. Ecology of fire in grasslands. *Adv. Ecol. Res.*, 5: 209–266.

Day, J.M. and Döbereiner, J., 1976. Physiological aspects of N_2 fixation by a *Spirillum* from *Digitaria* roots. *Soil. Biol. Biochem.*, 8: 45–50.

De Boissezon, P., 1961. *Contribution à l'étude de la microflore de quelques sols typiques du Congo*. ORSTOM, Paris, No. 110: 131 pp.

De Polli, H., Matsui, E., Döbereiner, J. and Salati, E., 1977. Confirmation of nitrogen fixation in two tropical grasses by $^{15}N_2$ incorporation. *Soil Biol. Biochem.*, 9: 119–123.

De Rham, P., 1970. *L'azote dans quelques fôrets, savanes et terrains de culture d'Afrique tropicale humide*. ETH, Stiftung Rübel, Zürich, No. 45: 127 pp.

De Rham, P., 1973. Recherches sur la minéralisations de l'azote dans les sols de savane de Lamto (Côte d'Ivoire). *Rev. Ecol. Biol. Sol*, 10: 169–196.

Derx, H.G., 1953. Sur la cause de la distribution géographique limitée des *Beijerinckia*. *C.R. 6ème Congr. Int. Microbiologie, Rome*, VI: 354–355.

Döbereiner, J., 1974. Nitrogen fixing bacteria in the rhizosphere. In: A. Quispel (Editor), *The Biology of Nitrogen Fixation*.

North-Holland, Amsterdam, pp. 86–120.

Döbereiner, J., 1977. Biological nitrogen fixation in tropical grasses, possibilites for partial replacement of mineral N fertilizers. *Ambio, Spec. Rep.*, 6: 174–177.

Döbereiner, J. and Campelo, A.B., 1971. Non-symbiotic nitrogen fixing bacteria in tropical soils. *Plant Soil*, Spec. Vol., pp. 457–470.

Döbereiner, J. and Day, J.M., 1975. Nitrogen fixation in the rhizosphere of tropical grasses. In: W.D.P. Stewart (Editors), *Nitrogen Fixation by Free Living Microorganisms*, Cambridge University Press, Cambridge, pp. 39–56.

Döbereiner, J. and Day, J.M., 1976. Associative symbioses in tropical grasses. Characterization of microorganisms and dinitrogen fixing sites. In: W.E. Newton and C.J. Nyman (Editor), *Symposium on Nitrogen Fixation*, Washington State University Press, Pullman, Wash., pp. 518–538.

Döbereiner, J., Day, J.M. and Dart, P.J., 1972. Nitrogenase activity and oxygen sensitivity of the *Paspalum notatum, Azotobacter paspal*; association *J. Gen. Microbiol.*, 71: 103–116.

Döbereiner, J., Marriel, I.F. and Nery, M., 1976. Ecological distribution of *Spirillum lipoferum. Can. J. Microbiol.*, 22: 1464–1473.

Dommergues, Y., 1952a. L'analyse microbiologique des sols tropicaux acides. *Mém. Inst. Sci. Madagascar*, D 4: 169–181.

Dommergues, Y., 1952b. Influence du défrichement de forêt suivi d'incendie sur l'activité biologique du sol. *Mém. Inst. Sci. Madagascar*, D 5: 273–296.

Dommergues, Y., 1953. Note précisant la biologie d'*Azotobacter indicum* ainsi que sa répartition à Madagascar. *Mém. Inst. Sci. Madagascar*, D 5: 327–335.

Dommergues, Y., 1954. Action du feu sur la microflore des sols de prairie. *Mém. Inst. Sci. Madagascar*, D 6: 149–158.

Dommergues, Y., 1956. Etude de la biologie des sols de forêts tropicales sèches et de leur évolution après défrichement. In: *6ème Congr. Int. Science du Sol, Paris*, V 98: 605–610.

Dommergues, Y., 1959. Caractéristiques biologiques de quelques grands types de sol de l'Ouest africain. In: *IIIe Conf. Interafricaine Sols, Dalaba*, 1: 215–220.

Dommergues, Y., 1960. Un exemple d'utilisation des techniques biologiques dans la caractérisation des types pédologiques. *Agron. Trop.*, 15: 61–72.

Dommergues, Y., 1962. Contribution à l'étude de la dynamique microbienne des sols en zone semi-aride et en zone tropicale sèche. *Ann. Agron.*, 13: 265–324; 391–468.

Dommergues, Y., 1963. Distribution des *Azotobacter* et des *Beijerinckia* dans les principaux types de sol de l'Ouest africain. *Ann. Inst. Pasteur, Paris*, 105: 179–187.

Dommergues, Y., 1967. Nouvelles possibilités d'étude et d'amélioration de la fertilité des sols tropicaux offertes par les techniques biologiques. In: *Colloque Fertilité Sols tropicaux, Tananarive*, II: 1627–1635.

Dommergues, Y., 1968. Dégagement tellurique de CO_2. Mesure et signification. Rapport général. *Ann. Inst. Pasteur, Paris*, 115: 627–656.

Dommergues, Y., 1971. Effet litière. In: P. Pesson (Editor), *La vie dans les sols*. Gauthier-Villars, Paris, pp. 423–471.

Dommergues, Y. and Mangenot, F., 1970. *Ecologie microbienne du sol*. Masson, Paris, 796 pp.

Dommergues, Y., Balandreau, J., et al. 1973. Non symbiotic nitrogen fixation in the rhizosphere of rice, maize and different tropical grasses. *Soil Biol. Biochem.*, 5: 83–89.

Drouineau, G. and Lefèvre, G., 1949. Première contribution à l'étude de l'azote minéralisable dans les sols. *Ann. Agron., N.S.*, 19: 518–536.

Dwivedi, R.S., 1960. *Soil Fungi of Grasslands of Varanasi*. Thesis, Banaras Hindu University, Varanasi.

Dwivedi, R.S., 1966. Ecology of the soil fungi of some grasslands of Varanasi. 2. Distribution of soil mycoflora. *Trop. Ecol.*, 7: 84–89.

Eicker, A., 1974. The mycoflora of on alkaline soil of the open-savanna of the Transvaal. *Trans. Br. Mycol. Soc.*, 63: 281–288.

Garcia, J.L., 1974. Réduction de l'oxyde nitreux dans les sols de rizières du Sénégal.: mesure de l'activité dénitrifiante. *Soil Biol. Biochem.*, 6: 79–84.

Hilger, F., 1964. Comportement des bactéries fixatrices d'azote du genre *Beijerinckia* á l'égard du pH et du calcium. *Ann. Inst. Pasteur, paris*, 106: 279–291.

Jaubert, P., 1952. Deuxième étude sur la symbiose bactérienne des légumineuses du Sénégal. *Bull. Agron. (Ann. CRA Bombey)*, 3: 155–161.

Jung, G., 1967. *Influence de l'*Acacia albida (Del.) *sur la biologie des sols.* ORSTOM, Hann-Dakar, 63 pp.

Kauffmann, J. and Boquel, G., 1951. Nouvelle méthode de détermination du pouvoir nitrificateur d'une terre. *Ann. Inst. Pasteur, paris*, 81: 667–669.

Kauffmann, J. and Toussaint, P., 1951. Un nouveau germe fixateur de l'azote atmosphérique: *Azotobacter lacticogenes. Rev. Gén. Bot.*, 58: 553–561.

Khanna, P.K., 1970. Succession of fungi on decaying shoots of *Bothriochloa pertusa* A. Camus. *Trop. Ecol.*, 1: 201–208.

Kiss, S., Dragan-Bularda, M. and Radulescu, D., 1975. Biological significance of enzymes accumulated in soil. *Adv. Agron*, 27: 25–87.

Lamouroux, M., 1956. Etude de la fertilité et de l'utilisation des sols ferrugineux tropicaux du moyen Togo. In: *6ème Congr. Int. Science du Sol, Paris*, IV, 61: 423–426.

Laudelout, A. and Dubois, H., 1951. *Microbiologie des sols latéritiques de l'Uele. Publ. INEAC, Sér. Sci.*, No. 50: 36 pp.

Lavelle, P. and Schaefer, R., 1974. Les sources de nourriture des organismes du sol dans les savanes de Lamto. *Bull. Liaison Chercheurs Lamto*, Numéro Spéc., 5: 27–38.

Maggi, O., Bartoli, A., Puppi, G., Albonetti, S.G. and Rambelli, A., 1977. Première contribution à la connaissance de la microflore rhizophérique de *Loudetia simplex*, graminée typique de la savane de Lamto Côte d'Ivoire. *Rev. Ecol. Biol. Sol*, 14: 403–419.

Martin, G., 1967. Synthèse agropédologique sur les sols de la vallée du Niari. Evolution des sols sous culture. In: *Quinze ans de travaux et de recherches dans les pays du Niari*. P. Bory, Monaco, pp. 49–149.

Meiklejohn, J., 1955. Nitrogen problems in tropical soils. *Soils Fertil.*, 18: 459–463.

Meyer, J.A. and Laudelout, H., 1960. Biologie des sols tropicaux. *Agricultura (Louvain)*, 8: 567:–594.

Mills, W.R., 1953. Nitrate accumulation in Uganda soils. *E. Afr. Wildl. J.*, 19: 53–54.

Mishra, R.R., 1965. Seasonal distribution of fungi in four

different grass consociations of Varamasi, India. *Trop. Ecol.*, 6: 133–140.

Mishra, R.R., 1966. Seasonal variation in fungal flora of grasslands of Varanasi, India. *Trop. Ecol.*, 7: 100–112.

Mouraret, M., 1965. Contribution à l'étude de l'activité des enzymes du sol: l'asparaginase. *Cah. ORSTOM*, 9: 111 pp.

Moureaux, C., 1957a. Microbiologie de quelques sols de l'Ile Maurice. *Nat. Malgache*, 9: 11–27.

Moureaux, C., 1957b. Tests biochimiques de l'activité biologique de quelques sols malgaches. *Mém. Inst. Sci. Madagascar*, D 8: 225–241.

Moureaux, C., 1959a. Fixation de gaz carbonique par le sol. *Mém. Inst. Sci. Madagascar*, D IX: 109–120.

Moureaux, C., 1959b. L'activité microbiologique et ses variations dans l'année en divers sols des Hauts-Plateaux malgaches. *Mém. Inst. Sci. Madagascar*, D 9: 121–199.

Moureaux, C., 1959c. Observations microbiologiques sur quelques sols de la région de Morondava. *Mém. Inst. Sci. Madagascar*, D 9: 201–227.

Moureaux, C., 1962. Existence de sols noirs humifères en sommets de collines aux environs de Tananarive. *Bull. Acad. Malgache*, 38a: 47–49.

Moureaux, C., 1965. Glycolyse et activité microbiologique globale en divers sols ouest-africains. *Cah. ORSTOM, Sér. Pédol.*, 3: 43–78.

Moureaux, C., 1967. Influence de la température et de l'humidité sur les activités biologiques de quelques sols ouest-africains. *Cah. ORSTOM, Sér. Pédol.*, 5: 393–420.

Munro, P.E., 1966a. Inhibition of nitrite oxidizers by root of grass. *J. Appl. Ecol.*, 3: 227–229.

Munro, P.E., 1966. Inhibition of nitrifiers by grass root extracts. *J. Appl. Ecol.*, 3: 231–238.

Nguyen Cong Vien, 1953. Contribution à l'étude biologique des taches stériles en terres rouges. *Arch. Rech. Agron. Pastorales Vietnam (Saigon)*, No. 21: 29 pp.

Norris, D.O., 1972. Leguminous plants in tropical pastures. *Trop. Grassl.*, 6: 159–169.

Odu, C.T. and Vine, H., 1968. *Proc. Symp. IAEA.* FAO, Vienna, pp. 335–350.

Pandeya, S.C. and Sharma, S.C., 1973. *Autoecology and genecology of Anjan grass (Cenchrus ciliaris) complex.* Saurashtra University, Saurashtra, Report II, US PL 480 Project Anjan Grass, 191 pp.

Pandeya, S.C., Jain, H.K. and Krishnamurthy, L., 1975. A static model of the grazing lands at Khirasara near Rajkot. Paper presented at the *12th Int. Bot. Congr.*

Pochon, J. and Bacvarov, I., 1973. Données préliminaires sur l'activité microbiologique des sols de la savane de Lamto (Côte d'Ivoire). *Rev. Ecol. Biol. Sol.*, 10: 35–43.

Rai, P., 1968. *Succession of Fungi in Decaying Leaves of* Saccharum munja *Roxob.* Thesis, Banaras Hindu University, Varanasi.

Rambelli, A., 1971. Recherches mycologiques préliminaires dans les sols de forêt et de savane en Côte d'Ivoire. *Rev. Ecol. Biol. Sol*, 8: 219–226.

Rambelli, A. and Bartoli, A., 1972. Recherches sur la microflore fongique de terrains de Lamto en Côte d'Ivoire. *Rev. Ecol. Biol. Sol*, 9: 41–53.

Rambelli, A., Puppi, G., Bartoli, A. and Albonetti, S.G., 1973. Recherches sur la microflore fongique de terrains de Lamto

en Côte d'Ivoire. *Rev. Ecol. Biol. Sol*, 10: 13–18.

Raud, G., 1977. Données préliminaires sur les cyanophycées du sol fixatrices d'azote dans la savane de Lamto. *Rev. Ecol. Biol. Sol*, 14: 311–319.

Roy, R.Y. and Dwivedi, R.S., 1962. A comparison of soil fungal flora of three different grasslands. *Proc. Natl. Acad. Sci. India*, B 32: 421–428.

Rhoades, E.D., Locke, L.F., Taylor, H.M. and McIlvain, E.H., 1964. Water intake on a sandy range as affected by 20 years of differential cattle stocking rates. *J. Range Manage.*, 17: 185–190.

Roche, P. and Velly, J., 1961. Efficacité des cultures d'engrais verts dans le maintien de la fertilité de quelques types de sols de Madagascar. *Agron. Trop.*, 16: 7–51.

Rodin, L.E. and Bazilevich, N.I., 1967. *Production and Mineral Cycling in Terrestrial Vegetation.* Oliver and Boyd, Edinburgh.

Ruinen, J., 1971. The grass sheath as a site for nitrogen fixation. In: T.F. Preece and O.M. Dickinson (Editors), *Ecology of Leaf Surface Micro-organisms.* Academic Press, London, pp. 567–579.

Ruinen, J., 1974. Nitrogen fixation in the phyllosphere. In: A. Quispel (Editor), *The Biology of Nitrogen Fixation.* North-Holland, Amsterdam, pp. 121–167.

Saksena, S.B., 1955. Ecological factors governing the distribution of soil microfungi in some forest soils of Sagar. *J. Ind. Bot. Soc.*, 34: 262–298.

Sanogho, S.T., Sasson, A. and Renault, J., 1978. Contribution à l'étude des *Rhizobium* de quelques espèces de Légumineuses spontanées de la région de Bamako (Mali). *Rev. Ecol. Biol. Sol*, 15: 21–38.

Schaefer, R., 1974a. Le peuplement microbien du sol de la savane de Lamto. *Bull. Liaison Chercheurs Lamto*, Numéro Spéc., 5: 39–44.

Schaefer, R., 1974b. Activité métabolique du sol. Fonctions microbiennes et bilans biogéochimiques dans la savane de Lamto. *Bull. Liaison Chercheurs Lamto*, Numéro Spéc., 5: 167–184.

Sharma, P.D., 1967. *Succession of Fungi on Decaying* Setaria glauca Leauv. Thesis, Banaras Hindu University, Varanasi.

Shtina, E.A., 1960. Zonality in the distribution of soil algae communities. In: *7th Int. Congr. Soil. Sci., Madison*, III, 24: 630–634.

Smith, R.L., Bouton, J.H., Shank, S.C., Quesenberry, T.H., Tyler, M.E., Milan, J.R., Gárskine, M.H. and Littell, R.C., 1976. Nitrogen fixation in grasses inoculated with *Spirillum lipoferum. Science (Wash.)*, 193: 1003–1005.

Tchan, Y.T., 1953. Studies on N-fixing bacteria. V. Presence of *Beijerinckia* in Northern Australia and geographic distribution of non-symbiotic-N-fixing microorganisms *Proc. Linn. Soc. N.S.W.*, 78: 171–178.

Tchan, Y.T., 1955. Nitrogen economy in semi-arid plant communities. Part II: The non-symbiotic nitrogen fixing organisms. *Proc. Linn. Soc. N.S.W.*, 80: 97–104.

Tiwari, V.K. and Rai, B., 1977. Effect of soil burning on microfungi. *Plant Soil*, 47: 693–697.

Villecourt, P., 1973. Contribution à l'étude du bilan du carbone dans un sol de la savane de Lamto en Côte d'Ivoire. *Rev. Ecol. Biol. Sol*, 10: 19–23.

Chapter 26

THE PLACE OF PARASITES IN SAVANNA ECOSYSTEMS: INTRODUCTION

FRANÇOIS BOURLIERE

Like soil micro-organisms, parasites (broadly defined to include viruses, bacteria, fungi, protozoans, helminths and arthropods) do not represent an important biomass within savanna ecosystems. Yet, they may be of great functional importance in community dynamics. This is not readily apparent in undisturbed conditions, but becomes obvious when man attempts to modify the "balance" of natural communities for his own advantage. Suffice it to say that, in the African savannas alone, seven million square kilometres of grazeable land, capable of supporting 120 million head of cattle, remain largely unproductive, chiefly because of two parasitic diseases, trypanosomiasis and East Coast fever.

For a long time, the possible role of parasites in constraining the growth of plant and animal populations has not been given much attention in the ecological literature. Many prominent ecologists used to consider diseases as mere incidental factors in population regulation. Lack (1954), for instance, considered that disease is "a secondary not a primary factor" in the natural regulation of mammal, bird and fish populations, even in respect to population cycles. It was only in the case of plant-eating insects that parasites were considered capable of holding in check their increase in numbers. Even Andrewartha (1970) concurred with Lack's viewpoint when he pointed out that "in general, pathogens seem not to press heavily on their hosts as some predators press heavily on their prey". Among plants, Harper (1977), although devoting a whole chapter to pathogens, went as far as to say that "epidemic diseases (like epidemic outbreaks of pest populations) are not seen in natural vegetation — except after some major disturbance".

This last remark is very significant. It quite succinctly expresses the widely held view that, in nature, epidemics only take place in very special circumstances, whereas hosts and parasites can coexist peacefully in "normal" conditions. Odum (1971) stated "(1) that where parasites and predators have long been associated with their respective hosts and prey, the effect is moderate, neutral, or even beneficial from the long term view, and (2) that newly acquired parasites or predators are the most damaging". Burnet and White (1972) also declared that "in general terms, where two organisms have developed a host–parasite relationship, the survival of the parasite species is best served, not by destruction of the host, but by the development of a balanced condition in which sufficient of the substance of the host is consumed to allow the parasite's growth and multiplication, but not sufficient to kill the host".

The few reliable epidemiological surveys undertaken in populations of tropical animals also emphasize the fact that between epidemic outbreaks parasites and pathogens exist "silently" among some of their normal maintenance hosts, with little visible effects upon their health (Muul, 1970). This is the case, for instance, with the arboviruses of free-living wild animals (Simpson, 1968). Although virus multiplication and subsequent viraemia take place, the animal does not become ill overtly. Only in rare cases will infection cause the death of a natural host, as opposed to what happens when exotic hosts become infected. Therefore, to be parasitized is not necessarily a pathological condition but is a normal state of affairs in some situations, particularly in the tropics, where heavy parasite burdens have been discovered in apparently healthy game animals (Pester and Laurence, 1974; Horak, Ch. 27).

The two conditions which were, and still are, thought to increase the pathogenicity of the parasites are malnutrition of the host and severe disturbance of its habitat.

It has been well established (Scrimshaw et al., 1968) that poor nutrition of the host, especially protein deficiency, has a profound effect on resistance to infection in mammals. Malnutrition impairs the functioning of the host's immune system, since normal antibody response may be inhibited if protein deficiency is sufficiently severe. This causes parasites to become pathogenic. Infection, even when trivial, results in increased nitrogen loss, which further reduces the nutritional status of the host. The end-result of this process is the multiplication of the parasite and the appearance of a disease episode. Convincing evidence of the operation of such a mechanism in domestic animals is given by Reveron and Topps (1970), and in wild African ungulates by Sinclair (1977). A seasonal weakening of the host's condition can even occur in some savanna areas during the dry season, when the quality of food eaten by herbivores falls below the minimum necessary for animals to maintain their body weight. Decreased resistance of the host and increased virulence of the parasite have been suspected in such cases, leading to an increased mortality rate in the host population.

Severe habitat disturbance is another cause of imbalance between host and parasite populations. Large-scale development schemes, "bush" clearance, and introduction of non-native but taxonomically related species of domestic animals and plants, are often followed by severe disease outbreaks. In such circumstances it has often been noticed that the parasites which cause the most severe morbidity and mortality are those least adapted to their hosts in terms of long-term infection (Muul, 1970). Furthermore, the normal transmission cycle of the parasite is often altered by human interference, and vectors are compelled to seek new host species. The end-result of such environmental disturbances may therefore be a shift from the endemic to the epidemic phase of a disease. Many examples of severe disease outbreaks following environmental changes are known in tropical forest areas (Cooper and Tansley, 1978), and to a lesser extent in savanna areas. This has been attributed to the greater subdivision of the undisturbed rain forest into distinct vegetation layers, each of them harbouring a different community of parasites (Dunn et al., 1968). That a similar event can also take place in more open woodlands is illustrated by the Kyasanur forest disease (Boshell, 1969).

The traditional view that the role of parasites is only incidental in the regulation of animal and plant host populations is now openly questioned. Although well-planned epidemiological studies still have to be undertaken, especially in wild populations of native tropical species, the mathematical models which have been developed on the basis of available data definitely suggest that parasites are probably at least as important in regulating natural populations as the predators and insect parasitoids more usually studied (Anderson and May, 1978, 1979; May and Anderson, 1978, 1979; Anderson, 1979).

This more realistic approach to the relationships between parasites and hosts assumes that all population ecologists are well aware of the fact that the term "parasite" encompasses a great diversity of life-history strategies and patterns of association.

Two main categories of parasites may be distinguished, the micro- and the macroparasites. Microparasites (viruses, bacteria and protozoans) are characterized by small size, short generation times, extremely high rates of direct reproduction within the host, and a tendency to induce immunity to reinfection in those hosts that survive the initial onslaught. The duration of infection is typically short in relation to the expected life-span of the host, through some remarkable exceptions exist (the slow viruses, for instance). In contrast, macroparasites (parasitic helminths and arthropods) tend to have much longer generation times, and direct multiplication within the host is either absent or occurs at a low rate. The immune responses elicited by these metazoans generally depend on the number of parasites present in a given host, and tend to be of relatively short duration. Hence, macroparasitic infections tend to be of a persistent nature, with hosts being continually reinfected.

Both categories of parasites may pass from one host to the next, and thus complete their life-cycle either directly or indirectly via one or more intermediate host species. Direct transmission may be by contact between infected hosts, or by specialized or unspecialized transmission stages of the parasite that are either ingested or inhaled directly by the host, or actively enter its body through the tegument. Indirect transmission can involve biting by vectors that serve as intermediate hosts, or penetration by free-living transmission stages that are produced by intermediate hosts. A special case of direct transmission arises when the infection is conveyed by a parent to its unborn offspring, as is the case for many viral infections of arthropods.

Anderson and May (1978, 1979) have also called attention to the importance of some biological characteristics of an infection that determine its impact on host population growth, on the population consequences of immunological responses, and on the conditions which lead to endemic or to epidemic infections. Several general points emerge from their models, which they summarized as follows:

(1) For a disease to regulate the host population in a particular situation, the case mortality rate must be high in relation to the intrinsic growth rate of the disease-free host population; ability to achieve this degree of regulation is decreased by lasting immunity and high rates of recovery from infection.

(2) Diseases with long incubation periods have less impact on population growth.

(3) Diseases which affect the reproductive capacities of infected hosts are more liable to suppress population growth.

(4) Infection conveyed by a parent to its unborn offspring lowers the magnitude of the threshold population, that is the critical host density below which the infection cannot be maintained.

(5) This threshold density is set by the rate of loss of infected hosts divided by the rate of transmission; high threshold densities are therefore required for the maintenance of diseases with short duration of infection, long incubation periods and high case-mortality rate.

The same models also provide some insights into the population consequences of acquired immunity, clarifying its importance in determining the population consequences of a disease. However, it must be remembered that acquired immunity, in the sense discussed by Anderson and May, only exists in higher vertebrates. The defence mechanisms of invertebrates and plants are different, and perhaps less efficient (Whitfield, 1979). Encapsulation of the invading parasites by hae-mocytes or amoebocytes appears to be the rule in most insects and molluscs. Macromolecule-mediated defences analogous to the vertebrate immune system are far less common, although bactericidins can be produced in the haemolymph of some arthropods as a response to infections by bacteria. In invertebrates, and even in plants, glycoproteins called agglutinins (lectins) have been found which bind specifically to the growing tips of the hyphae of parasitic fungi. Some plant secondary compounds, such as some phenolic glucosides of dicotyledons, are also able to inhibit the development of rust fungi. It is clear, however, that caution must be exercised in extending the conclusions drawn from vertebrate models to invertebrates and plants.

The dynamic models proposed by Anderson and May (1979) also throw some light on conditions leading to epidemic disease patterns. As Anderson and May remark, they demonstrate the importance of "the rate at which new susceptible individuals appear in the population; hence the general correlation between endemicity and host population size and the observation that the host birth rate is of central importance". They also help to explain why infections of short duration which induce lasting immunity will tend to exhibit epidemic patterns, and how infectious agents of low pathogenicity persisting for long periods in the host can give rise to endemic diseases. Whether an infection will be endemic or epidemic depends on the interplay of many biological parameters.

May and Anderson (1979) have proposed models for infections caused by macroparasites with direct and indirect life cycles, and they have also discussed the mechanisms that can produce cyclic patterns, or multiple stable states in the levels of infection in the host population. Among these mechanisms that can generate multiple states, three are considered to be of principal importance: pairing of adult parasites for sexual reproduction in the primary host; non linearities associated with the transmission from primary to intermediate host, or vice versa; and parasite pathogenicity dependent on the nutritional state of the host.

The plausibility of the control of host populations by their parasites has been demonstrated in theory. It is now necessary for the field ecologist to obtain quantitative data to prove the operation of such a mechanism in natural conditions. In this perspective, opportunities should be taken to study the outbreaks of disease which occur relatively frequently during large-scale "development" projects; they may amount to large-scale experiments.

Furthermore, in order to understand better what takes place during these complex interactions, ecologists will need to keep themselves better informed on the many ways, ranging from natural host immunity to immunosuppression, by which parasites evade the defence mechanisms of their hosts (Bloom, 1979). This is a rapidly developing research field whose implications in ecology are many.

REFERENCES

Anderson, R.M., 1979. The influence of parasitic infection on the dynamics of host population growth. In: R.M. Anderson, B.D. Turner and L.R. Taylor (Editors), *Population Dynamics*. Blackwell, Oxford, pp. 245–281.

Anderson, R.M. and May, R.M., 1978. Regulation and stability of host–parasite population interactions. I. Regulatory processes. *J. Anim. Ecol.*, 47: 217–247.

Anderson, R.M. and May, R.M., 1979. Population biology of infectious diseases: Part I. *Nature*, 280: 361–367.

Andrewartha, H.G., 1970. *Introduction to the Study of Animal Populations*. Methuen, London, 2nd ed., 283 pp.

Barbehenn, K.R., 1969. Host–parasite relationships and species diversity in mammals: an hypothesis. *Biotropica*, 1(2): 29–36.

Bloom, B.R., 1979. Games parasites play: how parasites evade immune surveillance. *Nature*, 279: 21–26.

Boshell, J.M., 1968. Kyasanur forest disease: ecologic considerations. *Am. J. Trop. Med. Hyg.*, 18: 67–80.

Burnet, F.M. and White, D.O., 1972. *Natural History of Infectious Disease*. Cambridge University Press, Cambridge, 278 pp.

Cooper, J.I. and Tansley, T.W., 1978. Some epidemiological consequences of drastic ecosystem changes accompanying exploitation of tropical rain forest. *Terre Vie*, 32: 221–240.

Dunn, F.L., Lim, B.L. and Yap, L.F., 1968. Endoparasite patterns in mammals of the Malayan rain forest. *Ecology*, 49: 1179–1184.

Harper, J.L., 1977. *Population Biology of Plants*. Academic Press, London, 892 pp.

Lack, D., 1954. *The Natural Regulation of Animal Numbers*. Clarendon Press, Oxford, 343 pp.

May, R.M. and Anderson, R.M., 1978. Regulation and stability of host–parasite population interactions. II. Destabilizing processes. *J. Anim. Ecol.*, 47: 249–268.

May, R.M. and Anderson, R.M., 1979. Population biology of infectious diseases: Part II. *Nature*, 280: 455–461.

Muul, I., 1970. Mammalian ecology and epidemiology of zoonoses. *Science*, 170: 1275–1279.

Odum, E.P., 1971. *Fundamentals of Ecology*. Saunders, Philadelphia, Pa., 3rd ed., 574 pp.

Pester, F.R.N. and Laurence, B.R., 1974. The parasite load of some African game animals. *J. Zool. Lond.*, 174: 397–406.

Reveron, A.E. and Topps, J.H., 1970. Nutrition and gastro-intestinal parasitism in ruminants. *Outlook Agric.*, 6: 131–136.

Scrimshaw, N.S., Taylor, C.E. and Gordon, J.E., 1968. *Interactions of Nutrition and Infection*. Geneva, WHO, Publ. No. 57. [Quoted after Sinclair, 1977.]

Simpson, D.I.H., 1968. Arboviruses and free-living wild animals. In: A. McDiarmid (Editor), *Diseases in free-living wild animals. Symp. Zool. Soc. London*, 24: 13–28.

Sinclair, A.R.E., 1977. *The African Buffalo. A Study of Resource Limitation of Populations*. Chicago University Press, Chicago, Ill., 355 pp.

Whitfield, P.J., 1979. *The Biology of Parasitism: An Introduction to the Study of Associating Organisms*. Edward Arnold, London, 277 pp.

Chapter 27

HELMINTH, ARTHROPOD AND PROTOZOAN PARASITES OF MAMMALS IN AFRICAN SAVANNAS

I.G. HORAK

INTRODUCTION

Parasitism is an ecological relationship between two species populations in which the one, the parasite, is dependent upon the other, the host, for some of its physiological requirements, and may kill heavily infested hosts (Crofton, 1971). Both host and parasite are subject to controls exerted by the environment, and the parasite is also exposed to controls imposed by the host (for reviews see Gordon, 1973; Kennedy, 1975). In order to circumvent these controls and the accidental manner in which parasitic infestation is frequently acquired, the reproductive potential of parasites is generally high. Over-infestation of hosts, however, with their subsequent demise is not to the parasites' advantage, although a few heavily infested hosts within a population may largely be responsible for maintaining the parasite population.

Hosts are seldom infested with a single species of parasite; most harbour numerous helminth and arthropod species and may also host a number of protozoa, all of which may fluctuate seasonally. Because of the difficulty involved in studying such multispecific, changing populations little research has been conducted along these lines. However, several surveys of either helminths or ticks or the larvae of oestrid flies have been carried out in tropical savanna, while a few surveys have taken both endo- and ecto-parasites into account.

Of equal importance in the overall approach to parasitism is the fact that numerous host species may utilize the same habitat. They may be closely or distantly related, or not at all, and may live in sympatric or prey–predator relationships. They may harbour the same or widely differing parasites, and two or more host species may be essential for completion of certain parasitic life cycles.

Man frequently has a pronounced effect on the nature of parasitism within an ecosystem. His domestic animals may introduce new parasites and in turn be exposed to parasites of the inhabitants already present. His exploitation of his surroundings often leads to deterioration of the habitat, and an alteration in the degree and type of parasitic infestation because controls have been disrupted and equilibrium destroyed.

The lack of data from tropical savanna forces me to make use of some results obtained in studies in temperate regions and in laboratories. Whenever such findings are used I have striven to ensure that both the hosts and parasites involved, or closely related species, are also found in tropical savanna.

THE COMPLEXITY OF PARASITIC INFESTATION

Virtually all animals harbour more than one species of parasite simultaneously. A single host may harbour nematodes, cestodes, trematodes, the parasitic larvae of dipteran flies, haematophagous flies, fleas, lice, ticks, mites, intestinal and blood protozoa. Or these parasites may be dispersed within a host population in a particular habitat. In addition to the adult parasites which may be present, large numbers of immature helminths and arthropods frequently account for a major proportion of the total parasitic burden, and are often recovered from sites or organs other than the habitat of the adults. Both adult and immature parasites of all species are subject to fluctuations of seasonal or other origin.

Tropical savanna is particularly suitable as a habitat of numerous parasites; the complexity of infestation in this type of environment is apparent from Table 27.1, in which the parasites recovered

TABLE 27.1

Parasites recovered from blue wildebeest in the Kruger National Park, South Africa

Organ or system	Parasites recovered	
Hide	mites	*Sarcoptes* sp. (L,N,Ad)
	ticks	*Amblyomma hebraeum* (L,N,Ad), *Boophilus decoloratus* (L,N,Ad), *Rhipicephalus appendiculatus* (L,N), *Rhipicephalus evertsi evertsi* (Ad)
	lice	*Damalinia theileri* (N,Ad), *Linognathus* sp. (N,Ad)
Respiratory system		
nasal passages and sinuses	oestrid larvae	*Gedoelstia cristata* (1st, 2nd, 3rd), *Gedoelstia hässleri* (1st, 2nd, 3rd), *Kirkioestrus minutus* (1st, 2nd, 3rd), *Oestrus aureoargentatus* (1st, 2nd, 3rd), *Oestrus variolosus* (1st, 2nd, 3rd)
lungs	oestrid larvae	*Gedoelstia* spp. (1st)
	Pentastomida	*Linguatula nuttalli* (N)
	nematodes	*Dictyocaulus viviparus* (Ad)
Dura mater	oestrid larvae	*G. cristata* (1st), *G. hässleri* (1st)
Cardiovascular system		
heart	oestrid larvae	*Gedoelstia* spp. (1st)
	Pentastomida	*Linguatula nuttalli* (N)
blood vessels	trematodes	*Schistosoma mattheei* (Ad)
blood	Protozoa	*Theileria* spp.
Gastro-intestinal tract		
abomasum	nematodes	*Haemonchus bedfordi* (4th, Ad), *Trichostrongylus axei* (Ad), *Trichostrongylus thomasi* (4th, Ad)
small intestine	nematodes	*Cooperia* sp. (4th, Ad), *Gaigeria pachyscelis* (Ad), *Oesophagostomum* sp. (4th), *Strongyloides* sp. (Ad), *Trichostrongylus colubriformis* (Ad), *Trichostrongylus falculatus* (Ad)
	cestodes	*Avitellina centripunctata, Moniezia benedeni, Moniezia expansa*
large intestine	nematodes	*Agriostomum gorgonis* (4th, Ad), *Oesophagostomum multifoliatum* (4th, Ad), *Trichuris* sp. (Ad)
faeces	Protozoa	*Eimeria* spp. oocysts
Liver	oestrid larvae	*Gedoelstia* spp. (1st)
	Pentastomida	*Linguatula nuttalli* (N)
	cestodes	*Stilesia hepatica*
	trematodes	*Schistosoma mattheei* (Ad)
Mesenterium	cestodes	Cysticerci

[1]L = larvae; N = nymphae; Ad = adults; 1st, 2nd, 3rd, 4th = larval stages.

from blue wildebeest (*Connochaetes taurinus*) are listed against the sites from which they were recovered, during a seasonal prevalence survey conducted in the Kruger National Park, Republic of South Africa (Horak and De Vos, unpubl.).

A total of 33 parasitic species were recovered from the wildebeest. These consisted of a mite, four tick species, two lice, the larvae of five oestrid flies, the nymphae of a tongue worm, a larval cestode, four species of adult cestodes, a trematode, twelve nematodes and two protozoa. Each of these parasites has its own seasonal occurrence, preferred site in or on the host, nutritive requirements, and methods of satisfying these, and is subject to controls exercised by the host and the environment.

THE SEASONAL PREVALENCE OF INFESTATION

Many excellent works exist listing the parasites, their vertebrate hosts, and often their geographic distribution and the diseases they transmit. Thus, Hoogstraal (1956), Yeoman and Walker (1967) and Walker (1974) have listed the ticks, their distribution, their hosts and the diseases they transmit in the Sudan, Tanzania and Kenya respectively. Zumpt (1965) has described the myiasis-producing flies, their larvae, hosts and distribution in the Old World, as well as the parasitic insects (Zumpt, 1966) and parasitic mites (Zumpt, 1961) of vertebrates in Africa south of the Sahara. The fleas, their distribution and hosts in southern Africa have been described by De Meillon et al. (1961), while Round (1968) compiled a check-list of the helminth parasites of African mammals and Neitz (1965) a check-list and host-list of the zoonoses occurring in mammals and birds in South Africa and South-West Africa (Namibia).

Although invaluable as aids for the study of parasitic infestation, these lists do not necessarily reflect the dynamism of parasitism, which only really becomes apparent in the seasonal changes that occur within parasitic populations. These fluctuations reflect both environmental and host controls, are essential for survival of the parasite and the host, and forewarn of disease in the latter.

Of the numerous parasites spending any length of time in or on domestic or wild animals in tropical savanna, nematodes, ticks and the larvae of oestrid flies are those most commonly encountered, and they will constitute the major part of this discussion.

Nematodes

The seasonal prevalence of nematode parasites of blesbok (*Damaliscus dorcas phillipsi*), impala (*Aepyceros melampus*) and cattle has been determined in savanna in the northern Transvaal, by the regular slaughter and examination of animals (Horak, 1978b,c,d).

Two blesbok were shot at monthly intervals during a period of seventeen months. They were processed for helminth recovery and their worm burdens are summarized in Table 27.2.

The blesbok were infested with five nematode

TABLE 27.2

The mean monthly helminth burdens of blesbok in the northern Transvaal, South Africa (derived from Horak, 1978b)

Month	Mean numbers of helminths recovered								
	Haemonchus contortus		*Trichostrongylus*		*Impalaia nudicollis*		*Skrjabinema alata*		*Avitellina centripunctata*
			T. axei	*T. falculatus*					
	4th	adult	adult	adult	4th	adult	imm	adult	
January	0	191	1	10	3	7314	0	0	0
February	160	675	23	3	53	10 360	40	467	4
March	170	138	35	0	119	6817	0	0	0
April	0	25	5	0	3	9939	143	106	1
May	13	28	13	8	76	4221	1578	138	1
June	3	1	16	0	22	4094	32	60	0
July	64	1	9	0	30	2361	1	152	2
August	44	137	132	0	29	683	0	0	1
September	542	18	5	0	171	12 255	0	0	0
October	151	479	14	407	130	1536	211	72	1
November	27	228	0	200	0	480	2	31	0
December	20	223	6	486	6	721	211	732	2

4th = fourth-stage larvae; imm = immature worms.

TABLE 27.3

The mean monthly worm burdens of twelve-month and older impala in the northern Transvaal, South Africa (derived from Horak, 1978c)

Month	Mean numbers of helminths recovered															Mean number of eggs per gram of faeces
	Haemonchus placei		Longistrongylus sabie		Trichostrongylus			Impalaia tuberculata		Cooperia and C. hungi		Cooperoides		Oesophagostomum columbianum		
	4th	adult	4th	adult	T. axei adult	T. colubriformis adult	T. falculatus adult	4th	adult	4th	adult	C. hamiltoni adult	C. hepaticae adult	total		
February	41	103	12	245	3	475	37	1379	1706	427	624	425	38	6		3300
March	243	89	314	130	5	445	89	2676	1870	3405	874	1157	12	12		2567
April	715	35	183	15	0	797	82	6631	1737	1801	1230	432	2	31		1450
May	5040	13	1099	1	66	1562	0	10 354	421	3013	685	2436	1	32		200
June	1334	50	429	145	227	1697	251	8748	1967	3339	363	577	17	8		700
July	1159	10	2704	3	105	1568	45	9764	935	6112	256	461	125	23		50
August	739	0	538	1	259	1143	5	15 070	288	4545	1491	718	65	29		300
September	1588	40	3396	11	150	8058	0	7040	0	18 285	180	780	61	5		433
October	551	313	194	4	0	360	0	2927	1188	720	332	373	16	11		433
November	97	91	46	519	87	716	218	714	6019	569	1294	1280	96	34		3667
December	24	35	172	257	0	1846	55	6698	4083	3112	1045	2295	2	46		2300

Only six-week-old impala were shot during January and their burdens are not included in the table.
4th = fourth-stage larvae.

TABLE 27.4

The mean worm burdens of one- to twelve-months-old impala in the northern Transvaal, South Africa (derived from Horak, 1978c)

Age in months	Month	Haemonchus placei	Longistrongylus sabie	Trichostrongylus spp.	Impalaia tuberculata	Cooperia and Cooperioides spp.	Strongyloides sp.	Oesophagostomum columbianum	Moniezia expansa
1½	January	5	1	9	19	14	3	0	0
3	February	46	0	60	269	346	0	0	0
5	April	464	63	226	2379	747	0	0	0
6	May	496	183	799	2509	1267	0	0	0
7	June	472	86	600	3181	1100	0	0	2
10	September	375	105	859	4600	1455	0	13	1
12	November	188	565	1021	6733	3239	0	34	2

TABLE 27.5

The mean worm burdens of tracer cattle exposed to infestation in a habitat containing cattle and impala in the northern Transvaal, South Africa (derived from Horak, 1978d)

Month	Haemonchus placei		Longistrongylus sabie		Trichostrongylus			Cooperia			Impalaia tuberculata		Oesophagostomum radiatum	
	4th	adult	4th	adult	T. spp. 4th	T. colubriformis	T. falculatus	C. spp. 4th	C. pectinata	C. punctata	4th	adult	4th	adult
March 1976	394	50	41	39	50	80	96	521	63	25	288	0	2	5
April	560	0	0	1	25	0	38	1458	691	731	108	0	4	3
May	376	0	1	0	0	42	3	687	208	290	0	0	3	5
June	548	0	25	0	138	0	113	1689	325	364	552	0	40	12
July	190	0	0	0	0	0	0	513	1	100	0	0	50	13
August	0	0	0	0	0	0	0	25	0	0	25	0	169	199
September	0	0	0	0	0	2	2	0	0	0	0	0	70	9
October	0	0	0	0	0	0	0	0	0	0	0	0	41	8
November	25	125	1	0	75	204	41	238	363	50	138	0	50	12
December	600	800	225	38	263	2019	325	1425	1167	563	550	850	125	4
January 1977	250	965	2	14	0	302	0	75	138	325	0	0	86	114
February	316	188	0	0	13	2	0	214	53	27	0	0	3	17

4th = fourth-stage larvae.

species and a cestode. Of these, *Haemonchus contortus* is a parasite frequently encountered in sheep (Horak and Louw, 1977; Horak, 1978a); no sheep were, however, present in the immediate vicinity. The largest numbers of *H. contortus* were recovered from October to February and of *Trichostrongylus falculatus* from October to December, while peak numbers of *Impalaia nudicollis* were present from January to June and during September.

The survey on impala was conducted from February 1975 to February 1976, and during this survey some cattle grazed with these animals from February to May 1975, and during January and February, 1976. The survey on cattle commenced after that on impala had been completed, and lasted from March 1976 to February 1977. It was conducted at the same site as that on the impala, and impala were present in the habitat throughout the latter survey period.

Two to four impala were shot and examined each month, while twelve sets of cattle, each consisting of two worm-free, tracer animals, were exposed to infestation for periods of four or six weeks with a small group of continuously exposed cattle, and were then slaughtered for worm recovery. The worm burdens of the impala are summarized in Tables 27.3 and 27.4, and those of the cattle in Table 27.5.

In addition to the ten nematode species listed in Table 27.3, some of the impala harboured a further two nematode species, two trematodes and a cestode. Of the nematodes recovered, *Haemonchus placei* is considered a parasite of cattle, while *Trichostrongylus colubriformis*, *T. falculatus* and *Oesophagostomum columbianum* are parasites frequently found in sheep.

A striking feature of this survey was the large proportions of the total burdens constituted by *Cooperia hungi*, *Cooperioides* spp., *Haemonchus placei*, *Impalaia tuberculata* and *Longistrongylus sabie* that were in the fourth stage of larval development from approximately May to early September (late autumn to early spring). This phenomenon, known as arrested development, is characterized by a cessation of development in the early fourth larval stage, thus considerably prolonging the parasitic phase of the life cycle. It is a strategy acquired by many nematodes to ensure their survival in a stable environment within the host during regularly recurring seasons in which conditions in the pasture are unsuitable for the development or survival of their free-living stages either because of heat, cold or drought (for reviews see Michel, 1974; Schad, 1977).

The mean faecal worm-egg counts were also depressed from May to October, indicating reduced burdens of adult worms, and leading to a reduction in pasture contamination at a time when the free-living stages were unlikely to survive.

Although the total worm burdens during some months appear large, they consisted mainly of arrested fourth-stage larvae which are not as pathogenic as normally developing larvae or adult worms. Nevertheless, it is probable that the level of infestation was sufficient to depress host appetite and live weight.

The examination of young impala at various occasions during the course of the survey permitted a study of the acquisition of infestation during the first year of life of these antelope. Only the very young animals were infested with *Strongyloides* sp. This nematode is acquired by young animals of several species via the milk of the mother (Lyons et al., 1970; Moncol and Grice, 1974) and there is no apparent reason why this should not also occur in impala.

The level of infestation in the young impala rose from January to April and remained more or less static until September, to rise again thereafter and reach peak numbers in November. This infestation pattern confirmed that the climatic and environmental changes associated with winter (May–August) had an adverse effect on the development or survival of the free-living stages. The comparatively large adult-worm burdens of the 12-month-old impala shot during November indicate that the lambs of the previous year play a major role in comtaminating the habitat during the following summer.

The cattle were infested with eight nematode species and two animals harboured a ninth species, one a cestode and one a trematode. Two of the nematodes recovered, *Longistrongylus sabie* and *Impalaia tuberculata*, are parasites of impala.

With the exception of *Oesophagostomum radiatum*, which was acquired mainly from June to January, all other infestation was virtually exclusively picked up from November to July. Lack of rainfall coupled with the absence of cloud cover, depleted vegetation cover and warm temperatures

made the months July or August to October unfavourable for the survival of the free-living stages on pasture.

Nearly all *Haemonchus placei* and a large proportion of *Cooperia* spp. acquired from March to August failed to develop to adulthood, and were present as arrested fourth-stage larvae. thus ensuring the survival of these parasites in the host during winter.

Bindernagel and Todd (1972) studied the seasonal prevalence of the abomasal nematode *Ashworthius lerouxi* in African buffalo (*Syncerus caffer*) in Uganda. They found that infestation (presumably with adult worms) was greatest during and immediately after the long dry season. They suggested that this may reflect a continuously high level of larval intake during the preceding periods of rainfall. It seems likely that in order to survive during this dry season the development of these larvae was arrested, and that they matured towards the end of the dry period.

Ticks

The ixodid ticks which infest cattle and antelopes may require one, two or three hosts to complete their life cycles. The larval, nymphal and adult stages of the one host ticks all occur on the same host without the first two stages having to drop off to moult. In two host ticks the larvae and nymphae occur on one host and the nymphae then drop off to moult to the adult stage which infests the second host. The larvae, the nymphae and the adults of the three host ticks each occur on separate hosts. Of the ticks which will be mentioned in the following paragraphs *Boophilus decoloratus* is a one host tick, *Hyalomma marginatum rufipes*, *H. truncatum*, *Rhipicephalus evertsi evertsi* and *R. glabroscutatum* are two host ticks and *Amblyomma hebraeum*, *A. variegatum*, *Haemaphysalis silacea*, *Ixodes cavipalpus*, *Rhipicephalus appendiculatus* and *R. simus* are three host ticks.

The seasonal prevalence of ticks on cattle in savanna has been determined in Zambia (MacLeod, 1970; MacLeod et al., 1977), Uganda (Smith, 1969a), Zimbabwe (Matson and Norval, 1977) and in South Africa (Londt et al., 1979; Horak, unpubl. data) and on impala in South Africa (Horak, unpubl. data).

The seasonal fluctuations in numbers of the major

TABLE 27.6

The mean monthly numbers of adult ticks collected from cattle in Zambia (derived from MacLeod et al., 1977)

Month	Amblyomma variegatum	Boophilus decoloratus	Hyalomma marginatum rufipes	Hyalomma truncatum	Rhipicephalus appendiculatus	Rhipicephalus evertsi evertsi	Rhipicephalus simus
January	6.9	3.5	1.4	1.4	46.0	1.0	1.2
February	4.9	11.4	1.0	2.3	15.8	1.2	1.8
March	1.7	11.4	0.1	2.9	5.5	1.4	1.5
April	1.3	19.4	0.2	1.7	2.3	1.8	0.7
May	0.6	25.4	0.1	0.7	0.5	1.6	0.3
June	0.2	18.1	0.1	0.5	0.3	0.7	<0.1
July	0.3	15.1	0.1	0.5	0.1	0.8	<0.1
August	0.3	3.6	0.2	0.2	<0.1	0.6	<0.1
September	1.5	9.8	0.2	0.1	0.1	2.1	0.0
October	8.8	12.7	0.2	0.0	0.1	2.4	0.2
November	20.1	10.6	0.2	0.1	0.4	1.5	2.7
December	18.6	7.4	1.0	0.5	15.7	1.3	3.3

tick species encountered in Zambia by MacLeod et al. (1977) are summarized in Table 27.6. Peak numbers of adult Amblyomma variegatum were present from October to February, Rhipicephalus simus from November to March, R. appendiculatus and Hyalomma marginatum rufipes from December to February, and H. truncatum from January to April. Adult Boophilus decoloratus and Rhipicephalus evertsi evertsi were present throughout the year, the latter tick, however, exhibiting a slight preference for summer. In addition to the seven major species recovered, MacLeod et al. (1977) were also able to determine the seasonal occurrence of the adult ticks of six other species.

Macleod et al. (1977) also collected adult ticks from sheep and goats as well as from a number of wild ruminants. Sheep and goats and the smaller wild ruminants such as impala and bushbuck (Tragelaphus scriptus) were not efficient hosts of

adult ticks normally encountered on cattle, while buffalo, eland (Taurotragus oryx) and kudu (Tragelaphus strepsiceros) were efficient hosts for some of the species.

The mean tick counts recorded by Smith (1969a) on cattle at three localities below an altitude of 1200 m in Uganda are summarized in Table 27.7. The three localities were slightly north of the Equator, and mean monthly atmospheric temperature and humidity were fairly constant throughout the year. Rain fell in all months but precipitation was least during January and February (Smith, 1969b). The numbers of immature and adult ticks recovered exhibited no pronounced fluctuations and were a reflection of the uniform climate.

Londt et al. (1979) determined the seasonal prevalence of adult ticks on cattle in the northern Transvaal. The tracer cattle and the impala slaughtered for worm recovery (Tables 27.3, 27.4 and

TABLE 27.7

The mean numbers of ticks collected from cattle at three localities in Uganda (derived from Smith, 1969a)

Month	Amblyomma variegatum			Boophilus decoloratus (total)	Rhipicephalus appendiculatus			Rhipicephalus e. evertsi (total)
	L	N	adult		L	N	adult	
May	10	11	17	2	5	29	120	<1
June/July	18	18	3	2	3	33	90	<1
Sept/November	13	22	6	1	8	27	116	1
January	45	36	6	4	15	58	118	<1
April	2	9	19	2	2	17	76	1

L = larvae; N = nymphae.

TABLE 27.8

The mean monthly prevalence of adult ticks on cattle in the northern Transvaal, South Africa (derived from Londt et al., 1979)

Month	Amblyomma hebraeum	Hyalomma marginatum rufipes	Hyalomma truncatum	Rhipicephalus appendiculatus	Rhipicephalus evertsi evertsi	Rhipicephalus simus
February 1976	1	13	12	41	27	0
March	1	1	1	9	10	0
April	1	1	1	2	12	0
May	1	1	1	0	3	0
June	1	1	1	0	4	0
July	1	0	0	0	12	0
August		1	0		5	0
September	1	1	1	1	27	1
October	2	3	1	1	34	1
November	11	9	8	71	42	1
December	11	20	29	207	34	1
January 1977	9	22	15	228	25	1
February	15	11		290	48	1
March	7	6	11	130	32	0

TABLE 27.9

The mean monthly ecto-parasite burdens of impala in the northern Transvaal, South Africa

Month	Mean numbers of ticks recovered											Mean numbers of lice removed		
	Amblyomma hebraeum			Rhipicephalus appendiculatus			Rhipicephalus evertsi evertsi			Boophilus decoloratus (total)	Ixodes cavipalpus (total)	Damalinia aepycerus (total)	Damalinia elongata (total)	Linognathus spp. (total)
	L	N	adult	L	N	adult	L	N	adult					
February 1975	1	1	0	0	0	39	111	46	0	0	0	0	90	0
March	13	1	0	28	0	12	40	18	4	0	1	1	3	0
April	7	0	1	305	1	0	81	41	0	0	0	9	4	3
May	7	1	0	135	14	0	371	75	1	2	0	10	54	13
June	6	0	0	192	89	2	297	92	2	0	2	8	11	47
July	5	1	0	135	54	0	463	146	5	0	1	5	1	8
August	0	1	0	47	37	1	45	27	2	0	0	0	0	3
September	15	1	0	50	120	1	82	77	5	0	0	49	1	35
October	1	1	0	1	61	1	109	39	3	0	0	1	0	1
November	13	2	0	1	18	3	109	63	2	3	1	4	1	12
December	12	1	0	1	6	8	123	46	5	3	0	9	4	9
January 1976	33	2	0	0	1	6	40	53	5	1	0	9	3	135
February	32	8	1	1	1	27	31	41	3	1	2	9	40	5

L = larvae; N = nymphae.

27.5) were also processed for the recovery of ecto-parasites. These have been counted and identified in the case of the impala, and counted but not yet identified in the case of the cattle (Horak, unpubl. data). The results of these three surveys are summarized in Tables 27.8, 27.9 and 27.10.

In addition to the six species listed in Table 27.8, *Boophilus decoloratus* and *Ixodes cavipalpus* were also occasionally recovered from the cattle. Most of the species recovered were the same as those collected from cattle in Zambia by MacLeod et al. (1977). The seasonal prevalence in the two countries was also similar even in respect of the slightly later peak of activity of *Hyalomma truncatum* when compared with that of *H. marginatum rufipes*. Although *Amblyomma variegatum* was not recovered in South Africa, *A. hebraeum* exhibited a seasonal prevalence similar to that of the former species in Zambia.

Impala were good hosts of the larvae of *A. hebraeum*, and of the larvae, nymphae and to a lesser extent adults of *Rhipicephalus appendiculatus* and *R. evertsi evertsi*. The larvae of *R. appendiculatus* reached peak numbers from April to July, the nymphae from June to October and the adults from December to March. All stages of development of *R. evertsi evertsi* were present throughout the year, but the larvae and nymphae reached peak numbers from May to July.

The impala were also infested with the biting lice *Damalinia aepycerus* and *D. elongata*, and at least two sucking lice belong to the genus *Linognathus*. The heaviest infestations of sucking lice were encountered on the six-week-old impala shot during January 1976. Some animals were infested with the louse fly *Lipoptina binoculus*, and one harboured a single adult *Hyalomma marginatum rufipes*.

Table 27.10 summarizes the mean monthly total tick burdens of the impala and cattle slaughtered during the helminth surveys (Tables 27.3, 27.4 and 27.5). The ticks recovered from the cattle and impala belonged to the same species (Tables 27.8 and 27.9), and these findings indicate that impala and cattle are equally efficient as hosts of the larval and nymphal stages of these ticks. Only the cattle were, however, efficient hosts of adult ticks.

A seasonal incidence survey (Table 27.11) of ticks on kudu in the eastern Cape Province, South Africa (Knight and Rechav, 1978), although not carried out in tropical savanna, is worth mentioning as these animals frequently inhabit tropical savanna. *Rhipicephalus appendiculatus* and *Haemaphysalis silacea* were the ticks most commonly recovered. The larvae and nymphae of these ticks reached peak numbers from May to June, with the nymphae attaining a second peak during August. Peak numbers of adult *R. appendiculatus* were recovered from February to April, while compara-

TABLE 27.10

The mean monthly total tick burdens of impala and cattle in the northern Transvaal, South Africa

Month	Impala 1975/76			Cattle 1976/77		
	larvae	nymphae	adults	larvae	nymphae	adults
March	81	20	15	1491	27	96
April	392	42	1	139	25	40
May	513	91	2	116	25	12
June	496	182	4	921	142	73
July	603	200	5	215	149	36
August	91	65	2	10	183	60
September	147	197	5	2	426	17
October	111	101	3	17	182	60
November	122	82	9	15	51	107
December	135	54	15	200	7	301
January	79	57	11	3	10	409
February	64	50	30	149	20	1051
Mean	236	95	9	273	104	189

TABLE 27.11

The mean monthly tick burdens of kudu in the eastern Cape Province, South Africa (derived from Knight and Rechav, 1978)

Month	Amblyomma hebraeum			Rhipicephalus appendiculatus			Rhipicephalus glabroscutatum	Haemaphysalis silacea		
	L	N	adult	L	N	adult	adult	L	N	adult
June 1976	0	0	2	80	29	15	0	60	16	99
July	0	1	6	0	5	1	1	15	9	34
August	0	1	0	0	297	7	6	0	99	231
September	0	10	0	0	33	0	30	0	29	96
October	0	0	0	0	90	0	48	0	7	140
November	0	22	7	0	10	20	133	0	5	350
December	1	24	5	0	6	4	44	0	0	310
January 1977	5	9	7	0	2	66	26	2	6	195
February	7	0	3	0	0	273	9	9	3	103
March	0	0	11	2	1	228	2	12	0	85
April	46	19	19	0	30	192	2	29	2	164
May	2	5	13	85	107	60	1	92	51	235
June	4	3	4	167	186	46	0	109	81	121

L = larvae; N = nymphae.

tively large numbers of adult *H. silacea* were present throughout the year with a peak during November and December. Kudu also proved to be good hosts of *Amblyomma hebraeum* and of adult *Rhipicephalus glabroscutatum*, which according to Knight and Rechav (1978) is usually considered to be a rare tick.

From the foregoing it would appear that cattle and the larger wild ruminants, such as buffalo, eland and kudu, can sustain large tick populations within a particular habitat. Smaller species such as sheep, goats and impala cannot do this under normal circumstances because they are not efficient hosts of adult ticks. Because of their efficiency as hosts of the immature stages, however, they can contribute towards the maintenance of large tick populations in habitats in which cattle or large wild ruminants are already present.

The larvae of oestrid flies

The larvae of *Oestrus ovis*, the nasal bot fly, are parasites of the nasal passages and paranasal sinuses of sheep and goats in most countries, while many African antelope are parasitized by larvae of closely related species, and equids and elephants harbour the larvae of related flies in their gastrointestinal tracks (Zumpt, 1965).

The results of surveys conducted in the Transvaal, South Africa, in order to determine the seasonal prevalence of the larvae of *O. ovis* in sheep and goats (Horak, 1977; Horak and Butt, 1977a), although not carried out in savanna, are applicable to many savanna habitats, and are compared with those of a survey conducted on goats in tropical savanna (Unsworth, 1949).

The larval burdens of consecutive sets of three oestrid-free, tracer lambs exposed to natural infestation on the Transvaal Highveld, for periods of approximately 33 days (Horak, 1977), are summarized in Table 27.12. This table also summarizes the results obtained when sheep and goat heads obtained from two abattoirs were examined for larvae (Horak, 1977; Horak and Butt, 1977a). All mature, third-instar larvae collected from these heads were permitted to pupate and the lengths of the pupal periods before the flies hatched determined. The results may be compared with those obtained by Unsworth (1949) in a survey conducted on goats in tropical savanna in northern Nigeria (Table 27.13).

The larval burdens of the tracer lambs indicated

TABLE 27.12

The mean monthly burdens of *Oestrus ovis* larvae in tracer lambs and continuously exposed sheep and goats in South Africa (derived from Horak, 1977; Horak and Butt, 1977a)

Month slaughtered and mature larvae collected	Tracer lambs (mean total larval burdens)	Abattoir sheep mean numbers of larvae recovered				Abattoir goats (mean total larval burdens)	Length of pupal period in days
		1st	2nd	3rd	total		
January	12	14	2	3	19	4	21–30
February	7	13	3	1	17	–	25–34
March	8	13	2	2	17	6	28–63
April	4	14	2	1	17	3	failed to hatch
May	13	17	2	1	20	5	failed to hatch
June	17	19	2	1	22	3	failed to hatch
July	0	12	2	1	15	6	77
August	0	12	2	1	15	4	44–54
September	<1	3	1	2	6	4	33–55
October	5	4	1	2	7	4	28–36
November	17	13	2	2	17	4	23–53
December	21	6	2	2	10	4	21–30

1st, 2nd, 3rd = larval stages.

that infestation was acquired from September to June. In the continuously exposed sheep slaughtered at the abattoir, infestation reached a peak in May and June and was characterized by an accumulation of first-stage larvae. Larval burdens decreased thereafter, presumably because no new infestation had occurred during winter, and reached the lowest numbers in September. Fresh infestation in October and November would account for the increase in larval numbers during these months.

The pupal periods of the flies confirmed these infestation patterns. The last flies to hatch before

TABLE 27.13

Infestation of goats in Nigeria with *Oestrus ovis* larvae (derived from Unsworth, 1949)

Month	% infested
January	4
February	16
March	15
April	33
May	64
June	67
July	58
August	32
September	42
October	21
November	9
December	9

winter were those from pupae formed during March. These flies were presumably responsible for infestation that occurred until June. All pupae that formed during April, May and June were unable to survive the winter, and no flies hatched from them. The first flies to hatch after winter were invariably recovered during the first two weeks of October, and hatched from pupae which had formed during July and August. These flies were responsible for the new infestations which occurred after the winter months.

The fact that pupae formed from April to June were unable to survive the winter, and that all flies died after June, imply that, in order to survive, *O. ovis* had to overwinter in the heads of the sheep. This was accomplished by an arrest in larval development, as evidenced by an increase in the proportion of first-instar larvae in the total larval burdens from February to June.

The larval burdens of the goats from the abattoir were small and reasonably constant, so that no clear pattern of seasonal prevalence was obvious. An interesting observation results from a comparison of the larval burdens and infestation rates of the sheep and goats slaughtered at abattoirs. Of all sheep examined 73.4% were infested and the mean larval burden was 15.5, whereas 73.8% of goats were infested with a mean burden of 4.4 larvae (Horak, 1977; Horak and Butt, 1977a).

Thus, although both hosts were equally susceptible to infestation, the sheep harboured considerably more larvae and could be considered as better hosts.

In Nigeria mean minimum temperatures seldom fell below 16°C (Unsworth, 1949), the critical temperature for pupal survival (Rogers and Knapp, 1973). Consequently, the life cycle probably continued uninterrupted throughout the year. The lower percentage of goats infested from autumn to spring possibly reflected longer pupal periods occasioned by cooler temperatures.

The blesbok shot in savanna in the northern Transvaal for helminth recovery (Table 27.2) were also examined for oestrid larvae (Horak and Butt, 1977b). Their larval burdens and the pupal periods of flies hatched from mature third-instar larvae recovered from these buck are summarized in Table 27.14. The blesbok were infested with larvae of three oestrid flies; these were *Gedoelstia hässleri*, which lays its larvae on the cornea of the eyes, and *Oestrus macdonaldi* and *Oestrus variolosus*, which like *O. ovis*, lay their larvae in and around the nostrils.

The smallest numbers of *G. hässleri* larvae were present during August and September and of *Oestrus* spp. during October and November. These low larval numbers reflected the absence of flies during winter, which in turn was occasioned by the inability of the pupae to survive during this season. The earlier increase in the numbers of *G. hässleri* larvae after winter when compared with that of the *Oestrus* spp. was directly attributable to the consistently shorter pupal periods of the former fly, so that the first flies hatched sooner after winter. The seasonal prevalence and pupal periods of *G. hässleri* were similar to those of *O. ovis* in sheep. A difference was obvious, however, in the considerably larger numbers of third-stage larvae of the former species recovered when compared with the one or two larvae in this stage of development recovered from sheep (Tables 27.12 and 27.14).

Competition for space in the sinus cavities of the blesbok by the large third-instar larvae of the two *Oestrus* spp. was avoided by a difference in seasonality. The third-stage larvae of *O. macdonaldi* were recovered from May to September, and those of *O. variolosus* from July to February. The short period during which third-instar larvae of *O. macdonaldi* were present suggested that this fly only completed one generation a year, while more than one was probably completed by *O. variolosus*.

Four or five blue wildebeest, shot at monthly intervals over a period of thirteen months in

TABLE 27.14

The mean monthly numbers of oestrid larvae recovered from blesbok in the northern Transvaal, South Africa (derived from Horak and Butt, 1977b)

Month shot and mature larvae collected	Mean numbers of larvae recovered							Pupal periods in days	
	Gedoelstia hässleri			*Oestrus* spp.		*Oestrus macdonaldi*	*Oestrus variolosus*	*Gedoelstia hässleri*	*Oestrus variolosus*
	1st	2nd	3rd	1st	2nd	3rd	3rd		
January	37	17	17	72	10	0	4	23–27	failed to hatch
February	5	7	22	238	1	0	8	27	39
March	20	2	5	76	0	0	0	27–32	–
April	16	6	5	134	4	0	0	–	–
May	54	7	26	39	12	18	0	failed to hatch	–
June	41	9	10	143	5	9	0	failed to hatch	failed to hatch
July	12	13	18	165	51	6	40	failed to hatch	failed to hatch
August	3	3	6	91	28	1	47	–	67
September	0	2	7	37	17	8	23	46	failed to hatch
October	31	18	20	12	8	0	20	30–39	41–52
November	14	11	9	16	0	0	0	33–35	–
December	26	21	19	215	20	0	8	22–28	35

1st, 2nd, 3rd = larval stages.

TABLE 27.15

The mean numbers of oestrid larvae recovered monthly from blue wildebeest in the Kruger National Park, South Africa

| Month | Gedoelstia | | | | | Kirkioestrus minutus | | | Oestrus | | | | |
| | G. spp. (1st) | G. cristata | | G. hässleri | | 1st | 2nd | 3rd | O. spp. 1st | O. aureoargentatus | | O variolosus | |
		2nd	3rd	2nd	3rd					2nd	3rd	2nd	3rd
November 1977	331	2	12	9	13	3	2	25	1	8	8	0	0
December	293	12	10	4	8	3	1	23	8	8	6	2	2
January 1978	862	2	0	19	29	45	3	33	3	20	7	8	0
February	152	2	2	3	7	0	1	25	0	1	2	0	0
March	306	10	16	6	16	36	0	12	4	2	4	1	4
April	299	11	4	28	40	4	10	11	1	2	4	0	0
May	165	6	14	11	18	2	11	15	2	6	10	0	0
June	162	12	10	10	14	18	6	26	10	9	11	2	3
July	196	9	26	13	25	10	4	31	16	18	16	1	1
August	233	4	10	17	25	5	4	18	2	6	6	1	0
September	317	2	0	9	16	0	7	30	12	15	7	1	1
October	147	5	2	9	19	0	12	56	1	6	6	0	1
November	397	3	3	12	16	1	6	89	1	13	6	1	0

1st, 2nd, 3rd = larval stages.

tropical savanna in the Kruger National Park, South Africa, were processed for the recovery of endo- and ecto-parasites and the larvac of oestrid flies. Only the latter have to date been specifically identified (Horak and De Vos, unpubl. data); the mean larval burdens of all wildebeest older than four months of age are summarized in Table 27.15, and the pupal periods of the flies in Table 27.16.

The larvae of five oestrid fly species were recovered from the wildebeest, and the presence of mature larvae in virtually every month indicated that the life cycles probably continued throughout the year. The pupal periods of G. hässleri were always shorter than those of the other flies, particularly during autumn and winter when pupal periods increased considerably, and Gedoelstia cris-

TABLE 27.16

The pupal periods (in days) of oestrid flies recovered from blue wildebeest in the Kruger National Park, South Africa

Month mature larvae collected	Gedoelstia cristata	Gedoelstia hässleri	Kirkioestrus minutus	Oestrus aureo-argentatus
November 1977	27	21	–	24–25
December	27	21–22	–	27–28
January 1978	–	–	failed to hatch	failed to hatch
February	–	23	–	27
March	28–29	–	31–32	28–30
April	54–57	38–42	failed to hatch	50–58
May	86	49–57	failed to hatch	70–76
June	81–86	–	failed to hatch	70–73
July	failed to hatch	45–49	53–58	58
August	failed to hatch	35	44–46	43–44
September	–	28–32	37	–
October	33	25–29	33–38	32–35
November	–	20–21	–	–

tata and *Kirkioestrus minutus* appeared unable to survive mid-winter as pupae.

These surveys also made possible a comparison of the life cycles of *G. hässleri* in blesbok and blue wildebeest. In blesbok, first-instar larvae of *G. hässleri* were deposited on the cornea of the eyes, whence the majority appeared to migrate via the vascular system to the heart and then the lungs. After breaking through into the alveoli they migrated via the trachea and pharynx to the nasal passages and sinuses. A few were found on the dura mater of the brain, from where they would migrate through the foramina of the cribriform plate to the nasal cavity (Basson, 1966). Hundreds of first-stage larvae were found on the dura mater of wildebeest from where they migrated via the foramina of the cribriform plate (permitting the passage of nerves) to the nasal cavity. A few apparently utilized the route via the heart, lungs and trachea to the nasal cavity. However, in Zambia, Howard (1976) found few *Gedoelstia* spp. first-stage larvae on the brain surfaces of blue wildebeest despite the fact that large numbers of second- and third-stage larvae were present in the frontal sinuses, thus indicating that the alternative route to the nasal cavity may well be used. This is a remarkable life cycle, evidently causing no obvious clinical manifestations and few macroscopically visible lesions in the normal host animals (Basson, 1966).

Howard (1977) counted and identified the larvae of oestrid flies recovered from nine Lichtenstein's hartebeest (*Alcelaphus lichtensteini*) in Zambia. These animals were infested with larvae of six flies, all the species encountered in both blesbok and blue wildebeest in South Africa (Tables 27.14 and 27.15) being present.

Two zebra (*Equus burchelli*) were shot in tropical savanna in the Kruger National Park, and examined for oestrid larvae (Horak and De Vos, (unpubl. data). The sites from which larvae were recovered in the zebras and the larval burdens are summarized in Table 27.17. The nasal passages of the zebra harboured larvae of *Rhinoestrus* sp., while large numbers of larvae of six *Gasterophilus* spp. were recovered from several sites in their gastrointestinal tracts.

Protozoa

The seasonality of many protozoan infections is dependent upon the seasonal prevalence of their arthropod vectors. Thus, in Zimbabwe outbreaks of *Theileria lawrencei* in cattle occurred mainly from December to March, corresponding with the seasonal summer peak of the vector tick, *Rhipicephalus appendiculatus* (Matson, 1967).

TABLE 27.17

Oestrid larvae recovered from zebra in the Kruger National Park, South Africa

Site	Species	Numbers of larvae recovered					
		zebra 1			zebra 2		
		1st	2nd	3rd	1st	2nd	3rd
Nasal passages	*Rhinoestrus* sp.	7	0	0	56	0	0
Gums	*Gasterophilus nasalis*	0	2	0	0	0	0
	G. ternicinctus	6	2	0	7	9	0
Pharynx	*G. pecorum*	0	118	189	0	352	168
Stomach	*G. meridionalis*	0	0	3	0	0	0
	G. nasalis	0	8	2	0	3	1
	G. pecorum	0	0	34	0	1	42
	G. ternicinctus	0	37	218	0	56	670
Pylorus and duodenum	*G. meridionalis*	0	42	47	0	16	58
	G. nasalis	0	175	290	0	63	59
Colon	*G. haemorrhoidalis*	0	0	75	0	4	94
Rectum	*G. inermis*	0	0	0	0	0	28

1st, 2nd and third = larval stages.

CROSS-INFESTATION

The findings in the seasonal prevalence surveys indicated that many instances of cross-infestation between hosts were possible. None of these appeared to have a deleterious effect on the alternate hosts, and they had decided advantages for the parasites.

The blesbok were infested with *Haemonchus contortus*, a parasite usually encountered in sheep, and large numbers of this nematode were successfully transmitted artificially to sheep, while the other nematode parasites of the blesbok only became established in small numbers (Horak, 1979). The impala harboured *H. placei*, *Trichostrongylus colubriformis* and *T. falculatus* in common with the cattle in their habitat, as well as *Oesophagostomum columbianum*, a nematode often recovered from sheep. None of these worms or the other nematode parasites of the impala could be transmitted artificially in large numbers to sheep, goats or calves (Horak, 1979), indicating that they had probably become adapted to impala. Although the tracer cattle grazing with impala became infested with *Longistrongylus sabie* and *Impalaia tuberculata*, both parasites of impala, few of these worms developed to adulthood in the cattle (Table 27.5).

Considerable cross-infestation with immature ticks occurred between cattle and impala (Tables 27.9 and 27.10), while Knight and Rechav (1978) showed that kudu harbour adult and immature ticks of species usually encountered on cattle.

Oestrus ovis infested both goats and sheep (Tables 27.12 and 27.13), although sheep appeared the better hosts. The blesbok and blue wildebeest were infested with *Gedoelstia hässleri* and *Oestrus variolosus* (Tables 27.14 and 27.15), while *Gasterophilus haemorrhoidalis*, *G. nasalis* and *G. pecorum* found in the zebra are frequently recovered from domestic equids (Zumpt, 1965).

Certain cross-infestations do have a detrimental effect on the alternate hosts. The lungworm of cattle, *Dictylocaulus viviparus*, is found in many blue wildebeest in the Kruger National Park, and may cause fairly extensive lung lesions (Horak and De Vos, unpubl. data). The larvae of *Gedoelstia* spp. flies, which normally parasitize alcelaphine antelope (Zumpt, 1965), if deposited in the eyes of sheep can lead to severe ocular, vascular and nervous complications (Basson, 1969).

The stick-tight flea (*Echidnophaga larina*) of warthogs (*Phacochoerus aethiopicus*) can cause very severe infestations on domestic pigs (De Meillon et al., 1961).

The tick *Rhipicephalus appendiculatus* can transmit *Theileria lawrenci* from buffalo, in which it is apparently not pathogenic, to domestic cattle in which it can cause mortalities (Matson, 1967). The tsetse flies, *Glossina* spp., transmit *Trypanosoma* spp., which they may have acquired from wild animals which serve as reservoir hosts to domestic animals which may be severely affected (Levine, 1973). *Babesia equi* is a protozoan parasite of the red blood cells of zebras and horses, and may cause fatalities in the latter host (Levine, 1973).

The ability of parasites of wild animal origin to produce immunity to the same or related parasites in domestic animals is being investigated. It has been shown by Bigalke et al. (1974) that a tissue-culture vaccine prepared from a strain of the protozoan parasite *Besnoitia besnoitii* isolated from blue wildebeest can induce a durable immunity to the clinical form of besnoitiosis in vaccinated cattle.

INTERACTIONS BETWEEN PARASITES

The presence of a parasite in an organ may result in conditions which make that organ unsuitable for other parasites which inhabit it or migrate through it. On the other hand, prior infestation with one parasite may assist the establishment of another.

It has been demonstrated experimentally that infestation of sheep with *Cysticercus tenuicollis* may adversely affect the subsequent establishment of the liver fluke, *Fasciola hepatica* (Campbell et al., 1977). This cyticicersus is the larval stage of a dog tapeworm, *Taenia hydatigena*, and migrates through the livers of the intermediate hosts, which are usually sheep.

Similarly, previous infestation with the lungworm, *Dictyocaulus filaria*, can interfere with the establishment of a subsequent infestation of the hookworm, *Gaigeria pachyscelis* (Horak, 1971). The lungworms possibly elicit a non-specific immune response in the lungs through which the hookworm larvae have to migrate before reaching the intestines.

Prior infestation with the stomach bankrupt-worm, *Trichostrongylus axei*, can adversely affect the subsequent establishment of the wireworm, *Haemonchus contortus*, in the abomasum of sheep (Reinecke, 1974). A similar reaction to the detriment of *Haemonchus placei* possibly occurred between *H. placei* and *Longistrongylus sabie* in impala examined in the seasonal prevalence survey. These animals harboured the lowest total numbers of *H. placei* from November to February, and the largest numbers of adult *L. sabie* during these months (Table 27.3), while cattle, utilizing the same pastures and harbouring only small numbers of adult *L. sabie*, carried the greatest total numbers of *H. placei* at that time (Table 27.5).

The blowflies supply examples of advantageous interactions. The secondary blowfly, *Chrysomyia albiceps*, will deposit its eggs once the larvae of the primary blowflies, *Lucilia cuprina* or *Chrysomyia chloropyga*, have suitably liquified the tissues of the host animal (Howell et al., 1978). The Old World screwworm fly, *Chrysomyia benziana*, utilizes the wounds caused by the mouth parts of feeding ticks to deposit its eggs (Zumpt, 1965), thus affording a site of entry for the newly hatched larvae.

THE EFFECTS OF PARASITISM

The effects of mono-specific, artificial infestations on domestic livestock have been determined in numerous studies. However, naturally infested animals seldom if ever harbour a single species of parasite and it is the combined effect of the multitude of species they carry that can have a marked influence on productivity.

It has been demonstrated that helminth infestation in sheep caused decreased birth weights, slower rates of live-weight gain and lower wool production (Snijders et al., 1971). Infestation with the larvae of the nasal bot fly, *Oestrus ovis*, resulted in slower rates of live-weight gain (Horak and Snijders, 1974), while infestation with the sheep ked, *Melophagus ovinus*, resulted in reduced live-weight gains and wool yield (Nelson and Slen, 1968).

In cattle, helminth and tick infestation have caused decreased live-weight gains in Shorthorn × Hereford yearling cattle (Turner and Short, 1972), while in goats protozoan parasites of the genus *Eimeria*, parasitizing the epithelium of the intestinal tract, resulted in marked reductions in live-weight gain (Marlow, 1968).

It is practically impossible to determine the effects of parasitism on the productivity of free-ranging wild animals, but several studies on the pathology of parasitic infestations have been conducted. Thus, the lesions associated with parasites in hippopotami (*Hippopotamus amphibius*) (McCully et al., 1967b), African buffalo (*Syncerus caffer*) (Basson et al., 1970), African elephant (*Loxodonta africana*) (Basson et al., 1971) and baboons (*Papio ursinus*) (McConnell et al., 1974) have been described. Studies have also been conducted on the lesions caused in certain wild ruminants by the nematode *Cordophilus sagittus*, which lives in the chambers of the heart, coronary veins and branches of the pulmonary artery (McCulley et al., 1967a), and by the larvae of the oestrid flies *Gedoelstia cristata* and *G. hässleri* (Basson, 1966).

The overall effects of parasitism can be summarized as being a lack of appetite leading to poor utilization of food resources, frequently coupled with impaired absorption of one or several components and impaired live-weight gain. In certain instances death may supervene, in others reproduction may be prevented because of frank disease, or delayed because of the longer time required to reach optimal breeding weight, thus leading to a decrease in host numbers.

THE BENEFICIAL ASPECTS OF PARASITISM

Parasitism may have certain long-term advantages for the species parasitized and for the habitat. It serves as a selection screen, and, because young animals are particularly susceptible to infestation, the weaker ones succumb or are more easily caught by predators before they can contribute to the genetic pool.

Injured, sick and old animals may become heavily parasitized, either because they are unable to groom themselves properly, or because their movement may be restricted and they are forced to feed in habitats that they would normally avoid, and which may be heavily contaminated with parasites, or because their resistance to parasitism has broken down with age. The added effects of parasitism ensure the more rapid elimination of these animals from a habitat. Thus, a blue wildebeest calf with a broken leg harboured considerably more ticks than other healthy animals examined at the same time (Horak and De Vos, unpubl. data), and an old impala ewe had lice and nematode burdens greatly in excess of those of other impala examined in her vicinity (Horak, unpubl. data).

Parasitism may also prevent the introduction of, or invasion by, hosts foreign to a particular region, and so prevent over-exploitation of the habitat. Thus, for many years trypanosomiasis prevented the introduction of domestic stock into areas in which wild animals served as reservoirs of this protozoan parasite.

Overgrazing and overstocking can result in heavy parasitic burdens, and these coupled with the deterioration in nutrition may lead to numerous deaths, thus decreasing the population pressure and preventing further deterioration of the habitat.

THE IMPACT OF MAN

In Zimbabwe, Norval (1977a, b) considered factors such as density of human settlement, farming methods, conservation practices and host distribution, in relation to the distribution and abundance of the economically important tick species and prevalence of tick-borne disease. In regions in which overstocking was common the vegetation cover was poor, and the cutting of trees frequently led to these areas being poorly wooded. As a result the microhabitats in which the non-parasitic stages of the life cycle had to survive were unsuitable for several species. This applied particularly to the three-host ticks, some of which may spend 98% of their life cycles off the host (Norval, 1977c). The one-host tick *Boophilus decoloratus*, which spends about 30% of its life cycle on the host, could survive under these conditions, and diseases such as babesiosis and anaplasmosis transmitted by this tick were common in cattle in such areas if regular dipping ceased.

In regions where, because of scientific farming methods, the vegetation cover was good and stocking densities high, the single- and multi-host ticks were plentiful, with the three-host tick *Rhipicephalus appendiculatus* being the biggest problem (Norval, 1977a). In these areas, *Theileria lawrencei*, transmitted by *R. appendiculatus* (Matson, 1967) and tick toxicoses resulting from infestations with this tick, *Hyalomma truncatum*, *R. evertsi evertsi*

and *R. simus* could cause serious stock losses (Matson, 1966, Norval, 1977a).

Good vegetation cover and high host density, without the influence of chemical tick control, led to large burdens of *R. appendiculatus* on game in small game farms in Zimbabwe (Norval, 1977a). In a small game reserve in savanna in South Africa, blesbok grazed with two other antelope species, at a stocking rate of one animal per 2 ha, harboured five species of nematode and total mean burdens of approximately 5000 worms (Horak, 1978b). While on another reserve blesbok grazing with seven other antelope species, at a stocking rate of one animal per 0.8 ha, were infested with eleven nematode species and harboured total mean burdens of approximately 15 000 worms (Horak, unpubl. data).

Man can also by means of stock movement introduce parasites into habitats in which they did not previously occur. Thus, it is probable that the Old World screwworm fly. *Chrysomyia benziana*, was introduced into the eastern Cape Province, South Africa, with cattle returning from savanna in the northern Transvaal, where they had been grazing because of drought (Baker et al., 1968).

ACKNOWLEDGEMENTS

Part of the work reported here was undertaken within the South African Savanna Ecosystem Project, Nylsvley, and I wish to thank the South African Council for Scientific and Industrial Research for financial support of this research.

My thanks also to the National Parks Board, Republic of South Africa, for placing the blue wildebeest and zebras at my disposal.

REFERENCES

Baker, J.A.F., McHardy, W.M., Thorburn, J.A. and Thompson, G.E., 1968. *Chrysomya benziana* Villeneuve — Some observations on its occurrence and activity in the Eastern Cape Province. *J. S. Afr. Vet. Med. Assoc.*, 39: 3–11.

Basson, P.A., 1966. Gedoelstial myiasis in antelopes of southern Africa. *Onderstepoort J. Vet. Res.*, 33: 77–91.

Basson, P.A., 1969. Studies on specific oculo-vascular myiasis (uitpeuloog) in sheep. V. Histopathology. *Onderstepoort J. Vet. Res.*, 36: 217–231.

Basson, P.A., McCully, R.M., Kruger, S.P., Van Nierkerk, J.W., Young, E. and De Vos, V., 1970. Parasitic and other diseases of the African buffalo in the Krüger National Park. *Onderstepoort J. Vet. Res.*, 37: 11–28.

Basson, P.A., McCully, R.M., De Vos, V., Young, E. and Kruger, S.P., 1971. Some parasitic and other natural diseases of the African elephant in the Kruger National Park. *Onderstepoort J. Vet. Res.*, 38: 239–254.

Bigalke, R.C., Basson, P.A., McCully, R.M., Bosman, P.P. and Schoeman, J.H., 1974. Studies in cattle on the development of a live vaccine against bovine besnoitiosis. *J. S. Afr. Vet. Assoc.*, 45: 207 209.

Bindernagel, J.A. and Todd, A.C., 1972. The population dynamics of *Ashworthius lerouxi* (Nematoda: Trichostrongylidae) in African buffalo in Uganda. *Br. Vet. J.*, 128: 452–456.

Campbell, N.J., Kelly, J.D., Townsend, R.B. and Dineen, J.K., 1977. The stimulation of resistance in sheep to *Fasciola hepatica* by infection with *Cysticercus tenuicollis*. *Int. J. Parasitol.*, 7: 347–351.

Crofton, H.D., 1971. A quantitative approach to parasitism. *Parasitology*, 62: 179–193.

De Meillon, B., Davis, D.H.S. and Hardy, F., 1961. *Plague in Southern Africa, 1. The Siphonaptera (excluding Ischnopsyllidae)*. The Government Printer, Pretoria, 280 pp.

Gordon, H.McL., 1973. Epidemiology and control of gastrointestinal nematodes of ruminants. *Adv. Vet. Sci.*, 17: 395–437.

Hoogstraal, H., 1956. *African Ixodoidea. 1. Ticks of the Sudan (with Special Reference to Equatoria Province and with Preliminary Reviews of the Genera* Boophilus, Margaropus *and* Hyalomma). Department of the Navy, Bureau of Medicine and Surgery, Washington, D.C., 1101 pp.

Horak, I.G., 1971. Immunity in *Gaigeria pachyscelis* infestation. *J. S. Afr. Vet. Med. Assoc.*, 42: 149–153.

Horak, I.G., 1977. Parasites of domestic and wild animals in South Africa. I. *Oestrus ovis* in sheep. *Onderstepoort J. Vet. Res.*, 44: 55–63.

Horak, I.G., 1978a. Parasites of domestic and wild animals in South Africa. V. Helminths in sheep on dry-land pasture on the Transvaal Highveld. *Onderstepoort J. Vet. Res.*, 45: 1–6.

Horak, I.G., 1978b. Parasites of domestic and wild animals in South Africa. IX. Helminths in blesbok. *Onderstepoort J. Vet. Res.*, 45: 55–58.

Horak, I.G., 1978c. Parasites of domestic and wild animals in South Africa. X. Helminths in impala. *Onderstepoort J. Vet. Res.*, 45: 221–228.

Horak, I.G., 1978d. Parasites of domestic and wild animals in South Africa. XI. Helminths in cattle on natural pastures in the northern Transvaal. *Onderstepoort J. Vet. Res.*, 45: 229–234.

Horak, I.G., 1979. Parasites of domestic and wild animals in South Africa. XII. Artificial transmission of nematodes from blesbok and impala to sheep, goats and cattle. *Onderstepoort J. Vet. Res.*, 46: 27–30.

Horak, I.G. and Butt, M.J., 1977a. Parasites of domestic and wild animals in South Africa. II. *Oestrus ovis* in goats. *Onderstepoort J. Vet. Res.*, 44: 65–67.

Horak, I.G. and Butt. M.J., 1977b. Parasites of domestic and wild animals in South Africa. III. *Oestrus* spp. and *Gedoelstia hässleri* in the blesbok. *Onderstepoort J. Vet. Res.*, 44: 113–118.

Horak, I.G. and Louw, J.P., 1977. Parasites of domestic and wild animals in South Africa. IV. Helminths in sheep on irrigated pasture on the Transvaal Highveld. *Onderstepoort J. Vet. Res.*, 44: 261–270.

Horak, I.G. and Snijders, A.J., 1974. The effect of *Oestrus ovis* infestation on Merino lambs. *Vet. Rec.*, 94: 12–16.

Howard, G.W., 1976. Parasite and disease transmission from wildebeest to cattle in Zambia. In: *Proc. 4th Reg. Wildl. Conf. E. and Centr. Afr., Luangwa Valley, Zambia*, pp. 188–196.

Howard, G.W., 1977. Prevalence of nasal bots (Diptera: Oestridae) in some Zambian hartebeest. *J. Wildl. Dis.*, 13: 400–404.

Howell, C.J., Walker, J.B. and Nevill, E.M., 1978. Ticks, mites and insects infesting domestic animals in South Africa. Part 1. Descriptions and biology. *Dep. Agric. Tech. Serv. Rep. S. Afr. Sci. Bull.*, No. 393: 69 pp.

Kennedy, C.R., 1975. *Ecological Animal Parasitology*. Blackwell, Oxford, 163 pp.

Knight, M.M. and Rechav, Y., 1978. Ticks associated with kudu in the eastern Cape: Preliminary report. *J. S. Afr. Vet. Assoc.*, 49: 343–344.

Levine, N.D., 1973. *Protozoan Parasites of Domestic Animals and of Man*. Burgess, Minneapolis, Minn., 406 pp.

Londt, J.G.H., Horak, I.G. and De Villiers, I.L., 1979. Parasites of domestic and wild animals in South Africa. XIII. The seasonal incidence of adult ticks (Acarina: Ixodidae) on cattle in the northern Transvaal. *Onderstepoort J. Vet. Res.*, 46: 31–39.

Lyons, E.T., Drudge, J.H. and Tolliver, S.C., 1970. *Strongyloides* larvae in milk of sheep and cattle. *Mod. Vet. Pract.*, 51 (May): 65–68.

McConnell, E.E., Basson, P.A., De Vos, V., Myers, B.J. and Kuntz, R.E., 1974. A survey of diseases among 100 free-ranging baboons (*Papio ursinus*) from the Kruger National Park. *Onderstepoort J. Vet. Res.*, 41: 97–168.

McCully, R.M., Van Niekerk, J.W. and Basson, P.A., 1967a. The pathology of *Cordophilus sagittus* (v. Linstow, 1907) infestation in the kudu (*Tragelaphus strepsiceros* (Pallas, 1766)), bushbuck (*Tragelaphus scriptus (*Pallas, 1766)) and African buffalo (*Syncerus caffer* (Sparrman, 1779)) in South Africa. *Onderstepoort J. Vet. Res.*, 34: 137–159.

McCully, R.M., Van Niekerk, J.W. and Kruger, S.P., 1967b. Observations on the pathology of bilharziasis and other parasitic infestations of *Hippopotamus amphibius* Linnaeus, 1758, from the Kruger National Park. *Onderstepoort J. Vet. Res.*, 34: 563–617.

MacLeod, J., 1970. Tick infestation patterns in the southern province of Zambia. *Bull. Entomol. Res.*, 60: 253–274.

MacLeod, J., Colbo, M.H., Madbouly, M.H. and Mwanaumo, B., 1977. Ecological studies of ixodid ticks (Acari: Ixodidae) in Zambia. III. Seasonal activity and attachment sites on cattle, with notes on other hosts. *Bull. Entomol Res.*, 67: 161–173.

Marlow, C.H.B., 1968. Amprolium as a coccidiostat for Angora goats. *J. S. Afr. Vet. Med. Assoc.*, 39: 93.

Matson, B.A., 1966. Epizootiology and control of the tick borne diseases of cattle in Rhodesia. *Rhod. Agric. J.*, 63: 118–122.

Matson, B.A., 1967. Theileriosis in Rhodesia: I. A study of diagnostic specimens over two seasons. *J. S. Afr. Vet. Med. Assoc.*, 38: 93–102.

Matson, B.A. and Norval, R.A.I., 1977. The seasonal occurrence of adult ixodid ticks on cattle on a Rhodesian Highveld farm. *Rhod. Vet. J.*, 8: 2–6.

Michel, J.F., 1974. Arrested development of nematodes and some related phenomena. In: B. Dawes (Editor), *Advances in Parasitology*, 12: 279–366.

Moncol, D.J. and Grice, M.J., 1974. Transmammary passage of *Strongyloides papillosus* in the goat and sheep. *Proc. Helminthol. Soc. Wash.*, 41: 1–4.

Neitz, W.O., 1965. A checklist and hostlist of the zoonoses occurring in mammals and birds in South and South West Africa. *Onderstepoort J. Vet. Res.*, 32: 189–374.

Nelson, W.A. and Slen, S.B., 1968. Weight gains and wool growth in sheep infested with the sheep ked *Melophagus ovinus*. *Exp. Parasitol.*, 22: 223–226.

Norval, R.A.I., 1977a. Tick problems in relation to land utilization in Rhodesia. *Rhod. Vet. J.*, 8: 33–38.

Norval, R.A.I., 1977b. Ticks and tick-borne disease in Rhodesia's north-eastern operational area. *Rhod. Vet. J.*, 8: 60–66.

Norval, R.A.I., 1977c. Studies on the ecology of the tick *Amblyomma hebraeum* Koch in the Eastern Cape Province of South Africa. II. Survival and development. *J. Parasitol.*, 63: 740–747.

Reinecke, R.K., 1974. Studies on *Haemonchus contortus*. I. The influence of previous exposure to *Trichostrongylus axei* on infestation with *H. contortus*. *Onderstepoort J. Vet. Res.*, 41: 213–215.

Rogers, C.E. and Knapp, F.W., 1973. Bionomics of the sheep bot fly, *Oestrus ovis*. *Environ. Entomol.*, 2: 11–23.

Round, M.C., 1968. *Check List of the Helminth Parasites of African Mammals of the Orders Carnivora, Tubulidentata, Proboscidea, Hyracoidea, Artiodactyla and Perissodactyla*. Commonwealth Agricultural Bureaux, Tech. Commun., No. 38, 252 pp.

Schad, G.A., 1977. The role of arrested development in the regulation of nematode populations. In: G.W. Each (Editor), *Regulation of Parasite Populations*. Academic Press, New York, N.Y., pp. 111–167.

Smith, M.W., 1969a. Variations in tick species and populations in the Bugisu district of Uganda. Part I. The tick survey. *Bull. Epizoot. Dis. Afr.*, 17: 55–75.

Smith, M.W., 1969b. Variations in tick species and populations in the Bugisu district of Uganda. Part II. The effects of altitude, climate, vegetation and husbandry on tick species and populations. *Bull. Epizoot. Dis. Afr.*, 17: 77–105.

Snijders, A.J., Stapelberg, J.H. and Muller, G.L., 1971. Low level thiabendazole administration to sheep. I. Susceptibility of medicated sheep to natural infestations at Outeniqua. *J. S. Afr. Vet. Med. Assoc.*, 42: 155–167.

Turner, H.G. and Short, A.J., 1972. Effects of field infestations of gastrointestinal helminths and of the cattle tick (*Boophilus microplus*) on growth of three breeds of cattle. *Aust. J. Agric. Res.*, 23: 177–193.

Unsworth, K., 1949. Observations on the seasonal incidence of *Oestrus ovis* infection among goats in Nigeria. *Ann. Trop. Med. Parasitol.*, 43: 337–340.

Walker, J.B., 1974. *The Ixodid Ticks of Kenya*. Commonwealth Institute of Entomology, London, 220 pp.

Yeoman, G.H. and Walker, J.B., 1967. *The Ixodid Ticks of Tanzania*. Commonwealth Institute of Entomology, London, 215 pp.

Zumpt, F., 1961. The arthropod parasites of vertebrates in Africa south of the Sahara (Ethiopian region). Volume I. (Chelicerata). *Publ. S. Afr. Inst. Med. Res.*, 9(1): 457 pp.

Zumpt, F., 1965. *Myiasis in Man and Animals in the Old World*. Butterworths, London, 267 pp.

Zumpt, F., 1966. The arthropod parasites of vertebrates in Africa south of the Sahara (Ethiopian region). Volume III. (Insecta excl. Phthiraptera). *Publ. S. Afr. Inst. Med. Res.*, 13(52): 283 pp.

Chapter 28

ENERGY FLOW AND NUTRIENT CYCLING IN TROPICAL SAVANNAS

MAXIME LAMOTTE and FRANÇOIS BOURLIÈRE

INTRODUCTION

It is not easy to accurately describe and quantify the energy flow and transfer of nutrients in a tropical ecosystem. Although the number of species of micro-organisms, plants and animals in a savanna or woodland is not as great as in a rain forest, species richness is nevertheless high and, in some groups at least, the taxonomic status is far from clear. Furthermore, nutritional requirements or feeding habits are in many cases very poorly known. Any attempt to broadly outline the characteristic patterns of energy flow and nutrient transfer within a savanna ecosystem as a whole would therefore be premature at this stage. However, some long-term studies have been carried out during the past decade which throw some light on the structure and functioning of at least a significant part of a few savanna ecosystems. Whether the results of these studies can be extended to all tropical savannas is still open to question, but some of the tentative conclusions reached are worth reporting in order to stimulate further research.

Some methodological problems

The estimation of energy flow and nutrient transfer within a representative savanna sample implies reliable estimates of a number of population parameters of the major participating species. Numbers and biomass must be assessed. In addition, measurements are required of other demographic and physiological characteristics, such as age structure and sex ratio, population production, turnover and movements during a yearly cycle at least, as well as feeding habits, nutritional requirements and respiration. As indicated in many chapters of this book, the estimation of numbers and standing crops can be achieved under satisfactory conditions for some taxonomic groups, mostly vertebrates, and for some arthropod groups living in the litter and grass layers. Such estimations are much more difficult for soil micro-organisms, fungi and soil invertebrates, as well as for the arthropods of the shrub and tree layers for which no adequate sampling methods exist at present.

Once a given species has been reliably censured and its seasonal population dynamics ascertained, the study of its nutritional requirements, qualitative as well as quantitative, raises even more difficult problems. It is tempting to select a few abundant species in the most numerous taxonomic categories, study their diet and extrapolate the obtained results to all other more or less closely related species. This grossly oversimplifies the problem. In species-rich tropical communities, such as many savannas, the diet and feeding habits of sympatric species are remarkably varied. There are not only food specialists and "generalists", as in any other living community, but the same category of food resource is often partitioned in the most subtle ways to avoid or reduce inter-specific competition.

It is obviously impossible in a short time to identify and quantify the diet of thousands of species, at various stages of their life-cycle and during the different seasons of the year. A working compromise has therefore to be found. The best one is to allocate the various organisms to a number of major dietary categories and "functional groups", more or less sharing the same food habits and metabolic processes.

For instance, among the primary consumers of the tropical savannas, there are: **fresh-leaf eaters**, such as grazing and browsing mammals, short-horned and long-horned grasshoppers, crickets and caterpillars of butterflies and moths; **dry-leaf eaters**, such as termites; **seed eaters**, such as many rodents and passerine birds; **sap suckers**, such as most Homoptera and Heteroptera, Reduviidae excepted; **honey-dew eaters**, **pollen eaters** and **spore eaters**, such as a number of Diptera (Drosophilidae mainly), Hymenoptera and Collembola; **root eaters** such as many plant-parasitic nematodes and some beetle larvae.

Once dead and already partly decomposed by micro-organisms, savanna plant material is consumed by a variety of other organisms, sometimes called **detritivores** or **saprophages**. In actual fact, some of them feed more on the living micro-organisms themselves than on the dead plant material. Examples are provided by cockroaches and Diplopoda in the litter, earthworms, Symphyla, some beetle larvae and fungus-growing termites in the soil.

When decomposed, plant material is used up by **humus consumers** and **soil consumers**, such as many earthworms and termites, as well as fungi and bacteria.

Similar functional categories can be established among secondary consumers, whether invertebrates (mantids, reduvids and spiders in the grass layer, carabid beetles, many ants, blood-sucking mites, and Chilopoda in the litter and upper soil layers) or vertebrates. Ecto- and endo-parasites must not be neglected.

All organisms belonging to a single functional category do not necessarily use the same food in the same way and with the same efficiency. For instance, the ratio of assimilated energy to ingested energy can differ greatly from one species to another; the ratio is far smaller for an elephant than for an impala. A fair knowledge of the nutritional physiology of the various species studied is therefore necessary in order to avoid gross mistakes. Seasonal changes in the nutritive value of the diet must also be taken into consideration, especially for plant material which loses a large part of its proteins and other nutrients once dry.

The borderline between trophic levels is not always as clear-cut as it would appear at first sight. This is the case for the so-called omnivorous animals, which are far more numerous than commonly believed to be among mammals (from rodents to primates), birds (many granivorous passerines

are seasonally insectivorous) and insects (many ants, cockroaches, long-horned grasshoppers and crickets). This is also the case for many bacteria and fungi, often called decomposers, which can thrive on any organic material, either of plant or animal origin. Although their biomass is small, their role in the energy flow is most impressive; it is far more important than that of most large animals.

The energy cost of maintenance of the various kinds of organisms composing a savanna community cannot be estimated in the field. Such respiration measurements have to be carried out in the laboratory, in conditions as similar as possible to field conditions, and on individuals representative of the various age categories of the populations studied.

THE ENERGY BUDGET OF TROPICAL SAVANNAS

Incoming flux of solar energy

Most savannas are located in the inter-tropical zone, where the sun is directly overhead at its zenith for at least part of the year. These zones therefore benefit from a maximum amount of radiant solar energy, averaging at Lamto (Ivory Coast) 16.51 GJ m^{-2} yr^{-1}. Only some wavelengths, however, are of ecological significance at the ecosystem level — that is, those ranging from 290 to 5000 nm in length (Odum, 1959). However, a large part of this incoming radiation is reflected upwards by the cloud cover, and is therefore unusable for photosynthesis; indeed, less than half reaches the earth's surface.

For instance, the incoming radiation at ground level in northern Indian savannas averages only 6.37 GJ m^{-2} yr^{-1} in humid areas, exceeding 7 GJ m^{-2} yr^{-1} in dry zones where the cloud cover is reduced (Singh et al., 1975). Close to the rainforest edge, at Lamto, it averages 5.69 GJ m^{-2} yr^{-1}, ultraviolet representing 11.5 and infrared 3.11 GJ m^{-2} yr^{-1} (Bony, 1974).

In any case, incoming solar radiation cannot be considered as a limiting factor for primary production in tropical savannas; generally speaking, net production is greater in humid regions where there is more cloud cover than in drier, sun-parched areas.

Primary production[1]

Many estimates of standing crop biomass and primary production of tropical grasslands and savannas have been reported in the ecological literature. Unfortunately, the interpretation and comparison of these estimates are made difficult by the variety of the methods used, and by the very great variation in the comprehensiveness of the different studies. In many cases, only the above-ground component of the grass layer has been studied, and the shrub and tree layers have been neglected. Also the extent of grazing pressure, if any, is not mentioned for many sites. Our discussion will therefore be limited to a comparison of the most reliable figures for the "peak biomass" of the herbaceous cover, and to the results of the few production studies so far available.

Peak biomass values

A mere comparison of the maximum values of the standing crop of the herbaceous biomass is quite informative by itself, as it highlights both the role of the amount of yearly rainfall, and that of the floristic composition of the herb layer.

In West Africa, the value of the maximum *above-ground* biomass usually increases with the amount of rainfall. At Fété Olé, in the Sahelian savanna zone of northern Senegal, the peak living biomass did not exceed 0.5 t ha^{-1} in 1970 when the rainfall was 200 mm (Bille and Poupon, 1972; Bille, 1977). In a semi-arid ecosystem of Zimbabwe (rainfall 300 mm), Walker (1974) has found a maximum biomass of about 1 t ha^{-1}. The maximum standing crop of pastures in Mali ranges from 1 t ha^{-1} around Gao (rainfall 250 mm) to 1.5 t ha^{-1} around Mopti (rainfall 450 mm) and 2 t ha^{-1} at Niono (rainfall 500 mm), according to Breman (1975) and Breman and Cissé (1977). The same variability is also found in Tanzania. In the M'Komasi reserve, the peak biomass value (dead material included) averages 2 t ha^{-1} in a dry shrub savanna (rainfall 350 mm) and 2.5 t ha^{-1} in a short grass savanna (rainfall > 500 mm) (Harris, 1972). Large variations have also been observed by Braun (1973) in the Serengeti National Park, depending on the floristic structure of the grass cover, and on the variations in yearly rainfall: From 0.5 to 3 t ha^{-1} in *Kyllinga–Sporobolus* communities, 2 to 8.5 t ha^{-1} in *Andropogon* grasslands, and 1.5 to 11.5 t ha^{-1} in *Themeda*

savannas. In the *Burkea africana* savanna of Nylsvley, Transvaal (rainfall 695 mm), the peak biomass of the grass layer amounts to 4.8 t ha^{-1} above ground, 0.9 t ha^{-1} as ground litter, and 7.8 t ha^{-1} below ground (Huntley and Morris, 1979). Near Zaria, Nigeria, Haggar (1970) reported a peak biomass of 3.8 t ha^{-1} in an *Andropogon gayanus* community (rainfall 1000 mm).

In more humid regions, the maximum standing crop of the herbaceous layer usually reaches much higher values. In a northern Guinean savanna of Nigeria (rainfall 1200 mm), Egunjobi (1973) found a peak biomass value of 19.5 t ha^{-1}. Further south, near Lagos, (rainfall 1600 mm), Adegbola (1964) recorded maximum standing crops of 14.8 t ha^{-1} for an *Andropogon gayanus* community, and of 20.8 t ha^{-1} for a *Pennisetum purpureum* grassland. At Lamto, Ivory Coast, the peak living standing crop varies according to the facies of the forest savanna mosaic concerned (Menaut and César, 1979). It averages 6.9 t ha^{-1} in the *Loudetia simplex* grass savanna and 5.5 to 7 t ha^{-1} in the *Hyparrhenia* open shrub savanna. The total biomass (above- and below-ground) of these two facies averages 23.5 t ha^{-1} and 14.2 t ha^{-1}, respectively.

A number of figures are also available for various grassland types in northern India (see Table 28.1). Peak values are low (0.7 t ha^{-1}) in the *Dactyloctenium–Cenchrus* pastures at Pilani (rainfall 390 mm), higher (4.6 t ha^{-1}) in the *Dichanthium–Cymbopogon* savannas of Ujjain (rainfall: 900 mm), and exceed 20 t ha^{-1} in *Cynodon–Dichanthium* grasslands around Varanasi (rainfall 1200 mm).

Unfortunately there is very little information on most Latin-American savannas. San Jose and Medina (1976) and Medina (1980) reported that the maximum values for the above-ground herbaceous biomass ranged from 2 to 6 t ha^{-1} in the *Trachypogon* savannas of the Venezuelan *llanos*; the corresponding below-ground peak biomass reached 2.3 t ha^{-1}. In the flooded savannas of Venezuela, the maximum above-ground living biomass varied from 4.3 t ha^{-1} in the *banco* to 9.1 t ha^{-1} in the *estero*, the total peak biomass ranging from 7 to 9.2 t ha^{-1} (Escobar and Gonzalez-Jimenez, 1979).

---

[1]All figures are dry weight, unless otherwise stated.

TABLE 28.1

Maximum and minimum values of above-ground live biomass (t ha^{-1}) for some Indian grasslands (after Singh and Joshi, 1979)

Site	Grassland type	Minimum biomass	Maximum biomass
Rajkot	*Cenchrus ciliaris*	0	2.28
Jodhpur[1]	mixed grass	0.07	1.64
Pilani	mixed grass	0.035	0.76
Kurukshetra	mixed grass	1.05	19.74
Delhi[1]	*Heteropogon*	0	7.71
Varanasi[1]	*Eragrostis–*	8.71	32.96
	Desmostachya	5.73	23.60
Jhansi[1]	*Sehima–Heteropogon*	4.96	14.08
Ujjain	*Dichanthium*	0.24	4.57
Ratlam	*Sehima*	0.01	3.63
Ambikapur[1]	mixed grass	1.23	4.23
Sagar	*Heteropogon–*	0.14	5.72
	Dichanthium	0.11	3.37

[1]Includes standing dead shoots also.

Net annual herbaceous production

The harvest techniques used by various investigators to estimate the above-ground production of the grass cover are very diverse. They range from a simple harvest of the grass cover at the end of its growing period to a precise measurement of the seasonal changes in biomass, or crop growth-rates, of every species belonging to the community. As shown by Singh et al. (1975), some of these techniques tend to underestimate the actual net production while others tend to overestimate it. Comparison of published data must therefore be approached with caution.

Some data have been obtained from studies designed to investigate the effects of grazing by wild or domestic ungulates upon the vegetation (e.g., César, 1978; Strugnell and Pigott, 1978). Here, the heights of cutting were based on the actual heights to which particular animal species reduce the sward, e.g. 20 cm above the ground for elephants (*Loxodonta africana*), and 5 cm for kob (*Kobus kob*), hippopotamus (*Hippopotamus amphibius*) or buffalo (*Syncerus caffer*). The extent to which such monthly clippings stimulate or reduce net grass production is not yet clear; however, it does allow precise knowledge to be gained of the monthly variations in growth rates (Fig. 28.1).

Estimation of net below-ground primary production is even more difficult than measurement of root biomass. The more precise techniques now available cannot be used in field conditions, and it is quite unrealistic to rely upon assumed fixed ratios between standing-crop and production figures. Many published values are actually no more than sensible guesswork.

So far, most of the more comprehensive studies on above-ground primary production in tropical savannas and grasslands have been carried out in Africa and India. Their results are, on the whole reasonably consistent.

At Fété Olé, in the Sahel (rainfall 200 mm) primary production only occured during a 70-day period, from August to early October 1970 (Bille and Poupon, 1972). However, production differed greatly from one community to the other: 0.8 t ha^{-1} yr^{-1} in *Aristida mutabilis* grasslands on the top of former dunes; 2.6 t ha^{-1} yr^{-1} in *Chloris prieurii* shrub savannas on slopes; and 4.8 t ha^{-1} yr^{-1} in *Panicum humile* communities, on low-lying wet areas. The estimated above-ground production, in this case, was about 20% greater than peak biomass values. Total net production averaged 3.6 t ha^{-1} yr^{-1} in the study area as a whole.

In a semi-arid East African savanna, Harris (1972) recorded production figures of c. 1.5 t ha^{-1} yr^{-1} for *Chloris roxburghiana* communities in dry areas (rainfall 150 mm), and c. 3 t ha^{-1} yr^{-1} in *Themeda triandra* savannas under more humid conditions (rainfall 300 mm). A similar situation is found in the

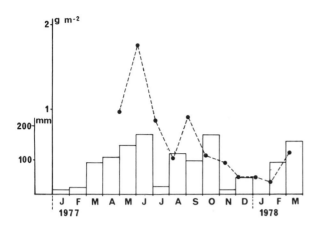

Fig. 28.1. Mean daily rates of production of grass (dry weight) in the Lamto savanna, measured by analysis of successive cuts at 5 cm above the ground, at the plateau site. Monthly rainfall is shown on histograms. (After César, 1978.)

Serengeti National Park (Braun, 1973) where grass production ranges from 0.5 to 3.0 t ha^{-1} yr^{-1} in the drier areas to over 12 t ha^{-1} yr^{-1} in the more humid parts.

The influence of increased rainfall is also apparent in the northern Sudanian savannas of West Africa. According to Breman (1975), herbaceous primary production around Bamako, Mali averages 2.6 t ha^{-1} yr^{-1} in the drier areas (rainfall 800 mm) and rises to 3.2 t ha^{-1} yr^{-1} in areas where rainfall reaches 1100 mm.

Production figures are also greatly influenced by rainfall in southern Guinean savannas. At Lamto, Ivory Coast (rainfall 1300 mm) above-ground production reaches 8.3 t ha^{-1} yr^{-1} in *Loudetia simplex* communities and 12.8 t ha^{-1} yr^{-1} in *Hyparrhenia* communities. In both cases the growth of grasses continues during most of the year, and the grass cover is burnt in January. Total net production (above- and below ground) ranges from 21.5 to 35.8 t ha^{-1} yr^{-1} (Menaut and César, 1979). Working near Mokwa, Nigeria (rainfall 1175 mm) Ohiagu and Wood (1979) have found an annual above-ground production by grass of only 2.7 t ha^{-1} yr^{-1} in ungrazed plots, and 3.1 t ha^{-1} yr^{-1} in grazed ones in an *Andropogon–Hyparrhenia* savanna woodland.

Available estimates of the yearly production of the herbaceous layer of southern African savannas and managed grasslands have been reviewed by Rutherford (1978). In areas where the mean annual rainfall averages less than 500 mm, very low production figures are obtained: 0.2 to 0.6 t ha^{-1} yr^{-1} in various communities of the Etosha National Park in South-West Africa (Namibia); 0.4 to 0.8 t ha^{-1} yr^{-1} in northeastern Botswana; 0.6 to 1.1 t ha^{-1} yr^{-1} in the Kalahari thornveld, South-West Africa; 0.7 to 0.8 t ha^{-1} yr^{-1} in the arid parts of western Transvaal, South Africa.

Most of the areas with an herbaceous production falling roughly between 1.0 and 2.0 t ha^{-1} yr^{-1} lie in the northerly parts of the savanna woodland zone of southern Africa. Production values over 20 t ha^{-1} yr^{-1} have been reported in Zimbabwe, Malawi, Zambia, southern Angola and Zaire. In the *miombo* woodland near Lumumbashi, Zaire, Fresson (1973) has reported a herbaceous production of 2.2 t ha^{-1}during one growing season. In southern Angola, the production of open grasslands with some *Themeda* and *Loudetia* spp. ranged from 3.3 to 4.8 t ha^{-1} yr^{-1}.

Net annual primary production has been measured in a number of Indian savannas (Singh and Misra, 1978; Singh and Joshi, 1979; see also Misra, Ch. 7). In the more arid areas (rainfall <400 mm) the above-ground production is low, 1.6 t ha^{-1} yr^{-1} at Jodhpur, 2.2 t ha^{-1} yr^{-1} at Pilani, and 2.4 t ha^{-1} yr^{-1} at Rajkot for instance. But the below-ground production is higher, c. 2.9 t ha^{-1} yr^{-1} at Pilani. In more humid areas, there is a steady increase in the primary production values. Total net production reaches 35.4 t ha^{-1} yr^{-1} at Kurukshetra (rainfall 790 mm), and ranges from 22 to 45.5 t ha^{-1} yr^{-1} around Varanasi depending on the grassland type and location (upland versus lowland grasslands). In both localities, the above-ground component of the net primary production exceeds the below-ground component: 24.1 versus 11.3 t ha^{-1} yr^{-1} at Kurukshetra, 33.9 versus 11.6 t ha^{-1} yr^{-1} and 22.2 versus 13.8 t ha^{-1} yr^{-1} at two locations around Varanasi.

Very few figures are available for Neotropical savannas. According to Medina (1980), the above-ground net grass production ranges from 2 to 5.7 t ha^{-1} yr^{-1} in *Trachypogon* savannas of the cerrado type, and from 4.3 to 9.1 t ha^{-1} yr^{-1} in the flooded savannas of the banco–estero type. In the flooded *Paspalum fasciculatum* grasslands, the above-ground grass production is much higher, from 10.1 to 25.4 t ha^{-1} yr^{-1}.

Net annual woody production

Available data are very scanty. The net annual production of the woody component of the Fété Olé study area in northern Senegal has been studied by Bille and Poupon (1972) and Poupon (1979). Whereas the total woody biomass of this Sahelian savanna averages 5.55 t ha^{-1} (ranging from 2.35 t ha^{-1} on top of the old dunes to 24 t ha^{-1} in low-lying areas), the mean annual production of woody material (above- and below-ground) is only 0.42 t ha^{-1} yr^{-1}, and that of leaves and fruits is 0.12 t ha^{-1} yr^{-1}. This is very low compared with the total grass production (3.63 t ha^{-1} yr^{-1}).

In the southern African savannas studied by Rutherford (1979) and in some of those studied by Kelly (1973), both receiving roughly 500 mm mean annual rainfall, the production of the leaf and current twig component of woody species is of the order of 1.5 t ha^{-1} yr^{-1}, and a girth increment of trunk and branches averaging 0.6 t ha^{-1} yr^{-1}.

Woody production is much higher in the southern Guinean savanna at Lamto (César and Menaut, 1974; Menaut, 1977). The estimates of net annual woody production in an open shrub savanna reach 3 t ha^{-1} yr^{-1}, out of which 2.3 t ha^{-1} yr^{-1} consists of deciduous organs (leaves,flowers and fruits), and 0.2 t ha^{-1} yr^{-1} are roots.

In the Chandraprabha sanctuary, India (rainfall 1057 mm), Singh and Misra (1978) and Misra (Ch. 7) reported an annual woody production of 2.6 t ha^{-1} yr^{-1}, of which 1.5 t ha^{-1} yr^{-1} were deciduous organs, and 1 t ha^{-1} yr^{-1} were roots.

Seasonal variations of primary production

This is one of the most striking characteristics of the savanna ecosystem. This is not to say that primary production is not seasonal in rain-forest environments; some seasonality can be traced everywhere, even in the most apparently stable climates. But it is in savannas that seasonality has the most far-reaching implications.

As shown in Fig. 28.2 seasonality in rainfall not only affects the amount of grass available to herbivores, but also influences the quality of the available food — in this case, in Uganda. both the digestible fraction of the food and the proportion of crude protein (Strugnell and Pigott, 1978). These seasonal changes in quantity and quality of food compel the animal consumers to adjust their life-cycles, and particularly their breeding activities, to

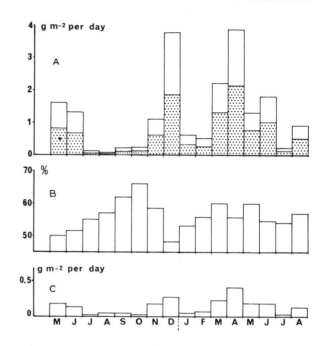

Fig. 28.2. A. Mean daily rates of production of organic matter (g m^{-2} day^{-1}) in ungrazed grassland of *Sporobolus pyramidalis* and *Chloris gayana* in Uganda, measured by analysis of successive cuts at 5 cm above the ground (digestible fraction stippled, residue unshaded). B. Percentage digestibility. C. Crude protein production. (After Strugnell and Pigott, 1978.)

the local seasonal pattern of primary production.

As mentioned earlier, it is in the most arid savannas that seasonality in primary production is the most extreme; in the Sahel it can be restricted to less than three months (Bille and Poupon, 1972). Seasonal changes in production are far less pronounced in more humid savannas, such as those of Lamto — although they are quite noticeable.

Too little is known at present about seasonal variations in root production to allow any conclusions to be reached. Root growth is apparently synchronous with shoot growth in the drier areas, whereas it takes place mostly at the end of the rains in more humid environments, the underground reserves being actively used up at the beginning of the rains. However, the regular occurrence of burning can blur the whole picture (see Ch. 30).

Compartmental transfers

The transfer of dry matter through the various compartments of the grass layer has been estimated for a few semi-arid, dry subhumid and moist subhumid grasslands in India (Singh and Joshi,

1979). While the net accumulation in the root compartment was twice as great as in the shoot compartment in the semi-arid Pilani grassland, the reverse was true for the dry subhumid Kurukshetra grassland; in the moist subhumid Sagar grassland, on the other hand, the share of roots and shoots was almost equal.

System transfer functions have also been calculated on the basis of annual production values for some of the Indian sites. While in comparatively dry grasslands such as Pilani or Rajkot, 65 to 97% of total net production was channelled underground, in more humid grasslands, only 26 to 48% of total production was used for root growth

Biomass turnover

Biomass turnover, or $P/\bar{B}$ coefficient (Petrusewicz and Macfadyen, 1970) represents the fraction of biomass which is exchanged during the time interval T. When all the above-ground standing crop of grass is burnt once a year in a tropical savanna, the yearly rate of above-ground turnover obviously exceeds unity. At Lamto, for instance, it ranges from 1.3 to 2.0 depending upon the facies of the vegetation; the lower value corresponds to the *Loudetia* savanna, and the upper value to the tree- and shrub-facies of the *Hyparrhenia* savanna. The turnover rate of the below-ground grass biomass quite likely approximates unity. This means that the turnover rate for the total grass biomass is probably close to 1.5.

The reciprocal of the $P/\bar{B}$ ratio can also be estimated. It represents the turnover time for the standing crop. At Lamto it requires approximately eight months for the grass compartment of the savanna to renew itself. Obviously the tree and shrub compartments take much longer (10–15 years?), though the leaves shed annually represent a significant part of the yearly production.

According to Singh and Joshi (1979), the turnover rate of underground plant material in tropical grasslands seems to be quite rapid. An average annual rate of 93% has been reported for the semi-arid site of Pilani, and another of 97% for the dry subhumid grassland of Kurukshetra. This is ascribed to the presence of a large number of rainy season ephemerals, strong seasonality of climate in these study sites, and high temperatures.

Efficiency of energy capture

The proportion of the incident solar radiation at the earth's surface (half of that received from the sun) which is fixed in net primary production (Singh and Misra, 1968) has been calculated for a number of Indian sites. The values range from a minimum of 0.23% (for the growing season only) for the semi-arid grassland at Pilani, to a maximum of 1.66% for the dry subhumid grassland at Kurukshetra (Singh and Joshi, 1979). Most tropical grasslands show a higher efficiency of energy capture than temperate grasslands (Sims and Singh, 1971). This is due to several reasons. Most of the sun-loving tropical plants presumably possess the C^4-pathway of carbon-dioxide assimilation, and have higher water-use efficiency, little or no light saturation, a higher range of optimal temperatures for photosynthesis, and no photorespiration (Singh and Joshi, 1979).

Water-use efficiency

There is a general lack of data concerning the actual amounts of water used by tropical grasslands. A crude index of water-use efficiency can be derived by comparing annual above-ground net production (g m^{-2}) with annual rainfall (mm). The ratio of these two parameters ranges from 0.1 to 4.0 in tropical savannas and grasslands. The efficiency of water use has some tendency to decline in areas of high rainfall (Singh and Joshi, 1979).

On the whole, primary production in tropical savannas is strongly correlated with rainfall (Fig. 28.3). For example, in South-West Africa (Namibia) Walter (1964) has found a linear correlation between annual rainfall and primary production; production increases by one ton for each extra 100 mm of rainfall. Braun (1973) also shows that the average grass yield per site, in the Serengeti National Park, and the variation in yield as well, increase with amount of rainfall. However, every vegetation type has its particular relationship to rainfall.

Nevertheless, such a correlation mostly holds true for lower rainfall values, that is in the more arid areas. In most places it is apparently no longer valid for annual rainfall values exceeding 600–800 mm. In the Ivory Coast, for instance, Talineau (1970) did not find any significant relationship between grass production and rainfall in his experimental grass plots. This is well in line

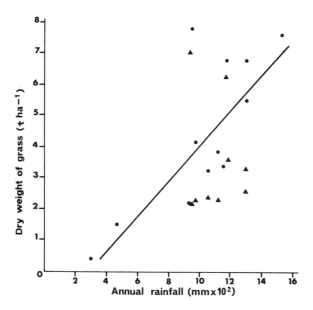

Fig. 28.3. Relationship between rainfall, maximum grass biomass and grass production in Central and West Africa. (After Ohiagu and Wood, 1979.) Black circles = peak biomass values; black triangles = estimated annual production.

with the tendency of water-use efficiency to decline in areas of high rainfall.

Furthermore, the time at which the rain falls must also be taken into consideration, as well as the micro-relief and the water-holding capacity of the soil. Widely spaced heavy showers can be much less productive than the same amount of water falling steadily over a more limited period. The production figures at Fété Olé tend to substantiate these points.

The influence of soil characteristics

For a given value of annual rainfall, and a given species of plant, the yield can vary considerably with the structure and nutrient content of the soil. On the Serengeti Plains, for instance, Schmidt (1975) has shown that the various grass communities are closely associated with the carbon and nitrogen content of the soil. Furthermore, the above-ground grass production along a soil catena is largely independent of the local climatic gradient (Anderson and Talbot, 1965; Braun, 1973).

The effects of system perturbations

Burning and grazing have a considerable effect on primary production of tropical savannas and grasslands. Chapters 30 and 31 deal with these problems in detail. Suffice it to say that these

effects vary a great deal, depending upon the intensity and periodicity of the perturbations, as well as on the length of time during which they took place in the areas under study.

In general, grazing creates a more open herbaceous layer and makes the invasion of annuals and aliens possible, thus increasing species diversity. When grazing is heavy, unpalatable plants often invade the grasslands; if it continues for a long time, it decreases the basal cover of grasses and triggers soil erosion. This leads to a decrease in net primary production. On a number of sites in the Ivory Coast, César (1978) has shown that repeated clipping of the above-ground herbaceous vegetation throughout the year leads to a diminution of total annual grass production to only half the normal values.

Conversely, moderate grazing in humid climates often increases the above-ground net primary production through enhanced tillering and increased diversity (Singh, 1968).

The effect of fire on herbaceous productivity has often been discussed, though reliable quantitative data are few. In some cases, it appears to be rather favourable, at least on a short term basis.

The consumers trophic level

To estimate the energy flow through the animal component of a savanna ecosystem, one has to measure: (1) the size of the animal populations and their seasonal variations throughout the annual cycle; (2) the intake of plant material by herbivores, and their assimilation efficiency; (3) the proportion of this intake that is transferred to carnivores, broadly speaking, and from all consumers to the detritus food web as excreta and dead organic material; and (4) the losses from the ecosystem through the respiration of consumers.

The production of consumers must also be evaluated — that is, the amount of animal tissue synthesized during the year, both through growth of the parental generations and through the production of new individuals.

The large number of species involved, particularly in invertebrates, necessitates the selection of a few key species, representative of the major "life-styles" in each trophic category, and the arbitrary application of the findings to each category as a whole.

Population numbers and biomass

As described in many chapters of this book, it is possible to get reliable estimates of numbers and standing crop values for most of the savanna vertebrates and some of the larger arthropods and other invertebrates. This is unfortunately not the case for the soil micro-organisms, which play such a dominant role in energy flow. The numbers of cells detected by traditional plating methods are often much smaller than the total numbers actually present, as determined by direct-counting techniques. An effort has been made to separate micro-organisms into "functional groups", but this has proved hard to achieve in practice, as some of these groups are difficult to separate. Therefore, the activity rate of micro-organisms in their role as decomposers has often been measured by carbon-dioxide output. In field conditions, however, this gas is also emitted by plants and animals, and the validity of the estimates of decomposition rates from soil carbon-dioxide emission depends on the rates of respiration by roots and soil animals, which are seldom measured. Hence, decomposition rates have often been estimated from the rate of weight loss of buried portions of litter or filter paper.

Trophic ecology of main consumer groups

The first step is to establish the energy budget of individual populations representative of the most numerous trophic groups — herbivores, carnivores and detritivores. To achieve this aim, one has to determine as accurately as possible the relevant components of the energy budget: consumption (I), assimilation (A), maintenance cost or respiration (R), as well as production (P), which includes production due to reproduction (P_r) and production ascribable to body growth (P_g). This is necessary in order to compare the efficiencies of different species belonging to different taxonomic categories and to different trophic groups (Figs. 28.4 and 28.5). The indices of efficiency most commonly used for such comparisons are the assimilation efficiency (A/I), the production efficiency (P/A), the ecological efficiency (P/I), and the index of efficiency of biomass production or biomass turnover rate ($P/\bar{B}$). In this latter index, $\bar{B}$ represents the average biomass during the year.

The **assimilation efficiency** varies a great deal among species, depending both upon their taxonomic status and physiological characteristics, and

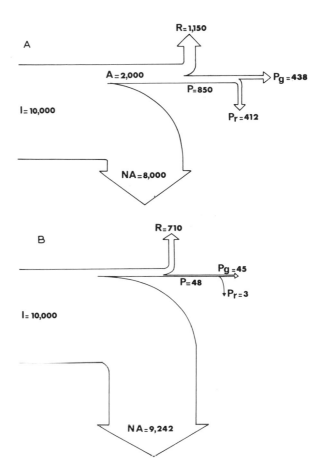

Fig. 28.4. Diagrammatic representation of the energy flow through populations of two invertebrates at Lamto (Ivory Coast). A. An insect herbivore, the grasshopper *Orthochtha brachycnemis*. B. A soil detritus feeder, the earthworm *Millsonia anomala*. (After Gillon, 1973; Lavelle, 1978.) NA = non-assimilated energy; for other symbols, see text.

also upon the trophic category to which they belong. Even within a given species it can vary according to the developmental stage of the animal, and to the seasonal changes in its food habits and in the nutritive value of its staple diet. Furthermore, many free-living animals, particularly vertebrates, may exercise a considerable degree of dietary selection. Preferences for particular plant species, parts of plants and for living versus dead herbage have been described for many herbivorous species, from grasshoppers to domestic ungulates. Therefore, the intake and assimilation values obtained in laboratory feeding experiments, where dietary selection is restricted, can only represent approximations to what actually occurs under field conditions. It is all the more important to bear in mind the point that

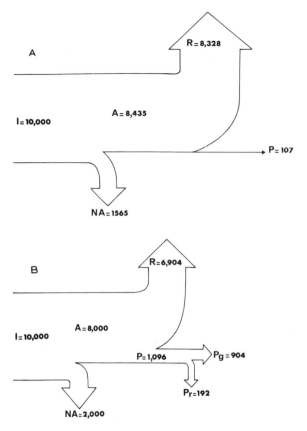

Fig. 28.5. Diagrammatic representation of the energy flow through populations of two savanna vertebrates. A. A grazing ungulate, the kob, *Kobus kob*. B. A carnivorous lizard from Lamto. *Mabuya buettneri*. (After Buechner and Golley, 1967; and Barbault, 1974.) Symbols as in Fig. 28.4.

laboratory estimations of food intake and respiration do not include the energy cost of food search, which may be an important factor in the energy balance of certain species.

Estimation of the average A/I ratio in natural populations is consequently a most delicate task. The figures obtained in the laboratory on individuals belonging to different developmental stages must be extrapolated to the population as a whole, taking into account the variation in age structure of the consumer population during the year, as well as seasonal changes in its diet. Some examples for a few savanna animals are given in Table 28.2. They are very similar to the figures published by Hutchinson and King (1979) for temperate grassland animals. The mean assimilation efficiency is highest for carnivores, still very high in herbivores, and much lower in detritivores.

The **production efficiency** values for savanna animals listed in Table 28.2 also correspond quite well with those calculated by authors working on temperate-zone species (McNeill and Lawton, 1970; Humphreys, 1979; Hutchinson and King, 1979). Production efficiency is very high in invertebrate and cold-blooded vertebrate carnivores, high in invertebrate herbivores and very low in warm-blooded herbivores and invertebrate detritus feeders. The difference between homoeothermic and poikilothermic herbivores is particularly striking. As May (1979) has put it, warm-blooded animals, relative to cold-blooded ones, pay a marked cost in order to keep their metabolic machinery ticking over at a constant rate.

The **ecological efficiencies** in Table 28.2 range from 0.005 to 0.50. Again, these figures are consistent with the values given by Hutchinson and King (1979) for temperate grassland animals. The lowest P/I ratio so far observed in the animal kingdom is that of the savanna earthworm *Millsonia anomala* (Lavelle, 1978). This ratio can be explained both by the very low nutritive value of the soil ingested by this worm — 376 J g^{-1} as against 2561 J g^{-1} for the litter ingested by the temperate *Allobophora rosea* (Philippson and Bolton, 1976) — and by the higher temperatures of tropical savanna soils, which increase the respiration of the animals. Moreover, mucus production has not been estimated; if it had been estimated, then the production figure would probably be twice as great as that given.

Among vertebrates, ecological efficiency is again higher in poikilothermic than in homoeothermic herbivores. In the latter category, it mostly depends on the body size of the animals. Values reported for African wild ungulates range from 0.01 to 0.05 (Petrusewicz and Macfadyen, 1970).

The **index of efficiency of biomass production**, or biomass turnover rate per year, is also very variable, ranging from 0.27 in the Uganda kob to 9.7 in fungus-growing termites. In a balanced system, it can be a helpful short-cut in compiling energy budget, if a relation between P and $\bar{B}$ can be confidently established for a given population, as only one of the two quantities needs to be measured directly. Generally speaking, the turnover rate within a given taxonomic group is all the more rapid as the number of generations per year is higher. For example, the rate ranges from 3 to 4 in

TABLE 28.2

Energy conversion parameters, at population level, for some tropical savanna animals

	Assimilation efficiency A/I	Production efficiency P/A	Ecological efficiency P/I	Turnover rate $P/\bar{B}$ year	Source
HERBIVORES					
Grasshoppers					
Burkea savanna, various spp.	0.32	0.19	0.057		Gandar (1979)
Acacia savanna, various spp.	0.32	0.21	0.068		Gandar (1979)
Orthochtha brachycnemis	0.20	0.42	0.085	9.6	Gillon (1973)
Caterpillars					
Cirina forda	0.43	0.15	0.064		Gandar (1979)
Herbivorous termites					
Trinervitermes geminatus			0.09	1.04	Josens (Ch. 23)
Ancistotermes cavithorax (fungus growing)			0.018	9.7	Josens (1973)
Hodotermes mossambicus	0.61				Nel et al. (1970)
Ungulates					
Uganda kob (*Kobus kob*)	0.84	0.01	0.01	0.27	Buechner and Golley (1967)
impala (*Aepyceros melampus*)[1]	0.59	0.04	0.024		Gandar (1979)
domestic cattle (*Bos taurus*), Transvaal	0.57	0.05	0.023		
african elephant (*Loxodonta africana*)	0.30	0.02	0.005		Petrides and Swank (1965)
CARNIVORES					
Spiders					
Orinocosa celerierae[1]	0.95	0.53	0.50		Celerier (pers. comm.)
Afropisaura valida				6	Blandin (1979)
Amphibian					
Nectophrynoides occidentalis	0.83	0.21	0.18	1.3	Lamotte (1972)
Lizard					
Mabuya buettneri			0.11	3	Barbault (1974)
DETRITIVORE					
Earthworm					
Millsonia anomala	0.09	0.04	0.006	2	Lavelle (1978)

[1]Values based upon the study of a single individual.

monovoltine grasshopper species and reaches 10 to 12 in those having three breeding seasons per year. For invertebrates, the rate also varies seasonally and from year to year, depending upon the environmental conditions.

Energy flow through successive trophic levels

So far we have attempted to trace energy flow through individual populations of consumers, and to assess the relative importance of different species within a given trophic level. The next step in ecosystem analysis is to quantify the flow of energy between different compartments, from green plant producers to top animal predators. This allows one not only to estimate the amount of energy that is routed through above-ground invertebrate and vertebrate consumers, but also to partition more accurately the relative activity of micro-organisms and invertebrates in the release of energy below the soil surface.

Very few attempts have been made to estimate the energy flow through the various levels of animal consumers in tropical savannas. To date, the most elaborate study has been carried out in the Lamto savanna by Lamotte (1977, 1978) and his associates.

Lamto is located within the southern Guinean savanna belt, in the forest savanna mosaic of the Ivory Coast. As mentioned earlier, the herbaceous cover mostly consists of perennial grasses (*Hyparrhenia* and *Loudetia* facies), interspersed with a number of trees and shrubs and also some tall *Borassus* palms. The grass is burnt yearly during the dry season, except in a few protected study areas. Average daily temperatures are always high, ranging from 24 to 30°C. Annual rainfall averages 1250 mm, but can vary from 900 to 1700 mm. The main dry season, four months long, extends from November to March, with a short dry season regularly occurring during a few weeks in July and August.

The average *total* plant biomass is about 66 t ha^{-1}, out of which the biomass of herbaceous plants and palms each accounts for 17.1 t ha^{-1}, and the tree and shrub standing crop for 31.8 t ha^{-1}. The above-ground plant biomass amounts to 42.6 t ha^{-1} and the below-ground compartment to 23.4 t ha^{-1}, of which 10.1 t ha^{-1} is composed of grass roots.

The total net production, below-ground as well as above-ground, of the palm savanna averages $476 \times 10^6 \text{ kJ ha}^{-1} \text{ yr}^{-1}$. This represents a photosynthetic efficiency of 0.67%, which compares with the rates commonly found in temperate grasslands (Coupland, 1979).

Of this annual primary production 6 to 8 t are destroyed by bush fires. This is about half of the above-ground primary production.

Grasshoppers, Homoptera, foraging termites and rodents are the most important primary consumers in this kind of savanna; it is also likely that certain ant species and caterpillars play a significant role. As shown in Table 28.3 these herbivores do not consume more than $260 \times 10^6 \text{ kJ ha}^{-1} \text{ yr}^{-1}$. Therefore the proportion of energy that is routed through above-ground consumers is very low.

On the other hand, the role of litter and soil invertebrates and micro-organisms is very important. Two groups of invertebrates are particularly active in the Lamto savanna, the earthworms and the fungus-growing termites. Earthworms alone consume about $230 \times 10^6 \text{ kJ ha}^{-1} \text{ yr}^{-1}$ and fungus-

growing termites more than $16 \times 10^6 \text{ kJ ha}^{-1} \text{ yr}^{-1}$. However, at this trophic level, production remains low in all the above-mentioned taxonomic groups ($2.51 \times 10^6 \text{ kJ ha}^{-1} \text{ yr}^{-1}$). Their ecological efficiency is also low, particularly that of the earthworms whose P/I ratio averages 0.006. This is due to the fact that most of the ingested plant material is not assimilated; some $226 \times 10^6 \text{ kJ ha}^{-1} \text{ yr}^{-1}$ is excreted and thus remains available for other detritus feeders.

The amount of energy lost by respiration (R), and estimated by the difference between A and P, is low: $c. 31 \times 10^6 \text{ kJ ha}^{-1} \text{ yr}^{-1}$. This is due to the fact that most primary consumers in the Lamto savanna ecosystem are poikilothermic invertebrates, with low maintenance costs.

Utilization of the energy produced by the plant compartment of the ecosystem can therefore be summarized in a diagrammatic way as follows (figures in $\times 10^6 \text{ kJ ha}^{-1} \text{ yr}^{-1}$):

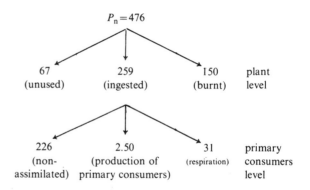

$$P_n = 476$$

| 67 | 259 | 150 | plant |
| (unused) | (ingested) | (burnt) | level |

| 226 | 2.50 | 31 | primary |
| (non-assimilated) | (production of primary consumers) | (respiration) | consumers level |

The biomass of secondary consumers amounts to $217 \times 10^6 \text{ kJ ha}^{-1} \text{ yr}^{-1}$, with most of the standing crop accounted for by driver ants (Table 28.4). The various levels of "carnivores" consume $c. 5.4 \times 10^6 \text{ kJ ha}^{-1} \text{ yr}^{-1}$ and produce $1.45 \times 10^6 \text{ kJ ha}^{-1} \text{ yr}^{-1}$. Respiratory losses are low (ca. $2.3 \times 10^6 \text{ kJ ha}^{-1} \text{ yr}^{-1}$), most of the species being poikilotherms. The amount of rejecta they produce has been estimated at $1.7 \times 10^6 \text{ kJ ha}^{-1} \text{ yr}^{-1}$. Assimilation efficiency and ecological efficiency are high, contrasting with the situation observed for herbivores.

In the Lamto savanna, there are at least two levels of secondary consumers, and one can assume that the flow of energy is partitioned among them in the following way (figures in $\times 10^6 \text{ kJ ha}^{-1} \text{ yr}^{-1}$):

TABLE 28.3

Average biomass ($\bar{B}$), annual production (P), consumption (I), and energy budget of some major groups of primary consumers in the Lamto savanna (after Lamotte, 1977), estimates in kJ ha^{-1} for biomass, and kJ ha^{-1} yr^{-1} for production and consumption

	$\bar{B}$	P	I	P/I	$P/\bar{B}$	$I/\bar{B}$
Rodents	5225	14 600	710 000	0.02	2	100
Grasshoppers	11 700	83 600	1 254 000	0.06	7	104
Termites (fungus growing)	23 400	234 000	16 720 000	0.014	10	714
Earthworms	752 400	1 567 000	2 299 000 000	0.005	1.8	450

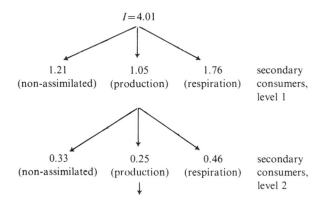

The importance of the production of the first level of secondary consumers is at first sight surprising (1.05 × 10^6 kJ ha^{-1} yr^{-1} as opposed to 2.50 × 10^6 kJ ha^{-1} yr^{-1} for primary consumers). The material ingested (4.01 kJ ha^{-1} yr^{-1}) is also larger than the production of sympatric herbivores (2.50 kJ ha^{-1} yr^{-1}). This is probably due to the fact that many "carnivorous" invertebrates feed upon the excreta of the herbivores, and also on many microscopic detritus feeders living on these rejecta.

The importance of "wastage" must not be forgotten either: many herbivorous animals, from grasshoppers to monkeys, destroy as much plant material as they actually consume, and often more.

At Lamto, for instance, the grasshopper *Rhabdoplea klaptoczi* ingests less than half the grass blades that are actually cut with its mandibles (Gillon, 1968). The rest is immediately incorporated into the litter, and consequently made directly available to detritus feeders and decomposers without passing through the other categories of microscopic consumers. The energy equivalent of the plant material which is not used by herbivores has been estimated at 67 × 10^6 kJ ha^{-1} yr^{-1}, and that of the unassimilated organic material available in the excreta of various herbivores and detritus feeders at 226 × 10^6 kJ ha^{-1}yr^{-1}. This makes a grand total of 293 × 10^6 kJ ha^{-1}yr^{-1} which is left at the disposal of micro-organisms, together with the bodies of animals that die.

The study of soil actinomycetes, fungi and bacteria is still in progress at Lamto, but some estimates of their overall metabolism have been made (Schaefer, 1974a,b), using a soil respiration method. The respiration of all soil detritus feeders and decomposers has thus been estimated to reach 267 × 10^6 kJ ha^{-1} yr^{-1}, out of which only 25 × 10^6 kJ ha^{-1} yr^{-1} is due to earthworms. The difference (242 × 10 kJ ha^{-1} yr^{-1}) is most likely ascribable to micro-organisms.

As there is about 293 × 10^6 kJ ha^{-1} yr^{-1} of or-

TABLE 28.4

Average annual biomass($\bar{B}$), production (P), consumption (I) and energy budget of a few major groups of secondary consumers in the Lamto savanna (after Lamotte, 1977); estimates are in kJ ha^{-1} yr^{-1}

	$\bar{B}$	P	I	P/I	$P/\bar{B}$	$I/\bar{B}$
Birds	2341	920	52 250	0.02	0.4	22
Lizards	585	1840	18 390	0.10	2.8	30
Mantodea	2424	14 630	43 890	0.33	6.0	18
Spiders	13 376	96 140	313 500	0.30	7.0	23
Driver ants	200 640	1 254 000	5 016 000	0.25	6.2	25

ganic matter which remains unused by the various animal groups present in the system, and as this system is apparently stable, it would appear that about 51×10^6 kJ ha^{-1} yr^{-1} is decomposed above the ground level — that is, that some decomposition of leaves and stems takes place *before* this dead plant material drops to the soil surface. This proportion of energy that does not appear in the litter layer amounts to about 16% at Lamto. This figure is well in line with those found in other grassland studies, which range from 4% in some Indian sites to 39% in natural temperate humid grasslands (Coupland, 1979).

Tropical savannas, especially in Africa, are rightly famous for their richness in "game" species. This is unfortunately not the case in the southern Guinean savanna of Ivory Coast, Lamto included, where large wild ungulates are scarce and domestic livestock have not yet been successfully introduced. The average yearly biomass of wild ungulates in some East African national parks and game reserves ranges from about 1 t km^{-2} to 31 t km^{-2} (live weight). Even the llanos of Venezuela, reputedly poor in large wild animals, can harbour an important biomass of mammals. For instance, Eisenberg et al. (1979) report mammalian biomass values of 1.4 to 2.2 t km^{-2} for their two Maseguaral study sites in Venezuela, out of which 0.4 and 1.5 t km^{-2} respectively corresponded to terrestrial herbivores. The impact of such a large standing crop of herbivores on plant resources can be expected to be important, since it reduces the amount of food left at the disposal of other vertebrate and invertebrate primary consumers.

This situation has been carefully studied in the Serengeti National Park in Tanzania (Sinclair, 1975). The total annual production of the three savanna categories present in the study area, as well as the production and off-take of the major categories of primary consumers are shown in Table 28.5. This table indicates the ways in which the primary production of grass is partitioned between the various consumer groups. In the long grasslands and *kopjes*, grass fire remove over 50% of the total above-ground production. In the long grasslands, herbivores (mammals and grasshoppers) remove an additional 27%, leaving at least 20% to detritus feeders and decomposers. In the short-grass area, the same categories of herbivores take a larger proportion (38%) of the primary

production; burning is relatively unimportant and about half the plant production seems to pass directly to detritivores, mostly termites.

However, such values, computed over the whole year for available food and off-take, are somewhat misleading in that they imply that in all cases there is an excess of total food produced over total herbivore requirements. This is certainly not the case, as shown by Sinclair's (1977) study of the Serengeti buffalo.

Monthly estimates of mean food availability and population requirements are far more meaningful. In this way, it has been possible to show, in long-grass areas, that the mean total food requirement for herbivores was always in excess of the food available during the dry season (Fig. 28.6). At that time of the year, some 95% of this requirement was due to the ungulates, and the invertebrates contributed very little to the total requirement. On the other hand, the invertebrates build up in numbers during the wet season, and their impact on the grass layer must have been at least equal to that of the resident ungulates, if not greater. Small mammals remained low in numbers throughout the year. On the short-grass area, there was practically no green food available during the dry season, and the invertebrate consumers required more than all the mammals. However, during the wet period the migrant ungulates utilized the plains and accounted for over 90% of the total off-take.

The Serengeti study also highlights the key importance of the decline in the quantity of good quality food at the beginning of the dry season. Most ungulates require on the average from 4 to 5% crude protein to maintain their body weight. This can only be achieved if they selectively ingest the small quantity of green material that is available. Rodents and grasshoppers also prefer the green grass. Sinclair (1975) concludes that primary production cannot simply be equated with food, and that in the Serengeti savanna green grass is the limiting resource for all invertebrate and vertebrate grazers.

NUTRIENT CYCLING IN TROPICAL SAVANNAS

The study of nutrient cycling in tropical savannas is still in its infancy. Data have been collected in a number of tropical grasslands, mostly in India, concerning the quantity of certain nutrients in the

TABLE 28.5

Total annual grass production and herbivore off-take in the three grassland types of the Serengeti (after Sinclair, 1975)

	Long grassland		Short grassland		Kopjes	
	kg ha^{-1} yr^{-1}	% of total	kg ha^{-1} yr^{-1}	% of total	kg ha^{-1} yr^{-1}	% of total
Annual grass						
above-ground production	5978		4703		5978[a]	
Total herbivore off-take	1647	27.6	1795	38.2	865	14.4
ungulate off-take	1122	18.8	1597	34.0	122	2.0
small mammal off-take	69	1.2	4	0.1	259	4.3
grasshopper off-take	456	7.6	194	4.1	484	8.1
Removed by burning	3185	53.3	586	12.5	3430	57.4
Removed by detritivores	1146	19.2	2322	49.5	1683	28.2
Biomass (dry wt, kg ha^{-1})						
ungulates	22.90		17.90		3.39	
small mammals	0.42		0.07		2.44	
grasshoppers	0.79		0.34		0.81	
detritivores	6.80		13.8		10.0	

[a]Estimated value. No actual measurement was made and the grass production was assumed to be the same as that of the surrounding long-grass areas, the grass species and the height of the grass cover being the same (A.R.E. Sinclair, pers. comm., 1980).

soil, grass, and sometimes the shrub compartments of these ecosystems, as well as its changes throughout the year (Singh et al. 1975; Singh and Misra, 1978; Singh and Joshi, 1979; Misra, Ch. 7).

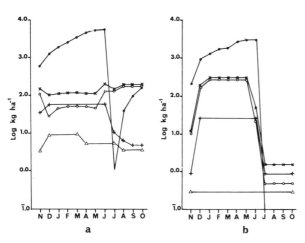

Fig. 28.6. The mean monthly available food, and food requirement by herbivores, in the long grasslands (a) and the short grasslands (b) of the Serengeti National Park, Kenya (log scale): ● = available food; ○ = ungulate requirement; + = invertebrate requirement; △ = small mammals' requirement; × = total herbivore requirement. In the long grass, requirements are in excess of available food for the dry season months (July to September), but are nearly equal in October. In the short grass, there is no measurable growth of food during the dry season and, consequently, a very low density of herbivores. (After Sinclair, 1975.)

Unfortunately, there is very little information on the situation in other continents, and it is still too early to judge whether the static models of nutrient flow which have been constructed for Indian grasslands can be considered valid for other tropical savannas. Caution is all the more necessary since Misra (Ch. 7) considers that some of the figures on nitrogen uptake for dry subhumid Indian savannas reported by Singh et al. (1975) are probably underestimates. Furthermore, the nutrients which have received the most attention are nitrogen and phosphorus; next to nothing is known about the others.

Nitrogen cycling

Nitrogen enters the soil in three different ways: through rainfall and dust, and by fixation of nitrogen.

Nitrogen is carried into the soil in rain water and is believed to be converted to nitrate and ammonia, via the action of lightning. This input of nitrogen from rain is far from negligible: it has been estimated as c. 2.3 kg ha^{-1} yr^{-1} at Fété Olé, northern Senegal (Bille, 1977), as 4 kg ha^{-1} yr^{-1} at Lamto (Villecourt, 1973), and from 2 to 5 kg ha^{-1} yr^{-1} in the Burkea savanna of southern Africa (Bate, 1979).

Dust is collected by wind from the topsoil which can be very rich in nitrogen. There is at present no indication of the possible contribution of dust deposition to nitrogen input in tropical savannas, but it might not be negligible in some situations.

Nitrogen fixation occurs by legume nodules, by intracellular association of *Azospirillum* with the roots of many tropical grasses, and by blue-green algae. More details are given in Chapter 25. Suffice it to say here, that nitrogen fixation has seldom been evaluated at the plant community level in tropical savannas. In the *Loudetia* savanna of Lamto, it averages 13 kg ha^{-1} yr^{-1}, and only 9 kg ha^{-1} yr^{-1} in the *Hyparrhenia* savanna (Balandreau, 1976). In the *Burkea* savanna of Nylsvley, Transvaal (South Africa), it has been estimated by Grobbelaar and Rösch (in Bate, 1979) to reach between 30 and 90 kg ha^{-1} yr^{-1}.

The contribution of the possible methods of fixation has not yet been estimated. Apparently legumes do not play a major role in the few African savannas studied. They are scarce at Lamto. Even in southern Africa, where many of the trees and shrubs, and a number of herbs, are leguminous, only a limited number of them have been shown to be nodulated (Bate, 1979). The role of blue-green algal mats is certainly important, as shown both in Kenya (Smith, 1977) and in the Ivory Coast (Raud, 1976). In *kikuyu* grass in Kenya the amount fixed reached 61 mg N m^{-2} day^{-1} (Smith, 1977). At Lamto, their role in nitrogen fixation is more important on wet soils, at the beginning of the rainy season. Fixation rates by *Azospirillum* have not yet been reported for natural savannas.

The **amount of nitrogen in soil organic matter** ranges from 64 g m^{-2} in Indian semi-arid grasslands, to 452 g m^{-2} in subhumid grasslands. Values reported for temperate regions vary between 300 and 500 g m^{-2}, except for a mesohalophytic meadow in the U.S.S.R. where it reaches a peak value of 1400 g m^{-2} (Coupland, 1979).

The **annual flow of nitrogen from soil to plants** ranges from 3 g m^{-2} yr^{-1} to 26 g m^{-2} yr^{-1} in the International Biological Programme (I.B.P.) and Man and the Biosphere (M.A.B.) sites in India. On the average, it is higher in dry subhumid grasslands, and lower in the semi-arid zone. In the latter area, as well as in humid grasslands, more than half (53–66%) of the nitrogen uptake is retained in the underground net production, while in the moist subhumid grasslands most (53–73%) of the total uptake is associated with the above-ground net production. The rate of nitrogen uptake by plants is highest during the rainy season in semi-arid and subhumid Indian grasslands, but this does not appear to be the case in the humid zone.

The **mean amount of nitrogen in plants and litter** was 1 g m^{-2} in the semi-arid Indian site, ranged from 12 to 17 g m^{-2} in the subhumid grasslands and reached 20 g m^{-2} in the humid grasslands. Corresponding values for wet temperate meadows vary from 12 to 36 g m^{-2} (Coupland, 1979).

The **amount of nitrogen that returns to the soil each year** obviously depends a great deal on the management of the grass cover by cattle herders. In India, it ranges from 2 g m^{-2} yr^{-1} in semi-arid grasslands, to between 10 and 15 g m^{-2} yr^{-1} in subhumid and humid grasslands, reaching a maximum of 18 g m^{-2} yr^{-1} in the fenced savanna of the Chandraprabha sanctuary. The rate of release back to the soil is highest during the winter.

Phosphorus cycling

This has also been studied on the same I.B.P. and M.A.B. sites in India as nitrogen cycling. The amount of phosphorus in the soil organic matter reaches 11.4 g m^{-2} in the fenced savanna of the Chandraprabha sanctuary, as compared to 119.8 g m^{-2} in the nearby natural forest. The annual uptake by plants ranges from 0.49 g m^{-2} yr^{-1} in semi-arid grasslands, to 4.96 g m^{-2} yr^{-1} in dry subhumid grasslands. Most of the phosphorus uptake is associated with the above-ground net production in subhumid and humid sites. Conversely, in the semi-arid grasslands, 65% of the phosphorus uptake is directed to the below-ground net production. The rate of phosphorus uptake is highest during the rainy season in all Indian eco-climatic zones. The mean amount of phosphorus in plants and litter ranges from 0.15 g m^{-2} in semi-arid grasslands to 1.88 g m^{-2} in dry subhumid grasslands; it is low (0.18 g m^{-2}) in humid grasslands. The amount of phosphorus that returns from plants to soil varies from 0.32 g m^{-2} yr^{-1} to 3.03 g m^{-2} yr^{-1}, the mimimum value corresponding to the humid grassland and the maximum figure to the dry subhumid site. In the more humid situations phosphorus tends to be locked up in dead plant material. The rate of phosphorus release from plants to soil is greater during the winter than during the monsoon.

Potassium cycling

This has been studied at Chandraprabha; its characteristics are summarized by Misra (Table 7.6).

Comparing nutrient cycling in adjacent natural forests and savannas at Chandraprabha, Singh and Misra (1978) reached the following conclusions: While the storage of nutrients in the soil is comparable in natural forests and protected savannas (except for phosphorus), the storage of nitrogen in the forest is almost eight times that of the savanna. Whereas about 80% of the total nitrogen is stored in the forest soil and 20% in plant material, in the savanna over 92% of the nitrogen is stored in the soil.

For phosphorus the situation is quite different. Most of this nutrient is stored within the soil (c. 70% in the forest, and c. 94% in the adjacent savanna). For potassium, the proportions stored in the soil are 44% in the forest and 68% in the savanna.

Nutrient uptake in the protected savanna is considerably greater than in the forest. Generally speaking, nutrient cycling is also much faster and the turnover rate in the savanna is double that in nearby natural forest for nitrogen, almost three times for phosphorus, and nearly seven times for potassium.

In the Sahelian savannas of Senegal and Mali, Bille (1977) and Penning de Vries et al. (1980) have found very low amounts of both nitrogen and phosphorus in ungrazed savanna as well as in overgrazed pastures. The amount of nitrogen available in the soil ranges from 1 to 3.5 g m^{-2} in the Mali sites, and from 12 to 144 g m^{-2} at Fété Olé, Senegal. However, this scarcity of nitrogen does not appear to be as important as the deficiency in phosphorus (7–33 g m^{-2} at Fété Olé). It is the low availability of this nutrient which was found to limit nitrogen absorption and plant yield.

CONCLUDING REMARKS

Although quantitative data on energy flow and nutrient cycling in tropical savannas are deplorably few, some preliminary conclusions can tentatively be drawn.

First of all, there is no such thing as a "typical" savanna ecosystem. Rather, there is a gradient of related ecosystems ranging from open woodlands to almost treeless "steppes". However, these systems all share a number of common features which make them similar to each other rather than different.

Although the incoming radiation is always high, primary production varies a great deal from site to site. This is largely due to differences in the amount and distribution of the annual rainfall, and also to differences in water availability resulting from local topography. In a given area, short grass will usually be found on high ground and long grass on low ground. When enough water is available, soil quality becomes a major variable; it depends not only on the nature of the parent material but also on other factors such as leaching or erosion.

Above-ground herbaceous production is sometimes very high, in humid savannas exceeding 20 to 25 t ha^{-1} yr^{-1}, and occasionally exceeding that of forests growing in similar ecological conditions. In general, such highly productive savannas are not stable formations; once protected from fire and human interference, they revert to forests.

Primary production decreases as one enters drier areas, to become almost zero at the edge of the desert. In large parts of eastern and southern Africa, altitude can partially alleviate the consequences of low rainfall. In northern India, the cool winter temperatures that coincide with the beginning of the dry season slow down herbaceous production at this time of the year.

Whatever these variations in different savanna areas, primary production is always highly seasonal, the growth of most plants stopping entirely during the long dry season in the most xeric environments. In many areas, this extreme seasonality is made even more marked by the traditional practice of grass fires, which also destroy part of the production of the previous year.

Primary production can therefore be considered as highly variable within the savanna biome, both regionally and seasonally. The staple food of most primary consumers being "green grass", the regional variations in its average biomass, as well as its seasonal changes in availability, have far reaching consequences upon all kinds of herbivores.

The most conspicuous savanna herbivores are the ungulates, wild and/or domestic, the former being particularly numerous in Africa. However spectacular their numbers and standing crops may be, they do not necessarily play the leading role in energy transfer or nutrient cycling within the savanna ecosystem, as illustrated by several contributors to this book. Wild ungulates are particularly numerous, in species as well as in individuals, in dry African savannas; the long grass of more humid regions is mostly used by a few larger species such as the African buffalo and the African elephant. Medium- or small-sized grazing animals cannot properly exploit tall species of grass, except for a short time after burning. Domestic ungulates are also far more numerous in the Sahelian and Sudanian vegetation zones than in the more humid Guinean savannas.

The grasshoppers (mostly Acrididae) are another major group of primary consumers in the savannas of both hemispheres; in some areas they may compete, at least seasonally, with ungulates. In Africa, foraging termites also use a noticeable part of the grass production. A similar role may be played by foraging ants in the Neotropics.

Another important kind of plant resource, also seasonally produced, is seeds, which constitute the bulk of the diet of a number of invertebrates (e.g. some harvesting ants) and vertebrates (mostly graminivorous birds and rodents). However, seed production is more important in dry savannas than in the more humid ones, where conditions are less favourable for temporary storage during the rainy season.

Grass which is not eaten when fresh loses most of its nutritive value as it dries out. In areas where the grazing pressure is slight, such as in Lamto, a large amount of dead plant material remains unused at the beginning of the dry season and half, if not more, of it is usually destroyed by the annual grass fire. In dry savannas, the amount of grass fuel consumed by burning can be even greater, as the peak biomass of grass is reached close to the onset of the dry season. In more arid areas, grass fires are generally less destructive due to the scarcity of fuel and to the frequency of bare areas which act as natural fire-breaks.

Dry grass which has escaped burning can no longer be eaten by most herbivores, from grasshoppers to many large mammals. Dead grass can only be digested by those animals (termites and ruminants) that harbour symbiotic bacteria or protozoa able to break down cellulose in their stomach or gut. Termites are particularly active in this respect, greatly reducing the amount of litter on the soil surface in most savannas.

A large part of the organic material absorbed by primary consumers is not assimilated by them and passes out in faeces, to be taken up again by dung-feeding arthropods, many invertebrate detritus feeders and bacterial decomposers. All this helps one to understand why large ungulates do not consume more than 10% of the net total primary production (i.e. underground compartment included) in many tropical savannas.

A few rhizomes and bulbs excepted, the root system of the savanna grass layer is very seldom consumed alive. It is only when roots and rootlets die off that detritus feeders and decomposers make use of them. In this way, actinomycetes, fungi and bacteria become responsible for the major fraction of the energy flow through the savanna ecosystem. This is particularly true in the more humid situations, where earthworms and termites are the main consumers of cellulose and lignin. In dry savannas, the role of earthworms is less important, but termites are still very numerous.

There is a similar pattern of energy flow through the various categories of secondary consumers and decomposers of animal tissue. In each case, the diet is very rich in easily digestible animal proteins, with a high production efficiency — about 25% on average for poikilothermic consumers which do not ingest any cellulose. Unfortunately, very little is known about the energy exchange characteristics of the various functional groups of bacteria.

It must be stressed that annual energy budgets only provide an overall view of the functioning of a savanna ecosystem. The seasonality of production is a major characteristic of all savannas, and it is necessary to establish seasonal energy budgets in order to understand more fully the workings of the system. Only in this way can the many adaptive characteristics of savanna organisms be put in a true perspective.

In effect, seasonality of climate and the resulting seasonality of primary production affect all categories of plants and animals (and quite likely micro-organisms too) that live in tropical savannas. Many morphological, physiological and behavioural adaptations to seasonality, at the individual level, have been mentioned in various chapters of this book. But adaptive strategies at the population level are also very important. Some species are short-lived and have a very rapid population turnover; they can therefore adapt very quickly to even unpredictable changes in environmental conditions, provided they have the physiological capacity to survive drought conditions at the spore, egg or pupal stage. Other species with a longer life-span have to adapt their life-cycles, particularly their breeding periods, to the locally prevailing patterns of rainfall and production. The longest-lived species have to develop some kind of "escape strategy" in order to survive the lean period of the year. Those which have the ability to move away can undertake seasonal long-range migration to more congenial regions, and those which can enter diapause or can aestivate avoid seasonal food shortages by becoming inactive and living on their metabolic reserves. In any case, and even in the dry Sahelian savannas, the underground environment always remains more stable than the grass layer. This is obviously why so many animals take refuge in the soil or termite mounds during the dry season; it also explains why the root compartment of so many xeric savanna plants is as large, if not larger, than the above-ground compartment.

REFERENCES

Adegbola, A.A., 1964. Forage crop research and development in Nigeria. *Nigerian Agric. J.*, 1: 34–39.

Anderson, G.D. and Talbot, L.M., 1965. Soil factors affecting the distribution of the grassland types and their utilization by wild animals on the Serengeti plains, Tanganyika. *J. Ecol.*, 53: 33–56.

Balandreau, J., 1976. Fixation rhizosphérique de l'azote en savane de Lamto. *Rev. Ecol. Biol. Sol*, 13: 529–544.

Barbault, R., 1974. Structure et dynamique d'un peuplement de lézards: les Scincidae de la savane de Lamto (Côte d'Ivoire). *Terre Vie*, 28: 352–428.

Bate, G.C., 1979. Nitrogen in South African savannas. In: *Workshop on Dynamic Changes in Savanna Ecosystems, Kruger National Park, May 1979*, in press.

Bille, J.C., 1977. Etude de la production primaire nette d'un ecosystème sahélien. *Trav. Docum. ORSTOM*, 65: 1–82.

Bille, J.C. and Poupon, H., 1972. Recherches écologiques sur une savane sahélienne du Ferlo septentrional, Sénégal: biomasse végétale et production primaire nette. *Terre Vie*, 26: 366–382.

Blandin, P., 1979. Cycle biologique et production de l'araignée *Afropisaura valida* (Simon, 1885) (Araneae–Pisauridae) dans une savane d'Afrique occidentale (Lamto, Côte d'Ivoire). *Trop. Ecol.*, 20: 78–93.

Bony, J.P., 1974. Le rayonnement solaire à Lamto. *Bull. Liaison Chercheurs Lamto*, Numéro Spec., 1: 105–119.

Braun, H.M.H., 1973. Primary production in the Serengeti; purpose, methods and some results of research. *Ann. Univ. Abidjan*, E 6: 171–188.

Breman, H., 1975. *Les pâturages maliens*. Ecole Normale Supérieure, Bamako, 39 pp.

Breman, H. and Cissé, A.M., 1977. Dynamics of sahelian pastures in relation to drought and grazing. *Oecologia*, 28: 301–315.

Buechner, H.K. and Golley, F.B., 1967. Preliminary estimation of energy flow in Uganda Kob (*Adenota kob thomasi* Neumann). In: K. Petrusewicz (Editor), *Secondary Productivity of Terrestrial Ecosystems*. Institute of Ecology, Warsaw, pp. 243–254.

César, J., 1978. *Cycles de la biomasse herbacée et des repousses après fauche dans quelques savanes de Côte d'Ivoire*. Centre de Recherches Zootechniques, Minankro–Bouaké. Rapp. No. 16: 45 pp.

César, J. and Menaut, J.C., 1974. Le peuplement végétal. *Bull. Liaison Chercheurs Lamto*, Numéro, Spéc., 2: 1–161.

Coupland, R.T., 1979. Conclusions. In: R.T. Coupland (Editor), *Grasslands Ecosystems of the World*. Cambridge University Press, Cambridge, pp. 335–355.

Egunjobi, J.K., 1973. Studies on the primary productivity of a regularly burnt tropical savanna. *Ann. Univ. Abidjan*, E 6: 157–169.

Eisenberg, J.F., O'Connell, M.A. and August, P.V., 1979. Density, productivity, and distribution of mammals in two Venezuelan habitats. In: J.F. Eisenberg (Editor), *Vertebrate Ecology in the Northern Neotropics*. Smithsonian Institution Press. Washington, D.C., pp. 187–207.

Escobar, A. and Gonzalez-Jimenez, E., 1979. La production primaire de la savane inondable d'Apure (Venezuela). *GeoEcoTrop.*, 3: 53–70.

Fresson, R., 1973. Contribution à l'étude de l'écosystème forêt claire (Miombo). Note 13, Aperçu de la biomasse et de la productivité de la strate herbacée au miombo de la Luiswishi. *Ann. Univ. Abidjan*, E 6: 265–277.

Gandar, M.V., 1979. Trophic ecology of selected consumers in a South African savanna. In: *Workshop on Dynamic Changes in Savanna Ecosystems, Kruger National Park, May 1979*, in press.

Gillon, Y., 1968. Caractéristiques quantitatives du développement et de l'alimentation de *Rhabdoplea klaptoczi* (Karny, 1915) (Orthoptera: Acridinae). *Ann. Univ. Abidjan*, E 1: 101–112.

Gillon, Y., 1973. Bilan énergétique de la population d'*Orthochtha brachycnemis* Karsh., principale expèce acridienne de la savane de Lamto (Côte d'Ivoire). *Ann. Univ. Abidjan*, E 6: 105–125.

Haggar, R.J., 1970. Seasonal production of *Andropogon gayanus*. 1. Seasonal changes in field components and chemical composition. *J. Agric. Sci.*, 74: 487–494.

Harris, L.D., 1972. An ecological description of a semi-arid East African ecosystem. *Colo. State Univ., Range Sci. Dep., Publ.*, 11: 80 pp.

Humphreys, W.F., 1979. Production and respiration in animal populations. *J. Anim. Ecol.*, 48: 427–453.

Huntley, B.J. and Morris, J.W., 1979. Paper presented at the Savanna Symposium, Pretoria.

Hutchinson, K.J. and King, K.L., 1979. Arable grasslands. Consumers. In: R.T. Coupland (Editor), *Grasslands Ecosystems of the World*. Cambridge University Press, Cambridge, pp. 259–265.

Josens, G., 1973. Observations sur les bilans énergétiques dans deux populations de termites à Lamto (Côte d'Ivoire). *Ann. Soc. R. Zool. Belg.*, 103: 169–176.

Kelly, R.D., 1973. *A Comparative Study of Primary Productivity under Different Kinds of Land Use in Southeastern Rhodesia.* Thesis, University of London, London. [Quoted after Rutherford, 1978.]

Lamotte, M., 1972. Bilan énergétique de la croissance du mâle *Nectophrynoides occidentalis* Angel, Amphibien Anoure. *C.R. Acad. Sci. Paris*, 274: 2074–2076.

Lamotte, M., 1977. Observations préliminaires sur les flux d'énergie dans un écosystème herbacé tropical, la savane de Lamto (Côte d'Ivoire). *GeoEcoTrop.*, 1: 45–63.

Lamotte, M., 1978. La savane préforestière de Lamto, Côte d'Ivoire. In: M. Lamotte and F. Bourlière (Editors), *Structure et Fonctionnement des écosystèmes terrestres.* Problèmes d'Ecologie, 5. Masson, Paris, pp. 231–311.

Lavelle, P., 1978. *Lers vers de terre de la savane de Lamto (Côte d'Ivoire): peuplements, populations et fonctions dans l'écosystème.* Thesis, Université de Paris VI, Paris, 301 pp.

McNeill, S. and Lawton, J.H., 1970. Annual production and respiration in animal populations. *Nature*, 225: 472–474.

May, R.M., 1979. Production and respiration in animal communities. *Nature*, 282: 443–444.

Medina, E., 1980. Ecology of tropical American savannas: an ecophysiological approach. In: D.R. Harris (Editor), *Human Ecology in Savanna Environments.* Academic Press, New York, N.Y., pp. 297–319.

Menaut, J.C., 1977. Analyse quantitative des ligneux dans une savane arbustive préforestière de Côte d'Ivoire. *GeoEcoTrop.*, 1: 77–94.

Menaut, J.C. and César, J., 1979. Structure and primary productivity of Lamto savannas, Ivory Coast. *Ecology*, 60: 1197–1210.

Nel, J.J.C., Hewitt, P.H. and Joubert, L., 1970. The collection and utilization of redgrass *Themeda triandra* (Forsk,) by laboratory colonies of the harvester termite *Hodotermes mossambicus* (Hagen) and its relation to population density. *J. Entomol. Soc. S. Afr.*, 33: 331–340.

Odum, E.P., 1959. *Fundamentals of Ecology.* Saunders, Philadelphia, Pa., 2nd ed., 546 pp.

Ohiagu, C.E. and Wood, T.G., 1979. Grass production and decomposition in Southern Guinea savanna, Nigeria. *Oecologia*, 40: 155–165.

Penning de Vries, F.W.T., Krul, J.M. and Van Keulen, H., 1980. Productivity of Sahelian rangelands in relation to the availability of nitrogen and phosphorus from the soil. In: T. Rosswall (Editor), *Nitrogen Cycling in West African Ecosystems.* Royal Swedish Academy, Stockholm, pp. 95–113.

Petrides, G.A. and Swank, W.G., 1965. Estimating the productivity and energy relations of an African elephant population. In: *Proc. IXth Int. Grassland Congr.*, São Paulo, pp. 831–842.

Petrusewicz, K. and Macfadyen, A., 1970. *Productivity of Terrestrial Animals. Principles and Methods.* I.B.P. Handbook, 13. Blackwell, Oxford, 190 pp.

Phillipson, J. and Bolton, P.J., 1976. The respiratory metabolism of selected Lumbricidae. *Oecologia*, 22: 135–152.

Poupon, H., 1979. *Structure et dynamique de la strate ligneuse d'une steppe sahélienne au nord du Sénégal.* Thesis, Université de Paris-Sud, Paris, 351 pp.

Raud, G., 1976. Etude des algues du sol fixatrices d'azote dans la savane de Lamto (Côte d'Ivoire). Elaboration d'une nouvelle technique. *Ann. Univ. Abidjan*, F 8: 306–335.

Rutherford, M.C., 1975. *Aspects of Ecosystem Function in a Savanna Woodland in South West Africa.* Thesis, University of Stellenbosch, Stellenbosch. [Quoted after Rutherford, 1978.]

Rutherford, M.C., 1978. Primary production ecology in southern Africa. In: M.J.A. Werger (Editor), *Biogeography and Ecology of Southern Africa.* Junk, The Hague, 2nd ed., pp. 621–659.

San Jose, J.J. and Medina, E., 1976. Organic matter production in the *Tachypogon* savannas in Venezuela. *Trop. Ecol.*, 17: 113–124.

Schaefer, R., 1974a. Le peuplement microbien du sol de la savane de Lamto. *Bull. Liaison Chercheurs Lamto*, Numéro Spéc., 5: 39–44.

Schaefer, R., 1974b. Activité métabolique de sol: fonctions microbiennes et bilans biogéochimiques dans la savane de Lamto. *Bull. Liaison Chercheurs Lamto*, Numéro Spéc., 5: 167–184.

Schmidt, W., 1975. Plant communities on permanent plots of the Serengeti plains. *Vegetatio*, 30: 133–145.

Sims, P.L. and Singh, J.S., 1971. Herbage dynamics and net primary production in certain ungrazed and grazed grasslands in North America. In: N.R. French (Editor), *Preliminary Analysis of Structure and Function in Grassland.* Colorado State University, Fort Collins, Colo., pp. 59–174.

Sinclair, A.R.E., 1975. The resource limitation of trophic levels in tropical grassland ecosystems. *J. Anim. Ecol.*, 44: 497–520.

Sinclair, A.R.E., 1977. *The African Buffalo. A Study of Resource Limitation of Populations.* The University of Chicago Press, Chicago, Ill., 355 pp.

Singh, J.S., 1968. Net aboveground community productivity in the grassland at Varanasi. In: R. Misra and B. Gopal (Editors), *Proceedings of a Symposium on Recent Advances in Tropical Ecology. Part II.* International Society for Tropical Ecology, pp. 631–654.

Singh, J.S. and Joshi, M.C., 1979. Tropical grasslands. Primary production. In: R.T. Coupland (Editor), *Grasslands Ecosystems of the World.* Cambridge University Press, Cambridge, pp. 197–218.

Singh, J.S. and Misra, R., 1968. Efficiency of energy capture by the grassland vegetation at Varanasi. *Curr. Sci.*, 37: 636–637.

Singh, J.S., Lauenroth, W.K. and Steinhorst, R.K., 1975. Review and assessment of various techniques for estimating net aerial primary production in grasslands from harvest data. *Bot. Rev.*, 41: 181–232.

Singh, K.P. and Misra, R., 1978. *Structure and Functioning of Natural, Modified and Silvicultural Ecosystems of Eastern Uttar Pradesh.* Banaras Hindu University, Varanasi, 161 pp.

Smith, K., 1977. Acetylene reduction by Blue-green algae in subtropical grassland. *New Phytol.*, 78: 421–426.

Strugnell, R.G. and Pigott, C.D., 1978. Biomass, shoot-production and grazing of two grasslands in the Ruwenzori National Park, Uganda. *J. Ecol.*, 66: 73–96.

Talineau, J.C., 1970. Action des facteurs climatiques sur la production fourragère en Côte d'Ivoire. *Cah. ORSTOM, Sér. Biol.*, 14: 51–76.

Villecourt, P., 1973. Premiers résultats du bilan d'azote dans la savane à roniers de Lamto-Pacobo (Côte d'Ivoire). *Ann. Univ. Abidjan*, E 6: 33–34.

Villecourt, P., 1975. Apports d'azote minéral (nitrique et ammoniacal) par la pluie dans la savane de Lamto. *Rev. Ecol. Biol. Sol*, 12: 667–680.

Walker, B.H., 1974. Ecological considerations in the management of semi-arid ecosystems in south central Africa. In: *Proc. 1st Int. Congr. Ecol., The Hague*. PUDOC, Wageningen, pp. 124–129.

Walter, H., 1964. Productivity of vegetation in arid countries, the savanna problem and bush encroachment after overgrazing. In: *L'écologie de l'homme dans le milieu tropical. IUCN Publ., N. S.*, 4: 221–229.

Chapter 29

SUCCESSIONAL PROCESSES

BRIAN HOPKINS

INTRODUCTION

Savanna is a dynamic ecosystem. Long- and short-term changes are constantly modifying its physiognomy, composition and ecological processes. Some of these changes are predictable; others are random fluctuations. Succession is a predictable cumulative change in vegetation. It is accompanied by associated changes in the animal populations, but these have been insufficiently studied.

The major causes of these changes are variations in the climate, burning, populations of large herbivores, and man's activities. Climate varies in long-term trends (such as have occurred during the Pleistocene), seasonally (with alternating wet and dry periods), and in year-to-year fluctuations; these were discussed in the first two chapters. The other causes are directly or indirectly related to human activities. Fire is the most important and is almost always started by man; it will be discussed in the next chapter. The present-day populations of large wild herbivores, if their distribution is included, are also under man's control; their effects are considered in this chapter (pp. 607–608), whilst those of domesticated livestock will be described in Chapter 31. The effects of man as a cultivator will also be considered here.

Primary successions are on new land and are consequently somewhat rare. Secondary successions are changes from man-induced simpler ecosystems towards the more complex ones which existed previously. They occur when the intensity of the particular anthropogenic factor (cultivation, fire, grazing) is reduced, and result in a more complex physiognomy with increases particularly in the cover and species of the woody plant strata.

These changes parallel those taking place with increasing rainfall — or rather, decreasing dry-season severity — and with increasing soil-water availability. Where these factors vary in space there is a zonation of vegetation types, which may be on a continental scale with respect to climate, or on a topographical scale with respect to soil type. When the variation is in time, the man-induced effects and the rainfall fluctuations may reinforce each other, as when rainfall decreases and cattle density increases, or tend to cancel each other out, as when both rainfall and cattle density increase simultaneously.

So great is the dynamism caused by these various factors that it is often difficult or impossible to decide if a particular ecosystem is a climax, or a stage in either a succession or a fluctuation. Their impact is thus to mask successions and, in little-known vegetation types, to make their study and elucidation more difficult.

Consequently, little is known about savanna successions, although the literature contains considerable speculation. I shall not deal further with those aspects already mentioned and covered elsewhere in this volume (particularly in Chapters 30 and 31). I shall first outline knowledge of primary successions and then consider three other aspects: the effects of large mammals, and of cultivation, and forest/savanna relations. (Savanna–desert relations are covered in Chapter 31.) Successions in Africa, where most tropical savanna occurs, are the best known and will be taken as a model, whilst examples from other continents will be given for comparison.

PRIMARY SUCCESSIONS

There are three main types of primary succession: marine, fresh-water and xeric. All are rare, especially in tropical savannas.

Probably the two most characteristic types of vegetation of tropical shores are mangroves (especially *Rhizophora* spp.) and coconut palms (*Cocos nucifera*); neither is relevant to savanna successions. Mangroves are virtually the only intertidal flowering plants on tropical shores, although there are many strand species which can withstand sea spray, including the widespread *Ipomoea pescaprae*. Vegetation zones have been described for several localities. Coastal savannas may occur around brackish lagoons where they too are zoned; they are common along parts of the coast of West Africa. However these, like the strand zones, appear to be related to edaphic factors — often the nature of the substratum, but including water depth and salinity (Brand and Brammer, 1956; Adjanohoun, 1962, 1964, 1965; Adejuwon, 1970; Guillaumet and Adjanohoun, 1971; Schnell, 1971). There is no critical evidence for any of these, or similar types in other continents, being successional, although some man-induced coastal savannas will revert by a secondary succession to mangroves (Hills and Randall, 1968).

There is much more information on fresh-water swamps which abut onto savannas. As in still or slowly flowing waters throughout the world, the vegetation zones are related to water depth, with submerged species (e.g. *Ceratophyllum demersum*) in the deeper water, a zone of rooted plants with floating leaves (such as *Nymphaea* spp.) and, in progressively shallower water, zones of semi-aquatic vegetation especially consisting of grasses and sedges. Where the water is virtually still, free-floating vascular plants (such as *Azolla* spp., *Eichornia crassipes*, *Pistia stratioides*, *Salvinia* spp. and members of the Lemnaceae) occur mixed with the rooted species. Often there is a zone of free-floating grasses and sedges which may or may not be anchored to the bank; in Africa species such as *Cyperus papyrus* and *Vossia cuspidata* are common here. In some areas there are swamps containing woody species.

References to the literature on these communities are given by Beard (1953), Champion and Seth (1968), Walter (1971), Graham (1973) and Schnell (1977). This literature contains some detailed accounts of the zonation in particular localities and suggests that many of these zonations represent successions, but extremely few have been critically evaluated. Indeed, Vesey-FitzGerald (1970) in a paper terminating his long and detailed work on the Rukwa Valley in southern East Africa concluded that its floodplain grasslands were edaphic climaxes determined mainly by flooding.

A major difficulty to accepting such zonations as representing successions, as undoubtedly they would do in many other biomes, is the seasonal climate and the corresponding large differences in water level; many swamps are seasonally dry. Further, there is often considerable between-year variation in rainfall and consequent fluctuation in water level. A changing water level is not conducive to succession, although Misra (1946) described depressions in India with allogenic successions due to the monsoon rains washing in eroded soil and so causing silting.

Xeric successions occur on new volcanic deposits. Inevitably most are in mountain ecosystems rather than lowland savannas. Between 1926 and 1931, vegetation at the intermediate height of *c.* 1450 m on lava from the 1912 eruption of Rumoka Volcano in Central Africa mainly consisted of mosses, although 26 vascular species were present — a very slow rate of succession (Robyns, 1932). Taylor (1963) described volcanic successions to derived savanna in Nicaragua.

Volcanism too may result in vegetation zonations which do not represent successions, as is the case on *c.* 5000 km^2 of the eastern Serengeti (Tanzania) where a lithosequence of volcanic tuff deposits occurs (Anderson and Talbot, 1965; Lamprey, 1979).

SECONDARY SUCCESSIONS

In contrast to primary successions, and as pointed out by Goodall (1977), secondary successions are common and very important in applied ecology. As stated above, only those due to large herbivorous mammals and cultivation will be considered in this chapter; other types are treated elsewhere in the volume.

Effects of large mammals

High densities of large mammals have a considerable effect on vegetation due to the consumption of plant food, trampling, and other activities, such as elephants debarking trees and pushing them over. The effects on the woody strata are more dramatic than those on the herbs; they result in a decrease in woody plants and may cause woodland to change into grassland. Although many areas are in dynamic equilibrium, a decrease in animals generally results in a more woody vegetation.

Thus, these changes may be considered as successions, although the climax vegetation is by no means always clear. Much depends on the extent to which the high mammal densities are considered as a natural part of the ecosystem; many are largely, if indirectly, due to man. However, this was not always the case, and Kortlandt (1972) considered that large mammals created a forest/savanna mosaic from forest during the Miocene. Changes in density of large mammals were considered earlier in this volume; here their main effects will be discussed irrespective of whether they are truly successional. The discussion will be limited to Africa; compared to Africa, with its wealth of large mammal species, the effects on vegetation of large mammals in other continents are almost negligible.

Elephants have the greatest effect on vegetation. Buechner and Dawkins (1961) showed that they halved the tree density in 24 years in parts of Kabalega (formerly Murchison Falls) National Park, Uganda. Even greater tree mortalities occurred later (Laws et al., 1970, 1975). In many East African national parks annual tree mortalities of from 2 to 20% have been reported and, for some areas or species, values >40% (Lamprey et al., 1967; Field, 1971; Harrington and Ross, 1974; Laws et al., 1975; Leuthold, 1977; Lamprey, 1979). Elephants have been the prime cause, but fire normally plays a secondary role in these changes towards grassland.

These physiognomic changes are not only due to the death of mature trees. Jarman and Thomas (1968–1969) showed that it was only in areas where large mammals were absent that *Colophospermum mopane* had high seedling densities (103 m^{-2} compared with a maximum of 15 m^{-2} elsewhere) and survival (65% compared with 48%). Experimentally simulated grazing of *Acacia senegal* seedlings showed

that maximum mortality occurred when they were c. 40 days old (Seif El Din and Obeid, 1971). Exclosure experiments are very recent, but preliminary results indicate considerable increases in the density and height of woody plants (Harrington and Ross, 1974; Lock, 1977).

Grazing animals often cause considerable decreases in herb height, cover and biomass, although these are not always obvious. They have, however, been clearly demonstrated in Ruwenzori (formerly Queen Elizabeth) National Park (Uganda) where exclosure plots in three vegetation types had higher biomasses than unenclosed control plots, although only in two were the differences statistically significant (Lock, 1972); the removal of hippopotamus from the Mweya Peninsula caused a decrease in bare ground from c. 80% to c. 40% in eight years, with a corresponding increase in litter (1–43%) (Laws, 1968; Thornton, 1971).

By stimulating new shoot production, grazing may lead to a more nutritive pasture and so attract more herbivores. Further, the decrease in herb height caused by very large mammals (elephant, hippopotamus, buffalo) may make the grassland available for species such as zebra, wildebeest and topi which, in turn, may make it available for gazelles. Such a grazing succession has been described for several localities (Vesey-FitzGerald, 1960, 1964b, 1965).

Less is known about herbivore feeding preferences causing changes in floristic composition. Changes due to browsing have been demonstrated by Agnew (1968), Field (1971), Vesey-FitzGerald (1973a, b, 1974), Laws et al. (1975) and Leuthold (1977). The effects of grazers are more complex, and the time of grazing in relation to the physiological state of the plants appears to be very important. Certainly, grazing of dry standing hay has virtually no effect on floristic composition. Observations on grazing at other seasons are not all in agreement (Rattray, 1960; Lamprey, 1963; Heady, 1966; Thornton, 1971; Lock, 1972; Ndawula-Senyimba, 1972; Eltringham, 1973; and others including Moore, 1973, in Australia). The diversity of results appears to be due to the interplay of other factors, especially the type of savanna, fire, rainfall and shade, as well as the species of herbivore concerned.

The major effect of large mammalian herbivores, particularly when one species predominates, is thus towards a more open vegetation with a simpler physiognomy. Such changes are, however, more likely to be fluctuations than successions.

Effects of cultivation

Secondary successions commence when cultivation ceases, and are thus common where the agricultural system consists of a period of farming alternating with one of fallow. This is typical of much peasant agriculture in the tropical savanna biome and is known as bush fallow, shifting cultivation, slash and burn, etc. A site is cleared by fire and chopping down the unwanted trees (trees of economic or other importance are retained). This usually occurs late in the dry season. Crops are then grown, but the yields of successive harvests decline so that, after a few years, the land is not worth the effort of recultivating. The farm is abandoned and a new farming site is cleared. The vegetation on the abandoned farm undergoes a secondary succession during its period of fallow, when the nutrient content of both the soil and vegetation increases.

There are many variations on this simple pattern due to differences in climate, vegetation, soils, crops and the traditions and cultures of the farming peoples. Nevertheless, in many areas, two or three years of cultivation are followed by twenty or thirty years of fallow.

The succession

To a considerable extent the succession on an abandoned savanna farm consists of the regeneration of the woody plants which survived the period of farming either as protected trees or, more usually, as underground rootstocks. These rootstocks are very common (Ch. 5) and are able to withstand regular burning (Ch. 30) and coppicing. As soon as cultivation ceases, their shoots grow unchecked, and so the same individual woody plants as were there before farming re-establish themselves. Further, some of the more resistant grass tufts may also survive cultivation. Of course, seeds of both woody and herbaceous species do invade and become established, but these rarely form a major part of the secondary succession of the woody strata.

The initial stage of the succession consists of a fairly dense shrub layer with scattered, previously protected, trees and a rather heterogeneous herb layer largely consisting of agricultural weeds and savanna herbs — both categories being predominantly composed of grasses, such as *Imperata cylindrica*. Regular burning ensures that all fire-tender species are eliminated and tall grasses are favoured. These shade out some smaller herbs, and gradually the savanna becomes more mature and more woody.

There is considerable variation on this generalized theme related to the severity of the dry season, soil moisture relations, the degree of burning and grazing, and other human influences. Knowledge of this process comes from many observations on many areas, but in no case have all the stages been observed on the same abandoned farm. Examples illustrating this variation will be selected from savannas of increasing humidity.

Arid savannas are near or beyond the limit of rain-fed cultivation, and their irrigated areas are usually permanently cropped. Consequently, post-cultural successions are rare and unstudied. They will be profoundly modified by grazing — the major agricultural system of these regions — and are better considered under the heading of pastoralism (Ch. 31).

Mesic savannas, with an annual rainfall between 500 and 1200 mm, are subject to both grazing and cultivation. In some areas, as in West Africa, their human population is high so that large areas have a sparse tree population, and there is often a concentric series of vegetation zones around the larger towns.

Fairbairn (1939) distinguished the following five stages on abandoned farms in the Sudan zone of Nigeria: after 1 year, grassland with weeds and shrubs 1.2 to 1.5 m tall; after 3 to 6 years, grassland with shrubs 2 to 3 m tall; after *c.* 10 years, grassland with scattered woody plants up to 4.5 m tall; grassland with many trees 6 to 9 m tall; and a wooded grassland climatic climax with trees 11 to 14 m tall.

Abandoned farms in northern Ghana initially have abundant *Combretum glutinosum* and little *Acacia* spp., but the proportions of these two taxa are later reversed (Ramsay and Rose-Innes, 1963).

Miège et al. (1966) followed the changes during six years of fallow following three years of farming in Senegal. In control (unburnt) plots of·1 m², there was a decrease in mean number of flowering plant species from 15 to 7 over the fallow period, and three-quarters of the between-year comparisons of species numbers during this period showed significant differences ($P < 0.05$). Most woody plants belonged to the Combretaceae of which *Guiera senegalensis* was the most abundant. The herbs showed large between-year variations due to differences in rainfall (522–1113 mm yr^{-1} during the period of observations), so that it was difficult to determine species trends, which were also often different on burnt and unburnt plots. The biomass (dry weight) of herbaceous shoots tended to increase, especially on the unburnt plots, as the following figures (in g m^{-2}) show:

years of fallow	1 and 2	5 and 6
unburnt plots	177	338
burnt plots	194	285

Humid savannas are influenced by cultivation rather than pastoralism, and their secondary successions have been described from many places. In West Africa, during the first year annual and ruderal species predominate, whilst the perennial grasses become established during the following two years. These include *Pennisetum* spp. and members of the Andropogoneae, especially *Hyparrhenia* spp. and, on more degraded areas, *Imperata cylindrica* (Clayton, 1958; Letouzey, 1968). Gradually the *I. cylindrica* declines and is replaced by *Hyparrhenia* spp. (Dundas, 1942; Letouzey, 1968); this is sometimes taken as an indication that the area is ready to be farmed again. The succession on Puerto Rico described by Molinari (1949) is basically similar. Adjanohoun (1962) recorded the changes in life-form spectrum over ten years on a cleared plot in Ivory Coast coastal savanna; chamaephytes and geophytes declined whereas hemicryptophytes gradually increased. Similar changes occur in southern Zaïre (Mullenders, 1954).

The changes in woody plants are much more variable, depending on the number of root stocks removed or killed during the farming period and the degree of burning afterwards (Keay, 1960; Adjanohoun, 1964). In *miombo*, Delevoy (1938) found that the shoot density of woody plants initially increased but soon declined rapidly. Similar results have been observed elsewhere. They are probably related to the increase in fire severity as the density of the herb layer increases over the

first five years (Leay, 1960; Boaler and Sciwale, 1966). In miombo, Boaler and Sciwale (1966) found that, between five and fifteen years, fire decreased as the woody plant density increased, then it slowly increased due to competitive thinning of the woody plants until an equilibrium was achieved in mature miombo. They also found that the basal area of trees ($\geqslant 5$ cm diameter at 130 cm above ground level) increased as follows:

years since cultivation:	10	20	25	50	100	Mature miombo
basal area (%)	0.01	0.02	0.02	0.08	0.12	0.14

Similar results (0.14% after c. 50 years) were obtained by Strang (1974).

Floristic changes accompanied these changes in density and basal area. Initially *Terminalia sericea* was the most abundant canopy species, for it could withstand the farming practices. The shrub *Hymenocardia acida* became abundant about seven years after cultivation, whilst the main canopy dominants (*Brachystegia* spp. and *Julbernardia* spp.) did not become prominent until about twenty years had elapsed.

In very humid areas near forests the successions may involve changes to or from forest; these are considered below (pp. 610–613).

Nutrient aspects

Nye and Greenland (1960) summarized the information on nutrients in vegetation and soil, and showed that both increased during the fallow period. Many of their data were for forest ecosystems, but savannas were included and most were humid savannas — especially from Ejura in Ghana (Nye, 1958). Because the amount of vegetation is less than in forest, the amounts of nutrients are also less, and those in the soil are consequently of greater importance. These effects are reinforced in savannas with fewer trees.

A typical humid wooded grassland type of savanna fallow increases its humus carbon by from 8 to 19 g m^{-2} yr^{-1}. The nitrogen increase parallels this; c. 4 g m^{-2} yr^{-1} is fixed but c. 3 g m^{-2} yr^{-1} is lost on burning, so that the net increase is only c. 1 g m^{-2} yr^{-1}. Organic phosphorus increases by from 0.03 to 0.1 g m^{-2} yr^{-1}, and the increases in potassium, calcium and magnesium in the soil exchange complex are *c.* 2, 4.5 and 1 g m^{-2} yr^{-1}, respectively.

Little intensive work has been carried out since Nye and Greenland's (1960) account; most of it has confirmed their general conclusions.

FOREST–SAVANNA RELATIONS

Introduction

Information on tropical forests is given in Volume 14 of this series. Structurally they are the most complex type of vegetation in the world and mainly consist of woody plants. Their great physiognomic difference from savannas may be expressed quantitatively in life-form spectra. Table 29.1 presents the life-form spectra for forest and savanna plots only 5.5 km apart, and shows that, whereas phanerophytes are the predominant forest life-form, hemicryptophytes, geophytes and therophytes are also abundant in savanna.

In general, it is the driest ·type of forest, often called moist semi-deciduous forest, which abuts onto the savanna. This forest has an incomplete upper tree stratum, and its taller trees are usually deciduous whilst the smaller trees and other plants are evergreen. Similarly, it is a humid type of savanna which is generally adjacent to the forest. This is often called derived savanna (Jones, 1945) and is also known as peri-, post-, pre- and subforest savanna; periforest savanna, *savane périforestière* of Letouzey (1968), is probably the most satisfactory term. This savanna may contain a high density of trees, or be virtually treeless. Amongst its

TABLE 29.1

Forest and savanna life-form spectra; percentages of the species on a 50×50 m plot in the Olokemeji Forest Reserve, southwestern Nigeria, belonging to the various life forms (from Hopkins, 1962)

Life-form	Forest	Savanna
Phanerophytes	93	30
Chamaephytes	1	0
Hemicryptophytes	2	23
Geophytes	5	21
Therophytes	0	25

trees are a number of species which have close relatives in the forest; lists of pairs of such African species are given by Chevalier (1900), Aubréville (1949a, b), Adjanohoun (1964), Richards (1952), Keay (1959) and Schnell (1970).

The effect of farming

Abandoned peasant farms in forest revert towards forest by a secondary succession (Ewel, 1982). However, such farms which are near savanna will almost certainly be invaded by savanna grasses, for these have efficient means of dispersal. Once cultivation ceases these may become established and, if so, they will probably be burnt. Savanna grasses are tolerant of burning (Chapter 30), whereas forest plants are not. Thus, the area will be changed into a grassland. If, as is normally the case, these patches are burnt annually, they will increase and form a savanna.

This explains how much of the savanna adjacent to the forest has arisen and how the forest has receded in the past — by as much as 500 km in some places in Africa, but only to a small extent in New Guinea (Taylor in Hills and Randall, 1968); this process has been reviewed by Aubréville (1949a, b), Schnell (1950), West (1956) and Hills and Randall (1968). The advance of the savanna has not been uniform; it has depended on the human population pressure, and was initially on soils which were reasonably fertile and not too difficult to clear or work. Consequently, heavy clays and rocky areas often remain as forest in a zone consisting of a mosaic of patches of forest, savanna, farms and fallow.

There is considerable floristic heterogeneity in the early successional stages. *Imperata cylindrica* is often abundant and, in Africa, species of *Andropogon* and *Hyparrhenia* are often common. The woody plant flora is more variable, for it takes time for their seeds to arrive and to become established; the degree of burning is crucial in regulating their rate of progress.

The forest/savanna boundary

The forest/savanna boundary is generally abrupt; it is a fire boundary for, whereas savanna is regularly burnt, forest is normally resistant to fire.

The boundary at Olokemeji, in southwestern Nigeria, was studied in detail by Clayton (1958). His profile diagram of the transition is reproduced as Fig. 29.1. The whole transition occurs in 50 m. Clayton recognized five zones: (1) "open savanna" of typical humid savanna species; (2) "shaded savanna" which was similar but with more trees and the shade-tolerant species *Andropogon tectorum* and *Afromomum latifolium*; (3) "Anogeissus woodland", a transition zone 25 m wide of vegetation 15 to 18 m tall, intermediate between forest and savanna, with a sparse herb layer and with *Anogeissus leiocarpus* accounting for about half the trees; and (4) a narrow zone of "thicket" adjacent to (5) the "forest". This boundary may be regarded as typical. Fosberg (in Hills and Randall, 1968) considered the boundary to be an ecosystem rather than an ecotone because of its importance for the animal species which inhabit it.

Advance of the forest into the savanna has been described for many places in Africa (Lebrun, 1936; Eggeling, 1947; Sillans, 1958; Hopkins, 1962; Miége, 1966; Letouzey, 1968) and other continents (Hills and Randall, 1968). Fires generally die out at the savanna edge of the transition. Only occasionally do feeble fires enter the woodland, be-cause its sparse herb layer provides an inadequate amount of fuel, and so it is able to support both fire-tolerant and fire-tender trees. The woodland canopy gradually shades out the herbaceous layer in the adjacent savanna and so changes it to woodland. Within the woodland there is a sequence of both forest and savanna trees, with their older members towards the forest edge and their younger ones mainly towards the savanna edge; the forest is encroaching into the woodland, and savanna trees are unable to become established or survive under heavy shade. Thus, the whole transition is a succession moving slowly into the savanna at a rate depending on the degree of burning and, of course, dependent on the absence of other agencies destructive to forest such as cultivation and high densities of large mammals. [Robyns (1936) and Buechner and Dawkins (1961) described the creation of savanna from forest by the action of African elephants and fire.] The forest advance is very erratic, being determined by the severity of each annual fire at each point along its boundary. Average rates, over a decade or more, of from 2.5 to 3.5 m yr^{-1} have been recorded in West Africa (Hopkins, 1962; Miège, 1966).

Where there is cultivation the savanna will ad-

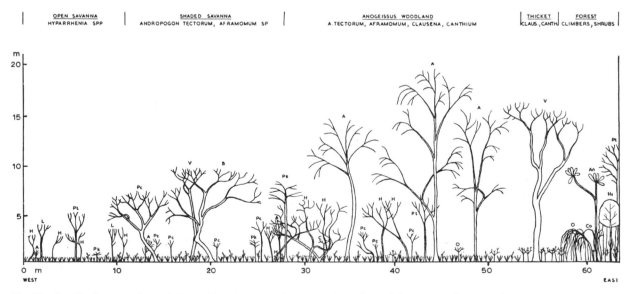

Fig. 29.1. Profile diagram of a strip 9 m wide taken near the western boundary of the Olokemeji Forest Reserve, and passing from savanna, through a fringe of *Anogeissus* woodland, to forest. Vegetation less than 1.5 m high has been shown diagrammatically: the trees were in leaf at the time. The following symbols have been used to indicate the species: *A = Anogeissus leiocarpus*; *An = Anthocleista* sp.; *B = Butyrospermum parkii*; *C = Cussonia barteri*; *Co = Combretum platypterum*; *H = Hymenocardia acida*; *Hi = Hildegardia barteri*; *L = Lophira lanceolata*; *O = Ochna* sp.; *Pc = Parinari curatellifolia*; *Pk = Parkia clappertoniana*; *Pt = Pterocarpus erinaceus*; *V = Vitex doniana*.

vance into the forest. Here the boundary is different; it is even more abrupt and has no zone of transition woodland; normally grassland abuts directly onto forest. Further, these boundaries tend to be straight, indicating the limits of old farms. This type of boundary generally occurs where there is a high human population density (Adejuwon, 1971).

Distribution of the vegetation types

Fig. 29.2 diagrammatically illustrates the distribution of forest and savanna in and near the boundary; for simplicity farms and fallows have been excluded. There are two types of forest outlier in the savanna: fringing (or gallery) forest along the watercourses; and relict patches which have escaped being changed to savanna. These relicts are abundant near the main forest/savanna boundary and gradually decrease away from it as the climate becomes less humid. Their distribution is also dependent on soil type and on social and historical factors. Those furthest from the main boundary may mark the climatic limit of forest. Some may remain as forest although they are farmed periodically, possibly due to low human pressure or high

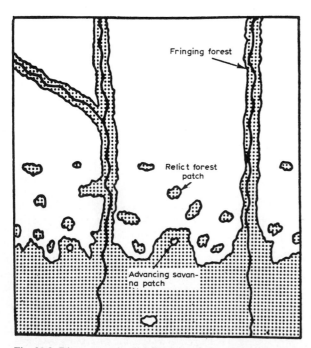

Fig. 29.2. Diagrammatic distribution of the vegetation types near the forest/savanna boundary. Forest stippled; savanna blank. Width of figure is c. 280 km. (From Hopkins, 1965.)

soil nutrients (Morgan and Moss, 1965). Due to the rapid uphill spread of fire, savanna usually occupies the ridges and crests in hilly areas, whilst forest remains in the less severely burnt valleys.

The relict forest patches have boundaries very similar to those already described, whereas the fringing forest boundaries are often more abrupt and more stable because they are reinforced by the topographic soil-moisture sequences; typical examples are illustrated by Keay (1960) and Bonvallot et al. (1970). Beyond their immediate valley, however, they behave as relict forest/savanna boundaries and the forest may advance into the savanna (Jones, 1963).

Savanna enclaves too are fairly common near the main boundary due to rapid advance of either forest or savanna; in both cases they rapidly disappear — either to forest or by coalescence respectively.

There are also savanna enclaves within the forest zone, and occasionally under an annual rainfall of >2000 mm in Africa or almost twice this in South America (Robyns, 1936; Keay and Onochie, 1947; Aubréville, 1949a, 1966; Beard, 1953; Devred, 1956; Crosbie, 1957; Sillans, 1958; Ahn, 1959; Asare, 1962; Richardson, 1963; Adjanohoun, 1964; Hills and Randall, 1968; Letouzey, 1968; Cole and Jarrett, 1969). Most are in the size range from 1 ha to 1 km^2. Almost all are annually burnt and some are grazed by cattle. Most are treeless, due either to their distance from parent savanna (i.e. fire-tolerant) trees or to soil factors. Their herbaceous flora is also impoverished.

Whilst many, like the periforest savanna, are fire climaxes, others are not. Some of these have soils too shallow for tree growth and are edaphic climaxes (Ahn, 1959; Denevan, in Hills and Randall, 1968; Cole and Jarrett, 1969; Adejuwon, 1971). In other cases it has been claimed that their soils are too waterlogged or too deficient in nutrients for tree growth; these may seem unlikely factors, for forest may occur on equally waterlogged and deficient soils, but they do seem to be the controlling factors in parts of South America (Beard, 1953; Wilhelmy, 1957; Cole, 1960; Ferri, 1963; Whyte, 1974; and others). Some are on sites of ancient cultivation or other types of forest destruction, which may have altered the soil so that forest cannot recolonize (Aubréville, 1949a; Schnell, 1952; Ferri, 1963). Others may be relics of a dry climate earlier in the

Pleistocene (Portères, 1950; Schnell, 1952; Leneuf and Aubert, 1956; Adjanohoun, 1964; Aubréville, 1966; Adjanohoun and Assi, 1968; Bellier et al., 1969). Edaphic factors, especially dry and shallow soils, may prevent or slow down their recolonization by forest, but the major factor maintaining most of them as savanna is undoubtedly fire (Thomas, 1951; Taylor, 1963; Aubréville, 1966; Letouzey, 1968; Whyte, 1974).

Experimental work

Considerable knowledge of the dynamics of forest/savanna relations comes from experimental work. Most of these experiments consist of plots protected from fire compared with other plots burnt early or late during the dry season.

At Olokemeji, in southwestern Nigeria, is one of the first ecological experiments. It commenced in 1929 and consists of three plots, each of 1700 m², which were originally burnt and coppiced (MacGregor, 1937; Charter and Keay, 1960). Tree densities (m⁻²) after 6 and 28 years were:

Years	6	28
protected from fire	0.23	0.25
annual early burn	0.13	0.10
annual late burn	0.04	0.06

After 6 years 35% of the trees on the protected plot belonged to forest species; after 28 years the figure was 64%, at which time only one species of savanna grass (the shade-tolerant *Andropogon tectorum*) was present.

The plots of Kokondékro, in the central Ivory Coast, had the same three treatments but were larger (2 ha) and established on a seven-year fallow without cutting the woody plants (Ramsay and Rose-Innes, 1963; Adjanohoun, 1964). Despite the different beginnings, the result after 24 years' protection from fire was very similar to that at Olokemeji; 0.40 trees m⁻², of which 67% were forest species.

The early-burn plots of these experiments had considerably greater tree densities than the late-burn plots and a higher proportion of forest tree species. Thus, even with annual burning, provided it was not severe, there is a slow succession towards the climatic climax of forest, as has been observed

along the forest/savanna boundary as already described.

Adjanohoun (1964) and Adjanohoun and Assi (1968) attempted to create a savanna under an annual rainfall of c. 1950 mm at Adiopodoumé in the southern Ivory Coast. A plot of forest measuring 2000 m² was cleared and burnt, and large quantities of seeds of 38 savanna species, including 19 grasses, were introduced. In the following year more seeds were added, and eight of the same species were introduced vegetatively. By the third year a grassland had been produced after which there was little change; despite the annual burning, in the eighth dry season the grassland, then 2 to 3 m tall, contained only fifteen of the introduced savanna species. Forest species still accounted for half the total, and weeds for c. 40%. A savanna-like vegetation had been produced, but only by enormous and continuing inputs.

Conclusions

Fire has caused woodlands in Uganda to change to savanna (Dawkins, 1954), but it is doubtful if, as Aubréville (1966) claimed, fire alone can change moist forest to savanna. However, it is likely that in the past dry forest and woodland were much more widespread in Africa, forming part of a continuum from moist evergreen forest to desert, but that these have been largely eliminated by fire. Nevertheless, the experiment of Adjanohoun (1964) and Adjanohoun and Assi (1968) clearly shows that more than fire is required to create savanna under an annual rainfall of c. 2000 mm. Aubréville (1949a, 1966) agreed that fire usually acts in association with cultivation, as stated by Adam (1947, 1948), Devred (1956), Sillans (1958), Germain (1965) and Hopkins (1965).

Periforest savanna now occupies much of the moist semi-deciduous forest zone. It is maintained by fire and was caused, at least partly, by it. The elimination of fire would result in most non-cultivated periforest savanna changing back to forest by a secondary succession.

ACKNOWLEDGEMENTS

I am grateful to R. Misra and R.M. Moore for supplying information on Indian and Australian savannas.

REFERENCES

Adam, J.G., 1947. La végétation de la source du Niger. *Ann. Géogr.* 56: 192–200.

Adam, J.G., 1948. Les reliques boisées et les essences des savanes dans la zone préforestière en Guinée française. *Bull. Soc. Bot. Fr.*, 95: 22–26.

Adejuwon, J.O., 1970. The ecological status of coastal savannas in Nigeria. *J. Trop. Geogr., Singapore*, 30: 1–10.

Adejuwon, J.O., 1971. The ecological status of savannas associated with inselbergs in the forest areas of Nigeria. *Trop. Ecol.*, 12: 51–65.

Adjanohoun, E., 1962. Etude phytosociologique des savanes de Basse Côte d'Ivoire (savanes lagunaires). *Vegetatio*, 11: 1–38.

Adjanohoun, E., 1964. Végétation des savanes et des rochers découverts en Côte d'Ivoire Centrale. *Mém. ORSTOM*, No. 7: 178 pp.

Adjanohoun, E., 1965. Comparaison entre les savanes côtières de Côte d'Ivoire et du Dahomey. *Ann. Univ. Abidjan, Sci.*, 1: 41–62.

Adjanohoun, E. and Assi, L.A., 1968. Experimentation on the creation of savannas inside forest area of Ivory Coast. In: *VI West Africa Sci. Assoc. Conf. Commun.*, B 70.

Agnew, A.D.Q., 1968. Observations on the changing vegetation of Tsavo National Park (East). *East Afr. Wildl. J.*, 6: 75–80.

Ahn, P., 1959. The savanna patches of Nzima, south-western Ghana. *J. West Afr. Sci. Assoc.*, 5: 10–25.

Anderson, G.D. and Talbot, L.M., 1965. Soil factors affecting the distribution of the grassland types and their utilization by wild animals on the Serengeti Plains, Tanganyika. *J. Ecol.*, 53: 33–56.

Asare, E.O., 1962. A note on the vegetation of the transition zone of the Tain Basin in Ghana. *Ghana J. Sci.*, 2: 60–73.

Aubréville, A., 1949a. *Climats, forêts et désertification de l'Afrique tropicale*. Société d'Editions géographiques, maritimes et coloniales, Paris, 351 pp.

Aubréville, A., 1949b. *Contribution à la paléohistoire des forêts de l'Afrique tropicale*. Société d'Editions géographiques, maritimes et coloniales, Paris, 98 pp.

Aubréville, A., 1966. Les lisières forêt-savane des régions tropicales. *Adansonia*, 6: 175–187.

Beard, J.S., 1953. The savanna vegetation of northern tropical America. *Ecol. Monogr.*, 23: 149–215.

Bellier, L., Gillon, D., Gillon, Y., Guillaumet, J.-L. and Perraud, A., 1969. Recherches sur l'origine d'une savane incluse dans le bloc forestier du Bas-Cavally (Côte d'Ivoire) par l'étude des sols et de la biocoenose. *Cah. ORSTOM Sér. Biol.*, 10: 65–94.

Boaler, S.B. and Sciwale, K.S., 1966. Ecology of a miombo site, Lupa North Forest Reserve Tanzania. III. Effects on the vegetation of local cultivation practices. *J. Ecol.*, 54: 577–587.

Bonvallot, J., Dugerdil, M. and Duviard, D., 1970. Recherches écologiques dans la savane de Lamto (Côte d'Ivoire): répartition de la végétation dans la savane préforestière. *Terre Vie*, 24: 3–21.

Brand, B. and Brammer, H., 1956. Provisional grassland associations of the interior and coastal savannah zones of the Gold Coast. *Gold Coast Dep. Soil Land Use Surv., Techn. Rep.*, No. 21: 17 pp.

Buechner, H.K. and Dawkins, H.C., 1961. Vegetation changes induced by elephants and fire in Murchison Falls National Park, Uganda. *Ecology*, 42: 752–766.

Champion, H.G. and Seth, S.K., 1968. *A Revised Survey of the Forest Types of India*. Government of India Press, Delhi, 404 pp.

Charter, J.R. and Keay, R.W.J., 1960. Assessment of the Olokemeji fire-control experiment (investigation 254) 28 years after institution. *Nigerian For. Inform. Bull.*, 3: 1–32.

Chevalier, A., 1900. Les zones et les provinces botaniques de l'Afrique occidentale française. *C. R. Acad. Sci., Paris*, 130: 1205–1208.

Clayton, W.D., 1958. Secondary vegetation and the transition to savanna near Ibadan, Nigeria. *J. Ecol.*, 46: 217–238.

Cole, M.M., 1960. Cerrado, caatinga and pantunal: the distribution and origin of the savanna vegetation of Brazil. *Geogr. J.*, 126: 166–179.

Cole, N.H.A. and Jarrett, H.O., 1969. Tropical plant communities of limited occurrence. *J. West Afr. Sci. Assoc.*, 14: 95–102.

Crosbie, A.J., 1957. *The Yinahin Range*. Ghana Department of Agriculture, Division of Soil and Land Use Survey, Accra (Mimeograph).

Dawkins, H.C., 1954. Timu, and the vanishing forests of North-East Karamoja. *East Afr. Agric. J.*, 29: 164–167.

Delevoy, G., 1938. A propos de la végétation des savanes boisées. *Bull. Séances, Inst. R. Col. Belge, Bruxelles*, 9: 363–379.

Devred, R., 1956. Les savanes herbeuses de la région de Mvuazi (Bas-Congo). *Inst. Nat. Etude Agron. Congo, Bull. Inform.*, 65: 115 pp.

Dundas, J., 1942. Farming conditions and grass succession near Ilorin. *Farm Forest*, 3: 19.

Eggeling, W.J., 1947. Observations on the ecology of the Budongo rain forest, Uganda. *J. Ecol.*, 34: 20–87.

Eltringham, S.K., 1973. Fluctuations in the numbers of wildfowl on an equatorial hippo wallow. *Wildfowl*, 24: 81–87.

Ewel, J., 1982. Succession. In: F.B. Golley (Editor), *Tropical Rain Forest Ecosystems, A. Structure and Function*. Ecosystems of the World, 14A. Elsevier, Amsterdam, pp. 217–223.

Fairbairn, W.A., 1939. Ecological succession due to biotic factors in northern Kano and Katsina Provinces of Northern Nigeria. *Inst. Pap., Imp. For. Inst.*, 22: 32 pp.

Ferri, M.G. (Editor), 1963. *Simposio sobre o Cerrado*. University of São Paulo, São Paulo.

Field, C.R., 1971. Elephant ecology in the Queen Elizabeth National Park, Uganda. *East Afr. Wildl. J.*, 9: 99–123.

Germain, R., 1965. Les biotypes alluvionnaires herbeux et les savanes intercalaires du Congo équatorial. *Mém. Acad. R. Sci. Outre-Mer*, 15: 399 pp.

Goodall, D.W., 1977. Dynamic changes in ecosystems and their study: the roles of induction and deduction. *J. Environ. Manage.*, 5: 309–317.

Graham, A. (Editor), 1973. *Vegetation and Vegetation History of Northern Latin America*. Elsevier, Amsterdam, 393 pp.

Guillaumet, J.L. and Adjanohoun, E., 1971. Vegetation. In: J.M. Avenard et al. (Editors), *Les Milieu Naturel de la Côte d'Ivoire. Mém. ORSTOM*, pp. 157–263.

Harrington, G.N. and Ross, I.C., 1974. The savanna ecology of Kidepo Valley National Park I. The effects of burning and browsing on the vegetation. *East Afr. Wildl. J.*, 12: 93–105.

Heady, H.F., 1966. Influence of grazing on the composition of *Themeda triandra* grassland, East Africa. *J. Ecol.*, 54: 705–727.

Hills, T.L. and Randall, R.E., 1968. *The Ecology of the Forest/Savanna Boundary*. McGill University Savanna Research Project. Savanna Research Series No. 13. Department of Geography McGill University, Montreal, Que., 128 pp.

Hopkins, B., 1962. Vegetation of the Olokemeji Forest Reserve Nigeria I. General features and the research sites. *J. Ecol.*, 50: 559–598.

Hopkins, B., 1965. *Forest and Savanna. An Introduction to Tropical Plant Ecology with Special Reference to West Africa*. Heinemann, Ibadan, 100 pp.

Jarman, P.J. and Thomas, P.I., 1968/69. Observations on the distribution and survival of mopane (*Cochlospermum mopane* (Kirk ex Benth.) Kirk ex J. Leonard) seeds. *Kirkia*, 7: 103–107.

Jones, A.P.D., 1945. Notes on terms for use in vegetation description in Southern Nigeria. *Farm For.*, 6: 130–136.

Jones, E.W., 1963. The forest outliers in the guinea zone of Northern Nigeria. *J. Ecol.*, 51: 415–434.

Keay, R.W.J., 1959. *An Outline of Nigerian Vegetation*. Government Printer, Lagos, 46 pp.

Keay, R.W.J., 1960. An example of Northern Guinea Zone vegetation in Nigeria. *Nigerian For. Inform. Bull.*, 1: 1–46.

Keay, R.W.J. and Onochie, C.F., 1947. Some observations on the Sobo plains. *Farm For.*, 8: 71.

Kortlandt, A., 1972. *New Perspectives on Ape and Human Evolution*. Stichting voor Psychobiologie, Amsterdam, 100 pp.

Lamprey, H.F., 1963. Ecological separation of the large mammal species in the Tarangire Game Reserve, Tanganyika. *East Afr. Wildl. J.*, 1: 63–92.

Lamprey, H.F., 1979. Structure and function of the semi-arid grazing land ecosystem of the Serengeti Region (Tanzania). In: *Tropical Grazing Land Ecosystems, a State of Knowledge Report Prepared by UNESCO/UNEP/FAO. Nat. Resour. Res.*, 16: 562–601.

Lamprey, H.F., Glover, P.E., Turner, M.I.M. and Bell, R.H.V., 1967. Invasion of the Serengeti National Park by elephants. *East Afr. Wildl. J.*, 5: 151–166.

Laws, R.M., 1968. Interactions between elephant and hippopotamus populations and their environments. *East Afr. Agric. For. J., Spec. Iss.*, 33: 140–147.

Laws, R.M., Parker, I.S.C. and Johnstone, R.C.B., 1970. Elephants and habitats in North Bunyoro, Uganda. *East Afr. Wildl. J.*, 8: 163–180.

Laws, R.M., Parker, I.S.C. and Johnstone, R.C.B., 1975. *Elephants and Their Habitats: the Ecology of Elephants in North Bunyoro, Uganda*. Oxford University Press, Oxford, 376 pp.

Lebrun, J., 1936. La forêt équatoriale congolaise. *Bull. Agric. Congo Belge*, 27: 163–192.

Leneuf, N. and Aubert, G., 1956. Sur l'origine des savanes de la basse Côte d'Ivoire. *C. R. Acad. Sci., Paris*, 243: 859–860.

Letouzey, R., 1968. *Etude phytogéographique du Cameroun*. Lechevalier, Paris, 511 pp.

Leuthold, W., 1977. Changes in tree populations of Tsavo East National Park, Kenya. *East Afr. Wildl. J.*, 15: 61–69.

Lock, J.M., 1972. The effects of hippopotamus grazing on grasslands. *J. Ecol.*, 60: 445–467.

Lock, J.M., 1977. Preliminary results from fire and elephant exclusion plots in Kabalenga National Park, Uganda. *East Afr. Wild. J.*, 15: 229–232.

MacGregor, W.D., 1937. Forest types and succession in Nigeria. *Emp. For. J.*, 16: 234–242.

Miège, J., 1965. Les savanes et forêts claires de Côte d'Ivoire. *Etud. Edurnéenes*, 4: 62–81.

Miège, J., 1966. Observations sur les fluctuations des limits savanes-forêts en basse Côte d'Ivoire. *Ann. Fac. Sci. Univ. Dakar*, 19: 149–166.

Miège, J., Bodard, M. and Carrère, P., 1966. Evolution floristique des végétations de jachère en fonction des méthodes culturales à Darau (Senegal). *Trav. Fac. Sci., Univ. Dakar*, 58 pp.

Misra, R., 1946. A study in the ecology of low-lying lands. *Indian Ecol.*, 1: 27–47.

Molinari, O.G., 1949. Succession of grasses in Puerto Rico. *Rev. Agric., Puerto Rico*, 39: 199–217.

Moore, R.M., 1973. Australian Grasslands. In: R.M. Moore (Editor), *Australian Grasslands*. Australian National University Press, Canberra, A.C.T., pp. 87–110.

Morgan, W.B. and Moss, R.P., 1965. Savanna and forest in Western Nigeria. *Africa*, 35: 286–294.

Mullenders, W., 1954. La végétation de Kaniama (Entre-Lubishi-Lubilash, Congo Belge). *Inst. Natl. Etude Agron. Congo, Ser. Sci.*, No. 61: 499 pp.

Ndawula-Senyimba, M., 1972. Some aspects of the ecology of *Themeda triandra*. *East Afr. Agric. For. J.*, 38: 83–93.

Nye, P.H., 1958. The mineral composition of some shrubs and trees in Ghana. *J. West Afr. Sci. Assoc.*, 4: 91–98.

Nye, P.H. and Greenland, D.J., 1960. *The soil under shifting cultivation. Commonw. Agric. Bur., Tech. Commun.*, 51: 156 pp.

Portères, R., 1950. Problèmes sur la végétation de la Basse Côte d'Ivoire. *Bull. Soc. Bot. Fr.*, 97: 153–156.

Ramsay, J.M. and Rose-Innes, R., 1963. Some quantitative observations on the effects of fire on the guinea savanna vegetation of northern Ghana over a period of eleven years. *Afr. Soils*, 8: 41–86.

Rattray, J.M., 1960. *The grass cover of Africa. FAO Agric. Stud.*, No. 49: 168 pp.

Richards, P.W., 1952. *The Tropical Rain Forest*. Cambridge University Press, Cambridge, 450 pp.

Richardson, W.D., 1963. Observations on the vegetation and ecology of the Aripo savannas, Trinidad. *J. Ecol.*, 51: 295–313.

Robyns, W., 1932. Contribution à l'étude de la végétation du Parc National Albert. La colonisation végétale des lavas récentes du volcan Rumoka (laves de Kateruzi). *Inst. R. Col. Belge, Sect. Sci. Nat. Med.*, 1: 3–33.

Robyns, W., 1936. Contribution à l'étude des formations herbeuses du district forestier central du Congo belge. *Inst. R. Col. Belge, Sect. Sci. Nat. Med.*, 5: 151 pp.

Schnell, R., 1950. *La Forêt Dense*. Lechevalier, Paris, 330 pp.

Schnell, R., 1952. Végétation et flore de la région montagneuse du Nimba (A.O.F.). *Mém. Inst. Fr. Afr. Noire*, 22: 604 pp.

Schnell, R., 1970, 1971. *Introduction à la Phytogéographie des Pays Tropicaux. Vol. I. Les flores et les structures. Vol. II, Les milieux. Les groupements végétaux.* Gauthier-Villars, Paris, 2 vols., 980 pp.

Schnell, R., 1977. *Introduction à la Phytogéographie des Pays Tropicaux. Vol. 4. La Flore et la Végétation de l'Afrique tropicale.* Gauthier-Villars, Paris, 378 pp.

Seif El Din, A. and Obeid, M., 1971. Ecological studies of the vegetation of the Sudan. IV. The effect of simulated grazing on the growth of *Acacia senegal* (L.) Wild seedlings. *J. Appl. Ecol.*, 8: 211–216.

Sillans, R., 1958. *Les Savanes de l'Afrique Centrale.* Encyclopedia Biologique, Lechevalier, Paris, Vol. 55, 424 pp.

Strang, R.M., 1974. Some man-made changes in successional trends on the Rhodesian highveld. *J. Appl. Ecol.*, 11: 249–263.

Taylor, B.W., 1963. An outline of the vegetation of Nicaragua. *J. Ecol.*, 52: 27–54.

Thomas, A.S., 1951. The vegetation of the Sese Islands, Uganda. *J. Ecol.*, 29: 330–353.

Thornton, D.D., 1971. The effect of complete removal of hippopotamus on grassland in the Queen Elizabeth National Park, Uganda. *East Afr. Wildl. J.*, 9: 47–55.

Vesey-FitzGerald, D.F., 1960. Grazing succession among East African game animals. *J. Mammal.*, 41: 161–172.

Vesey-FitzGerald, D.F., 1964a. Ecology of the red locust. In: Davis, D.H.S. (Editor), *Ecological Studies in Southern Africa.* Junk, The Hague, pp. 255–268.

Vesey-FitzGerald, D.F., 1964b. Grasslands. *IUCN Publ.*, 4: 111–115.

Vesey-FitzGerald, D.F., 1965. The utilization of natural pastures by wild animals in the Rukwa Valley, Tanganyika. *East Afr. Wildl. J.*, 3: 38–48.

Vesey-FitzGerald, D.F., 1970. The origin and distribution of valley grasslands in East Africa. *J. Ecol.*, 58: 51–75.

Vesey-FitzGerald, D.F., 1973a. Animal impact on vegetation and plant succession in Lake Manyara National Park, Tanzania. *Oikos*, 24: 314–325.

Vesey-FitzGerald, D.F., 1973b. Browse production and utilization in Tarangire National Park. *East Afr. Wildl. J.*, 11: 291–305.

Vesey-FitzGerald, D.F., 1974. The changing state of *Acacia xanthophloea* groves in Arusha National Park, Tanzania. *Biol. Conserv.*, 6: 40–47.

Walter, H., 1971. *Ecology of Tropical and Subtropical Vegetation.* Oliver and Boyd, Edinburgh, 539 pp.

West, O., 1965. Fire in vegetation and its use in pasture management with special reference to tropical and subtropical Africa. *Commonw. Bur. Pastures Field Crops, Bull.*, No. 1: 53 pp.

Whyte, R.O., 1974. *Tropical Grazing Lands.* Junk, The Hague, 222 pp.

Wilhelmy, H., 1957. Das grosse Pantanal in Mato Grosso. In: *Geographentag Würzburg*, pp. 45–71.

Chapter 30

THE FIRE PROBLEM IN TROPICAL SAVANNAS

DOMINIQUE GILLON

INTRODUCTION

Grass fires are so intimately associated with savanna landscapes that many people still consider all tropical savannas as fire-climaxes, particularly in Africa. This is obviously a gross over simplification of a complex problem, as shown by other contributors to this volume. However, fire exerts such an important influence on the soil, the vegetation and the fauna of tropical savannas, that its ecological role needs to be reviewed in more detail. In this chapter I shall successively discuss the location and origin of grass fires, their characteristics, and their impact upon plants, soils and animals. This contribution is based upon my own field experience in West Africa as well as on the data from the literature.

LOCATION AND ORIGIN OF GRASS FIRES

Large-scale grass fires are more likely to take place in areas having a climate moist enough to permit the production of a large amount of grass, and seasonally dry enough to allow the dried material to catch fire and burn quickly. It is therefore no surprise that the tropical areas where grass fires are the most frequent and widespread are those where the rainy and dry seasons alternate most regularly. Batchelder (1967) has mapped the areas of the tropical world where fires occur regularly, giving the times at which they are most frequently observed. Phillips (1968) has also compared the frequency of fires in the various geographical and bioclimatic zones of tropical Africa. The savannas which burn most frequently are those whose seasonal rainfall is relatively high — that is,

the derived savannas (or forest/savanna mosaic), the moist wooded savannas [or Guinea(n) savannas], and to a lesser extent, the dry wooded savannas [or Sudan(ian) savannas]. Perennial grass species, by far the most numerous here, quickly start growing after burning as a result of the high atmospheric moisture. On the other hand, fires are far less numerous in Sahelian savannas where the sparse ground vegetation limits the spread of fires; in such savannas, however, the grass cover takes a longer time to recover after burning, since most grass species are annuals.

Grass fires may be natural or man-induced. Natural fires have undoubtedly occurred since the earliest appearance of land vegetation, and at present they continue to occur in all tropical areas (Budowski, 1958; Komarek, 1965, 1971; Lacey et al., 1979), their most frequent cause being lightning. Their frequency, however, is far less than man-made fires, man having learned at a very early date some advantages of the use of vegetation fires.

In African savannas, intentional burning has been practiced for at least 50 000 years (Clark, 1959; Rose-Innes, 1971). In America, it has long been thought that man began to set fire to tropical grasslands at a much later date. More recent archaeological research has shown that Amerindians were already living in central Brazil more than 10 000 years ago (Coutinho, 1979). In the *llanos* of Columbia and the Guyana savannas, modern palynological studies have nonetheless established that the impact of fire on these grasslands was still almost negligible 5000 years ago (Batchelder, 1967). The Australian aborigines entered their island continent about 40 000 years ago and traditionally used grass burning as a form of incipient animal husbandry (Mulvaney, 1975;

Lacey et al., 1979). Thus, man-made fires have a very long history, in most of the Palaeotropics at least.

At the present time, man is obviously responsible for the induction of most grass fires in tropical savannas — especially in areas of high human population density. Seasonal burning has even become a tradition in many places. Deliberate grass burning serves many functions. It is used to drive out game animals, to encourage new succulent growth to attract game, to clear ground for agriculture, to produce or maintain good grazing land for domesticated herds, or even to establish fire breaks around permanent settlements and to improve communication.

THE CHARACTERISTICS OF GRASS FIRES

Temperatures during vegetation fires

The rise in temperature during the advance of the burning front is very brief and does not last for more than four to five minutes, the highest temperatures not persisting for more than a few seconds (Fig. 30.1).

At the soil surface, the temperatures can become very high, but vary a great deal from place to place, even between areas a few metres apart: 90 to 140°C (Pitot and Masson, 1951); 220 to 340°C (Rains, 1963); 32 to 538°C (Hopkins, 1965); 75 to 350°C (Gillon and Pernès, 1968). Inside a burning grass tussock itself, the temperature rises much less: to 52 or 56°C according to Adam and Jaeger (1967) or 65°C according to Gillon and Pernès (1968).

Above soil level, temperatures rise very quickly with height and reach their maximum in the densest part of the grass cover — that is, at a height of about 50 cm, when grass is 1.20 to 1.50 m high. Here the maximum temperature reaches 500 to 600°C (Pitot and Masson, 1951; Rains, 1963; Hopkins, 1965; Viani et al., 1973). Above this temperatures decrease, but remain high up to 3 m above soil surface.

Within the soil, the temperature rise under the burning front is much smaller and decreases rapidly with increasing depth. Just under the soil surface it did not exceed 60°C in the Ivory Coast (Gillon and Pernès, 1968). An increase in temperature of from 3 to 14°C was recorded 2 cm below the soil surface in Senegal (Masson, 1948; Pitot and Masson, 1951). This increase did not exceed 3°C at a depth of 3 cm in Venezuela (Vareschi, 1962), and no change in temperature was recorded 5 cm below soil surface in the Brazilian *cerrado* (Coutinho, 1979).

The thermal effect of the blaze is therefore of short duration and limited in height to the layer of the grass cover where the amount of fuel is greatest. The temperature rise inside grass tussocks is moderate, and almost nil underground. This explains why so many invertebrates and soil organisms, as well as the underground parts of savanna plants, can withstand the passage of the narrow zone of flames.

The severity of burning

The amount of heat released during a grass fire depends on a number of factors.

First of all, the time of the year at which it occurs. In the Olokemeji Forest reserve, Hopkins (1965) has observed that the temperatures recorded during fires were lower at the beginning of the dry season, when the grass was still green, than in the middle or the end of the dry season when most of the above-ground grass biomass consisted of dead material.

Second, the amount of "grass fuel" available. In the northern Nigerian savannas, Rains (1963) has shown that the height of the flames of the burning front increased with the available grass biomass.

Third, the time elapsed since the last fire. If grass fires occur infrequently, they are far more destructive than if they take place annually.

Other factors to be taken into account are the velocity and direction of the wind (headfires versus backfires), the slope, the atmospheric moisture, and even the time of the day.

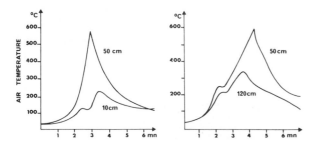

Fig. 30.1. The rise in temperature at different heights (10, 50 and 120 cm) during the passage of the burning front, in the course of a grass fire at Lamto, Ivory Coast. The average height of the grass cover was 70 cm. (After Viani et al., 1973.)

In a given area, the intensity of the fire and the speed at which it spreads (500 m h^{-1} on the average in derived savanna in the Ivory Coast) can vary depending on the place and time. At Lamto, fires can burn many hectares of grass in a single day and can last for several days in succession; on the other hand, they can also stop short quickly, when they reach previously burned areas, gallery forests or marshy ground. At least in this part of Africa the end-result is a patchwork landscape of burnt and unburnt areas which increases the heterogeneity of the environment, especially since fires do not all occur at the same time and with the same periodicity. The severity of grass fires in similar savannas of southern Nigeria has been estimated by Hopkins (1965) by calculating the percentage of ground vegetation destroyed by fire. An early burning can destroy as little as 25% of the grass cover, whereas late burnings can destroy 64 to 96% of the grass cover (84% on the average). In the Ruwenzori National Park, Uganda, when fires were uncontrolled, the areas covered by grassland burns have been recorded in six dry seasons, from July 1970 to July 1973 (Eltringham, 1976). During these three years 55% of the Park was burnt on at least one occasion, and on the average 13.4% was burnt annually.

In the Columbian llanos the burning front seldom exceeds 2 to 5 km, being stopped by various obstacles ranging from cattle trails to drainage ways (Blydenstein, 1968). In northern Australian savannas, fires occurring towards the end of the dry season are much more extensive and destructive than those taking place earlier (Lacey et al., 1979).

THE ENVIRONMENTAL EFFECTS OF GRASS FIRES

Effects of burning on the grass layer

Phenology

The impact of fire upon the annual growth cycle of the grass cover differs greatly between various climatic zones. New shoots on a fresh burn appear more quickly when the moisture content of the soil and atmosphere is high.

In the derived savanna of the Ivory Coast (average annual rainfall 1300 mm), the above-ground biomass of the grass layer reaches its peak values during the dry season. However, this standing crop is largely made up of dead material and most of the nutrients have already been translocated underground, where the root biomass reaches its maximum value precisely at this time of the year (Figs. 30.2 and 30.3). Once the grass has been burnt, new shoots appear with surprising promptness, and the above-ground grass biomass averages 1 t ha^{-1} (dry weight) just one month after the fire, despite the fact that very little rain (or no rain at all) has fallen in the interim. As mentioned earlier, the root biomass decreases continuously from December to June, then reaches relatively stable values until the next October when it rapidly increases, attaining a maximum again in December. At this time of the year (i.e. the dry season) when root growth is apparently nil, the increase in below-ground biomass is probably due to storage of food within the root system of perennial herbs (Monnier, 1968; César and Menaut, 1974). Such a rapid post-fire regrowth takes place in all wet savannas. Neal (1970) has described a similar course of events in western Uganda where, four months after the fire, the grass cover of burnt areas was quite similar to that of unburnt ones.

The situation is very different in dry savannas, where burning retards the development of most grass species (Brockington, 1961; Skovlin, 1971). Taking place at a time when the grass has stopped growing and is in a dormant stage, fire accentuates the effects of the seasonal drought. However, it can

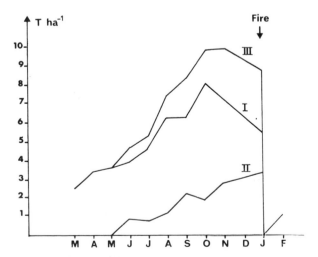

Fig. 30.2. The seasonal variations in above-ground grass biomass (in t ha^{-1}) in the Lamto savanna, Ivory Coast: *I* = live grass; *II* = dead grass material; *III* = total grass biomass (after César and Menaut, 1974).

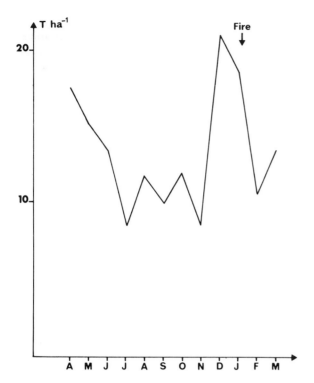

Fig. 30.3. Seasonal variations of the root biomass (in t ha^{-1}) of grass in the Lamto savanna, Ivory Coast (after César and Menaut, 1974).

also stimulate sprouting, the high temperatures acting either as a "sprouting promoter" or as a suppressor of a "sprouting inhibitor" (Hopkins, 1963). According to Granier and Cabanis (1976), this stimulation might also be due to the "heat shock" of the fire which, once growth inhibition is removed by the disappearance of the vegetative organs of the plant and of its terminal buds, provides in a short time the same amount of energy as is needed in the spring for the sprouting of buds. In this way, the "heat shock" would mobilize the nutrients stored in the root system, as well as the water retained in the tissues.

Primary production

The percentage of the annual grass production burnt during a grass fire has been estimated on a few occasions. In the derived savanna of Olokemeji, Nigeria, Hopkins (1965) has made the following calculations. The dry weight of the herb layer in the area studied reached its maximum in November and December, when it amounted to about 6.8 t ha^{-1}. Later in the dry season the weight fell by

almost one quarter, as seeds and fruits were dispersed, leaves fell from the forbs, and food reserves were translocated to the root system. At the time of burning the dry weight of the herb layer averaged 5.2 t ha^{-1}. Only 0.6 t ha^{-1} remained after burning. The heat produced by a "typical" fire averages 1850 kcal m^{-2} (7733 kJ m^{-2}) and the calorific value of the remaining ash layer was only 23 kcal m^{-2} (96 kJ m^{-2}). Hence in this case at least three quarters of the annual production of the aerial shoots of herbs was lost to the biological system when the savanna was burnt after December. Hopkins' figures, however, clearly demonstrate the importance of the time of burning in determining the percentage of the above-ground standing crop destroyed: early in the dry season it was about 25%, but it averaged 84% from January onwards. In derived savanna in the Ivory Coast, César and Menaut (1974) estimated that the amount of grass layer material destroyed by the January burnings ranged from 6.0 to 8.9 t ha^{-1}, depending upon the savanna facies concerned.

On a long-term basis it is difficult to appreciate the impact of repeated burnings on above-ground grass production. The data published so far are contradictory and comparison is made even more difficult by the climatic differences among the areas involved. In the llanos of Columbia, for instance, San Jose and Medina (1975) recorded an above-ground biomass of 4.15 t ha^{-1} in a burnt area, as against a biomass of only 3.25 t ha^{-1} in the protected area. In the cerrado of central Brazil, the action of fire is apparently negligible according to Coutinho (1979). On the other hand, burning savanna grasses in northern Australia reduces their productivity. The later the burn occurs in the wet season the greater the reduction (Smith, 1960). Similar results have been reported from Zambia (Brockington, 1961).

In a single area the impact of fire on subsequent primary production also depends a great deal on the time of burning and the amount of grass fuel available. In the Venezuelian llanos, burning of the *Trachypogon* savanna at the start of the dry season (November–December) resulted in an increase of shoot production during the first post-burn season of about 50%, whereas a later burning reduced the production (Blydenstein, 1963). The same author has also shown that burning in November increased shoot production and plant height more than did clipping on the same date, whereas the comparison between the same treatments was reversed during the wet season. Also working in the llanos, Aristeguieta and Medina (1966) have noticed a much smaller regrowth of the grass after a fire if the savanna had previously been unburnt, probably because of the accumulation of dead material which increased the severity of the burning.

The quality of the grass

For a long time, traditional cattle herders have recognized the attraction new grass appearing after a fire has for grazing mammals, wild or domestic. The inferred changes in the nutritive quality of the plant shoots during the post-fire period have received much attention.

Working in Malawi, Lemon (1968b) has shown that the protein content of the mature grass was much higher (16%, five weeks after a late season burning) in burnt areas than in grasslands which were protected from fire for three and eleven years (8.7% and 7.8% of crude protein respectively). Quite on the contrary, the protein content of the legumes was the same in both grasslands. The mineral content of the forage samples was not affected by the fire either, with the exception of potassium and manganese which showed some increase in the burnt areas, possibly because of the release of ash materials.

Though not carried out in tropical savannas, the study of Allen et al. (1976) gives an insight into the ways in which burning and the presence of ashes can influence the chemical composition of grass. In the Kansas grassland they studied, range burning increased the nutritive value of carbohydrates and increased the nitrogen, protein and mineral content of the plants. It also diminished their content of crude fibre and cell-wall constituents such as cellulose and lignin — all constituents whose percentage is generally low in high-quality grass.

It can therefore be concluded that burning increases the nutritive value of the plants, but it must be kept in mind that in experiments of this kind the two categories of grass compared are of a different biological age. It has been known for a long time that young grass leaves are richer in protein, calcium and phosphorus than older, drying leaves whose contents of nitrogen, calcium and phosphorus steadily decrease while their indigestible material increases. This has, for instance, been convincingly established for *Themeda triandra* and *Cynodon dactylon* in Kenya by Dougall and Glover (1964).

To establish the true effects of fire on grass in burnt areas, Christensen (1977) compared the nutrient content of shoots of the same age, after burning and clipping. He found that the concentrations of nitrogen, calcium and magnesium were higher in plant tissues from the burnt area than in clipped plants, while concentrations of potassium and phosphorus were not significantly different between the two areas. Thus, increased tissue content of potassium and phosphorus following burning may be a reflexion of tissue age rather than increased availability of these nutrients due to fire. Leaf concentrations of nitrogen and magnesium in the clipped plots were intermediate between those of the unburned and burned areas, suggesting that at least a part of the increase in concentration of these nutrients in burned areas may have been due to tissue age. Six to twelve months after burning, tissue concentrations of all nutrients in leaves from burned areas were significantly lower than in those from the unburned areas. Christensen' study was carried out in a pine–wiregrass (*Pinus palustris–Aristida stricta*) savanna of North Carolina, and it is likely that his conclusions also apply to tropical grasslands.

During a grass fire, part of the minerals present in the vegetation are lost to the atmosphere (nitrogen, sulphur and phosphorus), or dispersed in very fine atmospheric particles (calcium, potassium and magnesium). The remaining minerals stay in the ashes, on the ground. In North Carolina the atmospheric loss of nitrogen has been estimated at 70% by Christensen (1977). Part of the nutrients lost to the atmosphere return to the soil by gravity. or in solution as rain water. Coutinho (1979) measured the amount of this return month by month in a Brazilian cerrado. He found that there was a normal monthly nutrient import correlated with the amount of rain, and an occasional import correlated with the deposition of ash fragments derived from fires a short distance away. In the derived savanna of Lamto (Ivory Coast), Villecourt et al. (1979) reported a substantial decrease in the amounts of nitrogen, phosphorus and potassium (by a factor of 1.6 to 1.8) in the aerial portion of grasses, and showed that the loss by fire is partly (potassium) or entirely (nitrogen and phosphorus) compensated by rain water.

During a grass fire, therefore, a very large loss of both energy and nutrients takes place, but organic matter is also mineralized and part of the minerals and nitrogen is quickly recycled. To study the influence of the ashes on the post-fire regrowth, Coutinho (1979) removed the ashes after burning on an experimental plot and studied the primary production during the subsequent months on this plot with the ash removed and on a control area. The removal of ashes decreased the above-ground primary production after the fire. Thus, the role of ashes, though transitory, is very important in regularly burnt savanna.

Community composition

Over the millennia, most savanna plants have become adapted to burning, through natural selection. This is particularly true of the grass species which are adapted to both fire and drought.

Definite life forms predominate in fire-maintained communities: annuals (therophytes), and hemicryptophytes and geophytes among perennial herbaceous plants. The buds buried in the soil or protected by the bases of tillers and leaf sheaths, are fire-resistant, and the buried seeds can withstand a temporary and moderate rise in temperature (West, 1971; Granier and Cabanis, 1976).

In the derived savanna at Lamto, Ivory Coast, the plant community includes many geophytes and most grasses are hemicryptophytes. The basal buds are well protected at the base of the tussocks, where temperatures seldom reach a lethal level during the fire (see p. 618). Seed-reproducing therophytes are scarce (César and Menaut, 1974).

Repeated burning also favours rapidly growing species — that is, those which quickly provide a new grass cover. Through litter destruction and humus mineralization it also increases the dryness of the habitat and favours xeric species. Prevention of burning also puts caespitose species to a disadvantage — plants no longer competing for room in an horizontal plane, but for space in a vertical plane. Species well adapted to such competition being scarce among perennial grasses, suffrutescent plants will progressively become dominant, among them many legumes.

The prolonged absence of burning therefore results in a change in plant community structure, a change in vegetation stratification, and the replacement of fire-adapted life forms by others less resistant to burning (Granier and Cabanis, 1976). In derived savannas in Ghana, Ramsay and Rose-Innes (1963) have also recorded a decrease in grass cover and a reduction of the area occupied by short-grass species in fire-protected plots; tall-grass species increased in the mean time. In derived savanna at Lamto, protection from burning first causes changes in the community composition of the grass layer, then bush encroachment, and finally, an increase in the number of woody species — the end-result of all these changes being important micro-climatic changes in the savanna itself (Vuattoux, 1970, 1976; Menaut, 1977). The role of fire at the forest/savanna boundary is further discussed in Chapter 28.

Flowering activity

The flowering of many savanna species shortly after burning is a well-known phenomenon (West, 1971). Adam and Jaeger (1976) reported that some species of Poaceae and Cyperaceae of the Guinean savannas no longer flowered in the absence of grass fires, and that the flowering activity of other herbaceous species was markedly decreased. They concluded that fire is necessary to stimulate the reproductive organs of such plants. In the absence of burning, the vegetative activity of the plants would be favoured at the expense of reproductive activity. In Brazil, Coutinho (1979) studied the influence of fire on the flowering activity of many grasses. Ten days after burning, the morphology of dormant buds began to change, and they started to differentiate into reproductive organs. Thus, the post-fire bloom of the savannas would not result from the activation of already differentiated structures, but from morphogenetic processes inducing the appearance of reproductive organs. In the same way as drought, fire destroys the aging parts of plants and possibly the growth-inhibiting factors they might contain. Burning also acts as a synchronizer of flowering at the population level, and thus facilitates sexual reproduction in many species, increasing their vitality in the meantime.

Seed survival and germination

Since seeds are generally buried in the soil and dormant during the dry season, burning does not harm them much, especially when they are specifically adapted to enter the soil surface. For instance, the awns of seeds of Andropogonae, once humidified by atmospheric moisture, describe a rotatory movement which helps them enter the soil (Granier and Cabanis, 1976). When burning takes place earlier in the season, before the seeds have dropped, or later on when they are germinating, most of them are killed. Seeds not killed by fire, however, may germinate earlier as a result of it. Thus, the germination of fresh seeds of *Themeda triandra* can be induced by treatment with dry heat (West, 1965). Granier and Cabanis (1976) have also noticed a higher germination percentage of seeds in burnt West African savannas. The same has also been recorded by Lacey et al. (1979) in Australia, although they were uncertain whether this effect was due to a stimulation of the germination processes by fire, or to reduced competition.

The effect of burning upon woody vegetation in savannas

Phenology

In the derived savanna of Lamto, Ivory Coast, there is a definite phenological cycle for the woody component of the community. Buds start to appear at the end of the dry season, flowering and fruiting take place during the rainy season, and fruits and seeds mature at the end of the rains. Fires therefore take place during the "rest" phase of the woody vegetation.

Leaf fall occurs mostly during the dry season, but it is spread over a period which varies from year to year, whether the savanna burns or not. If the leaves start falling early during the dry season, the litter has no time to decompose and is burnt by the next grass fire. On the other hand, when the leaves fall after burning they form a fresh litter which plays a crucial role in protecting the soil and ground fauna (César and Menaut, 1974).

Tree density

Fire damage to savanna trees varies greatly with the time of burning, the intensity of the fire, the species concerned, and the size and phenological stage of the tree. In the derived savanna of Olokemeji, Nigeria, an area subjected to late burning in five consecutive years suffered a 32% reduction of the tree population (Hopkins, 1965) — an average loss of 7.9% per year. The loss was far

from uniform. The chance of a tree's survival increased considerably with increasing height and basal area. There were two reasons for the greater mortality of small individuals. First, their shoots were entirely within the zone of high fire temperatures, so that apical buds were likely to be damaged. Second, their bark was thinner than that of larger individuals, so that the cambium was less protected (Fig. 30.4). In East Africa, Glover (1968) also found that trees 1 to 4 m high were the most vulnerable to fire. Working in Zambia, Trapnell (1959) found average annual mortality rates of 0.38, 0.64 and 1.58% for trees of over 18-cm^2 basal area in his fire-protected, early-burnt and late-burnt plots respectively. These low mortality rates, as compared to those of Olokemeji, are possibly due to a less dense tree population and thicker herb layer at Olokemeji. Trapnell, however, also emphasized the drastic effects of late burning on regeneration, most notably on the canopy dominants.

Since young trees suffer most from burning, it is no wonder that tree regeneration is especially active when a savanna area is protected from fires. This was very obvious in the derived savanna of Lamto. Vuattoux (1970, 1976) and Menaut (1977) have shown that both the number of woody species and the number of individuals increased very quickly

Fig. 30.4. The burning of small trees during a grass fire in the derived savanna at Lamto, Ivory Coast (photo by Yves Gillon).

during the first three years of fire protection, and continued to increase regularly (Fig. 30.5). The number of individuals increased 6-fold in 3 years, and 13-fold in 13 years of fire protection. This increase in density was due to the existing savanna species, since the number of species present did not change much during the first six years following cessation of burning (Fig. 30.6). Later on, the situation changed. The increased shrub cover and its resulting shade inhibited the growth of the grass layer, and the habitat then became favourable for settlement by forest species (Menaut, 1977). Grass burning at Lamto not only keeps the density of

savanna trees at a given level, but also prevents the settlement of the area by forest trees.

Fire resistance

According to Glover (1968), the ability of some savanna trees to survive repeated burning is due to their ability to take to a geophytic habit, while their aerial parts are repeatedly destroyed by fire. He noticed, both in Rhodesia and in the Serengeti National Park (Kenya), that most of the woody species of small size had stout root stocks at the ground level or just below, showing numerous fire scars. In derived savannas, most of the woody species also assume hemicryptophytic or geophytic forms which last as long as the individual can grow shoots large enough to resist burning (César and Menaut, 1974). Other fire adaptations of woody plants are the development of a thick bark, and the ability to grow shoots from undergound parts of the plant (see also Ch. 5). In the savannas of northern Australia, some eucalypts can produce lignotubers, sometimes completely buried in the soil, when environmental conditions become harsh. If the aerial parts of the tree are destroyed by fire, the lignotubers can grow shoots again (Lacey et al., 1979).

Woody plants in the savanna, in general have the option of surviving underground, when burning occurs repeatedly. They can also quickly regenerate their aerial parts when given respite from grass fires for a few years. This is why it is so difficult to prevent "bush encroachment" in fire-protected savanna areas (Glover, 1968).

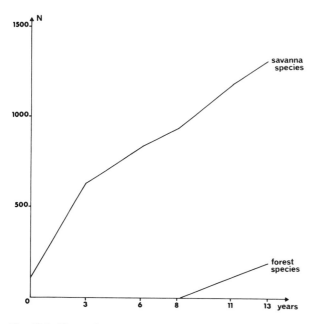

Fig. 30.5. Change in tree density (number of trees per hectare) following complete protection from burning in the derived savanna of Lamto, Ivory Coast (after Menaut, 1977).

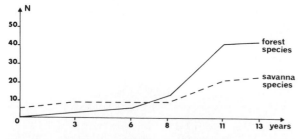

Fig. 30.6. Change in numbers of woody species per hectare following complete protection from burning in the derived savanna of Lamto, Ivory Coast. (after Menaut, 1977).

Effects of fire on the soil

Burning not only substantially reduces the amount of organic material which would otherwise be incorporated into the soil, but also leaves the unprotected soil surface exposed to sunlight and rain.

Microclimatic consequences

As shown previously, the rise in temperature at the soil surface during the passage of the burning front is slight. During the following weeks or even months, however, the lack of grass cover causes prolonged microclimatic changes at the soil level. In derived savanna in the Ivory Coast it takes about six months for the grass cover to be completely restored, and for the litter to re-accumulate.

Soil temperature. Just after a fire the temperature of the bare, black, ash-covered, soil surface is much higher than that of the unburnt areas; the thermal amplitudes are also much greater (Fig. 30.7). At Lamto, the daily maxima of temperature in the top 10 cm of the soil exceeded the daily atmospheric maxima for six months after the January fire. The situation changed only in July, when the grass cover was dense enough to prevent sunshine from reaching ground level. During the rest of the year, the situation within the grass layer was similar in the two plots; the daily temperature maxima of the upper soil layer were always smaller than the atmospheric maxima, whereas the daily soil minima exceeded the atmospheric minima (Athias et al., 1975).

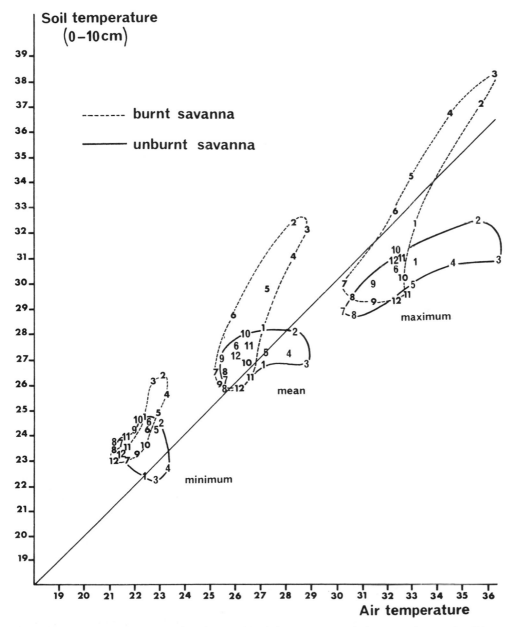

Fig. 30.7. The minimum, average and maximum soil and air temperatures, during successive months of the year, in a regularly burnt plot and in a plot protected from fire for eight years in the derived savanna at Lamto, Ivory Coast. The successive months are numbered from *1* to *12*. (after Athias et al., 1975.)

Soil moisture. Burning also greatly influences soil moisture. In the derived savanna at Lamto the lowest moisture values were attained during the weeks immediately following the fire, pF values often reaching 4.7 in the top 10 cm of the soil (Athias et al., 1975). Burning therefore exaggerates the effects of the dry season.

In the Venezuelian llanos San Jose and Medina (1975) have compared the seasonal changes in soil-water content in two plots of pure grassland, one of which was burnt near the middle of the dry season. Following burning, the soil-water content in the burnt plot was higher than in the control area, due to reduced transpiration. This difference disappeared after the first rains, vegetation on the burnt plot having developed significantly by this time. The maximum values on the burnt plot remained lower than expected from the water retention capacity of the soil. During the rainy season, all soil layers in the control plot were moistened homogeneously, the deepest levels remaining near saturation until the end of the rains. During two short dry periods which occurred toward the end of the rainy season, the upper soil layer in both plots showed strong oscillations in water content. In the burnt plot, the upper 5 cm never reached saturation, probably because of the absence of a complete soil cover, resulting in increased evaporation and run-off. The deepest layers in this plot did not reach saturation either, possibly due to higher root activity.

Erosion

After a grass fire very little material is left on the soil surface, except some plant debris, and the bases of tillers and leaf sheaths. However, in derived savannas and open woodlands where tree and shrub density is high, the leaf fall which follows burning can be important, and compensates somewhat for the loss of the grass litter. At Lamto, for instance, the leaf fall can reach 600 kg ha^{-1} dry weight (Menaut, 1971).

Rains falling upon bare soil cause erosional losses, with ashes and plant debris washed off first. In the Ivory Coast savannas, Roose (1971) estimated the amount of surface soil lost by erosion after burning 150 kg ha^{-1} yr^{-1}, as against 50 kg ha^{-1} yr^{-1} in an unburnt control plot. Granier and Cabanis (1976) recorded a loss of 610 kg ha^{-1} yr^{-1} in a grassland burnt yearly, as against

only 13 kg ha^{-1} yr^{-1} in a similar grassland (7% slope) mown annually at the same time of the year. Obviously the effects of burning on erosion and run-off depend greatly on the relief and grazing pressures. This is noticeable even on rather flat land and is catastrophic on steep slopes (Du Plessis and Mostert, 1965; West, 1965).

In Zambia, Brockington (1961) and Trapnell et al. (1976) have also observed an increased compaction of the surface soil in burnt areas, as compared with unburnt ones.

Soil chemistry

In most cases savanna soils are temporarily enriched in mineral nutrients after burning.

Coutinho (1979) has shown, for instance, that the concentration of mineral nutrients in the soil, mainly calcium. potassium and magnesium, increased in the more superficial soil layers of the Brazilian cerrado immediately after burning; this enrichment, however, was of short duration — about four months.

In the deeper soil layers, no increase of nutrient concentration was noticed, implying that most nutrients were taken up by grass roots at the expense of more deeply rooted trees. In this way mineral nutrients from tree branches, twigs, leaves, flowers and fruits are transferred from the tree layer to the grass layer by burning. Christensen (1976, 1977) observed the same phenomenon in North American prairies after grass fires. Burning apparently effected an abrupt release of the nutrients stored in dead leaves and straw, which would otherwise have only gradually become available by the slow decay of plant litter. In this way fire speeds up the recycling process by accelerating mineralization.

Meiklejohn (1955) ascribed the post-burn increase in nitrogen content in the top 25 mm of savanna soils in Kenya to a temporary disturbance of the soil microflora. Anaerobic nitrogen fixers (*Clostridium*), though poorly represented, persisted through the burning. Eleven weeks later, and again five months after the fire, total nitrogen was higher, probably a result of the destruction of nitrifiers which otherwise would have converted much nitrogen into the soluble forms that could leach deeper.

On a long-term basis, burning does not appear to decrease savanna soil fertility either. In the Olokemeji derived savanna, Nigeria, Moore (1960) compared the nutrient content of the top 20 cm of soil in three experimental plots subjected during eight consecutive years to three different fire regimes: total protection from fire; mild fires coming at the start of the dry season; and hot fires coming late in the dry season. Bush encroachment took place in the unburnt plot, some encroachment occurred in the second plot, whereas the third plot remained as it was at the beginning of the experiment. Chemical determinations performed showed that: (a) the amount of organic matter, total nitrogen, and nitrates was always greater in the soil of the plot burnt at the start of the dry season; (b) in the fire-protected plot, the nutrient values were intermediate between those of the two other plots, except that phosphorus was decreased; and (c) the value of the C/N ratio of the fire-protected plot (11.9) was similar to that of a forest soil, whereas in the two burnt plots (13.1 and 13.3 respectively) it resembled that of a savanna soil. Repeated burning can nevertheless decrease fertility on a long-term basis, as suggested by Moore's observations of his "hot fire" plot, where both nitrification and the amount of humus were decreased. Mild fires started early in the dry season had opposite effects, and the grasses were invigorated.

Other factors have to be considered, however. The long-term effects of burning on a deciduous *Brachystegia–Julbernardia* woodland have been studied by Trapnell et al. (1976) in Zambia. The effects of early and late burning and of complete fire protection on the top 15 cm of soil were compared after a period of 23 years. The soils of all experimental plots had a very low organic content and extremely low base saturation. The pH, exchangeable calcium, and phosphate status were markedly improved by the burning treatments, particularly late burning, when compared with the effects of protection. A notable result was the complete lack of effect that fire protection had on the nitrogen and organic matter content of the soil. Here, this lack of effect on organic matter was due to the disposal of litter by large populations of both humus-feeding and wood- and litter-feeding termites. Small mounds of the humus-feeding *Cubitermes* had three to four times the organic carbon concentration of the surface soils, and a marked increase in exchangeable bases with high base saturation. Large mounds of wood-feeding Macrotermitinae showed corresponding increases

in organic carbon over normal subsoil levels and still greater increases in the bases. Trapnell et al. (1976) suggested that fire and termites exert complementary effects on the base status of the soil: the fires promote the return of bases brought down in the form of litter from the woodland canopy, while the termites deplete this return by consuming unburnt litter and thus gradually concentrate the derived bases in their mounds. In either case, the organic content of the soil is kept at a very low level.

Generally speaking, savanna soils are poorer in mineral nitrogen than nearby forest soils, as shown by De Rham (1973) in the Ivory Coast. This has been ascribed to the destruction of the litter by fire and the resulting changes in soil temperature and soil moisture. It must be remembered, however, that at least some savanna grasses can fix atmospheric nitrogen through their symbiosis with N-fixing micro-organisms (Balandreau, 1971; see also Ch. 25 of this volume).

Soil animals

Very little is known about the effects of fire on the soil animals of the savannas. The following statements are based upon the observations of Athias et al. (1975) in the derived savanna of the Ivory Coast. Burning exerts its effects mainly through a decrease in soil moisture and food sources.

As shown previously, dry-season grass fires exaggerate the seasonal decrease in soil moisture which normally takes place after the cessation of the rains. However, this drying out of the soil — which already begins in December at Lamto — proceeds progressively and many soil animals have time to adapt slowly to the new situation. This helps them to withstand the temporary stress of fire.

The various categories of soil animals react differently to a drop in soil moisture. Earthworms first move deeper into the soil, and then enter a phase of inactivity when the pF value exceeds 3. Soil arthropods are less sensitive to drought and cannot become torpid. An abrupt increase in mortality rate of most groups was observed after burning, but recolonization took place very quickly, in Collembola particularly. Some migration into deeper layers of the soil occurred in most groups, except the oribatid mites.

The almost complete disappearance of the litter slows down the recolonization of the first top

TABLE 30.1

Average biomasses (mg m^{-2}, fresh weight) of soil animals in the derived savanna of Lamto, Ivory Coast: comparison of a regularly burnt plot (BS) with a plot protected from burning (UBS) (after Athias et al., 1975)

Primary consumers	BS	UBS
Symphyla	50	20
Protura	4	4
Diplura	130	190
Coccoidea	7	10
Foraging termites	425	340
Ants	400	400
Beetle larvae	200	80?
Total	1216	1044

Detritus eaters	BS	UBS
Earthworms	3000	7600
Pauropoda and Polyxenidae	5	10
Diplopoda	250	750
Acari	70	170
Collembola	38	80
Foraging termites	425	340
Fungus-growing termites	490	490
Ants	200	200
Beetle larvae	70	40?
Total	4548	9690

Soil eaters	Burned	Protected
Enchytreidae	10	—
Earthworms	27 000	15 000
Humivorous termites	500	500
Total	27 510	15 500

Predators	Burned	Protected
Chilopoda	25	75
Pseudoscorpiones	50	620
Acari	70	120
Spiders	25	100?
Ants	1400	1600
Beetle larvae	20	10
Total	1590	2525

centimetre of the soil by microarthropods and diplopods. Population numbers of many saprophagous animals, like oribatid and tydeid mites, are kept at a very low level as long as the litter is not thick enough, and competition for the remaining plant debris is high. Fungus-growing and foraging termites which are capable of storing food in their nests apparently do not suffer much from the temporary food shortage. Saprophagous earthworms temporarily become geophagous, and the breeding season of *Millsonia anomala* is postponed from March to June in burnt savanna.

Therefore, in most savannas the soil microfauna is probably composed of highly adaptable species, which cope with the difficult situation caused by burning by means of a wide range of adaptations: migration to deeper soil layers, change in diet, storage of food, and postponement of their reproductive season.

The major differences between the soil animal community of an annually burnt savanna at Lamto, and that of an area protected from fire for nine years, can be seen in Table 30.1. There is little difference between the biomasses of primary consumers of the two plots, except for Diplura which are more numerous in the unburnt savanna, and Symphyla and foraging termites, two sun-loving groups which are put at a disadvantage by bush encroachment.

Saprophagous soil invertebrates, on the other hand, are much more numerous in the unburnt savanna, particularly earthworms, diplopods and microarthropods. Fungus-growing termites remain remarkably unaffected by the difference in fire regime. Geophagous soil invertebrates, mostly earthworms, are more numerous in the burnt plot than in the unburnt one. The biomass of the secondary consumers, on the other hand, is greater in the fire-protected plot, like the saprophagous soil animals on which they mostly prey.

On a long-term basis, therefore, fire decreases the number of saprophagous soil animals and their invertebrate predators, mostly by reduction of the litter; on the other hand, it favours soil primary consumers and soil eaters.

Effects of fire upon the above-ground fauna

Arthropods of the grass layer

Data on the effect of fire upon the arthropods of

the grass layer are scanty. Early publications dealt with the use of fire for controlling populations of tsetse flies. Bush clearing by burning has been used extensively to decrease tsetse numbers in many parts of Africa (Swynnerton, 1925, 1934, 1936; Lloyd et al., 1927, 1933; Nash, 1944; Buxton, 1955; Glasgow, 1963). Some attention has also been given to the mechanisms of the spectacular colour changes of a number of insects, acridids particularly, after burning (Poulton, 1926; Burtt, 1951; Hocking, 1964).

So far, quantitative studies on the effects of grass fires upon the arthropod community of the grass layer have only been carried out in the derived savanna of Lamto, Ivory Coast (Gillon and Pernès, 1968; Daget and Lecordier, 1969; D. Gillon, 1970, 1972; Y. Gillon, 1972). The immediate, short-term and long-term effects will be discussed in turn.

Instant effects of fire. On the day following the fire in the annually burnt savanna studied, from 64 to 88% of the arthropods present the day before burning, and 32 to 60% of their biomass, were still present in the burnt area. It is important to note that the drop in biomass was greater than that in numbers, which implies that the survivors were mostly of a small size. Among insects, the strong flyers such as acridids, tettigonids, Homoptera, reduvids and coreids moved away from the burning front (Table 30.2). Just after the fire, the only arthropods left at ground level were poor flyers, such as carabid beetles, lygaeids and cockroaches, or non-flying forms such as spiders. Among acridids, 88% were estimated to have flown away, 4% to have died, while only 8% were found again in the sampled area the day after the fire.

Quantitatively, the carabid beetles community was midly affected by fire. These ground-living beetles easily take shelter in the soil and the impact of fire on them is therefore relatively the same for all species. Pentatomid bugs maintain an intermediate position between acridids and carabid beetles: great differences in flying ability exist between species, the flightless nymphs being much more vulnerable than the winged adults. Drosophilids, being very poor flyers, pay a heavy tribute to burning (Lachaise, 1974).

Fire therefore affects the various arthropod dwellers of the grass layer in very different proportions, depending largely on their unequal abil-

TABLE 30.2

Number of individuals (per m^2) of the major arthropod groups in the grass layer of the derived savanna of Lamto (Ivory Coast), the day before burning, the day following the fire, and one month later; the percentages of individuals "surviving" the grass fire are given in parentheses (after D. Gillon, 1970)

	The day before burning	The day following the fire	One month later
Arachnida	13.41	12.73 (95%)	5.33 (39%)
Myriapoda	0.10	0.07 (70%)	0.04 (40%)
Lepidoptera, caterpillars	0.83	0.23 (28%)	0.30 (36%)
Blattaria	1.04	0.86 (83%)	0.29 (28%)
Mantodea	0.84	0.39 (46%)	0.36 (43%)
Acrididae	1.35	0.09 (7%)	2.00 (48%)
Tetrigidae	0.12	0.08 (67%)	0.02 (17%)
Gryllidae	1.44	0.64 (44%)	0.19 (13%)
Tettigoniidae	0.22	0.00 (0%)	0.01 (5%)
Pentatomidae	0.82	0.21 (26%)	0.18 (22%)
Coreidae	0.42	0.01 (2%)	0.01 (2%)
Lygeidae	0.40	0.24 (60%)	0.32 (80%)
Reduvidae	0.37	0.07 (19%)	0.13 (35%)
Homoptera	3.31	0.11 (3%)	0.17 (5%)
Carabidae	0.30	0.13 (43%)	0.02 (7%)

ities to move away from the burning front. Most of those which manage to escape the flame curtain settle temporarily in previously burnt areas and at the edges of gallery forests (Pollet, 1972, 1974).

Short-term effects of fire. During the month which followed the fire, the total number of arthropods continued to decrease, despite a vigorous growth of young grass. The total arthropod population averaged only 35 to 39% of that present the day before burning; thus, in one month's time it had been reduced by half. An increase in population numbers did not become apparent until the second month after the fire.

In the mean time, the arthropod community structure changed drastically. The arthropod groups living in the lower strata of the grass layer which has best withstood burning, such as carabid beetles. cockroaches, tetrigids and spiders (Table 30.2), became less numerous, and those which had flown away started to come back. In two months the acridids, first to arrive, became even more abundant than they were before burning. Such changes in community structure can be explained by the radical change in the structure of the new and more open grass layer, and its microclimatic and nutritional implications. Shade-seeking and hygrophilous species disappear, while sun-loving ones make their appearance, and the fresh, nutrient-rich grass, attracts many primary consumers. When it started to rain again, two months after the fire, the arthropod community was becoming more similar to its pre-fire conditions. Most groups, however, continued to increase their numbers during the following months, many of the insects which had previously taken refuge in forest edges resettling in the savanna at that time (Pollet, 1972).

Thus, every major taxonomic group has its own way of reacting to the burning of the savanna, but there are also very interesting intra-group differences. Among acridids, for instance, some sun-loving and ground-living species quickly return to the burnt areas apparently being more attracted by the new physical characteristics of the environment than by food. Following the heavy mortality of "wild" species of drosophilids after the fire, anthropic species such as *Drosophila melanogaster* and *D. ananassae* temporarily invade the burnt area, coming from nearby plantations and villages (Lachaise, 1974). The specific preferences of the various species therefore play a major role in the resettlement of the burnt savanna.

Long-term effects of fire. Whatever the season, there are always fewer arthropods per unit of surface area in a regularly burnt savanna than in an unburnt one. On the average, burning of the derived savanna at Lamto, whether it takes place early or late in the dry season, decreases both the arthropod numbers and biomasses by about 30% (Figs. 30.8 and 30.9).

At Lamto, the qualitative and quantitative differences in the arthropod community of the grass layer in regularly burnt and unburnt savannas, are shown in Figs. 30.10 and 30.11:

(a) Alone among all the other arthropod groups, acridids and tettigonids are always more numerous in regularly burnt areas.

(b) Although less abundant in burnt savannas, spiders nevertheless make up a higher percentage of the total arthropod fauna in burnt areas.

(c) In contrast, cockroaches, lygaeids, and, to a lesser extent, pentatomid bugs and carabid beetles, are relatively less important in burnt savannas.

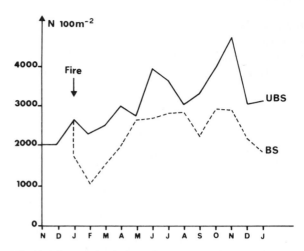

Fig. 30.8. Seasonal variations of the number of arthropods in the grass layer of annually burnt plots (*BS*) and unburnt plots (*UBS*) in the derived savanna at Lamto, Ivory Coast (after D. Gillon, 1970).

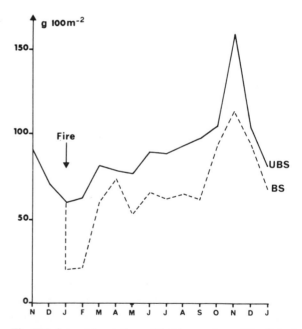

Fig. 30.9. Seasonal variations of the biomass (g per 100 m², fresh weight) of arthropods in the grass layer of annually burnt plots (*BS*) and unburnt plots (*UBS*) in the derived savanna at Lamto, Ivory Coast (after D. Gillon, 1970).

(d) All the other taxonomic groups, though less numerous in burnt plots as compared with unburnt ones, are present in similar proportions in both kinds of savannas.

When considered as groups, acridids and tettigonids are the only arthropods to benefit from

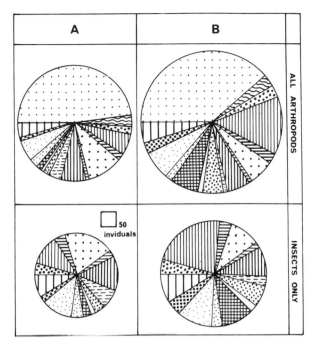

Fig. 30.10. Biomass (fresh weight) of the major taxonomic categories of arthropods captured from March to December 1965 on a 100-m² plot of a savanna burnt less than a year earlier (A), and on a 100-m² plot of a savanna burnt more than a year before (B) (after Gillon and Pernès, 1968.)

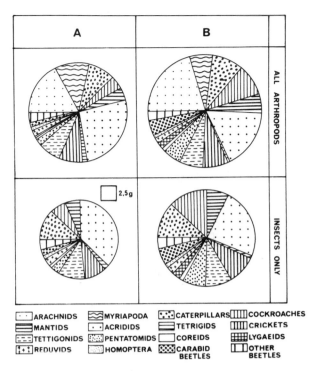

ARACHNIDS	MYRIAPODA	CATERPILLARS	COCKROACHES
MANTIDS	ACRIDIDS	TETRIGIDS	CRICKETS
TETTIGONIDS	PENTATOMIDS	COREIDS	LYGAEIDS
REDUVIDS	HOMOPTERA	CARABID BEETLES	OTHER BEETLES

Fig. 30.11. Numbers of individuals of the major taxonomic categories of arthropods captured from March to December 1965 on a 100-m² plot of a savanna burnt less than a year earlier (A), and a 100-m² plot of a savanna burnt more than a year before. Same symbols as on Fig. 30.10. (After Gillon and Pernès, 1968.)

periodic savanna burning, but there are nonetheless important interspecific differences. Carabid beetles, on the other hand, suffer from burning. For most of the year their numbers in burnt areas are only half those in fire-protected ones. This is quite probably due to the fact that most species are definitely hygrophilous and suffer greatly from the reduction of the grass cover. During the last two months of the year, however, their numbers and community structure are about the same in both kinds of savannas (Lecordier, 1975). Burning also decreases the number of pentatomids by half and their community structure is very different in burnt and unburnt savannas in all seasons.

The resettlement of burnt areas by "wild" species of drosophilids takes from two weeks to five months, depending upon the species concerned. Their flight range being restricted, much depends on the local conditions and prevailing winds. Such a situation might eventually lead to "founder effects" and even genetic drift at the population level (Lachaise, 1974).

Burning leads to a temporary change in diet for the ant *Camponotus acvapimensis*. Although their nest are located underground, these ants feed mostly upon the honeydew of Homoptera in the grass layer. After the fire they turn to predatory habits until new generations of plant-lice have developed on the fresh grass (Lévieux, 1967). Fire is also detrimental to species building aerial carton "sheds" and nests in the grass layer, such as *Acrocoelia impressa*. It also has an indirect influence upon their overall population density; since bush encroachment is prevented, it reduces the availability of nest sites (Delage-Darchen, 1971).

Annual burning is also responsible for the presence of some sun-loving forbs in the savanna, such as *Vernonia guineensis* in the derived savanna of the Ivory Coast. This is the most common composite at Lamto, which progressively disappears when fires are prevented. As shown by Duviard (1970), many insects depend upon this common and fast growing forb, which provides a variety of foods for them at a time of seasonal food shortage. This is a good example of one of the many ways by which grass burning can influence the community structure of the savanna arthropods.

Amphibians and reptiles

As shown by Barbault (1972, 1976, 1977) and Lamotte in Chapter 12, the distribution and abundance of populations of savanna-dwelling amphibians very closely depend on the rainfall regime of the areas concerned, as well as on the availability of permanent or long-lasting water bodies.

The seasonal cycle of abundance for amphibians in the derived savanna of Lamto, Ivory Coast, is shown in Fig. 30.12. Their numbers increase slowly at first during the rainy season, and more rapidly later, before quickly diminishing from November to February; this means that the population density of amphibians within the savanna itself is at its lowest level when burning occurs, most adults having taken refuge for the dry season in nearby marshy areas, residual pools, rodent burrows or termite mounds. The direct effect of fire upon adult amphibian populations is therefore negligible.

Its indirect role, however, is much more important. The microclimate of an unburnt derived savanna is more attractive to adult amphibians than that of one which has been annually burnt. Consequently, populations reach higher densities, and year-to-year fluctuations in their numbers are far less important than in periodically burnt areas. The structure of amphibian communities is also different in the two cases. At Lamto, for instance,

Arthroleptis poecilonotus is mostly found in fire-protected savannas, and *Bufo regularis* in those which are annually burnt.

In the same savannas of the Ivory Coast, burning has little direct effect upon the lizard populations, but does influence their density through its effect on vegetation structure (Barbault, 1974, 1975, 1977). Populations of the skink *Mabuya buettneri* were twice as numerous in unburnt plots as in annually burnt ones. This increase in population density took place very rapidly, as early as the second year following the fire. It was due both to an improved survival rate of the young of the post-fire cohorts, and to some migration of animals coming from surrounding burnt areas. However, the increased population density quickly attracted predators, and population growth was checked. Five years after the last fire, the *M. buettneri* population of unburnt plots was still at the same level as four years earlier (Fig. 30.13). This skink can be considered as a true "annual" species, laying its eggs at the end of the rains, before the traditional time for burning of the savanna. The eggs being buried in the soil, and the adult population being already much reduced at that time of the year, the fire causes little damage to the species. If burning is delayed, mortality is no longer negligible, and the population takes longer to recover. Obviously, the impact of grass fires would be very different on populations of long-lived lizard species. Unfortunately, none of them has so far been studied in tropical savannas.

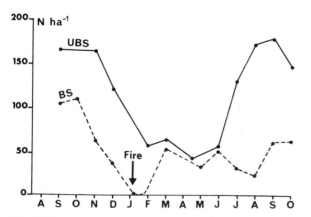

Fig. 30.12. Seasonal variations in the density per hectare of amphibians in the grass layer of annually burnt plots (*BS*) and unburnt plots (*UBS*), in the derived savanna at Lamto, Ivory Coast (after Barbault, 1976).

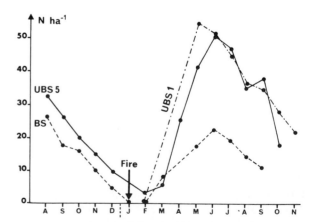

Fig. 30.13. Seasonal variations in the density per hectare of the skink *Mabuya buettneri* in the grass layer of annually burnt savanna plots (*BS*), plots unburnt for one year (*UBS 1*), and plots unburnt for five years (*UBS 5*), in the derived savanna of Lamto, Ivory Coast (after Barbault, 1977).

Again, at Lamto, snakes were on the average three times more numerous in unburnt savannas than in those which were annually burnt (Fig. 30.14). As shown by Barbault (1970, 1977), this was mostly due to the sharp increase in numbers of the frog-eating species whose favourite prey is more abundant in fire-protected savannas, as shown previously. This helps to explain why "batrachophagous" snakes can reach densities ten times as high as in annually burnt areas, whereas species feeding on other prey are only 1.5 times as abundant. The overall seasonal pattern of abundance, however, remains the same whether burnt or unburnt. About 90% of the individuals are replaced each year, fire being directly or indirectly responsible for the death of half of them.

Birds

Grass fires are detrimental to bird populations only when they destroy nests or kill the brood of ground-nesting species, such as the ostrich or the francolins in Africa, or the eggs or young of birds nesting in high grass (Phillips, 1930, 1965; Brynard, 1964). However, the breeding season seldom corresponds to the burning season, and the damage, if any, is generally limited. This was the case in the derived savanna at Lamto. The hole- or tree-nesters suffer very little from the blaze (Thiollay, 1974).

Burning can temporarily provide extra food resources for a number of species. This is apparently the case for birds of prey on all continents

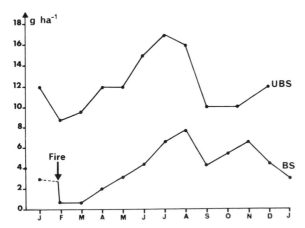

Fig. 30.14. Seasonal variations of the snake density per hectare in the grass layer of annually burnt savanna plots (*BS*), and unburnt plots (*UBS*) at Lamto, Ivory Coast (after Barbault, 1970).

(Komarek, 1969). For example, the resident black kite (*Milvus migrans parasitus*) actively looks for grass fires in West Africa. and will congregate in numbers around them (Fig. 30.15). Many other species join the kites looking for small prey driven out by the burning front. Three different categories have been distinguished by Thiollay (1971):

(1) Those which fly over the flame curtain, such as the black kite already mentioned, the grasshopper buzzard (*Butastur rufipennis*), various eagles and hawks, coucals, swifts and swallows.

(2) Those which follow the burning front, either walking on the ground, like the cattle egret (*Ardeola ibis*), or looking from nearby trees, such as rollers and bee-eaters.

(3) Those which busily investigate the hot ashes behind the burning front such as the marabou stork (*Leptoptilos crumeniferus*) and the pied crow (*Corvus albus*).

Some of these birds do not stay in the burnt area for more than a day (hawks and storks), or a week (black kites and pied crows), but others (such as the grasshopper buzzard) can linger there for one month. Other birds settle in the burnt areas as long as the new grass remains short, for example, Abyssinian roller (*Coracias abyssinica*), palm-nut vulture (*Gypohierax angolensis*), harrier hawk (*Polyboroides radiatus*), West African cuckoo falcon (*Aviceda cuculoides*).

Among these "grass-fire followers", the most numerous are those diurnal birds of prey which are able to hunt insects in flight. They are also gregarious and to a large extent migratory. These roving habits enable them to take advantage of the temporary insect mannas made available to them by the grass fires. In this way, burning can be considered as one of the major factors determining the timing of local movements of a number of intertropical African migrants (Thiollay, 1971).

As a rule, the first birds to arrive (the black kites in West Africa) are quickly attracted by the rising column of smoke, and others are attracted by the activity of the first comers. The number of individuals belonging to "large" species is variable, usually ranging from 20 to 80 in the Ivory Coast, but sometimes reaching many hundreds. Acridids are the main prey consumed by these birds, as well as tettigonids, mantids and crickets to a lesser extent. From 50 to 100 large insects per day can be consumed by a single bird during a fire (Thiollay, 1971).

Fig. 30.15. Black kites (*Milvus migrans parasitus*) hunting insects over a grass fire in Lamto, Ivory Coast (photo by Y. Gillon).

The long-term impact of regular burning on the structure of bird communities in the derived savannas of the Ivory Coast is hard to evaluate. A fire starting early during the dry season only leads to temporary changes in community structure, and apparently has little influence on the resident populations. Undoubtedly, this would be different if burning occurred during the major breeding season, although during the rainy season a fire would not last very long or extend very far.

Small mammals

The effect of grass fires upon rodents and lagomorphs has already been discussed in Chapter 17. Most rodents react to the advance of the burning front either by taking refuge in their burrows or by moving away and quickly entering any available shelter. This is so well known to local hunters that, in some parts of Africa at least, fires may be started purposefully to catch large numbers of edible species. On the Batéké plateau, in the Congo

Republic, for instance, heaps of stones are quickly built ahead of the leading edge of the fire, where "rats" take a temporary refuge before being caught by the native hunters once burning is over (F. Petter, pers. comm., 1979). The observations made by Cheeseman and Delany (1979) in Ruwenzori National Park (Uganda) provide an insight into the complexity of rodent responses to fires. Some animals move away from the fire apparently in response to the noise of combustion and/or the smell of smoke; their absence from the advancing front of the fire suggests the existence of a warning stimulus other than the heat of the fire itself. Some animals become stranded on islands of vegetation such as bush clumps and unburnt grass patches. For some of them this is apparently no more than a temporary displacement as they subsequently return to their original home area. Some other rodents do not move during a burn, but escape by retreating down their burrows or other subterranean refuges such as ventilation shafts of termite mounds. Mortality from fire is usually rare. Some species, however, may be more vulnerable than others. Such is the case of *Trichomys apereoides* in the *certão* of northeastern Brazil. F. Petter (pers. comm., 1979) has found numerous dead bodies of these rodents, close to the entrance of their burrows, after a fire. This species lives in very shallow tunnels, with a single broad entrance, which probably affords very little protection against burning.

Accidental, fierce fires occurring at long and irregular intervals can be much more destructive than fires occurring annually, because of the greater amount of grass fuel and the ensuing higher temperatures during the blaze. Yet, this was apparently not the case on a sheep range at the edge of the Namib Desert, where the effects of a fire started by a lightning strike without accompanying precipitation after a luxuriant crop of dry grass, was studied by Christian (1977). No differences in breeding percentages, survivorship, and movement distances of *Desmodillus auricularis* and *Gerbillurus paeba* were found during the twelve weeks following the fire, as compared with nearby control populations. However, this lack of demographic response was probably more a function of an adaptation of the two species concerned to desert existence rather than to habitats altered by burning. On the other hand, *Rhabdomys pumilio*, a largely diurnal rodent, only found in dense grass and bushes, was conspicuously absent from the burnt trapping grid.

The influence of burning on rodent populations during the months following a fire also varies from species to species. In their study area in the Ruwenzori National Park, Cheeseman and Delany (1979) observed that *Mus triton* and *Lemniscomys striatus* increased their numbers in the five months following a burn to levels not previously attained, whereas *Mylomys dybowskyi* and *Lophuromys sikapusi* were less frequent than usual.

In the long term, unburnt savanna areas generally have larger populations of rodents than regularly burnt areas (Bellier, 1967; Neal, 1970).

Larger mammals

Extensive grass fires can also kill larger mammals, but this seldom happens. However, this was the case during a very severe accidental fire which raged through the northern territories of the Kruger National Park (South Africa) in 1954; forty animals representing ten species were burnt to death, including lions and elephants (Brynard, 1964, 1971). In such a situation, young individuals generally suffer more than adults (Phillips, 1930, 1965). Extensive and repeated burnings may also temporarily disturb the social organization of the game herds, although this has never been convincingly proved.

In the long run, however, fire protection is rather detrimental to the populations of most savanna ungulates. At the time when regular *veld* burning was stopped in the Kruger National Park, larger mammal populations steadily declined in numbers, and moved into the adjoining areas where grass was still burnt regularly and where they were slaughtered by the hundreds. Only a few species, like the reedbuck and, to a lesser extent, the sable antelope, benefitted by the absence of burning. The decline in numbers of most ungulate species has been attributed to the encroachment of *Hyparrhenia dissoluta*, a fire-resistant grass, which is to a large extent stimulated by excessive burning (Brynard, 1971). It is likely, however, that the game were also attracted by the better nutritional quality of the grass in the regularly burnt areas. Some experiments have indeed shown that burning may lead to a more rapid weight increase in cattle which feed on regularly burnt pastures. During the same rainy season, the weight gain of domestic zebus averaged 43 kg, when feeding on regularly burnt grasslands, as against 8 kg on unburnt pastures. In another experiment, adult cattle grazing on a grassland protected from burning for four years, lost 66.5 kg on the average, with a 40% mortality rate (Granier and Cabanis, 1976). All this suggests a poor nutritional quality of unburnt tropical pastures for domestic as well as wild ungulates.

Fire may also benefit larger mammals by reducing the number of biting arthropods, from tsetse flies to ticks, which are often vectors of serious diseases.

A number of larger ungulates have also been observed on freshly burnt areas, perhaps attracted by the ashes themselves (Komarek, 1969).

Fire and overstocking with some large game species can sometimes join forces to favour the extension of derived savannas at the expense of the forest edge (see Ch. 29). In the Kabalega National Park, northwestern Uganda, elephants destroy the fire-resistant woody species along the forest edge; this eliminates the shade produced by the canopy and allows grass to develop. This grass will be maintained by the annual burning regime, and by continued elephant destruction of tree and shrub regeneration; in this way grasslands are taking over the forest (Buechner and Dawkins, 1961; Laws et al., 1975). According to Lock (1977), the lack of regeneration of woody species in such areas of relatively high rainfall appears to be the result of a combination of both burning and elephant browsing.

Fire as a management tool

It has been pointed out correctly that the savanna regions can be regarded as areas of natural or semi-natural vegetation which provide an ideal habitat for many wild and most domestic ungulates in the tropics. It is little wonder then, that for many decades cattle breeders as well as managers of national parks have been concerned with the use of fire as a practical way of controlling bush encroachment on their open pastures or savannas, as well as for improving grass vigour. Intensive grazing, especially by domestic stock, tends to promote the spread of unpalatable plant species, many of them woody, through continuous selection of the most palatable kinds of grass and forbs. Absolute fire

protection, on the other hand, also favours the accumulation of litter material and forest regeneration in most savanna habitats. A combination of overgrazing and prevention of burning in savanna landscapes, can therefore lead to a progressive disappearance of the grass cover, the habitat becoming unsuitable, for a time at least, both for cattle raising and forest exploitation. Periodic burning has thus become one of the more popular tools for controlling bush encroachment in many parts of the tropics. Its use, however, is somewhat tricky. To a certain extent it can indeed balance the effects of selective grazing, but the subsequent reduction of grass fuel can also limit the spread and intensity of the fire and prevent the destruction of the woody species unpalatable for cattle (Budowski, 1966). In East Africa, for instance, overstocking combined with the total control of grass fires has led to extensive bush encroachment in many countries (Skovlin, 1972).

Moreover, when burning is prevented, litter accumulates and this accumulation of dead grass fuel increases the risk of accidental fires. Such fierce fires even become unavoidable, statistically speaking, on very large areas, and their results can be disastrous.

All African game reserves and national parks have therefore been faced with the necessity of adopting a definite fire policy adapted to their goals. As Russel (quoted by Owen, 1972) put it, "In most parks, uncontrolled fires come through every dry season and the management decision is either to let them come through, or to prevent any fire coming through, or to control the fire so that it comes through at a predetermined time at a predetermined interval of years.".

Indeed, many authors (Phillips, 1965; and Daubenmire, 1968; among others) feel that burning is not only an "inescapable evil" better to master than to suffer from, but that fire is also an efficient and economic way to maintain the quality of pastures. As suggested in the title of Phillips (1936) paper, fire is a bad master, but a good servant. Its advantages are many, if the management objective is to maintain a habitat suitable for large grazing ungulates. This view is supported by many observations made in Kenya (Edwards, 1942; Bogdan, 1954), Uganda (Masefield, 1948), Tanzania (Van Rensburg, 1952) and Central Africa (Kennan et al., 1955; Trapnell, 1959). Burning destroys unpalatable plant species, stimulates grass growth and increases the nutritional quality of the grass, at least for some months. It also provides fresh grass out of season, kills many ectoparasites, and reduces the likelihood of accidental fierce fires which may destroy trees as well as large animals.

Thus the basic facts on which range management procedures must be based concern the differential response of various grass and tree species to burning at different times of the year, and the needs of the grazing animals. In concluding his study of the effects of fire on the Guinean and Sudanian savannas of West Africa, Rose-Innes (1972) summarized his views in the following way:

(1) Early burning at the end of the growing season tends to damage perennial grasses which are still partially green and have not yet returned all their food reserves from leaves to storage in the roots. Should they be induced by fire to sprout again, out of season, they must do so at the expense of partially replenished root reserves. Finally, if the resulting regrowth is grazed, root reserves are even further depleted, and the most vigorous plants must die in the course of a very few years of such treatment.

(2) Burning of dormant grass early in the dry season causes minimal damage but, if the regrowth is then grazed the plants are seriously weakened.

(3) Savanna trees usually sprout well before the grasses and often long before the beginning of the rains: they are most sensitive to fire damage at this time, whereas grasses are still dormant and escape harm.

In the savannas of northern Nigeria, Rains (1963) indeed found that annual fires started early during the growing season just after the first rains, when grass is still dormant and trees are in full growth, can turn an "uncleared bush" into an open grassland in four to five years. Similar results can be reached more slowly if the range is alternately burnt and grazed every year. The amount of dry grass fuel necessary to control shrubs 50 to 100 cm high is estimated to range from 800 to 1000 kg ha^{-1}; to control larger bushes and small trees, 1.7 to 2.8 t ha^{-1} are needed.

In eastern, central and southern Africa, Van Rensburg (1972) advocated late burning which results in denser, higher and more vigorous grass. He condemned fires started in the middle of the dry season which deprive animals of their food re-sources at a critical period of the yearly cycle when regrowth is made difficult by the lack of moisture, and when the depletion of root reserves makes the grass less vigorous. Burning of pasture early in the dry season does little harm to woody vegetation and is easy to control, but the resulting bush development often results in the progressive disappearance of the grass cover. Early mild fires can only be advocated in the more humid areas, such as Zambia, according to the same author.

In Zimbabwe, the experiments reported by Kennan (1972) tend to show that where grazing is withheld for a full growing season before burning, and for six to eight weeks afterwards, and where burning occurs very late in the dry season, at intervals of not less than about four years, good bush control can be achieved without lasting damage to the grasses. Despite this, however, if a run of adverse weather occurs, damage can result which may take many years to repair, even if the most lenient burning treatments are followed.

In keeping with their different objectives, and taking into consideration the variety of rainfall patterns in savanna areas, different burning procedures have been adopted by cattle ranchers and managers of national parks. Such a pragmatic approach is exemplified by the case of the Kruger National Park, as reported by Brynard (1972). At the time of the proclamation of the Park (1926) veld fires used to sweep the Transvaal low veld annually, and no official burning policy existed during the early years of the Park. At the beginning of the dry season, the warden and his staff only used to burn the old grass which escaped accidental fires, in order to provide short green grass for game during the winter months and to prevent the disastrous effects of accidental late fires. A new warden having been appointed in 1946, a new burning policy was adopted. The time of burning was switched from autumn to spring after the first good rains, and the period between burns was extended from once every year to "not more often than once every five years". Such a policy quickly led to a tremendous accumulation of old, decaying material all over the Park. This had a very serious adverse effect on the game populations, many animals moving out of the protected area in search of short grazing. The lack of regular burning was also responsible for the onset of a period of bush encroachment, and for the occurrence of disastrous conflagrations killing a number of animals. In 1954 this led to the formulation of a new burning policy, involving the division of the Park into sections, separated by properly constructed fire breaks, and burnt once every three years on rotation, after the first rains. On the whole this policy has proved to be sound and workable; even on the small burning plots where the grazing intensity is abnormally high there seems to be very little or no qualitative or quantitative changes in the grass stratum in the triennially burnt plots (Van Wyk, 1972). Since 1975, the burning policy has again been made more flexible; time and frequency of burning are now fixed according to the particular needs of the plant and animal communities concerned in order to preserve the present desirable mosaic of grassland, scrub and woodland (Joubert, 1979).

A very similar policy has also been adopted in Uganda national parks. In the Kabalega National Park (formerly Murchison Falls National Park), the fires are started as early in the dry season as possible (Wheater, 1972). In the Ruwenzori National Park (formerly Queen Elizabeth National Park), burning also takes place during dry seasons, but Eltringham (1976) has called attention to the necessity of also protecting the savanna trees and the animals dependent upon them, especially where tree destruction by elephants is taking place at a rate of 6.4% per year. In such areas, the protection of saplings which do not develop resistance to fire for some time, certainly not within three years, may necessitate a burning policy different from that of open grasslands.

On the Nyika Plateau (Malawi), burning is carried out annually during the dry season, on one third of the area at a time, over a period of five to six months (Lemon, 1968a,b). In the Wankie National Park (Zimbabwe) after a period of regular early burning which ultimately resulted in scrub encroachment detrimental to the continued existence of pure grasslands, a policy of hot fires during the peak of the dry season, just before the rains, was adopted. Such late burns carried out every two or three years, gave very good results according to Austen (1972).

The above examples clearly demonstrate that fire as a management tool must be handled cautiously. The optimal timing of the fires depends as much on the objectives pursued as on the rainfall regime and the vegetation structure of the area.

REFERENCES

Adam, J.G. and Jaeger, P., 1976. Suppression de la floraison consécutive à la suppression des feux dans les savanes et prairies de la Guinée (Afrique Occidentale). *C.R. Acad. Sci. Paris*, D 282: 637–639.

Allen, L.J., Harbers, L.H., Schalles, R.R., Owensby, C.E. and Smith, E.F., 1976. Range burning and fertilizing related to nutritive value of bluestem grass. *J. Range Manage.*, 29: 306–308.

Aristeguita, L. and Medina, E., 1966. Proteccion y quema de la sabana llanera. *Bol. Soc. Venez. Cienc. Nat.*, 26(109): 129–139.

Athias, F., Josens, G. and Lavelle, P., 1975. Influence du feu de brousse annuel sur le peuplement endogé de la savane de Lamto (Côte d'Ivoire). In: *Proc. 5th Int. Coll. Soil Zoology, Prague, 1973*, pp. 389–397.

Austen, B., 1972. The history of veld burning in the Wankie National Park, Rhodesia. *Proc. Tall Timbers Fire Ecol. Conf.*, 11: 277–296.

Balandreau, J., 1971. *Méthode d'étude in situ de la fixation microbienne d'azote dans le sol.* CNRS, Centre de Pédologie biologique. Document No. 16.

Barbault, R., 1970. Recherches écologiques dans la savane de Lamto (Côte d'Ivoire): les traits quantitatifs du peuplement des Ophidiens. *Terre Vie*, 24: 94–107.

Barbault, R., 1972. Les peuplements d'Amphibiens des savanes de Lamto (Côte d'Ivoire). *Ann. Univ. Abidjan*, E 5: 59–142.

Barbault, R., 1974. Structure et dynamique des populations naturelles du Lézard *Mabuya buettneri* dans la savane de Lamto (Côte d'Ivoire). *Bull. Ecol.*, 5: 105–121.

Barbault, R., 1975. Les peuplements de Lézards des savanes de Lamto (Côte d'Ivoire). *Ann. Univ. Abidjan*, E 8: 147–221.

Barbault, R., 1976. Structure et dynamique d'un peuplement d'Amphibiens en savane protégée du feu (Lamto, Côte d'Ivoire). 30: 246–263.

Barbault, R., 1977. Structure et dynamique d'une herpétocénose de savane (Lamto, Côte d'Ivoire). *Geo-Eco-Trop*, 4: 309–334.

Batchelder, R.B., 1967. Spatial and temporal patterns of fire in the tropical world. *Proc. Tall Timbers Fire Ecol. Conf.*, 6: 171–208.

Bellier, L., 1967. Recherches écologiques dans la savane de Lamto (Côte d'Ivoire): densités et biomasses des petits Mammifères. *Terre Vie*, 21: 319–329.

Blydenstein, J., 1963. Cambios en la vegetacion despues de proteccion contra el fuego. Parte I: El aumento anual en materia vegetal en varios sitios quemados y no quemados en la Estacion Biologica. *Bol. Soc. Venez. Cienc. Nat.*, 103: 223–238.

Blydenstein, J., 1968. Burning and tropical American savannas. *Proc. Tall Timbers Fire Ecol. Conf.*, 8: 1–14.

Bogdan, A.V., 1954. Bush clearing and grazing trial at Kisokon, Kenya. *East Afr. Agric. J.*, 19: 253–259.

Brockington, N.R., 1961. Studies on the growth of *Hyparrhenia*-dominant grassland in Northern Rhodesia. I. Fertilizer response, II. The effect of fire. *J. Br. Grassl. Soc.*, 16: 54–64.

Brynard, A.M., 1964. The influence of veld burning on the vegetation and game of the Kruger National Park. In:

D.H.S. Davis (Editor) *Ecological Studies in Southern Africa*, Junk, The Hague, pp. 371–393.

Brynard, A.M., 1972. Controlled burning in the Kruger National Park — History and development. *Proc. Tall. Timbers Fire Ecol. Conf.*, 11: 219–231.

Budowski, G., 1956. Tropical savannas, a sequence of forest felling and repeated burnings. *Turrialba*, 6: 22–33.

Bodowski, G., 1958. The ecological status of fire in tropical American lowlands. *Cong. Int. Americanistas, Costa Rica, Actas*, 33: 264–278.

Budowski, G., 1966. Fire in tropical American lowland areas. *Proc. Tall Timbers Fire Ecol. Conf.*, 5: 5–22.

Buechner, H.K. and Dawkins, H.C., 1961. Vegetation change induced by elephants and fire in Murchinson Falls National Park, Uganda, *Ecology*, 42: 752–766.

Burtt, E., 1951. The ability of adult grasshoppers to change colour on burnt ground. *Proc. R. Entomol. Soc. Lond.*, A 26: 45–48.

Buxton, P.A., 1955. *The Natural History of Tsetse Flies*. School of Hygiene and Tropical Medicine, London, Mem. No. 10, 816 pp.

César, J. and Menaut, J.C., 1974. Le peuplement végétal. In: *Analyse d'un écosystème tropical humide: la savane de Lamto. Bull. Liaison Chercheurs Lamto*, Numéro Spéc., 2: 161 pp.

Cheeseman, C.L. and Delany, M.L., 1979. The population dynamics of small rodents in a tropical African grassland. *J. Zool. Lond.*, 188: 451–475.

Christensen, N.L., 1976. Short-term effects of mowing and burning on soil nutrients in big meadows, Shenandoah National Park. *J. Range Manage.*, 29: 508–509.

Christensen, N.L., 1977. Fire and soil–plant nutrient relations in a pine–wiregrass savanna on the coastal plain of north Carolina. *Oecologia*, 31: 27–44.

Christian, D.P., 1977. Effects of fire on small mammal populations in a desert grassland. *J. Mammal.*, 58: 423–427.

Clark, J.D., 1959. *The Prehistory of Southern Africa*. Penguin Books, London, 341 pp.

Coutinho, L.M., 1979. The ecological effects of fire in Brazilian cerrado. Paper presented at the *Workshop on Dynamic Changes in Savanna Ecosystems, Kruger National Park, May 1979*, in press.

Daget, J. and Lecordier, C., 1969. Influence du feu sur les peuplements de Carabiques dans la savane de Lamto (Côte d'Ivoire). *Ann. Soc. Entomol. Fr., N.S.*, 5: 315–327.

Daubenmire, R., 1968. Ecology of fire in grasslands. *Adv. Ecol. Res.*, 5: 209–266.

Delage-Darchen, B., 1971. Contribution à l'étude écologique d'une savane de Côte d'Ivoire (Lamto). Les Fourmis des strates herbacée et arborée. *Biol. Gabon.*, 7: 461–496.

De Rham, P., 1973. Recherches sur la minéralisation de l'azote dans les sols des savanes de Lamto (Côte d'Ivoire). *Rev. Ecol. Biol. Sol.*, 10: 169–196.

Dougall, H.W. and Glover, P.E., 1964. On the chemical composition of *Themeda triandra* and *Cynodon dactylon*. *East Afr. Wildl. J.*, 2: 67–70.

Du Plessis, M.C.F. and Mostert, J.W.C., 1965. Run-off and soil losses at the Agricultural Research Institute, Glen. *S. Afr. J. Sci.*, 8: 1051–1060.

Duviard, D., 1970. Place de *Vernonia guineensis* Benth.

(Composées) dans la biocénose d'une savane préforestière de Côte d'Ivoire. *Ann. Univ. Abidjan*, E 3: 7–174.

Edwards, D.C., 1942. Grass burning. *Emp. J. Exper. Agric.*, 10: 219–231.

Eltringham, S.K., 1976. The frequency and extent of uncontrolled grass fires in the Ruwenzori National Park, Uganda. *East Afr. Wildl. J.*, 14: 215–222.

Gillon, D., 1970. Recherches écologiques dans la savane de Lamto (Côte d'Ivoire): les effets du feu sur les Arthropodes de la savane. *Terre Vie*, 24: 80–93.

Gillon, D., 1972. The effect of bush fire on the principal Pentatomid bugs (Hemiptera) of an Ivory Coast savanna. *Proc. Tall Timbers Fire Ecol. Conf.*, 11: 377–417.

Gillon, D. and Pernès, J., 1968. Etude de l'effet du feu de brousse sur certains groupes d'Arthropodes dans une savane préforestière de Côte d'Ivoire. *Ann. Univ. Abidjan*, E 1: 113–197.

Gillon, Y., 1972. The effect of bush fire on the principal Acridid species of an Ivory Coast savanna. *Proc. Tall Timbers Fire Ecol. Conf.*, 11: 419–471.

Glasgow, J.P., 1963. *The Distribution and Abundance of Tsetse*. Pergamon Press, Oxford.

Glover, P.E., 1968. The role of fire and other influences on the savannah habitat, with suggestions for further research. *East Afr. Wildl. J.*, 6: 131–137.

Granier, P. and Cabanis, Y., 1976. Les feux courants et l'élevage en savane soudanienne. *Rev. Elev. Méd. Vét. Pays Trop.*, 29: 267–275.

Hocking, B. 1964. Fire melanism in some African grasshoppers. *Evolution*, 18: 332–335.

Hopkins, B., 1963. The role of fire in promoting the sprouting of some savanna species. *J. West Afr. Sci. Assoc.*, 7: 154–162.

Hopkins, B., 1965. Observations on savanna burning in the Olokemeji Forest Reserve, Nigeria. *J. Appl. Ecol.*, 2: 367–381.

Jaeger, P. and Adam, J.G., 1967. Sur le mécanisme d'action des feux de brousse en prairie d'altitude (Monts Loma, Sierra Leone). Observations et expériences. *C.R. Acad. Sci. Paris*, 264: 1428–1430.

Joubert, S.C.J., 1979. The implementation of a research programme to determine the response of savannas to varied fire regimes and large indigeneous herbivore populations. Paper presented at the *Workshop on Dynamic Changes in Savanna Ecosystems, Kruger National Park, May 1979*, in press.

Kennan, T.C.D., 1972. The effects of fire on two vegetation types of Matapos, Rhodesia. *Proc. Tall Timbers Fire Ecol. Conf.*, 11: 53–98.

Kennan, T.C.D., Staples, R.R. and West, O., 1955. Veld management in southern Rhodesia. *Rhod. Agric. J.*, 52: 4–21.

Komarek, E.V., 1965. Fire ecology — Grassland and man. *Proc. Tall Timbers Fire Ecol. Conf.*, 4: 169–220.

Komarek, E.V., 1969. Fire and animal behaviour. *Proc. Tall Timbers Fire Ecol. Conf.*, 9: 161–207.

Komarek, E.V., 1972. Lightning and fire ecology in Africa. *Proc. Tall Timbers Fire Ecol. Conf.*, 11: 473–512.

Lacey, C.J., Walker, J. and Noble, I.R., 1979. Fire in Australian savannas. Paper presented at the *Workshop on Dynamic Changes in Savanna Ecosystems, Kruger National Park, May 1979*, in press.

Lachaise, D., 1974. Les Drosophilidae des savanes préforestières de la région tropicale de Lamto (Côte d'Ivoire). I. Isolement écologique des espèces affines et sympatriques; rhythmes d'activité saisonnière et circadienne; rôle des feux de brousse. *Ann. Univ. Abidjan*, E 7: 7–152.

Laws, R.M., Parker, I.S.C. and Johnstone, R.C.B., 1975. *Elephants and Their Habitats. The Ecology of Elephants in North Bunyoro, Uganda*. Clarendon Press, Oxford, 376 pp.

Lecordier, C., 1975. *Les peuplements de Carabiques (Coléoptères) dans la savane de Lamto (Côte d'Ivoire)*. Thesis, University of Paris, Paris, 234 pp.

Lemon, P.C., 1968a. Effects of fire on an African plateau grassland. *Ecology*, 49: 316–322.

Lemon, P.C., 1968b. Fire and wildlife grazing on an African Plateau. *Proc. Tall Timbers Fire Ecol. Conf.*, 8: 71–88.

Lévieux, J., 1967. La place de *Camponotus acvapimensis* Mayr (Hyménoptère Formicidae) dans la chaîne alimentaire d'une savane de Côte d'Ivoire. *Insectes Soc.*, 14: 313–322.

Lloyd, L.L. Johnson, W.B. and Rawson, P.H., 1927. Experiments in the control of tsetse fly. *Bull. Entomol. Res.*, 17: 423–455.

Lloyd, L.L., Lester, H.M., Taylor, A.W. and Thornewill, A.S., 1933. Experiments in the control of tsetse fly. Pt II. *Bull. Entomol. Res.*, 24: 233–245.

Lock, J.M., 1977. Preliminary results from fire and elephant exclusion plots in Kabalega National Park, Uganda. *East Africa. Wildl. J.*, 15: 229–232.

Masefield, G.B., 1948. Grass burning: some Uganda experience. *East Afr. Agric. J.*, 13: 135–138.

Masson, H. 1948. La température du sol au cours d'un feu de brousse au Sénégal. *Agron. Trop.*, 3: 174–179.

Meiklejohn, J., 1955. The effect of bush burning on the microflora of a Kenya upland soil. *J. Soil Sci.*, 6: 111–118.

Menaut, J.C., 1971. *Etude de quelques peuplements ligneux d'une savane guinéenne de Côte d'Ivoire*. Thesis, University of Paris, Paris, 141 pp.

Menaut, J.C., 1977. Evolution of plots protected from fire since 13 years in a Guinea savanna of Ivory Coast. In: *Actas del IV Symposium Internacional de Ecologia Tropical, Panama

Monnier, Y., 1968. Les effets des feux de brousse sur une savane préforestière de Côte d'Ivoire. *Etud. Eburnéennes*, 9: 260 pp.

Moore, A.W., 1960. The influence of annual burning on a soil in the derived savanna zone of Nigeria. In: *Trans. Int. Congr. Soil Sci.*, 7th Sess., Madison, Wisc., pp. 257–264.

Mulvaney, D.J., 1975. *The Prehistory of Australia*. Ringwood, Penguin Books, Ringwood, Vic., revised ed., 327 pp.

Nash, T.A.M., 1944. The probable effect of the exclusion of fire upon the distribution of Tsetse in Northern Nigeria. *Farm For.*, April 1944.

Neal, B.R., 1970. The habitat distribution and activity of a rodent population in Western Uganda, with particular reference to the effects of burning. *Rev. Zool. Bot. Afr.*, 81: 29–50.

Owen, J.S., 1972. Fire and management in the Tanzania National Parks. *Proc. Tall Timbers Fire Ecol. Conf.*, 11: 233–242.

Phillips, J.F.V., 1930. Fire. Its influence on biotic communities and physical factors in South and East Africa. *S. Afr. J. Sci.*, 27: 352–367.

Phillips, J.F.V., 1936. Fire in vegetation: a bad master, a good servant, and a national problem. *J. S. Afr. Bot.*, 2: 35–45.

Phillips, J.F.V., 1965. Fire — as master and servant; its influence in the bio-climatic regions of trans-Saharan Africa. *Proc. Tall Timbers Fire Ecol. Conf.*, 4: 7–109.

Phillips, J.F.V., 1968. The influence of fire in trans-Saharan Africa. *Acta Phytogeogr. Suec.*, 54: 13–20.

Pitot, A. and Masson, H., 1951. Quelques données sur la température au cours des feux de brousse aux environs de Dakar. *Bull. IFAN*, A 13: 711–732.

Pollet, A., 1972. Contribution à l'étude du peuplement d'insectes d'une lisière entre forêt galerie et savane éburnéenne. I. Données générales sur les phénomènes. *Ann. Univ. Abidjan*, E 5: 395–473.

Pollet, A., 1974. Contribution à l'étude des peuplements d'insectes d'une lisière entre savane et forêt galerie éburnéennes. II. Données écologiques sur les principales espèces constitutives de quelques grands groupes taxonomiques. *Ann. Univ. Abidjan*, E 7: 315–357.

Poulton, E.B., 1926. Protective resemblance borne by certain African insects to the blackened areas caused by grass fires. In: *Proc. III Int. Congr. Entomol. Zürich, July 1925*, 2: 433–451.

Rains, A.B., 1963. Grassland research in northern Nigeria (1952–62). *Misc. Pap. Inst. Agric. Res. Samaru*, 11: 1–67.

Ramsay, J.M. and Rose-Innes, R., 1963. Some quantitative observations on the effects of fire on the Guinea savanna vegetation of Northern Ghana over a period of eleven years. *Afr. Soils.*, 8: 41–120.

Roose, E., 1971. *Influence de la modification du milieu sur l'érosion, le ruissellement, le bilan hydrique et chimique. Suite à la mise en culture. Résultats sous pluie naturelle*. Comité Technique ORSTOM, Abidjan, 20 pp.

Rose-Innes, R., 1972. Fire in West African vegetation. *Proc. Tall Timbers Fire Ecol. Conf.*, 11: 147–173.

San Jose, J.J. and Medina, E., 1975. Effect of fire on organic matter production in a tropical savanna. In: F.B. Golley and E. Medina (Editors), *Tropical Ecological Systems*. Ecological Studies, Springer-Verlag, Berlin, pp. 251–264.

Skovlin, J.M., 1972. The influence of fire on important range grasses of East Africa. *Proc. Tall Timbers Fire Ecol. Conf.*, 11: 201–218.

Smith, E.L., 1960. Effects of burning and clipping at various times during the wet season on tropical tall grass range in Northern Australia. *J. Range Manage.*, 13: 197–203.

Swynnerton, C.F.M., 1925. An experiment in control of tsetse flies at Shinyanga, Tanganyika territory. *Bull. Entomol. Res.*, 15: 313–337.

Swynnerton, C.F.M., 1934. Protection of vegetation against grass fires as a possible solution for tsetse problems. *Bull. Entomol. Res.*, 25: 415–438.

Swynnerton, C.F.M., 1936. The tsetse flies of East Africa. *Trans. R. Entomol. Soc. Lond.*, 84: 579 pp.

Thiollay, J.M., 1970. Recherches écologiques dans la savane de Lamto (Côte d'Ivoire): le peuplement avien. Essai d'étude quantitative. *Terre Vie*, 24: 108–144.

Thiollay, J.M., 1971. L'exploitation des feux de brousse par les Oiseaux en Afrique Occidentale. *Alauda*, 39: 54–72.

Thiollay, J.M., 1974. Le peuplement avien de la savane de *1977*, pp. 541–558.

Lamto. *Bull. Liaison Chercheurs Lamto*, Numéro Spec., 4: 39–68.

Trapnell, C.G., 1959. Ecological results of woodland burning experiments in Northern Rhodesia. *J. Ecol.*, 47: 129–168.

Trapnell, C.G., Friend, M.T., Chamberlain, G.T. and Birch, H.F., 1976. The effects of fire and termites on a Zambian woodland soil. *J. Ecol.*, 64: 577–588.

Van Rensburg, H.J., 1952. Grass burning experiments on the Msima River Stock Farm, Southern Highlands, Tanganyika. *East Afr. Agric. J.*, 17: 119–129.

Van Rensburg, H.J., 1972. Fire: its effect on grasslands, including swamps — Southern, Central and Eastern Africa. *Proc. Tall Timbers Fire Ecol. Conf.*, 11: 175–200.

Van Wyk, P., 1972. Veld burning in the Kruger National Park. *Proc. Tall Timbers Fire Ecol. Conf.*, 11: 9–32.

Vareschi, V., 1962. La quema como factor ecologico en los llanos. *Bol. Soc. Venez. Cienc. Nat.*, 23: 9–26.

Viani, R., Baudet, J. and Marchant, J., 1973. Réalisation d'un appareil portatif d'enregistrement magnétique de mesures. Application à l'étude de la température lors du passage d'un feu de brousse. *Ann. Univ. Abidhan*, E 6: 295–304.

Villecourt, P., Schmidt, W. and César, J., 1979. Recherche sur le composition chimique (N, P, K) de la strate herbacée de la savane de Lamto (Côte d'Ivoire). *Rev. Ecol. Biol. Sol.*, 16: 9–15.

Vuattoux, R., 1970. Observations sur l'évolution des strates arborée et arbustive dans la savane de Lamto (Côte d'Ivoire). *Ann. Univ. Abidjan*, E 3: 285–315.

Vuattoux, R., 1976. Contribution à l'étude de l'évolution des strates arborée et arbustive dans la savane de Lamto (Côte d'Ivoire). *Ann. Univ. Abidjan*, C 12: 35–63.

West, O. 1965. *Fre in Vegetation and its Use in Pasture Management, with Special Reference to Tropical and Subtropical Africa.* Commonw. Bur. Pastures and Crops, Farnham Royal.

West, O., 1972. Fire, man and wildlife as interacting factors limiting the development of vegetation in Rhodesia. *Proc. Tall Timbers Fire Ecol. Conf.*, 11: 121–146.

Wheater, R.J., 1972. Problem of controlling fires in Uganda National Parks. *Proc. Tall Timbers Fire Ecol. Conf.*, 11: 259–276.

Chapter 31

PASTORALISM YESTERDAY AND TODAY:
THE OVER-GRAZING PROBLEM

H.F. LAMPREY

INTRODUCTION

Published knowledge on pastoralism, of which there is a large amount, is mainly anthropological in its approach, consisting of descriptions of the history, structure, organization, attitudes and inter-relationships of pastoral societies and the adaptive significance of their economies. By comparison there has been a surprising neglect of the role of pastoral regimes in the functioning of the rangeland ecosystems of which they form part. Hitherto, there has been an almost total absence of measurements in accounts of the interaction of pastoralists and their livestock with their habitats.

In the last two decades a small number of authors has obtained and analyzed data which relate quantitatively the livestock populations of pastoral economies with their environments on the one hand and with the resource requirements of the human populations which subsist upon them on the other. Several authors have collated data which provide the bases for preliminary ecological models linking the gross mean productivity of arid and semi-arid savanna rangelands to the production of livestock populations (Coe et al., 1976; Le Houérou and Hoste, 1977) and thence to their capacity to support pastoral peoples (Dyson-Hudson, 1962; Brown, 1971; Western, 1974; Dahl and Hjort, 1976).

The recent widespread recognition that livestock populations are an important factor in the ecological degradation prevailing in many of the arid and semi-arid grazing lands of the world (UNEP, 1977) has drawn attention to the lack of detailed knowledge of pastoral ecosystems in general, and of the nature and degree of animal and human impact upon their soils and vegetation in particular. While it is evident that ecological degeneration of grazing lands, particularly in semi-arid regions, is indeed resulting from their over-exploitation by pastoral societies, the tendency to ascribe the problem simply to "over-grazing" directs attention only to an immediate and obvious cause, and obscures to a great extent the complex of indirect historical, socio-economic, political and administrative factors which have greatly accentuated it (Fig. 31.1).

Pastoralism may be defined as human subsistence economy based wholly or partly upon domestic animals. Baxter (1975) has defined three intergrading categories of stock-keeping people:

(1) "Pure pastoralists" who do not cultivate. They may be either (a) specialist producers in a wider economy, as for example are the Basseri (of southern Iran), or (b) they may be only marginally involved in the wider economy as are some of the peoples of northern Kenya, e.g. the Rendille.

(2) Those people who perceive themselves as being primarily pastoral but cannot subsist by their stock alone and are frequently transhumant. Such peoples, for example the Karimojong (of north-eastern Uganda), may maintain both permanent villages and cattle camps.

(3) Those people who are primarily agriculturalists, but maintain, at least in their own conception, strong pastoral values, such as the Jie (Uganda), the Barabaig or the "semi-pastoral" Gogo (both of Tanzania).

In the tropics pastoralism is practised mainly in the drier areas (with less than 600 mm mean annual rainfall and commonly with less than 400 mm). In general the driest country is inhabited by the more purely pastoral societies, which are virtually all nomadic or transhumant. To a great extent the distributions of pastoral societies coincides with eco-climatic conditions which are unsuitable for

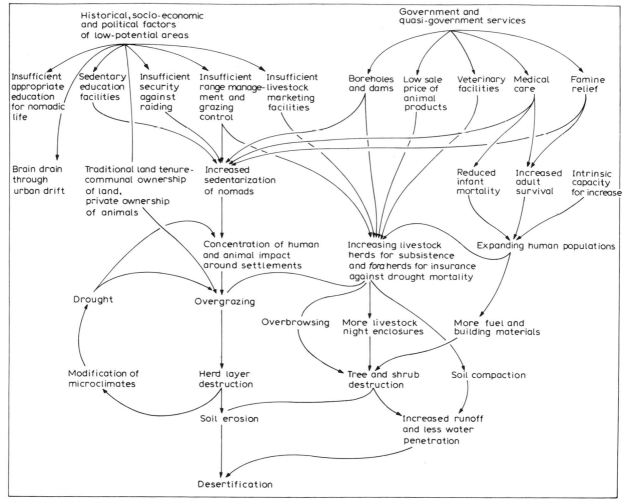

Fig. 31.1. Some causal factors in desert encroachment in northern Kenya.

agriculture; but, where this is not so, it indicates that a traditionally pastoral tribe has, by virtue of early occupation of the region, excluded agriculturalists. In several such areas, the Masai people of Kenya and Tanzania have recently been persuaded to convert extensive former grazing lands to wheat, maize and bean farming. All such areas have a mean annual rainfall of at least 600 mm. It should be noted that land which is manifestly unsuitable for rain-fed agriculture (less than 300 mm mean annual rainfall) is traditionally cultivated in many areas in North Africa and the Sahel (Le Houérou, 1970). In such areas, the rapid desert encroachment, which is frequently attributed to "overgrazing" is in fact primarily due to the cultivation of millet and sesame in the Sahel and wheat in

North Africa. The subsequent occupation of the degraded land by pastoralists may result in further depletion of the soils and vegetation.

Only through mobility is it possible for pastoral societies to subsist for long periods upon the low-density vegetation of extensive arid and semi-arid rangelands. The margins of the Sahara Desert and the Rub al Khali in the Arabian Peninsula in areas with mean annual rainfall as low as 25 mm, where rain may fall at a particular site at intervals of several years, support nomadic pastoralists who must move their camels, sheep and goats over long distances to take advantage of infrequent rain and ephemeral vegetation.

Nomadic pastoralism is today the primary land use over about one eighth of the land surface of the earth. It was formerly more extensive, particularly in Asia and Africa. It is now limited to small areas in the Indian Peninsula but continues to be widespread in the Middle East, northern, western and eastern Africa. In the Kalahari, nomadic pastoralism had ceased by 1920 (Devitt, 1971), but a limited form of transhumance continues locally. Elsewhere in Africa, in response to a variety of modern influences and incentives, the nomadic way of life is practised by fewer stock-keepers each year as pastoralist societies become increasingly sedentarized. Western (1976) remarked, "In the non-arable, or rangeland areas of Kenya, subsistence pastoralism is the dominant form of land use, having progressively displaced hunter-gatherers over the centuries. In turn, pastoralism is being modified and will almost certainly be displaced by commercial ranching in the future."

This prediction is probably justified, at least for the semi-arid savanna regions of Africa, but may not be correct for the arid zone, with its very low livestock-carrying capacities and limited commercial potential. The future of pastoralism is further discussed in the conclusion to this chapter (pp. 663–665).

ORIGINS AND HISTORY OF PASTORALISM

There can be little doubt that the history of pastoralism is closely bound up with the gross-ecological changes that have taken place during the last 6000 years, notably in northern Africa and southwestern Asia, and which are continuing today at an accelerating pace. Seen in perspective against the time scale of ecosystem evolution, it is a relatively short time ago that pastoralism was rather suddenly superimposed as an alien component upon the long established semi-arid savanna ecosystem that had occupied large parts of the Middle East and Sahara regions for several million years (albeit experiencing several long-term climatic fluctuations during that period). In the short time that pastoralism has existed, it has become the dominant biotic factor across these regions, and has extended its influence into eastern Africa during the last 2000 years and into southern Africa in the last 600 years.

In the course of their evolution since the Neolithic period, human populations have exhibited a remarkable adaptive radiation which has enabled them to exploit, as their regular habitats, most of the eco-climatic zones and land biomes of the world. Since the geographic dispersal of man in the later Pleistocene, from his probable centre of origin in Africa, each of the main ethnic divisions has undergone its separate development, isolated in varying degrees until recent historical times, and has invented or adopted a life supporting economy adapted to its particular habitat. Several peoples independently discovered pastoralism and others adopted it from neighbouring societies. They became the most numerous and the dominant inhabitants of the grasslands and savannas, environments where they could subsist only by virtue of the capacity of their domestic animals to convert vegetation into milk and meat.

Nomadic herding, like agriculture, developed as a sequel to the hunting, fishing and food gathering which were the only means of livelihood until the Neolithic period. Although an approximate period can be adduced from archaeological evidence in the Middle East, for the earliest known existence of domestic sheep, goats, cattle and pigs, there is very little information on the time and place of the early domestication of livestock species elsewhere. The techniques of domestication, which have also been unrecorded by history, are the subject of speculation. Payne (1970) pointed out that, in the case of each of the cattle species, domestication could have occurred at one locality at one or several times, or at several places at approximately the same time. He mentioned the theory that the domestication of cattle took place where man and wild cattle lived in close proximity, as they must have done on the flood plains in the semi-arid areas. The discovery of some of the earliest remains of domestic cattle in the valleys of the Nile, the Tigris and the Euphrates supports this suggestion. Baskin (1974) described the recent successful herding of wild reindeer with the help of domestic reindeer in Kamchatka and suggested that the resemblance between the herding methods of different tribes in the region indicates that they are derived from a common origin, namely from ancient methods used by hunters to follow the wild herds upon which they preyed. He wrote, "It is probable that when man changed from being a hunter to becoming a herder he retained stock in order to insure production and the availability of meat at a desired time. About 9000 to 11 000 years ago man discovered the possibility of managing ungulates that were kept in compact herds. This represented a neolithic economic revolution".

Payne (1970) has described an existing practice in Assam which provides evidence of how domestication could have occurred at some localities. "The Naga owners of gayal cattle provide salt licks in the forest. These attract wild gaur and then the female domesticated gayal are driven into the forest to breed with the wild bulls. This practice may have been used in the past to create types of cattle that originated from *Bos primigenius* and *Bos namadicus*". The current experimental domestication of oryx (*Oryx gazella*) in Kenya has shown that with this species, captured as young animals and herded in their original habitat, the process of domestication is likely to be intermittent because individual animals frequently return to the wild and rejoin the domestic herd several weeks later (M. Stanley-Price, pers. comm., 1979).

The earliest domestication of ungulate animals, like the first cultivation of plant crops, took place before recorded history and, from archaeological evidence, is generally understood to have occurred, in the Old World at least, between 10 000 and 11 000 years ago. The earliest known remains of sheep and goats with recognizable characteristics of domestication occurred together in a settlement dated between 9000 and 10 000 B.P., excavated at Jarmo in the alluvium of the southern Euphrates Valley (Hawkes and Wooley, 1963). At the archaeological site of the Belt Cave in northern Iran, sheep and goat remains occurred in a Neolithic occupation preceding the appearance of pottery and dated by the ^{14}C method as the first half of the eighth millennium B.P. At the same site, the earliest evidence of domestic pigs and cattle appeared in deposits from the second half of the same millennium, tending to confirm the generally accepted theory that sheep and goats were the first ungulates to be domesticated.

Cattle

Payne (1970), Epstein (1971) and Williamson and Payne (1971) have given accounts of the origins and history of the many present-day breeds which are herded by pastoralists, paying most attention to the cattle. Domestic cattle belong to three genera and sub-genera, all classified within the sub-family Bovinae:

Bos, including all domestic cattle of the temperate breeds (*Bos taurus*); the tropical humped breeds or zebu (*Bos indicus*); the yak of the Tibetan plateau (*Bos mutus*) and crossbreeds of these species.

Bos (sub-genus *Bibos*), including the gaur [*B.(B.) gaurus*], indigenous in the Indian Peninsula and southeastern Asia; the banteng [*B.(B.) banteng*] from the mainland of southeastern Asia and Indonesia; the kouprey [*B.(B) sauveli*], indigenous in Cambodia; the gayal [*B.(B.) frontalis*] from Assam and Burma; and crossbreeds between these species and *Bos taurus* and *Bos indicus* breeds:

Bubalus bubalis, the Asiatic buffalo of the Middle East, the Indian Peninsula and southeastern Asia; *B. mindorensis*, the buffalo of the Philippines and *B. depressicornis* of Celebes, are all important domestic animals. In the wild they are endangered and close to extinction. Among the cattle, the ancestral species of *Bos taurus* and *Bos indicus* are extinct.

The *Bibos* breeds of Burma, India and southeastern Asia, while becoming numerous locally, especially in Indonesia, have not spread far from their centres of origin. The breeds that belong to the *Bos taurus* and *Bos indicus* species have spread widely in Neolithic and historical times and the many distinctive varieties herded by pastoralists today have resulted from a complex background of diverse ancestry, different times of dispersal from the centres of origin, interbreeding, adaptive breeding for different environments and, most recently, intensive selection of specialized breeds in the developed cattle industry.

Payne's (1970) account of the origins and history of cattle breeds is somewhat detailed and, in places, unavoidably speculative. The history of the *Bos taurus* and *Bos indicus* breeds is closely bound up with their dispersal through the greater part of the world from their supposed centres of origin in the Middle East. This dispersal has taken place mainly through the outward migrations of pastoral peoples who took their livestock with them as they spread through Asia, Europe and Africa. A few salient points from Payne's account of this dispersal can be mentioned here.

Two wild species, both of which survived into recent historical times, appear to have been ancestral to modern cattle of the *Bos taurus* and *Bos indicus* types. They were the aurochs (*Bos primigenius*), the large, long-horned species which inhabited the forests of North Africa, Asia and Europe, and the urus (*Bos namadicus*), from the Indian Peninsula and elsewhere in Asia. The latter species, remains of which were found at an archaeological site at Anau in southern Turkestan, appear to have been domesticated by *c.* 10 000 B.P. (Duerst, 1908), although this very early estimate is not easily reconciled with more recent dating of livestock remains from other sites. Payne (1970) has suggested that *Bos namadicus* gave rise to the Hamitic longhorn type of cattle which is found locally in Central and West Africa today, but that this type possibly derived its longhorns from some *Bos primigenius* ancestry.

By 8000 B.P. a small species, *Bos brachyceros* (shorthorn-type cattle) had apparently replaced *Bos namadicus* in the Middle East and led subsequently to the widespread dispersal of a small shorthorned type into Europe, Asia and Africa. Several existing breeds in these continents are predominantly derived from this type. Epstein (1934) believed that *B. brachyceros* could have been a mutant form of *B. namadicus*, but Payne (1970) suggested that it could have arisen through the selective effects of domestication which might have favoured small size.

Both the Hamitic longhorn and the shorthorn types, together classified as *Bos taurus*, are humpless cattle. The humped zebu species (*Bos indicus*) also appears to have originated in the Middle East and the first archaeological evidence of it was found at Arpachiyah near Mosul in Iraq, where it had apparently been domesticated by 6500 B.P. in company with the Hamitic longhorn type. However, the ancestral species of the zebu is not known for certain, and the alternative possibility exists that it was first domesticated in the Indus Valley and later taken to the Middle East by traders or migrants. As no ancestral humped cattle have been found in archaeological sites, the origin of the hump of the zebu is difficult to explain, unless it is a feature which, like the fat tail of some sheep breeds, has arisen under domestication.

During the late Neolithic period it appears that there were dispersal migrations of several ethnic groups. Semitic and Hamitic peoples inhabiting the Middle East region migrated westwards and southwards with their livestock towards the eastern Mediterranean Basin and Africa. Vedic Aryans moved north from the region into Europe and western Asia and also into the Indian Peninsula. Ural Altaic people moved northwards and eastwards into Asia.

It is notable that the main dispersal of cattle, together with sheep and goats, took place during a minor wet phase in the climate which lasted approximately from 7000 to 4350 years B.P. (Butzer, 1961). It is probable that all three species of livestock first arrived in Africa, herded by Proto-Hamites, during that period, when the Sahara region was less arid than it is today (Payne, 1970). North Africa was sub-humid to semi-arid, and the semi-arid zone with its savanna vegetation provided a relatively favourable environment for the herding of livestock. The highland areas of Tibesti and Ahaggar, now desert, were likely to have been good habitats for pastoralists at that time and rock engravings of cattle found at Wadi Zirmei in Tibesti lend support to this suggestion.

Although it is not known whether any cattle were domesticated in North Africa it is clear that Hamitic longhorn, shorthorn and zebu types were brought into northeastern Africa by nomadic peoples and thence spread throughout the continent into suitable areas. In some regions the original types remain unchanged. For instance, a small number of shorhorn-type cattle are to be found in the Nuba mountains of the Sudan (Faulkner and Epstein, 1957). Over the greater part of eastern Africa the zebu type is common, while further west from Uganda and Chad through the Sahelian zone the Hamitic longhorn type, typified by the Ankole cattle of Uganda, is locally dominant. However, it is evident that wherever the cattle breeds have come together in Africa they have interbred. The type that is intermediate between the humped and humpless cattle, and generally carries a small hump, is known as the sanga. This is the form that reached southern Africa about 600 years ago with the Hottentot people who were moving ahead of the waves of migrating Bantu people from the southern Congo forests (Epstein, 1933). Payne (1970) elaborated his account of cattle in Africa to explain the present distribution of each of the important breeds and he discussed the barriers which have limited cattled dispersal. The most important have been the increasingly arid Sahara region and the huge area of forest and woodland of West, Central and East Africa occupied by tsetse (*Glossina* spp.), the secondary host of the protozoan blood parasites *Trypanosoma* spp., which cause "sleeping sickness" in mammals. Although a few cattle breeds that have been longest in Central Africa have apparently developed some tolerance to trypanosomiasis, in general cattle distribution is limited to the open savannas where tsetse are absent. Very few cattle occur in approximately half the area of the continent south of the Sahara which is occupied by tsetse.

Sheep and goats

The origins of domesticated sheep and goats are obscure. As already mentioned, the remains 'of what appear to have been both species have been found together in alluvial deposits in the Euphrates Valley estimated to be more than 9000 years old. Their skeletons are so similar that only when the relatively distinctive horn cores of mature males are found is it possible to distinguish them in archaeological sites. It is generally accepted that, like cattle, sheep and goats were first domesticated in southwestern Asia and were subsequently taken by migrating pastoralists into Asia, Africa and Europe.

Sheep and goats are classified in the sub-family Caprinae. Williamson and Payne (1971) suggest that modern domestic sheep, (*Ovis aries*) have descended from three closely related wild species, the argali (*Ovis ammon*) of Central Asia, the urial (*O. vignei*) of southwestern Asia and the mouflon (*O. musimon*) of Asia Minor and Europe. The wild sheep of Africa (*Ammotragus lervia*) apparently has not been domesticated. In the period since sheep were first domesticated, they have diverged morphologically into a variety of breeds adapted to several eco-climatic zones. Those breeds which have been subjected to seasonal food shortage have developed reserves of fat around the rump or tail. Those subjected to cold conditions have grown wool. Williamson and Payne (1971) have described the main distinctive breeds which are herded by pastoralists, distinguishing the white coarse-wooled, thin-tailed lohi of western Punjab, the hairy coated nellore of arid regions in India, the fat-tailed, hairy hejazi, capable of withstanding the most extreme arid tropical conditions of Arabia and the Horn of Africa; the fat-rumped blackhead

Persian and the similar fat-rumped Somali sheep. The Masai sheep of Kenya and Tanzania are fat-tailed, with coarse dark-brown wool. In Sudan the desert breed is large with a brown or pied hairy coat and a long fat tail. Other distinctive breeds are those of the Zaghawa nomadic desert Arabs of Darfur and the Fulani pastoralists of northern Nigeria. The West African dwarf sheep is widely distributed throughout the region outside the forest zone.

Williamson and Payne (1971) have suggested that domestic goats (*Capra hircus*) have descended from the species *Capra aegagrus*, the ibex of the mountains of southwestern Asia. As with the sheep, goats have spread out with their nomadic herdsmen from the Middle East into Asia, Africa and Europe and, in the course of their dispersal, have adapted morphologically and physiologically to a variety of environments. Three basic types that are herded by pastoralists in the tropics may be mentioned here: the long-legged, long-haired breeds of the Middle East and northern Africa; the short-haired compact type of East Africa; and the dwarf type of West Africa.

Camels

The domestic camels of the cold regions of Central and Eastern Asia are the two-humped *Camelus bactrianus*, descended from the wild species which survives in very small numbers in Mongolia. The ancestral species of the single-humped *Camelus dromedarius* of the tropics is extinct in the wild. From archaeological evidence (Bibby, 1970) it appears to have been domesticated at approximately 3500 B.P. from a wild population living in the northeast of the Arabian Peninsula, and was thus the last of the important ungulates to have been domesticated in the arid tropics. Archaeological excavation at sites along the northern Arabian coast has provided evidence of several vigorous city states which existed between 7000 and 2000 B.P. (Bibby, 1970). The evidence also indicates that, during that period, the climate of the region, originally semi-arid, became arid and the former savanna vegetation changed to sub-desert. With the drying up of the fresh-water sources the settlements died out.

It seems probable that the single-humped camel was indigenous in the semi-arid savanna habitat of parts of the Arabian Peninsula. As the region became progressively arid and degraded, wild camels going to water were probably forced to come into close contact with pastoral people at the few remaining oases and would have been vulnerable to capture for domestication. With increasing desert conditions, the region would have become unsuitable for cattle, and their replacement by camels would have been a logical step for pastoralists who were forced to adapt to a more arid environment.

This hypothesis is supported by recent evidence of changes in livestock populations in northern Kenya. With the current progressive loss of the herb layer vegetation due to over-grazing in their tribal areas, the Gabra and Rendille peoples have been abandoning cattle herding to concentrate increasingly on camels.

Interest in the origins and history of camel herding has not compared with that of cattle herding, and references are very few. There are estimated to be some nine million camels in the world of which about six million are kept in tropical areas. The herders of camels are mostly nomadic pastoralists of the more arid regions of the Middle East and Africa and are normally found in areas having less than 300 mm mean annual rainfall, where cattle can only be kept in small numbers. However, camel nomads can exploit desert regions with mean annual rainfall as low as 25 mm, by virtue of their ability to travel long distances between water sources and to use sparse and ephemeral desert vegetation. In winter conditions, in the Sahara region and in the rainy season in East Africa, camels that have access to green vegetation do not need to drink. Under dry conditions, camels normally drink every six days but can survive for twelve days between watering points.

One-humped camels are herded by pastoralists throughout the arid regions of the Middle East, Africa north of the Equator, Pakistan, and western India. Two distinct types occur; the very specialized riding camel and the common herd camel or baggager. Among the latter there is considerable variation from one region to another, mainly in size. The camels of the Rendille and Gabra of northern Kenya are unusually small, whereas those of the neighbouring Somali people tend to be large.

Horses and donkeys

In general, horses and donkeys are of relatively little significance in pastoral ecosystem compared with other livestock species owing to their small numbers, although they may perform important roles as riding and transport animals. Horses are ridden by several camel nomad tribes who use their camels to carry their tents and other baggage. Cattle herders, such as the Masai, use donkeys as pack animals for moving their camps. The Turkana of northern Kenya keep large herds of donkeys, numbering as many as a hundred, because these animals have the ability to withstand severe drought conditions and can serve as an emergency food supply, enabling the Turkana people to retain their more highly valued camels, sheep and goats.

The time and place of the first domestication of horses and donkeys is uncertain. It is likely that *Equus przewalskii*, now close to extinction in its original habitat on the Mongolia/China border, is one of two or three species ancestral to domestic horses. The others are extinct. The domestic donkey appears to have descended from the Nubian wild ass (*Equus asinus*) of the eastern Sahara and the Red Sea coast, which probably became extinct in recent years. One cause of its extinction has been interbreeding with feral animals, of which many are to be found in its original habitat.

THE PASTORAL ECONOMY

Pastoral economy has been described with various emphases and approaches by several authors. The following summarized account is derived mainly from Brown (1971), Dahl and Hjort (1976) and Western (1974).

Pastoral peoples subsist wholly or mainly upon the products of their livestock: milk, meat, blood and hides. Their dependence on domestic stock may be from choice or necessity and they may indulge in some trade or exchange to obtain other commodities. Where the rainfall is insufficient to support reliable agriculture (below 650 mm mean per annum in East Africa and below 500 mm in the Sahel) pastoralists must rely upon their livestock, but in wetter areas they may have the choice of cultivating.

In tropical Africa an average family consists of about eight persons including children, and is estimated (Brown, 1971) to have dietary requirements equivalent to 6.5 adults. The calorific requirements for pastoralists, judged by standard scales for "active individuals doing no heavy work", are 2300 kcal (*c.* 9626 kJ) per day, or for a family of 6.5 adult equivalents, about 14 950 kcal (*c.* 62 573 kJ), although Brown (1971) suggests that this may be too high an estimate for most pastoralists who tend to be slim, light people for their height.

The most important item in the diet of pastoralists is milk, which is likely to be available in approximately regular quantities every day. Brown gives three possible diets theoretically sufficient for a family of 6.5 adult equivalents for one day:

pure milk: 21 l
$\frac{3}{4}$ milk, $\frac{1}{4}$ meat: 16 l milk and 2.41 kg meat
$\frac{1}{2}$ milk, $\frac{1}{2}$ meat: 10.5 l milk and 4.82 kg meat

In practice pastoralists' diets will vary from one society to another and from year to year depending on the climatic conditions and the plant production. They will also vary seasonally, Under dry conditions less milk is produced and more animals are killed or die, so that meat becomes more important in the diet. However, Brown (1971) suggests that, over the year in most pastoral societies in Africa, a diet of 75% milk and 25% meat is a good estimate, and that with such a diet an average pastoralist family would consume 6606 l of milk and 704 kg of meat in a year.

Brown (1971) estimated the size of the family livestock herd which would be necessary to supply these requirements, allowing for a variety of species composition for the herd. His Standard Stock Unit (SSU) of 1000 lb (453.6 kg) live weight is given as the weight of one mature camel, or two average East African cattle or ten sheep or goats. If the goats and sheep are not considered, the family's daily requirement of 16 l of milk can be produced by 7 cows or 4 camels. Since the lactation period of cows is usually less than six months in semi-arid savanna grasslands, the family must keep at least twice as many cows — that is, 14 or 15 — to ensure sufficient milk. The individual milk yield of camels is higher than that of cows and their lactation period up to eighteen months, so that fewer may be kept to obtain the daily amount of milk.

The biomass of 14 to 15 cows is approximately 3000 kg, and for 4 to 5 camels 1800 kg. In the drier regions where camels are normally herded, a pastoralist family can live with a lower demand on the habitat if they subsist on camels rather than cattle.

For several other reasons mentioned later, camels have less impact on the environment than cattle in supporting a pastoralist population of a given size.

In order to maintain 14 milking cows, of which about 7 will be lactating at a time, at a calving rate of 70% and allowing for an annual loss of 10% of all age and sex classes, with four years to the birth of the first calf, all heifers born would have to be kept; these will add a further 15 to 18 animals. At least 2 or 3 adult bulls will be kept by each family to insure against the loss of one and about four young males will be raised to replace them. Thus, an average pastoralist family will be obliged to keep 35 to 40 cattle to ensure its milk requirements. With camels less than half this number would be needed.

The theoretical meat requirement of 704 kg yr^{-1} for the average African pastoral family is commonly met by the consumption of animals which die together with an additional quantity which would be supplied by 3 or 4 young males. This probably necessitates the retention of all young males born to a herd of 14 or 15 milking cows each year, since an additional four are likely to be needed for the replacement of bulls.

Brown (1971) points out that, in practice, pastoralists obtain a proportion of their meat requirements from sheep and goats, which are of a more convenient size than cattle. One animal provides from 7 to 9 kg of meat which can be eaten by a family in two or three meals. A camel or a steer must be shared with other families or must be dried for storage (not practical under nomadic conditions) otherwise much of it will rot in a hot climate. The role of small stock in pastoral economies will be further discussed below (pp. 651, 662 and 663).

Brown's (1971) estimate of the minimum cattle numbers required for the subsistence of a pastoralist family has been substantially confirmed by the observations of several field workers, as quoted by Dahl and Hjort (1976). However, in general, actual observations of cattle numbers indicate that pastoral societies maintain herds that are considerably larger than the minimum needed for subsistence in an average year. Thus, the Boran in Ethiopia with small families of about four persons have herds averaging 50 cattle (Asmaron, 1975). P. Spencer (pers. comm., 1979) noted that the average Samburu family in 1958–59 consisted of 6.75 persons; for each person, including children, there were 13.4 cattle. Dyson-Hudson (1966) gave from 50 to 150 cattle as the normal "unit size" among the Karimojong, and Huntingford (1969) found 70 to 80 cattle per stock owner among the Barabaig. These observations tend to show that actual cattle numbers are approximately twice to three times the number theoretically necessary for subsistence. However, as Brown (1971) observed, to provide for the possibility of disaster, such as drought, disease and cattle-raiding (all of which are prevalent), pastoralists would be expected to keep at least 50% more animals than the minimal requirement for an average year.

The role of small stock in the economy of pastoral societies in Africa has been discussed in considerable detail by Dahl and Hjort (1976). In general, sheep and goats are regarded as less important than cattle or camels. However, virtually all pastoralists keep small stock as an essential contribution to their economy. While the main purpose of cattle and camels is normally to provide milk, sheep and goats are usually kept for meat, though also producing milk when required. In Iran the normal practice of nomadic pastoralists is to herd only small stock. In East Africa it is also a common practice among pastoral peoples that normally herd cattle and camels for some individuals to keep only small stock. Nevertheless, the normal practice among East African pastoralists is for a stock owner to possess at least as many small stock as cattle or camels. A "moderately prosperous" Karimojong family would have 100 sheep and goats in addition to some 100 or 150 cattle (Dyson-Hudson and Dyson-Hudson, 1970). As a means of subsistence by themselves, Hunt (1951) stated that a Somali mother in the pastoral areas would need as few as 50 sheep and goats to exist with three children, and with 100 animals they would subsist "reasonably" with meals every day. There is general agreement that, under average conditions, a minimum flock of small stock for the subsistence of a family is at least 50 or 60 animals, while 150 may be regarded as the average number per married woman and 200 are common in Somalia.

Dahl and Hjort (1976) quoted several sources on milk production by sheep and goats. Williamson and Payne (1971) stated that the most likely yield for tropical goats subsisting on "diligent foraging" would be in the range from 57 to 85 g per day during a short lactation period.

Livestock numbers and densities in tropical savannas

In general there is a lack of reliable data on livestock population sizes and densities in tropical ecosystems. Some estimates of total animal numbers for particular countries and regions have been published, commonly without explanations of how they were obtained. Ground census carried out by government livestock authorities was the method normally used in the colonial period in Africa. In the last ten years systematic aerial sample surveys have been undertaken in several countries in Africa, notably in Kenya and the Sudan. This method, now adopted as standard practice in East Africa for censusing wildlife and livestock populations, has been described by Jolly (1969), Watson (1969) and Norton-Griffiths (1978). The technique consists of counting animals visually and photographically from a light aircraft flying at a constant low height (commonly 100 m) above the ground using a radar altimeter. By following the parallel lines of the map grid (usually at intervals of 10 km) a ground transect of selected width (commonly between 100 and 300 m depending on the density of the vegetation) can be observed on either side of the aircraft, defined for each observer by horizontal rods mounted on the wing struts. Following a grid pattern with lines 10 km apart, a total transect width of 600 m gives a 6% sample which can be expanded to a population estimate with measures of distribution and standard deviation. Herd sizes are normally verified by photography. The aerial survey technique can be combined with the assessment of other variables such as surface water, the condition of vegetation, nomad settlements and wildlife numbers and distribution.

MacGillivray (1967) gave tables showing the regional distribution of livestock in Kenya, Tanzania and Uganda in 1964, estimating the total cattle population of East Africa as 19 354 000, of which between three and four million were owned by nomadic and transhumant pastoralists. He gave estimates of the numbers of other livestock species in East Africa: camels, 1 266 000 (virtually all owned by nomad pastoralists in Kenya); donkeys, 287 000 (mostly in Kenya and Tanzania); goats, 11 153 000 and sheep, 6 985 000.

Boudet (1976) gave livestock population estimates for Mali with a regional breakdown in which he separated the Sudanian and Sahelian zones. Cattle numbers for the Sahelian zone in Mali decreased from c. 3 200 000 (density 6.4 km^{-2}) in 1970 to 1 877 000 (3.8 km^{-2}). Watson (1972) obtained livestock biomass density estimates from systematic aerial sampling in six arid and semi-arid districts in Kenya which ranged from 1151 kg km^{-2} (4.6 TLU* km^{-2}) in Wajir district (mean annual rainfall 250 mm) to 7884 kg km^{-2} (31.5 TLU km^{-2}) in Kaputei district (mean annual rainfall 650 mm).

Field (1979) gave camel density estimates, obtained in nine periodic aerial surveys between 1976 and 1979 in the 22 000 km^2 of arid savanna of the UNEP–MAB study area in Marsabit District, Kenya (mean annual rainfall 250 mm); these estimates varied from 0.5 km^{-2} to 3.1 km^{-2} owing to nomadic movement, with a mean density for the period of 1.4 km^{-2}. From the same aerial surveys the density of sheep and goats varied between 4.5 and 19.9 km^{-2} with an average of 11.5 km^{-2}, an estimate which is close to that of Boudet for the same rainfall zone in Mali in 1972. Hunting Technical Services (1977), using data from the same series of surveys, gave cattle density estimates with an average of 1.4 km^{-2}. During the period covered by these surveys the total mean biomass density of the four livestock species together was 1150 kg km^{-2}, which agrees with the estimate of Watson (1972) for Wajir district, Kenya, in the same eco-climatic zone, and is close to that of Boudet (1976) in the Sahelian zone near the beginning of the long drought.

It is noteworthy that the biomass densities of the wild grazing ungulates occupying the same area in northern Kenya, oryx (*Oryx gazella*) and Grevy's zebra (*Equus grevyi*), were 8.5 and 12.0 kg km^{-2} respectively and that their combined biomass densities amounted to 1.8% of that of the domestic ungulates in the region. Unfortunately, it is not possible to determine how much the wildlife populations have been depressed by competition with livestock for food, and by illegal hunting.

The human population of the region, pastoralists of the Rendille and Gabra tribes, are almost completely dependent upon their herds for subsistence. Approximately 20 000 people occupy the area under study, giving a density of c. 1 km^{-2} and a ratio of one person to 4.6 TLU (1150 kg live weight). For an average family of eight, including children, the average family herd would thus consist of the equivalent of 37 TLU (c. 51 cattle or 30 camels or 370 sheep and goats).

During the drought from 1968 to 1976 cattle populations in the Sahelian zone and in northeastern Africa were reduced by 50% in several countries (UNEP, 1977). Normal irregularity in rainfall in the arid zone, with the resultant fluctuation in the production of the vegetation, causes livestock populations to fluctuate. During a series of good rainfall years animal numbers increase in response to increased food availability. Under favourable conditions sheep and goat herds can increase by 30% per year. Between 1960 and 1966 cattle numbers in Chad increased from c. 1 million to 4.5 million (SEDES, 1975). With the onset of a series of dry years, livestock numbers may remain high for the first two years. In the third year of drought, as consumption by livestock continues to exceed productivity and the standing crop of dry plant material is consumed and broken up, animal impact upon the vegetation is likely to be at its highest level. If drought conditions continue beyond three years, as happened in 1972 in the Sahelian zone, the shortage of grazing will lead to heavy mortality and a decline in livestock populations. With the fourth year of drought, as in 1973 in the Sahelian countries and northeastern Africa, a precipitate die-off is likely to occur. Under these circumstances the affected pastoral communities experience famine, and many of the poorer families lose their whole herds and become destitute. The existence of pastoralist societies is governed by the exigencies of periodic drought and by adaptive strategies which have enabled the successful to survive.

Carrying capacity

The carrying capacity of an area is usually understood to be the maximum biomass density of animals that it can support on a continuing basis. In the absence of gross natural disturbance and human interference, ecosystems possess dynamic equilibrium. Component plant and animal populations may fluctuate numerically under the influence of normal environmental variability which, in terrestial ecosystems, is mainly climatic. As suggested previously (p. 645), pastoralism may be regarded as a major consumer component

---

*TLU: Tropical Livestock Unit (250 kg live weight).

newly imposed upon the plant communities of long established savanna ecosystems which formerly included their own finely adapted communities of indigenous ungulates (and, in all probability in Africa and Asia, man as a hunter-gatherer) existing at characteristic biomass densities which fluctuated within limits determined by density-dependent regulating mechanisms. In the case of wild bovids living today in relatively undisturbed East African national parks, the most important of these regulating mechanisms is mortality caused by the decreasing quality and quantity of the food available during periodic and seasonal scarcity periods (Sinclair, 1974). In tropical savannas such periods occur in dry seasons and recurrent droughts. At times when animals are weakened by poor nutrition they become relatively vulnerable to disease, predation and accident, which act as secondary regulating factors.

In combination these limiting influences tend to maintain wild ungulate populations at density levels which are within the long-term carrying capacity of their habitats. Thus, the grazing ecosystem as a whole is self-regulating and ungulate populations tend to increase to equilibrium levels determined by primary production and by limited competition with other herbivorous species with which they overlap in their resource requirements (Sinclair, 1973). Domestic ungulate populations are evidently not regulated in this manner and frequently increase to levels which are greater than the carrying capacity of their habitats. The reasons for their relative lack of vulnerability to natural regulating mechanisms and the consequences of "over-stocking" by domestic animal populations are discussed below.

The productivity of savanna ecosystems depends on several environmental factors including climate, soil type, botanical composition and patterns and intensities of exploitation by animals. Le Houérou and Hoste (1977) have synthesized the available data on the relationship between annual rainfall and range production in the Mediterranean Basin and the Sudano–Sahelian zone of tropical Africa and have shown a close relationship between average primary production and average rainfall over large geographical areas. Although mean annual rainfall is one of several factors influencing range production, it is correlated with other climatic factors, including annual rainfall variability, num-

ber of rainy days, length of dry and rainy seasons, and potential evapo-transpiration. These authors have used a large number of data to derive theoretical livestock carrying capacities over a range of climatic zones in Africa, based upon measured plant production in relation to rainfall and upon known average fodder consumption figures for the local domestic ungulates. For a given amount of rain it has been found that net primary production is higher in the Mediterranean Basin than it is in the Sahel. (For reasons given below the rainfall production relationship in East Africa resembles the Mediterranean situation rather than that of the Sahel.) Le Houérou and Hoste (1977) have calculated regressions which show that, on average, each millimetre of rainfall produces approximately 2 kg ha^{-1} of consumable dry matter in the Mediterranean Basin and 1 kg ha^{-1} in the Sudan and Sahel zones. The differences in estimated rainfall efficiency between the two geographical regions are ascribed to several factors: shrub and tree production, which are relatively important in the Sahelian zone, were not included in measurements of productivity; rainfall distribution differs in the two regions, the growing season being longer in the Mediterranean and occurring in the winter when plant water requirements are lower, whereas the short rainy season in the Sahel coincides with the summer when water requirements are high; potential evaporation for a given amount of rain is about 50% higher in the Sahel than in the Mediterranean; soils in the Sahel have relatively low nitrogen and phosphorus content; the character of the vegetation of the two regions differs markedly, that of the Mediterranean consisting mainly of perennial subshrubs with deep roots and that of the Sahelian zone mainly of short-lived annuals with shallow root systems. In most respects conditions in East African savanna rangelands resemble those of the Mediterranean rather than the Sahel, with the relatively long growing season in East Africa, resulting from a bimodal annual rainfall distribution, having a marked effect on productivity.

Estimates of average range production are given in Table 31.1, which is derived mainly from Le Houérou and Hoste (1977). Productivity, expressed as total dry matter (DM) and consumable dry matter, is shown for five eco-climatic zones between 125 and 1000 mm annual rainfall, giving the average annual production for each rainfall level in the

TABLE 31.1

Estimates of livestock carrying capacity and maximum human density under subsistence pastoralism, in relation to eco-climatic conditions; mean wildlife biomass densities for comparison

| Eco-climatic zone | Mean annual rainfall (mm) | Range production[a] | | Livestock[a] carrying capacity based on 3105 kg DM TLU^{-1} yr^{-1} intake (ha TLU^{-1}) | Maximum sustainable densities | | Estimated[a] livestock carrying capacities (kg km^{-2}) | Mean[c] wildlife biomass densities (kg km^{-2}) |
		total dry matter (kg ha^{-1} yr^{-1})	consumable dry matter (kg ha^{-1} yr^{-1})		livestock[a] (TLU km^{-2} yr^{-1})	people[b] (adult equiv./ km^{-2} yr^{-1})		
Sub-humid savanna	1000	4000[d] (2500)[e]	2000 (1000)	6.35 (12.72)	63.00 (31.50)	14 (7)	15 570 (7787)	10 000
Dry sub-humid to semi-arid savanna	750	3000 (1850)	1500 (750)	9.52 (19.04)	42.00 (21.00)	9 (4.5)	10 500 (5250)	6600
Semi-arid savanna: dry thornbush	500	2000 (1250)	1000 (500)	12.72 (25.44)	31.50 (15.75)	7 (3.5)	7787 (3937)	2820
Arid sub-desert scrub	250	1000 (625)	500 (250)	25.44 (50.88)	15.75 (7.85)	3.5 (1.75)	3937 (1968)	954
Very arid desert	125	500 (312)	250 (125)	50.88 (101.76)	7.85 (3.95)	1.75 (0.85)	1968 (987)	269

[a]Le Houérou and Hoste (1977). [b]Brown (1971). [c]Coe et al. (1976). [d]Unbracketed figures: Mediterranean and East Africa. [e]Bracketed figures: Sahelian zone.

Mediterranean (and East African) region and the Sudano-Sahelian region (bracketed in the Table). Table 31.1 also gives estimated livestock carrying capacities in the two regions for each of the climatic zones, based upon two main assumptions: (1) that forage consumption can be as high as 50% of total production in the Mediterranean region and 40% in the Sudano-Sahelian zone; (2) that the annual intake of one tropical livestock unit (TLU) of 250 kg liveweight is 3105 kg DM, which includes a maintenance diet of 2300 kg and an additional 805 kg to give a weight gain of 100 kg ha^{-1} or a production of 1000 kg of milk. Estimates of subsistence herd size (30 TLU) for a standard pastoral family of 6.5 adult equivalents (Brown, 1971; Dahl and Hjort, 1976; Pratt and Gwynne, 1977) have been used to indicate maximum sustainable population densities of livestock and pastoral people at each rainfall level.

Coe et al. (1976) suggested that biomass densities[1] of wild ungulate communities in relatively undisturbed ecosystems in national parks could provide an indication of the true carrying capacity in pastoral ecosystems in the same eco-climatic conditions. These authors found close correlation between biomass densities of large herbivores and mean annual rainfall from data obtained in 24 African wildlife areas. In Table 31.1 mean wildlife biomass densities for each of the five rainfall levels between 125 and 1000 mm obtained from the regression of Coe et al. (1976) are compared with the predicted livestock carrying capacities given by Le Houérou and Hoste (1977) for the same rainfall levels. The comparison is of the greatest interest and brings to light a significant anomaly in the two sets of data. The two regressions differ considerably in their slope (Fig. 31.2). While there is approximate agreement between them at the 1000 and 750 mm rainfall levels, there is increasing divergence between them below the 750-mm rainfall level. At the 500-mm level the predicted livestock carrying capacities of the Mediterranean and Sahelian regions are respectively 2.67 and 1.40 times the corresponding mean wildlife biomass densities shown by Coe et al. (1976). At 250 mm predicted carrying capacities are 4.13 and 2.06 times higher than observed wildlife biomass densities and at 125 mm they are 7.31 and 3.67 times as high.

The discrepancy between the carrying capacity levels predicted by Le Houérou and Hoste (1977)

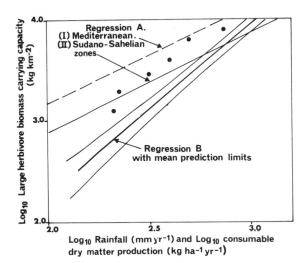

Fig. 31.2. A comparison between two possible regressions relating savanna carrying capacities (in terms of large herbivore biomass density) to average annual rainfall and primary productivity: *Regression A*. Derived from Le Houérou and Hoste (1977); calculated from rainfall/productivity data obtained at 45 sites in the Mediterranean Basin and 44 in the Sudano-Sahelian Zone; assuming an offtake of 50% of total above-ground phytomass in the Mediterranean region and 40% in the Sudano-Sahelian region, and an annual intake of 3105 kg dry matter per tropical livestock unit (1 TLU = 250 kg). *Regression B*. Derived from Coe et al. (1976); from wild ungulate biomass density data obtained in 24 wildlife areas in Africa. The authors postulate that existing (undisturbed) wildlife biomass densities indicate the approximate carrying capacities of the climatic zone in which they occur. The possible significance of the wide divergence of the regressions at the lower rainfall levels is discussed in the text. Black dots indicate biomass density estimates of six pastoral livestock populations in Kenya (from Watson, 1972).

and those given by Coe et al. (1976) on the basis of observed wildlife biomass densities is large at the low rainfall levels. In the absence of prolonged stocking rate trials which, in the arid zone rangelands, might have to extend over a period of at least twenty years to include the full range of rainfall variability, and to show up long-term responses in the vegetation, there is no known absolute measure of the livestock carrying capacity of a tropical savanna ecosystem. The estimates of range management specialists are based upon observed livestock consumption levels in relation to mean productivity for the eco-climatic conditions in question and upon practised observation on the long-term effects of known stocking rates. There is general agreement between the estimates of carry-

---

[1]Biomass per unit area.

ing capacity made by range specialists (Pratt and Gwynne, 1976; Boudet, 1976; Le Houérou and Hoste, 1977), but in every case these authors have expressed caution in the interpretation of the values they give, which can only be gross approximations of average carrying capacities for relatively large areas, and generally based upon mean annual rainfall. Such estimates cannot take into account the heterogeneity of the savanna habitats nor the high degree of variability of rainfall in the arid regions, both in space and time.

It seems likely that carrying capacity estimates based upon broad averages over time and space will be subject to several important qualifications, the most important of which is the uncertainty as to whether the long-term sustainable stocking rate will be related to the mean annual rainfall, or to some value nearer to the minimum annual rainfall. The fact that wildlife populations appear to be regulated at particularly low levels in the driest regions suggests that their population densities are more closely related to minimum rather than average values of rainfall and range production. Assessment of carrying capacity is complicated by a number of highly variable factors, of which the climate is one. There will be the additional factors of irregular grass fires (where grass production is high enough to provide the necessary fuel); consumption by insects, particularly termites and grasshoppers; and the breaking up of dry material through trampling and its removal by wind erosion. Under certain circumstances and in some vegetation communities, the percentage of the standing crop or of the total annual production that can be removed by grazing without causing reduction in the viability of the plants may be less than the 40 or 50% normally allowed for.

Pratt and Gwynne (1977) commented (in the context of range management) that there is a tendency, when using quantitative methods and mathematical relationships, to set carrying capacity too high. They wrote, "Too often it is overlooked that as rainfall and yield decrease, so carrying capacity decreases geometrically towards infinity". This observation appears to confirm the suggestion of Coe et al. (1977) that the true carrying capacity of dry savanna rangelands may be at the relatively low levels of existing wildlife densities. If it is true that, under managed ranching conditions, rangeland carrying capacities are apt to be over-

estimated, it seems even more likely that most pastoralists, whose objective and strategy is to encourage the greatest possible increase in their herds, and who do not appear to understand the concept of carrying capacity, will tend to overexploit their savanna habitats. The brief account of prevailing livestock numbers and densities given in the previous section indicates that, in general, the livestock populations of pastoralists, where these are known, are considerably above the estimated carrying capacities of the savanna grazing lands. Widespread damage to semi-arid and arid zone grasslands in eastern Africa and the Sahelian zone provides overwhelming evidence of over-grazing by the domestic livestock of pastoralists. The effects of pastoralism upon savanna ecosystems are discussed below (pp. 657–660).

As mentioned earlier, it is evident that the livestock populations of pastoral societies are generally less vulnerable to factors limiting their populations than wild ungulates, and are not subject to the regulating mechanisms which tend to maintain wild species at numbers which are within the carrying capacity of their habitats. Domestication confers some disadvantages and some advantages, but it is apparent that the latter normally outweigh the former in the struggle for survival. Most obviously, domestic ungulates receive almost complete protection from predation, so that not only are healthy animals given a better chance of survival but so are the mildly diseased and injured which, in wild populations, are especially vulnerable. However, it seems probable that the greatest advantage gained by domestic animals is in the protection given to their young, with the result that the relatively heavy mortality normally experienced by the young of wild ungulates does not occur in domestic populations, except during drought periods and epidemics.

The disadvantages of domestication include the inability to graze at night, periodic increased exposure to disease through crowding in night enclosures and, probably, several inherited characteristics selected as adjuncts to domestication such as the fat tails of certain drought-adapted sheep breeds. In balance it seems that the symbiosis of pastoral man and his domestic animals has been very successful if viewed as a survival strategy in the short term. In the long term it appears less successful since it tends to destory its own habitat.

THE IMPACT OF PASTORALISM ON TROPICAL SAVANNAS

Many African national parks and reserves, particularly those which contain tsetse populations, where pastoralism has had little or no impact, provide examples of normally functioning savanna ecosystems in which the indigenous ungulate populations are subject to natural regulation (Sinclair, 1975), and generally remain within the long-term carrying capacity of their habitats. Studies of the Serengeti grasslands in Tanzania have shown the nature of the dynamic equilibrium which exists in the interaction of a community of grazing animals and the herb-layer communities (Sinclair, 1977; McNaughton, 1978), resulting from a long period of plant–herbivore co-evolution. Embodied in this equilibrium are numerous adaptive strategies evolved by the herbivores towards the exploitation of the vegetation, and others evolved by the plant species to protect themselves in the face of intense grazing pressures. Among the protective devices evolved by savanna plants are the "plant defence guilds" (Atsatt and O'Dowd, 1976; McNaughton, 1978) by which, under conditions of heavy grazing, palatable species may be protected by their spatial association in mixed stands with unpalatable species. The most important aspect of the natural regulation mechanisms affecting wild ungulates is the density-dependent limiting action of the combined effects of under-nutrition, disease and predation which, in an animal population approaching habitat capacity level, becomes critical in drought periods when the quantity and quality of the grazing available are at their lowest (Sinclair, 1975).

In such "natural" grazing ecosystems several characteristics of the herb communities are strongly influenced by grazing: phenology, plant size, annual productivity, protein content, and species composition. Variations in grazing intensity and frequency and in ungulate grazing succession produce related variations in the composition and growth patterns of grassland communities (Vesey-Fitzgerald, 1965a,b). It is evident that tropical savanna ecosystems have the flexibility to adjust to such variations within certain limits and to do this under variable climatic conditions. They are also adapted to periodic burning, and their characteristics will vary according to the frequency and intensity of fires. In view of the considerable adaptive flexibility of tropical grassland ecosystems, it might be expected that partial or total replacement of an indigenous ungulate fauna by exotic ungulate species, introduced by pastoral peoples, need not adversely affect the viability of the plant communities, provided that the level of exploitation remains within their adaptive limits. The experience of range management specialists and ranchers has shown that the productivity of tropical savannas grazed by domestic livestock can be maintained through the management of stocking rates and grazing rotation. Pratt and Gwynne (1977) make the following fundamental observation, "In the grazing animal is the means both for the economic exploitation of rangeland and for its destruction. In grazing management is the means for manipulating the grazing animal so as to effect use without misuse. The objective in grazing management, therefore, is to achieve the highest level of animal production commensurate with maintaining or improving range condition."

Sound range management applied to tropical savanna ecosystems is a relatively new technology. History records numerous examples of extensive over-grazing of tropical and subtropical grazing lands under cattle ranching in America, Australia and southern Africa, when the principles of range management were unknown or ignored. Two primary principles of range management appear to be generally accepted. Firstly, stocking rates must be maintained within the carrying capacity of the rangeland in question, taking into account year-to-year variability in primary production due to fluctuating climate: secondly, grazing land should be rested from use for periods of several weeks, particularly while the vegetation is making new growth. If continuous grazing takes place, without resting periods, the stocking rate must be held at a relatively low level to permit adequate regrowth. For any particular savanna type, appropriate management will consist of the manipulation of stock rotations and levels consistent with the potential of that type under the prevailing climatic conditions.

Subsistence pastoralists are no less subject to the "rules" and constraints of savanna range management than ranchers if they are to maintain the viability of their habitats. It is frequently pointed out in anthropological literature that the practices of traditional pastoralism represent highly adaptive strategies which enable pastoralist societies to exploit grazing lands and to survive periods of extreme stress during drought. It is undoubtedly true that, under most circumstances, pastoralists behave rationally in realizing or attempting to realize their immediate, and sometimes critical, goal of survival, and their less immediate objective of maintaining known standards of living (Hjort, 1976). [The strategies which have evolved in such societies by which these vital objectives have been achieved, are discussed elsewhere (pp. 660–663).] However, there remains the crucial question "To what extent have pastoralists' strategies incorporated measures to ensure the continued viability of their savanna habitats, and has traditional pastoralism been sufficiently adaptive in the past to ensure, not only the survival of the pastoral societies themselves, but also that of their habitats?." Although there are known examples of traditional behaviour directed towards the conservation of the pastoralists' grazing lands (discussed below, pp. 660–663), the overwhelming evidence of continuing extensive ecological degradation in these areas indicates that such adaptive behaviour has been infrequent, ineffective or absent over the greater part of the arid and semi-arid savannas of Africa and southwestern Asia for long periods. The view that the present phase of desert formation is being greatly accelerated or caused by the effects of human activities upon soils and vegetation is supported by ecologically oriented assessments (UNEP, 1977), although it is generally understood that the current dry phase in the climate of the Sahelian zone (Jackson, 1957; Wickens, 1975) and periodic droughts are contributory factors.

Whereas communities of indigenous herbivores are normally regulated in their numbers by density-dependent limiting factors in their environment, the livestock populations of pastoralists are evidently less vulnerable to such limitations by virtue of advantages gained through their domesticated existence. Their numbers often give stocking rates considerably above the carrying capacity of the grazing lands. In one sense, pastoralism has been too successful as a consumer community. It has enabled the domestic consumer populations to overcome the natural defences of the savanna vegetation against excessive exploitation. In a "natural" ecosystem gross modifications of a habitat would be expected to be accompanied either by the reduction or extinction of its animal inhabitants, since their limited environmental tolerance prevents them from extending their ranges into neighbouring habitats. In the case of pastoralism, the biological advantages provided by domestication have given domestic stock the tolerance to move into new regions after their former habitats have been rendered unproductive by over-grazing.

It was noted above that the vegetation of savanna ecosystems is normally tolerant to grazing and browsing by ungulate animals within certain limits of feeding intensity. Moderate grazing pressure produces a characteristic "grazing climax" community for any particular eco-climatic situation which will exhibit successional changes if the grazing is increased or decreased. Although the details of a plant community succession in response to changes in grazing intensity will vary very greatly from one savanna ecosystem to another, a general description can be given of the changes which occur as a result of over-grazing.

Grazing ungulates select the most palatable plant species, and under light grazing conditions may take only the most palatable parts of each plant that they graze, normally the young foliage, leaving the stems untouched together with a proportion of the leaf growth. Under these conditions a large enough proportion of the plant is likely to remain to enable it to survive and produce seed or, in the case of a perennial species, to transfer sufficient nutrient material to its storage roots and rhizomes for the following year's growth. Each species is able to tolerate the removal of a limited proportion of its foliage, which may vary from 20 to 75% and is also likely to depend on the climatic conditions prevailing. Under intensive grazing pressure the palatable species may have higher proportions of their foliage eaten than they can tolerate and they may fail to form seed or, in perennials, to transfer nutrients to their underground storage organs.

In addition to the process of intensive grazing, there is the mechanical damage done by the trampling of the hooves of densely herded livestock. The hooves of cattle in particular, and of sheep and goats to a lesser extent, break up the surface layers of the soil, and heavy trampling tends to destroy the roots and rhizomes of perennial grasses.

Under heavy grazing, as the palatable species are reduced, livestock may be forced to feed upon progressively less palatable plants and, in the case of the grazing species, to browse upon the foliage of shrubs which they normally avoid. Continued heavy grazing may completely remove several species from a community, or may reduce them to a low density. In the latter case a reduction in grazing pressure or a particularly wet year may provide the opportunity for them to recover. On the other hand, several drought years may hasten the loss of a palatable plant species from a heavily grazed area.

Thus, heavy grazing tends to reduce species diversity in plant communities and, if permitted to continue, may progress towards threshold levels beyond which recovery is unlikely or, in the event of subsequent protection, would be greatly delayed. As the process of degradation continues with the removal of vegetation, denudation exposes the soil surface to increased wind and water erosion, which commonly lead to the removal of the top soil layer. When this happens, the former plant communities are prevented from re-establishing themselves and a threshold of virtual irreversibility may have been passed. The resulting bare subsoil surfaces are often colonized by a small number of sub-shrub species which constitute a specialized plant community adapted to the new soil conditions, but which may be palatable only to goats and camels. Several thousand square kilometres of the Elbata plains and the Samburu grazing lands in northern Kenya have been changed from grassland to eroded dwarf shrubland in forty years by over-grazing. As grasslands are replaced by shrublands, they cease to support cattle and tend to be occupied by camels, sheep and goats. Sobania (1979) has recorded the evidence for a gradual change in the Marsabit District of northern Kenya as cattle herding has been replaced by camel herding over the last sixty years.

The above brief and over-simplified account describes only one of many successional processes caused by over-grazing in tropical savannas. In sandy soil areas the reduction of plant cover by over-grazing may lead to the formation of loose sand surfaces and mobile dunes which may be blown across productive agricultural and grazing lands (Sudan Government, 1979).

Tropical savanna wooded grasslands are adapted to periodic burning. Frequent grass fires tend to reduce tree and shrub growth, and to strengthen certain perennial grasses (e.g. *Themeda triandra*) in semi-arid savannas in Africa. Fires usually occur towards the end of the dry season when the herb vegetation is dry and much of the perennial standing grass has become dead hay and has transferred its nutrient reserves to storage roots. Under these circumstances even intense fire does no harm to the grass plants. It does, however, burn young trees back to ground level without necessarily killing them and it inhibits woodland growth.

It is a very common practice for pastoralists to start grass fires. They normally do so to stimulate the early growth of new green shoots from the roots of the perennial species so that grazing is made available as soon as possible at the end of the dry season and at the beginning of the following rains. Other benefits which the pastoralists obtain from grass burning can be mentioned. In addition to controlling bush encroachment in grazing lands, fires kill parasites, primarily ticks; they also remove cover for predators.

Where heavy grazing occurs the grass may be kept so short that it cannot be burnt. When this happens in wooded savanna, tree and shrub growth normally increases, and the savanna may turn into thicket. Serious over-grazing, which destroys the herb layer, may lead to thicket development in the semi-desert thorn scrub of Africa; this has happened over very large areas of northern Tanzania and northern and eastern Kenya during the past fifty years.

In the semi-arid and arid savannas of eastern Africa pastoralists normally corral their animals at night in fenced enclosures made of thorn trees (mainly the small *Acacia* species), to prevent them from wandering, and to protect them from human raiders and animal predators. Observations on the livestock *manyattas* of nomadic Rendille and Samburu pastoralists indicate that each family cuts (on the average) six trees of *Acacia reficiens* or other scrub *Acacia* species as its contribution to the group livestock enclosure every time it moves to a new site (Hunting Technical Services, 1977). At least eight moves per year are normal. As a rough estimate, at least 50 000 trees are felled annually for fence construction by the 9000 Rendille people of the southwestern Marsabit District in northern Kenya, in an area of approximately 9000 km². Over very large areas within the District, the rate of tree felling is well within the recruitment rate of the tree populations. In the more favoured settlement areas, especially near wells and springs, denudation of the land both by over-grazing and tree felling spreads outwards in concentric circles. Rapp (1974) studied the pattern of exploitation of arid and semi-arid grazing lands by the livestock of pastoralists in relation to water distribution. In the wet season, the herds can graze the greater part of the land. When vegetation is green camels, sheep and goats need to drink at infrequent intervals and normally find free water in rain pools. In the dry season, the herds have to be brought back to permanent water sources at regular intervals: every two (occasionally three) days for cattle, every two days for sheep and goats, and every six to ten days for camels. The foraging distance from water will be approximately 8 km for cattle, sheep and goats and about 30 km for camels. The distribution of water sources will determine the total area of land which can be reached by the herds during the dry season. If wells are close together the whole rangeland area may be accessible. Rapp (1974) suggested that the optimum spacing for wells is such that, during the dry season, two-thirds of the total rangeland area is inaccessible, so that it is bound to be rested during the whole of the dry season. The carrying capacity of the region as a whole is limited by the area within foraging range of water in the dry season (Lamprey, 1964; Western, 1974). Where water sources are widely distributed the total numbers of livestock will be proportionately low and, except for the areas accessible to water throughout the year, the region may be preserved from over-grazing. Examples of undamaged savanna wooded grasslands are to be seen in the *qoz* stabilized sand-dune areas of the northern Sudan where there is no surface water and where no livestock can be taken for several months in the year.

ENVIRONMENTAL CONSTRAINTS AND ADAPTIVE STRATEGIES

Pastoralism in tropical savannas, seen as a form of economy, has been successful in so far as it has permitted some human populations to exploit the arid and semi-arid regions of southwestern Asia and Africa as habitats and to survive there under tolerable conditions during the last ten millennia. On the other hand, seen as a biological system, pastoralism has been unsuccessful to the extent that it has persisted and spread at the expense of the

progressive degradation of these regions. In any attempt to evaluate the adaptive nature of pastoral strategies, it should be borne in mind that they fall far short of the degree of mutual adaptation which is characteristic of the indigenous communities in "natural" ecosystems. While it is evident that pastoralists are, for the most part, rational in the adaptive behaviour which enables them to achieve their primary goal of survival, it is apparent that survival at subsistence levels is commonly achieved without regard to the maintenance of the savanna habitats. There is no suggestion that this failing is peculiar to the pastoral economy. It is an unprofitable philosophical question as to whether subsistence agriculturalists and pastoralists are to be blamed for their failure to conserve the land resources of their habitats. In the last fifty years numerous modern influences have undoubtedly exacerbated destructive land use by pastoralists in the savanna rangelands of developing countries (see Fig. 31.1) and to this extent they are victims of circumstances beyond their control.

The first requirement in the existence of a pastoralist family is a subsistence herd with sufficient animals to ensure the production of milk and meat amounting to approximately 2300 kcal per person per day (although survival near starvation levels may occur at half this intake for several months). As mentioned above (pp. 650–651), the likelihood of loss through drought, disease, raiding, predation or accident makes it necessary for a pastoral family to keep herds which exceed the minimum subsistence number (approximately 30 TLU for eight people) and generally to keep at least twice as many. While most pastoralist families live close to the subsistence level, some are relatively rich and possess many times the number of animals that are needed for subsistence. Although rich herdsmen contribute disproportionately to the grazing pressure upon savanna pastures, they are nevertheless an important factor in the society's adaptation to survival under marginal conditions and, indeed, may be an integral element in the society's organization. A pastoralist who is starting to build up a herd or whose herd is recovering from disaster may borrow animals from his richer relatives and friends. A young or impoverished pastoralist may serve as herdsman to his more affluent neighbour and, in payment, be allowed to borrow or keep some of the calves born in the herd as well as to use the milk from the herd for his own subsistence. Hjort (1976) describes the *stock friend* system common among pastoralists, as a mutually beneficial arrangement between two adult males whereby one who may, at the time, have animals that he can spare, will agree to help a less fortunate friend or relation with the loan of an animal. In this way a "mutual insurance network" is established which tends to protect pastoralists from final destitution, but can only operate under limited drought conditions when the total number of animals in the society is not seriously reduced.

Hjort (1976) mentioned three other traditional forms of redistribution of livestock wealth which make it possible for individuals to build up subsistent herds. A prevalent but drastic method of obtaining animals is by raiding neighbouring tribes. Understandably, this approach is restricted through law enforcement by governments, but nevertheless continues. In some societies ritual leaders receive gifts of animals and then distribute them to poor families in their communities, a practice of the Boran people of southern Ethiopia and northern Kenya. The fourth method is by bartering with neighbouring tribes. A family may take advantage of the rapid rate of increase of its sheep and goats and may exchange several for a cow or a camel.

When a serious drought occurs or when a disease, such as rinderpest, greatly depletes livestock populations, such redistribution systems will fail to provide for the subsistence of all families, and some inevitably become destitute. In the past, the loss of livestock during disasters has led to increased mortality from famine among pastoral peoples. Their numbers would tend to be limited by the reduced productivity of the savanna ecosystem during drought periods in a manner similar to that in which animal populations are limited by their food supplies in natural ecosystems. While it is questionable whether pastoral systems have ever been *regulated* at levels within the carrying capacity of their habitats, it seems probable that there was formerly a much closer approach to an equilibrium between pastoralist populations and their habitats than exists today. The present rate of population increase among pastoralists (over 2% per annum in northern Kenya), a recent development influenced primarily by the introduction of public health and medical care facilities, is the main factor in the recently accelerated over-exploitation of the graz-

ing lands. For normal subsistence the increasing human populations have depended upon increased livestock populations, and this has been made possible in the greater part of the dry savannas of Africa by the introduction of veterinary facilities and artificial water sources, mainly boreholes. It is acknowledged by critical observers of pastoralist societies (Western, 1976; Campbell, 1979) that some populations are too large to be supported by their traditional pastoralist economies.

As Hjort (1976) has pointed out, traditional systems of land use in arid regions are based upon the assumption that land is not a scarce resource, although for some pastoral societies, whose numbers have increased in recent years, it has become scarce. In general, a well-defined area of land is owned or claimed by a tribe or by a clan, and its members have the right to herd their animals there. Occasionally tribal lands are claimed by adjacent tribes and are the subject of periodic hostilities, although neighbouring tribes commonly respect boundaries.

Pastoralism in the arid and semi-arid zones of Africa and southwestern Asia depends to varying degrees upon mobility. Pastoral societies migrate to enable them to exploit the availability of grazing and temporary water of their savanna rangelands. During the seasonal rains livestock are normally dispersed over the greater part of the country, taking advantage of green forage and water in the rain pools. In the dry season, the herds are forced to concentrate within foraging distance of permanent water. Depending on the degree of aridity of the country and the distribution of water sources, the seasonal movements may be very short and consist of a shift from the valley grasslands in the dry season to the neighbouring hills in the rains. The Masai and the Samburu of Kenya frequently make such short movements with their cattle. A second type of movement is referred to as transhumance and consists of a regularly repeated migration between traditional wet- and dry-season pastures, such as is found among several pastoral tribes in the Sahelian zone who move north in the rains and southwards in the dry season. The ecology of such transhumant movements is described by Breman et al. (1978).

Nomadic pastoralists commonly separate their livestock into two kinds of herds. The home or subsistence herd consists mainly of the lactating cows with their calves to provide milk for the family. The *fora* herd, as it is called in northeastern Africa, consists of the dry or barren females, weaned calves, castrates and the bulls that are not needed for breeding at the time. The home herd is kept close to the family settlement and the fora herd is normally taken by the young men as far afield as possible, following traditional nomadic practice and thereby relieving the grazing pressure upon the areas near the family settlements.

Fora herds are sometimes referred to as reserve herds kept mainly as insurance against loss. Dahl and Hjort (1976) suggested that the implication of this term that the animals in the fora herds are, for most of the time, surplus to the pastoralists' requirements, is misleading, since these animals are almost invariably needed in the economy at some time. Nevertheless, Field (1979) has pointed out that the Rendille and Gabra people of northern Kenya frequently keep at least twice as many male camels as are needed for breeding, and suggested that the fora herds in that tribe may include surplus animals. The question is of great importance in view of the probable need to reduce livestock numbers in many areas to reduce the over-stocking of the rangelands.

Traditionally, the family herds are also nomadic, and tend to remain so today among most of the tribes in the arid zone in Africa. However, there is an increasing incidence of sedentarization by nomads near the small settlements which grow up around boreholes and wells. The rate of such settlement has greatly increased in the arid zone of northern Kenya during and since the drought from 1968 to 1976 when many destitute pastoralists became dependent upon famine relief food. Other incentives to settle or to spend more time than formerly near the settlements are the presence of medical facilities, schools and shops and a degree of security provided by police protection against raiding.

Traditional pastoralism embodies a great deal of understanding on the part of the herdsmen of the skills of animal husbandry, on the characteristics of the rangeland habitats and on the physiological limitations of their animals. Traditional knowledge and practices, which are passed down from father to son include the customs of using particular vegetation and water for their curative properties during the occurrence of mild diseases such as

mange or camel pox and to keep their animals in the best possible condition. The Kababish people of Northern Kordofan (Sudan) take their camels large distances across the Libyan Desert to take advantage of the *gizu* ephemeral vegetation which grows irregularly in desert areas which have received winter rainfall (Lamprey, 1975; Wilson, 1978). The Kababish believe, probably correctly, that the fertility of their camels, is greatly increased when they can feed upon the gizu vegetation, which has a high protein content.

There is considerable geographical variation in pastoral systems, which can be interpreted partly as adaptive response to the various eco-climatic conditions of their environments. A major basis for this adaptation is the species composition of herds, a subject that has been well reviewed by Dahl and Hjort (1976). In general terms the composition of domestic herds reflects two important requirements in the pastoral economy. Firstly, it enables the pastoralist to exploit the different elements of the vegetation by using an appropriately balanced "spectrum" of livestock species in a manner similar to that of a community of wild ungulate species (Lamprey, 1964). Secondly, it enables the pastoralist to spread the risks due to drought and disease so that heavy mortality in one species will not deprive him of all his animals. A third advantage that has been mentioned earlier is that, while cattle and camels are normally kept for milking, sheep and goats are kept mainly to provide meat.

Although the number of species herded by one pastoralist family is commonly two, three or four and may be as high as five, it is evident that this assemblage of species does not have the adaptive precision to be seen in the ecological separation of the species of an indigenous ungulate community, which in semi-arid African savannas commonly exceeds fifteen. Nevertheless, the principle of ecological complementarity applies to the domestic animal community. Thus, an Ariaal Rendille family in northern Kenya may keep cattle and donkeys as the large grazers of the community (herb-layer feeders), sheep as small grazers, camels as large browsers (tree and shrub feeders) and goats as small browsers. This classification of the domestic species into feeding types is over-simplified, but indicates their feeding preferences. Each species is capable of eating categories of plants that it normally avoids. In times of scarcity, after heavy grazing has removed the grass layer, particularly in arid regions where little grass is available, the grazing species commonly browse the foliage of woody plants (Field, 1978), but revert to grazing after rain when grass becomes available again.

In most sub-humid and semi-arid savanna grasslands pastoralists normally prefer to keep cattle. Some pastoral societies, such as the Tswana of Botswana tend to keep cattle only. Other pastoralists keep sheep and goats to supplement their economy. In northern Kenya and across the Sahelian zone cattle are kept in areas with as little rainfall as 300 mm yr^{-1}. In drier areas, although cattle may be kept in relatively small numbers, they tend to be replaced by camels as the main milk producers. In rainfall zones of less than 150 mm yr^{-1} cattle are not normally kept but certain specialized breeds are capable of surviving in such dry areas; of these the Borero longhorn breed of West Africa is the most drought-adapted. In dry regions, pastoralists keep sheep and goats of certain varieties which have evidently been bred, perhaps through a process more akin to natural selection than to conscious selection on the part of the pastoralists, for their great tolerance to food shortage and to drinking at two- or three-day intervals. These are the fat-tail and fat-rumped sheep of northeastern Africa and southwestern Asia, which apparently use their fat reserves to enable them to survive several months with little food intake.

It is of considerable interest that the numbers of cattle kept by some pastoral tribes in northeastern Africa, notably the Turkana of Kenya, reflect the prevailing and recent climate. Several years of above-average rainfall, as have occurred from 1977 to 1979, have made it possible for the Turkana to become cattle owners on country that does not normally support cattle. With recurrent droughts the Turkana are obliged to rely again on their camels and their sheep and goats. Development schemes, particularly fishing on Lake Turkana, have enabled the Turkana people to earn money. This money has tended to be exchanged for the recognized currency of the Turkana people — livestock, particularly cattle. Unfortunately, the cattle herds will inevitably contribute to the overgrazing of already degraded land before they die or are sold in the next drought.

In the absence of such modern influences as introduced "development" schemes, pastoralists continue to practise their traditional methods and skills, of which very little has been recorded. Pastoral systems incorporate traditional techniques that are likely to be essential adjuncts in the continuation of animal husbandry in the savanna grazing lands in Africa and southwestern Asia, even if pastoralism as we know it today is replaced by other forms of livestock production, such as co-operative ranching. There is an urgent need to study the livestock management methods of pastoralists to ensure that valuable skills are not lost (Schwartz, 1978). At the same time such studies may be expected to indicate other practices, perhaps based upon superstition, which may not be advantageous to livestock management. It seems probable, however, that thousands of years of experience of animal husbandry under marginal conditions have produced systems in which very few aspects of the pastoralists' beliefs, social behaviour and structure are not of adaptive significance, at least in relation to the survival of pastoralism itself. Unfortunately, the adaptive nature of pastoralism has not extended as far as inducing motivation to preserve its own habitats.

DISCUSSION AND CONCLUSIONS

Subsistence pastoralism will probably remain the predominant land use in the greater part of the arid and semi-arid savanna regions of Africa and southwestern Asia for many years to come, but may be expected to be increasingly modified by contact with the "developing" world. Although some societies living in areas remote from the influence of governments may continue to practise traditional pastoralism, it seems inevitable that it will eventually be superseded by other systems of animal husbandry in most arid regions. There are many reasons for this, not the least of which is the urgent necessity for governments to introduce systems of controlled grazing in order to prevent further overgrazing and the continuing degradation of productive pastures. Pastoralism has reached a turning point, and its future remains to be determined.

Over thousands of years pastoral societies have been able to adapt to the gradual processes of ecological decline through two major strategies.

Either they could move on from degraded lands into new territories, or they could adapt their pastoral practices to increasingly marginal conditions (for instance by herding camels instead of cattle) and remain where they were. Under twentieth century conditions in which the pastoralists' physical and political environments are changing at unprecedented rates, such long-standing stratagems are no longer appropriate. Restrictions imposed by modern boundaries, together with the fact that virtually all available grazing lands are now occupied, prevent further migration to new territories. Adaptation to the rapidly changing environmental conditions of the developing world seems to call for adaptability beyond the reach of normally conservative pastoral societies. It requires a new insight into the basic principles of ecology and range management, radical changes in the functioning of the pastoral economy and the adoption of new authoritarian systems under which the supervision of grazing management can take place. If such changes are not initiated from within the pastoralist societies themselves, it seems inevitable that they will be imposed by governments. The introduction of the changes that are necessary will present difficulties which should not be underestimated.

The form that grazing control system should take is very far from certain. In one North African country pastoralists have been settled by the Government in highly organized and disciplined village communities, with the animal husbandry run on a communal basis and the numbers of livestock regulated to conform to the carrying capacity of the rangelands. The system is recently introduced and it remains to be seen how well it will function in the long term. In an East African country the land of one pastoral tribe has been demarcated into group ranches, so that the responsibility for the use and conservation of limited areas is well defined. Obvious difficulties arise over the allocation of ranches of different productivity, and over the policy to be followed in drought years when rain may fall on some ranches and not on others. The political and administrative problems inherent in such schemes are proving to be very great, but may be overcome with many years of experience.

The basic and most difficult problem is that, over very considerable areas, it is no longer possible for the increasing numbers of people who live there to subsist upon the products of the pastoral economy without over-stocking their grazing lands. Neither settlement schemes nor group ranches can solve this problem unless some means is found to diversify the economy. Agriculture is the most obvious basis for such diversification but is only possible in areas of relatively high rainfall. In the arid zone there are very few possibilities of introducing additional sources of livelihood, but such industries as bee-keeping, arid zone forestry and gum arabic production offer some hope of supplementing the pastoral economy.

In many African countries tourism is a very valuable adjunct to the pastoral economy, especially where wildlife is abundant, and where the pastoralists themselves are involved in the management of the wildlife parks and their local councils receive direct financial benefit from them. However, it is evident that, although free-ranging wild ungulate populations of low to moderate density are reasonably compatible with traditional pastoral practices, wild grazing ungulates, such as wildebeest (*Connochaetes taurinus*), African buffalo (*Syncerus caffer*) and zebra (*Equus burchelli*), may occur locally at densities high enough to be serious competitors for the available grazing and are unlikely to be tolerated indefinitely except where they contribute substantially to the local economy through tourism. Where traditional pastoral systems have been replaced by group ranches, especially where the land is demarcated by fences, wildlife may become a serious liability. Under these circumstances, a compromise has to be reached in which a proportion of the rangeland remains unfenced, especially in the vicinity of parks and reserves. Although the coexistence of wildlife conservation areas and animal husbandry is likely to present problems for several African governments, it seems likely that the difficulties need not be insuperable and that these governments will continue to set aside areas — not only the parks themselves, but also the dispersal zones around the parks — where wild ungulates will inevitably compete with the herds of the pastoralists who occupy these areas. As the demand for grazing land becomes critical so will the economic contribution of wildlife conservation in pastoral areas be more critically assessed.

Over the greater part of the Sahelian zone and southwestern Asia, in the absence of government intervention, pastoralism has proved to be imcompatible with wildlife survival, and the wildlife populations of former years have been almost eliminated by armed people, largely belonging to pastoral tribes (Lamprey, 1975). The destruction of wild ungulate populations in the arid zone in Africa has run parallel with the destruction of the vegetation, particularly in recent years. Although the most immediate prospect for slowing up and preventing desert encroachment in the arid and semi-arid regions depends upon firm government action, the success of such action in the long term must depend on acceptance by the pastoralists themselves of the measures being taken. This acceptance will depend on an understanding of the issues and principles involved in rational land use, and that understanding will depend on extensive education programmes in which the emphasis is put upon practical resource management. The extent of the problem of over-grazing and the rate at which productive land is being irreversibly destroyed, necessitates international action on an unprecedented scale.

REFERENCES

Asmarom, L., 1975. *Ecological Stress and Pastoralism in Africa.*

Atsatt, P.R. and O'Dowd, D.J., 1976. Plant defence guilds. *Science*, 193: 24.

Baskin, L.M., 1974. Management of ungulate herds in relation to domestication.In: *The Behaviour of Ungulates and its Relation to Management.* Symposium, University of Calgary, Alta., 1971. *IUCN Publ. N. S.*, No. 24: 530–541.

Baxter, P., 1975. Some consequences of sedentarization for social relationships. In: T. Monod (Editor), *Pastoralism in Tropical Africa.* Oxford University Press, pp. 206–228.

Bibby, G., 1970. *Looking for Dilmun.* Penguin Books, London, 240 pp.

Boudet, G., 1976. Mali. Regional studies and proposals for development. In: A. Rapp, H.N. Le Houérou and B. Lundholm (Editors), *Can Desert Encroachment be Stopped? Ecol. Bull.*, 24: 137–153.

Bourn, D., 1978. Cattle rainfall and tsetse in Africa. *J. Arid Environ.*, 1: 49–61.

Breman, H., Diallo, A., Traore, G. and Djiteye, M.M., 1978. *The Ecology of the Annual Migrations of Cattle in the Sahel.* Unpublished report of the Netherlands project "Primary Production Sahel": Mali; Wageningen, 21 pp.

Brown, L.H., 1971. The biology of pastoral man as a factor in conservation. *Biol. Conserv.*, 3: 93–100.

Butzer, K.W., 1961. In: L. Dudley Stamp (Editor), *A History of Land Use in Arid Regions.* UNESCO, Paris, No. 17: 31.

Campbell, D.J., 1979. *Development or Decline: Resources. Land Use and Population Growth in Kajiado District.* Working Paper 352. Institute for Development Studies, University of Nairobi, Nairobi.

Coe, M.J., Cummings, D.H. and Phillipson, J., 1976. Biomass and production of large African herbivores in relation to rainfall and primary production. *Oecologia*, 22: 341–354.

Dahl, G. and Hjort, A., 1976. *Having Herds.* University of Stockholm, Stockholm, 335 pp.

Devitt, P., 1971. Man and his environment in the western Kalahari. *Bostswana Not. Rec.*, Spec. Edi., No. 1: 50–56.

Duerst, J., 1908. Animal remains from the excavations at Anau. In: R. Pumpelly, *Explorations in Turkestan. Expedition of 1904.* Carnegie Institution, Washington, D.C., pp. 341–393.

Dyson-Hudson, N., 1962. Factors inhibiting change in an African pastoral society: the Karimojong of North-east Uganda. *Trans. N.Y. Acad. Sci.*, 2: 771–801.

Dyson-Hudson, N., 1966. *Karimojong politics.* Clarendon Press, Oxford, 280 pp.

Dyson-Hudson, N. and Dyson-Hudson, R., 1970. The food production system of a semi-nomadic society, the Karimojong, Uganda. In: P. McLoughlin (Editor), *African Food Production Systems, Cases and Theory.* Johns Hopkins University Press, Baltimore, pp. 91–123.

Eckholm, E.P., 1979. *Losing Ground.* Norton, New York, N.Y., 223 pp.

Epstein, H., 1933. Descent and origin of the Afrikander cattle. *J. Heredity*, 24: 448–462.

Epstein, H., 1934. Studies in native animal husbandry. 9. The west African shorthorn. *J. S. Afr. Vet. Med. Assoc.*, 5: 187–201.

Epstein, H., 1971. *The Origin of the Domestic Animals of Africa.* Africana Publications, New York, N.Y., 2 vols., 573 + 719 pp.

Faulkner, D.E. and Epstein, H., 1957. The indigenous cattle of the British dependent territories in Africa with material on certain other countries. *Publ. Col. Adv. Comm. Agric. Anim. Health, For.*, No. 5.

Field, A.C., 1978. The impact of sheep and goats on the vegetation in the arid zone. O.D.M. Sheep and Goat Project. *UNEP–MAB IPAL, Tech. Rep.*, No. E-3.

Field, C.R., 1979. Ecology and management of camels, sheep and goats in northern Kenya. *UNEP–MAB IPAL, Tech. Rep.*, No. E-1.

Hawkes, J. and Wooley, L., 1963. *Prehistory and the Beginnings of Civilization. Vol. I Pt. I. The Domestication of Animals.* UNESCO, Paris, pp. 279–283.

Hjort, A., 1976. Constraints on pastoralism in drylands. In: A. Rapp, H.N. Le Houérou and B. Lundholm (Editors), *Can Desert Encroachment be stopped? Ecol. Bull.*, 24: 71–81.

Hunt, J.A., 1951. *A General Survey of the Somaliland Protectorate 1944–1950.* Hargeisa, 203 pp.

Hunting Technical Services (Lewis, J.G.), 1977. Report of a short term technical consultancy on the grazing ecosystem of the Mt. Kulal region, northern Kenya. *UNEP–MAB IPAL Tech. Rep.*, No. E-4.

Jackson, J.K., 1957. Changes in the climate and vegetation of the Sudan. *Sudan Not. Rec.*, 38: 47–66.

Jolly, G.M., 1969. Sampling methods for aerial censuses of wildlife populations. *East Afr. Agric. For. J.*, 34: 50–55.

Lamprey, H.F., 1963. Ecological separation of the large mammal species in the Tarangire Game Reserve, Tanganyika, *East Afr. Wildl. J.*, 1: 63–92.

Lamprey, H.F., 1964. Estimation of the large mammal densities, biomass and energy exchange in the Tarangire Game Reserve and the Masai Steppe in Tanganyika. *East Afr. Wildl. J.*, 2: 1–46.

Lamprey, H.F., 1975. *Report on the Desert Encroachment Reconnaissance in Northern Sudan.* United Nations Environment Programme, Nairobi, 14 pp. (mimeo).

Le Houérou, H.N., 1970. North Africa: past, present and future. In: *Arid Lands in Transition. Am. Assoc. Adv. Sci.*, pp. 227–278.

Le Houérou, H.N. and Hoste, C.H., 1977. Rangeland production and annual rainfall relations in the Mediterranean Basin and in the African Sahelo-Sudanian Zone. *J. Range Manage.*, 30: 183–189.

MacGillivray, D., 1967. *East African Livestock Regional Survey — Kenya, Tanzania, Uganda.* FAO/SF, Rome, 3 vols.

McNaughton, S.J., 1978. Serengeti ungulates: feeding selectivity influences the effectiveness of plant defense guilds. *Science*, 199: 806–807.

Norton-Griffiths, M., 1978. *Counting Animals. Handbook No. 1.* African Wildlife Leadership Foundation, Nairobi, 139 pp.

Payne, W.J.A., 1970. *Cattle Production in the Tropics. Breeds and Breeding.* Longman, London, 336 pp.

Pratt, D.J. and Gwynne, M.D., 1977. *Rangeland Management and Ecology in East Africa.* Hodder and Stroughton, London, 310 pp.

Rapp, A., 1974. *A Review of Desertization in Africa. Water, Vegetation and Man.* Secretariat for International Ecology, Rep. No. 1, Stockholm, 77 pp.

Schwartz, H.J., 1978. *Report on a Five Month Consultancy Mission Within the Framework of the Funds-in-Trust Project, The Efficiency of Livestock Management in Traditional Nomadic Pastoralism.* Unpublished report to UNESCO.

SEDES, 1975. *Recueil statistique de la production animale.* Republ. Fr., Min. Coop., 1201 pp.

Sinclair, A.R.E., 1973. Regulation and population models for a tropical ruminant. *East Afr. Wildl. J.*, 11: 307–316.

Sinclair, A.R.E., 1974. The natural regulation of buffalo populations in Africa. IV. The food supply as a regulating factor, and competition. *East Afr. Wildl. J.*, 12: 291–311.

Sinclair, A.R.E., 1975. The resource limitation of trophic levels in tropical grassland ecosystems. *J. Anim. Ecol.*, 44: 497–520.

Sinclair, A.R.E., 1977. *The African Buffalo.* University of Chicago Press, Chicago, Ill., 355 pp.

Sobania, N., 1979. Background history of the Mt. Kulal region of Marsabit District, northern Kenya. *IPAL Tech. Rep.*, No. A-2.

Spencer, P., 1973. *Nomads in Alliance. Symbiosis and Growth among the Rendille and Samburu of Kenya.* Clarendon Press, Oxford, 230 pp.

Sudan Government, 1979. *Sudan's Desert Encroachment Control and Rehabilitation Programme* (DECARP). Khartoum, 227 pp.

UNEP, 1977. *Ecological Change and Desertification.* UN Conference on Desertification, Background Document, Nairobi, 124 pp.

Vesey-Fitzgerald, D.F., 1960. Grazing succession among East African game animals. *J. Mammal.*, 41: 161–172.

Vesey-Fitzgerald, D.F., 1965a. The utilization of natural pastures by wild animals in the Rukwa Valley, Tanganyika. *East Afr. Wildl. J.*, 3: 38–48.

Vesey-Fitzgerald, D.F., 1965b. Lechwe pastures. *Puku*, 3: 143–147.

Vesey-Fitzgerald, D.F., 1970. The origin and distribution of valley grasslands in East Africa. *J. Ecol.*, 58: 51–75.

Watson, R.M., 1969. Aerial photographic methods in censuses of animals. *East Afr. Agric. For. J.*, 34: 32–37.

Watson, R.M., 1972. *Results of Aerial Livestock Surveys of Kaputei Division, Samburu District and North Eastern Province.* Ministry of Finance and Planning, Statistics Division, Nairobi.

Western, D., 1974. *The Environment and Ecology of Pastoralists in Arid Savannas.* New York Zoological Society, N.Y., (Typescript report).

Western, D., 1976. The environment of the range area of Kenya and the impact of man and animals on the ecosystem. *J. Anim. Prod. Soc. Kenya*, 9: 35–44.

Wickens, G.E., 1975. Changes in the climate and vegetation of the Sudan since 20 000 B.P. *Boissieria*, 24: 43–65.

Williamson, G. and Payne, W.J.A., 1971. *An Introduction to Animal Husbandry in the Tropics.* Longmans, London, 2nd ed., 435 pp.

Wilson, R.T., 1978. The 'gizu': winter grazing in the south Libyan Desert, 1. *J. Arid Environ.*, 1: 291–310.

AUTHOR INDEX[1]

---

[1]Page references to text are in Roman type, to bibliographical entries in italics.

667

SYSTEMATIC INDEX

GENERAL INDEX

nutrient(s) *(continued)*

- cycling, 162, 163, 583–603, 609, 610
- in rain, 163
- in soil, *see* soil nutrients
- pools, 14
nutrition, 450–454
nyala *(Tragelaphus angasi)*, 404, 410, 422
–, mountain *(T. buxtoni)*, 404, 414, 422
Nyika Plateau (Malawi), 638
Nylsvley Nature Reserve (South Africa), 317, 585, 598

oaks (tropical) *(Quercus)*, 176, 273, 274, 279
Oaxaca (Mexico), 273
Oceania, 43, 55
oestrid flies *(Oestrus)*, 563–565, 572–576, 578
Ogle bridge (Guiana), 23
Okavango Swamp (Botswana), 411
Old World Tropics, 29–33, 306, 325, 359
Oligocene Epoch, 33, 227, 402
oligohumic earthworms, 486, 492, 493, 496, 500
Olokemeji (Nigeria), 81, 366, 381, 382, 389, 610, 611, 613, 618, 620, 623, 627
omnivores, 346, 350, 372, 373, 389, 392, 469, 535, 536
Omo Valley (Ethiopia), 29
ootheca (egg case), 283
open forest, 167–181
- formations, 129–140
opossum, short bare-tailed *(Monodelphis brevicaudata)*, 464
opportunistic species, 94, 96, 97, 103, 106, 107, 410, 454, 457, 458
orchard savannas, 134
orchids (Orchidaceae), 92, 117, 143, 265, 270, 281, 282
Ord River (N.T., Australia), 200, 201, 236
oribatid mites, 628
oribi *(Ourebia ourebi)*, 405, 418, 422
Oriental Region and domain, 123, 506, 525
Orinoco River (Venezuela), 27, 39, 61, 70, 90, 253, 254, 257, 260, 269, 283, 285, 432
Oriomo Plateau (Papua New Guinea), 222
Oro Bay (Papua New Guinea), 224
orthopteroid insects (Orthopteroidea), 295, 298, 321, 467, 534
oryx(es), 401, 407, 408, 410, 418, 646
–, beisa (gemsbuck) *(Oryx gazella)*, 404, 410, 422, 646, 652
–, scimitar-horned *(O. dammah)*, 404, 422
ostrich (Struthionidae) *(Struthio camelus)*, 338, 339, 340, 345, 347, 354, 467
overgrazing, 75, 76, 120, 129, 135, 175, 273, 636, 644, 657
overstocking, 122, 651–656
Owen Stanley Range (Papua New Guinea), 224
owls, 338, 342, 349, 396
oxic horizon, 64
Oxisols, 64–67, 69, 70, 72, 73, 254
oxpeckers *(Buphagus)*, 347

P to *A* ratio, *see* production efficiency
P to *B* ratio, *see* turnover rate
P to *I* ratio, *see* ecological efficiency
Pacific Islands *(see also* Southwest Pacific), 183–243, 526
Pacific Ocean, 273

pajonales, 272
Pakaraima Highlands (Guyana), 265, 267
Pakistan, 649
Palaearctic Region, 351–353, 486
palaeoclimatology of savannas, 402, 403
palaeoecology of savannas, 19–35, 276
palaeogeography of savannas, 19–35, 276
Palau District, Western Caroline Islands (U.S. Trust Territory of the Pacific Islands), 232, 234
Paleocene epoch, 281
Paleotropics, 29–32, 101, 111
palm(s) (Arecaceae), 25, 82, 86, 87, 116, 144, 176, 205, 207, 218, 221, 251, 258, 265, 267, 270, 273, 274, 279–281, 304, 393, 517, 528, 594
–, buriti *(Mauritia vinifera)*, 253, 273
–, Moriche *(M. minor)*, 261, 263
- savannas, 134–136, 154, 220, 253, 260, 272, 273, 316
palynology, 19–35
pampas, 282, 428, 464
- deer *(Ozotoceros bezoarticus)*, 427–432, 436
Panama, 274, 327, 331, 490, 491, 492
pandan savannas, 221, 227
pandans (screw pines) *(Pandanus)*, 191, 193, 196, 208–210, 214, 217, 218, 220, 221, 228, 231–234
Pangola grass *(Digitaria decumbens)*, 237, 553
pangolin(s) (Pholidota), 464, 465
–, ground *(Manis temmincki)*, 465
Pantanal *(see also* Gran Pantanal), 73, 272
Pantar Island (Indonesia), 234, 235
paperbarks *(Melaleuca)*, 191, 193, 205, 207, 209–211, 214–216, 218, 219, 221, 228, 238, 451
Papua *(see also* Papua New Guinea), 186, 218
Papua New Guinea *(see also* New Guinea), 183–190, 193, 194, 196–200, 211, 217–229, 232, 236–238
Paraguay, 245, 271
Paraguay River, 271
paramos, 464
Parana [State] (Brazil), 252
parasites, 456–458, 559–584
parasitic plants, 117
park savannas, 134
parrots (Psittacidae), 338, 341, 342, 345
partitioning of nest sites, 528–530
- of resources, *see* resource partitioning
Pasoh (West Malaysia), 10
passerines, 10, 584
pastoral economy, 649–656
pastoralism, 643–666
pathogens, 559–581
peak biomass values, *see* biomass, plant
peccary(ies), 427, 428
–, white-lipped *(Tayassu pecari)*, 427, 430
pentatomids (Pentatomidae), 289, 291, 292, 294, 297, 302, 303, 629–631
perennial plants (or perennials), 14, 94, 97, 100, 102, 103–107, 117, 119, 120, 154, 155, 246, 260, 617, 623
Perija Cordillera (Venezuela), 263
perissodactyls (Perissodactyla), 401, 402, 421, 428
Permian Era, 141
Pernambuco (Brazil), 247

Tropi